QE 1495

David Craft
student # 144323
c/o Bama Towers
1209½ Univ. Blvd.
Tusc., Ala., 35401
phone - 758-0302

BIOLOGY

BIOLOGY

HELENA CURTIS

SECOND EDITION

This book is dedicated to C. P. Rhoads, M.D.

BIOLOGY, SECOND EDITION

ILLUSTRATOR: SHIRLEY BATY

PICTURE EDITOR: ANNE FELDMAN

DESIGN: MALCOLM GREAR DESIGNERS, INC.

PRINTED IN THE UNITED STATES OF AMERICA

LIBRARY OF CONGRESS CATALOG CARD NO. 74–27183

ISBN: 0-87901-040-1

FIRST PRINTING, JANUARY 1975

WORTH PUBLISHERS, INC.

444 PARK AVENUE SOUTH

NEW YORK, NEW YORK 10016

PREFACE TO THE SECOND EDITION

Since the first edition of *Biology* appeared seven years ago, there has been a notable change in biology textbooks in general. They are better looking, with more color, better photographs and drawings, and above all the writing is livelier, clearer, and more interesting. A book, by virtue of its very existence, has an obligation to engage its readers, and more and more textbooks are recognizing and meeting this obligation. Moreover, with only a few exceptions, the quality of these books as teaching instruments has not been sacrificed; generally speaking, they are more instructive and also far more representative of modern biological research than were their predecessors. In short, biology books are better than ever. We hope that the existence of the first edition of *Biology* played some role in bringing about these changes.

This new edition has been completely rewritten and also, to some degree, reorganized. The Introduction, as before, deals with evolution, the major unifying principle of the life sciences. The rest of the text is divided into three parts, corresponding to levels of biological organization: cells, organisms, and populations. The first part, comprising Sections 1 and 2, deals with life on the cellular level and contains an introduction to cell chemistry that presupposes no prior knowledge of chemistry. In reviewing many current texts, including our own, we came to the conclusion that shorter does not always mean simpler or easier; therefore, in this second edition, we deliberately spend more time on particular subjects, such as glycolysis, so that the student really understands what goes on and doesn't simply memorize a few key words in order to pass a multiple-choice test. Failing to involve the student in this kind of intellectual interaction is dishonest, it seems to me. One says, in effect, "I'll pretend I taught you this and you pretend you learned."

In Section 1, The Organization of Living Things, we try to avoid the static catalog of organelles and molecules that one sometimes gets trapped into by the nature of the material. We concentrate instead on the dynamics of the cell—the cell in action, so to speak.

Section 2 tries to strike a balance between so-called classical genetics and molecular genetics. As with many sections of the book, it is open-ended, concluding on the note of what we don't yet understand. Throughout, we would like to encourage students to stretch their minds and their imaginations, instead of

presenting the subject wrapped up in a tidy package like the bland, artificial products offered as substitutes for real food, which has to be chewed before swallowing.

Part II, which deals with organisms, is divided into three sections. The first, Section 3, introduces the student to the vast diversity of living things and shows him (or her, of course) how taxonomy and evolutionary theory provide patterns for this diversity. We have attempted to make this section not only instructive but readable, rather than following the well-trod path of providing merely a brief reference encyclopedia. Nor have we tried to make it "relevant" in the sense of talking only about bacteria that cause human diseases, or fungi that make us itch between the toes, or plants that we eat or build our houses with. For many students, this will be both the first and last glimpse of the enormous complexity, variety, and beauty of the living world, and we would not want anyone to lose his chance for such a glimpse.

The second section of Part II, Section 4, is plant anatomy and physiology; it is concentrated almost exclusively on the angiosperms. (Other members of the plant family are discussed in Chapter 21 of the diversity section.) In presenting plant physiology separately in this way, we are moving against the current fashion of seeking unity between plants and animals and emphasizing principles. In all the books that present a "unified" treatment, we think the plants get short shrift indeed. Of all organisms, relationships between form and function are the easiest to see in the modern land plants, and yet, in the unified approach, one never sees the plant as a dynamic, functioning biological unit but only as something that performs more slowly and less efficiently than an animal.

Part II concludes with a section on animal physiology. The section is devoted primarily to man, although comparisons between man and invertebrates or other vertebrates are included wherever they are illuminating. (The student is introduced to invertebrate physiology in Section 3.) We have here reduced the section on development to a single chapter, Chapter 31, again focusing on man and using much comparative material. The reason for concentrating on man is partly one of student motivation, but more, it is a matter of seeking a focus for what can otherwise be a very large and diffuse subject.

Most of the topics that were included in the behavior section of the first edition now appear elsewhere in this book: invertebrate behavior is integrated with invertebrate anatomy and physiology in the section on diversity; there is a great deal of material, much of it new, on social behavior in the part of the text that deals with populations; and the section on animal physiology concludes with a chapter on sensory perception and the brain. Again we close the animal physiology section by focusing on the frontiers of scientific knowledge, stressing once more that biology is not a static collection of facts but, in fact, a process of discovery.

In Part III of the book, Populations, the emphasis is on the relationships among the organisms that populate our planet and with the environment in which they live. In addition to the more traditional subjects, such as the flow of energy, the cycling of nutrients, mimicry, coevolution, territoriality, and social dominance, which are covered to some degree—or should be—in all introductory biology texts, we have included some other topics we believe will be of special

student interest. Chief among these are two analyses of social behavior in mammals, one of wolves and the other of baboons. These are not only fascinating in their own right—at least, to me—but they also provide extraordinarily cogent summaries of the ecological principles presented in a more abstract form in earlier chapters. Environmental resistance, competition, survival curves, and predator-prey relationships become vividly real through these examples.

The final chapters of the book begin with the evolution of man, returning in effect to a variation on the theme with which we began. Human evolution is not a traditional part of the biology curriculum; rather it has become the domain of physical anthropologists and, for better or worse, ethologists, both professional and amateur. Just as the social behavior of wolves provides a dynamic synthesis of traditional ecological concepts, the history of the hominids serves as a provocative summation of the principles of evolution and, perhaps more important, of man's place in nature.

Man's evolution leads quite naturally to the ecology of modern man, with which the book ends. Our purpose here is not so much to present "ecological problems"; students are already so bombarded with such material that ecology has come to be practically synonymous with aluminum beer cans. By attempting to analyze how we arrived where we are today, we try to give a larger perspective—both broader and deeper—and so provide for a more fundamental understanding both of today's problems and, more important, of tomorrow's.

Acknowledgments

The second edition of *Biology* has been reviewed even more extensively than the first edition. The manuscript was read by more than 50 experts in various fields. Then, in the spring of 1974, it was set in type, printed, and bound, with illustrations and captions, in a limited preliminary edition of 100 copies. These copies were reviewed, in whole or part, by teachers and other specialists in order to ferret out any remaining inaccuracies or ambiguities and to ensure that all data were the most current available. It is impossible to acknowledge adequately those whose comments were of the greatest value in this strenuous process, but I must give special thanks to David Shappirio, who reviewed the entire book, sharing with me not only his encyclopedic knowledge of biology and his broad teaching experience, but also his deft command of the English language and his sense of humor. Others to whom I owe particular gratitude are Elmer Palmatier, for his scrupulous attention to detail and his patience with our many telephone calls, Dorothy Luciano, for her important assistance with the section on animal physiology, and Stephen Vogel for his suggested chapter-end questions for Part II.

I am listing first those advisors whose criticisms and suggestions played a large part in shaping a particular section of the book.

Part I: E. J. DuPraw, Stanford University; Charles E. Holt, Massachusetts Institute of Technology; James D. Jamieson, Yale University; Albert L. Lehninger, Johns Hopkins School of Medicine; David G. Shappirio, University of Michigan.

Part II: Antonie W. Blackler, Cornell University; Paul B. Green, Stanford University; Earl D. Hanson, Wesleyan University; Dorothy S. Luciano; Elmer A.

Palmatier; University of Rhode Island; Thomas Roos, Dartmouth College; David G. Shappirio, University of Michigan; Robert M. Thornton, University of California, Davis; Stephen Vogel, Duke University.

Part III: Thomas C. Emmel, University of Florida; Terrill Hamilton, University of Texas; Stephen P. Hubbell, University of Michigan; Richard C. Lewontin, Harvard University.

Next I wish to thank those others who reviewed the second edition, working either from the manuscript or the preliminary edition, or both.

Glenn D. Aumann, University of Houston; George T. Barthalmus, North Carolina State University; Jack Burk, California State University, Fullerton; Edwin Burling, De Anza College; Irven DeVore, Harvard University; Michael F. Duff, Indiana University; Paul R. Ehrlich, Stanford University; Susan E. Eichhorn, University of Wisconsin; Abraham S. Flexer, University of Colorado; Arthur Forer, York University; Esther M. Goudsmit, Oakland University; Govindjee, University of Illinois, Urbana-Champaign; Michael J. Greenberg, Florida State University; Burton S. Guttman, Evergreen State College; Calvin S. Hall, University of California, Santa Cruz; Jean B. Harrison, Wellesley College; Mildred Harry, University of Houston; James D. Haynes, State University of New York, Buffalo; Alvin E. Hixon, Daytona Beach Junior College; Robert K. Josephson, University of California, Irvine; Kenneth A. R. Kennedy, Cornell University; W. R. Klemm, Texas A & M University; C. C. Lamberg-Karlovsky, Harvard University; Richard N. Keogh, Rhode Island College; Paul Licht, University of California, Berkeley; Sheldon Lustick, Ohio State University; J. Geoffrey Magnus, University of Georgia; Robert McNally; L. David Mech, U.S. Bureau of Sport Fisheries and Wildlife, Minnesota; Gerald Myers, South Dakota State University; Cyril Ponnamperuma, University of Maryland; David Pratt, University of California, Davis; Peter H. Raven, Missouri Botanical Garden; William R. Rayburn, Washington State University; Michael Salmon, University of Illinois; Clifford L. Schmidt, San Jose State University; Bodil Schmidt-Nielsen, The Mount Desert Island Biological Laboratory; Mary Jane Sherfey; Jerry Smolsky, North Adams State College; Gunther S. Stent, University of California, Berkeley; Richard F. Thompson, Harvard University; John H. Troughton, Carnegie Institution of Washington; Eugene D. Weinberg, Indiana University; H. W. Woolhouse, University of Leeds.

Finally, I would like also to express my gratitude to my friend Marguerite Sheffield, who assisted with the copyediting, and to the staff of the East Hampton Post Office, who made it all possible.

HELENA CURTIS

East Hampton, New York
January, 1975

PREFACE TO THE FIRST EDITION

This book is the result of an unusually broad cooperative venture. There were several essential factors in this venture. My own contribution was my long experience as a professional writer in the field of biology and a desire to communicate my interests and enthusiasms about the subject. Second, and more important, was the arrival on the publishing scene of a vigorous young company with fresh ideas and initiative. The third and by far the most important factor was the willingness of a large number of teachers and experts in all fields of biology to lend their assistance at every stage of the planning, writing, and innumerable revisions of the manuscript and illustrations.

Our purpose was to write a book for students who are taking biology as their first science course in college. This is a particularly interesting group of students from the point of view of one concerned with science and with education. For a great many of them, it will be their only exposure to science in their academic career. What is it we want them to learn within this short period of time, and how can we best achieve this purpose? I personally feel that there is no reason to present an "easier" or less challenging course in science to the student who will major in the humanities or the social sciences than to the student who will go on to be a biology or physics major. In fact, I could present a case for the opposite approach. I think we all agree, however, that the emphasis should be somewhat different. There is no point, for example, in stressing the acquisition of a large amount of detailed data or, in particular, an exhaustive technical vocabulary unless the student is going to continue to use this material, since it will be, at best, memorized and forgotten. On the other hand, new ideas and new ways of looking at things tend to endure, and nothing tugs so persistently at the strings of memory as an interesting unanswered question. So, therefore, we have tried to deal with broad ideas, to remind the student, and ourselves, of the fundamental questions and concerns of biology, and also to show them how biologists themselves ask questions and seek answers.

The unifying theme of the book is evolution, which is, of course, a unifying concept of all biology. The Introduction, which discusses the background and the essential ideas of Darwinian evolution, provides a perspective that will be useful to the student throughout his study of biology. The rest of the book is designed as a series of movements from past to present, from simpler to more complex, from the cell considered in isolation to the biosphere.

Each teacher will adapt the book to his own particular needs and interests and those of his class. Although it is true that high schools are offering more and more in the way of science teaching, most students now entering college have not had a thorough grounding in either biology or chemistry. The first half of the book, the three sections covering Cells, Organisms, and Genetics, was planned with such students in mind. For students with a better background in the sciences, the chapters on cell chemistry, for instance, or taxonomy, or classical genetics, may be used as reviews, with teaching emphasis placed upon the more recent material.

In the second half of the book, we have covered areas of biology in which much new and exciting work is being done: Development, Behavior, and Population Biology and Ecology. These draw upon information offered from a different point of view in the early chapters, and particularly in the last section, we have attempted to bring together many of the previous themes of the story and to leave the student with a sense of his place in the world of living things. The teacher who has limited time at his disposal may want to select only one of these sections, or isolated chapters from each, to give a flavor of the work in these different fields. I would hope that at least one of them could be pursued in depth.

We have tried throughout to illustrate and illuminate the ideas of biology with concrete examples, not only in words but in the many drawings and photographs that accompany the text. Some topics such as cell chemistry, concepts of molecular biology, mitosis and meiosis, and alternation of generations can be particularly difficult to understand and so have been given special attention and have been illustrated more extensively for additional clarity. Many illustrations are accompanied by measurement scales in order to awaken the student's realization that he is looking at a living organism which exists outside of a biology text.

Although we have not hesitated to make use of new and sometimes complicated material when it seemed pertinent, we have tried in such cases to give the student special assistance with it. For example, each electron micrograph is accompanied by a labeled diagram to aid the student in picking out the elements in the micrograph. Unfamiliar terms and, in particular, new and crucial concepts are defined in the unusually ample glossary. At the end of each section, we have provided a bibliography which includes both sources of additional information on special subjects for the more ambitious student and books that can be read for pleasure by any interested amateur. We have not included *Scientific American* reprints in the bibliographies, since complete lists of articles are generally available, but we do recommend them highly.

In addition to the principal consultants to the book [Keith R. Porter, Harvard University; Albert L. Lehninger, Johns Hopkins School of Medicine; Peter H. Raven, Stanford University; Walter F. Bodmer, Stanford University; Clifford L. Grobstein, University of California, La Jolla; Edward O. Wilson, Harvard University; Richard C. Lewontin, University of Chicago; Robert H. MacArthur, Princeton University] and to whom my indebtedness is unlimited, I express my thanks to the many teachers throughout the country who gave their opinions about what they considered the most important topics to be studied by general biology students and what features should be emphasized, thus helping greatly

in the formulation of the original plan for the book. I also record my particular appreciation to the following teachers, writers, and investigators, who helped me by reviewing various sections of the book during its many stages of development.

Jay Martin Anderson, Bryn Mawr College; Daniel Arnon, University of California, Berkeley; Andrew S. Bajer, University of Oregon; Edwin M. Banks, University of Illinois; Alan P. Brockway, University of Colorado, Denver; John Buettner-Janusch, Duke University School of Medicine; Allison L. Burnett, Case Western Reserve University; A. G. W. Cameron, Institute for Space Studies; Sir Wilfrid Le Gros Clark, Oxford University; Mary E. Clutter, Yale University; Carleton S. Coon, University of Pennsylvania; Thomas Eisner, Cornell University; B. E. Frye, University of Michigan; J. Woodland Hastings, Harvard University; Harold Hempling, Cornell University School of Medicine; W. W. Howells, Harvard University; Lionel Jaffe, University of Pennsylvania; Ursula Johnson, Harvard University; Bostwick H. Ketchum, Woods Hole Oceanographic Institute; Donald L. Kimmel, Jr., Brown University; John H. Law, University of Chicago; Monte Lloyd, University of Chicago; Charles F. Lytle, Pennsylvania State University; Everett Mendelsohn, Harvard University; Thomas R. Mertens, Ball State University; Roger Milkman, Syracuse University; Ingirth Deyrup-Olsen, University of Washington, Seattle; Lee D. Peachey, University of Pennsylvania; Shepherd K. Roberts, Temple University; Arnold Ross, American Museum of Natural History; Roberts Rugh, Columbia Presbyterian Medical Center; Joseph Schwab, University of Chicago; Malcolm S. Steinberg, Princeton University; Mary E. Townes, North Carolina College; Charles A. West, University of California, Los Angeles, John S. Willis, University of Illinois; Emil Witschi, Universität Basel.

I would like to acknowledge my debt to the staff of Worth Publishers, who organized this large undertaking, and in particular, to Walter Meagher, who aided and encouraged me through every stage of the book. I was greatly assisted by Mrs. Jean Ely, who provided editorial assistance and supplied captions and summaries.

I feel particularly privileged to be writing for young people; I like their curiosity, their energies, their imaginativeness, and their dislike of the pompous and the pedantic. I hope I serve them well.

HELENA CURTIS

Woods Hole, Massachusetts
December, 1967

CONTENTS IN BRIEF

xiii   *Contents in Brief*

CONTENTS

I–1 *A Galapagos tortoise. The bird is a*
ground finch, one of the thirteen species
of finches Darwin observed on the
Galapagos archipelago. The birds eat
ticks that they remove from their
amiable hosts.

INTRODUCTION

In 1831, a young Englishman, Charles Darwin, sailed from Devonport on what was to prove the most consequential voyage in the history of biology. Not yet 23, Darwin had already abandoned a proposed career in medicine—he describes himself as fleeing a surgical theater in which an operation was being performed on an unanesthetized child—and was a reluctant candidate for the clergy, a profession deemed more suitable for the younger son of an English gentleman. An indifferent student, Darwin was an ardent hunter and fisherman, a collector of beetles, mollusks, and shells, and an amateur botanist and geologist. When the captain of the surveying ship *Beagle*, himself only a little older than Darwin, offered passage and a berth in his own cabin to any young man who would volunteer to go without pay as a naturalist, Darwin eagerly seized the opportunity to escape from Cambridge. This voyage, which lasted five years, shaped the course of Darwin's future work. He returned to an inherited fortune, an estate in the English countryside, and a lifetime of work and study.

THE ROAD TO EVOLUTIONARY THEORY

That Darwin was the founder of the modern theory of evolution is well known. In order to understand the meaning of his theory, however, it is useful to look briefly at the intellectual climate in which it was formulated.

The Ladder of Life

Aristotle, the first great biologist, believed that all living things could be arranged in a hierarchy. This hierarchy became known as the *Scala Naturae*, or ladder of nature, in which the simplest creatures had a humble position on the bottommost rung, man occupied the top, and all other organisms had their proper places between. Up until the end of the last century, many European biologists believed in such a natural hierarchy. But whereas to Aristotle living organisms had always existed, the later Europeans, in harmony with the teachings of the Scriptures, believed that all living things were the products of a divine creation. In either case, the concept that prevailed for 2,000 years was that all the kinds, or species (*species* simply means "kinds"), of animals had come into existence in their present form. Even those who believed in spontaneous generation (toads forming from the mud and snakes from a lady's hair dropped in a rain barrel) did not believe that any species had an historical relationship to any other one—any common ancestry, so to speak.

I–2 *"Afterwards, on becoming very intimate with FitzRoy (the captain of* Beagle*), I heard that I had run a very narrow risk of being rejected on account of the shape of my nose! He . . . was convinced that he could judge of a man's character by the outline of his features; and he doubted whether anyone with my nose could possess sufficient energy and determination for the voyage. But I think he was afterwards well satisfied that my nose had spoken falsely."* (*Charles Darwin,* The Voyage of the Beagle.)

One of the more famous biologists who believed in original creation was Carolus Linnaeus of Sweden, the great eighteenth century systematist who established a method of classifying the plants and animals. All the time that Linnaeus was at work on his encyclopedic *Systema Naturae* and *Species Plantarum*, explorers of the New World were continuing to return to Europe with new species of plants and animals and even new kinds of human beings. Linnaeus revised edition after edition to accommodate these findings, but he did not change his opinions of the fixity of species. He was sometimes criticized by his contemporaries, who felt that his classifications were too artificial. Few, however, took issue with his belief in a special creation.

Preevolutionists

Among the first to suggest that species might undergo some changes in the course of time was the French scientist Georges-Louis Leclerc de Buffon (1707–1788). Buffon believed that these changes took place by a process of degeneration. He suggested that, in addition to the numerous creatures that were the product of the special creation at the beginning of the world, "There are lesser families conceived by Nature and produced by Time." In fact, as he summed it up, ". . . improvement and degeneration are the same thing, for both imply an alteration of the original constitution." If you are familiar with classical philosophy, you will see that Buffon was influenced by the Platonic concept of the ideal, or true form, of which all worldly expressions are merely imperfect copies.

Another early doubter of the fixity of species was Erasmus Darwin (1731–1802), Charles Darwin's grandfather. Erasmus Darwin was a physician, a gentleman naturalist, and a prolific and discursive writer, often in verse, on both botany and zoology. Erasmus Darwin suggested, largely in asides and footnotes, that species have historical connections with one another, that competition plays a role in the development of different species, that animals may change in response to their environment, and that their offspring may inherit these changes. For instance, a polar bear, he maintained, is an "ordinary" bear which has become modified by living in the Arctic and which passes these modifications along to its cubs. These ideas were never clearly formulated; what Erasmus Darwin thought is of interest largely because of its possible effects on Charles Darwin, although the latter, who was born after his grandfather died, did not hold his grandfather's views in high esteem.

The Age of the Earth

It was the geologists far more than the biologists who paved the way for evolutionary theory. One of the most influential of these was James Hutton (1726–1797). Hutton proposed that the Earth had been molded not by sudden, violent events but by slow and gradual processes—wind, weather, and the flow of water—the same processes that can be seen at work in the world today. This theory of Hutton's, which was known as uniformitarianism, was important for two reasons. First, it implied that the Earth has a living history. This was a new idea. Christian theologians, by counting the successive generations since Adam (as recorded in the Bible), had calculated the maximum life span of the world at about 6,000 years. No one had ever thought in terms of a longer period. And 6,000 years is not enough time for such major evolutionary changes as the formation of new species to have taken place. Second, the theory of uniformi-

I–3 Trilobites were abundant in the Cambrian epoch, 500 to 600 million years ago. They flourished for about 300 million years and then became extinct, leaving only their fossils, such as those shown here. Most species are about an inch long. Trilobites were a type of arthropod; modern arthropods include crabs, lobsters, spiders, and insects.

I–4 The first ideas about evolution came from the work of geologists who discovered that different layers, or strata, of the Earth's surface contain different types of fossils. This view of the Grand Canyon clearly shows such strata, now seen as chapters in evolutionary history.

tarianism stated that change is the *normal* course of events, as opposed to the concept of a normally static system interrupted by an occasional unusual event.

The Fossil Record

During the late eighteenth century, there was a revival of interest in fossils. In previous centuries, fossils had been collected as curiosities, but they had generally been regarded either as accidents of nature—stones that somehow looked like shells—or as evidence of great natural catastrophes, such as proof of Noah's Flood. The English surveyor William Smith (1769–1839) was the first to make a systematic study of fossils. Whenever his work took him down into a mine or along canals or across country, he carefully noted the order of the different layers of rock, which are called strata, and collected the fossils from each layer. He eventually established that each stratum, no matter where he came across it in England, contained a characteristic group of specimens and that these fossils were actually the best means of identifying a particular stratum. (The use of fossils to identify strata is still widely practiced.) Smith did not interpret his findings, but the implication that the present surface of the Earth had been formed layer by layer over the course of time was an unavoidable one.

Like Hutton's world, the world seen and reported by William Smith was clearly a very ancient one. A revolution in geology was beginning; earth science was becoming a study of time and change rather than a mere cataloging of types of rocks. As a consequence, the history of the Earth became inseparable from the history of living organisms, as revealed in the fossil record.

I–5 *A drawing by Georges Cuvier of a mastodon. Although Cuvier was one of the world's experts in reconstructing extinct animals from their fossil remains, he was a powerful opponent of evolutionary theories.*

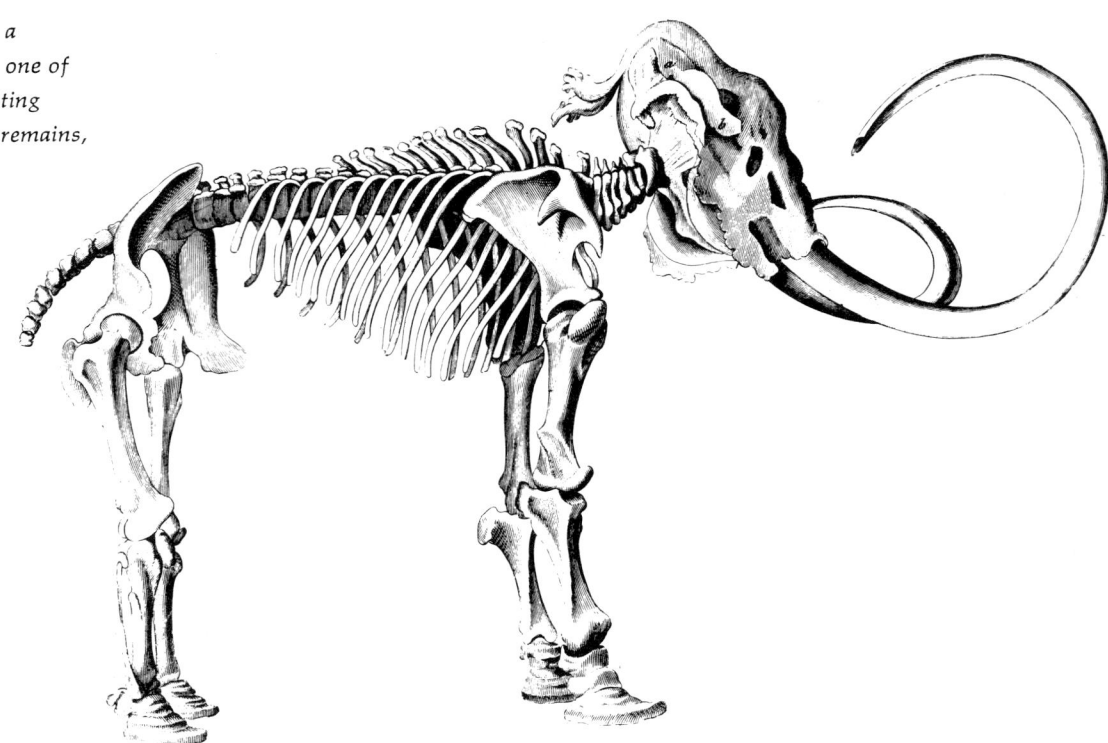

Catastrophism

Although the way was being prepared, the time was not yet ripe for a parallel revolution in biology. The dominating force in European science in the early nineteenth century was Georges Cuvier (1769–1832). Cuvier was the founder of paleontology, the scientific study of the fossil record. An expert in anatomy and zoology, he applied his special knowledge of the way in which animals are constructed to the study of fossil animals, and he was able to make brilliant deductions about the form of an entire animal from a few fragments of bone. We think of paleontology and evolution as so closely connected that it is surprising to learn that Cuvier was a staunch and powerful opponent of evolutionary theories. He recognized the fact that many species that had once existed no longer did. (In fact, according to modern estimates, considerably less than 1 percent of all species that have ever lived are represented on the Earth today.) He explained the extinction of species by postulating a series of catastrophes. After each catastrophe, the most recent of which was the Flood, the species that remained alive repopulated the world.

The proponents of catastrophism were of two schools of thought: the deluvianists held that all the great upheavals which had destroyed extinct species were floods, and the vulcanists believed that the world had periodically been inundated with lava. According to both theories, however, the fossil record is simply the remains of those species of once-living things which the violence of nature had eliminated. Time and nature, the catastrophists believed, act **not** to create but to eliminate.

The first scientist to work out a systematic theory of evolution was Jean Baptiste Lamarck (1744–1829). "This justly celebrated naturalist," as Darwin himself referred to him, boldly proposed in 1801 that all species, including man, are descended from other species. Lamarck, unlike most of the other zoologists of his time, was particularly interested in the one-celled organisms and other invertebrates (animals without backbones), and it is undoubtedly his long study of these "simpler" forms of life that led him to think of living things in terms of constantly increasing complexity, each form derived from an earlier, simpler form.

Like Cuvier and others, Lamarck noted that the older the rocks, the simpler the forms of life they contain, but unlike Cuvier, he interpreted this as meaning that the higher forms had risen from the simpler forms by a kind of progression. According to his hypothesis, this progression, or "evolution," to use the modern term, is dependent on two main forces. The first is the inheritance of acquired characteristics. Organs in animals become stronger or weaker, more or less important, through use or disuse, and these changes, according to Lamarck's theory, are transmitted from the parents to the progeny. His most famous example, and the one that Cuvier used most often to ridicule him, was that of the giraffe, which stretched its neck longer and longer to reach leaves on higher branches and transmitted this longer neck to its offspring, which again stretched its neck, and so on.

The second important factor in Lamarck's theory of evolution was a universal creative principle, an unconscious striving upward on the *Scala Naturae* that moved every living creature toward greater complexity. Every amoeba was on its way to man. Some might get waylaid—the orangutan, for instance, by being caught in an unfavorable environment had been diverted off its course—but the will was always present. Life in its simplest forms was constantly emerging by spontaneous generation to fill the void left at the bottom the ladder. In Lamarck's formulation, the ladder of life of the ancients had been transformed into a steadily ascending escalator powered by a universal will.

Lamarck's concept of the inheritance of acquired characteristics is an attractive one. Darwin borrowed from it consciously and unconsciously, and in fact, successive editions of *The Origin of Species* show that he was maneuvered into this particular Lamarckian point of view by some of his critics and by his own inability (owing to the primitive state of the science of genetics at that time) to explain some of the ways in which animals changed. Belief in the inheritance of acquired characteristics persisted into the twentieth century in the work of the Russian biologist Lysenko.

Lamarck's contemporaries did not object to his ideas about the inheritance of acquired characteristics, as we would today with our more advanced knowledge of the mechanics of inheritance, nor did they criticize his belief in a metaphysical force, which was actually a common element in many of the theories of the time. But these vague, unprovable postulates provided a very shaky foundation for so radical a proposal. Lamarck's championship of evolution was damaging not only to his own career but to the concept of evolution itself. The result of his theories was that both scientists and the public became even less prepared for an evolutionary doctrine.

I–6 *According to Lamarck's hypothesis, the necks of giraffes became longer when they stretched them to reach high branches, and this acquired characteristic was transmitted to their offspring.*

DARWIN'S THEORY

The Earth Has a History

The person who most influenced Darwin, it is generally agreed, was Charles Lyell (1797–1875), a geologist who was Darwin's senior by 10 years. One of the few books that Darwin took with him on his voyage was the first volume of Lyell's newly published *Principles of Geology*, and the second volume was sent to him while he was on the *Beagle*. On the basis of his own observations and those of his predecessors, Lyell opposed the theory of catastrophes. Instead, he produced new evidence in support of Hutton's earlier theory of uniformitarianism. According to Lyell, the slow, steady, and cumulative effect of natural forces had produced continuous change in the course of the Earth's history. Since this process is demonstrably slow, its results being barely visible in a single lifetime, it must have been going on for a very long time. In his earlier works, Lyell did not discuss the biological implications of his theory, but apparently they were clear to Darwin. If the Earth had a long continuous history and if no forces other than outside physical agencies were needed to explain the events as they were recorded in the geologic record, might not living organisms have had a similar history? What Darwin's theory needed, as he knew very well, was time, and it was time that Lyell gave him.

The Voyage of the Beagle

This, then, was the intellectual equipment with which Charles Darwin set sail from Devonport. As the *Beagle* moved down the Atlantic coast of South America, through the Straits of Magellan, and up the Pacific coast, Darwin traveled

I–7 The *Beagle*, *at anchor in Sydney Harbor. Only 90 feet in length, this "good little vessel" set sail on its five-year voyage with 74 people aboard. Darwin shared the poop cabin with a midshipman and 22 chronometers belonging to Captain FitzRoy, who had a passion for exactness. His sleeping space was so confined that he had to remove a drawer from a locker to make room for his feet.*

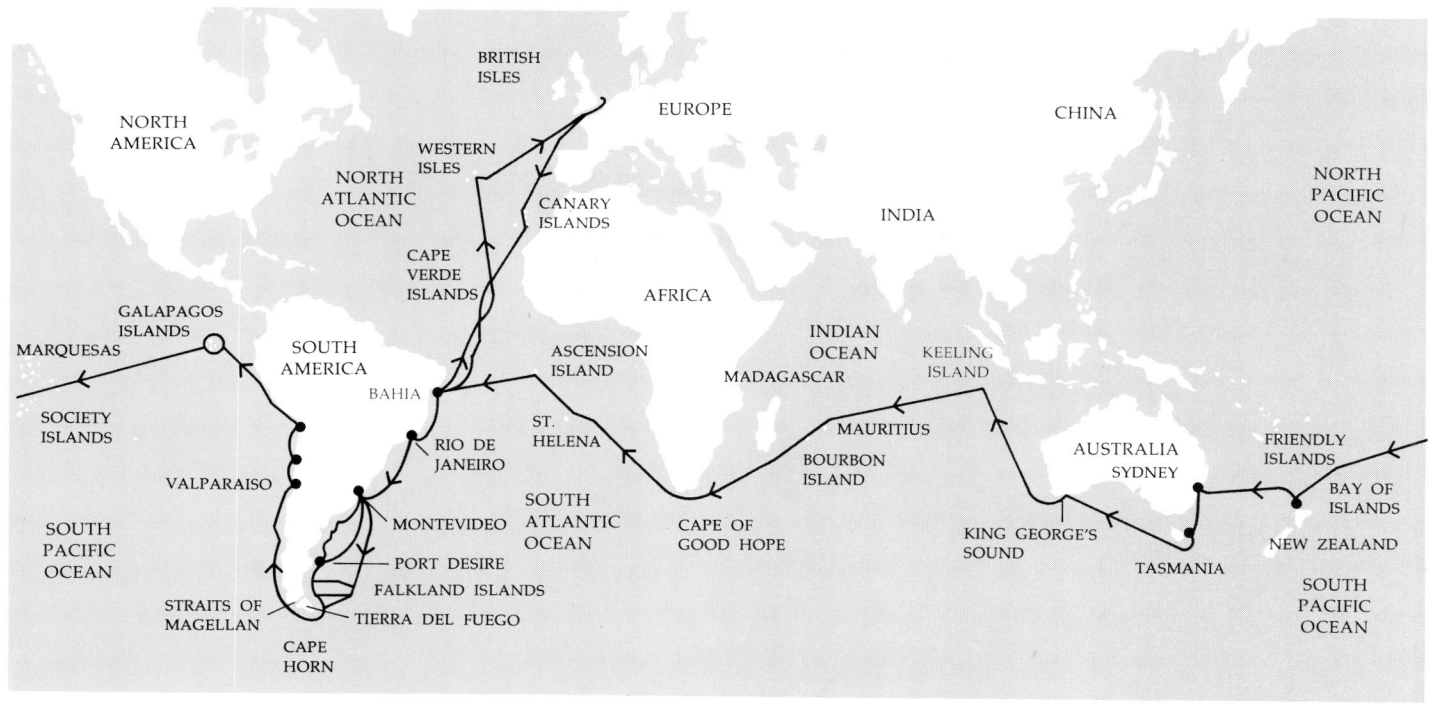

I–8 *The* Beagle's *voyage.*

the interior, fished, hunted, and rode horseback. He explored the rich fossil beds of South America (with the theories of Lyell fresh in his mind) and collected specimens of the many new kinds of plant and animal life he encountered. He was impressed most strongly during his long, slow trip down the coast and up again by the constantly changing varieties of organisms he encountered. The birds and other animals on the west coast, for example, were very different from those on the east coast, and even as he moved slowly up the western coast, one species would give way to another one.

Most interesting to Darwin were the animals and plants that inhabited a small barren group of islands, the Galapagos, that lie some 580 miles off the coast of Ecuador. The Galapagos were named after the islands' most striking inhabitants, the tortoises (*galápagos* in Spanish), some of which weigh two hundred pounds or more. Each island has its own type of tortoise; the fishermen who frequented the islands and hunted the tortoises for food could readily tell which island any particular tortoise had come from. Then there was a group of finchlike birds, 13 species in all, which differed from one another in body size, and particularly in the type of food they ate. In fact, although still clearly finches, they had taken on many characteristics seen only in completely different types of birds on the mainland. The woodpeckerlike finch, for example, had taken on the woodpecker's role in routing insects out of the bark of trees. It is not fully equipped for this, however, lacking the long tongue with which the woodpecker flicks out insects from under the bark. Instead, the woodpecker finch carries with it a small stick to pry the insects loose.

I–9 *A geyser, near Alcedo Volcano on Isabela, one of the Galapagos. The archipelago consists of 13 volcanic islands which pushed up from the sea more than a million years ago. Craters still covered with black basaltic lava rise to 3,000 or 4,000 feet. The major vegetation is a dreary grayish-brown thornbush, making up miles and miles of dense leafless thicket, and a few tall tree cactuses—"what we might imagine the cultivated parts of the Infernal regions to be," young Charles Darwin wrote in his diary.*

From his knowledge of geology. Darwin knew that these islands, clearly of volcanic origin, were much younger than the mainland. Yet the plants and animals of the islands were different from those of the mainland, and in fact the inhabitants of different islands in the archipelago differed from one another. Were the living things on each island the product of a separate special creation? "One might really fancy," Darwin mused at a later date, "that from an original paucity of birds in this archipelago one species had been taken and modified for different ends." For years after his return, this problem continued, in his own word, to "haunt" him.

Development of the Theory

Not long after Darwin's return, he came across a book by the Reverend Thomas Malthus that had first appeared in 1798. In this book, Malthus warned, as economists have warned frequently ever since, that the human population was increasing so rapidly that it not only would soon outstrip the food supply but would leave "standing room only" on Earth. Darwin saw that this conclusion—that food supply and other factors hold populations in check—is true for all species, not just the human one. For example, a single breeding pair of elephants, which are the slowest breeders of all animals, could produce 19 million elephants in 750 years, yet the average number of elephants generally remains the same over the years. Where there might have been 19 million elephants in theory, there are, in fact, only two. The process by which the two survivors are "chosen" was termed by Darwin *natural selection*. He saw it as a process analogous to the type of selection exercised by breeders of cattle, horses, or dogs—which, as a country squire, he was very familiar with. In the case of artificial selection, man chooses variants for breeding on the basis of characteristics which seem to him to be desirable. In the case of natural selection, environmental conditions are the principal active force, operating on the variations continually produced in all species to "favor" some variants and "discourage" or eliminate others.

Where do the variations come from? According to Darwin's theory, variations occur absolutely at random. They are not produced by the environment, by a "creative force," or by the unconscious striving of the organism. In themselves, they have no direction; *direction is imposed entirely by natural selection*. A variation that gives an animal even a slight advantage makes that animal more likely to leave surviving offspring. Thus, to return to Lamarck's giraffe, an animal with a slightly longer neck has an advantage in feeding and so is apt to leave more offspring than one with a shorter neck. If the longer neck is an inherited trait, some of these offspring will also have long necks, and since the animals with longer necks are always favored, the next generation will have even longer necks. Finally, the population of short-necked giraffes will give way to a population of long-necked ones.

As you can see, the essential difference between Darwin's formulation and that of any of his predecessors is the central role he gave to the process of variation. Others had thought of variations as mere disturbances in the overall design, whereas Darwin saw that variations are the real fabric of the evolutionary process. Species arise, he saw, when differences among individuals within a group are gradually converted into differences between groups as the groups become separated in space and time.

On the Origin of Species, which Darwin pondered for more than 20 years before its publication in 1859, is, in his own words, "one long argument." No experiments are performed. No new information is revealed. Fact after fact, observation after observation, culled from the most remote Pacific island to a neighbor's pasture, is recorded, analyzed, and commented upon. Every objection is anticipated and countered. Because the process of evolution is so slow, Darwin did not believe that direct proof of his theory was possible. However, as we shall see, the twentieth century has produced clear evidence of evolution in progress. No scientist now doubts that species have originated in the past and are still originating, that species have become extinct in the past and are still becoming extinct, and that all living things today have an ancestral species in the past.

Importance of the Theory in Modern Biology

What is the importance of this theory to modern biologists, many of whom are concerned with such areas of investigation as the chemistry of heredity—a phrase that would have been meaningless in Darwin's time—or the interpretation of subcellular structures newly revealed by the electron microscope or the tracing of a radioactive substance through the components of an ecosystem? In the words of Ernst Mayr of Harvard University:*

The theory of evolution is quite rightly called the greatest unifying theory in biology. The diversity of organisms, similarities and differences between kinds of organisms, patterns of distribution and behavior, adaptation and interaction, all this was merely a bewildering chaos of facts until given meaning by the evolutionary theory. There is no area of biology in which that theory has not served as an ordering principle.

Because they are of particular importance in this book, we shall single out for mention three of the many ways in which Darwin's theory has "served as an ordering principle" in biology. The first involves the kinds of living organisms that fill the world about us. On the basis of evolutionary theory, we can understand, as Linnaeus could not, both the resemblances and differences among the various groups of plants and animals and other living things and correlate these differences, to some extent, with where and how these organisms live and with their history.

Second, we begin to understand the seeming purposefulness of living things and their activities. Why do flowers often have sweet odors and bright colors? Why does our heart pump blood through our lungs? If we answer that the heart pumps blood through our lungs in order to oxygenate it, we imply that the heart has a purpose, that it knows what it is doing—which is, of course, not true. Yet in another sense, this sort of explanation is the correct one. Flowers and their insect pollinators evolved together; brightly colored flowers were more apt to be pollinated and, therefore, left more offspring. As you can see, within the framework of evolutionary theory, "why" and "what for" questions become interesting and meaningful.

The third point, closely related to the first two, is that evolutionary theory emphasizes the dynamic relationship between structure and function. Anatomy,

* Ernst Mayr, *Animal Species and Evolution*, Harvard University Press, Cambridge, Mass., 1963.

I–10 *Darwin accurately predicted the discovery of an insect with a tongue 11 inches long in Madagascar because this length would be required to reach the nectar of a species of orchid that blooms there. A similar, though slightly less startling, relationship exists between the spinx moth and the tobacco blossom, shown here.*

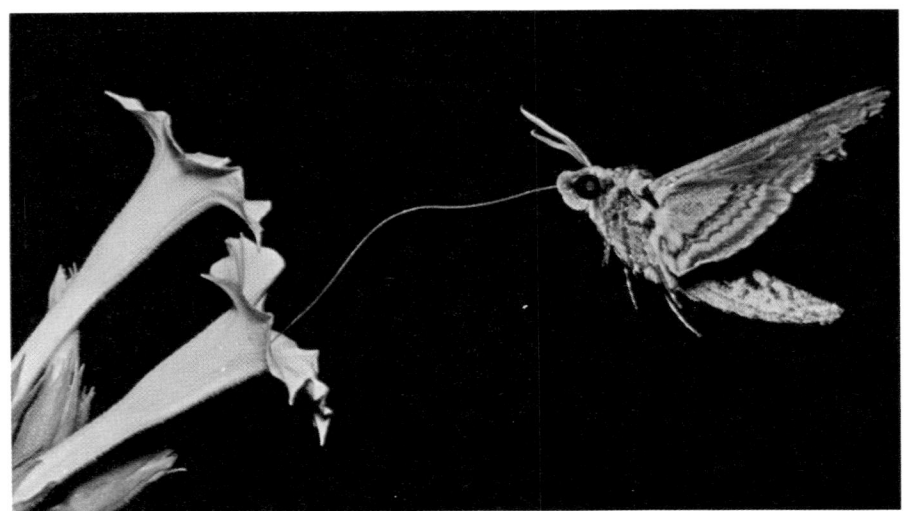

on the one hand, and physiology or behavior, on the other, cannot logically be studied or understood apart from one another. The two have necessarily evolved together, the function shaping the structure and the structure directing and guiding the activity.

As you will see, we can say with equal truthfulness that man can use his hands for grasping because he has an opposable thumb and that man has an opposable thumb because he uses his hands for grasping. This does not imply perfection; the wisdom of nature is a sentimental notion. Your body will automatically reject a skin graft from another person—unless the person happens to be your identical twin—even if you are dying of an extensive burn and the graft could save your life. Evolution has imposed a number of clumsy burdens. The human vertebral column is a good example. An engineer could devise a better structural support for modern man if he were permitted to start from the beginning. The point is, however, that our backbone and our pelvis have been molded by time—by variation and natural selection—into their present form, and their structure and function are inseparable.

LEVELS OF ORGANIZATION

As we look at evolution through a modern perspective, we can see it not merely as a slow pattern of gradual change but also as a process of ever-increasing organization.

During the seventeenth century, a school of biologists called mechanists arose, of which the French philosopher René Descartes (1596–1650) was an exponent. The mechanists set about proving that the body worked essentially like a machine; the arms and legs moved like levers and pulleys, the heart like a pump, the lungs like a bellows, and the stomach like a mortar and pestle. Opposing them was a group known as the vitalists, who contended that living things were qualitatively different from inanimate objects, containing within them a "vital spirit" which enabled them to perform activities that could not be carried out outside of the living organism or fully described by chemistry or physics.

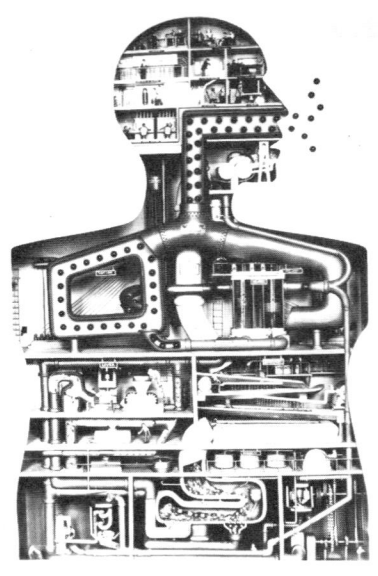

I–11 *Mechanistic view of the human body.*

By the nineteenth century, simple mechanical models of living organisms had been abandoned, and the argument now centered around whether or not the chemistry of living organisms was governed by the same principles as the chemistry performed in the laboratory by man. The vitalists claimed that the chemical operations performed by living tissues could not be carried out experimentally in the laboratory. The reductionists, as their opponents were now called (since they believed that the complex operations of living systems could be reduced to simpler and more readily understandable ones), achieved a partial victory when the German chemist Friedrich Wöhler (1800–1882) converted an "inorganic" substance (ammonium cyanate) into a familiar organic substance (urea). The claims of the vitalists were supported by the fact that as chemical knowledge improved, many new compounds were found in living tissues that were never seen in the nonliving, or inorganic, world. In spite of many advances in chemistry, this phase of the controversy lasted until 1898. Until this time, the chemical processes involved in turning fruit juice to alcohol had been carried out only in the presence of living yeast cells. Finally, however, the German chemist Eduard Büchner (1860–1917) showed that a substance extracted from the yeast cells could produce fermentation outside of the living cell. This substance was given the name enzyme, from *zyme*, the Greek word meaning "yeast" or "ferment."

Today it is clear that the vitalists were wrong; the chemistry of living systems "obeys the rules" of inorganic chemistry. The mechanists were wrong, too, of course—the body is not merely a simple machine—just as the reductionists were somewhat foolish when they claimed in "Believe It or Not" columns that the human body is nothing more than 98 cents worth of chemicals. It is, of course, much more, as water is more than its component parts—hydrogen and oxygen atoms do not have any of the special properties by which we identify water. What then is the difference between 98 cents worth of chemicals and the human body, or between free hydrogen and oxygen, and water? The difference

I–12 *The laboratory at Giessen where Wöhler was a student. This laboratory, which belonged to Wöhler's friend and teacher Justus von Liebig, was one of the first where practical work in chemistry could be done. The drawing was made in 1842.*

is one of organization. The long, slow movement of evolution can be described as the gradual imposition of order and of organization. And, as is the case with hydrogen, oxygen, and water, when we move from one level of organization to another, we move to something qualitatively different. An atom is one level of organization, a molecule another.

We know now that living things are made up of a relatively few elements, all of which can be found in simple forms in the thin film of earth and atmosphere in which we live. But when these elements become organized into molecules, we have an entity with properties quite different from those of the elements alone. Similarly, although cells are composed of "nothing but" molecules, they have properties very different from the molecules. We progress from single cells to many-celled organisms. This occurred in the course of evolution, and it occurs over and over again as each many-celled plant or animal develops from a single cell, the fertilized egg, and in so doing becomes something more than and different from the egg. Finally, the organisms, in turn, are organized into populations, communities and societies, in which the behavior of any individual organism is not the same as the behavior of the organism in isolation. All these different levels of organization, from cells to societies, are the subjects of biology. In this book, we shall move along the various levels of organization, beginning with atoms and molecules—as life began—then examine the living cell and the processes that take place within it. Next we shall look at the great variety of organized groups of cells, organisms, including plants and animals, with an emphasis on man. Finally, we shall look at communities of organisms and try to understand man's place in the complex web of life.

THE NATURE OF SCIENCE

Before we start to examine the nature of living systems, however, let us take a brief look at the nature of science itself, which will tell us something about the methods we shall be using in our observations. Science is a way of seeking principles of order in the natural world. (Art is another, and religion yet another.) Science has two separate components: (1) objective evidence based on observation or experiment or a combination of the two—the data—and (2) the structuring of data by making meaningful connections among them. The great discoveries in science are not the addition of new facts but the perceiving of new relationships among them, the revelation of form, order, and meaning among the seeming chaos of facts.

However, this is not to say that the facts are not important. It is the facts that endure, that are passed from one worker to the next, from one scientific generation to its successor. It is for this reason that scientists stress objectivity; in all science, observations and experiments must always be reported in such a way that they can be repeated and verified, and must always be repeated and verified before they are incorporated into the body of knowledge. The data are the unyielding building blocks, the structural elements, stubborn and concrete, with which theories are erected and against which dreams are shattered.

Most texts, and this one is no exception, tend to stress what is known at the present time, rather than what is not known or how we came to know what we do. This tendency, although understandable, somewhat distorts the nature of biology, and indeed of science in general. A modern science is not a static accumula-

Indeed, scientists are in the position of a primitive tribe which has undertaken to duplicate the Empire State Building, room for room, without ever seeing the original building or even a photograph. Their own working plans, of necessity, are only a crude approximation of the real thing, conceived on the basis of miscellaneous reports volunteered by interested travelers and often in apparent conflict on points of detail. In order to start the building at all, some information must be ignored as erroneous or impossible, and the first constructions are little more than large grass shacks. Increasing sophistication, combined with methodical accumulation of data, make it necessary to tear down the earlier replicas (each time after violent arguments), replacing them successively with more up-to-date versions. We may easily doubt that the version current after only 300 years of effort is a very adequate restoration of the Empire State Building; yet, in the absence of clear knowledge to the contrary, the tribe must regard it as such (and ignore odd travelers' tales that cannot be made to fit).

E. J. DuPraw: Cell and Molecular Biology,
Academic Press, Inc., New York, 1968

tion of facts organized in a particular way, but a somewhat amorphous body of knowledge that not only constantly grows, developing new bulges and unpredictable appendages, but also may suddenly change its entire shape (as biology did in the nineteenth century with the acceptance of the theory of evolution). Science is dynamic, not static; consequently it cannot be contained within textbooks or libraries or information retrieval centers, but rather it is a process taking place in the minds of living scientists.

One final word: In our enthusiasm for telling you all that biology has discovered, do not let us convince you that all is known. Many questions are still unanswered. More important, many good questions have not yet been asked. Perhaps you may be the one to ask them.

QUESTIONS

1. What is the essential difference between Darwin's theory of evolution and that of Lamarck?
2. The chief predator of an English species of snail is the song thrush. Snails that inhabit woodland floors have dark shells, whereas those that live on grass have yellow shells, which are less clearly visible against the lighter background. Explain, in terms of Darwinian principles.
3. The phrase "chance and necessity" has been applied to the process by which species develop. Relate this to the fact that snails living on grass do not have green shells but there are, for example, green frogs and green insects.
4. In general terms, how would a modern reductionist explain love, religion, the creation of a great work of art? Do you consider yourself a vitalist or a reductionist? Would it be possible to have a science based on a form of vitalism?

SUGGESTIONS FOR FURTHER READING

BAKER, JEFFREY J. W., and GARLAND E. ALLEN: *Hypothesis, Prediction, and Implication in Biology*, Addison-Wesley Publishing Company, Inc., Reading, Mass., 1971.*

An introducion, with interesting examples, of the nature of scientific inquiry in the biological sciences.

DARWIN, CHARLES: *The Origin of Species by Means of Natural Selection, or The Preservation of Favored Races in the Struggle for Life,* Doubleday & Company, Inc., Garden City, N.Y., 1960.*

Darwin's "long argument." Every student of biology should, at the very least, browse through this book to catch its special flavor and to begin to understand its extraordinary force.

DARWIN, CHARLES: *The Voyage of the Beagle,* Natural History Library, Doubleday & Company, Inc., Garden City, N.Y., 1962.*

Darwin's own chronicle of the expedition on which he made the discoveries and observations that eventually led him to his theory of evolution. The sensitive, eager young Darwin that emerges from these pages is very unlike the image many of us have formed of him from his later portraits.

MOOREHEAD, ALAN: *Darwin and the Beagle,* Harper & Row, Publishers, Inc., New York, 1969.

A delightful narrative of Darwin's journey, beautifully illustrated with contemporary or near contemporary drawings, paintings, and lithographs.

TOULMIN, STEPHEN, and JUNE GOODFIELD: *The Discovery of Time,* Harper & Row, Publishers, Inc., New York, 1965.*

The historical development of our concepts of time as they relate to nature, human nature, and human society.

* Available in paperback.

PART I

Cells

SECTION 1

THE ORGANIZATION OF LIVING THINGS

Chapter 1

Life: Its Origin and Definition

Sagittarius, the archer, is a constellation that is visible in our summer skies. If you look toward Sagittarius on a clear moonless night, you are looking toward the center of the Milky Way, the galaxy of which our planet is a part. There are 250 billion stars in our own galaxy, and within the range of our most powerful telescopes there are about 10 billion more such galaxies.

The Milky Way was perhaps 10 billion years old, the astronomers calculate, when the star that is our sun came into being. According to current hypotheses, it formed, like other stars, from a condensation of particles of dust and hydrogen and helium gases whirling in space among the older stars. Now there are fewer such clouds of particles and gases in the Milky Way, but you can still sometimes see them as dark patches in the starlit night.

The immense cloud that was to become the sun condensed gradually as the hydrogen and helium atoms were pulled toward one another by the force of gravity, falling into the center of the cloud and gathering speed as they fell. As the cluster grew denser, the atoms moved faster and faster. More and more atoms collided with each other, and the gas in the cloud became hotter and hotter. The collisions became increasingly violent as the temperature rose, until the hydrogen atoms collided with such force that their nuclei fused, forming helium and releasing nuclear energy (in a process like the explosive reaction in an H-bomb). Energy from this thermonuclear reaction, still going on at the heart of the sun, is radiated from its glowing surface. It is this energy, captured in the cells of green plants, on which all life on Earth depends.

The planets, according to current theory, formed from the remaining gas and dust moving around the new-formed star. At first, in the planets-to-be, particles were collected merely at random, but as each mass grew larger, other particles began to be attracted by the gravity of the largest masses. The whirling dust and forming spheres continued to revolve around the sun until finally each planet had swept its own path clean, picking up loose matter like a giant snowball. The orbit nearest the sun was swept clean by Mercury, the next by Venus, the third by Earth, the fourth by Mars, and so on out to Neptune and Pluto, the most distant of the planets. The solar system, including Earth, is calculated to have come into being about 4.5 billion years ago.

1-1 *Our sun is near the outer edge of a disk-shaped galaxy, the Milky Way, containing 250 billion stars. These stars, like the sun, formed from nebulae— clouds of gas and dust—such as this nebula in the constellation Sagittarius.*

1–2 *Photograph of Earth taken from a NASA satellite 22,000 miles out in space. The continent of South America is visible, as are portions of North America and Africa. Antarctica is blanketed under a heavy cloud cover. The clouds that stretch over central North America from the Great Lakes to Mexico represent a cold front moving eastward.*

THE COMPOSITION OF THE EARTH

During the time Earth and the other planets were being formed, the release of energy from radioactive materials kept their interiors very hot. When Earth was still so hot that it was mostly liquid, the heavier materials collected in a dense core whose diameter is about half that of the planet. As soon as the supply of stellar dust, stones, and larger rocks was exhausted, the planet ceased to grow and began to cool. As Earth's surface cooled, an outer crust, a skin as thin by comparison as the skin of an apple, was formed. The oldest known rocks are about 3.98 billions years old.*

Between the thin crust of the Earth and its dense interior is a thick layer, known as the <u>mantle</u>, made of hot compressed rock. The mantle is slightly plastic, so it "gives" a little. The portions of the mantle on which the continents rest are depressed slightly under their weight. Only 30 miles below us, the Earth is still hot—a small fraction of it is even still molten. We see evidence of this in the occasional volcanic eruption that forces lava, which is molten rock, through weak points in the Earth's skin or in the geyser which spews up boiling water that has trickled down to the Earth's interior. Very little is known about what lies beneath the surface of the Earth.

The biosphere is the part of the planet within which life exists. It forms a thin film on the crustal layer, extending about 5 or 6 miles up into the atmosphere and about equally far down into the depths of the sea.

* The process by which these dates are estimated is described on page 39.

1–3 *Because man has never been able to penetrate more than a small fraction of the Earth's crust, theories about the interior of our planet are based largely upon the analysis of how earthquake (seismic) waves travel. According to these data, the outer layer, the crust, is rarely more than 40 miles thick. It is in the form of plates on which the continents rest. The boundary between the crust and mantle is called the moho, after Andrija Mohorovičić, the Yugoslav geologist who was its discoverer. The main ingredient of the core is iron. The inner core is compressed solid by the immense weight of all the materials above it. At the bottom of the mantle, for example, where the temperature is 7,000°F, the pressure is believed to be about 17 million pounds per square inch.*

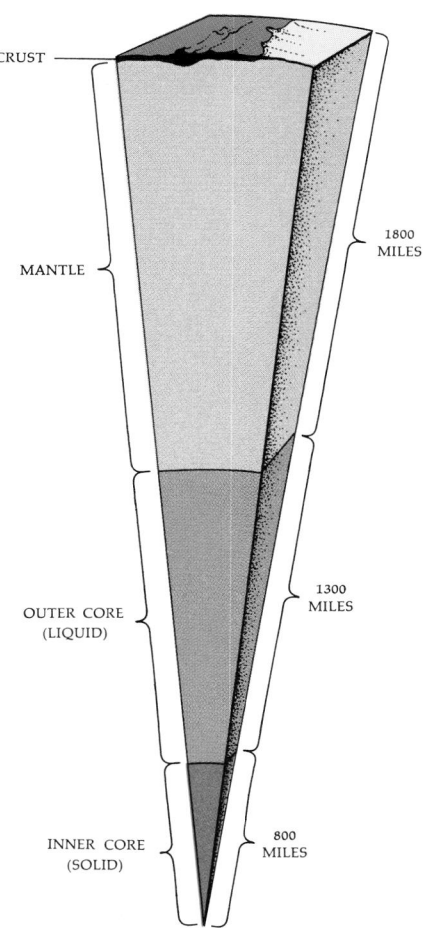

CRUST

MANTLE

1800 MILES

OUTER CORE (LIQUID)

1300 MILES

INNER CORE (SOLID)

800 MILES

WHY ON EARTH?

In our solar system, Earth among all the planets is most favored for the production of life. For one thing, Earth is neither too close nor too distant from the sun. At very low temperatures, the chemical reactions on which life—any form of life—depends must virtually cease. At high temperatures, compounds are too unstable for life to form or survive.

Earth's size and density are also greatly in its favor. Planets much smaller than Earth do not have enough gravitational pull to hold an adequate atmosphere, and any planet much larger than Earth may hold so dense an atmosphere that radiations from the sun cannot reach its surface.

As recently as the late 1960s, serious students of exobiology (the science of life outside our own planet) were pessimistic about the discovery of life within our solar system. However, the wealth of data obtained from the Mariner 9 spacecraft, which made nearly 700 orbits around Mars in the early 1970s, has reawakened interest in the possibility of finding some primitive, perhaps dormant form of life on that planet. It seems likely that some of the canals and other surface irregularities that have long intrigued observers of the red planet were probably made by liquid water and in the not too distant past. The frozen polar caps, once thought to be carbon dioxide, are now believed to be largely water, and several cloud systems containing water vapor were detected in the Martian atmosphere. The Viking spacecraft is scheduled to land on Mars in 1976 and collect samples that may settle the question of the presence of life on that planet.

Venus is another possibility, although a more remote one. The Soviet space probes of 1969 reported temperatures of 400° to 530°C on Venus's surface, but the planet has a very dense atmosphere, some 15 miles thick (mostly carbon dioxide) and temperatures in regions of the Venusian atmosphere are likely to be more moderate. Jupiter is surrounded by a film of clouds of ammonia crystals ranging from −140°C on the outer surface to about +2,000°C at the innermost surface. In between is a layer of water droplets at temperatures at which biochemical reactions could occur. Most recently (at this writing), as-

1–4 *Mars, the red planet, as seen from Earth, shows bright and dark areas and white polar caps, now believed to be largely water. Surface features change with seasons.*

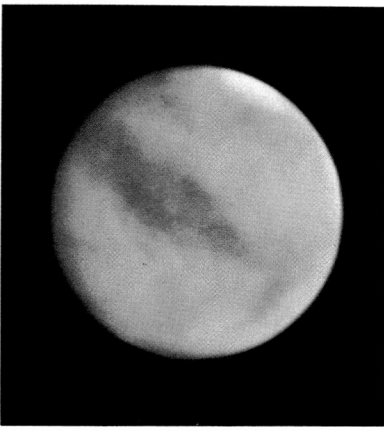

tronomers report that Titan, the largest of the 10 moons of the giant planet Saturn, probably has atmospheric conditions similar to those of the primitive Earth. Two unmanned spacecrafts will fly by Jupiter in 1977 and on out past Saturn in 1981, passing perhaps within 100 miles of Titan.

Some biologists contend that given certain conditions—such as an energy source, water, and a temperature range in which the water can exist in liquid form—plus a long enough time, the evolution of some forms of life is inevitable. The discovery on another planet of a living organism, no matter how primitive, whose origin was independent of Earth, would strongly support this hypothesis. There are approximately 10^{20} (the number 1 followed by 20 zeroes) stars like our own sun in the universe, which can provide energy for living things. At least 10 percent of these, according to astronomers, are likely to be surrounded by planetary systems such as our own. If only 1 percent were to have planets with environments roughly similar to those of Earth, that would offer some 10^{18} possibilities for the existence of life in other solar systems.

Until comparatively recently, astronomers believed that the stars and planets that surround the Earth in the vastness of space whirled in orbit around us. As a result of the revolution in astronomy of the sixteenth and seventeenth centuries (Copernicus, Kepler, Galileo, and Newton), mankind learned that Earth was not the center of the universe, nor indeed even of our own solar system. Not long after, in the nineteenth century, with the acceptance of Darwin's theory, we also learned to accept the fact that man is the product of the same blind evolutionary process as other species and not, as far as science can show, created for any special purpose or as part of any universal design. We now appear to be on the verge of discovering that the presence on this planet of living systems, a phenomenon which once seemed so special and unique, is perhaps only one of a billion such fleeting experiments now under way on so many distant planets. And we have no reason not to believe that some of the

Table 1–1   *Some Data on the Planets*

Planet	Sun distance (miles)	(ratio)	Diameter (miles)	(ratio)	Mass (m/m_e)	Rotation period (days)	Revolution period (years)	Surface temperature (°C)
Mercury	0.36×10^8	0.39	3,170	0.40	0.055	59	0.24	340
Venus	0.68×10^8	0.73	7,750	0.98	0.81	243	0.62	530
Earth	0.93×10^8	1.00	7,910	1.00	1.00	1.00	1.00	15
Mars	1.42×10^8	1.53	4,270	0.54	0.11	1.02	1.88	− 20
Jupiter	$4.8 \ \times 10^8$	5.2	89,000	11.3	314	0.41	11.9	− 140*
Saturn	$8.9 \ \times 10^8$	9.6	74,000	9.4	94	0.43	29.5	− 180*
Uranus	$17.8 \ \times 10^8$	19.2	33,000	4.2	14	0.47	84	− 210*
Neptune	$28 \ \ \times 10^8$	30	31,000	3.9	17	0.60	165	− 220*
Pluto	$37 \ \ \times 10^8$	40	∼ 4,000	∼ 0.5	< 0.2	6.6	248	− 230(?)

* These temperatures apply to some level in a thick atmosphere rather than to some surface.

1–5 (a) *Nicolas Copernicus, in* De Revolutionibus, *published in 1543 (the year of his death), set forth the new concept that the sun, not the Earth, is the center of the solar system. His theory was supported by the German astronomer Johannes Kepler (1571–1630), who discovered the laws of planetary motion, and by the Italian Galileo Galilei (1564–1642). The latter spent the last ten years of his life confined to his home for heresy because of his advocacy of Copernican beliefs. (b) Building on the work of Galileo, who had studied the movements of falling bodies, Isaac Newton (1642–1727) was the first to describe the interrelations of force and motion, in what are now known as Newton's laws of motion. It was Newton who formulated the concepts of momentum, force, and mass, invented the mathematical methods of calculus, demonstrated that white light contains all the colors of the rainbow, and proposed the theory of universal gravitation, thus explaining the attractive forces among objects in space. Generally considered the greatest genius that ever existed, he completed work on most of his greatest contributions by the time he was 30.*

(a)

(b)

living systems are more complex, more intelligent, and more technologically advanced than ourselves. Thus an increase in man's knowledge has brought about a precipitous decline in his sense of his own significance.

THE ORIGIN OF LIFE ON EARTH

Sometime between the time Earth formed and the date of the earliest fossils discovered so far—an interval of about a billion years—life began. The chief raw materials for life were to be found in the atmosphere of the young Earth. The principal component of the sun and hence of the solar system is hydrogen, and Earth, when it formed, was probably surrounded by a cloud of hydrogen gas. Oxygen was present mainly in the form of water (H_2O). Under conditions such as an abundance of hydrogen, the available nitrogen and carbon and oxygen atoms also present would tend to combine with hydrogen to form ammonia (NH_3), methane (CH_4) and other carbon–hydrogen gases, and water. These were the raw materials for living systems. Combinations of these four elements available in the primitive atmosphere—hydrogen, oxygen, carbon, and nitrogen—make up more than 95 percent of the tissues of all living things.

In order to break apart the simple gases of the atmosphere and reform them into organic* molecules, energy was required. And energy abounded on the

* The complex molecules of living things are generally referred to as organic molecules. The term "organic chemistry" was first applied to substances made by plants and animals, and by extension it came to include all compounds that contain both carbon and hydrogen, since these are the chief constituents of such compounds. Now many such compounds are made routinely in the laboratory, and some of them, like synthetic fibers and plastics, are never found in organisms. The laboratories in which these compounds are made are still referred to as laboratories of organic chemistry. The term "organic molecules" will be used in this book to mean the relatively complex carbon-containing molecules found in living systems.

young Earth. First there was heat, both boiling (moist) heat and baking (dry) heat. Water vapor spewed out of the primitive seas, cooled in the upper atmosphere, collected into clouds, fell back on the crust of the Earth, and steamed up again. Violent rainstorms were accompanied by lightning, which provided electrical energy. The sun bombarded the Earth's surface with high-energy particles and ultraviolet light. Radioactive elements within the Earth released their energy into the atmosphere. These conditions can be simulated in the laboratory, and scientists have now show that under such conditions, organic molecules are produced. Among these molecules are some of the amino acids, the important building blocks of proteins, which are basic components of living matter.

In the next step in the sequence of events which led to life, as biochemists reconstruct the events, these compounds were washed out of the atmosphere by the driving rains and began to collect in the oceans, which grew larger as the Earth cooled. As a consequence of winds and tides, increasingly rich mixtures of organic molecules would have concentrated in certain areas of the ocean. Some organic molecules have a tendency to aggregate in groups; in the primitive ocean, these groups probably took the form of droplets, similar to the droplets formed by oil in water. Such droplets may have been the forerunners of the first primitive cells.

1–6   *Bolts of lightning in the steam boiling up from a volcanic crater. Such sources of energy would have been present on the primitive Earth and might have contributed to the formation of organic molecules. This photograph, taken in 1963, shows the birth of the island of Surtsey off the coast of Iceland.*

1–7 *Conditions believed to exist on the primitive Earth are simulated in this apparatus. Methane and ammonia are continuously circulated between a lower "ocean," which is heated, and an upper "atmosphere," through which an electric discharge is transmitted. At the end of 24 hours, 45 percent of the carbon originally present in the methane gas is converted to amino acids, the building blocks of proteins, and other organic molecules.*

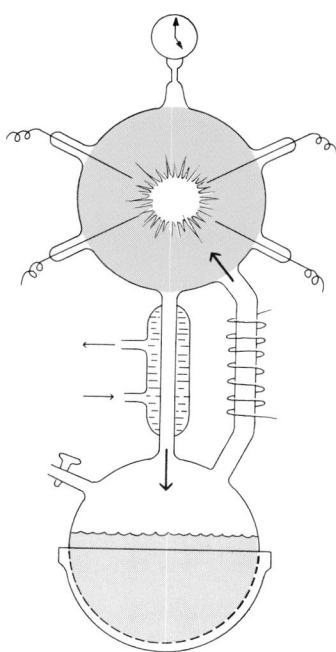

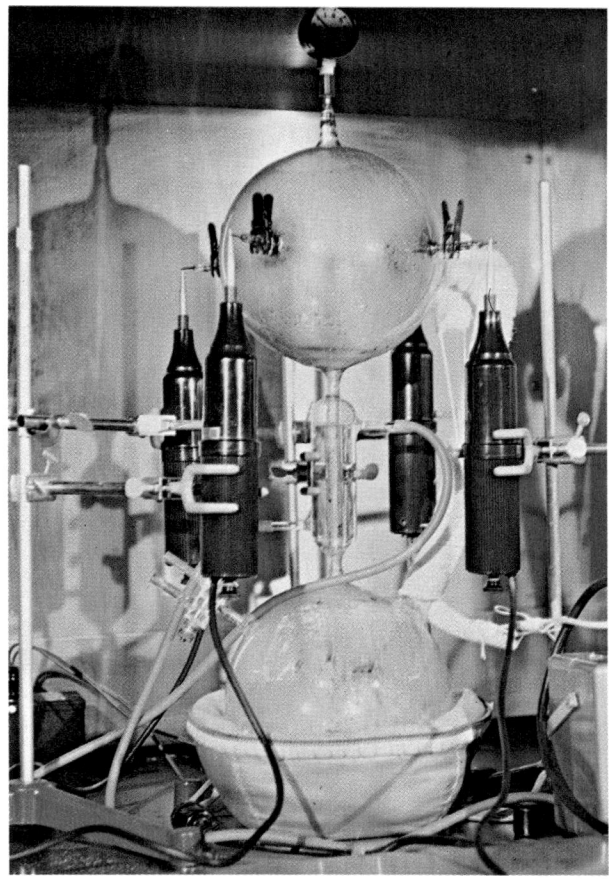

The exact moment at which living systems first came into being is not known. The earliest fossils found so far, which are about 3.5 billion years old, are sufficiently complex so that it is clear that some little aggregation of chemicals moved through that twilight zone separating the living and nonliving some millions of years before.

We have, however, a tantalizing clue about the origin of life. One of the outstanding capacities of living systems is self-replication, the ability to make copies of themselves. As we shall see in Section 2, every single living thing now present on this planet possesses the same basic self-replicating mechanisms, the same molecule (DNA), organized in the same way. Does this mean that the process of self-replication developed only once? Did all life as it now exists begin only once? Was it some single chance event? Does everything that is—or was ever—alive have the same single molecule for an ancestor? Or did the same system come into being many different times and in many different places?

Perhaps several forms of replicating systems once existed but one was superior to the others and so continued at their expense. Here again, the discovery of living systems on another planet might illuminate unresolved questions about our own.

1-8 Fossil algae 2 billion years old (a) and
living algae (b). Note the similarities in
appearance. Interspersed among the living algae are chains of rod-shaped
bacteria.

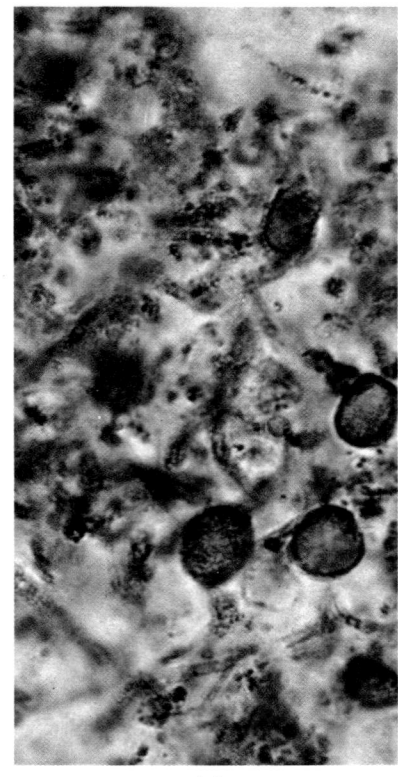
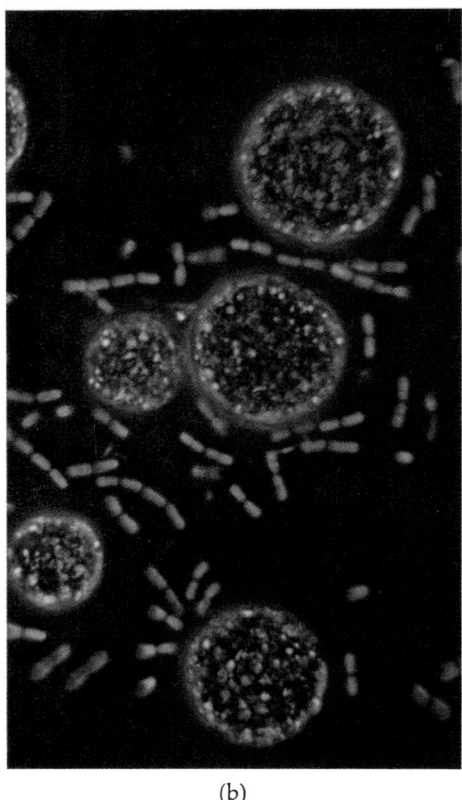

(a)   (b)

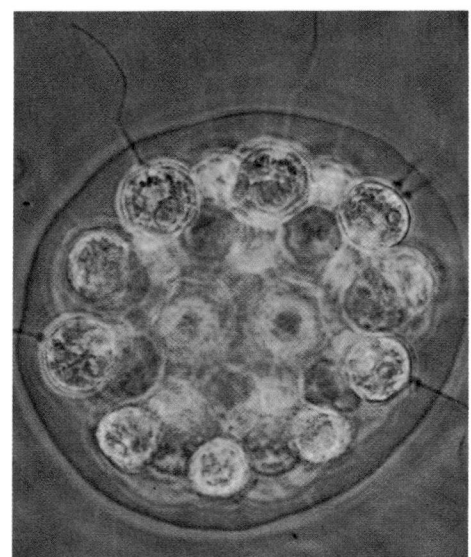

1-9 A modern autotroph. This simple organism, called Pandorina, is made up of
32 cells, many of which are visible here,
held together by a jellylike substance,
whose outlines you can also see. Each of
these cells is photosynthetic and can
survive independent of the others. Each
cell has two flagella (protruding whiplike
structures), several of which are visible
in this micrograph. The flagella all point
outward and their beat is coordinated
so that the colony rolls through the water
like a ball.

THE EARLIEST FORMS OF LIVING SYSTEMS

The earliest fossils are representatives of two types of living organisms—bacteria and blue-green algae. Both are single cells. As cells, they are surrounded
by an outer membrane that separates them from their external environment.
Such very simple cells are known as *prokaryotes* (from *pro* meaning "before"
and *karyote* meaning "nucleus"). Larger, more complex cells have a nucleus separated from the rest of the cell interior by a special membrane; such cells
are known as *eukaryotes* (*eu* meaning "true"). Eukaryotes also characteristically
have a number of specialized structures (organelles) not present in prokaryotes.
The cells that make up the body tissues of plants and animals are all eukaryotic,
as are many of the one-celled organisms. Prokaryotes still exist in abundance in
the form of modern bacteria and blue-green algae, not fundamentally different
from their fossil ancestors.

The fossil bacterial cell differs from the fossil algal cell in that the algal cell,
on the basis of comparisons with its modern counterparts, clearly appears to
have been photosynthetic. In other words, it could synthesize its own energy-
containing foodstuffs from simple elements (in this case, carbon dioxide and
water) using the light of the sun as a power source. This kind of an organism
is known as an *autotroph*, or "self-feeder."

Most modern bacteria require an outside source of organic compounds. This
kind of organism is known as a *heterotroph*, "other feeder."

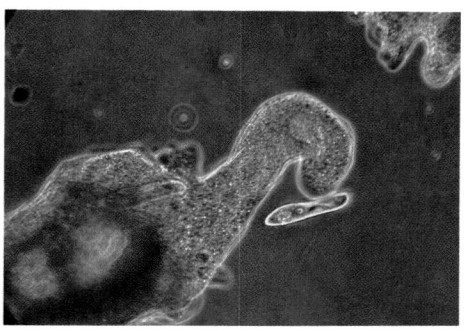

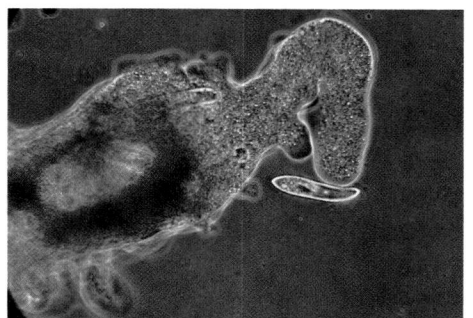

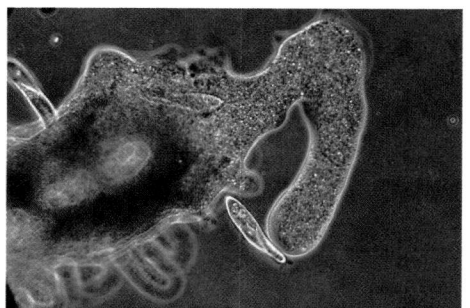

These oldest fossils are approximately the same age, so there is no way of telling which type came first. It might seem logical to assume that the self-feeders, autotrophs, were first to evolve, but photosynthesis is such a complex process (as you will see in Chapter 9) that most biologists now believe that photosynthetic organisms represent a more advanced form of life. As the past is now reconstructed, the first primitive cells or cell-like structures were probably heterotrophs. As we noted earlier, a heterotrophic organism is one that cannot make all the complex organic molecules needed by living things from simple molecules like H_2O, CO_2, and NH_3 as autotrophs can, but must receive them from other sources. Modern heterotrophs, including man, get such organic molecules from other living things. The earliest heterotrophs, however, could not depend on other organisms, since these were, for all practical purposes, nonexistent. They must have received their nourishment from the rich organic soup in which they arose. At first, presumably, these cells used the most complex molecules, the ones most like themselves and requiring the least modification. These molecules, which had taken millions of years to accumulate, would soon have been used up, however. The primitive cells that required the most complex organic substances could then no longer multiply, while cells able to make more complex chemicals out of slightly simpler organic ones had a selective advantage in this changing environment. As the medium in which they lived became increasingly depleted, the less competent cell-like systems were weeded out. So in the course of time there evolved cells—autotrophs—that were able to make organic molecules out of very simple nonorganic materials.

By this time in the history of the planet, the basic pattern of life as it is today had been established. Organisms existed that were able to convert the radiant energy of the sun directly into the chemical energy of organic compounds. Other organisms survived by eating these photosynthetic organisms or eating organisms that had eaten them. Although these photosynthetic autotrophs and the heterotrophs that ate them were only single cells, they were, as far as the fossil record reveals, remarkably similar both in chemical composition and in physical structure not only to modern single-celled organisms but to the cells that make up the bodies of the complex, many-celled plants and animals that are the familiar organisms of our biosphere.

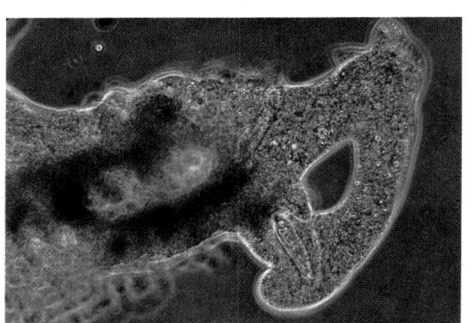

1–10 *A modern heterotroph, the giant amoeba*
Chaos chaos, *captures a paramecium.*
Although seemingly disorganized, as its
name would indicate, Chaos chaos *is able*
to sense its prey, move toward it, and
send out a pseudopod ("false foot") of
the right size and shape to envelop it.
Heterotrophs, unlike autotrophs, are
dependent on other organisms for their
sources of chemical energy.

WHAT IS LIFE?

What do we mean when we speak of "the evolution of life" or "life on other planets" or "when life begins"? If we were to be transported through time or space in the search for "life," what would we look for? Scientists concerned with this question, whether for practical or philosophical reasons, appear to agree that there are no single, simple answers.

Life does not exist in the abstract. There is, in fact, no "life." What exists and what can be examined and studied are living systems, individual living organisms. Even within the confines of our biosphere, such organisms show astonishing variety. Are these highly varied living organisms distinguishable from nonliving systems? Can we make distinctions that would apply to unknown living forms as well as to known ones? To focus this question even more sharply, we might imagine that we are programming a computer for an unmanned space expedition to detect signs of living things.

What would we instruct it to look for? Here again, there seems to be some area of general agreement. Although life is defined by no single criterion, there are certain properties shared by living systems or that are more common, collectively, in living than in nonliving systems.

The list that follows was compiled from various sources. Before you read on, you might wish to formulate your own list of the attributes that distinguish living systems from nonliving, and we invite your criticisms of ours.

1–11   *The clockface of biological time. Life first appears relatively early in the Earth's history, before 7:00* A.M. *on a 24-hour time scale. The first multicellular organisms do not appear until the twilight of that 24-hour day, and man himself is a late arrival—at about 20 seconds to midnight.*

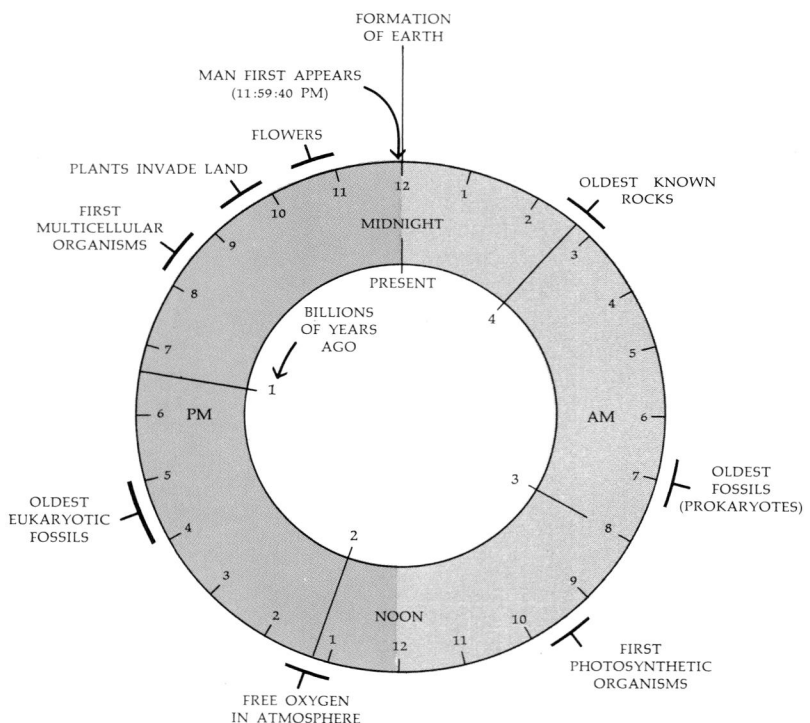

SCALE: 1 SECOND = 52,000 YEARS
1 MINUTE = 1,125,000 YEARS
1 HOUR = 187,500,000 YEARS

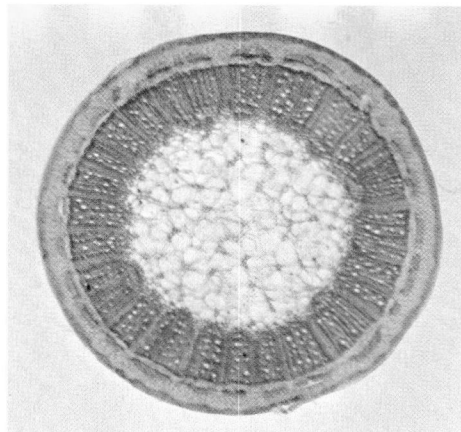

1–12 *Cross section of a rose stem. The existence of order and form—organization—is one of the signs of life.*

THE SIGNS OF LIFE

1. *Living organisms are complex and highly organized.* In nonliving things—soil, water, air—atoms and molecules are seldom arranged in a precise and ordered way. They are generally random mixtures of fairly simple chemical compounds.

Even crystals, which consist of one or a few atoms arranged in a regular, repeating pattern, are quite simple compared to a living cell. The same atoms and molecules are found in living systems as in nonliving systems, as we have noted, but in the former they are put together in very specific and complex ways. One of the principal ways in which a biologist distinguishes a microfossil, such as that shown in Figure 1–8, from the rock in which it is embedded, is by observation of its structure. The complexity of even a very small and simple living thing stands out in contrast to the relative simplicity of the rock.

In living things, atoms and molecules are organized in such a way that they form cells and the many special structures found within cells. Virtually the same structures are found within all cells. The fossil cells of 3.2 billion years ago are remarkably similar to modern prokaryotes. In higher (higher meaning more complex) organisms, cells are organized into tissues, tissues into organs, and so on.

Of course, not everything that is organized is alive. A beehive is a complex, organized structure. An automobile is highly organized, yet not living. The letters on this page are highly organized (into words), as distinct from random (and so meaningless). Yet neither the beehive, nor the automobile, nor the words on the page are living. On the other hand, they are all produced by living organisms. So if we programmed that imaginary computer to look for signs of organization, we could be reasonably sure it would pick up traces of living things if not the things themselves.

2. *Living things take energy from their environment and change it from one form to another.* The original source of energy for living systems on this planet is the energy of the sun. Green plants and algae transform this radiant energy to chemical energy by the process of photosynthesis. Animals and other non-photosynthetic organisms derive their energy from the chemical energy stored by plants. Energy is used to create and maintain the organization characteristic of living things. (This relationship between energy and order is one of the most interesting principles of modern biology, and one which we shall discuss at greater length in subsequent chapters.) Energy is used to make the molecules of which living systems are composed. Organisms transform the chemical energy of their nutrient molecules to mechanical energy, the energy of motion and muscle contraction. Some cells convert it largely to electric energy; our nerve impulses are the results of such a conversion. Some luminescent organisms turn the chemical energy back into light energy. For some, it is the source of body heat.

Energy conversion is one of the most characteristic and striking properties of living things, although it is not an exclusive property of living systems. Nonliving things also can convert energy, as, for example, conversion from light energy to heat energy (as in a rock absorbing sunlight), or from chemical energy to light and heat (as in paper burning), or electrical energy to light and kinetic energy (as in a bolt of lightning striking a tree). Machines, nonliving things made by living things, are characteristically able to convert chemical

1-13 *Corn snake converting chemical energy to mechanical energy. The prey is a white-footed mouse.*

energy (fuel) to motion, heat, or other forms. Notice, however, that whereas in living systems, energy input can result in an increase in orderliness—as in a growing child or a plant—in nonliving systems, inputs of energy are more likely to result in a decrease in order.

3. *Living things are homeostatic.* Homeostatic means simply "staying the same." Perhaps the most familiar example of homeostasis is the remarkably constant body temperature we and other mammals are able to maintain despite temperature changes in the outside environment. Although not all living things maintain a constant temperature, all are homeostatic in their chemical composition. Living things, as we have seen, are made of the same atoms as nonliving things, and they exchange materials with their environment. In fact, any single atom may be present only briefly—a fraction of a second—in a living structure. Yet living things maintain an internal environment quite different from that of the external environment that surrounds them. Homeostasis is, of course, closely related to the first attribute, organization.

4. *Living things respond to stimuli.* Plant seedlings bend toward light; mealworms congregate in dampness; cats pounce on small moving objects; even certain bacteria move toward or away from particular chemicals. Although different organisms respond to widely varying stimuli, the capacity to respond is a fundamental and almost universal characteristic of life. Moreover, the responses of living things to their external environment can often be distinguished from those of nonliving things by the fact that the living responses can be interpreted as generally useful to the organism responding.

5. *Living things reproduce themselves.* They make more of themselves, copy after copy after copy, with astonishing fidelity (and yet, as we shall see, with just enough variation to provide the raw material for evolution). Collections of inanimate matter show no apparent capacity to reproduce themselves in nearly identical size, shape, and internal structure through "generation" after "generation."

1–14 *Living things, although they exchange materials with their environment, maintain a relatively stable internal environment quite unlike that of their surroundings. Even this tiny, apparently fragile animal, a rotifer, has a constant chemical composition that, as long as the organism is alive, does not change with changes in the environment. This capacity for staying the same is known as homeostasis.*

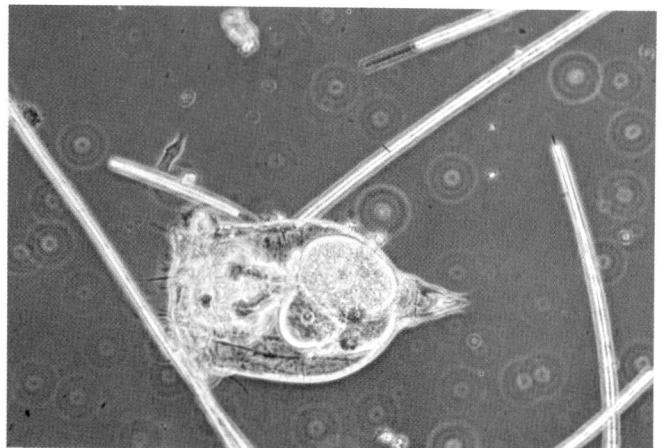

1–15 *Scallops, sensing an approaching starfish, leap to safety. They move by jet propulsion, expelling water forcibly by snapping shut their shells. The stimulus to which the scallops are responding is probably not the sight of the starfish—despite the scallop's numerous blue eyes—but its smell. Extracts of starfish tissue produce the same response.*

1–16 *Reproduction is another of the criteria by which we recognize living systems.*

1–17   *Adaptations. (a) Moles, which seldom venture above ground, have small, almost sightless eyes. For sensing their prey—consisting largely of worms and other underground invertebrates—they rely almost entirely on their sense of smell, hence their fleshy noses. Notice also the large shovel-like forepaws. (b) Tarsiers, primitive primates which live entirely in trees, have hands and feet with enlarged skin pads, specialized for grasping. Like other treetop dwellers, they depend mostly on vision. Their huge eyes are related to their nocturnal existence.*

(a)

(b)

6. *Living things grow and develop.* Growth and development are the processes by which, for example, a single living cell, the fertilized egg, becomes a tree, or an elephant, or a human infant. Development can be viewed as part of the process of reproduction—the way by which one or two complex adults produce another, similar adult. It is, however, such a spectacular and unusual (although not universal) capacity of living systems that we have chosen to list it separately.

7. *Living things are adapted.* Living organisms and the component parts of living organisms are suited to the environments in which the organisms live and the way they function in these environments. This is a familiar concept. By looking at a polar bear or a cactus plant or a beaver or a giraffe, you can make a number of assumptions about where they live and what they do to sustain life.

Because organisms are so exquisitely suited to their roles in the biosphere—to their ecological niche, to use a more precise term—it was long believed that each type of organism must have been specially created. There is, however, no objective evidence to support this concept of special creation. According to the scientific explanation, organisms have become suited to their environment by a gradual process of adaptation. Adaptation takes place in two steps. The first involves the appearance at random of inheritable differences among similar individuals. The second involves differences in the number of offspring left by individuals whose differences make them better suited to the environment as compared to other individuals of the same type. For example, among primitive horses, some members of each generation had slightly longer than average legs and some had slightly shorter. Those with slightly longer legs, provided these longer legs appeared in chance combination with other characteristics, such as good lung capacity, could run faster. Faster runners, who could escape from predators, left more offspring in general than slower runners, who were more likely to die young. To the extent that long-leggedness was hereditary, the offspring of the long-legged horses would have longer legs than the offspring of the short-legged horses. Thus, very slowly, by a fraction of a centimeter a century, horses, on the average, would come to have longer legs.

In biology, adaptation also has a second meaning. In the previous example, adaptation refers to gradual changes in a group of organisms. The second meaning of adaptation refers to changes in an individual that take place in response to particular environmental factors. Our skins darken on exposure to the sun. Or, to take a slightly less familiar example, people who live at high altitudes, where oxygen is scarce, develop additional red blood cells (which carry oxygen).

When we speak of adaptation in this text, unless specifically noted otherwise, we shall be using the word in the first sense. Note that the second type of adaptation—the capacity to change under environmental stress—is actually the result of an adaptation in the first sense. The end result of adaptation is evolution.

As a consequence of adaptation, living things have an appearance, as we previously noted, of having been designed purposefully. For this reason, the study of biology is quite different from that of chemistry or of physics. As

Albert Lehninger* points out, "In living organisms it is quite legitimate to ask what the function of a given molecule is. However, to ask such questions about molecules in collections of inanimate matter is irrelevant and meaningless."

Of course, this appearance of having been designed for a purpose is a property that, like other properties we have discussed, is shared by man-made objects. Every component of a man-made object, even one that is merely decorative or amusing, has a purpose. Moreover, it is at least theoretically possible to design machines that possess the other properties which we think of as belonging to living things. We can imagine, for example, a spaceship that runs on solar energy, replaces its component parts by using materials from the atmosphere, and even reproduces itself! However, here it is easy to make an essential distinction between the living and the nonliving. The information required to construct such a machine must come from outside the machine—that is, at least as far as our planet is concerned, from the mind of man. This leads us to the final characteristic in our list.

8. *The information by which living things organize their purposeful structures and functions, maintain homeostasis, convert energy, respond to stimuli, reproduce, and develop is all contained within the individual organism itself.*

In the following chapters of this section, we shall examine the "lifeless" materials of which living systems are composed and attempt to explore, within the present limits of biological knowledge, why, how, and under what circumstances those nonliving components come together and, on coming together, exhibit the signs of life.

SUMMARY

The sun and its planets were formed 4.5 billion years ago—the sun probably from the condensation and contraction of a hydrogen gas cloud, and the planets as accumulations of interstellar debris. Of the nine planets in this solar system, only Earth is known to support life, but there are likely to be other planets in the galaxy with some form of life.

The primitive atmosphere of Earth held the chief raw materials of living matter—hydrogen, oxygen, carbon, and nitrogen—combined in water vapor and gases. The energy required to break apart the simple gases in the atmosphere and re-form them into more complex molecules was supplied by heat, lightning, radioactive elements, and high-energy radiation from the sun. Laboratory experiments have shown that the types of molecules characteristic of living organisms—that is, organic molecules—can be formed under such conditions. These molecules gradually accumulated to form cells, the first living things.

The earliest fossil cells resemble modern bacteria and blue-green algae. Such cells are known as prokaryotes. They are distinguished from the larger, more complex cells of plants and animals—known as eukaryotic cells—chiefly by the fact that eukaryotes have a membrane-bounded nucleus and many specialized structures (organelles) not found in prokaryotes.

The earliest cells were probably heterotrophs, that is, organisms that depend on outside sources of organic molecules for their energy. Organisms that can

* A. L. Lehninger, *Biochemistry*, 2d ed., Worth Publishers, New York, 1975.

use the sun's energy directly to make organic molecules are known as auto-trophs. The most familiar of the modern autotrophs are the green plants.

Living systems are characterized by their possession of certain properties:

1. Living matter is highly organized and so is much more complex in its structure than nonliving matter.
2. Living things take energy from their environment and change it from one form to another.
3. Living things are homeostatic.
4. Living things respond to stimuli.
5. Living things reproduce themselves.
6. Living things grow and develop.
7. Living things are adapted.
8. Living things contain within themselves the information by which they create their own organization and by which they carry out the other functions characteristic of living things.

QUESTIONS

1. Define the following terms: fossil, autotroph, heterotroph, photosynthesis, biosphere. Check your definitions with those in Glossary at the back of the book.
2. Among the experiments planned for the Viking mission to Mars are several designed to detect replication among microorganisms in Martian soil. Such experiments are to take place on the surface, with results transmitted to Earth. How would you design such an experiment?

Chapter 2

Atoms and Molecules: Some Basic Principles

Almost a hundred different elements are present in the Earth's crust, in the waters that cover much of the Earth's surface, and in the thin veil of atmosphere surrounding it. Of these many elements, only four make up more than 95 percent of all living matter. These four are carbon, hydrogen, nitrogen, and oxygen. In combination with only two other elements, phosphorus and sulfur, these four make up the principal molecules of living systems. These six elements are conveniently remembered as CHNOPS.

Although the basic six elements can be combined in many different ways to form a wide variety of chemical compounds, only a few different kinds of compounds are found in large quantities in living systems—five, to be exact. One of these is water. The other four are (1) sugars and combinations of sugars, such as starch and cellulose; (2) lipids, including fats and waxes; (3) amino acids, which combine to form proteins; and (4) nucleotides, which play important roles in energy transfer and which combine to form nucleic acids, the self-replicating molecules that carry the hereditary information.

This chapter and the three that follow it will be concerned with these five compounds and the atoms that constitute them. At first, it may seem to you as if electrons, protons, covalent bonds, and similar subjects have little to do with biology which is, after all, by definition, the science of life. Bear with us, however, and you will come to see that, in fact, these lifeless molecules are the very fabric of which life is constructed.

Notice that by beginning the study of biology with an examination of how atoms and molecules are put together and how they interact, we are making a very fundamental assumption. We are saying, in effect, that living things are made up of the same chemical and physical components as nonliving things and that they obey the same chemical and physical laws.* Yet, there are recognizable differences between living and nonliving systems. What then is the basis of these

---

* Notice that the word "law," like the word "theory," has a somewhat different meaning in biology than in ordinary usage. A man-made law can be broken yet remain intact. If a scientific law is broken, it simply ceases to exist.

2–1 *This is a single living cell from a pea plant. Made up mostly (more than 95 percent) of only four kinds of atoms, it is nevertheless capable of carrying out a wide variety of intricate chemical reactions. The sum total of these reactions is manifested in this cell's capacity to exhibit properties, such as energy conversion, homeostasis, reproduction, and growth, that are characteristic of living organisms.*

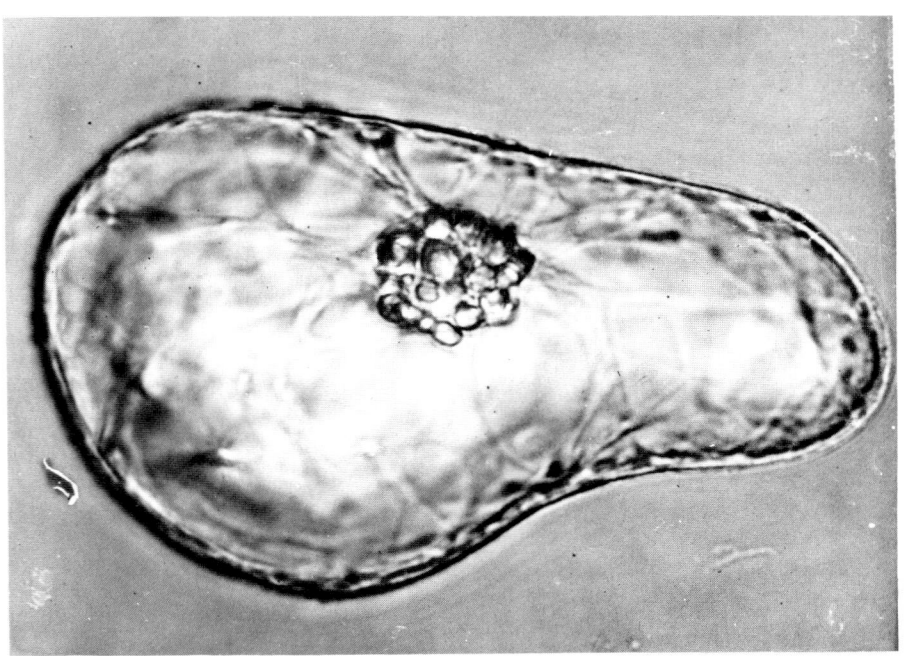

differences? As we noted in Chapter 1, the difference between a collection of chemicals and a living organism, with all its remarkable capacities, is not a matter of materials but one of organization. When we move from one level of organization to another, we move to something qualitatively different.

ATOMS

Atoms are the irreducible units of chemical elements. This does not mean that atoms cannot be broken apart but rather that, once broken apart, the atoms no longer have the chemical properties of the elements. The core, or *nucleus,* of an atom contains one or more positively charged particles, called *protons.* The number of protons in an atom determines its chemical characteristics. Thus, the number of protons is unique for each element: A hydrogen atom has one proton; a helium atom has two; carbon, six; oxygen, eight; all the way up to uranium, which has 92, the largest number of protons found in any naturally occurring atom. (Atoms with more than 92 protons exist, as far as is known, only in the laboratory.) The number of protons in an atom is its *atomic number.*

In addition to protons, most atoms contain particles in their nucleus called *neutrons,* which, as their name implies, have no charge, although they do have mass. The mass of a neutron is approximately equal to that of a proton, and the atomic mass of an atom is about equal to the combined numbers of protons and neutrons. Atoms that contain the same number of protons but different numbers of neutrons differ, therefore, in their atomic mass. However, they are very similar chemically; that is, each can substitute for the other in chemical compounds or chemical reactions. Such chemically interchangable atoms are called *isotopes.*

$E = mc^2$

Protons and neutrons, as we mention in the text, are arbitrarily assigned an atomic mass of one. One would, therefore, expect that an element with, for example, twice as many protons and neutrons as another element would weigh twice as much. This assumption is true—almost. If the weights of nuclei are measured with great accuracy, as they can be by instruments developed by modern physics, small nuclei always have proportionately slightly greater masses than larger nuclei. For example, the most common isotope of carbon has a combined total of 12 protons and neutrons. If we assign carbon an atomic mass of 12, we find the hydrogen atom has an atomic mass not of 1, but of 1.008. Helium has two protons and two neutrons. It does not have a mass of exactly 4, however, or of 4.032 (four times the weight of hydrogen); it weighs, in relation to carbon, 4.0026. Similarly, oxygen, with a combined total of 16 protons and neutrons, has an atomic mass, in relation to that of carbon, of 15.995.

In short, when protons and neutrons are assembled into an atomic nucleus, there are slight changes in mass. One of the oldest and most fundamental concepts of chemistry is the law of conservation of mass—that mass is never created or destroyed. Yet under conditions of extremely high temperature, atomic nuclei fuse to make new elements. What happens to the mass "lost" in the course of this fusion? This is the question answered by Einstein's fateful equation $E = mc^2$. E stands for energy, m stands for mass, and c is a constant equal to the speed of light. Einstein's equation means simply that under certain extreme and unusual conditions, mass is turned to energy.

The sun consists largely of hydrogen nuclei. At the extremely high temperature at the core of the sun, hydrogen nuclei strike each other with enough velocity to fuse. In a series of steps, four hydrogen nuclei fuse to form one helium nucleus. In the course of these reactions, energy is released, enough to keep the fusion reaction going and to emit tremendous amounts of radiant energy into space.

We may seem to have moved a long way from biology in the course of this discussion of mass and energy, but, as we shall see in future chapters, life on this planet depends on energy emitted by the sun in the course of this reaction. This same reaction provides, of course—as Einstein foresaw—the energy of the hydrogen bomb.

Albert Einstein in 1905, the year he published his paper on the theory of relativity. He was 26 years old and working at the Swiss Patent Office in Bern as a technical expert third class.

2–2 *Atomic models of hydrogen, deuterium, tritium, and helium. Note that hydrogen, deuterium, and tritium each have only a single proton and a single electron, and so they behave similarly chemically although they differ in the number of neutrons. Helium, which has one more proton and one more electron, is chemically very different from hydrogen and its isotopes.*

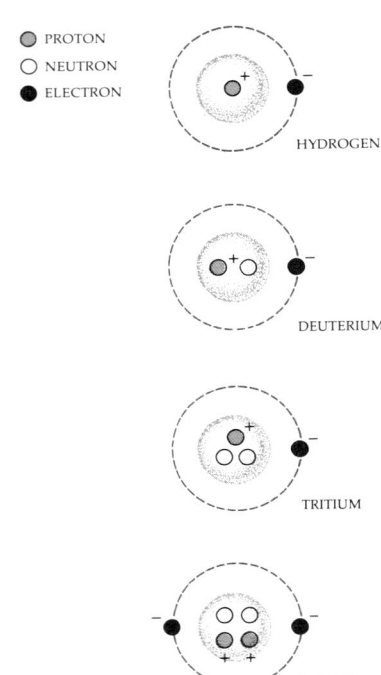

○ PROTON
○ NEUTRON
● ELECTRON

HYDROGEN

DEUTERIUM

TRITIUM

HELIUM

2–3 *Another representation of the helium atom, with the electron orbits indicated in three dimensions.*

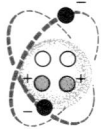

The ordinary isotope of hydrogen (H), for example, contains a single proton in its nucleus. One fairly common isotope of hydrogen, heavy hydrogen, or deuterium, contains one proton and one neutron and so has an atomic mass of about two (^{2}H). Another isotope of hydrogen, tritium, has one proton and two neutrons and so has an atomic mass of about three (^{3}H). As indicated previously, chemically each of the isotopes is similar. Physically, however, they behave differently. Deuterium and tritium are heavier than hydrogen. Moreover, tritium, like a number of other isotopes, is unstable, or radioactive. Tritium atoms become stable by emitting high-energy particles. As long as the nucleus contains only one proton, the atom is still a hydrogen atom—that is, it behaves chemically like hydrogen—even though the number of neutrons increases. If, however, the nucleus gains a proton, a different element is formed —helium. The common isotope of helium contains two protons and two neutrons, and so has an atomic mass of four.

Because protons are positively charged, the nucleus of any atom always has a positive charge, the strength of which is indicated by the atomic number (the number of protons). For example, the nucleus of a hydrogen atom with one proton has a positive charge of one, the helium atom nucleus has a positive charge of two, the carbon atom nucleus a positive charge of six, and so forth. The positive charge (+) of the nucleus is offset by negatively charged particles, called *electrons,* which move around the atomic nucleus at varying distances from it, traveling at high velocity. The negative charge (−) of one electron is sufficient to balance out the positive charge of one proton; therefore an atom is neutral in charge if the number of electrons orbiting the nucleus equals the number of protons within it.

Although electrons have a charge equal to that of protons, they are so much smaller and lighter that they contribute virtually nothing to the mass of an atom. When you weigh yourself on a scale, only 1 ounce approximately of your total weight is made up of electrons; all the rest is protons and neutrons.

Structure of the Atom

The drawings in Figure 2–2 are based upon the model of atomic structure suggested by the physicist Niels Bohr. In Bohr's model, negatively charged electrons are pictured as moving in planetary orbits around the nucleus. The relative sizes of electrons and nucleus are not meant to be accurately represented. Electrons are much, much lighter than atomic nuclei. Moreover, most of an atom is empty space. If a picture of a hydrogen atom were drawn to scale and the nucleus of the atom were depicted as 1 inch in diameter, the electron, in proportion, would be almost a mile away.

In the Bohr model, electrons move at fixed distances from the nucleus in what are called *electron shells.* These shells represent different energy levels for the electrons. An input of energy—in the form of heat or light—can boost an electron up to a higher energy level. When it drops down again—as it does almost instantaneously—the energy is reemitted. Bohr's essential clue to the existence of electron shells was the fact that the amount of energy necessary to boost an electron of a particular kind of atom from one shell to the next was always the same and that it always equaled the amount reemitted.

The largest atoms may have as many as seven concentric electron shells.

THE NATURE OF MODELS

Many hypotheses in science are presented in the form of models. A model is a way of assembling a collection of data. It is not necessarily a copy of a physical reality, nor does it attempt to fully duplicate it. For example, a model airplane constructed of balsa wood could be instructive about how an airplane looks and, perhaps, about how it is assembled. Suitably painted and viewed from appropriate angles, it could be helpful in learning to identify such an airplane from a distance. It would be no use at all, however, in studies of structural requirements for air flight because of the tremendous weight differences between the model and a real plane. To study problems of aerodynamic stress, it would be necessary to construct another sort of model. This model might have little resemblance to the first model, and it might not even look much like the plane itself, but it would be more useful for some purposes.

In explaining chemical and biochemical phenomena, it is common to use models of atoms and molecules. The models used in this text are based on those proposed by the physicist Niels Bohr. Although newer and more complex models have been formulated, Bohr's are still the most useful for interpreting most chemical events.

Niels Bohr in 1956. Bohr's theory of the atom, proposed in 1913, was one of the turning points of modern science.

The atoms important in living systems, which are among the lighter atoms, have one, two, three, or rarely four.

Only a limited number of electrons can orbit at each energy level. The first shell of any atom can contain only two electron orbits. Additional outer shells can contain up to eight, and eight are necessary for a stable outer shell. Atoms tend to complete their outer shells, a fact which, as we shall see, leads to the bonding together of atoms into molecules.

The attractive force between the negatively charged electrons and the positively charged nucleus is greater for electrons in orbit in the first electron shell and successively less at orbits more distant from the nucleus. Electrons can be boosted from one energy level to another by an input of energy. The electron then is a source of *potential energy* that can be used in living systems.

In order to understand what potential energy is and how an electron can have potential energy, an analogy may be useful. A rock on flat ground has no useful energy. If you push it up a hill, it gains energy—potential energy. So long as it sits on the peak of the hill, it neither gains nor loses energy. If it rolls down the hill, however, it loses its potential energy as it rolls toward level ground (Figure 2–4). Similarly, a rocket poised on the launching pad has no useful energy. The chemical energy of rocket fuel is required to blast it away from the Earth, but as long as it remains within the Earth's gravitational pull, no input energy is required for its return to Earth. In the fall back to Earth, the rocket releases its potential energy. Similarly, water that has been pumped up to a water tank for storage has potential energy that will be released when the water runs out the faucet.

The electron is like the boulder, or the rocket, or the water in that an input of energy can raise it to a higher level—farther away from the nucleus. So long as it remains in this higher level, it possesses potential energy. When it returns to a lower level, its potential energy is released. Electrons that have been forced to a higher energy level possess potential energy.

The flow of energy through the living world has its origin, as we shall see in Chapter 9, in electrons boosted to a higher energy level by light striking a chlorophyll molecule in a green plant.

2–4 (a) *The energy used to push a boulder to the top of a hill (less the heat energy produced by friction between the boulder and hill) becomes potential energy, stored in the boulder as it rests at the top of the hill. This potential energy is converted to kinetic energy (energy of motion) as the boulder rolls downhill.* (b) *When an input of energy—such as light energy—boosts an electron to a high energy level, the electron, like the boulder, possesses potential energy which is released when the electron returns to its previous energy level.*

Table 2–1 *The Relative Abundance of Some Chemical Elements in the Earth's Crust*

Element (chemical symbol)	Relative abundance
Oxygen (O)	62.5 %
Silicon (Si)	21.2
Aluminum (Al)	6.47
Sodium (Na)	2.64
Calcium (Ca)	1.94
Iron (Fe)	1.92
Magnesium (Mg)	1.84
Phosphorus (P)	1.42
Carbon (C)	0.08
Nitrogen (N)	0.0001

Table 2–2 *Atomic Composition of Three Representative Organisms*

Element	Man	Alfalfa	Bacterium
Oxygen	62.81%	77.90%	73.68%
Carbon	19.37	11.34	12.14
Hydrogen	9.31	8.72	9.94
Nitrogen	5.14	0.83	3.04
Phosphorus	0.63	0.71	0.60
Sulfur	0.64	0.10	0.32
Total	97.90	99.60	99.72

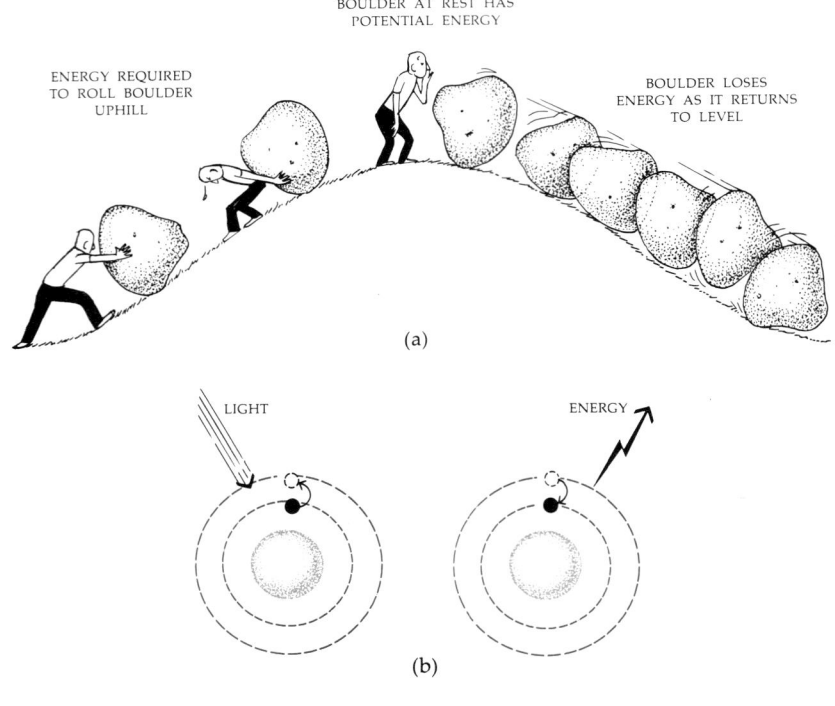

(a)

(b)

Common Atoms in Living Matter

As we mentioned at the beginning of the chapter, hydrogen, oxygen, carbon, and nitrogen are the most common atoms in living systems. It is no accident that living systems are built of these particular atoms rather than other elements also present in the Earth's surface and atmosphere. First, these atoms are light, so that all of them are capable of existing either alone or, as in the case of carbon, in combination with others as simple gases. In addition, they can be dissolved in water. Second, also because of their internal structure, they are capable of combining with one another in a variety of stable combinations.

Hydrogen

The hydrogen atom is the simplest of all atoms. Its nucleus contains a single proton and outside the nucleus, as you can see in Figure 2–5, is a single electron. In this state, the atom is electrically neutral. If, however, the hydrogen atom loses its single orbiting electron, it will have a charge of +1, designated H^+. Such charged atoms or groups of atoms are called _ions._ A hydrogen ion is often referred to simply as a proton, which is what it is.

Hydrogen gas is made up of two hydrogen atoms sharing their electrons and is designated H_2. (The number of atoms in a molecule is indicated by a

Many naturally occurring isotopes are radioactive. All the heavier elements—atoms that have 84 or more protons in their nuclei—are unstable and, therefore, radioactive. All radioactive elements emit radiation (as particles or rays) at a fixed rate; this process is known as radioactive "decay." The rate of decay is measured in terms of half-life: The half-life of a radioactive element is defined as the time in which half the atoms in a sample of the element lose their radioactivity and become stable. Since the half-life of an element is constant, it is possible to calculate the fraction of decay that will take place for a given isotope in a given period of time.

Half-lives vary widely, depending on the isotope. The radioactive nitrogen isotope ^{13}N has a half-life of 10 minutes, and tritium has a half-life of 12¼ years. The most common isotope of uranium (^{238}U) has a half-life of 4½ billion years. The uranium atom undergoes a series of decays, eventually being transformed to an isotope of lead (^{206}Pb).

Isotopes have a number of important uses in biological research. One of these is in dating the age of fossils and of the rocks in which fossils are found. The proportion of ^{238}U to ^{206}Pb in a given rock sample, for example, is a good indication of how long ago that rock was formed. (The lead formed as a result of the decay of uranium is not the same as the lead commonly present in the original rock, ^{204}Pb.) A second use of isotopes is as radioactive tracers. A plant, for example, that is exposed to radioactive carbon dioxide ($^{14}CO_2$) will use the radioactive carbon to make sugar just as it would the more common isotope. The use of radioactive carbon dioxide has played an important role in enabling plant physiologists to trace the path of carbon in photosynthesis, as described in Chapter 9. A third use is autoradiography, a technique in which a sample of material containing a radioactive isotope is placed on a photographic film. Energy emitted from the isotope leaves traces on the film and so reveals the exact location of the isotope within the specimen.

Autoradiograph of a fossil clam shell. The linear black marks are tracks of high-energy particles emitted from ^{238}U which was incorporated into the shell as it formed from seawater and from mineral particles in the sea floor.

2–5 (a) *A hydrogen atom has one electron and so tends to gain another electron and complete its outer shell.* (b) *Hydrogen atoms combine by sharing their two electrons. The bond thus formed is known as a covalent bond. A covalent bond always involves the sharing of two electrons.*

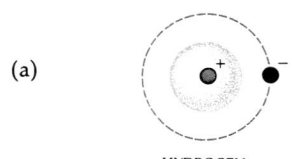

(a)

HYDROGEN

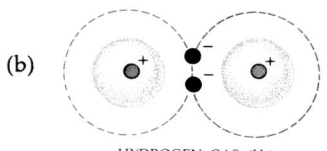

(b)

HYDROGEN GAS (H₂)

subscript.) Bonds formed in this way, by the sharing of pairs of electrons, are known as *covalent bonds*. By forming covalent bonds, each atom attains a stable outer shell. Helium, which already has two electrons in its outer shell, does not combine with other atoms and so is chemically unreactive, or inert.

The H_2 molecules are so light that Earth, which is a relatively small planet, does not have enough gravitational force to hold them. However, the atmosphere of Jupiter, a much larger planet, contains a high proportion of gaseous hydrogen.

Oxygen

The oxygen atom, as you can see in Figure 2–6, is larger and a little more complicated than the hydrogen atom. In its nucleus are eight protons and eight neutrons (giving it a positive charge of eight), and orbiting its nucleus are eight electrons. (An inner shell never has more than two electrons, according to the laws of atomic structure.) There are six electrons in the outer shell when oxygen is electrically neutral, but eight electrons are required for a stable conformation. Therefore oxygen requires two more electrons to complete its outer shell. Thus when hydrogen and oxygen combine, oxygen combines with two hydrogen atoms—H_2O—sharing a pair of electrons with each of the two

2–6 *An oxygen atom has eight protons in its nucleus and so is electrically neutral with eight orbiting electrons, as shown here. Two of these are in the innermost shell and six are in the outer shell, leaving the atom with an unstable outer shell. What is the atomic number of oxygen? The atomic mass?*

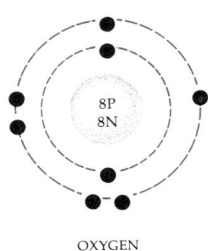

OXYGEN

2–7 *(a) The carbon atom has four electrons in its outer shell. (b) It can complete its shell by sharing electron pairs with four hydrogen atoms thus forming methane. (c) A carbon atom requires four electrons to complete its outer shell, and an oxygen atom requires two; both requirements are satisfied by carbon's sharing of two pairs of electrons with each of two oxygen atoms to form carbon dioxide. Bonds involving two pairs of electrons are known as double bonds.*

hydrogen atoms. The fact that when the two elements combine they combine in definite proportions—one oxygen with two hydrogens, for example—was known long before any theoretical explanation for the fact could be found. The Bohr model, as you can see, accounts for the proportions' being definite and constant.*

Carbon

Carbon, you will note, has four electrons in its outer shell and therefore needs four more to complete its shell. It can combine with four hydrogen atoms to produce the gas methane, CH_4, or with two oxygen atoms to produce carbon dioxide, CO_2. As you can see in Figure 2–7, when carbon combines with oxygen to produce carbon dioxide, it shares two pairs of electrons with each oxygen atom, thus completing the outer shells of each atom. A bond formed by the sharing of two pairs of electrons is known as a double covalent bond or, more simply, as a double bond. Carbon's capacity to form so many covalent bonds means that carbon atoms can form the points of attachment—the backbones—for a great variety of complex compounds. This capacity, along with its ability to combine with other atoms to form gases, makes carbon uniquely fit for its role in living systems.

Nitrogen

The nitrogen atom contains seven protons, seven neutrons, and seven electrons, five of which are in its outer shell. Since three electrons are needed to complete its outer shell, nitrogen, like oxygen and carbon, can form double as well as single covalent bonds.

* In combining elements, chemists use a convenient shortcut which involves a quantity known as a mole. Mole is an abbreviation for gram molecular weight, which is the sum of the masses of all the atoms of a molecule expressed in grams. A mole of hydrogen (H_2), for example, is 2 grams of hydrogen. A mole of oxygen (O_2) is 32 grams. To make water, you would combine 2 moles of hydrogen (4 grams) with 32 grams of oxygen, and you would get 2 moles of water, each weighing 18 grams. A mole of any substance contains the same number of particles (whether atoms or molecules) as a mole of any other substance. Can you see why? This number turns out to be 6.023×10^{23}.

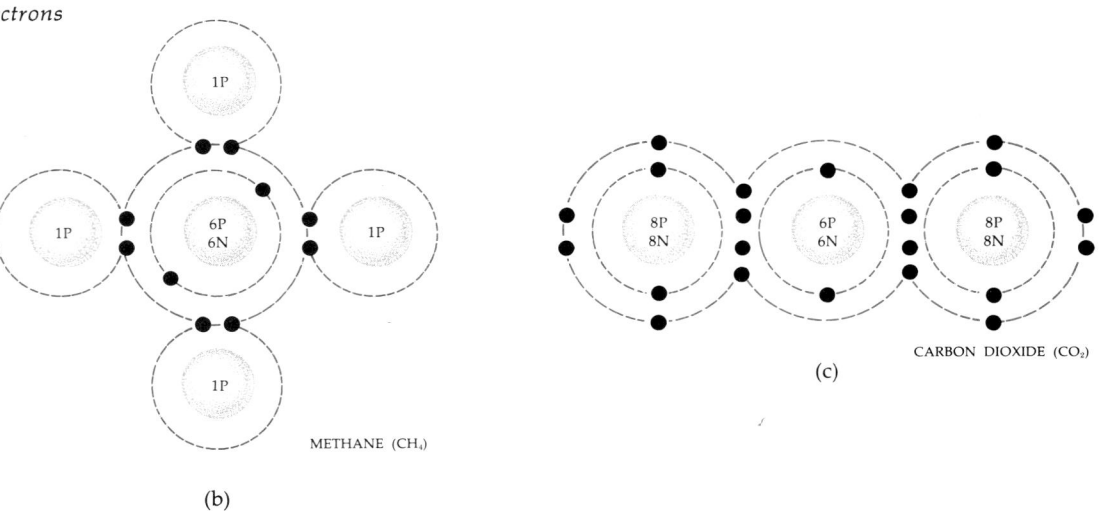

CARBON

(a)

METHANE (CH_4)

(b)

CARBON DIOXIDE (CO_2)

(c)

2–8 *Nitrogen, with five electrons in its outer shell, forms covalent bonds with three hydrogen atoms. The product is ammonia.*

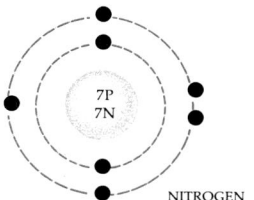

NITROGEN

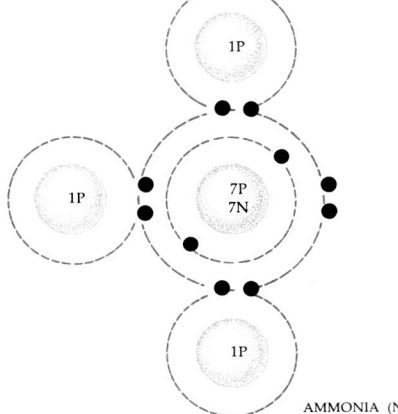

AMMONIA (NH₃)

In general, the lighter the elements, the stronger the covalent bonds formed between them. Hydrogen, oxygen, carbon, and nitrogen are the lightest elements capable of forming covalent bonds. Thus, living systems are based on atoms with capacities for forming strong bonds, resulting in stable combinations.

MOLECULES

Ions and Ionic Bonds

As we just saw, the formation of one or more covalent bonds is one way in which electron shells can become stabilized. Another way is by losing or gaining electrons. The sodium atom, for example, has 11 protons and 12 neutrons in its nucleus and so is electrically neutral when it has 11 orbiting electrons; two of these are in its first shell, eight in its second, and one in a third shell. As a consequence, the sodium atom has a tendency to lose the one electron, thereby stabilizing its outer shell. The loss of one electron gives the atom a net positive charge of 1. The sodium atom is Na in chemical shorthand, and the sodium ion, the atom minus the one electron, is represented as Na^+.

Chlorine (Cl), on the other hand, has an atomic number of 17, with seven electrons in its outer shell, and so tends to gain one, thereby forming a stable outer shell of eight electrons. The chlorine ion is represented as Cl^-.

When sodium and chlorine interact, they become ionized—that is, sodium tends to lose its single outermost electron and chlorine gains it, completing its outermost shell. These oppositely charged ions attract one another. Table salt (NaCl) is a latticework of alternating Na^+ and Cl^- ions held together by their opposite charges. Such bonds between ions with opposite charges are known as *ionic bonds.*

Many common atoms must gain or lose more than one electron to form a stable outer shell. Calcium, for instance, has an atomic number of 20 (20 protons) and so has 2 electrons in its outermost shell. Thus it can combine with two atoms of an element such as chlorine. The combining power of an element

2–9 (a) *Sodium (Na), which has only one electron in its outer shell, becomes stable if it loses the electron. Chlorine (Cl), which has seven electrons in its outer shell, becomes stable if it gains one. When sodium and chlorine interact,* *sodium loses its single electron and chlorine gains it; following this transaction, sodium has a positive charge and chlorine a negative one. Such charged atoms are called ions. (b) Oppositely charged ions attract one another. Table* *salt is crystalline NaCl, a latticework of alternating Na⁺ and Cl⁻ ions held together by their opposite charges. Such bonds between oppositely charged ions are known as ionic bonds.*

(a)

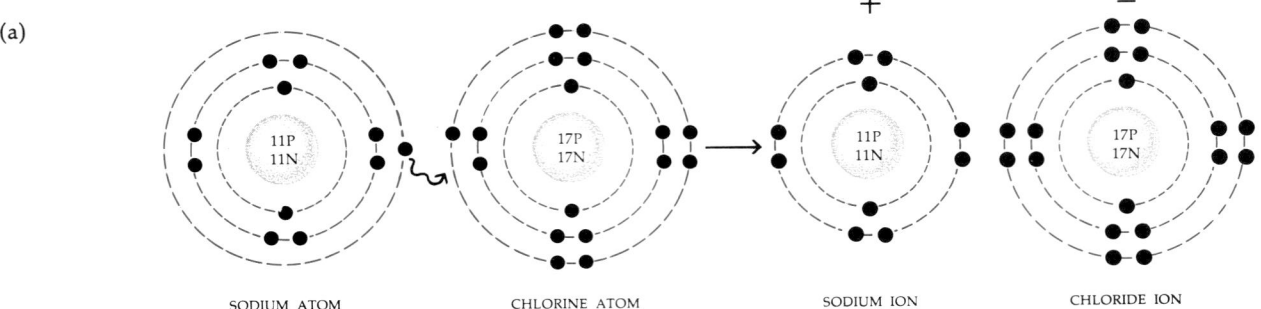

SODIUM ATOM   CHLORINE ATOM   SODIUM ION   CHLORIDE ION

(b)

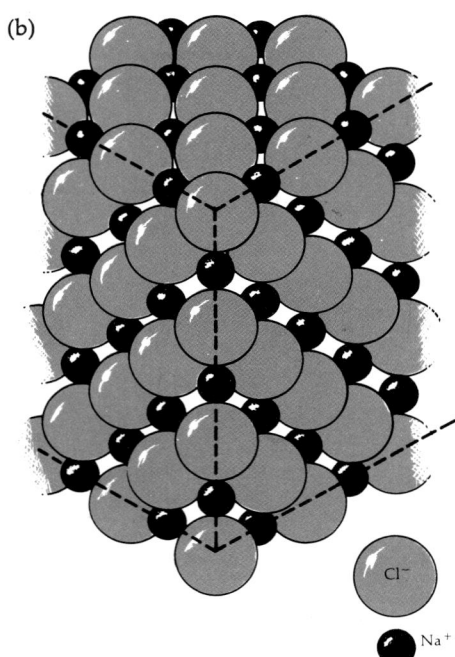

Cl⁻

Na⁺

is called its *valence*. Sodium has a combining power of +1, calcium has a combining power of +2, and chlorine has a combining power of −1.

An ion may be a single charged atom, such as Na⁺ or Cl⁻, or a combination of two or more atoms with a total positive or negative charge. Substances wholly or partially ionized in water play many important roles in the functioning of biological systems. Also water itself has a slight tendency to ionize—that is, to separate into positively charged hydrogen ions (H⁺) and negatively charged hydroxyl ions (OH⁻). The ionization of water is discussed in Chapter 3.

Oxidation-Reduction Reactions

The passing of electrons from one atom or one molecule to another, as in the interaction between sodium and chloride, is an *oxidation-reduction reaction.* The loss of an electron is known as *oxidation,* and the compound that loses the electron is said to be oxidized.

Reduction is, conversely, the gain of an electron. Oxidation and reduction take place simultaneously because an electron that is lost by one atom is accepted by another.

When carbon-containing compounds are oxidized and reduced, the electron often travels in company with a proton—that is, as a hydrogen atom. When carbon is oxidized—that is, when it loses a hydrogen atom—the hydrogen atom is usually accepted by an oxygen atom, forming water. It is because of oxygen's role as an acceptor in many biological reactions that electron loss came to be known as oxidation.

Molecules and Compounds

Molecules are simply two or more atoms held together by covalent or ionic bonds. The atoms in a molecule may be the same, as in H₂ and O₂, or they may be different, as in H₂O or NaCl. Chemical compounds are made up of molecules composed of two or more different kinds of atoms in definite proportions.

CHEMICAL REACTIONS

A chemical reaction always involves a change in the molecular structure of one or more substances. Matter is changed from one substance with its own characteristic properties to another substance with new, different properties. The atoms of which matter is composed are neither created nor destroyed (although, as we noted previously, matter may, under extraordinary circumstances, be changed to energy). Because atoms are neither gained nor lost in chemical reactions, such reactions are written in the form of equations which "balance"; that is, there are the same number of atoms in the products of the reaction as in the original reactants. The equation for a chemical reaction thus tells both the kinds of atoms that are present and their proportions.

Types of Reactions

Chemical reactions can be classified into a few general types. One type can be represented by the formula $A + B \longrightarrow AB$. An example of this sort of reaction is the combination of hydrogen gas with oxygen gas to produce water. Hydrogen gas is H_2, and oxygen gas is O_2. However, we know that each molecule of water contains two atoms of hydrogen and one of oxygen, and therefore the proportions must be two to one:

$$2H_2 + O_2 \longrightarrow 2H_2O$$

A reaction may also take the form $AB \longrightarrow A + B$. For example, the equation above showing the formation of water can be reversed:

$$2H_2O \longrightarrow 2H_2 + O_2$$

This means that from water you can obtain hydrogen and oxygen gases.

A reaction may also involve an exchange, taking the form

$$AB + CD \longrightarrow AD + CB$$

An example of such a reaction is the combination of hydrochloric acid (HCl) and sodium hydroxide (NaOH) to make table salt and water:

$$HCl + NaOH \longrightarrow NaCl + H_2O$$

As the example of the formation and splitting of the water molecule indicates, all chemical reactions are theoretically reversible. In fact, in any system we are studying, any chemical reaction that is taking place is taking place in both directions. What an equation indicates is not what is happening to every atom in the system, but what is happening on the average. To indicate that a reversible reaction is taking place, equations can be written with double arrows:

$$AB + CD \rightleftharpoons AD + CB$$

When the arrows are the same length, as they are in the above equation, it indicates that the reaction has an equal tendency to go in either direction and that one would find some molecules of all four types.

This situation is easy to comprehend when one looks at the chemists' model of chemical reactions. According to this concept, chemical reactions take place as a result of collisions of molecules. When more AB and CD are present, it is more likely that collisions will take place between these particular molecules, thus increasing the rate of formation of AD and CB molecules. As these latter molecules increase, the possibilities of collisions between them increase in

number, and so the mixture reaches an equilibrium at which collisions are equally as likely to occur between AB and CD as between AD and CB. At the equilibrium point, the two reactions take place at the same rate and the amounts of the four kinds of molecules stay constant.

The addition of energy in the form of heat increases the rate of reactions because it increases the rate at which the individual molecules move. A rise in temperature is, by definition, an increase in the rate of movement of molecules. Living systems cannot tolerate the great increases in temperature such as those applied in the chemistry laboratory. Enzymes, the structure and function of which we shall examine in Chapter 5, provide the means by which living systems increase the rates of chemical reactions without an increase in temperature.

Notice that if AD and CB were to be continuously removed from the system, the reaction would never reach equilibrium but would continue to move to the right. Similarly, if AB and CD were to be removed, the only collisions that could occur would be between AD and CB molecules, and thus the reaction would have to move to the left.

Energy Changes in Chemical Reactions

As we saw earlier, energy changes in atoms come about when electrons move from one energy level to another. Energy changes in molecules come about when bonds between atoms are broken and re-formed. Bonds between atoms have characteristic strengths. A carbon–oxygen bond has a different strength from a carbon–hydrogen bond, for example. The strength of a bond is generally calculated by measuring the amount of heat that will break the bond. Bond energies are expressed in terms of calories required to break the bonds of a given amount of the substances involved (see Figure 2–10). A *calorie* is, by definition, the amount of heat that will raise the temperature of 1 gram of water 1°C. A kilocalorie is the amount of heat that will raise 1 kilogram of water 1°C. Kilocalories, which are often termed simply Calories (with a capital C), are used to measure dietary requirements.

If the total bonding energy of the products of the reaction is less than that of the reactants, energy is released in the course of a reaction. Such a reaction is known as *exergonic*. Exergonic reactions are downhill reactions; although they require an input of energy to get started—so-called activation energy— once they are under way, they are self-sustaining. A reaction in which the bond energies of the product are greater than those of the reactants is an *endergonic*, or uphill, reaction, requiring an input of energy.

If a reaction is strongly exergonic, it will proceed rapidly to completion. If the energy differences are not great, however, there will still be some tendency for the reaction to go in either direction, although it will be more likely to move downhill. Such an equation is written

$$AB + CD \rightleftharpoons AC + BD$$

The longer arrow indicates the exergonic direction. Every reaction which does not go to completion has a characteristic point of equilibrium. Notice here again that if we remove the reactant molecules or the products as they are formed, we can force the reaction to move in one direction or the other.

2–10 *A chemical bond is a force holding atoms together. The strength of the bond is measured in terms of the energy required to break it. The figures at the left indicate the number of kilocalories that will break the bonds between the pairs of atoms shown. The lines connecting the atoms represent the characteristic center-to-center distances between them. Double lines indicate double bonds, which, as you can see, hold the atoms closer together and are stronger. An angstrom (Å) is 10^{-8} centimeters, the diameter of a hydrogen atom.*

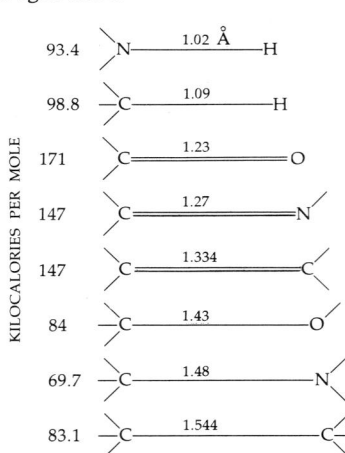

Reactions in Sequence

Compared to chemists in a laboratory, a living organism carries out its chemical activities with remarkable efficiency, not only with comparatively little loss of energy in the form of heat but also with a remarkably small accumulation of by-products and waste materials. One of the chief reasons for this efficiency is the fact that chemical reactions in living systems characteristically occur in sequence:

$$\text{Reactants} \longrightarrow \text{Product 1} \longrightarrow \text{Product 2} \longrightarrow \text{Product 3} \longrightarrow \text{end product}$$

In living systems, such sequences may involve as many as 20 steps, each of which uses the product of a previous reaction. The importance of such sequential reactions can be understood readily if you remember the principle of reversible reactions. If Product 2, for example, is used up (by being assimilated into Product 3) almost as rapidly as it is formed, the reaction Product 1 $\longrightarrow$ Product 2 can never reach equilibrium, and the whole series of reactions will move toward completion. We shall see an example of such a series in Chapter 8.

In short, although the details of chemical reactions may be complex, the principles that govern them are relatively simple. They involve the bonding capacities (valences) of the elements involved, the energies of the bonds linking one atom to another, and the proportions of the products to the reactants. Chemical reactions in living systems, although they are governed by the same principles as apply to reactions in nonliving systems, are different from the latter in some important ways. One important difference is that they do not involve large or sudden changes in heat. Another is that they are remarkably efficient, with the products of one often becoming the reactants of the next. Finally, as we noted at the beginning of the chapter, the great majority of all reactions in living systems, complicated as these systems may be, involve only a remarkably few kinds of atoms—CHNOPS.

SUMMARY

Living and nonliving matter are composed of the same fundamental particles and are subject to the same chemical and physical laws. The unusual properties associated with living systems are a consequence of their organization.

Atoms are the smallest units of the chemical elements. The nucleus of any atom contains one or more protons, which are positively charged, and it usually contains neutrons, which have no charge. The number of protons determines the chemical properties of an atom.

The neutrons and protons combined make up the atomic mass of the atom. Each neutron or proton has a mass of about one. Atoms with the same numbers of protons but different numbers of neutrons—i.e., with different atomic masses —are known as isotopes. The net positive charge of the nucleus (which is the same as the atomic number of that element) is offset by the negatively charged electrons that surround it. For an atom to be electrically neutral, the number of electrons surrounding the nucleus must equal the number of protons within it.

In the Bohr model of the atom, electrons orbit the nucleus at fixed distances from it. These distances constitute electron shells, which have different energy levels. When an electron moves from an inner to an outer shell, it acquires

potential energy which is released when it returns to its previous energy level. Each shell can accommodate only a fixed number of electrons—the first shell, two; and each outer shell, no more than eight. Atoms tend to stabilize, or complete their shells, either by sharing, gaining, or losing electrons. When an atom gains or loses an electron, it acquires a positive or negative charge. Electrically charged atoms or groups of atoms are called ions, and many of the inorganic chemical components of the living organism are in this state.

Transfers of electrons from one atom or molecule to another are oxidation-reduction reactions. Oxidation is the loss of an electron (or hydrogen atom) and reduction is the gain of an electron (or hydrogen atom).

Molecules are formed of two or more atoms bound together by chemical bonds. The two general types of chemical bonds that bind atoms into molecules are covalent bonds, which result from the sharing of pairs of electrons, and ionic bonds, which are formed by the attraction between ions of opposite electrical charge.

Matter is neither lost nor gained in a chemical reaction. This principle is expressed in chemical equations in which the kinds and numbers of atoms on either side of the equation are the same although the reactants and the products are different.

All chemical reactions involve the breaking of bonds and the formation of new bonds, and all involve changes in energy. An endergonic reaction is one in which the energies of the products are greater than those of the reactants; energy must be added to cause the reaction to proceed. An exergonic reaction liberates energy, and the energies of the products are less than those of the reactants. Reversible reactions may reach a point of equilibrium, at which the rates of forward and reverse reaction are equal and no further net chemical change takes place. In most biological systems, however, the products of any one reaction are removed by a further reaction, and so reactions proceed in sequence.

QUESTIONS

1. Define the following: ion, proton, nucleus, neutron, electron, oxidation, reduction. Check your definitions in the Glossary at the back of the book.
2. Construct Bohr models of the following: oxygen gas (O_2), the element neon (10 protons), coal gas or methane (CH_4).
3. Argon, which has an atomic number of 18, is essentially nonreactive. Can you explain why?
4. Magnesium has an atomic number of 12. How would you expect magnesium and chlorine to interact? Write the formula for magnesium chloride.

Chapter 3

Water

Life on this planet began in water, and today, almost wherever water is found, life is also present. There are one-celled organisms that eke out their entire existence in no more water than that which can cling to a grain of sand. Some species of algae are found only on the melting undersurfaces of polar ice floes. Certain species of bacteria and certain blue-green algae can tolerate the near-boiling water of hot springs. In the desert, plants race through an entire life cycle—seed to flower to seed—following a single rainfall. In the jungle, the water cupped in the leaves of a tropical plant forms a microcosm in which a myriad of small organisms are born, spawn, and die. We are interested in whether the soil of Mars and the dense atmosphere surrounding Venus contain water principally because we want to know whether life is there. On our planet, and probably on others where life exists, life and water have been companions since life first began.

3–1 *Leaves of clover after rain.*

3-2 *Water. (a) A single molecule. Note the four-cornered distribution of positive and negative charges. (b) As a result of these positive and negative charges, each water molecule can form hydrogen bonds with four other water molecules. (c) All the water molecules in a sample of liquid water are linked together by shifting hydrogen bonds. This is a diagram of water as "seen" by a computer.*

(a)

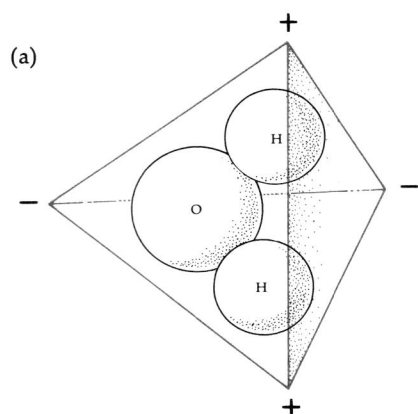

(b)

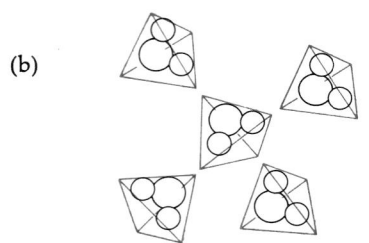

(c)

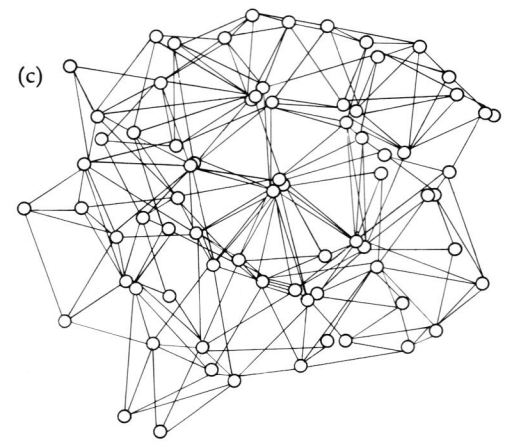

In the previous chapter, we discussed some of the physical and chemical principles that govern the behavior of the atoms and molecules of which all matter, including living matter, is composed. In this chapter, we are going to look more closely at one particular combination of atoms and molecules: water.

Water is a common liquid, by far the most common. Three-quarters of the Earth's surface is covered by water. In fact, if all the Earth's surface were absolutely smooth, all of it would be 1½ miles under water. Water makes up 50 to 95 percent of the weight of any functioning living system. If you weigh 145 pounds, about 100 pounds of this is water.

But "common" is not the same as "ordinary." Water is not in the least an ordinary liquid. Compared with other liquids it is, in fact, quite extraordinary. If it were not, it is highly probable that life could never have evolved.

WATER AND THE HYDROGEN BOND

In order to understand why water is so extraordinary and how, as a consequence, it can play its unique and crucial role in relation to living systems, we have to look again at its atomic structure. As we noted in the previous chapter, each water molecule is made up of two atoms of hydrogen and one of oxygen held together by covalent bonds (bonds formed by the sharing of two electrons). The water molecule as a whole is neutral in charge, having an equal number of electrons and protons. However, as is shown in the model, the single electrons of the two hydrogen atoms, which are shared with the oxygen atom, are more strongly attracted to the oxygen nucleus, which, with its eight protons, has a stronger positive charge than the hydrogen nucleus. As a consequence, the oxygen atom has two weak local negative charges and the hydrogen atoms each have a weak local positive charge. Thus, the water molecule, in terms of its electrical charges, is four-cornered, with two positive "corners" and two negative ones.

A molecule such as the water molecule which has zones of negative and positive charges is known as *polar*, by analogy with a magnet, which has a negative pole and a positive pole.

When a positively charged hydrogen atom of a water molecule comes into juxtaposition with an atom carrying a sufficiently strong negative charge—such as, in this case, the oxygen atom of another water molecule—the force of the attraction forms a bond between them, which is known as a *hydrogen bond*. A hydrogen bond is a relatively weak type of bond, linking a hydrogen atom that is covalently bonded to another atom—usually oxygen or nitrogen—to the oxygen or nitrogen atom of another molecule. In water, hydrogen bonds form between the negative "corners" of one water molecule and the positive "corners" of another. As a consequence, every water molecule can establish hydrogen bonds with four other water molecules. Liquid water is made up of water molecules bound together in this way, as shown in Figure 3-2. Any single hydrogen bond has only an exceedingly short lifetime; on an average, each such bond lasts about 1/100,000,000,000th of a second. All together, however, the hydrogen bonds have considerable strength, making water both fluid and stable under ordinary conditions of pressure and temperature.

Now let us look at some of the consequences of these attractions among water molecules, especially as they affect living systems.

3-3 Ammonia is very similar to water in its chemical structure, and biologists have speculated about whether it might substitute for water in life processes. The ammonia molecule, like the water molecule, is made up of hydrogen atoms covalently bonded to nitrogen, which, like the oxygen in the water molecule, retains a slight negative charge. But because there are three hydrogens with slight positive charges to one nitrogen, ammonia does not have the cohesive power of water and evaporates much more quickly. Perhaps this is why no form of life based on ammonia has been found, although NH_3 was very common in the primitive atmosphere.

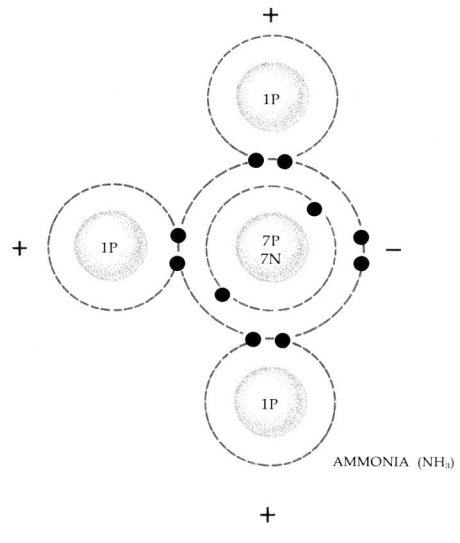

AMMONIA (NH_3)

"Ammonia! Ammonia!"

[Drawing by R. Grossman; © 1962 The New Yorker Magazine, Inc.]

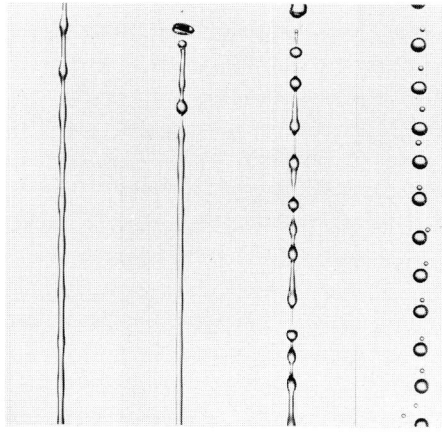

3-4 The formation of water droplets.

SURFACE TENSION

Look, for example, at water dripping from a faucet. Each drop clings to the rim and dangles for a moment by a thread of water; then just as the tug of gravity breaks it loose, its outer surface is drawn taut, enclosing the entire sphere as it falls free. Take a needle or a razor blade and gently place it flat on the surface of the water in a glass. Although the metal is denser than water, it floats! Look at a pond in spring or summer; you will see water boatmen and other insects walking on its surface as if it were almost solid.

The phenomena just described are all the result of *surface tension.* Surface tension is the result of the cohesion, clinging together, of the water molecules. (Cohesion is, by definition, the holding together of like substances. Adhesion is the holding together of unlike substances.) The only liquid with a greater surface tension than that of water is mercury. Atoms of mercury are so greatly attracted to one another that they will not adhere to anything else.

Water, because of its negative and positive charges, adheres strongly to any other charged molecules and to charged surfaces. The "wetting" capacity of water—that is, its ability to coat a surface—depends on its capacity to adhere.

3–5 *Because of hydrogen bonds, water molecules on the water's surface form a tough elastic film. The water strider, which lives on this surface, has specialized hairs on its first and third pairs of legs that enable it to rest upon the surface film, depressing it but not penetrating. The second pair of legs, which penetrate the film, serve as propelling oars.*

3–6 *The germination of seeds begins with changes in the seed coat that permit a massive uptake of water. In these acorns, photographed on the forest floor, the embryonic roots have penetrated the seed coat and emerged through the tough outer layer of the fruit.*

CAPILLARY ACTION

If you hold two dry glass slides together and dip one corner in water, the interaction of cohesion and adhesion will cause water to spread between the two slides. This is *capillary action.* Capillary action similarly causes water to rise in glass tubes of very fine bore, to creep up a piece of blotting paper, or to move slowly through the micropores of the soil and so become available to the roots of plants.

IMBIBITION

Imbibition ("drinking up") is the movement of water molecules into substances such as wood or gelatin, which swell or increase in volume as a result of interaction between them and the water molecules. The pressures developed by imbibition can be astonishingly great. It is said that stone for the ancient Egyptian pyramids was quarried by driving wooden pegs into holes drilled in the rock face and then soaking the pegs with water. The swelling of the wood created a force great enough to break the stone slab free. Seeds imbibe water as they begin to germinate, often swelling to many times their original size as a result.

SPECIFIC HEAT OF WATER

The amount of heat a given amount of a substance requires for a given increase in temperature is its *specific heat.* One calorie of heat, by definition, is the amount of heat that will raise the temperature of 1 gram (1 milliliter) of water 1°C. The specific heat of water is about twice the specific heat of oil or alcohol; that is, approximately 0.5 calorie will raise 1 gram of oil or alcohol 1°C. It is four times the specific heat of air or aluminum and nine times that of iron. Only

Table 3–1 *Comparative Specific Heats (The quantity of heat, in calories, required to raise the temperature of 1 gram through 1°C)*

Substance	Specific heat
Water	1.00
Lead	0.03
Iron	0.10
Salt (NaCl)	0.21
Glass	0.20
Sugar (sucrose)	0.30
Liquid ammonia	1.23
Chloroform	0.24
Ethyl alcohol	0.60

liquid ammonia has a higher specific heat. In other words, it requires a high input of energy to raise the temperature of water.

The high specific heat of water is also a consequence of hydrogen bonding as is water's surface tension and cohesiveness. Heat is a form of energy, the kinetic energy, or energy of movement, of molecules. In order for the kinetic energy of water molecules to increase sufficiently for the temperature to rise 1°C, it is necessary to rupture a number of the hydrogen bonds holding the molecules together. If you put an iron skillet over a gas flame, the skillet will soon glow red because the heat energy produced by the flame is transferred to the metal molecules. However, if you put a pan of water over the same gas flame, it will take much longer, and therefore more heat energy, to heat the water than the skillet. The reason for this difference is that most of the heat energy added to the water is used in breaking the hydrogen bonds holding the water molecules together, leaving only a relatively small amount to increase molecular movement. (Note that heat and temperature are not the same. A lake may have a lower temperature than that of a bird flying over it, but the lake contains more heat because it comprises many more molecules than the bird and therefore has more molecular motion. Temperature is measured in degrees and reflects average energy of the molecules. Heat is measured in calories and reflects both molecular movement and volume.)

What does this high specific heat mean in biological terms? It means that for a given rate of heat input, the temperature of water will rise more slowly than the temperature of almost any other material. Conversely, the temperature will drop more slowly as heat is released. Because so much heat input or heat loss is required to raise or lower the temperature of water, organisms that live in the oceans live in an environment where the temperature is relatively constant. Similarly, land plants and animals, because of their high water content, are more easily able to maintain a relatively constant internal temperature. This constancy of temperature is important because, as we shall see, biological reactions take place only within a narrow temperature range.*

HEAT OF VAPORIZATION

Water has a high heat of vaporization. Vaporization—or evaporation, as it is more commonly called—is the change from a liquid to a gas. It takes more than 500 calories to change a gram of liquid water into vapor, five times as much as for ether and almost twice as much as for ammonia.

Hydrogen bonding is also responsible for water's high heat of vaporization. Vaporization comes about because the rapidly moving molecules of a liquid break loose out of the surface and enter the air. The hotter the liquid, the more rapid the movement of its molecules and, hence, the more rapid the rate of evaporation; but, whatever the temperature, so long as a liquid is exposed to air that is less than 100 percent saturated, evaporation will take place and will continue to take place right down to the last drop.

In order for a water molecule to break loose from its fellow molecules—that is, to vaporize—the hydrogen bonds have to be broken. This requires heat

* Another important property of water—that it expands as it freezes—is discussed on page 759.

Table 3–2 *Comparative Latent Heats of Vaporization (The quantity of heat, in calories, required to convert 1 gram of liquid to 1 gram of gas)*

Liquid	Latent heat of vaporization
Water (at 0°C)	596
Water (at 100°C)	540
Ammonia	295
Chlorine	67.4
Hydrofluoric acid	360
Nitric acid	115
Carbon dioxide	72.2
Ethyl alcohol	236.5
Ether	9.4

3–7 *Because of the polarity of water molecules, water can serve as a solvent for ions and other polar molecules. This diagram shows table salt (NaCl) dissolving in water as the water molecules cluster around the individual ions, separating them from each other.*

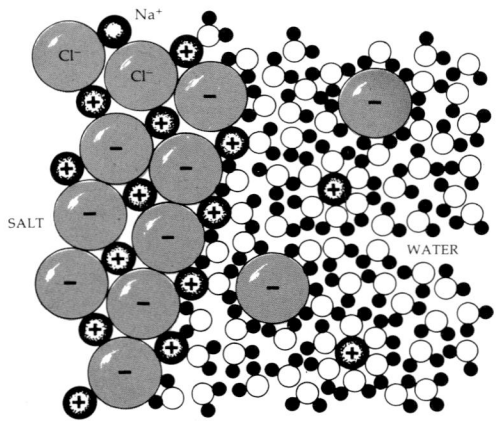

energy. As we noted previously, more than 500 calories are needed to change 1 gram of liquid water into vapor (the exact amount depends on the initial temperature of the water). The same amount of heat would raise the temperature of 500 grams of water 1°C. As a consequence, when water evaporates, as from the surface of your skin or a leaf, it absorbs a great deal of heat from the immediate environment. Thus evaporation has a cooling effect. Evaporation from the surface of a land-dwelling plant or animal is one of the principal ways in which these organisms "unload" excess heat.

WATER AS A SOLVENT

Many substances within living systems are found in solution. (A solution is a uniform mixture of the molecules of two or more substances in which one—the solvent—is a liquid.) The polarity of the water molecules is responsible for water's usefulness as a solvent. The charged water molecules tend to pull apart molecules such as NaCl into their constituent ions. Then, as shown in Figure 3–7, the water molecules cluster around and so segregate the charged ions. Many of the small molecules important in living systems also bear areas of positive and negative charge and so attract water molecules in the same way. Because of water's tendency to dissolve and to cluster around other molecules, virtually all of the chemical reactions of the cell take place in water. Moreover, as we shall see in the next chapter, water molecules participate directly in many of these reactions.

IONIZATION OF WATER

In liquid water, there is a slight tendency for a hydrogen atom to jump from the oxygen atom to which it is covalently bonded to the oxygen atom to which it is hydrogen-bonded. In this reaction, two ions are produced: the hydronium ion (H_3O^+) and the hydroxyl ion (OH^-).* In any given volume of pure water, a small but constant number of water molecules will be ionized in this way. The number is constant because the tendency of water to ionize is exactly offset by the tendency of the ions to reunite; thus even as some are ionizing, an equal number of others are forming bonds.

Although the hydrogen ion does not exist in a separate form, by convention the ionization of water is expressed by the equation:

$$HOH \rightleftharpoons H^+ + OH^-$$

The arrows indicate that the reaction can go in either direction. The fact that the arrow pointing toward HOH is longer indicates that the reaction is more likely to go in that direction. As a consequence, in any sample of water, only a small fraction exists in ionized form.

In pure water, the number of H^+ ions exactly equals the number of OH^- ions. This is necessarily the case since neither ion can be formed without the other when only H_2O molecules are present. A solution acquires the properties we recognize as acid when the number of H^+ ions exceeds the number of OH^- ions; conversely, a solution is alkaline (basic) when the OH^- ions exceed the H^+ ions.

* This is more correctly termed an hydroxide ion, but we are retaining this older term to conform with present common usage.

THE pH SCALE

Chemists define degrees of acidity by means of the pH scale. In the expression "pH," the p stands for "power" and the H stands for the hydrogen ion.

In a liter of pure water, 1/10,000,000 mole* of hydrogen ions can be detected. For convenience, this is written in terms of a power, 10^{-7}; and in terms of the pH scale, this is referred to simply as pH 7 (see Table 3–3). At pH 7, the concentrations of free H^+ and OH^- are exactly the same, and thus pure water is "neutral." Any pH below 7 is acidic, and any pH above 7 is basic. The lower the pH number, the higher the concentration of hydrogen ions. Thus pH 2 means 10^{-2} mole of hydrogen ions per liter of water, or 1/100 mole per liter— which is, of course, a much larger figure than 1/10,000,000.

We can now define "acid" and "base" more exactly:

1. An _acid_ is a substance that donates H^+ ions to a solution, i.e., a hydrogen ion donor. A solution with a pH less than 7 (with more than 10^{-7} mole of H^+ ions per liter) is acidic.
2. A _base_ is a substance that decreases the number of H^+ ions, i.e., a hydrogen ion acceptor. A solution with a pH more than 7 (with less than 10^{-7} mole of H^+ ions per liter) is basic.

* For a definition of mole see page 40.

Table 3–3   *The pH Scale*

	Concentration of H^+ ions (moles per liter)		pH	Concentration of OH^- ions (moles per liter)	
Acidic	1.0	$= 10^0$	0	10^{-14}	
	0.1	$= 10^{-1}$	1	10^{-13}	
	0.01	$= 10^{-2}$	2	10^{-12}	
	0.001	$= 10^{-3}$	3	10^{-11}	
	0.0001	$= 10^{-4}$	4	10^{-10}	
	0.00001	$= 10^{-5}$	5	10^{-9}	
	0.000001	$= 10^{-6}$	6	10^{-8}	
Neutral	0.0000001	$= 10^{-7}$	7	$10^{-7} =$	0.0000001
		10^{-8}	8	$10^{-6} =$	0.000001
		10^{-9}	9	$10^{-5} =$	0.00001
		10^{-10}	10	$10^{-4} =$	0.0001
Basic		10^{-11}	11	$10^{-3} =$	0.001
		10^{-12}	12	$10^{-2} =$	0.01
		10^{-13}	13	$10^{-1} =$	0.1
		10^{-14}	14	$10^0 =$	1.0

SOME WEAK ACIDS (HYDROGEN ION DONORS):

$$CH_3COOH \rightleftharpoons CH_3COO^- + H^+$$

| ACETIC ACID | ACETATE ION | HYDROGEN ION |

$$CH_3C \underset{OH}{|} HCOOH \rightleftharpoons CH_3CHCOO^- \underset{OH}{|} + H^+$$

| LACTIC ACID | LACTATE ION | HYDROGEN ION |

A WEAK BASE (HYDROGEN ION ACCEPTOR):

$$NH_3 + H^+ \rightleftharpoons NH_4^+$$

| AMMONIA | HYDROGEN ION | AMMONIUM ION |

Table 3–4 *pH Values for Various Solutions*

Solution	pH
Gastric secretions	0.9
Cola drink	2.5–3.0
Vinegar	3.0
Grapefruit juice	3.2
Orange juice	2.6–4.4
Tomato juice	4.3
Urine	4.8–7.5
Saliva	6.6
Milk	6.6–6.9
Pure (distilled) water	7.0
Human blood	7.4
Tears	7.4
Intestinal secretions	7.0–8.0
Pancreatic secretions	7.5–8.0
Seawater	8.0
Egg white	8.0

STRONG AND WEAK ACIDS AND BASES

Hydrochloric acid (HCl) is an example of a common acid. It is a strong acid, meaning that it tends to be almost completely ionized. Sodium hydroxide (NaOH) is a common strong base; it exists entirely as Na^+ and OH^-. Weak acids and weak bases are those which ionize only slightly.

Organic acids all contain the carboxyl group (—COOH). The carboxyl group dissociates to yield hydrogen ions when dissolved in water:

$$—COOH \rightleftharpoons —COO^- + H^+$$

Thus any compound containing this group is a hydrogen donor, or an acid. It is a weak acid, however, because, as you can see by the arrows, the —COOH ionizes only slightly.

Compounds that contain the amino group (—NH$_2$) act as weak bases since the —NH$_2$ has a weak tendency to accept hydrogen ions or protons, thereby forming —NH$_3^+$:

$$—NH_2 + H^+ \rightleftharpoons —NH_3^+$$

Because of the strong tendency of H^+ and OH^- to combine and the weak tendency of water to ionize, the concentration of H^+ ions will always decrease as the concentration of OH^- increases, and vice versa. If HCl is added to a solution containing NaOH, the following reaction will take place:

$$Na^+ + OH^- + H^+ + Cl^- \longrightarrow H_2O + Na^+ + Cl^-$$

In other words, if an acid and a base are added in equivalent amounts, the solution will once more be neutral. (Chemical interaction of an acid and a base always produces a salt and water.)

BUFFERS

Solutions more acidic than pH 1 or more basic than pH 14 are possible, but these are not included in the scale because they are almost never encountered in biological systems. In fact, almost all the chemistry of living things takes place at pH's of between 6 and 8. A notable exception is the chemical processes in the stomach of man and other animals, which take place at a pH of about 2. Human blood, for instance, maintains an almost constant pH of 7.4 despite the fact that it is the vehicle for a large number and variety of nutrients and other chemicals being delivered to the cells, as well as for the removal of wastes, many of which are acids or bases.

The maintaining of a constant pH—an example of homeostasis—is important because the pH greatly influences the rate of chemical reactions. Organisms resist strong sudden changes in the pH of blood and other fluids by means of *buffers.*

A buffer is a combination of H^+-donor and H^+-acceptor forms of weak acids or bases. The capacity of a buffer system to resist changes in pH is greatest when the concentrations of H^+ donor and H^+ acceptor are equal. As the ratio changes in either direction, the buffer becomes less effective.

The major buffer system in the human bloodstream is the acid-base pair H_2CO_3–HCO_3^-. The weak acid H_2CO_3 (carbonic acid) dissociates into H^+ and bicarbonate ions as follows:

$$H_2CO_3 \text{ (H}^+\text{ donor)} \rightleftharpoons H^+ + HCO_3^- \text{ (H}^+\text{ acceptor)}$$

The H_2CO_3–HCO_3^- buffer system resists the changes in pH that might result from the addition of small amounts of acid or base by "soaking" up the acid or base. For example, if a small amount of H^+ in the form of HCl is added to the system, it combines with the H^+ acceptor HCO_3^- to form H_2CO_3. If a small amount of OH^- is added, it combines with the H^+ to form H_2O; more H_2CO_3 tends to ionize to replace the H^+ as it is used.

Control of the pH of the blood is rendered even "tighter" by the fact that the H_2CO_3 is in equilibrium with dissolved CO_2 in the blood:

$$H_2O + CO_2 \rightleftharpoons H_2CO_3$$

As the arrows indicate, the two reactions are in equilibrium and the equilibrium favors the formation of CO_2; in fact, the ratio is about 100 to 1 in favor of CO_2 formation.

Dissolved CO_2 in the blood is, in turn, in equilibrium with the CO_2 of the air in the lungs. Changes in the rate of loss of the CO_2 in the lungs, by changes in the rate of breathing, can change the HCO_3^- concentration in the blood and thus help control pH.

Obviously, if the blood should be flooded with a very large excess of acid or base, the buffer would fail, but normally it is able to adjust continuously and instantaneously to the constant small additions of acid or base which normally occur in body fluids.

SUMMARY

One of the chief components of living systems is water. Water, which is the most common liquid in the biosphere, has a number of remarkable properties which are responsible for its "fitness" for its role in living systems. These properties are a consequence of the molecular structure of water.

Water is made up of two hydrogen atoms and one oxygen atom held together by covalent bonds. Because the electrons shared by the molecules tend to cluster nearer the oxygen molecule, which has a stronger positive charge than the hydrogen molecule, the water molecule is polar, carrying two weak negative charges and two weak positive charges. As a consequence, weak bonds form between water molecules. Such bonds, which link a positively charged hydrogen atom that is part of one molecule to a negatively charged oxygen atom that is part of another, are known as hydrogen bonds. Each water molecule can form hydrogen bonds with four other molecules, and although each bond is weak and constantly shifting, the total strength of the bonds holding the molecules is very great.

Because of the hydrogen bonds holding the water molecules together (cohesion), water has a high surface tension and a high specific heat. (The specific heat of a substance is the amount of heat that a given amount of the substance requires for a given increase in temperature.) It also has a high heat of vaporization (the heat required to change it from a liquid to a gas).

The polarity of the water molecule (having zones of positive and negative charge) is responsible for water's capacity for adhesion which makes possible capillary movement and imbibition. Similarly, water's polarity makes it a good solvent for ions and other charged molecules.

Water has a slight tendency to ionize, that is, to separate into H^+ (actually H_3O^+, hydronium ions) and OH^- ions. In pure water, the number of H^+ and

OH⁻ ions is equal. An acid is a substance that donates H^+ ions to a solution. The pH scale reflects the proportion of H^+ ions to OH⁻ ions. An acid is a solution with more H^+ than OH⁻ ions; a base has more OH⁻ than H^+. Almost all of the chemical reactions of living systems take place within a narrow range of pH. Organisms maintain this narrow pH range by means of buffers.

QUESTIONS

1. Define the following terms: polarity, surface tension, hydrogen bond, acid, alkaline. Check your definitions in the Glossary in the back of the book.
2. What is vaporization? Describe the changes which take place in water as it vaporizes. What is heat of vaporization? Why does water have an unusually high heat of vaporization?
3. Sodium bicarbonate ($NaHCO_3$) is a common buffer. What reactions would take place if sodium bicarbonate was in a solution to which hydrogen chloride was added?

Chapter 4

4–1 *A diamond is pure carbon. Each atom bonds to four others in a stable lattice whose geometry is mirrored in the crystal shape and whose stability is shown by the diamond's unrivaled hardness.*

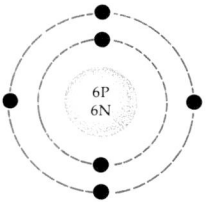

4–2 *The carbon atom, surrounded by its electrons. Carbon atoms are the skeleton on which living organisms are constructed, forming the basic framework of all organic molecules.*

The Carbon Atom and Molecules for Energy

As we mentioned in Chapter 2, all living systems are largely composed of only a very few different kinds of molecules: sugars, lipids, amino acids (which form proteins), and nucleotides (which form nucleic acids). All of these molecules contain both hydrogen and oxygen atoms and all also contain carbon atoms. The modern definition of organic chemistry is, in fact, the chemistry of complex compounds containing carbon. (The term "complex" excludes only very simple compounds, such as CO_2.)

This chapter is primarily concerned with molecules involved in supplying energy to living systems. We are going to introduce these molecules with a discussion of the carbon atoms they all contain, taking a look at the special properties that make carbon uniquely suited for its functions in these molecules. We are then going to look more closely at two types of carbon-containing, energy-storing molecules: sugars and fats (lipids). Finally we are going to examine the molecule that serves as the intermediary between energy-containing storage molecules and the energy-requiring activities of living systems. The molecule that plays this role in all living cells, including prokaryotic bacteria and blue-green algae and every cell in your own body, is adenosine triphosphate, ATP.

THE CARBON ATOM

Just as knowledge of the structure of the water molecule helps us understand water's unusual characteristics, so we can better understand the central role of the carbon atom in the chemistry of living matter if we look at the way it is put together. As you will recall from Chapter 2, a carbon atom has six protons and six electrons, two in the inner shell and four in the outer shell. As a result of having four electrons in its outer shell, carbon does not tend to gain or to lose electrons and become ionized. Instead, it usually forms covalent bonds. Also, because of the four electrons in its outer shell, each carbon atom can form four covalent bonds with as many as four different atoms. Coal gas, or methane (CH_4), is a simple example.

Even more important, in terms of carbon's biological role, carbon atoms can form bonds with each other. Ethane contains two carbons; propane, three; butane, four; and so on. In these gases, every carbon bond that is not occupied by another carbon atom is taken up by a hydrogen atom. Compounds consisting of only carbon and hydrogen are known as *hydrocarbons*.

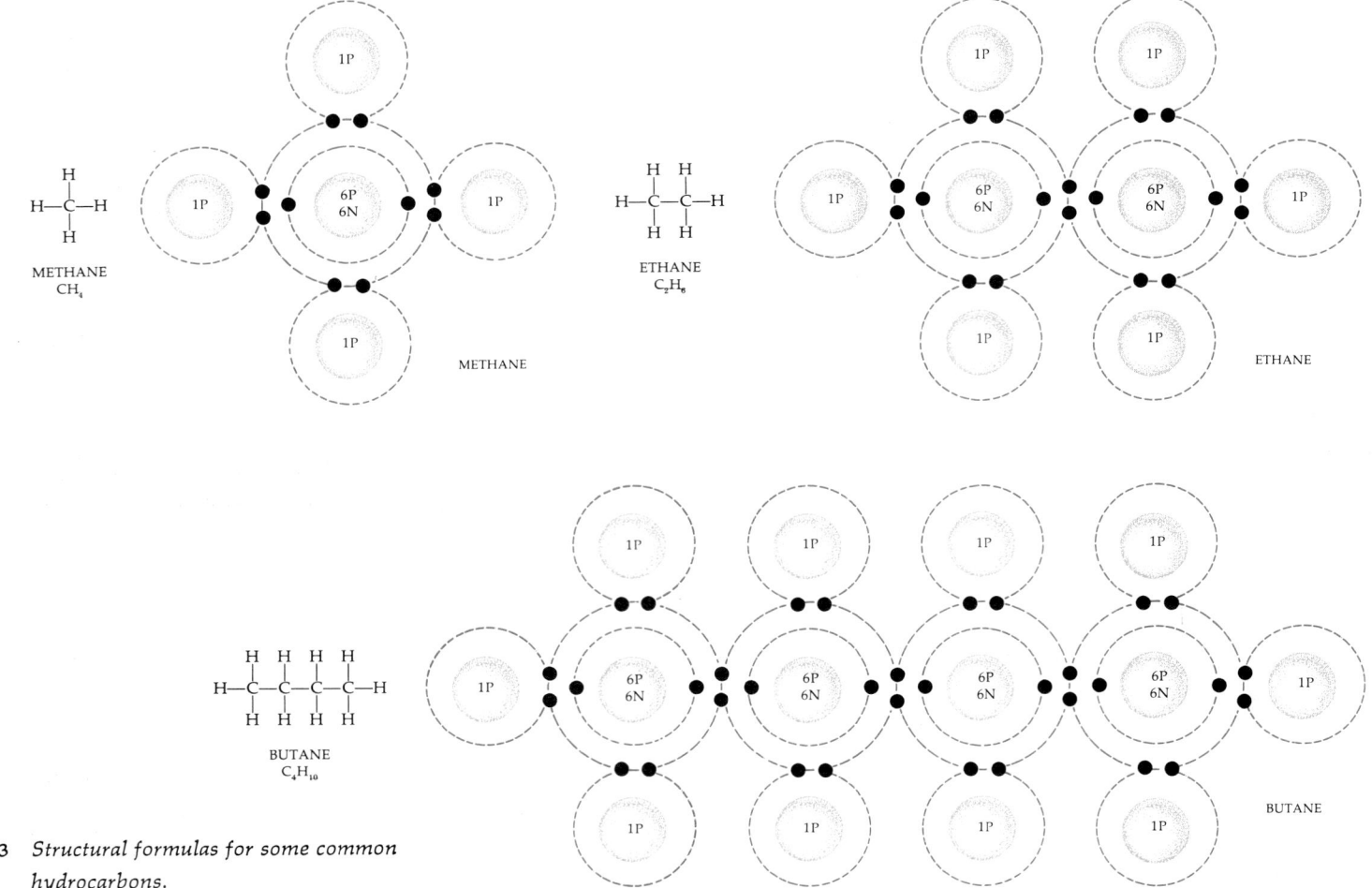

H   H
 \ /
  C
 / \
H   H

METHANE
CH$_4$

METHANE

H   H
| |
H—C—C—H
| |
H   H

ETHANE
C$_2$H$_6$

ETHANE

H H H H
| | | |
H—C—C—C—C—H
| | | |
H H H H

BUTANE
C$_4$H$_{10}$

BUTANE

4–3 *Structural formulas for some common hydrocarbons.*

4–4 *Some functional groups which play important roles in organic compounds.*

—OH   HYDROXYL GROUP

—NH$_2$ AMINO GROUP

$$-\overset{\displaystyle}{\underset{\displaystyle O}{C}}-CH_3$$ ACETYL GROUP

$$-\overset{\displaystyle}{\underset{\displaystyle O}{C}}-OH$$ CARBOXYL GROUP

$$-O-\overset{\displaystyle O^-}{\underset{\displaystyle O}{P}}-O^-$$ PHOSPHATE GROUP

Carbon atoms can bond not only to each other and to hydrogen but to a wide variety of other atoms and groups of atoms. Bonding to an OH (hydroxyl) group is an important example. When one hydrogen and one oxygen are bonded covalently, one electron is left over for sharing. If a hydroxyl group is substituted for one or more of the hydrogens in a hydrocarbon, the compound takes on the properties of an alcohol. Methane thus becomes methyl alcohol, or wood alcohol (CH$_3$OH), a pleasant-smelling, poisonous compound noted for its ability to cause blindness or death. Ethane becomes ethyl alcohol, or grain alcohol (C$_2$H$_5$OH), which is present in all alcoholic beverages. Glycerol, C$_3$H$_5$(OH$_3$), contains, as its formula indicates, three carbon atoms and three hydroxyl groups. OH is what is known as a functional group. Functional groups are combinations of atoms that occur repeatedly and that confer similar properties on otherwise dissimilar compounds. The carboxyl group (COOH), mentioned in the previous chapter, is a functional group that gives a compound the properties of an acid. OH is the functional group of the alcohols.

FRIEDRICH KEKULÉ

Friedrich Kekulé, Professor of Chemistry in Ghent, Belgium, was the first to deduce that carbon has a valence (combining power) of four, although nothing was known in his time of the nature of atoms or electrons. Another of Kekulé's important discoveries was that carbon compounds can form rings. For some time, he had been pondering the nature of the structure of benzene, but to no avail. One afternoon in 1865, he turned away from his work:

I turned my chair to the fire and dozed. Again the atoms were gambolling before my eyes. This time the smaller groups kept modestly in the background. My mental eye, rendered more acute by repeated visions of this kind, could now distinguish larger structures, of manifold conformations; long rows, sometimes more closely fitted together; all twining and twisting in snakelike motion. But look! What was that? One of the snakes had seized hold of its own tail, and the form whirled mockingly before my eyes. As if by a flash of lightning I awoke and this time also I spent the rest of the night working out the consequences of the hypothesis.

The snake grasping its own tail was Kekulé's clue to the benzene ring. The story of the professor's dream is often recounted as an example of the role of the creative subconscious in scientific discovery.

4–5 (a) *Ethylene, because of its double bond, is a highly reactive compound. Letting go one of the double bonds leaves each carbon atom with an electron available for sharing at either end of the molecule, as in polyethylene. (b) Polyethylene is very unreactive and so is widely used for tubing and containers which hold reactive chemicals. It is a conspicuous component of the pollution problem.*

(a) ETHYLENE MOLECULES

(b) POLYETHYLENE MOLECULE

4–6 *Benzene ring.*

Carbon atoms can form not only single but also double (Figure 4–5a) or even triple bonds. A compound that contains one or more double bonds is said to be unsaturated because more atoms or groups of atoms can be added at the location of the double bond. Polyethylene, a plastic, for instance, can be produced from ethylene by breaking the double bonds and sharing them between adjacent molecules (Figure 4–5b).

Chains of carbon atoms can unite with one another to form a ring. Benzene is an example of such a ring compound. Notice the double lines in Figure 4–6 indicating a double bond or two shared pairs of electrons.

Carbon and Combustion

A common feature of these hydrocarbons, as you have probably noticed, is that they burn readily. For example, methane (CH_4), mentioned previously, is often the gas responsible for fires and explosions in coal mines. Most of our common fuels are carbon–hydrogen combinations. Natural gas consists mostly of methane (CH_4) with some ethane (C_2H_6) and propane (C_3H_8). Bottled gas is usually propane or butane (C_4H_{10}). Gasoline consists of combinations of five-carbon, six-carbon, seven-carbon, eight-carbon, and nine-carbon hydrocarbons. All of these substances combine readily with oxygen, producing carbon dioxide and water. The total bond energy of the products is considerably less than the total bond energy of the reactants; in other words, they are highly exergonic.

$$\text{Fuel molecules} + O_2 \longrightarrow CO_2 + H_2O + \text{energy}$$

Notice that this is an example of an oxidation-reduction reaction, in which the carbon is oxidized and oxygen, the acceptor molecules for the hydrogen ion and electron, is reduced. This same type of reaction takes place when food molecules—all of which contain carbon—are oxidized in living systems.

THE SWEETNESS OF SUGAR

Beauty, as we know, is in the eye of the beholder. Similarly, sweetness resides in perception systems that detect these valuable food-energy sources.

Houseflies, for instance, have sugar detectors in their feet. If a housefly lights on a droplet of weak sugar solution, its proboscis will automatically be extended. This useful response evolved over the millennia as a food-detecting mechanism. We similarly have sensory receptors especially tuned for the detection of sugars, although our receptors, in keeping with our eating habits, are in our tongues. As you might expect, because of its value as a food-energy source, many other organisms have sugar-detecting mechanisms. Houseflies react positively to about the same range of sugars that we do, although its detection mechanism, as gauged by the proboscis extension test, is about 10 million times more sensitive.

The sweet taste of sugar molecules presumably depends upon the three-dimensional shape of the sugar molecule and the orientation of its functional groups, although the exact relationship between configuration and taste is not known. The recent discovery that eating certain foods, such as artichokes, confers a sweet taste on foods eaten subsequently may provide a clue both to our taste mechanisms and to a source of low-calorie sugar substitutes.

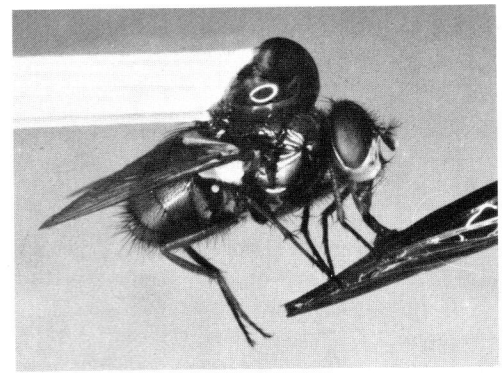

This housefly is glued to a wooden kitchen match. Its front feet are touching a small paintbrush, which has been dipped in a weak solution of sugar water. Because its feet have "tasted" the sugar, its proboscis is extended.

SUGARS

Sugars are a principal chemical energy source for living systems. There are three general classes: (1) monosaccharides ("single sugars") such as glyceraldehyde, glucose, and fructose; (2) disaccharides ("two sugars") such as sucrose, maltose, and lactose; and (3) polysaccharides ("many sugars") such as glycogen and starch. The disaccharides and polysaccharides are built up from monosaccharides by living systems.

Monosaccharides: Ready Energy for Living Systems

In each monosaccharide, the ratio of carbon to hydrogen to oxygen atoms is 1:2:1, as indicated by the shorthand symbol $n(CH_2O)$, in which n may be as small as three—$C_3H_6O_3$—or as large as eight—$C_8H_{16}O_8$. In every monosaccharide molecule, whether it has three, four, five, or more carbons, the carbon atoms are bonded to each other in a chain. The functional group of sugar is either an aldehyde group—a carbon atom double-bonded to oxygen and single-bonded to a hydrogen atom (Figure 4–7a) or a keto group—a central carbon atom double-bonded to oxygen and single-bonded to two carbon atoms (Figure 4–7b). Because of the characteristic CH_2O ratio, simple sugars and combinations of them are known as *carbohydrates* (carbon plus water).

There are many different ways to describe a molecule. Glucose, for example, has 6 carbon atoms, 12 hydrogen atoms, and 6 oxygen atoms, so $C_6H_{12}O_6$, its chemical formula, is one way to describe it. If we want to emphasize the fact that it is a carbohydrate, we can make this point by describing glucose as $6(CH_2O)$, which, of course, says just the same thing in a slightly different way. However, fructose, another simple sugar, also contains 6 carbons, 12 hydrogens, and 6 oxygens and has a similar structure—a chain of carbon atoms to which hydrogen and oxygen atoms are attached. Despite this similarity, glucose and fructose have different chemical characteristics; for instance, fructose tastes

4–7 *The encircled atoms in these three-carbon sugars represent the functional groups for sugar. One (a) is known as an aldehyde group; the other (b), as a keto group.*

(a)   (b)

4-8 (a) *In glucose, the functional group is an aldehyde group, and in fructose it is a keto group. Yet the total numbers of atoms in glucose and fructose are the same and so, therefore, are their chemical formulas. (b) In water, the molecules tend to assume a ring form, which also involves some rearrangements in the positions of hydrogen and oxygen atoms. In the ring forms, carbon atoms are not usually indicated.*

GLUCOSE
$C_6H_{12}O_6$

FRUCTOSE
$C_6H_{12}O_6$

(a)

GLUCOSE

FRUCTOSE

(b)

4-9 *Although carbon atoms are not usually indicated in these structural formulas, each has a number. The numbering of the carbon atoms in glucose and fructose is shown.*

GLUCOSE
$C_6H_{12}O_6$

FRUCTOSE
$C_6H_{12}O_6$

much sweeter than glucose. How is it that two compounds with the same formula are different? The reason is that the hydrogen and oxygen atoms are arranged differently in glucose than in fructose. The molecules can therefore be described more precisely by drawing structural formulas. (See Figure 4–8a.)

When a monosaccharide containing five or more carbon atoms is dissolved in water (as it is in living systems), it tends to take on a ring formation in which the electrons of the characteristic double-bonded oxygen atom are shared between two of the carbon atoms. In glucose, the first and fifth carbons are linked to the oxygen atom, producing a six-sided ring. In fructose, the oxygen atom is linked between the second and fifth carbons, producing a five-sided ring. The lower edge of the ring is made darker and thicker to indicate the three-dimensional nature of the molecule. (See Figure 4–8b.)

By convention, the carbon atoms are "understood" and not labeled. However, each carbon atom in such a diagram has a number (Figure 4–9) which makes it possible to refer to particular carbon atoms. Thus, for example, we can say that changing from chain to ring forms of glucose involves changes in H atoms and OH groups attached to carbon atom 1 and carbon atom 5.

Carbohydrates and Energy

Carbohydrates react with oxygen to produce carbon dioxide and water:

$$n(CH_2O) + n(O_2) \longrightarrow n(CO_2) + n(H_2O)$$

This reaction is energy-yielding. This energy can be released by burning in the form of heat, as with the energy of fuel oil molecules. Glucose, when burned, yields carbon dioxide, water, and 686 kilocalories per mole. In a living cell, comparatively little energy is converted to heat and a comparatively large amount is recovered in a form in which it can do work for the cell.

A principal energy source of man and other vertebrates is the monosaccharide glucose. It is in this form that sugar is generally transported in the animal body. A patient receiving an intravenous feeding in a hospital is getting glucose dissolved in water. This glucose is carried through the bloodstream to the cells of the body where the energy-yielding reactions are carried out.

Disaccharides: Transport Forms

Disaccharides consist of two simple sugar molecules (monosaccharides) coupled together. Although glucose is the common transport sugar for vertebrates, sugars are often transported in other organisms as disaccharides. Sucrose, commonly called cane sugar, is the form in which sugar is transported in plants from the photosynthetic cells (mostly in the leaf), where it is produced, to other parts of the plant body. Sucrose is composed of the monosaccharides glucose and fructose. As is the case with all disaccharides, the bond of the two monosaccharides is formed by removal of a molecule of water. In sucrose, the bond forms between the carbon in the one position on the glucose molecule and the carbon in the five position on the fructose molecule. (See Figure 4–10.)

When the sucrose molecule is split into glucose and fructose, as it is when it is used as an energy source, the molecule of water is added again. This splitting involving the addition of a water molecule is known as *hydrolysis*, from *hydro*, meaning "water," and *lysis*, meaning "breaking apart." It is a "downhill" reaction, releasing about 5.5 kilocalories per mole. Conversely, the

4–10 *Sucrose is a disaccharide made up of two monosaccharides, glucose and fructose. The formation of sucrose, shown here, involves the removal of a molecule of water. Splitting sucrose requires, conversely, the addition of a water molecule (hydrolysis).*

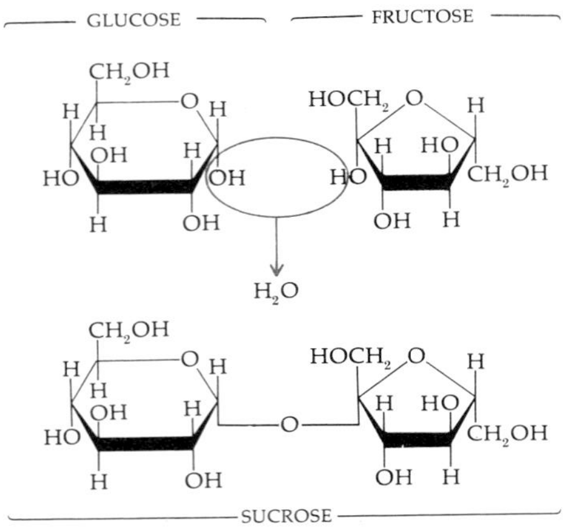

formation of sucrose from the two monosaccharides is an "uphill" reaction, requiring an input of about 5.5 kilocalories.

Another common disaccharide is lactose, a sugar which occurs only in milk. Lactose is made up of glucose combined with another monosaccharide, galactose. Sugar is transported through the blood of many insects in the form of another disaccharide, trehalose, which consists of two glucose units linked together.

Polysaccharides: Sugars in Storage

Polysaccharides are made up of monosaccharides linked together in long chains. Unlike monosaccharides and disaccharides, polysaccharides are often insoluble. They constitute a storage form for sugar. When there is a surplus of sugar, polysaccharides are formed. When there is a shortage, the polysaccharides are hydrolyzed (broken apart again).

Starch is the principal storage form for sugar in higher plants. A potato, for example, contains starch produced from the sugar formed in the green leaves of the potato, transported underground and accumulated there in a form suitable for winter storage, after which it will provide for new growth in the spring. Starch occurs in two forms, amylose and amylopectin. Both consist of glucose units linked together. In amylose, the first carbon of one glucose ring is linked to the fourth carbon of the next in a long unbranched chain. Amylopectin also contains chains of glucose units linked in this way but also has many branches, formed by links between the first and sixth carbons on adjacent glucose units.

4–11 *In plants, sugars are stored in the form of starch. Starch is composed of two different types of polysaccharides, amylose (a) and amylopectin (b). A single molecule of amylose may contain 1,000 or more glucose units in a long unbranched chain, which winds to form a helix (c). A molecule of amylopectin is made up of about 48 to 60 glucose units arranged in shorter, branched chains. Starch molecules, perhaps because of their helical nature, tend to cluster into granules.*

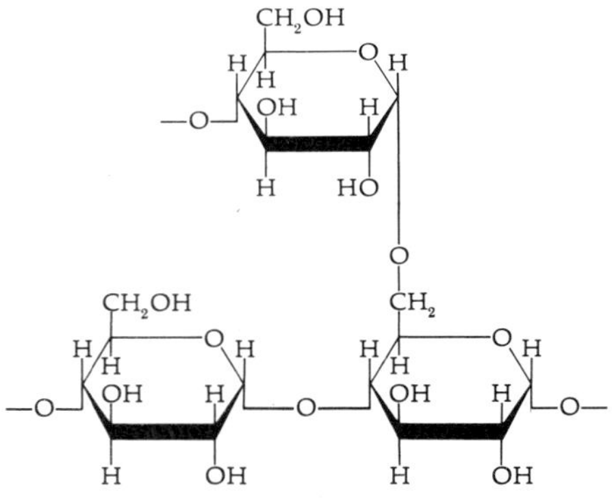

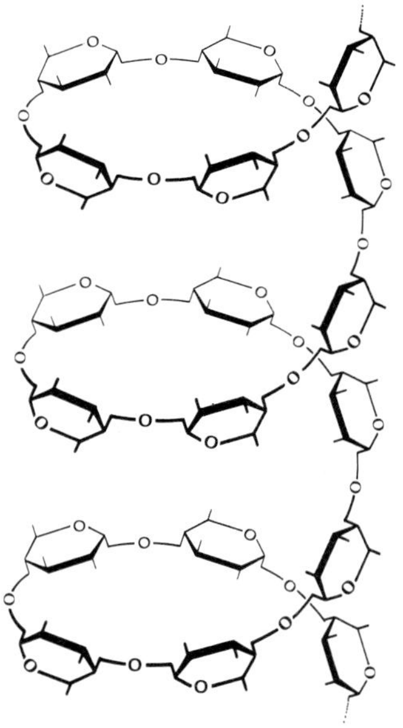

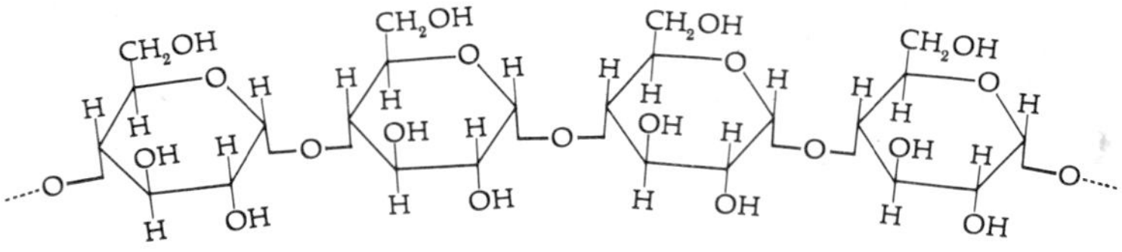

4–12 *In this electron micrograph of a portion of a cell in a tomato leaf, the large pale objects are starch grains. They are within the chloroplasts, the structures in which simple sugars are synthesized from carbon dioxide and water, using the light of the sun as an energy source.*

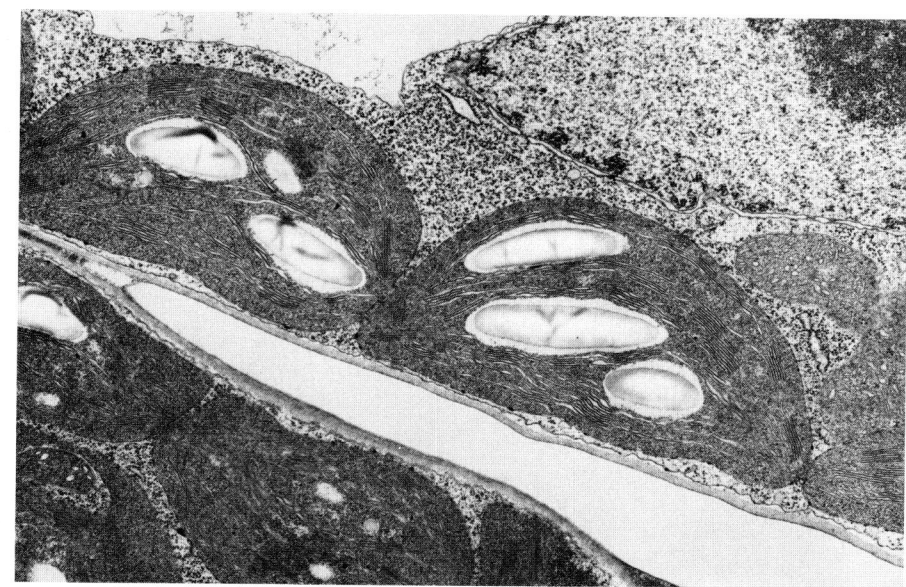

4–13 *Glycogen, which is the common storage form for sugar in vertebrates, resembles amylopectin in its general structure except that each molecule contains only 16 to 24 glucose units. The dark granules surrounding the central fat droplet in this liver cell are glycogen. When needed, the glycogen can be converted to glucose. The large body in the lower right is the nucleus.*

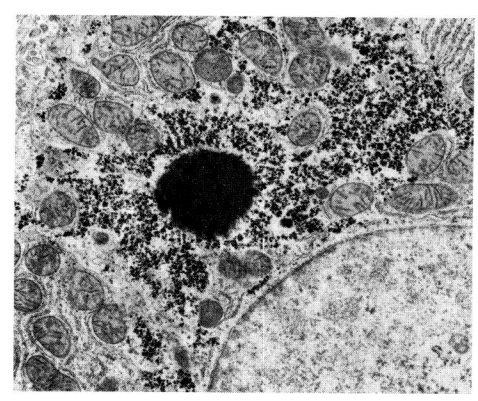

4–14 *Glyceraldehyde is a sugar, as you can see from the characteristic aldehyde group. In glycerol the oxygen atom characteristic of sugar is replaced by the hydroxyl (OH) group characteristic of an alcohol.*

```
        H                    H
        |                    |
        C=O              H—C—OH
        |                    |
    H—C—OH             H—C—OH
        |                    |
    H—C—OH             H—C—OH
        |                    |
        H                    H
  GLYCERALDEHYDE         GLYCEROL
```

In amylopectin, the branches occur at about every twelfth glucose unit. In plant cells, starch is often synthesized from sugar in special organelles known as plastids.

Glycogen is the principal storage form for sugar in higher animals. Glycogen has a structure much like that of amylopectin except that it is more highly branched, with branches occurring every eight to ten glucose units. In man, glycogen is stored principally in the liver and in muscle tissue. When there is an excess of glucose in the bloodstream, the liver forms glycogen. When the concentration of glucose in the blood drops, the hormone glucagon, produced by the pancreas, is released into the bloodstream; glucagon stimulates the liver to hydrolyze glycogen to glucose, which then enters the bloodstream.

FATS: ENERGY IN STORAGE

Unlike many higher plants, such as the potato plant, higher animals have only a limited capacity to store carbohydrates. In higher animals, sugars in excess of what can be stored as glycogen are converted into fats. Some plants also store food energy as fats (oils), especially in seeds and fruits. Fats contain a higher proportion of energy-rich carbon–hydrogen bonds than carbohydrates do and, as a consequence, contain more chemical energy. On the average, fats yield about 9.3 kilocalories per gram* as compared to 3.79 per gram of carbohydrate or 3.12 per gram of protein. They are insoluble in water.

A fat molecule consists of three molecules of fatty acids joined to one glycerol molecule. Fatty acids, which are seldom found in cells in a free state, typically consist of chains between 14 and 22 carbon atoms long. About 70 different fatty acids are known which differ in their chain length and in whether the chain

---

* 1,000 grams = 1 kilogram = 2.2 pounds, so oxidation of a pound of fat would yield about 4,700 kilocalories (remember that 1 kilocalorie = 1 Calorie), more than the 24-hour requirement for moderately active adults.

4–15 *A fat molecule consists of three fatty acids joined to a glycerol molecule. The long hydrocarbon chains of which the fatty acids are composed terminate in carboxyl groups, which become covalently bonded to the glycerol molecule. The*

bonds are formed by the removal of a molecule of water. The physical properties of the fat—such as its melting point—are determined by the lengths of the chains and by whether its component fatty acids are saturated or unsaturated.

Three different fatty acids are shown here. Stearic acid and palmitic acid are saturated, and oleic acid is unsaturated, as you can see by the double bond in its structure.

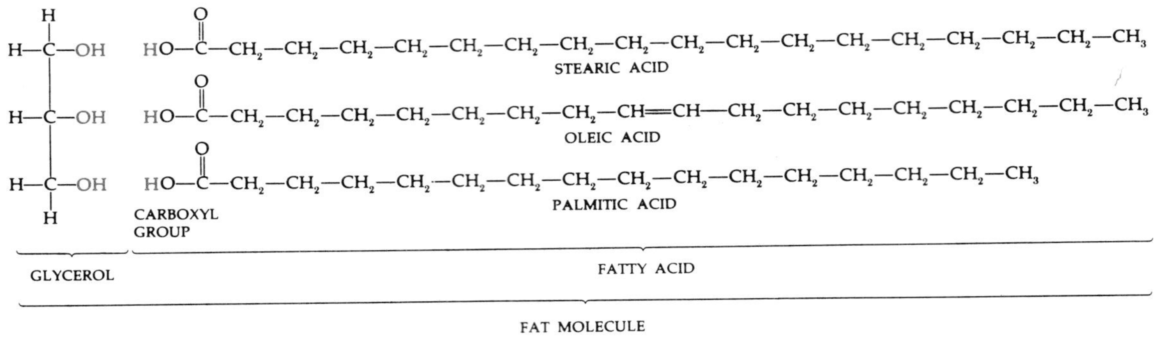

STEARIC ACID

OLEIC ACID

PALMITIC ACID

CARBOXYL GROUP

GLYCEROL

FATTY ACID

FAT MOLECULE

4–16 *Diets high in saturated fats, which are largely animal fats, are associated with heart disease. Studies of different populations in seven countries revealed that, with little variation, the coronary death rate rose with the amount of saturated fats in the daily diet. In the United States, railroad employees made up the study group.*

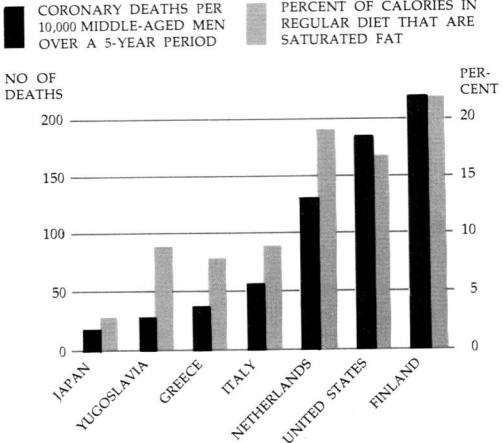

CORONARY DEATHS PER 10,000 MIDDLE-AGED MEN OVER A 5-YEAR PERIOD

PERCENT OF CALORIES IN REGULAR DIET THAT ARE SATURATED FAT

contains any double bonds (as in oleic acid) or not (as in stearic acid), and in the position in the chain of these double bonds. (See Figure 4–15.) A fatty acid such as stearic acid in which there are no double bonds is known as a saturated fatty acid. A fatty acid such as oleic acid which contains carbon atoms joined by double bonds is said to be unsaturated. Unsaturated fats, which tend to be oily liquids, are more common in plants than in animals; examples are olive oil, peanut oil, and corn oil. Animal fats, such as lard, contain saturated fatty acids and usually have higher melting temperatures. For reasons which are not clear, diets high in saturated fats are associated with heart disease. (See Figure 4–16.)

Sugars, Fats, and Calories

As we noted earlier, when carbohydrates are taken into the body in excess of the body's energy requirements, they are stored temporarily as glycogen or, more permanently, as fats. Conversely, when the energy requirements of the body are not met by its immediate intake of food, glycogen and, subsequently, fat are broken down to fill these requirements. Whether or not the body uses up its own storage molecules has nothing to do with the molecular form in which the energy comes into the body. It is simply a matter of whether these molecules, as they are broken down, release sufficient numbers of calories. The only scientific basis for various claims of extraordinary weight loss on diets which limit the intakes of particular kinds of foods is that, because they usually tend to be so monotonous, the dieter eats less.

SOAP AND WATER

Homemade soap is produced by boiling animal fat with lye (sodium hydroxide). The boiling hydrolyzes the bonds linking the fatty acids to the glycerol molecule, and the sodium hydroxide then reacts with the fatty acid:

$$NaOH \quad + \quad HC_{18}H_{33}O_2 \quad \longrightarrow \quad NaC_{18}H_{33}O_2 \quad + \quad HOH$$

sodium hydroxide $\qquad$ stearic acid $\qquad\qquad$ sodium stearate $\qquad$ water

(soap)

Because the sodium end of the molecule is soluble in water and the fatty acid chain is not, soap molecules tend to form clusters (called "micelles") in water. In

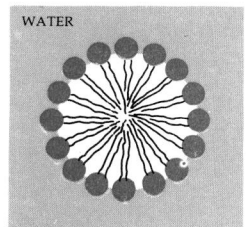

WATER

these clusters, the soluble ends of the soap molecules point outward, and the insoluble, fatty acid ends point inward. Soaps remove grease because the grease particles, not soluble in water, are trapped within these fatty acid clusters. Other dirt particles are usually charged so that, once released from the grease, they go into solution, like the sodium and chloride ions shown in the drawing on page 52, and can be washed away. The clustering of soap molecules is an example of a hydrophobic (water-hating) interaction. Such interactions play an important role in the self-assembly of cell membranes and other structures.

4–17 (a) *Adenine, a nitrogen base.* (b) *Ribose, a five-carbon sugar.* (c) *A phosphate group.* (d) *ATP, adenine plus ribose plus three phosphate groups. The wavy line ~ linking each of the last two phosphates to the molecule indicates a high-energy bond.*

(a) ADENINE

(b) RIBOSE

(c) PHOSPHATE

(d) ADENINE / RIBOSE / PHOSPHATES

ATP: THE ENERGY CARRIER MOLECULE

When a compound such as glucose is broken down in living systems, some of the energy released is caught and packaged in adenosine triphosphate, or ATP, molecules. These molecules are present in all living cells and participate in virtually every series of biochemical reactions.

To understand how ATP performs this function, we must look at its structure. ATP is composed of three types of components. One of these is a nitrogen-containing compound (a nitrogen base) known as adenine (Figure 4–17a). Adenine is a compound which we shall be mentioning frequently in Section 2, because it is one of the principal components of the genetic material.

A second component is a five-carbon sugar, ribose (Figure 4–17b).

The third component is phosphate, which is an atom of phosphorus combined with four oxygen atoms. This combination of a nitrogen base plus a sugar plus one phosphate group makes up a compound known as a *nucleotide*, which is one of the basic and important many-purpose chemical combinations in the cell. As the name implies, ATP (adenosine triphosphate) contains three phosphates (Figure 4–17d).

The third phosphate can be removed from ATP by hydrolysis, leaving ADP (adenosine diphosphate) and a phosphate:

$$ATP + H_2O \longrightarrow ADP + phosphate$$

In the course of this reaction, about 7,000 calories per mole of ATP are released, a relatively large amount of chemical energy. Removal of the second phosphate produces AMP (adenosine monophosphate) and releases a slightly larger amount of chemical energy. To indicate that relatively large amounts of chemical energy are involved, the bonds linking these two phosphates to the rest of the molecule are often called high-energy bonds, symbolized by ~. This term is somewhat misleading, however, because the energy released during this reaction does not arise entirely from the bond. The products of these reactions, ADP or AMP and phosphate, contain 7,000 calories less than the reactants, ATP or ADP and H_2O. This difference in energy between reactants and products is due only in part to bond energy. It is also a result of the internal structure of the ATP or ADP molecules. The phosphate groups each carry negative charges and so tend to repel each other.

In most reactions that take place within a cell, the terminal phosphate groups of ATP or ADP are not simply removed but transferred to another molecule. The amounts of energy released in such transfers are of a magnitude well suited to the step-by-step activities of living cells. (A corollary to this is that the activities of living systems have evolved in such a way as to take advantage of the energy made available by hydrolysis of the "high-energy" bonds of ATP and ADP.)

The ADP (or AMP) molecule is subsequently "recharged"—regaining its phosphate group(s) and again becoming ATP—with the energy obtained from the oxidation of a carbon-containing compound. (In photosynthetic cells, as we shall see, the energy of sunlight is also used directly to form ATP from ADP.) We shall discuss the formation of ATP at greater length in the final chapters of this section.

Let us look at a simple example of an energy exchange involving ATP. As we mentioned previously, sucrose is formed from the monosaccharides glucose

and fructose; this is an energy-requiring (endergonic) reaction. The energy for the reaction can be supplied by coupling the synthesis of sucrose to the removal of a phosphate group from the ATP molecule.

First, the terminal phosphate group of ATP is transferred to the glucose molecule.

$$\text{ATP} + \text{glucose} \longrightarrow \text{glucose phosphate} + \text{ADP}$$

In this reaction, some of the 7,000 calories available from the hydrolysis of ATP are conserved by the transfer of the phosphate group to the glucose molecule, which thus becomes "energized."

Next glucose phosphate reacts with fructose to form sucrose:

$$\text{Glucose phosphate} + \text{fructose} \longrightarrow \text{sucrose} + \text{phosphate}$$

In this second step, the phosphate group is released from the glucose and most of the energy made available by its release (energy originally derived from the ATP) is used to form the bond between glucose and fructose. The free phosphate formed is then available, with an input of energy, to recharge an ADP molecule to ATP.

As we noted on page 62, the formation of sucrose requires 5.5 kilocalories per mole. The conversion of ATP to ADP + phosphate releases 7 kilocalories per mole. Thus most of the energy in the phosphate bond has been effectively utilized. The ATP–ADP molecule serves as a universal energy carrier, shuttling between energy-releasing reactions, such as the breakdown of glucose, and energy-requiring ones.

SUMMARY

The chemistry of living organisms is, in essence, the chemistry of compounds containing carbon. Carbon is uniquely suited to this central position by the fact that it is the lightest atom capable of forming four covalent bonds. Because of this capacity, carbon can combine with carbon and other atoms to form strong and stable chain and ring compounds with numerous small variations in structure that are reflected by variations in chemical properties. Compounds of carbon with hydrogen, such as methane and propane, are known as hydrocarbons. Combinations of carbon, hydrogen, and oxygen in the proportions of 1:2:1 are carbohydrates. When carbon–hydrogen compounds react with oxygen, as in burning, carbon dioxide and water are formed, and energy is released.

Sugars serve as a primary source of chemical energy for living systems. The simplest sugars are the monosaccharides ("single sugars"), such as glucose and fructose. Monosaccharides can be combined to form disaccharides ("two sugars"), such as sucrose, which is the form in which sugars are transported through the plant body. Polysaccharides (chains of many monosaccharides), such as starch and glycogen, are storage forms for sugars. These molecules can be broken apart again by hydrolysis, the addition of a water molecule.

Fats, which belong to a general group of compounds known as lipids, are used for longer-term storage of food energy by animals and some plants. A fat molecule consists of one molecule of glycerol bonded to three molecules of fatty acids. Fats are designated as saturated or unsaturated depending on whether or not their fatty acids contain any double bonds. Unsaturated fats, which tend to be oily liquids, are more commonly found in plants.

Sugars, fats, and other energy-storing molecules are broken down by the cell and used to form ATP from ADP. ATP supplies the energy for most of the energy-requiring activities of the cells. The ATP molecule consists of a nitrogen base, adenine; a ribose sugar; and three phosphate groups. Two of the groups are linked to the molecule by high-energy bonds—bonds that release a relatively large amount of energy when the terminal phosphate is transferred to another compound. ATP participates as an energy carrier in most series of reactions that take place in living systems.

QUESTIONS
1. Define the following terms: carbohydrate, hydrocarbon, polysaccharide, glycogen, hydrolysis, unsaturated bond.
2. Draw a Bohr model of the carbon atom.
3. Explain the advantages to living systems of having a single energy-yielding compound (ATP) that participates in a wide variety of reactions.
4. Both disaccharides, such as sucrose, and polysaccharides, such as starch and glycogen, are broken down by hydrolysis. Considering the function of these reactions in terms of the life of a cell, what are two advantages of hydrolysis?

Chapter 5

Informational and Structural Macromolecules

In this chapter, we are going to discuss two additional ways in which molecules function in living systems. One is the carrying out of a variety of highly specific chemical activities, including facilitating chemical reactions (enzymes), carrying oxygen (hemoglobin), serving as chemical messengers (hormones), and transmitting hereditary information (nucleic acids). The other is structural, making up the special walls, membranes, organelles, tubules, fibers, and other structures that are part of the cell's visible organization. The molecules involved in these kinds of activities are often called macromolecules. (*Macro* means "large.")

INFORMATIONAL MACROMOLECULES

The two types of molecules primarily involved in controlling the cell's chemical activities are proteins and nucleic acids. All enzymes are proteins, hemoglobin is a protein, and many hormones also are proteins. The chemicals concerned with storage and transmission of genetic information are nucleic acids.

Like polysaccharides, proteins and nucleic acids are made up of a large number of smaller components joined together by covalent bonds so that they form long chains. They differ from polysaccharides in important ways. First, they differ in their components: The building blocks of proteins are called amino acid residues and those of nucleic acids are called nucleotides. Secondly, whereas polysaccharides are composed of one or a few different kinds of components repeated over and over again, the proteins and nucleic acids are composed of different kinds of components arranged in a very precise order. The order of the components enables the particular protein or nucleic acid to perform a highly specific function.

These macromolecules are sometimes referred to as informational, by analogy with speech or written language. Just as a sentence can carry information by virtue of the order and organization of its subunits—the letters, syllables, and words—the functions of these molecules depend on the order of their subunits, the amino acids or nucleotides. Without the information contained in these macromolecules, living systems would not be able to perform the large number of complex activities characteristic of living matter.

5–1 (a) *Every amino acid contains an amino group (NH$_2$) and a carboxyl group (COOH) bonded to a central carbon atom. A hydrogen atom and a side group are also bonded to the same carbon atom. This basic structure is the same in all amino acids. The "R" stands for the side group, which is different in each kind of amino acid. (b) The 20 kinds of amino acids found in proteins. As you can see, their basic structures are the same, but they differ in their side groups. Because of differences in their side groups, amino acids may be nonpolar (with no difference in charge between one part of the molecule and another), polar but with the two charges balancing one another out so that the amino acid as a whole is uncharged, positively charged (acidic), or negatively charged (basic). The nonpolar molecules are not soluble in water, whereas the charged and polar molecules are.*

(a)

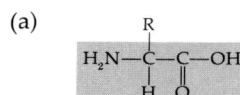

(b)   NONPOLAR

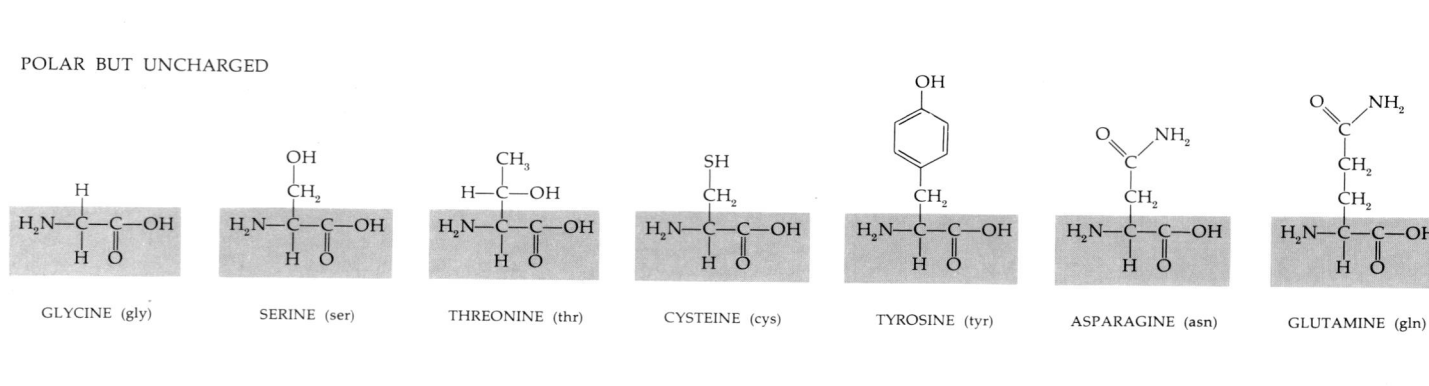

ALANINE (ala)   VALINE (val)   LEUCINE (leu)   ISOLEUCINE (ile)

PROLINE (pro)   PHENYLALANINE (phe)   TRYPTOPHAN (trp)   METHIONINE (met)

POLAR BUT UNCHARGED

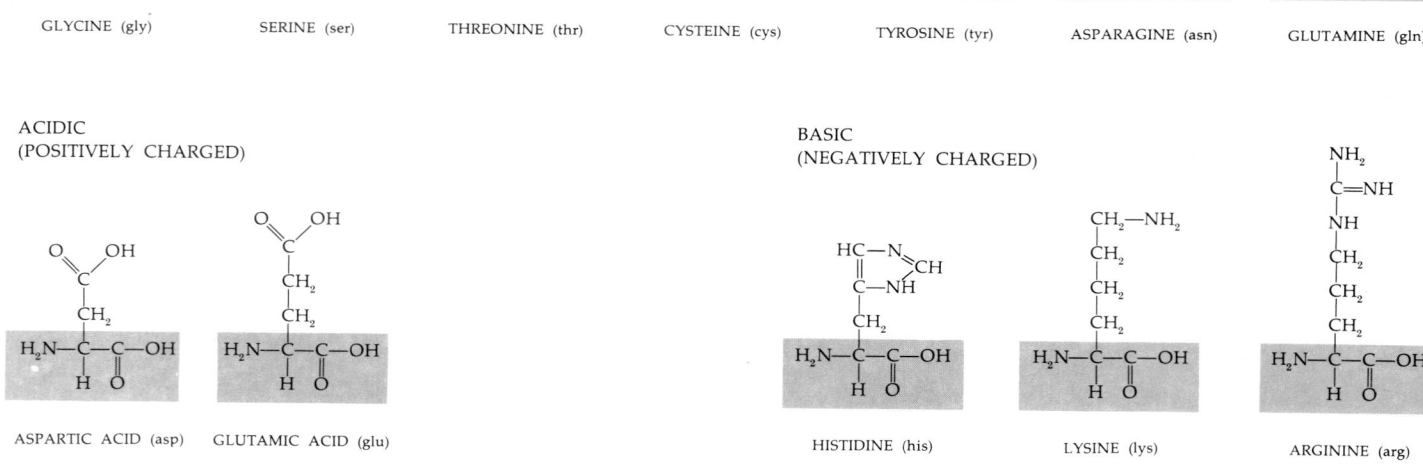

GLYCINE (gly)   SERINE (ser)   THREONINE (thr)   CYSTEINE (cys)   TYROSINE (tyr)   ASPARAGINE (asn)   GLUTAMINE (gln)

ACIDIC
(POSITIVELY CHARGED)

ASPARTIC ACID (asp)   GLUTAMIC ACID (glu)

BASIC
(NEGATIVELY CHARGED)

HISTIDINE (his)   LYSINE (lys)   ARGININE (arg)

AMINO ACIDS: THE BUILDING BLOCKS OF PROTEINS

All proteins are made up of one or more chains of relatively simple molecules called amino acids. Amino acids contain carbon, hydrogen, and oxygen, as do sugars. All of them also contain nitrogen. Every amino acid has the same "backbone" structure, which consists of a central carbon atom bonded to an amino group ($-NH_2$) and a carboxyl group ($-COOH$). In every amino acid there is also another atom or group of atoms bonded to the central carbon. As shown in Figure 5–1, this side (R) group can be a hydrogen atom, in which case the amino acid is glycine; a CH_3 group, in which case the amino acid is alanine; and so on. The side group, depending on the atom or atoms that compose it, may have a positive charge, have a negative charge, be polar (with a negative and positive zone), or have no charge at all (in which case it is insoluble in water).

A large variety of different amino acids is theoretically possible, but only about 20 are found in proteins. And it is always the same 20, whether in a bacterial cell, a plant cell, or a cell in your own body.

Amino acids are linked together by peptide bonds (see Figure 5–2) to form proteins. The sequence of amino acids in these chains determines the biological character of the protein molecule; even one small variation in this sequence may alter or destroy the way in which it functions. Although only 20 different kinds of amino acids are found in proteins, protein molecules are large, often containing several hundred amino acids. Thus the number of different amino acid sequences, and therefore the possible variety of protein molecules, is enormous—about as enormous as the number of different sentences that can be written with our own 26-letter alphabet. The single-celled bacterium *Escherichia coli*, for example, contains 600 to 800 different kinds of proteins at any one time, and the cell of a higher plant or animal has several times that number. In a complex organism such as man, there are at least several thousand different proteins, each with a special function and each, by its unique chemical nature, specifically fitted for that function.

Amino Acids and Nitrogen

Like fats and carbohydrates, amino acids can be taken into the body as food or can be formed within living cells using sugars as starter materials. But whereas fats and carbohydrates are made up only of carbon, hydrogen, and

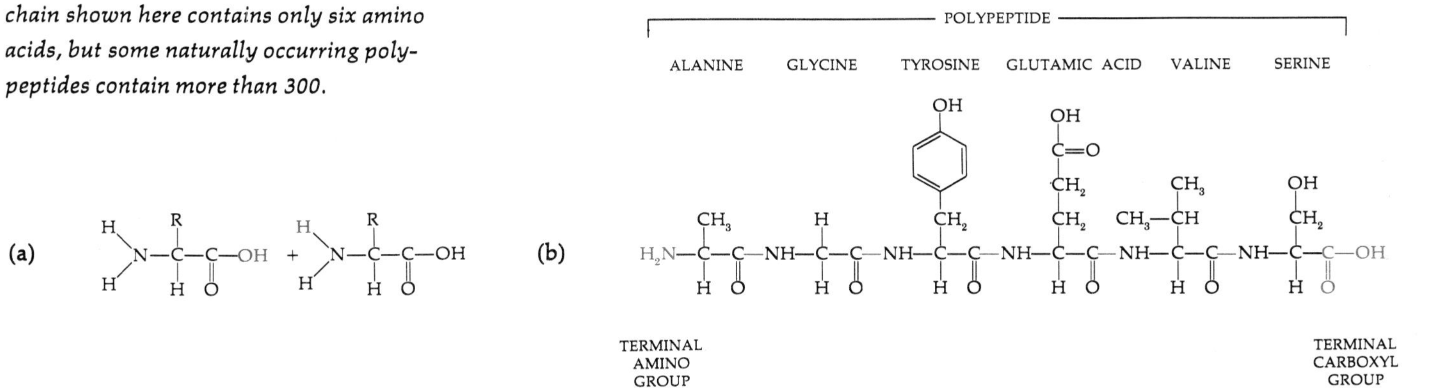

5–2 (a) *Formation of a peptide bond involves removal of a molecule of water. The amino acid thus becomes an amino acid residue. The bond is broken by hydrolysis (addition of a molecule of water).* (b) *Polypeptides are chains of amino acids linked together by peptide bonds (shown in color), with the nitrogen atom from the amino group of one acid covalently bonded to the carbon atom of the carboxyl group of the next. The polypeptide chain shown here contains only six amino acids, but some naturally occurring polypeptides contain more than 300.*

(a)

(b)

5-3 *Primary structure of the enzyme lysozyme. This enzyme, which contains 129 amino acids, is found in many different types of animal cells. The sequence of the amino acids constitutes its primary structure. One consequential feature of the primary structure is the occurrence of the sulfur-containing amino acid cysteine. Covalent bonds form between the sulfur atoms, linking parts of the molecule together. Lysozyme breaks down the cell walls of many types of bacteria. It was first detected by Alexander Fleming, better known for his discovery of penicillin, as a consequence of his having a bad cold. A droplet fell from his nose onto a glass slide containing bacteria, and Sir Alexander, as he was to become, noticed that the bacteria in the vicinity of the droplet were destroyed. Lysozyme was subsequently discovered in tears, saliva, milk, and various other body fluids of many different animals. The molecule shown here was extracted from egg white. Lysozymes from other organisms might differ in some of their amino acids.*

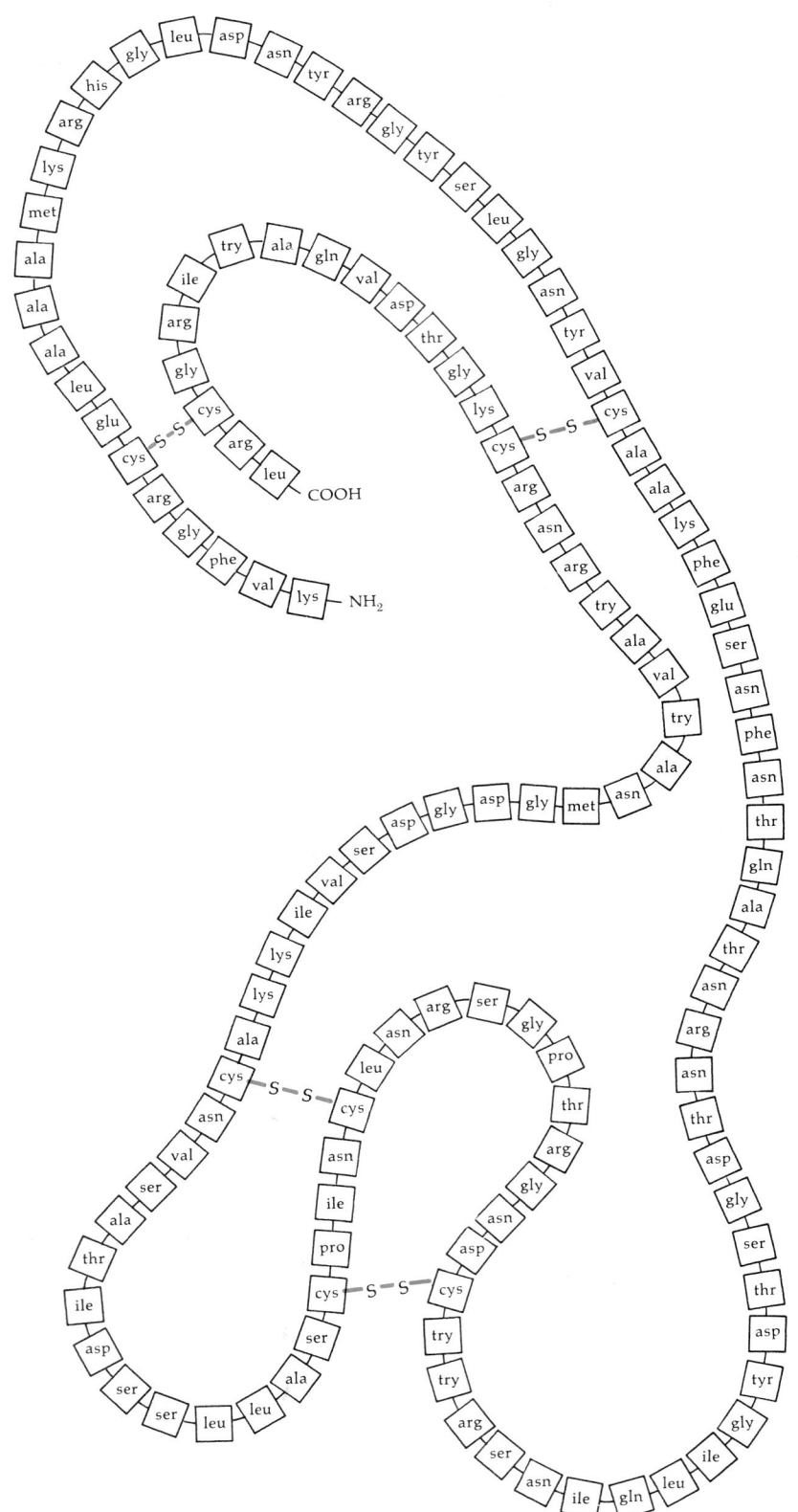

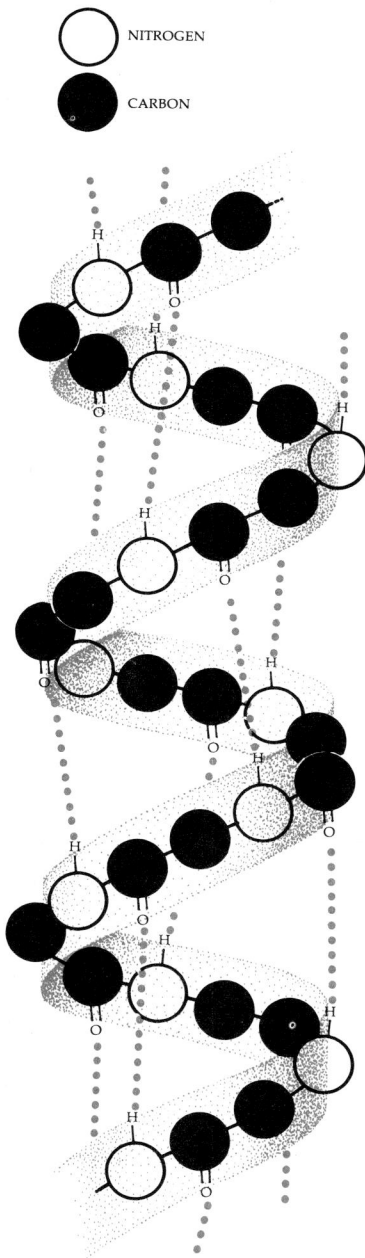

5–4 *Secondary structure of a protein. The helix is held in shape by hydrogen bonds, indicated by the dotted lines. In this particular helix (the alpha helix), bonds form between every fourth amino acid.*

○ NITROGEN

● CARBON

oxygen atoms, all available in the sugar and water of the cell, amino acids also contain nitrogen. Most of the nitrogen in the biosphere exists as a gas in the atmosphere. Only a few organisms, all microscopic, are able to incorporate gaseous nitrogen from the air into inorganic nitrogen compounds that can be assimilated by plants. Hence the amount of nitrogen available to the living world is limited.

Plants obtain nitrogen in the form of ammonium nitrites and nitrates present in the soil, convert these to ammonia, and incorporate the ammonia into amino acids. Animals obtain some of their amino acids by eating plants. They can also synthesize some of their amino acids by combining sugars with amino groups from dietary proteins. They can also interconvert some amino acids, changing one to another. Some, however, they cannot either synthesize or make from other amino acids. These, the so-called essential amino acids, must be obtained either directly or indirectly from plants. For adult human beings, the essential amino acids are lysine, tryptophan, threonine, methionine, phenylalanine, leucine, valine, and isoleucine.

Until recently, agricultural scientists concerned with the world's hungry people concentrated on developing plants with a high caloric yield. Ironically, these new strains, although richer in carbohydrates, were usually poorer in proteins than the wild strains from which they were developed. Now, increasing recognition of the role of plants in supplying amino acids to the animal world has led to emphasis on the development of strains of food plants that contain not only more proteins but, in particular, proteins that include the amino acids often in short supply in vegetable diets.

The latter is important because, as you can see, in order to assemble amino acids into proteins a cell must have available not only a large enough quantity of amino acids but also all the different kinds—just as a typesetter, even though he may have a large supply of letters, cannot set a sentence if he does not have enough *e*'s.

We shall discuss the pathways of nitrogen through living systems in greater detail in Chapter 41.

THE LEVELS OF PROTEIN ORGANIZATION

Proteins are assembled in living systems with the amino groups of one amino acid linked to the carboxyl of another, like a line of boxcars. The bond between amino acids is known as a peptide bond, and chains of amino acids may be called *polypeptides*.* This linear sequence is known as the *primary structure* of the protein. Each different protein has a different primary structure. The primary structure of one protein is shown in Figure 5–3.

As the chain is assembled, it takes on a secondary structure, which is determined by the sequence of amino acids in the primary structure. Secondary structure refers to the arrangement of the polypeptide chain along one axis. A common secondary structure is a helix (see Figure 5–4). The helix is held in place by hydrogen bonds that form between the amino acids, linking oxygen and hydrogen atoms at regular intervals. Another common secondary structure is

* The distinction between a polypeptide and a protein is not always clear-cut. A polypeptide, however, can be any chain composed of many amino acids, whereas a protein is always a naturally occurring, biologically active one.

5–5 *Linus Pauling, who discovered the helical structure of proteins and the role of the hydrogen bond in maintaining it.*

the "pleated sheet," which is formed when polypeptide chains line up side by side, held together (as in the helix) by hydrogen bonds.

In informational proteins—proteins that perform specific activities in the cell—the chains fold back on one another into compact globular arrangements which are referred to as *tertiary* structures (see Figure 5–6). Proteins composed of such folded chains are called *globular proteins*. This folding-back process often alters the secondary, helical structure in portions of the long molecules.

Globular proteins assume their special configurations spontaneously as a result of interactions between their amino acids. Wherever two cysteine units come together, for example, the sulfur atoms on their side groups form covalent disulfide bonds, which serve as bridges from one part of the molecule to the other. Other amino acids on the long protein chain are attracted to or repelled from one another by their positive or negative charges, and others are drawn together by hydrophobic interactions. Changes in temperature or pH can rupture these relatively fragile bonds based on electric attractions or hydrophobic interactions (as in soap molecules) and so cause the protein to lose its special configuration. A protein that has lost its tertiary structure in this way is said to be denatured. When a denatured protein is restored to its normal temperature and pH, it will sometimes spontaneously resume its normal tertiary structure.

Different proteins lose their biological activity at different temperatures. For example, the protein controlling the synthesis of the darker colored pigment in Siamese cats is active only in the cooler, peripheral parts of the body, such as the ears, nose, paws, and tip of the tail.

5–6 *The primary structure of a protein is the linear sequence of its amino acids. Because of interactions among these amino acids, the molecule spontaneously coils into a secondary structure, such as the alpha helix shown here, and into a tertiary structure, such as a globule. Many globular proteins, including hemoglobin and some enzymes, are made up of more than one amino acid chain. This structure is known as a quaternary structure.*

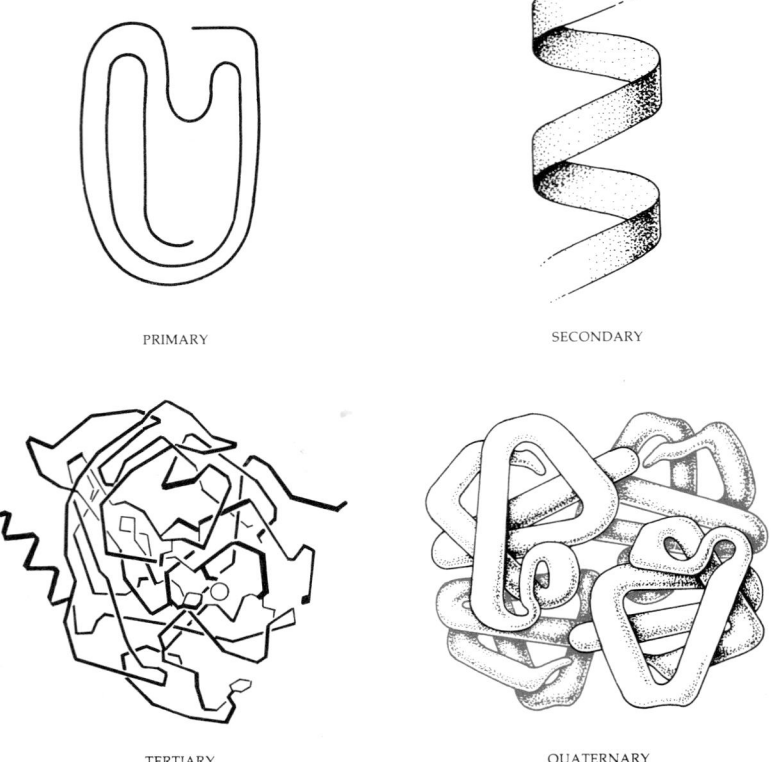

PRIMARY

SECONDARY

TERTIARY

QUATERNARY

5–7 *Chemical reactions require an input of energy to get them started—analogous to the spark which lights a fire or the push which sends the boulder on its downhill course. An uncatalyzed reaction requires more activation ("input") energy than a catalyzed one, such as an enzymatic reaction. Note, however, that the overall energy change from the initial state to the final state is the same.*

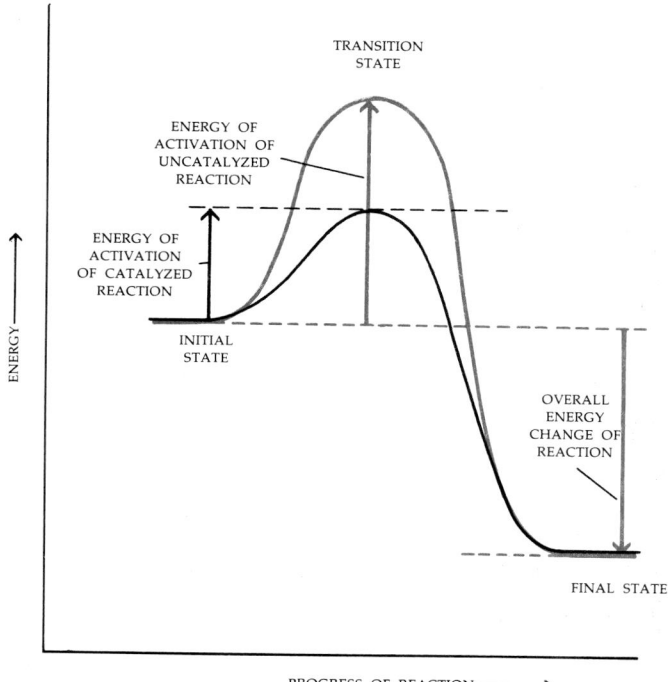

5–8 *The effect of temperature on the rate of an enzyme-controlled reaction. The concentrations of enzyme and reacting molecules (substrate) were kept constant. As you can see, the rate of the reaction approximately doubles for every 10°C rise in temperature up to about 40°C (104°F). Above this temperature, it decreases, and at about 60°C (140°F) it stops altogether, presumably because the enzyme is denatured.*

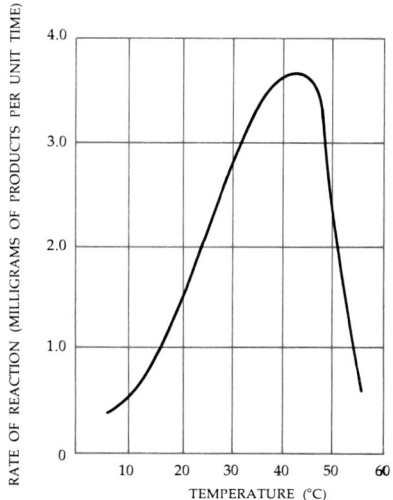

ENZYMES

Enzymes are globular proteins. They are responsible for virtually all of the chemical activities of living systems. They act as *catalysts,* a term used by chemists to describe any substance which regulates the rate of a chemical reaction and itself remains unchanged. Because each enzyme catalyst can be used over and over again, enzymes are typically effective in very small quantities. A single enzyme molecule may catalyze several thousand reactions in a second. Because of enzymes, cells can carry out chemical reactions that otherwise would require greater heat. Heat, as we noted in Chapter 2, speeds chemical reactions by increasing molecular motion, thus making the potential reactants move more quickly and so increasing the probability of their bumping into one another with sufficient energy to cause a reaction between them. Enzymes similarly speed the occurrence of reactions that, if left to occur spontaneously, would occur so rarely that they would serve no useful function. However, catalysts do not cause reactions to take place which would not otherwise be possible—that is, which are not in accord with the principles of energy requirements and equilibrium discussed in Chapter 2.

The exact way in which enzymes work is not fully understood. It is known, however, that they provide a site to which the reacting chemical or chemicals adhere and on which the reactions take place. This portion of the enzyme is known as the *active site.* The reacting molecules (called the *substrate*) fit into the active site. The relationship between the active site and the substrate is very precise, like that of a lock and a key. The active site is a result of the very precise folding of the protein chain into a highly specific tertiary, globular structure.

5–9 *Model of an enzyme. This enzyme (the digestive enzyme chymotrypsin) is composed of three polypeptide chains. The amino (—NH₂) and carboxyl (—COOH) ends of each are labeled. The numbers represent the position of particular amino acids in the chains. Five disulfide bridges connect residues 1 and 122, 42 and 58, 136 and 201, 168 and 182, and 191 and 220. The three-dimensional shape of the molecule is a result of a combination of disulfide bonds and of interactions among the chains and between the chains and the surrounding water molecules, based on the plus or minus charges or the polarity of the various amino acids. As a result of this bending and twisting of the polypeptide chains, particular amino acids come together in a highly specific configuration to form the active site of the enzyme. Two amino acids known to be part of the active site are shown in color.*

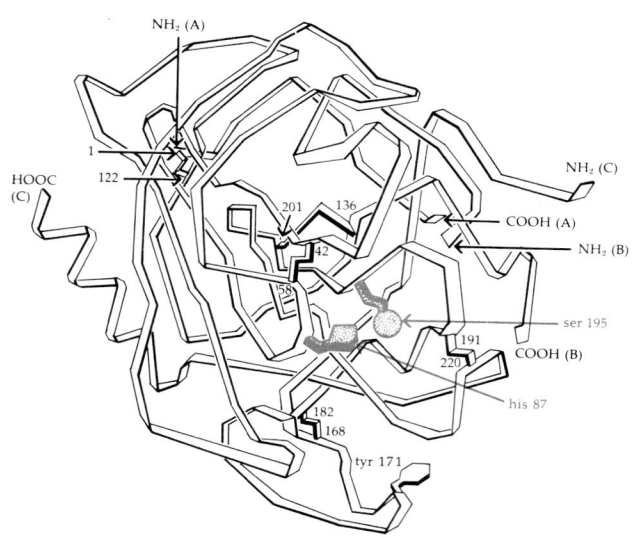

As you can see in Figure 5–9, the active site often involves amino acids from different segments of the protein chain. Variations in primary structure —that is, in the sequence of amino acid units—can render the enzyme nonfunctional even if they do not involve the active site because of their effect on the folding of the molecule. The different chemical reactions taking place in living systems are each regulated by a specific enzyme with a specific primary and so a specific tertiary structure, which provides an active site exactly conforming to the substrate involved.

Enzymes characteristically work in sequence, carrying out series of reactions of the A ⟶ B ⟶ C ⟶ D type described in Chapter 2. Each step in the reaction is catalyzed by a separate enzyme.

Feedback Inhibition

How does a living cell with its hundreds of metabolic pathways control the rates and amounts of products biosynthesized by each? This is a central question in modern cell biology and one that we shall explore further in Section 2. One of the simplest and most direct means, however, is by an "allosteric (other shape) interaction." One of the products of an enzyme series—usually the final product

5–10 *The first step in the formation of sucrose: ATP + glucose ⟶ glucose phosphate + ADP. This reaction, as the diagram shows, is regulated by a specific enzyme. As the arrows indicate, it can theoretically go in either direction. Why is it more likely to proceed in the direction indicated?*

5–11 *An allosteric ("other shape") effector can bind to an enzyme and, by disrupting the bonds determining its tertiary structure, alter the active site. The altered enzyme cannot interact with its normal substrate. The end product of an enzyme sequence may act as an allosteric effector, and so, as the end product accumulates, the reactions become slower.*

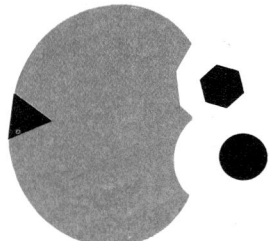

ALLOSTERIC
EFFECTOR

—combines with one of the enzymes—usually the first enzyme—at a site other than the active site. When the product, the effector, and the enzyme interact, the shape of the active site is altered and the enzyme is no longer functional, and so the sequence stops. (See Figure 5–11.) This is one way cells effect a tight regulatory control over their biosynthetic machinery. The effector fits its site as exactly as the substrate fits the active site.

Coenzymes

Many enzymes work in conjunction with other compounds known as coenzymes. These coenzymes often function as carrier or acceptor molecules for substances involved in enzymatic reactions. Analysis of various coenzymes has shown that they all either are vitamins or contain vitamins as components of their structure. Vitamins, by definition, are substances that animals require in small amounts but cannot synthesize within their bodies and so must supplement in their diets. The vitamins were all originally discovered as a consequence of trying to understand and correct deficiency diseases, such as scurvy and rickets, caused by lack of particular vitamins. Nucleotides are components of many coenzymes. (Vitamins are discussed further on pages 645 and 646.)

PROTEIN HORMONES

A number of the important hormones of the human body, including growth hormone and insulin, are complex polypeptide chains. Hormones are substances which are synthesized in one type or one group of cells and released into the bloodstream (or other body fluids) where they travel to another group of cells on which they exert their effects. Obviously, the activity of hormones depends both on the hormone itself and on a specific receptor site on the target cell that makes it susceptible to the hormone (since all cells are exposed to any substance circulating in the blood). We shall discuss the effects of various hormones and the way they act in more detail in Chapter 32.

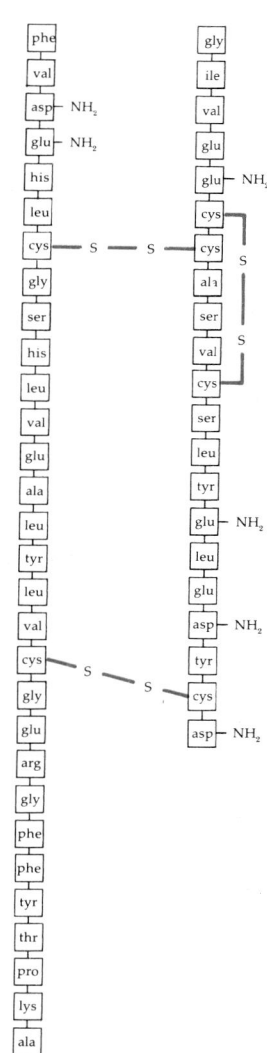

-12 *Insulin is a hormone consisting of two polypeptide chains held together by disulfide bonds.*

5–13 *The heme group of hemoglobin. It contains an iron atom (Fe) held in a porphyrin ring. The porphyrin ring consists of four nitrogen-containing rings, which are numbered in the diagram. The heme group is attached to a long polypeptide chain which wraps around it. The oxygen atom is held flat against the heme.*

HEMOGLOBIN

Sometimes globular proteins consist of two or more polypeptide chains that cluster together as a consequence of a variety of electrical attractions and other weak bonds such as are responsible for the folding of a chain into its tertiary structure. Two or more polypeptide chains clustered into one globular protein constitute what is known as *quaternary structure*. Hemoglobin, the molecule which carries oxygen in the human blood, is a protein with such quaternary structure.

Hemoglobin consists of four chains, each of which is combined with an iron-containing substance known as *heme*. In heme, an iron atom is held in the center of a ring of nitrogen-containing atoms known as a porphyrin ring. Hemoglobin has two identical alpha chains and two identical beta chains, each about 150 amino acids long, for a total of about 600 amino acids in all. In man, hemoglobin molecules are manufactured and carried in the red blood cells. A mature red blood cell contains 265 million molecules of hemoglobin. These molecules possess the special property of being able to combine loosely with oxygen in such a way that they can collect oxygen in the lungs and release it in the tissues.

Sickle cell anemia is a disease in which the hemoglobin molecules are defective. When oxygen is removed, the defective molecules change shape and interact with one another, forming stiffened rodlike structures. Red cells containing large proportions of these molecules become stiff and deformed, taking on the characteristic sickle shape. The deformed cells may clog the smallest blood vessels (capillaries) cutting off the local blood supply. Also, they have a shorter lifespan than normal red blood cells, resulting in anemia. The disease is usually painful and often fatal.

Analysis of the hemoglobin molecules of patients with sickle cell anemia reveals that the only difference between normal and sickle hemoglobin is that in a precise location in each beta chain, one glutamic acid is replaced by one valine.

5–14 *The hemoglobin molecule consists of four heme groups with their polypeptide chains intertwined in a quaternary structure. The outside of the molecule and the hole through the middle are lined by charged amino acids, and the uncharged amino acids are packed inside. Each molecule can hold up to four oxygen atoms. (Adapted with permission from R. E. Dickerson and I. Geis, The Structure and Action of Proteins, W. A. Benjamin, Inc., New York, 1969. Copyright 1969 by Dickerson and Geis.)*

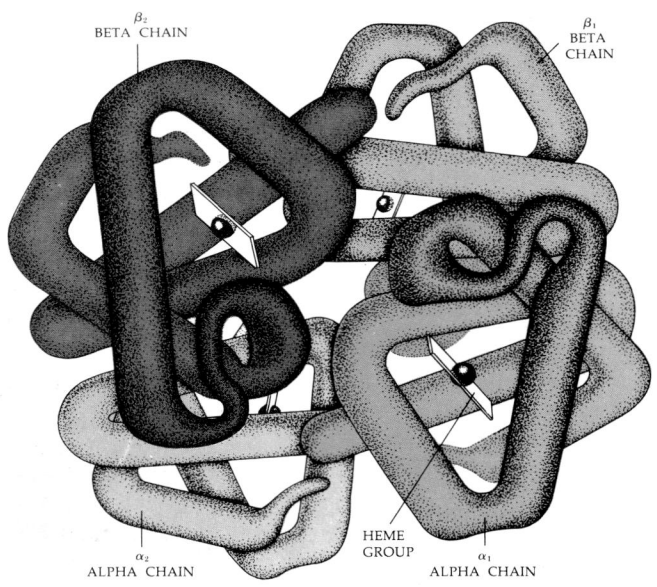

-15 *Scanning electron micrograph of normal and sickle cells. Persons homozygous for the sickle gene have a severe anemia. In order for a child to be born with sickle cell anemia, each parent must carry at least one gene for the disease.*

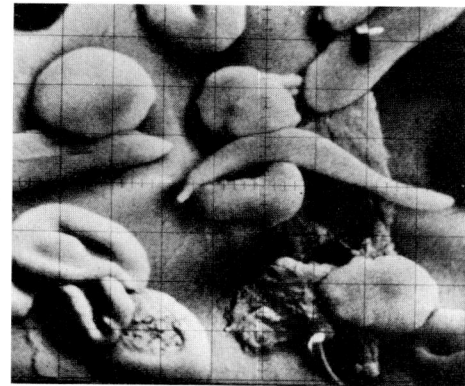

5-16 *An example of the remarkable precision of the "language" of proteins. Portions of the beta chains of the hemoglobin A (normal) molecule and the hemoglobin S (sickle cell) molecule are shown. The hemoglobin molecule is composed of two identical alpha chains and two identical beta chains, each chain consisting of about 150 amino acids, or a total of 600 amino acids in the molecule. The entire structural difference between the normal molecule and the sickle cell molecule (literally, a life-and-death difference) consists of one change in the sequence of each beta chain: one glutamic acid is replaced by one valine.*

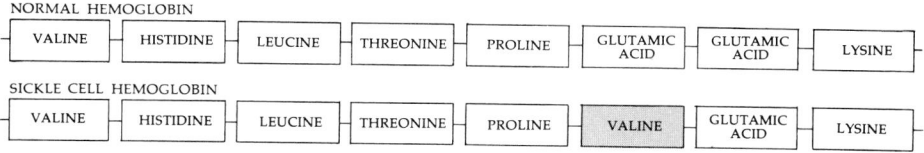

NORMAL HEMOGLOBIN

| VALINE | HISTIDINE | LEUCINE | THREONINE | PROLINE | GLUTAMIC ACID | GLUTAMIC ACID | LYSINE |

SICKLE CELL HEMOGLOBIN

| VALINE | HISTIDINE | LEUCINE | THREONINE | PROLINE | VALINE | GLUTAMIC ACID | LYSINE |

When one considers that this difference of two amino acids in a total of almost 600 is actually the difference between life and death, one begins to get an idea of the precision and the importance of the "information" carried by these macromolecules.

NUCLEIC ACIDS

Proteins, in the form of enzymes, carry out the moment-to-moment chemical activities of the cell. Nucleic acids direct the synthesis of the enzymes and so control and specify the chemical activities of which the cell is capable. These directions, contained in particular nucleic acid molecules, are passed from cell to cell and organism to organism from generation to generation; they constitute the hereditary material. In Chapter 17, we shall describe the way in which information is transferred to the proteins from the nucleic acids.

Just as proteins are composed of long chains of amino acids, nucleic acids consist of sequences of a different type of building block known as nucleotides. A nucleotide, as we saw in the preceding chapter, is a more complex molecule than a monosaccharide, a fatty acid, or an amino acid. It is made up of three subunits: a phosphate group (PO_4), a five-carbon sugar, and a nitrogen base, so-called because the ring structure of the base contains nitrogen as well as carbon. Two types of nitrogen bases are found in nucleotides: pyrimidines, which have a single nitrogen–carbon ring, and purines, which have two fused nitrogen–carbon rings. The three pyrimidines found in nucleic acids are thymine, cytosine, and uracil, and the two purines are adenine and guanine.

There are two types of nucleic acid, each composed of chains of nucleotides: deoxyribonucleic acid, or DNA, and ribonucleic acid, or RNA. Chemically, the differences between them are slight. *Deoxy* means "minus one oxygen," and as you can see in Figure 5–19, this is the only difference between the two five-carbon sugars. DNA bases are adenine, guanine, *thymine*, and cytosine; RNA bases are adenine, guanine, *uracil*, and cytosine. (You need not remember these at this time. You will find, however, that you need to know the difference between DNA and RNA when you come to Chapter 17.)

-17 *The nitrogen bases found in nucleic acids. Thymine is found only in DNA; uracil only in RNA.*

MAJOR PURINES

ADENINE

GUANINE

MAJOR PYRIMIDINES

CYTOSINE

THYMINE

URACIL

5–18 (a) *A nucleotide is made up of three different parts: a nitrogen base, a sugar, and a phosphate. (b) Each nucleotide in DNA contains a deoxyribose sugar and one of the four possible nitrogen bases.*

(a)

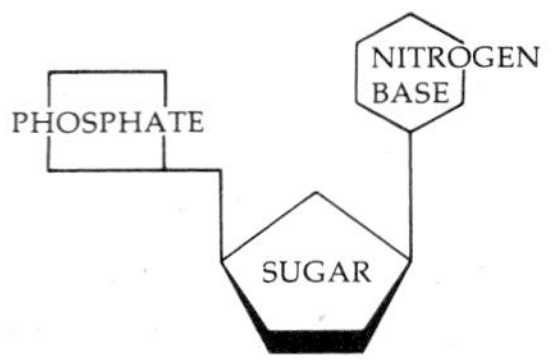

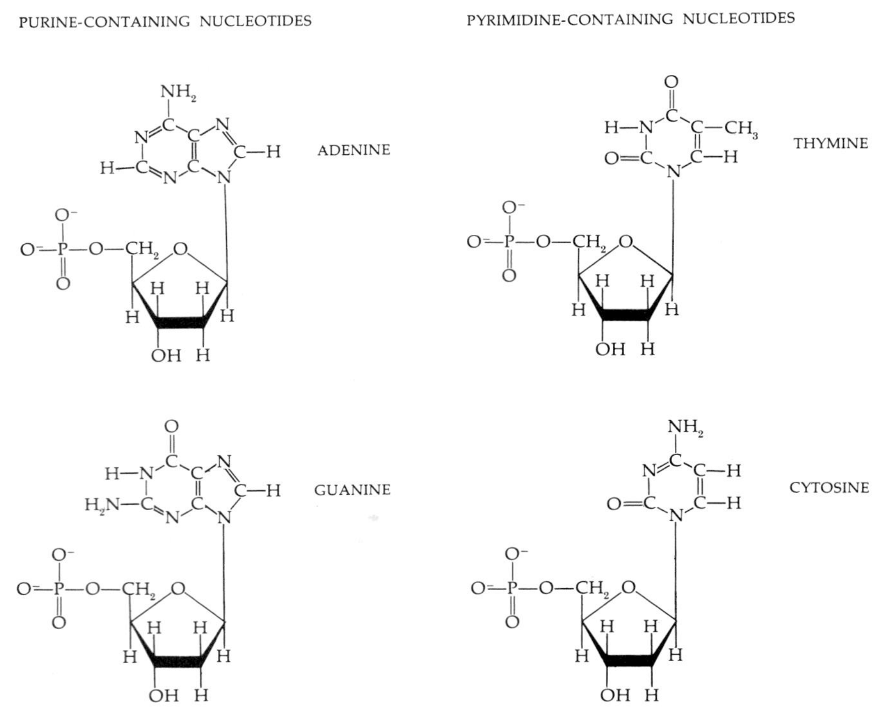

5–19 *Chemically, RNA is very similar to DNA, but there are two differences in its chemical groups. One difference is in the sugar component; instead of deoxyribose, RNA contains ribose, which has an additional oxygen atom. The other difference is that instead of thymine, RNA contains the closely related pyrimidine uracil. (A third, and very important, difference between the two is that most RNA does not possess a regular helical structure and is usually single-stranded.) In these structural models, the carbon atoms are "understood," as in the sugar molecules shown previously.*

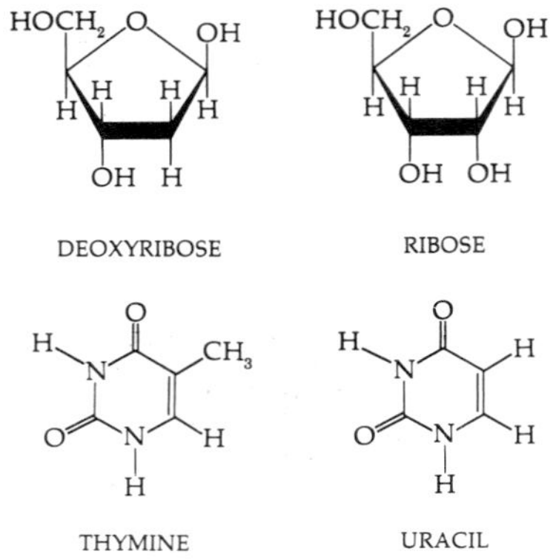

(b)

PURINE-CONTAINING NUCLEOTIDES   PYRIMIDINE-CONTAINING NUCLEOTIDES

Although their chemical components are very similar, DNA and RNA play very different biological roles. DNA is the nucleic acid found in the chromosomes of the cell, the carrier of the genetic message. The function of RNA is to transcribe the genetic message present in DNA and translate it to proteins. The discovery of the structure of these molecules and how they perform their functions was one of the great triumphs of twentieth-century science. We are going to discuss it in the section on genetics because we believe you will better appreciate the magnitude of this discovery after you have learned something about its background and the work that went before it.

STRUCTURAL MACROMOLECULES

A major function of molecules in living systems is to form the structural components of cells and tissues. The principal structural molecule in plants is cellulose. In fact, half of all of the organic carbon in the biosphere is incorporated into cellulose. Wood is about 50 percent cellulose, and cotton, for instance, is nearly pure cellulose.

Cellulose molecules are synthesized inside plant cells and transported outside the cell membrane where they form the fibrous part of the cell wall. The cellulose fibers are cemented together by other kinds of polysaccharides. Thus plant cells, in effect, live in little cellulose boxes. When the cells are young, these boxes are flexible and stretch as the cell grows. They grow thicker and more rigid as the cell matures. In some plant tissues, such as the tissues that form wood and bark, the cells eventually die, leaving only their tough outer cellulose containers.

Cellulose is a macromolecule composed of glucose units, just as starch and glycogen are. Starch and glycogen can be readily utilized as fuel by almost all kinds of living systems, but cellulose can be hydrolyzed by only a few micro-organisms (such as fungi and bacteria) and a very few animals, silverfish, for example. Cows and other ruminants, termites, and cockroaches are able to use cellulose for energy only because the necessary enzyme is produced by micro-organisms that inhabit their digestive tracts.

To understand the structural differences between cellulose and starch or glycogen, we have to look briefly once again at the glucose molecule. You will remember that the molecule can exist as a chain of carbon atoms and that when it is in solution, as it is in the cell, it assumes a ring form. The ring may close in either of two ways (see Figure 5–20a). One ring form is known as alpha, and the other as beta. The numbers of alpha and beta forms are in equilibrium, with a certain number of the molecules changing from one to the other all of the time, going through the chain structure to reach the other form. Starch and glycogen are both made up entirely of alpha units. In cellulose, alpha units alternate with beta units (see Figure 5–20b). Because of this slight difference

–20 (a) In living systems, alpha and beta forms of the ring-structured glucose molecule are in equilibrium. The molecules pass through the straight-chain form to get from one ring structure to another. (b) Cellulose consists of alternating alpha and beta glucose units. The OH groups, which project from both sides of the chain, form hydrogen bonds with neighboring OH groups resulting in the formation of bundles of cross-linked parallel chains (c). In the starch molecule (d), composed of alpha glucose units, most of the OH groups capable of forming cross-linkages project into the interior of the coiled molecule (e).

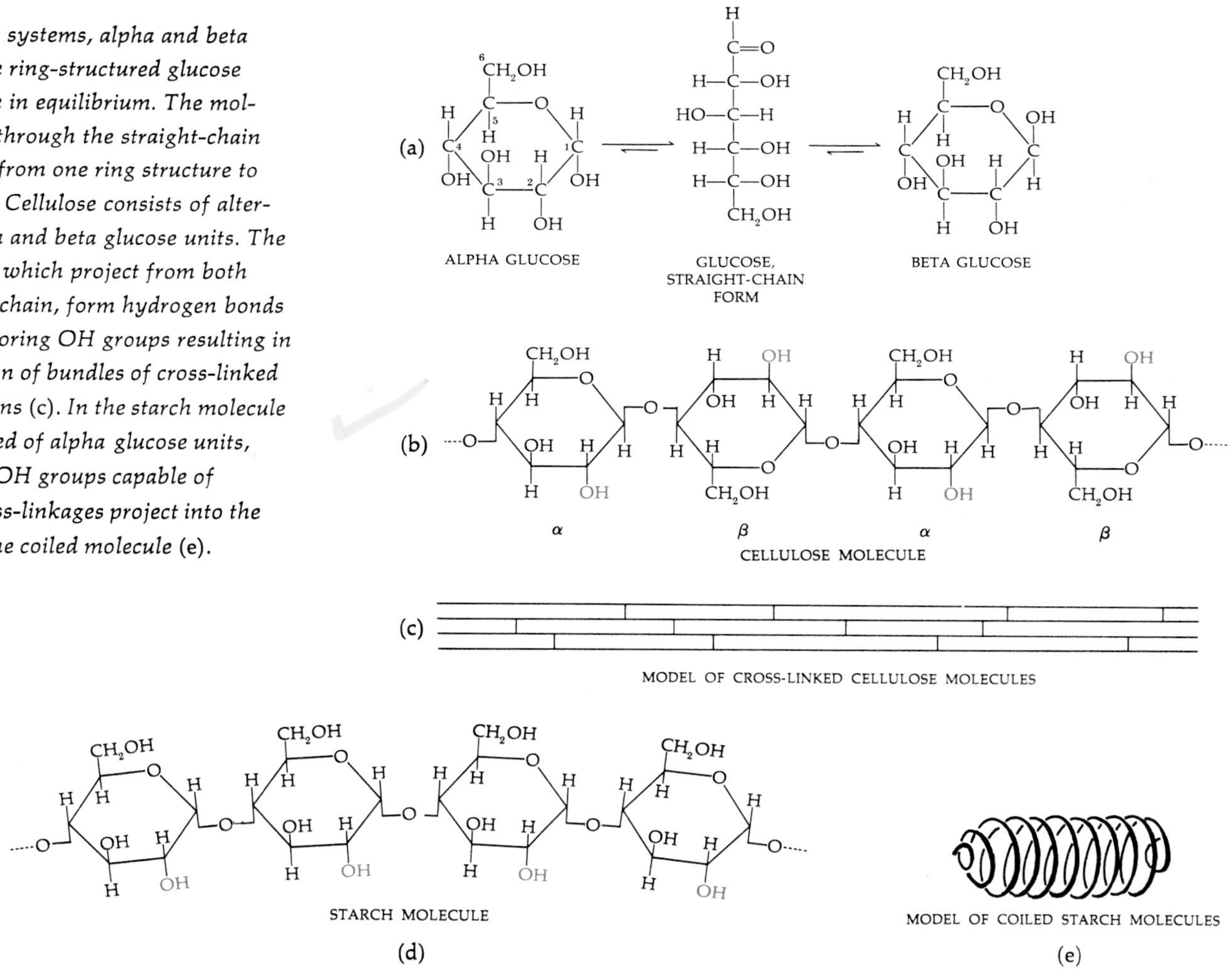

ALPHA GLUCOSE   GLUCOSE, STRAIGHT-CHAIN FORM   BETA GLUCOSE

CELLULOSE MOLECULE

(c) MODEL OF CROSS-LINKED CELLULOSE MOLECULES

STARCH MOLECULE
(d)

MODEL OF COILED STARCH MOLECULES
(e)

5-21　*The plant cell wall is composed largely of cellulose. Each of the microfibrils you can see here is a bundle of hundreds of cellulose strands, and each strand is a chain of glucose units (Figure 5-20b). The microfibrils, as strong as an equivalent amount of steel, are embedded in other polysaccharides, one of which is pectin.*

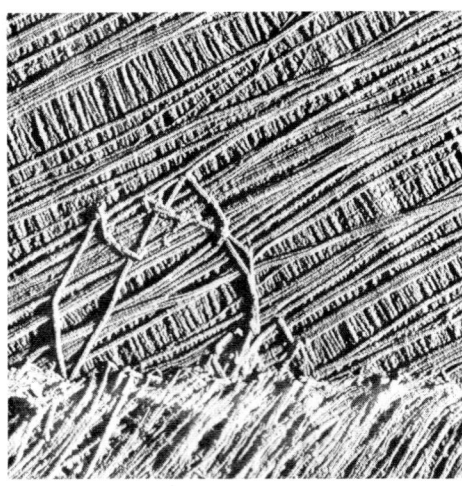

in structure, cellulose is impervious to the enzymes that so successfully break down the storage polysaccharides.

Chitin, which forms the exoskeletons of insects and is a major component of the cell walls of many fungi, is also a tough, resistant, common polysaccharide.

LIPIDS AND MEMBRANES

Lipids also play extremely important structural roles. Lipids, you will recall, are a general group of substances of which the fats, discussed in the previous chapter, are a particular example. The lipids most important for structural purposes are phospholipids. Like fats, the phospholipids are composed of fatty acid chains attached to glycerol. In the case of the phospholipids, the third carbon of the glycerol molecule is occupied not by a fatty acid but by a phosphate group (of the same sort as the phosphate groups in the ATP and ADP molecules and in the nucleotides of the nucleic acids). Phosphate groups are electrically charged; as a consequence the phosphate end of the molecule is soluble in water and the fatty acid portions are not. The result is that phospholipids tend to behave like soap molecules, with their fatty acid portions clustering together in hydrophobic interactions.

Cells, as we shall see in the next chapter, are the units of living matter, and membranes are among the most important constituents of cells. Cell membranes, on chemical analysis, are found to be made up largely of lipids and proteins. The phospholipid molecules may be the same or of only a few different kinds. It is likely that at least some of the protein molecules that form part of the cell membrane are globular enzymes and other informational molecules. Cell membranes are described in greater detail in Chapter 7.

5-22　*Green darner molting. The shells, or exoskeletons, of insects are made of chitin, a nitrogen-containing polysaccharide. Some types of insects, after molting, recycle their sugars and nitrogen by thriftily eating their discarded exoskeletons.*

5-23　*A phospholipid molecule consists of two fatty acids linked to a glycerol molecule, as in a fat, and a phosphate group (indicated by colored screen) linked to the glycerol's third carbon. It also usually contains an additional chemical group, indicated by the letter R. The fatty acid tails are nonpolar and therefore insoluble in water; the phosphate and R groups are soluble.*

POLAR HEAD   NONPOLAR TAILS

$$R—O—P—O—^3CH_2$$

$$H—^2C—O—C—CH_2CH_2CH_2CH_2CH_2CH_2CH_2CH=CHCH_2CH_2CH_2CH_2CH_2CH_2CH_3$$

$$H—^1C—O—C—CH_2CH_2CH_2CH_2CH_2CH_2CH_2CH_2CH_2CH_2CH_2CH_2CH_2CH_2CH_3$$

H

GLYCEROL

5–24 (a) *Because phospholipids have water-soluble heads and water-insoluble tails, they tend to form a thin film on a water surface with their tails extending above the water. (b) Surrounded by water, they spontaneously arrange themselves in two layers with their "heads" extending outward and their hydrophobic (water-hating) "tails" inward. Organic molecules enter cells at a rate proportional to their solubilities in lipid. This was one of the first clues that an arrangement of lipid molecules, such as (b), might form a major component of the cell membrane.*

(a)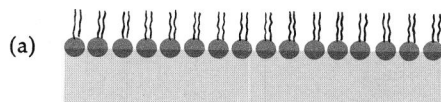

(b)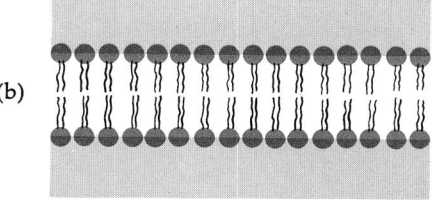

5–25 *The cholesterol molecule consists of four carbon rings and a hydrocarbon chain. Cholesterol is also a component of the membranes of eukaryotic cells.*

Cholesterol

Cholesterol is also found in some cell membranes. It belongs to a group of compounds known as the steroids. (This group also includes the sex hormones and the hormones of the adrenal cortex, which we shall discuss in Chapter 32). All the steroids have four linked carbon rings, like cholesterol, and several of them, like cholesterol, have a tail. In addition, many of them have the OH functional group which classifies them as alcohols. Although steroids do not resemble the other lipids structurally, they are grouped with the lipids because they are soluble in water.

About 25 percent (by dry weight) of the membranes of red blood cells is cholesterol; it is also a major component of the myelin sheath, the fatty membrane that wraps around and insulates fast-conducting nerve fibers (see Figure 5–26).

In some older persons, cholesterol forms fatty, waxy deposits on the inner lining of the blood vessels, blocking the vessels and decreasing their elasticity and so making such persons more susceptible to high blood pressure, heart attacks, and strokes. Attempts have been made to reduce the incidence and extent of such deposits by reducing the dietary intake of high-cholesterol foods, such as eggs and cheese, in the diet, but since many cells of the human body are able to synthesize cholesterol, it is not clear that a reduction of dietary cholesterol will solve the problem.

5–26 *Cross section of a nerve fiber (axon) wrapped in a myelin sheath. The sheath, which is composed of the membranes of specialized cells, appears here as a dark border surrounding the axon.*

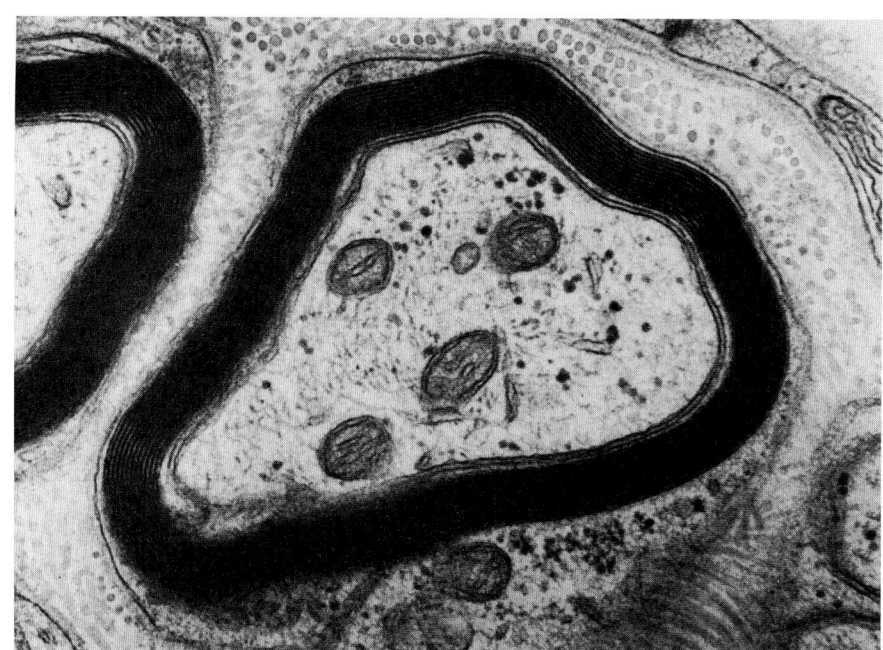

5-27 *Waxes are also made of fatty acids. This electron micrograph shows waxy deposits on the upper surface of a eucalyptus leaf. Biosynthesis of waxes, which protect exposed plant surfaces from water loss, is a property of all groups of higher plants.*

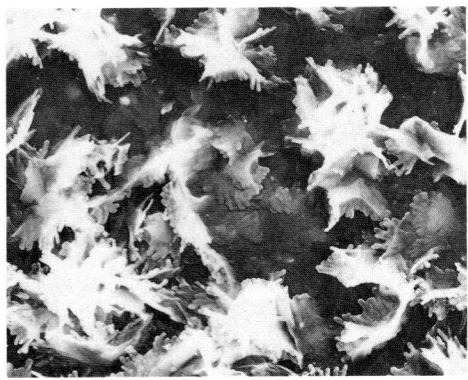

5-28 *Steps in the self-assembly of tobacco mosaic virus (TMV) from its nucleic acid core and protein subunits. Each of the subunits is an identical globular protein, each composed of 158 amino acids. There are 2,200 subunits in all. These assemble themselves spontaneously to form the outer coat of the virus particle. The inner core of the virus—the long threadlike structure in the diagrams—is a single molecule of nucleic acid. TMV, because of its simplicity and because it was the first example of self-assembly to be observed, is considered a model system for other biological self-assemblies, including those involved in cell structures. The electron micrograph (j) shows the virus particles, each 0.3 micrometer in length. (A micrometer is 1/10,000th of a centimeter.)*

Waxes

Waxes are also a form of lipid. They are found as protective coatings on skin, fur, feathers, on the leaves and fruits of higher plants, and on the exoskeletons of many insects.

STRUCTURAL PROTEINS

Proteins are the major structural components, by weight, of the animal world.

The tobacco mosaic virus (TMV), because of its simplicity, provides a useful example of the structural use of protein. Like all viruses, TMV consists essentially of a nucleic acid core surrounded by a protein coat. The coat is made up of 2,200 identical protein molecules, each containing 158 amino acids. The molecules coil up into football-shaped units (a conformation dictated by weak interactions among the amino acids), and the units in turn, because of attractions between them, assemble themselves into a helix.

Microtubules

Microtubules, which are found widely throughout living cells, resemble the protein coat of TMV in their general structural plan. Microtubules are very slender—they can be seen only under a high-magnification electron microscope

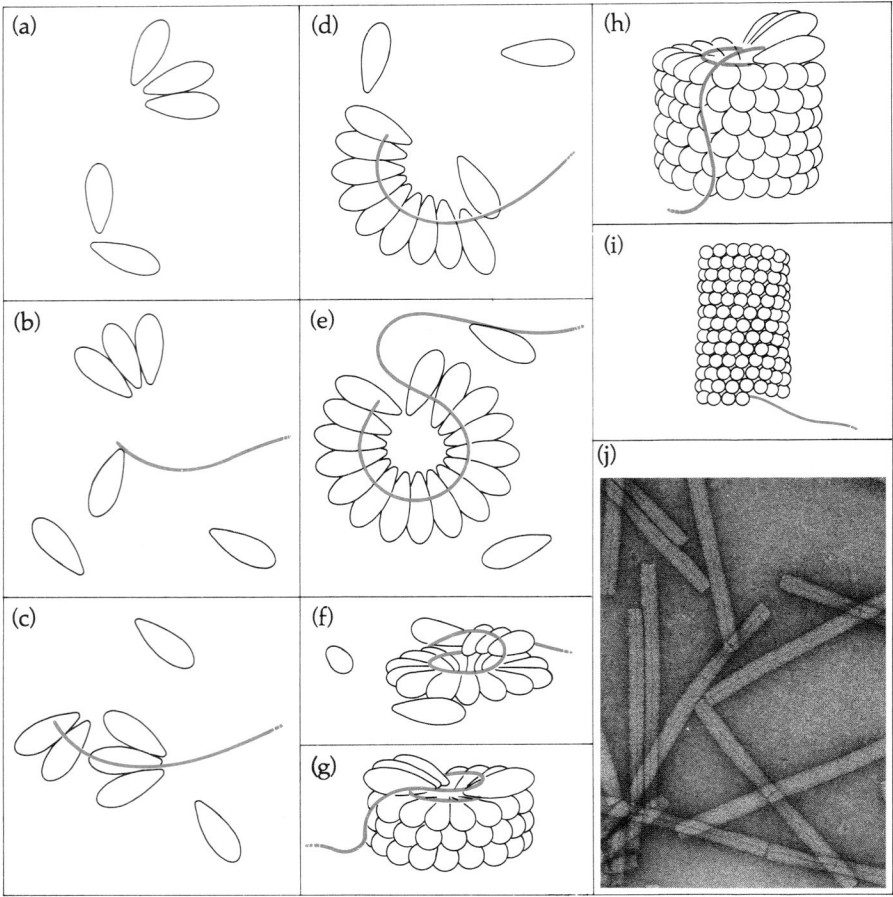

5–29 *Microtubules are tiny tubes, only about 0.02 micrometer in diameter. Like the tobacco mosaic virus, microtubules are capable of self-assembly from their sub-units, which are also globular proteins. Unlike TMV, microtubules do not have a nucleic acid core.*

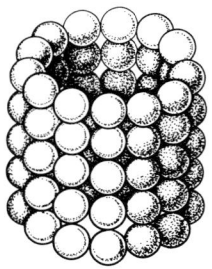

—but so long that their entire length can seldom be traced in a single microscopic section. They function in a variety of ways inside the cells. They apparently act as internal skeletons, stiffening parts of the cell body. They also may serve as tracks along which substances may move inside the cell. The formation of a new cellulose cell wall in a plant can be predicted by the appearance at that site of large numbers of microtubules, and when cellulose fibrils are being laid down outside a plant cell membrane, as a cell wall forms or grows, it is possible to detect microtubules inside the cell aligned in the same direction as the fibrils outside.

Microtubules also form the major components of cilia and flagella and of the spindle fibers that appear in dividing cells.

Fibrous Proteins

In addition to these microscopic structures, proteins make up a large number of familiar and clearly visible structures. Hair, silk, wool, nails, horns, and feathers are all structural proteins, for example. They are all made inside cells, transported to the cell surface, and released. These proteins are known as fibrous proteins (as distinct from globular). Fibrous proteins usually have a regularly repeated primary structure and, often, secondary structure that takes the form of a regular helix. The molecules instead of turning back on themselves are attracted to other similar molecules so that they form sheets or cables. Collagen, which is the major protein constituent of skin, tendon, ligament, and bone, is formed of long protein molecules that wrap around each other in threes, forming a cablelike fiber. Collagen makes up about one-third of the protein in the human body. (See Figure 5–30a.)

5–30 (a) *The structural protein collagen makes up about one-third of the protein in the human body and is probably the most abundant protein in the animal kingdom. Collagen is composed of long polypeptide chains woven together to form fibrils, such as those shown here. These fibers are a major constituent of skin, tendon, ligament, cartilage, and bone. (b) The capacity to produce the structural protein keratin is characteristic of all vertebrates. It is found in scales, horns, wool, nails, and feathers. These are cross sections of human hairs as seen by the scanning electron microscope. (c) Keratin also is the chief component of the formidable armor of this member of the class Reptilia.*

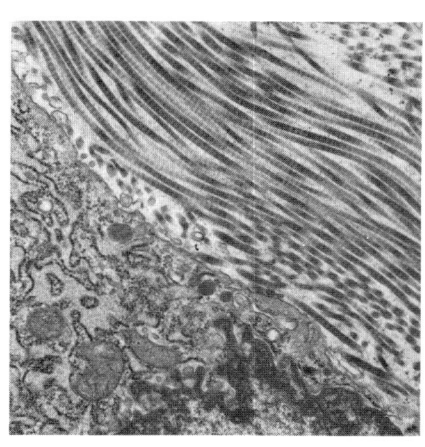

(a)

(b)

(c)

LEVELS OF ORGANIZATION

Notice that in these last few chapters we have progressed through a number of levels of organization, describing the ways in which neutrons and protons and electrons combine to form atoms, and atoms combine to form simple molecules, and simple molecules combine to form larger, more complex molecules which, in turn, combine to form the structures of which living systems are composed. We are now prepared to look at some of these living systems, both in terms of how they are put together and how they operate.

SUMMARY

Proteins are very large molecules composed of long chains of amino acids which are nitrogen-containing substances. There are 20 different amino acids in proteins, and from these an extremely large variety of different kinds of protein molecules can be synthesized, each of which has a highly specific activity and function in living systems. Informational proteins serve as enzymes, as hemoglobin, and as hormones. Some proteins serve as major structural components of living systems.

The sequence of the amino acids is known as the primary structure of the protein. Depending on the amino acid sequence, the molecule may coil into a helix, held together by hydrogen bonds; this is one form of the secondary structure. In globular proteins, the long helix folds back on itself, in a tertiary structure. The way it folds is determined by interactions among its various amino acids. In fibrous proteins, the long molecules interact with other long molecules to form cables or sheets. The interaction of two or more protein molecules to form a composite molecule is called quaternary structure.

Enzymes are globular proteins that control the rates of reactions in living systems. These organic catalysts are able to perform this function because the molecule contains an active site onto which the reactants, or substrate, are temporarily bound. The specificity of enzymes is due to the lock-and-key fit of the active site on the enzyme with the substrate molecule.

Hemoglobin, the oxygen-carrying molecule of the blood, is composed of four (two pairs of) protein chains each attached to an iron-containing heme group. Substitution of one amino acid for another in one of the pair of chains results in a serious and sometimes fatal form of anemia known as sickle cell anemia.

Nucleic acids are composed of units known as nucleotides, each of which is made up of a nitrogen base, a five-carbon sugar, and a phosphate group. DNA (deoxyribonucleic acid) is found principally in the chromosomes of cells and is the carrier of the genetic information.

Cellulose is the principal structural polysaccharide of plants. It is formed within plant cells and deposited outside of the cell membrane. Like starch and glycogen, cellulose is composed of units of glucose, but because of slight differences in molecular configuration, cellulose cannot be broken down by the enzymes that hydrolyze the storage polysaccharides.

Phospholipids are major structural components of cellular membranes. Phospholipids consist of one unit of glycerol, two fatty acids (instead of the three fatty acids present in fats), and a phosphate group. Because of their soluble "heads" and hydrophobic "tails," phospholipids spontaneously form

films and clusters which are the basis of membrane structure. Eukaryotic cell membranes also include cholesterol, a lipidlike substance, and proteins, at least some of which are informational.

Both globular and fibrous proteins serve as structural molecules. Microtubules, which are important cell components, are composed of globular protein molecules assembled into a very long helix. Collagen, hair, silk, wool, horns, nails, and feathers are all fibrous structural proteins.

QUESTIONS

1. Define the following terms: hydrolysis, secondary structure, enzyme, amino group, active site, substrate, energy of activation, allosteric effect.
2. What is the basis of the specificity of enzyme action? What is the advantage to the cell of such specificity? The disadvantage?
3. Sketch the arrangement of phospholipids in water.
4. Sketch the structure of a microtubule.

| Chapter 6 | # How Cells Are Organized |

As we noted earlier and have demonstrated in the preceding chapters, the properties of a group of atoms or molecules or combinations of molecules are different from, greater than, and not predictable by the sum total of the properties of the individual members of the group. In Chapter 1, we listed a number of properties of living systems, the signs of life. In this chapter, these two trains of thought come together. The properties of living systems do not emerge gradually as the degree of biochemical organization increases. They appear quite suddenly and specifically, in the form of the living cell.

The cell theory, like the theory of evolution, is a major unifying concept in biology. The modern cell theory states simply that (1) all living matter is composed of cells; (2) all cells arise from other cells; and (3) all the metabolic* reactions of a living organism, including all energy exchanges and all biosynthetic processes, take place within cells. It also states that cells contain the hereditary information of the organisms of which they are a part, and that this information is passed from parent cell to daughter cell.

ORGANIZATION OF CELLS

There are many, many different kinds of cells. Within our own bodies are more than 100 different and distinct cell types. In a teaspoon of pond water, you are likely to find several different kinds of one-celled organisms, and in a whole pond, there are probably several hundred clearly different kinds. Plants are composed of cells superficially quite different from those of our own body, and insects have many cells of kinds not found either in plants or in vertebrates. Thus, the first remarkable fact about cells is their diversity.

The second, even more remarkable fact is their similarity. Every cell is a self-contained and at least partially self-sufficient unit, surrounded by an outer membrane that controls the passage of materials in and out of the cell and so makes it possible for the cell to differ biochemically and structurally from its surroundings. All cells also, at least at some time in their existence, contain control centers, or nuclei. Many have a variety of internal structures, the organelles, which are similar or identical from one cell to another throughout a wide range of cell types. And, as we have seen in the preceding pages, all are composed of the same remarkably few kinds of atoms and molecules.

* Metabolism is simply the sum of all the chemical reactions taking place in a living system.

6–1 *The single 4-inch cube, the eight 2-inch cubes, and the sixty-four 1-inch cubes all have the same total volume; however, as the volume is divided up into smaller units, the amount of surface area increases (e.g., the single 4-inch cube has only one-fourth the total surface area of the sixty-four 1-inch cubes). Similarly, smaller cells have more surface area (more membrane surface through which materials can diffuse into or out of the cell) and less volume (shorter diffusion distances within the cell).*

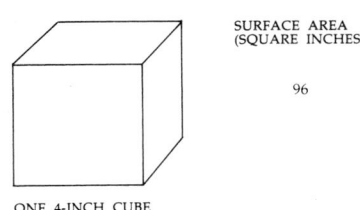

SURFACE AREA
(SQUARE INCHES)

96

ONE 4-INCH CUBE

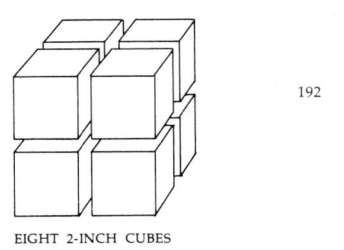

192

EIGHT 2-INCH CUBES

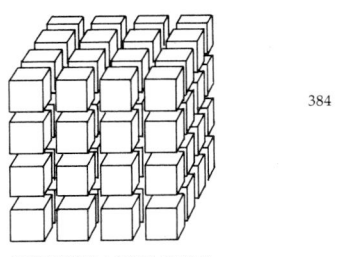

384

SIXTY-FOUR 1-INCH CUBES

In this chapter, we are going to be concerned particularly with eukaryotic cells, which, as we noted on page 24, are larger, more complex, and presumed to have evolved more recently than prokaryotic cells. Prokaryotic cells are dealt with in considerably more detail in Chapter 19.

CELL SIZE AND SHAPE

Most of the cells that make up a plant or animal body are within a size range of between 10 and 30 micrometers in diameter. A principal restriction on cell size seems to be the relationship between volume and surface area. As you can see in Figure 6–1, as volume increases, surface area decreases rapidly in proportion to volume. Materials entering and leaving the cell have to move through the surface, and the more active a cell is, the more rapidly these materials must pass through. Also, oxygen, carbon dioxide, and other metabolically important molecules move in and out of the cell by diffusion, which, as we shall see in the next chapter, is effective only over short distances. Materials can move faster into, out of, and through small cells.

It is not surprising, therefore, that the most metabolically active cells are usually small. The relationship between cell size and metabolic activity is nicely illustrated by egg cells. Many egg cells are very large. A frog's egg, for instance, is 1,500 micrometers in diameter. Some egg cells are several inches across. (Most of this is storage material, of course.) When the egg cell is fertilized and begins to be active metabolically, it first divides several times before there is any increase in volume or mass, thus cutting each of its cellular units down to an efficient metabolic size.

A second limitation on cell size appears to involve the capacity of the nucleus, the cell's control center, to regulate the cellular activities of a large, metabolically active cell. Again, the exceptions seem to "prove" the rule. In certain large, complex one-celled animals—the ciliates, of which *Paramecium* is an example—each cell has two or more nuclei, the additional ones apparently copies of the original.

Like drops of water and soap bubbles, cells tend to be spherical. Cells take other shapes because of cell walls, found in most plant cells and in many one-celled organisms, or because of attachments to and pressure from other, neighboring cells or surfaces.

Table 6–1   *Measurements Used in Microscopy*

1 centimeter (cm) = 1/100 meter = 0.4 inch
1 millimeter (mm) = 1/1,000 meter = 1/10 cm
1 micrometer (μm)* = 1/1,000,000 meter = 1/10,000 cm
1 nanometer (nm) = 1/1,000,000,000 meter = 1/10,000,000 cm
1 angstrom (Å) = 1/10,000,000,000 meter = 1/100,000,000 cm
or
1 meter = 10^2 cm = 10^3 mm = 10^6 μm = 10^9 nm = 10^{10} Å

* Micrometers were formerly known as microns, indicated by the Greek letter μ, which corresponds to our *m* and is pronounced "mew."

ORIGIN OF THE CELL THEORY

In the seventeenth century, Robert Hooke, using a microscope of his own construction, noticed that cork and other plant tissues are made up of small cavities separated by walls. He called these cavities "cells," meaning "little rooms." The word did not take on its present meaning, however, for more than 150 years.

In 1838, Matthias Schleiden, a German botanist, came to the conclusion that all plant tissues were organized in the form of cells. In the following year, zoologist Theodor Schwann extended Schleiden's observation to animal tissues and proposed a cellular basis for all life. The cell theory is of tremendous and central importance to biology because it emphasized the basic sameness of all living systems and so brought an underlying unity to widely varied studies involving many different kinds of organisms.

Schleiden, although so wise in his first generalization, erred on a second. He had been studying cell division in plant endosperm, the tissue that nourishes the embryo of flowering plants. This tissue is unusual in that all the nuclei of the cells divide first and then cell walls are laid down around the nuclei. Observing this process, Schleiden concluded that cells formed like crystals, condensing around a central core.

In 1858, this misconception was swept aside, and the cell theory took on an even broader significance when the great pathologist Rudolf Virchow generalized that cells can arise only from preexisting cells: "Where a cell exists, there must have been a preexisting cell, just as the animal arises only from an animal and the plant only from a plant.... Throughout the whole series of living forms, whether entire animal or plant organisms or their component parts, there rules an eternal law of continuous development."

In the broad perspective of evolution Virchow's concept takes on an even larger significance. There is an unbroken continuity between modern cells—and the organisms which they compose—and the primitive cells that first appeared on Earth more than three billion years ago.

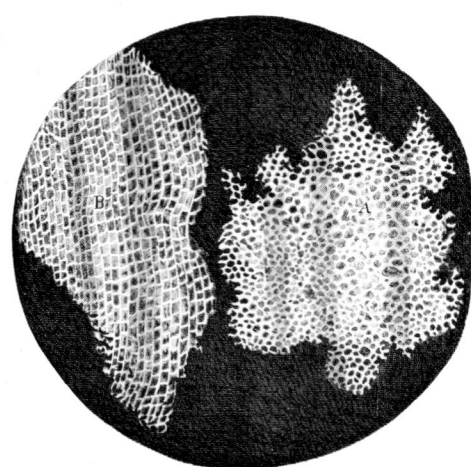

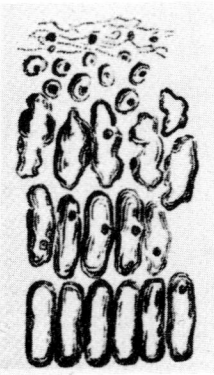

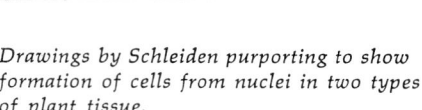

Robert Hooke's drawings of two slices of a piece of cork, reproduced from his Micrographia (1665). Hooke was the first to use the word "cells" to describe these tiny compartments into which living organisms are organized.

Drawings by Schleiden purporting to show formation of cells from nuclei in two types of plant tissue.

6–2 *Cell from a barley root as seen using an electron microscope. The cell wall is in the upper left-hand corner of the picture. Notice the many different kinds of structures within the cell and the variety and intricacy of their forms. For an explanation of the size marker in the lower right-hand corner of the micrograph, refer to page 93.*

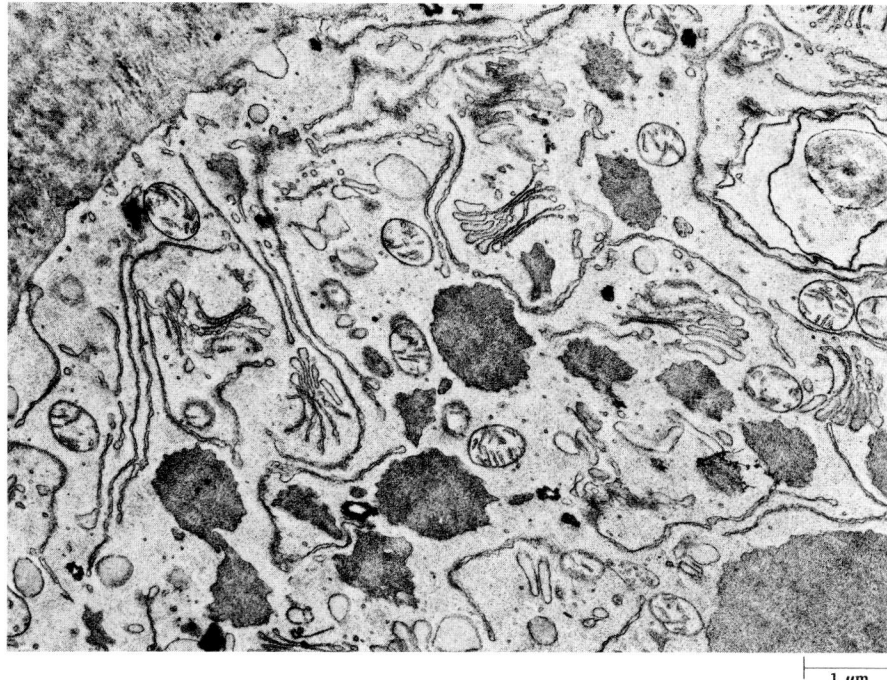

1 μm

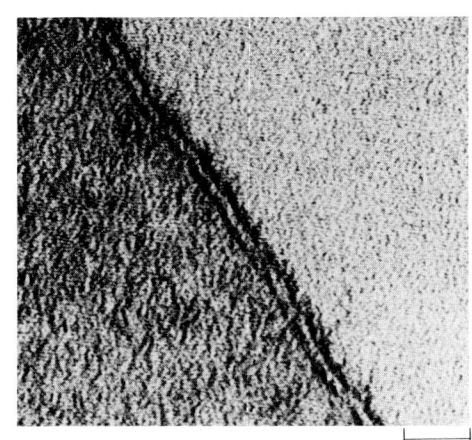

25 nm

6–3 *Electron micrograph of the cell membrane of a red blood cell. The "molecular sandwich" structure of the membrane is believed to consist of two electron-dense (dark) layers of phospholipid and cholesterol molecules arranged with their hydrophobic tails pointing inward, forming the electron-transparent (light) "filling," with globular proteins embedded throughout. (See page 115 for diagram.)*

CELL STRUCTURE

All cells, as we stated previously, are basically very similar, containing many of the same structures, the same types of enzyme systems, and the same kind of genetic material. An outer membrane, apparently conforming to the same biochemical design in all cells, surrounds the cell. The living materials bounded by the membrane, sometimes called the protoplasm, consist of the cytoplasm and the nucleus.

The Cell Membrane

Cell membranes, which are only about 90 Å in thickness, are not visible in the light microscope. Before electron microscopy became available, cytologists presumed that the cell was surrounded by an invisible film, because if this hypothetical film was ruptured, the contents of the cell could be seen to spill out. Now, with the electron microscope, the membrane can be visualized as a continuous thin double line. According to current hypotheses, the membrane consists essentially of the phospholipid and cholesterol molecules arranged so that their hydrophobic tails are pointed inward, forming the filling of a molecular sandwich, with globular proteins embedded at intervals (see Figure 7–8, page 115).

At one time this lipoprotein membrane was referred to as the "unit membrane," emphasizing its similarity from cell to cell and from structure to structure within the cell. More detailed chemical analyses reveal that, although the basic structure is indeed similar throughout living systems, there are local differences in the types of lipids and in the number and types of proteins. These differences undoubtedly represent differences in membrane function.

VIEWING THE MICROSCOPIC WORLD

Unaided, the human eye has a resolving power of about 1/10 millimeter, or 100 micrometers. This means that if you look at two lines that are less than 100 micrometers apart, they merge into a single line. Similarly, two dots less than 100 micrometers apart look like a single blurry dot. To separate structures closer than this, optical instruments such as microscopes are used. The best light microscope has a resolving power of 0.2 micrometer, or 200 nanometers, or 2,000 Å, and so improves on the naked eye about 500 times. It is theoretically impossible to build a light microscope that will do better than this.

Notice that resolving power and magnification are two different things; if you take a picture through the best light microscope of two lines that are less than 0.2 micrometer, or 200 nanometers, apart, you can enlarge that photograph indefinitely, but the two lines will continue to blur together. By using more powerful lenses, you can increase magnification, but this will not improve resolution. The limiting factor is the wavelength of light.

With the electron microscope, resolving power has been increased almost 1,000 times over that provided by the light microscope. This is achieved by using "illumination" consisting of electron beams instead of light rays. Areas in the specimen that permit the transmission of more electrons—"electron-transparent" regions—show up bright, and areas that absorb or deflect electrons—"electron-dense" regions—are dark. Electron microscopy at present, under the very best conditions, affords a resolving power of

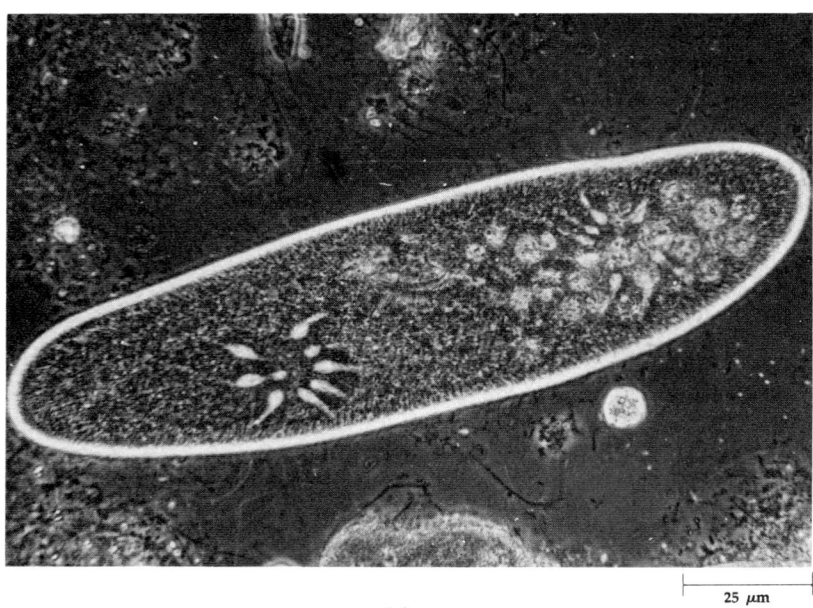

25 µm

(a)

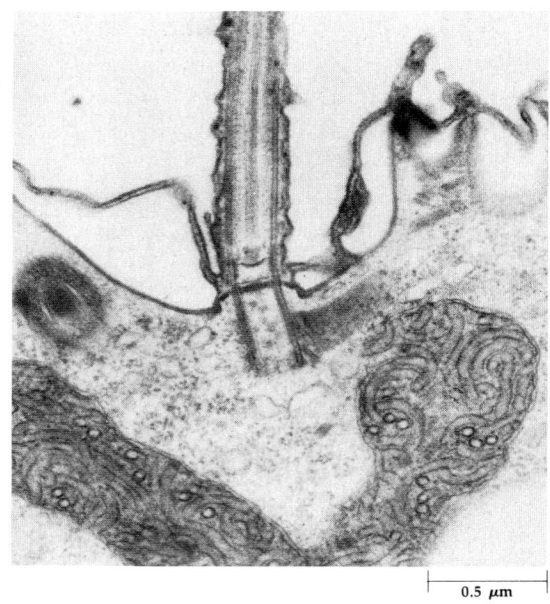

0.5 µm

(b)

A paramecium as seen using (a) a light microscope, (b) a conventional electron microscope, and (c) a scanning electron microscope. In the photomicrograph (a), light penetrating the living cell reveals some of its internal structures (those in the plane of focus), whereas in the scanning electron micrograph (c), electrons are reflected from the surface of the cell, making it possible to see the numerous hairlike cilia. A paramecium is a large cell, about 150 micrometers long. The conventional electron micrograph (b) shows only a small portion (an area about 2 micrometers square) of the cell, which has been sliced thin to permit passage of the electrons. You can see a cilium in longitudinal section and the internal structure of two specialized organelles (mitochondria). These details would not be visible in a photomicrograph,

about 2 Å, roughly 500,000 times greater than that of the human eye. (A hydrogen atom is about 1 Å in diameter.)

Electrons, which have a very small mass, can pass through specimens only when the specimens are exceedingly thin. As a consequence, living matter has to be fixed, dehydrated, embedded in hard materials, and sliced by special cutting instruments before it can be examined by transmission electron microscopy. Moreover, most specimens are electron-transparent and have to be stained with heavy metals that bind differentially to different subcellular components. Thus only dead cells can be studied, and, moreover, it is sometimes difficult to determine whether or not the specimens have been changed by the preparation methods.

Recently, electron microscopy has been expanded in scope by the development of the scanning electron microscope. In scanning electron microscopy, the electrons whose imprints are recorded come from the surface of the specimen. The electron beam is focused into a fine probe, which is rapidly passed back and forth over the specimen. Complete scanning from top to bottom usually takes a few seconds. As a result of the electron bombardment from the probe, the specimen emits low-energy secondary electrons. Variations in the surface of the specimen alter the number of secondary electrons emitted. Holes and fissures appear dark, and knobs and ridges are light. Electrons scattered from the surface plus secondary electrons are amplified and transmitted to a screen, which is scanned in synchrony with the electron probe.

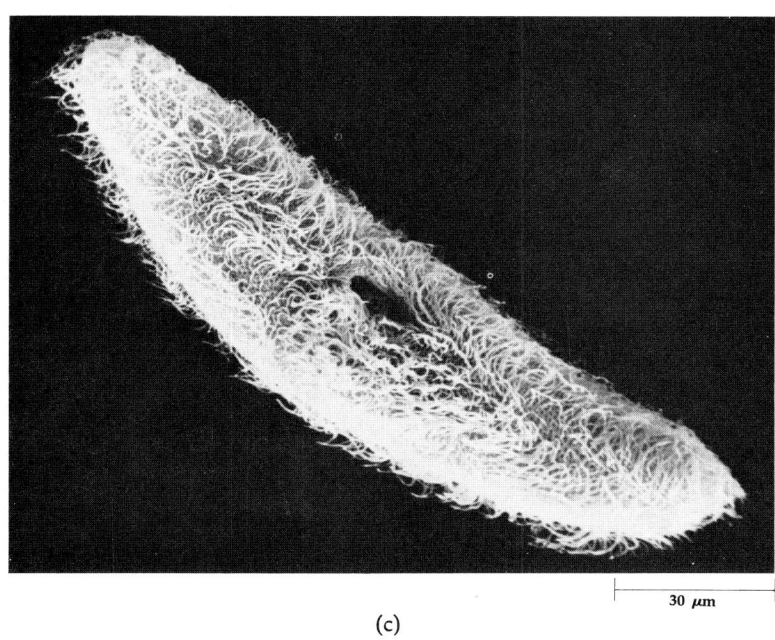

30 μm

(c)

no matter how great the magnification. The short straight line at the bottom of each picture provides a reference for size; a micrometer, abbreviated μm, is 1/10,000 centimeter. The same system is used to indicate distances on a road map.

The Cell Wall

Most plant cells are surrounded by an outer cell wall, composed chiefly of cellulose. This wall is built, molecule by molecule, by the cell as it grows. In young cells, the wall is thin and flexible, but as the plant cell matures, it becomes more rigid. In some types of cells, such as those which form the woody tissue of plants, the cellulose walls are impregnated with a macromolecule known as lignin, which has stiffening properties. In such cells, the protoplast (the cell itself, within the wall) often dies, leaving only the cell wall. Between walls of adjacent cells is a thin gluey layer called the *middle lamella*; it is composed of pectins and other noncellulose polysaccharides and serves to hold the cells in place. Adjacent cells in the plant body are often interconnected by plasmodesmata (singular, *plasmodesma*), which are fine strands of cytoplasm crossing through the cell walls and the middle lamella. These are believed to integrate the plant body by forming passageways for the exchange of materials between cells, although there is little direct evidence to support this reasonable hypothesis.

The Nucleus

In eukaryotic cells, the nucleus is a large, spherical body, usually the most prominent structure within the cell. It is surrounded by two "unit membranes," which together make up the nuclear envelope (see Figure 6–5). Penetrating the surface of the envelope are nuclear pores, which are covered by a thin single membrane. These pores apparently permit only specific large molecules to go through, thus keeping the chemical composition of the nuclear material different from that of the cytoplasm.

6–4 (a) *Diagram of plant cell walls. The cells are separated from one another by the middle lamella. On each side of the middle lamella is the primary wall of an individual cell. In plants that have secondary walls, this wall is deposited inside the primary wall. In some plant cells, the protoplast—the membrane-bounded cell within the cell wall—is dead at maturity. (b) Electron micrograph of two adjacent cell walls of vessel elements, the cells that form the tubes through which water is conducted. You can see the middle lamella and the layered, heavily lignified primary and secondary walls. The cells are from the wood of a black locust tree. The protoplasts are no longer present.*

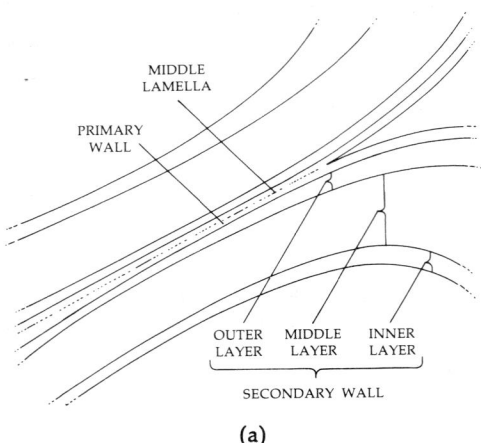

(a)

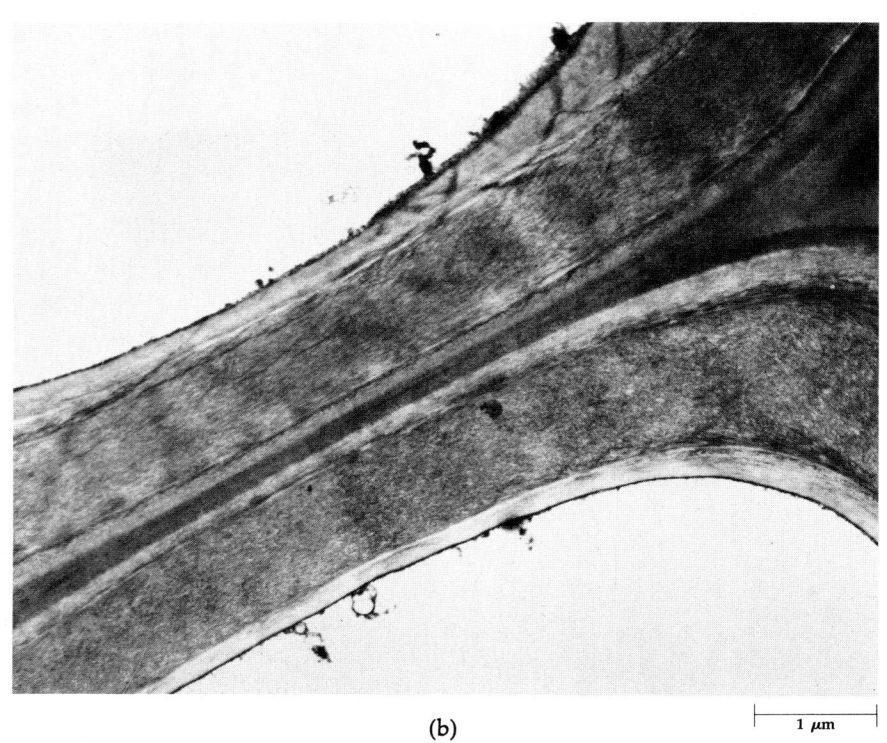

1 μm

(b)

6–5 (a) *The cell nucleus fills the left half of this electron micrograph. What you see here is the surface of the nuclear envelope. Clearly visible on this surface are pores, through which, it is believed, the nucleus and the cytoplasm are able to "communicate." (b) Diagram of a nuclear pore, as interpreted from electron micrographs. Each pore appears to contain eight spheroid elements. There is a thin central membrane across the pore opening which seems to disperse during times of rapid synthesis of material in the nucleus. It has been suggested that the pores can open and close, thus regulating the passage of materials between nucleus and cytoplasm.*

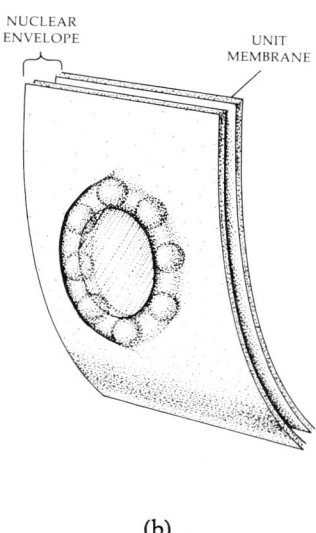

(a) |———— 1 μm ————| (b)

6–6 *Electron micrograph showing the large central nucleolus in the nucleus of a plant cell. The small, dense patches near the nucleolus are areas of chromatin. The pillow-shaped objects outside the nucleus are chloroplasts, the organelles in which photosynthesis takes place.*

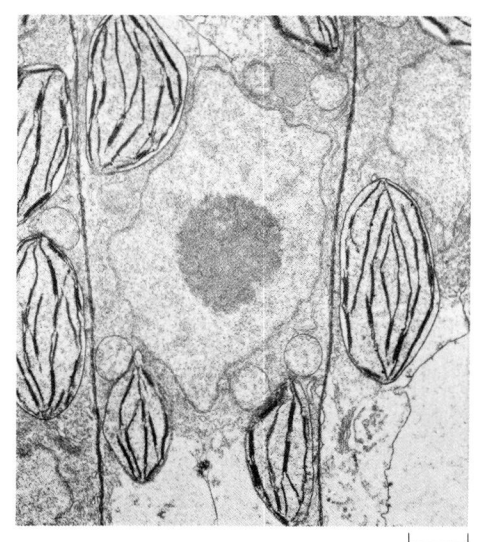

|— 1 μm —|

The chromosomes, which are composed of DNA and protein, are found within the nucleus. When the cell is not dividing, the chromosomes are visible only as a tangle of fine threads, called *chromatin*. The most conspicuous body within the nucleus is the *nucleolus*. The nucleolus, which is composed of DNA and protein, like the chromosomes, and is formed from a portion of a chromosome, is the site at which a particular kind of RNA—the ribosomal RNA—is formed.

The Functions of the Nucleus

The nucleus performs two crucial functions for the cell. First, it carries the hereditary information for the cell, the instructions that determine whether a particular organism will develop to become a paramecium, oak tree, or human being—and not just any paramecium, oak tree, or human being, but one that resembles the parent or parents of that particular, unique organism. Second, the nucleus directs the ongoing activities of the cell, ensuring that the complex molecules the cell requires are synthesized in the number and of the kind needed. These molecules are involved in carrying out the cell's various activities and also in forming organelles and other structures. The way in which the nucleus performs these functions will be described in Section 2.

The Cytoplasm

The cytoplasm includes all the functioning substances of the cell (the protoplasm) except the nucleus. It consists of an aqueous (watery) solution of enzymes and other macromolecules, ATP, electron carriers, amino acids, nucleotides, and inorganic substances, such as phosphates, sodium, and potassium, largely in the form of ions. Depending on the type of cell, it may also contain specialized materials, such as hemoglobin, starch granules, pigments, or oil droplets. It also includes the cellular organelles.

Endoplasmic Reticulum

The cytoplasm is partitioned by a network of membranes called the *endoplasmic reticulum,* which is continuous with the outer membrane of the nuclear envelope. It has the same basic structure as the cell membrane and the nuclear membranes. The endoplasmic reticulum divides the cell into separate compartments, making it possible for the cell's different chemical products and activities to be segregated from each other. Many of the enzymes that carry out these activities form a part of the lipoprotein structure of the membrane forming the endoplasmic reticulum. It is believed that enzymes that work in sequence are lined up in order on a membrane surface, like factory workers on an assembly line. The many folds of the endoplasmic material thus form a large work surface on which many of the cell's biochemical activities take place. (See Figure 6–7.)

Ribosomes

Both eukaryotic and prokaryotic cells contain *ribosomes,* which are the sites where amino acids are assembled into enzymes and other proteins. Ribosomes are composed of relatively small molecules of RNA and of protein. Each ribosome consists of two subunits, one slightly larger than the other. In cells that are synthesizing proteins "for export"—such as digestive enzymes that are secreted into the stomach or intestine—most of the ribosomes are attached to membranes of the endoplasmic reticulum. Endoplasmic reticulum with ribosomes attached is known as rough endoplasmic reticulum. In cells with rough endoplasmic reticulum, the membrane network forms a pathway by which substances are channeled in and out of cells. By contrast, in cells that are growing or making proteins for internal use, such as hemoglobin, large numbers of ribosomes are found unattached to other structures in the cytoplasm.

6–7 *Groups of ribosomes (polyribosomes) on the surface of endoplasmic reticulum. Endoplasmic reticulum is a network of membranes that fills the cytoplasm of the eukaryotic cell, dividing it into channels and compartments and providing surfaces on which chemical reactions take place. Ribosomes are the sites at which amino acids are assembled into proteins. This electron micrograph shows a portion of an epidermal cell of a radish root.*

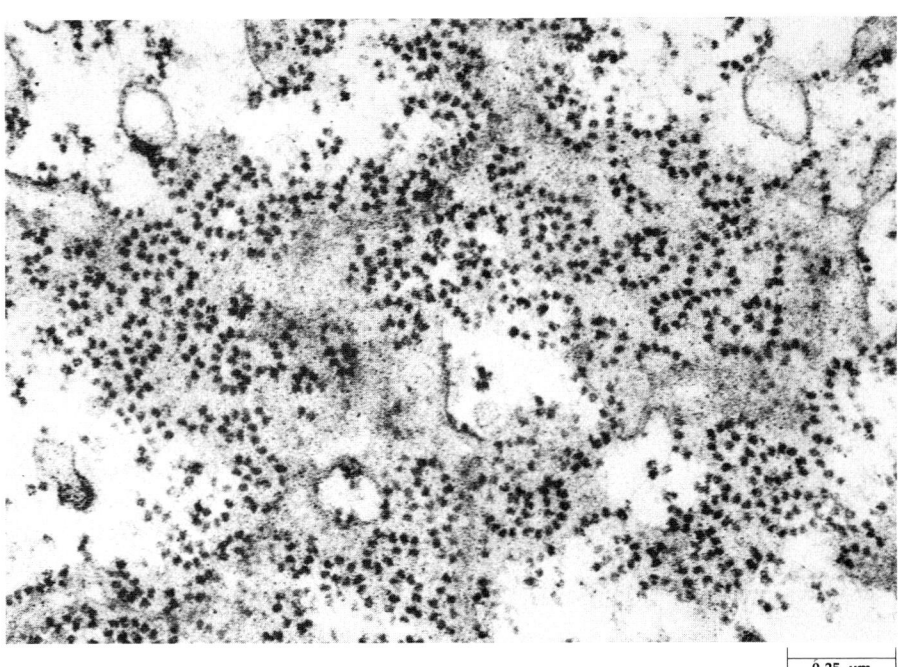

0.25 μm

6–8 *Golgi bodies are also composed of special arrangements of membranes. Compounds formed within the cell are packaged within membrane-enclosed vesicles at the Golgi bodies. The vesicles then move to the cell surface and release their contents. For instance, in the case of plant cells, the contents of the vesicles are used in the construction of the plant cell walls. Note the numerous Golgi bodies in the cell shown in Figure 6–2 on page 91.*

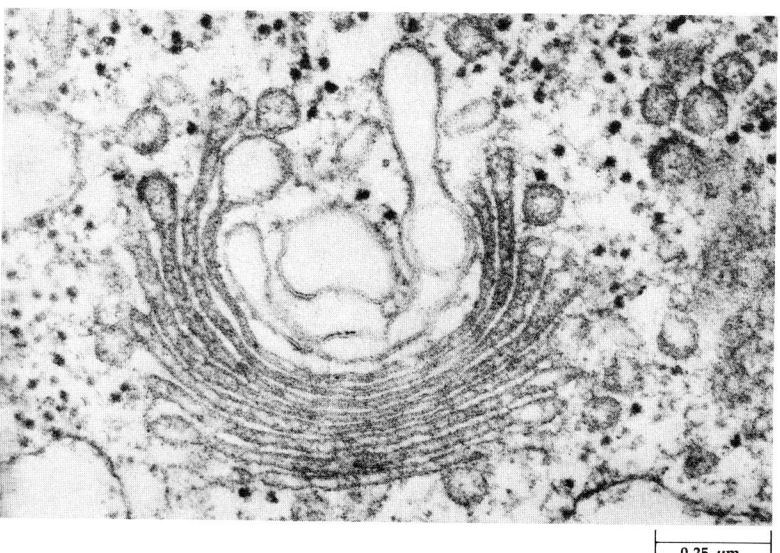

0.25 μm

Golgi Bodies

Golgi bodies, which function as packaging centers, are found in almost all types of eukaryotic cells. Animal cells usually contain one; plant cells may have as many as several hundred. Each consists of a group of flattened sacs, composed of membranes, stacked loosely on each other, surrounded by tubules and small spherical vesicles. *Vesicles*—little membrane-surrounded packages—bud off from the larger sacs. Evidence suggests that proteins synthesized for export, formed on ribosomes on the surface of the endoplasmic reticulum, are channeled into Golgi bodies where they accumulate and are packaged in vesicles in which they travel to the outer cell membrane and are discharged to the exterior of the cell. Components of the plant cell wall are also assembled within the cell and transported from it within the vesicles of Golgi bodies. (See Figure 6–8.) A cell's total set of Golgi bodies is called the Golgi complex.

Lysosomes

Important products of the Golgi body, especially in animal cells, are *lysosomes*, which are bags of hydrolytic (from "hydrolysis") enzymes enclosed in membranes which separate them from the rest of the cell. These enzymes are capable of digesting practically any cellular constituent. Lysosomes can be regarded as the cell's internal digestive system. They function primarily in two ways. First, in single-celled organisms, such as paramecia, they digest bacteria and other food particles taken up by the cell. In specialized cells of the human body, such as certain white blood cells, this digestion of bacteria and other disease-causing organisms forms the body's first line of defense against infection. Second, lysosomes are involved in the breakdown of worn-out cell organelles by a process known as autophagy (self-eating). Sometimes entire cells are destroyed, as in the resorption of the tadpole tail which occurs in the course of its metamorphosis into a frog. In plant cells, lysosomes may be involved in the breakdown of protoplasts that occurs in some types of cells as they mature.

Table 6–2 *Some Representative Sizes*

Human egg cell	100 μm
RESOLVING POWER OF UNAIDED VISION	1/394 inch = 0.1 mm = 100 μm
Body cell of plant (average)	30 μm (diameter)
Body cell of animal (average)	20 μm (diameter)
Chloroplast (in higher plants)	about 3 $\times$ 5 μm
Mitochondrion	0.2–1 $\times$ 3–10 μm
Bacterium (*Escherichia coli*)	1 $\times$ 2 μm
RESOLVING POWER OF LIGHT MICROSCOPE	0.2 μm = 200 nm = 2000 Å
Ribosomes	250 Å (diameter)
Tobacco mosaic virus	180 $\times$ 3000 Å
Hemoglobin molecule	64 $\times$ 55 $\times$ 50 Å
RESOLVING POWER OF CONVENTIONAL ELECTRON MICROSCOPE	2 Å
Hydrogen atom	1 Å

The products of lysosomal digestion are small molecules (amino acids, sugars, nucleotides, etc.), which diffuse from the site of digestion out into the cytoplasm where they can be reutilized. Defects in lysosomal function have been implicated in several degenerative diseases, especially of the heart and brain.

Mitochondria

Mitochondria are the organelles in which most of the ATP of the eukaryotic cell is produced. They are present in almost all kinds of cells, including both plant and animal. Mitochondria are surrounded by two membranes and have a complex inner membrane structure. We shall examine their structure in detail in Chapter 8, when we discuss their function.

Plastids

Plastids, like mitochondria, are bound by two outer membranes and have a system of internal membranes. Unlike mitochondria, they are present only in plant cells. They are of three general types: leucoplasts, chromoplasts, and chloroplasts. Leucoplasts, which are colorless, are the sites at which glucose is converted to starch and where lipids or proteins may be stored. Chromoplasts synthesize and retain pigments and are responsible for the bright colors of many kinds of fruits, vegetables, flowers, and leaves. Chloroplasts, the organelles in which photosynthesis takes place, will be discussed in detail in Chapter 9.

6–9 *Animal cell, as interpreted from electron micrographs.*

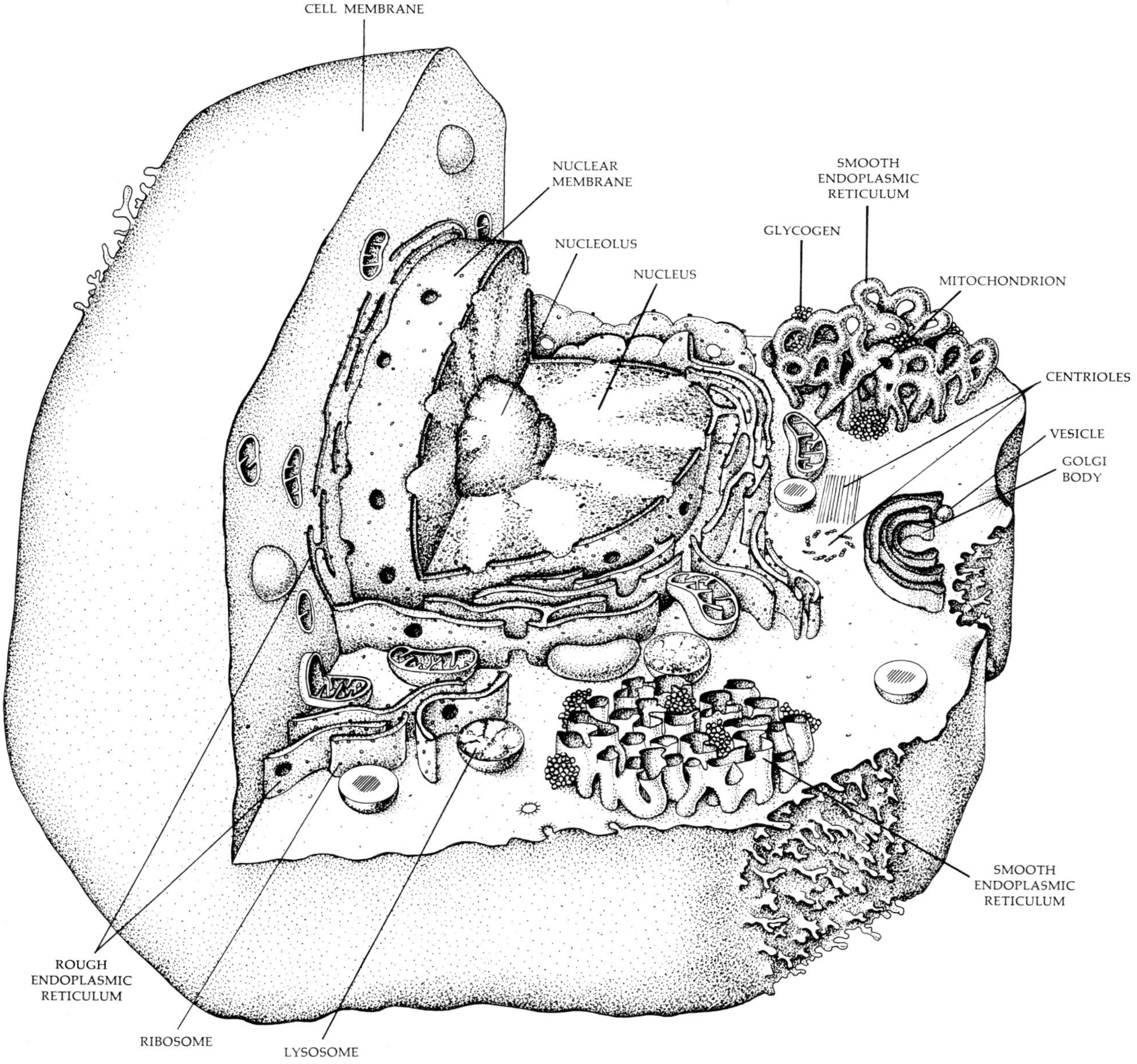

CELL MEMBRANE

NUCLEAR
MEMBRANE

NUCLEOLUS

NUCLEUS

SMOOTH
ENDOPLASMIC
RETICULUM

GLYCOGEN

MITOCHONDRION

CENTRIOLES

VESICLE

GOLGI
BODY

SMOOTH
ENDOPLASMIC
RETICULUM

ROUGH
ENDOPLASMIC
RETICULUM

RIBOSOME

LYSOSOME

6–10 (a) *Electron micrograph of a food vacuole of a paramecium. The vacuole is full of bacteria. You can clearly see the surrounding membrane. (b) Edge of a vacuole. The membrane runs across the center of the picture. The small, dense vesicles outside (below) the membrane, indicated by arrows, are lysosomes. They contain digestive enzymes which are discharged into the food vacuole (the vesicle merges with the vacuole). In the lower right-hand corner of the picture are two mitochondria. Inside the vacuole, digestion of the bacteria has begun.*

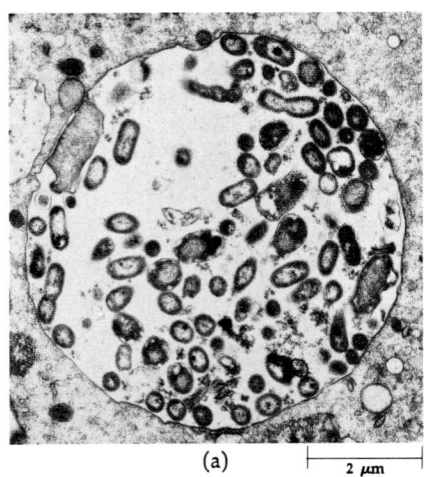

(a) |———— 2 µm ————|

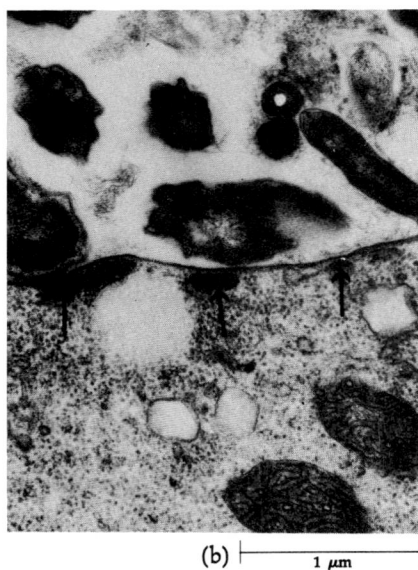

(b) |——— 1 µm ———|

Vacuoles

Many cells also contain *vacuoles*.* A vacuole is a space in the cytoplasm filled with water and solutes, and surrounded by a single membrane. Immature plant cells characteristically have many vacuoles, and as the plant cells mature, the numerous smaller vacuoles coalesce into one large, central, fluid-filled vacuole that then becomes a major supporting element of the cell. In plant cells, such vacuoles also may serve the functions of storing sucrose or of accumulating waste products.

Microfilaments

Microfilaments, as their name implies, are fine threadlike fibrils; they can be detected by the electron microscope in many types of cells. These filaments, only about 60 Å in diameter, are believed to be capable of contracting. Elaborate arrangements of contractile filaments are present in skeletal muscle cells, as we shall see in Chapter 32, and microfilaments are believed to be made of one of the major protein components (actin) of muscle cells. More recently, these microfilaments have been shown to be associated with movements of the cytoplasm, called amoeboid movement, or cyclosis, or cell streaming, depending on what type of cell they are observed in. This cytoplasmic motion, which is an energy-requiring process, is associated with locomotion in cells such as amoebas and the white blood cells found in our bloodstreams. It is readily visible under the light microscope in many types of plant cells, and typically occurs in animal cells during developmental processes in which cells change shape or develop long extensions; in the case of a nerve cell, such extensions may be as long as 3 feet.

* Both vesicles and vacuoles are membrane-surrounded bubbles in the cytoplasm. They are distinguished by size: vesicles are usually less than 100 nanometers in diameter, whereas vacuoles are larger.

6–11 *When animal cells, cultured in a medium outside the body, are placed on solid substrate, such as a glass dish, the cells attach and begin to spread over the substrate. The spreading portion of the cell characteristically includes microfilaments (mf), which are threadlike structures associated with various other kinds of cell movements, such as cytoplasmic streaming and ameboid motion. They are frequently found in association with microtubules (mt), as they are here. The microfilaments and microtubules are characteristically extended in the direction in which the cytoplasmic movement is taking place.*

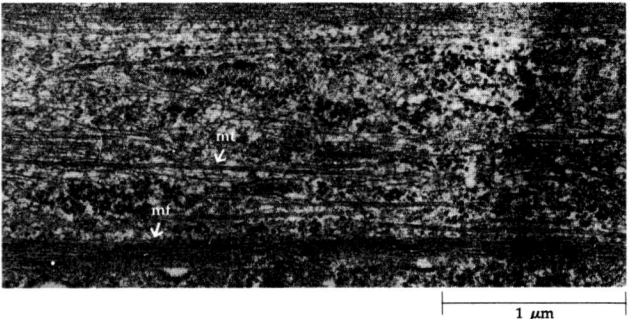

|——— 1 µm ———|

-12 *Photosynthetic cell from the leaf of a tobacco plant. A young plant cell usually contains several vacuoles in different parts of the cell. These vacuoles will* swell *as the cell reaches maturity, stretching the flexible cell wall until most of the interior of the cell is taken up by a single large central vacuole. In the mature cell, the cellulose cell wall becomes thick and inflexible. Within the chloroplasts, the largest bodies in the cytoplasm, you can see several starch grains.*

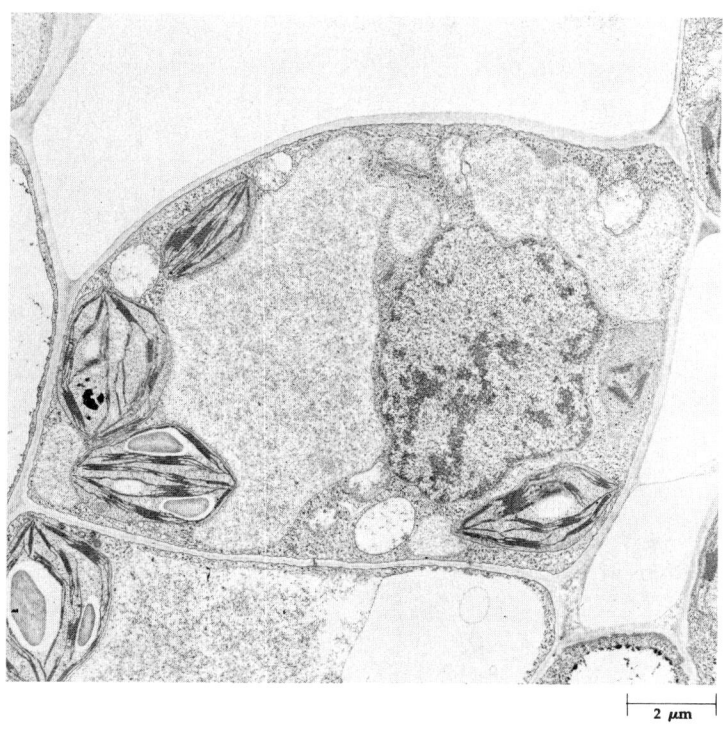

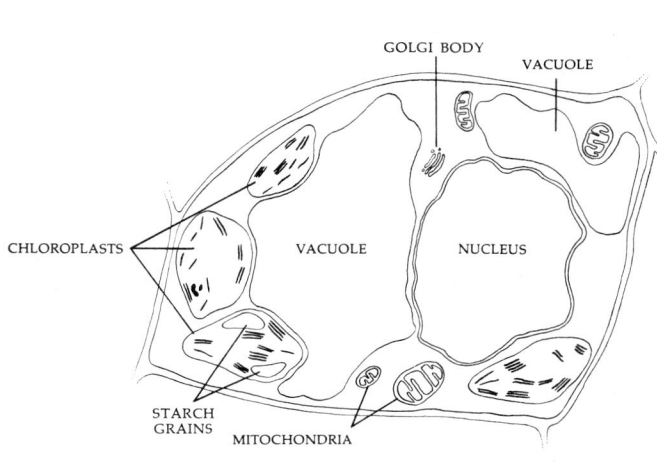

2 μm

-13 *These are bundles of microtubules from the cytoplasm of a paramecium (the arrow points to one of the bundles). Such bundles are characteristically found near the gullet, the area in the cell membrane where food particles are taken in. It has been suggested that they help channel the food vacuoles from this area deeper into the cytoplasm.*

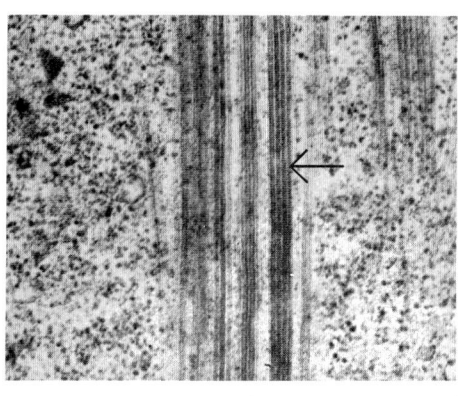

1 μm

Microtubules

Microtubules, the structure of which was described on page 85, serve a variety of functions in cells, as we noted previously. They are found in the spindle at cell division, as we shall see in Chapter 10, and they make up the internal structures of cilia and flagella, which we shall discuss in the next paragraph. They also seem to act as a scaffolding or internal skeleton in cells, as, for example, when plant cell walls are being laid down, or when cells are changing shape in the course of embryonic development.

Cilia and Flagella

Cilia and *flagella* are long, thin (0.2 micrometer) structures extending from the surface of many types of eukaryotic cells. They are the same except for length; flagella are longer. (Prokaryotic cells also have flagella, but these are different in structure.) In one-celled organisms and some small animals (such as flatworms), cilia and flagella are associated with cell movement. For example, one type of paramecium has approximately 17,000 cilia, each about 10 micrometers long, which propel it through water by beating in a coordinated fashion. Other one-celled organisms, such as the members of the genus *Chlamydomonas*, have only

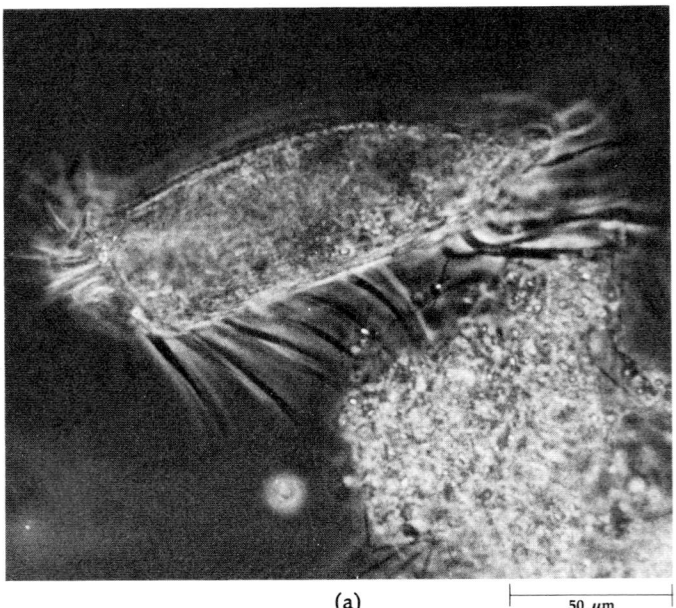

(a)

|— 50 μm —|

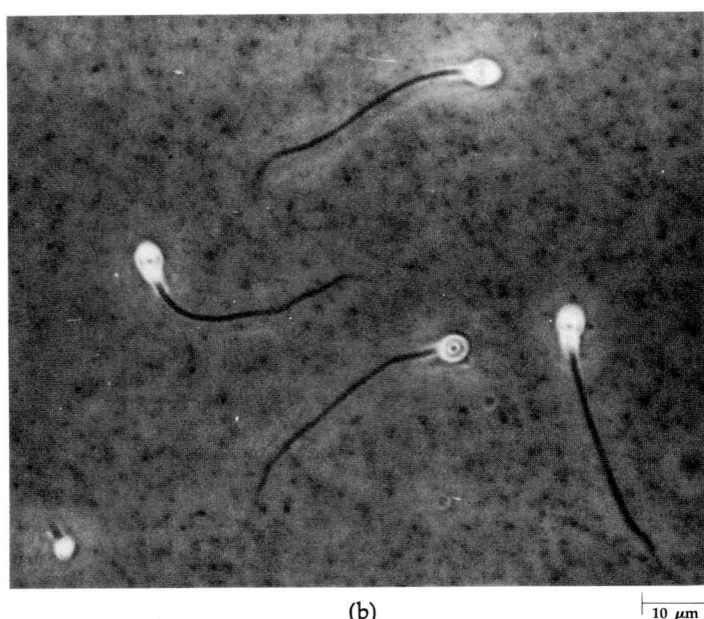

(b)

|— 10 μm —|

6–14   (a) Euplotes patella, *a single-celled
organism. The cilia on the protist's lower
surface are fused together in groups to
form heavy cirri. These cirri do not beat
rhythmically like ordinary cilia but
instead move in rapid jerks. They are used
by* Euplotes *as legs, enabling it to crawl
about on vegetation. (b) Human sperm.
The flagella of these sperm cells have
the same basic 9 + 2 microtubular
structure as the cilia of* Euplotes.

two whiplike flagella, protruding from the anterior end of the organism, which
move them through the water. The motile power of the human sperm cell comes
from its single powerful flagellum, or "tail."

Many of the cells which form the tissues of our bodies are also ciliated. These
cilia do not move the cells, but rather serve to sweep substances across the cell
surface. For example, cilia on the surface of cells of the respiratory tract beat
upward, propelling bits of soot, dust, pollen, tobacco tar—whatever foreign
substances we have inhaled either accidentally or on purpose—to the backs of
our throats, where they can be removed by swallowing. A very long single
flagellum propels the human sperm cell through the female reproductive tract,
and the human egg cell is propelled down the Fallopian tubes by the beating of
cilia that line the inner surfaces of these tubes. Cilia or flagella are found
extensively throughout the living world, on the cells of invertebrates, vertebrates, the sex cells of ferns and other plants, as well as on one-celled organisms.
Only a few large groups of organisms, such as the true fungi and the flowering
plants, have no cilia or flagella in any cells.

All cilia and flagella, whether on a paramecium or a human sperm cell, have
the same structure: nine pairs of microtubules form a ring which surrounds
two additional, solitary microtubules in the center. (The microtubules themselves, as you will recall from the previous chapter, are composed of identical
globular protein units assembled in a long helix.) The movement of cilia and
flagella comes from within the structures themselves; cilia removed from cells
and exposed to ATP can move by themselves. The movement, according to one
hypothesis, is caused by one tubule of an outer pair moving tractor fashion
along its partner. The little "arms" that you can see on one of each pair of
outer tubules (Figure 6–16) are believed to be enzymes involved in the splitting
of ATP to ADP within the cilium, and so converting chemical to kinetic energy.

6–15 *Cross section of flagella. These are from* Trichonympha, *a one-celled organism that lives in the guts of termites where it breaks down cellulose.*

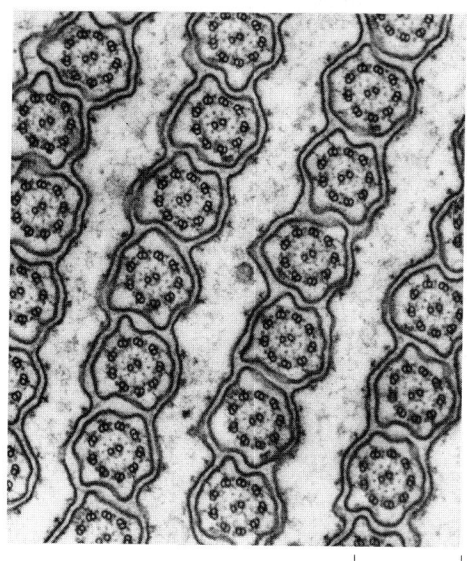

0.2 μm

6–16 *Diagram of a cilium with its underlying basal body. All cilia and flagella, whether they are found on one-celled organisms or on the surfaces of cells within our own bodies, have this same basic structure, which consists of an outer ring of nine pairs of microtubules surrounding two additional microtubules in the center. The basal bodies from which they arise have nine outer triplets, with no microtubules in the center. The structural basis of the wheel-like formation in the basal body is not known. The "hub" is not a microtubule, although it has about the same diameter.*

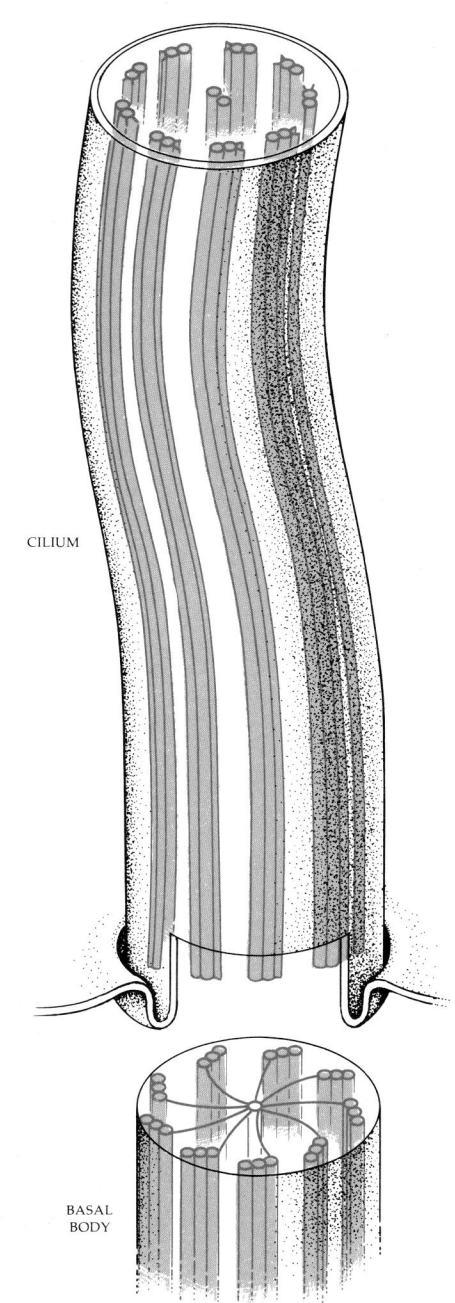

CILIUM

BASAL BODY

Underlying each cilium is a structure known as a basal body, which has the same diameter as a cilium, about 0.2 micrometer. It consists of microtubules arranged in nine triplets around the periphery and, unlike the cilium, has no tubules in the center. Cilia and flagella arise from basal bodies. For instance, as a sperm cell takes form, a basal body moves near the cell membrane, and the sperm's flagellum arises from it. Basal bodies also apparently serve to transmit ATP and perhaps other nutrients to cilia and flagella. You will notice that none of the microtubules in the basal body has arms.

Centrioles

Many types of eukaryotic cells contain _centrioles_. Centrioles, which typically occur in pairs, are small cylinders, about 0.2 micrometer in diameter, containing nine microtubule triplets (Figure 6–17). Their structure is identical to that of basal bodies; however, their distribution and function in the cell are different, which is why they have continued to be called by different names even though electron microscopy has revealed their identical structures. Centriole pairs usually lie at right angles to one another in the cytoplasm, near the nuclear membrane. They are found only in those groups of organisms which also have cilia or flagella (and, therefore, basal bodies). There is evidence that centrioles are important in organizing a structure known as the _spindle_ which appears at the time of cell division and is involved in chromosome movements. However, despite this evidence, which we shall present in Chapter 10, cells without centrioles can, paradoxically, form spindles.

THE UNITY OF LIFE

In this chapter, we have concentrated on those structures common to many or all cells. Like the sugars, lipids, proteins, and nucleic acids found in all cells, the presence of these same structures, over and over again, throughout the living world, serves to remind us once more of the essential unity of living systems.

6–17 _Centrioles are structurally identical to basal bodies. The one shown here in cross section is from a mouse embryo cell._

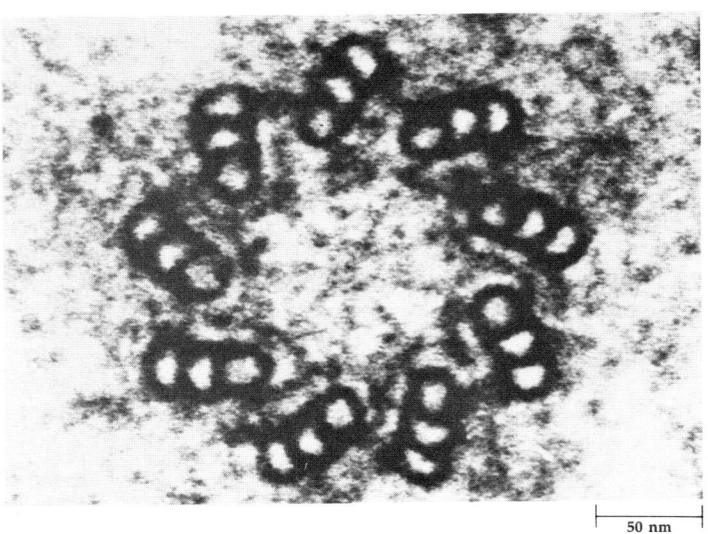

50 nm

Table 6–3 *Comparison of Animal, Plant, and Prokaryotic Cells*

	Animal	Plant	Prokaryote
Cell membrane	Present	Present	Present
Cell wall	Absent	Present (cellulose)	Present (noncellulose polysaccharide plus protein)
Nucleus	Surrounded by nuclear envelope	Surrounded by nuclear envelope	No nuclear envelope
Chromosomes	Multiple, containing DNA and protein	Multiple, containing DNA and protein	Single, containing only DNA
Endoplasmic reticulum	Usually present	Usually present	Absent
Mitochondria	Present	Present	Absent
Plastids	Absent	Present in many cell types; chloroplasts in photosynthetic cells	Absent
Ribosomes	Present	Present	Present (smaller)
Golgi bodies	Present	Present	Absent
Lysosomes	Often present	Usually absent	Absent
Vacuoles	Small or absent	Usually large single vacuole in mature cell	Absent
9 + 2 cilia or flagella	Often present	Absent (in higher plants)	Absent
Centrioles	Present	Absent (in higher plants)	Absent

SUMMARY

All living matter is composed of cells. All cells arise from other cells. The metabolism of living systems—the sum total of all their biochemical activities—takes place within cells and as a result of cellular activities.

Cells are units of protoplasm—living matter—separated from their environment by an outer cell membrane which restricts the passage of materials in and out of the cell and so protects the cell's structural and functional integrity. The size of cells is limited by proportions of surface to volume; the smaller a cell's volume in proportion to its surface area, the more readily substances can diffuse in and out of the cell. Cells which are active metabolically are likely to be small.

In eukaryotic cells, the hereditary material is separated from the cytoplasm by the nuclear envelope, which contains pores through which molecules pass to the cytoplasm. The nucleus contains the chromosomes, which are composed of DNA and protein and which, when the cell is not dividing, exist in an extended form called chromatin. The nucleolus, visible within the nucleus, is the site of formation of the ribosomal RNA.

The cytoplasm is surrounded by a membrane composed of two layers of phospholipid and cholesterol molecules arranged with their hydrophobic tails pointing inward, forming a molecular sandwich, with globular proteins embedded throughout. Outside the cell membrane in plant cells is a cellulose cell wall.

The cytoplasm is subdivided by a network of membranes known as the endoplasmic reticulum, which serves as a work surface for many of the cell's biochemical activities, a way of segregating different activities from one another and also, perhaps, a system of channels for moving materials through the cell. Golgi bodies, also composed of membranes, are packaging centers for materials being moved through and out of the cell.

All metabolically active cells contain ribosomes, the organelles on which amino acids are assembled into proteins. Ribosomes may be bound to the surface of the endoplasmic reticulum or free in the cytoplasm. Bound ribosomes are found characteristically in cells producing proteins for export.

Mitochondria are bounded by two membranes, the inner one of which folds to form a complex internal membrane structure. They are found in virtually all eukaryotic cells and are the chief centers for the formation of ATP from ADP.

Lysosomes, which usually arise from Golgi bodies, are sacs of digestive enzymes. They may fuse with other vesicles or vacuoles, such as food vacuoles, and digest their contents, or they may burst and, by releasing the content of the lysosome into the cell, destroy it.

Plastids are present in many types of plant cells. Leucoplasts are involved in synthesizing starch from sugars. Chromoplasts synthesize and retain many of the pigments found in plants. Chloroplasts are the sites of photosynthesis. Like mitochondria, chloroplasts are surrounded by two membranes and have a complex internal membrane system.

Vacuoles and vesicles are spaces within cells that are enclosed by a single membrane. They contain water, enzymes, stored sugars, food particles, waste substances, etc. Immature plant cells characteristically have numerous small vacuoles that coalesce into a large single vacuole as the plant cell matures.

Microfilaments are contractile, threadlike protein filaments associated with cytoplasmic movements in a variety of types of cells.

Microtubules, which are long, thin assemblies of globular proteins, act as skeletal elements within cells and also are the components of cilia and flagella.

Cilia and flagella are hairlike appendages found on the surface of many cell types and are associated with the movement of cells or the movement of materials across cell structures. They have a highly characteristic 9 + 2 structure, with nine pairs of microtubules forming a ring surrounding two central microtubules. One of each pair of outer microtubules contains enzymes which split ATP and provide the energy for ciliary motion.

Cilia and flagella arise from basal bodies, which are cylindrical structures containing nine microtubule triplets with no inner pair. Centrioles have the same

internal structure as basal bodies and are found only in those groups of organisms that have cilia or flagella. They typically occur in pairs, lying near the nuclear membrane, and appear to play a role in the separation of the chromosomes at cell division.

QUESTIONS

1. Sketch a cell, either plant or animal. Include the principal organelles and label them.
2. What are the functions of the following: endoplasmic reticulum, Golgi body, chromoplast, vesicle?
3. Sketch a cross section of a cilium.

Chapter 7

How Things Get Into and Out of Cells

Cells are able to regulate the passage of materials across cell membranes. This is an important capacity. One of the criteria by which we identify living systems, as we noted in the first chapter, is that living matter, although surrounded on all sides by nonliving matter, is different from it in the kinds and amounts of chemical substances it contains. Without this difference, of course, living matter would be unable to maintain the organization and structure on which its existence depends. The cell membrane is not simply an impenetrable barrier, however. Living matter constantly exchanges substances with the non-living world around it. Control of these exchanges is essential in order to protect the cell's integrity and to maintain those very narrow conditions of pH and salt concentrations, common to all cells, at which enzyme activity can take place. The cell membrane thus has a complex double function of keeping things out and letting things in. Moreover, cell membranes not only control the passage of material from outside the cell, but internal membranes, such as those surround-

7–1 *Electron micrograph of a cell from the root tip of a corn plant. The cell has a large, central nucleus with scattered chromatin, several clearly delineated Golgi bodies, many mitochondria, endo-plasmic reticulum, small vacuoles, and numerous other cellular organelles. Not only is the cell itself surrounded by a membrane and the nucleus by a double membrane system (the nuclear enve-lope), but many of its organelles are also surrounded by membranes, and the membranes of endoplasmic reticulum further divide the cell into membrane-bounded compartments. These mem-branes regulate the movements of sub-stances in and out of cells and restrict their passage from one part of the cell to another.*

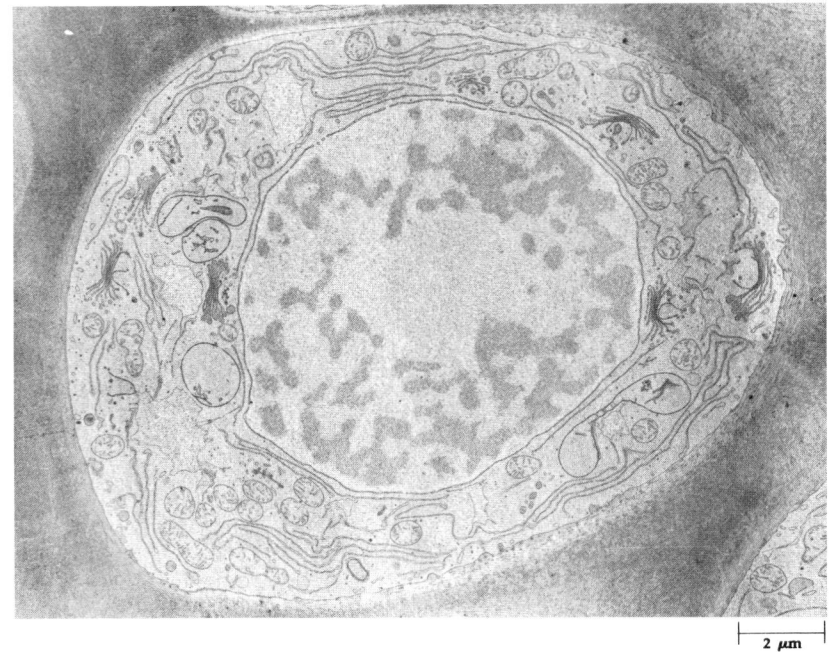

2 μm

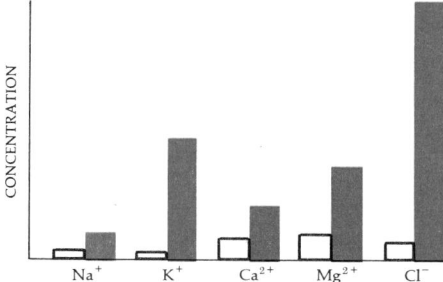

7–2 *Diagram showing the relative concentrations of different ions in pond water (clear boxes) and in the cytoplasm of the green alga* Nitella *(colored boxes). Differences such as these between cells and their surroundings demonstrate that cells regulate the passage of materials across their membranes.*

ing mitochondria, chloroplasts, and the nucleus, regulate the passage of materials between intracellular compartments and so regulate their internal environment.

The regulation of substances moving across membranes depends on interactions between the physical and chemical properties of the membrane and those of the molecules that move through them. Of the many kinds of molecules moving in and out of cells, by far the most important is water. Let us therefore look again at water, focusing our attention this time on how water moves.

WATER MOVEMENT

Three principles govern the movement of water: bulk flow, diffusion, and osmosis.

Bulk Flow

Bulk flow is the overall movement of water (or some other liquid). It occurs in response to differences in the potential energy of water, usually referred to as *water potential.*

A simple example of water that has potential energy is water at the top of a hill. As this water runs downhill, its potential energy can be converted to mechanical energy by a watermill or to electrical energy by a hydroelectric turbine.

Pressure is another source of water potential. If we put water into a rubber bulb and squeeze the bulb, this water, like the water at the top of a hill, also has water potential, and it will move to an area of less water potential. Can we make the water that is running downhill run uphill by means of pressure? Obviously we can. But only so long as the water potential produced by the pressure exceeds the water potential produced by gravity. Water moves from an area where water potential is greater to an area where water potential is less, regardless of the reason for the water potential. Your heart pumps blood to your brain against gravity by creating a greater water (blood) potential.

The concept of water potential is a useful one because it enables physiologists to predict the way in which water will move under various combinations of circumstances. Measurements of water potential are usually made in terms of the pressure required to stop the movement of water—that is, the hydrostatic (water-stopping) pressure—under the particular circumstances. This pressure is usually expressed in units of atmosphere. One atmosphere is the average pressure of the air at sea level.

Diffusion

Diffusion is a familiar phenomenon. If you sprinkle a few drops of perfume in one corner of a room, the scent will eventually permeate the entire room even if the air is still. If you put a few drops of dye in one end of a glass tank full of water, the dye molecules will slowly distribute themselves evenly throughout the tank. (The process may take a day or more, depending on the size of the tank.) Why do the dye molecules move apart?

Most of us, being human, tend to explain these movements in human terms, that is, in terms of members of a crowded population who, in order to find "elbowroom," move out of the city into the suburbs. This is not true of diffusion

7–3 *Water at the top of a falls, like a boulder on a hilltop, has potential energy. The movement of water from one energy level to another, as from the top of the falls to the bottom, is referred to as bulk flow.*

processes. If you could observe the individual dye molecules in the tank (see Figure 7–4), you would see that each one of them moves individually and at random; looking at any single molecule—at either its rate of motion or its direction of motion—gives you no clue at all as to where the molecule is located with respect to the others. So how do the molecules get from one side of the tank to the other? In your imagination, take a thin cross section of the tank, running from top to bottom. Dye molecules will move in and out of the section, some moving in one direction, some moving in the other. But you will see more dye molecules moving from the side of greater concentration. Why? Simply because there are more dye molecules at that end of the tank. Since there are more dye molecules on the left, more dye molecules, moving at random, will move to the right, even though there is an equal probability that any one molecule of dye will move from right to left. Consequently, the overall (net) movement of dye molecules will be from left to right. Similarly, if you could see the movement of the individual water molecules in the tank, you would see their overall movement is from right to left.

What happens when all the molecules are distributed evenly throughout the tank? The even distribution does not affect the behavior of the molecules as individuals; they still move at random. And, since the movements are random, just as many molecules go to the left as to the right. But because there are now as many molecules of dye and as many molecules of water on one side of the tank as on the other, there is no overall direction of motion. There is, however, just as much overall motion as before, provided the temperature has not changed.

7-4 *Diagram of the diffusion process. Diffusion is the random movement of molecules from a more concentrated to a less concentrated area. Notice that as one type of molecule (indicated by color) diffuses to the right, the other diffuses in the opposite direction. The result will be an even distribution of both types of molecules. Can you see why the net movement of molecules will slow down as equilibrium is reached?*

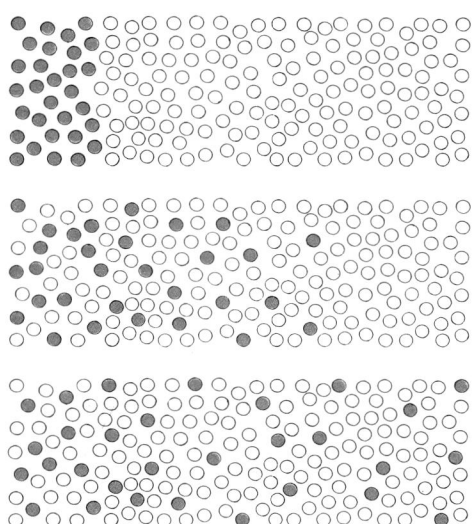

Substances that are moving from a region of greater concentration of their own molecules to a region of lesser concentration are said to be moving *down a gradient*. Diffusion only occurs along a gradient. A substance moving in the opposite direction, toward a greater concentration of its own molecules, moves *against a gradient*, which is analogous to pushing something uphill. The steeper the downhill gradient—that is, the larger the differences in concentration—the more rapid the diffusion process. Also, diffusion is more rapid in gases than in liquids and at higher rather than at lower temperatures. Can you explain why?

Notice that, in our imaginary tank, there are two gradients; the dye molecules are moving along one of them, and the water molecules are moving along the other in the opposite direction. When the molecules have reached a state of equal distribution, that is, when there are no more gradients, they continue to move, but there is no net movement in either direction. This situation is similar to that which occurs in a reversible chemical reaction when equilibrium has been reached. (See page 44, Chapter 2.)

A system in which diffusion is occurring is a system that possesses water potential. Since water, like other substances, will move from a region of greater concentration to a region of lesser concentration, the area of the tank in which there is pure water has a greater water potential than the area containing water plus dye or some other dissolved substance (solute).

The essential characteristics of diffusion are (1) that each molecule moves independently of the others and (2) that these movements are random. The net result of diffusion is that the diffusing substance becomes evenly distributed.

Cells and Diffusion

Most organic molecules cannot freely diffuse through the lipid barrier of the cell membrane. Carbon dioxide and oxygen, however, which are soluble in lipids, move freely through the membrane. Water also moves in and out freely. Water is not soluble in lipids (which is, of course, just another way of saying that lipids are not soluble in water), and so the fact that water moves freely has led biologists to postulate the presence of small pores in the membrane, which permit the passage of water molecules (and also of some small ions). Within a cell, materials may be produced at one place and used at another place. Thus a concentration gradient is established between the two areas, and the material diffuses down the gradient from the site of production to the area of use.

Diffusion is essentially a slow process, except over very short distances. It is efficient only if the concentration gradient is steep, the surface area is relatively large, and the volume relatively small. For instance, the rapid spread of a scent, such as perfume, through the air is due not primarily to diffusion but rather to the circulation of air currents. Similarly, in many cells, the transport of materials is speeded by active streaming of the cytoplasm (see page 100). Also cells hasten diffusion by their own metabolic activities. For example, oxygen is used up within the cell almost as rapidly as it enters, thereby keeping the gradient steep. Carbon dioxide is produced by the cell, and so a gradient from inside to outside is maintained.

Osmosis

As we noted, water moves freely through membranes. However, the membranes, while permitting the passage of water, block the passage of most materials dissolved in it. Such a membrane is known as a semipermeable or, preferably, a selectively permeable membrane, and the movement of water through such a membrane is known as *osmosis*. In the absence of other factors that influence water potential (such as pressure), the movement of the water in osmosis will typically be from a region of lesser solute concentration (and therefore of greater water concentration) into a region of greater solute concentration (lower water concentration). The presence of solute decreases the water potential and so causes the movement of water from a region of greater to a region of lesser water potential.

Osmotic pressure can be measured by a device known as an osmometer. The principle is illustrated in Figure 7–5. The beaker contains distilled water, and within the tube is water plus a solute such as salt or sugar. Across the mouth of the tube is the selectively permeable membrane; such a membrane is freely permeable to water, but not permeable to the solute.

Water moving from the beaker through the membrane into the solution causes the solution level in the tube to rise until equilibrium is reached—that is, until the water potential on both sides of the membrane is equal. The amount of hydrostatic pressure—in this case, the force of gravity—exerted on the rising column of liquid in the tube eventually equals the osmotic pressure exerted by the decreasing concentration of the solution in the tube.

7–5   *Osmosis and osmotic pressure. (a) The tube contains a solution and the beaker contains distilled water. (b) The selectively permeable membrane permits the passage of water but not of solute. The movement of water into the solution causes the solution to rise in the tube until the osmotic pressure resulting from the tendency of water to move into a region of lower water concentration is counterbalanced by the height, h, of the column of solution. (c) The force that must be applied to the piston to oppose the rise in the tube of the solution is a measurement of the osmotic pressure. It is proportional to* h.

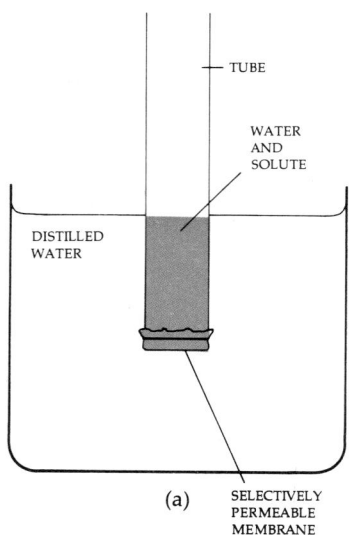

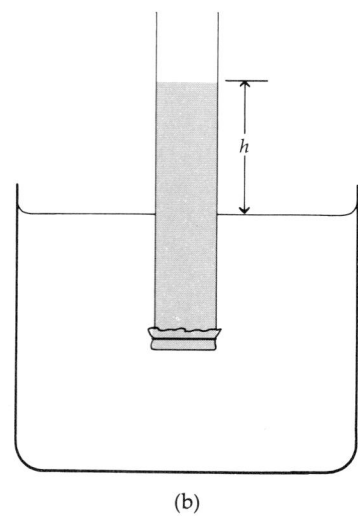

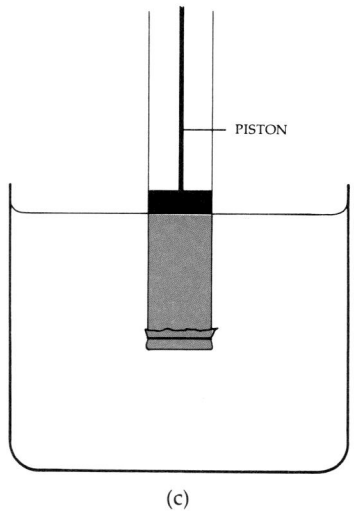

The movement of water is not affected by what is dissolved in the water, only by how much solute it contains—how many molecules or ions. The word *isotonic* was coined to describe solutions that have equal numbers of particles dissolved in them and therefore equal osmotic pressures. There is no net movement of water across a membrane separating two solutions that are isotonic to one another, unless, of course, hydrostatic pressure is exerted on one side. In comparing solutions of different concentration, the solution that has less solute and therefore a lower osmotic concentration is known as *hypotonic,* and the one that has more solute and more osmotic concentration is known as *hypertonic.* (Note that *iso* means "the same"; *hyper* means "more"—in this case, more molecules of solute; and *hypo* means "less"—in this case, less molecules of solute.)

Since solutes decrease water potential, a hypotonic solution has a higher water potential than a hypertonic one. In osmosis, water molecules move through a selectively permeable membrane into a hypertonic solution until the water potential is equal on both sides of the membrane.

Osmosis and Living Organisms

The movement of water across the cell membrane from a hypotonic to a hypertonic solution causes some crucial problems for living systems. These problems vary according to whether the cell or organism is hypotonic, isotonic, or hypertonic in relation to its environment. One-celled organisms that live in salt water, for example, are usually isotonic with the medium they inhabit,

–6 *A paramecium is hypertonic in relation to its environment, and hence water tends to move into the cell by osmosis. Excess water is expelled through a contractile vacuole. (a) Collecting tubules converge toward the vacuole, filling it. (b) Then it contracts, emptying outside the cell membrane by way of a small central pore.*

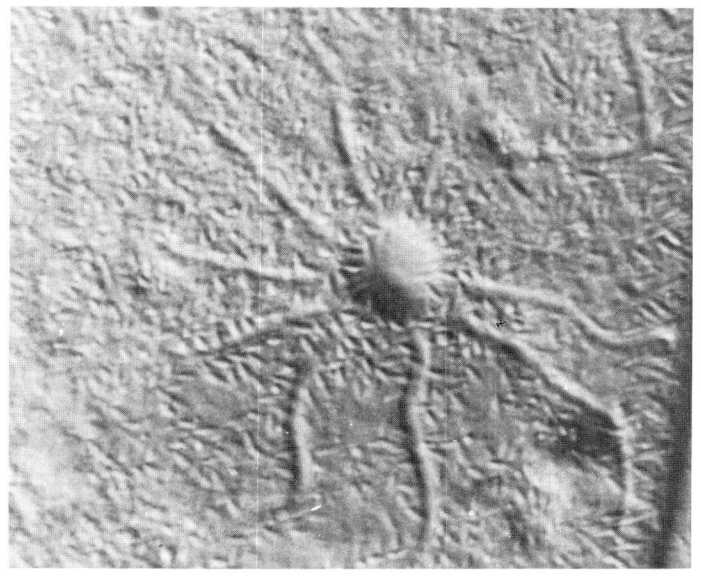

(a)

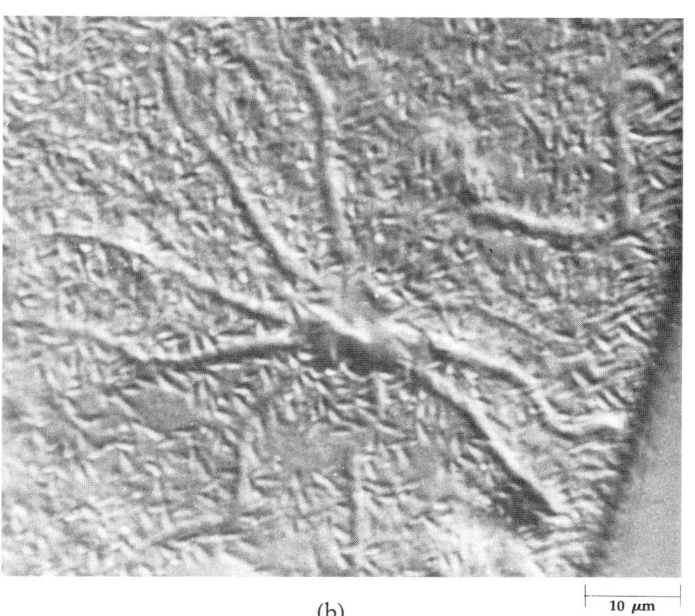

(b)

$\vdash\!\!\!\dashv$ 10 μm

which is one way of solving the problem. Similarly, the cells of higher animals are isotonic with the blood and lymph that constitute the watery medium in which they live.

Many types of cells live in a hypotonic environment. In all single-celled organisms that live in fresh water, such as paramecia, the interior of the cell is hypertonic to the surrounding water; consequently water tends to move into the cell by osmosis. If too much water were to move into the cell, it could dilute the cell contents to the point of interfering with function and could even eventually rupture the cell membrane. This is prevented by a specialized organelle known as a contractile vacuole, which collects water from various parts of the cell body and pumps it out with rhythmic contractions.

Turgor

Plant cells expand as a result of the water pressure created by osmosis. Because plant cells are usually hypertonic to their surrounding solution, water tends to move into them. This movement of water into the cell creates pressure within the cell against the cell wall. The pressure causes the cell wall to expand and the cell to enlarge. Some 90 percent of the growth of a plant cell is a direct result of water uptake.

If the water potential on both sides of the plant cell membrane becomes equal, the net movement of water ceases. (We say "net movement" here because water molecules continue to diffuse back and forth across the membrane. However, these movements are in equilibrium; that is, as many water molecules are going in as are coming out.)

As the plant cell matures, the cell wall stops growing. Moreover, mature plant cells, as we saw in Chapter 6, typically have large central vacuoles. These vacuoles often contain solutions of salts and other materials. (In citrus fruits, for example, they contain the acids which give the fruits their characteristic sour taste.) Because of these concentrated solutions, water continues to "try"

7–7 (a) *A turgid plant cell. The central vacuole is hypertonic in relation to the fluid surrounding it and so gains water. The expansion of the cell is held in check by the cell wall. (b) A plant cell begins to wilt if it is placed in an isotonic solution, so that water pressure no longer builds up within the vacuole. (c) A plant cell in a hypertonic solution loses water to the surrounding fluid and so collapses, with its membrane pulling away from the cell wall. Such a cell is said to be plasmolyzed.*

NET MOVEMENT OF WATER
SOLUTES

CELL WALL   CYTOPLASM
VACUOLE

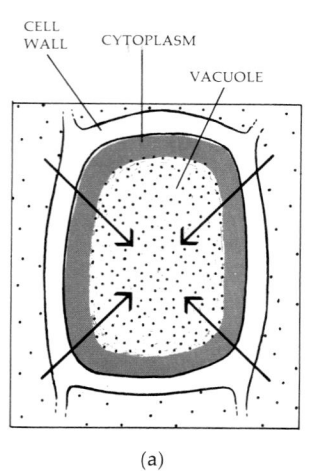
(a)

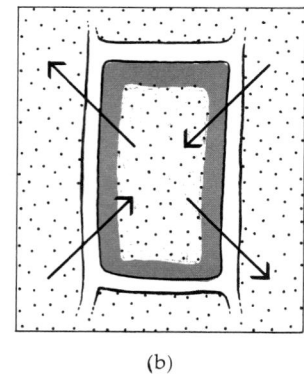
(b)

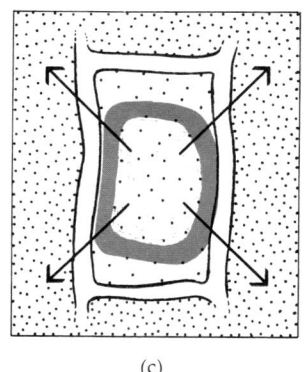
(c)

to move into the cells. In the mature cell, however, the cell wall does not expand further, and equilibrium of salt concentration is not reached. As a consequence, the cell wall remains under constant pressure. This water pressure exerted on the cell wall from inside is known as _turgor_. Turgor keeps the cell wall stiff and the plant body crisp. When turgor pressure is reduced, as a consequence of water loss, a plant wilts.

Osmotic pressure also plays an extremely important part in determining the composition of the urine excreted by the kidney, as we shall see in Section 5.

MEMBRANE TRANSPORT

The double layer of lipid molecules of the cell membrane forms a barrier to sugars, amino acids, nucleotides, and many other essential cell nutrients. How do they get in? There are apparently several different kinds of mechanisms. One of these, known as facilitated diffusion, involves a sort of shuttle system by which a protein molecule that can move through the cell membrane binds with some other specific molecule that cannot, such as glucose, presumably in the same way an enzyme binds with its substrate. The combined molecules then diffuse through the membrane to the inner surface where the escorted molecule is released. Facilitated diffusion, like simple diffusion, moves only along a concentration gradient and does not require an expenditure of energy on the part of the cell.

7–8 *Model of a cell membrane. The membrane is composed of phospholipid and cholesterol molecules with their hydrophobic "tails" forming the inner layer, and of protein molecules (indicated in color). According to this model, the proteins vary from membrane to membrane, according to cell function, and also from place to place on the same membrane. Some are enzymes and some are probably involved in active transport. The membrane also contains pores which permit the passage of water and other small molecules.*

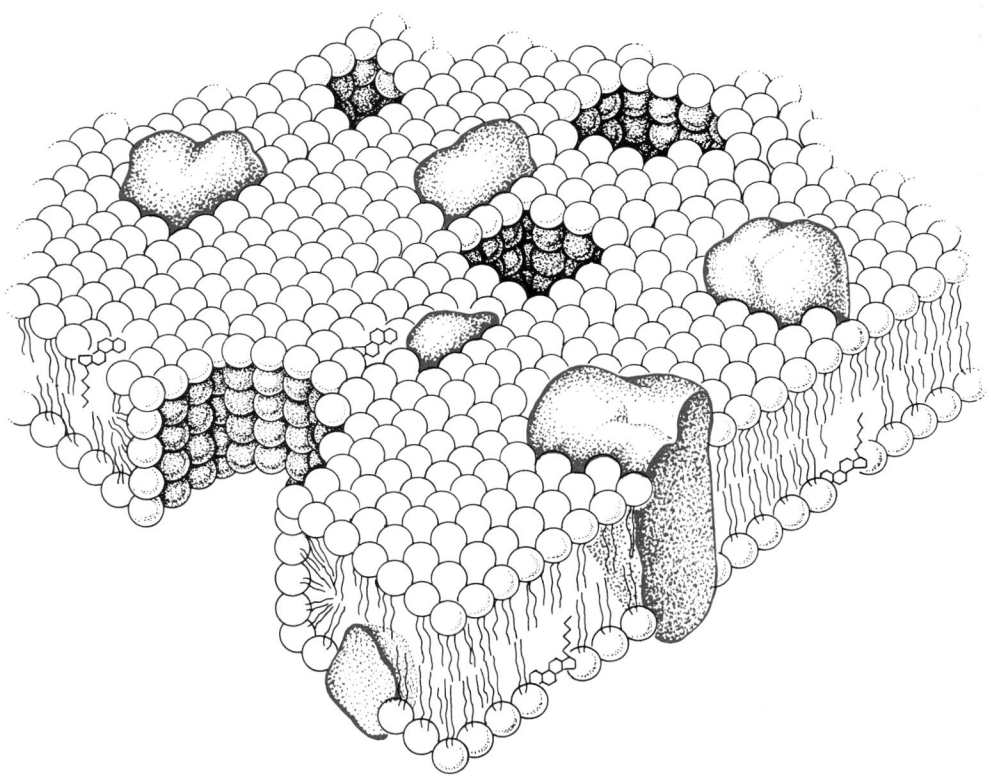

7–9 *The "revolving door" model of active transport. The molecule to be transported locks into the carrier molecule in much the same way a substrate locks into its enzyme. The carrier molecule rotates, releasing the passenger molecule on the other side of the membrane, and then returns to its former position. Sodium ions are pumped out of cells in this way.*

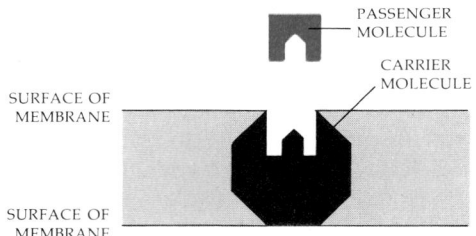

PASSENGER MOLECULE

CARRIER MOLECULE

SURFACE OF MEMBRANE

SURFACE OF MEMBRANE

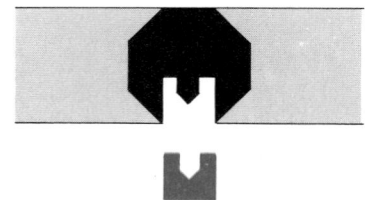

7–10 *Mitochondria clustered near the surface of kidney cells. These cells are concerned with pumping out sodium against the concentration gradient. The mitochondria, presumably, provide the energy for this active transport process.*

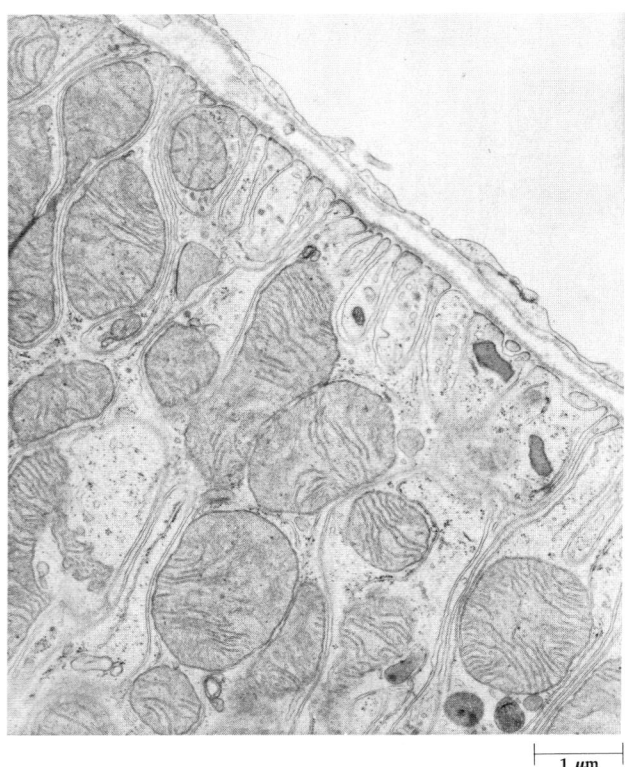

1 μm

Cells can also move substances against a concentration gradient. For example, the cells lining your intestine can continue to take in food molecules even though their concentration within the cell is a hundred times greater than it is outside the cell membrane. This process, known as *active transport*, also appears to involve carrier molecules. Unlike facilitated diffusion, however, it requires a great deal of ATP energy; in cells involved in active transport, mitochondria are often located near the membrane across which transport is occurring.

Substances can also be moved out of cells. The sodium ion is an example. Cells living in blood or seawater characteristically contain far less Na^+ than their surrounding medium, although sodium, being a small ion, diffuses in constantly along the gradient. The sodium is exported by what is known as a sodium pump, which may be a sort of revolving door, run by ATP energy. In some cells, this same pump that moves Na^+ out apparently pumps potassium ions (K^+) into the cell. Some of the consequences of the sodium pump will be examined in Section 5, when we discuss the movement of electrical impulses along the membrane of a nerve cell, a phenomenon which is dependent on this pumping mechanism.

ENDOCYTOSIS AND EXOCYTOSIS

In *endocytosis*, material to be taken into the cell attaches to special places on the cell membrane and induces the membrane to slide or flow inward, producing a little pouch or vacuole enclosing the substance. This vacuole is released into the cytoplasm.

When the substance to be taken in is a solid, such as a bacterial cell, the process is usually called *phagocytosis*, from the Greek word *phage*, "to eat." (See Figure 7–11.) Many one-celled organisms, such as paramecia, feed in this way, and white blood cells in our own bloodstreams engulf bacteria and other invaders in phagocytic vacuoles. Often lysosomes fuse with these vacuoles, emptying their enzymes into them and so digesting or destroying their contents.

The taking in of dissolved molecules, as distinct from particulate matter, is sometimes given the special name of *pinocytosis*, although it is the same in principle as phagocytosis.

Pinocytosis occurs not only in single-celled organisms but also in multicellular animals. One type of cell in which it has been frequently observed is the human egg cell. As the egg cell matures in the ovary of the female, it is surrounded by "nurse cells," which apparently transmit nutrients to the egg cell, which takes them in by pinocytosis.

Phagocytosis and pinocytosis can also work in reverse. Substances are characteristically exported from cells in vacuoles. The vacuoles move to the surface of the cell. When the vacuole reaches the cell surface, its membrane fuses with the membrane of the cell, thus expelling its contents to the outside. This process is sometimes referred to as *exocytosis*.

–11 *Solid materials or dissolved molecules may induce certain types of membranes to form vacuoles around them and so transport them into the cytoplasm. Phagocytosis involves the incorporation of solid materials, and pinocytosis, dissolved molecules. Exocytosis is endocytosis in reverse; the membrane enclosing the exported material fuses with the cell membrane as the vacuole expels its contents.*

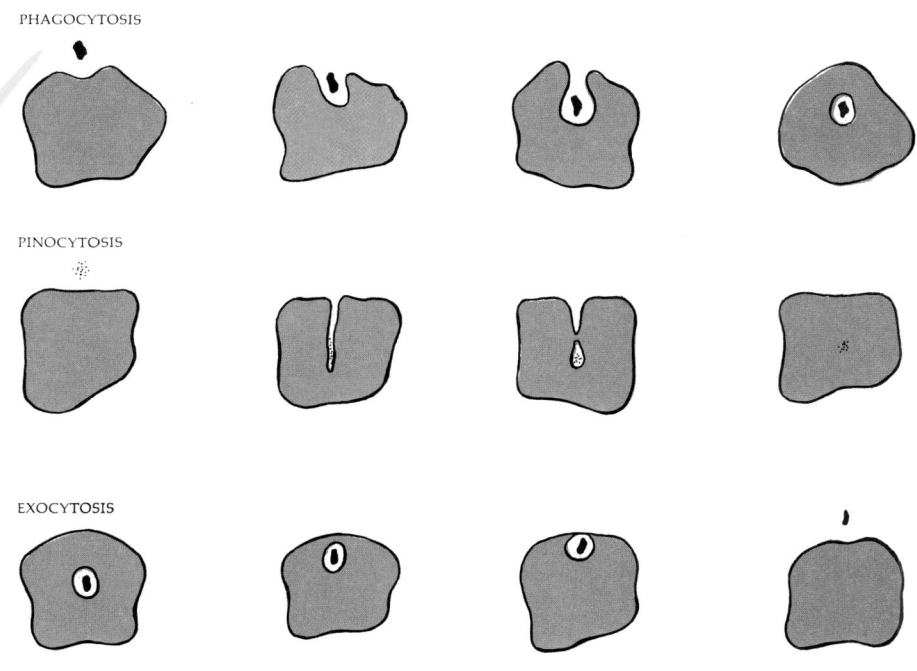

PHAGOCYTOSIS

PINOCYTOSIS

EXOCYTOSIS

7–12 *Phagocytosis of a paramecium by a didinium. Both are one-celled organisms (protists). (a) The barrel-shaped didinium has made contact with its prey, a paramecium which has discharged a defensive barrage of tiny barbs (visible at the top of the micrograph). From its anterior proboscis, the didinium is about to eject a bundle of slender, poisonous strands (not visible), which will paralyze the paramecium in a matter of seconds. (b) Ingestion of the inactivated paramecium has begun. The concave area just above the oral rim of the didinium is the oral groove of the paramecium. Paramecia, also heterotrophs, feed largely on bacteria. (c) Because the paramecium is larger than the didinium, folding helps. (d) The paramecium is half swallowed and the process of compression has begun, as you can see in the posterior tip protruding from the oral groove.* **This compression is largely a matter of squeezing out water.** *(e) The paramecium has almost disappeared. The part that is within the didinium is surrounded by a membrane composed of the cell membrane of the didinium. Once the paramecium is completely inside, the didinium's cell membrane will fuse over it forming a food vacuole. A didinium can eat a dozen paramecia, each larger than itself, in a single day. Moreover, the paramecium must also provide the means for its own demolition: the didinium apparently lacks the digestive enzyme dipeptidase which the paramecium supplies.*

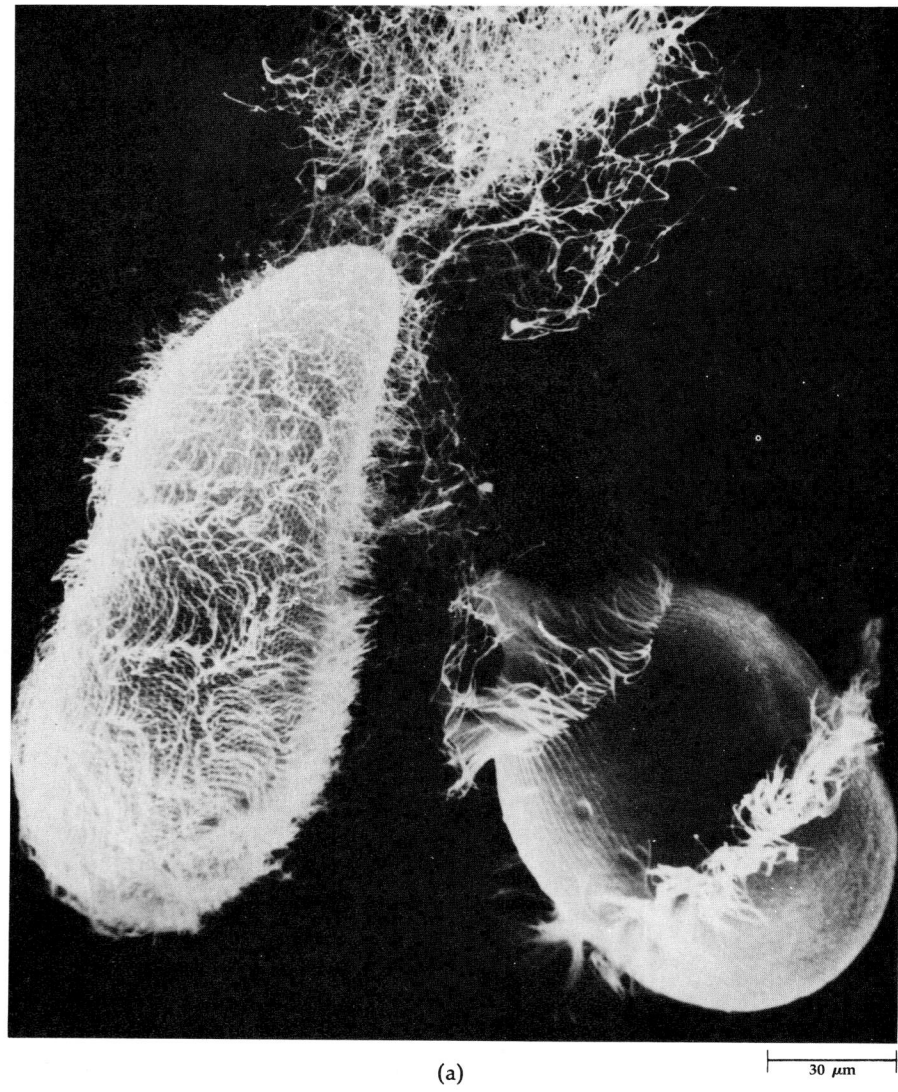

(a)

30 μm

Phagocytosis and pinocytosis appear superficially different from membrane transport systems involving carrier molecules. They are fundamentally similar, however, in that all depend on the capacity of the membrane to "recognize" particular molecules. This capacity is, of course, the result of billions of years of an evolutionary process that began, as far as we are able to discern, with the formation of a fragile film around a few organic molecules, thus separating the molecules from their external environments and permitting them to maintain the particular kind of organization that we recognize as life.

SUMMARY

The cell membrane regulates the passage of materials into and out of the cell, a function which makes it possible for the cell to maintain its structural and functional integrity. This regulation depends on interaction between the structure of the membrane and the structure of the materials that pass through it.

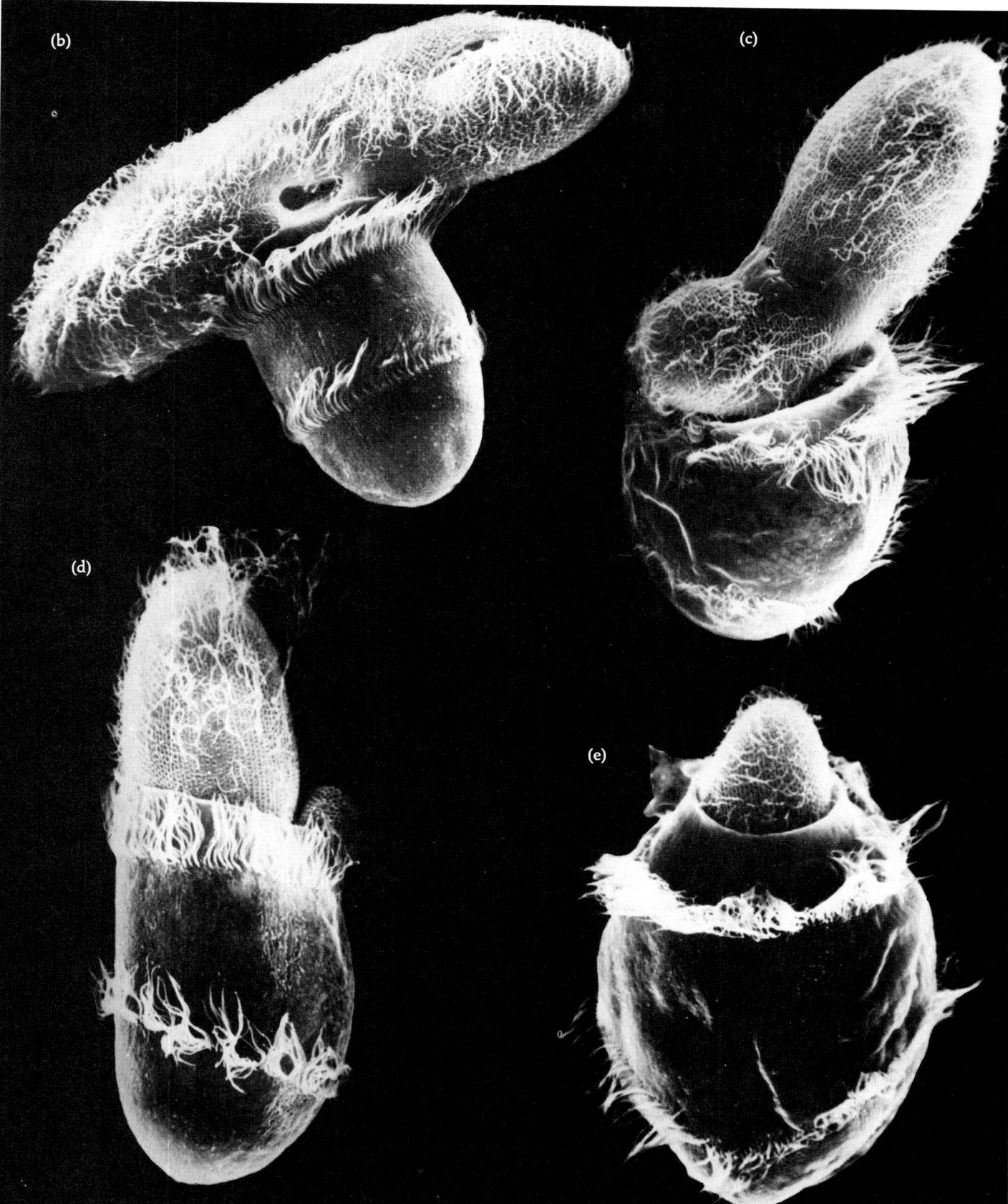

One of the principal materials passing into and out of cells is water. Water moves by bulk flow, diffusion, and osmosis. Water potential determines the direction the water moves; that is, water movement takes place from where the water potential is greater to where it is less. Bulk flow is the overall movement of water, as when water flows downhill.

Diffusion involves the random movement of molecules and results in net movement down a concentration gradient. Carbon dioxide and oxygen are two important molecules that move into and out of cells by diffusion across the membrane. Diffusion is most efficient when the surface area is large in relation to volume and when the distance involved is small. The rate of movement of substances within cells is increased by cytoplasmic streaming.

Osmosis is the movement of water through a membrane that permits the passage of water but inhibits the movement of solutes; such a membrane is called a selectively permeable membrane. In the absence of other forces, the movement of water in osmosis is from a region of lesser solute concentration (a hypotonic medium), and therefore of higher water potential, to one of greater solute concentration (a hypertonic medium), and so of lower water potential. Turgor in plant cells is a consequence of osmosis.

Active transport is the movement of molecules across the cell membrane against a concentration gradient. It requires energy and probably involves the presence of specific carrier molecules on the membrane. Controlled movement in and out may also occur by endocytosis or exocytosis, in which substances are transported in vacuoles composed of portions of cell membrane.

QUESTIONS

1. Define water potential, turgor, osmosis, and diffusion. At this point, we are going to stop reminding you that you can check your definitions in the Glossary, but it still might not be a bad idea.
2. Imagine a pouch with a selectively permeable membrane containing a saltwater solution. It is immersed in a dish of fresh water. Which way will the water move? If you add salt to the water in the dish, how will this affect the water movement? What living systems exist under analogous conditions? How do you think they maintain water balance?
3. Three funnels have been placed in a beaker containing a solution (see figure). What is the concentration of that solution? Explain your answer.

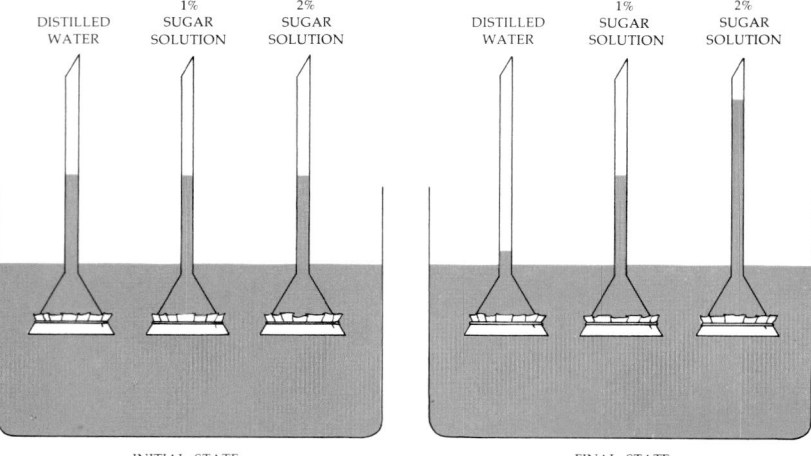

DISTILLED WATER   1% SUGAR SOLUTION   2% SUGAR SOLUTION   DISTILLED WATER   1% SUGAR SOLUTION   2% SUGAR SOLUTION

INITIAL STATE   FINAL STATE

How Cells Make ATP

Chapter 8

ATP is the principal energy carrier in living systems. As we have noted in previous chapters, it participates in a great variety of cellular events, from the uniting of glucose and fructose to make sucrose, to the flick of a cilium, or the active transport of a molecule across a cell membrane. In the following pages, we shall show in some detail how a cell breaks down carbohydrates and uses the released energy to recharge ADP to ATP, thereby storing energy in the terminal "high-energy" phosphate bond of that molecule. We describe the process in detail, not because we believe it should be memorized but because it provides an excellent illustration both of chemical principles described in previous chapters and also of the way in which cells perform biochemical work.

THE LAWS OF THERMODYNAMICS

Before we begin a detailed discussion, however, let us place it in a larger context. Certain laws apply to all energy exchanges in biological as well as in physical systems. These laws are usually referred to as the laws of thermo-dynamics because laboratory measurements of energy changes are usually made in terms of heat (*thermo*). The *first law of thermodynamics* states: *Energy can be changed from one form to another, but it cannot be created or destroyed.* The total energy of any closed system remains constant, regardless of the physical or chemical changes it may undergo. Light is a form of energy, as is electricity. Light can be changed to electrical energy, and electrical energy can be changed to light (for example, by letting it flow through the tungsten wire in a light bulb).

Motion is another kind of energy. A gasoline engine, for example, produces energy in the form of motion. In order to do this, it must use up energy in another form—the chemical energy stored in the gasoline. By burning gasoline as fuel, the engine changes the chemical energy to heat and then changes the heat to mechanical movements.

Energy can be stored in various other forms. A boulder on a hill is an example of stored energy. The energy required to push the boulder up the hill is "stored" in the boulder as potential energy. When the boulder rolls down-hill, the energy is released as kinetic (motion) energy.

8–1   *The flow of energy in an ecological system. Radiant energy from the sun is transformed to chemical energy by the grasses of this African savanna. The zebras, eating the grasses, use this chemical energy for growth and convert some of it to kinetic energy, which may* *help to keep them from participating in another energy transfer involving the waiting lions. Here we have a biological example of the second law of thermodynamics. Of the radiant energy falling on the grass, less than 10 percent is converted to chemical energy. Less than* *10 percent of the chemical energy stored in the grass is converted to chemical energy stored in the zebra, and less than 10 percent of the zebra's stored energy is transferred to its predator, the lion.*

Some useful energy is converted to heat in the transaction. When a boulder rolls downhill, the potential energy it received on its push up is converted to kinetic energy and also to heat energy caused by friction. In a gasoline engine, about 75 percent of the energy originally present in the fuel is dissipated to the surroundings in the form of heat. The rest of the energy is converted to the organized energy of motion as, similarly, an electric motor converts only 25 to 50 percent of the electrical energy into mechanical energy, the rest going into heat. Heat, you will recall, is simply the random motion of atoms or molecules. The higher the heat, the greater the motion. Heat can be used for work, as in a steam engine, but only if a heat gradient exists. In engines that run on heat gradients, the exhaust area must be cooler than the inside of the engine. A steam engine in a room full of steam cannot run at all.

When we say that energy is "lost" as heat, what we actually mean is that it is no longer available to do work, because the heat is evenly distributed. Energy stored in a stick of dynamite can do work; that part of the energy that is released, or randomized, as heat at the time of the explosion is no longer available for work. The energy in the boulder at the top of the hill is also available to do work. If it is difficult to visualize a rolling boulder as having the capacity to do work, it is probably because boulders are not conventionally harnessed for this purpose. Substitute water, for which the first law of thermodynamics applies equally well. Suppose that, instead of the boulder, we transport water to the top of our imaginary hill and let it rush down again. We know from experience that the water could be used to turn a series of paddle wheels and that machines powered by such wheels could be used to grind corn or do other work useful in human terms.

Energy changes also take place, as we noted previously, during chemical reactions. In the course of these changes some or all of the chemical energy may be converted to heat energy, as when we burn wood or gas. Living systems are organized, however, so that heat changes of this sort are minimized. In their studies of living systems, biochemists are primarily interested in a form of energy called free energy. *Free energy* is the energy available to do work under conditions of constant temperature and pressure. In particular, biochemists are concerned with free energy that is released during metabolic reactions. In an exergonic (energy-yielding) reaction, the energy of the reactants is greater than the energy of the products. As the first law of thermodynamics reminds us, this energy is not lost. A part of it is dissipated to the surroundings. The remainder is in the form of free energy, which is available to do work in the living systems.

Entropy

This brings us to the *second law of thermodynamics*. The law states that *all natural processes tend to proceed in such a direction that the disorder or randomness of the system increases*. This disorder is given the special name of *entropy*. The concept that systems tend to increase in disorder is in keeping with our observations of the world around us. For example, in our model of diffusion (page 111), we would never expect the process to reverse itself— that is, once the dye molecules are distributed randomly throughout the tank,

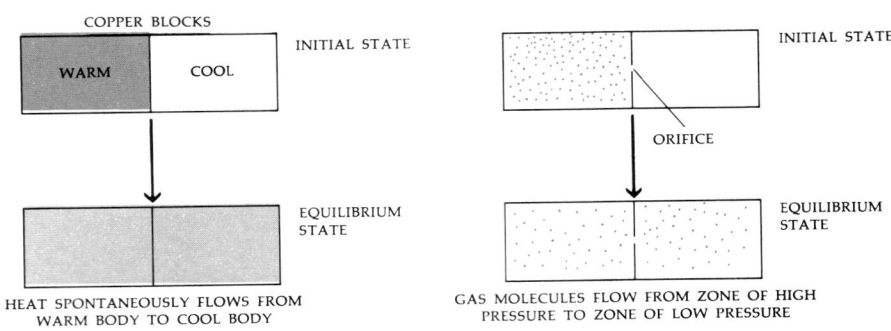

COPPER BLOCKS | INITIAL STATE

WARM | COOL

EQUILIBRIUM STATE

HEAT SPONTANEOUSLY FLOWS FROM WARM BODY TO COOL BODY

INITIAL STATE

ORIFICE

EQUILIBRIUM STATE

GAS MOLECULES FLOW FROM ZONE OF HIGH PRESSURE TO ZONE OF LOW PRESSURE

we would not expect them, by any process of chance, to reaggregate themselves at one end of the tank. Nor would we expect water to run uphill. When we say that water moves from a position of higher water potential to a position of lower water potential, we are stating an example of the second law of thermodynamics.

A highly organized collection of boards and bricks—a building—eventually weathers and falls apart. A random collection of boards and bricks, no matter how much time is available, will not, in our experience, assemble itself into a building. In terms of chemical energy, more highly organized molecules such as glucose tend to break apart into smaller, less organized groupings of carbon dioxide and water. Free energy—that is, useful energy—declines.

Free energy and entropy have an inverse relationship. When chemical reactions take place in living systems, as we noted previously, the energy of the products is less than the energy of the reactants. As the first law of thermodynamics indicates, however, none of this energy is lost. If there is no change in heat, the total energy change that takes place under biological conditions is the sum of the change in free energy and the change in entropy. Thus free energy and entropy have an inverse relationship. The greater the change in entropy, the less the change in free energy; as entropy increases, free energy declines.

Energy Flow in Living Systems

When the concept of entropy was first formulated, some biologists believed that living systems violated the second law of thermodynamics, because in living systems order increases and randomness, or entropy, decreases. This objection is countered, however, if one looks at the universe as a whole. Living systems take in useful energy from outside sources and release it into the environment in a less useful form. If one measures not just the living system alone but the living system and its environment, the total entropy increases.

The energy flow through living systems takes place in three stages. The first stage, which we will discuss in the following chapter, is photosynthesis, the capture of the energy of sunlight and its conversion to chemical form. The second stage is the conversion of this chemical energy produced by radiant energy to energy in forms that can participate in cellular transactions. This second stage is the subject of this chapter. The third stage is the utilization of

8–3 *The flow of biological energy. The radiant energy of sunlight is produced by the fusion of hydrogen atoms to form helium. Chloroplasts of green plant cells capture this energy and use it to convert water and carbon dioxide into glucose, starch, and other foodstuff molecules. Oxygen is released into the air as a product of the photosynthetic reactions. Mitochondria break down these compounds and capture their stored energy in ATP molecules. This process, cellular respiration, consumes oxygen and produces carbon dioxide and water, completing the cycle.*

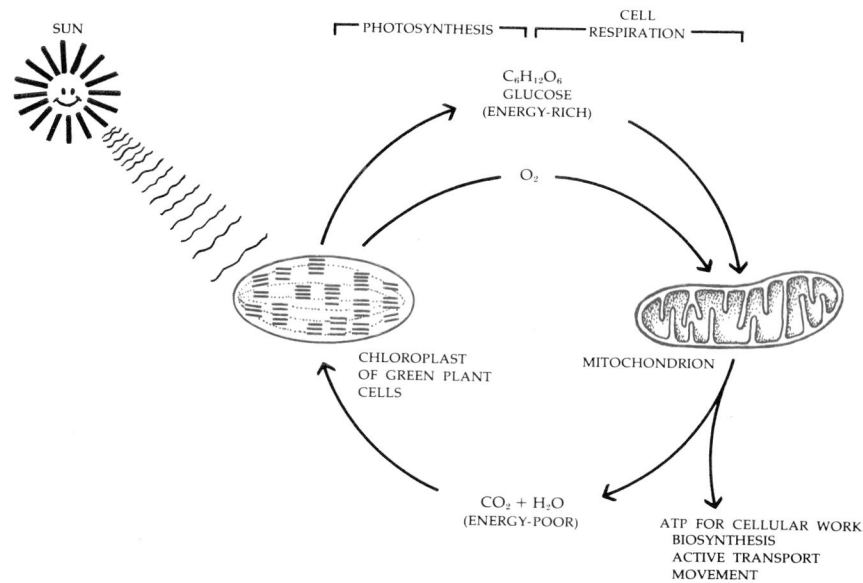

this energy to do the cellular work of metabolism, motion, active transport, and so forth. Examples of this final stage have been noted in previous chapters and will be seen in chapters to come. The consequences of the second law of thermodynamics, in terms of the operation of the biosphere as a whole, will be discussed in Chapter 41.

GLYCOLYSIS

The process by which cells break down food molecules is complicated in detail —it involves more than 70 sequential chemical steps—but it is simple in overall design. It is the oxidation of an energy-rich carbon-containing molecule, resulting in energy-poorer molecules plus free energy, some of which is used to convert ADP to ATP. As we noted in Chapter 4, any energy-rich molecule— sugar, fat, amino acid, nucleotide—can be broken down and used as an energy source. In this chapter we shall focus, however, on the oxidation of glucose, which is the most common fuel molecule for the animal cell. The overall equation for this reaction is

Glucose + oxygen $\longrightarrow$ carbon dioxide + water + energy
$C_6H_{12}O_6 + 6O_2 \longrightarrow 6CO_2 + 6H_2O + 683$ kilocalories

The oxidation of glucose takes place in two stages. The first is known as *glycolysis*; it is also sometimes called fermentation. The second is *respiration*. In glycolysis, organisms convert fuel molecules to ATP in the absence of oxygen. Since living organisms probably first arose in an atmosphere lacking oxygen, glycolysis is believed to have evolved early. This hypothesis is supported by the fact that almost all modern organisms—from bacteria to human beings— not only carry out glycolysis but do so using almost identical enzyme systems.

Some few organisms (anaerobes) that live without oxygen employ only glycolysis. In most cells, however, glycolysis is only a preparatory stage for respiration, and it is in the latter stage that most of the cell's useful chemical energy is obtained.

Electron Acceptor Molecules

Before we begin our discussion of glycolysis, however, we need to introduce one more type of molecule involved in biological energy exchanges—the *electron acceptor molecule*. As we mentioned previously (page 42), the addition of an electron to an atom or molecule is reduction; the removal of an electron is oxidation. In photosynthesis, carbon atoms (present in carbon dioxide) are reduced to sugar. When sugar is used for energy, the carbon atoms are oxidized; that is, electrons are removed from them. Reduction requires an input of energy; in photosynthesis, the energy comes from the sun. The breakdown of sugar releases energy; that is, it is a downhill reaction, during which electrons are passed from a higher energy level to a lower one. This passage involves the assistance of electron acceptor molecules of which NAD is an example. NAD is a coenzyme. It is composed of several subunits including the nucleotide adenine (also present in ATP), attached by phosphate groups to a ribose sugar. The portion of the molecule that accepts and releases the electrons is the vitamin nicotinamide (Figure 8–4). NAD in the reduced form (with the electrons) is designated as NAD_{red}; NAD in the oxidized form (with the electrons removed) is NAD_{ox}.

NADP and the flavoproteins (FP) are other electron acceptors closely related both structurally and functionally to NAD. They also play roles in the oxidation and reduction of carbon in biological systems.

Steps of Glycolysis

The first stage in the oxidation of glucose, glycolysis, takes place in the cytoplasm. In glycolysis, the six-carbon glucose molecule is split into 2 three-carbon molecules of a compound known as pyruvate (see Figure 8–5). This is carried out in a series of reactions that release about 143 kilocalories. The next three pages describe the nine steps by which one glucose molecule is broken down (oxidized) into two pyruvate molecules.

Do not try to memorize these steps, but follow them closely. Note especially the formation of ATP from ADP and of NAD_{red} from NAD_{ox}. These represent the cell's net gain from this energy transaction.

8–4 The nicotinamide component of nicotinamide adenine dinucleotide (NAD) in its oxidized form (a) and reduced form (b). In the reduced form (NAD_{red}), the nicotinamide ring has accepted two hydrogen atoms, at the points shown by the colored dots. One of the hydrogen atoms is then released and becomes an H^+ ion. The rest of the molecule does not change.

(a)

NAD_{ox}

NAD_{red}

(b)

8–5 In glycolysis, the six-carbon glucose molecule is split into 2 three-carbon molecules of a compound known as pyruvate.

GLUCOSE ⟶ 2 PYRUVATE

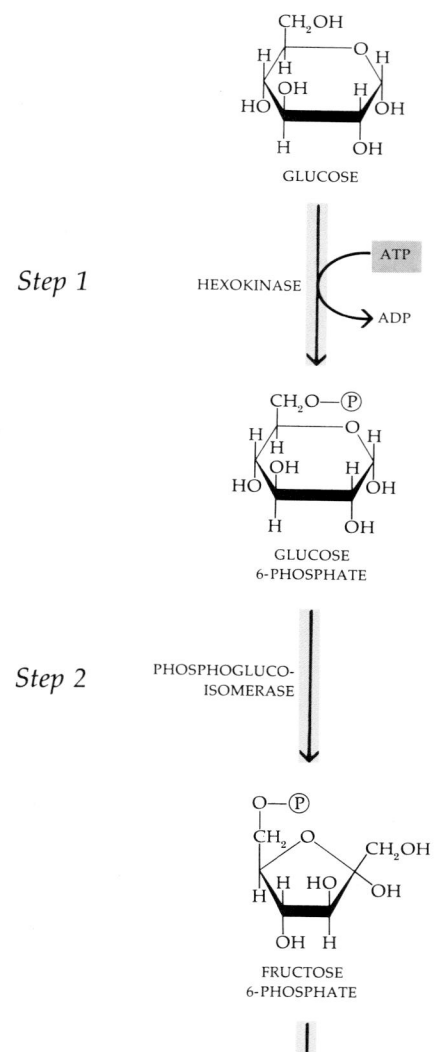

Step 1

HEXOKINASE

ATP

ADP

GLUCOSE

CH₂OH

GLUCOSE
6-PHOSPHATE

CH₂O—Ⓟ

Step 2

PHOSPHOGLUCO-
ISOMERASE

FRUCTOSE
6-PHOSPHATE

Step 3

PHOSPHO-
FRUCTOKINASE

ATP

ADP

FRUCTOSE
1,6-DIPHOSPHATE

Step 1. The first steps in glycolysis require an input of energy. This activation energy is supplied by the hydrolysis of ATP to ADP. The terminal phosphate group is transferred from an ATP molecule to the glucose molecule, to make glucose 6-phosphate. (This is also the first step, you may recall, in the biosynthesis of sucrose, page 67.) The combining of ATP with glucose to produce glucose 6-phosphate and ADP is an energy-yielding reaction, as we saw previously (page 67). Some of the energy released from the ATP is conserved in the chemical bond linking the phosphate to the sugar molecule. This reaction is catalyzed by a specific enzyme (hexokinase), and each of the reactions that follows is similarly regulated by a specific enzyme.

Step 2. The molecule is reorganized, again with the help of a particular enzyme. The six-sided ring characteristic of glucose becomes a five-sided fructose ring. As you know, glucose and fructose both have the same number of atoms—$C_6H_{12}O_6$—and differ only in the arrangements of these atoms. This reaction can proceed in either direction; it is pushed forward by the accumulation of glucose 6-phosphate and the disappearance of fructose 6-phosphate as the latter enters Step 3. (You may wish to review the discussion of chemical equilibria on pages 43–45.)

Step 3. This step, which is similar to Step 1, results in the attachment of a phosphate to the first carbon of the fructose molecule, producing fructose 1,6-diphosphate, that is, fructose with phosphates in the 1 and 6 positions. Note that in the course of the reactions thus far two molecules of ATP have been converted to ADP and that no energy has been recovered.

Step 4. The molecule is split into 2 three-carbon molecules. The two are interconvertible. However, because the glyceraldehyde phosphate is used up in subsequent reactions, all of the dihydroxyacetone phosphate is eventually converted to glyceraldehyde phosphate. Thus, all subsequent steps must be counted twice to account for the fate of one glucose molecule. With the completion of Step 4, the preparatory reactions that require an input of ATP energy are complete.

Step 4

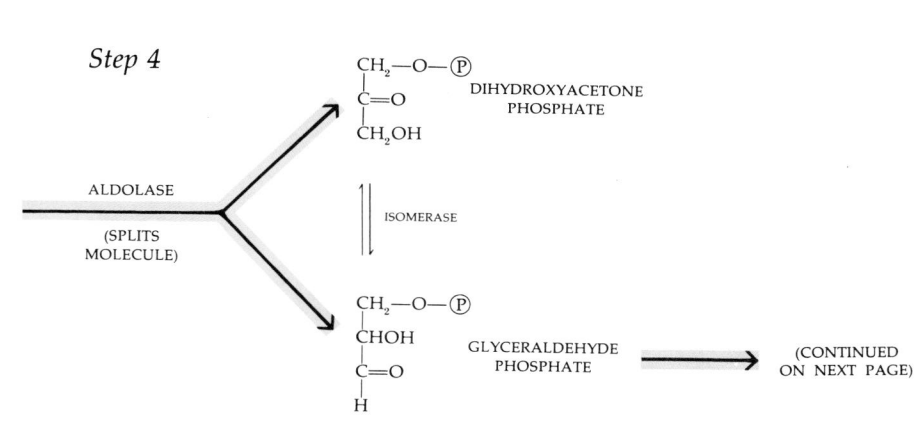

ALDOLASE

(SPLITS
MOLECULE)

ISOMERASE

CH_2—O—Ⓟ
C=O
CH_2OH

DIHYDROXYACETONE
PHOSPHATE

CH_2—O—Ⓟ
CHOH
C=O
H

GLYCERALDEHYDE
PHOSPHATE

(CONTINUED
ON NEXT PAGE)

Step 5. Glyceraldehyde phosphate molecules are oxidized—that is, hydrogen atoms with their electrons are removed—and NAD_{ox} becomes NAD_{red}. This is the first reaction from which the cell gains energy. Energy from this oxidation reaction is also used to attach phosphate groups to what is now the 1 position of each of the glyceraldehyde molecules. (The designation P_i indicates inorganic phosphate available as a phosphate ion in solution in the cytoplasm.) Note that a high-energy bond is formed.

Step 6. The high-energy phosphate is released from the diphosphoglycerate molecule and used to recharge a molecule of ADP (a total of two molecules of ATP per molecule of glucose). This is a highly exergonic reaction and so pulls all the previous reactions forward.

Step 7. The remaining phosphate group is transferred from the 3 position to the 2 position.

Step 8. In this step, a molecule of water is removed from the three-carbon compound, and as a consequence of this internal rearrangement of the molecule, a high-energy phosphate bond is formed.

Step 9. The high-energy phosphate is transferred to a molecule of ADP, forming another molecule of ATP (again, a total of two molecules of ATP per molecule of glucose). This is also a highly exergonic reaction, and so the sequence runs downhill with accelerating force as it ends.

In summary, the complete sequence begins with one molecule of glucose. Energy is put into the sequence at Steps 1 and 3 by the transfer of a phosphate group from an ATP molecule—one at each step—to the sugar molecule. The six-carbon molecule splits at Step 4, and from this point onward, the sequence is exergonic. At Step 5, a molecule of NAD_{ox} takes energy from the system and becomes NAD_{red}. At Steps 6 and 9, molecules of ADP take energy from the system, form additional phosphate bonds, and become ATP. Thus one glucose molecule, using energy from the phosphate bonds of two ATP molecules to initiate the glycolytic reaction, produces two NAD_{red} from two NAD_{ox} molecules and four ATP from four ADP molecules.

$$\text{Glucose} + 2\text{ATP} + 4\text{ADP} + 2\,\text{NAD}_{ox} \longrightarrow 2\text{ pyruvate} + 2\text{ADP} + 4\text{ATP} + 2\text{NAD}_{red}$$

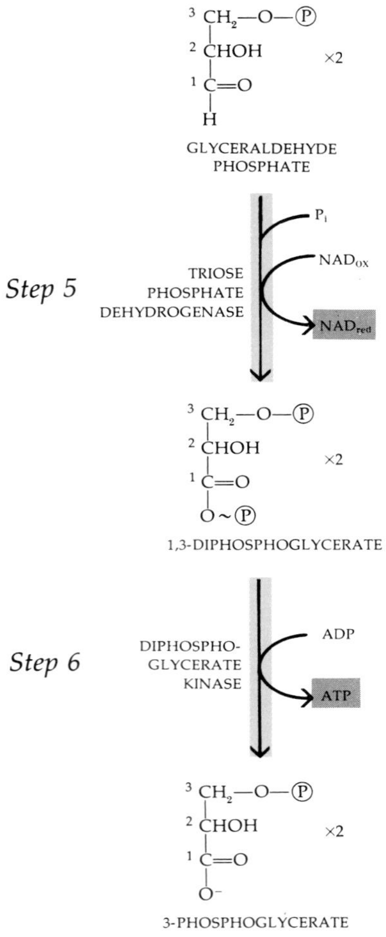
GLYCERALDEHYDE PHOSPHATE
Step 5
TRIOSE PHOSPHATE DEHYDROGENASE
NADox
NADred
1,3-DIPHOSPHOGLYCERATE
Step 6
DIPHOSPHO-GLYCERATE KINASE
ADP
ATP
3-PHOSPHOGLYCERATE

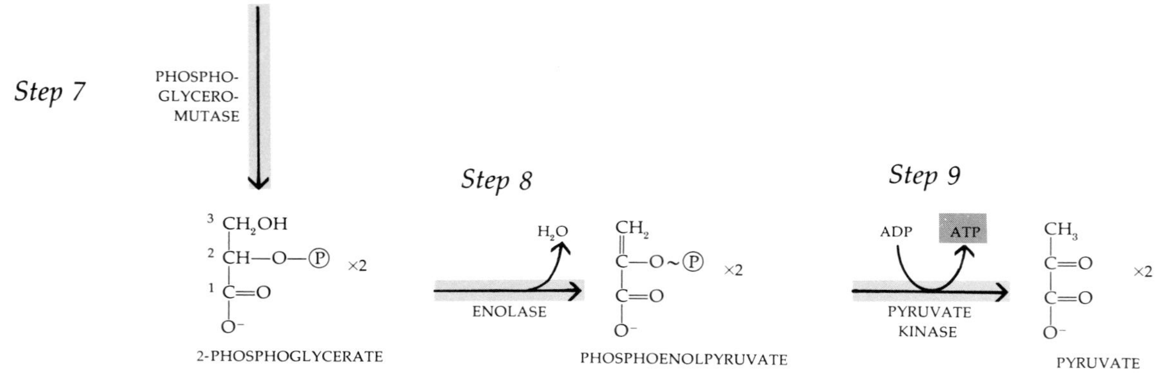
Step 7
PHOSPHO-GLYCERO-MUTASE
2-PHOSPHOGLYCERATE
Step 8
ENOLASE
PHOSPHOENOLPYRUVATE
Step 9
ADP
ATP
PYRUVATE KINASE
PYRUVATE

Phosphates and water (formed by the splitting of ATP) also participate in the reaction.

Two molecules of pyruvate* remain, and these two molecules still contain a large amount of the energy stored in the original glucose molecule. This series of reactions is carried out by virtually all living cells—from prokaryotes to the eukaryotic cells of man.

* Pyruvic acid dissociates, producing pyruvate and a hydrogen ion (proton). The two forms (pyruvic acid and pyruvate) exist in dynamic equilibrium and the two terms are used interchangeably.

8–7   *The two stages of glycolysis. Compounds other than glucose, such as galactose, mannose, pentoses, glycogen, and starch, can undergo glycolysis once they have been converted to glucose 6-phosphate.*

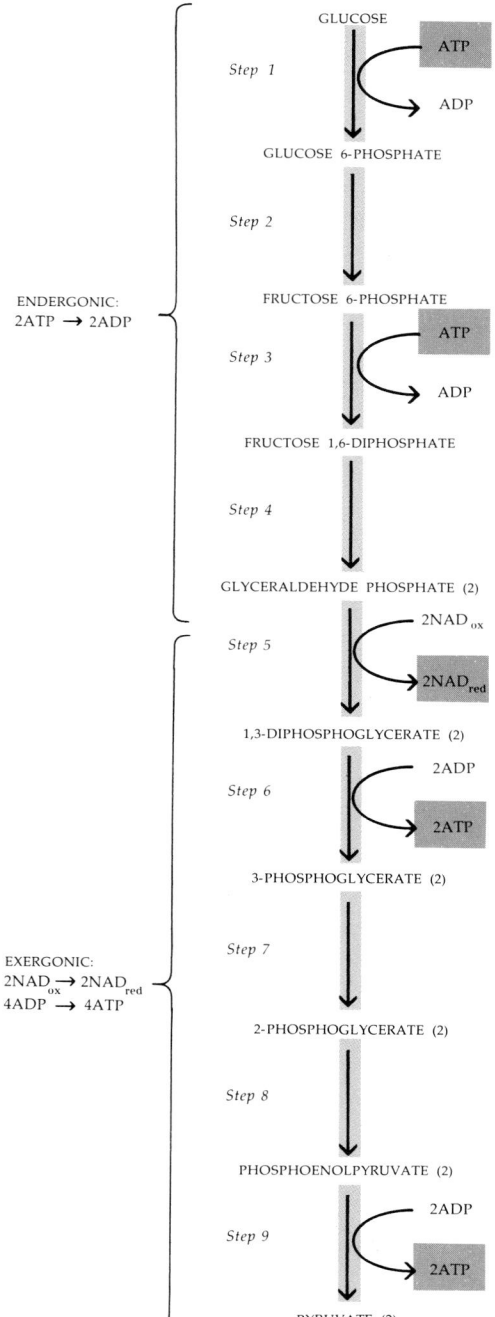

GLUCOSE

Step 1

ATP
ADP

GLUCOSE 6-PHOSPHATE

Step 2

FRUCTOSE 6-PHOSPHATE

ENDERGONIC:
$2ATP \rightarrow 2ADP$

Step 3

ATP
ADP

FRUCTOSE 1,6-DIPHOSPHATE

Step 4

GLYCERALDEHYDE PHOSPHATE (2)

Step 5

$2NAD_{ox}$
$2NAD_{red}$

1,3-DIPHOSPHOGLYCERATE (2)

Step 6

2ADP
2ATP

3-PHOSPHOGLYCERATE (2)

Step 7

EXERGONIC:
$2NAD_{ox} \rightarrow 2NAD_{red}$
$4ADP \rightarrow 4ATP$

2-PHOSPHOGLYCERATE (2)

Step 8

PHOSPHOENOLPYRUVATE (2)

Step 9

2ADP
2ATP

PYRUVATE (2)

THE ANAEROBIC PATHWAY

Most of the rest of this chapter will concentrate on what happens to the pyruvate molecules in cells such as those of higher plants and animals in the presence of oxygen. The fact that glycolysis does not require oxygen, however, suggests that the glycolytic sequence evolved early, before free oxygen was present in the atmosphere, and that primitive one-celled organisms used glycolysis or something very much like it to extract energy from organic compounds which they absorbed from their watery surroundings. Moreover, certain modern cells are able to extract their energy from glycolysis alone if oxygen is not available to them. Yeast cells, for example, can live without oxygen. Under anaerobic (oxygenless) conditions, they convert glucose to pyruvic acid by the glycolysis sequence. The pyruvic acid so formed is then converted to alcohol by the steps shown in Figure 8–8.

8–8 *The steps by which pyruvic acid, formed by glycolysis, is converted to ethyl alcohol (ethanol). In the first step, carbon dioxide is released. In the second, NAD is oxidized, and acetaldehyde is reduced. Most of the energy of the glucose remains in the alcohol, which is the end product of the sequence.*

$$CO_2 \qquad NAD_{red} \quad NAD_{ox}$$

$$
\begin{array}{ccc}
CH_3 & CH_3 & CH_3 \\
C{=}O & C{=}O & H{-}C{-}OH \\
C{=}O & H & H \\
OH & &
\end{array}
$$

PYRUVIC ACID (FROM GLYCOLYSIS) — ACETALDEHYDE — ETHANOL

8–9 *Tapestry depicting two methods of extracting grape juice for winemaking: wooden press and foot. As the grapes are crushed, yeast cells present on the grape skins mix with the grape juice. Storing the mixture under anaerobic conditions causes the yeast to break down the glucose in the grape juice to alcohol.*

8–10 *The enzymatic reaction that produces lactic acid from pyruvic acid in muscle cells. In the course of this reaction, NAD is oxidized and pyruvic acid is reduced. The NAD_{ox} molecules produced in this reaction and the one shown in Figure 8–8 are recycled in the glycolytic sequence. Lactic acid accumulation results in muscle soreness and fatigue.*

$$
\begin{array}{ccc}
CH_3 & & CH_3 \\
| & & | \\
C=O & \xrightarrow{NAD_{red} \quad NAD_{ox}} & H-C-OH \\
| & & | \\
C=O & & C=O \\
| & & | \\
OH & & OH \\
\end{array}
$$

PYRUVIC ACID
(FROM GLYCOLYSIS)   LACTIC ACID

8–11 *Repaying the oxygen debt.*

When the glucose-filled juices of grapes and other fruit are extracted and stored in airtight kegs, the yeast cells, present as a bloom on the skin of the fruit, turn the fruit juice to wine by converting glucose into ethyl alcohol. Yeast, like all living things, have a limited tolerance for alcohol, and when a certain concentration (about 12 percent) is reached, the yeast cells die and fermentation ceases.

Because of the economic importance of the wine industry, alcoholic fermentation was the first enzymatic process to be intensively studied. Louis Pasteur was the first to recognize the role of yeast cells in the process, and this process was the final setting for the vitalism-reductionism controversy (page 11) that came to an end around the beginning of the twentieth century. The formation of alcohol from sugar is called "fermentation," and for many years enzymes were commonly referred to as "ferments," although the scope of their activities is now known to be so broad that this word is inappropriate in the light of present-day knowledge.

Cells of higher animals do not form alcohol from pyruvate in the absence of oxygen, but rather produce another substance known as lactic acid. Lactic acid is produced, for example, in muscle cells during bursts of extra hard work, as by an athlete in a sprint. We breathe hard when we run fast, thereby increasing the supply of oxygen to the muscle cells, but the capacity of the muscles to do work is not limited by the amount of oxygen available to them at that moment. They can accumulate what is known as an oxygen debt by producing lactic acid from glucose (which is stored in the muscles in the form of glycogen). The accumulated lactic acid may be one of the substances that produces the sensations of muscle fatigue. After a sprint, we continue to breathe hard until we have paid off the oxygen debt and the accumulated lactic acid has been oxidized.

RESPIRATION AND THE MITOCHONDRION

Respiration has two meanings in biology. One is the breathing in of oxygen and breathing out of carbon dioxide; this is also the ordinary, nontechnical meaning of the word. The second meaning of respiration is the oxidation of food molecules by cells. This process, sometimes qualified as cellular respiration, is what we are concerned with here. Cellular respiration takes place in two stages: the Krebs cycle and the electron transport chain. They both require oxygen, unlike glycolysis, and in eukaryotic cells they take place in the specialized organelle known as the mitochondrion. Mitochondria are surrounded by two membranes. The outer one is smooth and the inner one folds inward. The folds are called cristae. The more active a cell, the more numerous are both its mitochondria and the cristae within them. Within the inner compartment, surrounding the cristae, is a dense solution containing enzymes, coenzymes, water, phosphates, and other molecules involved in respiration. The mitochondrion is a self-contained chemical plant. The outer membrane lets most small molecules in or out freely, but the inner one permits the passage only of certain molecules, such as pyruvate and ATP, and restrains the passage of others. The enzymes of the Krebs cycle are in solution in the inner compartment. The enzymes and other components of the electron transport chain are built into the

8–12 *Mitochondria are surrounded by two membranes. The inner membrane folds inward to make a series of shelves, or cristae. The enzymes and electron carriers involved in the final stage of cellular respiration are built into these internal membranes.*

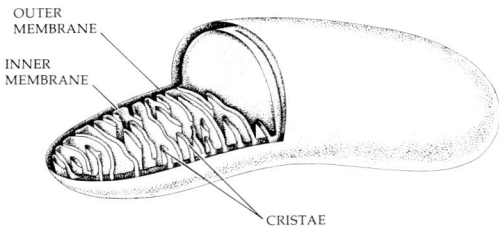

OUTER MEMBRANE

INNER MEMBRANE

CRISTAE

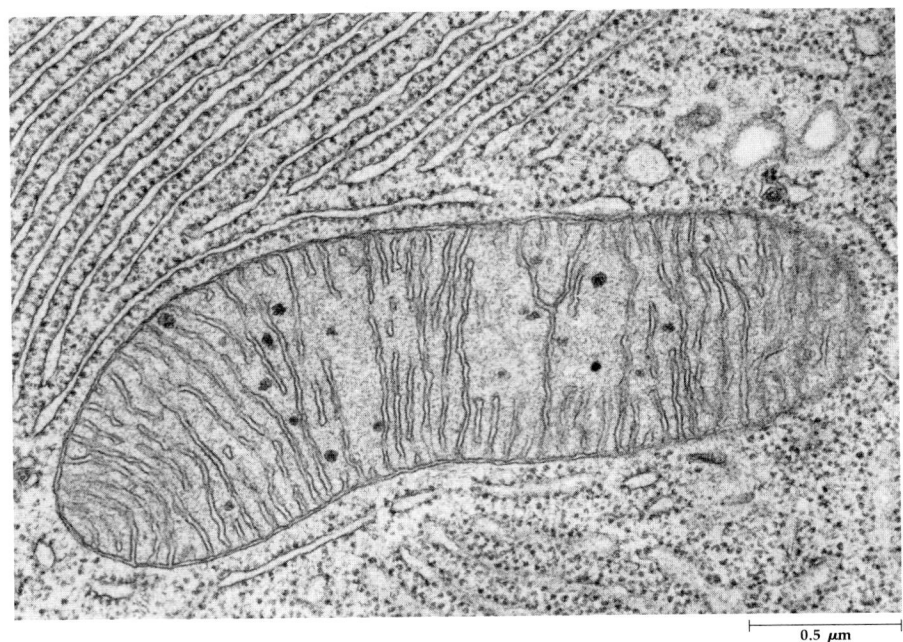

0.5 μm

surfaces of the cristae. In the mitochondria, pyruvic acid from glycolysis is oxidized to carbon dioxide and water, completing the breakdown of the glucose molecule.

The Krebs Cycle

Before entering the Krebs cycle, each of the three-carbon pyruvate molecules is oxidized. The third carbon (the one in the carboxyl group) is removed, forming carbon dioxide. In the course of this exergonic reaction, a molecule of NAD_{red} is produced from NAD_{ox}. The glucose molecule has now been oxidized to two acetyl (CH_3CO) groups. These groups are momentarily accepted by a coenzyme known as coenzyme A. Like the coenzymes we have examined previously, it is a large molecule, a portion of which is a nucleotide and a portion of which is a vitamin, pantothenic acid, one of the B complex vitamins. The combination of the acetyl group and CoA is known simply as acetyl CoA.

Fats and amino acids can also be converted to acetyl CoA and enter the respiratory sequence at this point. A fat molecule is first hydrolyzed to glycerol and three fatty acids. Then successive two-carbon groups are removed, beginning at the carboxyl end. A molecule such as palmitic acid, which contains 16 carbon atoms, yields eight molecules of acetyl CoA. Amino acids also can be converted, by various reactions, to acetyl CoA.

In the Krebs cycle, the two-carbon acetyl group is combined with a four-carbon compound (oxaloacetic acid) to produce a six-carbon compound (citric acid). In the course of the cycle, two of the six carbons are oxidized to CO_2, and oxaloacetic acid is regenerated—thus making this series literally a cycle. Each turn around the cycle uses up one acetyl group and regenerates a molecule

8–13 *The three-carbon pyruvic acid molecule is oxidized to the two-carbon acetyl group which is combined with coenzyme A to form acetyl CoA. The oxidation of the pyruvic acid molecule is coupled to the reduction of NAD_{ox}. Acetyl CoA enters the Krebs cycle.*

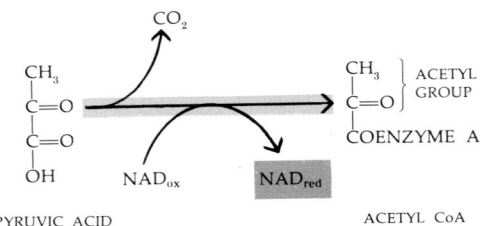

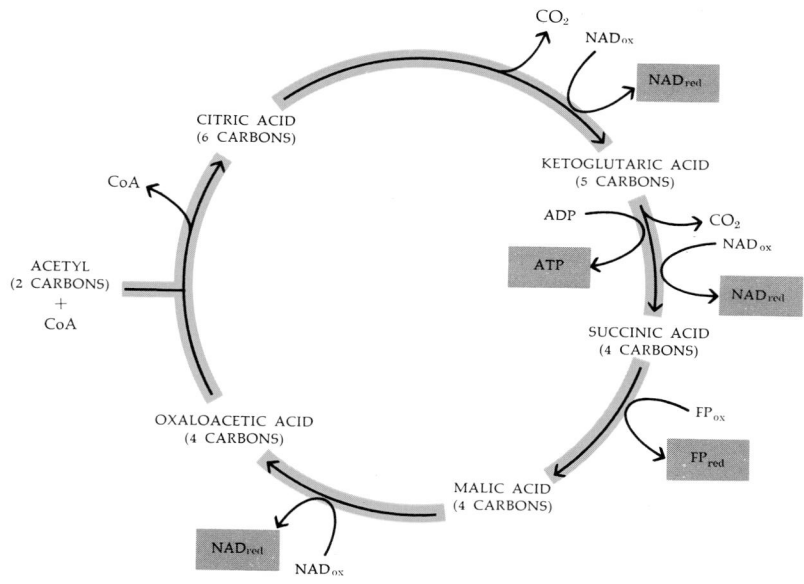

8–14 Summary of the Krebs cycle. In the course of the cycle, the carbons donated by the acetyl group are oxidized to carbon dioxide and the hydrogen atoms are passed to electron carriers. As in glycolysis, a specific enzyme is involved at each step. One molecule of ATP, three molecules of NAD_{red}, and one molecule of FP_{red} represent the energy yield of the cycle.

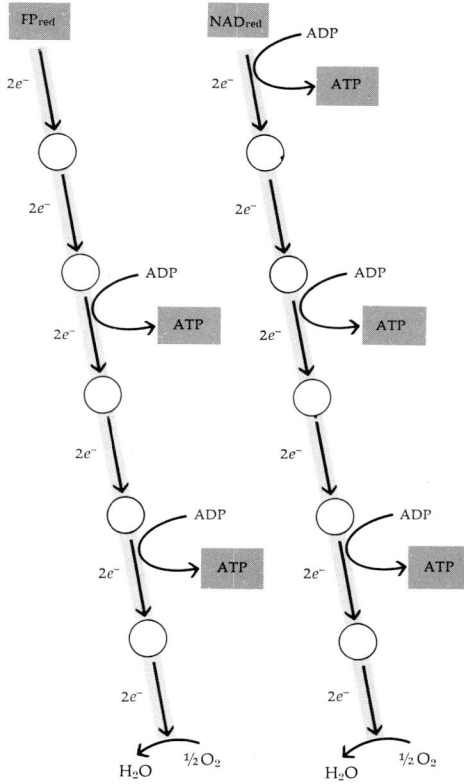

–15 As electrons flow in pairs along the electron transport chain from a point of higher energy to a point of lower energy, the released energy is harnessed by the cytochromes, represented by circles, and their coupling enzymes to produce ATP from ADP. The ATP is then used for the work of the cell.

of oxaloacetic acid, which is then ready to begin the sequence again. In the course of these steps, some of the energy released by the oxidation of the carbon atoms is used to convert ADP to ATP (one molecule per cycle), some is used to produce NAD_{red} from NAD_{ox} (three molecules per cycle), and some is used to reduce a second electron carrier, a flavoprotein, abbreviated FP (one molecule of FP_{red} from FP_{ox} per cycle).

$$\text{Oxaloacetic acid} + \text{acetyl CoA} + \text{ADP} + 3NAD_{ox} + FP_{ox} \longrightarrow$$
$$\text{oxaloacetic acid} + 2CO_2 + \text{CoA} + \text{ATP} + 3NAD_{red} + FP_{red}$$

Electron Transport

The glucose molecule is now completely oxidized. Some of its energy has been used to produce ATP from ADP. Most of it, however, still remains in electrons removed from the carbon atoms as they were oxidized and passed to the electron carriers NAD and FP. These electrons are at a high energy level. In the course of the electron transport chain, they are passed "downhill" and the energy released is used to form ATP molecules from ADP. This process is known as *oxidative phosphorylation.*

The electron carriers of the electron transport chain of the mitochondria differ from NAD and FP in their chemical structure. Most of them belong to a class of compounds known as *cytochromes.* Cytochromes contain an atom of iron, in a porphyrin ring, like the heme portion of the hemoglobin molecule, surrounded by a protein containing about 100 amino acids. Each cytochrome differs in its protein chain and also in the energy level at which it holds the electrons. They thus work in sequence somewhat like a series of water-

Cytochrome c. The model shows only the heme (iron-porphyrin group) and parts of the protein chain closest to it.

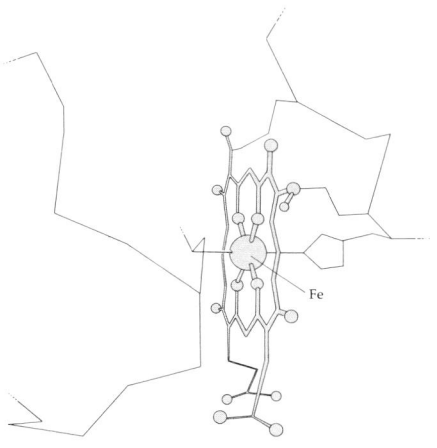

Fe

wheels. Water, as it flows from a point of higher water potential to one of lower water potential, releases energy which can be harnessed by waterwheels to do work. As electrons flow along the electron transport chain from a higher to a lower energy level, the cytochromes harness the released energy and use it to convert ADP to ATP. At the end of the chain, the electrons are accepted by oxygen and combine with protons (hydrogen ions) to produce water. Each time one pair of electrons passes from NAD_{red} to oxygen, three molecules of ATP are formed from ADP and phosphate. Each time a pair of electrons passes from FP_{red}, which holds them at a slightly lower energy level than NAD_{red}, two molecules of ATP are formed.

We are now in a position to see how much of the energy originally present in the glucose molecule has been recovered in the form of ATP.*

Glycolysis yielded two molecules of ATP directly and two molecules of NAD_{red}, for a total net gain of 8 ATP.

The conversion of pyruvic acid to acetyl CoA yields two molecules of NAD_{red} for each molecule of glucose and so produces six molecules of ATP.

The Krebs cycle yields, for each molecule of glucose, two molecules of ATP, six of NAD_{red}, and two of FP_{red}, or a total of 24 ATP.

As the balance sheet, Table 8–1, shows, the complete yield from a single molecule of glucose is 38 molecules of ATP. Note that all but two of the 38

* These figures are based on recent research using animal liver cells. The total yield from glycolysis may differ somewhat according to the cell type used.

8–17 *Energy of motion.*

8–18 *Summary of glycolysis and respiration. Glucose is first broken down to pyruvic acid, with a yield of two ATP molecules and the reduction (dashed arrows) of two NAD molecules. Pyruvic acid* is oxidized to acetyl CoA, and one molecule of NAD is reduced (note that this and subsequent reactions occur twice for each glucose molecule; this electron passage is indicated by solid arrows). In the Krebs cycle, the acetyl group is oxidized and the electron acceptors, NAD and FP, are reduced. NAD and FP then transfer their electrons to the series of cytochromes that make up the electron transport chain. As these cytochromes pass the electrons "downhill," the energy released is used to make ATP from ADP.

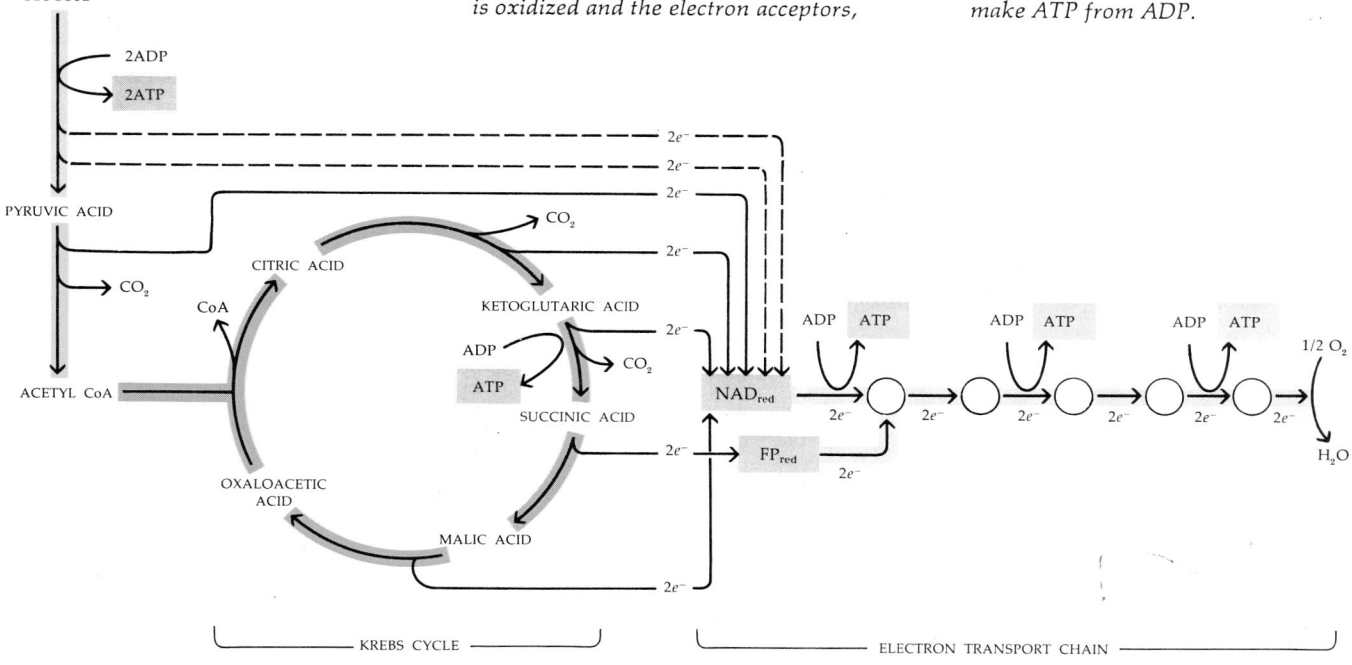

8–19 *Energy changes in the oxidation of glucose. The complete respiratory sequence (glucose $+ 6O_2 \longrightarrow 6CO_2 + 6H_2O$) proceeds with an energy drop of 686 kilocalories. Of this, about 40 percent (266 kilocalories) is conserved in 38 ATP molecules. In anaerobic glycolysis (glucose $\longrightarrow$ lactic acid), by contrast, only 2 ATP molecules are produced, representing only about 2 percent of the available energy of glucose.*

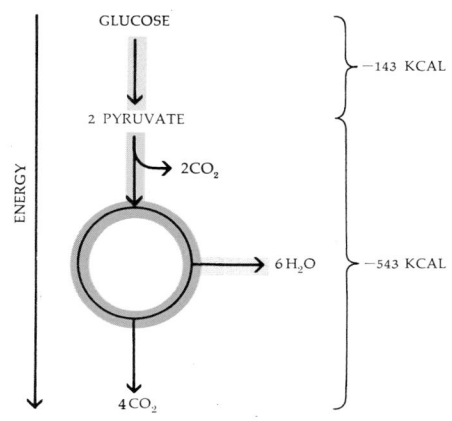

molecules of ATP have come from reactions taking place in the mitochondrion, and all but four involve the oxidation of NAD_{red} or FP_{red} along the electron transport chain.

The total difference in free energy between the reactants (glucose and oxygen) and the products (carbon dioxide and water) is 686 kilocalories. Almost 40 percent of this, about 266 (7×38) kilocalories, has been captured in the "high-energy" bonds of the 38 ATP molecules.

Table 8–1   *Summary of Energy Yield from One Molecule of Glucose*

Glycolysis:	$2\ NAD_{red} \longrightarrow$ $\begin{array}{l}2\ ATP \\ 6\ ATP\end{array}\Big\}$	$\longrightarrow$ 8 ATP
Respiration:		
Pyruvate $\longrightarrow$ acetyl CoA:	$1\ NAD_{red} \longrightarrow 3\ ATP$ ($\times 2$)	$\longrightarrow$ 6 ATP
Krebs cycle:	$\begin{array}{l}3\ NAD_{red} \longrightarrow 9\ ATP \\ 1\ FP_{red} \longrightarrow 2\ ATP\end{array}\Big\}$ $\begin{array}{l}1\ ATP\end{array}$ ($\times 2$)	$\longrightarrow$ 24 ATP

Table 8–2 *Energy Production from Foodstuffs*

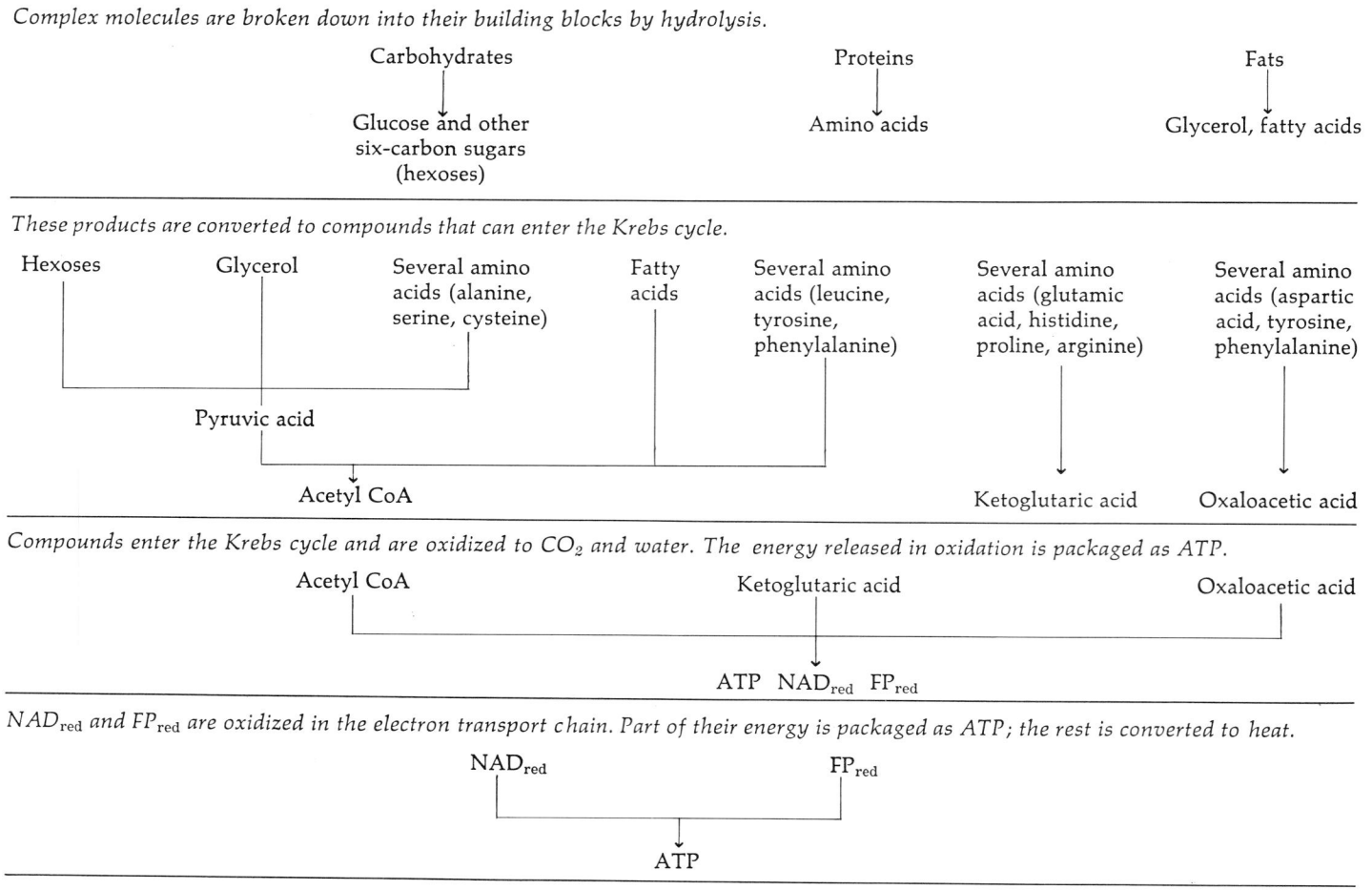

Complex molecules are broken down into their building blocks by hydrolysis.

Carbohydrates	Proteins	Fats
Glucose and other six-carbon sugars (hexoses)	Amino acids	Glycerol, fatty acids

These products are converted to compounds that can enter the Krebs cycle.

Hexoses   Glycerol   Several amino acids (alanine, serine, cysteine)   Fatty acids   Several amino acids (leucine, tyrosine, phenylalanine)   Several amino acids (glutamic acid, histidine, proline, arginine)   Several amino acids (aspartic acid, tyrosine, phenylalanine)

Pyruvic acid

Acetyl CoA   Ketoglutaric acid   Oxaloacetic acid

Compounds enter the Krebs cycle and are oxidized to CO_2 and water. The energy released in oxidation is packaged as ATP.

Acetyl CoA   Ketoglutaric acid   Oxaloacetic acid

ATP NAD_{red} FP_{red}

NAD_{red} and FP_{red} are oxidized in the electron transport chain. Part of their energy is packaged as ATP; the rest is converted to heat.

NAD_{red} FP_{red}

ATP

SUMMARY

Energy exchanges in living and nonliving systems are governed by physical laws known as the laws of thermodynamics. The first law states that energy can be changed from one form to another but cannot be created or destroyed. The second law states that all natural processes tend to proceed in such a direction that the entropy (disorder or randomness) of the system increases. Living systems are able to maintain their high degree of organization and order because they are not energetically isolated from their environment. They have an abundant outside source of energy (ultimately the sun), and although in the course of their energy conversions they decrease their own entropy, they greatly increase the entropy of their environment.

The oxidation of glucose is a chief source of energy in most cells. As the glucose is broken down in a series of small enzymatic steps, the energy in the molecule is packaged in the form of "high-energy" bonds in molecules of ATP.

The first phase in the breakdown of glucose is glycolysis, in which the six-carbon glucose molecule is split into 2 three-carbon molecules of pyruvate, and two new molecules of ATP and two of NAD_{red} are formed. This reaction takes place in the cytoplasm of the cell.

The second phase in the breakdown of glucose and other fuel molecules is respiration. It requires oxygen and, in eukaryotic cells, takes place in the mitochondria. It occurs in two stages: the Krebs cycle and the electron transport chain.

In the course of respiration, the three-carbon pyruvate molecules are broken down to two-carbon acetyl groups, which then enter the Krebs cycle. In the Krebs cycle, the two-carbon acetyl group is broken apart in a series of reactions to carbon dioxide. In the course of the oxidation of each acetyl group, four electron acceptors (three NAD and one FP) are reduced, and another molecule of ATP is formed.

The final stage of breakdown of the fuel molecule is the electron transport chain, which involves a series of electron carriers and enzymes embedded in the inner membranes of the mitochondrion. Along this series of electron carriers, the high-energy electrons accepted by NAD_{red} and FP_{red} during the Krebs cycle pass downhill to oxygen. Each time a pair of electrons passes down the electron transport chain, ATP molecules are formed from ADP and phosphate. In the course of the breakdown of the glucose molecule, 38 molecules of ATP are formed, most of them in the mitochondrion.

QUESTIONS

1. Define the following terms: free energy, entropy, cellular respiration, and glycolysis.
2. What are the two laws of thermodynamics? Give illustrations of each.
3. In general terms, describe the events that take place during cellular respiration. Compare your description with the chart on page 135.

9–1 *The energy on which life depends enters*
the biosphere in the form of light.
Photosynthesis is the process by which
this light energy is converted to chemical
energy.

Chapter 9

How Cells Capture the Energy of the Sun

In the history of scientific ideas, one concept or theory is formulated, endures for a period of time, stimulates further research that often eventually brings the concept itself into question, and so is revised or replaced. The history of the study of photosynthesis is a good example of the way in which scientific theories grow and change as well as a useful introduction to our present, still incomplete knowledge of the process.

HISTORICAL BACKGROUND

Until about 300 years ago, observers of the biological world, noting that the life processes of animals were dependent on the food they ate, thought that plants derived their food from the soil in a similar way.

This concept was widely accepted by those who thought about the problem until the Belgian physician Jan Baptista van Helmont (1577–1644) offered the first experimental evidence that soil is not the major source of plant nourishment. Van Helmont grew a small willow tree in an earthenware pot for five years, adding only water to the pot. At the end of five years, the willow had increased in weight by 164 pounds, while the earth had decreased in weight by only 2 ounces. On the basis of these results, van Helmont concluded that all the substance of the plant was produced from the water and none from the soil.

Priestley's "Planetary Ventilation"

Fire, one of the "elements" of the ancient Greeks, intrigued the early alchemists and the Renaissance chemists who were their successors. What happens when an element is consumed by fire was similarly a major focus of study and debate among the chemists of the eighteenth century, and led to a revolution in the science of chemistry. The study of plant physiology played an important part in this revolution.

One of those concerned with the problem of the changes produced by fire was Joseph Priestley, an English clergyman and chemist. On August 17, 1771, Priestley "put a sprig of mint into air in which a wax candle had burned out and found that, on the 27th of the same month, another candle could be burned in this same air." Priestley believed, as he reported, that he had "accidentally hit upon a method of restoring air that has been injured by the

"I took an earthen vessel in which I put 200 pounds of earth that had been dried in a furnace, which I moistened with rainwater, and I implanted therein the trunk or stem of a willow tree, weighing five pounds. And at length, five years being finished, the tree sprung from thence did weigh 169 pounds and about three ounces. When there was need, I always moistened the earthen vessel with rainwater or distilled water, and the vessel was large and implanted in the earth. Lest the dust that flew about should be co-mingled with the earth, I covered the lip or mouth of the vessel with an iron plate covered with tin and easily passable with many holes. I computed not the weight of the leaves that fell off in the four autumns. At length, I again dried the earth of the vessel, and there was found the same 200 pounds, wanting about two ounces. Therefore 164 pounds of wood, bark and roots arose out of water only." (Jan Baptista van Helmont.)

burning of candles." The "restorative which nature employs for this purpose," he stated, "is vegetation." Priestley extended his observations and soon showed that air "restored" by vegetation was not "at all inconvenient to a mouse." Priestley's experiments offered the first logical explanation of how the air remained "pure" and able to support life despite the burning of countless fires and the breathing of many animals. When he was presented with a medal for his discovery, the citation read in part: "For these discoveries we are assured that no vegetable grows in vain . . . but cleanses and purifies our atmosphere."

Only in the Light

Priestley's reports that plants purify the air were of great interest to his fellow chemists, but they soon attracted criticism because the experiments could not readily be confirmed. In fact, when Priestley tried to do the experiments again himself, he did not get the same results. Some historians think that he simply may have moved his equipment to a dark corner of his laboratory, not realizing that light was important to the effect. It was a Dutch physician, Jan Ingenhousz (1730–1799), who was finally able to confirm Priestley's work with an important addition. He found that the purification process takes place only in sunlight. Plants at night or in the shade, he reported, "contaminate the air which surrounds them, throwing out an air hurtful to animals." He also observed that only the green parts of plants restore the air and that "the sun by itself has no power to mend air without the concurrence of plants." While Ingenhousz was performing his experiments on plants, Antoine Lavoisier (1743–1794) was carrying out the experiments that put chemistry on an essentially modern basis.

Among his discoveries, those which are most directly connected with the story we are following here concern animal respiration. Working with the mathematician P. S. Laplace (1749–1827), Lavoisier confined a guinea pig in oxygen for 10 hours and measured the carbon dioxide produced. He also measured the amount of oxygen used by a man active and at rest. In these experiments, he was able to show that combustion of carbon compounds with oxygen to carbon dioxide and water is the true source of animal heat and that oxygen consumption increases during physical work. "Respiration is merely a slow combustion of carbon and hydrogen, which is similar in every respect to that which occurs in a lighted lamp or candle, and, from this point of view,

9–3 *Joseph Priestley—"... no vegetable grows in vain...."*

animals that breathe are really combustible bodies which burn and are consumed."

In modern language, this is expressed in the now familiar equation

$$(CH_2O) + O_2 \longrightarrow CO_2 + H_2O + \text{energy (heat)}$$

Lavoisier was guillotined on May 8, 1794 during the French Revolution. The judge presiding over the case is reported to have said, "The Republic has no need of savants."

The work of Ingenhousz spanned the prematurely terminated career of Lavoisier. Quick to adopt Lavoisier's ideas about gases, Ingenhousz hypothesized that the plant was not just exchanging "good air" for "bad" and so making the world habitable for animal life. In the sunshine, he suggested, a plant absorbs the carbon from carbon dioxide, "throwing out at that time the oxygen alone, and keeping the carbon to itself as nourishment."

Nicholas Theodore de Saussure (1767–1845) applied Lavoisier's principles of quantitative measurements and showed that equal volumes of CO_2 and O_2 are exchanged during photosynthesis and that the plant does indeed retain the carbon. He also showed that during photosynthesis the plant gained more weight than could be accounted for by the carbon taken in as carbon dioxide. In other words, the carbon in the dry matter of plants comes from carbon dioxide, but, equally important, the rest of the dry matter (with the exception of minerals from the soil) comes from water.

Thus, all the components were identified—carbon dioxide, water, and light —and it became possible to write the overall photosynthetic equation:

$$CO_2 + H_2O + \text{energy (light)} \longrightarrow (CH_2O) + O_2$$

In short, although it differs in detail, as we shall see, photosynthesis is actually respiration (or combustion) in reverse.

9–4 *Lavoisier conducting an experiment on respiration. This drawing was made by Mme. Lavoisier, who pictured herself as an observer.*

LIGHT AND LIFE

Almost 300 years ago, the English physicist Sir Isaac Newton (1642–1727) separated visible light into a spectrum of colors by letting it pass through a prism. By this experiment, Newton showed that white light is actually made up of a number of different colors, ranging from violet at one end of the spectrum to red at the other. Their separation is possible because light of different colors is bent at different angles in passing through the prism. Newton believed that light was a stream of particles (or, as he termed them, "corpuscles") because, in part, of its tendency to travel in a straight line.

In the nineteenth century, through the genius of James Clerk Maxwell (1831–1879), it came to be known that what we experience as light is in truth a very small part of a vast continuous spectrum of radiation, the electromagnetic spectrum. As Maxwell showed, all the radiations included in this spectrum travel in waves. The wavelengths— that is, the distances from one peak to the next—vary from those of x-rays, which are measured in angstroms, to those of low-frequency radio waves, which are measured in miles. The shorter the wavelength, the greater the energy. Within the spectrum of visible light, red light has the longest wavelength, violet the shortest. Another feature that these radiations have in common is that, in a vacuum, they all travel at the same speed—186,300 miles (300,000 kilometers) per second.

By 1900 it had become clear, however, that the wave theory of light was not adequate. The key observation, a very simple one, was made in 1888: When a zinc plate is exposed to ultraviolet light, it acquires a positive charge. The metal, it was soon deduced, becomes positively charged because the light energy dislodges electrons, forcing them out of the metal atoms. Subsequently it was discovered that this photo-electric effect, as it is known, can be produced in all metals. Every metal has a wavelength critical for the effect; the light (visible or invisible) must be of that wavelength or a shorter (more energetic) wavelength for the effect to occur. (The hypothesis that electrons orbit atoms at particular energy levels, as formulated by Bohr and others, is based on these observations.)

With some metals, such as sodium, potassium, and selenium, the critical wavelength is within the spectrum of visible light, and as a consequence, visible light striking the metal can set up a moving stream of electrons (such a stream is, of course, an electric current). The electric eyes that open doors for you at supermarkets or airline terminals, exposure meters, and television cameras all operate on this principle of turning light energy into electrical energy.

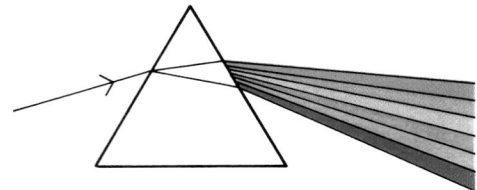

When white light passes through a prism, as Sir Isaac Newton showed almost 300 years ago, it is sorted into a rainbowlike spectrum. By this experiment, Newton showed that white light is actually a mixture of different colors, ranging from violet at one end of the spectrum to red at the other. It is now known that the spectrum of visible light represents only a very small portion of the vast electromagnetic spectrum.

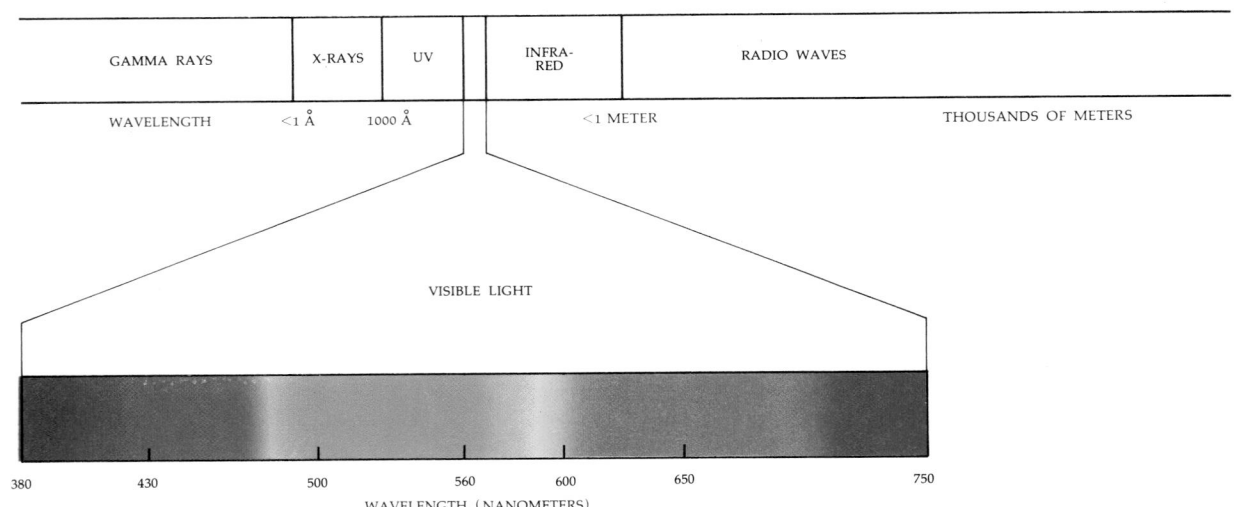

Wave or Particle?

Now here is the problem. The wave theory of light would lead you to predict that the brighter the light, the greater the force with which the electrons would be dislodged. But as we have already seen, whether or not light can eject the electrons of a particular metal depends not on the brightness of the light but on its wavelength. A very weak beam of the critical wavelength or a shorter wavelength is effective, while a stronger (brighter) beam of a longer wavelength is not. Furthermore, as was shown in 1902, increasing the brightness of the light increases the number of electrons dislodged, but not the velocity at which they are ejected from the metal. To increase the velocity, one must use a shorter wavelength of light. Nor is it necessary for energy to be accumulated in the metal. With even a dim beam of a critical wavelength, an electron may be emitted the instant the light hits the metal.

To explain such phenomena, the particle theory of light was proposed by Albert Einstein in 1905. According to this theory, light is composed of particles of energy called photons. The energy of a photon is not the same for all kinds of light but is, in fact, inversely proportional to the wavelength—the longer the wavelength, the lower the energy. Photons of violet light, for example, have almost twice the energy of photons of red light, the longest visible wavelength.

The wave theory of light permits physicists to describe certain aspects of its behavior mathematically, while the photon theory permits another set of mathematical calculations and predictions. These two models are no longer regarded as opposed to one another; rather, they are complementary, in the sense that both are required for a complete description of the phenomenon we know as light.

The coexistence of these two theories further illustrates the scientific method. If a scientist defines light as a wave and measures it as such, it behaves like a wave. Similarly if one defines it as a particle, one gets information about light as a particle. As Einstein once said, "It is the theory which determines what we can observe."

The Fitness of Light

Light, as Maxwell showed, is only a tiny band in a continuous spectrum. From the physicist's point of view, the difference between light and darkness—so dramatic to the human eye—is only a few nanometers of wavelength. There is no qualitative difference at all marking the borders of the light spectrum. Why does this particular small group of radiations, rather than some other, bathe our world in radiance, make the leaves grow and the flowers burst forth, cause the mating of fireflies and paolo worms, and, when reflecting off the surface of the moon, excite the imaginations of poets and lovers? Why is it that this tiny portion of the electromagnetic spectrum is responsible for vision, for phototropism (movement toward light), for photoperiodism (the changes that take place in an organism with the changing length of day and night in the changing seasons), and also for photosynthesis, on which all life depends? Is it an amazing coincidence that all these biological activities depend on these same wavelengths?

George Wald of Harvard, one of the greatest living experts on the subject of light and life, says no. He thinks that if life exists elsewhere in the universe, it is probably dependent on this same fragment of the vast spectrum. Wald bases this conjecture on two points. First, living things, as we have seen, are composed of large, complicated molecules held in special configurations and relationships to one another by hydrogen bonds and other weak bonds. Radiation of even slightly higher energies than the energy of violet light breaks these bonds and so disrupts the structure and function of the molecules. Radiations with wavelengths below 200 nanometers drive electrons out of atoms; hence they are called ionizing radiations. The energy of light of wavelengths longer than those of the visible band is largely absorbed by water, which makes up the

great bulk of all living things. When such light does reach organic molecules, its lower energy causes them to increase their motion (increasing heat) but does not trigger changes in their structure. Only those radiations within the range of visible light have the property of exciting molecules—that is, of raising electrons from one energy level to another—and so of producing biological changes.

The second reason that the visible band above all others of the electromagnetic spectrum was "chosen" by living things is that it, above all, is what is available. Most of the radiation reaching the Earth from the sun is within this range. Higher-energy wavelengths are screened out by the oxygen and ozone high in the atmosphere. Much infrared radiation is screened out by water vapor and carbon dioxide before it reaches the Earth's surface.

This is an example of what has been termed "the fitness of the environment"; the suitability of the environment for life and the suitability of life for the physical world are exquisitely interrelated. If they were not, life could not, of course, exist.

Van Niel's Hypothesis

For more than 100 years, it was generally assumed that in the equation

$$CO_2 + H_2O + light \longrightarrow (CH_2O) + O_2$$

the carbohydrate (CH_2O) resulted from the combination of the carbon and water molecules and that the oxygen was released from the carbon dioxide molecule. This entirely reasonable hypothesis was widely accepted. But, as it turned out, it was wrong.

The investigator who upset this long-held theory was C. B. van Niel of Stanford University. Van Niel, then a graduate student, was investigating photosynthesis in different types of photosynthetic bacteria. In their photosynthetic reactions, bacteria reduce carbon to carbohydrates, but they do not release oxygen. Among the types of bacteria van Niel was studying were the purple sulfur bacteria, which require hydrogen sulfide for photosynthesis. In the course of photosynthesis, globules of sulfur (S) are excreted or accumulated inside the bacterial cells. In these bacteria, van Niel found that this reaction takes place during photosynthesis:

$$CO_2 + 2H_2S \xrightarrow{light} (CH_2O) + H_2O + 2S$$

This finding was simple enough and did not attract much attention until van Niel made a bold extrapolation. He proposed that the generalized equation for photosynthesis is

$$CO_2 + 2H_2A \xrightarrow{light} (CH_2O) + H_2O + 2A$$

In this equation, H_2A stands for some oxidizable substance such as hydrogen sulfide, free hydrogen, or any one of several other compounds used by photosynthetic bacteria—or water. In the photosynthetic algae and higher green plants, H_2A is water. In short, van Niel proposed that it was the water that was split by photosynthesis, *not* the carbon dioxide.

Purple sulfur bacteria. In these cells, hydrogen sulfide plays the same role as water does in the photosynthetic process of plants. The hydrogen sulfide (H_2S) is split, and the sulfur that is released accumulates as globules, visible within the cells.

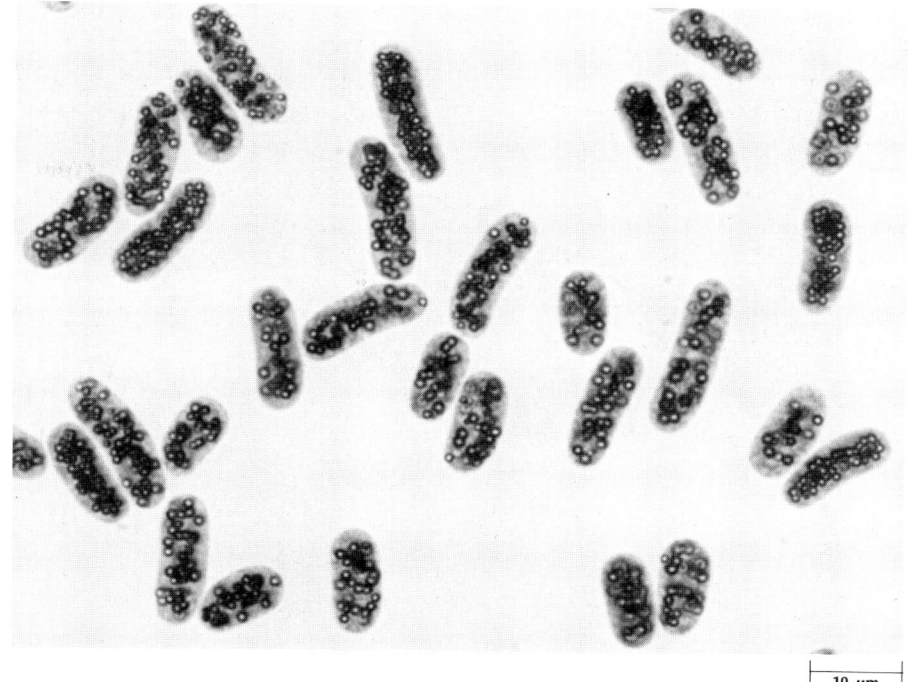

10 μm

This brilliant speculation, first proposed in the early 1930s, was not proved until many years later. Eventually, investigators, using a heavy isotope of oxygen ($^{18}O_2$), traced the oxygen from water to oxygen gas

$$CO_2 + 2H_2{}^{18}O \xrightarrow{\text{light}} (CH_2O) + H_2O + {}^{18}O_2$$

and confirmed van Niel's hypothesis. The overall concept of photosynthesis has remained unchanged from the time of van Niel's proposal. However, many of its details have subsequently been worked out and more are still under active investigation.

PHOTOSYNTHESIS: LIGHT AND DARK REACTIONS

The reactions of photosynthesis, it is now known, take place in two stages. In the first stage, light energy is used to form ATP from ADP and to reduce electron carrier molecules. These reactions require light and so are known as the light reactions. In the second stage of photosynthesis, the energy products of the first stage are used to reduce carbon from carbon dioxide to a simple sugar, thus converting the chemical energy of the carrier molecules to forms suitable for transport and storage and, at the same time, forming a carbon skeleton on which other organic molecules can be built. The reactions of this second stage, which do not require light (although they do not require dark, either), are known as the dark reactions. However, in the living plant (except in the course of laboratory experiments) they always take place in the daytime and in the light.

9–6 *(a) The chloroplast is the site of photosynthesis in all higher plants. Chloroplasts, like mitochondria, are surrounded by a double membrane and have a system of internal membranes, the lamellae. The light-capturing reactions* *take place in the lamellae where the chlorophylls and other pigments are found. The series of reactions by which this energy is used to produce carbon-containing compounds takes place in the stroma, the material surrounding the* *photosynthetic membranes. (b) At higher magnification the membrane can be seen to be organized in flattened sacs, the thylakoids. The stacks of thylakoids are called grana.*

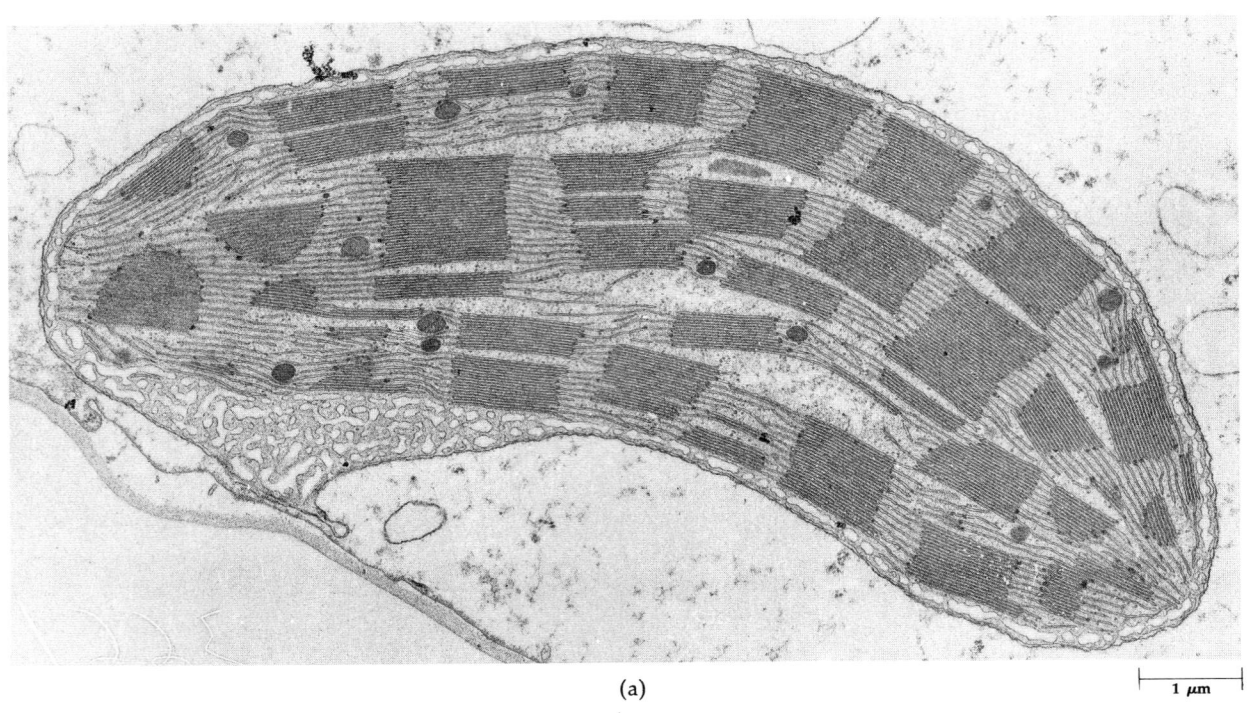

(a)

|—————| 1 μm

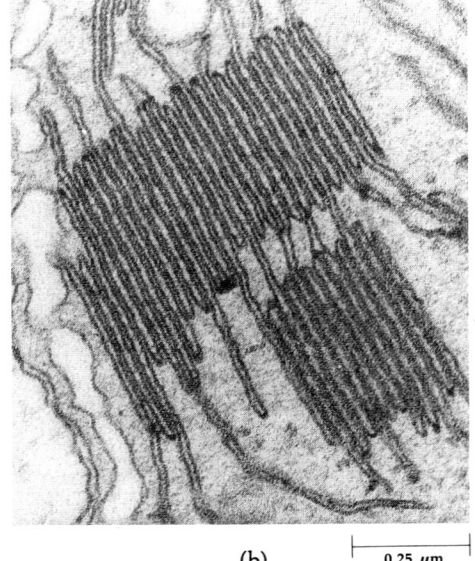

(b) |———| 0.25 μm

THE CHLOROPLAST: SITE OF PHOTOSYNTHESIS

The light and dark reactions of photosynthesis take place within the cellular organelles known as chloroplasts. A single leaf cell may contain 40 to 50 chloroplasts, and there are often 500,000 per square millimeter of leaf surface.

Chloroplasts, like mitochondria, are surrounded by two outer membranes. Within the inner membranes, or lamella, is a dense solution, the _stroma_, which is different in composition from the material surrounding the organelles in the cytoplasm. In the light microscope under high power, it is possible to see little spots of green within the chloroplasts of leaves. The early microscopists called these green specks _grana_ ("grains"), and this term is still in use.

Under the electron microscope, it can be seen that the grana are part of an elaborate system of membranes organized in parallel pairs. The pairs of membranes are joined at each end, forming a disk-shaped structure, or _thylakoid_ (from _thylakos_, the Greek word for "sac"). The grana are stacks of thylakoids. Some of the thylakoid membranes have extensions that interconnect the grana through the stroma that separates them.

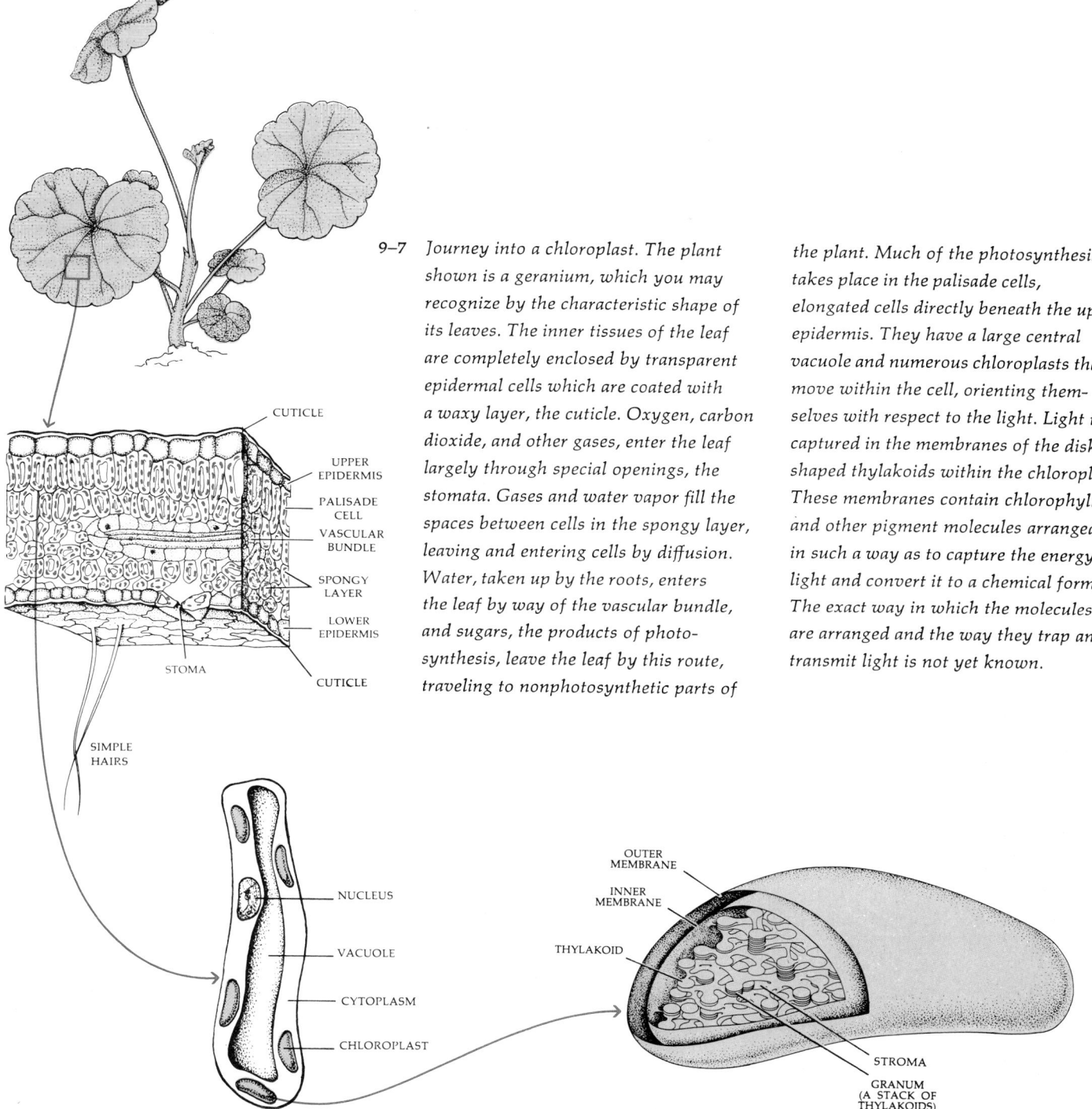

9–7 *Journey into a chloroplast. The plant shown is a geranium, which you may recognize by the characteristic shape of its leaves. The inner tissues of the leaf are completely enclosed by transparent epidermal cells which are coated with a waxy layer, the cuticle. Oxygen, carbon dioxide, and other gases, enter the leaf largely through special openings, the stomata. Gases and water vapor fill the spaces between cells in the spongy layer, leaving and entering cells by diffusion. Water, taken up by the roots, enters the leaf by way of the vascular bundle, and sugars, the products of photosynthesis, leave the leaf by this route, traveling to nonphotosynthetic parts of*

the plant. Much of the photosynthesis takes place in the palisade cells, elongated cells directly beneath the upper epidermis. They have a large central vacuole and numerous chloroplasts that move within the cell, orienting themselves with respect to the light. Light is captured in the membranes of the disk-shaped thylakoids within the chloroplast. These membranes contain chlorophyll and other pigment molecules arranged in such a way as to capture the energy of light and convert it to a chemical form. The exact way in which the molecules are arranged and the way they trap and transmit light is not yet known.

CUTICLE

UPPER EPIDERMIS

PALISADE CELL

VASCULAR BUNDLE

SPONGY LAYER

LOWER EPIDERMIS

STOMA

CUTICLE

SIMPLE HAIRS

NUCLEUS

VACUOLE

CYTOPLASM

CHLOROPLAST

OUTER MEMBRANE

INNER MEMBRANE

THYLAKOID

STROMA

GRANUM (A STACK OF THYLAKOIDS)

9–8 *The absorption spectrum of a pigment is measured by a spectrophotometer. This device directs a beam of light of each wavelength at the object to be analyzed and records what percentage of light of each wavelength is absorbed by the pigment sample as compared to a reference sample. Because the mirror is lightly (half) silvered, half of the light is reflected and half is transmitted. The photoelectric cell is connected to an electronic device that automatically records the percentage absorption at each wavelength.*

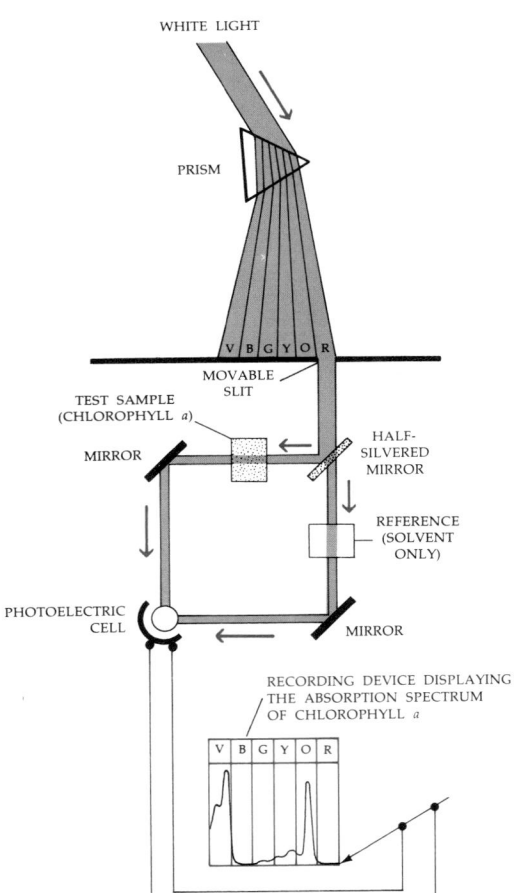

Light is captured in the chloroplasts by groups of pigments. A pigment is simply any substance that absorbs light. Some pigments absorb all wavelengths of light and so appear black. Some absorb only certain wavelengths and so transmit or reflect the wavelengths they do not absorb. Chlorophyll, the pigment which makes leaves green, absorbs light in the violet and blue wavelengths and also in the red; because it reflects green light, it appears green. Different pigments absorb light energy at different wavelengths. The absorption pattern of a pigment is known as the *absorption spectrum* of that substance. (See Figure 9–8.) There are several different kinds of chlorophyll that vary slightly in their molecular structure. The kinds found in the higher plants are chlorophylls *a* and *b*. Molecules with the same structure may have slightly different absorption spectra, depending on their position in the chloroplast in relation to other molecules.

Light is absorbed by pigment molecules. When pigments absorb light, electrons are boosted to a higher energy level with three possible consequences: (1) The energy may be converted to heat; (2) it may be remitted as light energy, a phenomenon known as "fluorescence"; (3) the energy may cause a chemical reaction, as happens in photosynthesis.

Different groups of plants and algae use various pigments in photosynthesis. However, with the exception only of the photosynthetic bacteria (which have their own special types of chlorophyll), all photosynthetic cells contain chlorophyll *a*. It is therefore generally believed that chlorophyll *a* is the pigment directly involved in the transformation of light energy to chemical energy. Most photosynthetic cells also contain a second type of chlorophyll—higher plants, as we noted, have chlorophyll *b*—and a representative of another group of pigments called the carotenoids. One of the carotenoids in higher plants is beta-carotene. The carotenoids are red, orange, or yellow pigments. In the green leaf, their color is masked by the chlorophylls, which are more abundant. In some tissues, however, such as those of a ripe tomato or the root of a carrot plant, the carotenoid colors predominate, as they do also when leaf cells stop synthesizing chlorophyll in the fall. The other chlorophylls and the carotenoids are able to absorb light at wavelengths different from that of the light absorbed by chlorophyll *a* and apparently can pass energy on to the chlorophyll *a*, thus extending the range of light available for photosynthesis. Whether or not a particular pigment can cause a chemical reaction depends not only on its physicochemical structure but also on its relationship with neighboring molecules.

If chlorophyll molecules are isolated in a test tube and light is permitted to strike them, they fluoresce. In other words, they absorb the energy of the sunlight momentarily and then bounce it back again at a different wavelength. None of the light is converted to any form of energy useful to living systems. However, chlorophyll molecules embedded in the fatty membrane of a thylakoid can trap this light energy in such a way that it can be converted to chemical energy. In the chloroplast, chlorophyll and other molecules are packed into what are known as photosynthetic units. Each unit contains from 250 to 400 molecules of pigment which can absorb light and pass it on to a reactive molecule. The energy absorbed by the reactive molecule is sufficient to force an electron

9–9 (a) *Chlorophyll* a *is a large molecule with a central core of magnesium held in a porphyrin ring. Attached to the ring is a long, insoluble carbon–hydrogen chain which serves to anchor the molecule in the internal membranes of the chloroplast. Chlorophyll* b *differs from chlorophyll* a *in having a CHO group in place of the encircled CH$_3$ group. Alternating single and double bonds, such as those in the chlorophylls, are common among pigments. (b) The estimated absorption spectra of chlorophyll* a *and chlorophyll* b. *(Prepared by Govindjee.)*

9–10 (a) *Related carotenoids. Cleavage of the beta-carotene molecule at the point shown yields two molecules of vitamin A. Oxidation of vitamin A yields retinene, the pigment involved in vision. All of this explains why you were told to eat your carrots. (b) The estimated absorption spectra of carotenoids in the chloroplast. (Prepared by Govindjee.)*

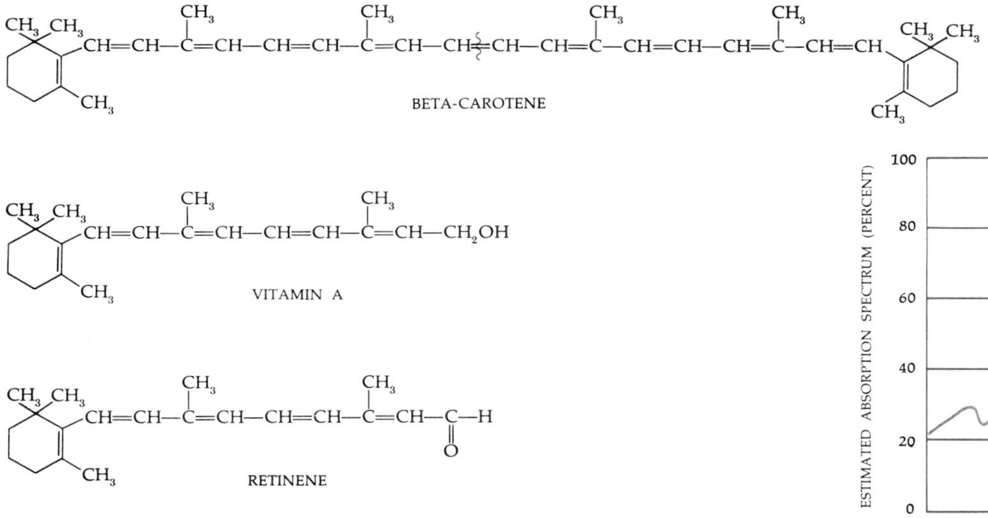

(a)

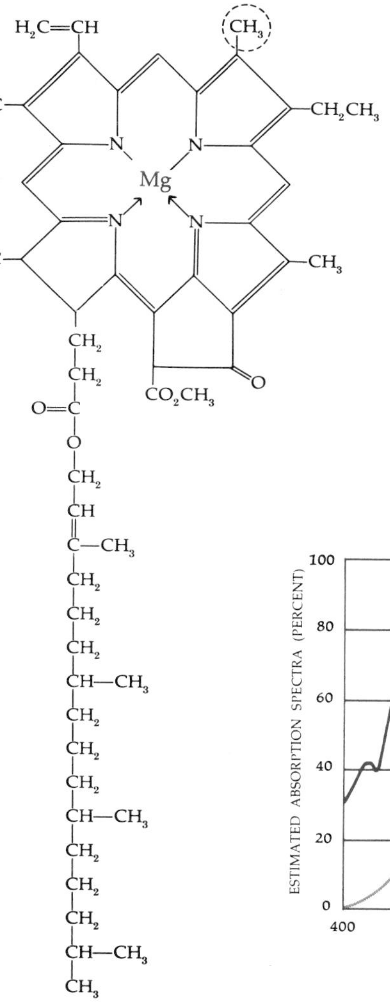

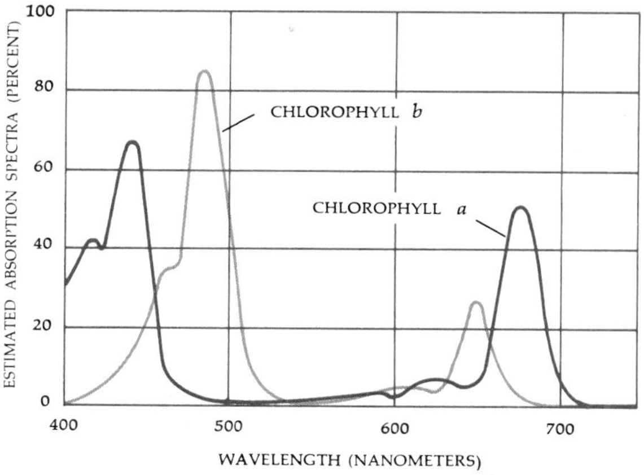

(b)

BETA-CAROTENE

VITAMIN A

RETINENE

(a)

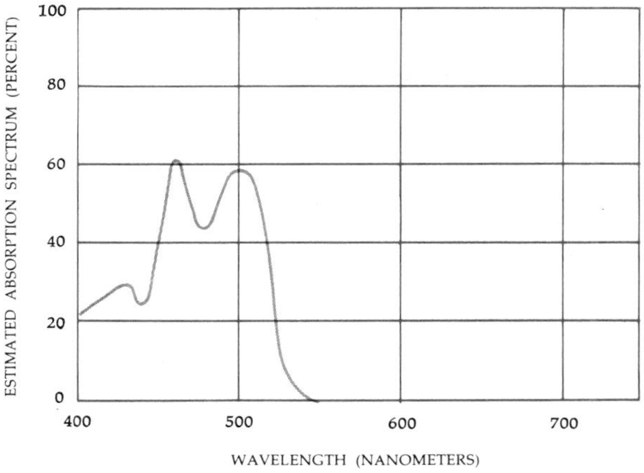

(b)

9–11 *Action spectrum for photosynthesis (upper curve). "Action spectrum" is a term that refers not to a pigment but to a light-requiring process, such as phototropism (the bending of a plant toward light), flowering, or photosynthesis. The action spectrum defines the relative effectiveness (per incident number of photons) of different wavelengths of light for a light-requiring process. Lower curves represent estimated absorption spectra for chlorophyll a, chlorophyll b, and carotenoids in the chloroplast. Note that the action spectrum of photosynthesis indicates that chlorophyll a, chlorophyll b, and the carotenoids all absorb light used in photosynthesis. (Prepared by Govindjee.)*

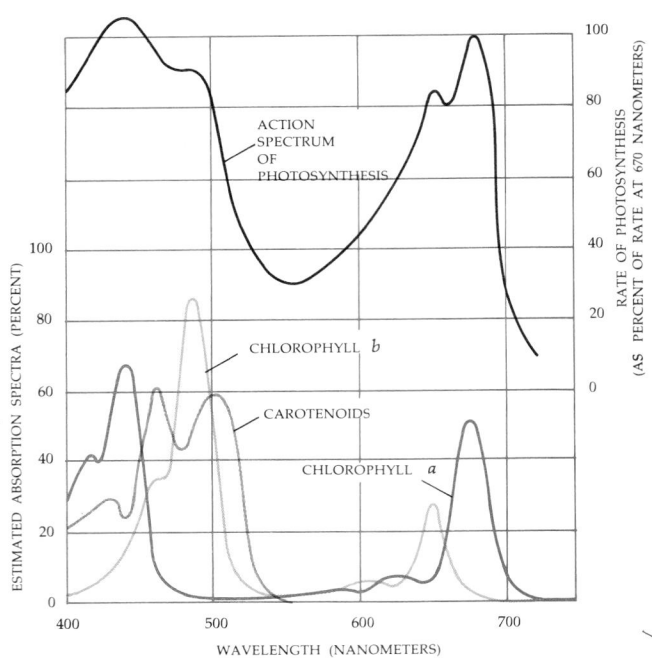

9–12 *Electron micrograph of the internal membrane surfaces of a thylakoid disk of a spinach chloroplast, following freeze-fracturing. In this process the sample is frozen and then fractured with a knife edge to yield cleavage surfaces, which are replicated by metal-casting. The image is that of the metal replica. The little particles have been called quantasomes, and it has been postulated, although not proved, that each is a functional unit containing a complete photochemical system.*

0.25 μm

not just to a higher energy level within the molecule but completely away from the molecule, like a spaceship going out of orbit and heading for the moon. The reactive molecule is thus oxidized and minus an electron, a situation sometimes referred to as having an electron "hole." (Remember the "hole"; it is important in the interpretation of subsequent events.)

The Pigment Systems

Recent work has revealed that the thylakoids of photosynthetic cells contain two different kinds of pigment systems. In green leaves, one pigment system, Pigment System I, contains a larger proportion of chlorophyll a to chlorophyll b than Pigment System II. Carotenoids of different types are present in both systems. These pigment systems can be separated from one another by centrifuging (which separates particles of different weights), and they are believed to be the particles visible on the inner surface of thylakoids under high-power electron microscopy (Figure 9–12).

In Pigment System I, the reactive molecule is a form of chlorophyll a known as P_{700} because one of the peaks of its absorption spectrum is at 700 nanometers, slightly to the right of the usual chlorophyll a peak. When P_{700} is oxidized, it bleaches, which is how it was detected. No one has managed to detect P_{700} in isolation. It therefore seems that P_{700} is not an unusual kind of chlorophyll but rather chlorophyll a which has unusual properties because of its arrangement in the membrane and its position in relation to other molecules. Pigment System II also contains a specialized chlorophyll molecule, which is capable of passing an electron on to an electron acceptor, although the identity of this reactive molecule is not yet established. Thus each pigment system forms part of a different photochemical system.

MODEL OF LIGHT REACTIONS

Figure 9–13 shows the present model of how the two photochemical systems work together in photosynthesis. Light energy first enters Pigment System II where it is responsible for two important events: (1) electrons are boosted uphill from a reactive molecule, and (2) water is split into protons, electrons, and oxygen gas. As the electrons are removed, they are passed uphill at increasingly higher energy levels to a primary electron acceptor. (The electron acceptors in these reactions, some of which are cytochromes, some of which are other electron-transfer molecules whose identities are known, and some of which are unidentified, are all indicated by circles.) The electrons then pass downhill along an electron transport chain similar to the electron transport chain in the mitochondria. In the course of this passage, ATP is formed from ADP. This process is known as _photophosphorylation_ and is comparable to the oxidative phosphorylation process in cellular respiration. These systems probably evolved separately (with Photochemical System I coming first), and Photochemical System I, as we shall see, can probably operate separately. In general, however, they work together.

9–13 *Light energy trapped in the reactive molecule of Pigment System II boosts electrons uphill from chlorophyll to a primary electron acceptor. These electrons are replaced by electrons pulled away from water molecules, which then fall apart, producing protons and oxygen gas. The electrons are passed from the electron acceptor along an electron transport chain to a lower energy level, the reaction center of Pigment System I. As they pass along the electron transport chain, some of their energy is packaged in the form of ATP. Light energy absorbed by Pigment System I boosts the electrons to another primary electron acceptor. From this acceptor, they are passed via other electron carriers to $NADP_{ox}$ to form $NADP_{red}$. The electrons removed from Pigment System I are replaced by those from Pigment System II. ATP and $NADP_{red}$ represent the net gain from the light reactions.*

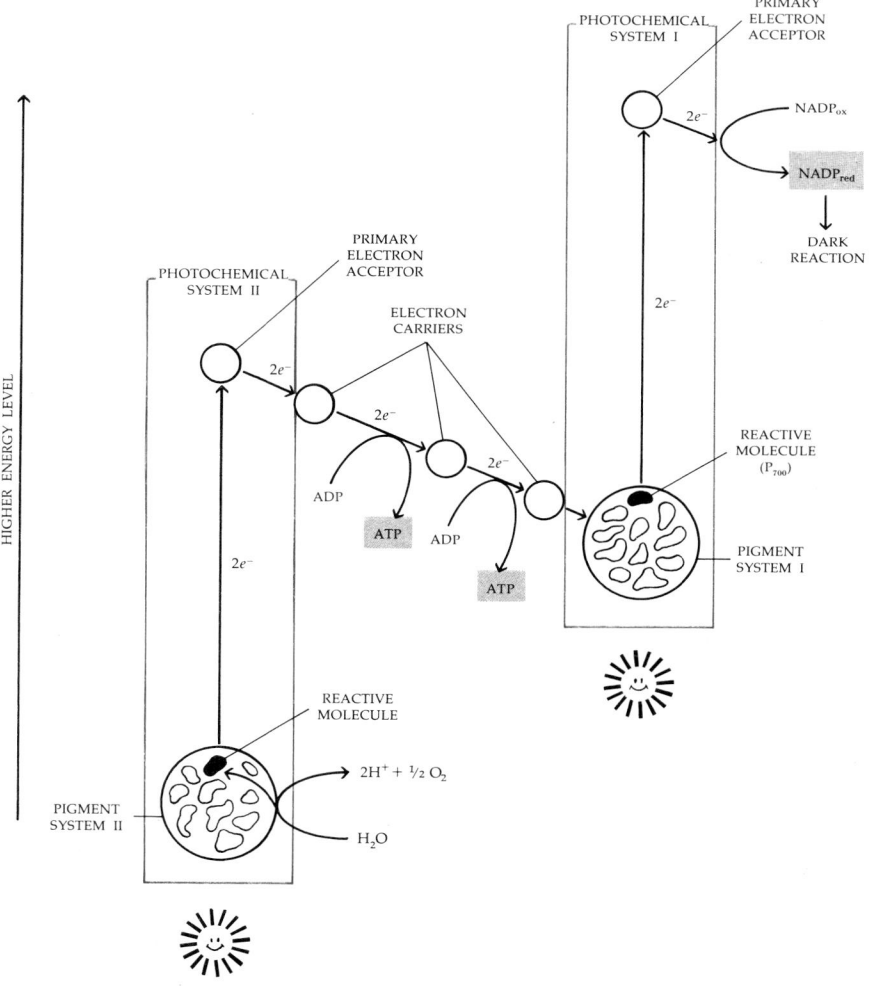

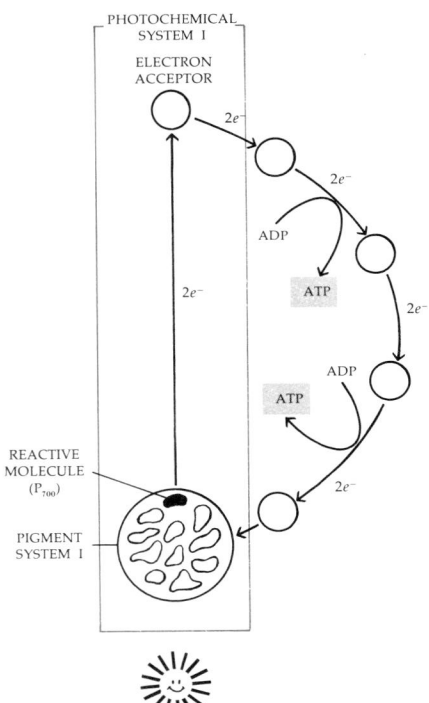

9–14 *Cyclic electron flow bypasses Photochemical System II and requires only Photochemical System I. ATP is produced from ADP but oxygen is not released and NADP is not reduced.*

PHOTOCHEMICAL SYSTEM I

ELECTRON ACCEPTOR

2e⁻

ADP

ATP

2e⁻

ATP

2e⁻

ADP

ATP

2e⁻

REACTIVE MOLECULE (P₇₀₀)

PIGMENT SYSTEM I

In Photochemical System I, light energy boosts electrons from P_{700} to an electron carrier from which they are passed downhill to an electron acceptor molecule NADP. NADP closely resembles the related coenzyme NAD, which functions in cellular respiration (it differs only in having an additional phosphate group). The gain from these two steps in photosynthesis is represented by the ATP molecules (the exact number is not certain) and the $NADP_{red}$. Oxygen is a by-product. The $NADP_{red}$ is the chief source of the reducing power used to reduce carbon to organic compounds.

Why are there two photochemical systems rather than one? The answer seems to lie in the problem of the electron "holes." Suppose for a moment that System I were isolated from System II. There would be no way to fill the "holes" left in the P_{700} molecule in System I. This function is now carried out by the electrons boosted up from System II. And what fills the holes in System II? They are filled by electrons from the water molecules. When the electrons pass from the water molecule to the chlorophyll molecule (which has a greater electron affinity), the water molecule falls apart, releasing oxygen gas.

Thus all the pieces fit together.

As we mentioned previously, there is also some evidence that System I can work independently. When this occurs, no $NADP_{red}$ is formed. In this process, called *cyclic electron flow*, electrons are boosted from P_{700} to an electron acceptor and from there pass downhill through a series of cytochromes back into the pigment system. ATP is produced in the course of the passage. It is believed that the most primitive photosynthetic mechanisms did, indeed, work in this way. These early photosynthetic mechanisms probably produced ATP; the fact that ATP is an energy carrier in every known kind of living system testifies to its very early appearance in the history of evolution. They did not give off oxygen, however, and it is not known if they reduced carbon to sugar.

Summary of Light Reactions

The reactions that we have just described are the "light reactions" of photosynthesis. In the course of these reactions, as we saw, light energy is converted to electrical energy—the flow of electrons—and the electrical energy is converted to chemical energy stored in the bonds of $NADP_{red}$ and ATP.

In the words of Nobel laureate Albert Szent-Györgi: "What drives life is . . . a little electric current, kept up by the sunshine."

THE DARK REACTIONS

In the second stage of photosynthesis, the energy generated by the light reactions is used to incorporate carbon into organic molecules. These reactions take place in the stroma of the chloroplasts and do not need light, and so they are often referred to as the "dark reactions" (although they can take place equally well in the light). The dark reactions involve a cycle called the Calvin cycle. It is analogous to the Krebs cycle (page 132) in that, in each turn of the cycle, the starting product is again regenerated. The dark reactions convert the chemical energy produced by the light reactions to forms more suitable for storage and transport and also build the basic carbon structures from which all of the other organic molecules of living systems are produced. (This last process is known as the *fixation of carbon*.)

PHOTOSYNTHESIS AND THE COMING OF OXYGEN

The first photosynthetic organism probably appeared almost 3 billion years ago. Before the evolution of photosynthesis, the physical characteristics of the planet itself were by far the most powerful forces in shaping the course of natural selection. But with the evolution of photosynthesis, organisms began to change the face of the planet, and they have continued to do so, at an ever-increasing rate, up to the present day.

The atmosphere in which the first cells evolved probably consisted largely, as we noted previously, of ammonia and methane and other carbon gases. Gradually, methane and ammonia were broken down, and their hydrogen atoms, which are very light, were released into space as hydrogen gas (H_2). These earliest forms of life were, of course, adjusted to living in an environment without free oxygen. Their energy probably came from glycolysis. Anaerobic glycolysis (page 130) might have resulted in the accumulation of carbon dioxide in the atmosphere. Thus carbon dioxide and nitrogen replaced methane and ammonia as significant components of the atmosphere.

Then, it is hypothesized, there slowly evolved photosynthetic organisms that used carbon dioxide as their carbon source and released oxygen, as do most modern photosynthetic forms. As photosynthetic organisms multiplied, free oxygen began to accumulate. In the struggle for survival under these changing conditions, cell species arose for which oxygen was not a poison but a requirement of existence.

From our knowledge of modern anaerobes, it would seem as though oxygen-consuming organisms have an advantage over those that do not use oxygen: as we saw in the previous chapter, a higher yield of energy can be extracted per molecule from the oxidation ("burning") of carbon-containing compounds than from anaerobic fermentation processes in which fuel molecules are not broken down completely. Energy resulting from the use of oxygen by cells made possible the development of increasingly active, increasingly complex organisms. Without oxygen, the higher forms of life that now exist on Earth could not have evolved.

The oxygen now present on the planet continues to depend on photosynthesis. In fact, all the oxygen of the biosphere is renewed about every 2,000 years by photosynthetic cells.

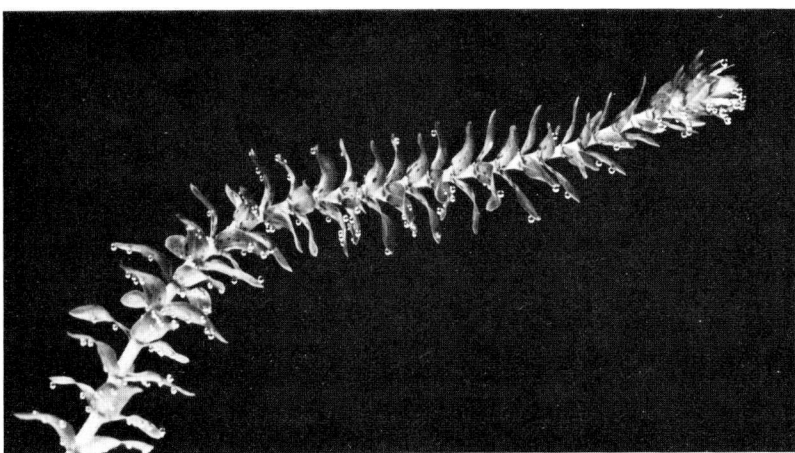

Oxygen produced by photosynthesis forms gas bubbles on this pondweed, Elodea, *growing under water.*

9–15 *Calvin and his collaborators briefly exposed photosynthesizing algae to radioactive carbon dioxide ($^{14}CO_2$). They found that the radioactive carbon is first incorporated into ribulose diphosphate (RuDP), which then immediately splits to form two molecules of phosphoglycerate (PGA). The radioactive carbon atom, indicated in color, next appears in one of the two molecules of PGA. This is the first step in the Calvin cycle.*

The Calvin Cycle

In the Calvin cycle (called after its discoverer Melvin Calvin of the University of California at Berkeley), the starting (and ending) compound is a five-carbon sugar with two phosphates attached, ribulose diphosphate (RuDP). The cycle begins when carbon dioxide enters the cycle and is incorporated into RuDP, which then splits to form two molecules of phosphoglycerate (Figure 9–15).

The complete cycle is diagramed in Figure 9–16. As in the Krebs cycle, each step is regulated by a specific enzyme. At each full turn of the cycle, a molecule of carbon dioxide enters the cycle, is reduced, and a molecule of RuDP is regenerated. Six revolutions of the cycle, with the introduction of six atoms of carbon, are necessary to produce a six-carbon sugar, such as glucose. The overall equation is

$$6RuDP + 6CO_2 + 18ATP + 12NADP_{red} \longrightarrow$$
$$6RuDP + glucose + 18P_i + 18ADP + 12NADP_{ox}$$

As you can see, the immediate product of the cycle itself is glyceraldehyde phosphate. This same sugar–phosphate is formed when the fructose diphosphate molecule is split at the fourth step in glycolysis (page 127). These same steps, using the energy of the phosphate bond, can be reversed to form glucose from glyceraldehyde phosphate.

Recent research indicates that the Calvin cycle is not the only pathway for the fixation of carbon. In some of the grasses, such as sugar cane and maize, carbon is incorporated into oxaloacetic acid, a four-carbon compound, which you will remember as one of the intermediates in the Krebs cycle. There seem to be a number of plants in the grass family that incorporate carbons into four-carbon rather than three-carbon compounds, and the high productivity of this group of plants indicates that this may be a more efficient biosynthetic

CH₂—O—(P)
|
C=O
|
CHOH   CO₂   H₂O
|
CHOH
|
CH₂—O—(P)
RIBULOSE
DIPHOSPHATE
(RuDP)

CH₂—O—(P)
|
CHOH
|
C=O
|
O⁻

CH₂—O—(P)
|
CHOH
|
C=O
|
O⁻
PHOSPHO-
GLYCERATE
(PGA)

9–16 *Summary of the Calvin cycle. Three molecules of ribulose diphosphate (RuDP), a five-carbon compound, are combined with three molecules of carbon dioxide, yielding six molecules of phosphoglycerate, a three-carbon compound. These are converted to six molecules of glyceraldehyde phosphate. Five of these three-carbon molecules are combined and rearranged to form three five-carbon molecules of RuDP. The "extra" molecule of glyceraldehyde phosphate represents the net gain from the Calvin cycle. The energy that "drives" the Calvin cycle is in the form of ATP and NADP$_{red}$, produced by the light reactions.*

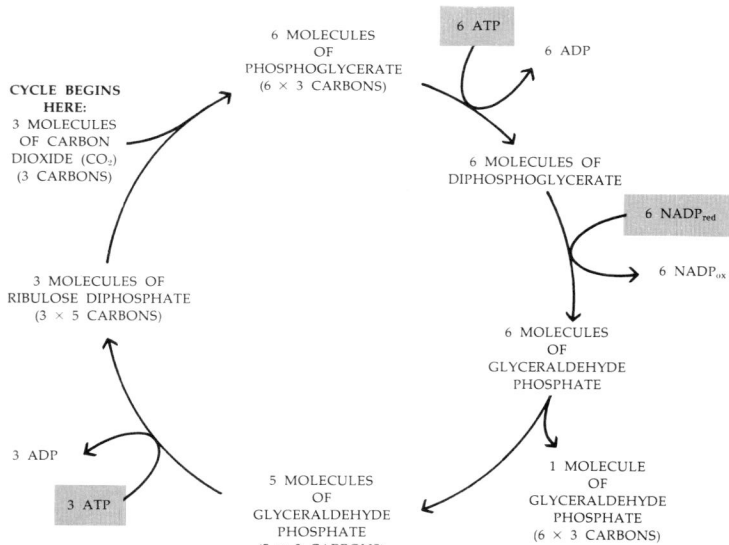

6 MOLECULES OF PHOSPHOGLYCERATE (6 × 3 CARBONS)

6 ATP

6 ADP

CYCLE BEGINS HERE: 3 MOLECULES OF CARBON DIOXIDE (CO₂) (3 CARBONS)

6 MOLECULES OF DIPHOSPHOGLYCERATE

6 NADP$_{red}$

6 NADP$_{ox}$

3 MOLECULES OF RIBULOSE DIPHOSPHATE (3 × 5 CARBONS)

6 MOLECULES OF GLYCERALDEHYDE PHOSPHATE

3 ADP

3 ATP

5 MOLECULES OF GLYCERALDEHYDE PHOSPHATE (5 × 3 CARBONS)

1 MOLECULE OF GLYCERALDEHYDE PHOSPHATE (6 × 3 CARBONS)

Table 9–1 *Summary of the Stages of Photosynthesis*

	Conditions	Where	What Appears to Happen	Results
Light reactions	Light	Thylakoids	Light striking Pigment System II boosts electrons uphill. (Electrons replaced by electrons from water molecules which are lysed, releasing oxygen gas.) Electrons then pass downhill along the electron transport chain, forming ATP, to Photochemical System I. Light hits Pigment System I, boosts electrons uphill to an electron acceptor from which they are passed to $NADP_{ox}$ to form $NADP_{red}$. Electrons are replaced in Pigment System I by electrons from Pigment System II.	Energy of light is converted to chemical energy stored in bonds of ATP + $NADP_{red}$.
Dark reactions	Do not require light	Stroma	Calvin cycle. $NADP_{red}$ and ATP formed in light reactions are used to reduce carbon dioxide. The cycle yields glyceraldehyde phosphate from which glucose or other organic compounds can be formed, or the glyceraldehyde phosphate can be oxidized to form ATP.	Chemical energy of light reactions stored in ATP + $NADP_{red}$ is used to incorporate carbon into organic molecules.

pathway. All known pathways, however, accomplish the same principal objective: the use of $NADP_{red}$ and ATP energy generated by photosynthesis to reduce carbon and fix it into organic compounds.

THE ROLE OF PHOTOSYNTHESIS

As the laws of thermodynamics instruct us, energy runs downhill, and the organization and activity characteristic of living systems can be maintained only by taking in energy from an outside source and releasing it in a less useful or less highly ordered form. The source of that energy for life here on Earth is the series of thermonuclear reactions taking place at the heart of the sun. This energy enters the biological world through the activities of individual photosynthetic cells.

SUMMARY

In photosynthesis, light energy is converted to chemical energy, and carbon is fixed into organic compounds. The generalized equation for this reaction is:

$$CO_2 + 2H_2A + \text{light energy} \longrightarrow (CH_2O) + H_2O + 2A$$

in which H_2A stands for water or some other substance from which electrons can be removed.

Photosynthesis takes place within cellular organelles known as chloroplasts. These organelles are surrounded by two membranes. Inside the second membrane is a solution of organic compounds and ions known as the stroma, and a complex internal membrane system consisting in higher plants of fused pairs of membranes that form sacs called thylakoids. The pigments and other molecules responsible for capturing light are located in and on these membranes.

Pigments are compounds that absorb light energy. The pigments involved in photosynthesis include the chlorophylls and the carotenoids. Light absorbed by pigments boosts their electrons to a higher energy level. Because of the way the pigments are packed into the membranes, they are able to transfer this energy to trapping molecules, probably chlorophyll packed in a particular way. Two pigment systems have been identified in thylakoids.

In the currently accepted model of the light reactions in photosynthesis, light energy first strikes Pigment System II, which contains several hundred molecules of chlorophyll a and chlorophyll b. Electrons are passed uphill to an electron acceptor from the reactive molecule. The electron holes are filled from the lysis and oxidation of water. The electrons then pass downhill to Pigment System I along an electron transport chain, in the course of which ATP is generated (photophosphorylation). Light absorbed in Pigment System I results in the ejection of an electron from the reactive molecule chlorophyll P_{700}. The resulting electron holes left in the reactive molecule are filled by the electrons from Pigment System II. The electrons are accepted by the electron carrier molecule, NADP. The energy yield from the light reactions is represented by molecules of $NADP_{red}$ and the ATP formed by photophosphorlylation.

In the dark reactions, which take place in the stroma, the $NADP_{red}$ and ATP produced in the light reactions are used to reduce carbon dioxide to organic carbon. This is accomplished by means of the Calvin cycle. In the Calvin cycle, a molecule of carbon dioxide is combined with the starting material, a five-carbon sugar called ribulose diphosphate. At each turn of the cycle, one carbon atom enters the cycle. Three turns of the cycle produce a three-carbon molecule, glyceraldehyde phosphate. Two molecules of gylceraldehyde phosphate (six turns of the cycle) can combine to form a glucose molecule. At each turn of the cycle, RuDP is regenerated. The glyceraldehyde can also be used as starting material for other organic compounds needed by the cell.

By this process, a flow of energy is created from the sun through living systems.

QUESTIONS

1. Define the following terms: anaerobe, absorption spectrum, action spectrum, and pigment.
2. Sketch a chloroplast and label its structures. Compare your sketch with the one on page 147.
3. Describe, in general terms, the events of photosynthesis. Compare your description with Table 9–1.
4. If one writes the equations for glycolysis and respiration starting with glucose, they appear to be merely the reverse of those for photosynthesis. In what ways are these processes *not* simply reversals of photosynthesis?

Chapter 10

How Cells Divide

The cell is the basic unit of life, and living systems grow by the multiplication of cells. In one-celled organisms, cells grow to a certain size by assimilating materials from the environment and synthesizing these materials into new structural and functional molecules. When such a cell reaches a certain size, it divides. The two daughter cells, each about half the size of the original mother cell, then begin growing again. In a one-celled organism, such as a paramecium, cell division may occur every day or even every few hours, producing a succession of organisms that are potentially immortal. In many-celled plants and animals, cell division is the means by which the organism grows and also by which injured or worn-out tissues are replaced and repaired. In all of these instances, the new cells produced are structurally and functionally similar both to the parent cell and to one another.

Cell division in eukaryotes consists of two overlapping stages, _mitosis_ and _cytokinesis._ In mitosis, the nuclear membrane breaks down, the chromosomes, which have previously been duplicated, are divided equally, and two new nuclei are formed. In cytokinesis, the cytoplasm of the parent cell is divided into two parts, each containing one of the nuclei.

At the end of cell division, the two new daughter cells will always contain exactly the same complement of genetic material. They will also usually be about the same size and have about the same numbers of the different organelles. The precise division of the genetic material is the result of mitosis; the division of the cytoplasm is the result of cytokinesis.

THE CELL CYCLE

Dividing cells pass through a regular sequence of events, known as the cell cycle. Completion of the cycle requires varying periods of time, depending both on the type of cell and external factors, such as temperature or available nutrients. Whether it lasts an hour or a day, however, the relative amount of time spent at each phase is about the same.

The S (synthesis) phase of the cell cycle is the period during which the genetic material (DNA) is duplicated. G (gap) phases precede and follow the S phase. The G_1 period occurs after mitosis and precedes the S phase; the G_2 period follows the S phase and occurs before mitosis. The G and S phases together are referred to as interphase. Mitosis, which occupies the rest of the cell cycle (amounting to only 5 to 10 percent of it), will be discussed in detail later in this chapter.

10–1 *A cell divides. The dark bodies are chromosomes, the carriers of the genetic information. The chromosomes have duplicated and moved apart. Each set of chromosomes is an exact copy of the other. Thus the two new cells, the daughter cells, will contain the same hereditary material.*

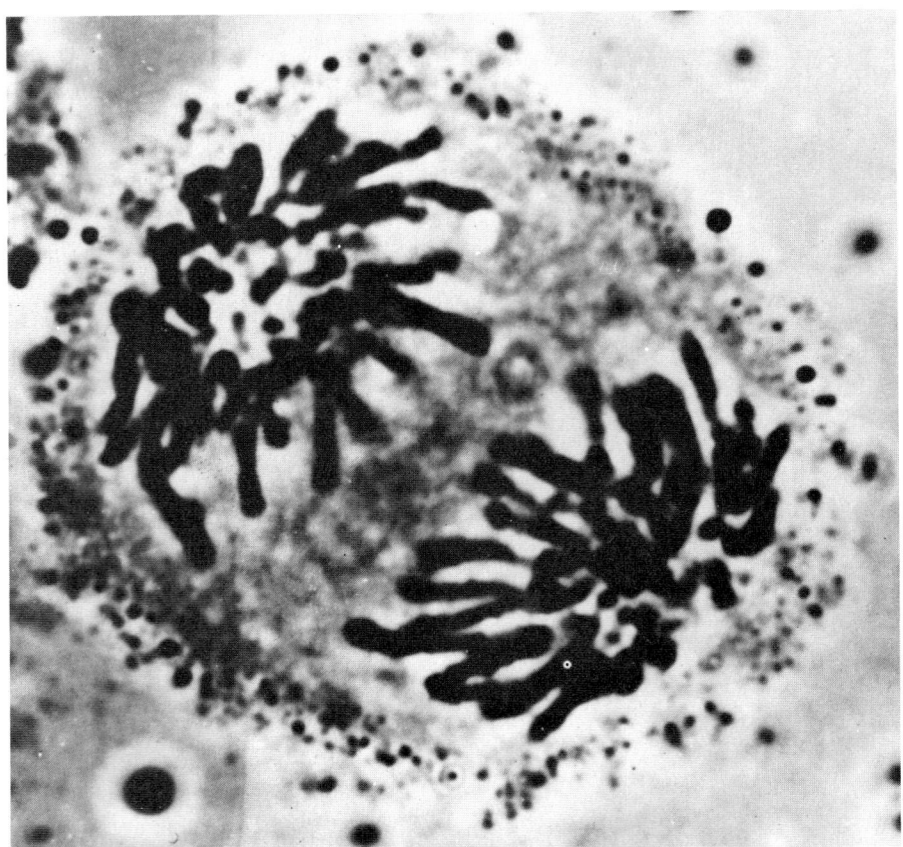

10–2 *The cell cycle. Dividing cells go through four principal phases, including the phases of mitosis (shown in color) and the S phase, during which the chromosomes are duplicated. Separating mitosis and the S phase are two G phases. The first of these (G_1) is a period of general growth and replication of cytoplasmic organelles. During the second (G_2), structures directly associated with mitosis, such as the spindle fibers, are synthesized. After the G_2 phase comes mitosis, which is, in turn, divided into four phases. Mitosis actually occupies only 5 to 10 percent of the cell cycle.*

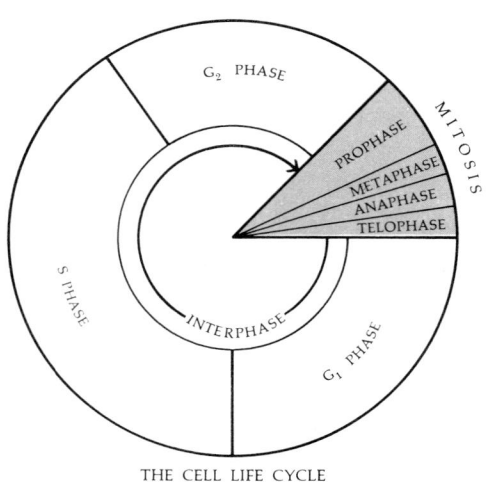

THE CELL LIFE CYCLE

The G_1 phase, between mitosis and chromosome synthesis, is principally a period of growth of the cytoplasmic material, including all the various organelles. Also, during this G_1 period, according to current hypothesis, substances are synthesized which either inhibit or stimulate the S stage and the rest of the cycle, thus determining whether or not cell division will occur. During the G_2 phase, structures involved directly with mitosis, such as the spindle fibers, are synthesized.

Some cells pass through successive cell cycles repeatedly. This group includes the one-celled organisms and certain cells in growth centers of both plants and animals. Some highly specialized cells, such as nerve cells, lose their capacity to replicate once they are mature. A third group of cells retains the capacity to divide but does so only under special circumstances. Cells in the human liver, for example, do not ordinarily divide, but if a portion of the liver is removed surgically, the remaining cells (even if only about a third of the total remain) continue to replicate themselves until the liver reaches its former size. Then they stop. Further knowledge of the control mechanisms involved would not only be interesting biologically but might be of great importance in the control of cancer. Cancer cells differ from normal cells largely in that they keep on dividing at the expense of the host tissues.

10–3　*Photomicrograph of dividing cells in the tip of an onion root. By comparing these to the drawings on the following pages, you should be able to identify the various phases of mitosis.*

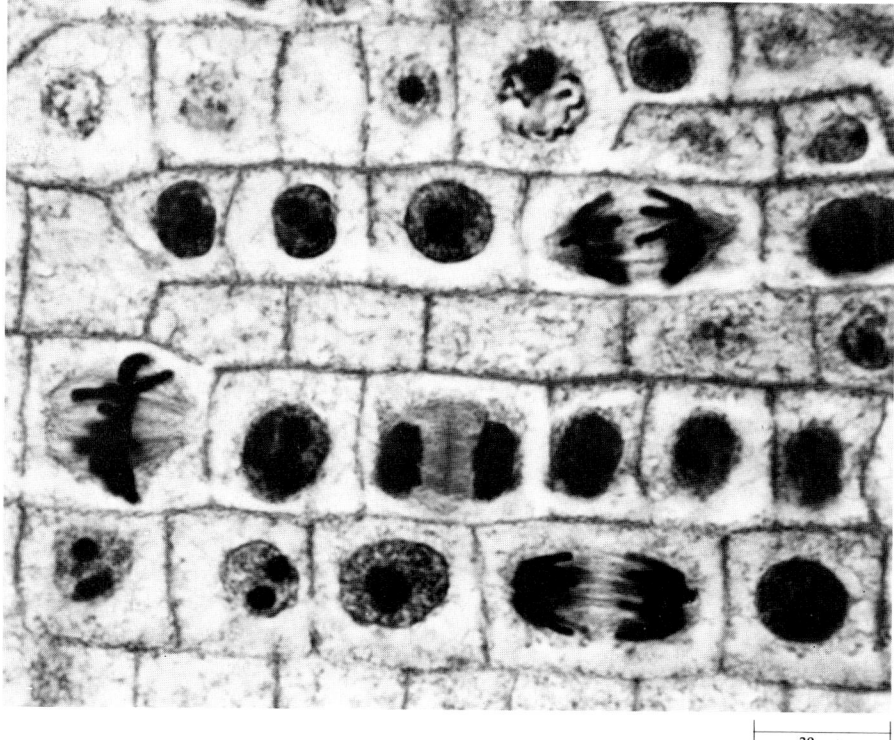

30 μm

10–4　*Diagram of a chromosome at the beginning of mitosis. The chromosomal material has replicated during interphase so that each chromosome now consists of two identical parts, called chromatids. The chromatids are joined together at their kinetochores.*

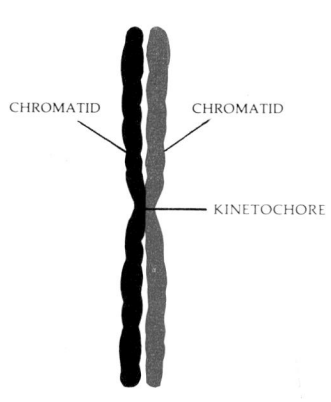

CHROMATID　　　CHROMATID

KINETOCHORE

CHROMOSOMES

The genetic material of eukaryotic cells is in the form of chromosomes.* Chromosomes are composed of nucleic acid (DNA) and proteins. The DNA is in the form of a long double helix, but all of the details of the way in which the helix is folded into the chromosome and the attachment of the proteins to it (and, indeed, the function of these proteins) are not known. When cells are not dividing, the chromosomal material is dispersed and is visible, if at all, only as threadlike strands, the *chromatin.* As we shall see in Section 2, it appears that the genetic material can play its role in controlling and determining cell functions only when it is in this extended form. The most prominent body in the nucleus at this time, as we have mentioned previously, is the nucleolus, which is the site of biosynthesis of RNA for the ribosomes.

At the beginning of mitosis, the chromosomes slowly coil and condense into a compact form; this condensation would appear to be necessary for the complex movements of the chromosomes at mitosis.

When the chromosomes become visible at mitosis, each chromosome consists of two replicas. Each replica is called a *chromatid,* and the two identical chromatids remain joined together like a pair of Siamese twins. Their point of attachment is a constricted area common to both chromatids known as the *kinetochore,* or centromere. (See Figure 10–4.)

* Chromosomes ("colored bodies") were given this name by the early microscopists because they took up certain kinds of stains used in preparing specimens for microscopy and so were first seen as brightly colored. Chromatin means simply "colored thread."

MITOSIS

The process of mitosis is conventionally divided into four phases: prophase, metaphase, anaphase, and telophase; of these, prophase is usually much the longest. If a mitotic division takes 10 minutes (which is about the minimum time required), during about six of these minutes the cell will be in prophase. The schematic drawings that follow show mitosis as it takes place in an animal cell.

In early prophase, the chromatin condenses, and the individual chromosomes begin to become visible. If you look carefully, you can see that each chromosome consists of two duplicate chromatids pressed closely together longitudinally and connected at the kinetochore (represented by a white dot). Two centriole pairs can be seen at one side of the nucleus, outside the nuclear envelope. Each pair consists of one mature centriole and a smaller daughter centriole lying at right angles to it.

10–5   *Early prophase in an onion cell. The chromosomes have begun to condense.*

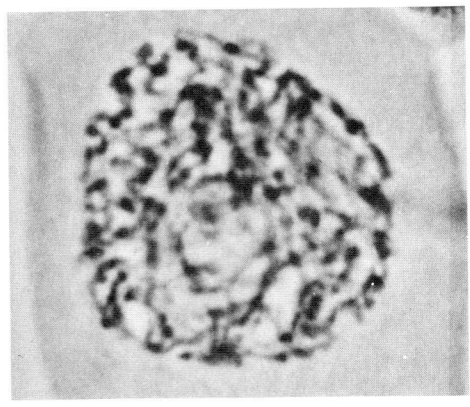

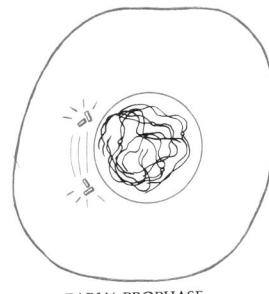

During prophase, the centriole pairs move apart. Between the centriole pairs, forming as they separate (or perhaps separating as they form), are spindle fibers made up of microtubules and other proteins. Additional fibers, known collectively as the *aster*, radiate outward from the centrioles. The nucleolus usually disperses, disappearing from view. The nuclear envelope breaks down, marking the end of prophase.

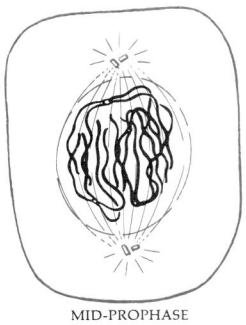

MID-PROPHASE

By the end of prophase, the centriole pairs are at opposite ends of the cell, and the members of each pair are of equal size. The spindle is fully formed. It is a three-dimensional football-shaped structure consisting of three groups of

protein fibers: (1) astral rays, (2) continuous fibers reaching from pole to pole, and (3) shorter fibers that are attached to the kinetochores (centromeres) of the individual chromosomes.*

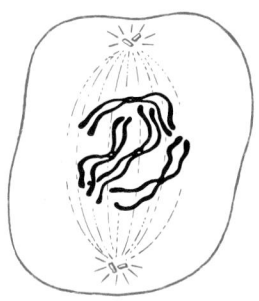

EARLY METAPHASE

During early metaphase the chromosomes move back and forth within the spindle, apparently maneuvered by the spindle fibers, as if they were being tugged first toward one pole and then the other. Finally they become arranged precisely at the midplane (equator) of the cell. This marks the end of metaphase.

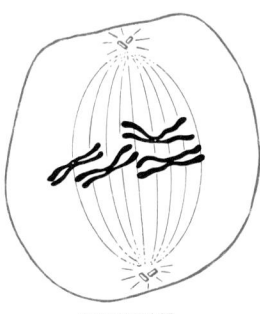

METAPHASE

At the beginning of anaphase, each kinetochore divides, and the chromatids (now chromosomes) separate. The two duplicate chromosomes pull apart slowly, apparently drawn toward the poles by the spindle fibers. The kinetochores move first, while the "arms" of the chromosomes drag behind.

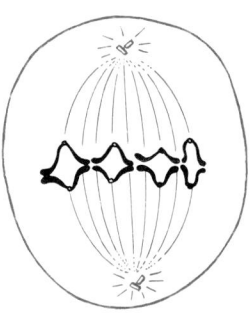

EARLY ANAPHASE

10–6   *Middle anaphase in a cell from the seed of a South African lily. The chromosomes have moved halfway toward the poles. The three-dimensional effect is a result of a special interference-contrast technique.*

| 20 μm |

* In counting, it is often difficult to know whether to count a chromosome that has duplicated but has not divided as 1 or 2. It is customary to count such a chromosome as 1. The trick is to count kinetochores.

161   *How Cells Divide*

During anaphase, the identical sets of chromosomes move toward the opposite poles of the spindle.

LATE ANAPHASE

10–7 *Early telophase in an onion cell.*

By the beginning of telophase, the chromosomes have reached the opposite poles and cytokinesis has begun. The spindle disperses. During this period a new daughter centriole arises from each mature centriole so that each new cell has two centriole pairs.

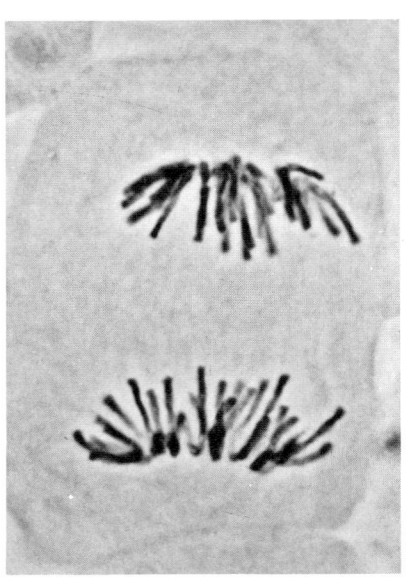

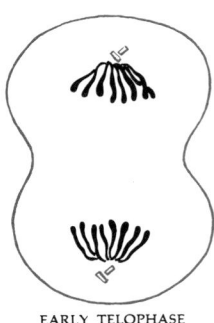

EARLY TELOPHASE

During late telophase, the nuclear envelopes re-form, and the chromosomes once more become diffuse.

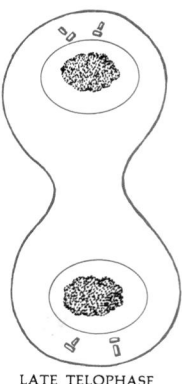

LATE TELOPHASE

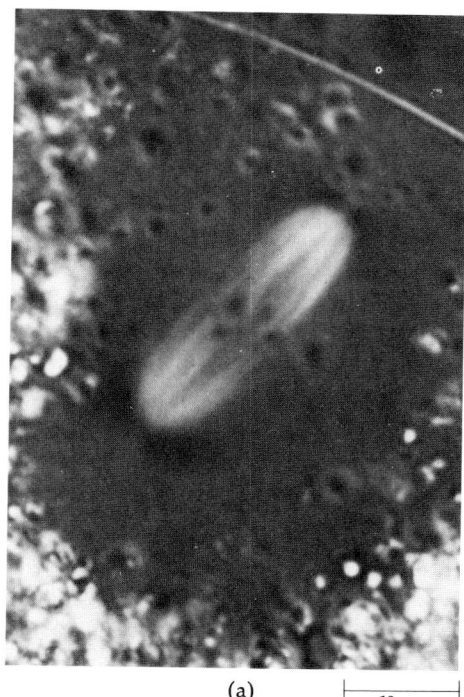

(a)

├─────── 10 μm ───────┤

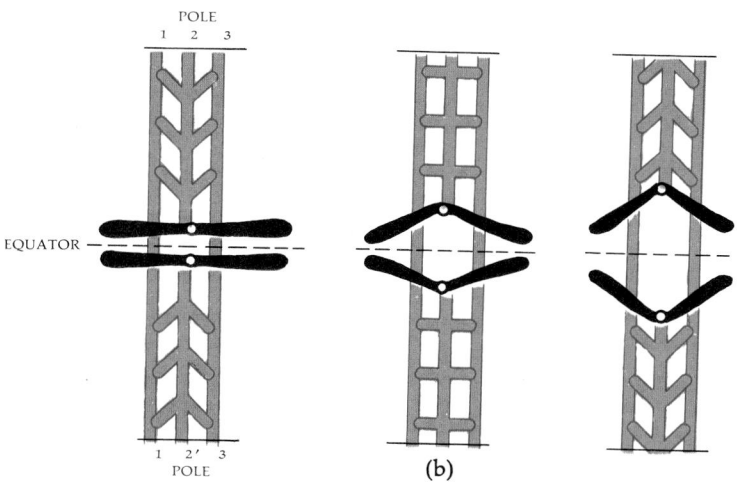

(b)

0–8   (a) *This unusual micrograph of a spindle in an egg cell of a marine worm emphasizes the spindle's three-dimensional qualities. The bright streaks are the spindle fibers. The chromosomes appear as indistinct gray bodies near the equator of the spindle. The curved line at the top is part of the cell membrane.*
(b) *A model of spindle fiber movement at mitosis. Fibers 1 and 3 reach pole to pole. The fibers marked 2 and 2' are attached to the kinetochores of duplicate chromosomes. During anaphase, the fibers move by "walking" along adjacent parallel fibers.*

The Spindle

The spindle, as we noted previously, contains microtubules and other protein fibers, some of which are attached to the kinetochores and some of which stretch from pole to pole. Although the fibers lengthen during prophase and then shorten during anaphase, they do not appear to get thinner or thicker. This suggests that they do not contract but that new material is removed from the fiber as the spindle changes shape. A recent model of spindle action has been proposed in which the microtubules have arms which walk along one another (Figure 10–8). (This is, of course, analogous to the proposed model of ciliary action described in Chapter 6, page 102.) This model of spindle action is still very speculative.

The Centrioles

It is tempting to think that the centrioles organize the microtubules of spindle fibers—spinning them out somewhat as a spider spins out silk. It appears, in fact, that basal bodies, which are structurally identical to centrioles, do organize the microtubules of flagella and cilia. However, much evidence contradicts the hypothesis that centrioles are the site of organization or production of spindle microtubules. For one thing, plant cells that do not have centrioles or basal bodies also form spindles with microtubules. No asters appear, but the spindles are as well formed as those in animal cells. Also, in some animal cells, it is possible to separate the centrioles from the spindles—yet cell division proceeds normally. It has been suggested that the spindles, instead of forming from the centrioles, serve to separate the centrioles, pushing them apart. But then what is the function of centrioles? This question remains to be answered.

Although the way in which the chromosomes move at mitosis is almost as mysterious to modern microscopists as it was to those who first observed it almost a hundred years ago, the results of mitosis are very clear. The duplicated chromosomes have been precisely divided among the two daughter cells, thus maintaining the long thread of genetic continuity.

CYTOKINESIS

Cytokinesis, the division of the cytoplasm, usually but not always accompanies nuclear division. It is preceded by an increase in the size of the cell, in the number of molecules that make up the cytoplasm, and in the number of organelles. Some of these organelles are synthesized from simpler materials by the cell. They include the membranous structures, such as the endoplasmic reticulum, Golgi bodies, ribosomes, vacuoles, and vesicles. Most of these new structures are produced during the G_1 period of the cell cycle, prior to replication of the genetic material.

Mitochondria and chloroplasts are produced only from previously existing mitochondria and chloroplasts. Both of these organelles also have their own DNA, which is organized much like the DNA of bacterial cells; that is, it is a "naked" molecule, rather than one occurring in close combination with protein. These are two of the reasons that many biologists hypothesize that these organelles originated as separate organisms that took up a symbiotic ("living together") relationship with early eukaryotes more than a billion years ago. This question will be discussed further in Chapter 19.

The visible process of cytokinesis usually begins during telophase of mitosis. It usually, but not always, divides the cell into two equal parts. The cleavage always occurs at the midline of the spindle, although the spindle does not seem to be involved in cytoplasm division. For instance, if a spindle is pushed out of position at late metaphase, or even if it is removed from the cell, cytokinesis proceeds normally, with the cell dividing along the previously established equator.

Cytokinesis differs in some respects in plant and animal cells. In animal cells, during late anaphase or early telophase, the cell membrane begins to pinch in along the circumference of the cell in the area where the equator of the spindle used to be. At first a furrow appears on the surface, and this gradually deepens into a groove. Eventually the connection between the daughter cells dwindles to a slender thread, which soon parts. Microfilaments, seen in large numbers near the furrows, are thought to play a role in the constriction (see page 100). The microfilaments are believed to act as a sort of "purse string" drawing the ends of the two daughter cells together and pinching them apart.

In plant cells, the cytoplasm is divided by the formation of a *cell plate,* which is formed from a line of vesicles produced from Golgi bodies. The vesicles form in the interior of the cell and move outward. They eventually fuse to form a flat, membrane-bound space. As more vesicles fuse, the edges of the growing plate fuse with the membrane of the cell. In this way, space is established between the daughter cells, completing the separation of the two daughter cells. Cellulose is laid down against the membranes, and the space itself eventually becomes impregnated with pectins and becomes the middle lamella (see page 94).

When cell division is complete, two daughter cells are produced, indistinguishable from one another and from the parent cell. The precise, ordered events of mitosis ensure the continuity of the genetic material and the preservation of the information that it carries.

10-9 *Mother-daughter centriole pair. Centrioles always appear in cells of organisms that, at some time in their life cycles, form cilia or flagella. Centrioles are also usually—but not always—formed from preexisting centrioles with the daughter appearing, mysteriously, at right angles to the mature centriole.*

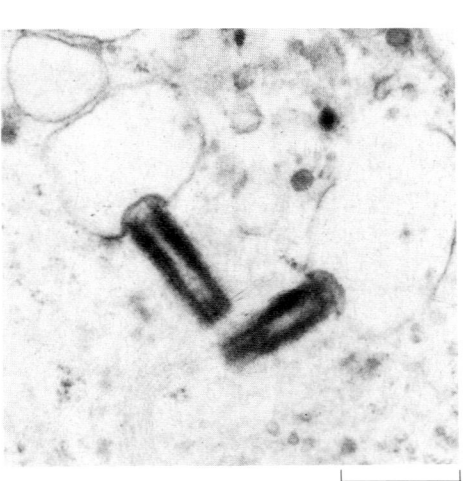

0.5 μm

0–10 *Mitosis in the embryonic cells of a whitefish. (a) Prophase. The chromosomes have become visible, the nuclear envelope has broken down, and the spindle apparatus has formed. (b) Metaphase. The chromosomes, guided by the spindle fibers, are lined up at the equator of the cell. Some of them appear to have begun to separate. (c) Anaphase. The two sets of chromosomes are moving apart. (d) Telophase. The chromosomes are completely separated, the spindle apparatus is disappearing, and a new cell membrane is forming which will complete the separation of the two daughter cells.*

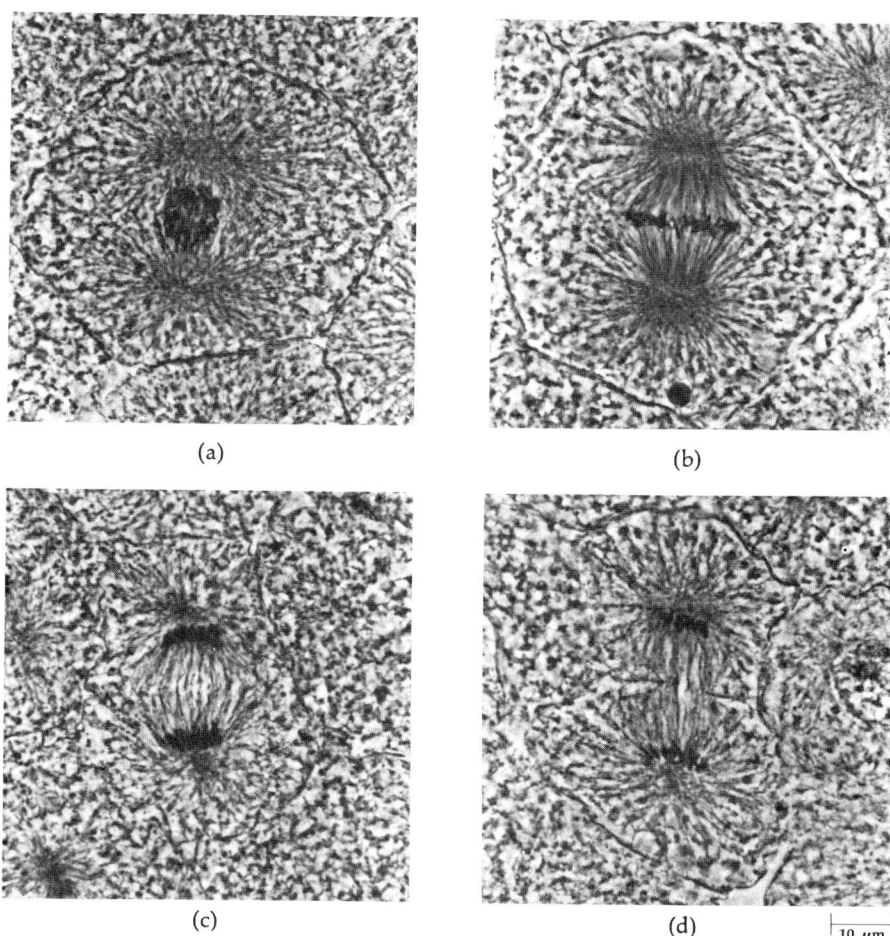

(a)　　　　　　　　　　　(b)

(c)　　　　　　　　　　　(d)　　10 μm

0–11 *In plants, the final separation of the two cells takes place by the formation of a structure known as a cell plate. Small droplets appear across the equatorial plane of the cell and gradually fuse, forming a flat membrane-bound space, the cell plate, which extends outward until it reaches the wall of the dividing cell. The dark forms are the chromosomes.*

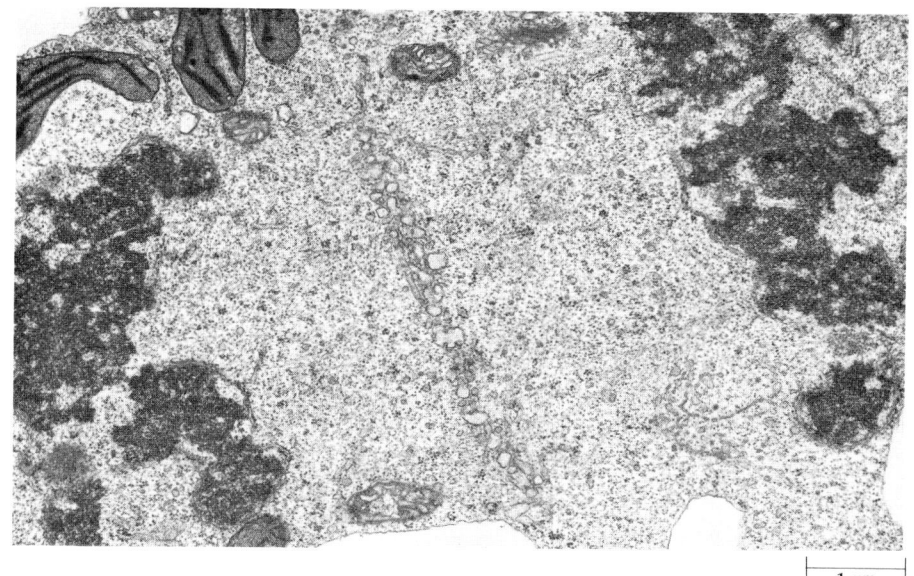

1 μm

The cell is the basic unit of life in all living systems. In the single-celled organism, the cell is the whole organism, capable of carrying out all the activities necessary for its survival. In the multicelled organism, the cell is a semiautonomous member of a society of cells; the capacities of the organism are the combined result of the activities and interactions of its individual cells. Thus the cell is the receptacle and the guardian of the particular organization of atoms, molecules, macromolecules, and organelles from which emerge the properties that are associated with, and are necessary to, life. The cellular basis of life is dependent upon the capacity of a cell to self-replicate.

SUMMARY

Cell division includes (1) division of the nucleus (mitosis), during which the chromosomes are separated into two equal groups, and (2) division of the cytoplasm (cytokinesis). When the cell is in interphase (not dividing), the chromosomes are visible only as thin strands of threadlike material (chromatin). During this period, if mitosis is to take place, the chromosomal material is duplicated. As mitosis begins, the chromosomal material condenses and each chromosome appears as two identical chromatids held together at the kinetochore (centromere). The spindle forms. In animal cells, it forms between the centrioles as they separate. In both animal and plant cells, some spindle fibers stretch from pole to pole and some are attached to the chromosomes at their kinetochores. Prophase ends with the breakdown of the nuclear membrane. During metaphase, the chromosomes, apparently maneuvered by the spindle fibers, move toward the center of the cell. At the end of metaphase, the chromosomes are arranged on the equatorial plane. During anaphase, each kinetochore divides, the chromatids separate, and each chromatid, now a daughter chromosome, moves to an opposite pole. During telophase, the nuclear membrane re-forms, the spindle disperses, and the chromosomes uncoil and once more become distended and diffuse.

Cytokinesis in animal cells results from constrictions in the cytoplasm between the two nuclei. In plant cells, the cytoplasm is divided by the coalescing of vesicles to form the cell plate. In both cases, the result is the production of two new, separate cells, each containing, as a result of mitosis, the same genetic material as the parent cell and, as a result of cytokinesis, approximately half of the cytoplasm and organelles.

QUESTIONS

1. Define the following terms: mitosis, cytokinesis, chromatin, chromatid, kinetochore.
2. Describe what is happening at each of the stages shown below.

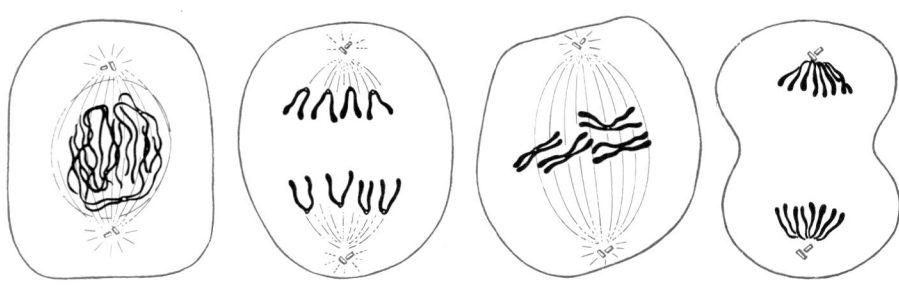

SUGGESTIONS FOR FURTHER READING

DUPRAW, E. J.: *Cell and Molecular Biology*, Academic Press, Inc., New York, 1968.

An excellent and provocative account of the frontiers of cell physiology. For the advanced student.

JASTROW, ROBERT: *Red Giants and White Dwarfs*, Harper & Row, Publishers, Incorporated, New York, 1967.

A short, handsomely illustrated book covering "the evolution of stars, planets and life."

LEDBETTER, M. C., and KEITH PORTER: *Introduction to the Fine Structure of Plant Cells*, Springer Publishing Co., Inc., New York, 1970.

An excellent atlas of electron micrographs of plant cells, with detailed explanations.

LEHNINGER, ALBERT L.: *Biochemistry*, 2d ed., Worth Publishers, Inc., New York, 1975.

This advanced text is outstanding both for its clarity and for its consistent focus on the living cell.

LEHNINGER, ALBERT L.: *Bioenergetics: The Molecular Basis of Biological Energy Transformation*, 2d ed., W. A. Benjamin, Inc., New York, 1971.

A brief, well-written account of the flow energy in living systems.

LOEWY, A. G., and P. SIEKEVITZ: *Cell Structure and Function*, 2d ed., Holt, Rinehart and Winston, Inc., New York, 1969.

An outstanding elementary text on cell structure and function.

MONOD, JACQUES: *Chance and Necessity*, Alfred A. Knopf, Inc., New York, 1971.

Chance and necessity refer to the two steps by which evolution takes place. You may not agree with Monod's bleak conclusions, but the process by which he arrives at them can be fascinating to all concerned with the implications of modern biology.

MOROWITZ, H. J.: *Energy Flow in Biology*, Academic Press, Inc., New York, 1968.

A stimulating analysis of the relationship between energy flow and biological organization.

OPARIN, A. I.: *Origin of Life*, Dover Publications, Inc., New York, 1938.*

Oparin, a Russian biochemist, was the first to argue that life arose spontaneously in the oceans of the primitive Earth. Although his concepts have been somewhat modified in detail, they form the basis for the present scientific theories on the origin of living things.

PONNAMPERUMA, CYRIL: *The Origins of Life*, E. P. Dutton & Co., Inc., New York, 1972.

A clear, authoritative, and up-to-date look at the origins of life on Earth and the possibilities of life on other planets. Written by one of the leading investigators in this field, with abundant and imaginative illustrations.

PORTER, KEITH R., and MARY A. BONNEVILLE: *An Introduction to the Fine Structure of Cells and Tissues*, 3d ed., Lea & Febiger, Philadelphia, 1968.

An atlas of electron micrographs of animal cells; detailed commentaries accompany each. These are magnificent micrographs, and the commentaries describe not only what the picture shows but also the experimental foundations of knowledge of cell ultrastructures.

RABINOWITCH, EUGENE, and GOVINDJEE: *Photosynthesis*, John Wiley & Sons, Inc., New York, 1969.*

A lucid introduction, suitable for undergraduate students, to the processes of photosynthesis and to related physical and chemical concepts, such as entropy and free energy.

*Available in paperback.

THOMAS, LEWIS: *The Lives of a Cell: Notes of a Biology Watcher*, Viking Press, New York, 1974.

Thomas, a physician and medical researcher, reveals the extent to which science can tune our intellectual antennae, broaden our perceptions, and extend our appreciation of ourselves and of the world around us. Anyone who wants to refute the contention that science destroys human values need look no further than these short, sensitive essays.

SECTION 2

THE CONTINUITY OF LIFE: GENETICS

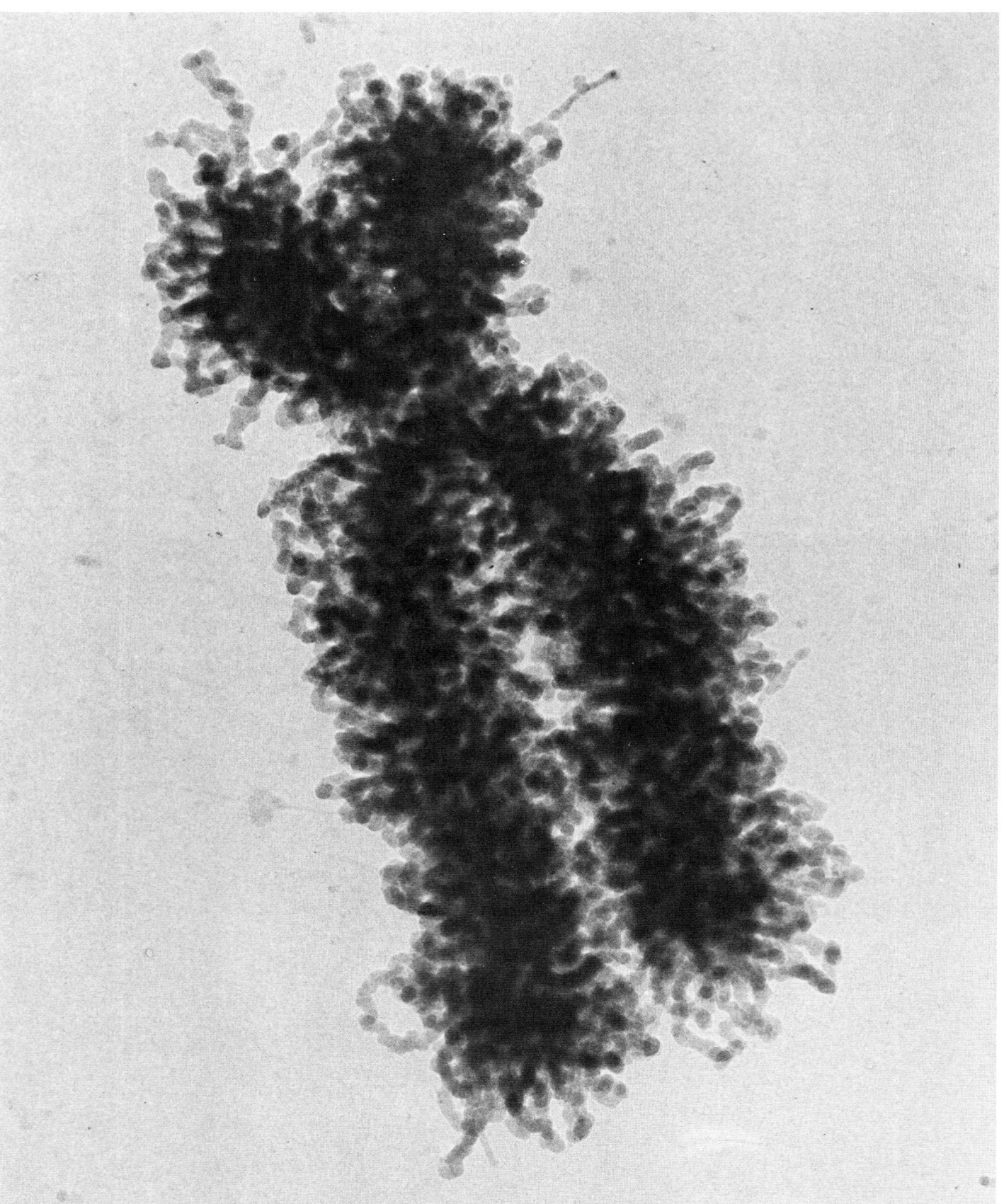

Chapter 11

11–1 *Human chromosome 12.*

11–2 *The protruding lip of the Hapsburgs is a famous example of an inherited trait. These portraits of members of the Hapsburg family encompass a period of about 300 years: (a) Rudolph I (1218–1291), King of Germany, (b) Maximilian I (1460–1519), Holy Roman Emperor, (c) Charles V (1500–1558), Holy Roman Emperor, (d) Ferdinand I (1503–1564), Holy Roman Emperor.*

Concepts of Heredity: Mendel

Among all the symbols in biology, perhaps the most widely used and most ancient are the hand mirror of Venus (♀) and the shield and spear of Mars (♂), the biologists' shorthand for female and male. Ideas about the nature of biological inheritance—the role of male and female—are even older than these famous symbols. Very early, men must have noticed that certain characteristics —hair color, for example, a large nose, or a small chin—were passed from parent to offspring. And throughout history, the concept of biological inheritance has been an important factor in the social organizations of men, determining the distribution of wealth, power, land, and royal privileges.

Sometimes the trait that is passed on is so distinctive that it can be traced through many generations. A famous example of such a characteristic is the Hapsburg lip (Figure 11–2), which has appeared in Hapsburg after Hapsburg, over and over again since at least the thirteenth century. Cases such as this have made it easy for men to accept the importance of inheritance in the formation of the individual, but it is only comparatively recently that we have begun to understand how this process works. In fact, the study of heredity as a science did not really begin until the second half of the nineteenth century. Yet the problems posed in this study are among the most fundamental in biology, since self-replication is the essence of the hereditary process and, as we noted in Chapter 1, one of the principal properties of living systems.

(a)

(b)

(c)

(d)

11–3 *Only fairly recently has mankind come to realize that living things come only from other living things of the same species and never from another species or from lifeless matter. This picture from an old Turkish history of India shows a wakwak tree, which bears human fruit. According to the account, the tree is to be found on an island in the South Pacific.*

11–4 *Fifteenth-century miniature showing the internal organs of the female sex.*

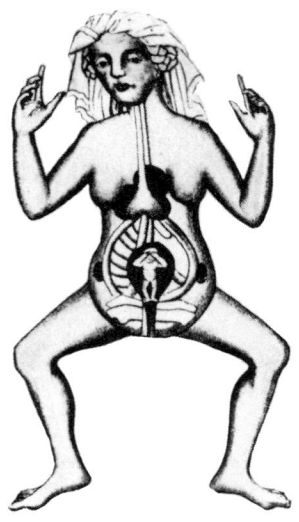

EARLY IDEAS ABOUT HEREDITY

Far back in human history, men learned to improve domestic animals and crops by inbreeding and crossbreeding. In the case of date palms, male and female flowers are found on different trees, and artificial fertilization of the palm was well known to the ancient Babylonians and Egyptians. The nature of the difference between the two flowers was understood by Theophrastus (380–287 B.C.). "The males should be brought to the females," he wrote, "for the male makes them ripen and persist." The mule, a man-made hybrid (a cross between a male donkey and a mare), was well known in the days of Homer. Both Plutarch and Lucretius noted in their writings that some children resemble their mothers, some resemble their fathers, and some even skip back a generation to resemble a grandparent; this fact, so easy to observe, continued to puzzle people for a very long time.

Since no genetic laws were known, bizarre crossbreeds were the subject of many legends. The wife of Minos, according to Greek mythology, mated with a bull and produced the Minotaur. Folk heroes of Russia and Scandinavia were traditionally the sons of women who had been captured by bears, from which these men derived their great strength and so enriched the national stock. The camel and the leopard also crossbred from time to time, according to the early naturalists, who were otherwise unable—and it is hard to blame them—to explain an animal as improbable as the giraffe. Thus folklore reflected early and imperfect glimpses of the nature of hereditary relationships.

The first scientist known to have pondered the nature of heredity was Aristotle. He postulated that the male semen was made up of a number of imperfectly blended ingredients and that, at fertilization, it mixed with the "female semen," the menstrual fluid, giving form and power (*dynamis*) to this amorphous substance. No one had a better idea—or indeed many ideas at all—for two thousand years. Seventeenth-century medical texts show various stages in the coagulation of the embryo from the mixture of maternal and paternal semens. Indeed, many scientists as well as laymen did not believe that such mixtures were even always necessary; they held that life, at least the "simpler" forms of life, could arise by spontaneous generation. Worms, flies, and various crawling things, it was commonly believed, took shape from putrid substances, ooze, and mud, and a lady's hair dropped in a rain barrel could turn into a snake. Jan Baptista van Helmont, who also carried out experiments on the growth of plants (see page 140), published his personal recipe for the production of mice: One need only place a dirty shirt in an open pot containing a few grains of wheat, and in 21 days mice would appear! He had performed the experiment himself, he said. The mice would be adults, both male and female, he added, and would be able to produce more mice by mating.

THE FIRST EXPERIMENTS

In 1677, the Dutch lens maker Anton van Leeuwenhoek discovered living sperm—"animalcules," he called them—in the seminal fluid of various animals, including man. Enthusiastic followers peered through Leeuwenhoek's "magic looking glass" (his homemade microscope) and believed they saw within each human sperm a tiny creature—the homunculus, or "little man." This little man was the future human being in miniature. Once implanted in the female womb,

SPONTANEOUS GENERATION

Most of the early biologists, from the time of Aristotle, believed that simple things such as worms, beetles, frogs, and salamanders could originate spontaneously in dust or mud, that rodents formed from moist grain, and that plant lice condensed from a dewdrop. In the seventeenth century, Francesco Redi performed a famous experiment in which he put out decaying meat in a group of wide-mouthed jars—some with lids, some covered by a fine veil, and some open—and proved that maggots arose only where flies were able to lay their eggs.

By the nineteenth century, no scientist continued to believe that complex organisms arise spontaneously. With the advent of microscopy, however, belief in spontaneous generation was vigorously revived. It was necessary only to put decomposing substances in a warm place for a short time and tiny "live beasts" appeared under the lens, before one's very eyes. By 1860, the controversy had become so spirited that the Paris Academy of Sciences offered a prize for experiments that would throw new light on the question. The prize was claimed in 1864 by Louis Pasteur, who devised experiments to show that microorganisms appeared only as contaminants from the air and not "spontaneously," as his opponents claimed. In his experiments he used swan-necked flasks which permitted the entrance of oxygen, thought to be necessary for life, but which, in their long, curving necks, trapped bacteria, fungal spores, and other microbial life and thereby protected the contents of the flask from contamination. He showed that if the liquid in the flask was boiled (which killed any microorganisms already present) and the neck of the flask was allowed to remain intact, no microorganism would appear. Some of his original flasks, still sterile, remain on display. Only if the neck of the flask was broken off, permitting air to enter the flask, did microorganisms appear.

"Life is a germ, and a germ is Life," Pasteur proclaimed at a brilliant "scientific evening" at the Sorbonne before the social elite of Paris. "Never will the doctrine of spontaneous generation recover from the mortal blow of this simple experiment!"

In retrospect, Pasteur's well-planned experiments were considered so decisive because the broad question of whether or not spontaneous generation ever occurred had been reduced to a simpler question of whether or not it occurred under the specific conditions claimed for it. Now, as we noted in Chapter 1, scientists generally believe that some form of "spontaneous generation" did indeed take place, although under quite different circumstances, when the Earth was young.

Pasteur's swan-necked flasks, which he used to counter the argument that spontaneous generation failed to occur in sealed vessels because air was excluded. These flasks permitted the entrance of oxygen, known to be essential for life, but their long curving necks trapped spores of microorganisms and thereby protected the liquid in the flask from contamination.

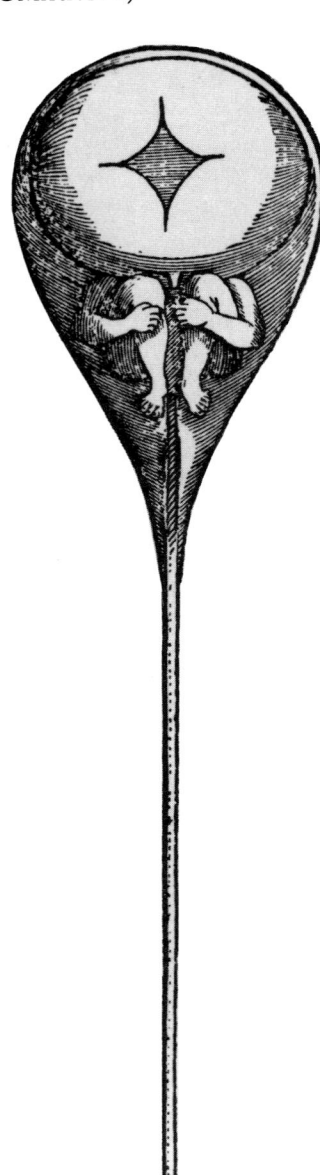

11–5 *What the animalculists, or spermists, of the seventeenth and eighteenth centuries believed they saw when they looked through a microscope at sperm cells. This is a homunculus ("little man"), a future human being in miniature, in a sperm cell. (OMIKRON.)*

the future human being was nurtured there; the only contribution that the mother made was to serve as an incubator for the growing fetus. Any resemblance a child might have to its mother, these theorists held, was because of "prenatal influences."

During the very same decade (the 1670s) that van Leeuwenhoek first saw human sperm cells, another Dutchman, Régnier de Graaf, described for the first time the ovarian follicle, the structure on the surface of the ovary in which the human egg cell forms. Although the actual human egg was not seen for another 150 years, the existence of a human egg was rapidly accepted. In fact, de Graaf attracted a school of followers, the ovists, who were as convinced of their opinions as the animalculists, or spermists, were of theirs and who soon contended openly with them. It was the female egg, the ovists said, that contained the future human being in miniature; the animalcules in the male seminal fluid merely stimulated the egg to grow. Ovists and animalculists alike carried the argument one logical step further. Each homunculus had within it another perfectly formed homunculus, and in that was still another one, and so on—children, grandchildren, and great-grandchildren, all stored away for future use. Some ovists even went so far as to say that Eve had contained within her body all the unborn generations yet to come, each egg fitting closely inside another like a child's hollow blocks. Each female generation since Eve has contained one less than the previous generation, they explained, and after 200 million generations, all the eggs will be spent and human life will come to an end.

BLENDING INHERITANCE

By the middle of the nineteenth century, the concepts of the ovists and spermists began to yield to new data. The facts that challenged these earlier hypotheses came not so much from scientific experiments as from practical attempts by master gardeners to produce new ornamental plants. Artificial crossings of such plants showed that, in general, regardless of which plant supplied the pollen (which contains the male cells, or sperm) and which plant contributed the egg cells, or ova, both contributed to the characteristics of the new variety. But this conclusion raised even more puzzling questions: What exactly did each parent plant contribute? How did all the hundreds of characteristics of each plant get combined and packed into a single seed?

The most widely held hypothesis of the nineteenth century was blending inheritance. According to this hypothesis, when the sex cells (or *gametes*, from the Greek word *gamos*, "to marry") combine, there is a mixing of hereditary material which results in a blend, analogous to a blend of two different-colored inks. On the basis of such a theory, one would predict that the offspring of a black animal and a white animal would be gray, and their offspring would also be gray because the black and white hereditary material, once blended, could never be separated again.

You can see why this concept was unsatisfactory. It ignored the phenomenon of traits skipping a generation, or even several generations, and then reappearing. To Charles Darwin and other proponents of the theory of evolution, it presented particular difficulties. Evolution, according to Darwin, as we noted in the Introduction (page 8), takes place as a result of random variations and natural selection. If the hypothesis of blending inheritance were valid, the

small hereditary variations would disappear, like a single drop of ink in the colored mixture. Sexual reproduction would eventually result in complete uniformity, natural selection would have no raw material on which to act, and evolution would never have occurred.

THE CONTRIBUTIONS OF MENDEL

Gregor Mendel, who was born into a peasant family in 1822, entered a monastery in Brünn (in what is now Czechoslovakia), where he was able to receive an education. He attended the University of Vienna for two years, pursuing studies in both mathematics and science. He failed his tests for the teaching certificate he was seeking and so retired to the monastery, of which he eventually became abbot. Mendel's work, carried on in a quiet monastery garden and ignored until after his death, marks the beginning of modern genetics.

Mendel's great contribution was to demonstrate that inherited characteristics are carried as discrete units, which are parceled out in different ways—reassorted—in each generation. These discrete units eventually came to be known as _genes_.

Mendel was successful where others were not, largely for three reasons. First, he planned his experiments carefully and imaginatively, choosing for study definite and measurable hereditary differences. Second, probably because of his background in physics and mathematics, he was one of the first to apply mathematics to the study of biology. Even though his mathematics was simple, the idea that statistics could be applied to living organisms was startlingly new. Finally, he carried out his experiments in a thoroughly scientific way (as his predecessors had not). He tested a very specific model in a series of logically designed experiments.

Mendel's choice of pea plants for his experiments was a very deliberate one. The plants were commercially available and easy to cultivate. Different varieties

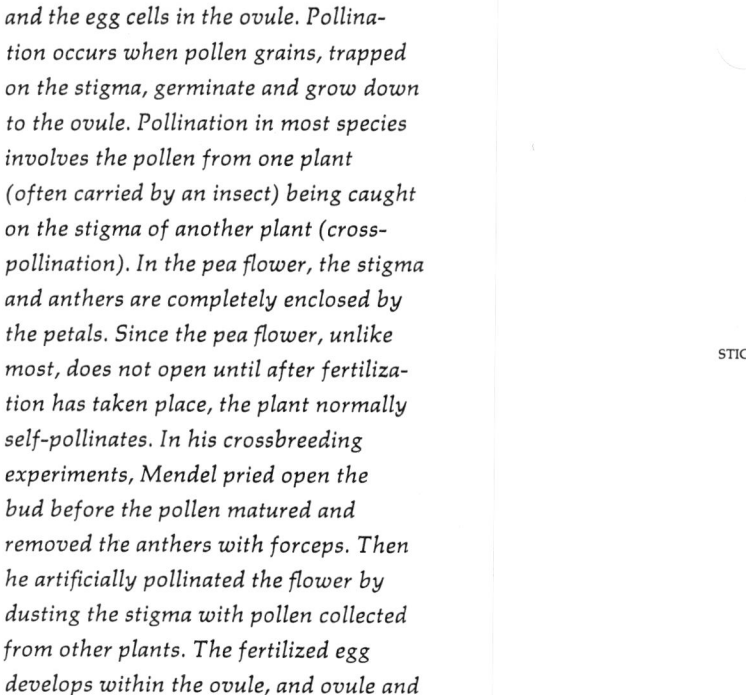

−6 In a flower, pollen develops in the anther and the egg cells in the ovule. Pollination occurs when pollen grains, trapped on the stigma, germinate and grow down to the ovule. Pollination in most species involves the pollen from one plant (often carried by an insect) being caught on the stigma of another plant (cross-pollination). In the pea flower, the stigma and anthers are completely enclosed by the petals. Since the pea flower, unlike most, does not open until after fertilization has taken place, the plant normally self-pollinates. In his crossbreeding experiments, Mendel pried open the bud before the pollen matured and removed the anthers with forceps. Then he artificially pollinated the flower by dusting the stigma with pollen collected from other plants. The fertilized egg develops within the ovule, and ovule and plant embryo form the pea (the seed).

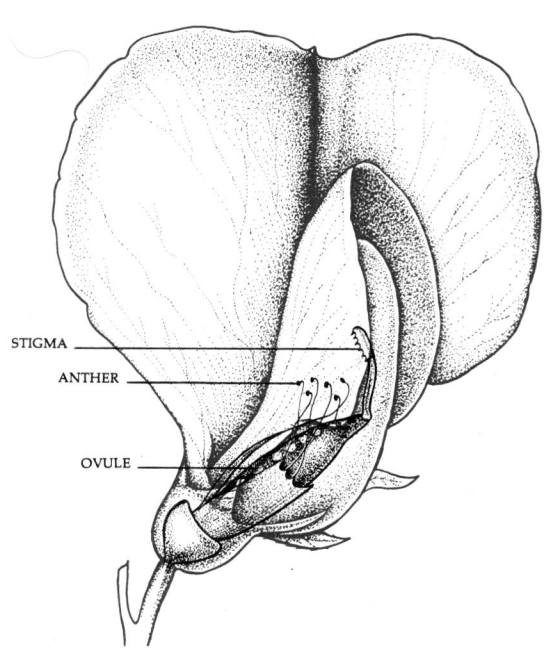

STIGMA

ANTHER

OVULE

had clearly different characteristics which "bred true," reappearing in crop after crop. Finally, the sexual organs of the pea flower are entirely enclosed by the petals, even when they are mature (see Figure 11–6). Consequently, the flower normally self-pollinates. Although the plants could be crossbred experimentally, accidental crossbreeding could not occur to confuse the experimental results. In Mendel's own words, quoted from his original paper, "The value and utility of any experiment are determined by the fitness of the material to the purpose for which it is used."

Principle of Segregation

Mendel began with 32 different types of pea plants, which he studied for two years to see which characteristics were clearly defined. As he said later in his report on this work, he did not want to experiment with traits in which the difference could be "of a 'more or less' nature, which is often difficult to define." As a result of these observations, he selected for study seven traits that appeared as conspicuously different characteristics in different types of plants. One variety of plant, for example, always produced yellow peas, or seeds, while another always produced green ones. In one variety, the seeds, when dried, had a wrinkled appearance, while in another variety they were smooth. The complete list of alternate traits is given in Table 11–1.

Mendel performed experimental crosses, removing the anthers from flowers and dusting the stigmas with pollen from a flower of another type. He found that in every case in the first generation (now known as F_1 in biological shorthand), one of the alternate traits disappeared completely without a sign. All the seeds produced as a result of a cross between yellow-seeded plants and green-seeded plants were as yellow-seeded as the yellow-seeded parent. All of the flowers produced by plants resulting from a cross between a red-flowered plant and a white-flowered plant were red. These traits and other such traits Mendel called *dominant*.

The interesting question was: What had happened to the second trait—the wrinkledness of the seed or the greenness of its color—which had been passed on so faithfully for generations by the parent stock? Mendel let the pea plant itself carry out the next stage of the experiment by permitting the F_1 to self-pollinate. The traits that had disappeared in the first generation reappeared in the second (F_2) generation. In Table 11–1 are the results of Mendel's actual counts. These traits, which originated in the parent generation and reappeared in the F_2 generation, must also have been present somehow in the F_1 generation, although not apparent there. Mendel called these traits *recessive*.

If you analyze the results in Table 11–1, as Mendel did, you will notice that the dominant and recessive traits appear in the second, or F_2, generation in ratios of about 3:1. How do the recessives disappear so completely and then appear again, and always in such constant proportions? It was in answering this question that Mendel made his greatest contribution. He saw that the appearance and disappearance of traits and their constant proportions could be explained only if hereditary characteristics are determined by discrete (separable) factors.* These factors, Mendel saw, must occur in the offspring as pairs, one factor inherited from each parent. These pairs of factors (genes)

11–7 *Garden pea plant with seeds (peas) in the approximate ratio of three yellow seeds for each green seed, as observed by Mendel.*

* Mendel called these factors *Elemente*. As we noted previously, they are now known as genes, and we shall use this term to refer to them.

A pea plant homozygous for red flowers is represented as WW in genetic short-hand. The gene for red flowers is designated W because of a convention by which geneticists, in indicating a pair of genes, use the first letter of the less common form (white). The capital indicates the dominant, the lowercase the recessive. A WW plant can produce egg cells or pollen grains with only a red-flower (W) gene. The female symbol ♀ indicates that this flower contributed the egg cells, or female gametes.

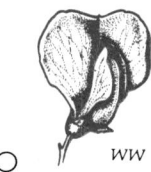

♀ ww

A white pea plant (ww) can produce egg cells or pollen grains with only a white-flower (w) gene. The male symbol ♂ indicates that this flower contributed the pollen, or male gametes.

♂ ww

When a w pollen grain fertilizes a W egg cell, the result is a Ww pea plant, which, since the W gene is dominant, will produce red flowers. However, this Ww plant can produce egg cells or pollen cells with either a W or a w allele.

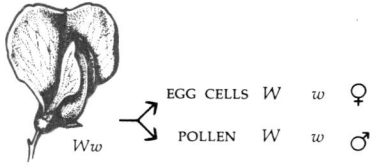

EGG CELLS W w ♀
Ww POLLEN W w ♂

And so, if the plant self-pollinates, four possible crosses can occur:

♀ W × ♂ W ⟶ red flowers
♀ W × ♂ w ⟶ red flowers
♀ w × ♂ W ⟶ red flowers
♀ w × ♂ w ⟶ white flowers

These results are summarized in Figure 11–9.

Table 11–1 *Mendel's Pea-plant Experiment*

| Trait | Original crosses | | F_2 generation | | |
	Dominant ×	Recessive	Dominant	Recessive	Total
Seed form	Round ×	Wrinkled	5,474	1,850	7,324
Seed color	Yellow ×	Green	6,022	2,001	8,023
Flower position	Axial ×	Terminal	651	207	858
Flower color	Red ×	White	705	224	929
Pod form	Inflated ×	Constricted	882	299	1,181
Pod color	Green ×	Yellow	428	152	580
Stem length	Tall ×	Dwarf	787	277	1,064

are separated again when the mature offspring produces the sex cells, resulting in two kinds of gametes, with one gene of the pair in each.

This hypothesis is known as Mendel's first law, or the *principle of segregation.*

The two factors in a pair might be the same, in which case the self-pollinating plant would breed true. Or the two factors might be different; such different, or alternative, forms came to be known as *alleles.* Yellow-seededness and green-seededness, for instance, are determined by alleles, different forms of the gene (factor) for seed color. When the genes of a gene pair are the same, the organism is said to be *homozygous* for that particular trait; when the genes of a gene pair are different, the organism is *heterozygous* for that trait.

When gametes are formed, genes are passed on to them, but each gamete contains only one of two possible alleles. When two gametes combine in the fertilized egg, the genes occur in matched pairs again. One allele may be dominant over another allele; in this case, the organism will appear as if it had only this gene. This outward appearance is known as its *phenotype.* However, in its genetic makeup, or *genotype,* each allele still exists independently and as a discrete unit, even though it is not visible in the phenotype, and the recessive allele will separate from its dominant partner when gametes are again formed. Only if two recessive alleles come together—one from the female gamete and one from the male—will the phenotype then show the recessive trait. When pea plants homozygous for red flowers are crossed with pea plants having white flowers, only pea plants with red flowers are produced, although each plant in the F_1 generation will carry a gene for red and a gene for white. Figure 11–8 shows what happens in the F_2 generation if the F_1 generation self-pollinates. Notice that the result would be the same if an F_1 individual is cross-fertilized with another F_1 individual, which is the way these experiments are performed with animals and with plants that are not self-pollinating.

In order to test his hypothesis (diagrammed in Figure 11–9), Mendel performed two additional experiments. He crossed white-flowering plants with white-flowering plants and confirmed that they bred true—that he got only white-flowering plants. Next he crossed one of his F_1 hybrids, the result of a cross between red- and white-flowering plants, with a true-breeding white-flowering plant. To the outside observer, it would appear as if Mendel were simply repeating his first experiment, crossing plants having red flowers with plants having white flowers. But he knew that if his hypothesis were correct, his results would be different from those of his first experiment. In fact, he actually predicted the results of such a cross before he made it. Can you? The easiest way is to diagram it, as in Figure 11–10. This experiment, which reveals the genotypes, is known as a testcross, or backcross. A testcross is an experimental cross between an F_1 and the parental homozygote recessive; it reveals the genotype of the F_1 organism.

11–9 *The use of a Punnett square for the determination of possible genotypes. A pea plant with two dominant genes for red flowers (WW) and one with two recessive genes for white flowers (ww) are crossed. The phenotype of the offspring in the F_1 generation is red, but note that the genotype is Ww. The F_1 heterozygote is self-pollinated, producing four kinds of gametes, ♀ W, ♂ W, ♀ w, and ♂ w, in equal proportions. The W and w pollen and eggs combine randomly to form, on the average, ¼ WW (red), ½ Ww (red), and ¼ ww (white) offspring. It is this underlying 1:2:1 genotypic ratio that accounts for the phenotypic ratio of 3 dominants (red) to 1 recessive (white). The distribution of traits in the F_2 generation is shown by a Punnett square, named after the English geneticist who first used this sort of checkerboard diagram for the analysis of genetic traits.*

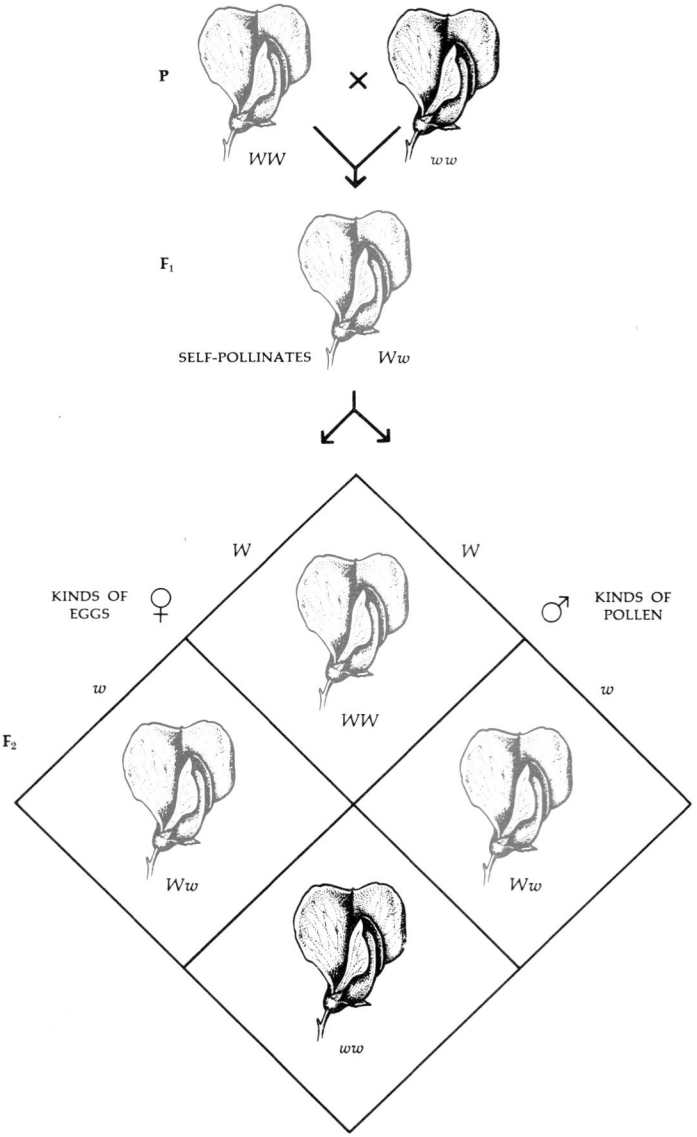

1–10 *A testcross. In order for a pea flower to be white, the plant must be homozygous for the recessive gene (ww). But a red pea flower can come from a plant with either a Ww or a WW genotype. How could you tell such plants apart? Mendel solved this problem by breeding such plants with homozygous recessives. As shown here, a phenotypic ratio in the F_1 generation of 2 red to 2 white indicates a heterozygous red-flowering parent. What would have been the result if the parent had been homozygous for red flowers?*

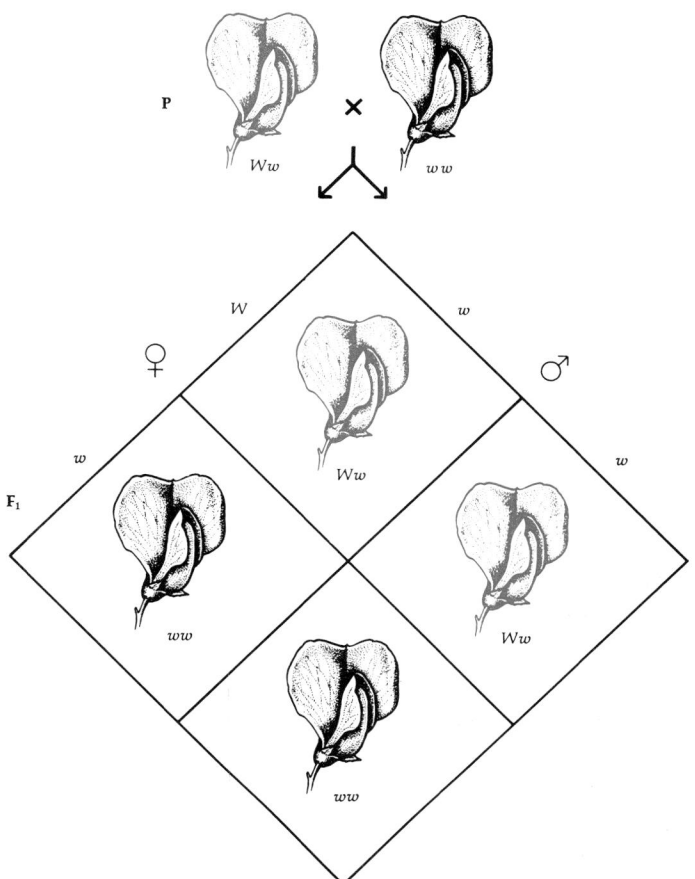

Principle of Independent Assortment

In a second series of experiments, Mendel studied crosses between pea plants that differed simultaneously in two characteristics; for example, one parent plant had peas that were round and yellow, and the other had peas that were wrinkled and green. The round and yellow traits, you will recall (see Table 11–1), are both dominant, and the wrinkled and green are recessive. As you would expect, all the seeds produced by a cross between the parental types were round and yellow. When these seeds were planted and the flowers allowed to self-pollinate, 556 seeds were produced. Of these, 315 showed the two dominant characteristics, round and yellow, but only 32 combined the recessive traits, green and wrinkled. All the rest of the seeds were unlike either parent; 101 were wrinkled and yellow, and 108 were round and green. Totally new combinations of characteristics had appeared. This experiment did not contradict Mendel's previous results. Round and wrinkled still appeared in the same 3:1 proportion (423 round to 133 wrinkled), and so did yellow and green (416 yellow to 140 green). But the round and the yellow traits and the wrinkled and the green ones, which had originally combined in one plant, behaved as if they were entirely independent of one another. From this, Mendel formulated his second law, the *principle of independent assortment.*

11-11 *One of the experiments from which Mendel derived his principle of independent assortment. A plant homozygous for round (RR) and yellow (YY) peas is crossed with a plant having wrinkled (rr) and green (yy) peas. The F₁ generation are all round and yellow, but notice how the traits will, on the average, appear in the F₂ generation. Of the 16 kinds of offspring, 9 show the two dominant traits (RY), 3 show one combination of dominant and recessive (Ry), 3 show the alternate combination (rY), and 1 shows the two recessives (ry). This 9:3:3:1 distribution is always the expected result from a cross involving two pairs of independent dominant-recessive alleles. (The letters R and Y are used because round and yellow are the less common forms in nature.)*

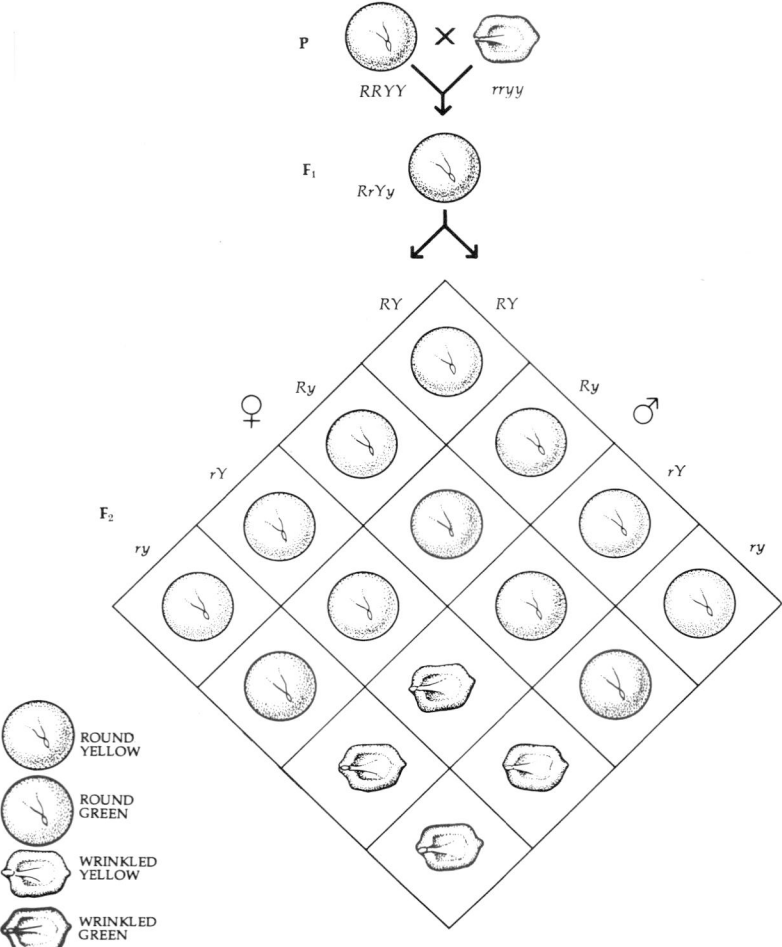

Figure 11–11 diagrams these results and shows why, in a cross involving two pairs of alleles, each pair with one dominant and one recessive allele, the ratio of distribution will be, on the average, 9:3:3:1, with 9 representing the proportion of F_2 progeny that will show the two dominant traits, 1 the proportion that will show the two recessive traits, and 3 and 3 the proportions of the two alternative combinations of dominants and recessives. This is true when one of the original parents is homozygous for both recessive traits and the other homozygous for both dominant ones, as in the experiment just described ($RRYY \times rryy$), as well as when each original parent is homozygous for one recessive and one dominant trait ($rrYY \times RRyy$).

A Testcross

Let us test the principle of independent assortment by performing a testcross, similar to that diagrammed in Figure 11–10, but involving two pairs of alleles instead of one. In such a cross, F_1 individuals, which are always heterozygous for the traits being studied, are crossed with individuals which are homozygous for the recessive traits. For simplicity, let us again study the distribution of

In applying statistics to the study of genetics, Mendel was stating that the laws of chance apply to biology as they do to the physical sciences. Toss an imaginary coin. The chance that it will turn up heads is fifty-fifty, or ½. The chance that it will turn up tails is also fifty-fifty, or ½. The chance that it will turn up one or the other is certain, or one chance in one. Now toss two imaginary coins. The chance that one will turn up heads is again ½. The chance that the second will turn up heads is also ½. The chance that both will turn up heads is ½ × ½, or ¼. The probability of two independent events occurring together is simply the probability of one occurring alone multiplied by the probability of the other occurring alone. The chance of both turning up tails is similarly ½ × ½. The chance of the first turning up tails and the second turning up heads is ½ × ½, and the chance of the second turning up tails and the first heads is ½ × ½. We can diagram this in a Punnett square:

The square indicates that the combination in each square has an equal chance of occurring. It was undoubtedly the observation that one-fourth of the offspring in the F_2 generation showed the recessive phenotype that indicated to Mendel that he was dealing with a simple case of the laws of probability.

If there were three coins involved, the probability of any given combination would be simply the product of all three fifty-fifty possibilities: ½ × ½ × ½, or ⅛. Similarly, with four coins, the probability of any given combination is ½ × ½ × ½ × ½, or ¹/₁₆. The Punnett square on page 182 expresses the probability (or chance) of each of any one of four possible combinations.

Notice that in planning his experiments, Mendel made two assumptions: (1) that an equal number of male and female gametes are produced and (2) that the gametes combined at random. Thus, the laws of probability could be followed.

If you toss two coins 4 times, it is unlikely that you will get the precise results diagrammed above. However, if you toss two coins 100 times, you will come close to the results predicted in the Punnett square, and if you toss two coins 1,000 times, you will be very close indeed. As Mendel knew, the relationship between dominants and recessives might well not have held true if he had been dealing with a small sample. The larger the sample, the more closely it will conform to results predicted by the laws of chance.

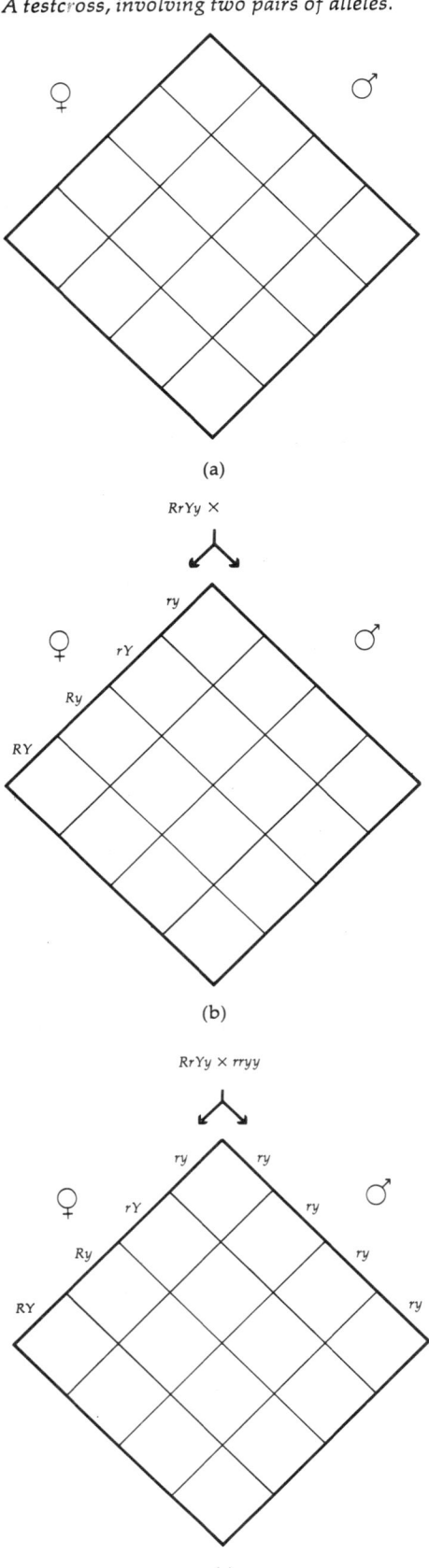

round vs. wrinkled (R vs. r) and green vs. yellow (Y vs. y). Draw a Punnett square with 16 squares (Figure 11–12a). Put the female symbol on one side and the male on the other. The F_1 heterozygote, with a genotype of $RrYy$, can produce four kinds of gametes: RY, Ry, rY, and ry. Let us assume that the heterozygote is female, that is, the contributor of the eggs. At the head of each column at the left, where the female symbol is, put one of these possible combinations (Figure 11–12b). (Notice that, in this step, we are assuming, as Mendel did, that each of the possible kinds of gamete is produced in equal numbers.)

The homozygous recessive can produce only one type of gamete in terms of the traits being studied: ry. Put ry at the head of each column on the right (Figure 11–12c). The advantage of using a Punnett square is that it makes it impossible to overlook any combination of gametes.

Now, starting with the column on the far left, begin to fill in the squares. You are less likely to make a mistake if you fill in all the female gametes first (or all male gametes), a column at a time, than if you try to work with both male and female gametes at the same time. When you are halfway through, your square will look like the one in Figure 11–12d.

Next fill in the symbols for the traits carried in the male gametes. Your square will then look like the one in Figure 11–12e.

Now count the phenotypes that this square predicts. Every capital R indicates a round seed (since R is dominant); there are eight capital R's, and hence eight round seeds. Conversely you can count eight wrinkled (rr) seeds. Similarly there are eight yellow (Y) seeds and eight green (yy) seeds.

Notice that each of the possible combinations of traits, round green, round yellow, wrinkled green, wrinkled yellow, appears in equal proportions.

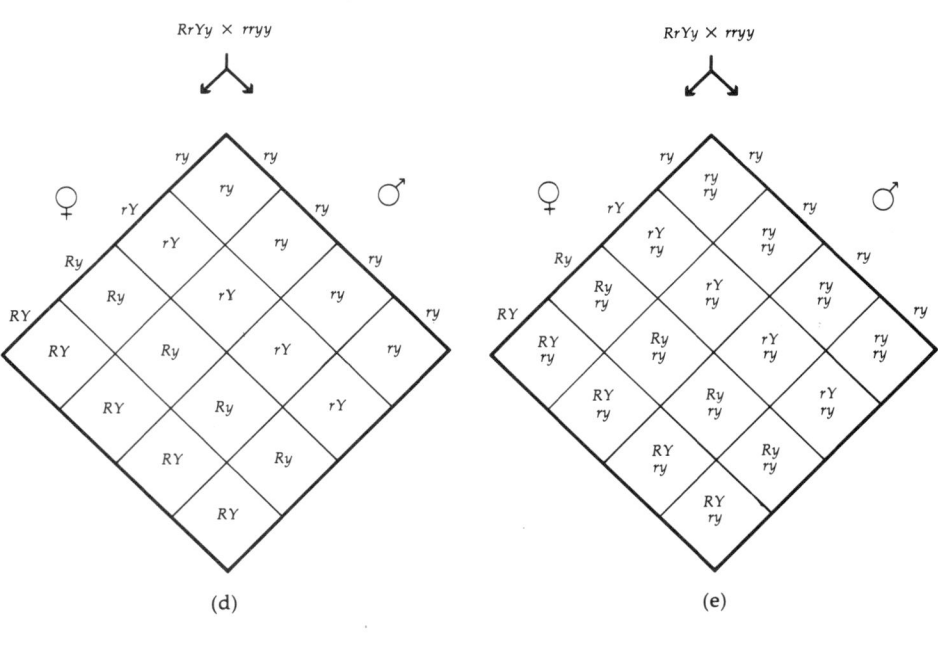

1–13 *Gregor Mendel, in experiments carried out in a monastery garden, showed that hereditary determinants are carried as separate units from generation to generation and explained how inherited variations can persist for generation after generation.*

Thus, the testcross reveals the genotype of the parent being tested (the female in this case) and, as it does so, makes both the segregation and the independent assortment of both gene pairs immediately apparent in the phenotypes of the individuals produced by the cross.

Influence of Mendel

Ironically, Mendel's experiments, although included today among the great triumphs of biological thinking, had almost no immediate effect on the course of scientific history. They were first reported in 1865 before a small group of people at a meeting of the Brünn Natural History Society, of which he was one of the founders. No one, apparently, understood what Mendel was talking about. But his paper was published the following year in the *Proceedings* of the Society, a journal which was circulated to libraries all over Europe. In addition, realizing the importance of his findings, Mendel himself continued for a number of years to send copies of his paper to other biologists.

By 1900, biology was finally prepared to accept Mendel's findings. Within a single year, his paper was rediscovered simultaneously and independently by three scientists working in three different European countries. Each of them had done similar experiments and was searching the scientific literature to seek confirmation of his results. And so, Mendel's brilliant analysis, written 35 years too soon, was rediscovered, but only after his death. Mendel's analysis of the questions he sought to answer, the design of his experiments, and the clarity of his results are so outstanding, however, that despite the fact he may not have significantly altered the course of genetic research, his name is permanently linked with the first principles of genetics.

SUMMARY

In this chapter, we started with man's earliest ideas about inheritance and traced the gradual development of these ideas into a science. The first question with which this new science was concerned was the mechanics of inheritance. How are hereditary characteristics passed from one generation to another?

By the middle of the nineteenth century, it was recognized that ova and sperm are specialized cells and that the ova and sperm both contribute to the hereditary characteristics of the new individual. But how were these special cells, called gametes, able to pass on the many hundreds of characteristics involved in inheritance? One answer to this question was the theory of blending inheritance, which held that the traits of the parents blended in the offspring, like a mixture of two fluids.

The revolution in genetics came when the blending theory was replaced by a unit theory. According to Mendel's principle of segregation, hereditary characteristics are always determined by discrete factors (now called genes), which appear in pairs, one of each pair inherited from each parent.

The genetic makeup of an organism is known as its genotype. Its appearance, or set of outward characteristics, is its phenotype. Both genes in a pair may be alike (a homozygous condition), or they may be different (heterozygous). Different genes of a gene pair are called alleles (alternative forms). In a heterozygous pair, only one allele may be detected in the phenotype. An allele that is expressed in the phenotype to the exclusion of the other is a dominant gene; one that is concealed in the phenotype is a recessive gene. When two

organisms that are homozygous for different alleles of the same gene are crossed, the ratio of dominant to recessive in the phenotype of the F_2 generation is 3:1.

Mendel's other great principle, the principle of independent assortment, applies to the behavior of two or more genes. This law states that the members of each pair of genes segregate independently. In crosses involving two independent pairs of alleles, the expected phenotypic ratio in the F_2 generation is 9:3:3:1. A testcross, in which an individual of the F_1 generation is crossed with a homozygous recessive, reveals the possible phenotypes in equal numbers.

QUESTIONS

1. *a.* In a pea plant that breeds true for tall, what possible gametes can be produced? (Use the symbol D for tall, d for dwarf.)
    b. In a pea plant that breeds true for dwarf, what possible gametes can be produced?
    c. What will be the genotype of an F_1 cross between these two types?
    d. What will be the phenotype of an F_1 cross between these two types?
    e. What will be the probable distribution of traits in the F_2 generation? Illustrate with a Punnett square.
2. You have just flipped a coin five times and it has turned up heads every time. What are the chances that the next time you flip it, it will be tails?
3. The ability to taste a bitter chemical, phenylthiocarbamide (PTC), is due to a dominant gene. In terms of the tasting gene, what are the possible phenotypes of a man both of whose parents are tasters? What are his possible genotypes? (This is a complex question.)
4. If such a man marries a woman who is a nontaster, what proportion of their children will probably be tasters? Suppose one of the children is a nontaster. What would you know about his genotype? Explain your results by drawing Punnett squares.
5. Suppose this couple had four children, all of whom could taste PTC. What is the father's probable genotype? Is there another possibility?
6. In chickens, two pairs of genes are involved in comb shape. One gene, R, results in rose comb, whereas its recessive allele, r, produces single comb. Gene P produces pea comb and its recessive allele, p, produces single comb. When R and P occur together, they produce a new type of comb: walnut. What would be the genotype of the F_1 cross resulting from $RRpp \times rrPP$? The phenotype? If F_1 hybrids were crossbred, what would be the probable distribution of genotypes? Of phenotypes? (Draw a Punnett square.)

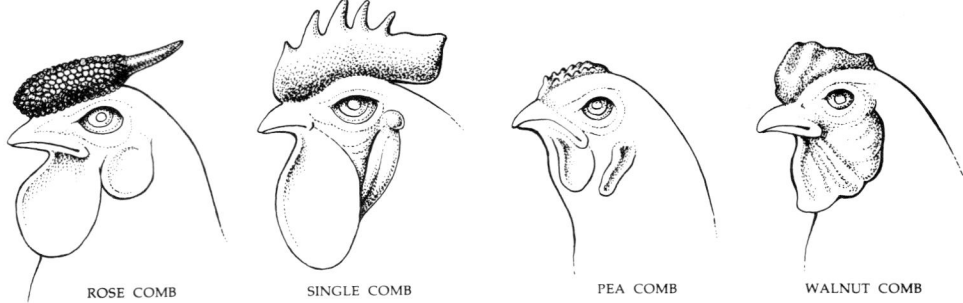

ROSE COMB   SINGLE COMB   PEA COMB   WALNUT COMB

Chapter 12

The Physical Basis of Heredity and Mendelian Inheritance in Man

Between the completion of Mendel's experiments in 1865 and their rediscovery in 1900, some important observations were made, largely as a result of improvements in the light microscope and in techniques of preparing materials for microscopic observation. By Mendel's time, you will recall, microscopists had seen sperm and egg cells, and had recognized them to be the hereditary links between parents and offspring. In 1875, the German zoologist Oscar Hertwig witnessed the fertilization of the egg of a sea urchin and noted that the fertilized egg contained two nuclei, one from the sperm and one from the egg. At about this time, as a result of new staining techniques, chromosomes were first observed within cells. In 1882, the German cytologist Walther Flemming pieced together the separate steps in mitosis, the process by which chromosomal material is divided exactly between two daughter nuclei. (See Chapter 10.)

As more and more cytologists focused their attention and their microscopes on these chromosomes, or "colored bodies," some new facts began to emerge.

12–1 *Despite the great differences in size of egg and sperm, both contribute equally to the hereditary characteristics of the individual. Since the nucleus is approximately the same size in both cells, the early microscopists postulated that this part of the cell must be the carrier of the hereditary determinants. The egg and sperm cells shown in (a) are from the sea urchin, which was used in many of these early studies because sea urchins are relatively easy to obtain and fertilization, which is external, can be easily observed in the laboratory. The electron micrograph at the right (b) shows the sperm penetrating the egg cell. The torpedo-shaped sperm nucleus fills most of the micrograph.*

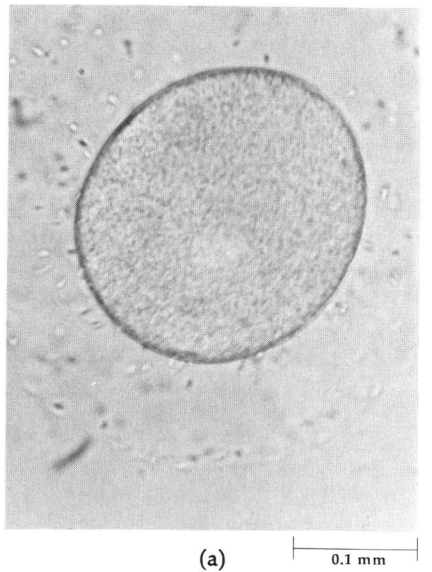 (a)   0.1 mm

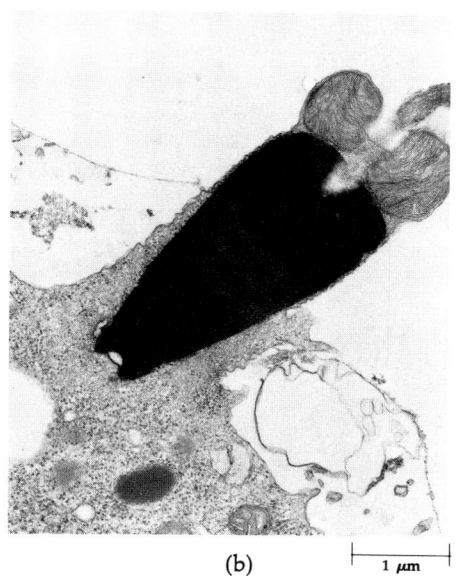 (b)   1 μm

First, it was noted that the number of chromosomes per cell in the tissues of any particular organism is generally the same from cell to cell. (One notable exception is found among the higher plants, in which many cells, for reasons which are not known, tend to have more than one set of chromosomes.) Moreover, individuals belonging to the same species have the same number of chromosomes per cell. Mosquitoes, for example, have 6 chromosomes per body cell; cabbages, 18; corn plants, 20; frogs, 26; sunflowers, 34; cats, 38; people, 46; plums, 48; dogs, 78; and goldfish, 94. When a cell divides by mitosis, as you will recall, each new cell receives the same number of chromosomes as the parent, dividing cell.

Second, the gametes of higher organisms—eggs and sperm—have exactly half the number of chromosomes that is characteristic for the somatic, or body, cells of the organism. The number of chromosomes in the sex cells is referred to as the haploid ("single") number, and the number in the somatic cells as the diploid ("double") number. For brevity, the haploid number is designated n and the diploid number $2n$. (Cells that have higher multiples of the haploid number of chromosomes, such as those in higher plants, are known as polyploid.) In man, for example, $n = 23$ and $2n = 46$. When a sperm fertilizes an egg, the two haploid nuclei fuse, and the diploid number is established.

The process by which the diploid number of chromosomes is reduced to the haploid number in the gametes is known as _meiosis_, from a Greek word meaning "to make smaller." Let us look first at the mechanics of meiosis, as observed by a microscopist. Subsequently, we shall look at the genetic interpretation of these events.

MEIOSIS VS. MITOSIS

The events that take place during meiosis somewhat resemble those that take place during mitosis, and meiosis, which is believed to have evolved from mitosis, uses much of the same cellular machinery. But there are a number of differences between them, of which three are of salient importance:

First, and most obvious, meiosis takes place in two stages involving two successive divisions and resulting in four new nuclei instead of two.

Second, whereas mitosis produces two nuclei that are identical to each other and to the parent nucleus from which they are formed, meiosis results in four nuclei that are not necessarily identical to each other and that have only half the number of chromosomes present in the parent nucleus.

Third, near the beginning of meiosis (but not mitosis), the chromosomes arrange themselves in homologous pairs; that is, each chromosome pairs up with another chromosome of the same size and shape, its homologue. The meaning of this crucial event in meiosis will be discussed later. (You may have guessed it already.)

THE STAGES OF MEIOSIS

In the following discussion, we shall describe meiosis in a plant cell in which the chromosome number is 8 ($n = 4$). Since meiosis consists of two successive divisions, it is convenient to refer to these divisions as meiosis I (the first division) and meiosis II (the second division).

12–2 _During prophase of meiosis, chromosomes become arranged in homologous pairs. Each homologous pair consists of four chromatids and is therefore also known as a tetrad (from the Greek_ tetra, _meaning "four")._

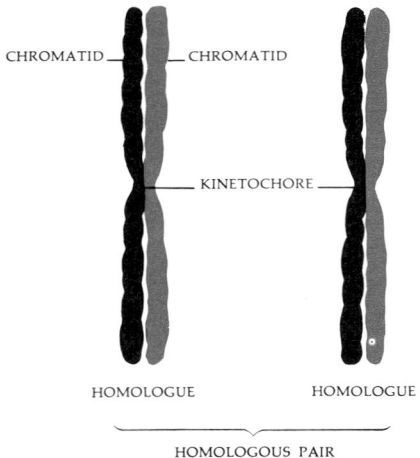

CHROMATID — CHROMATID

KINETOCHORE

HOMOLOGUE   HOMOLOGUE

HOMOLOGOUS PAIR
(TETRAD)

12–3 *Early prophase I in the formation of a sperm cell in the grasshopper. The homologous chromosomes are now paired; the individual chromatids are not visible, however, so each chromosome still appears as a single structure, and the tetrads appear double-stranded (rather than four-stranded). The dark area in the lower right is a sex chromosome, which is very prominent in the grasshopper.*

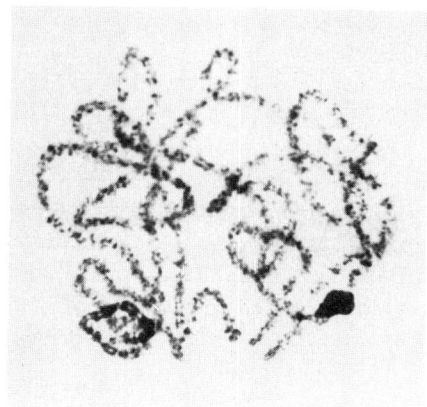

During interphase, before meiosis I, the chromosomes are duplicated (as they are before mitosis), so that by the beginning of meiosis each chromosome consists of two identical chromatids attached to each other at the kinetochore.

In prophase I (the prophase of meiosis I), the chromosomes come into view. During prophase, the homologous chromosomes (each consisting of two identical chromatids) come together in pairs. Since each homologue consists of two identical chromatids, the pairing process involves four chromatids; the resulting complex is known as a *tetrad*. At the time of pairing, some genetic material is generally exchanged between the homologues. This process, known as crossing over, is diagrammed in Figure 12–5 and will be discussed in Chapter 14.

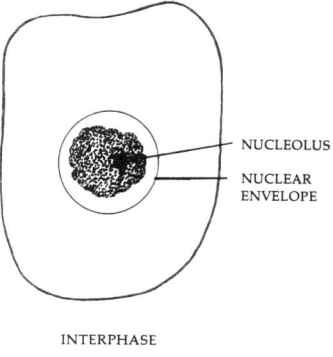

NUCLEOLUS

NUCLEAR ENVELOPE

INTERPHASE

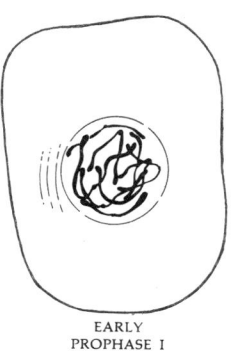

EARLY PROPHASE I

2–4 *Late prophase I. All four chromatids can be seen in most of the tetrads. The spindle fibers are not visible.*

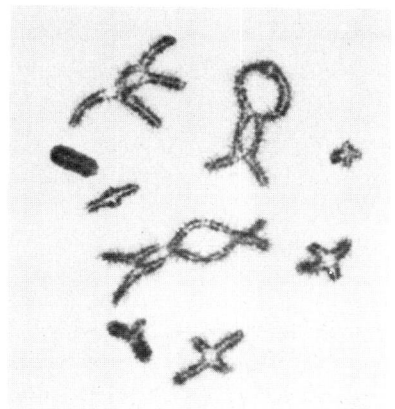

If you remember the arrangement of the homologous chromosomes at this stage of meiosis, you will be able to remember all the subsequent events with little difficulty.

Toward the end of prophase, the spindle apparatus is formed, the nucleolus disperses, and the nuclear envelope breaks down.

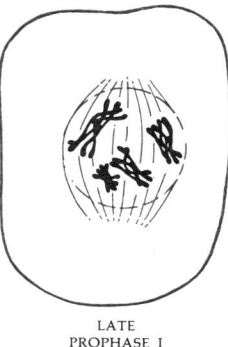

LATE PROPHASE I

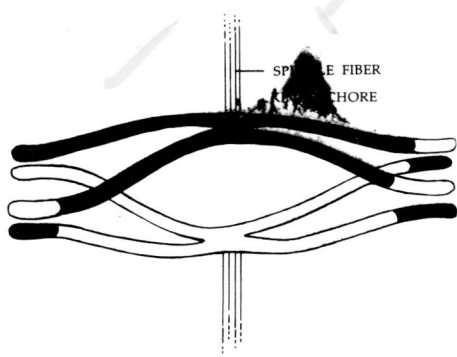

SPINDLE FIBER

KINETOCHORE

12–5   *Crossing over. During meiosis I, homologous chromosomes pair up to form tetrads, each consisting of four chromatids. Homologues connect at crossover points, where exchange of genetic material—crossing over—takes place. In this way, genetic material is redistributed among homologous chromosomes.*

In metaphase I, the homologous pairs (four, in this example) line up along the equatorial plane of the cell. Each pair consists of four chromatids (two chromosomes). The spindle has formed, and spindle fibers attach to each homologue in the area of the kinetochore. If this were an animal cell, centrioles and asters would be present, as in the mitosis series diagrammed on pages 160 to 162.

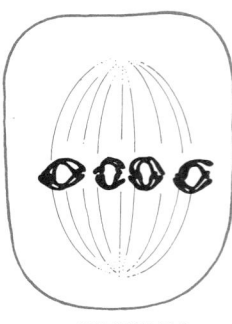

METAPHASE I

At anaphase I, the homologues, each consisting of two chromatids, separate, as if pulled apart by the spindle fibers attached to the kinetochores. The homologues cling to each other as they are pulled apart. The kinetochores do not divide, as they did in mitosis, and so the two chromatids of each chromosome do not separate.

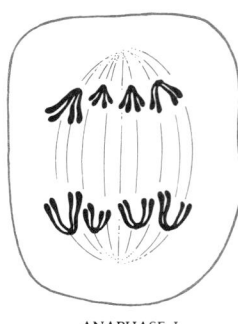

ANAPHASE I

By the end of the first meiotic division, the homologues have separated and nuclear membranes have formed around the two nuclei. Each nucleus now contains only half the number of chromosomes of the original nucleus. The cell may or may not divide into two cells; the diagrams here show an undivided cell.

TELOPHASE I

INTERPHASE II

Meiosis II resembles mitosis except that it is not preceded by duplication of the chromosomal material. At the beginning of the second meiotic division, the chromosomes, which may have dispersed somewhat, condense fully again. There are four in each nucleus (the haploid number), and they are still in the form of chromatids held together at the kinetochore.*

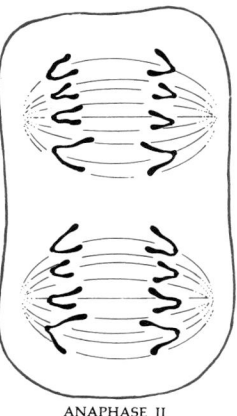

PROPHASE II

2–6 *Metaphase II in crested wheat grass. The chromosomes (chromatid pairs) are approaching the equatorial plane.*

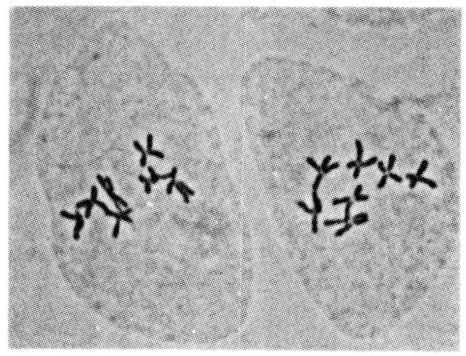

During prophase II, the nuclear envelopes, if present, dissolve, and new spindle fibers begin to appear. During the second metaphase, the four chromosomes in each nucleus line up on the equatorial plane. At anaphase II, as in mitosis, the kinetochores divide, the chromosomes separate, and each chromatid (now a single chromosome) moves toward one of the poles.

2–7 *Anaphase II in the royal fern* Osmunda regalis. *The kinetochores have divided, and the daughter chromosomes (previously each a chromatid) are moving to the opposite poles of the spindle.*

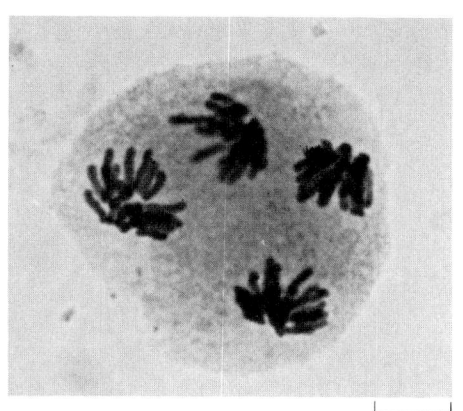

10 μm

METAPHASE II   ANAPHASE II

* In counting, it is often difficult to know whether to count a chromosome that has duplicated but has not divided as 1 or 2. It is customary to count such a chromosome as 1. The trick is to count kinetochores.

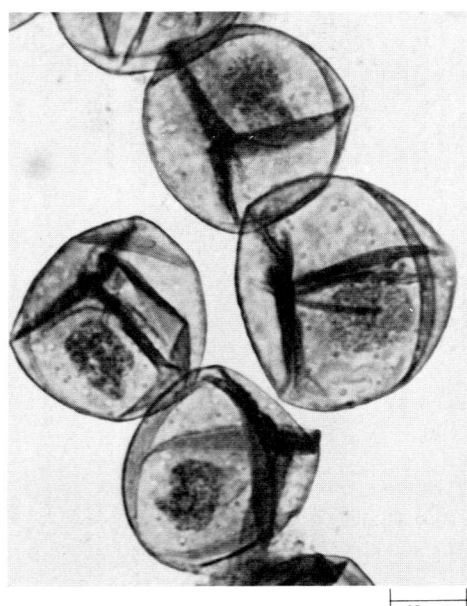

12–8 *The end of spore formation in* **Osmunda** *regalis. Each of these cells will develop into a pollen grain.*

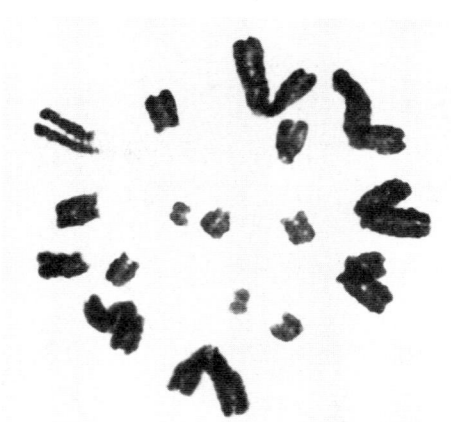

12–9 *Chromosomes from a diploid cell of a grasshopper. Note that even though these chromosomes are not paired, it is possible to pick out some of the homologues. The observation that chromosomes come in homologous pairs was one of Sutton's clues to the meaning of meiosis.*

During telophase, the spindles disappear and a nuclear envelope forms around each set of chromosomes. There are now four nuclei in all, each containing the haploid number of chromosomes. Cell division (cytokinesis) proceeds as it does following mitosis: Cell membranes and cell walls form, dividing the cytoplasm, and the cells begin to differentiate.

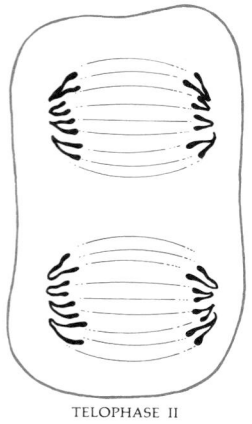

TELOPHASE II

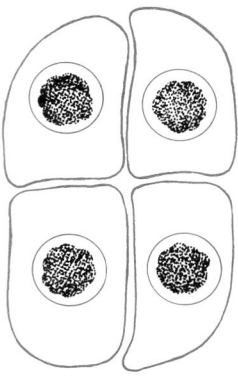

FOUR HAPLOID CELLS

CYTOLOGY AND GENETICS MEET: SUTTON'S HYPOTHESIS

In 1902 William S. Sutton, a graduate student at Columbia University, was studying the formation of sperm cells in male grasshoppers. In grasshoppers, as in other species, the male gametes are formed from a special group of cells in the testes, the *spermatogonia*. Like other cells in the grasshopper body, spermatogonia contain the diploid number of chromosomes. At the time the gametes are formed, the spermatogonia undergo two special cell divisions, the meiotic divisions, and then develop into sperm.

Observing the process of meiosis, Sutton was struck by the fact that the chromosomes that paired with one another at the beginning of the reduction divisions had physical resemblances to one another. In diploid cells, chromosomes apparently came in pairs. These pairs, or homologues, were only obvious at meiosis, but the discerning eye could find the homologues in the unpaired chromosomes when they became visible at the time of mitosis. Homologues, as we noted, are separated during the first meiotic division. When gametes—the newly formed sperm and egg—come together at fertilization of the egg, each chromosome from the sperm cell meets a new homologue, the corresponding chromosome in the egg. (The one exception is in the case of the sex chromosomes, which will be discussed in Chapter 14.)

Sutton postulated that the homologues that he could observe undergoing separation in the formation of the sperm cells of the grasshopper were replicas of the homologous chromosomes that had come together to make the fertilized egg, at the first moment in that particular grasshopper's biography. In other words, if one homologue of a chromosome pair came from the grasshopper's mother, the other homologue must have come from the father. The two had

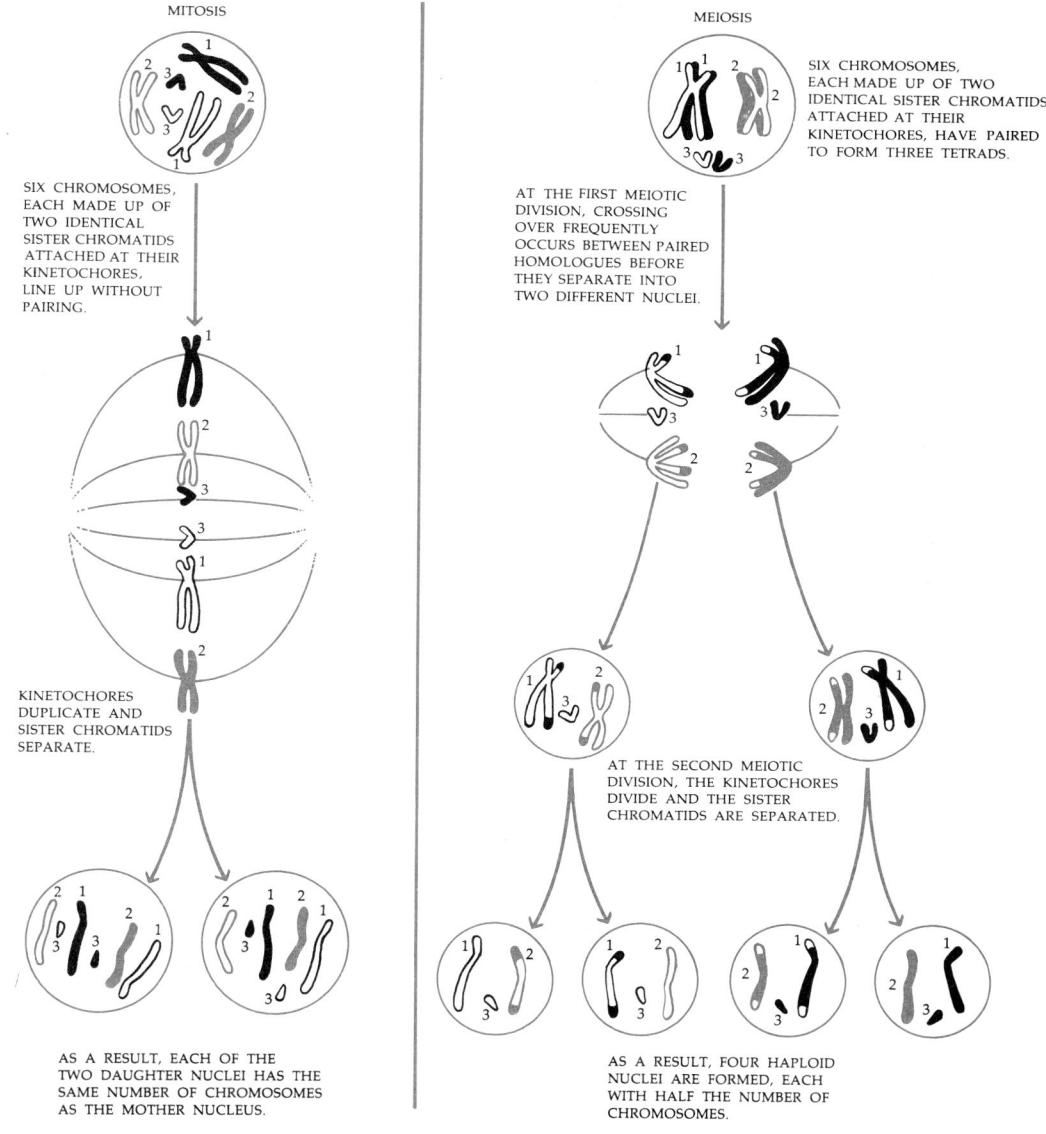

MITOSIS

SIX CHROMOSOMES,
EACH MADE UP OF
TWO IDENTICAL
SISTER CHROMATIDS
ATTACHED AT THEIR
KINETOCHORES,
LINE UP WITHOUT
PAIRING.

KINETOCHORES
DUPLICATE AND
SISTER CHROMATIDS
SEPARATE.

AS A RESULT, EACH OF THE
TWO DAUGHTER NUCLEI HAS THE
SAME NUMBER OF CHROMOSOMES
AS THE MOTHER NUCLEUS.

MEIOSIS

SIX CHROMOSOMES,
EACH MADE UP OF TWO
IDENTICAL SISTER CHROMATIDS
ATTACHED AT THEIR
KINETOCHORES, HAVE PAIRED
TO FORM THREE TETRADS.

AT THE FIRST MEIOTIC
DIVISION, CROSSING
OVER FREQUENTLY
OCCURS BETWEEN PAIRED
HOMOLOGUES BEFORE
THEY SEPARATE INTO
TWO DIFFERENT NUCLEI.

AT THE SECOND MEIOTIC
DIVISION, THE KINETOCHORES
DIVIDE AND THE SISTER
CHROMATIDS ARE SEPARATED.

AS A RESULT, FOUR HAPLOID
NUCLEI ARE FORMED, EACH
WITH HALF THE NUMBER OF
CHROMOSOMES.

been copied faithfully in cell division after cell division through the grasshopper's several stages of development until the time of meiosis, when they were once again separated.

Suddenly the facts fell into place. Suppose chromosomes carried genes, the *Elemente* described by Mendel. This idea does not seem very startling to us now, but remember that the gene was just an abstract idea or mathematical unit to the geneticist, and that the chromosome was just an unidentified colored body to the cytologist. Suppose, Sutton reasoned, alleles occurred on homologous chromosomes. Then the alleles could always remain independent and so could separate at meiosis, with new pairs of alleles forming when the gametes came together at fertilization. Mendel's law of the segregation of inherited traits could be explained by the segregation of the homologous chromosomes at meiosis.

Now, suppose that half of the chromosomes of the male grasshopper were inherited from his mother and half were inherited from his father. (See Figure 12–11.) At the time of meiosis, the maternal chromosomes could be separated from the paternal ones. Or in terms of the garden pea, cross-fertilization of a plant having smooth and yellow seeds with a plant having green and wrinkled seeds could produce, in the F_2 generation, plants with smooth and green seeds and plants with yellow and wrinkled seeds, although neither parent plant had had this phenotype. (See Figure 12–12.) Mendel's law of independent assortment could be explained on the basis of the independent movement of chromosomes at meiosis.

12–11 *An organism, such as the grasshopper, develops from a fertilized egg cell (zygote) containing two haploid sets of chromosomes, one from each parent. The maternal set is shown here in color;* *the paternal in black. Each cell in the grasshopper's body contains copies of these chromosomes, including the cells from which the sperm cells form. At meiosis, when the chromosomes are* *once again reduced to the haploid number, the chromosomes are reassorted so each gamete is likely to contain a mixture of maternal and paternal chromosomes.*

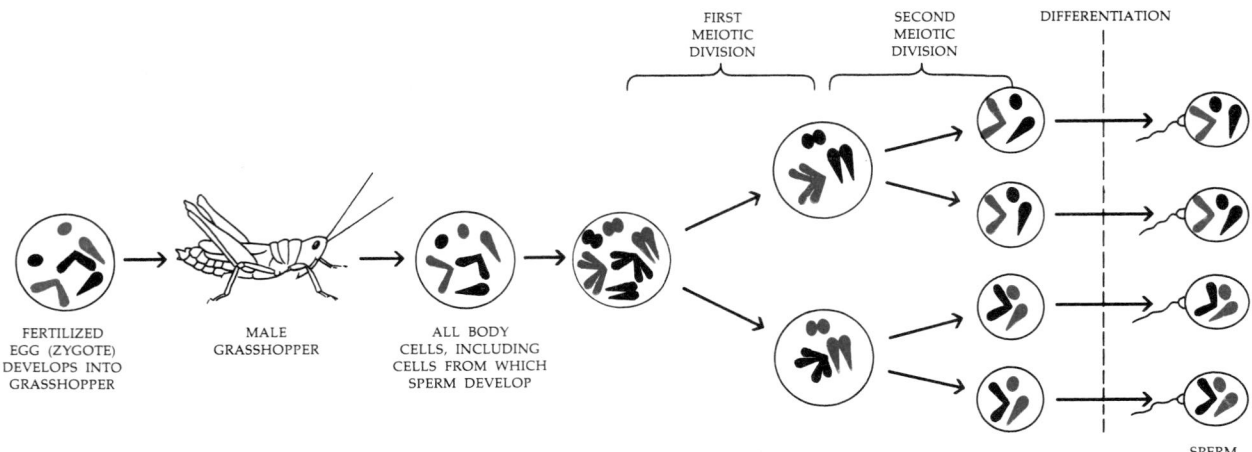

FERTILIZED EGG (ZYGOTE) DEVELOPS INTO GRASSHOPPER

MALE GRASSHOPPER

ALL BODY CELLS, INCLUDING CELLS FROM WHICH SPERM DEVELOP

FIRST MEIOTIC DIVISION

SECOND MEIOTIC DIVISION

DIFFERENTIATION

SPERM

2–12 *The chromosome distributions in Mendel's cross of round yellow and wrinkled green peas, according to Sutton's hypothesis. Although the pea has 14 chromosomes (n = 7), only 4 are shown here, the two carrying the genes for round or wrinkled and the two carrying the genes for yellow or green. (This is analogous to what Mendel did when he selected the two traits to study.) As you can see, one parent is homozygous for the recessives, one for the dominants. Therefore, the only gametes they can produce are RY and ry. (Remember, R now stands not just for the trait but for the chromosome carrying the trait, as do the other letters.) The F₁ generation, therefore, must be Rr and Yy. When a mother cell of this generation undergoes meiosis, R is separated from r and Y from y when the respective kinetochores divide in metaphase II. Four different types of n egg cells are possible, as the diagram reminds us, and also four different types of pollen nuclei. These can combine in 4 × 4, or 16, different ways; the ways in which they can combine are illustrated in the Punnett square at the bottom of the figure.*

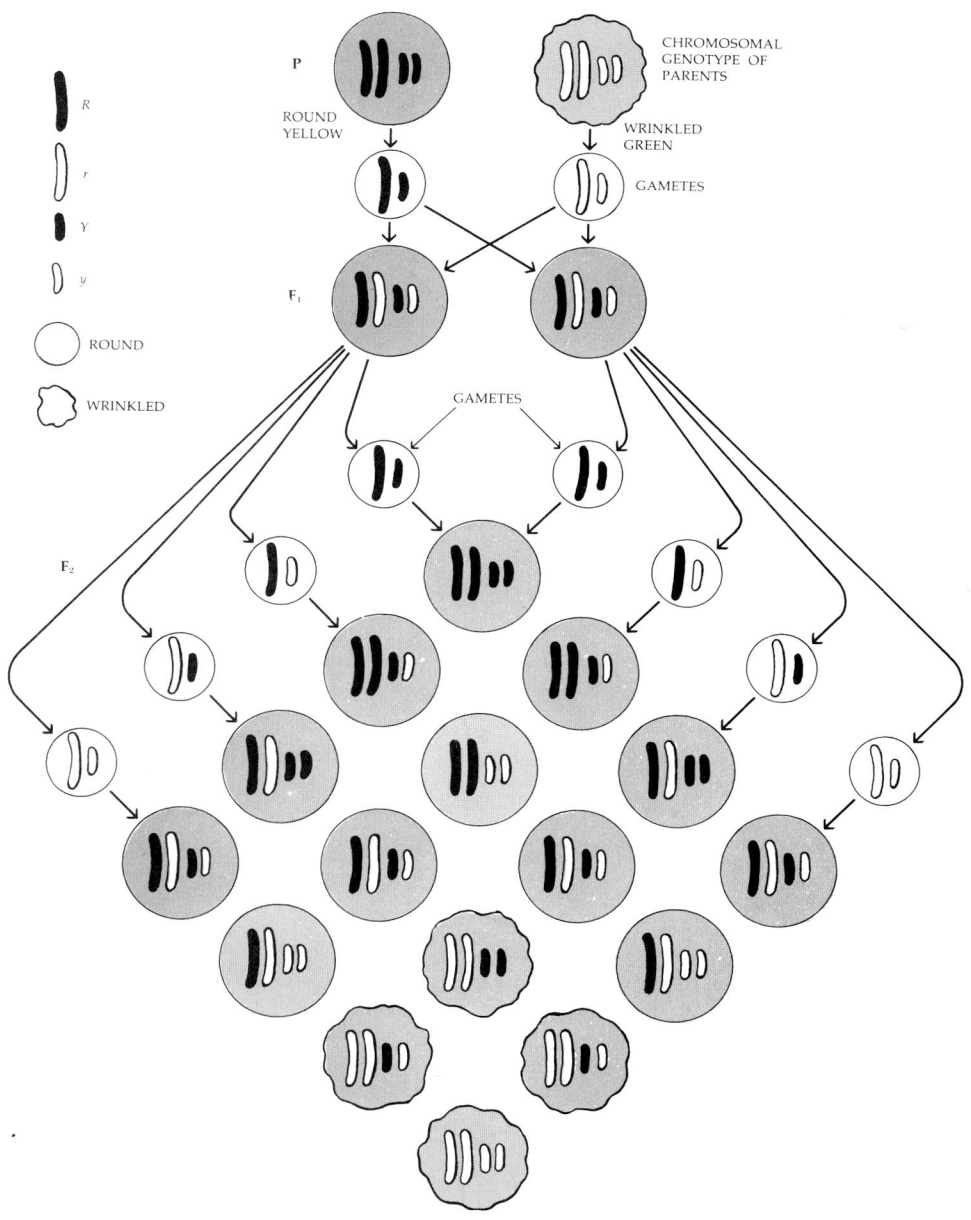

SEXUAL AND ASEXUAL REPRODUCTION

Sexual reproduction always involves two events: (1) the formation of gametes by meiosis and (2) the union of gametes, or fertilization. Man and other higher animals can reproduce only sexually, but many plants and some invertebrates can reproduce either sexually or asexually. (Asexual reproduction is so common among plants that it is often called vegetative reproduction, even when it occurs in animals.) Some few single-celled organisms, such as amoebas, reproduce only asexually. In a single-celled eukaryotic organism like the amoeba, asexual reproduction involves simple cell division with mitosis. In many-celled animals, asexual reproduction usually takes place by the production of a bud or the breaking off of a fragment of the parent animal; sea anemones, for example, break apart to form new sea anemones. Algae and plants may also break apart, or plants may produce roots or runners from which new individuals arise. Whether the new individual that is produced asexually is a single cell or a group of many cells, it is a product of mitosis and as such is genetically identical to its single parent.

Sexual reproduction demands a tremendous expenditure of energy on the part of the organisms involved. Male animals often produce thousands or even millions of sperm cells for each one that reaches an egg. Plants cover themselves with flowers. Birds invest in bright, improbable plumage. Human males write poems, fight duels, renounce kingdoms, even become gainfully employed, all in the cause of sexual reproduction.

As in the case for other adaptations, biologists can appropriately ask the meaning of the phenomena of sexual reproduction. The answer that biologists offer to this question is that sexual reproduction provides a source of variations in a population, and these variations provide the rich store of material upon which natural selection can operate. The fact that all higher animals—the most complex and advanced organisms, evolutionarily speaking—reproduce sexually appears to confirm the evolutionary advantages of this method of reproduction.

The arguments in support of the hypothesis that Mendel's factors, or genes, are carried on chromosomes can be summarized as follows:

1. All hereditary characteristics are carried in the sperm and egg cells since these cells are the only bridge from one generation to another.
2. Since sperm cells lose almost all their cytoplasm as they mature, the hereditary factors are probably carried in the nucleus.
3. The only visible parts of the nucleus which are accurately divided during cell division are the chromosomes. This suggests that the factors, or genes, must be carried on the chromosomes.
4. Chromosomes obey Mendel's laws.
   a. Chromosomes occur in pairs; so do Mendelian factors.
   b. Chromosomes segregate at meiosis; Mendelian factors segregate at the formation of the gamete.
   c. The members of a chromosome pair appear to segregate independently of other chromosome pairs; Mendelian factors segregate independently.

As occurs often in the history of science, two other biologists recognized the correlation between the behavior of Mendel's *Elemente* and the observed movements of the chromosomes, but young Sutton's paper appeared first, and his presentation was by far the most convincing.

A Test of the Hypothesis

Note that Sutton did not *prove* genes are carried in chromosomes. He offered a hypothetical explanation for the two sets of observed phenomena—Mendel's laws of inheritance and the behavior of chromosomes during cell division. On the basis of his hypothesis, it was possible to make certain predictions. One of the first and most obvious concerned independent assortment. For example, the gametes of a pea plant have only seven chromosomes ($n = 7$), and they clearly carry more than seven genes. Thus, one is led to predict that many genes will be present in the same chromosome and that such genes may not follow Mendel's law—that is, they may not assort independently. This prediction turned out to be true, and is an important exception to Mendel's second law. Genes on the same chromosome are said to exhibit "linkage," a subject we shall discuss more fully in Chapter 14.

-13 *Possible distributions of chromosomes at meiosis. The black chromosomes were originally of paternal origin, and the colored chromosomes of maternal origin. They were transmitted by replication and mitotic division to the cells from which the eggs and sperm were formed. In the course of meiosis, these chromosomes are sorted out among the haploid cells. As you can see, chromosomes of maternal or paternal origin do not stay together but are sorted independently. (a) If the original number of chromosomes is 4 (n = 2), the number of possible combinations of chromosomes is 2^2, or 4. (b) If the original number is 6, the number of possible combinations is 2^3, or 8, and (c) if there are 8 chromosomes, 16 different combinations (2^4) are possible. Because maternal and paternal chromosomes differ in some of their genetic material, each of these cells is genetically different. Man with his 46 chromosomes is capable of producing 2^{23} kinds of sperm cells—8,388,608 different combinations of chromosomes, equal in number to the population of New York City. And this does not take into account the additional variations introduced by crossing over, which we shall examine in Chapter 14.*

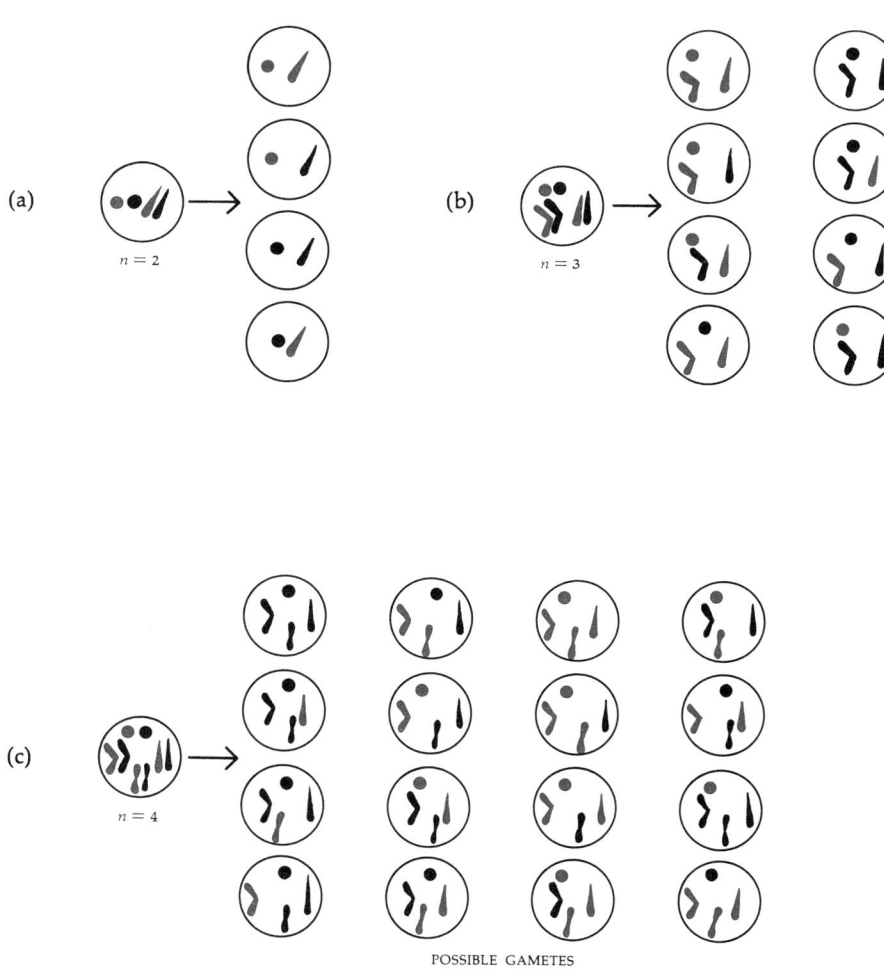

POSSIBLE GAMETES

LIFE CYCLES

Sexual reproduction is characterized by two events: the coming together of the sex cells (fertilization) and meiosis. Following meiosis, the number of chromosomes is haploid (n), represented by a single line. Following fertilization, the number is double, or diploid (2n), represented by a double line.

Fertilization and meiosis occur at different points in the life cycles of different organisms. In primitive eukaryotic cells (a), meiosis occurs immediately after fertilization and most of the life cycle is spent in the haploid state (signified by the single line). In higher animals (b), meiosis is typically soon followed by fertilization. As a consequence, during most of the life cycle the organism is diploid. In plants (c), fertilization and meiosis are separated and the organism characteristically has both a diploid and a haploid phase.

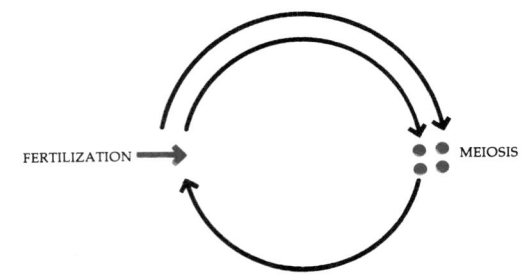

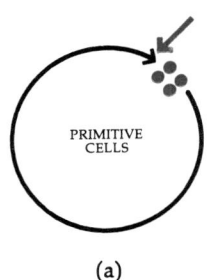

(a)

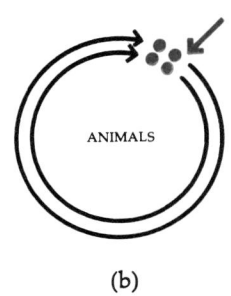

(b)

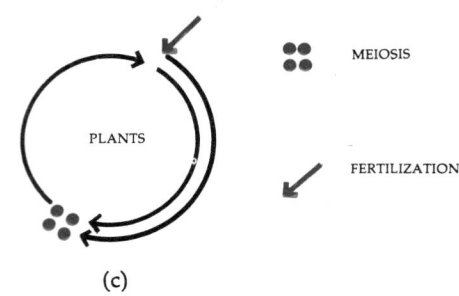

(c)

MEIOSIS

FERTILIZATION

MENDELIAN INHERITANCE IN MAN

A number of man's many inherited characteristics have been shown to follow the simple patterns of inheritance first observed by Mendel. One such characteristic is brachydactylism (see Figure 12–14a); this and other characteristics involving fingers and toes, including possession of extra ones, seem to involve simple dominants. Another simple dominant trait is tongue rolling (Figure 12–14b). Can you roll your tongue? Can your parents? What is your genotype for tongue rolling? If you cannot roll your tongue, does this mean that neither of your parents can? This may seem to be a very trivial sort of characteristic, one that neither natural selection nor society would favor, but oddly enough, very small differences such as this are often the reflection of more fundamental differences which may have considerable importance over an evolutionary time span.

Of more immediate consequence are a number of congenital diseases which are the result of the coming together of recessive genes. One such disease is sickle cell anemia. In persons homozygous for the sickling gene, a large proportion of the red blood cells "sickle"—that is, form a sickle shape—and then clog the small capillaries, causing blood clots and depriving vital organs of their full supply of blood. This produces continuous, painful illness and, usually, death at an early age. About 4 percent of the population in certain tropical regions in Africa are born with sickle cell anemia, and almost half of the members of some African tribes are known to carry the recessive gene. In this country, it is found almost exclusively among blacks.

(a) *A dominant gene is responsible for the trait known as brachydactylism (short fingers). In the brachydactylous hands shown here, the first bones of the fingers are of normal length but the second and third bones are abnormally short. (b) The tongue-rolling ability is transmitted by a dominant gene. Seven out of ten people have this ability. Do you? You can perform a simple experiment in genetics by checking to see if your parents and your brothers and sisters can do it.*

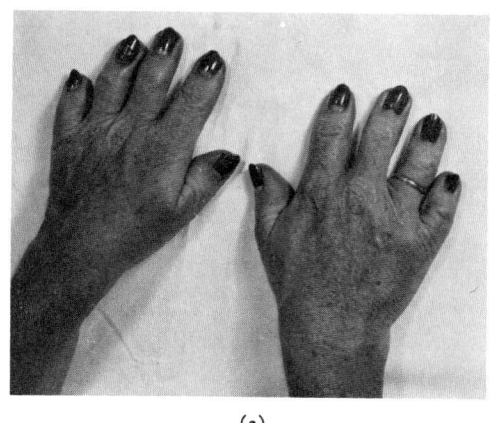

(a)

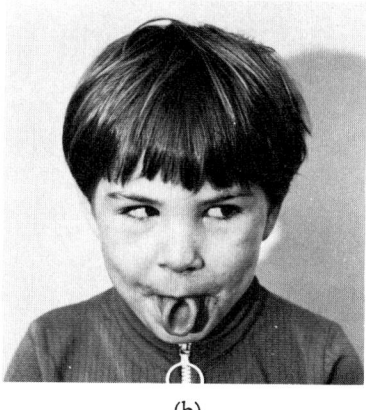

(b)

PKU (phenylketonuria), which produces mental deficiency in infants, is also the result of a "double dose" of a recessive gene; so is Tay-Sachs disease, which appears almost exclusively among Jews of Central European ancestry.

Blood Groups

Probably the most familiar characteristic in human beings that is determined by a single group of alleles is the ABO blood series. The existence of blood groups was discovered in 1900 by Karl Landsteiner. Mixing samples of blood taken from members of his laboratory staff, Landsteiner found that sometimes the red blood cells would clump together, or agglutinate, and sometimes they would not. From these experiments, he developed the theory that there were different categories of blood—mixing blood of different categories might often result in agglutination, while mixtures of blood of the same category would not agglutinate. He worked out four major blood groups: A, B, AB, and O. Before long, it was established that these blood types are inherited according to Mendelian laws.

The agglutination phenomenon is caused by antibodies, globular proteins that react against foreign substances in the blood. (We shall discuss antibodies at greater length in Chapter 33.)

If your blood type is A, this means that on the surface of your red blood cells is a specific polysaccharide, A, that is not found on the surface of blood cells of persons with type O or B. Persons with type B have polysaccharide B on their red blood cells; persons with type AB have the two types of polysaccharides; and persons with type O have neither A nor B polysaccharides. People of blood type A have in their blood antibodies to B. Similarly, type B have antibodies to A. Type O individuals have antibodies to both A and B, while type AB have neither. As a consequence, if you—still hypothetically blood type A—are given a transfusion of blood type B or blood type AB, your body's antibodies against the B or AB cells will agglutinate the donor B-type cells in your bloodstream. This reaction can be so violent that it is sometimes fatal. You can receive O cells safely, however, since they contain no polysaccharide that your body will recognize as foreign. (See Figure 12–15 and Table 12–1.)

12–15 *Severe and sometimes fatal reactions can occur following transfusions of blood of a different type from the recipient's. These reactions are the result of agglutination of the blood cells caused by antibodies present in the recipient's serum. Blood-group reactions to transfusions can be demonstrated equally well in test tubes, as shown here. The blood that is shown agglutinating has natural antibodies against the donor blood. Persons with type O blood used to be called* universal donors *and those with type AB blood,* universal recipients. *Now other factors are checked as well.*

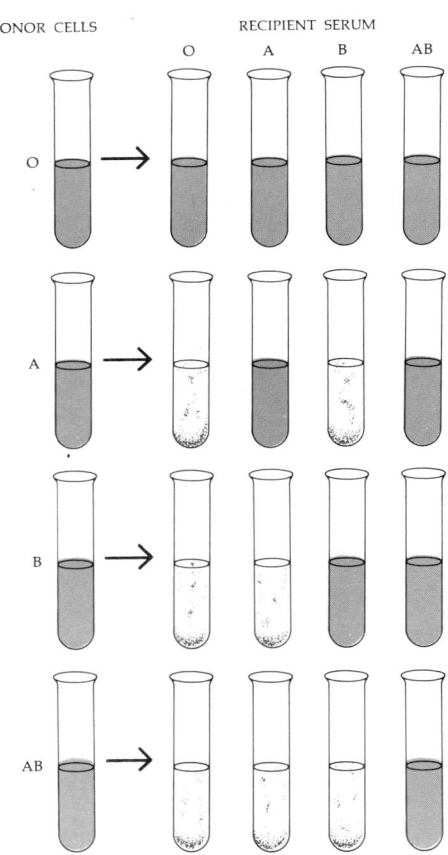

DONOR CELLS   RECIPIENT SERUM

Table 12–1 *Blood Groups*

Group	Genotype	Reaction with antibodies		Antibodies in blood plasma
		Antibody A	Antibody B	
O	O/O	—	—	Antibody A, antibody B
A	A/A, O/A	+	—	Antibody B
B	B/B, O/B	—	+	Antibody A
AB	A/B	+	+	None

Inheritance of Blood Types

Blood types are inherited, and the manner of inheritance can be determined from Table 12–1. If you have AB blood, it means that one of your parents is A or AB and the other is B or AB. If you have A blood, it means that you inherited an A gene from one parent and either an A gene or an O gene from the other. If you have O-type blood, you must have had parents who each carried one O gene, although they may have been, phenotypically, either A or B.

In the famous Charlie Chaplin paternity case in the 1940s, the baby's blood was B, the mother's A, and Chaplin's O. If you had been the judge, how would you have decided the case?* It can never be proved that someone *is* the father of a particular child, but, as you can see in Table 12–2, it is possible to prove that someone could *not* be the father.

The blood groups are examples of multiple alleles (A, B, and O). It is not unusual for genes to have more than two alternative forms, although, of course, only two alleles can be present in a diploid cell at one time.

Persons with O-type blood used to be known as universal donors and those with AB-type blood as universal recipients. However, a number of additional blood factors have been found since Landsteiner's time, and they are also checked in modern blood banks to determine blood compatibility.

Rh Factor

The Rh factor is named after the rhesus monkey, which is used as a test animal for it. If a woman is Rh-negative, meaning that she lacks the Rh factor, and her husband is homozygous for the Rh factor, any child that is conceived will be Rh-positive. The mother may form antibodies against the Rh substance on the red blood cells, and if these antibodies enter the infant's circulation across the placenta, an agglutination reaction may occur, endangering the infant's life. (Note that if the mother is Rh-positive, it does not matter whether the father is negative or positive, in terms of the Rh factor.) Transfu-

---

* As a matter of fact, Chaplin was judged guilty. Blood-group data are not admitted as evidence by some states in cases of disputed parentage.

Table 12-2 *Inheritance of Major Blood Types*

Phenotypes of parents		Children possible	Children not possible
A	A	A, O	AB, B
A	B	A, B, AB, O	—
A	AB	A, B, AB	O
A	O	A, O	AB, B
B	B	B, O	A, AB
B	AB	A, B, AB	O
B	O	B, O	A, AB
AB	AB	A, B, AB	O
AB	O	A, B	O, AB
O	O	O	A, B, AB

sion of the baby's blood at the time of birth—or even before, by new techniques —can flush the mother's antibodies out of the baby's system and greatly increase its chances of survival. Discovery of the ABO blood groups, which made such transfusions safe and practicable, and of the Rh system rank among the great medical advances of the twentieth century.

CONCLUSIONS

Thus Mendel's laws, based on observations of phenotypes and applications of mathematical laws, were corroborated by the observations of the behavior of chromosomes at meiosis. The rediscovery of Mendel's principles put genetics on a firm scientific basis. It also opened the way to the understanding of inherited factors in man, some of which are of great medical importance.

SUMMARY

In the late nineteenth century, rapid advances in the tools and methods of microscopy led to the discovery of mitosis and meiosis.

Mitosis (see Chapter 10) is the process by which the chromosome complement of a parent cell is passed on identically to the two daughter cells at the time of cell division. The chromosome complement of each daughter cell is identical to the other and to the parent.

Meiosis is the process by which gametes are formed. During meiosis, the chromosome number is reduced by half (to the haploid number) and chromosomes are reassorted. The diploid chromosome number (which is characteristic of the individual and the species) is restored at fertilization.

Before meiosis I, each chromosome duplicates into two chromatids joined at their kinetochore. At the start of meiosis, the chromosomes arrange themselves in pairs (homologues). Each homologous pair contains four chromatids; the

resulting complex is called a tetrad. One chromosome (two chromatids) is of paternal origin and the other is of maternal origin. Early in meiosis, crossing over occurs between homologues, resulting in exchanges of genetic material.

In the first meiotic division, the homologues are separated. Two nuclei are produced; each nucleus has the haploid number of chromosomes, each consisting of two chromatids. In the second division of meiosis, the kinetochores divide and the chromatids separate, as in mitosis. When the two new nuclei divide, four haploid cells result, each with a single copy of each of its chromosomes.

Meiosis provides a source of variation on which natural selection acts.

Sutton noted the analogy between the behavior of the chromosomes at meiosis and the assortment of genetic traits described by Mendel. On the basis of this observation, Sutton proposed that Mendelian factors (genes) are carried on chromosomes. Chromosomes appeared to come in pairs, and Sutton postulated that alleles occur on these homologous chromosomes. The chromosome pairs separate at meiosis, and new pairs form again when the egg is fertilized. Sutton's hypothesis led to the recognition of an important exception to Mendelian laws: Independent assortment may be modified if the genes involved are on the same chromosome.

A number of man's hereditary characteristics are inherited as simple Mendelian dominants or recessives. Among these are anatomical characteristics such as brachydactylism, certain muscular abilities (tongue rolling), as well as various diseases. Blood type is perhaps the most familiar example of inheritance from a single group of alleles. The four major blood groups are A, B, AB, and O.

QUESTIONS

1. Describe what is happening at each step in the illustrations which appear below.

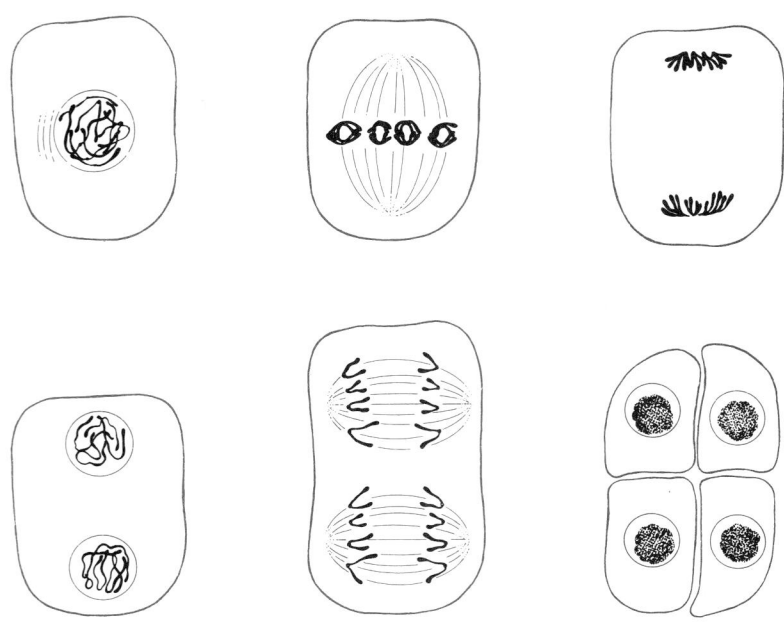

2. Draw a diagram of a cell with eight chromosomes ($n = 4$) at meiotic prophase I.

3. Diagram the possible gametes resulting from a single meiosis in an F_1 hybrid with six chromosomes ($n = 3$). If this were a self-fertilizing plant—a plant such as the garden pea, in which the same plant provides both pollen and ova—how many chromosome combinations would be possible in the fertilized egg? Diagram four of these combinations.

4. If two healthy parents have a child with sickle cell anemia, what are their genotypes with respect to this allele? Having had one such child, what are their chances of having another child with the same disease?

Chapter 13

More About Genes

During the decade that followed the rediscovery of Mendel's laws, many studies were carried out which, while confirming his work in principle, showed that the action of genes is more complex than it had at first appeared to be. This chapter will be concerned with these amplifications of Mendel's original observations.

MENDEL'S LAWS AND THE THEORY OF EVOLUTION

Mendel's work filled an important gap in Darwin's evolutionary theory by explaining why small variations persisted in populations and were not lost by blending. However, Mendelian principles presented new problems to the early evolutionists, because they appeared to offer no possibility for major changes in the genetic makeup of organisms. Segregation of characteristics explained how variations were maintained from generation to generation. Independent assortment explained how individuals could have combinations of characteristics not present in either parent and so be better adapted, in evolutionary terms, than either parent. But if all hereditary variations were to be explained by the reshuffling process proposed by Mendel, there would be little or no opportunity for major changes in organisms. How, for instance, could new species come about? In fact, these same principles were used by opponents of evolution to demonstrate that evolution could not possibly have occurred.

MUTATIONS

A solution to the problem was proposed by one of Mendel's rediscoverers, a Dutch botanist named Hugo de Vries. De Vries was studying genetics in the evening primrose. Heredity in the primrose, he found, was generally orderly and predictable, as in the garden pea, but occasionally a characteristic appeared that was not present in either parent or indeed anywhere in the lineage of that particular plant. De Vries hypothesized that this characteristic came about as the result of an abrupt change in a gene and that the trait embodied in the changed gene was then passed along like any other hereditary trait. De Vries spoke of this hereditary change as a *mutation* and of the organism that carried it as a *mutant*. Different alleles of the same gene, de Vries proposed, arise as a result of mutations. Mutations are the source of new genetic characteristics, and

13–1 *Hugo de Vries is shown standing next to* Amorphophallus titanum, *which has the largest flower of any of the flowering plants. De Vries, a Dutch botanist, was the first to recognize the nature of mutations and their role in hereditary processes.*

it is because of mutations, combined and recombined at meiosis, that species are able to adapt and evolve.

As it turns out, only about 2 of some 2,000 changes in the evening primrose observed by de Vries were actually mutations. The rest were due to new combinations of genes rather than to actual changes in any particular gene. However, de Vries's definition of a mutant and his recognition of the importance of the concept of mutation are still valid, although most of his examples are not.

How Mutations Are Produced

Mutations occur constantly. The average spontaneous mutation rate for a given gene has been estimated to be 1 or 2 new mutations per 100,000 genes per generation. This means that in every 100,000 sperm cells 1 or 2 can be expected to carry a new mutation for a particular gene.

In the 1920s it was found that exposure of gametes or gamete-forming cells to x-rays greatly increases the rate of mutation. Other radiations, such as ultraviolet light, and some chemicals can also act as mutagens—agents that produce mutations. Studies of mutations have shown that whether they are deliberately produced by radiations and chemicals or are "spontaneous,"* they are usually detrimental to the organism and, in fact, many of them are lethal. This is not surprising. If one were to change a word at random in a Shakespearean sonnet or a wire at random in a television set, an improvement would be unlikely, and the results might well be disastrous. It is reasonable to assume that the genetic makeup of an individual is as precise in its engineering and delicate in its balance as a sonnet or a television set.

* "Spontaneous," in this sense, means that we do not know why they happen.

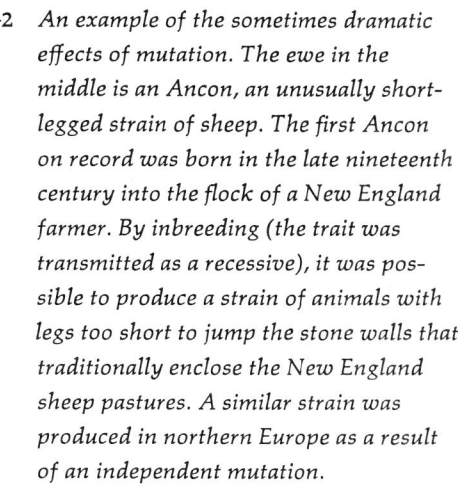

3–2 An example of the sometimes dramatic effects of mutation. The ewe in the middle is an Ancon, an unusually short-legged strain of sheep. The first Ancon on record was born in the late nineteenth century into the flock of a New England farmer. By inbreeding (the trait was transmitted as a recessive), it was possible to produce a strain of animals with legs too short to jump the stone walls that traditionally enclose the New England sheep pastures. A similar strain was produced in northern Europe as a result of an independent mutation.

Mutagens act, apparently, by altering the chromosomal material of the gametes. It is, in part, because of such possible effects on the sex cells—effects whose consequences will be concealed until the next generation—that many authorities are concerned about the increase in radiation exposure caused by atomic testing or by unwarranted or careless use of x-rays. (Radiation exposure can also produce harmful changes in somatic cells that may result in cancer—as shown, for example, by the increased incidence of leukemia among the "survivors" at Hiroshima.)

INCOMPLETE DOMINANCE

Another important modification of Mendel's work concerns the concept of dominance. Dominant and recessive traits are not always so clear-cut as those Mendel chose to study in the pea plant. Some traits do blend phenotypically. For instance, the cross between a red snapdragon and a white snapdragon produces a first generation that is pink (Figure 13–3). But when this generation is allowed to self-pollinate, the parental phenotypes reappear in the second (F_2) generation. Incomplete dominance is very common and a number of examples are known.

13–3 *A cross between a red snapdragon and a white snapdragon. This looks very much like the cross between a red- and a white-flowering pea plant shown in Figure 11–9, but there is a significant difference. Although the Mendelian genotypic ratio appears in the F_2 generation, the heterozygote is pink rather than red.*

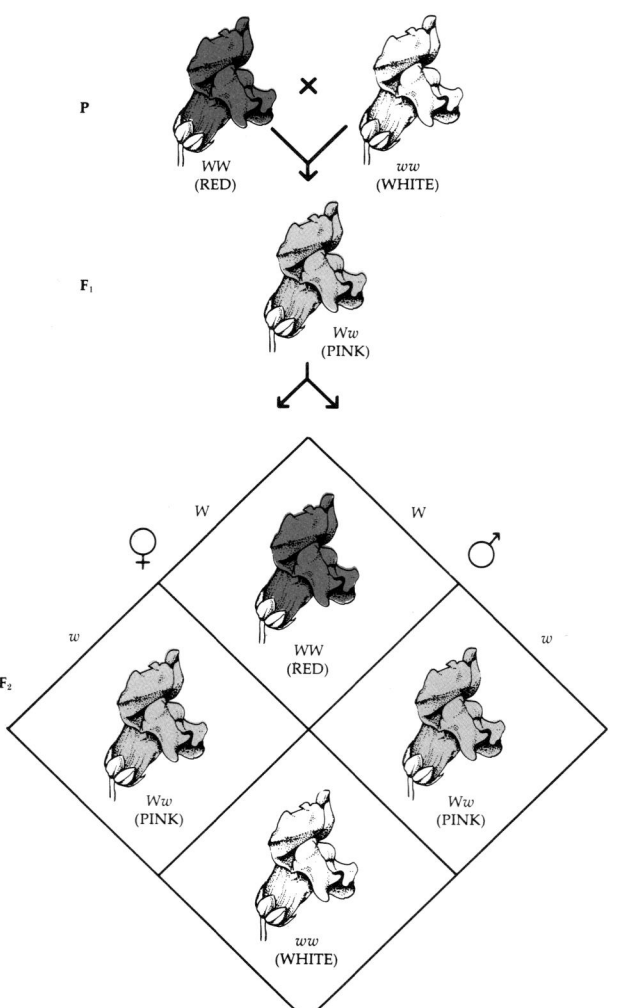

13-4 *A cross between any yellow mouse and an agouti mouse produces half yellow and half agouti. As the Punnett square indicates, this result shows that the yellow mouse is a heterozygote and that yellow is dominant over agouti.*

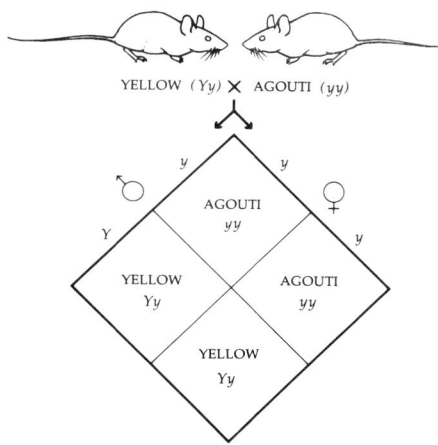

YELLOW (*Yy*) ✕ AGOUTI (*yy*)

13-5 *A cross between any two yellow mice produces yellow and agouti mice in a ratio of 2:1. The yellow mice are heterozygotes. Yellow homozygotes die before birth.*

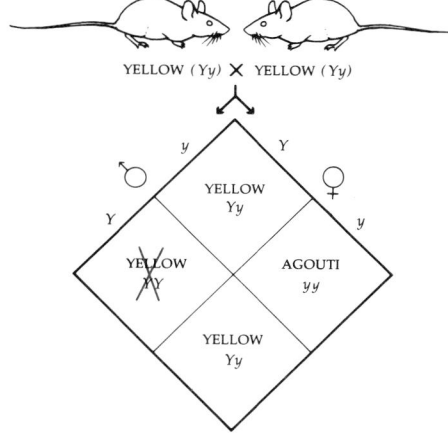

YELLOW (*Yy*) ✕ YELLOW (*Yy*)

LETHAL GENES

Another factor which may seem to alter Mendelian ratios is a lethal gene or a lethal combination of genes. Such a gene or combination of genes causes death of the offspring at conception or some time during development. Many spontaneous abortions in humans are the result of lethal genes.

An example of a lethal gene is found among the genes determining coat color in mice. The "mousy" color of wild mice is the result of a peculiar distribution of pigment in the individual hairs. Each hair is mostly black or dark brown, but, just below the tip, each has a yellow band. This color pattern is called agouti, after a wild animal (the agouti) that also has this color. When mice of a true-breeding agouti strain are crossed with true-breeding all-black animals (lacking the yellow bands), the F_1's are all agoutis, and in the F_2 (produced by a cross of F_1's) there are three agoutis to one black. In other words, in this cross, the agouti trait behaves like a simple Mendelian dominant.

Coat color in mice, like blood groups in man, is determined by multiple alleles. In addition to agouti and black, mice occasionally have yellow coats. If a yellow mouse is bred to a true-breeding agouti mouse, half of the mice of the first litter are yellow and half are agouti. This ratio can be explained if we hypothesize that the yellow mice are heterozygous and that, in this cross, the yellow is dominant over the agouti.

We should be able to test this hypothesis by crossing a yellow with a yellow. Theoretically, such a cross should produce litters that are three-fourths yellow and one-fourth agouti. In fact, what it produces are litters in which the ratio of yellow to agouti is approximately 2:1. Testcrosses of the yellows showed that they were always heterozygotes. Mice homozygous for the yellow gene were never detected in breeding experiments. There was another clue: Litters produced by crossing yellow mice were only about three-fourths the size of an average litter. It appeared that yellow homozygotes died before birth. Experimenters removed the uteri of yellow mice pregnant from a cross with yellow males and found, in confirmation, that about one in four of the embryos had died and was being reabsorbed. Yellow is an example of a lethal gene.

BROADENING THE CONCEPT OF THE GENE

Mendel's experiments seemed to suggest that each gene affects a single characteristic in a one-to-one relationship. It was soon discovered, however, that a single gene sometimes affects many traits in an organism (as indicated by the relationship between yellow coat color and mortality just described) and that, conversely, a single trait is often affected by many genes. The first property is known as *pleiotropy*, the second as *polygenic inheritance*.

Pleiotropy

Rats are sometimes born with a whole complex of congenital deformities, including thickened ribs, a narrowing of the trachea (windpipe), a loss of elasticity of the lungs, enlargement of the heart, blocked nostrils, a blunt snout, and needless to say, a greatly increased mortality.

All these changes, breeding experiments showed, are caused by a single mutation—that is, a mutation involving only one gene. This particular gene

governs production of a protein involved in the formation of cartilage, and since cartilage is one of the most common structural substances of the body, it is easy to see why one gene has such widespread effects.

Another example of pleiotropy is shown in Figure 13–6.

Polygenic Inheritance

A trait affected by a number of genes does not show a distinctive, clear-cut difference between groups—such as the differences tabulated by Mendel—but rather shows a gradation of small differences, which is known as _continuous variation_. If you make a chart of differences among individuals involving a single trait affected by a number of genes, you get a curve such as that shown in Figure 13–7.

Fifty years ago, the average height in the United States was less but the shape of the curve was the same; in other words, the great majority of men had heights within the middle range and the extremes in height were represented by only a few individuals. Some of these height variations are produced by environmental factors, such as diet, but even if all the men in a population were maintained from birth on the same type of diet, there would still probably

13–6 *The frizzle trait in fowls is an example of pleiotropy, the capacity of a gene to produce a variety of phenotypic effects. "Frizzle" is manifested primarily by differences in the feathers. (a) Under low magnification, feathers from normal birds show a closely interwebbed structure. (b) Frizzle feathers are weak and stringy and provide poor insulation. Some of the consequences of the manifestation of this single gene are shown in (c). Notice that the frizzle trait is a liability at low temperatures but may increase survival at high temperatures.*

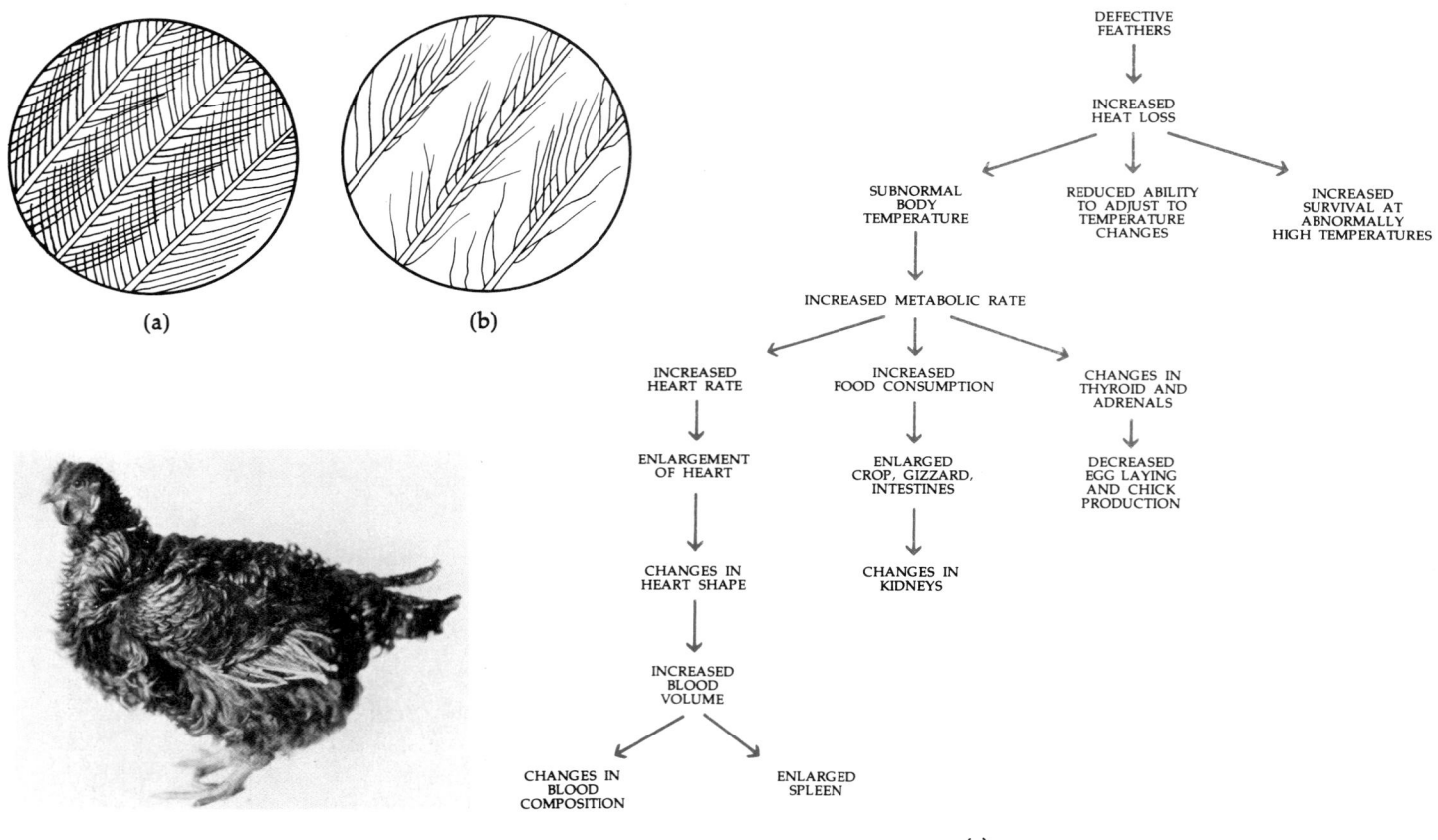

13-7 *Height distribution of males in the United States. Height and weight are examples of polygenic inheritance. These genetic traits are characterized by small gradations of difference. A graph of the distribution of such traits always takes the form of a bell-shaped curve, as shown, with the mean, or average, falling in the center of the curve. The larger the number of gene pairs involved, the smoother the curve.*

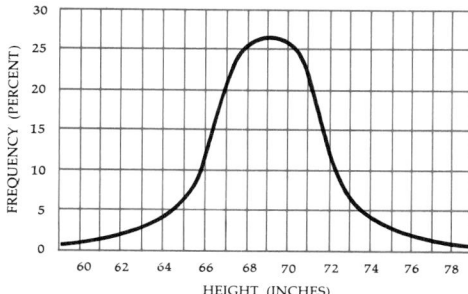

be a continuous variation in height in the population. This is due to genetic differences in hormone production, bone formation, and numerous other factors. Color in wheat kernels (Table 13–1) is an example of polygenic inheritance in which there is a blend in the phenotype, as in the snapdragon. A number of other characteristics in man are controlled by polygenic inheritance; among these is skin color, which may be the product of as many as five gene pairs.

GENOTYPE AND PHENOTYPE

From the moment of its conception, every organism is acted upon by the environment, and the expression of any gene is always the result of the interaction of gene and environment. To take a simple, familiar example, a seedling may have the genetic capacity to be green, to flower, and to fruit, but it will never turn green if it is kept in the dark, and it may not flower and fruit unless certain precise environmental requirements are met. Himalayan rabbits are all white if they are raised at high temperatures (above 95°F). However, rabbits of the same genotype, when raised at room temperature, have black ears, forepaws, noses, and tails. If the white hair is shaved off an area of the body and the shaved area is kept cold (with an ice pack, for instance), the hair grows back in black. Among humans, the gene for early baldness is a dominant gene expressed only in males; the determining factor here is the difference in internal environment, produced by male sex hormones.

In humans, also, similar genotypes are expressed quite differently in different environments. A simple example is found in height, which, as we have noted, is influenced by a group of genetic factors. This influence is indicated by the fact that tall parents generally have tall children and short parents tend to have short children.

13-8 *Height is influenced by both genetic and environmental factors. This chart shows, by age, median heights of boys of families in the United States with annual incomes of $3,000 or less and $10,000 or more, and of boys from India and the United Arab Republic. Which of these differences would you ascribe to "nature" and which to "nurture"? How could you test your hypothesis?*

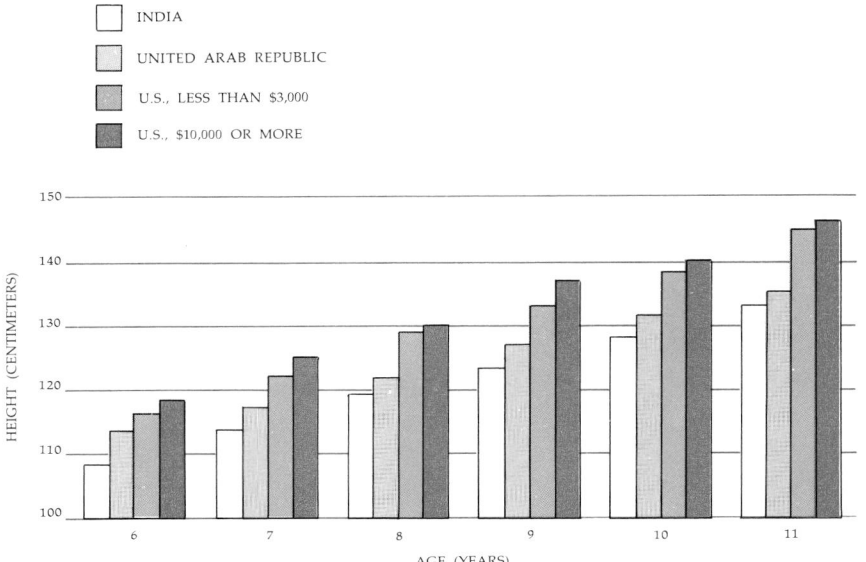

13–9 *The characteristic color pattern of the Siamese cat is a result of an interaction between genotype and environment. Because of a mutation in the gene affecting coat color, the gene is only expressed in the cooler, more peripheral areas of the body, such as the tips of the ears and tail, the muzzle, and the paws. As is often the case with animals with mutations resulting in reduced pigmentation (white tigers, pearl minks, and albino rats), Siamese cats are also frequently cross-eyed.*

Table 13–1 *The Genetic Control of Color in Wheat Kernals*

Parents:	$R_1R_1R_2R_2$ × $r_1r_1r_2r_2$		
	(dark red) (white)		

F_1:	$R_1r_1R_2r_2$ (medium red)		

F_2:	Genotype		Phenotype	
1	$R_1R_1R_2R_2$		Dark red	
2, 2 } 4	$R_1R_1R_2r_2$ $R_1r_1R_2R_2$		Medium-dark red Medium-dark red	
4, 1, 1 } 6	$R_1r_1R_2r_2$ $R_1R_1r_2r_2$ $r_1r_1R_2R_2$		Medium red Medium red Medium red	15 red to 1 white
2, 2 } 4	$R_1r_1r_2r_2$ $r_1r_1R_2r_2$		Light red Light red	
1	$r_1r_1r_2r_2$		White	

On the other hand, adult height is clearly also influenced by environmental factors in childhood, such as diet and incidence of disease. The population of the United States has grown taller with each generation for the past four generations, presumably as a result of general improvements in the standard of living and in the average diet. Similarly, Japanese teen-agers are taller than their parents, and the children raised in the kibbutzim of Israel come to tower over parents who grew up in the ghettos of Central and Eastern Europe.

As René Dubos reminds us,* not only are children growing taller, but they are growing faster; a boy now reaches full height at about 19 years, but 50 years ago, maximum stature was not usually attained until age 29. Of more social consequence is the fact that puberty is also being reached earlier. In Norway, for example, the mean age of the onset of menstruation fell from 17 in 1850 to 13 in 1960. Historical evidence indicates, however, that also in Imperial Rome and Western Europe in Shakespeare's time, teen-agers reached puberty at an early age (Juliet, remember, was not yet 14). Apparently, the slowing of the growth and maturation rate of the general population was a consequence of the industrial revolution and increasing urbanization, which reduced the standards of nutrition and health. Thus we have a situation in which the genetic potential has remained apparently unaltered over the cen-

*René Dubos, *So Human an Animal*, Charles Scribner's Sons, New York, 1968.

turies but the phenotypic expression has undergone fluctuations. A side effect of this change in phenotypic expression is that teen-agers are now reaching physical maturity early in a society in which childhood and dependency have become greatly prolonged.

CONCLUSIONS

Thus, continuing research on the principles of inheritance, while confirming Mendel's principles, has both broadened and refined knowledge about changes within genes, interactions among genes, and factors influencing relationships between genotypes and phenotypes.

SUMMARY

Mutations are abrupt changes in genotype. Together with the genetic recombinations that occur at fertilization and meiosis, they are the source of variations necessary for biological evolution.

Incomplete dominance is a situation in which the effects of the recessive allele are apparent in the heterozygote. Genes may affect two or more superficially unrelated characteristics; this property of the gene is known as pleiotropy.

Many characteristics are under the control of a number of separate genes and are said to be polygenetically inherited. Traits under the control of a number of genes typically show continuous variation, as represented by a bell-shaped curve.

Genes determine only potential capacities. Interactions of the genotype and the environment determine the phenotype.

QUESTIONS

1. The so-called "blue" (really gray) Andalusian variety of chicken is produced by a cross between the black and white varieties. What color chickens (and in what proportions) would you expect if you crossed two blues? If you crossed a blue and a black?
2. In one strain of mice, skin color is determined by five different pairs of alleles. The colors range from almost white to dark brown. Would it be possible for any given pair of mice to produce offspring darker or lighter than either parent? Explain.
3. Height and weight in animals follow a distribution similar to that shown in Figure 13–7. By inbreeding large animals, breeders are usually able to produce some increase in size among their stock. But after a few generations, increase in size characteristically stops. Why?
4. You and a geneticist are looking at a mahogany-colored Ayrshire cow with a newly born red calf. You wonder if it is male or female, and the geneticist says it is obvious from the color which sex the calf is. He explains that in Ayrshires the genotype *AA* is mahogany and *aa* is red, but the genotype *Aa* is mahogany in males and red in females. What is he trying to tell you —that is, what sex is the calf?

Chapter 14

Chromosomes and Genes

Sutton showed that there are similarities between the behavior of the chromosomes at mitosis and the Mendelian laws, but he did not actually prove that the genes, which were still abstract concepts, are really carried by the chromosomes. His closely argued thesis demonstrated only that whatever cellular organelles did carry the genes would have to behave like chromosomes. Proof of the physical location of the gene was to depend on another series of studies, which are the subject of this chapter.

SEX DETERMINATION

In the 1890s, microscopists noticed that male and female animals often show chromosomal differences, and they began to suspect that these differences were related to sex determination. One pair of chromosomes differs between the sexes, and these are known as the _sex chromosomes_; all the other chromosomes, which are the same whether the animal is male or female, are known as _autosomes_. In many animals, the two sex chromosomes are identical in the female but are dissimilar in the male, with one male sex chromosome usually smaller and of a different shape. The sex chromosome that is similar in the cells of both males and females is called the X chromosome, and the unlike chromosome characteristic of male cells is called the Y chromosome. Thus we can characterize the two sexes as XX (female) and XY (male).

However, in some insects, including the grasshopper, which Sutton studied, the Y chromosome is missing entirely. In this case, we usually speak of XX females and XO males. In birds, moths, and butterflies (and in occasional species in other groups), the chromosomes are reversed; the male has the two X chromosomes, and the female only one. The Y chromosome may or may not be present.

Human beings have 22 pairs of autosomes, which are structurally the same in both sexes. Women have a twenty-third matching pair, XX. Men, as their twenty-third pair, have one X and one Y. During meiosis, as each diploid spermatocyte undergoes reduction division into four haploid sperm cells, two of the sperm cells receive X chromosomes and two receive Y chromosomes. The ovum always contains an X chromosome, since a normal human female does not generally possess the Y in any of her cells. Thus the zygote will be-

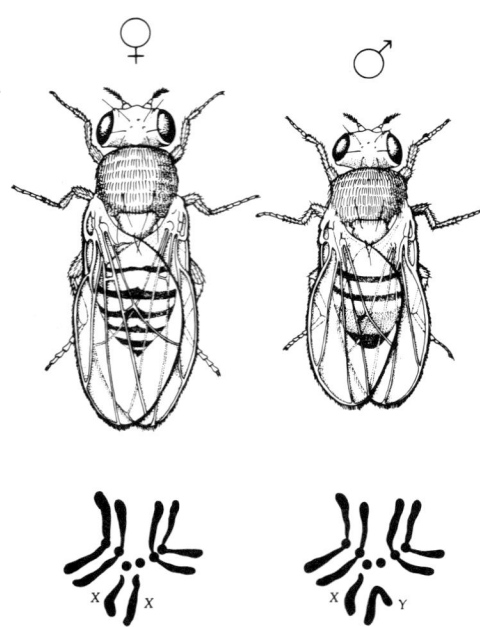

14–1 _The fruit fly and its chromosomes. This species has only four pairs of chromosomes, a fact that simplified Morgan's experiments. Six of the chromosomes (three pairs) are autosomes (including the two small spherical chromosomes in the center) and two are sex chromosomes._

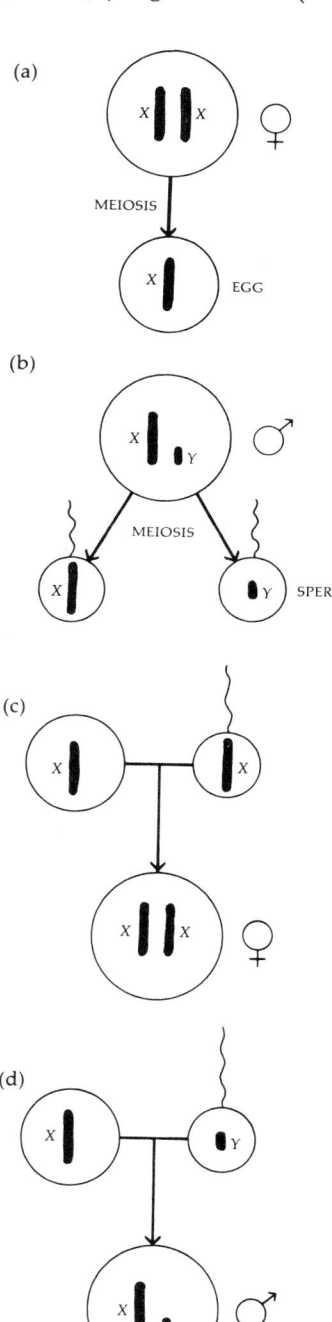

4–2 *How the sperm cell determines the sex of human offspring. (a) At meiosis, every egg cell receives an X chromosome from the mother. (b) A sperm cell may receive either an X chromosome or a Y chromosome. (c) If a sperm cell carrying an X chromosome fertilizes the egg, the offspring will be female (XX). (d) If a sperm cell carrying a Y chromosome fertilizes the egg, the offspring will be male (XY).*

(a)

MEIOSIS

EGG

(b)

MEIOSIS

SPERM

(c)

(d)

come XX or XY, depending on which of the sex chromosomes is carried in the sperm that fertilizes the egg. (See Figure 14–2.) It is in this way that the sperm cell determines the sex of the future offspring, and it is the process of meiosis that governs the almost equal production of male and female babies.*

As you can see, the correlation of the appearance of chromosomes with a particular characteristic—sex—was strong supporting evidence for Sutton's hypothesis. Stronger support was to come from a variety of studies carried out in what was known, because of the richness of the data it yielded, as "the Golden Age of Genetics." Much of the work of this Golden Age was carried out with the fruit fly, *Drosophila*.

DROSOPHILA

Early in the 1900s, Thomas Hunt Morgan began a study of genetics at Columbia University, founding what was to be the most important laboratory in the field for several decades. By a remarkable combination of foresight and good fortune, he selected *Drosophila* as his experimental material.†

Drosophila means "lover of dew," although actually this useful little fly is not attracted by dew but feeds on the fermenting yeast that it finds in rotting fruit. The fruit fly was a likely choice for a geneticist since it is easy to breed and maintain. These tiny flies, each only an eighth of an inch long, produce a new generation every two weeks. Each female lays hundreds of eggs at a time, and an entire population can be kept in a half-pint bottle, as they were in Morgan's laboratory. Also, *Drosophila* has only four pairs of chromosomes, a feature that turned out to be particularly useful, although Morgan could not have foreseen that. Three pairs of these are autosomes, and the fourth is an XX pair in the female and an XY pair in the male (Figure 14–1).

SEX-LINKED CHARACTERISTICS

The working procedure in Morgan's laboratory was, initially, to do breeding experiments similar to those Mendel had carried out in the pea plant. Such experiments involved examining under a magnifying lens hundreds—eventually thousands—of individual fruit flies.

The investigators, at first, were looking for genetic differences between individual flies that they could study by interbreeding experiments. Shortly after Morgan established his colony, such a difference appeared. One of the prominent and readily visible characteristics of the wild-type *Drosophila* is its brilliant red eyes. ("Wild-type" is a term frequently used by geneticists to describe an organism as it normally occurs in nature.) One day, a white-eyed fly, a mutant, appeared in the colony. The mutant, a male, was bred to a red-eyed female. All members of the first generation were red-eyed, indicating

* In actual fact, the ratio of human male to human female births is about 106 to 100. The reason for this is not known, but it has been suggested that the male-determining sperm may have an advantage in getting to the egg.

† Geneticists have often used for their experiments such "insignificant" little plants and animals—Mendel's pea plants, for instance, or Hertwig's sea urchins—organisms that seem to occupy very unimportant and out-of-the-way places in the natural order. Underlying this approach is the geneticist's assumption that genetic principles are universal, applying equally to all living things.

that the mutation was recessive. Then the members of the F₁ generation were interbred just as Mendel had done in his pea experiments, and this is what resulted:

Red-eyed females	2,459
White-eyed females	0
Red-eyed males	1,011
White-eyed males	782

Why were there no white-eyed females? Perhaps the gene for white eyes was carried only on the Y chromosome and not on the X. To test this hypothesis, Morgan crossed the original white-eyed male with one of the F₁ females. And the result was:

Red-eyed females	129
White-eyed females	88
Red-eyed males	132
White-eyed males	86

Morgan and his co-workers examined these figures and came to a second conclusion (the right one this time): the gene for eye color is carried only on the X chromosome. (In fact, as it was later shown, the Y chromosome carries very little genetic information.) The white allele is recessive. Thus a heterozygous female would never have white eyes—which is why there were no white-eyed females in the F₁ generation. However, a male that received an X chromosome carrying the allele for white eyes would always be white-eyed since no other allele would be present.

Further experimental crosses proved Morgan's hypothesis to be right. (See Figure 14–3.) They also showed that white-eyed fruit flies are more likely to die before they hatch than red-eyed fruit flies, which explains the lower-than-expected numbers in the F₁ generation and the testcross.

These experiments introduced the concept of sex-linked traits. An example of one such trait in man is shown in Figure 14–12. They also supported what Sutton had hypothesized some years before: Genes *are* on chromosomes.

LINKAGE

Mendel, you will remember, showed that certain pairs of alleles, such as round and wrinkled, assort independently of other pairs, such as yellow and green. However, as we noted previously, genes must assort independently only if they are on different chromosomes. If two genes are on the same chromosome, they will, of course, both be transmitted to the same sex cell at meiosis. The genes on any one chromosome are said to be in *linkage groups.*

As increasing numbers of mutants were found in Columbia University's *Drosophila* population, the mutations began to fall into four linkage groups, in accord with the four pairs of chromosomes visible in the cells. Indeed, in all organisms which have been studied in sufficient genetic detail, the number of linkage groups and the number of pairs of chromosomes have been the same, further supporting Sutton's hypothesis.

14–3 *Offspring of a cross between a white-eyed (w) female fruit fly and a red-eyed (W) male fruit fly, illustrating what happens when a recessive gene is carried on an X chromosome. The F₁ females, with one X chromosome from the mother and one from the father, are heterozygous (Ww) and so will be red-eyed. But the F₁ males, with their single chromosome received from the mother carrying the recessive (w) trait, will be white-eyed because the Y chromosome carries no gene for eye color and the recessive allele on the X chromosome inherited from the mother will be expressed.*

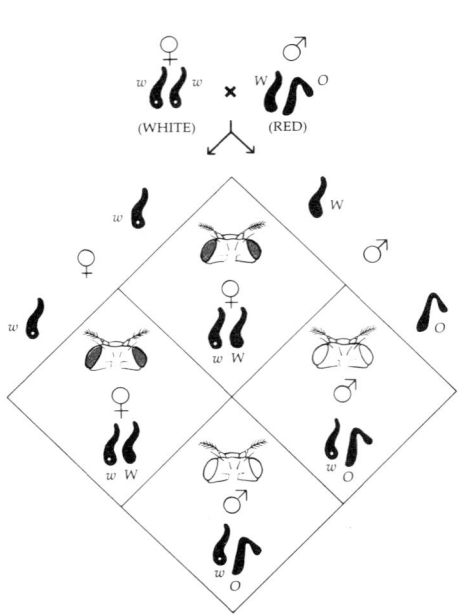

CROSSING OVER

Large-scale studies of linkage groups soon revealed some unexpected difficulties. For instance, wild-type fruit flies have gray bodies and long wings. When wild-type flies were bred with mutant fruit flies having black bodies and short wings (both recessive traits), all the progeny had gray bodies and long wings, as would be expected. Then the F_1 generation was inbred. Two outcomes seemed possible:

1. The two recessives would be assorted independently and would appear in the 9:3:3:1 ratio, indicating that they were on different chromosomes.
2. The two recessives would be linked. In this case, 75 percent of the flies would be gray with long wings, and 25 percent, homozygous for both recessives, would be black with short wings.

In the case of these particular traits, the results most closely resembled outcome 2, but they did not conform exactly. In a few of the offspring, the traits seemed to segregate independently, that is, some few flies appeared that were gray with short wings, and some that were black with long wings. How could this be? Somehow the alleles had recombined within linkage groups.

To find out what was happening, Morgan tried a testcross, similar to that diagrammed in Figure 11–10, breeding one of the F_1 generation with a homozygous recessive. If black and gray, long and short assorted independently—that is, if they were on different chromosomes—25 percent of the offspring of this cross should be black with long wings, 25 percent gray with long wings, 25 percent black with short wings, and 25 percent gray with short wings. On the other hand, if the two traits (color and wing size) were on the same chromosome and so moved together, half of the offspring of the testcross should be gray with long wings and half should be black with short wings. As it turned out, over and over, in counts of hundreds of fruit flies resulting from such crosses, 41.5 percent were gray with long wings, 41.5 percent were black with short wings, 8.5 percent were gray with short wings, and another 8.5 percent were black with long wings.

—4 *Crossing over takes place when breaks occur in chromatids at the beginning of meiosis, when the chromosomes are paired, and the broken end of each chromatid joins with the chromatid of an homologous chromosome. In this way, alleles are exchanged between chromosomes. The white circles symbolize kinetochores.*

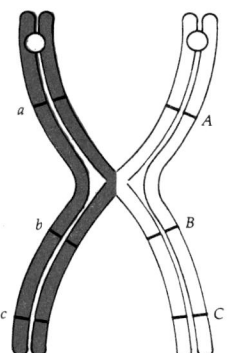

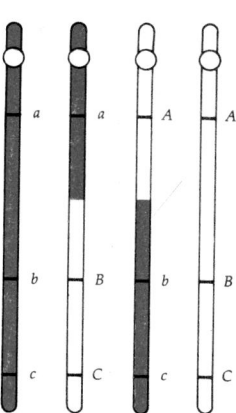

14–5 *Chiasmata in homologous chromosomes of a grasshopper.*

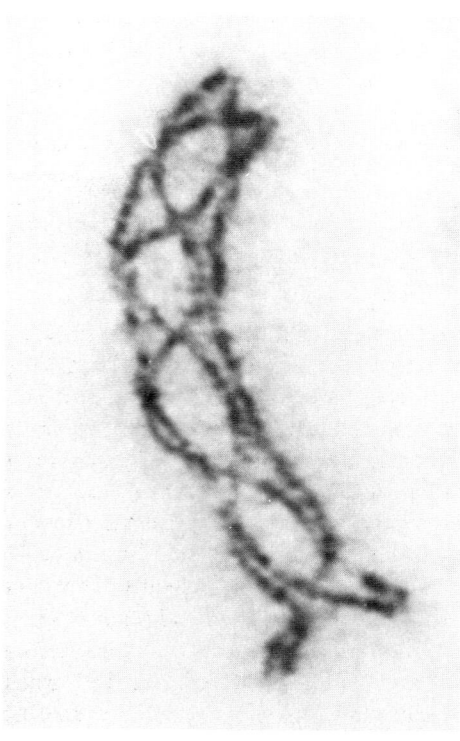

Morgan was convinced by this time that genes are located on chromosomes. It now seemed clear that the two traits were located on the same chromosome, since they did not show up in the 25:25:25:25 percentage ratios of separately assorted genes. The only way in which the observed figures could be explained was if one assumed (1) that the two genes were on one chromosome and their two alleles on the homologous chromosome, and (2) that sometimes genes could be exchanged between homologous chromosomes.

How can genes change chromosomes? Recall that at meiosis, the chromosomes pair before they divide. Consequently four homologous chromatids are lined up alongside each other before the first meiotic division. In 1909, the Belgian cytologist F. A. Janssens had observed that chromatids form crosses with one another and had suggested that these crosses, or *chiasmata* (singular, chiasma), as he called them, represented an exchange of chromosomal material. This exchange is difficult to see and interpret in the living cell, so that the suggestion was not so obvious as it now may seem. Furthermore, there is still no explanation for how homologous chromosomes manage to break in exactly corresponding sites and so exchange equal amounts of chromosomal material following each break. Nevertheless, vast accumulations of data have now confirmed that such exchanges, or *crossovers*, as they are called, do take place, and at virtually every meiosis.

MAPPING THE CHROMOSOME

With the discovery of crossovers, it began to seem clear not only that the genes are carried on the chromosomes, as Sutton had hypothesized, but that they must be arranged in a definite, fixed linear array along the length of the chromosome. Furthermore, all the alleles of one gene must be at corresponding sites on homologous chromosomes. If this were not true, exchange of sections of chromosomes could not possibly result in an exact exchange of alleles. In other words, genes behave as if they are arrayed along the chromosome like beads along a string; this simile was widely used in genetics for several decades.

14–6 *During meiosis, crossing over may occur at more than one point on adjacent chromatids. The result of a double crossover is shown.*

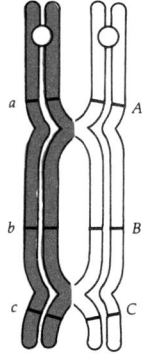

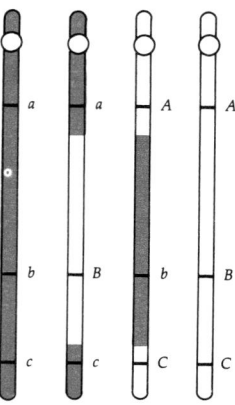

As other traits were studied, it became clear that the percentage of separations, or crossovers, between any two genes, such as gray body and long wing, was different from the percentage of crossovers between two other genes, such as gray body and long leg. In addition, as Morgan's experiments had shown, these percentages were very fixed and predictable. It occurred to A. H. Sturtevant, one of the many brilliant young geneticists attracted to Morgan's laboratory during those golden days of *Drosophila* genetics, that the percentage of crossovers probably had something to do with the distances between the genes, or in other words, with their spacing along the chromosome. (You can see in Figure 14–6, for example, that in a crossover, the chance of a strand's breaking and recombining with its homologous strand at the single junction between *B* and *C* is less than the chance of these events occurring between *A* and *C*.) This concept opened the way to the mapping of chromosomes.

Sturtevant reasoned (1) that genes are arranged in a linear series on chromosomes, (2) that genes which are close together will be separated by crossing over less frequently than genes which are further apart, and (3) that it should therefore be possible, by determining the frequencies of crossovers, to plot the sequence of the genes along the chromosome and the relative distances between them. (The distances cannot be absolute, because breaking and rejoining of chromosome segments may be more likely to occur at some sites on the chromosome than at others.) In 1913, Sturtevant began a series of crossover studies in fruit flies. As a standard unit of measure, he arbitrarily took the distance that would give (on the average) one crossover per 100 fertilized eggs. Thus genes with 10 percent crossover would be 10 units apart; those with 8 percent crossover would be 8 units apart. By this method, he and other geneticists constructed maps locating a wide variety of genes and their mutants in *Drosophila*.

As fixed locations were assigned on a chromosome to each gene and the maps began to develop, geneticists could begin to predict the amount of crossing over that should occur and what traits the offspring of any cross should exhibit. These predictions could then be tested by making the indicated crosses.

An Example

Suppose you discover that two traits—called *A* and *B* for simplicity—in the prolific fruit fly occur together in 92 percent of the offspring and recombine with their recessive counterparts *a* and *b* in 8 percent (Figure 14–8). Suppose that you then conduct breeding experiments to determine the percentage of recombinants between *A, a* and another pair of genes, *C, c*. These experiments show that *AC* and *ac* remain together in 75 percent of the offspring but recombine in 25 percent. Now draw a hypothetical chromosome—simply as a straight line. Using Sturtevant's arbitrary units, the number 25 can represent the distance between *A* and *C, a* and *c* (Figure 14–8a). Now *B* and *b* can be located, with 8 representing the distance between *A* and *B, a* and *b*. Indicate where *B* should go on the hypothetical chromosome. If you don't know how to do this, you're right. From the information available so far, there is no way to know whether *B* is 8 units to the left of *A* or 8 units to the right of *A*. However, it is not difficult to design an experiment to prove which is correct. One need only

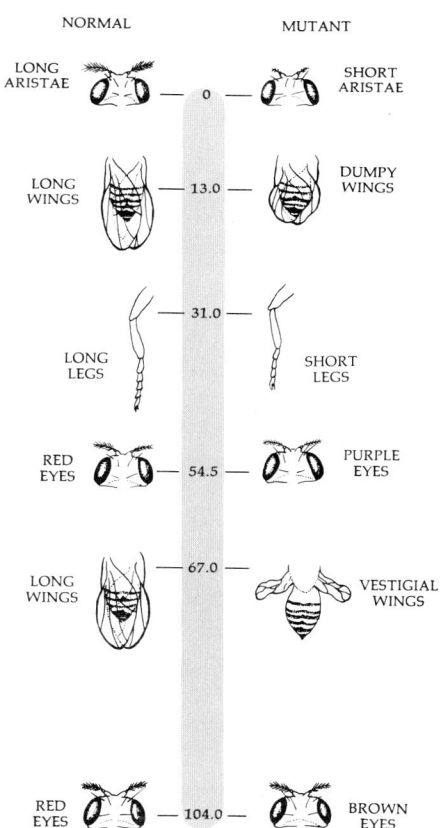

14–7 *A portion of the genetic map of* Drosophila melanogaster, *showing some of the genes on chromosome 2 and the distances between them, as calculated by the frequency of crossovers. As you can see, more than one gene may affect a single characteristic, such as eye color.*

NORMAL   MUTANT

LONG ARISTAE — 0 — SHORT ARISTAE

LONG WINGS — 13.0 — DUMPY WINGS

— 31.0 —

LONG LEGS   SHORT LEGS

RED EYES — 54.5 — PURPLE EYES

LONG WINGS — 67.0 — VESTIGIAL WINGS

RED EYES — 104.0 — BROWN EYES

14–8 *How to "map" a chromosome. Genes A and B recombine with a and b in 8 percent of the offspring and genes A and C with a and c in 25 percent of the offspring. Use as a unit of measure the distance that will give (on the average) one recombinant per 100 fertilized eggs. (a) Start with the largest number and establish the relative positions of A and C on the chromosome. (b) A and B, as you know from the data, are 8 units apart. Thus B could theoretically be either to the left of A or to the right. (c) However, if it were to the left, B and C would be 33 units apart, a distance that does not conform to the data. B must therefore be between A and C.*

CROSS	OFFSPRING		
AB × ab	AB + ab	(PARENTAL)	92%
	Ab + aB	(RECOMBINANT)	8%
AC × ac	AC + ac	(PARENTAL)	75%
	Ac + aC	(RECOMBINANT)	25%
BC × bc	BC + bc	(PARENTAL)	83%
	Bc + bC	(RECOMBINANT)	17%

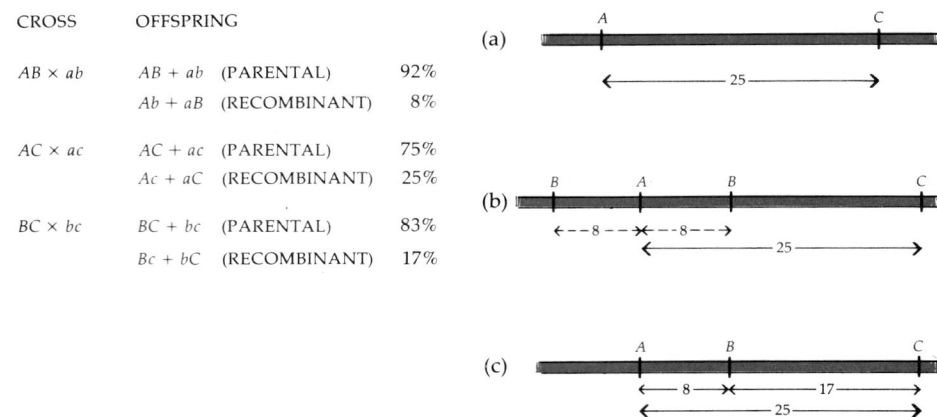

crossbreed a *Drosophila* which shows the traits *B* and *C* with one having traits *b* and *c*. If recombination occurs in 33 percent of the cases, *B* is to the left of *A* (8 + 25 = 33); if it occurs in 17 percent of the cases, *B* can be placed 8 units to the right of *A* and 17 units to the left of *C* (8 + 17 = 25). The data shown in Figure 14–8 dictate placement to the right (c). By similar experiments, any other trait linked to any of those already located can be placed on the line in relation to the others.

The figures used in the example are, of course, "ideal." In a case of actual mapping, the recombination "fractions" for a long-distance *A–C* would be smaller than those for *A–B* plus *B–C*. This is because more than one chiasma, or crossover, may occur along the length of the chromosome and so *A–C* may become separated by one break and then recombined by a second, and perhaps even separated again by a third. Since this happens frequently enough to distort the percentages for genes that are far apart from one another, the "units" are considered valid only for short distances. It is also true, as Sturtevant realized from the beginning, that the recombination fractions do not necessarily reflect distance but may equally well reflect something about the structure of the chromosome in particular areas that may make it more or less likely to break at certain points.

The Results of Mapping

By chromosome mapping, involving literally thousands and thousands of crossbreeding experiments, more than 500 different genes have been located in relationship to one another on the four chromosome pairs of the most commonly studied species of *Drosophila*.

Perhaps the most important conclusion of all from the mapping studies of *Drosophila* was the incontestable proof this work provided that the genes occur along the chromosome in an ordered linear array. Chromosome mapping has now been carried out in a variety of other organisms, including viruses, bacteria, fungi, plants, and animals, and this basic tenet has been found to hold true for them all.

4–9 *Chromosomes from the salivary gland of a* Drosophila *larva. These chromosomes are 100 times larger than the chromosomes in ordinary body cells, and their details are therefore much easier to see. (This micrograph was taken with a light microscope, not an electron microscope.) Because of the distinctive banding patterns, it was possible in some cases to assign genes to specific locations on particular chromosomes.*

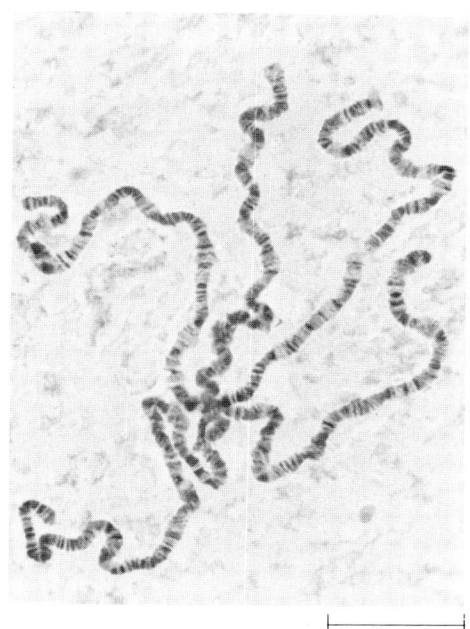

├─────┤ 20 µm ├─────┤

GIANT CHROMOSOMES

Further confirmation of Sturtevant's brilliant hypothesis came 20 years later from an unexpected source. In certain cells of the larva* of *Drosophila* (and of many other insects as well), the chromosomes are duplicated many times without separating. All the new duplicates line up beside the original chromosome until, eventually, several thousand copies are present. These closely aligned duplicates are easily visible in the light microscope. As you can see in Figure 14–9, these giant chromosomes are marked by distinctive dark and light bands. Changes in these banding patterns were found to correlate with observed genetic changes in the flies.

For example, in some chromosomes, *inversions* occurred (Figure 14–10), in which the order of the genes in a portion of the chromosome was reversed. Presumably such inversions are the result of a chromosome's breaking in two places and reattaching at the opposite ends. Morgan and his group had guessed that such inversions occurred, on the basis of mapping experiments (since crossovers do not take place between inverted portions of a chromosome and the corresponding portion of its normal homologue). Because of the distinctive banding patterns, it was actually possible to see some of these inversions, and therefore genes could be assigned quite accurately to specific parts of the chromosome.

Deletions, duplications, and translocations could also be observed. A deletion is the lack of a portion of a chromosome. If the circular chromosome segment *CB* shown reattaching in Figure 14–10 were not to reattach and the broken ends of the chromosome at *AD* were to mend, the result would be a deletion. A duplication occurs when a segment of a chromosome is repeated. A translocation occurs when a piece of one chromosome becomes attached to another place on the original chromosome or to another nonhomologous chromosome. Once chromosome maps are established, deletions and translocations can also be detected in breeding experiments.

* The wingless, often wormlike form of a newly hatched insect.

14–10 *Origin of a chromosomal inversion. An inversion is the reversal in sequence of some of the genes on a chromosome. Inversions can be detected by mapping studies and also by observations of* *pairing at meiosis. When a chromosome with an inversion (inversion heterozygote) replicates, it replicates the inversion, producing an inversion homozygote.*

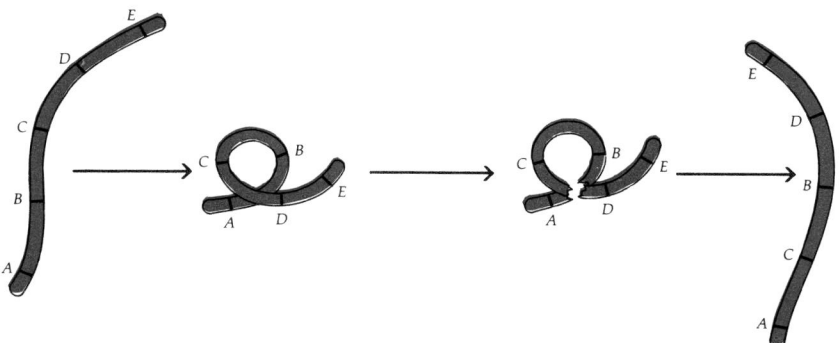

14–11 *Color blindness in humans is carried by a recessive allele on the X chromosome. In the chart, the mother has inherited one normal and one defective allele. The normal allele will be dominant, and she will have normal color vision. However, half her eggs (on the average) will carry the defective allele and half will carry the normal allele—and it is a matter of chance which kind is fertilized. Since her husband's Y chromosome, the one that determines a son rather than a daughter, carries no gene for color discrimination, the single gene the wife contributes (even though it is a recessive gene) will determine whether or not the son is color-blind. Therefore, half her sons (on the average) will be color-blind. Assuming that her children marry individuals with normal X chromosomes, the expected distribution of the trait among her grandchildren will be as shown on the chart.*

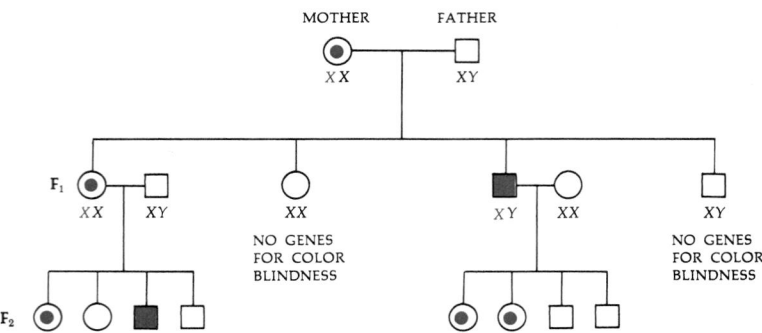

HUMAN GENETIC DISORDERS

As in *Drosophila*, the Y chromosome of man carries much less genetic information than the X chromosome. Genes for color vision, for example, are carried on the X chromosome in humans but not on the Y chromosome. Color blindness is produced by a recessive allele of the normal gene. The normal allele is dominant; a woman with one X chromosome with the normal allele and one X chromosome with the allele for color blindness will have normal color vision. If she transmits the X chromosome with the recessive allele to a daughter, the daughter also will have normal color vision if she receives a normal X chromosome from her father (that is, if he is not color-blind). If, however, the X chromosome with the recessive allele is transmitted to a son, he will be color-blind since, lacking a second X chromosome, he has only the recessive allele.

Hemophilia

A classic example of a recessive gene transmitted on the X chromosome is the hemophilia which has afflicted the royal families of Europe since the nineteenth century. Hemophilia is a disease in which the blood does not clot normally, so that even minor injuries carry the risk of the patient's bleeding to death. Queen Victoria was probably the original carrier in the family. Because none of her forebears or collateral relatives was affected, we conclude that the mutation may have occurred on an X chromosome in one of her parents or in the cell line from which her own eggs were formed. Prince Albert, Victoria's consort, could not have been responsible; male-to-male inheritance of the disease is impossible. (Why?) One of her sons, Leopold, Duke of Albany, died of hemophilia at the age of 31. At least two of Victoria's daughters were carriers, since a number of *their* descendants were hemophiliacs. And so,

–12 As this chart shows, Queen Victoria was the original carrier of the hemophilia that has afflicted male members of the royal families of Europe since the nineteenth century. The British royal family escaped the disease because King Edward VII, and consequently all his progeny, did not inherit the defective gene.

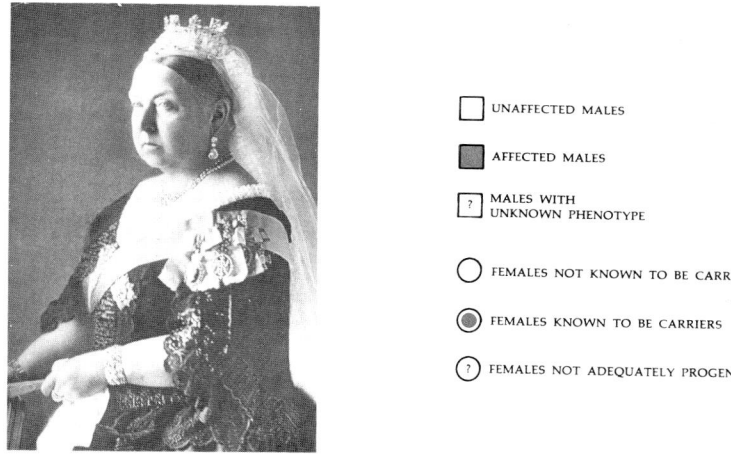

☐ UNAFFECTED MALES

▩ AFFECTED MALES

[?] MALES WITH UNKNOWN PHENOTYPE

◯ FEMALES NOT KNOWN TO BE CARRIERS

◉ FEMALES KNOWN TO BE CARRIERS

(?) FEMALES NOT ADEQUATELY PROGENY TESTED

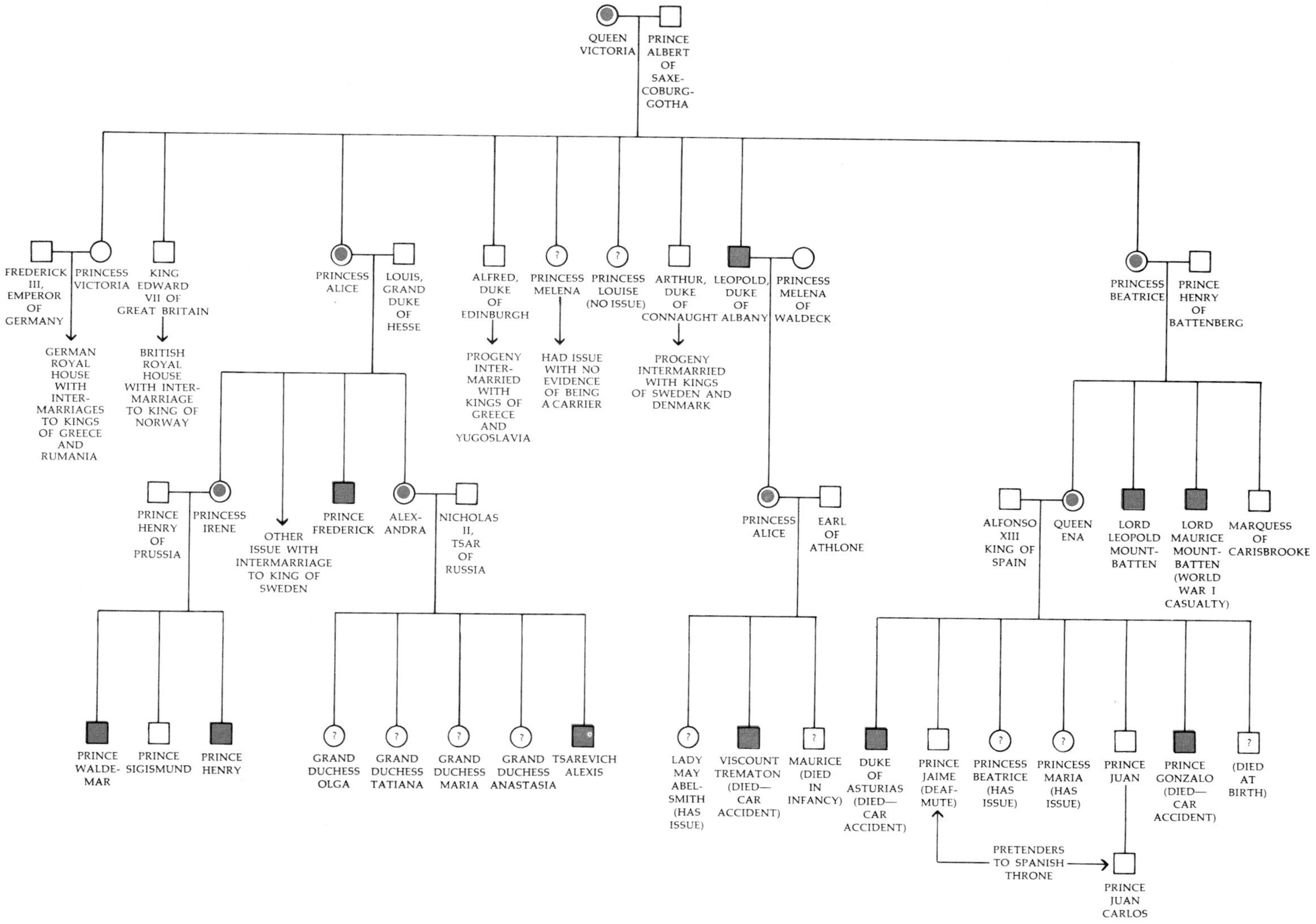

QUEEN VICTORIA — PRINCE ALBERT OF SAXE-COBURG-GOTHA

FREDERICK III, EMPEROR OF GERMANY — PRINCESS VICTORIA

KING EDWARD VII OF GREAT BRITAIN

PRINCESS ALICE — LOUIS, GRAND DUKE OF HESSE

ALFRED, DUKE OF EDINBURGH

PRINCESS MELENA (?)

PRINCESS LOUISE (NO ISSUE) (?)

ARTHUR, DUKE OF CONNAUGHT

LEOPOLD, DUKE OF ALBANY — PRINCESS MELENA OF WALDECK

PRINCESS BEATRICE — PRINCE HENRY OF BATTENBERG

GERMAN ROYAL HOUSE WITH INTERMARRIAGES TO KINGS OF GREECE AND RUMANIA

BRITISH ROYAL HOUSE WITH INTERMARRIAGE TO KING OF NORWAY

PROGENY INTERMARRIED WITH KINGS OF GREECE AND YUGOSLAVIA

HAD ISSUE WITH NO EVIDENCE OF BEING A CARRIER

PROGENY INTERMARRIED WITH KINGS OF SWEDEN AND DENMARK

PRINCE HENRY OF PRUSSIA — PRINCESS IRENE

OTHER ISSUE WITH INTERMARRIAGE TO KING OF SWEDEN

PRINCE FREDERICK

ALEXANDRA — NICHOLAS II, TSAR OF RUSSIA

PRINCESS ALICE — EARL OF ATHLONE

ALFONSO XIII KING OF SPAIN — QUEEN ENA

LORD LEOPOLD MOUNTBATTEN

LORD MAURICE MOUNTBATTEN (WORLD WAR I CASUALTY)

MARQUESS OF CARISBROOKE

PRINCE WALDEMAR

PRINCE SIGISMUND

PRINCE HENRY

GRAND DUCHESS OLGA (?)

GRAND DUCHESS TATIANA (?)

GRAND DUCHESS MARIA (?)

GRAND DUCHESS ANASTASIA (?)

TSAREVICH ALEXIS

LADY MAY ABEL-SMITH (HAS ISSUE) (?)

VISCOUNT TREMATON (DIED—CAR ACCIDENT)

MAURICE (DIED IN INFANCY) (?)

DUKE OF ASTURIAS (DIED—CAR ACCIDENT)

PRINCE JAIME (DEAF-MUTE)

PRINCESS BEATRICE (HAS ISSUE) (?)

PRINCESS MARIA (HAS ISSUE) (?)

PRINCE JUAN

PRINCE GONZALO (DIED—CAR ACCIDENT)

(DIED AT BIRTH) (?)

PRETENDERS TO SPANISH THRONE

PRINCE JUAN CARLOS

PREPARATION OF A KARYOTYPE

Chromosome typing for the identification of hereditary defects is being carried out at an increasing number of genetic counseling centers throughout the United States. The result of the procedure is known as a karyotype. The chromosomes shown in a karyotype are actually chromatid pairs held together by their kinetochores. All cells in the process of dividing have been interrupted at metaphase by the addition of colchicine, which prevents the subsequent steps of mitosis from taking place. After treating and staining, the chromosomes are photographed, enlarged, and arranged according to size. Certain abnormalities, such as an extra chromosome, can be detected. In some cases, karyotyping can help couples decide whether or not to have children. Karyotypes can also be made of fetuses still in the uterus and so can reveal the sex of the future child or the presence of genetic defects.

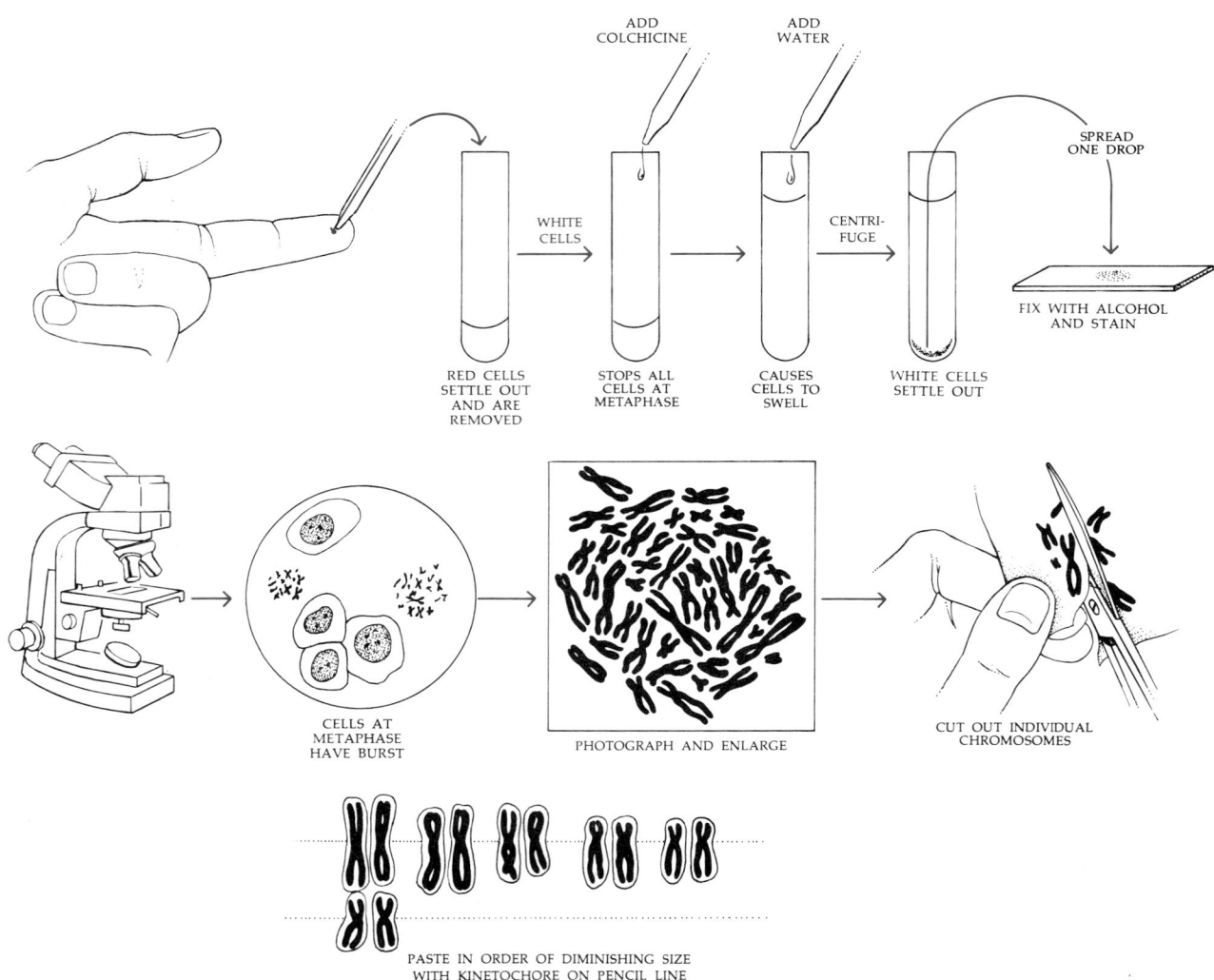

ADD COLCHICINE

ADD WATER

SPREAD ONE DROP

WHITE CELLS

CENTRI-FUGE

FIX WITH ALCOHOL AND STAIN

RED CELLS SETTLE OUT AND ARE REMOVED

STOPS ALL CELLS AT METAPHASE

CAUSES CELLS TO SWELL

WHITE CELLS SETTLE OUT

CELLS AT METAPHASE HAVE BURST

PHOTOGRAPH AND ENLARGE

CUT OUT INDIVIDUAL CHROMOSOMES

PASTE IN ORDER OF DIMINISHING SIZE WITH KINETOCHORE ON PENCIL LINE

-13 *The normal diploid chromosome number of a human being is 46, 22 pairs of autosomes and the 2 sex chromosomes. The arrangement here is called a karyotype. The autosomes are grouped by size (A, B, C, etc.), and then the probable homologues are paired. A normal woman has two X chromosomes and a normal man an X and a Y.*

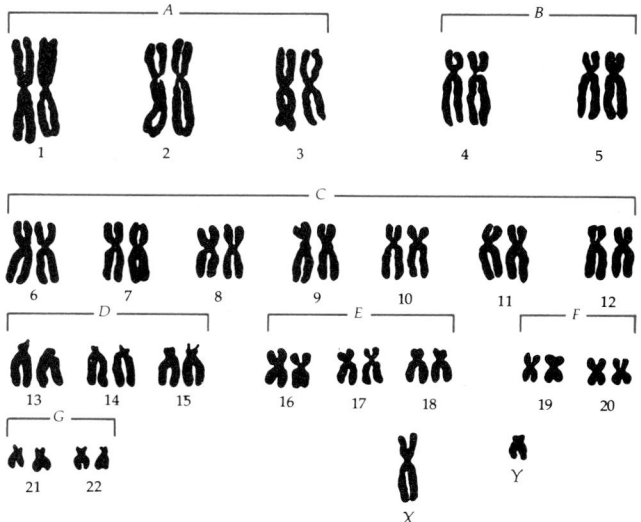

through various intermarriages, the disease spread from throne to throne across Europe. In the Tsarevitch, son of the last Tsar of Russia, and in the princes of Spain, the gene for hemophilia inherited from Victoria had considerable political consequences.

Nondisjunction

From time to time, usually because of "mistakes" at the time of meiosis, homologues may not separate. In this case, one of the sex cells has one too many chromosomes and the other one too few. This phenomenon is known as *nondisjunction.* The cell with one too few autosomes cannot produce a viable embryo, but the one with one too many sometimes can. The result is an individual with an extra chromosome in every cell of his body.

The presence of additional chromosomes often produces widespread abnormalities. Many such infants are stillborn. Among those who survive, many are mentally deficient. In fact, studies of abnormalities in human chromosome number among living subjects are usually carried out on patients in mental hospitals. These patients often have abnormalities of the heart and other organs as well.

Down's Syndrome

One of the most familiar chromosomal abnormalities is the form of mental deficiency known as *mongolism.* This name derives from the characteristic appearance of the eyefold in these patients, which makes them look "foreign," or mongoloid, to Europeans. Mongolism usually involves more than one defect and so is more precisely referred to as a syndrome, a group of disorders that occur together. In medical terminology it is usually referred to as Down's syndrome, after the physician who first described it. The syndrome includes, in most cases, not only mental deficiency but a short, stocky body type with a thick neck and, often, abnormalities of other organs, especially the heart.

221 *Chromosomes and Genes*

14–14 *The karyotype of a male patient with Down's syndrome caused by nondisjunction. Note that there are three chromosomes 21.*

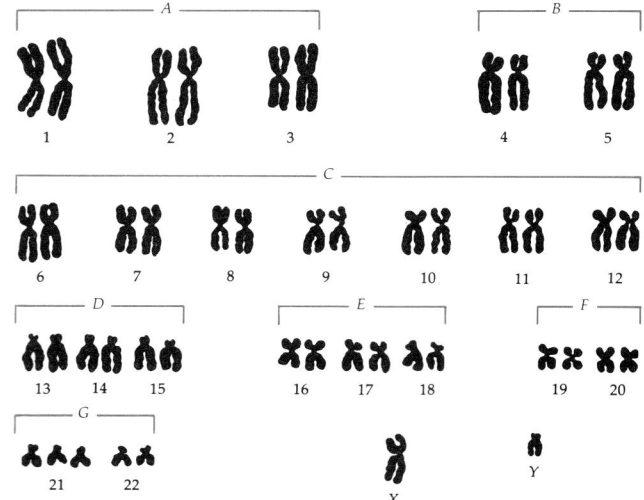

14–15 *The frequencies of mongoloid births in relation to the ages of the mothers. The number of cases shown for each age group represents the occurrence of mongolism in every 2,240 births by mothers in that group. As you can see, the risk of having a mongoloid child increases rapidly after the age of 40.*

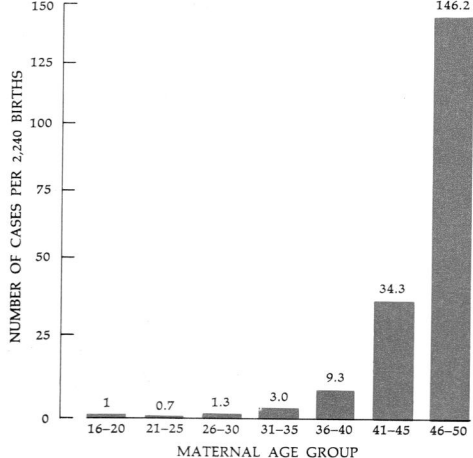

Down's syndrome and a number of other defects involving gross abnormalities of chromosomes are more likely to occur among infants born to older women. The reasons for this are not known, but the formation of the egg cells is well under way in the human female before she is born, so the increasing incidence of abnormalities may be correlated in some way with the aging of the mother's egg cells.

The most common cause of the genetic abnormality that produces Down's syndrome is nondisjunction involving chromosome 21. This results in an extra chromosome 21 (Figure 14–14) in the cells of the defective child.

Down's syndrome may also be the result of translocation of the chromosomes of one of the parents. Translocation, as we noted previously, occurs when a portion of a chromosome is broken off and becomes attached to another chromosome. The patient with translocation mongolism usually has a third chromosome 21 (or, at least, most of it) attached to a larger chromosome, such as 15. When cases of translocation mongolism are studied, it is usually found that one parent, although phenotypically normal, has only 45 separate chromosomes—one chromosome being composed of most of chromosomes 15 and 21 joined together. The possible genetic makeups of the offspring of this parent are diagrammed in Figure 14–16. Three out of the six possible combinations are lethal. One of the remaining three will produce Down's syndrome, one will be normal, and one will be a carrier. Not a very cheerful prognosis!

Thus parents who have had one abnormal child are faced with the terrible decision of whether or not to risk having another infant. Now there are special clinics throughout the country that can help them in this decision. In these clinics, cells from stillborn or abnormal infants are cultivated in a test tube for examination at the time of mitosis, when the chromosomes become visible. If an infant is found to have an abnormal karyotype, cells of the parents can be similarly tested. If the parents show an abnormality, they are warned that they are likely to transmit it to future infants through their gametes.

If, however, the karyotypes of both parents are normal, the parents are advised that the abnormalities in the child were probably the result of non-

4-16 *Transmission of translocation mongolism. The father, top row, has normal pairs of chromosomes 21 and 15, and each of his sperm cells will contain a normal 21 and a normal 15. The mother (more frequently, although not always, the translocation carrier) has one normal 15, one normal 21, and a translocation 15/21. She herself appears normal, but her chromosomes cannot pair normally at meiosis. There are six possibilities for the offspring of these parents: the infant will (1) die before birth (three of the six possibilities), (2) be mongoloid, (3) be a translocation carrier like the mother, or (4) be normal. Tests for the chromosomal abnormality can be made in prospective parents and in the fetus before birth.*

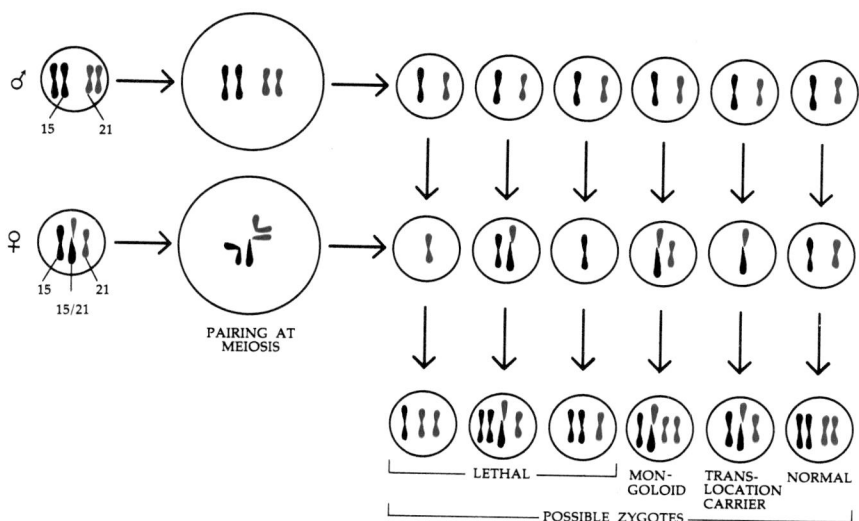

disjunction—a "mistake" that occurred during meiosis—and that they do not run a greater-than-normal risk for the mother's age group of having another congenitally ill child.

It is now possible to detect the presence of extra chromosomes in the fetus when it is 16 weeks old. This diagnosis can be made from a sample of the amniotic fluid surrounding the embryo. This fluid usually contains cells sloughed off by the embryo from which a karyotype can be made. There is no treatment, however. If the fetus is found to be abnormal, the parents may decide on abortion.

Abnormalities in the Sex Chromosomes

Nondisjunction may also produce individuals with extra sex chromosomes. An XY combination in the twenty-third pair, as you know, is associated with maleness, but so is XXY and XXXY and even XXXXY. These males, however are usually sexually underdeveloped and sterile. XXX combinations sometimes produce normal females, but many of the XXX women and all XO women (women with only one X chromosome) are sterile.

Many individuals with sex chromosome abnormalities are mentally retarded. More recently it has been discovered that white males with an extra Y chromosome (XYY) are found in a higher percentage in institutions for mentally disturbed or criminally insane than among males studied at random. (In the general population, on the basis of the rather limited studies done, about 1 in 1,000 males is XYY. Among the institutionalized males, the rate is about 2 per 100.) The XYY genotype is also associated with tallness, below-average intelligence, and severe acne. Some XYY males are of normal intelligence, however, and show no signs of mental disturbance. Consequently, it is difficult to know whether or not the institutionalized XYY males represent an unusual genotypically determined disorder or are just another example of a minority who, because of essentially harmless phenotypic variations, have been discriminated against by society.

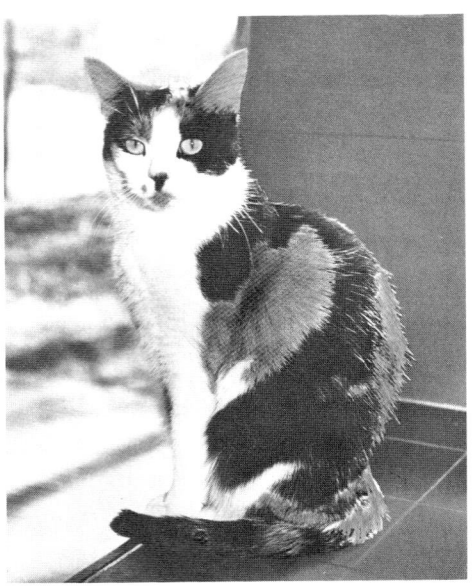

CALICO CATS, BARR BODIES, AND THE LYON HYPOTHESIS

Calico cats (also sometimes known as tortoise-shell cats) have coats that are both black and yellow (actually an orange-yellow) and are almost always female. Color distribution in these cats seems to support the Lyon hypothesis. In 1961, Mary Lyon hypothesized that the dark spot of chromatin observed at the periphery of the nucleus of female mammalian cells in interphase—the Barr body—represents an inactivated X chromosome. Early in embryonic life, "lyonization" of one or the other X chromosome occurs in each cell of the female, except for those cells from which egg cells will form. Thus all female somatic cells can be divided into two groups, in which one or the other X chromosome remains active. In cats, the alleles for black or yellow coat color are carried on the X chromosome, so calico cats neatly fit the Lyon hypothesis. There are, however, occasional male calicos. These are suspected of having an extra X chromosome, a supposition supported by the observation that they are almost always sterile.

SUMMARY

The developments which led to a final proof of Sutton's hypothesis that genes are located on the chromosomes began, in the late nineteenth century, with studies of chromosomes in animals. It was found that in many animals, one pair of chromosomes (the sex chromosomes) is different in the male and the female. All the other chromosomes (the autosomes) are the same in both sexes. Both sex chromosomes of females of most species are X; most males have one X and one Y. Men and women, for example, have 22 pairs of autosomes; women have, in addition, an XX pair, and men have an XY pair.

At the time of meiosis, the sex chromosomes are segregated. Each egg cell will receive an X chromosome, but half the sperm cells will receive X chromosomes and half will receive Y chromosomes. Thus in humans it is the sperm cell that determines the sex of the embryo.

In the early 1900s, experiments with mutations in the fruit fly *Drosophila* showed that certain characteristics are sex-linked, that is, carried on the sex chromosomes. Female *Drosophila* are XX, and males are XY. The sex-linked characteristics, it was found, are carried on the X chromosome; the Y chromosome contains very little genetic information. Therefore, while a female heterozygous for a sex-linked characteristic will show the dominant trait, a single recessive gene in the male, if carried on the X chromosome, will result in a recessive phenotype since no dominant allele is present. Recessive genes transmitted on the X chromosomes will thus appear in the phenotype far more often in males than in females. Examples of sex-linked characteristics of man are color blindness and hemophilia.

Further studies of *Drosophila* demonstrated that alleles are sometimes exchanged between homologous chromosomes at meiosis. Such crossovers could take place only if (1) the genes were arranged in a fixed linear array along the

length of the chromosome and (2) the alleles were at corresponding sites on homologous chromosomes. These assumptions were proved when chromosome maps, showing the relative locations of gene sites along the *Drosophila* chromosomes, were developed from crossover data provided by breeding experiments with *Drosophila*. Cytological confirmation of these findings was subsequently supplied by observations made on giant banded chromosomes in the salivary glands of *Drosophila*.

Chromosomal abnormalities include inversion, deletion, translocation, and nondisjunction. In man, nondisjunction—which is a failure of homologues to separate at meiosis—can result in Down's syndrome (mongolism) and other disorders, most of which involve mental deficiencies. Down's syndrome can also be produced by translocation, which involves the attachment of a portion of a chromosome to another nonhomologous chromosome.

QUESTIONS

1. Under what conditions would color blindness be found in a woman? If she married a man who was not color-blind, would her sons be color-blind? Her daughters?

2. A couple has three girls. What are the chances that the next child will be a boy?

3. Suppose you would like to have a family consisting of two girls and a boy. What are your chances, assuming you have no children now? If you already have one boy, what are your chances of completing your family as planned?

4. Construct a pedigree, similar to that shown in Figure 14–11, for a female calico cat, assuming she mates with a yellow male.

5. A man has 42 chromosomes. How many linkage groups does he have?

6. In a series of breeding experiments, a linkage group composed of genes *A, B, C, D,* and *E* was found to show the following crossover frequencies:

		A	*B*	*C*	*D*	*E*	
				Gene			
Gene	*A*	—	8	12	4	1	
	B	8	—	4	12	9	*Crossovers*
	C	12	4	—	16	13	*per 100*
	D	4	12	16	—	3	*fertilized eggs*
	E	1	9	13	3	—	

Using Sturtevant's standard unit of measure, map the chromosome.

7. Should infants be tested for their karyotypes? If a white male infant is discovered by a physician to have an *XYY* genotype, should the physician tell the parents? Should restrictions be placed on the freedom of an *XYY* adult? What sort of restrictions?

Chapter 15

Nature and Function of Genes: Molds and Microbes

By the early 1940s, the existence of genes and the fact that they were in chromosomes were no longer in doubt. But what were genes? What did they really do? What does biological inheritance mean? When we say, "She has her mother's eyes," we obviously mean something quite different from saying, "She has her mother's pocketbook." "He inherited his father's intelligence" is not at all the same as "He inherited his uncle's yacht." The microscopic fertilized egg contains neither eyes nor intelligence. All that can be inherited, biologically speaking, is a potentiality, an ability to develop in certain ways.

A turning point in genetics came when scientists began to focus on the question of how it was possible for these little lumps of matter—the chromosomes—to be the bearers of what they had come to realize must be an enormous amount of complex information. The chromosomes, like all the other parts of a living cell, are composed of atoms arranged into molecules. Some scientists, a number of them quite eminent in the field of genetics, thought it would be impossible to understand the complexities of genetics in terms of the structure of "lifeless" chemicals. (Although they did not use the term to describe themselves, they were, in fact, the vitalists* of the twentieth century.) Others thought that if the chemical structure of the chromosomes were understood, we could then come to understand how chromosomes could function as the bearers of genetic information. The work of the latter group is now termed "molecular genetics."

GENE-ENZYME RELATIONSHIPS

"Inborn Errors of Metabolism"

The first molecular geneticist—although he would not have understood the term—was an English physician, Sir Archibald Garrod. In 1908, Garrod presented a series of lectures in which he set forth a new concept of human diseases, one he called "inborn errors of metabolism." With a leap of the imagination that spanned almost half a century, Garrod postulated that certain diseases which were caused by the body's inability to perform certain chemical processes were hereditary in nature. In other words—although the terms would have been unfamiliar to him—he hypothesized that a change in the genetic material caused a change in an enzyme. This was the first suggestion of a direct link between genes and enzymes.

* See Introduction, page 11.

One of the most familiar examples of such a disease is phenylketonuria, or PKU, as it is commonly known. PKU is caused by lack of function of the enzyme that normally breaks down the amino acid phenylalanine. When this enzyme is missing or deficient, phenylalanine and its abnormal breakdown products accumulate in the bloodstream. Such an accumulation is particularly harmful to the developing cells of the brain and may result in mental retardation. Detection of PKU is relatively easy, involving only a simple urine or blood test, and tests for PKU in newborn infants are now required in many states. Phenylalanine is an "essential" amino acid, which means, you will recall, that it cannot be synthesized in the body and must be included in the diet. Therefore, infants with PKU are fed a special diet low in phenylalanine in an attempt to minimize the harmful accumulation of phenylpyruvic acid. PKU, like most of the diseases Garrod described, is inherited as a Mendelian recessive. About 1 in every 15,000 infants is homozygous for the recessive gene.

Garrod's work foreshadowed the beginning of modern molecular genetics. In postulating the existence of a relationship between genes and body chemistry, he anticipated, by almost 30 years, the coming together of genetics and chemistry. An essential feature of modern biology is that it attempts to explain genetic mechanisms in terms of the laws of physics and chemistry. It was not until the mid-1940s, however, that further research made it possible to begin to understand the role of genes in controlling enzyme production.

One Gene, One Enzyme

Based on observations such as Garrod's, the hypothesis—one of many—was formulated that genes act by influencing enzyme production. The problem was to devise a test for such a hypothesis. Traits such as wrinkledness in peas or eye color and wing shape in *Drosophila*—or indeed the visible features of any organism—are characteristically the end product of a vast number of chemical reactions, most of which are still unknown.

In 1941, George Beadle and Edward L. Tatum decided to turn the problem around, and in so doing they changed the course of genetics and won a Nobel prize. Instead of picking a genetic characteristic and working out its chemistry, they decided to begin with known chemical reactions—controlled by enzymes—and see how genetic changes affected these reactions. Their experiments are worth examining in some detail because of their influence on both the theories and the techniques of genetics in the next quarter century.

First, like Morgan and Mendel before them, they chose a suitable organism. For their purposes, it was the pink bread mold *Neurospora crassa*, which has since become almost as famous a research tool in genetics as the fruit fly. This organism has several obvious advantages for genetic research: (1) Its life cycle is brief; (2) it can be grown in vast quantities in the laboratory; and (3) unlike most higher organisms, it is haploid, that is, during most of its life cycle it has only one set of chromosomes (seven in number) rather than two sets. As a consequence, when a mutation occurs, its effects are detectable immediately. The lack of a homologous chromosome rules out masking or dominance of any mutation by the normal allele.

Another important feature of *Neurospora* from the point of view of the investigators was the fact that they could grow it on a very simple medium, containing any one of several sugars as a carbon and energy source, one vitamin (biotin), and a few ions. This undemanding mold is able to make for itself

5–1 *Hereditary variations, the raw material for evolution, also may be the source of congenital diseases. This little girl, age 1½, has phenylketonuria (PKU), which is caused by a defect in an enzyme that breaks down the amino acid phenylalanine. Accumulated phenylalanine may cause brain damage; about one-fourth of the patients with PKU are in mental institutions. This child has just had a blood test, part of a treatment program that includes a diet low in phenylalanine.*

15–2 *Asci of* Neurospora. *Because of the shape of these asci, the sexual spores contained within the ascus are immobilized in the order in which they are produced by meiosis.* Neurospora *also produces fine, dustlike asexual spores known as conidia.*

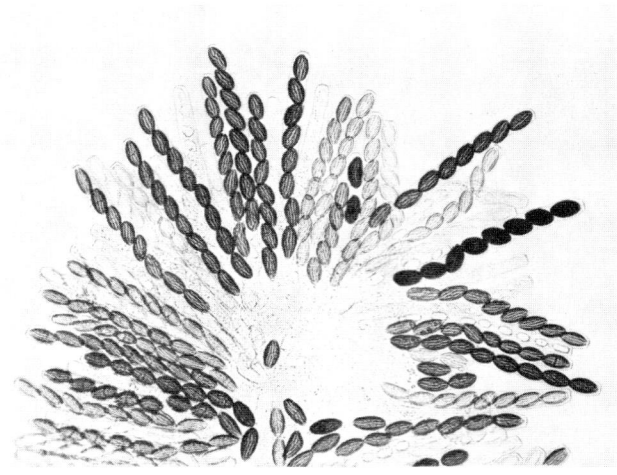

15–3 *Three enzymatic reactions involved in the biosynthesis of the amino acid arginine. Beadle and Tatum postulated that each enzyme is produced by a separate gene. To test this hypothesis, they sought mutant strains of* Neurospora *that could not carry out one particular biochemical step, such as the production of citrulline from ornithine, or arginine from citrulline. Such mutants could be analyzed by observing whether they could or could not grow in a particular medium. For example, a mold that could not perform step 2 could satisfy its arginine requirements with arginine or citrulline, but not with ornithine.*

Step 1

GENE 1 ⟶ ENZYME 1 ⟶ PRECURSOR

ORNITHINE

Step 2

GENE 2 ⟶ ENZYME 2 ⟶

CITRULLINE

Step 3

GENE 3 ⟶ ENZYME 3 ⟶

ARGININE

all the amino acids, other vitamins, polysaccharides, and other substances essential for its growth and functioning. On the other hand, if a substance—a particular amino acid, for example—is provided in the medium, *Neurospora* will use it "as is" rather than synthesize its own. The synthesis of an amino acid or a vitamin requires, as we know, a series of reactions, each of which is mediated by a particular enzyme. If, as a result of mutation, *Neurospora* were to lose any one of the enzymes involved, for example, in making the amino acid arginine, it could no longer grow in the simple minimal medium; it could, however, survive on a medium supplemented with arginine. With this new concept and this new tool, it became possible for geneticists to study not just the distant effects of gene activity, such as white eyes and long wings, but the most immediate products of gene function.

To understand the experiments, it is necessary first to know something about the life cycle of *Neurospora*. The mold produces tiny, dustlike spores known as conidia; these are asexual spores, meaning that they are not produced by any mating processes. They are carried in air currents and germinate when they land on a suitable medium, such as a loaf of bread. They grow to form a matlike fungus, the mycelium. Sometimes two of these fungal mycelia will encounter one another, and if they are of different mating strains, portions of the mycelia will fuse and form a zygote from which fruiting bodies are produced. Within the fruiting bodies are sacs, called asci (singular, ascus), and within the asci are sexual spores—that is, spores produced by meiosis (Figure 15–2). These, too, can germinate, producing new mycelia. The entire cycle from sexual spore to sexual spore takes only about 10 days, another advantage of *Neurospora*.

Beadle and Tatum collected asexual spores (conidia) of *Neurospora* and x-rayed them to increase the mutation rate. They then crossed these mutants with wild-type strains. The sexual spores thus produced were planted in culture media containing all the nutrients *Neurospora* normally needs plus amino acids. Cultures of those that grew were transplanted to a medium lacking the amino acids. This made it possible for the investigators to select mutants that could grow in the enriched medium but not in the minimal one. Subcultures of such molds were then tested individually to determine which function they lacked (Figure 15–4).

15–4 *How Beadle and Tatum tested the mutants of Neurospora. By these experiments, they were able to show that a change in a single gene results in a change in a single enzyme. (a) Asci are removed from fruiting bodies of* Neurospora *and the sexual spores dissected out. (b) Each spore is transferred to an enriched medium, containing all* Neurospora *normally needs for growth plus supplementary amino acids. (c) A fragment of the mycelium is tested for growth in the minimal medium. If no growth is observed on the minimal medium, it may mean that a mutation has occurred that renders this mutant incapable of making a particular amino acid, and so tests are continued. (d) Subcultures of mycelia that grow on the enriched medium but not on the minimal one are tested for their ability to grow in minimal media supplemented with only one of the amino acids. As in the example shown here, a mold that has lost its capacity to synthesize the amino acid proline is unable to survive in a medium that lacks that amino acid. Further tests are then made to discover, in each case, which enzymatic step has been impaired.*

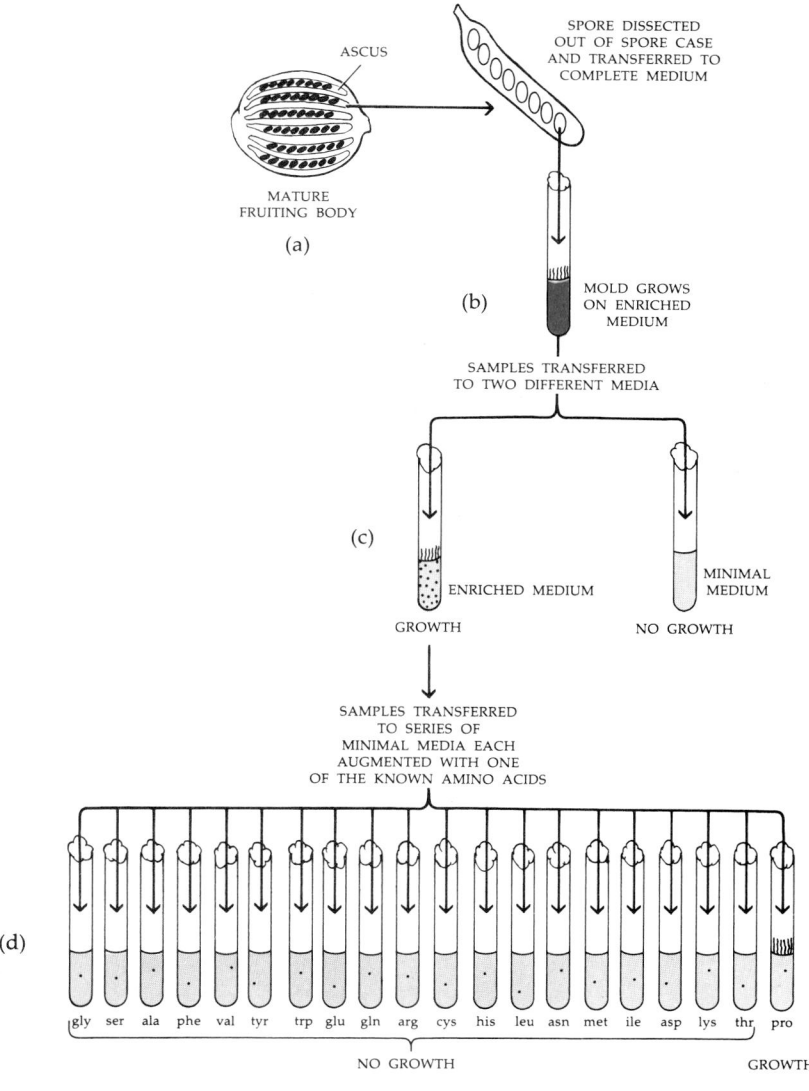

15–5 *Meiosis in* Neurospora, *showing only one of its seven chromosomes. The zygote represents a cross of a mutant strain (with a biochemical deficiency) with a wild-type strain (lacking the biochemical deficiency). As a result of meiosis and the mitotic division that follows it, eight spores are produced, lined up in a single, narrow spore case, the ascus. If the biochemical deficiency is the result of a single gene on a single chromosome, and no crossing over has occurred, the spores will be arranged as shown in the first ascus. If crossing over has occurred involving those particular alleles, the spores will be arranged as in the second ascus. By analyzing the products of meiosis, it is possible to demonstrate that a given hereditary change involves only a single gene.*

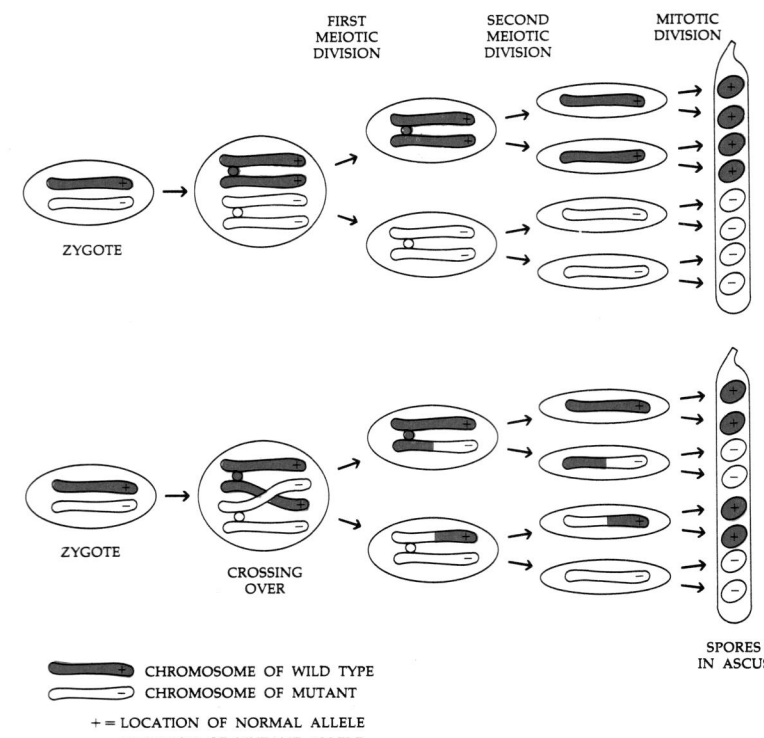

Notice the advantages of dealing with a haploid organism for genetic tests of this type (Figure 15-5). In diploid organisms, only a fraction of the haploid cells (the gametes) will ever develop into mature organisms whose phenotypes can be observed, whereas in haploid organisms, all of the products of meiosis can survive and be tested. Hence, in haploid organisms, one does not deal with statistical probabilities, as Mendel did, but with direct results. By mating *Neurospora* mutants with wild types, it was possible to prove that loss of a single enzyme function was the result of a mutation in a single gene.

At the time Beadle and Tatum's experiments were first performed, enzymatic pathways had been worked out for very few organic molecules in only a few species. Since that time, however, it has been found that a great variety of cells, including bacteria, yeast, and even the cells that make up the tissues of the human body, are very similar in their enzyme system and in their stepwise synthesis of the various basic cellular nutrients, underlining once more the remarkable biochemical unity of living systems.

One Gene, One Protein

The Beadle-Tatum hypothesis that a particular gene was responsible for a particular enzyme was quick to gain acceptance. That all enzymes are proteins had already been demonstrated in the 1930s. Not all proteins are enzymes, however; some, for instance, are hormones, like insulin. Others are structural proteins, like collagen. These proteins, too, are under gene control.

This expansion of the original concept did not modify it in principle: "One gene, one enzyme," as the theory was first abbreviated, was simply amended to the less memorable but more precise "one gene, one polypeptide chain." In other words, enzymes and other protein molecules are the direct product of genes.

Speculative thinkers in the field of biology were quick to see that the amino acids, the number of which was so provocatively close to the number of letters in our own alphabet, could be arranged in a variety of different ways and that these different arrangements might account for both the great diversity of enzymes and the great specificity of the biochemical reactions they mediate within the cell. The proteins were seen as making up a sort of language—"the language of life"—that spelled out the directions for all the many activities of the cell.

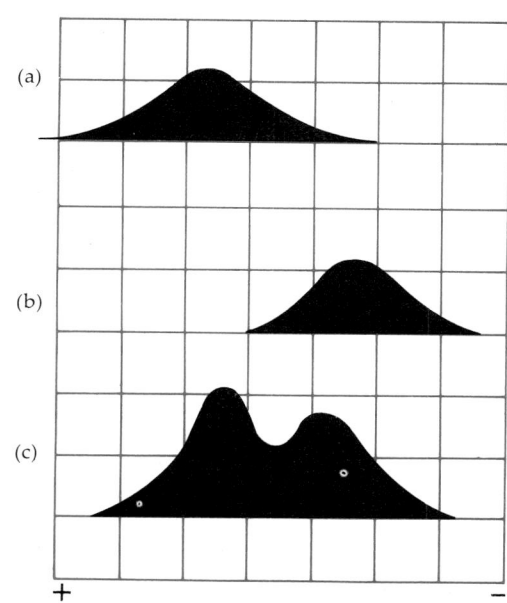

5–6 *Results from electrophoresis of the hemoglobin of (a) a normal person, (b) a person with sickle cell anemia, and (c) a heterozygote carrying the sickle cell trait. Because of slight differences in electric charge, the normal and sickle cell hemoglobins move differently in an electric field. The normal hemoglobin is more negatively charged; hence, it is closer to the positive pole than the sickle cell hemoglobin. The hemoglobin of the heterozygote separates into the two different types.*

THE STRUCTURE OF HEMOGLOBIN

Linus Pauling was one of the first to see some of the implications of these new ideas. Hemoglobin, as we mentioned previously, is also made up of chains of amino acids. Perhaps, Pauling reasoned, human diseases involving hemoglobin, such as sickle cell anemia, could be traced to a slight variation from normal in the protein structure of the hemoglobin molecule—a typographical error in one important sentence in the hypothetical language of the cells. To test this hypothesis—which, at this stage, was pure speculation—he took samples of hemoglobin from normal persons, from persons with sickle cell anemia, and from persons heterozygous for the disease. To study the differences in these proteins, he used a process known as electrophoresis, in which organic molecules are dissolved in a solution and exposed to a weak electric field.

As we noted in Chapter 5, individual amino acids may have a positive, a negative, or a neutral charge. Therefore, a mutation that causes the substitution of one amino acid for another may change the total charge of the protein molecule, and the normal and the mutant protein molecules will, as a result, move differently in an electric field.

Figure 15–6 shows the results of Pauling's experiment. A normal person makes one sort of hemoglobin, a person with sickle cell anemia makes a different sort, and a person who is heterozygous (carrying one copy of the recessive gene for sickling and one copy for normal hemoglobin) makes both, as you can see in the figure.

Although a heterozygote makes both kinds of hemoglobin molecules, he has enough "good" molecules to prevent anemia. (Notice that the terms "dominant" and "recessive" are beginning to take on a different meaning.)

A few years later, in 1959, it was found, as we noted previously (page 78), that the actual difference between the normal and the sickle hemoglobin molecules is 1 amino acid in 300. The hemoglobin molecule is composed of four polypeptide chains—two identical alpha chains and two identical beta chains. Each chain consists of about 150 amino acids. In a precise location in each beta chain, a glutamic acid is present in the normal hemoglobin, a valine in the sickle cell hemoglobin. The language of proteins is fantastically precise, and the consequences of even a small error or variation may be great.

THE CHEMISTRY OF THE GENE: DNA VS. PROTEIN

During the 1930s and 1940s, scientists became more and more concerned with the question "Exactly what is a gene?" The attempt to identify the chemical nature of the gene gave rise to two opposing schools of thought, and the controversy lasted for a number of years—until, as we shall see, it was conclusively resolved in 1953.

Chromosomes, as we mentioned previously, are composed of DNA and protein. Many prominent investigators, particularly those who had been studying proteins, believed that the genes themselves were proteins, that the chromosomes contained master models of all the proteins that would be required by the cell, and that enzymes and other proteins active in cellular life were copied from these master models. This was a logical hypothesis, but as it turned out, it was wrong.

The Transforming Factor

To trace the beginning of the other hypothesis—the one that proved right—it will be necessary to go back to 1928 and pick up an important thread in modern biological history. In that year, an experiment was performed which seemed at that time very remote from either biochemistry or genetics. Frederick Griffith, a public health bacteriologist, was studying the possibility of developing vaccines against the bacterial cells, pneumococci, that cause one kind of pneumonia. In those days, before the development of modern antibiotics, bacterial pneumonia was a serious disease, the grim "captain of the men of death." Pneumococci, as Griffith knew, come in either virulent (disease-causing) forms with capsules (gelatinous coats) or nonvirulent (harmless) forms without capsules. Griffith was interested in finding out whether injections of heat-killed virulent pneumococci, which do not cause disease, could be used to vaccinate against pneumonia. In the course of various experiments, he performed one that gave him very puzzling results. He injected mice simultaneously with heat-killed virulent bacteria and with living but nonvirulent bacteria, both of which were harmless—but all the mice died. When Griffith performed autopsies on them, he found their bodies filled with living encapsulated (and therefore virulent) bacteria. Had the dead virulent cells come back to life or had something been passed from them to the living nonvirulent cells which endowed the living cells with the capacity to make capsules and therefore to be virulent?

Within the next few years it was shown that the same phenomenon could be reproduced in the test tube and these questions could be answered. It was found that extracts from the killed encapsulated bacteria, when added to the living harmless bacteria, could transmit to them—and through them to their progeny—their virulent character and their ability to make capsules.

One of the laboratories that worked on the nature of this *transforming factor*, as it came to be called, was that of O. T. Avery at Rockefeller University. After almost a decade of patient chemical isolation and analysis, Avery and his co-workers were convinced that the chemical substance in the cellular extracts of killed bacteria that transmitted the new genetic quality was the molecule known as *deoxyribonucleic acid.* Subsequent experiments showed that a variety of genetic traits could be passed from one colony of bacterial cells to members of another, similar colony by means of isolated deoxyribonucleic acid, which soon took on the now familiar abbreviation of DNA.

5–7 Discovery of the transforming factor, a substance that can transmit genetic characteristics from one cell to another, resulted from studies of pneumococci, pneumonia-causing bacteria. One strain of these bacteria has capsules (protective outer sheaths); another does not. The capacity to make capsules and cause disease is an inherited characteristic, passed from one bacterial generation to another as the cells divide. (a) Injection into mice of encapsulated pneumococci killed the mice. (b) The nonencapsulated strain was harmless. (c) If the encapsulated strain was heat-killed before injection, it too was harmless. (d) If, however, heat-killed encapsulated bacteria were mixed with live nonencapsulated bacteria and the mixture was injected into mice, the mice died. (e) Blood samples from the dead mice revealed live encapsulated pneumococci. Something had been transferred from the dead bacteria to the live ones that endowed them with the capacity to make capsules and cause pneumonia. This "something" was later isolated and found to be DNA.

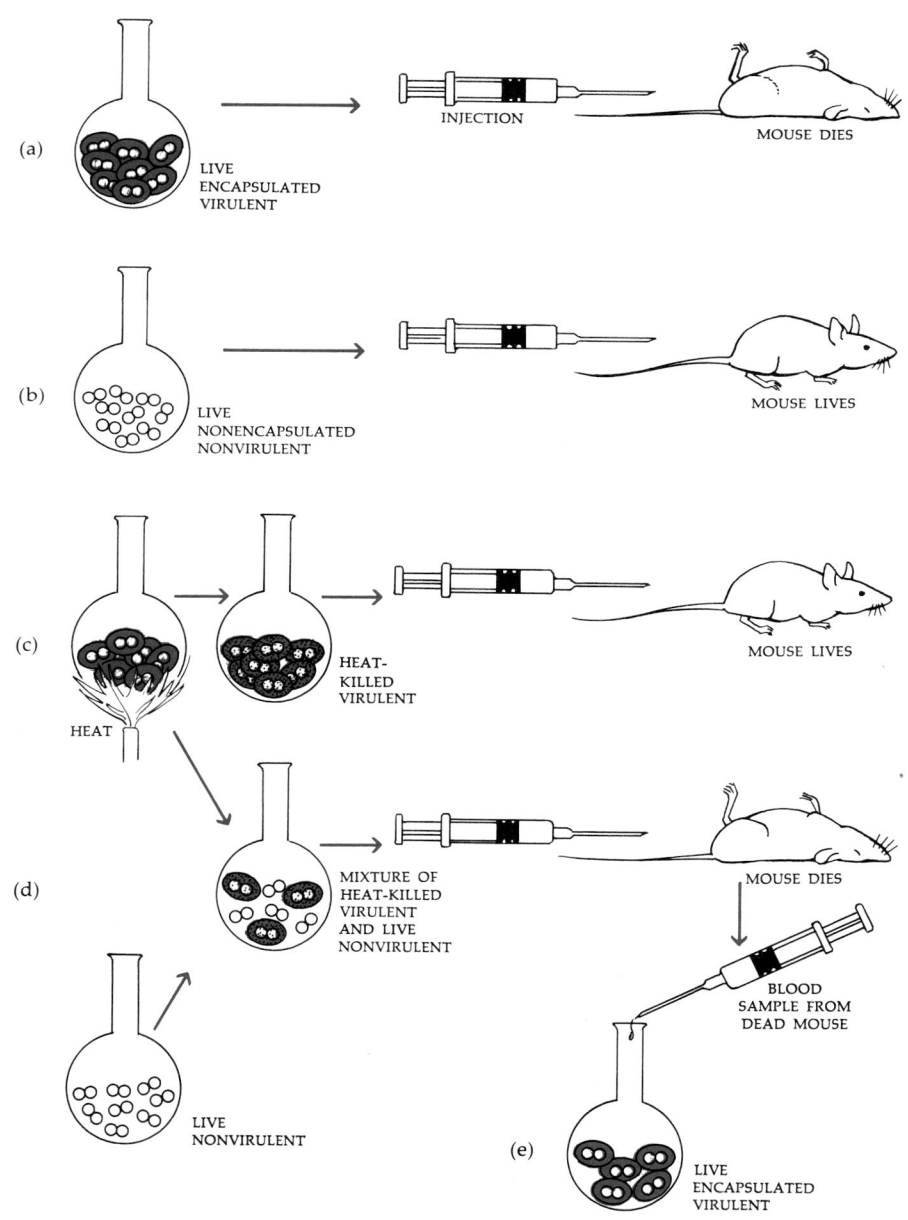

THE NATURE OF DNA

At this point, we shall examine DNA a little more closely and try to put its discovery into historical perspective—in order to understand why Avery's experiment, although beautifully designed and executed, did not receive full recognition for almost a decade.

DNA was first discovered in that same remarkable decade in which Darwin published *The Origin of Species* and Mendel presented his results to an audience of forty at the Natural History Society in Brünn. In 1869, a German chemist, Friedrich Miescher, had extracted a substance from the nuclei of cells which was white, slightly acid, and contained phosphorus. He called it nucleic acid, which was later amended to deoxyribonucleic acid to distinguish it from a closely related chemical, ribonucleic acid, which subsequently was also isolated from the cell.

In 1914, another German, Robert Feulgen, discovered that DNA had a remarkable attraction for fuchsin dye, but he considered this finding so unimportant that he did not trouble to report it for a decade. Feulgen staining, as it was called when it finally made its way into use, revealed that DNA was present in all cells and was characteristically located in the chromosomes.

There was no particular interest in DNA for several decades since no role had been postulated for it in cellular metabolism. Most of the work on its chemistry was carried out by the great biochemist P. A. Levene, who showed that DNA could be broken down into the four nitrogenous bases—adenine and guanine (the purines) and thymine and cytosine (the pyrimidines)—a five-

15–8 (a) *A nucleotide is made up of three different parts: a nitrogen base, a sugar, and a phosphate.* (b) *Each nucleotide in DNA contains one of the four possible nitrogen bases, a deoxyribose sugar, and a phosphate.*

carbon sugar, and a phosphate group. From the proportions of these components, he made two deductions, one correct and one incorrect:

1. Each nitrogenous base is attached to a molecule of sugar and to one of phosphate to form a single molecule, a *nucleotide*. This hypothesis was right.
2. Since in all the samples he measured, the proportions of the nitrogenous bases were approximately equal, all four nitrogenous bases must be present in nucleic acid in equal quantity. Furthermore, these molecules must be grouped in clusters of four—a tetranucleotide he called it—which repeated over and over again along the length of the molecule. This hypothesis, which was wrong, dominated scientific thinking about the nature of DNA for more than a decade.

Because of Levene's tetranucleotide theory, given great weight by his renown as a biochemist, biologists as a whole were slow to recognize the importance of Avery's experiment. In terms of DuPraw's analogy (page 13), Avery was a traveler bearing an odd tale that did not fit. How could a chemical whose structure was "known" to be simple and uniform be associated with something as complicated and various as heredity? For almost a decade, most influential biochemists continued to believe that proteins were the genetic material. In the next chapter, we shall review some of the evidence that made them change their minds.

SUMMARY

Classical genetics had been concerned with the mechanics of inheritance—how the units of hereditary material were passed from one generation to the next and how changes in the hereditary material were expressed in individual organisms. In the late 1930s, new questions arose and geneticists began to explore the nature of the gene—its structure, composition, and properties, and its role in the internal chemistry of living organisms.

In 1941, working with the bread mold *Neurospora crassa*, Beadle and Tatum established the "one-gene–one-enzyme" principle. They were able to trace biochemical defects in *Neurospora* to the nonfunction of specific enzymes and to relate each of these defects to a mutation in a specific gene.

It was soon discovered that some enzymes are made up of more than one protein chain and that other proteins, such as some hormones and hemoglobin, are also under genetic control, and the one-gene–one-enzyme principle was expanded to the "one-gene–one-polypeptide" principle.

Sickle cell anemia is inherited as a Mendelian recessive. Using electrophoresis, Linus Pauling demonstrated that the hemoglobin of a normal person is slightly different from the hemoglobin of a person with sickle cell anemia. Subsequently, it was shown that the actual difference lay in just two amino acid changes among the 600 amino acids making up the polypeptide chains in the molecule.

During the 1940s, many investigators believed that genes were proteins, but others were convinced that the hereditary material was deoxyribonucleic acid (DNA). Important, although not widely accepted evidence for the genetic role of DNA was presented by Avery in his experiments on the transforming factor of pneumococcus. Although DNA was first discovered in 1869, its role in genetics was not widely accepted and understood until the 1950s. In the next chapter, we shall discuss the development of the DNA theory.

QUESTIONS

1. Define the following terms: purine, pyrimidine, nucleotide, transforming factor.
2. Give two definitions of a gene, one in terms of Mendelian genetics, the other in terms of the Beadle-Tatum hypothesis.
3. In pea plants, a cross between a red-flowered plant and a white-flowered plant produces a red-flowered plant. In snapdragons, a cross between a red-flowered plant and a white-flowered plant produces a pink-flowered plant. Explain how the Beadle-Tatum hypothesis might account for these differences.
4. Although the person with sickle cell anemia is usually ill, the person with the sickle cell trait—that is, a person who is a carrier for sickle cell anemia—rarely has any symptoms of disease. Explain, in terms of the examples of flower color mentioned above.
5. In terms of survival of the organism, what is the advantage of diploidy?
6. A new mutant in *Neurospora* requires chemical X for growth and accumulates chemical Y. What is the order of these substances in the biosynthetic sequence and where in this sequence does the gene that is deficient in the mutant normally act?
7. In *Neurospora*, mutant *x* will grow only if provided with cystathionine, homocysteine, or cysteine; mutant *r* will grow only if provided with homocysteine, but accumulates cystathionine; mutant *w* will grow if provided with homocysteine or cystathionine, but not if only cysteine is available. Sketch the sequence of biosynthesis of these chemicals. What is the genetic defect in each of the mutants?

Chapter 16

Deoxyribonucleic Acid

About the time that Beadle and Tatum began their studies with *Neurospora,* another pair of scientists, Max Delbrück and S. E. Luria, initiated a series of studies with another "fit material," destined to become as important to genetic research as the garden pea, the fruit fly, and the pink bread mold. The fit material was a group of viruses that attacked bacterial cells and that were therefore known as *bacteriophages,* "bacteria eaters." Every known type of bacterial cell is preyed upon by its own type of bacterial virus, and many cells are host to many different kinds of virus. Delbrück, Luria, and the group that joined them in these studies agreed to concentrate on a series of seven related viruses that attacked *Escherichia coli,* a normal inhabitant of the human intestine. These were numbered T1 through T7, with the T standing simply for "type." As it turned out, most of the early work was done on T2 and T4, which came to be known as the T-even bacteriophages.

These viruses were inexpensive, readily available, and demanded little space or equipment. Furthermore, they were phenomenal at reproducing themselves. Only 20 minutes after the infection of one bacterial cell by one virus particle, the cell would burst open and hundreds of new viruses would be released, each an exact copy of the infecting particle. Moreover, although it was not known when the research was begun, this group of bacteriophages has a highly distinctive shape (see Figure 16–1), and so can be readily identified by the electron microscope.

THE INFECTION CYCLE

According to electron-microscope studies of infected *E. coli* cells (broken open at regular intervals after infection), the bacteriophages do not multiply like bacteria. They disappear the moment after infection, and for half of the total infectious cycle, not a single particle can be found within the bacterial cell. Then, depending on when the cell is opened during the course of the infection, increasing numbers of completed bacteriophage particles can be found and, mixed with them, odds and ends that resemble bits of incomplete phages.

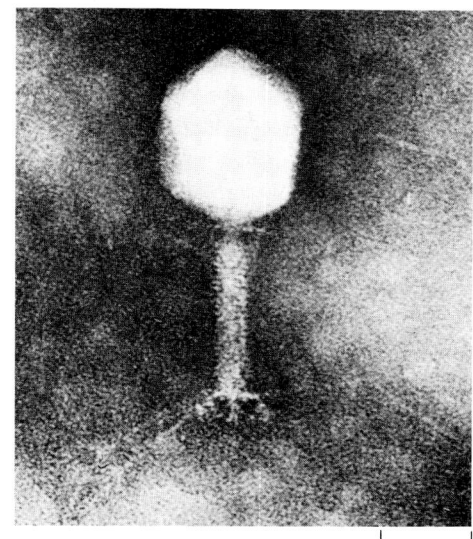

50 nm

1 *Electron micrograph of a T4 bacteriophage. Notice the highly distinctive "tadpole" shape. Each bacteriophage consists of a head, which appears hexagonal in electron micrographs, and a complex tail.*

The phages increase at a regular rate during this period, not geometrically—2, 4, 8, 16, 32, 64, like bacteria—but linearly, one after another, as if they were coming off an assembly line.

DNA vs. Protein

Chemical analysis of the bacteriophages revealed that they consist quite simply of DNA and of protein, the two leading contenders for the role of the genetic material. The chemical simplicity of the bacteriophage offered geneticists a remarkable opportunity. The viral genes—the hereditary material by which new viral particles are made within the cell—had to be carried either on the protein or on the DNA. If it could be determined which of the two it was, then the gene would be chemically identified. In 1952, a simple but ingenious experiment was carried out.

Alfred D. Hershey and his laboratory assistant, Martha Chase,* grew *E. coli* on a medium that contained radioactive phosphorus and radioactive sulfur and then infected the bacterial cells with T2. After a cycle of multiplication, the newly formed viruses all contained radioactive sulfur and radioactive phos-

*We are including the names of the scientists involved in these experiments, not only to give credit where it is due, but also because the names have become synonymous with the work. What we are describing now is the Hershey-Chase experiment.

16–2   *A summary of the Hershey-Chase experiments demonstrating that DNA is the hereditary material of a virus.*

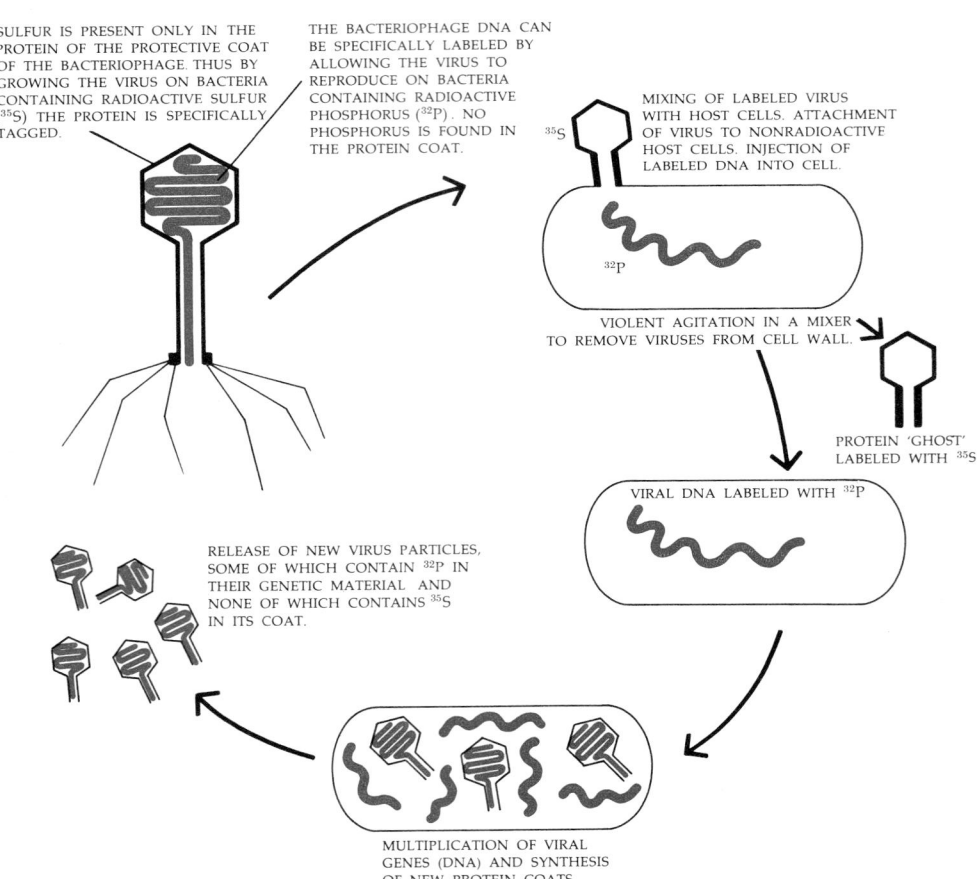

SULFUR IS PRESENT ONLY IN THE PROTEIN OF THE PROTECTIVE COAT OF THE BACTERIOPHAGE. THUS BY GROWING THE VIRUS ON BACTERIA CONTAINING RADIOACTIVE SULFUR (^{35}S) THE PROTEIN IS SPECIFICALLY TAGGED.

THE BACTERIOPHAGE DNA CAN BE SPECIFICALLY LABELED BY ALLOWING THE VIRUS TO REPRODUCE ON BACTERIA CONTAINING RADIOACTIVE PHOSPHORUS (^{32}P). NO PHOSPHORUS IS FOUND IN THE PROTEIN COAT.

MIXING OF LABELED VIRUS WITH HOST CELLS. ATTACHMENT OF VIRUS TO NONRADIOACTIVE HOST CELLS. INJECTION OF LABELED DNA INTO CELL.

^{35}S

^{32}P

VIOLENT AGITATION IN A MIXER TO REMOVE VIRUSES FROM CELL WALL.

PROTEIN 'GHOST' LABELED WITH ^{35}S

VIRAL DNA LABELED WITH ^{32}P

RELEASE OF NEW VIRUS PARTICLES, SOME OF WHICH CONTAIN ^{32}P IN THEIR GENETIC MATERIAL AND NONE OF WHICH CONTAINS ^{35}S IN ITS COAT.

MULTIPLICATION OF VIRAL GENES (DNA) AND SYNTHESIS OF NEW PROTEIN COATS.

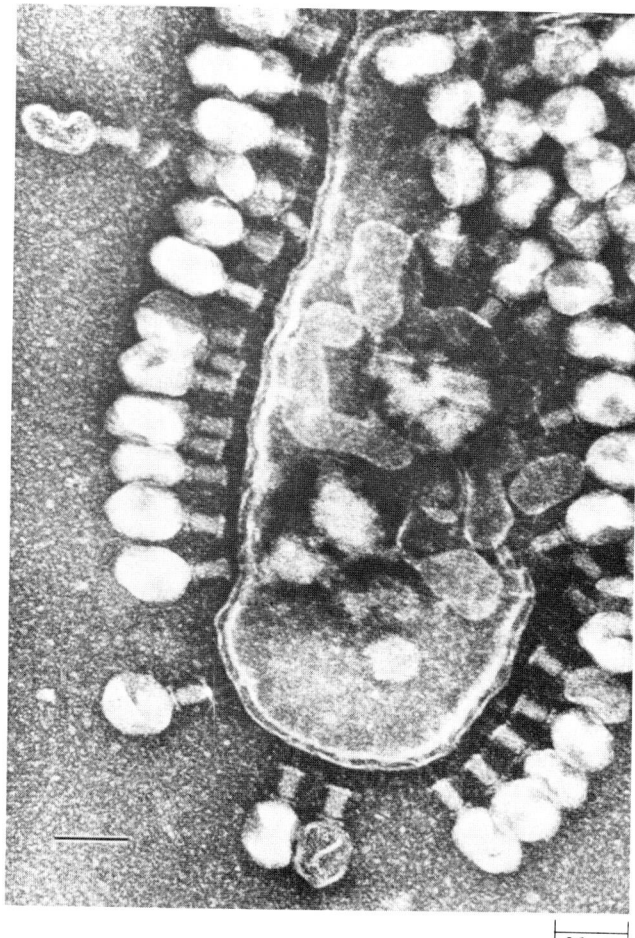

6–3 *Electron micrograph of bacteriophages attacking a cell of* E. coli. *The viruses are attached to the bacterial cell by their tails. Some of the viruses have injected their DNA into the cell, as revealed by their empty heads. A complete cycle of virus infection takes only about 20 minutes. At the end of that period, several hundred new virus particles are released from the cell.*

|‾ 0.1 μm ‾|

6–4 *Max Delbrück and Salvador Luria at Cold Spring Harbor in 1953. They shared the Nobel prize with A. D. Hershey in 1971 for "their discoveries concerning the replication mechanism and the genetic structure of viruses."*

phorus in place of the common isotopes of sulfur and phosphorus. Since proteins contain sulfur but no phosphorus, all the radioactive sulfur was confined to the protein. Conversely, since only the DNA contained phosphorus, only DNA was labeled with radioactive phosphorus. The labeled viruses were used to infect fresh *E. coli* cells growing in a medium with the ordinary isotopes of sulfur and phosphorus. Once infection was begun, the cells were agitated violently. This procedure led to the removal of most of the radioactive sulfur from the infected bacteria. However, the bacteria still yielded normal phage progeny, containing most of the radioactive phosphorus. This experiment showed that the phage DNA entered the cell while the protein "coat" stayed outside, attached to the bacterial surface (Figure 16–2).

Electron micrographs (Figure 16–3, for example) have now confirmed that the phage attaches to the bacterial cell wall by its tail and injects its DNA into the cell, leaving the empty protein coat on the outside. In short, the protein is just a container for the bacteriophage DNA. It is the DNA of the bacteriophages that enters the cell and carries the complete hereditary message of the virus particle, directing formation of new viral DNA and new viral protein.

16–5 *Electron micrograph of a bacteriophage surrounded by its single continuous molecule of DNA. The bacteriophage was burst open and the DNA released by osmotic shock, produced by placing the bacteriophage in distilled water and allowing the DNA to expand on the water surface. The DNA, which is normally circular, has been broken at one point—note the two free ends at the top and bottom.*

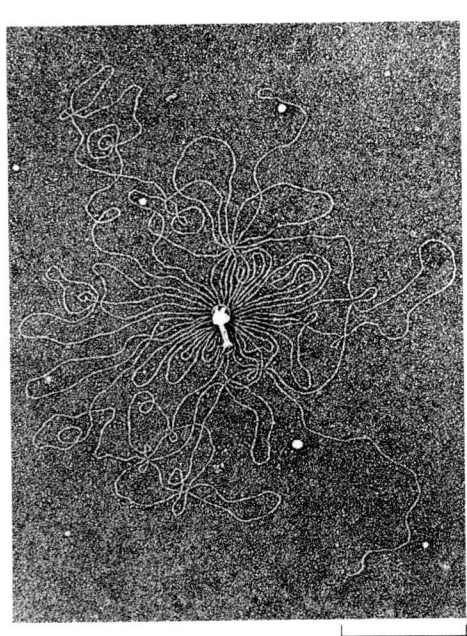

0.5 μm

FURTHER EVIDENCE FOR DNA

The role of DNA in transformation and in viral replication formed very convincing evidence for believing that DNA was the chemical basis of the gene. Two other lines of work also helped to lend weight to the argument. Alfred Mirsky, in a long series of careful studies conducted at the Rockefeller Institute, showed that all the tissue cells of any given species contain equal amounts of DNA. The only exceptions are the gametes, which regularly contain just half as much DNA as the other cells of the same species, and certain unusual cells, such as polyploid cells and cells with giant chromosomes (see page 217).

A second important series of contributions was made by Erwin Chargaff of Columbia University's College of Physicians and Surgeons. Chargaff analyzed the purine and pyrimidine content of the DNA of many different kinds of living things and found that, in contradiction to Levene's conclusions, the nitrogen bases do not occur in equal proportions. The proportion of nitrogen bases is the same in all cells of a given species but varies from one species to another. Therefore variations in base composition could very well provide a "language" in which the instructions controlling cell growth could be written. Some of Chargaff's results are reproduced in Table 16–1. Can you, by examining these figures, notice anything interesting about the proportions of purines and pyrimidines?

THE HYPOTHESIS IS CONFIRMED

The explanation of the way that the genetic information is contained in DNA was found in the structure of the DNA molecule. We have seen that the genetic material must meet at least four requirements:

1. It must carry genetic information from cell to cell and from generation to generation. Further, it must carry a great deal of information. Consider how many instructions must be contained in the set of genes that directs, for example, the development of an elephant, or a tree, or even a paramecium.
2. It must contain information for producing a copy of itself, for the chromosome is copied with every cell division. Moreover, the copying must take place with great precision; from mutation-rate data, we know that a human gene must on the average be copied for 100,000 years without a mistake.
3. On the other hand, it must sometimes mutate. When a gene changes, that is, when a "mistake" is made, the "mistake" must be copied as faithfully as was the original. This is a most important property, for without the capacity to replicate "errors," there could be no evolution by natural selection.
4. There must be some mechanism for decoding the stored information and translating it into action in the living individual.

It was when the DNA molecule was found to have the size, the configuration, and the complexity required to code the tremendous store of information needed by living things and to make exact copies of this code that DNA became widely accepted as the genetic material.

Table 16–1 *Composition of DNA in Several Species*

Source	Purines		Pyrimidines	
	Adenine	Guanine	Cytosine	Thymine
Man	30.4%	19.6%	19.9%	30.1%
Ox	29.0	21.2	21.2	28.7
Salmon sperm	29.7	20.8	20.4	29.1
Wheat germ	28.1	21.8	22.7	27.4
E. coli	26.0	24.9	25.2	23.9
Sheep liver	29.3	20.7	20.8	29.2

The scientists primarily responsible for working out the structure of the DNA molecule were James Watson and Francis Crick, and their feat is one of the milestones in the history of science.

THE WATSON-CRICK MODEL

In the early 1950s, a young American scientist, James D. Watson, went to Cambridge, England, on a research fellowship to study problems of molecular structure. There, at the Cavendish Laboratory, he met physicist Francis Crick. Both were interested in DNA, and they soon began to work together to solve the problem of its molecular structure. They did not do experiments in the usual sense but rather undertook to examine all the data about DNA and attempt to unify them into a meaningful whole.

The Known Data

By the time Watson and Crick began their studies, quite a lot of information on the subject had already accumulated:

1. The DNA molecule was known to be very large, and also very long and thin, and to be composed of nucleotides of adenine, guanine, thymine, and cytosine.
2. According to Levene's data, the nucleotides were assembled in repeating units of four.
3. X-ray diffraction studies of DNA from the laboratories of Maurice Wilkins at King's College, London, showed markings that almost certainly reflected the turns of a giant spiral, or helix.
4. Linus Pauling, in 1950, had shown that a protein's component chains of amino acids are often arranged in the shape of a helix and are held in that form by hydrogen bonds between successive turns in the helix. Pauling had suggested that the structure of DNA might be similar.

–6 *Watson and Crick with one of their models of DNA.*

16–7 *The double-stranded helical structure of DNA, as first presented in 1953 by Watson and Crick. The framework of the helix is composed of the sugar–phosphate units of the nucleotides. The rungs are formed by the four nitrogen bases adenine and guanine (the purines) and thymine and cytosine (the pyrimidines). Each rung consists of two bases. Knowledge of the distances between the atoms, determined from x-ray diffraction pictures, was crucial in establishing the structure of the molecule.*

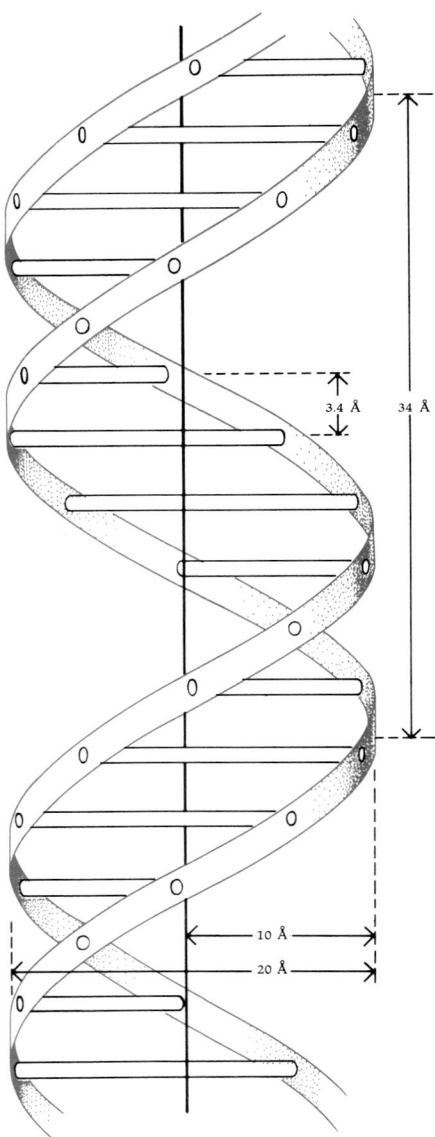

5. Also crucial (as you will see in the following paragraphs) were the data of Chargaff indicating, as you perhaps noticed in Table 16–1, that the ratio of nucleotides containing adenine to those containing thymine is 1:1, as is the ratio of nucleotides containing guanine to those containing cytosine.

Building the Model

From these data, some of them contradictory, Watson and Crick attempted to construct a model of DNA that would fit the known facts and explain the biological role of DNA. In order to carry such a vast amount of information, the molecules should be heterogeneous and varied. Also, there must be some way for them to replicate readily and with great precision in order that faithful copies could be passed from cell to cell and from parent to offspring, generation after generation.

On the other hand, they could not be sure that the chemical structure of DNA actually would reflect its biological function. After all, this idea had never really been tested rigorously. Perhaps DNA was merely some sort of biological clay on which some outside "vital force" operated. "In pessimistic moods," Watson recalled in a review of these investigations, "we often worried that the correct structure might be dull—that is, that it would suggest absolutely nothing."

It turned out, in fact, to be unbelievably "interesting." By piecing together the various data, they were able to deduce that DNA did not have a single-stranded helix structure, as do proteins, but was an exceedingly long entwined double helix.

The banister of a spiral staircase forms a single helix. If you take a ladder and twist it into the shape of a helix, keeping the rungs perpendicular, this would form a crude model of the molecule. (See Figure 16–7.) The two railings, or sides of the ladder, are made up of alternating sugar and phosphate molecules. The perpendicular rungs of the ladder are formed by the nitrogenous bases—adenine (A), thymine (T), guanine (G), and cytosine (C)—one base for each sugar–phosphate, as Levene had shown, and two bases forming each rung. The paired bases meet across the helix and are joined together by hydrogen bonds, the relatively weak, omnipresent chemical bonds that Pauling had demonstrated in his studies of the structures of proteins. (Hydrogen bonds are described in Chapter 3.)

The distance between the two sides, or railings, according to Wilkins' measurements, is 20 angstroms. Two purines in combination would take up more than 20 angstroms, and two pyrimidines would not reach all the way across. But if a purine paired in each case with a pyrimidine, there would be a perfect fit. The paired bases—the "rungs" of the ladder—would therefore always be purine-pyrimidine combinations.

As Watson and Crick worked their way through these data, they assembled actual tin and wire models of the molecule, seeing where each piece would fit into the three-dimensional puzzle. First, they noticed, the nucleotides along any one chain of the double helix could be assembled in any order: ATGCGTACA-TTGCCA, and so on. (See Figure 16–8.) Since a DNA molecule may be many thousands of nucleotides long, there is a possibility for great variety. The meaning of this variety, the molecular heterogeneity of DNA, will be explored more thoroughly in the next chapter.

16–8 (a) *The structure of a portion of one strand of a DNA molecule. Each nucleotide consists of a sugar, a phosphate group, and a purine or pyrimidine base. The sugar of each nucleotide is linked by a phosphate group to the sugar of an adjacent nucleotide. The sequence of nucleotides varies from one molecule to another. In the figure, the order of nucleotides is TTCAG. (b) The double-stranded structure of a portion of the DNA molecule. The strands are held together by hydrogen bonds between the bases. Because of bonding requirements, adenine can pair only with thymine and guanine only with cytosine. Thus the order of bases along one strand determines the order of bases along the other.*

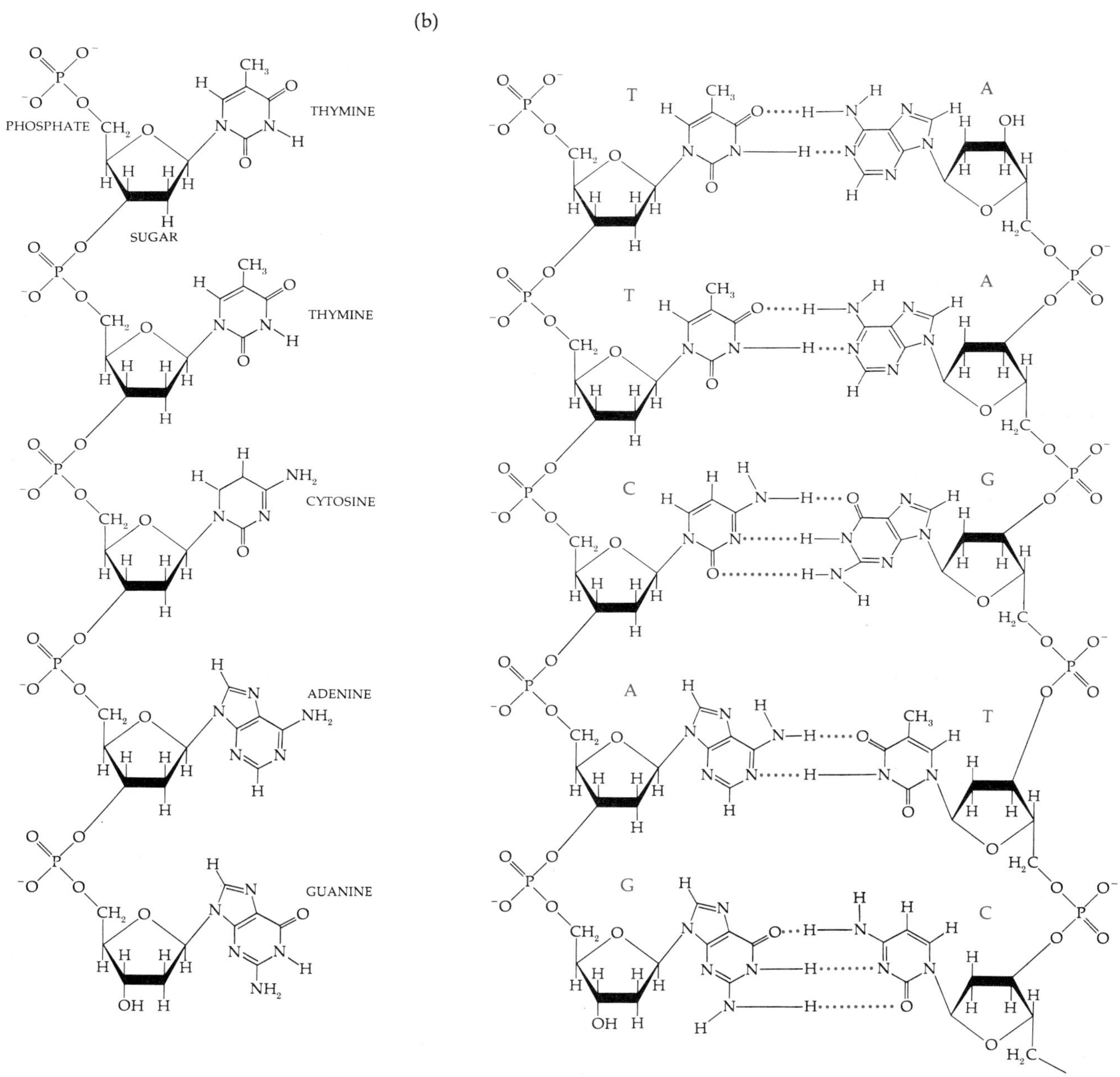

(a)

(b)

WHO MIGHT HAVE DISCOVERED IT?

Then there is the question, what would have happened if Watson and I had not put forward the DNA structure? This is "iffy" history which I am told is not in good repute with historians, though if a historian cannot give plausible answers to such questions I do not see what historical analysis is about. If Watson had been killed by a tennis ball I am reasonably sure I would not have solved the structure alone, but who would? Olby has recently addressed himself to this question. Watson and I always thought that Linus Pauling would be bound to have another shot at the structure once he had seen the King's College x-ray data, but he has recently stated that even though he immediately liked our structure it took him a little time to decide finally that his own was wrong. Without our model he might never have done so. Rosalind Franklin was only two steps away from the solution. She needed to realise that the two chains must run in opposite directions and that the bases, in their correct tautomeric forms, were paired together. She was, however, on the point of leaving King's College and DNA, to work instead on TMV with Bernal. Maurice Wilkins had announced to us, just before he knew of our structure, that he was going to work full time on the problem. Our persistent propaganda for model building had also had its effect (we had previously lent them our jigs to build models but they had not used them) and he proposed to give it a try. I doubt myself whether the discovery of the structure could have been delayed for more than two or three years.

There is a more general argument, however, recently proposed by Gunther Stent and supported by such a sophisticated thinker as Medawar. This is that if Watson and I had not discovered the structure, instead of being revealed with a flourish it would have trickled out and that its impact would have been far less. For this sort of reason Stent had argued that a scientific discovery is more akin to a work of art than is generally admitted. Style, he argues, is as important as content.

I am not completely convinced by this argument, at least in this case. Rather than believe that Watson and Crick made the DNA structure, I would rather stress that the structure made Watson and Crick. After all, I was almost totally unknown at the time and Watson was regarded, in most circles, as too bright to be really sound. But what I think is overlooked in such arguments is the intrinsic beauty of the DNA double helix. It is the molecule which has style, quite as much as scientists. The genetic code was not revealed all in one go but it did not lack for impact once it had been pieced together. I doubt if it made all that difference that it was Columbus who discovered America. What mattered much more was that people and money were available to exploit the discovery when it was made. It is this aspect of the history of the DNA structure which I think demands attention, rather than the personal elements in the act of discovery, however interesting they may be as an object lesson (good or bad) to other workers.

> Francis Crick: "The Double Helix: A Personal View,"
> Nature, **248**, 766–769, 1974.

HOW SCIENTISTS INVESTIGATE THE THREE-DIMENSIONAL STRUCTURE OF A MOLECULE

X-ray diffraction studies are used to investigate the physical structure of the molecule, that is, the spatial arrangements of the atoms within it. Such studies depend on the properties of crystals. When substances crystallize, their atoms line up in a latticework of regularly repeating units. These units will deflect x-rays in a regular pattern, and by studying these patterns one can sometimes determine the distances between various atoms in the latticework. An x-ray beam is projected through the crystal at various angles. The atoms in the crystal deflect the x-rays, and the deflection pattern is registered on a photographic plate. From this pattern, the way in which the atoms are arranged in the molecule is inferred by mathematical procedures. It was by this technique that investigators determined the structure of such simple crystals as those of ice and salt.

The methods of x-ray diffraction have been extended to the complex organic molecules of proteins and nucleic acids. But these giant molecules offer considerable difficulty since they contain a great many subgroupings and therefore so many planes for the x-rays to deflect from that it is sometimes impossible to interpret the photographic record. For this reason, when dealing with these complex molecules, scientists often build hypothetical models of the molecules. Once a model, consistent with available information on interatomic distances, bond angles, x-ray diffraction patterns, and so on, is constructed, additional x-ray diffraction patterns can be predicted. If these predictions are then confirmed, the evidence for the correctness of the structure is greatly strengthened.

The x-ray diffraction photograph of DNA shown here was taken by Rosalind Franklin in the laboratories of Maurice Wilkins, who shared the Nobel prize with Watson and Crick. The reflections crossing in the middle indicate that the molecule is a helix. The heavy, dark regions at the top and bottom are due to the closely stacked nitrogen bases perpendicular to the axis of the helix.

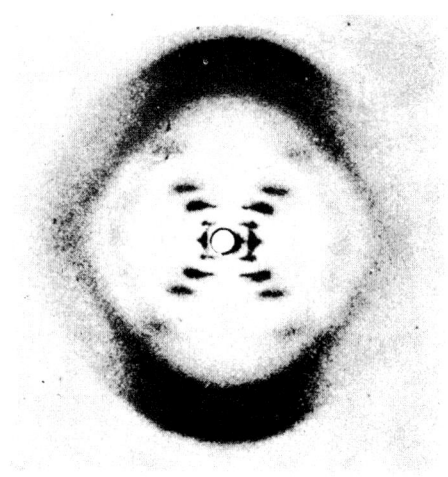

X-ray diffraction photo of DNA.

The most exciting discovery came, however, when they set out to construct the matching strand. They encountered an interesting and important restriction. Not only could purines not pair with purines and pyrimidines not pair with pyrimidines, but because of the configurations of the molecules, adenine could pair only with thymine and guanine only with cytosine. Look at Chargaff's table (Table 16–1, page 241) again and see how well these chemical requirements confirm his data.

DNA as a Carrier of Information

You will recall that a necessary property of the genetic material is the ability to carry genetic information. The Watson-Crick model shows clearly how the DNA molecule is able to do this. The information is carried in the sequence of the bases, and *any* sequence of the four pairs (AT, TA, CG, and GC) is possible. Since the number of paired bases ranges from about 5,000 for the simplest known virus up to an estimated 5 billion in the 46 chromosomes of man, the possible variations are astronomical. The DNA from a single human cell—which if extended in a single thread would be about 5 feet long—contains in-

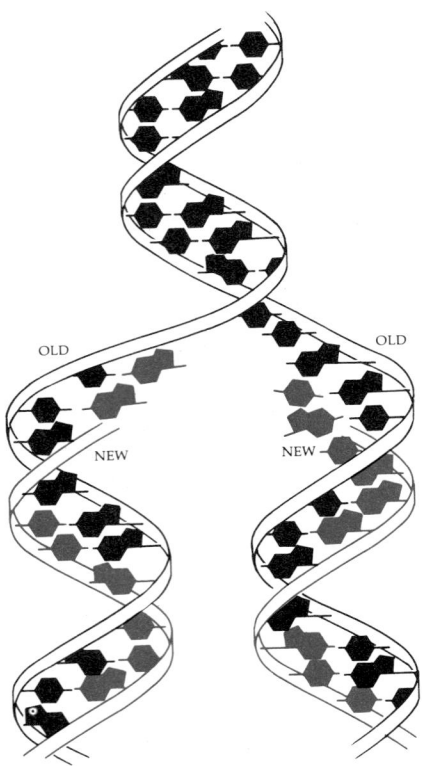

16–9 *The DNA molecule shown here is in the process of replication, separating down the middle as its base units separate at the hydrogen bonds. (For clarity, the bases are shown out of plane.) Each of the original strands then serves as a template along which a new, complementary strand forms from nucleotides available in the cell.*

OLD   OLD

NEW   NEW

formation equivalent to some 600,000 printed pages averaging 500 words each, or a library of about a thousand books. Obviously, the DNA structure can well explain the endless diversity among living things.

DNA REPLICATION

Another necessary property of the genetic material is the ability to provide for exact copies of itself. Does the Watson-Crick model satisfy this requirement? In their published account, Watson and Crick wrote, "It has not escaped our notice that the specific pairing we have postulated immediately suggests a possible copying mechanism for the genetic material." Implicit in the double and complementary structure of the DNA helix is the method by which it reproduces itself. The molecule "unzips" down the middle, the bases breaking apart at the hydrogen bonds. The two strands separate, and new strands form along each old one, using the raw materials in the cell. Each old strand forms a template, or guide, for the new one. If a T is present on the old strand, only an A can fit into place with the new strand, a G will pair only with a C, and so on. In this way, each strand forms a copy of the original partner strand, and two exact replicas of the molecule are produced. The age-old question of how hereditary information is duplicated and passed on, duplicated and passed on, for generation after generation, had in principle been answered.

A Confirmation

Matthew Meselson and Franklin W. Stahl, working at the California Institute of Technology, devised an ingenious test of the replication hypothesis, using the bacterial cell *E. coli*.

They grew *E. coli* for several generations in a medium containing the heavy isotope of nitrogen (^{15}N). At the end of this period virtually all of the ordinary isotope of nitrogen (^{14}N) had been replaced by heavy nitrogen. They isolated the DNA from some of these bacteria, put it in solution, and spun it in an ultracentrifuge with cesium chloride. (An ultracentrifuge operates on the same principle as a cream separator; it separates materials on the basis of their density.) The cesium chloride sediments partially to form a density gradient, and any mixture of molecules dissolved in the cesium chloride solution will simultaneously settle into bands according to their density. Each component will settle at a position where its density exactly equals that of the cesium chloride solution.

The investigators then placed a sample of cells containing heavy nitrogen in a medium containing ^{14}N and left them there long enough for the DNA to replicate once (as determined by a doubling of the number of cells). A sample of this DNA was spun in the ultracentrifuge. A second generation was then grown in the ^{14}N medium. This DNA was also ultracentrifuged.

Each sample of DNA contained more light DNA, as could be expected, because newly formed DNA had to incorporate the available ^{14}N (Figure 16–11c, d, and e). Moreover—and this was of crucial importance—the density of the first generation DNA was exactly halfway between that of the heavy parent DNA and that of ordinary light DNA, as it should be if each molecule contained one old (heavy) strand and one new (light) strand, as predicted by Watson and Crick (Figure 16–11d). The third generation contained one-fourth half-heavy DNA and three-fourths light DNA, which again, exactly and ingeniously, confirmed the Watson-Crick hypothesis.

–10 *Three possible mechanisms of replication of DNA. Newly replicated strands are shown in color. (a) Conservative replication. Each of the two strands of parent DNA is replicated, without strand separation. In the first generation, one daughter is all old DNA and one daughter is all new. The F$_2$ generation contains one helix composed of two old strands and three made up entirely of new strands. (b) Semiconservative replication of DNA. In the first generation, each daughter is half old and half new. The F$_2$ generation comprises two hybrid DNAs (half old, half new) and two new DNAs made up entirely of new strands. (c) Dispersive replication. During replication, parent chain breaks at intervals and replicated segments are combined into strands with segments from parent. All daughter helices are part old, part new. The Meselson-Stahl experiment (Figure 16–11) was undertaken to determine which of these three possibilities was correct. Watson and Crick had predicted (b).*

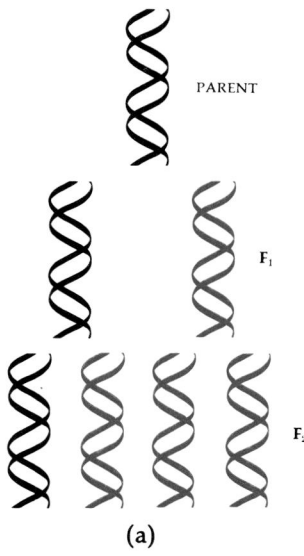

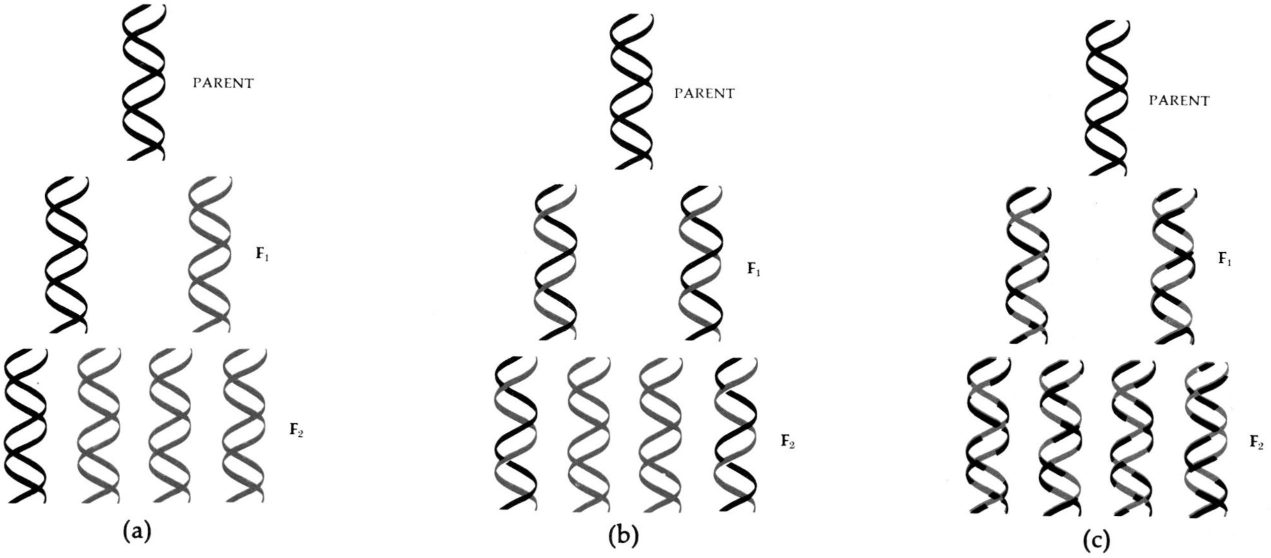

(a)   (b)   (c)

–11 *Meselson-Stahl experiment. Escherichia coli cultured in a medium containing heavy nitrogen (^{15}N) accumulated a "heavy" DNA. Then the cells were removed from the ^{15}N medium and permitted to multiply. Light and heavy DNAs were separated by ultracentrifuging the molecules in a cesium chloride gradient. The column on the left shows the results of the experiment; the column on the right shows the investigators' interpretation. As you can see, the experiment confirmed the Watson-Crick hypothesis. Where would the bands have appeared if DNA replication had been conservative (Figure 16–10a)? Dispersive (Figure 16–10c)?*

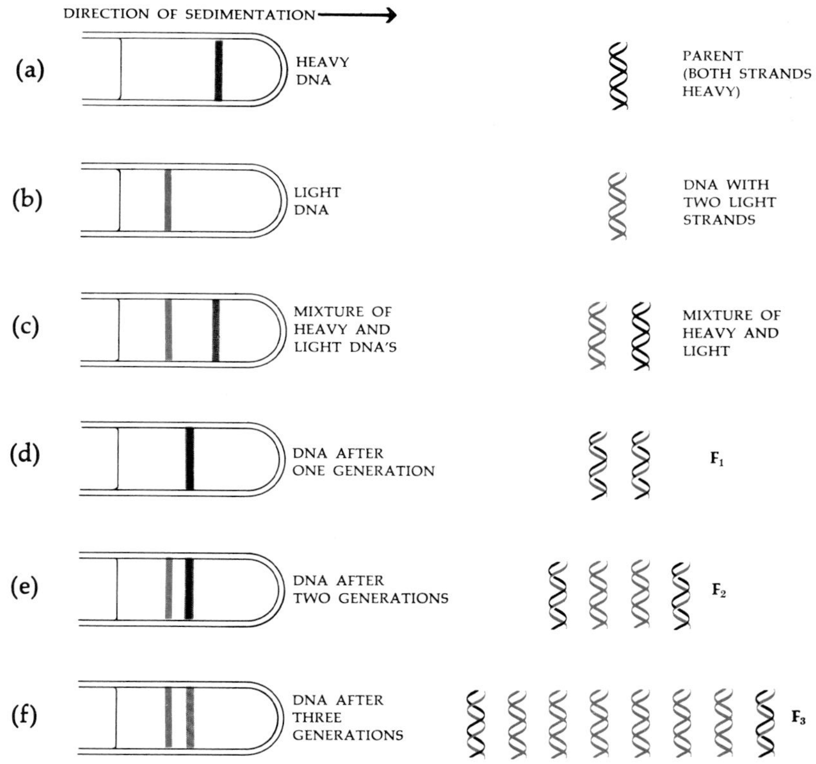

DIRECTION OF SEDIMENTATION ⟶

(a) HEAVY DNA   PARENT (BOTH STRANDS HEAVY)

(b) LIGHT DNA   DNA WITH TWO LIGHT STRANDS

(c) MIXTURE OF HEAVY AND LIGHT DNA'S   MIXTURE OF HEAVY AND LIGHT

(d) DNA AFTER ONE GENERATION   F$_1$

(e) DNA AFTER TWO GENERATIONS   F$_2$

(f) DNA AFTER THREE GENERATIONS   F$_3$

16–12 *DNA replicating. To make this autoradiograph, an* E. coli *cell was grown for a short time in a medium containing thymine labeled with a radioactive isotope (³H). The cell was broken open and the DNA, which was allowed to spread out, was placed on a photographic plate for two months. During this period, decay of the ³H left spots on the radioactive emulsion. The chromosome had been caught in the process of replication. A and B are believed to represent already replicated portions and C the unreplicated portion. This micrograph, taken by John Cairns, also established that the DNA of a bacterial cell is in the form of a circle—that is, of a single molecules with no end. The chromosome is about 1,100 micrometers long, or about 500 times as long as an* E coli *cell.*

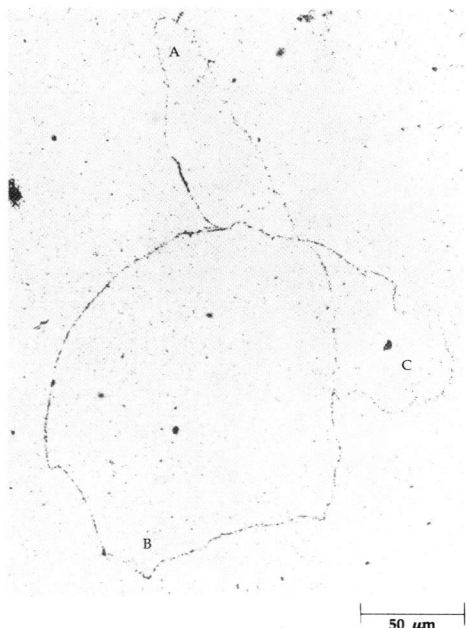

50 μm

Subsequently, it has become possible to visualize DNA replicating along the bacterial chromosomes. To produce the micrograph in Figure 16–12, the *E. coli* cell was grown in a medium containing thymine that had been labeled with tritium (³H), the radioactive isotope of hydrogen (see Chapter 2). When DNA from the cell was placed on a photographic plate, the emissions of high-energy particles from the isotope left tracks in the photographic emulsion, revealing the forming DNA. This is an example of autoradiography.

Similar studies of eukaryotic cells have revealed that DNA replication in these cells is also semiconservative—that is, old strands of DNA are conserved, and new strands are formed along them, using the old strands as a template.

The Mechanics of DNA Replication

Although DNA is often referred to as a self-duplicating molecule, this description is not precisely true. If DNA is placed in solution along with all the necessary components for new DNA, nothing happens. As in other biological reactions, a special enzyme is needed; this enzyme is known as DNA polymerase. The exact way in which DNA polymerase works is not known. It is known, however, that it requires a single strand of DNA as a template, in addition to the various raw materials.

Along with DNA polymerase, there are a dozen or more different enzymes that specialize in performing various operations on DNA. Some of them clip out nucleotides from just one strand, breaking the double helix; others fill in such breaks; and a third group seals together the broken ends of DNA strands. Probably some of these enzymes are important in repairing damaged DNA. Some, however, paradoxically, appear to initiate DNA damage. It is possible that such enzymes are important in helping to cause crossovers (see Chapter 14), thereby recombining DNA from different sources.

The Energetics of DNA Replication

With the help of DNA polymerase, the nucleotides are joined one by one to the growing DNA strand. The energy for this reaction comes from high-energy phosphate bonds. Each nucleotide, as it is synthesized, is coupled by a high-energy bond to a P~P (pyrophosphate) group. As the nucleotide is attached to the growing DNA strand, the two phosphates are removed and released. Almost immediately, another enzyme breaks the high-energy bond between the two phosphates, releasing them as inorganic phosphates.

Why does the cell do it this way? The release of pyrophosphate and the breaking of the high-energy bond seem like a waste of carefully stored chemical energy. Is the cell really so profligate?

Measurement of the energy changes involved reveals an interesting point. The reaction in which the activated nucleotide is attached to the growing DNA strand is only slightly exergonic. Therefore, it could tend to go in either direction (see page 44). Under certain equilibrium conditions, the DNA strand could come apart about as fast as it could go together, with the enzyme working both ways. However, with the release of the P~P fragment and its immediate degradation, the reaction becomes highly exergonic. Thus the reverse reaction—which would necessitate reforging the P~P group—becomes highly endergonic and so would almost never occur spontaneously. This is another example of the way in which cells exercise control over their biochemical activities.

SUMMARY

DNA (deoxyribonucleic acid) is the genetic material of the cell. Investigations showing that both the transforming factor in bacteria and the carrier of the genetic information in bacteriophages are DNA provided some of the first evidence for this hypothesis.

Further support for the genetic role of DNA came from two more findings: (1) Almost all tissue cells of any given species contain equal amounts of DNA and (2) the proportions of nitrogen bases are the same in the DNA of all cells of a given species, but they vary in different species.

In 1953, Watson and Crick proposed a structure of DNA. The DNA molecule, according to their model, is a double-stranded helix, shaped somewhat like a twisted ladder. The two sides of the ladder are composed of repeating groups of phosphate and a five-carbon sugar. The "rungs" are made up of paired bases, one purine base pairing with one pyrimidine base. There are four bases—adenine (A), guanine (G), thymine (T), and cytosine (C)—and A can pair only with T, and G only with C. The four bases are the four "letters" used to spell out the genetic message. The paired bases are joined by hydrogen bonds. On the basis of this structure, as revealed by Watson and Crick, the role of DNA as the carrier and transmitter of the genetic information became widely accepted.

When the DNA molecule replicates, the two strands come apart, breaking at the hydrogen bonds. Each strand forms a new complementary strand from nucleotides available in the cell. The semiconservative (one-half conserved) nature of this process was confirmed by studies using radioactive and heavy isotopes.

Replication of DNA is enzymatically mediated, and energy is supplied by high-energy phosphate bonds.

QUESTIONS

1. One of the chief arguments for the erroneous theory that proteins constitute the genetic material is that proteins are heterogeneous. Explain why the genetic material must have this property. What feature in the Watson-Crick DNA model is important in this respect?

2. What are the steps by which Griffith demonstrated the existence of the transforming principle? Can you think of any implications of Griffith's discovery for modern medicine?

3. Suppose you are talking to someone who has never heard of DNA. How would you support an argument that DNA is the genetic material? List at least five of the strong points in such an argument.

4. Eukaryotic cells are grown in a medium containing thymine labeled with ^3H. Then they are removed from the radioactive medium, placed in an ordinary medium, and allowed to divide. Autoradiographs taken after each generation show whether or not a chromatid contains radioactive material. Before they are placed in the nonradioactive medium, all the chromatids contain ^3H. After one generation in the nonradioactive medium, the ^3H is still divided evenly among the chromatids. Explain why. Does this confirm the Watson-Crick hypothesis? What would be the distribution of the ^3H after two divisions in the nonradioactive medium? Why?

Chapter 17

RNA and the Making of Proteins

Like most important scientific discoveries, the Watson-Crick model raised more questions than it answered.

Given that genes are made of DNA and that the products of genes are specific proteins, what is the link between them? How does DNA influence protein synthesis? One early hypothesis was that the DNA somehow formed a template for protein production. This was soon abandoned, however, because it was impossible to get a satisfactory physicochemical templatelike "fit." The relationship between DNA and protein had to be a more complicated one. If the proteins, with their 20 amino acids, were the "language of life," to extend the metaphor of the 1940s, the DNA molecule, with its four nitrogen bases, could be envisioned as a sort of code for this language. So the term "genetic code" came into being.

THE TRIPLET CODE

As it turned out, the idea of a "code of life" was useful not only as a dramatic metaphor but also as a working analogy. Scientists, seeking to understand how the DNA so artfully stored in the nucleus could order the quite dissimilar structures of protein molecules, approached the problem with methods used by cryptographers in deciphering codes. There are 20 biologically important amino acids, and there are four different nucleotides. If each nucleotide "coded" one amino acid, only four could be provided for. If two nucleotides specified one amino acid, there could be a maximum number, using all possible arrangements, of 4^2, or 16—still not quite enough. Therefore, following the code analogy, at least three nucleotides must specify each amino acid. This would provide for 4^3, or 64, possible combinations. This postulate, the triplet code, was widely and immediately adopted as a working hypothesis, although it was not actually proved until a decade after the Watson-Crick discovery. Proof depended on answering yet another question: How is the code translated?

THE RNAs

There were several clues indicating that RNA might play a role in the process of translating genetic information into a sequence of amino acids. As you will recall (page 79), RNA closely resembles DNA in its chemical characteristics. Unlike DNA, which is found primarily in the nucleus, most of the RNA is found in the cytoplasm, and it is there that protein manufacture takes place.

DEOXYRIBOSE

RIBOSE

THYMINE

URACIL

17–1 *Chemically, RNA is very similar to DNA, but there are two differences in its chemical groups. One difference is in the sugar component; instead of deoxyribose, RNA contains ribose, which has an additional oxygen atom. The other difference is that instead of thymine, RNA contains the closely related pyrimidine uracil (U). (A third, and very important, difference between the two is that most RNA does not possess a regular helical structure and is usually single-stranded.)*

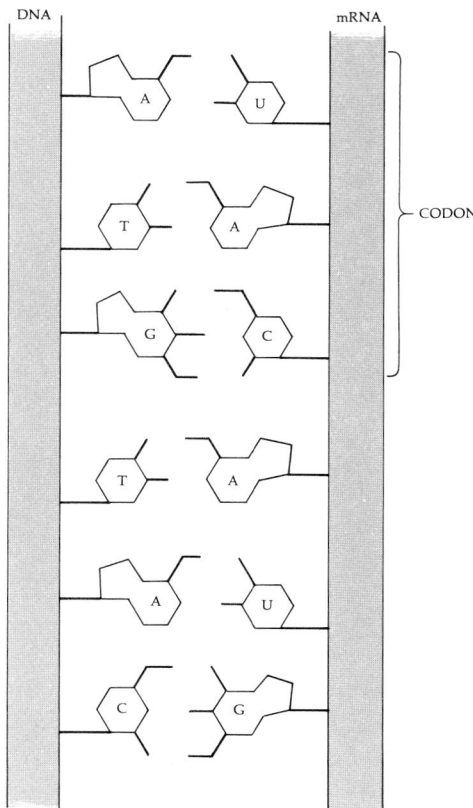

DNA   mRNA

CODON

7–2   *The beginning of the process of protein biosynthesis is the formation of mRNA on the DNA template. In the cell's nucleus, a strand of DNA that codes one sequence of amino acids for a protein forms a complementary strand of mRNA (instead of a partner DNA strand). The strand of mRNA is now a "negative print" of the sequence of nucleotides in the DNA. Transfer RNA and ribosomal RNA are also formed by this same copying process from chromosomal DNA. The copying of DNA by RNA is known as transcription.*

Cells making large amounts of protein have numerous ribosomes, and ribosomes are rich in RNA. When a bacterial cell is infected by a DNA-containing bacteriophage, a new RNA appears in the cell near the beginning of the infective cycle. Some viruses contain only RNA; tobacco mosaic virus (page 84) is an example. If the RNA is removed from TMV and rubbed onto a scratched tobacco leaf, infection results and new viruses are produced. In other words, RNA also contains information that can somehow determine both new RNA and proteins.

Many of the experiments to define the role of RNA in protein biosynthesis were carried out using cell-free extracts of *E. coli* and of animal liver cells. Cell-free in this context means simply that the cells have been broken apart. A principal advantage of using such extracts is that they can be separated into various fractions and the fractions studied separately.

As it turned out in the course of these studies, which occupied many different scientific groups for more than a decade, not one but three types of RNA are involved: messenger RNA, transfer RNA, and ribosomal RNA.

Messenger RNA

Messenger RNA is a long molecule consisting of a single strand of nucleotides. This molecule forms along the DNA helix, using one strand of the DNA as a template (Figure 17–2), in a way similar to that in which a new DNA strand forms. It is believed that the DNA helix opens up to facilitate RNA copying. The activation energy for the reaction is supplied by high-energy phosphate bonds. As in the replication of DNA, each nucleotide entering the reaction is coupled by a high-energy bond to a pyrophosphate (P~P) group. Such nucleotides are referred to as activated. (ATP is, in fact, the activated form of the adenine nucleotide of RNA.) The differences between the two processes are (1) the RNA strand contains ribose sugars rather than deoxyribose, (2) uracil substitutes for thymine (Figure 17–1), (3) a different enzyme—RNA polymerase—is involved, and (4) the RNA strand is released from the DNA template and becomes attached to a ribosome. Messenger RNA (mRNA) is the molecule that carries the genetic information from the DNA. It carries the message, as we shall see, in the form of *codons*, each codon consisting of a sequence of three nucleotides. The synthesis of RNA along a DNA template is known as transcription.

Transfer RNA

In the cytoplasm are amino acids, special enzymes, ATP molecules, ribosomes, and molecules of another kind of RNA, transfer RNA. Transfer RNA (tRNA) molecules are small, containing only about 80 nucleotides in a single strand. Like the other RNA molecules, tRNAs are formed in the nucleus along DNA templates.

All of the tRNA molecules are similar. One end always terminates in a guanine nucleotide, the other in a CCA sequence. The other nucleotides vary according to the particular type of RNA. There are more than 20 tRNA molecules—at least one for each amino acid found in proteins—and the sequence of nucleotides in a number of these types of molecules has now been analyzed. All tRNA molecules appear to have the cloverleaf shape shown in Figure 17–3; in addition, the cloverleaf is probably folded over on itself in some way to form a

17-3 (a) *Structure of a tRNA molecule. These molecules consist of about 80 nucleotides linked together in a single chain. One end of the chain always terminates in a guanine nucleotide, and the other in a CCA sequence. The amino acid is linked to the tRNA at the CCA end. The other nucleotides vary according to the particular tRNA. All tRNA molecules appear to have the configuration shown here; in some, however, there is an extra "arm." The "cloverleaf" is, in addition, folded in some way. Some of the bases are hydrogen-bonded to one another, following the DNA-type base pairing (A with U, G with C). The unpaired bases at the bottom of the diagram (indicated in color) serve as the anticodon and "plug in" the molecule to an mRNA codon. (b) Relationship between a specific mRNA codon and its corresponding tRNA anticodon.*

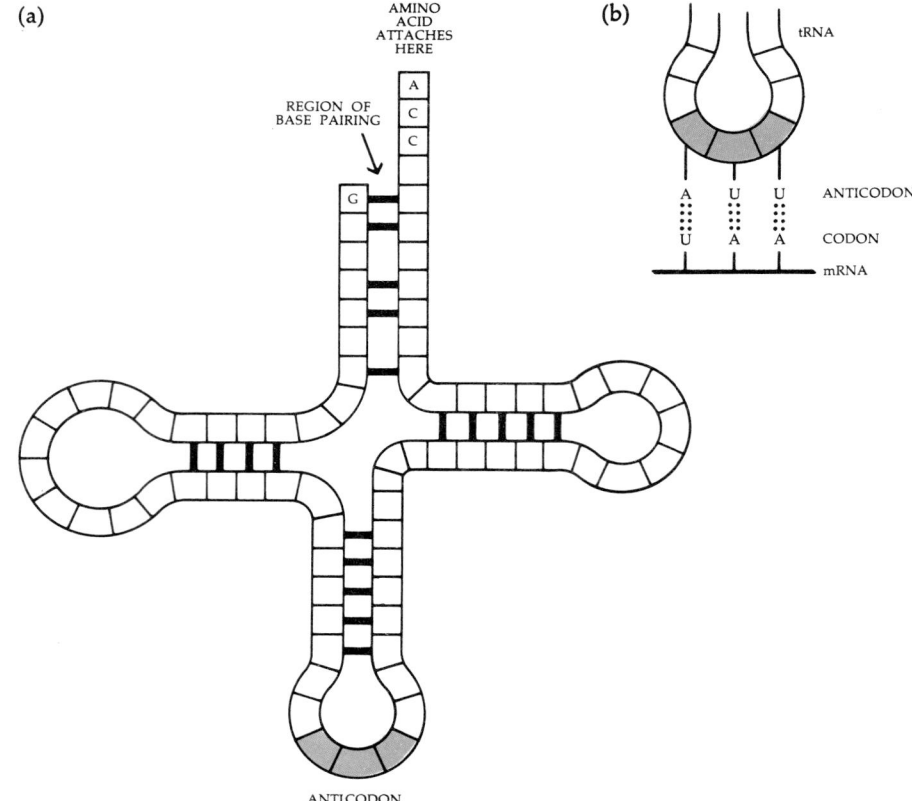

(a)

AMINO ACID ATTACHES HERE

REGION OF BASE PAIRING

ANTICODON

(b)

tRNA

ANTICODON

CODON

mRNA

three-dimensional structure. Some of the bases, as indicated in the diagram, are hydrogen-bonded to one another intramolecularly (A with U, G with C). Transfer RNA molecules characteristically contain a number of unusual bases. These unusual bases are formed after the molecule is assembled by the addition of extra chemical groups, such as the methyl (CH_3) group, mediated by special enzymes. The reason for these additional groups is not known, but it is hypothesized that their presence—by preventing base pairing within the molecule—maintains the tRNA in its particular three-dimensional configuration. Confirmation of this hypothesis depends on determination of the three-dimensional structure.

The function of the tRNAs is to serve as the links—the adapters—between amino acids and mRNA molecules. To follow our previous analogy, the tRNA molecules provide the genetic dictionary for the translation of mRNA to protein. The amino acid is bonded to the tRNA at the CCA end; this linking-up process is the function of a special enzyme, one for each tRNA, that recognizes both tRNA and amino acid. Once bonded to its amino acid, the tRNA molecule attaches to the mRNA molecule. Three unpaired bases in one of the loops of the tRNA molecule—the anticodon—specifically recognize and bond to three corresponding bases—the codon—on the mRNA molecule. In this way, only one kind of amino acid can be brought in at any one point. (See Figure 17–3b.) The attach-

ment of tRNA to mRNA occurs at the point at which the mRNA is attached to the ribosome.

All kinds of organisms use the same genetic code "dictionary." This fact is one of the strongest arguments in favor of the view that all modern organisms arose from a common ancestor.

Ribosomal RNA

Ribosomes are about half RNA and half protein. The ribosomal RNAs are formed on the DNA of the nucleolus. Each ribosome is composed of two subunits, each with its characteristic RNA and proteins. In the ribosomes of *E. coli*, which have been studied most extensively, the smaller subunit has one type of ribosomal RNA molecule, about 1,500 nucleotides in length, and about 20 different proteins. The larger subunit has two types of ribosomal RNA molecules, one about 100 nucleotides in length and the other about 3,000, and about 35 different proteins. Analysis of the ratios of the different bases shows that the RNA molecules do not exist in paired forms, like DNA in its double helix, but as single strands, perhaps with regions of internal base pairing. Messenger RNA attaches to the smaller subunit of the ribosome.

7–4 *Electron micrograph of the single-celled alga* Chlamydomonas. *The dark body in the center of the nucleus is the nucleolus, where the RNA of the ribosomes is assembled. Very fine RNA strands are barely visible, around which spherical, ribosomelike granules are grouped. There may be one or more nucleoli per nucleus, each a part of and attached to a specific area on a specific chromosome. Notice also the nuclear envelope, with its many clearly visible pores. There is a portion of a chloroplast to the left of the nucleus and a Golgi body below the nucleus to the left.*

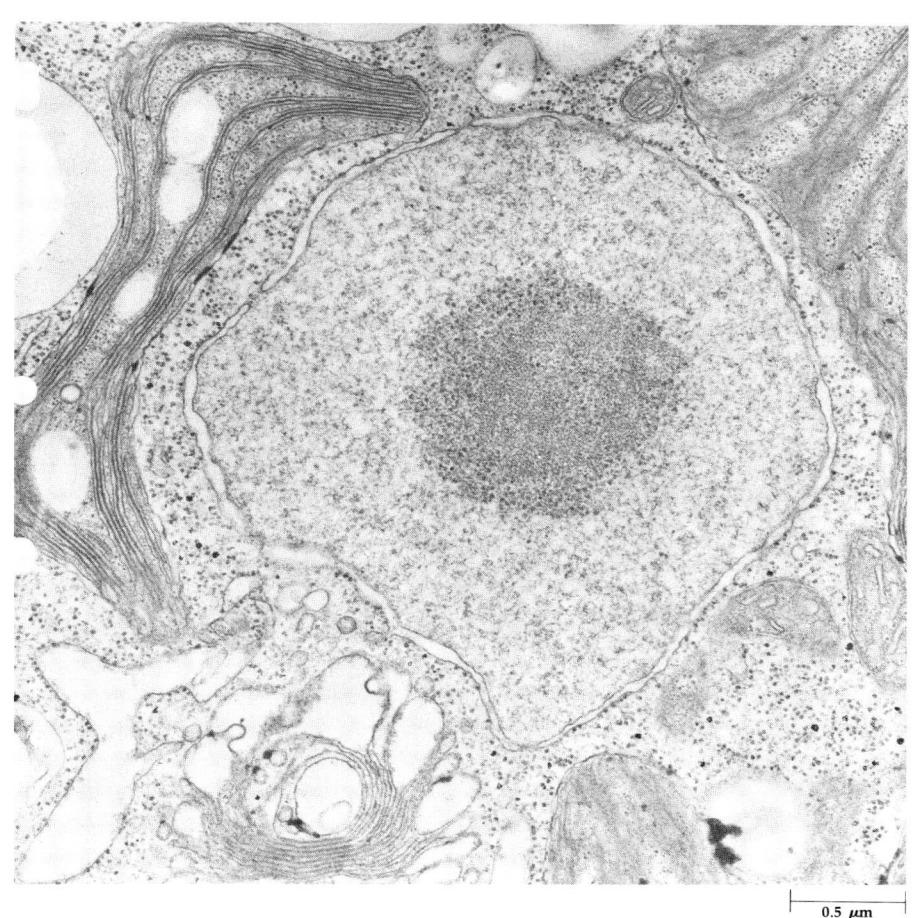

0.5 μm

17–5 *Groups of ribosomes, or polysomes. Each polysome is a group of ribosomes "reading" the same mRNA strand. These polysomes are from yeast cells.*

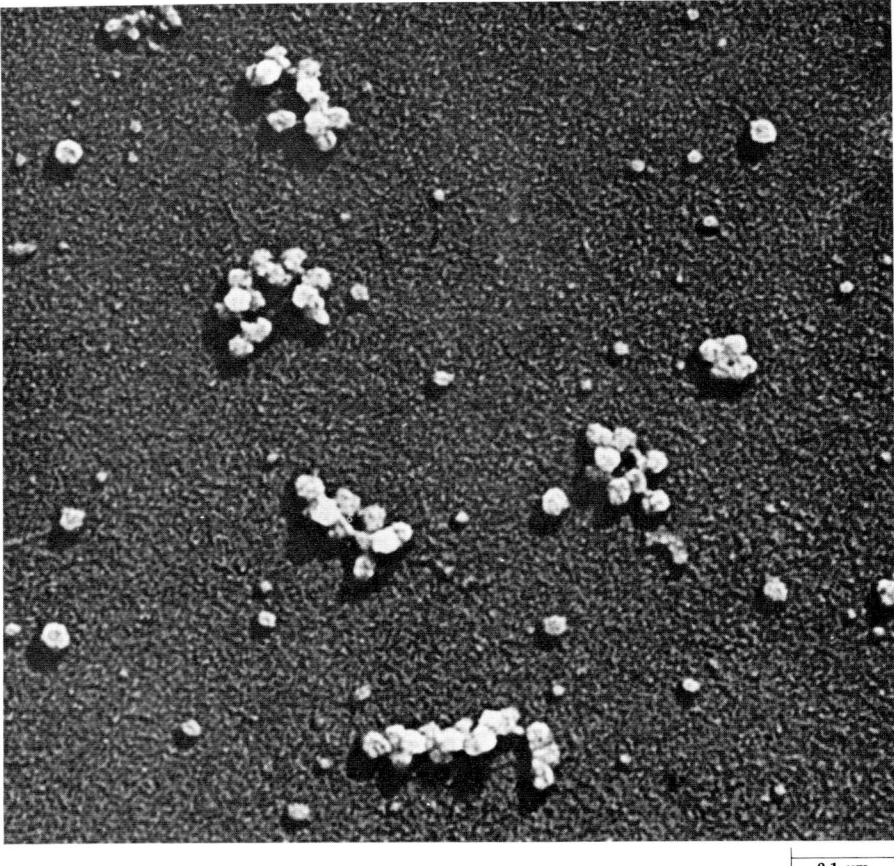

0.1 μm

17–6 *Diagram of a ribosome from a bacterial cell. As you can see, it consists of two subunits, one slightly larger than the other, and each composed of specific RNA and protein molecules.*

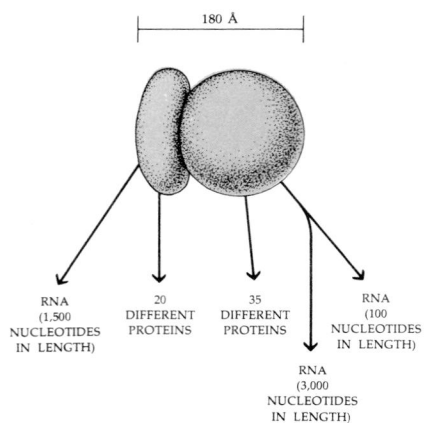

180 Å

RNA
(1,500
NUCLEOTIDES
IN LENGTH)

20
DIFFERENT
PROTEINS

35
DIFFERENT
PROTEINS

RNA
(100
NUCLEOTIDES
IN LENGTH)

RNA
(3,000
NUCLEOTIDES
IN LENGTH)

A rapidly growing *E. coli* cell contains about 15,000 ribosomes, constituting about one-fourth of the total mass of the cell. Ribosomes of eukaryotic cells are somewhat larger, but apparently structurally and functionally similar.

The function of the ribosome, presumably, is to orient mRNA, tRNA, amino acids, and the growing protein in a precise relation to one another. At this moment, however, there is no useful hypothesis as to how the ribosome carries out this function.

THE MAKING OF A PROTEIN

The ribosomes move along the mRNA strand, going from one codon to the next. As they move, the different tRNAs lock into place, one after another, and as each does so, its amino acid is transferred from the tRNA to the growing protein molecule (Figure 17–7). The protein coils into its secondary and tertiary structure (page 74) as it is formed. The tRNA molecules are used over and over again, recharged with new amino acids.

Characteristically, an mRNA molecule is "read" by a number of ribosomes simultaneously. The ribosomes always first attach to the mRNA at the end and apparently can attach only there, thus preventing reading backward or starting in the middle. The synthesis of a protein using the information coded in mRNA is often referred to as translation.

7-7 How a protein is made. At least 20 different kinds of tRNA molecules are formed on the DNA in the nucleus of the cell. These molecules are so structured that each can be attached (by a special enzyme) at one end to a specific amino acid. Each carries somewhere in the molecule an anticodon which fits an mRNA codon for that particular amino acid. The process of protein biosynthesis begins when an mRNA strand is formed on the DNA template in the nucleus and travels to the cytoplasm. A ribosome attaches to the strand, and, at the point of attachment, the matching tRNA molecule, with its amino acid, plugs in momentarily to the codon in the mRNA. As the ribosome moves along the mRNA strand, a tRNA linked to its particular amino acid fits into place and the first tRNA molecule is released, leaving behind its amino acid. As the process continues, the amino acids are brought into line one by one, following the exact order laid down by the DNA code, and are formed into a protein chain, which may be anywhere from fifty to hundreds of amino acids long.

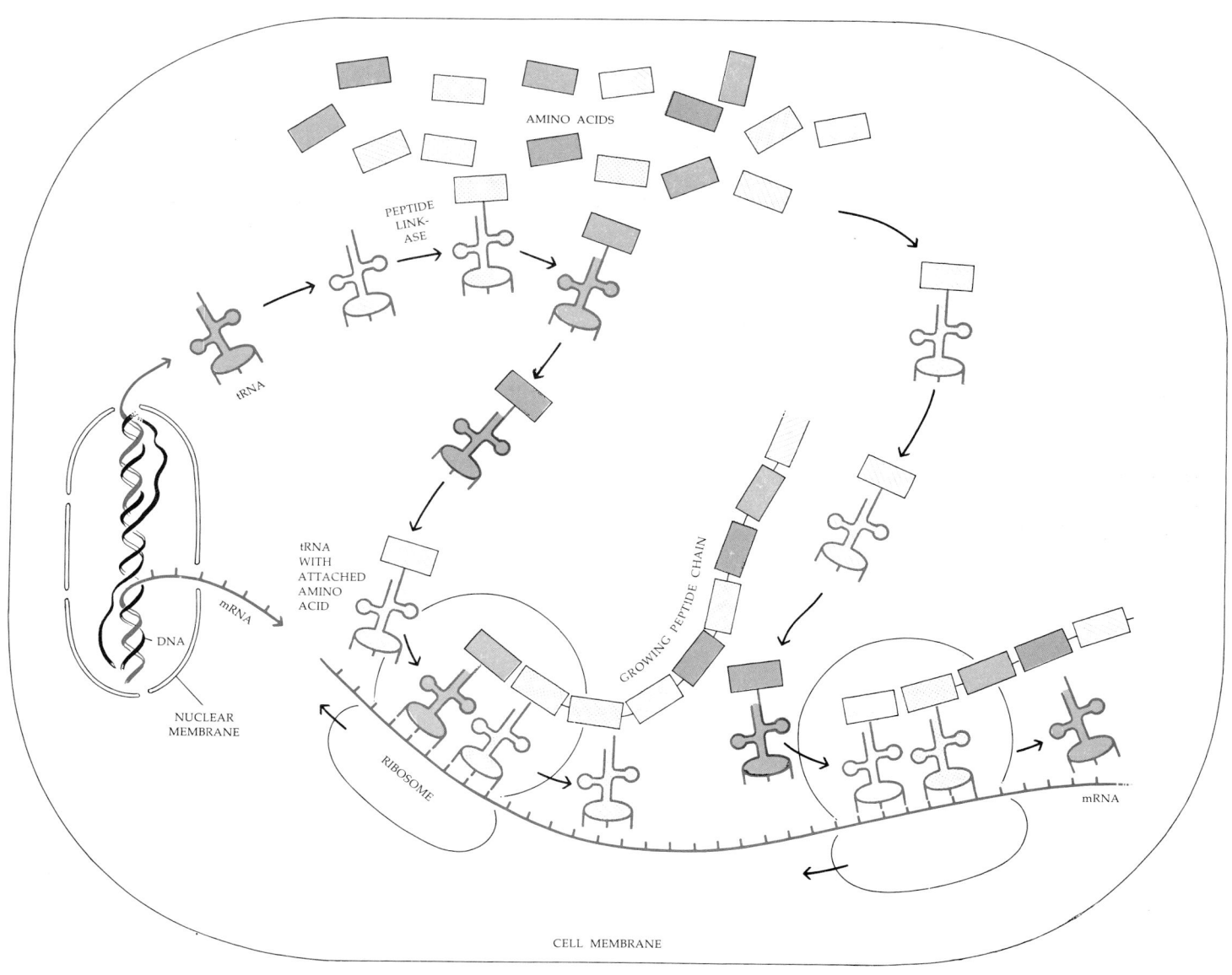

AMINO ACIDS

PEPTIDE LINK-ASE

tRNA

tRNA WITH ATTACHED AMINO ACID

mRNA

DNA

NUCLEAR MEMBRANE

GROWING PEPTIDE CHAIN

RIBOSOME

mRNA

CELL MEMBRANE

255 *RNA and the Making of Proteins*

The Energetics of Protein Biosynthesis

The activation energy for the synthesis of amino acids into proteins is supplied, again, by ATP, although the role of ATP in this transaction is slightly different from previous ones described. First, the amino acid reacts with an ATP molecule, and in the course of this reaction the adenine-containing portion of the ATP molecule, rather than the P~P, is linked to the amino acid:

$$\text{Amino acid} + \text{ATP} \rightleftharpoons \text{amino acid} \sim \text{AMP} + \text{P} \sim \text{P}$$

The P~P group is almost immediately hydrolyzed, thus making the overall reaction highly exergonic (see page 44). The enzymes that catalyze this reaction are called amino acid activating enzymes. There are at least 20 different kinds of activating enzymes, each one with specific binding sites for a particular amino acid and its matching tRNA molecule.

The amino acid~AMP complex remains bound to its activating enzyme until it reacts with the tRNA specific for that particular amino acid. Then the following reaction occurs, resulting in the transfer of the activated amino acid from AMP to its tRNA:

$$\text{Amino acid} \sim \text{AMP} + \text{tRNA} \rightleftharpoons \text{amino acid} \sim \text{tRNA} + \text{AMP}$$

The amino acid~tRNA molecule, held together by a high-energy bond, is then released from the enzyme.

When the amino acid is added to the growing protein chain, the energy of the bond linking the amino acid with its tRNA molecule is transferred to the peptide bond.

17–8 *Genes in action. This electron micrograph shows a section of DNA from the nucleolus of an amphibian egg cell. The fine fibrils are mRNAs that have formed along the DNA strand. The arrows indicate the direction in which the transcription is proceeding.*

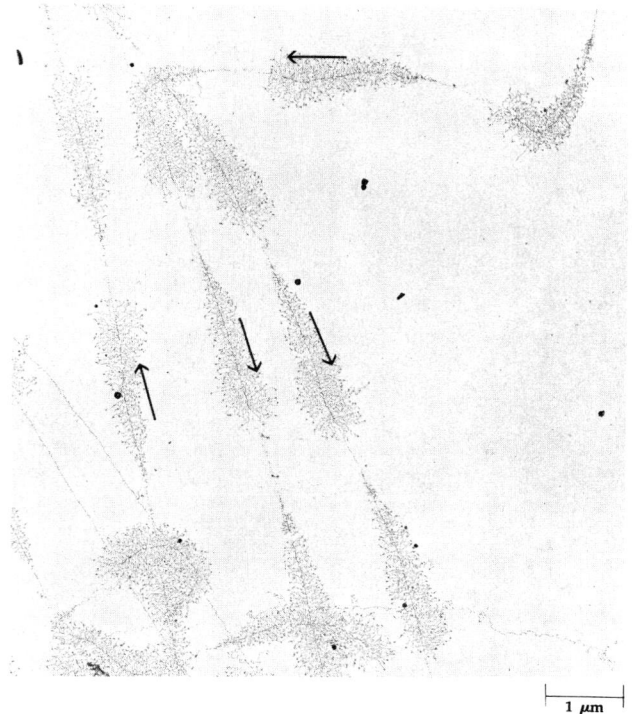

1 μm

The existence of mRNA was postulated in 1961 by the French scientists François Jacob and Jacques Monod. Almost immediately, Marshall Nirenberg of the National Institutes of Health set out to test their messenger RNA hypothesis. He added several crude extracts of RNA from a variety of cell sources and found that they all stimulated protein synthesis in cell-free extracts of *E. coli*. In other words, the cellular material would start producing protein molecules even when the RNA "orders" it received were from a "complete stranger." Even the tobacco mosaic virus, which naturally multiplies only in cells of the leaves of tobacco plants, could be "read" as an mRNA by the ribosomes and tRNA molecules of the bacterial cell. The "code" seemed to be a universal language.

Leaving others to ponder the evolutionary implications of his finding,* Nirenberg and Heinrich Matthaei raced ahead to give it practical application. Perhaps if the cell extracts could read a foreign message and translate it into protein, they could read a totally synthetic message, one dictated by the scientists themselves. Synthetic RNA was available; Severo Ochoa of New York University had developed an enzymatic process for linking ribonucleotides into a long strand of RNA. The trouble with the method, from Nirenberg's point of view, was that there was no way to control the order in which an assortment of ribonucleotides would be assembled. For Nirenberg's purposes, the order was of the utmost importance. He wanted to know the exact contents of any message that he dictated.

A simple solution for this seemingly perplexing problem presented itself: an RNA molecule that consisted of only one ribonucleotide, uracil, repeated over and over again. This molecule was known as Ochoa's synthetic poly-U. Nirenberg and Matthaei prepared 20 different test tubes, each of which contained cellular extracts of *E. coli* with ribosomes, tRNA, ATP, the necessary enzymes, and all the amino acids. In each test tube, one of the amino acids, and only one, carried a radioactive label. Synthetic poly-U was added to each test tube. In 19 of the test tubes, nothing detectable occurred, but in the twentieth one, to which radioactive phenylalanine had been added, the investigators were able to detect newly formed, radioactive polypeptide chains. When the polypeptide was analyzed, it was found to consist only of phenylalanines, one after another. Nirenberg and Matthaei had dictated the message "uracil . . . uracil . . . uracil . . .," and a clear answer had come back, "phenylalanine . . . phenylalanine . . . phenylalanine. . . ."

Within the year following Nirenberg's discovery, tentative codes were worked out by Nirenberg and Ochoa and their many co-workers for all the amino acids, by using synthetic mRNA. A synthetic polynucleotide made up entirely of adenine (poly-A), for instance, produces a peptide chain composed entirely of lysine. If two parts of guanine are combined with one part of uracil, the peptide that is dictated will be composed largely of valine, so it was presumed that the code of valine is GUU or UUG or UGU. As you can see, in the

---

*The genetic code may have been functional 3 billion years ago; almost certainly, it is more than 600 million years old. Some scientists have suggested that the code became frozen by the time organisms as complex as bacteria had evolved.

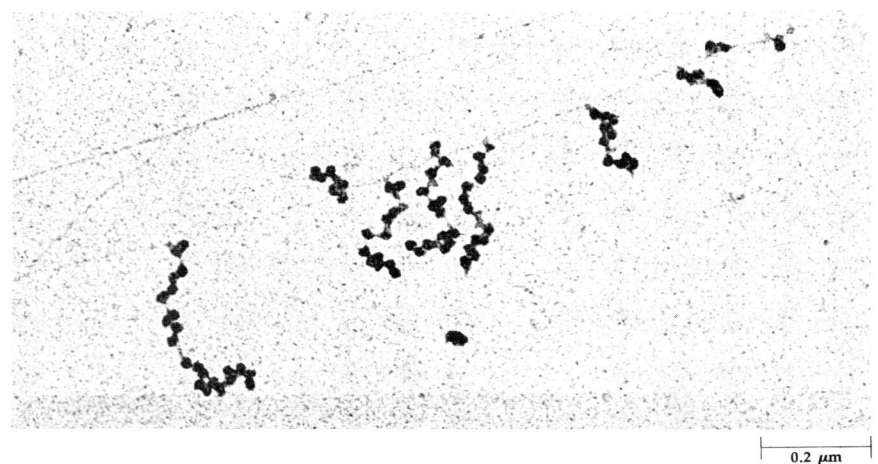

0.2 μm

17–9 *A bacterial gene in action. In the micrograph, you can see molecules of RNA polymerase, the enzyme that regulates the transcription of RNA from DNA. The one at the far right is approximately at the point where transcription begins. Several different mRNA strands (shown in color in the diagram) are being formed simultaneously. The longest one, at the left, was the first one synthesized. As each mRNA strand peels off the DNA molecule (active chromosome segment), ribosomes attach to the RNA, translating it into protein. The protein molecules are not visible in the micrograph.*

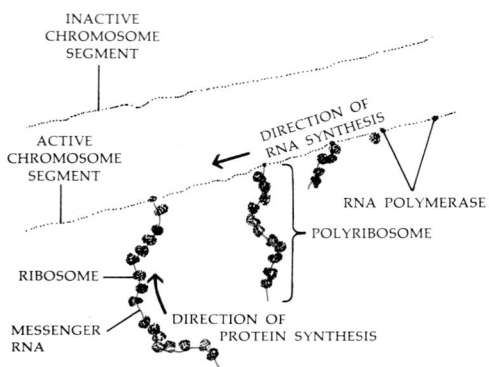

INACTIVE CHROMOSOME SEGMENT

ACTIVE CHROMOSOME SEGMENT

DIRECTION OF RNA SYNTHESIS

RNA POLYMERASE

POLYRIBOSOME

RIBOSOME

MESSENGER RNA

DIRECTION OF PROTEIN SYNTHESIS

synthetic messengers made of more than one type of ribonucleotide, there was then no way to tell the order of alignment of the bases.

By Nirenberg's method, it was possible to identify the composition of the codon that dictated a particular amino acid—that is, to determine which bases it contained. However, his method did not reveal in what order the bases occurred. With UUU there was no problem, of course, but adenine, guanine, and cytosine, for example, can be arranged in nine different ways in a triplet code.

The problem could, of course, be solved easily if one could determine the complete amino acid sequence of a protein and the base sequence of the piece of DNA or RNA that codes it. One could then simply compare the two. The amino acid sequence could be laboriously worked out, as we have seen, but, for technical reasons, determining the nucleotide sequence of a long nucleic acid molecule was then still impossible.

This problem, too, Nirenberg solved in an ingenious way. He found that, by adding nucleotides one at a time, it is possible to synthesize mRNA triplets— that is, codons—in which the bases appear in known predetermined order. This single triplet, although useless for the biochemical functions of the cell, will serve to bind the appropriate tRNA, along with its amino acid, to a ribosome. Unbound tRNA will slip right through a nitrocellulose filter, but tRNA attached to a ribosome will be caught in the filter. Nirenberg prepared various combinations of synthetic nucleotide triplets, tRNA, amino acids with radioactive labels (to facilitate their detection), and cell-free ribosome preparations. These combinations, after a brief period of exposure to one another, were passed through the cellulose filter. Then any radioactive material present in the filter was removed and analyzed. Knowing the triplet codon that he had put in the test tube, Nirenberg had only to identify the amino acid, bound by its tRNA to the ribosome, for an exact translation of his code triplet. As a result of the work of Nirenberg and others, RNA codes for all of the amino acids have been worked out (Figure 17–10).

Since 61 combinations code for 20 amino acids, you can see that there may be more than one codon for each amino acid. As you can see in Figure 17–10, codons specifying the same amino acid often differ only in the third nucleotide, leading

7–10 *The genetic code, consisting of 64 triplet combinations (codons) and their corresponding amino acids (see page 70). Since 61 triplets code 20 amino acids, there are "synonyms," as many as six for leucine, for example. Most of the synonyms, as you can see, differ only in the third nucleotide. Of the 64 codons, only 61 specify particular amino acids. The other three codons are stop signals which cause the chain to terminate. The code is shown here as it would appear in the mRNA molecule. How would you determine the corresponding DNA codes?*

SECOND LETTER

		U	C	A	G	
FIRST LETTER	U	UUU UUC } phe UUA UUG } leu	UCU UCC UCA UCG } ser	UAU UAC } tyr UAA stop UAG stop	UGU UGC } cys UGA stop UGG trp	U C A G
	C	CUU CUC CUA CUG } leu	CCU CCC CCA CCG } pro	CAU CAC } his CAA CAG } gln	CGU CGC CGA CGG } arg	U C A G
	A	AUU AUC AUA } ile AUG met	ACU ACC ACA ACG } thr	AAU AAC } asn AAA AAG } lys	AGU AGC } ser AGA AGG } arg	U C A G
	G	GUU GUC GUA GUG } val	GCU GCC GCA GCG } ala	GAU GAC } asp GAA GAG } glu	GGU GGC GGA GGG } gly	U C A G

THIRD LETTER

to the speculation that the first two may be sufficient to hold the tRNA in most instances.

Some of the biological implications of these findings are strikingly clear. Let us take another look at sickle cell anemia, for example, in the light of Figure 17–10. Normal hemoglobin contains glutamic acid; sickle cell hemoglobin contains valine. In mRNA, GAA or GAG specifies glutamic acid (glu), GUU or GUC, GUA or GUG specifies valine (val). So the difference between the two is merely the replacement of one adenine by one uracil in a molecule that, since it dictates a protein which contains more than 150 amino acids, must contain more than 450 bases. In other words, the tremendous functional difference —literally a matter of life and death—between the two hemoglobins can be traced to a single "misprint" in over 450 nucleotides.

MUTATIONS REDEFINED

One consequence of these studies was the formulation of a new definition for a mutation. A mutation, in terms of molecular genetics, is any change in the sequence of nucleotides in a DNA molecule. Such a change will result, when DNA replicates, in a corresponding complementary change in the new DNA strand. This change will, of course, result in a change in the mRNA produced along either DNA strand and so a change in the amino acid or amino acids specified. Such changes in the DNA molecule may occur as a result of mistakes in the copying process, or insertions or deletions taking place during crossing over. They may also be a result of physical damage to the DNA molecule by x-rays, ultraviolet radiation, or certain chemicals. Once the mutation—the change in the sequence of nucleotides—occurs, it is passed on by the normal copying process for the DNA molecule. Even a small alteration in base sequence may have drastic effects on the phenotype—as in the case of sickle cell anemia. However, other changes may be undetectable in the phenotype—such as changes affecting amino acids of an enzyme not involved in the active site.

An alternative approach to the study of the genetic code was developed in the laboratories of H. G. Khorana at the University of Wisconsin. Khorana and his co-workers devised a method for the synthesis of chains of DNA or RNA in which two or three nucleotides could be repeated over and over again in a known sequence. Thus he could make deoxyribose strands of TCTCTCTCTC, which would form a double helix with strands of AGAGAGAGAG, and strands of TGTGTGTGTG, which would form a double helix with ACACACACAC. Then, using RNA polymerase (the enzyme that makes mRNA, using a DNA template), he could obtain messenger strands of AGAGAGAGAG, UCUCUCUCUC, ACACACACAC, and UGUGUGUGUG. (You will recall that in RNA uracil is substituted for the thymine of DNA.)

Each of these RNA chains, when used as a messenger in the cell-free system, produced polypeptide chains of alternating amino acids. Poly-AG produced arginine and glutamic acid over and over again; poly-UC, serine and leucine; poly-AC, threonine and histidine; and poly-UG, cysteine and valine. This is, of course, what you would expect from a triplet code, since the message would be read AGA . . . GAG . . . AGA. . . . This was the first proof that mRNA is read sequentially (that is, one codon after another); it was also the first proof that the codon consists of an uneven number of nucleotides.

An artificial mRNA containing three different nucleotides in regular order could produce three different polypeptides, depending on where the reading process began; each would contain only one type of amino acid repeated over and over again. Ten years after the hypothesis was formulated, this experiment proved that the codon was a triplet.

mRNA BASE SEQUENCE	READ AS	AMINO ACID SEQUENCE OBTAINED
$(AG)_n$	· · · AGA GAG AGA GAG · · · ·	· · arg–glu–arg–glu · · ·
$(AGC)_n$	· · · AGC AGC AGC · · ·	· · · ser–ser–ser · · ·
	· · · GCA GCA GCA · · ·	· · · ala–ala–ala · · ·
	· · · CAG CAG CAG · · ·	· · · gln–gln–gln · · ·

REGULATING GENE ACTIVITY

Cells do not produce all enzymes and other proteins at the same time and the same rate. For example, cells of *E. coli* supplied with the disaccharide lactose as a carbon and energy source need the enzyme beta-galactosidase to split the disaccharide into glucose and galactose before they can use it (see Figure 17–11). In cells growing on lactose, approximately 3,000 molecules of beta-galactosidase are present in every normal *E. coli* cell. This represents about 3 percent of all the protein in the cell. In the absence of lactose, however, there is an average of only one molecule of this enzyme per cell. When the galactosidase is needed, it is produced from new mRNA.

Mutants of *E. coli* have been found that produce the enzyme even in the absence of lactose. These mutants are at a disadvantage compared to cells with

17–11 *Splitting of lactose to galactose and glucose requires the enzyme beta-galactosidase. Beta-galactosidase is an inducible enzyme; that is, its production is regulated by an inducer—in this case, lactose.*

LACTOSE   BETA-GALACTOSIDASE   GALACTOSE   GLUCOSE

a balanced protein synthesis, since by making an enzyme in the absence of its substrate they are using their energies and resources uneconomically. Substances such as lactose that increase the amount of enzyme produced by a cell are known as *inducers*, and the enzymes they influence are known as inducible enzymes.

Some substances act to repress enzyme production. For example, *E. coli* can make all its own amino acids from a carbon source and ammonia (NH_3). If a particular amino acid—histidine, for example—is present in the medium, the cell will then stop making all of the enzymes associated with the biosynthesis of histidine. Such enzymes are known as repressible enzymes. In bacterial cells, mRNA molecules are broken down very soon after they are produced. Therefore, control over mRNA production directly controls the rate of enzyme synthesis.

The Operon

To explain how bacterial cells regulate enzyme biosynthesis, Jacob and Monod developed the hypothesis of the *operon*. An operon is a group of related genes all aligned along a single segment of DNA. The operon consists of three different types of genes: the *promoter*, which is the site at which formation of the mRNA begins; the *operator*, which is the site of regulation; and one or more *structural genes*, which code for enzymes or other proteins. In the beta-galactosidase system, the operon includes the gene that codes for beta-galactosidase and two other genes, which also code for enzymes involved in lactose metabolism. These genes are adjacent to one another and are transcribed consecutively, one after the other along a single strand of DNA, forming a single mRNA molecule. The mRNA molecules produced are active for only a very short time, after which they are broken down by specific enzymes.

The activity of the operon is controlled by yet another gene, the *regulator*, which is not necessarily adjacent to the operon. The regulator codes for a protein, known as the *repressor*, which apparently binds to the DNA at the site of the operator gene. Since the operator gene is located between the promoter and the structural genes (Figure 17–13), this blocks the production of mRNA.

The repressor is controlled by another "signal" compound. In the case of inducible enzymes, this compound is the inducer. The inducer binds with the repressor molecule and changes its shape so that it can no longer attach itself

17–12 *According to the operon theory of Jacob and Monod, the synthesis of proteins may be regulated by interactions involving either a repressor and inducer or a repressor and corepressor. (a) In the case of repressible enzymes, such as beta-galactosidase, the repressor molecule is active until it combines with the inducer (in this case, lactose). It is then inactivated, and so the genes in the operon are no longer repressed. (b) In the case of a repressor molecule, the repressor is not active until it combines with the corepressor. Thus, in the absence of the corepressor, the genes in the operon are active.*

(a) ACTIVE REPRESSOR + INDUCER (LACTOSE) = INACTIVE REPRESSOR-INDUCER COMPLEX

(b) INACTIVE REPRESSOR + COREPRESSOR (HISTIDINE) = ACTIVE REPRESSOR

17–13 *The operon. An operon is a group of genes forming a functional unit. There are usually several structural genes, which code for different proteins, often a group of enzymes that work sequentially in a particular enzyme sequence. The transcription can begin only at the site marked "promoter." (a) In operons activated by inducers, the regulator gene codes for a protein that represses transcription of mRNA from the structural genes. The regulator, which need not be adjacent to the operon on the chromosome, directs production of a repressor protein. This repressor acts upon the operator gene, apparently by binding to the DNA at this site and blocking the formation of mRNA. In order to start transcription again, another compound, the inducer, is needed. The inducer counteracts the effects of the repressor, probably by binding to it and changing its shape. The repressor can no longer bind to the operator site, so synthesis of mRNA proceeds. (b) In operons regulated by the corepressors, transcription continues until the repressor and corepressor combined bind to the DNA of the operon. The repressor alone cannot halt transcription.*

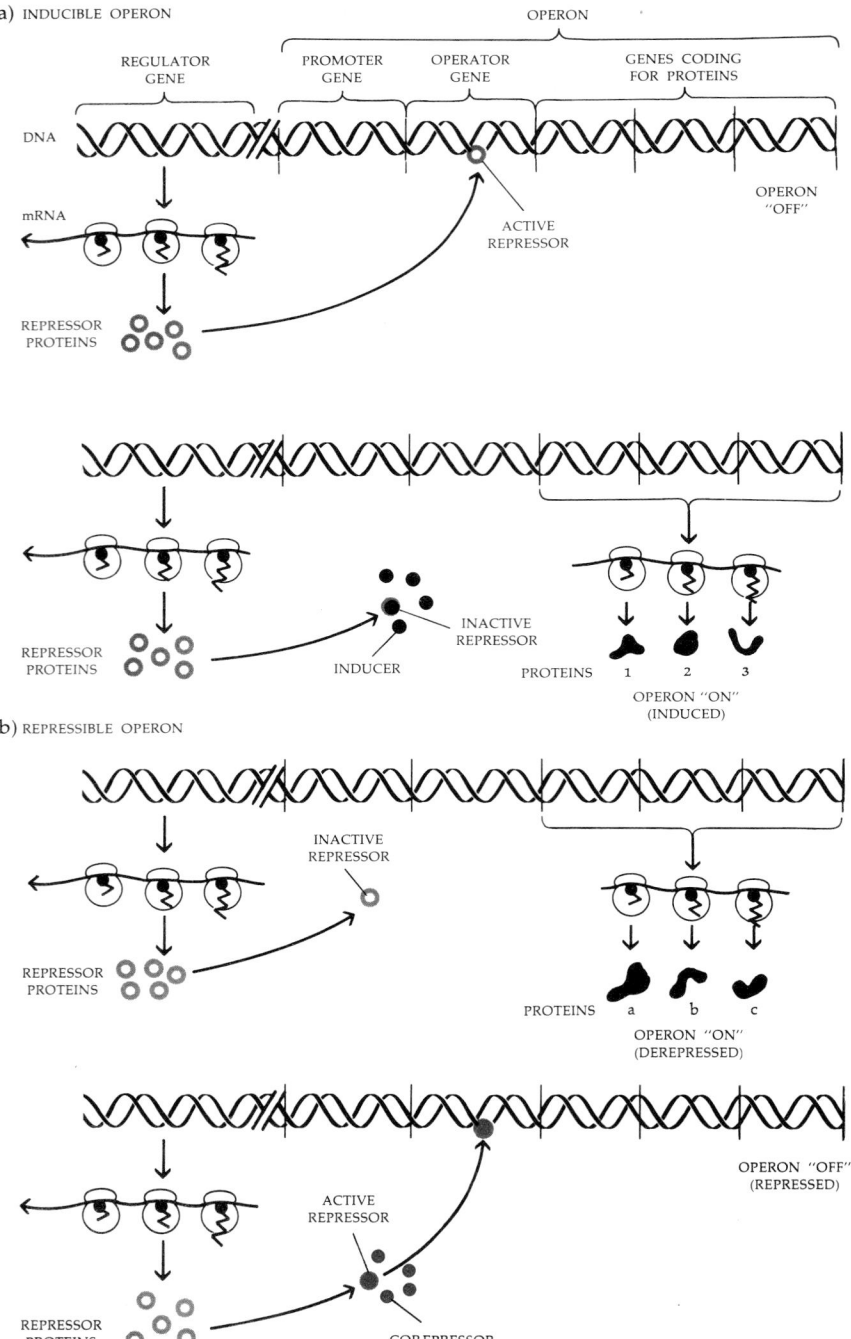

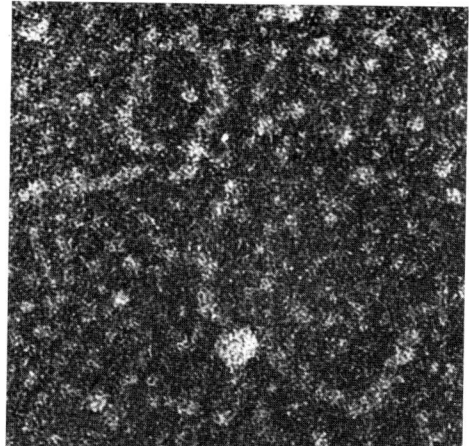

−14 *A repressor protein (the white spherical form below the center of the micrograph) attached to the lactose operon.*

├─── 50 nm ───┤

to the DNA. In the absence of the repressor, mRNA molecules are formed along the structural genes, and from these molecules proteins are produced. When the supply of the inducer is exhausted, the regulator once again assumes control, and mRNA production and protein formation cease. In the beta-galactosidase system, the inducer is lactose or a closely related compound derived from lactose. The lactose or related compound binds with the repressor and inactivates it, thus permitting enzyme biosynthesis to proceed.

In the case of repressible enzymes, the "signal" compound is a corepressor. The repressor is active only when bound with the corepressor. In the histidine system, which involves some 10 different enzymes, histidine combined with its tRNA is the corepressor. Histidine–tRNA$_{his}$ combines with the repressor and activates it, thus halting mRNA synthesis of histidine.

Allosteric Interactions

Allosteric interactions (page 76), which involve inactivation of an enzyme by an alteration in its shape, provide another means for regulation of the metabolic activity of the cell. Allosteric interactions provide a finer control system than the repressor-inducer system of the operon, but they do not serve the useful function of shutting off protein biosynthesis at its source—the production of mRNA.

Control Systems in Higher Organisms

There is little evidence that the operon system is a major regulatory factor in higher plants or animals. The chromosomes of these organisms are significantly different from the chromosomes of *E. coli* and other prokaryotes. Moreover, the gene control problems of these cells are also very different. As indicated by the mechanism of mitosis, each cell of a higher plant or animal contains all the genetic information present in the fertilized egg and in every other cell. (Further support for this sweeping generalization is offered in Chapter 31.) Therefore, most of the genes of any particular specialized cell will be turned off for the lifetime of the cell. The DNA of these higher cells is always found in association with protein, as we have mentioned previously, and it seems likely that this association of DNA and protein in eukaryotes is involved in gene repression. This subject will be discussed further in the chapter that follows.

SUMMARY

Genetic information is coded in molecules of DNA, and these, in turn, determine the sequence of amino acids in molecules of protein. A gene is a segment of a DNA molecule that specifies the complete sequence of one protein.

The way in which the gene directs the production of a protein, according to current theory, is as follows: Each series of three nucleotides along a DNA strand is the DNA code for a particular amino acid. The information is transferred from the DNA by means of a long, single strand of RNA (ribonucleic acid). This type of RNA molecule is known as messenger RNA, or mRNA. The mRNA forms along one of the strands of DNA, following the principles of base pairing first suggested by Watson and Crick, and therefore is complementary to it.

The mRNA strand leaves the cell's nucleus and attaches to a ribosome. At the point where the strand of mRNA is in contact with the ribosome, small molecules of another type of RNA, known as transfer RNA (tRNA), which serve as adapters between the mRNA and the amino acids, are bound temporarily to

the mRNA strand. This bonding is believed to take place by the same base-pairing principle as that which holds together the two strands of the double helix of DNA. Each tRNA molecule carries the specific amino acid called for by the mRNA codon to which the tRNA attaches. Thus, following the sequence dictated by the DNA, the amino acid units are brought into line one by one and are formed into a polypeptide chain.

Final proof of this hypothesis came when the DNA/RNA code was "broken," that is, when investigators were able to predict what amino acid would be formed from a given codon. Sixty-one of the sixty-four possible triplet combinations of the four-letter DNA code have been identified with one of the 20 amino acids that make up protein molecules. The other three triplets, it is believed, serve as "punctuation marks," terminating protein synthesis.

Cells have a variety of mechanisms by which they regulate the rate at which enzymes and other proteins are produced. One mechanism known to be important in prokaryotes is the operon. The operon consists of three different types of genes: the promoter, the site at which formation of the mRNA begins; the operator, which is the site of regulation; and one or more structural genes, which code for polypeptide chains. The operon is under the influence of another gene, not part of the operon, known as the regulator. The regulator codes for a protein, known as a repressor, which attaches to the DNA molecule at the operator and blocks mRNA transcription. The repressor does not act alone but rather in conjunction with another compound, often an end product of the enzyme sequence controlled by the operon. This compound may be an inducer, in which case it inactivates the repressor, or it may be a corepressor, in which case it makes it possible for the repressor to function. Enzymes produced as the result of the combined action of a repressor and an inducer are known as inducible; enzymes produced as the result of the combined action of the repressor and a corepressor are known as repressible.

QUESTIONS

1. Identify the following terms: repressible enzyme, operator gene, codon, anticodon, transcription, translation.
2. Explain the term "genetic code." In what ways is it a useful analogy?
3. In a hypothetical segment of DNA, the sequence of bases is AAGTTTGG-TTACTTG. What would be the sequence of bases in an mRNA strand transcribed from this DNA segment? What would be the amino acids coded by the mRNA? Does it matter where the transcription starts? Explain your answer.
4. Describe three ways in which a bacterial cell can control the activity of an enzymatic pathway.
5. Fill in the missing letters in the following:

T G T	_ _ _	_ _ _	
_ _ _	C _ _	_ _ _	} DNA
U _ _	_ C A	_ _ _	mRNA codon
_ _ _	_ _ _	G C A	tRNA anticodon

What amino acids will each triplet code for?

Chapter 18

Some Work in Progress

Research is still very active in the field of genetics. In this chapter, we are going to touch upon some of the current lines of investigation and interest, filling you in on some of the background work and leaving you to bring these stories up to date as new developments are reported.

ARRANGEMENT OF DNA IN CHROMOSOMES

One of the most active fields of current research and controversy concerns the arrangement of DNA in the chromosomes of eukaryotic cells. DNA is an "exquisitely thin filament," in the words of E. J. DuPraw, "so fine that a length sufficient to reach from the Earth to the sun would weigh a half a gram." The cells of a human body altogether contain between 10 and 20 billion miles of DNA double helix. DNA is tightly coiled, as we have noted, during mitosis and meiosis and dispersed during interphase. How does this packing take place?

This question takes on added significance because of a fairly considerable body of evidence that indicates that the packing and unpacking of DNA is crucial in making the DNA available for transcription by messenger RNA. In other words, the arrangement of DNA in the chromosomes may be a major control mechanism for eukaryotic cells.

DNA and mRNA Synthesis

The evidence comes from observations on three types of chromosomal material. The first is the mitotic chromosome. As we have mentioned previously, when chromosomes are condensed at the time of mitosis (or meiosis), no new DNA or RNA is produced. Evidently, the DNA must be unfolded as in the interphase cell, before it can be "read." The second is the giant chromosome of insects (see page 217). At various stages of larval growth in insects, it is possible to observe diffuse thickenings, or "puffs," in various regions of these chromosomes. The puffs are separated strands of DNA, and studies with radioactive isotopes indicate that these puffs are sites of rapid RNA synthesis. When ecdysone, a hormone that produces molting in the insects, is injected, the puffs occur in a definite sequence which can be related to the developmental stage of the animal. For example, in one species of *Drosophila*, ecdysone initiates three new puffs and

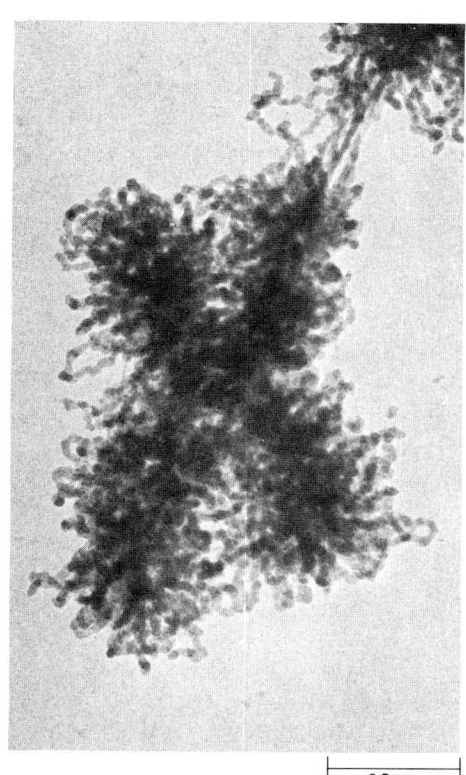

–1 *Electron micrograph of human chromosome 19-20. Note that at the tips, the fibers do not form free ends but loop back into the chromosomes.*

18–2 *Observations of chromosome puffs support the concept that the DNA is somehow unwound to make it available for mRNA transcription. These puffs were observed in chromosomes of the Brazilian gnat, which, like the fruit fly, has giant chromosomes in some of its cells. They occur normally but can also be induced experimentally. The gnat had previously been treated with a hormone that causes molting, and as the micrographs indicate, the puffs occurred sequentially along one chromosome.*

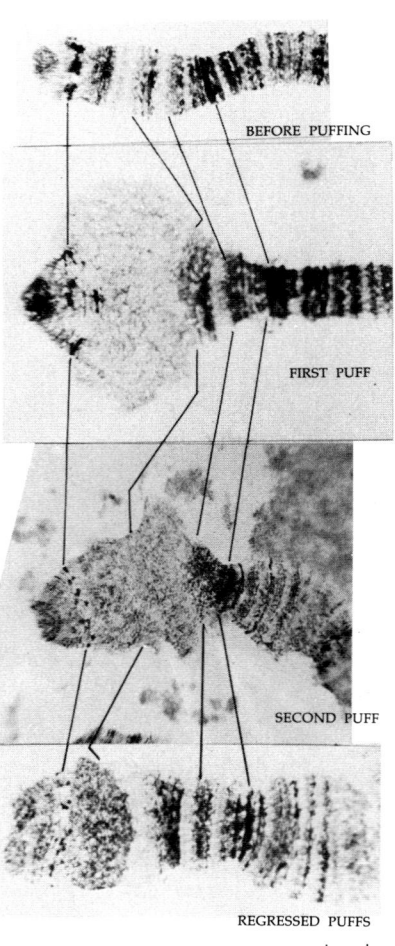

BEFORE PUFFING

FIRST PUFF

SECOND PUFF

REGRESSED PUFFS

|—————|
10 μm

causes increases in 18 other puffs within 20 minutes after it is injected; during this same period, 12 other puffs decrease in size. After four to six hours, five additional puffs can be seen. The looping out of the DNA occurs before RNA synthesis is initiated. The mechanism of this unraveling, or spinning out, of the chromosomal DNA is not known.

Somewhat similar observations have been made with a third type of chromosomal material, lampbrush chromosomes, so-called because they resemble the brushes used to clean kerosene lamps. They are found in the nuclei of egg cells of fish, birds, reptiles, and amphibians during the prophase of meiosis.* These cells are extremely actively engaged in the RNA and protein syntheses required for the rapid growth of the mature egg. Each lampbrush chromosome consists of two homologues held together by their kinetochores and chiasmata (page 214), each homologue consisting of two sister chromatids. As you can see in Figure 18–3, each chromatid has a number of lateral loops, which branch out from the main axis. These loops appear to be reeled in and out from the body of the chromosome. Each of these loops consists of a thin fiber. If this thin fiber is exposed to an enzyme that digests away the associated protein, all that remains is a thin filament about the diameter of a single DNA helix. These loops, like the chromosome puffs in giant chromosomes, are also the sites of active RNA synthesis. Thus, several different lines of evidence suggest that the tight coiling of DNA prevents the transcription of mRNA and that, conversely, the unfolding of DNA is related to mRNA transcription.

The Proteins of the Chromosomes

According to current interpretations of chromosome structure, the DNA in each chromatid is one long, continuous double helix. Within the chromatid, the DNA helix is tightly coiled and coated with several times its own weight in protein.

Among the proteins is a special group known as *histones.* These proteins are positively charged (basic) and so are attracted to DNA, which has negative charges. There are at least five distinct types of histones. One type is invariably associated with the nuclei of eukaryotic cells and is virtually identical from species to species; it consists of 102 amino acids. The molecules of this histone found in calf nuclei differ from those found in pea seedlings by only 2 amino acids out of the 102. This constancy of structure throughout billions of years of evolution strongly suggests that the histones play some important role in the chromosome, but as yet what this role is and how they play it remain speculative. According to a model proposed by DuPraw (Figure 18–4) the tight coiling of the DNA helix could be induced by wedge-shaped molecules composed of histones (or histones plus nonhistone proteins), which contain grooves providing specific binding sites for the DNA molecules. These hypothetical protein units self-assemble, like the protein molecules making up the coat of tobacco mosaic virus (page 84). Addition or subtraction of even small chemical sidegroups to either the DNA or the protein wedge could serve to disrupt their interaction and so free ("derepress") the DNA. At this moment, there are insufficient data to permit acceptance (or, for that matter, rejection) of this model, but it serves to bring the problem into clearer focus.

*For a review of meiosis, see page 186.

8–3 (a) *Two homologous lampbrush chromosomes held together by three chiasmata and a kinetochore (indicated by the arrow). (b) Model of a section of a lampbrush chromosome, showing individual loops. Experimental evidence indicates that the axis of each of the loops is a single double helix of DNA.*

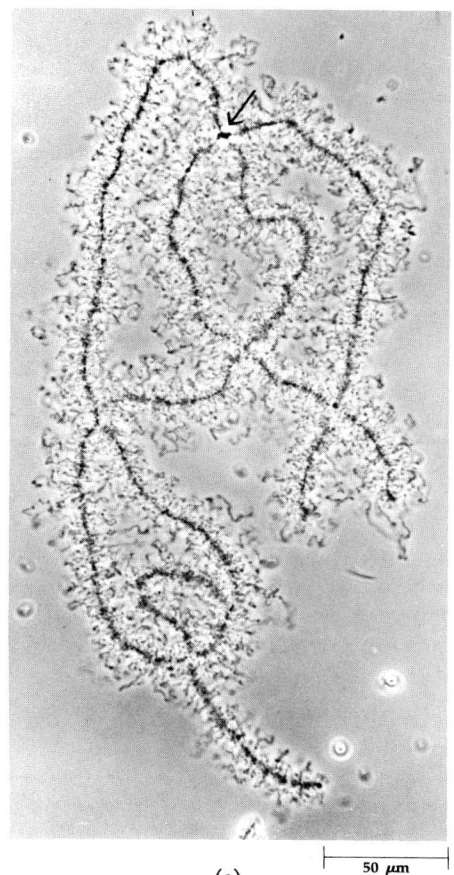

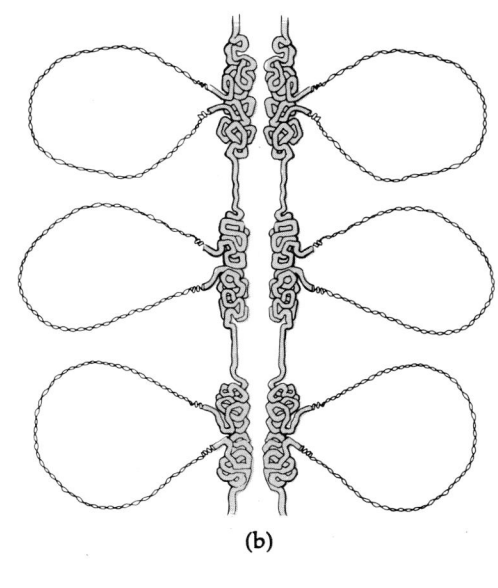

(b)

(a)

50 μm

8–4 *A model, by E. J. DuPraw, of how the DNA is packed into the chromosome. The DNA helix, tightly coiled, is held in place by histones and other proteins, shown hypothetically as wedge-shaped molecules with specific binding sites for the DNA coils.*

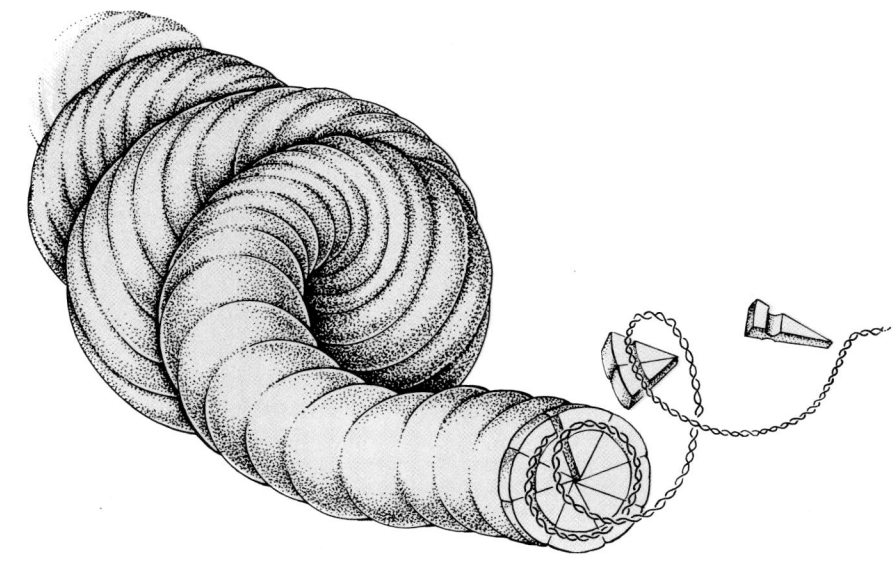

STUDIES OF GENE TRANSFER

Studies of the transfer of genetic information from one cell to another, which began some 30 years ago, are still an active area of research interest, in part because of the importance of the questions that are still unanswered—especially those concerning incorporation of genes into eukaryotic chromosomes—and in part because of their implications for various types of therapy in man. In Chapter 16, we mentioned two ways that new genetic information can be introduced into cells: (1) the release of DNA (transforming factor) into the environment, where it can be taken up by other cells, and (2) viral infection, in which DNA (or RNA) is injected into a host cell.

There are two additional means for transferring genes: conjugation and transduction.

Conjugation

Conjugation is a form of mating which takes place between prokaryotes. Like many other phenomena, it has been studied most extensively in *E. coli*. In the course of conjugation, donor cells pass some or all of their chromosome (which you will recall, is simply a continuous double helix of DNA) to recipient cells. Detection of such transfers is possible because of the existence of mutant strains of *E. coli* which, like the mutant strains of *Neurospora*, vary either in their capacities to make certain enzymes or in their sensitivity or resistance to drugs such as streptomycin and so can be isolated by their capacity to survive in particular media.

Whether or not an *E. coli* cell can function as a donor depends on its possessing an additional piece of DNA known as an F (fertility) factor. The F factor may exist independently, as a very small additional chromosome. (A gene or group of genes that can exist either free or as a part of a normal cellular chromosome is known as an episome). The F factor in its independent state can be transferred from one cell to another; when it is, the recipient cell becomes a donor cell.

Transfer of other genetic information occurs only when the F factor is part of the *E. coli* chromosome. The F factor always attaches to a point on the chromosome that is specific for the particular bacterial strain.

During conjugation, the chromosome opens and begins to replicate at the point of attachment of the F factor. As it replicates, the free end of the chromosome moves into the recipient cell (Figure 18–6). Usually only part of the chromosome is transferred. This fragment can then replace part of the chromosome of the recipient cell. The F factor, which would be the last portion of the chromosome to enter the cell, is seldom transferred in these cases.

The conjugation process takes about 90 minutes (at 37°C). By separating the cells at various points in the process (which can be accomplished by whirling them at high speed in an electric blender), it is possible to construct a map of the chromosome. (See Figure 18–7.) It was these studies that first gave a clue that the *E. coli* chromosome is circular (that is, that there is no end to it); no matter where the opening occurs in the donor chromosomes of the different mating strains, the genes always enter in the same order, like a freight train pulling a line of boxcars through a tunnel.

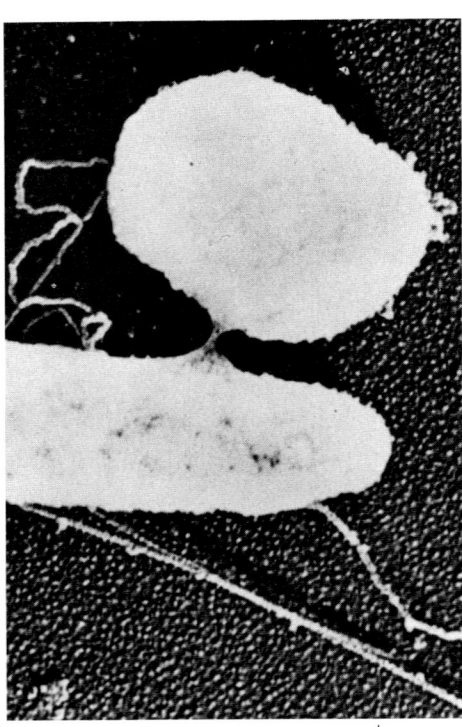

18–5   *Electron micrograph of two conjugating* E. coli *bacteria. A slender connecting bridge has been formed between an elongated donor cell and a rotund recipient cell.*

0.25 μm

18–6 (a) *In bacterial conjugation, a replica of the F (fertility) factor may be transferred from a donor (F⁺) cell to a recipient (F⁻) cell. The recipient then becomes a donor. (b) When the F factor is part of the chromosome (such cells are referred to as HFr, for high frequency of recombination), the chromosome itself, or a portion of it, enters the recipient cell. That portion may become part of the recipient's chromosome, replicating with it. The recipient cell usually remains an F⁻ cell.*

(a)

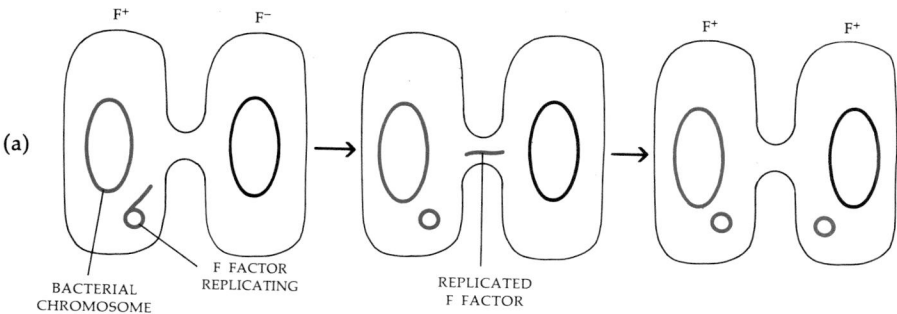

(b)

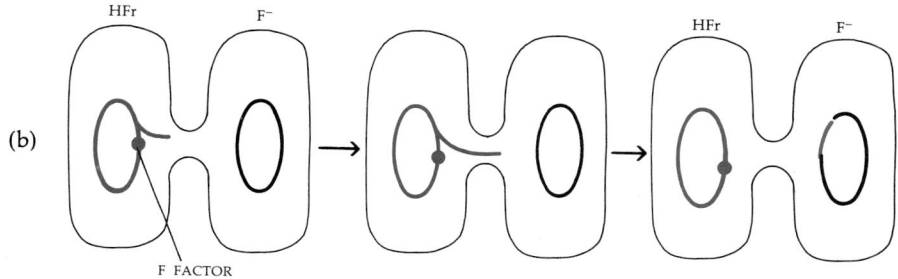

18–7 *The genes are transferred from one cell to another in a definite order and at a constant rate. By employing mating strains with a variety of different mutations, it was possible to map the bacterial chromosomes simply by observing the order in which genes entered the recipient cell. The symbols on the chromosome indicate specific genetic markers.*

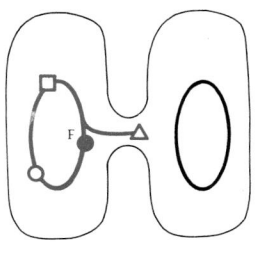

10 MIN

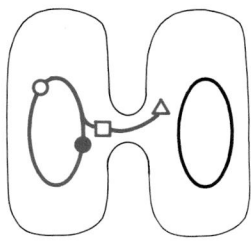

20 MIN

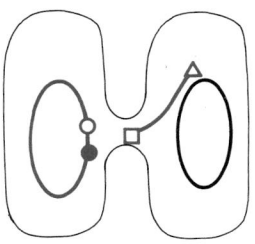

30 MIN

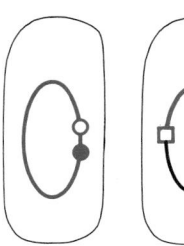

60 MIN

18–8 *When certain types of bacterial viruses infect bacterial cells, one of two events may occur. The virus DNA may enter the cell and set up an infection, such as we described in Chapter 16, or the DNA of the virus may simply lie latent in the cell. In the latter case it may become part of the bacterial chromosome, replicating with it. From time to time, such a virus becomes activated and sets up a new infective cycle. Bacteria harboring such viruses are known as lysogenic.*

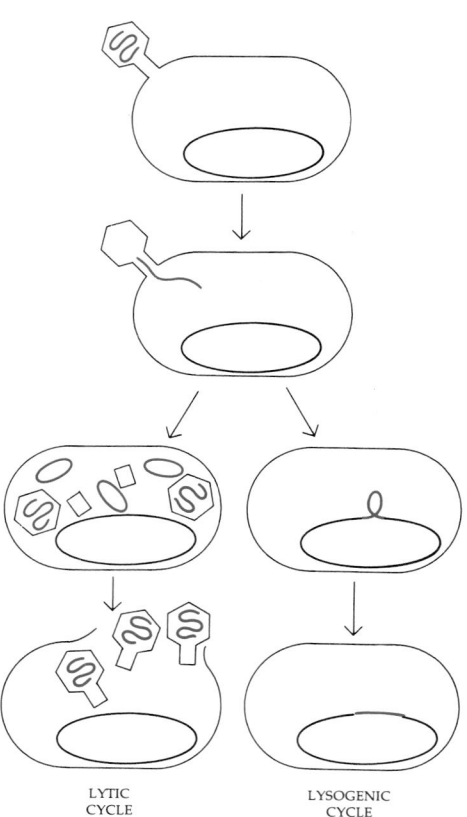

LYTIC
CYCLE

LYSOGENIC
CYCLE

Transduction

Two forms of transduction are known: general transduction and **restricted transduction**. General transduction occurs when fragments of the DNA of a bacterial cell, broken down in the course of infection, are caught up in a bacteriophage head and are carried to a new host cell. These particles carry little or no bacteriophage DNA and so they cannot set up an active infection in the recipient cell. Instead, the portion of the bacterial chromosome carried into the new host may become a part of the host chromosome, and if the chromosome donor differed from the host in any genetic capacities, such differences might be acquired by the new host cell.

Restricted transduction occurs only with certain types of bacterial viruses known as lysogenic bacteriophages. Such bacteriophages do not necessarily set up an infective cycle when they enter a host cell but may instead exist as episomes within the cell. Such a virus is known as a prophage or, more generally, a *provirus*. Like other episomes, they may become part of the cell chromosome. Once they are a part of the chromosome, they replicate with it, often for many generations. They also may produce mRNA and new proteins. It has been discovered recently that one such prophage, known as lambda, remains inactive because of the presence in this infected cell of a repressor protein that blocks expression of the viral genes.

From time to time the relationship between a bacterial cell and its resident virus may shift, and the virus breaks loose from the chromosome and sets up an infective cycle. Exposure to x-rays, ultraviolet, or certain chemicals (all generally the same agents that cause mutations) increases the rate of activation of viruses in host bacteria. (It is hypothesized that they may inactivate repressor proteins.) Bacteria harboring prophages are known as "lysogenic" bacteria. When a lysogenic bacterium releases its viruses, the viruses infect other nearby bacterial cells, either lysing or lysogenizing them.

When lysogenic viruses leave a host bacterial cell chromosome, they sometimes take a fragment of the chromosome with them (Figure 18–9). This fragment is replicated with them through any infectious cycle. If one of these viruses becomes a part of a chromosome of a new host, the genes from the previous host may be inserted into the new host's chromosome and become a part of its genetic equipment. For instance, the bacteriophage lambda has a chromosomal position near the bacterial genes that produce the enzymes concerned with utilization of galactose—in other words, the galactose operon. When a lambda provirus carrying the galactose genes becomes a provirus in a mutant bacterial cell that cannot synthesize one or more of these enzymes, the infected cells gain the capacity to utilize galactose for growth.

INFECTIOUS DRUG RESISTANCE

For many years, such transfers of information among bacterial cells have been regarded mainly as laboratory phenomena, chiefly of interest to research scientists engaged in studies such as gene mapping. However, it is now clear that such transfers also occur frequently in nature among certain groups of bacteria.

In every large bacterial population, there are a few cells which, as a result of mutation, are resistant to a particular drug—streptomycin, for example. If

18–9 (a) *Like the F factor, a virus may become part of a bacterial chromosome. (b) As a result of x-rays, chemical treatment, or "spontaneously," the virus may break loose from the chromosome, setting up a new infectious cycle. (c) Sometimes the virus takes a fragment of adjacent host chromosome with it. This fragment may be large enough to include several host genes. (d) When and if the virus combines with the chromosome of a new host cell, the cell acquires the new genetic information. This process is known as transduction.*

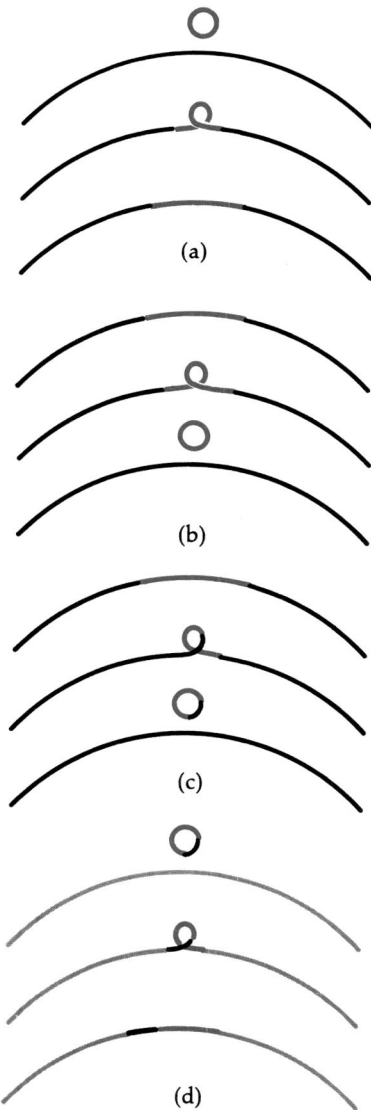

(a)

(b)

(c)

(d)

the bacterial population is exposed to streptomycin—as when a patient is being treated with a drug—the susceptible cells are destroyed and only the resistant ones remain. These multiply and produce a population of bacterial cells made up entirely of resistant individuals. This is the reason why responsible physicians advocate restraint in the use of antibiotics.

About 1959, a group of Japanese scientists discovered that bacteria can become drug-resistant by the transfer of genes from resistant to nonresistant cells. These genes, or resistance factors, as they are sometimes called, are in the cell in the form of isolated units of genetic material, like independent F factors. And like such F factors, they are readily transferable. Under experimental conditions, 100 percent of a population of sensitive cells can become resistant within an hour after being mixed with suitable resistant bacteria. This process, which was suggested by Griffith's experiments of nearly half a century ago, was the first discovered in a strain of *Shigella*, a bacterium that causes dysentery. Further studies showed that not only can *Shigella* transfer drug resistance to other *Shigella* organisms but also the innocuous *E. coli* can transfer resistance factors to this and other, unrelated groups of bacteria. Infectious resistance is now found among an increasing number of types of bacteria, including those that cause typhoid fever, gastroenteritis, plague, and undulant fever.

Fortunately for human populations, the majority of common infectious bacteria do not seem to possess this capacity for infectious drug resistance. However, almost half of the antibiotics produced in this country are given to livestock with their feed to promote growth. (In England, for example, it is no longer legal to give antibiotics to healthy livestock.) With constant exposure to such antibiotics, the ever-present harmless bacteria of such animals may all come to harbor resistance factors, which they may pass on to bacteria capable of causing disease in the animals themselves or to the people who eat them.

TRANSDUCING VIRUSES AND HUMAN CELLS

It has long been suspected that certain viruses can enter into relationships with eukaryotic cells similar to those relationships between the lysogenic bacteria and their viruses. Herpes simplex, the virus that causes fever blisters, is an example. If you suffer from fever blisters or know someone who does, you know that they tend to break out repeatedly in the same person and that they characteristically occur in times of stress, such as when that person has a fever or eats certain food or gets a sunburn (ultraviolet exposure). It is believed that certain cells constantly harbor the herpes simplex virus, that it multiplies when the cells multiply, and that it causes cell damage—the fever blister—only when it is activated.

Recently, it has been demonstrated that polyoma virus, a virus that causes cancer in some rodents, does become part of the host cell's chromosome. The implications of this relationship will be discussed at the end of this chapter.

Studies with human cells growing in tissue culture (that is, in the laboratory in special media, similar to the way in which bacteria are grown) suggest that cells can acquire new capacities when infected with particular viruses. For instance, in experiments as yet unconfirmed, cells were removed from a person with galactosemia, a fairly common "inborn error of metabolism" in which one

18–10 *SV40 (simian virus 40), originally isolated from monkeys, has been shown to produce cancers in baby hamsters and other laboratory animals. It is small, even for a virus; each particle is only about 45 nanometers (180 millionths of an inch) in diameter. It is composed of a molecule of DNA surrounded by an icosahedral (twenty-sided) protein coat. Although many viruses have now been shown to produce cancers in many different types of animals, there is no conclusive evidence that any cancer in man is caused by a virus.*

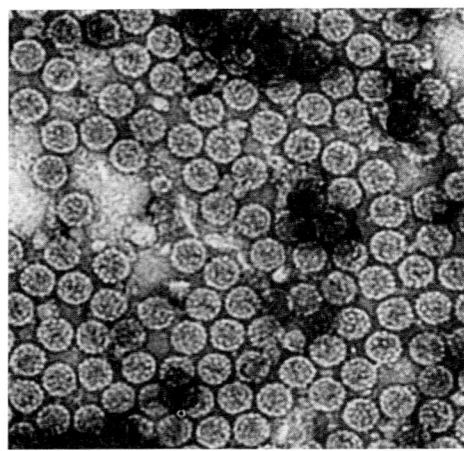

0.1 μm

of the genes involved in the breakdown of galactose is absent, owing to a genetic defect. This gene is one of those in the lactose operon. The cells, growing in tissue culture, were treated with lambda bacteriophages carrying lactose operons and also with the isolated DNA of such bacteriophages. In both types of experiments, the investigators claim that—the interpretation is still controversial—a small number of human cells apparently acquired the capacity to synthesize the missing enzyme.

The location of the viral DNA within the human cell is still unknown. But the experiments show clearly that bacteriophage DNA can enter human cells and that, once there, some of its genes can be replicated, transcribed, and translated.

GENETIC ENGINEERING

The discoveries that genetic information can be transferred from cell to cell and that "foreign" genes can operate in various host cells has quickened speculations about the possibility of someday inserting selected genes into human cells and so curing genetic diseases such as sickle cell anemia, PKU, and diabetes.

Some researchers, including Robert Sinsheimer of the California Institute of Technology, have suggested that viruses might be tailor-made to carry particular genes to particular cells. This would involve the synthesis of the necessary genes, a chain of perhaps 5,000 DNA nucleotides, the size of a small virus. (It is now possible to synthesize chains of about 75 nucleotides.) And Sinsheimer points out ". . . such a synthesis need only be done once. Once the DNA is available, nature provides the means to copy it with the highest fidelity." Secondly, the protein coat of normal virus could be assembled about the DNA strands. Scientists at the University of Maryland School of Medicine have already succeeded in assembling an artificial virus—a pseudovirion, as they call it—composed of polyoma protein and DNA from mouse embryo cells. This DNA, labeled with a radioactive tracer, was delivered into the nuclei of human embryo cells in tissue culture. As Sinsheimer says, "The technology needed here—for such a radically different approach to a major clinical problem—is in truth *almost* in reach."

Some observers hail these new possibilities as ways of lightening man's genetic burden. (See boxed essay, page 273.) Others, pointing out the irreversibility and the hereditary nature of such changes, have expressed concern that gene therapy might be used prematurely, before there is a real understanding of how genes are regulated naturally. An additional group, many of whose members compare these new discoveries with those that led to the development of atomic energy earlier in this century, voice the fear that gene insertion will be used to modify intelligence or behavior, producing changes desired by the state or some other influential group rather than changes intended to benefit the individual patient.

CLONING

The fact that all cells carry all the genetic information suggests that whole organisms can be generated from a single somatic cell. This procedure is known as cloning.

There has been much talk about the possibility of human genetic modification— of designed genetic change, specifically of mankind. I think there should be much discussion of this topic. I think this possibility—which we now glimpse for the first time—is potentially one of the most important concepts to arise in the history of mankind. I can think of none with greater long-range implications for the future of our species. Indeed, this concept marks a turning point in the whole evolution of life. For the first time in all time a living creature understands its origin and can undertake to design its future.

This is a fundamentally new concept. Even in the ancient myths man was constrained by his essence. He could not rise above his nature to chart his destiny. This day we can envision that chance and choice. . . .

It is a new horizon in the history of man. Some of you may smile and may feel that this is but a new version of the old dream of the perfection of man. It is that; but it is something more. The old dreams of the cultural perfection of man were always sharply constrained by his inherent, inherited imperfections and limitations. Man is all too clearly an imperfect, a flawed creature. Considering his evolution, it is hardly likely that he could be otherwise. And to foster his better traits and to curb his worse by cultural means alone has always been, while clearly not impossible, in many instances most difficult. It has been an Archimedean attempt to move the world but with the short arm of the lever. We now glimpse another route—the chance to ease the internal strains and heal the internal flaws directly—to carry on and consciously to perfect, far beyond our present vision, this remarkable product of two billion years of evolution. We are, it is true, very young for this task—young in skills, young in wisdom—but also, fortunately, young in heart.

Robert L. Sinsheimer: "The Prospect for Designed Genetic Change," American Scientist, **57**, 134–142, 1969.

A clone is a group of organisms that have been produced from a single parent by a series of mitotic divisions and that are (with the exception of possible mutations) genetically identical. Clones are found in nature only among one-celled organisms and in a few invertebrates and plants that reproduce asexually. About a decade ago, it was discovered, however, that it is possible to produce a whole carrot from an isolated carrot cell. Such work points to the possibility of producing strains of identical organisms, say domestic plants and animals, of proved and standard quality. It has even been suggested—although not, so far, by scientists in the field—that in this way a family could "reincarnate" a loved one simply by removing a cell from his body, a society could reproduce its leaders in politics, sciences, and the arts, and a government could produce large numbers of people of a proved useful type.

A way in which the cloning of human cells might be more practically applied is in the production of organs for transplant. There are two main obstacles to the widespread use of transplanted organs to replace diseased ones. One is the lack of suitable replacement organs, which need to be healthy, relatively young, and undamaged. The other, and at present more serious, obstacle is that an animal body reacts more or less strongly against any cells that are not genetically identical to its own cells. For this reason, tissues can be exchanged

Is a genetic counselor's responsibility first to the couple involved, or to society?

What constitutes a "defective" fetus?

When a mother being tested for mongolism in her unborn child turns out negative in that respect but the test shows the child has an extra Y chromosome—which might predispose him to antisocial behavior as an adult—should the doctor tell her? And if he does, what should she do?

Does an unborn baby have "rights"—including even the right not to be conceived (by cloning) as an identical twin of many others, or of his father? Can a parent ethically give "consent" on behalf of an unborn child—for gene manipulation, for example?

What will be the effect on society, the family, and the individual himself of being able to choose in advance which sex a child will be?

Mass prenatal screening can detect some 130 biochemical abnormalities. Who should do such screening? Which defects should have priority in the search? Should it be voluntary or compulsory?

Who should have access to the information from prenatal tests? How should it be used?

Could a child with genetic defects arising from a rubella infection early in pregnancy sue his mother's obstetrician—or his parents—for "wrongful life" in failing to abort him?

What public policy can and should be worked out to prevent abuses of current and future advances in genetics?

readily only between identical twins. Currently, transplants are done whenever possible between genetically similar persons, often members of the immediate family. Also, drugs are given to the patient to reduce his reactions to the foreign tissue, but since these drugs reduce his reactions to *all* foreign matter, they greatly increase the dangers of infection. One possible way to overcome these two obstacles might be the production of new organs from the patient's own cells growing in tissue culture; such organs, clearly, would be acceptable to his body. Needless to say, this possibility is a long way from materializing.

GENES, VIRUSES, AND CANCER

Cancer is a disease, or group of diseases, in which particular cells in the body cease to respond to whatever controls growth under normal conditions and multiply autonomously—crowding out, invading, and destroying other tissues. There have been many theories of cancer cause. For a number of years, the most widely held theory was that cancer is the result of a somatic mutation. A somatic mutation is simply a change in the genetic makeup of a body cell. It would not necessarily affect the germ cells (the cells that give rise to sperm or egg cells) but it would be passed on by mitosis to all the progeny of the malignant cell. Suppose, for example, that there is some sort of regulator gene that controls mitosis. A mutation that renders such a gene nonfunctional could result in the sort of wild growth characteristic of cancer. This hypothesis is supported

by the fact that many agents that cause cancer—x-rays, ultraviolet radiation, and certain chemicals—also cause mutations.

More recently, accumulating information about viruses has led many scientists to believe that viruses are the cause of at least some of the forms of cancer in man. One of the reasons for believing this is the fact that many different kinds of cancer in animals are caused by viruses, including cancers in frogs and chickens and in mice, rats, and other mammals. It has been impossible to test conclusively the hypothesis that human cancers are caused by virus. First the latent period between infection and the appearance of symptoms is so long that no infectious pattern emerges, as it does in influenza or measles, for example. Some of the animal viruses are passed from generation to generation—e.g., in the mother's milk—and may even skip a generation or two. Second, it is, of course, not permissible to inject human subjects with any virus or other substance suspected of causing cancer. Finally, although viruses can be detected in some electron micrographs of human cancer, there is no way to prove that such viruses are causative agents and not merely passengers.

THE THEORIES CONVERGE

With the discovery of viruses—and the proviruses that reside within their hosts as episomes (page 270)—the somatic theory of mutation and the virus theory appeared to merge. It has recently been discovered that at least one cancer-causing virus, the polyoma virus, does become part of the chromosome of cells growing in tissue culture. When the virus DNA is inserted into the host chromosome, the cell becomes cancerous and its offspring are also cancerous by all criteria, including the capacity to cause cancer in the animal or strain of animal from which the cells were originally taken.

One of the difficulties with the virus theory of cancer has been that many cancer viruses contain RNA instead of DNA. DNA viruses replicate themselves and produce mRNA, from which viral proteins are made. RNA viruses replicate themselves, presumably in a manner analogous to DNA replication, and also serve as mRNA for viral protein production. But RNA cannot serve as a genetic factor within a cell in the way that DNA can. So the theory that cancer viruses act by becoming part of the genetic apparatus of the cell did not seem reconcilable with the existence of cancer-causing RNA viruses. Recently, this problem was resolved in a totally unexpected way: A group of virologists discovered that some RNA viruses make DNA. When these viruses infect a cell, they produce, on the cell's ribosomes, a new enzyme that makes viral DNA on the template of the viral RNA. Moreover, most of the cancer-producing RNA viruses tested so far have been found to produce this new enzyme and to make DNA, whereas only one RNA virus not known to cause cancer has been found that makes the enzyme.

Thus, it is at least possible that the cancer-causing RNA viruses operate by the same mechanism as the cancer-causing DNA viruses: they become part of the host chromosome and, as they do so, endow the host cell with the murderous capacity to multiply, invade, and destroy.

Discovery of the cause of cancer, although a great intellectual achievement, would not necessarily mean control of the disease. We have, for instance, the example of sickle cell anemia, in which cause is known to the last nucleotide yet

for which, at this writing, there is no known cure. Nor are there any cures for virus diseases, although some of them can be prevented very effectively before infection occurs. Nevertheless, such research will surely hasten the day when a control will be found for what is probably the most dreaded of human diseases.

SUMMARY

The way in which DNA and protein are organized in the eukaryotic chromosome is not known. Several lines of evidence suggest that the combining of DNA with proteins (and especially with histone proteins) plays a role in repressing large segments of the chromosome during the life of the cell. Included in this evidence is the fact that mRNA transcription takes place only during interphase, when the chromosome is uncoiled, and that chromosome "puffs," observed in giant chromosomes, and the extended loops of lampbrush chromosomes are the sites of rapid mRNA synthesis.

Genetic material can be passed from one cell to another by means of (1) isolated DNA (transforming factors) as in pneumococcus; (2) viral infection; (3) conjugation; and (4) transduction.

In conjugation the DNA of a bacterial cell replicates, and as it replicates, some or all of the chromosome is passed to a recipient cell. Donor cells are characterized by the presence of the F (fertility) factor. The F factor is an episome; that is, it is one or more genes that can exist independently or as part of a chromosome. An isolated F factor can be passed from one conjugating cell to another; the recipient cell then becomes a donor cell. When the F factor is integrated in the chromosome, the chromosome breaks at that point, replicates, and all or part of one of the daughter chromosomes is passed to a recipient cell.

Transduction involves the carrying of a portion of a bacterial chromosome to another bacterial cell by a virus. Transducing viruses are those viruses that can exist in cells as proviruses (one type of episome). Bacterial cells harboring proviruses are known as lysogenic. In one form of transduction (restricted transduction), the transduction occurs when a provirus breaks loose from its bacterial chromosome and takes with it some of the genetic material of the host. The genetic material of the bacterial chromosome is replicated with the viral DNA and packaged into new viral particles which are able to infect new hosts. If such a virus becomes a provirus in a new host, the host cell acquires new genetic material and therefore new genetic capacities.

Infectious drug resistance involves the passage of genes conferring resistance to antibiotics from one strain of bacteria to another. The discovery of ways in which genes can be added to cells quickens hopes—and fears—that the day may be reached when genetic material can be inserted into the cells of human beings.

Cloning is the production of whole organisms from a single somatic cell removed from the body of an individual. It has been suggested as a method by which it would be possible to produce large numbers of individuals of a desirable and highly uniform genetic composition—food, plants, domestic animals, human beings.

Cancer, a disease caused by the uncontrolled multiplication of a group of cells, appears to involve genetic change in these cells. There is evidence that such

changes may be the result, at least in some cases, of infection by a provirus. Support for this concept comes from the discovery that the RNA of cancer-causing RNA viruses is transcribed to DNA, and so could exist within the cell in a provirus form.

QUESTION

1. How would you answer the questions in the boxed essay on page 274? Who do you think should determine such policies?

SUGGESTIONS FOR FURTHER READING

CROW, JAMES F.: *Genetic Notes*, 6th ed., Burgess Publishing Company, Minneapolis, Minnesota, 1966.*
Notable for its examples and good questions.

DUBOS, RENÉ: *So Human an Animal*, Charles Scribner's Sons, New York, 1968.*
In this mature, thoughtful book, which won the Pulitzer Prize in 1969, Dr. Dubos, a well-known microbiologist, analyzes the interactions between man's nature and his environment that produce that "unique, unprecedented, unrepeatable" creature, the human being.

DUPRAW, E. J.: *DNA and Chromosomes*, Holt, Rinehart and Winston, Inc., New York, 1970.
An excellent synthesis of information at the cellular and molecular levels.

JACOB, FRANÇOIS: *The Logic of Life: A History of Heredity*, Pantheon Books, New York 1973.
Jacob's principal theme concerns the changes in the way man has looked at the nature of living beings. These changes, which are part of man's total intellectual history, determine both the pace and direction of scientific investigation. Do not let this dry description dissuade you; The Logic of Life is brilliant, often witty, and always readable.

LURIA, S. E.: *Life: The Unfinished Experiment*, Charles Scribner's Sons, New York, 1973.
An interpretation of biology written for the layman by a Nobel laureate. A National Book Award winner.

PETERS, JAMES A. (ed.): *Classic Papers in Genetics*, Prentice-Hall, Inc., Englewood Cliffs, N.J., 1959.*
Includes papers by most of the scientists responsible for the important developments in genetics: Mendel, Sutton, Morgan, Beadle and Tatum, Watson and Crick, Benzer, etc. You should find this book very interesting; the authors are surprisingly readable, and the papers give a feeling of immediacy that no account can achieve.

SRB, A., R. OWEN, and R. EDGAR: *General Genetics*, 2d ed., W. H. Freeman and Company, San Francisco, 1965.
Although no longer up to date, this remains one of the best introductory texts. It is clear, well illustrated, and well balanced between molecular genetics and other aspects of the subject.

STENT, GUNTHER S.: *Molecular Genetics: An Introductory Narrative*, W. H. Freeman and Company, San Francisco, 1971.
A well-written, authoritative account of the genetics of bacteria and their viruses.

* Available in paperback.

PART II

Organisms

SECTION 3

THE DIVERSITY OF LIVING THINGS

19–1 *A confrontation between representatives of two prominent phyla. On the left, a tropical grasshopper; on the right, a South American insectivore (a marmosa). The grasshopper held its ground, and, after weighing the situation carefully, the marmosa withdrew.*

Chapter 19

The Classification of Living Things: Monera and Viruses

In the previous pages of this book, we discussed the organization of living systems at the cellular level, in terms of the organization of their atoms and molecules and organelles, and of the structures and functions of these various components. In Part II, we shall look at another level of biological organization —the organism.

As many as 10 million different kinds of living things share the biosphere of this planet. Guess the number of kinds of trees within a square mile of where you are right now. Roll over a decaying log or turn over a spadeful of rich soil; investigate the contents of a tidal pool. You will uncover a microcosm of small organisms. Take a microscope and look at a sample of your own saliva or feces. Another new world appears. Leap, in your imagination, to another continent. Are the trees the same? The contents of the soil?

Science, as we defined it in the introduction to this book, is a way that man seeks order in the universe that surrounds him. Is there any order to be found in the vast diversity of the living world?

Aristotle, whom many consider the first great biologist, pondered this question some 2,350 years ago. How do you group organisms in ways that make sense? He recognized, for example, that you could not choose any single characteristic as the basis for grouping. For instance, suppose you decided to group all the animals into those that fly and those that do not. You would end up with one category that included most insects, most birds, bats, and even an occasional squirrel, and you would have to separate some obviously close relations, such as the winged and the wingless ants which occur as members of the same colony. Aristotle recognized, as do modern taxonomists ("groupers"), that one had to look not just at a single feature but at the overall "plan" of a plant or animal. On the basis of overall similarities, an organism could be grouped with other forms in increasingly restricted categories: vertebrate, bird, duck, mallard, for example.

In the middle of the seventeenth century, an English clergyman, John Ray, set out to catalog all the organisms in the world and to arrange them systematically. He was the first to use the word "species" to describe a kind of organism. A species, according to Ray, was made up of organisms that were morphologically the same—from *morphe*, meaning "shape" in Greek—and that could reproduce their own kind.

19–2 *Strawberry plant from a French herbal dated about 550 A.D. During the Middle Ages, interest in science and in the natural world languished. The one exception was the plants, which were cataloged because of their use in medicine. In these catalogs, called herbals, plants were arranged alphabetically or, more often, by what they were supposed to be good for. The lists were accompanied by drawings, but as these drawings were copied from text to text and as the individual artists tended to lend their own embellishments, the drawings became more decorative and more stylized and bore less and less resemblance to any living organism. This picture, for example, has obviously been drawn from another picture and not from the plant itself. The leaves, for instance, are incorrectly shown in groups of four and five instead of groups of three.*

In other words, he tied the concept of species to the concept of heredity, which fits in, of course, with our experience about kinds of organisms—that a cat, for instance, may have an unusual or abnormal kitten but that it will never give birth to a rabbit.

Ray cataloged all the plants in the vicinity of Cambridge, the first complete catalog of one locality ever to be made. Then, traveling through England and the Continent, he worked his way, astonishingly, through all the plants that he could learn about (almost 19,000) and the birds, the fishes, and the four-footed animals.

At the time of Ray, and for a number of years thereafter, plants and animals were designated by cumbersome phrase names, or polynomials. These polynomials were brief descriptive phrases concerning the plant or animal to which they were applied. The first word in the polynomial had come, by the close of the seventeenth century, to designate the *genus* (plural, *genera*), an inclusive group of similar species. Thus, the numerous kinds of roses were grouped in the genus *Rosa,* many butterflies in the genus *Papilio,* and cats and catlike animals in the genus *Felis.* This grouping into genera simplified the system a great deal, since there are far fewer genera than species. Once a genus was described, its attributes could be expected to apply to all the included species. Thus the house cat and the lion share certain distinguishing features by which they are grouped together in the genus *Felis,* but they also have distinctive features which place them in different species of this genus.

THE BINOMIAL SYSTEM

The system of naming living things was simplified by the eighteenth-century Swedish professor, physician, and naturalist, Carolus Linnaeus (1707–1778). His ambition was to classify all the known kinds of plants and animals according to their genera. In 1753, he published a two-volume work, *Species Plantarum* ("the kinds of plants"), which contained brief analytical descriptions of every species of plant known to European science. Although he used the polynomial designations and regarded them as the proper names for species, he made an important innovation. In the margin of his book, opposite the "proper" name of each species, Linnaeus entered a single word, which, together with the generic name, formed a convenient "shorthand" designation for the species. In the book, catnip, which had previously been designated as *Nepeta floribus interrupte spicatus pedunculatis* ("*Nepeta* with flowers in an interrupted pedunculate spike"), was described under *Nepeta,* and *cataria* was put in the margin, making it *Nepeta cataria,* which is its name today.

The convenience of this system was obvious, and Linnaeus and subsequent authors soon replaced all "proper" names with "shorthand" ones. This binomial ("two-name") system is still used today.

A species name consists of two parts—the generic name and the specific epithet (adjective or modifier). However, a genus name may be written alone when one is referring to members of the entire group of species making up that genus, such as *Drosophila* or *Paramecium.*

A specific epithet, since it is merely a modifying adjective, is meaningless when written alone, however, because many different species in different genera may have the same specific epithet. The domestic dog is *Canis familiaris,* for

9–3 *At the time of the Renaissance, artists became interested in depicting the natural world. In* Primavera, *a section of which is shown here, over 30 species of plants, all realistically rendered, can be found.* Primavera *was painted by Botticelli in 1478.*

9–4 *Linnaeus, the inventor of the binomial system of classification. Linnaeus believed that each living thing corresponded more or less closely to some :deal model and that by classifying them, he was revealing the grand pattern of creation.*

instance; *familiaris* is a commonly used specific epithet and by itself would not identify any organism. For this reason, the specific epithet is always preceded by the genus name, or, in a context where no ambiguity is possible, the genus name may be abbreviated to its initial letter. Thus *Canis familiaris* may be designated *C. familiaris.*

WHAT IS A SPECIES?

Species in Latin simply means "kind," and so species, in the simplest sense, are the different kinds of organisms. A more technical definition of species is "a group of interbreeding organisms that do not ordinarily breed with members of other groups." This definition conforms to common sense. If a species interbred freely with other species, it would no longer be a distinctive kind of organism. This definition, although useful in animals, is not so useful in plant taxonomy, however, because fertile crossings can take place among very different kinds of plants. Moreover, very similar organisms often cannot breed with each other. Nor is it applicable to organisms that do not reproduce sexually.

In the light of evolutionary theory, a species is seen to change constantly in both time and space, rather than being the immutable, ideal form conceived by Linnaeus. However, even though a precise definition of species is elusive, the term is understandable in a practical way. For instance, anthropologists point

19–5 *Three members of the genus* Felis: *(a) a bobcat,* Felis rufus, *(b) a cougar,* Felis concolor, *and (c) a domestic cat,* Felis cattus.

(a)

(b)

(c)

out that primitive peoples, particularly hunter-gatherers, are able to recognize, distinguish, and name great numbers of the organisms with which they come in contact. What these people, operating on a practical level, recognize as a "kind" —something different enough to have its own name—almost always coincides with what scientists recognize as a "species." So despite the inability of experts always to agree upon a definition, the species is a clear and useful reality.

OTHER TAXONOMIC GROUPS

Linnaeus (and earlier scientists) recognized the plant, animal, and mineral kingdoms, and the *kingdom* is still the major unit used in biological classification. Between the level of genus and the level of kingdom, however, Linnaeus and subsequent taxonomists have added a number of categories. Thus, genera are grouped into *families*, families into *orders*, orders into *classes*, and classes into *phyla*. These categories (also known as taxa; singular, taxon) may be subdivided or aggregated into a number of less important ones, such as subgenera and superfamilies. By convention, generic and specific names are written in italics, while the names of families, orders, classes, and other taxa are not, although they are capitalized.

Table 19–1 shows the classification of two different types of organisms. Note how much a biologist knows about an organism if he knows its classification.

Taxonomy and Evolution

To Linnaeus and his immediate successors, taxonomic classification was a revelation of a grand, unchanging design. When evolutionary theory came to be the dominant ordering force in the biological sciences, taxonomy was seen to reflect evolutionary history. Species were groups that had diverged only recently; genera had more distant ancestors, and so on. However, even though the meaning of taxonomy changed, the actual classification of organisms, based almost entirely on morphological criteria (as are theories of evolutionary relationships) changed little.

A QUESTION OF KINGDOMS

In Linnaeus's time, as we mentioned, it was thought that there were three kingdoms: animals, plants, and minerals, and until very recently it was popular to classify every living thing as either an animal or a plant. Animals were organisms that moved, ate things, and breathed, and their limbs and organs and bodies grew to a certain size and then stopped growing. Plants were organisms that did not move, eat, or breathe, and that grew indefinitely. Thus the fungi, algae, and bacteria were grouped with the plants, and the protozoans—one-celled organisms that ate and moved—were classified with the animals.

In the twentieth century, problems began to arise, partly as a result of improvements in the light microscope and, subsequently, the development of the electron microscope, and partly because of the application of biochemical techniques to studies of differences and similarities among organisms. Not surprisingly, some major differences have been uncovered, and as a result, the number of groups recognized as constituting different kingdoms has increased.

THE NAMING OF ORGANISMS

The use of Latin names for organisms—a practice that makes biology in general, and taxonomy in particular, seem so formidable a subject in terms of language alone—came about in medieval times, when Latin was the language of scholarship. Although the terms seem cumbersome, they are a necessary tool of science. Common names are, at best, inadequate. We tend to give names only to what is of interest to us. For example, gauchos, the cowboys of Argentina who are famous for their horsemanship, have some two hundred names for different colors of horses but divide all the plants known to them into four groups: pasto, or fodder; paja, or bedding; cardo, wood; and yuyos, everything else. Moreover, even when common names do exist for a kind of organism, they may vary from place to place. A robin in North America is distinctly different from the English bird of the same name. A mole is a placental mammal in North America, a marsupial in Australia. A yam in the southern United States is a totally different vegetable from a yam a few hundred miles away in the West Indies. A sheepshead in Ohio is an ugly, unwanted, inedible cousin of the carp, but a sheepshead in Georgia is a delicate-fleshed saltwater fish. When different languages are involved, the problems of communication become virtually insurmountable. For this reason, biologists refer to organisms by Latin names, registered with and officially recognized by international organizations of botanists and zoologists.

Table 19–1 *Biological Classifications*

Red Maple

Category	Name	Characteristics
Kingdom	Plantae	Organisms that usually have rigid cell walls and usually possess chlorophyll
Sub-kingdom	Embryophyta	Plants forming embryos
Phylum	Tracheophyta	Vascular plants
Sub-phylum	Pteropsida	Generally large, conspicuous leaves, complex vascular pattern
Class	Angiospermae	Flowering plants, seed enclosed in ovary
Subclass	Dicotyledoneae	Embryo with two seed leaves (cotyledons)
Order	Sapindales	Soapberry order; trees and shrubs
Family	Aceraceae	Maple family; trees of temperate regions
Genus	*Acer*	Maples and box elder
Species	*Acer rubrum*	Red maple

Man

Category	Name	Characteristics
Kingdom	Animalia	Multicellular organisms requiring organic plant and animal substances for food
Phylum	Chordata	Animals with notochord, dorsal hollow nerve cord, gills in pharynx at some stage of life cycle
Subphylum	Vertebrata	Spinal cord enclosed in a vertebral column, body basically segmented, skull enclosing brain
Superclass	Tetrapoda	Land vertebrates, four-limbed
Class	Mammalia	Young nourished by milk glands, breathing by lungs, skin with hair or fur, body cavity divided by diaphragm, red corpuscles without nuclei, constant body temperature
Order	Primata	Tree dwellers or their descendents, usually with fingers and flat nails, sense of smell reduced
Family	Hominidae	Flat face, eyes forward, color vision, upright, bipedal locomotion, with hands and feet differently specialized
Genus	*Homo*	Large brain, speech, long childhood
Species	*Homo sapiens*	Prominent chin, high forehead, sparse body hair

19–6 *Three of the classification systems in common use. In (a), all organisms are classified as either plants or animals. In (b), the bacteria and blue-green algae, considered part of the plant kingdom in (a), constitute a third kingdom, the Monera. In (c), five separate kingdoms are recognized.*

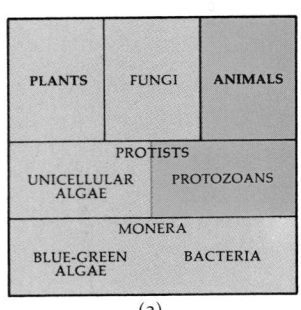

(a)

(b)

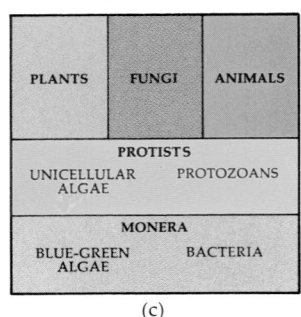

(c)

The most recent proposals recommend five kingdoms: Monera, Protista, Fungi, Plantae, and Animalia. In Appendix C, the major groups of organisms are classified in these five kingdoms. Other systems group all one-celled eukaryotes (cells with nuclear membranes) together as protists, and many group the fungi with the plants, ending up with three kingdoms: protists, plants, and animals. Some retain the two-kingdom classification system: plants and animals. The reason there is such a diversity is that no system is really satisfactory. For instance, at the one-celled level of life, there are no useful criteria for separating plants from animals. One finds two species of single-celled motile organisms almost identical in most respects except that one has chloroplasts and one does not. In some cases, the one that has chloroplasts can lose them from time to time and still continue to survive and reproduce indefinitely. Yet in a plant-animal division based on the presence of chloroplasts, these two closely related forms have to be separated. Either from the point of view of taxonomy based on morphology or one based on evolutionary relationships (see Figure 19–7), the two-kingdom division is unsatisfactory. On the other hand, as we shall see in Chapter 21, there is a clear evolutionary sequence, with modern, living representatives, leading from certain one-celled algae to the flowering plants. So if one groups all of the one-celled eukaryotes together—the protists—as many modern classification systems do, one has either to call some of the algae plants and some protists, as we do here, or to break up what would appear to be a clear evolutionary line leading to the higher plants.

Whether organisms are classified in two, three, or five kingdoms, however, their species designations are not affected, nor are the other taxa in which they are classified.

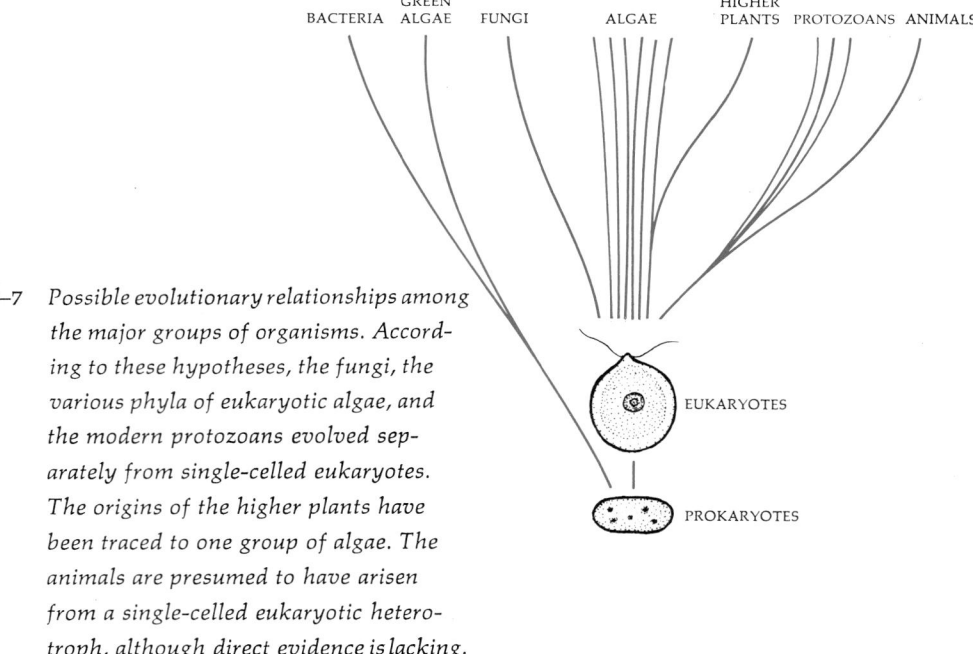

19–7 *Possible evolutionary relationships among the major groups of organisms. According to these hypotheses, the fungi, the various phyla of eukaryotic algae, and the modern protozoans evolved separately from single-celled eukaryotes. The origins of the higher plants have been traced to one group of algae. The animals are presumed to have arisen from a single-celled eukaryotic heterotroph, although direct evidence is lacking.*

19–8 *Representatives of the five kingdoms.*
(a) Monerans. Bacteria, Neisseria gonor-
rhoeae *(dark spheres), the causative
agent of gonorrhea, being ingested by
white blood cells. (b) Protists.* Vorticella
*is a single cell with a long stalk by which
it attaches itself. A contractile fiber,
visible in this micrograph, runs through
the stalk. (c) Fungi. Mushrooms of the
genus* Coprinus *are known as inky
caps because their gills autodestruct at
maturity, turning into a dark fluid mass.
(d) Plants. Marsh marigolds. The flowers
of angiosperms, attractive to pollinators, are among the principal reasons
for their evolutionary success. They are
by far the most abundant of land plants.
(e) Animals. A luna moth. A nervous
system with a variety of sense organs is
a chief characteristic of the animal
kingdom.*

(a) (b)

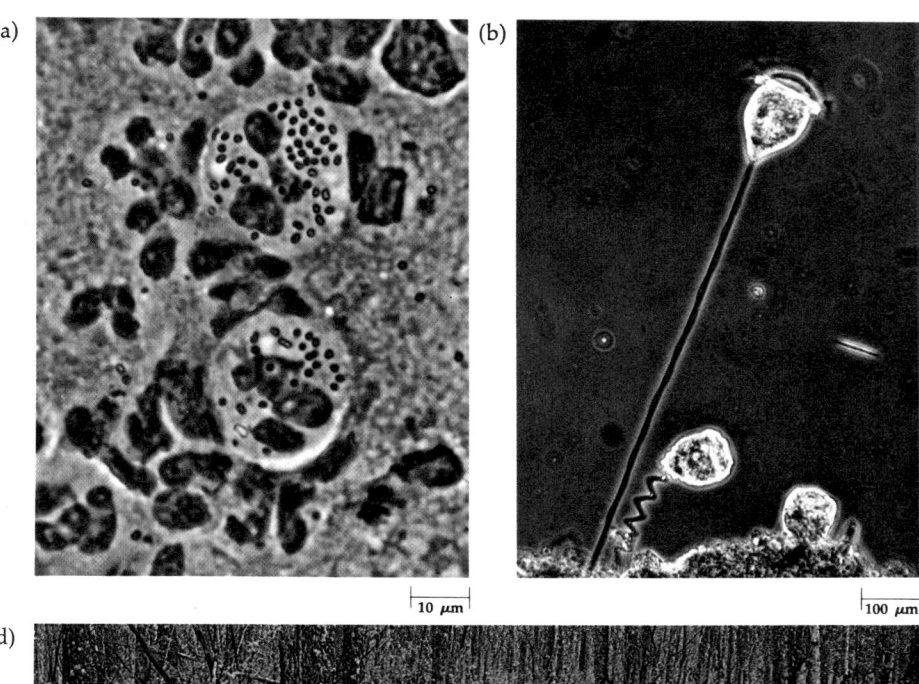

| 10 μm | | 100 μm |

(d)

(c)

(e)

Table 19–2 *Characteristics of Major Groups of Organisms*

	Monera	Protists	Fungi	Plants	Animals
Cell type	Prokaryotic	Eukaryotic	Eukaryotic	Eukaryotic	Eukaryotic
Chromosome	DNA (single circular molecule)	DNA plus protein	DNA plus protein	DNA plus protein	DNA plus protein
Nucleus	No nuclear envelope	Nuclear envelope present	Nuclear envelope present	Nuclear envelope present	Nuclear envelope present
Mitochondria	Absent	Present	Present	Present	Present
Chloroplasts	None (photosynthetic lamellae, chromatophores in some types)	Present (some forms)	Absent	Present	Absent
Cell wall	Noncellulose (polysaccharide plus amino acids)	Present in some forms, various types	Chitin and other noncellulose polysaccharides	Cellulose	Absent
Means of genetic recombination	Conjugation, transduction, transformation, or none	Fertilization (syngamy) and meiosis, conjugation, or none	Fertilization (syngamy) and meiosis, or none, dikaryosis (page 321)	Fertilization and meiosis	Fertilization and meiosis
Mode of nutrition	Autotrophic (chemosynthetic and photosynthetic) and heterotrophic (saprobic and parasitic)	Photosynthetic and heterotrophic or combination of these	Heterotrophic (saprobic and parasitic), by absorption	Photosynthetic mostly	Heterotrophic, by ingestion
Motility	Bacterial flagella, gliding, or non-motile	9 + 2 cilia and flagella, amoeboid, contractile fibrils	9 + 2 cilia and flagella in some forms, none in most forms	9 + 2 cilia and flagella in lower forms and in some gametes, none in most forms	9 + 2 cilia and flagella, contractile fibrils
Multicellularity	Absent	Absent	Present in most forms (multinucleate)	Present in higher forms	Present in all forms
Nervous system	None	Primitive mechanisms for conducting stimuli in some forms	None	None	Present

KINGDOM MONERA:
BACTERIA AND BLUE-GREEN ALGAE

The bacteria and blue-green algae are the simplest of all living organisms and probably bear the closest resemblance of any modern organism to the earliest forms of life on Earth. They resemble one another in basic architecture, being the two living groups of prokaryotic organisms. In many systems of classification, including the one followed here, they constitute the kingdom Monera.

Prokaryotes are much simpler in organization than eukaryotes. They lack organized nuclei with nuclear envelopes. They do not have complex chromosomes like those of eukaryotic organisms, nor do they have endoplasmic reticulum, mitochondria, chloroplasts or other plastids, or the 9 + 2 flagella. They do not reproduce sexually, although, as we saw in the previous chapter, some forms have mechanisms that lead to genetic recombination.

Nearly all prokaryotes have a rigid cell wall. Amino acids are incorporated into this basic wall structure, serving as links between the polysaccharide units. This unusual construction is evidence for the close relationship between bacteria and blue-green algae and also serves to distinguish both of these organisms from the eukaryotes.

No prokaryote is truly multicellular. Although some bacteria and most blue-green algae form filaments or masses of cells, these cells are connected only because their walls fail to separate completely following cell division or because they are held together within a common mucilaginous capsule (bacteria) or sheath (blue-green algae). Thus they are essentially separate cells.

–9 *The kingdom Monera is made up of the bacteria and the blue-green algae. These are the only modern prokaryotes.* (a) Lactobacillus acidophilus, *bacteria that sour milk.* (b) *A gelatinous colony of the blue-green alga* Nostoc. (c) Oscillatoria, *a filamentous blue-green alga.*

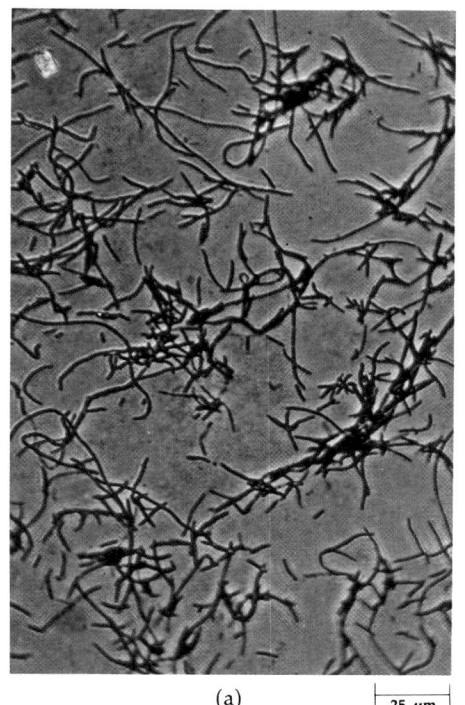

(a)   ⊢ 25 μm ⊣

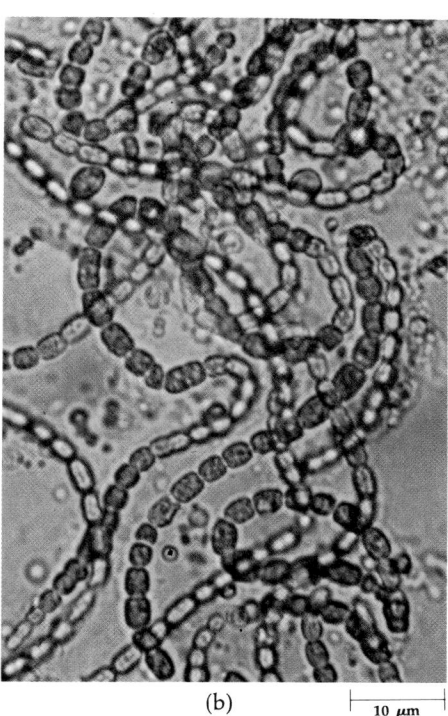

(b)   ⊢ 10 μm ⊣

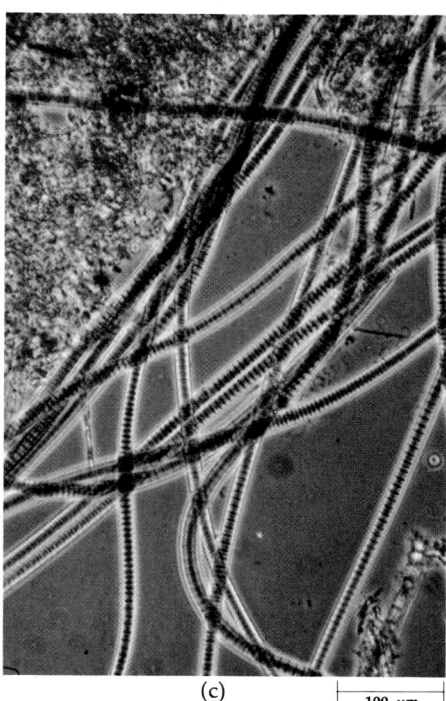

(c)   ⊢ 100 μm ⊣

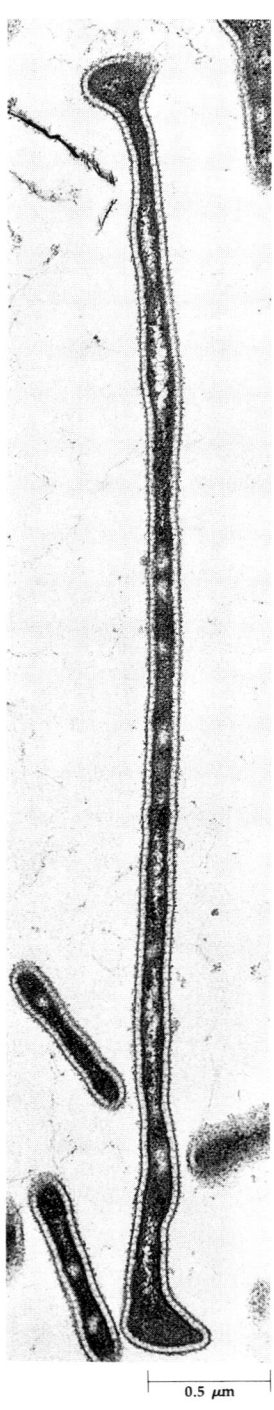

19–10 *Thermophilic bacteria thrive at 92°C, a temperature close to the boiling point of water. This electron micrograph shows a single long filamentous type and several short rodlike types.*

0.5 μm

THE BACTERIA: PHYLUM SCHIZOPHYTA

Bacteria are the oldest and most abundant group of organisms in the world; about 1,600 species have been described. They are the smallest cellular organisms. A single gram (about $\frac{1}{26}$ of an ounce) of fertile soil can contain as many as 2.5 billion bacteria.

The success of the bacteria, biologically speaking, is undoubtedly due to their rapid rate of cell division and their great metabolic versatility. A bacterial population, growing under optimum conditions, can double in size every 20 or 30 minutes. Bacteria can survive in many environments that support no other form of life. They have been found in the icy wastes of Antarctica, the near-boiling waters of natural hot springs, and even in the dark depths of the ocean. Some forms of bacteria are among the very few modern organisms that can survive without free oxygen, obtaining their energy by glycolysis (see page 127). Oxygen is lethal to some types (obligate anaerobes), whereas others can exist with or without oxygen (facultative anaerobes).

A few bacteria can form thick-walled, dry spores. Bacterial spores are resting forms that enable bacteria to survive for long periods of time without water or nutrients or in conditions of extreme heat or cold. They may stay dormant for years, and some remain viable even when boiled in water for as long as 2 hours.

From the ecological point of view, bacteria are most important as decomposers, breaking down organic material to a form in which it can be used by plants. Soil bacteria are also important in the process known as nitrogen fixation, by which nitrogen gas (N_2) is reduced to ammonia (NH_3) or ammonium (NH_4^+). Although nitrogen is abundant in the atmosphere, no eukaryotes are able to use nitrogen in its gaseous form, and so the crucial first step in the incorporation of nitrogen gas into organic compounds depends largely on a few species of blue-green algae and bacteria, some free-living and some in symbiotic association with plants.

A small number of bacterial species are important as causes of disease in man and other organisms. Others, living in the digestive tracts of animals, are essential to the survival of their hosts. Some bacteria are commercially valuable in the production of cheese, vinegar, and yogurt, and most of the antibiotics in medical use are synthesized by a group of bacteria called actinomycetes.

The Bacterial Cell

The most prominent structure within the bacterial cell is the nuclear material (chromosome). It is a circular (i.e., continuous) molecule of double-stranded DNA, about a thousand times as long as the cell in which it occurs. In a fast-growing culture, the DNA replicates more rapidly than the cells do, so each cell characteristically contains more than one DNA molecule, although a single molecule contains the full complement of genetic material.

In addition to the nuclear material, the cytoplasm contains a number of ribosomes, which are slightly smaller than the ribosomes of eukaryotes. The energy-yielding reactions that take place in the mitochondria of eukaryotes occur on the inner surface of the bacterial cell membrane.

Bacteria reproduce by fission (splitting), hence the designation "schizo" in the name of the phylum. When the bacterial cell divides, the DNA molecules attach

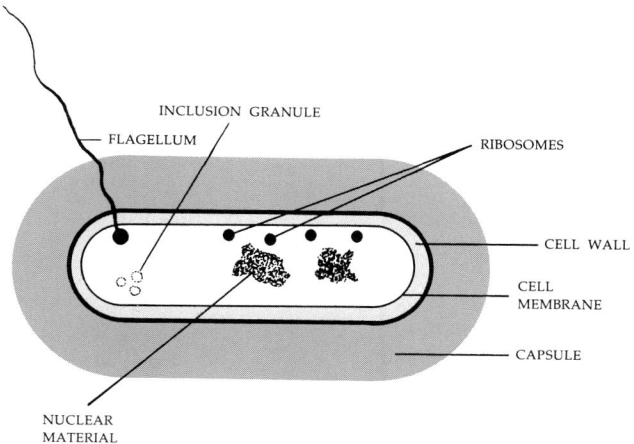

9–11 *Diagram showing the general structure of a bacterial cell. The relatively unstructured cytoplasm is surrounded by a cell membrane and then by a cell wall, which is a combination of sugars and amino acids woven together in a macromolecular mesh. Outside the cell wall of some bacteria is found a slimy capsule of polysaccharides (shown here in color) secreted by the cell. The ribosomes are somewhat smaller than those found in eukaryotes. Reserve foods and pigments are present in the inclusion granules. Two bacterial chromosomes (nuclear material) are present, as is often the case among cells in rapidly growing populations.*

to special areas in the cell membrane, which separate as the cell grows. Like mitosis in eukaryotes, this stratagem ensures that each new cell receives a chromosome. Fertilization and meiosis do not occur in bacteria, but bacterial conjugation (page 269) serves the same function of producing new combinations of genetic material.

The Bacterial Cell Wall

The bacterial cell wall is complex and, unlike the cell walls found in most plants, lacks cellulose. Instead, there is a network of molecules of another polysaccharide, also based on glucose, connected by cross-links of peptides, constituting a glycopeptide layer. Because most bacterial cells are hypertonic in relation to their environment, they would burst without their cell walls. The few kinds that do not have cell walls live as intracellular parasites in an isotonic environment.

The tough bacterial cell wall, which ranges from 10 to 80 nanometers in thickness, gives the different kinds of bacteria their characteristic shape.

In some bacteria, large molecules of lipopolysaccharide are deposited over the glycopeptide layer. Bacterial cell walls that lack the lipopolysaccharide layer combine firmly with such dyes as gentian violet, and those in which it is present do not. Those that combine with the dyes are known as gram-positive, whereas the others are gram-negative, after Hans Christian Gram, the Danish microbiologist who discovered this phenomenon. Gram staining is widely used as a basis for classifying bacteria, since it reflects a fundamental difference in the architecture of the cell wall. This architecture in turn affects various other characteristics of the bacteria, such as their patterns of susceptibility to antibiotics. Gram-positive bacteria are more susceptible to most antibiotics than are gram-negative bacteria. They are also more susceptible to lysozyme (page 72), an enzyme found in nasal secretions, saliva, and other body fluids that digests the cell walls of bacteria.

In certain bacteria, a mucilaginous polysaccharide capsule, which is secreted by the bacterium, is present outside the cell wall. The function of the capsule is

9–12 *A schematic diagram of the attachment of bacterial chromosomes to the cell membrane, which ensures that each daughter cell receives a chromosome.*

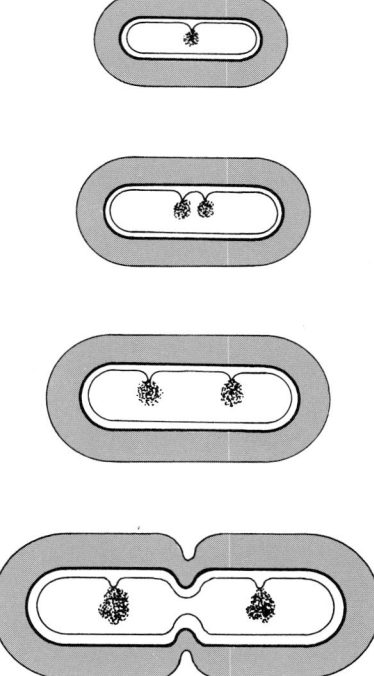

19-13 (a) *The structural formulas of the two amino (nitrogen-containing) sugars (NAM and NAG) that alternate to form the polysaccharide backbones of the bacterial cell wall. (b) A diagram of the cell wall structure of* Staphylococcus aureus. *The backbones of alternating NAM and NAG are cross-linked by five molecules of the amino acid glycine attached to the peptide side chains of the NAM molecules. Because its structure mimics that of a dipeptide (two amino acids), penicillin is able to bind the active site of (and thus inactivate) one of the specific enzymes involved in cross-linking. Without proper cross-links, the cell wall cannot hold the cell together and it subsequently lyses. Thus, penicillin acts specifically on growing bacterial cells and not on the eukaryotic cells of the infected animal.*

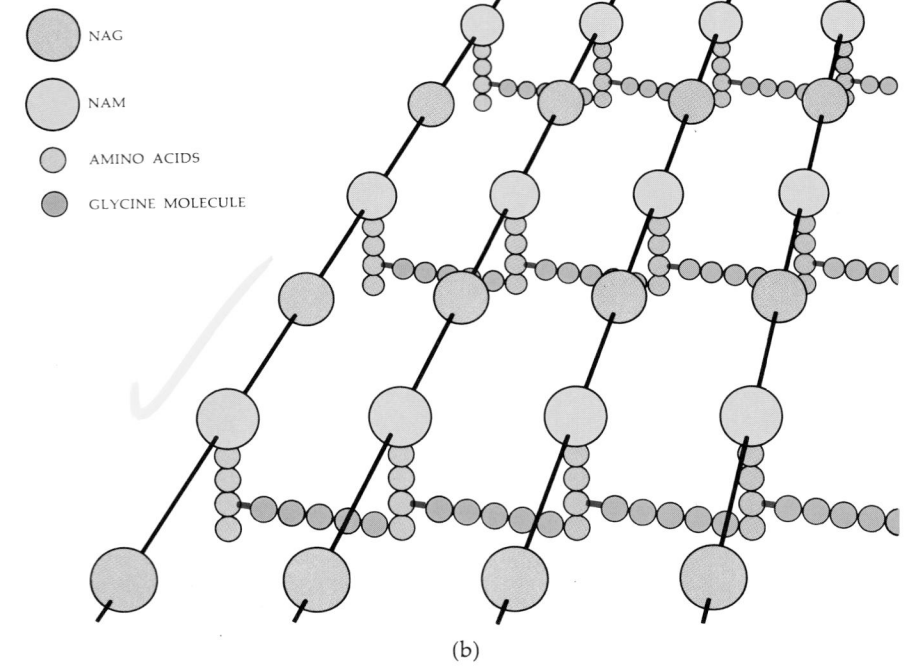

N-ACETYLMURAMIC ACID
(NAM)

N-ACETYLGLUCOSAMINE
(NAG)

(a)

NAG

NAM

AMINO ACIDS

GLYCINE MOLECULE

(b)

19-14 *Electron micrograph of conjugating cells of the bacterium* Escherichia coli, *an abundant and harmless inhabitant of the human intestine. The cells are connected by a long pilus, and numerous pili are visible on the upper cell.*

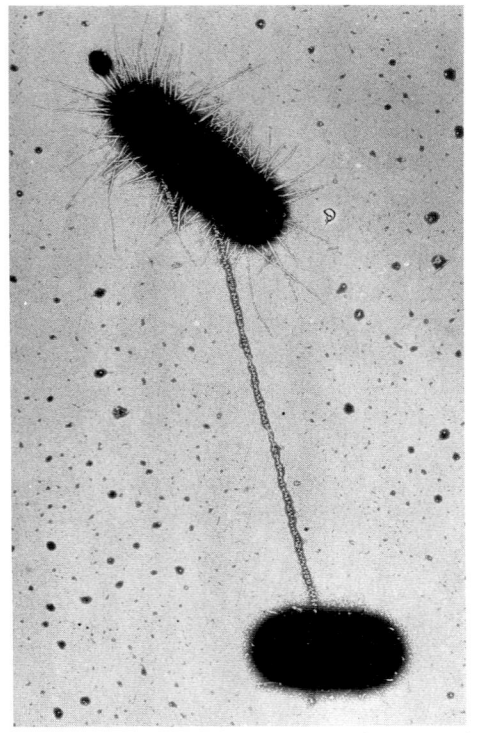

1 μm

not entirely clear, but its presence is associated with pathogenic activity in certain organisms. For example, the capsulated form of *Diplococcus pneumoniae* is virulent, whereas the noncapsulated form is generally nonvirulent. It appears that the capsule interferes with phagocytosis by host white blood cells.

Flagella and Pili

Some kinds of bacteria have long, slender flagella with which they swim, apparently by a longitudinal wavelike action. Since the flagella are only 10 to 20 nanometers in diameter, they are usually too fine to be seen by ordinary microscopic techniques. Bacterial flagella seem to consist entirely of protein of a single special type (flagellin). Pili are structures that somewhat resemble flagella, but they are even finer and usually shorter. Although they are proteins, their chemical composition is different from flagella. Some types of pili are formed only during bacterial conjugation. If they are removed, the bacteria cannot conjugate. Their exact role in conjugation is not known; however, they may serve as the bridge for the transfer of DNA, as the towline that pulls the conjugating cells together, or they may have some other, unknown function.

9–15 *Three of the four major form-groups of eubacteria: (a) bacilli, (b) cocci, and (c) spirilla. The rod-shaped bacteria (bacilli) include those microorganisms that cause lockjaw (Clostridium tetani), diphtheria (Corynebacterium diphtheriae), and tuberculosis (Mycobacterium tuberculosis), as well as the familiar E. coli. Among the cocci are Diplococcus pneumoniae, the cause of bacterial pneumonia, Streptococcus lactis, which is used in the commercial production of cheese, and Nitrosococcus, soil bacteria that oxidize ammonia to nitrates. The spirilla, which are less common, are helically coiled bacteria. Cell shape is a relatively constant feature in most species of bacteria. (Carolina Biological Supply Company.)*

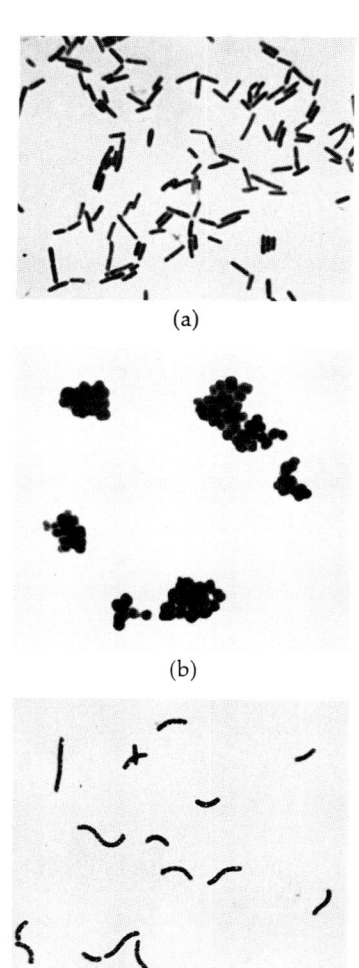

(a)

(b)

(c)

Classes of Bacteria

Eubacteria

The largest group of bacteria is the eubacteria, or true bacteria. There are several different kinds of eubacteria, distinguished principally by their shapes. Straight, rod-shaped forms like *E. coli* are known as bacilli, spherical ones are called cocci, long, spiral rods are called spirilli, and short, curved rods, probably incomplete spirals, are called vibrios. Cocci may stick together in pairs after division (diplococci), they may occur in clusters (staphylococci), or they may form chains (streptococci). The bacterium that causes pneumonia is a diplococcus, while the staphylococci are responsible for many serious infections characterized by boils or abscesses.

Rod-shaped bacilli usually separate after cell division. When they do remain together, they spread out end to end in filaments, since they always divide in the same plane. Because these filaments are funguslike in appearance, the combining form *myco-* (from the Greek word for "fungus") is often a part of the name of these organisms. *Mycobacterium tuberculosis*, for example, the cause of tuberculosis, is a rod-shaped bacillus that forms a filamentous, funguslike growth.

Myxobacteria

The myxobacteria (from *myxa*, or "slime," in Greek) are distinguished by their secretion of a slime track on which the cells glide and by the fact that some types form complex reproductive structures known as fruiting bodies. These fruiting bodies are treelike forms, usually less than a millimeter high. The main stem and branches are composed largely of hardened slime. On the tips of the branches are large, brightly colored cysts, each surrounded by a membrane and each containing hundreds or thousands of individual bacterial cells. These cysts are resting structures in which cells lie dormant during periods when temperature, moisture, or nutrition are unfavorable to growth. If favorable conditions are restored, the cysts break open and the bacteria are released. (See Figure 19–16a.)

Spirochetes

The spirochetes are distinguished by the fact that they are very long (from 5 to 500 micrometers) in relation to their width (usually about 0.5 micrometer) and by their axial filaments. The axial filament is a long, flexible fiber, apparently similar in composition to a bacterial flagellum. The cell body and the axial filament are wrapped around each other in a double helix. Contractions of the axial filament cause the spirochete to move with a whirling motion. (See Figure 19–16b.)

Rickettsia and Mycoplasma

Rickettsiae and mycoplasma are the smallest known prokaryotes. They were at one time believed to be evolutionary forerunners of the larger, more complex bacteria, but it is now generally agreed that they are derived forms—that is, organisms that once had certain structures and functions that have now been lost. This hypothesis is supported by the fact that the rickettsia and some of the mycoplasma are intracellular parasites, and they cannot be cultured, or can be cultured only with difficulty in the absence of the living cells on which they have become dependent.

19–16 *Examples of three classes of bacteria. (a) The myxobacteria are distinguished by their slime tracks. Also, some types form fruiting bodies, such as those in this micrograph of* Stigmatella aurantiaca. *(b) The spirochetes are spiral bacteria up to 500 micrometers long, which is an enormous size for bacteria. They move by means of a rapid spinning or whirling of the cell around its long axis.* Treponema pallidum, *shown here, is the causative agent of syphilis. (c)* Rickettsia *are the smallest known cells. Typhus is caused by* Rickettsia typhi, *shown here. It is spread from rats to man by fleas. It can then be transmitted by body lice and, under crowded conditions, large numbers of people can be infected in a very short time. More human lives have been taken by rickettsial diseases than by any other infection except malaria. At the siege of Granada in 1789, 17,000 Spanish soldiers were killed by typhus, 3,000 in combat. In the Thirty Years War, the Napoleonic campaigns, and the Serbian Campaign during World War I, typhus was also the decisive factor.*

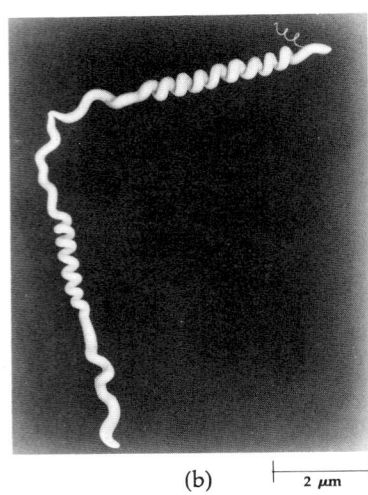

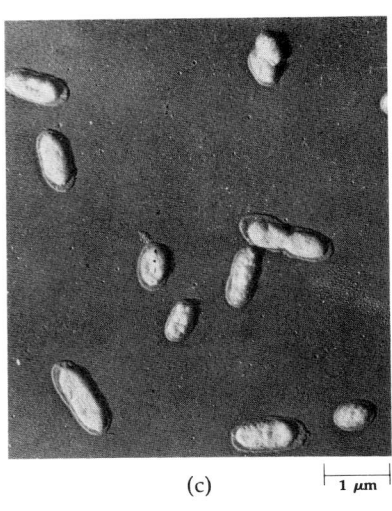

(a)   50 μm   (b)   2 μm   (c)   1 μm

19–17 *The red masses on the surface of this stagnant pond are made up largely of purple sulfur bacteria. Like all photosynthetic bacteria, these are anaerobes and so flourish in still waters in which the available oxygen has been used up by the respiration of aerobic organisms.*

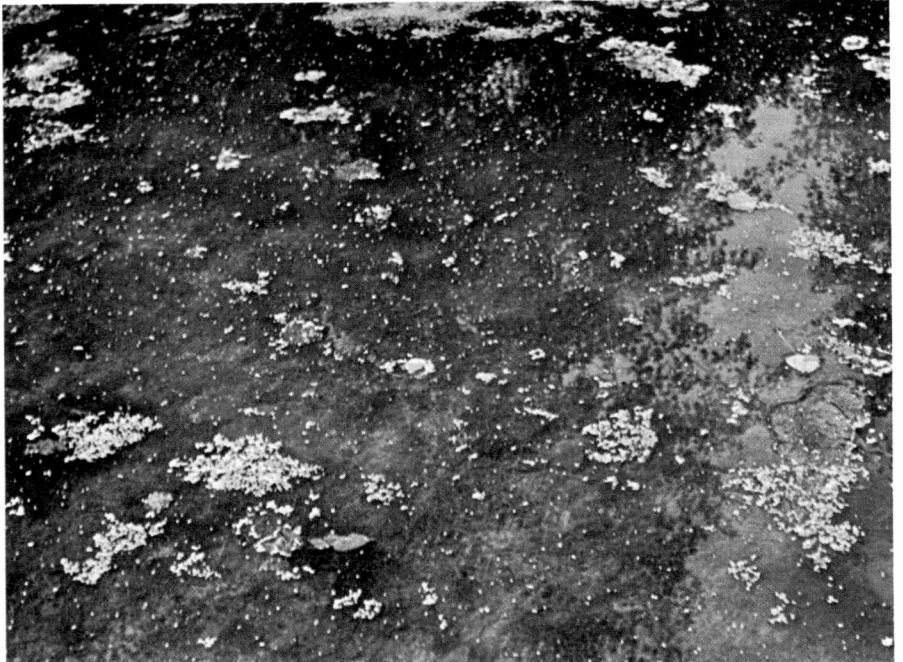

Bacterial Nutrition and Metabolism

Bacteria are extremely versatile metabolically. Some are *autotrophs*, organisms that can make their own organic compounds from inorganic elements or molecules, water, and an energy source. The most familiar autotrophic organisms are those that are photosynthetic, using light as their energy source, which is a characteristic of the plant kingdom (see Table 19–2 on page 288). Some bacteria are photosynthetic autotrophs. Some are also chemosynthetic autotrophs, organisms that derive their energy from inorganic molecules. This type of autotrophy is found only among bacteria.

Most bacteria are *heterotrophs*, using organic compounds that have been synthesized by other organisms. Heterotrophs may ingest other organisms, as most members of the animal kingdom do; they may be saprobes, procuring organic molecules from dead organic matter, or they may be parasites, using organic molecules from living organisms.

Photosynthetic Bacteria

Among the eubacteria are three photosynthetic forms: the green sulfur bacteria, the purple sulfur bacteria, and the purple nonsulfur bacteria. (The colors of the third group may actually range from purple to red or brown.) Like the green plants, photosynthetic bacteria contain chlorophyll molecules composed of a magnesium-containing porphyrin ring and a lipid tail. The chlorophyll found in the green sulfur bacteria, chlorobium chlorophyll, is chemically very similar to the chlorophyll *a* of higher plants. The chlorophyll found in the two groups of purple bacteria is bacteriochlorophyll, which differs chemically in several details from chlorophyll *a* and is a pale green in color. The colors of the purple bacteria are due to the presence of several different yellow and red carotenoids, which function as accessory pigments in photosynthesis.

In the photosynthetic sulfur bacteria, the sulfur compounds are the electron donors, playing the same role in bacterial photosynthesis that water does in plant photosynthesis.

$$CO_2 + 2H_2S \xrightarrow{\text{light}} (CH_2O) + H_2O + 2S$$

As we saw in Chapter 9, an understanding of the course of photosynthesis in the sulfur bacteria was the key that led van Niel to propose the generalized equation for photosynthesis:

$$CO_2 + 2H_2A \xrightarrow{\text{light}} (CH_2O) + H_2O + 2A$$

All bacterial photosynthesis is carried out anaerobically, and it never results in the production of molecular oxygen (O_2).

In the nonsulfur photosynthetic bacteria, other compounds, including alcohols, fatty acids, and a variety of other organic substances, serve as electron donors for the photosynthetic reaction.

Chemosynthetic Bacteria

Chemosynthetic bacteria obtain their energy from the oxidation of inorganic compounds. Certain chemosynthetic bacteria are essential components of the nitrogen cycle, the process by which nitrogen compounds are cycled and re-

19–18 *Beggiatoa, gliding myxobacteria, are relatively large cells that form filaments and move in a "gliding" fashion similar to blue-green algae. Waves of contraction cause periodic changes in the form of the cell and bring about its movements. Beggiatoa is a chemosynthetic autotroph. The conspicuous granules in the cells are sulfur, produced by the oxidation of hydrogen sulfide—a process utilized by the bacteria for the production of energy. The gliding bacteria are essentially like colorless blue-green algae, a fact that underscores the fundamental similarity between these prokaryotes.*

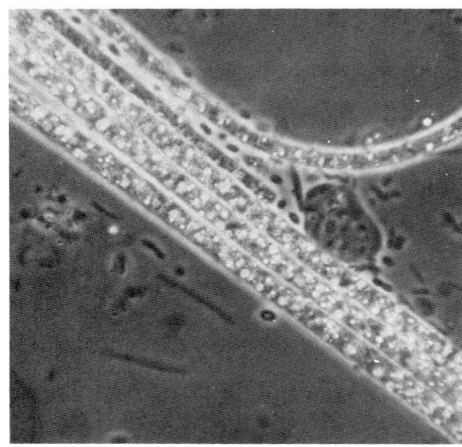

$\vdash$100 μm$\dashv$

19–19 *A sneeze.*

cycled through ecosystems. One group oxidizes ammonia or ammonium (derived from the breakdown of organic materials, the activities of nitrogen-fixing prokaryotes, or, to a minor extent, from lightning or volcanic activities). The products of this reaction are nitrite (NO_2^-) and energy. Another group oxidizes nitrites, producing nitrate (NO_3^-) and energy. Nitrate is the form in which nitrogen moves from the soil into the roots of plants.

Sulfur is also required by plants for amino acid synthesis. Like nitrogen it is converted to the form in which it is taken up by plant roots by the activities of chemosynthetic bacteria that oxidize elemental sulfur to sulfate:

$$2S + 2H_2O + 3O_2 \longrightarrow 2H_2SO_4$$

Other sulfur bacteria, such as *Beggiatoa*, obtain energy by oxidizing hydrogen sulfide.

These important biogeological cycles and the roles of bacteria in them will be discussed further in Chapter 41.

Heterotrophic Bacteria

Most bacteria are heterotrophs. The largest group of heterotrophic bacteria obtain their nourishment from dead organic matter. Bacteria and fungi are responsible for the decay and recycling of organic material in the soil. Typically, different groups of bacteria each play specific roles—such as the digestion of cellulose, starches, or other polysaccharides, or the hydrolysis of specific peptide bonds, or the breakdown of amino acids. These bacterial activities release the nutrients and make them available to plants and, through plants, to animals. Thus, they are an essential part of all ecological systems.

Some heterotrophic bacteria are parasites that break down organic material in the bodies of living organisms. The disease-causing (pathogenic) bacteria belong to this group, as do a number of other nonpathogenic forms. Some of the parasitic bacteria have little effect on their hosts, and some are actually beneficial. Cows and other ruminants can digest cellulose only because their stomachs contain bacteria and certain symbiotic* protozoans. Our own intestines contain a number of types of generally harmless bacteria (including *E. coli*). Some supply vitamin K, which is necessary for blood clotting. Others prevent us from developing serious infections. When the normal bacterial inhabitants of the human intestinal tract are destroyed—as can happen, for example, following prolonged antibiotic therapy—our tissues are much more vulnerable to disease-causing bacteria and fungi, and symptoms of vitamin deficiency may develop.

Pathogenic Bacteria

Bacteria may destroy tissues of their host. Generally, however, bacteria cause disease when their metabolism results in the production of substances toxic to their host.

Pathogenic bacteria are transmitted in a number of ways. Some of man's diseases are caused by airborne bacteria. One such disease is diphtheria, which is caused by the bacterium *Corynebacterium diphtheriae*. This organism, under the influence of a provirus (see page 270), produces a powerful toxin that circu-

*Symbiosis—"living together"—is defined as a close and permanent association between organisms of different species. If one species benefits and the other is harmed, the relationship is known as parasitism. Sometimes it is difficult to define an exact boundary between the two.

20   *Colony of* Clostridium botulinum. *Botulism, the type of food poisoning suffered by consumers of inadequately sterilized foods, is associated with the survival of the spores of this bacterium. The spores are extremely resistant to heat.* Clostridium, *which grows only in the absence of oxygen, produces its powerful toxin inside the can or jar. The toxin itself is destroyed by boiling for 15 minutes but can be fatal even in trace amounts.*

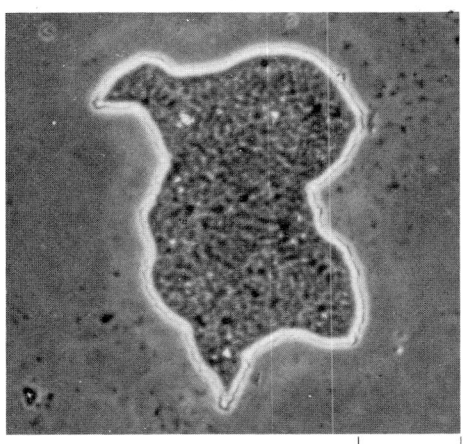

100 μm

lates throughout the body rapidly and causes serious damage to the heart muscle, nervous tissue, and kidneys. Other serious diseases, including scarlet fever and rheumatic fever, are caused by airborne bacteria of the genus *Streptococcus*. Tuberculosis and whooping cough are also airborne bacterial diseases.

A number of other diseases of bacterial origin are spread in food or water. Among these are typhoid fever and paratyphoid, caused by bacteria of the genus *Salmonella*; bacillary dysentery, caused by *Shigella dysenteriae*; and cholera, caused by *Vibrio cholerae*. Undulant fever, brucellosis, which affects both cattle and man, is usually contracted through drinking milk or eating milk products from an infected cow. Pasteurization of milk destroys the causative agent, *Brucella*, and the disease can be controlled in animals by immunization.

Bacteria play an enormous role in spoiling food and other stored organic products, and some organisms of this type are pathogenic. Food poisoning by *Clostridium botulinum*, an obligate anaerobe, is rare but extremely dangerous; botulism is fatal about 65 percent of the time. Spores of this pathogen may survive heating to temperatures as high as 100°C, which is why they may not be killed if the process of home or commercial canning is faulty. The bacteria can then grow and produce toxin in the rich, oxygenless (anaerobic) environment. *Staphylococcus* food poisoning is fairly common, but it is fortunately much less serious than botulism. Food poisoning can also be caused by eating raw or partially cooked eggs and meat from animals infected by *Salmonella* bacteria.

Syphilis and gonorrhea, caused by a spirochete and a diplococcus, respectively, are transmitted by sexual intercourse. Syphilis may also be transmitted by an infected mother to her unborn child (congenital syphilis), and gonorrhea, if present in the genital tract of the mother, can infect the eyes of an infant as it passes through the vagina at birth.

The most dangerous wound-infecting bacteria are members of the genus *Clostridium*. They are soil dwellers, form spores, and are ubiquitous, so any

wound exposed to soil is probably exposed to clostridia. Obligate anaerobes, these organisms thrive in deep puncture wounds, where there is an abundance of dead and dying (necrotic) tissue with an impaired blood supply and an absence of oxygen. *Clostridium tetani* produces a toxin that affects the nervous system and causes muscle spasms (lockjaw). *Clostridium perfringens* creates severe local damage, with extensive destruction of tissue and, often, the accumulation of gases as the result of bacterial activity (gas gangrene).

Most bacterial diseases are not usually fatal, even if untreated. This is logical. Strains that destroyed host after host would have a difficult time surviving and so would be selected against in favor of less harmful varieties. However, even infections that are ordinarily very mild may be fatal in some persons, particularly the very young, the very old, and patients debilitated from other causes.

Many economically important diseases of plants are also associated with bacteria. Some of the more important diseases of plants are called soft rots, blights, or wilts. Sometimes death is associated with an invasion and plugging of the vascular system by large numbers of bacterial cells and the slime they

Table 19–3   *Major Groups of Bacteria*

	Examples	Form	Motility	Mode of nutrition	Distribution	Ecological role	Diseases
Eubacteria	*Escherichia coli, Streptococcus, Staphylococcus, Mycobacterium tuberculosis*	Rod-shaped (bacillus), spherical (coccus), spirillum (many curves), vibrio (one curve)	Gliding, flagella	Chemoautotrophs, photosynthetic autotrophs, heterotrophs	Soil, water, parasites	Decomposers, symbionts, pathogens	Lockjaw, diphtheria, tuberculosis
Myxobacteria	*Myxococcus, Cytophage, Chondromyces*	Rod-shaped, flexible, in slime	Gliding	Heterotrophs	Soil, some aquatic	Decomposers, especially of complex polysaccharides	None
Spirochetes	*Spirocheta, Treponema, Leptospira*	Extremely long, helical	Twisting (axial filament)	Heterotrophs	Aquatic (polluted water), parasites	Symbionts (in mollusks), pathogens, decomposers	Syphilis, infectious jaundice, relapsing fever
Rickettsiae	Rickettsia	Small (0.3 × 0.5 micrometers)	None	Heterotrophs (parasites)	Intracellular parasites	Pathogens	Typhus, spotted fever
Mycoplasma	PPLO	Smallest free-living cells, no cell walls	None	Heterotrophs (many parasites)	Intracellular parasites, soil	Pathogens	Mycoplasmal pneumonia

-21 Staphylococcus aureus (a) *before and* (b) *after treatment with penicillin. Note the collapse of the cell walls in* (b). Staphylococcus aureus *causes boils, wound infections, and many other diseases.*

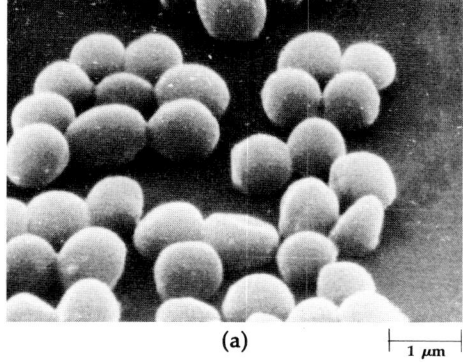

(a) ├─────┤ 1 μm

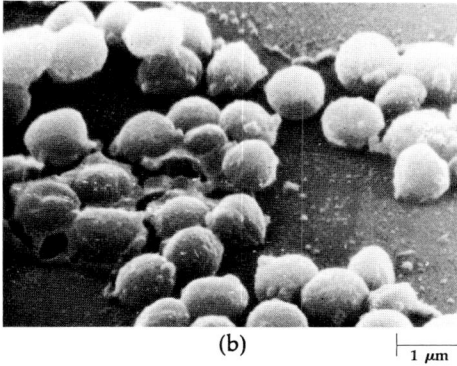

(b) ├─────┤ 1 μm

-22 *Sulfanilamide is effective against bacteria because of its close resemblance to PABA, a growth factor required by many pathogenic bacteria. Sulfanilamide is incorporated into folic acid, one of the B-complex vitamins, in place of PABA, but because of its structural differences, renders the folic acid inactive. Mammalian cells cannot make this vitamin and so are not susceptible to sulfanilamide poisoning.*

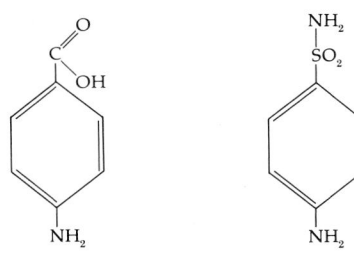

PARA-AMINOBENZOIC ACID (PABA)   SULFANILAMIDE

produce. In other instances, the diseases are associated with the production of toxic substances, like many of the diseases of man and animals.

Prevention and Control of Bacterial Disease

Although bacteria were first seen and depicted with remarkable accuracy by Antony van Leeuwenhoek in the late seventeenth century, they were not generally recognized as a cause of disease until 100 years ago. The relationship was made clear by the work of Louis Pasteur on disease in silkworms and by Robert Koch, who studied anthrax in sheep and tuberculosis in man.

Recognition of microbes as disease agents led directly and swiftly to control measures. Among the agents spreading bacterial infection were physicians, surgeons, nurses, and hospitals. Childbed fever, for example, once a leading cause of sickness and death among young women, was transmitted almost exclusively by physicians who carried streptococci from patient to patient on their hands or on unsterilized instruments.

The battlefields of World War II were the proving grounds for new antimicrobial drugs, such as sulfa and penicillin. These drugs made possible not only treatment of wounds but the widespread and often life-saving use of major surgery as a treatment for cancer and other diseases. Ironically, with the advent of strains of bacteria resistant to these drugs and endemic in hospitals, the latter are once more becoming reservoirs of serious bacterial diseases.

Even more important than improvement in medical practices was the institution of public health eradication of ticks, fleas, lice, mosquitoes, and other agents that carry disease, disposal of sewage and other wastes, protection of public water supplies, pasteurization of milk, quarantine, and other procedures. Not long ago, for instance, infant mortality during the first years of life was often as high as 50 percent in some localities owing to infant diarrhea caused by contaminated milk and water.

Immunization, which involves the stimulation of the body's natural defenses against a disease-causing agent, is also responsible for the prevention of many bacterial diseases, including diphtheria, whooping cough, and tetanus. (Immunity is discussed further at the end of this chapter and also in Chapter 33.)

In 1935, sulfanilamide, the first of the new "wonder drugs" for the control of bacterial infection, was discovered in Germany, and in 1940 the effects of penicillin (first noted by Alexander Fleming in 1929) were reported from England. Penicillin was the first-discovered antibiotic—by definition, a chemical that is produced by a living organism and is capable of inhibiting the growth of microorganisms. Sulfanilamide and penicillin received early and extensive trials on the battlefields of World War II.

Most antibiotics are produced by bacteria, especially the actinomycetes; a few are formed by fungi. Many, including penicillin, can now be synthesized in the laboratory. These antibacterial agents act by interfering with metabolic reactions essential to the bacteria but not to the host cells. Penicillin, for example, inhibits bacterial growth by interfering with the cross-linking of the peptide chains in the bacterial cell walls. (See Figure 19–13, page 292.) Because eukaryotes lack such walls, penicillin does not affect the cells of the host in this way.

The tremendous decrease of deaths from infectious disease over the last decades is the chief cause of the present population explosion.

The blue-green algae are prokaryotes and are organized much like the other prokaryotes, the bacteria. They are photosynthetic, but, unlike any photosynthetic bacteria, they contain chlorophyll *a*, which is also found in all photosynthetic eukaryotes. They have several kinds of accessory pigments, including xanthophyll, which is a yellow carotenoid, and several other carotenoids. (Carotenoids can also be found in photosynthetic eukaryotes and in some of the bacteria.) The cells of blue-green algae may also contain one or two pigments known as phycobilins: phycocyanin, a blue pigment, which is always present, and phycoerythrin, a red one, which is often present. Chlorophyll and the accessory pigments are not enclosed in chloroplasts, as they are in plant cells, but are scattered in a membrane system distributed in the peripheral portion of the cell.

The blue-green algae have a cell wall that does not contain cellulose, but is made up of the same sorts of polysaccharides linked with polypeptides that occur in the bacteria.

The cells lack cilia, flagella, or any other type of locomotive organelles, yet some filamentous blue-green algae are capable of gliding motion. Reproduction is by simple fragmentation or cell division. Like some of the bacteria, many species are able to form thick-walled spores in which they can lie dormant during periods unfavorable to growth.

Cells of the blue-green algae have an outer mucilaginous sheath, or coating. The outer sheath is often deeply pigmented, particularly in species that spread up onto the land, and their colors include a light golden yellow, brown, red, emerald green, blue, violet, and blue-black. In addition, the carotenoids and phycobilins modify the color of the cells in which they occur. Thus, despite their name, only about half of the blue-green algae are actually blue-green in

19–23 *Diagram and electron micrograph of the blue-green alga* Anabaena cylindrica. *Blue-green algae have no chloroplasts and no membrane-bound nucleus, such as are found in eukaryotic algae and plant cells. Photosynthesis takes place in chlorophyll-containing membranes scattered throughout the cell, and the nucleus is a single molecule of DNA. The three-dimensional quality of this electron micrograph is due to a technique known as freeze-etching. The specimen is frozen rapidly and split apart, and a replica is made of its surface. The electron micrograph is made of the replica.*

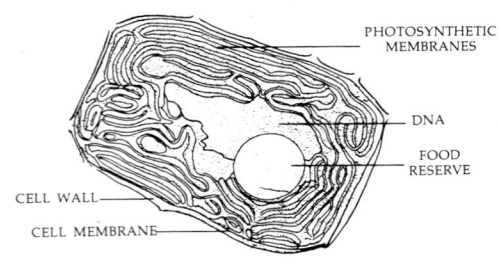

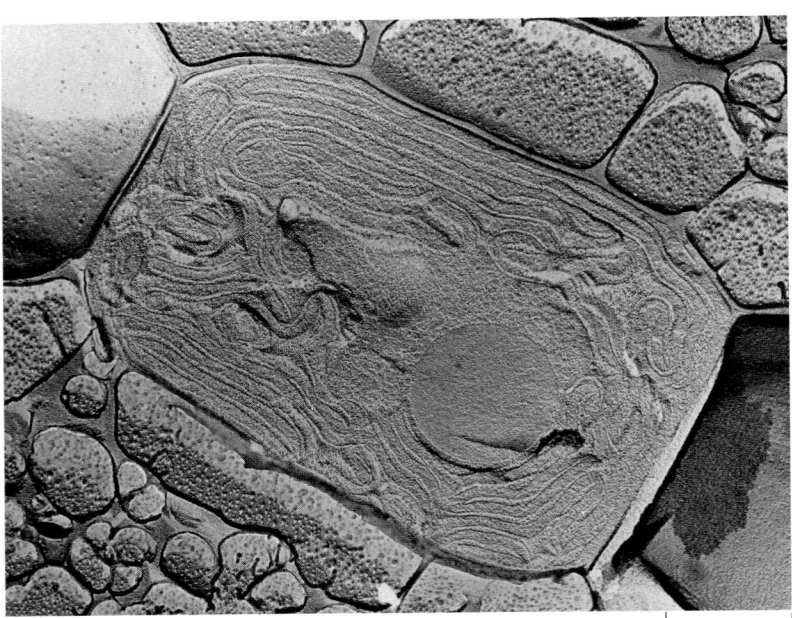

1 μm

color. Indeed, the Red Sea was named because of the dense concentrations, or "blooms," of red-pigmented blue-green algae that float on its surface.

Individual blue-green algae are microscopic, but they often grow in large masses as much as 1 meter or more in length. Some blue-green algae are unicellular, others are filamentous, a few form branched filaments, and a very few form plates or colonies.

Nutrition of Blue-Green Algae

Almost all species are photosynthetic. Many are also capable of nitrogen fixation. The photosynthetic, nitrogen-fixing blue-green algae have the simplest nutritional requirements of any living thing, needing only N_2 and CO_2, which are always present in the atmosphere, a few minerals, and water.

The ecological importance of the blue-green algae appears to be less than that of the nitrogen-fixing bacteria, at least for agriculture. However, in Southeast Asia, rice can be grown on the same land for years without the addition of fertilizers because of the rich growth of nitrogen-fixing blue-green algae in the rice paddies.

Because of their nutritional independence, the blue-green algae are able to colonize bare areas of rock and soil. A dramatic example of such colonization was seen on the island of Krakatoa in Indonesia, which was denuded of all visible plant life by its cataclysmic volcanic explosion of 1883. Filamentous blue-green algae were the first living things to appear on the pumice and volcanic ash; within a few years they had formed a dark-green gelatinous growth. The layer of blue-green algae eventually became thick enough to provide a substrate for the growth of higher plants. It is very probable that the blue-green algae were similarly the first colonizers of land in the course of biological evolution.

–24 *Two common genera of blue-green algae.* (a) Anabaena, *composed of barrel-shaped cells held in a gelatinous sheath, is capable of forming spores.* (b) Thio-thrix, *a genus that lacks chlorophyll. This species obtains energy by the oxidation of hydrogen sulfide. The filaments, which are attached to a substrate at one end, forming a characteristic rosette, are filled with sulfur droplets.*

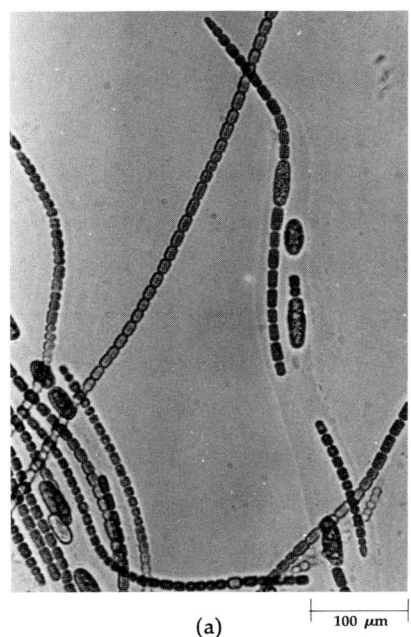

(a)   100 μm

(b)   50 μm

VIRUSES

Viruses do not fit easily into any of the traditional categories by which living organisms are classified, and the problem of categorizing them is made even more difficult by the fact that there is considerable doubt about whether or not they should be considered living. They are clearly not Monera. We discuss them in this chapter because they are traditionally studied with the bacteria, and, like bacteria, they are responsible for a number of diseases in animals and plants. Virus diseases of man include smallpox, chicken pox, measles, rubella (German measles), mumps, influenza, colds (often complicated by secondary infections caused by bacteria), infectious hepatitis, yellow fever, polio, and rabies. Other diseases, including cancer (see page 274) and multiple sclerosis, are suspected of being viral in origin. As we noted in the previous section, viruses are also important as research tools in molecular genetics.

Composition of Viruses

In size, viruses range from about 17 nanometers (a hemoglobin molecule is 6.4 nanometers in diameter) to about 300 nanometers, larger than small bacteria. The larger ones are at the limits of resolution of the light microscope.

Viruses are made up of nucleic acid, either DNA or RNA, enclosed in a protein coat. The protein coat determines the specificity of the virus; a cell can be infected by a virus only if that type of cell has a receptor site for the virus protein. Thus cold viruses infect cells in the mucous membranes of the respiratory tract; measles and chicken pox infect skin cells; polio infects the upper respiratory tract and intestinal and sometimes nerve cells. Even bacteria, as we noted in Section 2, have their own set of specific viruses, the bacteriophages. Fungi also have viruses, and recently a virus, termed a cyanophage, has been discovered that attacks blue-green algae.

In some virus infections, the protein coat is left outside the cell (see Figure 16–2, page 238); in others, the intact virus enters the cell, but once inside, the protein is destroyed by enzymes. The DNA of a DNA-containing virus serves as a template for more viral DNA and also for messenger RNA, which codes for viral enzymes, viral coat protein, and probably, at least in some cases, repressors and other regulatory chemicals. The virus uses the equipment of the host cell, including ribosomes, transfer RNA molecules, amino acids, and nucleotides (see page 255). Many viruses use host enzymes as well as their own, and some break up host DNA and recycle the nucleotides as viral DNA. In the case of the RNA viruses, the viral RNA not only serves as a template for more RNA but also acts directly as messenger RNA. As we noted in Chapter 18, viral RNA can also serve as a template for viral DNA. This phenomenon of reverse transcription is observed almost exclusively with cancer-causing viruses.

As with bacterial infections, virus infections do not usually kill their hosts. Rabies is the only virus infection of man that is regularly fatal if untreated.

Virus particles are assembled within the host cell. They then leave the cell, often lysing the membrane. Some viruses, such as influenza virus, bud off from the host cell membrane and, in so doing, become wrapped in fragments of it. Each new viral particle is capable of setting up a new infective cycle in an uninfected cell.

In animals, the infection cycle is usually checked by the appearance of anti-

19–25 (a) Adenovirus, one of the many viruses that cause colds in humans. This virus is an icosahedron. Each of its 20 sides is an equilateral triangle composed of identical protein subunits. Many viruses, and also Buckminster Fuller's geodesic domes, are constructed on this principle. There are 252 subunits in all. (b) A model of the adenovirus, made up of 252 tennis balls.

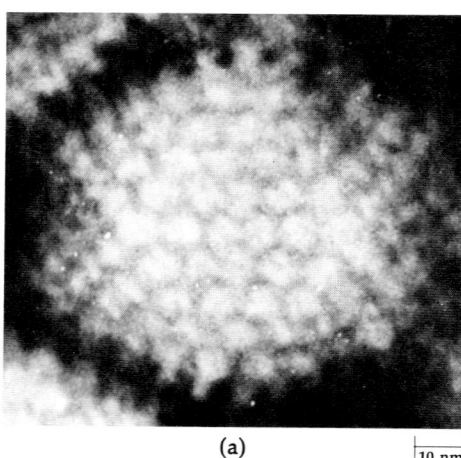

(a)

|10 nm|

(b)

bodies in the bloodstream of the host. Antibodies are specific protein molecules that combine with the virus particles and inactivate them. (The production of antibodies is discussed in greater detail in Chapter 33.) Because the antibody-producing cells respond very quickly to a second infection by the same agent, a virus infection usually confers immunity to reinfection for a varying length of time.

Control of Virus Infections

The most effective means for control of viral infections is immunization (or vaccination). The first vaccination was performed in 1796 by Edward Jenner. Jenner observed that persons who had had cowpox, a mild disease, did not develop smallpox, and he boldly undertook to inoculate well persons with cowpox (vaccinia). We know now that Jenner was successful because the protein of the vaccinia virus is similar enough to that of the smallpox virus to stimulate production of antibodies effective against both viruses.

Vaccines have now been produced against a number of other kinds of virus, using either killed viruses of the disease-causing strain or closely related but harmless live viruses. Viruses are not always susceptible to control by vaccination, however. The influenza virus, for reasons which are not understood, mutates frequently. The changes in its nucleic acid alter its protein coat and hence previously formed antibodies no longer "recognize" it. New strains of influenza virus are likely to arise more rapidly than new vaccines can be produced to combat them. The Pasteur Institute in Paris has recently announced a program to predict the changes that will take place in the virus in order to prepare vaccines against these new strains, thus staying one jump ahead of the virus.

Although immunization against viruses has been more widely successful than immunization against bacterial infections, chemotherapy (treatment by chemical means) has been much less so. No agents such as penicillin or sulfanilamide or any of the other antibiotics are effective against virus diseases. The reason for this lack of effectiveness is undoubtedly because the virus and the host cell are so intimately associated that a drug that affects one affects the other.

–26 (a) *Influenza virus* (A_2, *Hong Kong 1/68*). *The virus is surrounded by a lipoprotein envelope through which protrude stubby protein spikes.* (b) *A bacterial virus* (*T5 bacteriophage*), *showing the head, tail, and short tail fibers.*

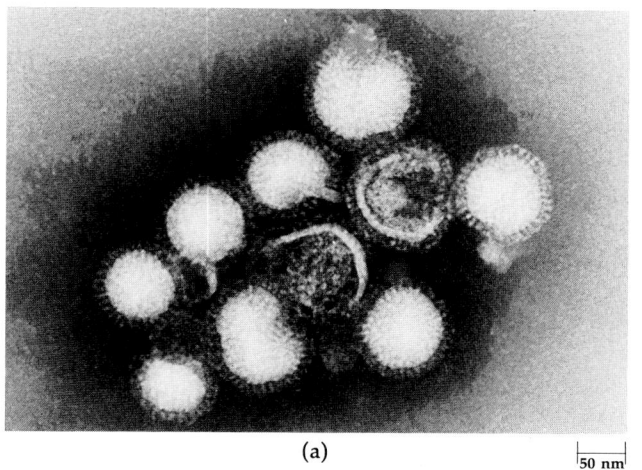

(a)

|50 nm|

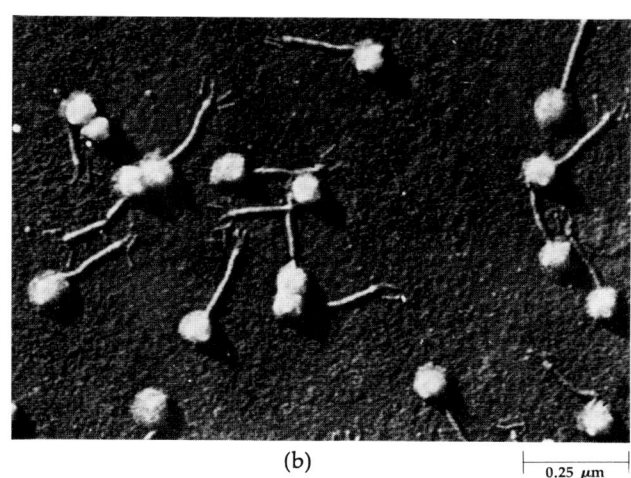

(b)

|0.25 μm|

19–27 *Until the 1950s, poliomyelitis was one of the most feared of all childhood diseases. It was brought under control first by a killed-virus vaccine (Salk vaccine), which was subsequently replaced by a live-virus vaccine (Sabin vaccine), the one now in wide general use. The possible after-effects of polio are depicted in this Egyptian hieroglyph, dating from the nineteenth dynasty (1320–1200 B.C.).*

Origin of Viruses

Viruses were once thought to be very simple forms of living things, and some biologists at one time believed that the earliest form of life on Earth may have been viruslike. More recently, knowledge of the architecture of viruses and of the capacity of viruses to become part of the host cell's chromosomes (page 270) make it seem more likely that viruses are cellular fragments that have set up a partially independent existence. S. E. Luria, who received the Nobel prize for his work with bacteriophages, has called viruses "bits of heredity looking for a chromosome."

SUMMARY

There are more than 10 million kinds of organisms in the biosphere. Scientists seek order in this vast diversity of living things by classifying them, that is, by grouping them in meaningful ways. A species is the basic unit of classification; species are made up of individuals that breed freely with members of the same species and not with those of others. Species are grouped into genera, genera into families, families into orders, orders into classes, classes into phyla, and phyla into kingdoms. Depending on the classification system used, the phyla may be grouped into as few as two kingdoms (plants and animals) or as many as five (Monera, Protista, Fungi, Plantae, and Animalia). The latter system is followed in this text.

Bacteria and blue-green algae, the two living groups of prokaryotic organisms, constitute the kingdom Monera. Prokaryotes lack an organized nucleus and cellular organelles, and do not undergo mitosis or meiosis. Their genetic material is contained in a single closed loop of DNA. Prokaryotes have rigid cell walls, which, unlike the cell walls of any other organisms, contain amino acids.

The bacteria share with the fungi the role of decomposers in the world ecosystem. They are versatile metabolically. Most bacteria are heterotrophic, but some are photosynthetic autotrophs, and others are chemosynthetic autotrophs. Some are aerobic, others are obligate anaerobes, and still others are facultative anaerobes. A number of genera play important roles in the cycling of nitrogen, sulfur, and other minerals.

Bacterial cells may be spherical (cocci), rod-shaped (bacilli), curved (vibrio), or spiral (spirilli). They may adhere in groups or filamentous masses if the cells do not separate after cell division. Some are nonmotile, some move by gliding, and others have flagella.

Bacteria reproduce by fission. Some form spores, which are resting stages. They do not reproduce sexually—that is, by meiosis and fertilization—but sometimes acquire new genotypes by exchanging genetic information (conjugation).

Photosynthetic bacteria, which are anaerobic, utilize hydrogen sulfide or another electron donor other than water and do not liberate oxygen. Chemosynthetic bacteria derive their energy from the oxidation of inorganic molecules. Some bacteria are important as pathogens.

Control of bacterial infection has improved greatly in the last few decades owing to improvements in sanitation and immunization procedures and to the development of antibiotics and other drugs. The effectiveness of chemo-

therapy (drug treatment) depends on metabolic differences between bacterial and normal cells.

The blue-green algae are photosynthetic prokaryotes in which chlorophyll *a* is the photosynthetic pigment, just as in the photosynthetic eukaryotes. They also have accessory pigments in the form of carotenoids and phycobilins, which aid in photosynthesis. These pigments, in various combinations, give the blue-green algae their characteristic colors. Their cell walls resemble those of the bacteria, but they lack pili or flagella. Like the bacteria, they are able to adapt to inhospitable environments, and some forms are able to fix nitrogen. A few blue-green algae are anaerobes.

Viruses are infectious agents composed of an inner core of nucleic acid, either RNA or DNA, and an outer protective coating of protein. They cannot reproduce themselves outside of living cells. In DNA viruses, the viral DNA serves as a template for more viral DNA and also for messenger RNA. In RNA viruses, the RNA, which is usually single-stranded, acts both as a template for viral RNA and as messenger RNA in the host cell. Both DNA and RNA viruses use host ribosomes, and new viral nucleic acid and protein are made from molecules available in the host cell.

QUESTIONS

1. Define the following terms: Monera, kingdom, antibiotic, virus, species.
2. Name the principal differences between prokaryotes and eukaryotes.
3. In what way is the species a different kind of a unit of classification from, say, a genus, order, or kingdom? How is the species a less arbitrarily assigned designation than any of the latter?
4. Before the structure of any virus was known, Crick and Watson predicted that the protein coats of viruses would prove to be made up of a large number of identical subunits. Can you explain the basis of their prediction?
5. Parasites often lack many of the structures characteristic of their nearest nonparasitic relatives. Why might this be so? How might this observation be used to support the notion that viruses or rickettsia, rather than being ancestral to more complex organisms, are instead derived from them?

Chapter 20

Kingdom Protista and Kingdom Fungi

In this chapter, we are going to discuss two large groups of organisms, the protists and the fungi. According to the classification system followed here, these constitute, respectively, the kingdom Protista and the kingdom Fungi. They are both eukaryotes, as are all plants and animals; the only extant prokaryotes are placed in the kingdom Monera, which consists of the bacteria and blue-green algae.

KINGDOM PROTISTA

The protists are all eukaryotic and unicellular; otherwise, they are an extremely heterogeneous group of organisms. The kingdom includes heterotrophs, among them a number of parasitic forms; photosynthetic autotrophs; and a few versatile organisms that are both heterotrophic and photosynthetic.

Protists usually reproduce asexually, by mitosis. Many also have sexual cycles, involving meiosis and the fusion of gametes, or *syngamy*. (Syngamy in animals is usually referred to as fertilization.) The fusion of gametes results in a diploid ($2n$) *zygote*. Among protists, the zygote is often a thick-walled, resistant resting cell, formed during periods of cold or drought. Some protists, notably the ciliated protozoans, have a special type of sexuality involving exchanges of genetic material between individuals.

Although protists are single cells, they are far from simple; in fact, many of them are extremely complex, with an astonishing variety of highly specialized structures found in no other types of living organisms. The kingdom as a whole might be viewed as representing a vast testing laboratory of evolutionary possibilities. In fact, it would be possible to set forth arguments that the kingdom Protista actually represents several kingdoms rather than a single one.

However, as their complexity reminds us, modern protists are not beginnings. They are in themselves the end products of millions of years of evolution. Those that have not changed much in that time—amoebas might be an example—have survived over the millennia because they are exquisitely adapted to the environ-

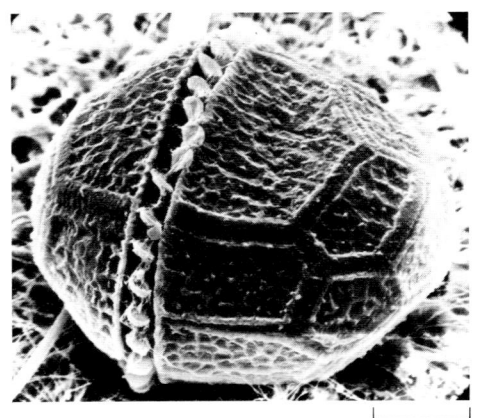

20–1 *Scanning electron micrograph of a photosynthetic protist, the dinoflagellate* Peridinium. *The stiff cellulose wall has two grooves in which two flagella beat, causing the cell to spin through the water. A coiled flagellum is visible here in the central groove.*

10 µm

20–2 *Comparison between a prokaryotic cell*
(a) and a eukaryotic cell (b). The eukary-
ote contains a membrane-bound nucleus;
complex membrane-bound organelles,
such as mitochondria and chloroplasts;
endoplasmic reticulum; and Golgi bodies.
Its chromosomes are a complex of DNA
and special proteins. Flagella, when pres-
ent, have a characteristic 9 + 2 struc-
ture. All these features are lacking in
prokaryotes.

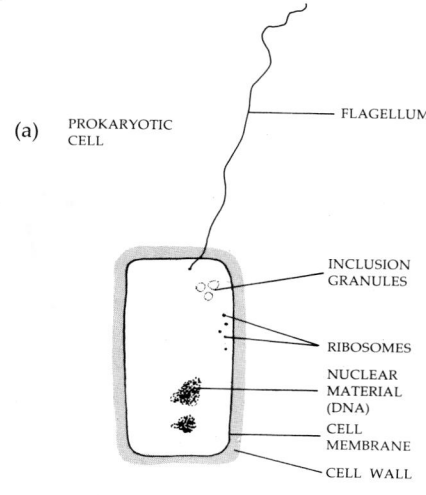

(a) PROKARYOTIC CELL
FLAGELLUM
INCLUSION GRANULES
RIBOSOMES
NUCLEAR MATERIAL (DNA)
CELL MEMBRANE
CELL WALL

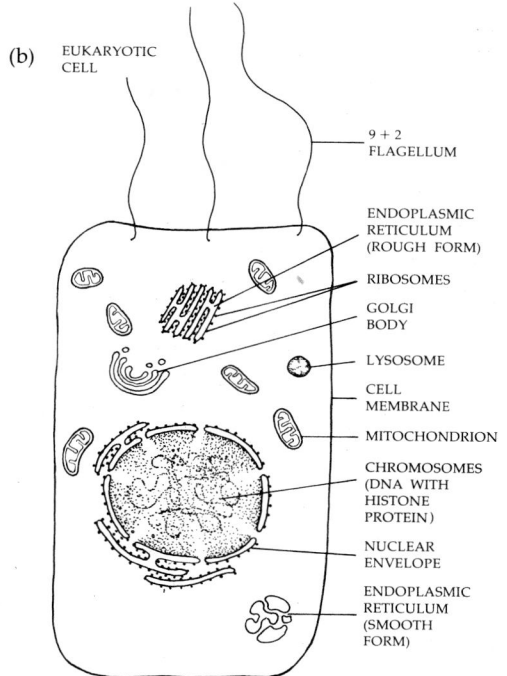

(b) EUKARYOTIC CELL
9 + 2 FLAGELLUM
ENDOPLASMIC RETICULUM (ROUGH FORM)
RIBOSOMES
GOLGI BODY
LYSOSOME
CELL MEMBRANE
MITOCHONDRION
CHROMOSOMES (DNA WITH HISTONE PROTEIN)
NUCLEAR ENVELOPE
ENDOPLASMIC RETICULUM (SMOOTH FORM)

ment in which they live. It is all a question of how one measures evolutionary success.

EVOLUTION OF THE EUKARYOTE

As we noted in Chapter 1, the earliest fossils, which are prokaryotes, have been dated at some 3½ billion years ago. The earliest forms that can be identified as eukaryotes appeared, again according to the fossil record, between 1.2 and 1.4 billion years ago—considerably later. There are many reasons to believe that the eukaryotes evolved from the prokaryotes. Perhaps the most compelling of these is that the genetic code and the way it is translated and transcribed is so strikingly similar in the two groups. Man and *Escherichia coli* even share some of the same enzyme systems. However, despite these similarities, the evolutionary jump from prokaryotic to eukaryotic cells is a large one—the largest such transition in the history of living organisms. It involved, for one thing, the development of complex chromosomes, in which the DNA is associated with special proteins (histones), and a nuclear envelope within which the chromosomes and their activities are segregated from the rest of the cell. The endoplasmic reticulum, which is continuous with the nuclear envelope in modern eukaryotes, probably evolved simultaneously. The evolution of mitosis and of the spindle apparatus and other machinery associated with it must also have taken place during this same period. The nuclear apparatus of certain protists is different in some details from that of higher organisms and will perhaps throw some light on these evolutionary steps.

Prokaryotes do not have mitochondria, as do virtually all eukaryotes, nor do they have chloroplasts or other plastids. In fact, the modern prokaryotes are often no larger than mitochondria and chloroplasts. One interesting current hypothesis (not universally accepted) suggests that these organelles originated as symbiotic prokaryotes. There is a body of evidence that supports this idea. Mitochondria and chloroplasts contain their own DNA, and this DNA, like the DNA of bacteria, is not associated with protein. Moreover, many of the same enzymes contained in the cell membranes of bacteria are found only in the mitochondrial membranes of the eukaryotes. Finally, mitochondria and chloroplasts are not synthesized and assembled entirely by the cell's own biosynthetic equipment, as are ribosomes (page 251), for example. Rather they appear to be self-replicating, arising only from the division of other mitochondria and chloroplasts. The hypothesis also gains support from the fact that there are numerous instances of blue-green algae and bacteria living as symbionts in modern eukaryotes. An alternative hypothesis is presented in Figure 20–3.

PHYLUM PROTOZOA

The Protozoa—"first animals"—are one-celled heterotrophs. Classification of the Protozoa into three of their four major groups is based upon their characteristically different methods of locomotion: (1) by flagellar movement (the zooflagellates–"animal flagellates"—or mastigophores), (2) by pseudopodia (the sarcodines), and (3) by ciliary movements (the ciliates). The fourth major group, the sporozoans, are nonmotile during the major phases of their lives and all are parasites.

20–3 *An hypothesis for the evolution of the eukaryotic cell. Invaginating (inward-turning) cell membranes enclose duplicates of a prokaryotic chromosome within double membranes. The structures so formed then, in the course of evolution, slowly take on separate specialized functions, becoming the eukaryotic nucleus, mitochondrion, and chloroplast and other plastids. An alternative hypothesis is presented in the text.*

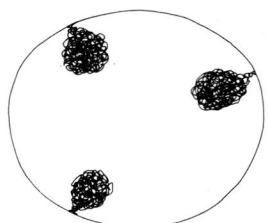

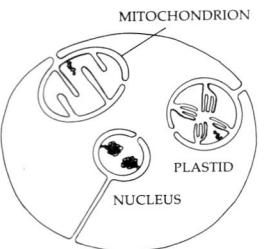

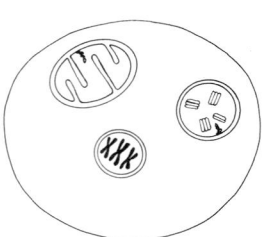

20–4 *Protists respond to chemical stimulus, as shown in these photographs of an amoeba in pursuit of prey. The initial stimulus produced by the "prey"—a fragment of* Hydra *tentacle—induces pseudopodia from the amoeba and causes the amoeba to move toward the* Hydra *fragment. As the piece of tentacle is moved away, the amoeba moves after it and will remain in pursuit so long as the prey is close enough for the stimulus to continue.*

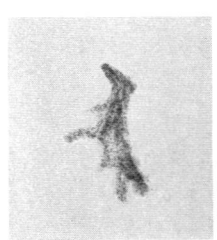 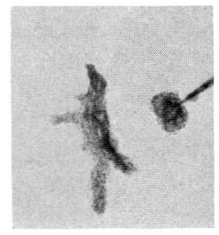

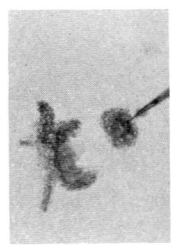 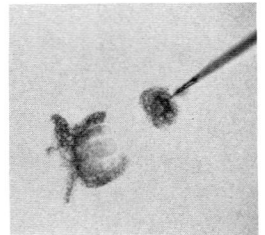

Class Mastigophora

The zooflagellates are regarded as the most primitive of the Protozoa and also as a link between the algae and the protozoans. They are thought to have been derived from photosynthetic forms, such as *Euglena* (page 313), which lost their chloroplasts. Almost all have one or two flagella, but some have more and some have none. These flagella have the characteristic 9 + 2 structure (page 102). The zooflagellates multiply asexually by mitosis and cell division. They generally have no outer wall, and some are able to form pseudopodia. Most are free-living, but some are parasitic. The latter include *Trypanosoma gambiense*, a flagellate that causes African sleeping sickness, and members of the genus *Trichonympha*, complex and beautiful flagellates (see Figure 6–14) that live as symbionts in the digestive tracts of termites, where they digest the wood ingested by the termite.

Class Sarcodina

The sarcodines are amoebalike organisms which have no coat or wall outside their cell membranes and which generally move and feed by the formation of pseudopodia. They take their name from the word "sarcode," coined in the early nineteenth century to describe the "simple, glutinous, and homogeneous jelly" of which, at one time, simple cells were thought to be composed. Despite their uncomplicated appearance, they are complex cells and are even capable of some complex behavior patterns, as when sensing and pursuing prey organisms.

Sarcodines are closely related to the mastigophores; some sarcodines may develop flagella during particular stages of their life cycle or under particular environmental conditions. They are found in both fresh water and salt water. Some are parasites, such as those that cause amoebic dysentery.

Reproduction may be asexual or sexual. Asexual reproduction takes place by cell division accompanied by mitosis in which the nuclear envelope usually does not break down. In sexual reproduction, the cells, which are diploid, undergo meiosis, forming gametes, which then fuse to form zygotes.

0–5 *Two sarcodines. (a) Actinosphaerium has numerous thin pseudopods (called axopods), which extend into the surrounding water. The pointed axopods function as a trap. When a food particle touches one of them, it becomes stuck in an adhesive secretion that coats the surface. The axopod is then drawn into the cell, particles and all. The axopods contain longitudinal arrays of microtubules (page 85) that disassemble and re-form as the axopod is withdrawn and reextended. (b) Foraminiferans begin life as amoebas and then secrete chalky shells around themselves, enlarging the shells with new chambers as they grow. When they die, their shells form an important part of the ooze that covers the ocean floor. Much of the limestone and chalk deposits in the earth are formed from foraminiferal shells.*

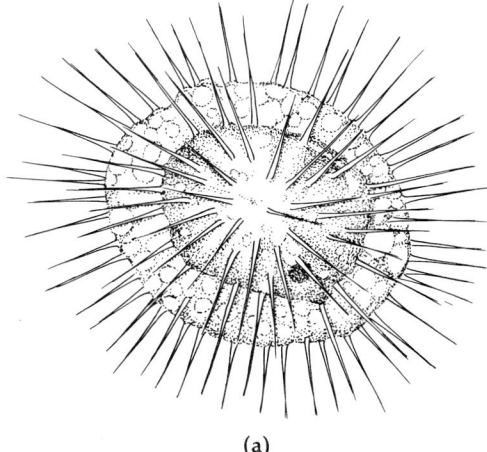

(a)

(b) |‾100 μm‾|

Table 20–1   *The Kingdom Protista*

Phylum Protozoa	Unicellular heterotrophs, including flagellates (mastigophores), amoebalike organisms (sarcodines), ciliates, and sporozoans
Phylum Euglenophyta	*Euglena* and related unicellular algae (some heterotrophs)
Phylum Chrysophyta	Diatoms and related unicellular algae (some heterotrophs)
Phylum Pyrrophyta	Dinoflagellates and related unicellular algae (some heterotrophs)
Phylum Gymnomycota	Slime molds

Many of the sarcodines cover their amoebalike characteristics with firm bright shells. Some, like *Arcella*, secrete a hard, keratinlike material, while others, like *Difflugia*, exude a sticky organic substance upon which forms a sort of haphazard mosaic made up of specks of sand and bits of the discarded shells of other minute creatures. Some, the "sun animals," resemble pincushions, with sharp silica spicules radiating from their soft bodies. *Actinosphaerium*, the protozoan shown in Figure 20–5, may reach as much as a millimeter in diameter and may sometimes be seen as a tiny white speck floating on the surface of a pond. Other sarcodines, the Foraminifera, have snail-like shells and live in the sea. Their shells are made of calcium carbonate, extracted from the seawater. The white cliffs of Dover and similar chalky deposits throughout the world are the result of the long accumulation of these discarded shells. The shells of the Foraminifera have been accumulating on the ocean bottom for millions of years, and in many areas, as a result of geologic changes, thick deposits of their skeletons (the foraminiferan ooze) can be found on the surface of the land or under later rock formations. Since the skeletons have evolved over this long period of time, it is possible to date a particular stratum by the type of Foraminifera that it contains, a fact that has proved of immense practical value in locating oil-bearing strata in Texas, Oklahoma, and many other oil-rich areas.

Class Ciliophora

The ciliophores, or ciliates, are the most highly specialized and complicated of the Protozoa, and indeed probably represent the most complex of all living cells. They are characterized by cilia. In some species, the cilia, all of which have the typical 9 + 2 structure, adhere to each other in rows, forming brushlike structures called membranelles or clumps of cilia called cirri, which can be used for walking or jumping. Cilia, membranelles, and cirri move in a coordinated fashion, although the way in which they are coordinated is not understood. Some ciliates also have myonemes, contractile threads. All have a complex skin, the cortex, which includes the cell membrane. In some groups, the cortex contains small barbs known as trichocysts, which are discharged when the cell is stimulated in certain ways.

The ciliates have another unusual feature: They have two kinds of nuclei, micronuclei and macronuclei. One or more of each kind is present in all cells.

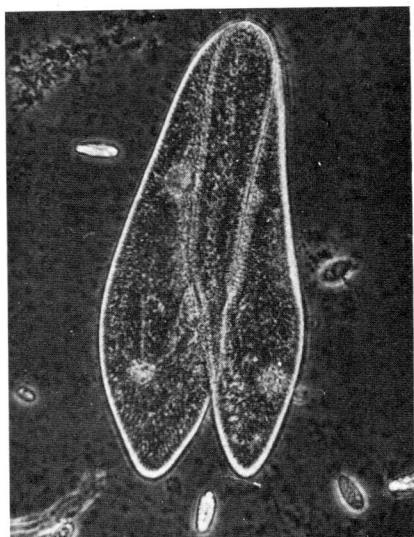

20-6 *Paramecia conjugating.*

|――― 100 μm ―――|

They also have a complex system for exchange of genetic information, in which cells conjugate and the micronuclei undergo meiosis. Cells then exchange haploid micronuclei which fuse, so that each cell has a new diploid micronucleus, which then divides. The old macronucleus dissolves, and a new macronucleus develops from one of the new micronuclei. The macronucleus in certain ciliates contains 50 to 100 times as much DNA as the micronucleus and so is believed to represent multiple copies of it. This view is supported by the fact that, in many ciliates, a cell can survive indefinitely without a micronucleus if even a small portion of a macronucleus is left, although it cannot conjugate. However, it cannot live without a macronucleus, even if it has a micronucleus. The macronucleus does not divide mitotically but is apportioned approximately equally between dividing cells as they constrict and separate.

The complexity of the conjugation process among ciliates reminds us again of the expenditure of energy and other resources involved in effecting exchanges of genetic information among individuals of the same species, from bacteria to man. The biological cost of these activities is an indication of the survival value to a species of the genetic variability produced by these exchanges.

About 6,000 species of ciliates are known, both freshwater and saltwater forms. Almost all are free-living (nonparasitic).

20-7 *Drawing of a paramecium, a ciliate. The body of this protist is completely covered by 9 + 2 cilia, although only a relative few are shown here. Like other ciliates, a paramecium feeds largely on bacteria, smaller microorganisms, and other particulate matter. The beating of specialized cilia drive particles into the gullet, where they are formed into food vacuoles. The food is digested in the vacuoles, and the undigested matter, still in vacuoles, is emptied out through the anal pore. As the contractile vacuoles indicate (page 113), members of the genus* Paramecium *live in fresh water.*

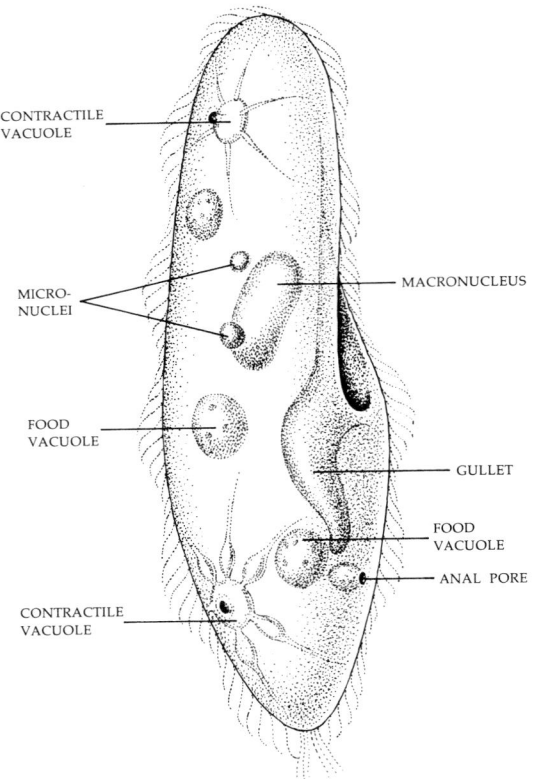

CONTRACTILE VACUOLE
MICRO- NUCLEI
MACRONUCLEUS
FOOD VACUOLE
GULLET
FOOD VACUOLE
ANAL PORE
CONTRACTILE VACUOLE

20–8 Stentor, *a ciliate. Extended, the cell looks like a trumpet (this genus is named after "bronze-voiced" Stentor of the Iliad, "who could cry out in as great a voice as 50 other men"). Stentor is crowned with a wreath of membranelles. These beat in rhythm, creating a powerful vortex that draws edible particles up to and into the funnel-like groove leading to the cytostome ("mouth"). Elastic protein threads, myonemes, run the length of the body. When contracted, Stentor is an almost perfect sphere. The macronuclei are visible in this micrograph.*

20–9 *Some examples of behavior in protists. (a) If a Stentor is annoyed, by a cloud of ink particles, for example, it will first bend away, perhaps repeatedly. If the offensive stimulus persists, the Stentor will reverse its cilia and try to sweep the particles away. If bending and sweeping are not successful, it contracts and waits. Once it has contracted, it does not bend or sweep again, but it may reach out to sample the water several times before it finally swims away. The length of time it tolerates the noxious stimulus apparently depends on whether or not its site had previously proved a good feeding area. Thus, even ciliates show some flexibility in behavior. (b) Avoidance behavior in the paramecium. The dark area at the top of the figure represents a drop of a toxic substance and the arrows indicate the direction of movement. The paramecium tests the substance, backs up, turns 30°, and starts forward again in a new direction. Many protists are capable of this sort of simple behavior.*

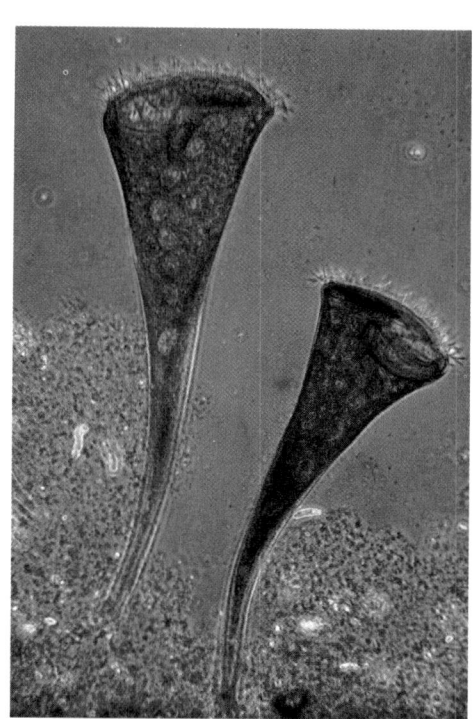

100 μm

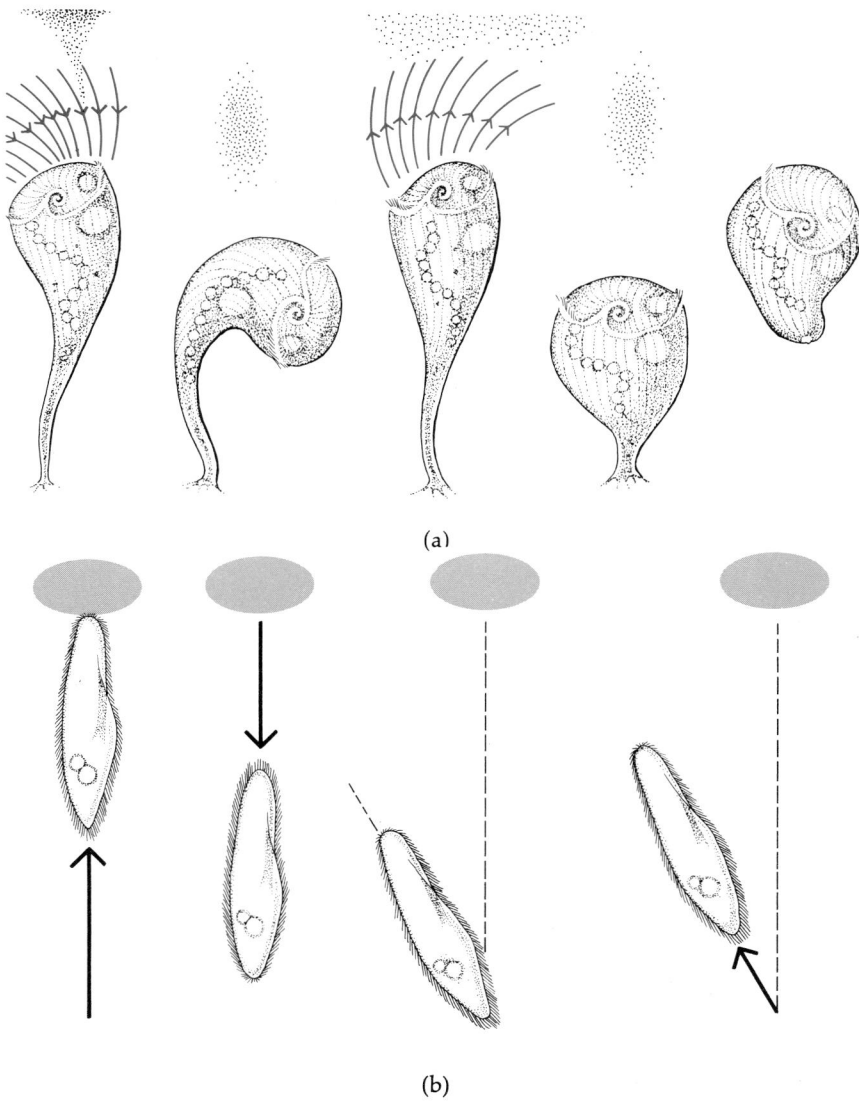

(a)

(b)

Class Sporozoa

The fourth class of Protozoa is the Sporozoa, all of which are parasitic. They are characterized by the lack of cilia or flagella in adult forms and by their complex life cycle. A single sporozoan undergoes multiple fission, dividing into numerous smaller cells (the spores) simultaneously.

The best known sporozoans are members of the genus *Plasmodium*, which cause malaria. *Plasmodium* is passed back and forth between man and the *Anopheles* mosquito. The female mosquito requires blood for the development of her eggs—the male lives on a more sybaritic diet of nectar. If the female draws her blood from a person with malaria, she will pick up *Plasmodium* cells which multiply within her body, although they do not harm her, and travel to her salivary glands. When she bites her next victim, she injects a droplet of salivary fluid under his skin. This fluid seems to anesthetize the victim against the bite; it also keeps the blood from clotting so that it will flow freely through her fine proboscis. Ultimately, through an inflammatory reaction, it raises the familiar welt. This drop of salivary fluid also carries *Plasmodium*, which eventually enter the blood cells and begin to multiply. The *Plasmodium* break out of the blood cells at regular intervals—usually every 48 or 72 hours, depending on the species—which is why malaria is characterized by recurrent bouts of chills and fever. The life cycle of the parasite is shown in Figure 20–10.

20–10 *Life cycle of* Plasmodium vivax, *the sporozoan that causes malaria in man. The cycle begins* (a) *when a female Anopheles mosquito bites a person with malaria and, along with his blood, sucks up gametes* (b) *of the protozoan. In the mosquito's body, the gametes unite* (c) *and form a zygote* (d). *From the zygotes, multinucleate structures called oocysts develop* (e), *which, within a few days, divide into thousands of very small, spindle-shaped cells, sporozoites* (f), *which then migrate to the mosquito's salivary glands. When the mosquito bites another victim* (g), *she infects him with the sporozoites. These first enter liver cells* (h), *where they undergo multiple division* (i). *The products of these divisions (merozoites) enter the red blood cells* (j), *where again they divide repeatedly* (k). *They break out of the blood cells* (l) *at regular intervals of about 48 or 72 hours, producing the recurring episodes of fever characteristic of the disease. After a period of asexual reproduction, some of these merozoites become gametes and, if they are ingested by a mosquito at this stage, the cycle begins anew.*

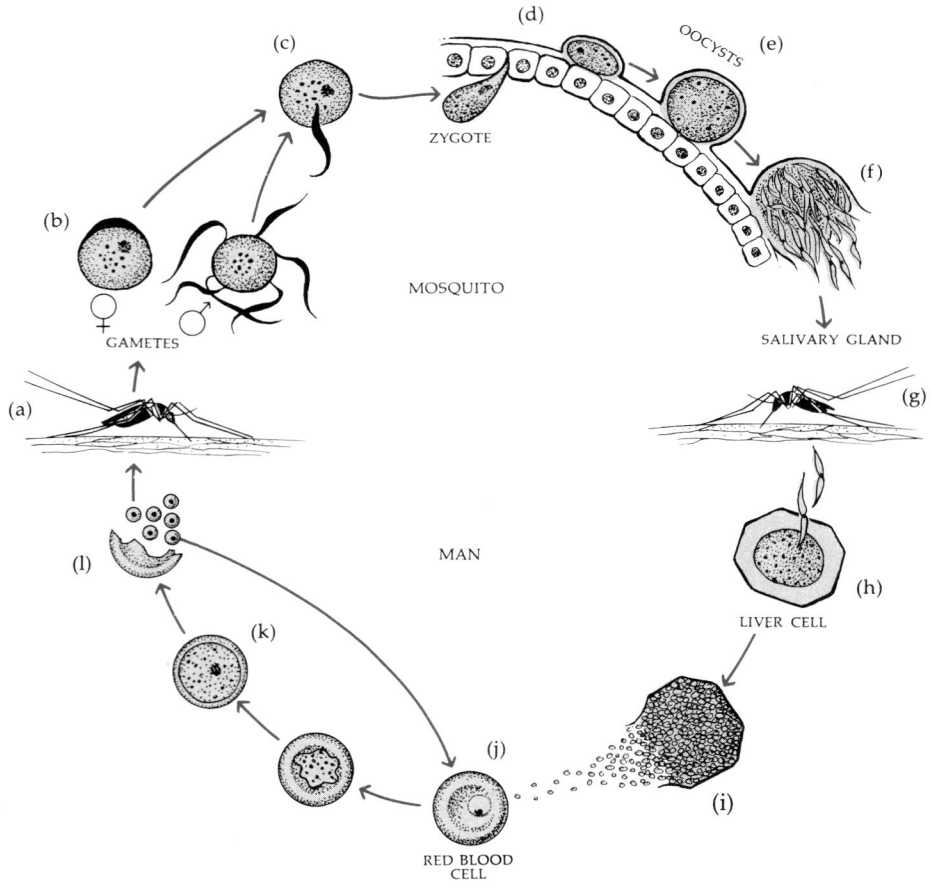

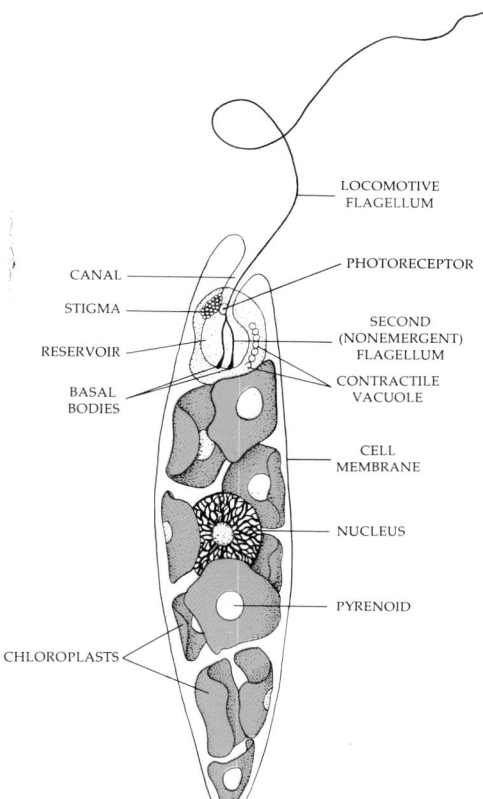

0–11   Euglena *is one of the most versatile of all one-celled organisms. Containing numerous chloroplasts, it is photosynthetic, but it can also absorb organic nutrients from the medium in which it lives. The pyrenoid is believed to be a food storage body.* Euglena *moves by the whiplike motion of a single flagellum.*

LOCOMOTIVE FLAGELLUM

CANAL

STIGMA

RESERVOIR

BASAL BODIES

CHLOROPLASTS

PHOTORECEPTOR

SECOND (NONEMERGENT) FLAGELLUM

CONTRACTILE VACUOLE

CELL MEMBRANE

NUCLEUS

PYRENOID

–12   *A cell of* Euglena, *broken open, showing the flexible protein strips that make up the pellicle. The large organelle at the bottom right is a chloroplast. The spongy network to the left is endoplasmic reticulum. At the top left is a broken mitochondrion, identifiable by its characteristic cristae.*

THE ONE-CELLED ALGAE

Algae are photosynthetic organisms that are either unicellular or have relatively simple multicellular structures (compared to land plants) and are adapted to life in water. The more than 20,000 species of eukaryotic algae are grouped into six phyla, which are considered to represent a series of parallel evolutionary lines. There are three phyla of unicellular forms—the Euglenophyta, the Chrysophyta, and the Pyrrophyta—and these are included in the kingdom Protista.

Phylum Euglenophyta

Euglenophyta is a small group of unicellular organisms (about 450 species), most of which occur in fresh water. They are named for the genus *Euglena*, the most common of the group. Characteristically, they have one or two flagella. Their chloroplasts contain chlorophylls *a* and *b* like those of higher plants. They store their food as paramylum, a polysaccharide not found in any other group of organisms. The cells lack a cell wall but have a flexible series of protein strips, which make up the pellicle, inside the cell membrane. They reproduce asexually, dividing longitudinally to form two new cells that are mirror images of one another. Superficially they resemble the Mastigophora.

The cells are complex. *Euglena* is characteristically an elongated cell with a single nucleus, two flagella (one of which is rudimentary), and numerous small emerald-green chloroplasts that give the cell its bright color. The flagella are attached at the base of the flask-shaped opening, the reservoir, at the anterior end of the cell. Emptying into the reservoir is the contractile vacuole, which collects excess water from all parts of the cell and discharges it into the reservoir. Unlike the stiff walls of the cells of higher plants, the flexible pellicle

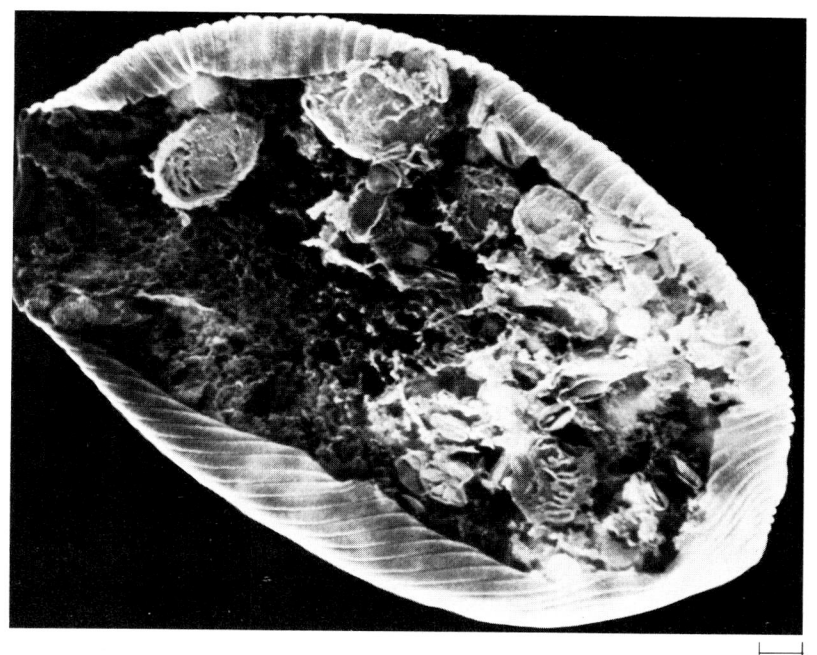

1 μm

permits *Euglena* to change its shape, providing an alternative means of loco-motion for mud-dwelling forms. No other phylum of photosynthetic organisms has a contractile vacuole or a pellicle, which are features that are common among the protozoans.

If one leaves a culture of *Euglena* near a sunny window, a clearly visible green cloud will form in the water, and this will move as the light changes. *Euglena* is probably able to orient with respect to light because of a pair of special structures: the stigma, or eyespot, which is a patch of pigment, and an underlying photoreceptor. The exact way in which these function is not known.

Some nonphotosynthetic protists—members of the genus *Peranema*, for ex-ample—closely resemble *Euglena* but lack chloroplasts. In fact, if some strains of *Euglena* are kept at an appropriate temperature and in a rich medium, the cells may replicate faster than the chloroplasts, producing nonphotosynthetic cells which can survive indefinitely in a suitable medium containing a carbon source. It is tempting to speculate that some modern heterotrophs arose from autotrophs in an analogous manner early in evolutionary history.

Phylum Chrysophyta: "Golden" Algae

The chrysophytes are the "golden" algae. There are 6,000 to 10,000 species, most of which are diatoms. Diatoms are the principal component of phyto-plankton, the photosyntheic cells that float near the surface of inland and oceanic waters and that are the primary source of food for all water-dwelling animals and for some land animals as well. Members of this group have several identifying characteristics: (1) their photosynthetic pigments are chlorophyll *a* and chlorophyll *c*, which closely resembles the chlorophyll *b* of higher plants; (2) they contain a yellow-brown carotenoid, fucoxanthin, which functions as an accessory pigment and gives them their characteristic color; (3) their cell walls, which contain no cellulose, are often impregnated with silicon compounds and thus are very rigid; (4) they store food in the form of oil rather than as starch. Because of these oil reserves, fresh water containing large amounts of these algae may have an unpleasant oily taste.

20-13 *Diatoms, chrysophytes. (a) Side view showing the characteristic intricately marked shell. (b) Pillbox type seen from above and from the side. Notice that one cell is dividing.*

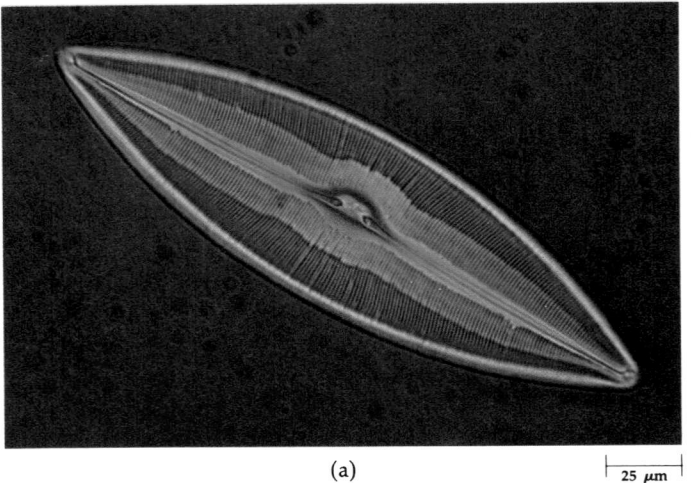

(a)   25 μm

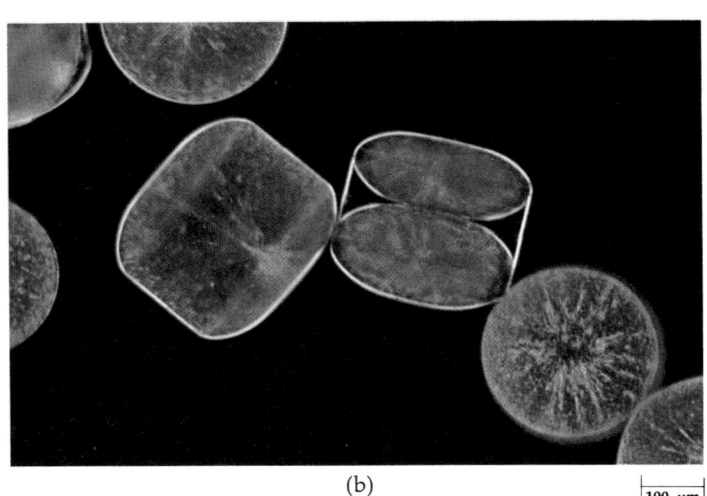

(b)   100 μm

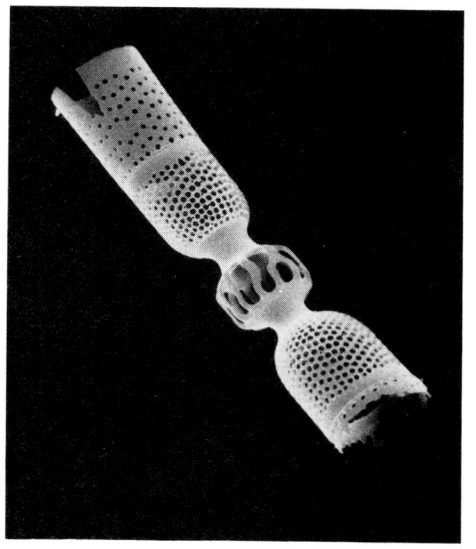 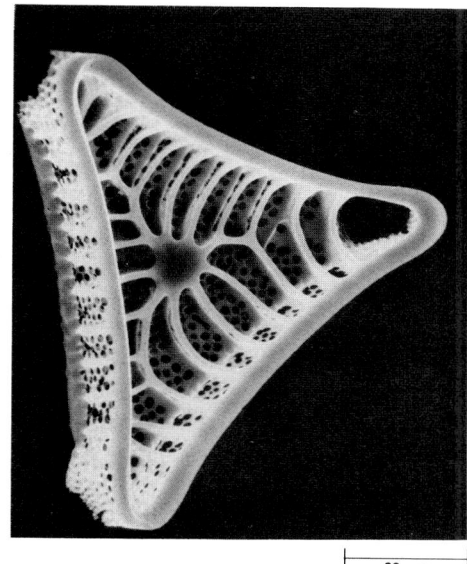

0–14 *The silicon-containing walls of diatoms as shown by the scanning electron microscope. Each species has its characteristic pattern of perforations in the walls.*

30 μm   30 μm

Diatoms are enclosed in a fine double shell, the two halves of which fit together, one on top of the other, like a carved pillbox. The delicate markings of these shells, by which the species are identified, were traditionally used by microscopists to test the resolving power of their lenses. Electron microscopy has shown that these fine tracings in diatom shells actually represent minute, intricately shaped pores or passageways connecting the protoplasm within the cell to the exterior environment. The piled-up silicon shells of diatoms, which have collected over millions of years, form the fine, crumbly substance known as "diatomaceous earth," used as an abrasive in silver polish and toothpaste and for filtering and insulating materials.

Some other members of the phylum lack cell walls and are amoeboid. Except for the presence of chloroplasts, the amoeboid cells are indistinguishable from amoeboid protozoans (the sarcodines), and the two groups may be closely related.

Phylum Pyrrophyta: "Fire" Algae

The Pyrrophyta is also largely composed of single-celled algae, the dinoflagellates ("spinning flagellates"), of which about 1,000 different species are known, almost all of them marine forms. Like the diatoms, they are important components of the phytoplankton. Other members of the phylum are heterotrophs, clearly related to the photosynthetic forms.

The dinoflagellates usually have two flagella that beat within grooves: one encircling the body like a belt and a second lying perpendicular to the first. The beating of the flagella in their respective grooves causes the cells to spin like tops as they move through the water.

Many of the dinoflagellates are bizarre in appearance, with a stiff cellulose wall (theca) which often looks like a strange helmet or an ancient coat of armor.

These dinoflagellates are often red in color (hence the name of the phylum), and the infamous red tides, in which thousands of fish die, are caused by great blooms of red dinoflagellates. The poison in these red tides has been traced to one species of armored dinoflagellate, *Gonyaulax catanella*. It is such an extraordinarily powerful nerve toxin that 1 gram of it would be enough to kill 5 million mice in 15 minutes. Blooms of *Gonyaulax catanella* appear regularly on the Pacific Coast and the Gulf of Mexico and have recently been reported off the New England coast. Mussels ingest the algae and concentrate the poison; the mussels then become dangerous for consumption by vertebrates, including man. Many dinoflagellates are bioluminescent, converting chemical energy into flashes of light that dance along the crests of the waves.

Dinoflagellates have a unique type of mitosis which appears to combine features of both eukaryotes and prokaryotes. Their cellular organization is eukaryotic, but their nuclei retain a number of prokaryote features. For example, the chromosomes of dinoflagellates are always condensed (and so visible) and undergo no additional condensation prior to mitosis. They are not associated with histone proteins. The nuclear membrane persists during mitosis, and apparently the chromosomes become attached to the membrane. At the time of mitosis, cytoplasmic channels invade the dividing nuclei. These channels contain bundles of microtubules, similar to the microtubules of the eukaryotic spindle. The microtubules, which remain entirely outside the nuclear membrane, may regulate the separation of the portions of nuclear membrane with the attached chromosomes. Thus, according to these recent findings by Donna Kubai and Hans Ris of the University of Wisconsin, the dinoflagellates appear to be a sort of living fossil, an intermediate step in evolution between the prokaryotic and the eukaryotic nucleus.

20–15 *Dinoflagellates.* (a) Ceratium tripos, *an armored dinoflagellate. You can see one flagellum in motion.* (b) Noctiluca scintillans, *a bioluminescent marine dinoflagellate. Yellow-brown diatoms that have been ingested can be seen inside the cell.*

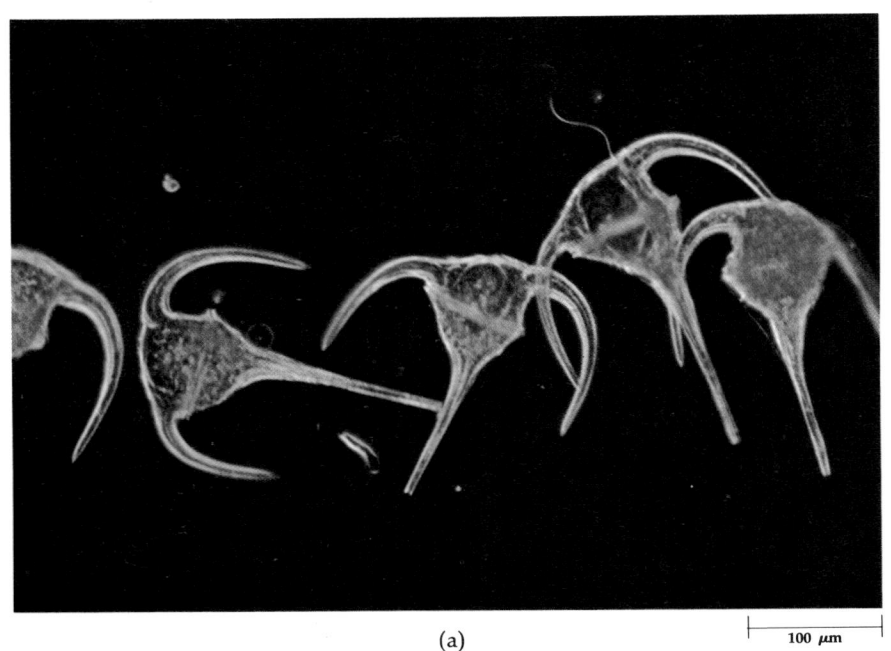

(a) |⎯ 100 μm ⎯|

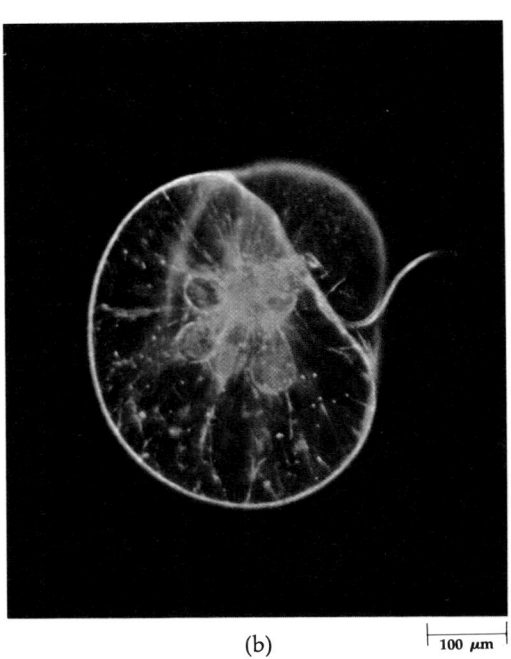

(b) |⎯ 100 μm ⎯|

−16 Mitosis in dinoflagellates appears to combine features of both eukaryotes and prokaryotes. Their cellular organization is eukaryotic, but their nuclei retain several prokaryote features. For example, as shown in (a), their chromosomes are condensed in interphase. Also, the nuclear membrane persists during cell division, and apparently the chromosomes become attached to it (b), as do the chromosomes of bacteria. During mitosis, cytoplasmic channels, as shown in the middle of (b), containing bundles of microtubules (similar to the microtubules of the eukaryotic spindle apparatus), invade the dividing nuclei. These microtubules are all oriented in one direction; they may direct the flow of cytoplasm through the channels and, in this way, regulate the separation of the portions of nuclear membrane with the attached chromosomes. The organism shown here is Gyrodinium cohnii.

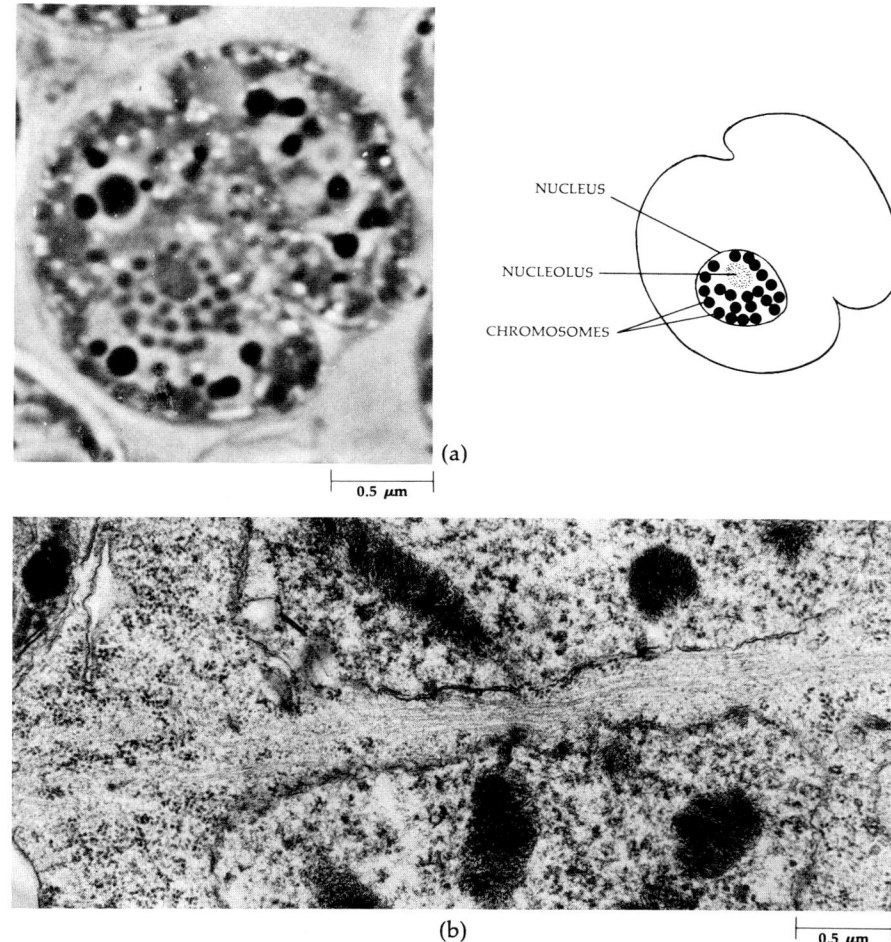

(a)

0.5 μm

NUCLEUS

NUCLEOLUS

CHROMOSOMES

(b)

0.5 μm

Table 20–2 Photosynthetic Protists

Phylum	No. of species	Photosynthetic pigments	Storage food	Flagella (9 + 2 structure)	Cell wall	Sexual reproduction	Distribution
Euglenophyta (euglenoids)	450	Chlorophylls *a* and *b*, carotenoids	Paramylum (a polysaccharide)	1, 2, or 3, at one end	No cell wall; protein pellicle	Unknown	Mostly freshwater
Chrysophyta ("golden" algae, mostly diatoms)	6–10,000	Chlorophylls *a* and *c*, carotenoids, fucoxanthin	Oils and leucosin (a polysaccharide)	1 or 2, at one end	Pectic compounds, often impregnated with silicon	Rare	Marine, freshwater, soil
Pyrrophyta ("fire" algae, mostly dino-flagellates)	1,100	Chlorophylls *a* and *c*, fucoxanthin	Starch and oils	2, lateral	Cellulose	Rare	Marine and freshwater

Phylum Gymnomycota: Slime Molds

The slime molds are a group of curious organisms classified with the protists—although some of them are multicellular in some stages—because of their similarity to protozoans, especially the amoebas. Two main groups (orders) are known: the plasmodial slime molds (Myxomycetes) and the cellular slime molds (Acrasiomycetes).

Most slime molds live in cool, shady, moist places in the woods—on decaying logs, dead leaves, or other damp organic matter. One of the common species (*Physarum cinereum*), however, is sometimes found creeping across city lawns. The plasmodia come in a variety of colors and can be spectacularly beautiful. The function of the pigments is not known with certainty, but they are probably photoreceptors because only slime molds with pigmented plasmodia require light for spore production.

During their nonreproductive stages, the Myxomycetes, or "true" slime molds, are thin streaming masses of protoplasm which creep along in amoeba fashion. As one of these plasmodia travels, it engulfs bacteria, yeast, fungal spores, and small particles of decayed plant and animal matter, which it digests. It may grow to weigh as much as several ounces, and, since slime molds are spread thinly, several ounces can cover an area several feet in diameter. The plasmodium contains many nuclei but is not partitioned by cell walls. As the plasmodium grows, the nuclei divide repeatedly and, in the early stages, synchronously.

Plasmodial growth continues as long as an adequate food supply and moisture are available. When either of these are in short supply, the plasmodium separates into many mounds of protoplasm, each of which develops into a mature sporangium (a structure in which spores develop) borne at the tip of a stalk. Meiosis takes place and cell walls form around the individual haploid nuclei to produce spores, which are resistant resting forms.

The spores germinate under favorable conditions, and each spore, depending on the species, produces one to four haploid flagellated cells. Some of these cells fuse to form a zygote, from which a new plasmodium develops.

20–17   (a) *Plasmodium of a slime mold. Such a plasmodium can pass through a piece of silk or filter paper and come out the other side apparently unchanged.*
(b) *Sporangia of a plasmodial slime mold on a rotting log.*

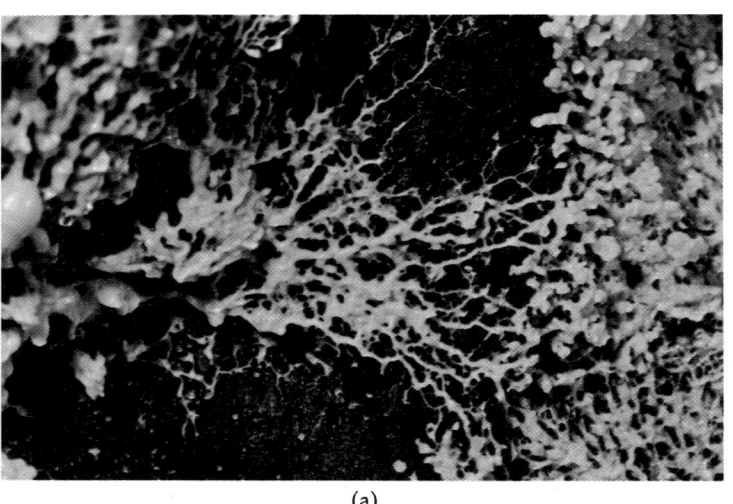

(a)

(b)

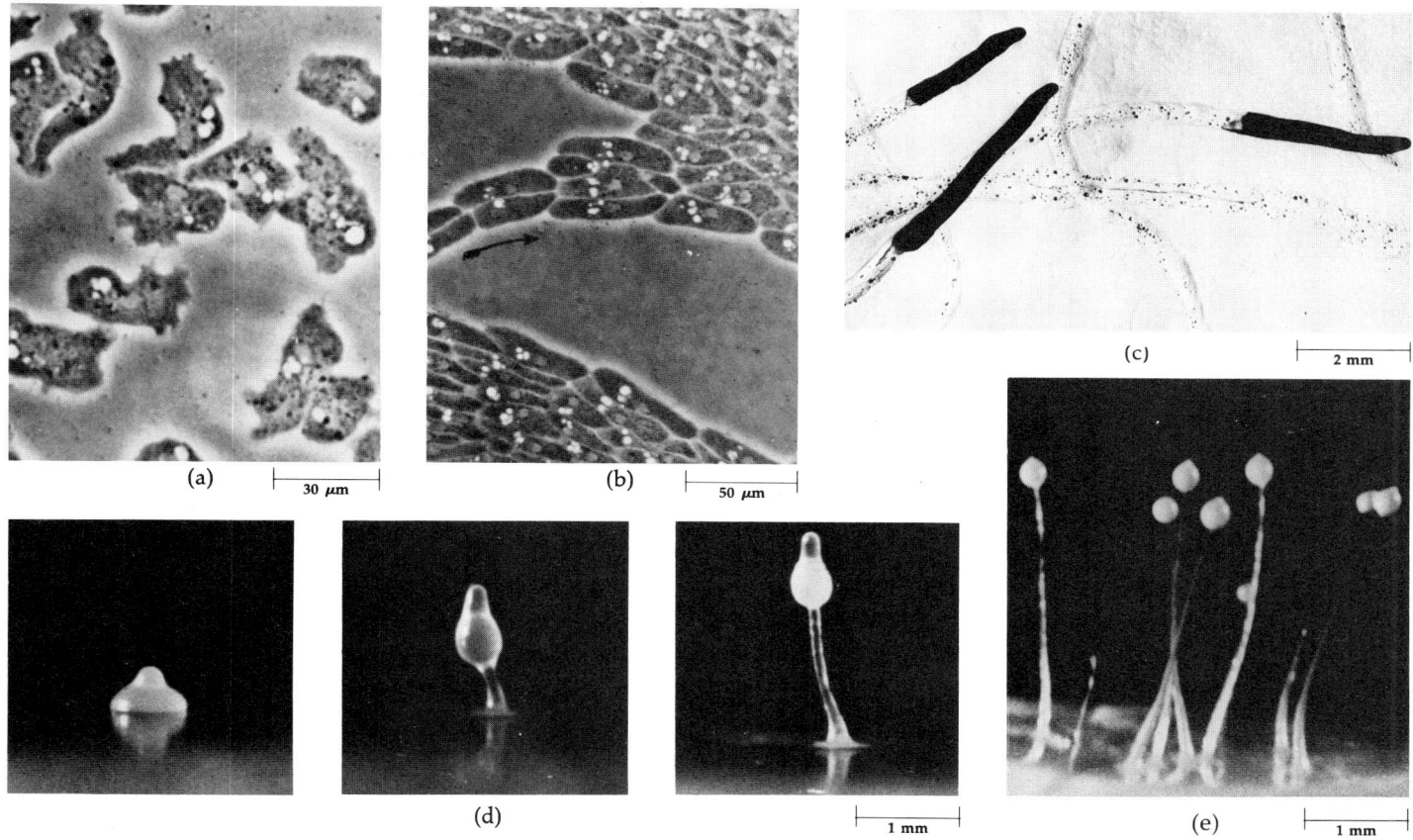

(a)   30 μm

(b)   50 μm

(c)   2 mm

(d)   1 mm

(e)   1 mm

20–18   *Life history of a cellular slime mold, Dictyostelium discoideum. (a) Amoebas in feeding stage. The light-gray area in the center of each cell is the nucleus, and the white areas are contractile vacuoles. (b) Amoebas aggregating. The direction in which the stream is moving is indicated by the arrow. (c) Migrating pseudoplasmodia. Each "slug" deposits a thick slime sheath which collapses behind it. (d) At the end of the migration, the pseudoplasmodium gathers together and begins to rise vertically, differentiating into a stalk and sporangium (e).*

The Acrasiomycetes, or cellular slime molds, have a similar life cycle. They also begin as amoebalike organisms but differ from the plasmodial slime molds in that the amoebas, on swarming together, do not lose their cell membranes but retain their identity as individual cells. The life history of this interesting group of organisms is shown in Figure 20–18.

KINGDOM FUNGI

The fungi are a group of organisms so unlike any others that, although they were long classified with the plants, it has come to seem appropriate to assign them to a separate kingdom. Except for some one-celled forms, such as the yeasts, the fungi are basically composed of masses of filaments. A fungal filament is called a *hypha*, and all the hyphae of a single organism are collectively called a *mycelium*. The mycelium may appear as a mass on the surface of the nutrient or may be hidden beneath the surface. The complex, spore-producing structures of fungi, such as mushrooms, are composed of tightly packed hyphae. In most groups of fungi, the cell walls are composed primarily of chitin, a polysaccharide that is never found in the kingdom Plantae (it is, however, the principal component of the exoskeletons, the hard outer coverings, of insects). Protoplasm containing the fungal nuclei and cytoplasmic organelles flows within the hyphae.

20–19 *Fungal cell walls characteristically contain chitin (a) rather than cellulose (b), the polysaccharide found in the cell walls of plants. Chitin resembles cellulose in that it is tough, inflexible, and insoluble in water. As you can see, they are structurally very similar; in chitin, the hydroxyl (OH) group in the 2 position is replaced by a nitrogen-containing group.*

(a)

(b)

In some groups, the mycelia are septate—divided by cell walls—but the walls, or septa, are perforated, and the cytoplasm and anything contained in it may be able to pass through the septa. Only the reproductive structures are completely separated. Thus the body of a fungus is multinucleate, without distinct cell boundaries.

All fungi are heterotrophs. Because of their filamentous form, each fungal cell is no more than a few micrometers from the earth or water or other substance in which the fungus lives, and is separated from it only by a thin cell wall. The fungi obtain food passively by absorption of organic compounds in a dissolved form. They characteristically digest organic materials first by means of extracellular enzymes, which they secrete into the food mass. Growth is their only form of motility (except for sex cells or spores, which may travel through water or air). They are capable of astonishingly rapid growth, as evidenced in the overnight appearance of a lawnful of mushrooms. (There is no scientific distinction between mushrooms and toadstools, although the former is sometimes used to refer to edible forms and the latter to inedible or poisonous ones. We shall call them all mushrooms—which should not be taken to mean that all are edible.)

The fungi, together with the bacteria, are the principal decomposers of the world, and as we shall see in Section 6, their activities are as vital to the continued survival of higher forms of life as are those of the food producers. Some are also destructive; they may interfere with the activities of man by attacking his foodstuffs, his domestic plants and animals, his shelter, his clothing, and even his person.

20–20 *Fungal mycelium growing on pine bark on the underside of a log.*

REPRODUCTION IN THE FUNGI

Fungi reproduce both asexually and sexually. Asexual reproduction takes place either by the fragmentation of the mycelium (with each fragment becoming a new individual) or by the production of spores. In some of the fungi, spores are produced in reproductive structures called *sporangia*, and are borne on specialized hyphae called *sporangiophores* or, sometimes, fruiting bodies. A *spore*, by definition, is a cell which is capable of developing into an adult organism without fusion with another cell; thus it contrasts with a gamete, which must unite with another gamete to form a zygote, which then develops into an adult individual. Spores are also often, but not necessarily, resting forms, surrounded by a tough resistant wall and able to survive during periods of lack of water and extreme temperatures. Some airborne spores are very small and so can

remain suspended in the air for long periods and be carried for great distances. Often the sporangia are raised above the mycelium; thus the spores are easily caught up and transported by air currents. The bright colors and powdery textures associated with many particular types of molds are produced by the spores. Usually, the mycelium is located beneath the surface of the material upon which the fungus is growing.

[Sexual reproduction is often initiated by the coming together of hyphae of different mating strains. Either before or after they come in contact, the touching tips of the hyphae may develop into specialized reproductive structures, the *gametangia* (see Figure 20–22). The gametangia fuse, the nuclei unite, forming a zygote, and meiosis takes place.]

Sometimes the fusion of hyphae is not followed immediately by the fusion of nuclei. Thus strains of fungi may exist with two genetically distinct kinds of nuclei operating simultaneously. Such a combination, known as *dikaryosis*, is found uniquely among the higher fungi (Ascomycetes and Basidiomycetes).

PHYLUM MYCOTA

The kingdom Fungi consists of only one phylum, Mycota, which is divided into five principal classes: Oomycetes, Zygomycetes, Ascomycetes, Basidiomycetes, and Fungi Imperfecti.

Class Oomycetes

The Oomycetes are also known as the water molds because many of them are aquatic. They are the only group of fungi that produce flagellated, swimming spores. Even the terrestrial forms produce swimming spores that require free water.

Most oomycetes are saprobes, living on dead organic matter. Some forms are parasitic and pathogenic, however, and, as mycologist C. J. Alexopoulos has said, "At least two of them have had a hand—or should we say a hypha!—in shaping the economic history of an important portion of mankind." The first of these is *Phytophthora infestans* (*phytophthora* literally means "plant destroyer"), the cause of the "late blight" of potatoes, which produced the great potato famines in Ireland. The second economically important member of this group is *Plasmopara viticola*, the cause of downy mildew of grapes. This mildew threatened the entire French wine industry during the latter part of the nineteenth century.

The oomycetes derive their name from *oion*, the Greek word for "egg." They are the only group of fungi in which the gametes are clearly male (sperm) and female (egg). They are also the only group of fungi with cellulose rather than chitin in their cell walls. Sperm and eggs, each of which are borne in their own type of gametangia, fuse to produce a zygote (see Figure 20–22).

Class Zygomycetes

The Zygomycetes are terrestrial fungi most of which live in the soil, feeding on dead plant or animal matter. Some are parasites of plants, insects, or small soil animals. Unlike oomycetes, they produce no flagellated spores at any stage of the life cycle, and, in sexual reproduction, they produce zygospores, which are thick-walled, resistant spores that develop from a zygote.

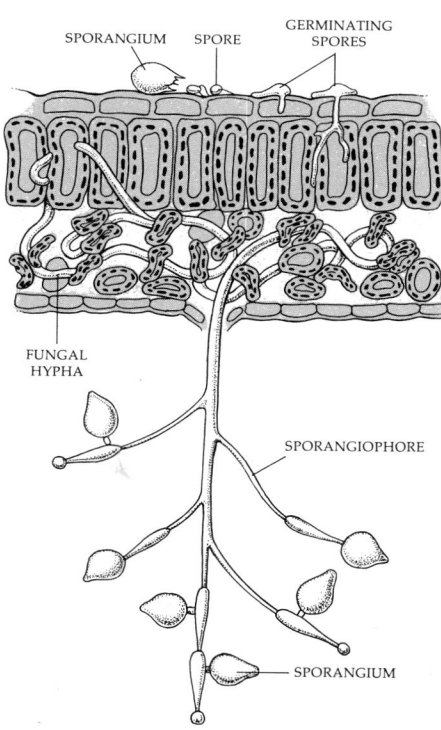

–21 Phytophthora infestans, *cause of potato blight. Infection begins when an airborne sporangium alights on a leaf, releasing spores that move about in the film of water on the leaf's surface. These spores germinate, producing hyphae that penetrate the epidermis and attack the mesophyll cells. Eventually aerial hyphae —sporangiophores—bear a sporangium from which a new generation of spores are released.*

SPORANGIUM   SPORE   GERMINATING SPORES

FUNGAL HYPHA

SPORANGIOPHORE

SPORANGIUM

20–22 *Mating in the fungus* Achyla ambi-sexualis, *an oomycete. The large spherical structure is the female gametangium. The dark bodies within it contain eggs. Encircling the female gametangium is the male gametangium. Fertilization tubes extending from the male into the female gametangium are barely visible. Male nuclei pass through these tubes to the egg nuclei. Development of the male gametangia and their attraction to the female are controlled by the production of a steroid hormone remarkably similar in structure to the human sex hormones.*

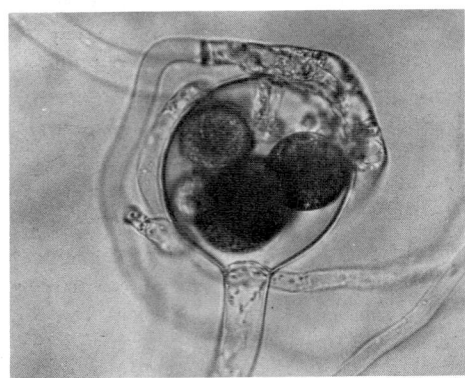

├────── 30 μm ──────┤

20–23 *Asexual and sexual reproduction in the common black bread mold* Rhizopus. *The mold consists of branched hyphae, including rhizoids, which anchor the mycelium; stolons which run aboveground; and sporangiophores. At maturity, the fragile wall of the sporangium disintegrates, releasing the asexual spores, which are carried away by air currents. Under suitable conditions of warmth and moisture, the spores germinate, giving rise to new masses of hyphae. Sexual reproduction occurs when two hyphae from different mating strains come together, forming gametangia, which fuse to form a thick-walled, resistant zygote, commonly called a zygospore. After a period of dormancy, the zygote undergoes meiosis and germinates, producing a new sporangium.*

One of the most common members of this class is *Rhizopus stolonifer*, the black bread mold. Infection begins when a spore germinates on the surface of bread, fruit, or some other organic matter and forms hyphae. Some of the hyphae extend rootlike anchoring structures, called *rhizoids*. The rhizoids secrete digestive enzymes and absorb dissolved organic materials. Other specialized hyphae, the sporangiophores, push up into the air, and sporangia form on their tips. As the sporangia mature, they become black, giving the mold its characteristic color. They eventually break open, releasing numerous airborne spores, each of which can germinate to produce a new mycelium.

Sexual reproduction in *Rhizopus* occurs when the specialized hyphae of two different mating strains meet and fuse, attracted toward one another by hormones that diffuse in the form of gases. The two strains are designated plus (+) and minus (−), since there are no morphological differences between them on which to base male and female designations. Septa, or cross walls, form behind the tips of the touching hyphae; the two tip cells thus formed are gametangia, one containing numerous + nuclei, the other containing numerous − nuclei. Two gametangia fuse, and the two types of nuclei then fuse, producing a diploid nucleus. The resulting multinucleate cell, the zygote, then forms a hard, warty wall and becomes dormant; during this dormant stage, it can survive periods of extreme heat or cold or desiccation. At the end of dormancy, only one diploid nucleus remains, and this undergoes meiosis when the zygote germinates. Only one of the four nuclei produced by meiosis generally survives. It commonly gives rise to a sporangiophore, from which airborne spores are released.

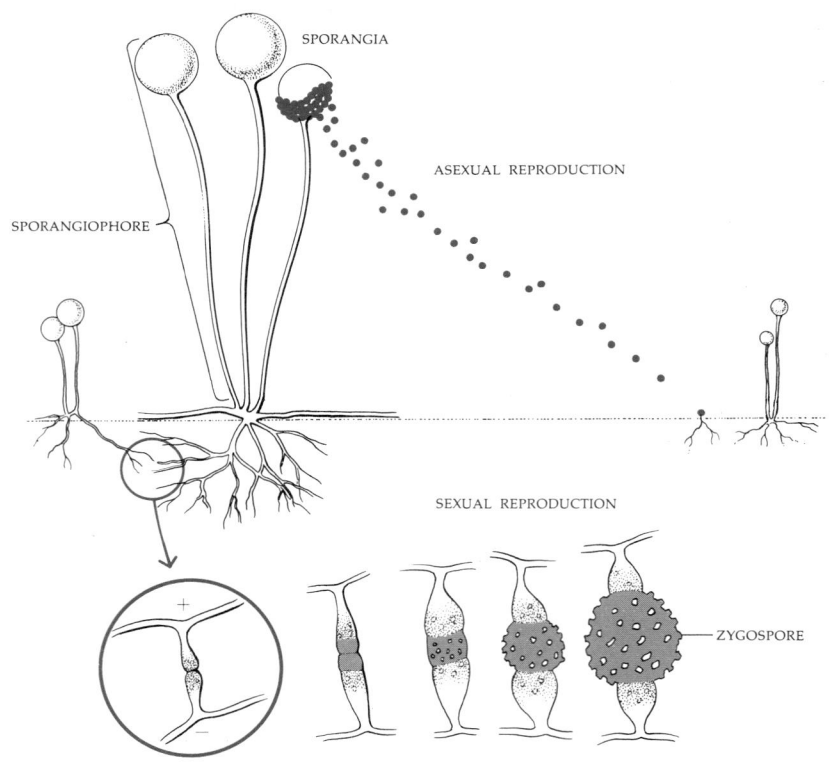

SPORANGIA

ASEXUAL REPRODUCTION

SPORANGIOPHORE

SEXUAL REPRODUCTION

ZYGOSPORE

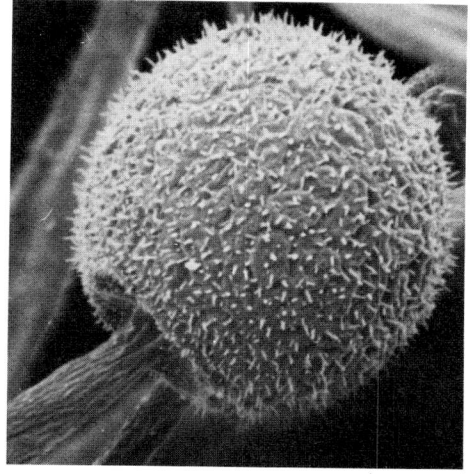

)–24 *Scanning electron micrograph of bread mold sporangium.*

10 μm

Class Ascomycetes

The Ascomycetes are the largest class of fungi (about 30,000 species), including the yeasts and powdery mildews, many of the common black and blue-green molds, and the morels and truffles prized by gourmets. Members of this group of fungi are the cause of many plant diseases, such as chestnut blight and Dutch elm disease, and are the source of many of the antibiotics. The red bread mold *Neurospora*, which played a major role in the history of modern genetics (see Chapter 15), is an ascomycete.

In ascomycetes, the hyphae are divided by cross walls, or septa, unlike the hyphae of the oomycetes and most zygomycetes. Each compartment generally contains a separate nucleus, but the septa have pores in them through which the cytoplasm and the nuclei can move. Spores are formed sexually and asexually. Asexual spores are formed either singly or in chains at the tip of a specialized hypha. They are characteristically very fine and so are often called *conidia*, from the Greek word for "dust."

Sexual reproduction always involves the formation of an *ascus* ("little sac"), a structure that is characteristic of the group. Ascus formation is preceded by the fusion of hyphae of different mating strains to form a dikaryon. The nuclei form pairs that divide synchronously as the hypha grows. Eventually some of the nuclei fuse; this is the only truly diploid stage in the life cycle. The diploid nuclei immediately undergo meiosis, producing four haploid nuclei, and then each of these four nuclei commonly divide mitotically, producing eight haploid nuclei. Each of these nuclei becomes surrounded by a tough wall; each mature ascus contains eight of these spores (ascospores). In most ascomycetes, the ascus becomes turgid at maturity and finally bursts, squirting its ascospores explosively into the air.

·25 *Two ascomycetes. (a) A common morel, edible and choice. Mushroom gatherers look for it when the oak leaves are "the size of a mouse's ear." (b) Scarlet cup, a harbinger of spring in hardwood forests throughout the United States. It is usually found arising from a fallen branch.*

(a)

(b)

20–26 *Electron micrograph of two asci. The closed ascus within which the sexually produced spores develop is the "trademark" of the ascomycete.*

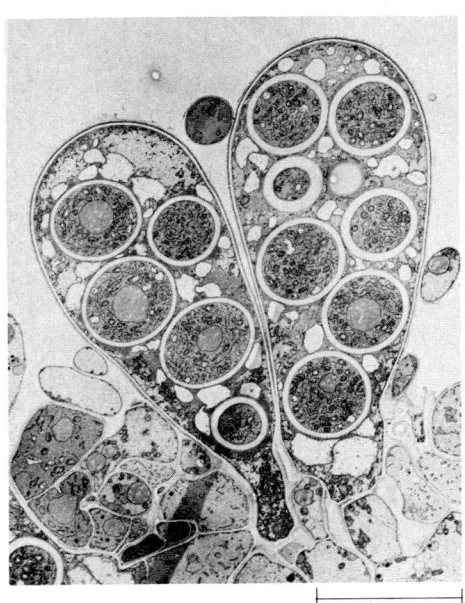

10 µm

20–27 *The basidium is the specialized reproductive hypha of the basidiomycetes, in which fusion of nuclei and meiosis takes place. In the common mushroom, basidia line the gills.*

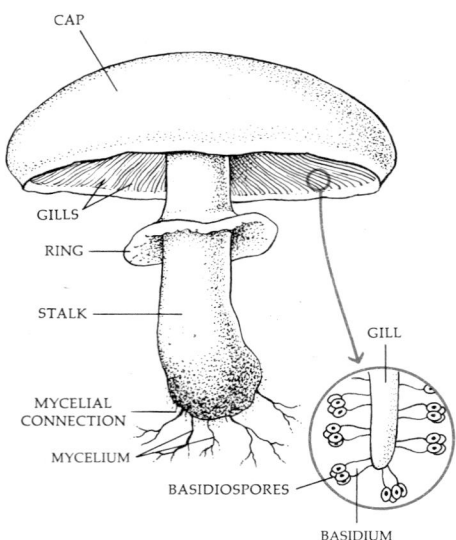

CAP

GILLS

RING

STALK

GILL

MYCELIAL CONNECTION

MYCELIUM

BASIDIOSPORES

BASIDIUM

Many members of this class are parasitic on higher plants. Ergot, which is one of the most famous fungus-produced diseases, is caused by *Claviceps purpurea*, a parasite of rye. Although ergot seldom causes serious damage to the crop of rye, it is dangerous because a small amount mixed with rye grains is enough to cause severe illness among domestic animals or among the people who eat bread made with the flour. Ergotism is often accompanied by gangrene, nervous spasms, psychotic delusions, and convulsions. It occurred frequently during the Middle Ages, when it was known as St. Anthony's fire. In one epidemic in 994, more than 40,000 people died. Ergot, which causes muscles to contract and blood vessels to constrict, has various medical uses. It is also the initial source for the psychedelic drug lysergic acid diethylamide (LSD).

Class Basidiomycetes

The most familiar basidiomycetes are mushrooms. The mushroom, which is the spore-producing body, is composed of masses of tightly packed hyphae. The mycelium from which the mushrooms are produced forms a diffuse mat which may grow as large as 100 feet in diameter. In an open area, the mycelium expands evenly in all directions. Mushrooms usually form at the outer edges, where the mycelium grows most actively, since this is the area in which there is the most fresh nutritive material. As a consequence, the mushrooms will appear in rings, and as the mycelium grows, the ring becomes larger and larger in diameter. Such circles of mushrooms, which might appear in a meadow overnight, were known in European folk legends as "fairy rings." They can make such a rapid appearance because most of the production of new protoplasm takes place underground, in the mycelium. The protoplasm then streams into the new hyphae of the fruiting body as it forms above ground.

The basidiomycetes have hyphae subdivided by perforated septa, like those found in the ascomycetes. Sexual reproduction is initiated by the fusion of hyphae to form a dikaryon. The dikaryon may persist for years, forming an elaborate mycelium. Eventually, some of the nuclei fuse to form diploid nuclei that immediately undergo meiosis. The spores (basidiospores) are formed externally, on a specialized hypha (called a *basidium*, from the Greek word for "base" or "club"). Many of the larger basidiomycetes seem to have lost the capacity to produce asexual spores.

The best-known mushrooms belong to the group known as the gill fungi, which includes *Agaricus campestris*, the common field mushroom. Varieties of this species are the only mushrooms it has been possible to cultivate commercially. The gill fungi also include most of the known poisonous mushrooms. Mushrooms of the genus *Amanita* are the most highly poisonous of all mushrooms; even one bite of the white *Amanita verna*, the "destroying angel," can be fatal. Some species of toxic mushrooms, such as *Psilocibe mexicana* (the source of psilocybin), are eaten by Mexican and Central American Indians and others for their hallucinogenic effects.

The spores of the gill mushrooms are found in the furrows, or "gills," under the cap. If you take a mature mushroom and place it on a piece of white paper, it will release fine spores that will trace out a negative copy of the gill structure.

20–28 *Basidiomycetes.* (a) *White coral mushroom.* (b) Amanita muscaria. *The ring on the stem is characteristic of the genus* Amanita, *which includes most of the highly poisonous mushrooms.* (c) *Mushrooms growing under reindeer lichen.* (d) *Puffballs,* Lycoperdon. *Spores line the inside of the puffball and are released in a "puff" when the fungus is touched.* (e) *Bracket fungi on a dead elm. The sudden appearance and rapid growth characteristic of these fungi is the result of cytoplasmic streaming from a large mass of underlying hyphae.*

(a)

(b)

(c)

(d)

(e)

The spores, which come in a variety of colors, are a useful means of identifying various mushrooms. Other types of basidiomycetes include puffballs (a few of which have a diameter of 3 feet), earthstars, stinkhorns, and the parasitic rusts and smuts, some of which cause severe losses among cereal crops.

The Fungi Imperfecti

This class comprises fungi whose complete reproductive cycles are unknown, either because they have not been sufficiently studied or because some part of the process has been lost in the course of evolution—hence the adjective "imperfect."

Among the Fungi Imperfecti are parasites that cause diseases of plants and animals. The most common human diseases caused by this group are infections of the skin and mucous membranes, such as ringworm (including "athlete's foot") and thrush. A few are of great economic importance in the production of certain cheeses (Roquefort and Camembert) and of antibiotics, including penicillin.

Table 20–3   *Major Classes of Fungi*

Class	No. of species	Examples	Distinctive characteristics	Diseases	Economic uses
Oomycetes	Several hundred	Potato blight fungus	Some aquatic; flagellated spores; formation of eggs and sperm in special gametangia; cell walls contain cellulose	Blights and mildews of plants, fish infections	None
Zygomycetes	Several hundred	Black bread mold	Formation of zygospores (tough, resistant spores resulting from a fusion of gametes); no flagellated cells	Few	None
Ascomycetes	30,000	*Neurospora,* yeasts, morels, truffles	Formation of fine asexual spores (conidia); sexual spores in asci; hyphae divided by perforated septa; dikaryons; no flagellated cells	Powdery mildews of fruits, chestnut blight, Dutch elm disease, ergot	Food (morels, truffles), wine, beer, bread-making (yeasts)
Basidiomycetes	25,000	Toadstools, mushrooms	Sexual spores in basidia; hyphae divided by perforated septa; dikaryons, no flagellated cells	Rusts, smuts	Food (mushrooms)
Fungi Imperfecti	25,000	*Penicillium*	Fungi with no known sexual cycles; no flagellated cells	Ringworm, thrush	Cheeses, antibiotics

(a)　　　　　　　(b)　　　　　　　(c)

-29　(a) *Foliose lichen growing on the face of a rock.* (b) *Two common lichens growing on a tree trunk.* (c) *Reindeer "moss," actually a lichen, is a principal food source for large herbivores—namely, reindeer and caribou—of the treeless Arctic plains.* (d) *British soldier lichen (Cladonia). Each soldier (so called because of the scarlet color) is about 3 millimeters tall.*

(d)

THE LICHENS

A lichen is a specific combination of a fungus and an alga. The organisms resulting from these combinations are completely different from those of either the alga or the fungus growing alone, as are the physiological conditions under which the lichen can survive. The lichens are widespread in nature; they occur from arid desert regions to the arctic, and grow on bare soil, tree trunks, sun-baked rocks, fence posts, and windswept alpine peaks all over the world. They are often the first colonists of bare rocky areas.

Lichens do not need an organic food source, as do their component fungi, and unlike free-living algae, they can remain alive even when very desiccated. They require only light, air, and a few minerals. They apparently absorb some minerals from their substrate (this is suggested by the fact that particular species are characteristically found on specific kinds of rocks or soil or tree trunks), but most often minerals enter the plant through the air and in rainfall.

Because lichens rapidly absorb substances from rainwater, they are particularly susceptible and sensitive to airborne toxic compounds; thus, the presence or absence of lichens is a sensitive index of air pollution.

The algae found in lichens also occur as free-living species, whereas the lichen fungi are generally found in nature only in the lichens. For these reasons, lichens are generally classified according to the fungus species with which they are associated.

There are about 17,000 species of lichens, that is, of lichen fungi. Most are ascomycetes or Fungi Imperfecti, although a few are basidiomycetes. Algae from some 30 different genera are found in symbiotic combination with these fungi.

Lichens reproduce most commonly by the breaking off of fragments containing both fungal hyphae and algae. New lichens are formed, according to some recent research, by the capture of an alga by a fungus. Sometimes the alga is destroyed by the fungus, in which case the fungus also dies. If it survives a lichen is produced.

SUMMARY

The kingdom Protista includes the phylum Protozoa, three phyla of one-celled algae, and the slime molds.

The Protozoa, the nonphotosynthetic protists, are subdivided into four classes: the Mastigophora (flagellates), the Sarcodina (amoebas), the Ciliophora (ciliates), and the Sporozoa. The first three classes may be identified on the basis of their locomotor structures. Members of the latter group, which is composed largely of parasitic forms, have no locomotor organelles. Among the protists are some of the largest known cells and also the most complex.

Euglenophyta is a small group of unicellular algae, mostly found in fresh water. They contain chlorophylls *a* and *b* and store carbohydrates in an unusual starchlike substance, paramylum. The cells lack a cell wall but have a flexible series of protein strips, which make up the pellicle, inside the cell membrane. The cells are highly differentiated, containing chloroplasts, a contractile vacuole, eyespot, or stigma, and flagella. No sexual cycle is known. The group also contains nonphotosynthetic forms.

The Chrysophyta are important components of freshwater and marine phytoplankton. Most of the known species are unicellular organisms known as diatoms. Diatoms are characterized by fine, double silicon-containing shells. Abundant fossil records of these shells have been found.

The Pyrrophyta are unicellular biflagellates, many of which are marine. This division includes the dinoflagellates, which are characterized by two flagella that beat in different planes, causing the organism to spin; dinoflagellates often have stiff, bizarrely shaped cellulose walls.

The slime molds, or Gymnomycota, are amoebalike organisms. They resemble fungi in that they reproduce by the formation of spores and are heterotrophic. Unlike the fungi, however, they lack a cell wall and they ingest their food as particles. There are two principal orders: Myxomycetes (plasmodial slime molds) and Acrasiomycetes (cellular slime molds).

The Fungi, considered here as a separate kingdom, are heterotrophs, which

derive their nutrition by absorption of organic compounds digested extracellularly by secreted enzymes. Their bodies are composed of masses of filaments called hyphae, sometimes subdivided by perforated septa (walls). The principal component of their cell walls is usually chitin. They form both asexual and sexual spores, although not all fungi form both kinds. The spores may be formed in sporangia. The sexual cycle is initiated by the fusion of hyphae of different mating strains. In some groups of fungi, the nuclei in the fused hyphae immediately combine and a zygote is formed. In others (the Ascomycetes and Basidiomycetes), the two genetically distinct nuclei remain separate, forming pairs that divide synchronously, sometimes over prolonged periods. A cell or organism with such paired, genetically distinct nuclei is known as a dikaryon. Once the nuclei fuse, meiosis always follows immediately. There are five principal classes: the Oomycetes, or water molds, which form flagellated spores; the Zygomycetes, which have nonmotile spores and whose zygotes become thick-walled zygospores; the Ascomycetes, in which the sexual spores develop within an ascus (sac); the Basidiomycetes, in which the sexual spores develop on a basidium (club); and the Fungi Imperfecti, with no known sexual cycle.

Fungi have an important ecological role as decomposers of organic material. They also are major plant pathogens. (disease)

Lichens are combinations of fungi and algae which are morphologically and physiologically different from either organism as it exists separately. They are able to survive under adverse environmental conditions where neither partner could exist independently. The lichen represents a symbiotic relationship in which the fungus encloses the algal cells and is dependent on the algal cells for nourishment.

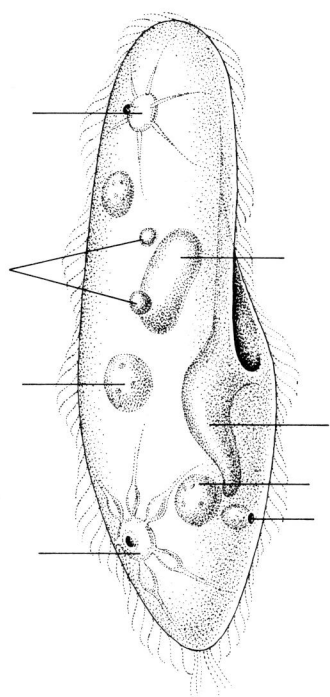

QUESTIONS

1. Define the following terms: chitin, dinoflagellate, hypha, gametangium, mycelium, lichen.
2. Label the drawing at the left.
3. Name the four classes of phylum Protozoa and give the distinguishing characteristics of each.
4. Name the five classes of kingdom Fungi and give the distinguishing characteristics of each.
5. It has been hypothesized that the prokaryote may be derived from the eukaryotes. What evidence can you think of in support of this hypothesis? Against?
6. Defend the view that it is misleading to consider ciliates as cells in the same sense as the cells of our bodies.
7. Consider the life cycle of *Plasmodium*. At what stages in the cycle do its numbers increase? Why might a parasite that requires several hosts find it necessary to evolve a life cycle in which its numbers increase at several stages?
8. Classify the following organisms: *Didinium* (pages 118 to 119), *Chaos chaos* (page 25), *Euplotes patella* (page 102).

Chapter 21

The Plant Kingdom

For most of Earth's history, the land was bare. A billion years ago, seaweeds may have clung to the shores at low tide and perhaps some patches of gray-green lichens added a bit of dull color to a few inland rocks, but had anyone been there to observe it, the planet's surface would generally have appeared as barren and forbidding as the moon's does today. According to the fossil records, plants first began to invade the land a mere 400 million years ago, and not until then did the Earth truly come to life. As a film of green spread from the edges of the waters, other forms of life, the heterotrophs, were able to follow. The shapes of these new forms and the ways in which they lived were determined by the plant life that preceded them, because, in freeing themselves from the water, they became increasingly dependent on the land plants not only for their food—their chemical energy and their molecular building blocks—but also for their nesting, hiding, stalking, and breeding places. In all communities except those created by man, the character of the land plants still determines the character of the animals and other forms of life that inhabit that particular area. Even man, who has seemingly freed himself from the life of the land and even, on occasion, from the surface of the Earth, is still dependent on the photosynthetic events that take place in the green leaves of the land plants.

In the first half of this chapter, we will examine three phyla of algae, including the green algae (Chlorophyta), whose ancestors, as far as the experts are able to determine, took part in this invasion of the land. These algae differ from the photosynthetic prokaryotes and eukaryotes discussed in the previous chapters in that these phyla consist of or include multicellular forms. The second half of the chapter traces the evolution of land plants and describes some of the modern forms, the green plants, that dominate today's landscapes.

EVOLUTION OF PLANTS

The story of plant evolution begins with one-celled algae floating on or near the surface of the water in the open seas. Such organisms have left little trace in the fossil record, but we assume that their way of life was much like that of modern members of the phytoplankton, although they may have been less complex and specialized. In the open water, light was abundant, and the oxygen, hydrogen, and carbon in the air and water were available to every floating cell, as they are today. Although groups of cells might form long filaments (as a

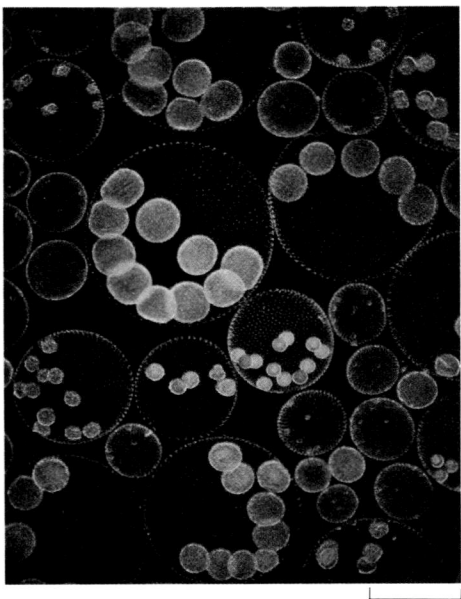

21–1 *Volvox, a colonial alga. Each colony is made up of hundreds or thousands (depending on the species) of individual bright green biflagellate cells, attached to one another by fine threads of protoplasm. Each colony forms a hollow sphere that spins through the water as a result of coordinated beating of the flagella. Daughter colonies form inside the mother colony. Many such mother-daughter combinations are visible in this micrograph.*

300 µm

Table 21–1 *The Plant Kingdom*

Phylum	Common name	No. of species
Phaeophyta	Brown algae	1,100
Rhodophyta	Red algae	4,000
Chlorophyta	Green algae	7,000
Bryophyta	Mosses, liverworts	23,000
Tracheophyta	Vascular plants	260,000

consequence of not separating after cell division), each remained a separate functional entity and lived an independent existence.

As the cellular colonies multiplied, they probably started to exhaust the supplies of nitrogen, phosphorus, sulfur, and other minerals available in the open ocean. (In fact, this shortage of essential elements is the limiting factor in any current plan to farm the seas.) Life was probably more abundant near the shores, where the waters were rich in minerals washed from the land by rivers and streams, scraped from the coasts by the action of the waves, and brought up from the bottom by upwellings of coastal waters. Here, along the coasts, in a much more challenging environment than that of the open sea, complex plant life evolved.

Among the challenges presented by life along the shore was the turbulent shore itself. Under evolutionary pressures, groups of plants developed anchoring structures, holdfasts, which adhered to the rocks. These were often below the surface of the water, where less light penetrated. Over millions of years, the upper portions of the plant body, which were near the light, became thinner and more spread out, increasing the surface area. The upper cells became specialists in photosynthesis, producing enough sugars to nourish the overshadowed lower cells that anchored them. Some of these multicellular algae developed specialized conducting tissues that transport the products of photosynthesis downward. None, however, developed complex tissues that enclose and protect the gamete and embryo as do the cones and flowers of higher land plants.

Not only did specialized shapes and structures evolve among these marine algae, the seaweeds, but specialized colors as well. Water filters light, removing first the longer wavelengths, the reds and oranges; at the deeper levels, only a faint blue light penetrates. As a consequence of the evolution of specialized pigments, the various seaweeds came to use the entire spectrum of light as well as the living space along the rocky shore.

Of the multicellular algae, three groups persist as modern representatives: the brown algae (Phaeophyta), the red algae (Rhodophyta), and the green algae (Chlorophyta).

21–2 *Sargassum, a brown alga. Notice the differentiation of the plant body into a number of specialized parts.*

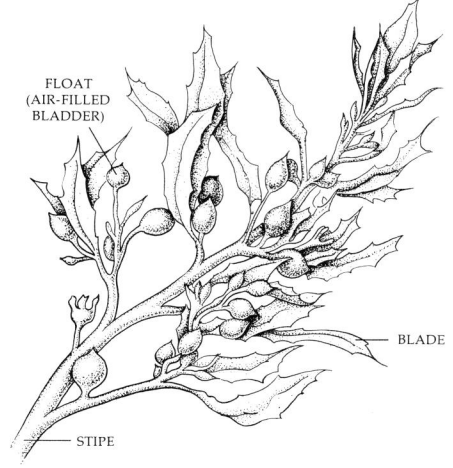

FLOAT
(AIR-FILLED
BLADDER)

BLADE

STIPE

The brown algae are the principal seaweeds of temperate and polar regions. An almost exclusively marine group, they dominate the rocky shores throughout the cooler regions of the world, and some, like the kelps, often form extensive beds offshore.

One species of brown alga, *Sargassum nitans*, is a major component of the dense floating masses of seaweeds that fill the Sargasso Sea, an area of more than 2 million square miles, reaching from the Bahamas to the Azores.

The brown algae contain chlorophylls *a* and *c* and fucoxanthin, as do the Chlorophyta. They differ from most other plants in that they store their food as an unusual polysaccharide (laminarin), or sometimes as oil, but never as starch, as do green algae and higher plants. Their cell walls contain cellulose.

The brown algae are often very large, and many have a variety of specialized tissues. Some of the giant kelps are nearly 200 feet long; these are annuals, reaching their full size in a single season. In some kelps, the plant body is well differentiated into *holdfast* ("root"), *stipe* ("stalk"), and *blade* ("leaf"). (The words are put into quotation marks because, although the corresponding parts of the algae superficially resemble the organs of higher plants, they are not really comparable in their internal organization.)

There are no modern unicellular forms in this group, except for the gametes. Sperm and spore cells are often flagellated.

21–3 (a) *A brown alga, rockweed* (Fucus vesiculosus), *covers many rocky shores that are exposed at low tide. The air-filled* bladders on the blades carry them up toward the light. The exposed plant parts are thick and leathery and moderately drought-resistant. (b) *A brown alga,* Laminaria, *showing holdfast, stipe, and the lower part of the blade.*

(a)

(b)

(a)

(b)

(c)

1–4 *Red algae. (a) In* Plenosporium dasy-oides, *the basically filamentous structure of the red algae is clearly evident. Notice the delicate, lacy plant body, as compared with the sturdier* Fucus, *which is adapted to life in the pounding surf rather than in the quieter ocean depths. Multicellularity in the red algae evolved separately from multicellularity in the brown and green algae, where no similar patterns of organization are found. (b) Reef-building red alga,* Porolithon craspedium, *from a reef in the Marshall Islands. These coralline algae, as they are called, are heavily encrusted with calcium carbonates and are an important part of reef communities. (c) Irish moss* (Chondrus crispus), *an important source of agar, a mucilagenous material with many commercial uses, including the culture of bacteria and the making of dental impressions.*

PHYLUM RHODOPHYTA: RED ALGAE

The red algae are most commonly found in warm marine waters; they make up the other principal group of large seaweeds. The red algae contain chlorophylls *a* and *d*, carotenoids, and also certain phycobilins, which give them their distinctive colors. (Phycobilins and chlorophyll *a* are also found in the blue-green algae, as you may recall, and it is believed that the blue-green algae and red algae are related.)

Most of the seaweeds of the world are red algae, of which there are some 4,000 species. Less than 2 percent are freshwater forms. Red algae usually grow attached to rocks and other algae; there are no large red algae capable of prolonged life in the floating state, like the sargasso weed of the Sargasso Sea. (They die when floating because the tidal surge is required for adequate gas exchange; without it, they suffocate.) As you would expect from their red color, which indicates that they absorb blue light, some grow at greater depths than other algae; they have been found attached 600 feet below the ocean surface in the clear water of the tropics. Although some are several yards long, red algae never attain the size of the largest of the brown algae. They have a reproductive pattern in which neither of the gametes is motile; the male is carried to the fixed female reproductive cell by the movement of the water. None of their cells are flagellated, and none of them contain centrioles.

PHYLUM CHLOROPHYTA: GREEN ALGAE

The brown and red algae, although they achieved considerable size and complexity, never, as far as is known, produced any land forms. All the land plants, it is believed, arose from members of a third group, the Chlorophyta. Evidence for this theory includes the fact that both the chlorophytes and all the land plants, but no other groups of algae, contain chlorophylls *a* and *b* and carotenoids as their photosynthetic pigments and store their foods as starch. All also have plant cell walls of cellulose.

The green algae are the most diverse of all the algae, comprising at least 7,000 species. Although most green algae are aquatic, others occur in a wide variety of habitats, including the melting surface of snow, as green patches on tree trunks, and as symbionts in lichens, protozoans, and invertebrates. Of the aquatic species, a few groups are entirely marine, but the great majority are

found in fresh water. Many green algae are microscopic, but some of the marine forms are large; *Codium magnum* in the Gulf of Mexico, for example, sometimes attains a breadth of 10 inches and a length of more than 25 feet.

The green algae are of special interest to biologists because living members of the group exemplify a wide variety both of forms and of life cycles. Thus they offer many provocative examples of possible evolutionary pathways in this important group of organisms.

A Single-celled Green Alga: Chlamydomonas

Members of the genus *Chlamydomonas*, which are among the most common freshwater green algae, are small (usually less than 25 micrometers long), grass green, and round or pear-shaped (Figure 21–5). They move very rapidly, with a characteristic darting motion caused by the beating of the two equal-sized flagella that protrude from their smaller, anterior end. In many species, a single chloroplast fills almost the entire cell. In the area of the chloroplast is a red pigment body, the eyespot or stigma, and a light-colored spherical body, the pyrenoid, which seems to be associated with the conversion of sugars to starch, the form in which carbohydrates are stored. The cell body is bounded by a membrane and is surrounded by a thin cellulose wall.

21–5 *Electron micrograph and diagram of* Chlamydomonas. *At the top, on the right, you can see a portion of one of the organism's two flagella. A single chloroplast fills most of the cell and gives the cell its bright green color. Note the mitochondria at the base of the flagellum.*

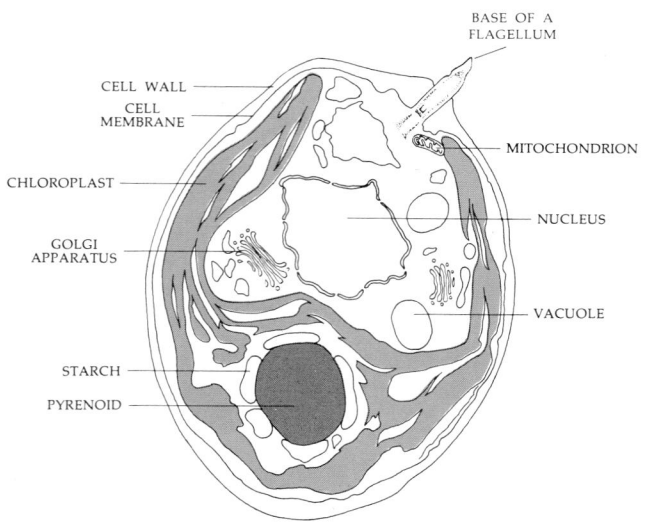

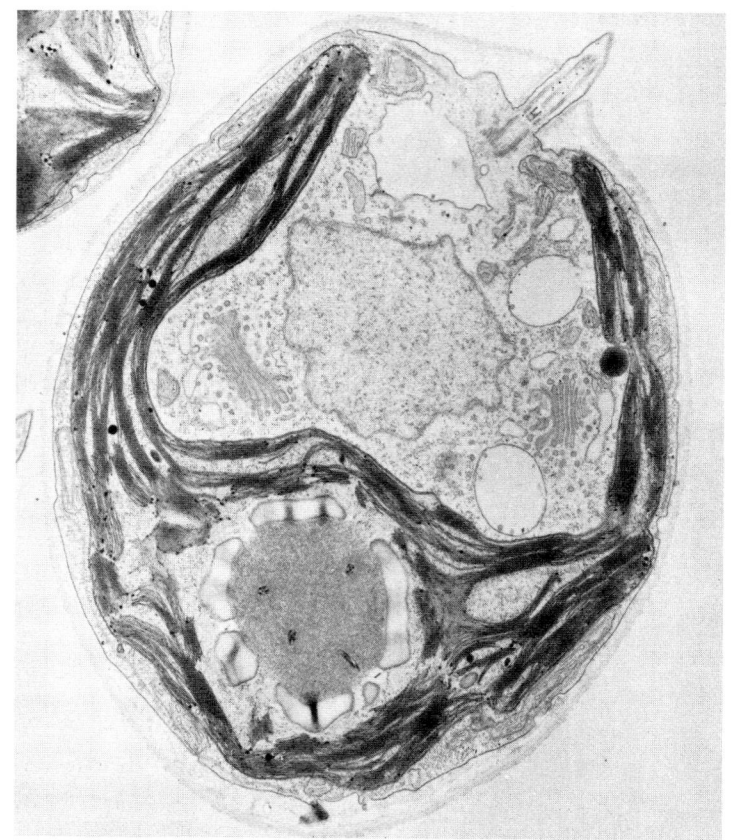

1 μm

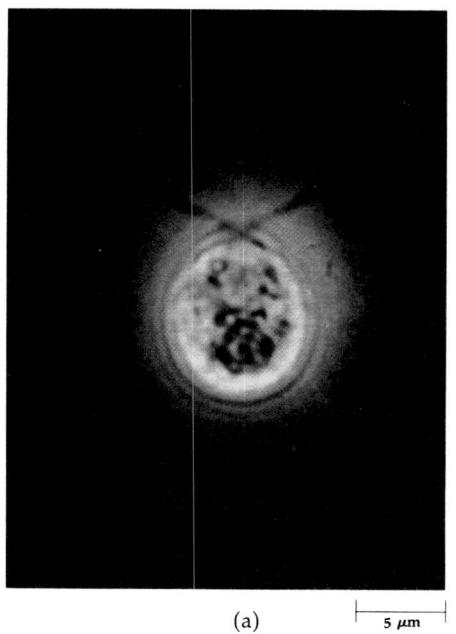

(a) ⊢——⊣ 5 μm

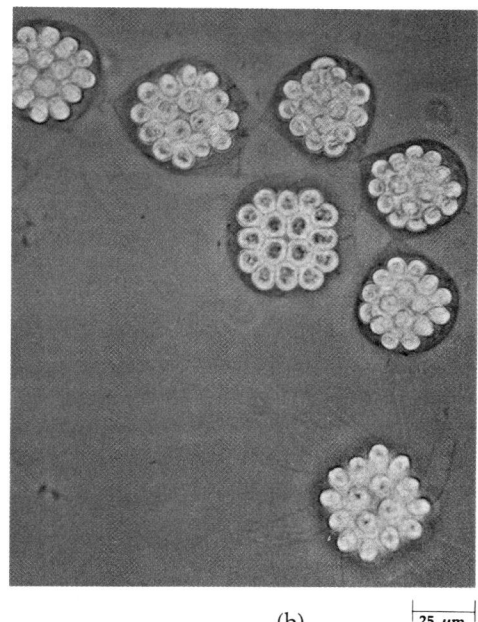

(b) ⊢——⊣ 25 μm

(c) ⊢——⊣ 5 μm

1–6   *The volvocine line. (a) Chlamydomonas, a single-celled alga. (b) Gonium colonies. Each is composed of a shield-shaped mass of Chlamydomonas-like cells, held together in a gelatinous matrix. (c) Pandorina, a larger colony in which the individual cells form an egg-shaped colony. Volvox, Figure 21–1, is the largest and most complex member of this evolutionary line.*

Evolutionary Trends

Beginning with unicellular organisms like *Chlamydomonas,* it is possible to trace several evolutionary lines.

One pathway leading to increased complexity is called the volvocine line, after one of its more spectacular members, *Volvox* (Figure 21–1). It is based on the cohesion of *Chlamydomonas*-like cells in motile colonies. Cellular colonies differ from multicellular organisms in that in colonies, the individual cells preserve a high degree of independent function. The cells are connected by cytoplasmic strands, which integrate the colony sufficiently so that it may be regarded as a single organism. The simplest member of the volvocine line is *Gonium.* The *Gonium* colony consists of 4 to 32 separate cells (depending on the species) arranged in a shield-shaped disk. The flagella of each cell beat separately, pulling the entire colony forward. Each cell in *Gonium* divides to produce an entire new colony.

A closely related colonial organism is *Pandorina,* which consists of 16 or 32 cells in a tightly packed ovoid or ellipsoid shape. The colony is polar, that is, one end is different from the other: the eyespots are larger in the cells at one pole of the colony. Each cell has two flagella, and because all the flagella point outward, *Pandorina* rolls through the water like a ball. When the cells attain their maximum size, the colony sinks to the bottom, and each of the cells divides to form a daughter colony. The parent colony then breaks open like Pandora's box (which suggested its name), releasing new daughter colonies.

Volvox, from which the group gets its name, is a hollow sphere that is made up, according to species, of a single layer of 500 to 50,000 tiny biflagellate cells. The flagella of each cell beat in such a way as to spin the entire colony around its axis, spinning it through the water. *Volvox* orients its anterior end toward the light, but moves away from light that is too strong. Most of the cells are

335   *The Plant Kingdom*

21–7 *Three other evolutionary lines among the Chlorophyta are represented by the three types shown here. (a) Valonia, which is about the size of a hen's egg, with many nuclei but no partitions separating them. (b) Spirogyra, in which* the cells all elongate and then are divided by transverse cross walls so that the cells in the plant body are strung together in long, fine filaments, in which the chloroplasts form spirals that look like strips of green tape within each cell. (c) Ulva, or sea lettuce, in which the cells divide both longitudinally and laterally, with a single division in the third plane, producing a broad plant body two cells thick.

(a)

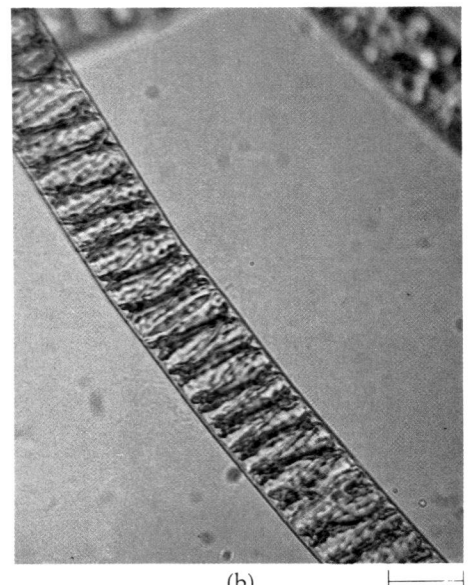

(b) |— 25 μm —|

(c)

21–8 *Origins of multicellularity. (a) Cell divisions in one plane produce a filament, such as Spirogyra. (b) Divisions in two planes produce a one-celled layer. (c) Divisions in three planes produce a three-dimensional solid body, such as that found in the higher plants and animals. The arrows indicate the direction of growth.*

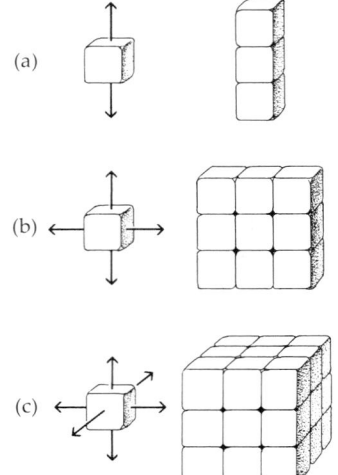

(a)

(b)

(c)

vegetative and do not take part in reproduction. For reasons that are not entirely understood, only some of the cells of the lower hemisphere can form daughter colonies; in other words, there is some specialization of function. These cells, which appear identical to others in the young colony, become larger, greener, and morphologically distinct as the colony matures. Then they enlarge and divide, forming a new sphere of hundreds or thousands of cells. Daughter colonies remain inside the mother colony until the mother colony eventually breaks apart.

Thus, in these four extant genera, *Chlamydomonas, Gonium, Pandorina,* and *Volvox,* there is a steady progression in size and complexity and also a trend toward specialization of function.

Sexual Reproduction

Chlamydomonas

Chlamydomonas reproduces both sexually and asexually. During asexual reproduction, the nucleus usually divides twice mitotically, resulting in the production of four daughter cells within the parent cell wall. Such aggregations of daughter cells may represent the forerunners of colonial organisms, such as *Gonium* and *Pandorina*. Eventually, each cell secretes a wall about itself, develops flagella, and escapes.

Sexual reproduction in some species of *Chlamydomonas* involves the fusion

1–9 Chlamydomonas *leads a solitary existence, except when it reproduces sexually. At this time, cells of different mating strains (indicated here by + and −) are attracted to one another. They fuse, their nuclei unite (syngamy), and the result is a single cell, the diploid zygote. The* zygote *loses its flagella, sinks to the bottom, and develops a thick protective coat in which it can survive through periods of drought or cold. When conditions are favorable once again, the zygote germinates and divides meiotically, producing four new haploid cells, which will probably divide by mitosis but which might enter into another sexual cycle. As in previous life cycles, the haploid phase of the cycle is indicated by a single arrow, the diploid phase by a double arrow.*

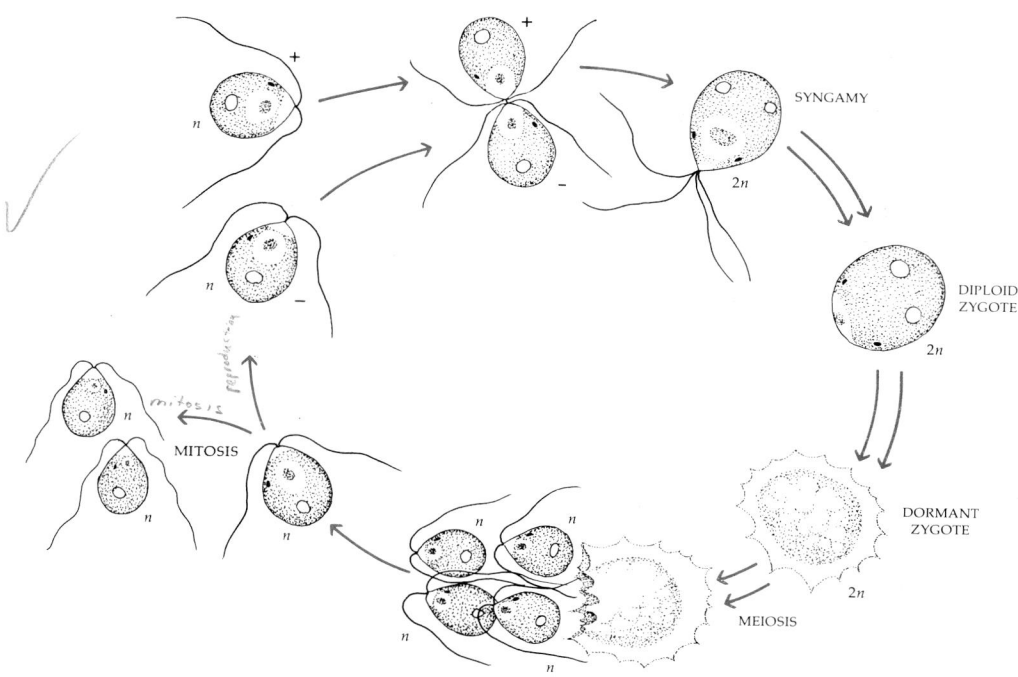

10 *Relative size of gametes. (a) Isogamy: the gametes are equal in size and shape, as in Figure 21–9. (b) Anisogamy: one gamete, conventionally termed "male," is smaller than the other. (c) Oogamy. There is a trend from (a) to (c) in the volvocine line.*

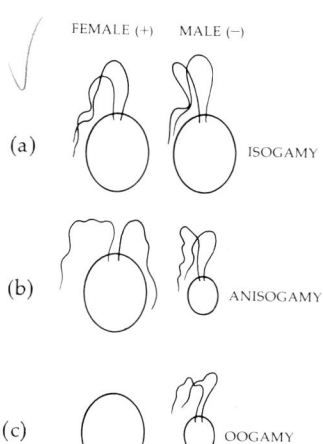

of individuals belonging to different mating strains. The cells first stick together by their flagella and then fuse completely, forming a zygote. A thick cell wall forms around the zygote. This thick-walled, resistant zygote is then dormant for a time. (Sexual reproduction in freshwater algae often results in the formation of a dormant, thick-walled zygote in which the organism can survive during unfavorable conditions, as when ponds or streams dry up. In marine representatives of the group, which are not exposed to such environmental hazards, the zygotes are generally thin-walled.)

Meiosis occurs at the end of the dormant period, resulting in the production of four haploid cells, each of which develops two flagella and a cell wall. These cells can either divide asexually (mitotically) or fuse with a cell of another mating strain to produce a new zygote.

In most species of *Chlamydomonas*, the cells of the two mating types, conventionally designated + and −, are identical in size and structure (*isogamy*). In addition to these isogamous species, there are other species in which *anisogamy* (large, motile female gametes) and *oogamy* (immotile female gametes) occur. Thus the single genus *Chlamydomonas* exhibits in its various species all the types of gametes that occur in all phyla of algae.

337 *The Plant Kingdom*

Ulva: *Alternation of Generations*

The sea lettuce, *Ulva*, has a reproductive pattern characteristic of many of the algae and of all the higher plants. This pattern is known as *alternation of generations*, in which one generation reproduces sexually and the next one reproduces asexually. The plant that produces gametes is known as the *gametophyte*, the gamete-producing plant. The individual cells of the plant are haploid (n) and so are the gametes it produces. The gametes fuse to form a diploid ($2n$) zygote. The zygote develops into a plant in which all the cells are diploid. This plant is known as the *sporophyte*, the spore-producing plant. Spores that are produced by meiosis are haploid.

As we noted previously, spores differ from gametes in that when a spore germinates it forms a new organism, while a gamete must first unite with another gamete before further development occurs. The gametophyte and the sporophyte are always genetically different since one is composed of haploid cells and the other of diploid cells.

Sometimes the two generations look alike, as in *Ulva*. In other chlorophytes, the sporophyte and the gametophyte do not resemble one another. In fact, in some cases, the two generations of the same plant were considered to be two entirely different species until the life cycles were studied in the laboratory. Figure 21–12 shows such an alga—a large multinucleate alga like *Valonia*—which was called *Derbesia* in its sporophyte ($2n$) form and *Halicystis* in its gametophyte (n) form.

As we shall see, with variations, this theme of alternation of generations runs through the life cycles of all the many groups of land plants.

21–11 *In the sea lettuce,* Ulva, *we can see the reproductive pattern known as alternation of generations, in which one generation produces spores, the other gametes. The haploid (n) gametophyte produces haploid isogametes, and the gametes fuse to form a diploid (2n) zygote. A sporophyte, a plant in which all the cells are diploid, develops from the zygote. This plant produces haploid spores by meiosis. The haploid spores develop into haploid gametophytes and the cycle begins again.*

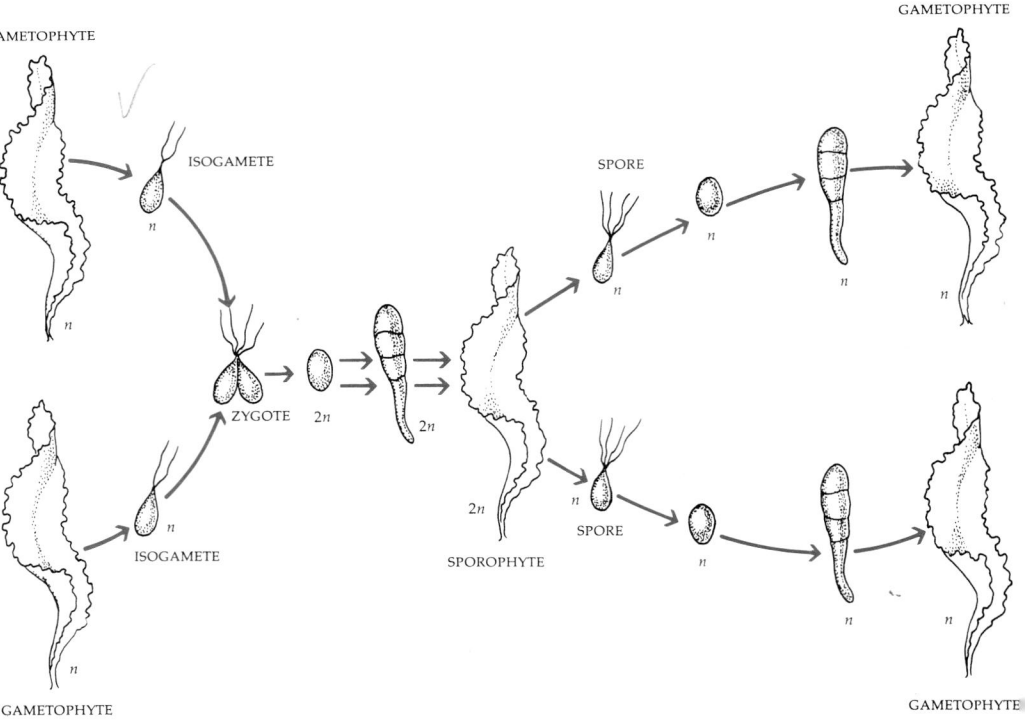

–12 *In some plants the alternating genera-tions are so different that they once were believed to be of different genera. At one time, the gametophyte of the Valonia-like algae shown here was called Hali-cystis and the sporophyte Derbesia.*

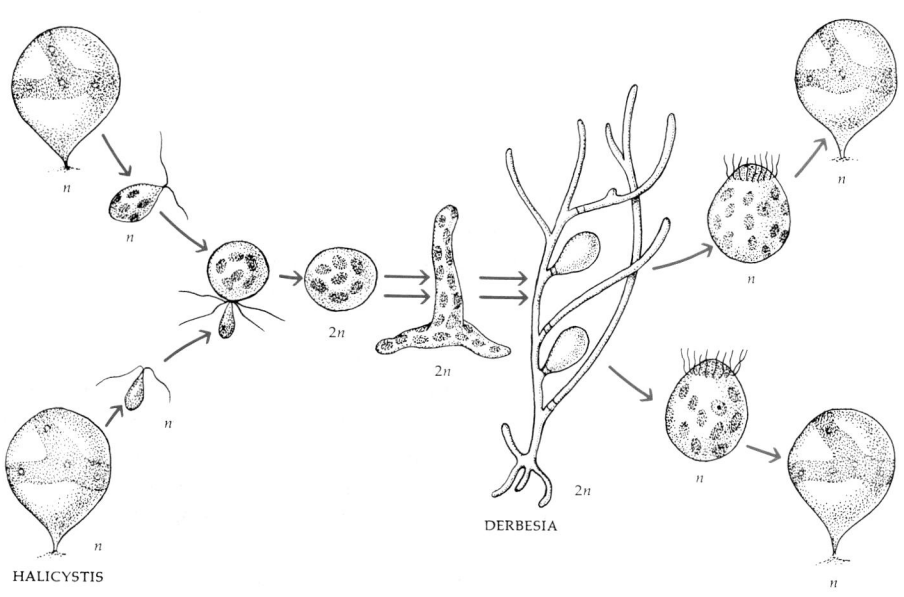

HALICYSTIS

DERBESIA

√ Table 21–2 *Multicellular Algae*

Phylum	No. of species	Photosynthetic pigments	Food reserve	Flagella	Cell wall components	Distribution
Phaeophyta (Brown algae)	1,100	Chlorophylls *a* and *c*, carotenoids (especially fucoxanthin)	Laminarin	2, lateral; in reproduc-tive cells only	Cellulose with alginic acids (poly-saccharides)	Almost all marine, mostly temperate
Rhodophyta (Red algae)	4,000	Chlorophylls *a* and *d*, carotenoids, phycobilins	Floridean starch	None	Cellulose, pectin, various mucilages (algin)	Mostly marine, but some freshwater; many species tropical
Chlorophyta (Green algae)	7,000	Chlorophylls *a* and *b*, carotenoids (especially beta-carotene)	Starch	2 (or more), apical	Cellulose, pectin	Mostly fresh-water, but some marine

THE LAND PLANTS

Water was the first life zone for living things. Land was the next life zone, and air was the third. We are now beginning the invasion of a possible fourth life zone, the zone of outer space.

The second life zone, the land, had a tremendous wealth to offer plants. Light was abundant from daylight to dusk, unfiltered by the turbulent water. Carbon dioxide, needed for photosynthesis, was plentiful in the atmosphere, as was oxygen by this time, and both circulated much more freely than in the water. And there was a great abundance of unoccupied space, as compared with the now-crowded shorelines.

The land presented some difficulties, however. First and foremost was the problem of water. Water is freely available to every cell of a *Volvox* or a sea lettuce, but the major source of water for land plants is beneath the surface of the soil, sometimes many feet beneath. The conquest of dry land was dependent on the evolution of roots with specialized conducting tissues that would supply water to the parts of the plant that are above ground. Conversely, the plant cells below the ground came to receive the organic molecules they required from the photosynthetic cells of these aboveground parts.

As events are reconstructed, only one group of plants, descendants of some unknown line of green algae, successfully made the transition to land. Even among the plants that did achieve a foothold on the land, most kinds are now extinct and are known to us only by their fossil record.

The two plant phyla with modern representatives are (1) Bryophyta, the mosses, hornworts, and liverworts, a small phylum, and (2) Tracheophyta,

21–13   *Chart of evolution of land plants.*

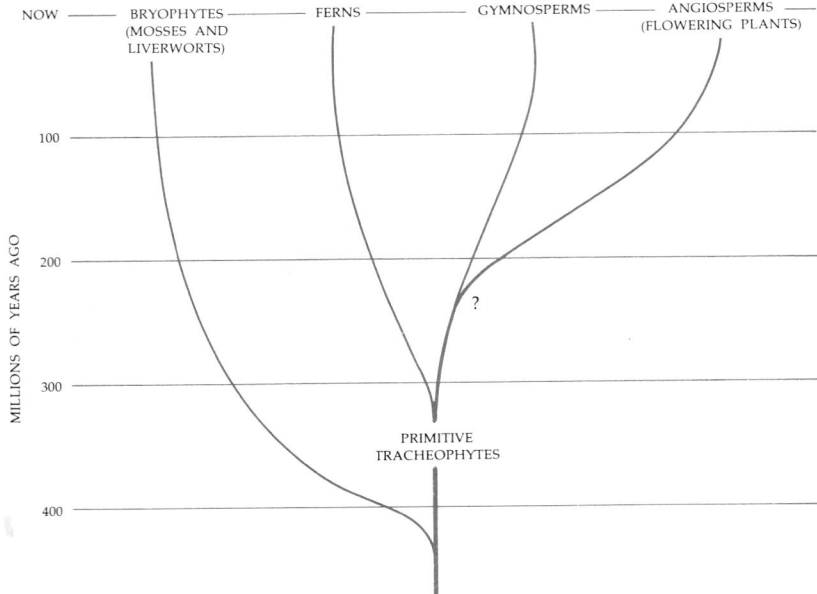

10–14 *The vascular plants, Tracheophyta, are characterized by specialized conducting tissues such as these tracheids, which carry water from the soil through stems and trunks to branches, leaves, and reproductive structures.*

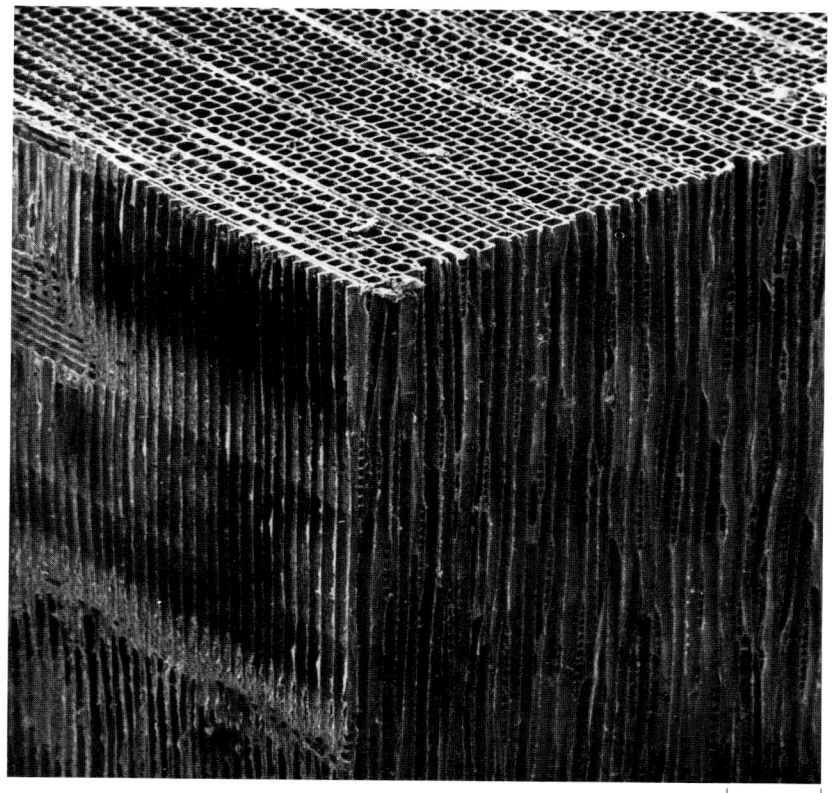

0.25 mm

the vascular plants, a large phylum that includes all of the dominant modern land plants.

The bryophytes and tracheophytes differ from one another in that the tracheophytes have a well-developed vascular system that transports water, minerals, sugars, and other nutrients. They also have a number of similarities, however. The tissues of members of both phyla are more specialized than the algae; this trend toward tissue specialization, with differentiation of the parts of the plant body and the development of special organs, reaches its fullest expression in the flowering plants, the angiosperms. In all the land plants, however (as distinct from the algae), the egg cells develop in the body of the parent plant in a specialized multicellular receptacle. Alternation of generations is also found in all of them, although the pattern differs from group to group. Cell division in all these groups involves the formation of a cell plate (Figure 10–11), which is almost never found among the algae.

Furthermore, all except the liverworts have *stomata* (singular, stoma), which are openings surrounded by specialized pairs of cells that admit gases (carbon dioxide and oxygen) into leaves and other parts of the plant body. Some of the mosses also have a *cuticle,* a waxy protective outer layer found on leaves and other surfaces of tracheophytes.

These similarities are taken as evidence that the bryophytes and tracheophytes had a common land-dwelling ancestor.

341 *The Plant Kingdom*

A haircap moss with spore capsules. The lower green structures are female gametophytes; the stalks and capsules are the sporophytes.

PHYLUM BRYOPHYTA: MOSSES, HORNWORTS, AND LIVERWORTS

The bryophytes, like frogs and toads, are essentially amphibious. Lacking specialized vascular tissues, they absorb moisture through their leaves and stems. They grow largely in moist, shady places and in bogs. Some of them, like sphagnum (peat moss), are capable of absorbing and holding large amounts of water, so that they maintain a watery existence even on dry land. Most bryophytes are tropical plants, but some species occur in temperate regions, and some even reach the Arctic and Antarctic.

In the damp environments frequented by the bryophytes, individual cells can, with relative ease, absorb water and nutrients directly from the air or by diffusion from nearby cells. The plants do not have any stiff supporting structures to hold them erect, and, in general, they have no conducting cells to transport water up the plant body. As a consequence, many of them are very small, and even the largest can reach little more than a foot above the ground.

Bryophytes have motile sperm, which develop on the gametophyte in *antheridia*. Egg cells develop on the gametophyte in *archegonia*. Free water is necessary for the completion of their sexual cycle, because the sperm must swim to the egg. Thus the bryophytes are incompletely adapted to land.

Table 21–3 *The Bryophytes (Land plants without true vascular systems)*

Class	Common name	No. of species
Hepaticae	Liverworts	9,000
Anthocerotae	Hornworts	100
Musci	Mosses	14,500

The bryophytes, however, have some special structural adaptations to a terrestrial existence. They have small leaflike structures in which photosynthesis takes place, and, as in the higher plants, the body of the plant is specialized for support and food storage. Thus the bryophytes resemble other land plants far more than they do the algae, although they are not usually classified as "higher plants" and do not seem to have been their ancestors.

The bryophytes do not have roots. In the mosses, the individual plants grow up like branches from a network of horizontal filaments known as protonemata (singular, protonema). A single moss plant, interconnected by protonemata, may grow to a conspicuous size, but most liverworts are so small that they are noticeable only to a keen observer. In both mosses and liverworts, small leaflike structures lack the specialized tissues of the leaves of higher plants and are only one or a few cell layers thick; hence moss leaves are believed to have evolved separately from "true" leaves.

Bryophyte Reproduction

In the bryophytes, unlike any tracheophytes, the gametophyte (the haploid form, you will recall) is larger than the sporophyte (see Figure 21–15). The antheridia and archegonia are borne on the gametophyte. The sperm, which are biflagellate, are released when sufficient moisture is present, enabling them to swim to the archegonium, to which they are attracted chemically. Fusion takes place within the archegonium.

Inside the archegonium, the zygote develops into a sporophyte, which remains attached to the gametophyte and is often nutritionally dependent on it. Typically the sporophyte has a single, large sporangium (often in the form of a capsule) elevated on a stalk, from which the spores are discharged.

Asexual reproduction, often by fragmentation, is also common.

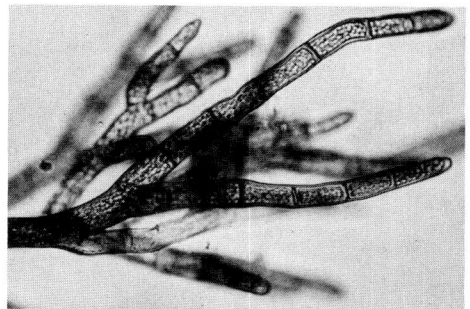

-16 *Protonema of a moss. Protonemata, which are young gametophytes, are characteristic of some bryophytes. They resemble filamentous green algae.*

-17 (a) *Young liverwort growing on a rock. In the liverwort* Marchantia, *the archegonia and the antheridia are formed on different plants, the female (b) and the male (c) gametophytes. The zygote, formed in the archegonium, develops into the sporophyte, which remains attached to the female gametophyte. The cuplike gemmae visible in (c) participate in asexual reproduction.*

(a)

(b)

(c)

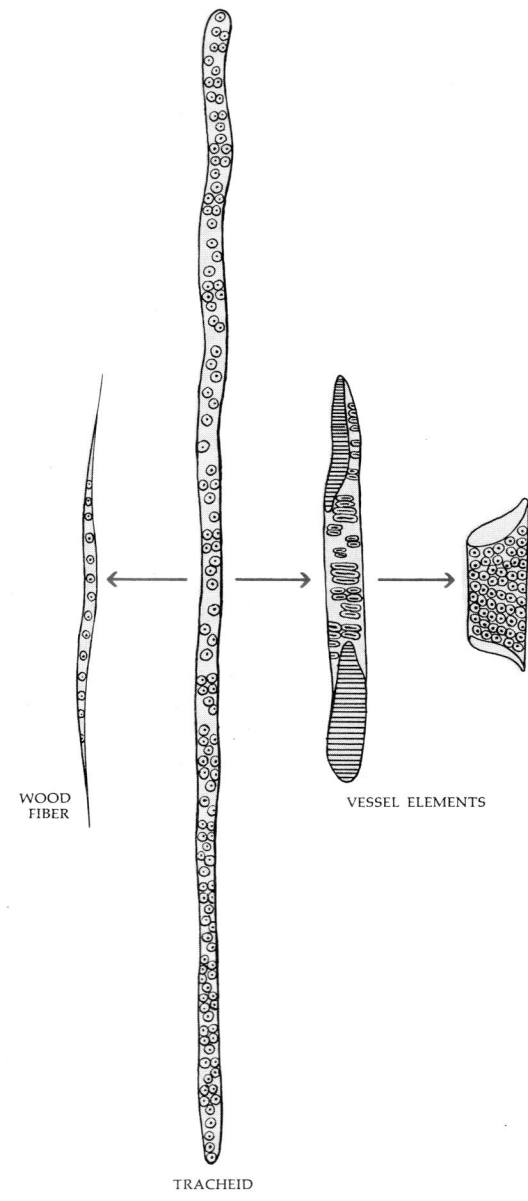

WOOD
FIBER

VESSEL ELEMENTS

TRACHEID

PHYLUM TRACHEOPHYTA: THE VASCULAR PLANTS

The bryophytes never managed to solve completely the problems of independent existence on dry land. Instead, they generally circumvent these problems by growing only in moist environments or during the moist, cool times of the year and reproducing sexually only when and if free water is available. A chief difference between the bryophytes and the tracheophytes is that all members of the latter group, although to varying degrees, have vascular systems made up of specialized conducting tissues. These tissues conduct water and minerals up to the leaves, and sugar from the leaves to the other parts of the plant. In addition, since the vascular systems also serve a supporting function, tracheophytes are able to grow taller than the more weakly supported bryophytes, thereby shading them.

The vascular system consists of two distinct tissues: the _phloem,_ which conducts sugars and other soluble organic molecules from the leaves to other parts of the plant body, and the _xylem,_ which conducts water and minerals up from the roots. The conducting elements of the phloem are _sieve cells_ or _sieve-tube elements,_ and the conducting elements of the xylem are _tracheids_ and _vessel elements._ In stems, the strands of xylem and phloem are side by side, either in vascular bundles or arranged in two concentric layers (cylinders) in which one tissue (typically the phloem) appears outside the other.

Tracheophytes also have true leaves, specialized photosynthetic organs that contain vascular tissues (veins) continuous with the vascular tissue of the stem. These leaves are covered with the cuticle, a waxy layer that conserves water in the leaf tissue. The plant body is also clearly differentiated into shoot and root.

Three groups of vascular plants still have living representatives. Two of the groups are small: the Lycophytina (club mosses) and the Sphenophytina (horsetails). Some examples are shown in Figure 21–19.

Members of the third group, the Pterophytina, differ from the others in that their vascular systems are more highly specialized. Also they have large leaves with complex venation (megaphylls) rather than the small leaves with single veins (microphylls) of other tracheophytes.

The Pterophytina, a very large subphylum, includes all the major groups of land plants: ferns, gymnosperms, and flowering plants.

21–18 _Evolutionary relationships among some cells of the tracheophytes. Tracheids, which are water-conducting cells of the conifers, are believed to resemble the more primitive (that is, earlier evolved) cells. Tracheids are composed of cells with thin areas (pits) in their lateral walls through which water moves from one tracheid to another up the trunk from the roots. Tracheids also serve as mechanical supporting structures. From tracheids evolved wood fibers, specialized for support, and vessel elements, specialized for conducting water. In the most highly evolved vessels, the end walls of the individual cells (vessel elements) disintegrate during development and the elements are stacked on top of one another, leaving a continuous tube. Both vessels and wood fibers are presumed to have been derived from primitive tracheids, the water-conducting tissues of primitive vascular plants._

1–19　(a) *In the club mosses,* Lycopodium, *the sporangia are borne on specialized leaves, sporophylls, which are aggregated into a cone at the apex of the branches, as shown in this running ground pine,* Lycopodium complanatum. *The airborne, waxy spores give rise to small, indepen-* dent, subterranean gametophytes. The sperm, which are biflagellated, swim to the archegonium, and the plant embryo develops there. (b) Tree club moss, Lycopodium obscurum. (c) *The horsetails, of which there is only one living genus* (Equisetum), *are easily recognized by their jointed, finely ribbed stems, which contain silicon. At each node, there is a circle of small, scalelike leaves. Spore-bearing leaves (sporophylls) are clustered into a cone at the apex of the stem. The gametophytes are independent, and the sperm are coiled, with numerous flagella.*

(a)

(b)

(c)

1–20　*Reconstruction of a Devonian forest showing western New York as it is believed to have looked 370 million years ago. The tall plant at the right is a relative of the club mosses. Although modern club mosses are typically only a few inches high, these primitive trees often grew to heights of 100 feet. The group of plants at the left are tree ferns. In the foreground at the far right is a primitive leafless plant,* Psilophyton, *related to the ferns. To the left are several clumps of horsetails, identifiable by their whorled branches.*

345　*The Plant Kingdom*

Table 21–4   *Vascular Plants: Phylum Tracheophyta*

	No. of species	Leaves	Gametophytes	Sperm	Seed
Subphylum Lycophytina (Club mosses)	1,000	Microphylls, in spirals	Independent	Biflagellate	Not present in modern forms
Subphylum Sphenophytina (Horsetails)	12	Scalelike, in whorls	Independent	Multiflagellate	Not present
Subphylum Pterophytina	More than 260,000	Megaphylls	Small, usually microscopic	Flagellated in some	Present in most forms
Class Filicinae (Ferns)	11,000	Megaphylls	Independent	Flagellated	Not present in modern forms
Class Coniferinae	550	Megaphylls, often reduced to needles or scales	Dependent, reduced	Not flagellated, carried in pollen tubes	Present, naked seed
Class Cycadinae	100	Megaphylls, large and fanlike	Dependent, reduced	Flagellated but carried in pollen tubes	Present, naked seed
Class Ginkgoidae	1	Megaphylls, fanlike	Dependent, reduced	Flagellated but carried in pollen tubes	Present, naked seed
Class Gnetinae	70	Megaphylls, often scalelike	Dependent, reduced	Not flagellated, carried in pollen tubes	Present, naked seed
Class Angiospermae (Flowering plants)	About 250,000	Megaphylls	Greatly reduced	Not flagellated, carried in pollen tubes	Present, enclosed in mature ovary (fruit)

Gymnosperms brackets Class Coniferinae, Class Cycadinae, Class Ginkgoidae, and Class Gnetinae.

Class Filicinae: Ferns

Ferns are vascular plants that are generally distinguishable from most other plants by their large feathery leaves that, in most species, unroll from base to tip during growth. The sporangia are on the undersurface of the leaves or, sometimes, on specialized leaves. Such spore-bearing leaves are called *sporophylls*. As we shall see, the carpels and stamens of flowers are also sporophylls.

According to the fossil record, ferns first appeared almost 400 million years ago, and they have remained relatively abundant until the present time. Most of the 11,000 living species in this class are found in tropical regions, but some occur in temperate and even arid portions of the globe. Because they have flagellated sperm and need free water for fertilization, those species growing in arid regions must, like the bryophytes, exploit the seasonal occurrence of water for sexual reproduction.

Ferns have anatomically simple stems compared with those of gymnosperms and angiosperms, and these stems are often reduced to a creeping

(a)

(b)

–21   (a) *The immature sporophyte of many common ferns develops as a "fiddle head," which uncoils and spreads as it elongates. (b) A cinnamon fern. The sporangia are borne on separate stalks.*

–22   *Fern spores develop on the sporophyte in sporangia, which are usually found in clusters (sori) on the underside of a leaf (sporophyll), as in the* Woodsia obtusa *shown here.*

underground stem called a *rhizome.* Although ferns do not exhibit secondary growth—the type of growth that results in increase of girth and formation of bark and woody tissue—some grow very tall. For instance, *Cyathea australis,* a tree fern found on Norfolk Island in the South Pacific, sometimes reaches 85 feet in height.

The leaves of ferns are usually finely divided into pinnae; these divided leaves spread widely and so collect more light, and apparently they are thus adapted to growing on the forest floor in diffused light. The sporangia develop on the lower surface of sporophylls, which may resemble the other green leaves of the plant or may be nonphotosynthetic stalks (modified leaves). The sporangia commonly occur in small clusters known as sori (singular, sorus).

In the ferns, as in all the vascular plants, the predominant plant form is the sporophyte generation. The gametophyte begins development as a small alga-like chain of cells, each filled with chloroplasts, and then develops into a flat, membranous structure. Although the gametophyte is small and sometimes microscopic, it is nutritionally independent (autotrophic), as is the sporophyte. In all but a few genera, it produces both antheridia and archegonia. The sperm are coiled and multiflagellate.

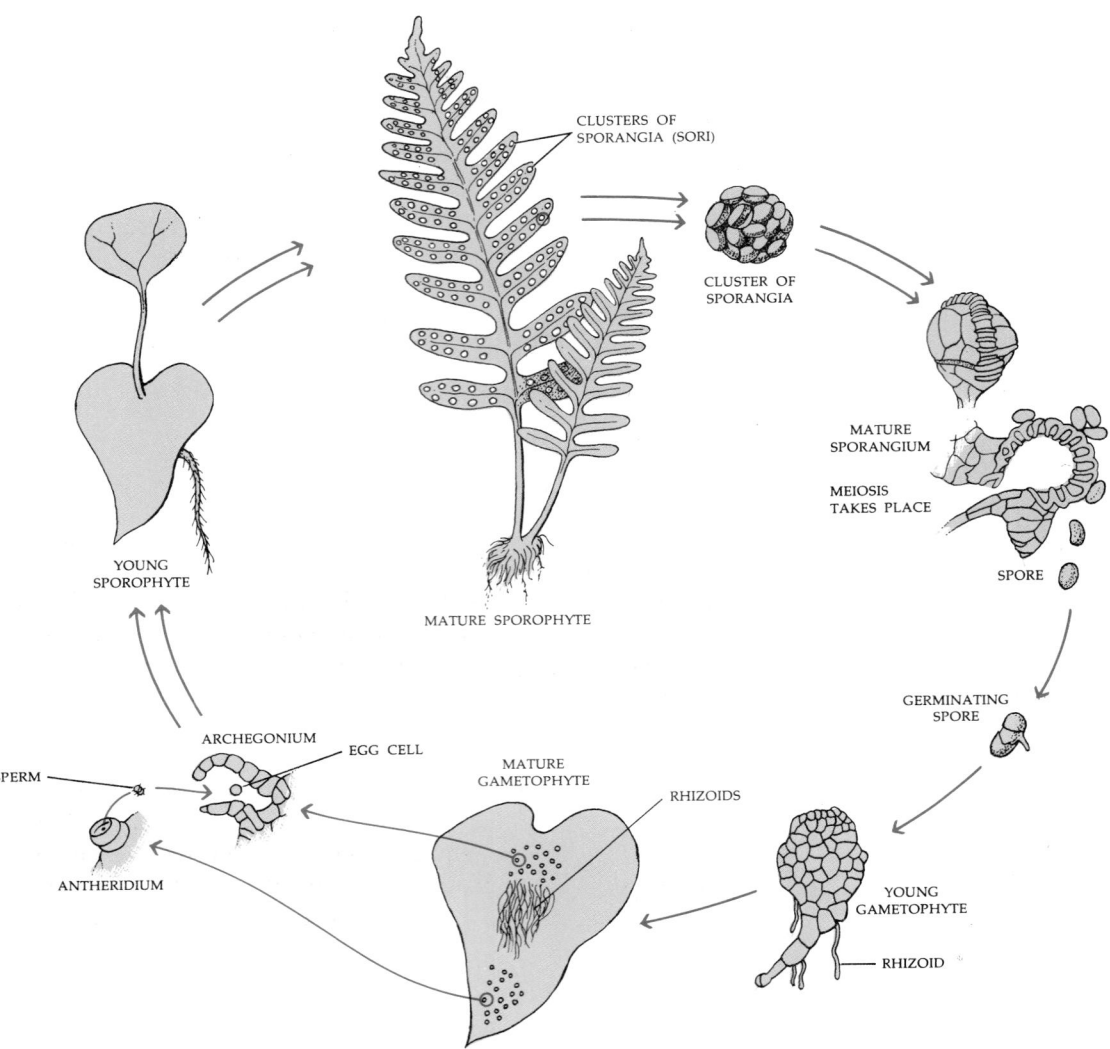

CLUSTERS OF
SPORANGIA (SORI)

CLUSTER OF
SPORANGIA

MATURE
SPORANGIUM

MEIOSIS
TAKES PLACE

SPORE

MATURE SPOROPHYTE

YOUNG
SPOROPHYTE

GERMINATING
SPORE

ARCHEGONIUM   EGG CELL

MATURE
GAMETOPHYTE

RHIZOIDS

SPERM

ANTHERIDIUM

YOUNG
GAMETOPHYTE

RHIZOID

21–23 *The life cycle of a fern. The spores, which are haploid, are produced by meiosis in sporangia on the lower surface of a leaf (sporophyll) and are shed. The spores develop into haploid gametophytes which are only one layer of cells thick and are somewhat heart-shaped, as shown here (bottom). From the lower surface of the gametophyte, rootlike filaments, the rhizoids, extend downward into the soil.*

On the lower surface of the gametophyte are the flask-shaped archegonia, which enclose the egg cells, and the antheridia, *which enclose the multiflagellate sperm. When the sperm are mature and there is an adequate supply of water, the antheridia burst and the sperm cells swim to the archegonia and fertilize the eggs. From the 2n zygote, the 2n sporophyte grows out of the archegonium within the gametophyte. After the young sporophyte becomes rooted in the soil, the gametophyte disintegrates. When the sporophyte becomes mature, certain of its leaves (sporophylls) develop sporangia, in which meiosis occurs, and the cycle repeats again.*

−24　*The cedars of Lebanon, great conifers with a life span of about a thousand years, were used to build the palaces of Solomon, the fleets of Alexander the Great, and the cross on which Jesus was put to death. Now in Lebanon, only a few hundred remain.*

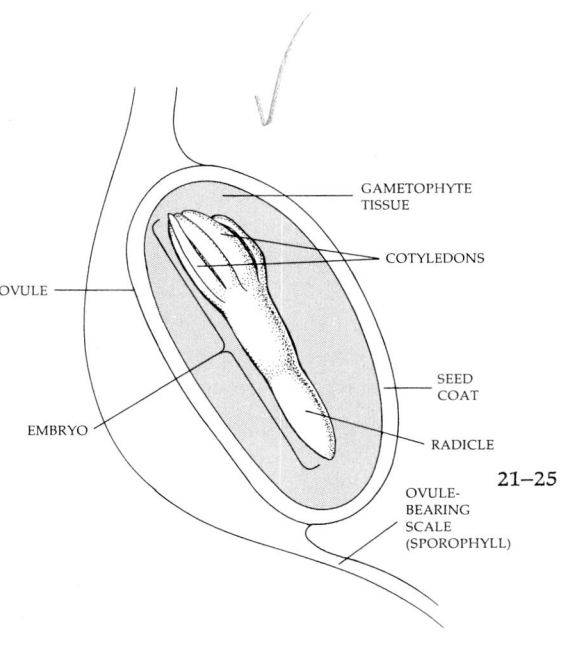

GAMETOPHYTE TISSUE

COTYLEDONS

OVULE

SEED COAT

EMBRYO

RADICLE

OVULE-BEARING SCALE (SPOROPHYLL)

21–25　*Pine seed. The ovule has hardened into a seed coat, enclosing the female gametophyte and the embryo, which now consists of an embryonic root, the radicle, and a number of embryonic leaves, the cotyledons. When the seed germinates, the radicle will emerge from the seed coat and penetrate the soil. When the root takes in water, the tightly packed cotyledons will elongate and swell with the moisture, rising above ground on the lengthening stem and forcing off the seed coat.*

Gymnosperms

The humid Paleozoic era was the age when our coal deposits were formed from lush vegetation that sank so swiftly into the warm, marshy soil that there was no chance for it to decompose. At the close of this era, in the Permian period (see Table 21–5, page 358), there were worldwide changes of climate, with the advent of widespread glaciers and drought. The amphibians gave way to the scaly-skinned reptiles, which were better suited to the harsher climate. Under similar selection pressures, plants began to evolve that, because of specialized structures, could maintain life during periods when no water was available.

It was during the Permian period that the gymnosperms—the naked-seed plants—evolved. There are four groups of gymnosperms with living representatives, three small (the cycads, the ginkgos, and the gnetinae) and one large and familiar (the conifers).

The Seed

The seed is a protective structure in which the embryonic plant can lie dormant until conditions become favorable for its survival. Thus, in its function, it parallels the spores of bacteria or the resistant zygotes of the freshwater algae. In structure it is far more elaborate, however. A seed includes the embryonic plant (the young, dormant sporophyte), a store of nutritive tissue, and an outer protective coat.

To understand the evolution of the seed it is necessary to return for a moment to the principle of alternation of generations found in all the vascular plants. In the green algae, the two generations, sporophyte and gametophyte, are independent (Figures 21–11 and 21–12). In the fern life cycle shown in Figure 21–23, the gametophytes, although still independent, are much smaller than the sporophytes. In the seed plants, the gametophyte generation is reduced still further and is totally dependent on the sporophyte. The female gametophyte develops on the mother sporophyte within a structure known as the *ovule*. Within the female gametophyte, one or more egg cells are formed. Sperm cells develop in another gametophyte, also protected and nourished by a sporophyte. They are carried to the female gametophyte and fertilize the eggs developing there. The zygote develops into an embryo. Following fertilization, the ovule enlarges and its outer surface hardens into a protective seed coat around the embryo and the gametophyte tissue in which the embryo is embedded. This complex is released as the seed.

Let us look at a specific example, the formation of a pine seed. A pine tree has two types of cones. The small "male" cones resemble those of club mosses, but on the large "female" cones, the scales (sporophylls) that bear the ovules

Male cones of jack pine (Pinus bank-siana) shedding pollen. The pollen grains are male gametophytes, which complete their maturation when they reach the ovules, embedded in the female cones. There they produce the naked (wall-less) sperm cells that fuse with the egg cells.

are much thicker and tougher. In the male cones, specialized cells inside the sporangia undergo meiosis to produce spores. Each spore differentiates into a microscopic, windborne pollen grain, a young male gametophyte. The wind is an unreliable messenger, disseminating the male gametophytes at random; therefore, natural selection has favored those plants that produce pollen in great excess (Figure 21–26).

Fertilization

Within the ovules of the female cones, a second type of spore is formed by meiosis. (Spores that give rise to female gametophytes are often referred to as megaspores, as distinct from microspores, from which male gametophytes arise.) Of the four spore cells produced by each meiotic sequence, three disintegrate and the remaining one forms a tiny gametophyte. This gametophyte grows within the cone and develops two or more archegonia, each of which contain a single egg cell. The development from the spore to the gametophyte with its egg cells may take many months—slightly more than a year, for example, in some common pines. The ripening ovule secretes a sticky liquid. When the cones become dusted with pollen, some of it sifts down between the cone scales and comes in contact with the liquid. Pollen grains are caught in the liquid, and as it dries, they are drawn into the neck of the ovule. Here the pollen grain continues its development. The male gametophyte produces two nonmotile nuclei, the male gametes. These travel toward the egg through the pollen tube, which is produced by the male gametophyte and which grows through the tissue of the ovule. During this period, both male and female gametophytes draw their nourishment from the tissues of the ovule-bearing sporophyte.

Because the drought-resistant pollen is blown to the ovule of the female cones by the wind and the sperm nuclei are carried to the egg by the pollen tube, the pines and other conifers are not dependent on free water for fertilization. Thus they are able to reproduce sexually when ferns and bryophytes cannot.

Following fertilization, the zygote begins to divide and form the embryo. As the ovule matures, its outer walls harden into a seed coat, enclosing both the embryo and the female gametophyte (the latter provides food for the embryo when the seed germinates). After the cone matures, it releases its seeds. As in most conifers, the seeds are winged and are spread widely by the wind.

Figure 21–25 shows cross sections of a pine seed. The seed coat and the wing on which the seed is carried arise from the hardened outer layers of the ovule, derived from the mother sporophyte. The next layer represents the body of the gametophyte; swollen and packed with stored food reserves, it grows and crushes out most of the original sporophyte tissue. The inner core is the embryo with its many *cotyledons,* the embryonic leaves which will appear as the first leaves of the shoot of the new sporophyte when the seed germinates. The lower part of the embryo, the radicle, will form the root.

The complete life cycle of a pine is shown in Figure 21–27.

The seed was not a gymnosperm invention; seeds have been found in fossil club mosses. However, there are no survivors among these more ancient lines of seed plants.

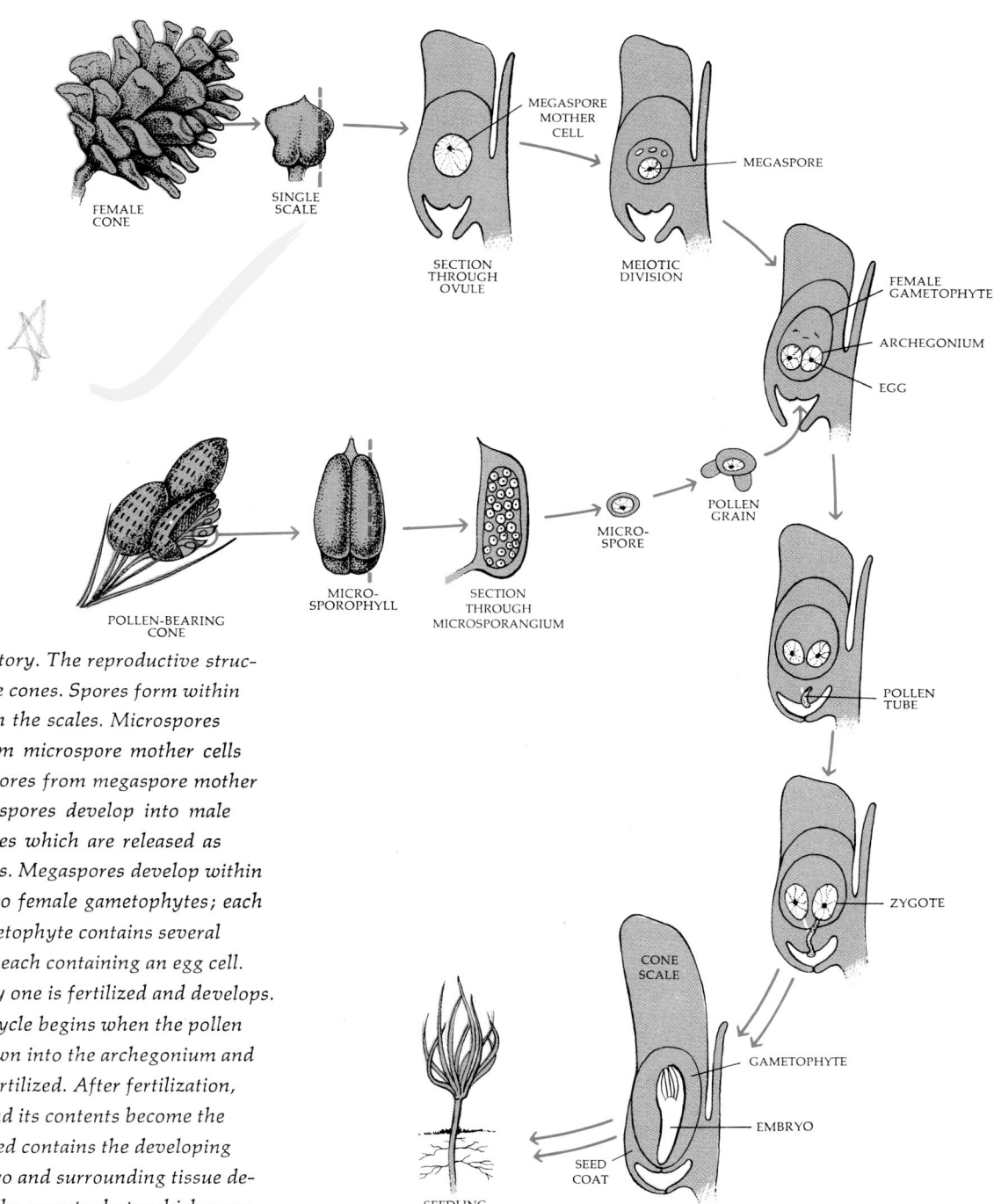

FEMALE CONE

SINGLE SCALE

SECTION THROUGH OVULE

MEGASPORE MOTHER CELL

MEIOTIC DIVISION

MEGASPORE

FEMALE GAMETOPHYTE

ARCHEGONIUM

EGG

POLLEN-BEARING CONE

MICRO-SPOROPHYLL

SECTION THROUGH MICROSPORANGIUM

MICRO-SPORE

POLLEN GRAIN

POLLEN TUBE

ZYGOTE

CONE SCALE

GAMETOPHYTE

EMBRYO

SEED COAT

SEEDLING (SPOROPHYTE)

MATURE SEED ON CONE SCALE

4–27   Pine life history. The reproductive structures are the cones. Spores form within
the cones on the scales. Microspores
develop from microspore mother cells
and megaspores from megaspore mother
cells. Microspores develop into male
gametophytes which are released as
pollen grains. Megaspores develop within
an ovule into female gametophytes; each
female gametophyte contains several
archegonia, each containing an egg cell.
Usually only one is fertilized and develops.

The life cycle begins when the pollen
grain is drawn into the archegonium and
the egg is fertilized. After fertilization,
the ovule and its contents become the
seed; the seed contains the developing
plant embryo and surrounding tissue derived from the gametophyte which nourishes the embryo. As the seed matures,
the cone opens, releasing the winged
seeds, which germinate, producing the
sporophyte. Both types of cones develop
on the sporophyte.

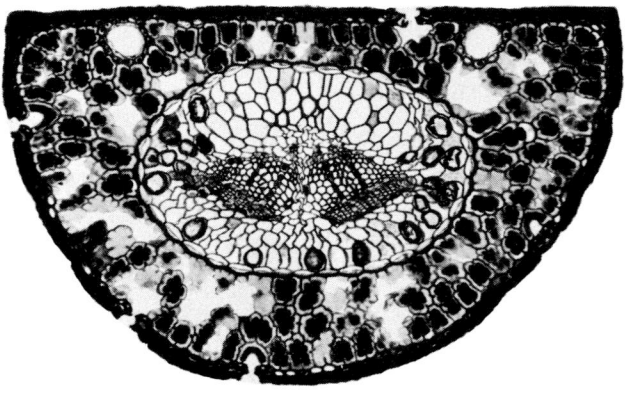

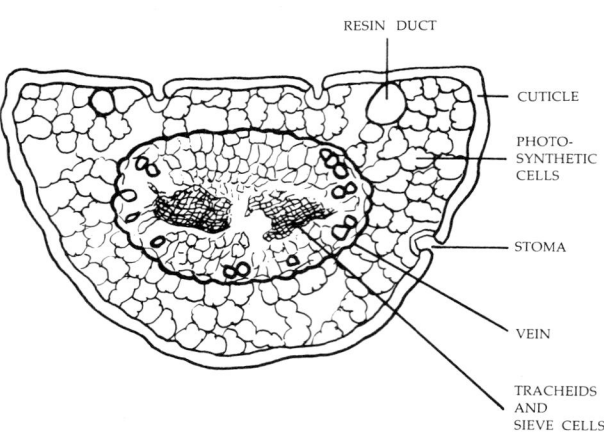

RESIN DUCT

CUTICLE

PHOTO-
SYNTHETIC
CELLS

STOMA

VEIN

TRACHEIDS
AND
SIEVE CELLS

21–28   *Cross section of a pine leaf. The hard outer covering and compact shape protect the leaf from water loss, an essential factor in the survival of these trees in areas in which there is little rainfall, or in which the water is locked in the ground as ice during many months of the year.*

In two of the other groups of living gymnosperms, the cycads and ginkgos, the male gametophytes are carried to the ovules in the form of pollen, as in the conifers. The male gametophytes do not complete their development until they have reached the ovule. Then particular cells of the gametophytes divide to produce microscopic (about 80 micrometers long) multiflagellated sperm that tumble a short distance to the egg cells and fuse with them. Thus these gymnosperms carry the evolutionary traces of their ancestors, which were dependent on free water for fertilization.

The Needle

Another feature commonly associated with conifers, although not with all gymnosperms, is the needlelike leaf. Figure 21–28 shows a cross section of a pine needle, which may be 4 or more inches long but only $\frac{1}{16}$ to $\frac{1}{8}$ inch in diameter. In the center you can see the single vein, the vascular transport system, which carries water in one set of conducting tubes (the tracheids) and sugars in another (the sieve cells). Outside the vein are parenchyma cells in which photosynthesis takes place. The ducts on the facing sides of the needles carry resin, a substance that is released if the plant is wounded and serves to close the break. The outside layer of cells is porous but very hard.

The needlelike leaf, despite its small size, is a megaphyll and is a special adaptation to the long periods of drought characteristic of many regions in which modern conifers are abundant. (In areas with more precipitation and more moderate temperature, they cannot usually compete with angiosperms.)

Class Angiospermae: Flowering Plants

The angiosperms are generally agreed to have arisen from a now-extinct group of gymnosperms. They appear in the fossil record suddenly and in abundance during the Cretaceous period, about 100 million years ago as the dinosaurs were vanishing and small rodentlike mammals were making their first appearance. Numerous genera appeared suddenly at this time, and many of these seem to have been very similar to our modern plants.

According to Daniel Axelrod, a paleobotanist at the University of California, angiosperms probably arose long before this time during the Permian period. They are likely to have originated on the less fertile hills and uplands of tropical areas, the richer lowlands being crowded with ferns, gymnosperms, and lycopods. Once established, they spread into the lowlands, where they soon became the dominant plant forms and were deposited as fossils. During this mid-Cretaceous period, the climate of the Earth was warmer and more uniform than it is at present, and by the end of the Cretaceous period, much of the land was covered with a rich forest of angiosperms, reaching almost as far north as the Arctic Circle.

About 250,000 different species of angiosperms are known. Just listing them, one after another, would require a book twice the length of this one. They dominate the tropical and temperate regions of the world. The angiosperms include not only the plants with conspicuous flowers but also most of the great trees, the oak, the willow, the elm, the maple, and the birch; all the fruits, vegetables, nuts, and herbs; the cactus and the coconut; and all the corn, wheat, rice, and other grains and grasses that are the staples of man's diet and the basis of his agricultural economy all over the world.

Angiosperms, like other vascular plants, contain chlorophylls *a* and *b* and beta-carotene, and have megaphylls, stomata, and a cuticle impervious to water. The modern forms have a more highly evolved vascular system than is found in other groups (Figure 21–18). They also have two new, interrelated features that distinguish them from all other plants: the flower and the fruit.

–29 *Angiosperms, the most recently evolved of the tracheophytes, now dominate most landscapes. Here fleabane, also known as Coulter's daisy, carpets an alpine meadow in Colorado in mid summer.*

21–30 *The generalized structure of a flower. The reproductive structures are shown in color. In the center is the carpel, consisting of the ovary, the style, and the pollen-catching stigma. Two stamens are shown, each consisting of a slender filament and a pollen-producing anther. Carpels and stamens are surrounded by petals that attract pollinators to the flower. At the base of the flower are the sepals, which once enclosed the entire structure when it was a bud.*

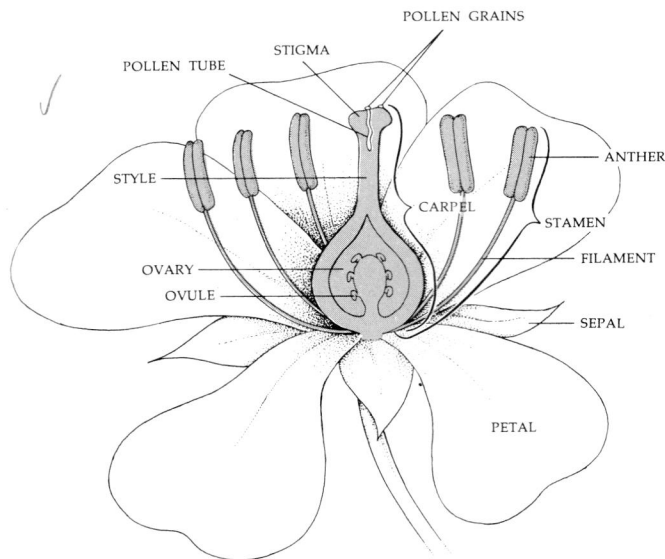

21–31 *Presumed evolutionary development of simple and compound ovaries. The three-lobed carpel with submarginal ovules (upper left) folded in on itself and the margins fused to form a simple ovary (upper right). Compound ovaries (lower) were formed by the fusing of separate infolded carpels.*

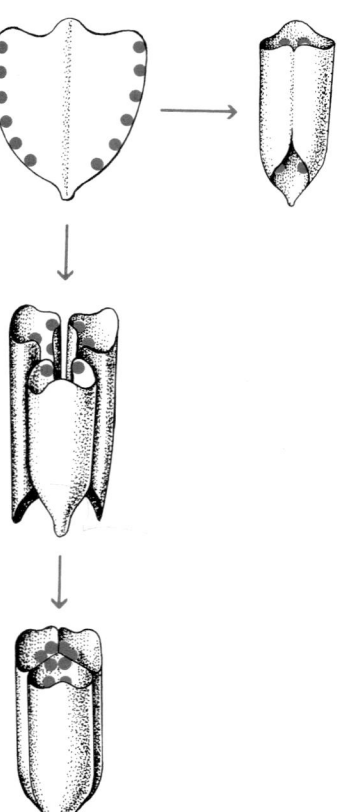

The Flower

Figure 21–30 is a diagram of a simple flower. The central structure is the *carpel,* the female reproductive structure.* The carpel is believed to be a modified leaf, which, in the course of evolution, enfolded in such a way that the ovule is attached to its inner surface (see Figure 21–31). A single carpel may contain one or more ovules, and a single flower, one or more carpels. The carpels may be separate (blackberry) or fused (tomato and apple). The swollen base of the carpel or carpels is the ovary. Within the ovary is the ovule, or ovules, in which the female gametophyte develops. The tip of the carpel has become specialized as a *stigma,* a sticky surface to which pollen grains adhere.

The pollen grains (male gametophytes) develop in the *stamen,* which, like the carpel, is the evolutionary descendant of a sporophyll. It consists of the *anther,* which is a group of sporangia, and a *filament.*

The other parts of the flower are specialized to attract insects and other pollinators. Pollen grains, which are male gametophytes produced in the anthers, are carried to the stigma, usually of another flower, where they germinate, developing pollen tubes that grow down through the style toward the ovule. Flowers often have special strategies that prevent self-pollination; for example, the pollen may mature only after the stigma of that flower has been pollinated by pollen from another flower. The sperm cells are carried by pollen tubes to the female gametophyte within the ovule. This gametophyte characteristically consists of only eight cells, one of which is the egg cell.

Two sperm nuclei are carried through the pollen tube into the ovule to the female gametophyte. One fuses with the nucleus of the egg cell and another fuses with two nuclei of the gametophyte to produce a triploid (3*n*) tissue.

*A single carpel or a group of fused carpels used to be known as the pistil because of their resemblance to an apothecary's pestle. This older term is now being abandoned, however.

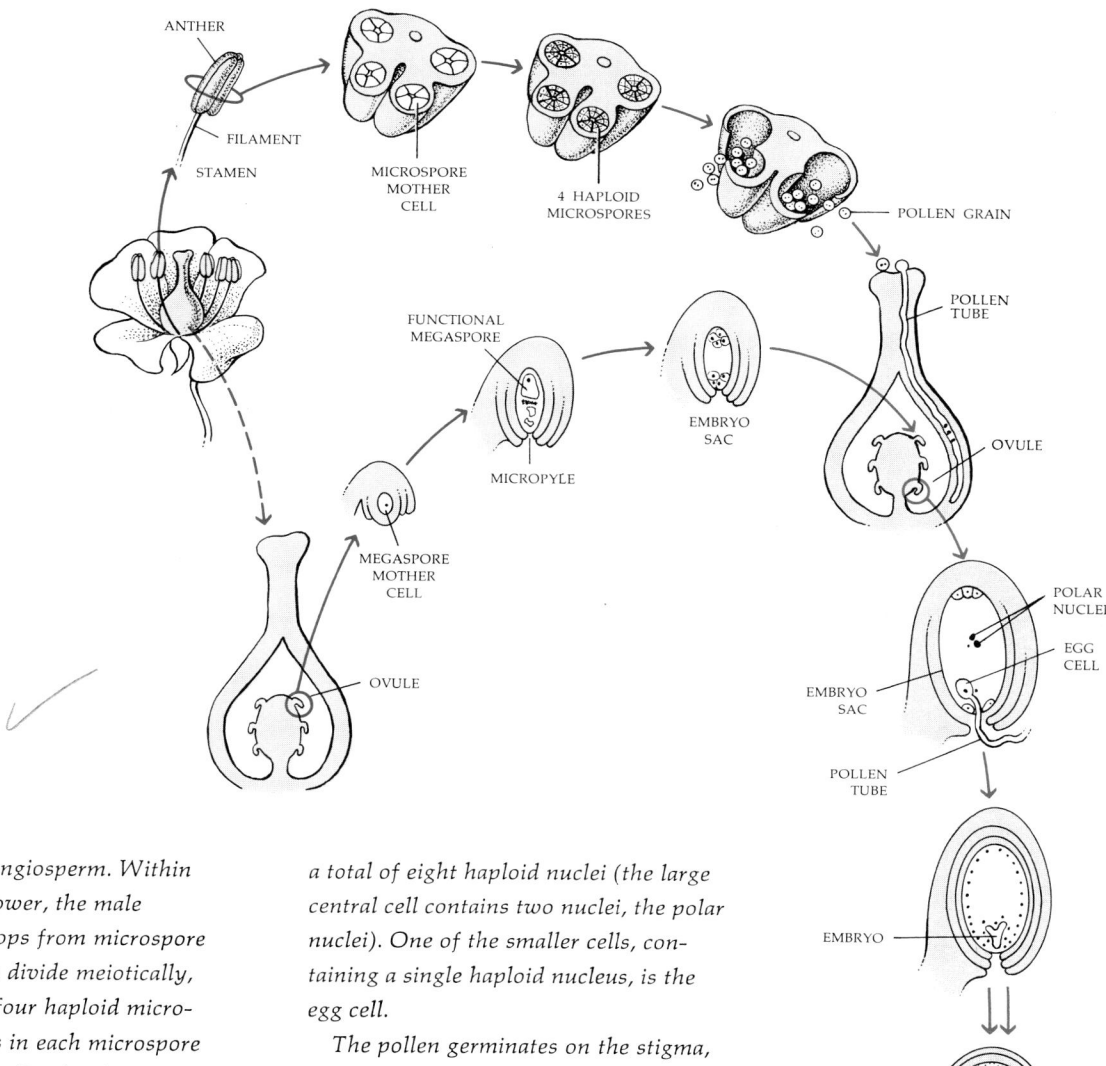

ANTHER

FILAMENT

STAMEN

MICROSPORE MOTHER CELL

4 HAPLOID MICROSPORES

POLLEN GRAIN

POLLEN TUBE

FUNCTIONAL MEGASPORE

OVULE

EMBRYO SAC

MICROPYLE

MEGASPORE MOTHER CELL

OVULE

POLAR NUCLEI

EGG CELL

EMBRYO SAC

POLLEN TUBE

EMBRYO

SEED

ENDOSPERM

SEEDLING (SPOROPHYTE)

−32 Life history of an angiosperm. Within the anther of the flower, the male gametophyte develops from microspore mother cells, which divide meiotically, each giving rise to four haploid microspores. The nucleus in each microspore then divides mitotically, developing into a pollen grain containing two nuclei. One of the nuclei subsequently divides again, usually upon germination, resulting in three nuclei per pollen grain: two sperm nuclei and one vegetative nucleus.

Within the ovule, the female gametophyte develops from a megaspore mother cell, which divides meiotically to produce four haploid megaspores. Three of the megaspores disintegrate; the fourth divides mitotically, developing into an embryo sac consisting of seven cells with a total of eight haploid nuclei (the large central cell contains two nuclei, the polar nuclei). One of the smaller cells, containing a single haploid nucleus, is the egg cell.

The pollen germinates on the stigma, producing a pollen tube that grows down the style into the ovary. The two sperm nuclei enter the embryo sac through the tube; one nucleus fertilizes the egg cell, the other merges with the polar nuclei, forming the triploid (3n) endosperm. The embryo undergoes its first stages of development while still within the ovary of the flower, and the ovary itself matures to become a fruit. The seed, released from the mother sporophyte in a dormant form, eventually germinates, forming a seedling.

21–33 *Angiosperms are characterized by fruits. A fruit is a mature ovary, including the seed or seeds and, often, accessory parts of the flower. The fruit aids in dispersal of the seed. Some fruits are borne on the wind, some are carried from one place to another by animals, some float on water, and some are even forcibly ejected by the parent plant. (a) Flowers and fruit of a dandelion. (b) In milkweed, the fruit is the seed pod, which when ripe bursts open, releasing seeds with tufts of silky hair that aid in their dispersal. (c) The seeds of fleshy fruits are usually dispersed by vertebrates that eat the fruits and later excrete the seeds. The female evening grosbreak shown here is eating the bright red fruits of winterberry, a member of the holly family. (d) Chipmunk stuffing his cheek pouch with an acorn. (e) In the tumbleweed the whole plant breaks off in the wind and is tumbled across open country, scattering the seeds.*

(a)

(b)

(c)

(d)

(e)

In the entire spectrum of living things, this phenomenon of triple fusion has been observed only among angiosperms. Neither its evolutionary history nor its adaptive value are understood. From the extraordinary 3n nucleus, a 3n tissue, the *endosperm*, is produced, which nourishes the young embryo. In some angiosperms, such as peas, the endosperm is used up in the course of development and is absent in the mature seed. In others, such as corn and wheat, it is present at maturity and nourishes the germinating sporophyte and, indeed, most of the human population of the world.

The Fruit

Following fertilization in the angiosperm, the *ovary* (the enlarged base of the carpel or carpels) develops into the fruit. As in the gymnosperms, the ovules develop into the seeds, each containing its embryo sporophyte. Fruits may develop from a single carpel or from fused carpels, and each carpel may contain one or more ovules. In the peach, for instance, the skin, the fleshy, edible portion of the fruit, and the stone are three layers of the matured ovary wall. The seed is inside the stone. This is an example of a simple seed formed from a single ovule. In the pea, the pod is the mature ovary wall, and the peas themselves are the seeds. The raspberry is an aggregate of many tiny, fleshy fruits, each formed from a separate carpel.

Fruits provide new and ingenious ways for seed dispersal. In some plants, the fruit itself carries wings, as in many of our common trees; in others, the fruit bursts open, shooting out the seeds. Some species of geranium send forth their seeds by a sort of slingshot. Often, the fruits are edible and brightly colored, tempting birds and animals to eat them. The seeds within the fruit pass unharmed through the digestive tract hours later and often miles away. Burrs adhere to fur, feathers, or one's trouser legs, to be carried by unwilling messengers to far-off fields and meadows. In the tumbleweed, the whole plant is blown across the open country, scattering seeds as it goes. Angiosperm seeds travel much farther and faster than the simpler seeds of the gymnosperms.

−34 *Development and structure of the apple. (a) Flower of apple. (b) Older flower, after petals have fallen. (c, d) Longitudinal and cross sections of the mature fruit. The core of the apple is the ripened ovary wall. The fleshy edible part develops from the floral tube.*

(a)

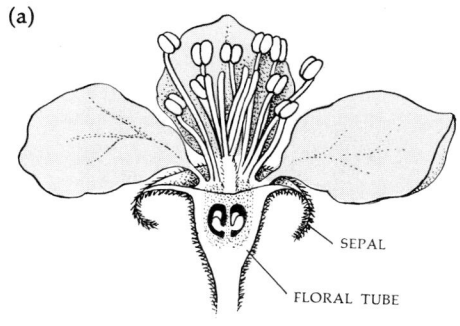

SEPAL

FLORAL TUBE

(b)

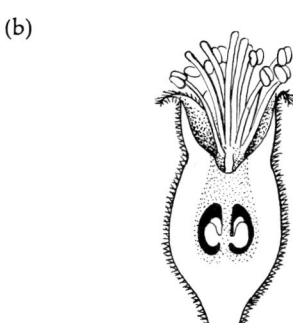

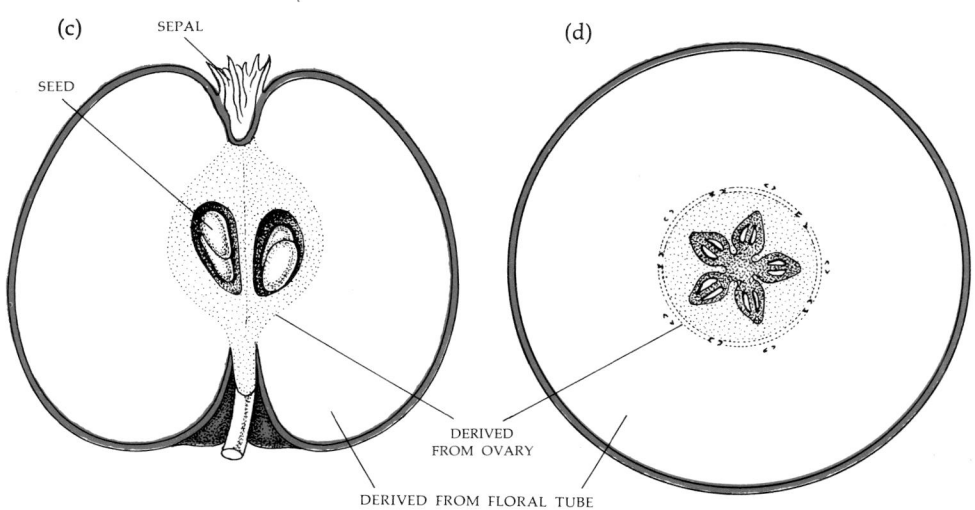

(c)

SEPAL

SEED

DERIVED FROM OVARY

DERIVED FROM FLORAL TUBE

(d)

Table 21–5 *Evolution of Land Plants: Major Physical and Biological Events in Geologic Time*

Millions of years ago	Era	Period	Life forms	Climates and major physical events
65–136	MESOZOIC	Cretaceous	Angiosperms become abundant by close of period.	Lands low and extensive. Last widespread oceans. Elevation of Rockies at end of period.
136–195		Jurassic	Gymnosperms, especially cycads and ferns. Dinosaurs' zenith.	Mild. Continents low. Large areas in Europe covered by seas. Mountains rise from Alaska to Mexico.
195–225		Triassic	Forests of gymnosperms and ferns.	Continents mountainous. Large areas arid. Eruptions in eastern North America. Appalachians uplifted and broken into basins.
230–280	PALEOZOIC	Permian	Origin of conifers and possible origin of angiosperms; earlier forest types wane.	Extensive glaciation in Southern Hemisphere. Appalachians formed by end of Paleozoic; most of seas drain from continent.
280–345		Carboniferous Pennsylvanian Mississippian	Great swamp forests of ferns, club mosses, gymnosperms, and horsetails. Insects become abundant.	Warm. Lands low, covered by shallow seas or great coral swamps. Mountain building in eastern U.S., Texas, Colorado. Moist, equable climate, conditions like those in temperate or subtropical zones, little seasonal variation, root patterns indicate water plentiful.
345–395		Devonian	Rise of land plants. Extinction of primitive vascular plants. Origin of modern subclasses of vascular plants. Amphibians appear.	Europe mountainous with arid basins. Mountains and volcanoes in eastern U.S. and Canada. Rest of North America low and flat. Sea covers most of land.
395–440		Silurian	First terrestrial plants. Modern groups of algae and fungi. Rise of fish and reef-building corals. Shell-forming sea animals abundant.	Mild. Continents generally flat. Mountain building in Europe. Again flooded.
440–500		Ordovician	First primitive fish. Invertebrates dominant. Microscopic plant life thriving.	Mild. Shallow seas, continents low; sea covers U.S. Limestone deposits.
500–600		Cambrian	Age of marine invertebrates. Shell animals.	Mild. Extensive seas. Seas spill over continents.
	PRECAMBRIAN		Earliest known fossils.	Dry and cold to warm and moist. Planet cools. Formation of Earth's crust. Extensive mountain building. Shallow seas.

The angiosperms differ from the other vascular plants by their more efficient conducting systems, their flowers, which increase the efficiency of fertilization, and their fruits, which increase the efficiency of distribution. They represent the most successful of all the land plants in terms of numbers of individuals, numbers of species, and their effects on the existence of other organisms. The anatomy and physiology of this dominant group of plants will be explored further in Section 4. Their ecological role will be described in Section 6.

SUMMARY

The plant kingdom comprises the multicellular algae (Phaeophyta, Rhodophyta, and Chlorophyta), the Bryophyta (mosses, liverworts, and hornworts), and the Tracheophyta (vascular plants).

The Phaeophyta (brown algae) and Rhodophyta (red algae) are the principal seaweeds. The brown algae, which include the kelps, are found more commonly in cooler water; the red algae, in the tropics. Both groups include forms differentiated into holdfast, stipe, and blade, analogous to the root, stem, and leaf of higher plants.

The Chlorophyta, the green algae, are hypothesized to be the ancestors of the land plants, on the basis of similarities between the groups. These similarities include the fact that both groups have chlorophylls *a* and *b* and beta-carotene as their photosynthetic pigments. Both store food reserves in the form of starch, and the cells of both have cellulose cell walls. Forms range from the single-celled *Chlamydomonas* to a variety of multicelled forms belonging to several different evolutionary lines represented, for example, by *Valonia, Spirogyra, Ulva,* and *Volvox.*

Alternation of generations is found among many algae and all land plants. The term designates a reproductive cycle in which sexual reproduction (by the fusion of gametes) and asexual reproduction (by means of spores) occur in alternating generations. In one generation, the plant body, known as the gametophyte, which is haploid (*n*), produces haploid gametes, the sperm and egg cells. The gametes fuse to form the zygote, which develops into a diploid (2*n*) sporophyte. The sporophyte produces spores by meiotic division. A spore is a single cell that, unlike a gamete, can develop into an adult organism asexually, that is, without combining with another cell. In alternation of generations, the spore, which is haploid, produces the haploid gametophyte.

Modern bryophytes are small plants found usually in moist locations Most lack specialized vascular tissues and true leaves, although the plant body is differentiated into photosynthetic, food storing, and anchoring tissues, as in higher plants. Unlike the algae but like the higher plants, they have multicellular reproductive structures in which the gametes develop. The female gamete (egg) develops in the archegonium; the male gamete (sperm) develops in the antheridium. The sperm, which are flagellated, swim to the archegonium and fertilize the egg cell. The zygote develops within the archegonium. In the bryophytes—and in no other members of the plant kingdom—the gametophyte (*n*) is dominant and the sporophyte (2*n*) smaller, attached, and often nutritionally dependent.

The important subdivisions of the true vascular plants, the tracheophytes, include the ferns, the gymnosperms, and the angiosperms, or flowering plants. All these groups have megaphylls (leaves with complex venation), stomata, cuticles, and specialized vascular tissue.

Ferns are characterized by large, often finely dissected leaves called fronds. The sporophyte is the dominant generation, but the gametophytes are independent. The sperm are flagellated, and free water is needed for fertilization. Sporangia are characteristically formed on the underside of leaves (sporophylls) of the sporophyte.

The conifers are the largest group of gymnosperms—naked-seed plants—so called in contrast to the angiosperms in which the seed is enclosed in an ovary which develops to form a fruit. Among conifers, the male gametophyte is formed on modified leaves (scales); these sporophylls form a cone. The gametophytes are released in the form of windblown pollen. The female gametophyte develops on the scales of a separate, larger cone within an ovule composed of the tissue of the parent sporophyte. Within the female gametophyte, an archegonium forms. The male gametophyte germinates and produces a pollen tube through which nuclei (the sperm) enter the archegonium and fertilize the egg cell. The seed consists of the ovule and its contents, including the young embryo and nutritive tissue derived from the gametophyte. The seed, which is shed from the female cone, can remain dormant for long periods of time and hence is adapted to withstanding cold and drought.

The angiosperms, of which there are about 250,000 species, are the dominant modern land plants. They are characterized by the flower and the fruit. The reproductive structures of the flower are the stamens, composed of filament and anther, and the carpels, composed of ovary, style, and stigma. The stamens and the carpels are highly specialized sporophylls. The ovule or ovules are enclosed in the ovary at the base of the carpel. Pollen grains, the male gametophytes, are formed in the anthers and germinate on the sticky surface of the stigma. The sperm cells are carried through a pollen tube which grows down the style into the ovary and enters the ovule. As the zygote begins to develop, the ovary enlarges and the other parts of the flower fall away. The ovary becomes the fruit, enclosing the mature ovules, the seeds. Flowers attract pollinators, and fruits aid in dispersal of the seeds.

QUESTIONS

1. Define the following terms: holdfast, volvocine line, zygote, alternation of generations.
2. Distinguish the following: moss (Musci), club moss (Lycophytina); megaphyll, microphyll; anisogamy, oogamy; sporophyte, gametophyte.
3. Define sporophyll. Describe the sporophylls of a club moss, a horsetail, a fern, an angiosperm.
4. Sketch a flower and label the structures.
5. Sketch a pine seed, label it, and indicate the origin of its components.
6. Bryophytes, among the plants, and amphibia, among the animals, often live in habitats intermediate between fresh water and dry land, rather than between the salt water and land. Propose a physiological argument (refer-

ring back to Chapter 7) to explain why invasions of the land were more likely by organisms previously adapted for life in fresh water.

7. Tall trees today are not appreciably taller than plants of the Devonian forests. What factors select for tallness in trees? What factors, by contrast, select against tallness? What kind of evolutionary innovation might be required to alter the optimal balance among these various factors?

8. In many areas on Earth, large gymnosperms are either dominant or they peacefully coexist with angiosperms. Try to imagine several advantages to a big land plant of being a gymnosperm.

9. Label these drawings and indicate the origin of each tissue.

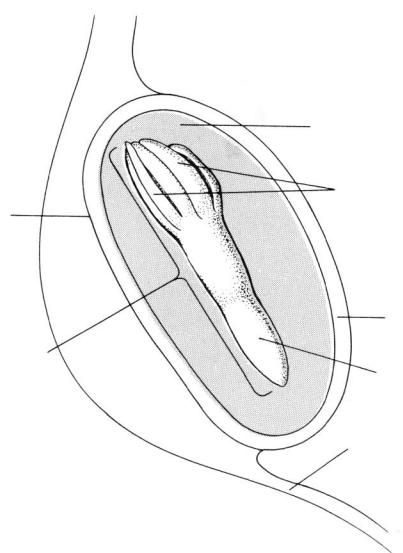

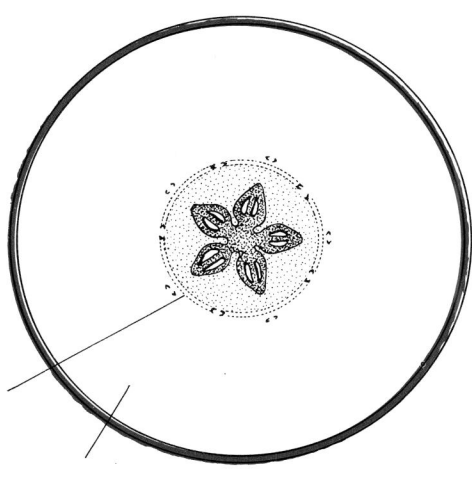

Chapter 22

The Animal Kingdom I:
Introducing the Invertebrates

Animals are many-celled heterotrophs. They depend directly or indirectly for their nourishment on land plants or algae. Most digest their food in an internal cavity, and most store food as glycogen or fat. Their cells do not have walls. Most move by means of contractile cells (muscle cells) containing characteristic proteins. Reproduction is usually sexual. Most are fixed in adult size and shape, in contrast to the plants, in which growth often continues for the lifetime of the organism. The higher animals—the arthropods and the vertebrates—are the most complex of all organisms, with many kinds of specialized tissues, including elaborate sensory and neuromotor mechanisms not found in any of the other kingdoms.

For most of us, animal means mammal, and mammals are, in fact, the chief focus of attention in Section 5. However, the mammals, or even the vertebrates as a whole, represent only a small fraction of the animal kingdom. More than 90 percent of the different species of animals are invertebrates, that is, animals without backbones, and most of these are insects.

Many of the great biologists, from Aristotle through Lamarck and Darwin and right up to the present day, have been intrigued by the invertebrates. One obvious reason is their fascinating variety; collecting beetles and butterflies rivals stamp collecting in aesthetic appeal and the joys of discovery. The invertebrates are also of great ecological importance, particularly the insects, which challenge man in his dominance of the globe. Although evolutionary relationships between us and the invertebrates are obscure, we can read into them traces of our own biological beginnings. In the twitch of a tiny segment of an earthworm's artery, we sense the echo of man's own heartbeat. Finally, and perhaps most importantly, the invertebrates, confronted with the same biological problems as we are, demonstrate a spectrum of dazzling solutions. In this way, they illuminate the essential nature of these problems and so help us to understand and evaluate our own solutions.

What are these problems? Let us begin to answer this question by recalling that an organism is a cell or group of cells. A primary need of the organism is to supply the cell or cells with the materials it requires. In the case of most heterotrophic cells, these are oxygen, water, and a nitrogen source plus a carbon

22–1 *Animals are characterized by their mobility, usually a result of the contraction of assemblies of protein fibers within specialized (muscle) cells.* Gonionemus murbachi, *the medusa (jellyfish) form of a coelenterate is shown here swimming actively, its bell contracted by muscles around its margin.*

2 mm

22–2 *Animals are heterotrophs. These goose-neck barnacles, which are small crustaceans, sweep food particles into their mouths with their six pairs of hairy appendages that protrude from the shell.*

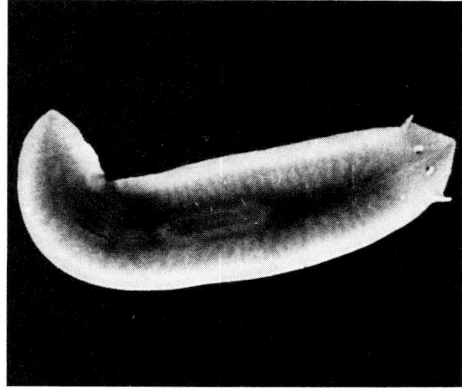

22–3 *Animals have a variety of sensory equipment. Even this small and relatively simple organism, a planarian, has two light-sensitive eyespots (ocelli) and a variety of chemoreceptors in the head region.*

source and an energy supply (usually the same molecules), and a few ions. Another requirement is to eliminate wastes, including excess carbon dioxide, nitrogenous wastes from the breakdown of amino acids, and, in some cases, excess water. Cells that live individual or colonial lives in a watery environment can solve these problems in relatively simple ways, but as organisms get larger, thicker, and more complex, the problem of servicing each individual cell becomes correspondingly complicated.

Another set of problems that an organism must solve in order to exist arises from the fact that it is more than just a group of cells. It is, in fact, a complex society, in which the needs of each individual cell or cell type are subordinated to the needs of the society. In a population of paramecia, the organisms may have common requirements, but each is in competition with the others. In a society of a few thousand cells—a small crustacean, for instance—the individual cells are dependent on the existence of the group and are organized in a system of mutual cooperation. The second group of problems faced by organisms, therefore, relates to the organization or integration of activities. Hormones are one of the chief means of integration in both plants and animals. In the animals, another, more rapid integrating mechanism has evolved: the nervous system, by which the organism keeps in touch with its environment and coordinates its own activities.

A QUESTION OF SIZE

At this point, one might well ask: Considering the problems faced by larger organisms, why did larger animals evolve? What selective advantages are there to the multicellular, more complex animals as compared with the smaller ones? Some answers to these questions are obvious and simple. Larger animals are, in general, more likely to eat than be eaten. Larger organisms, especially those that live underwater or on land, are generally able to travel faster and farther than small ones, and this is an advantage. A small ciliate, for instance, might starve only a few centimeters from a food supply.

Perhaps even more important, however, than mobility and edibility is what the famous French physiologist Claude Bernard called *milieu interieur*, the internal environment of the animal, as distinct from the external environment that surrounds it. A single-celled organism is as cold or as hot and as wet or as dry as its surroundings, whereas a larger animal is more independent and, to some extent, controls the environment in which its cellular society lives. Control of the internal environment is more readily achieved by the many-celled animal because of the simple surface-to-volume geometry we noted in Chapter 6. Exchanges between a cell and its surroundings take place across a cell's available surface area. This is a principal reason why a cell, which depends for its existence on the exchange of substances with its environment, cannot be very large. On the other hand, since it may be advantageous to conserve certain substances, such as water and heat, an organism may be better off, within limits of weight and mobility, if its relative surface area is reduced. One-celled animals can live successfully only in water or as parasites in the bodies of other organisms, which amounts to the same thing. Many-celled animals can live not only in water but on land, in the sky, and even, as we are now beginning to discover, in outer space—which is a very logical extension of an old evolutionary trend.

22–4　*A breadcrumb sponge showing some of the many openings (oscula) through which water leaves the animal. The oscula usually protrude above the rest of the animal. This arrangement allows the natural water flow in the habitat to draw water through the sponge.*

In this chapter, we shall discuss the so-called lower invertebrates and some that must clearly be considered higher, such as the clever and highly emotional octopus. The chapter that follows deals with the insects and other arthropods and, briefly, with the vertebrates.

There are almost thirty phyla of invertebrates. Of these, we are going to discuss relatively few, concentrating on the largest phyla and those of particular biological interest. A complete listing can be found in Appendix C.

PHYLUM PORIFERA: SPONGES

Sponges seem to have had a different origin from other members of the animal kingdom and to have traveled a solitary evolutionary route. For this reason, they are often classified in a subkingdom of their own, the Parazoa ("alongside of animals"). In fact, until the eighteenth century, the sponges were classified as plant-animals ("zoophytes") since they are all sessile (permanently attached to a substrate). Sponges are found on ocean floors throughout the world. Most live along the coasts in shallow water, but some, such as the fragile glass sponges, are found at great depths, where the water is almost motionless. A few types are found in fresh water.

The sponges are made up of a relatively few different cell types, the most characteristic of which are the choanocytes, or collar cells, which line the interior cavity of the sponge (Figure 22–5). Similar cells, the choanoflagellates,

22–5　*The body of a simple sponge is dotted with tiny pores, from which the phylum derives its name (Porifera, or "pore bearer"). Water containing food particles is drawn into the internal cavity of the sponge through these pores and is exhaled out the osculum. The water is moved by the beating of the flagella protruding from the collars of choanocytes and by the sucking effect of flow of the local currents across the osculum. Each collar is made up of about 20 filaments, each of which is retractile. The lashing of the flagella directs a current of water through the filaments. Minute particles are filtered out and cling to one or more filaments and are then drawn into the cell and digested. The digested food is then shared by diffusion with other sponge cells. A sponge filters as much as a ton of water to gain an ounce of body weight.*

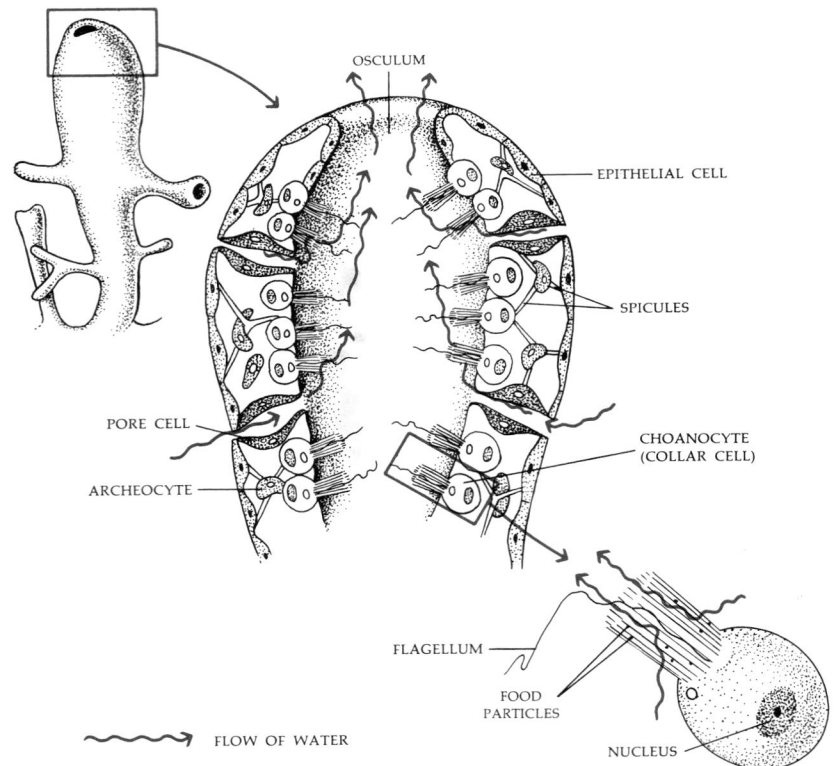

OSCULUM

EPITHELIAL CELL

SPICULES

PORE CELL

ARCHEOCYTE

CHOANOCYTE (COLLAR CELL)

FLAGELLUM

FOOD PARTICLES

NUCLEUS

FLOW OF WATER

2–6 *A cluster of sponges.*

2–7 *The skeleton of* Euplectella speciosis-
sima, *a species of glass sponge pictur-
esquely known as Venus's flower basket.
These fragile sponges, with their delicate
silica-containing skeletons, are usually
found at great depths. According to the
fossil record, glass sponges were present
in the Ordovician period, which began
500 million years ago.*

are found as either solitary or colonial protists, but the evolutionary relationships between these and the sponges are a matter of dispute. All of a sponge's
digestive processes are carried out by the individual choanocytes; hence, even a
giant sponge—and some stand taller than a man—can eat nothing larger than
microscopic food particles.

The outer surface of the sponge is covered with epithelial cells. Among these
epithelial cells are cells that are contractile. These cells contract in response to
touch or to irritating chemicals, and in so doing, close up pores and channels.
Each cell acts as an individual, however; there is no coordination among them.
Between the epithelial cells and the choanocytes is a middle, jellylike layer, and
in this layer are found largely undifferentiated cells, known as archeocytes,
which may develop into a number of other cell types. Some of these cells form
amoebocytes, which are amoebalike cells that carry digested food particles from
the choanocytes to the epithelial cells. Others secrete stiffening skeletal structures, or spicules, which may be horny and fibrous, as in the bath sponges;
calcareous (of calcium carbonate), as in the chalk sponges; or of silica, as in the
delicately beautiful glass sponges.

The sponge shown in Figure 22–5 is a small and simple one. In larger
sponges, the body plan, although it is essentially the same, looks far more
complex. These sponges have greatly increased feeding and filtering services,
owing to their highly folded body walls. We have already encountered this
evolutionary strategem for increasing biological work surfaces at the cellular
level—as in the inner membrane of the mitochondrion—and we shall be encountering it again as we examine the structure of gills and lungs.

Sponges are somewhere between a colony of cells and a true multicellular
organism. The cells are not organized into tissues or organs; each leads an

independent existence. Yet there is some attractive force among the cells that holds them together and organizes them. If the sponge *Microciona prolifera* is squeezed through a fine sieve or a piece of cheesecloth, the body of the sponge is separated into individual cells and small clumps of cells. Within an hour, the isolated sponge cells begin to reaggregate, and as these aggregations get larger, canals and flagellated chambers and other characteristics of the body organization of the sponge begin to appear. This phenomenon has been used as a model for the analysis of cell adhesion, recognition, and differentiation, all of which are basic biological features of development in higher organisms.

Sponges may reproduce sexually or asexually. In sexual reproduction, archeocytes divide and differentiate to produce eggs or sperm. Sponges reproduce asexually, either by branches that break off from the parent animal, or by gemmules, aggregations of archeocytes within a hard, protective outer layer. Production of such resistant forms is found, in general, only among freshwater organisms. In the ocean, conditions are relatively unchanging, but the freshwater environment is much harsher. Invertebrates that live in fresh water have protected embryonic forms, whereas even closely related marine species generally do not.

PHYLUM COELENTERATA: POLYPS AND JELLYFISH

The coelenterates are a large and often strikingly beautiful group of aquatic organisms. Their adult form is generally radially symmetrical; that is, their body parts are arranged around a central axis, like spokes around a hub (Figure 22–8). As you can see in Figure 22–9, the basic body plan is a simple one: the animal is essentially a hollow container, which may be either vase-shaped, the *polyp*, or bowl-shaped, the *medusa*. The polyp is usually sessile; the medusa, motile. Both consist of two layers of tissue: ectoderm and endoderm. Between the two layers is a gelatinous filling, the mesoglea ("middle jelly"), which is made of a collagenlike material. In the polyp form, the mesoglea is sometimes very thin, but in the medusa, it often accounts for the major portion of the body substance.

One distinctive feature of the animals in this phylum is the *coelenteron*, a digestive cavity with only one opening. Within this cavity, enzymes are released that break down food, partially digesting it extracellularly, as our own food is digested within the stomach and intestinal tract. The food particles are then taken up by the cells lining the cavity; they complete the digestive process and pass the products on to the other cells of the animal. Inedible remains are ejected from the single opening. The water-filled coelenteron also serves as a hydraulic support (like the water in a fire hose) for the soft-bodied animals.

The second distinctive feature is the *cnidoblast*. Coelenterates are carnivores. They capture their prey by means of tentacles that form a circle around the "mouth." These tentacles are armed with cnidoblasts, special cells that contain nematocysts (thread capsules). Nematocysts are discharged in response to chemical stimulus or touch. The nematocyst threads, which are often poisonous and may be sticky or barbed, can lasso prey, harpoon it, or paralyze it—or some useful combination of all three. The toxin apparently produces paralysis by attacking the lipoproteins of the nerve cell membrane of the prey.

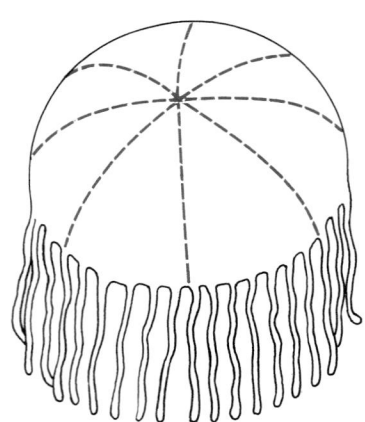

22–8 *In organisms with radial symmetry, any plane through the animal that passes through the central axis divides the body into halves that are mirror images of one another.*

22–9　Among coelenterates, there are two basic body plans: the vase-shaped polyp (left) and the bowl-shaped medusa (right). The coelenteron, characteristic of the phylum, is a digestive cavity with a single opening. The coelenterate body has two tissue layers, ectoderm and endoderm, with gelatinous mesoglea between them.

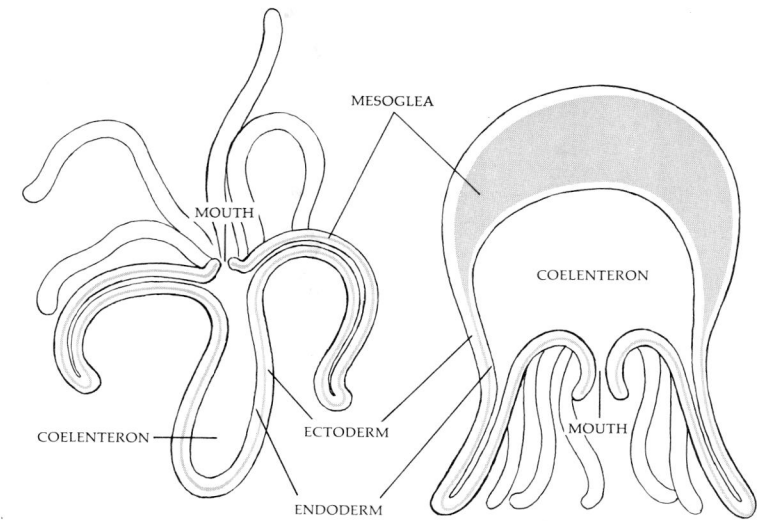

–10　Cnidoblasts, specialized cells located in the tentacles and body wall, are a distinguishing feature of coelenterates. The interior of the cnidoblast is filled by a nematocyst, which consists of a capsule containing a coiled tube, as shown on the left. A trigger on the cnidoblast, responding to chemical or mechanical stimuli, causes the tube to shoot out, as shown on the right. The capsule is forced open and the tube turns inside out, exploding to the outside. Four of the many different types of nematocysts are shown here. In (a) and (b), the tube is sticky and is used by the animal to fasten its tentacles to solid objects as it moves. The tube in (c) is used to wrap around and entangle prey. The nematocyst in (d) is armed with barbs and spines; the tube penetrates the tissues of the prey and injects a paralyzing poison.

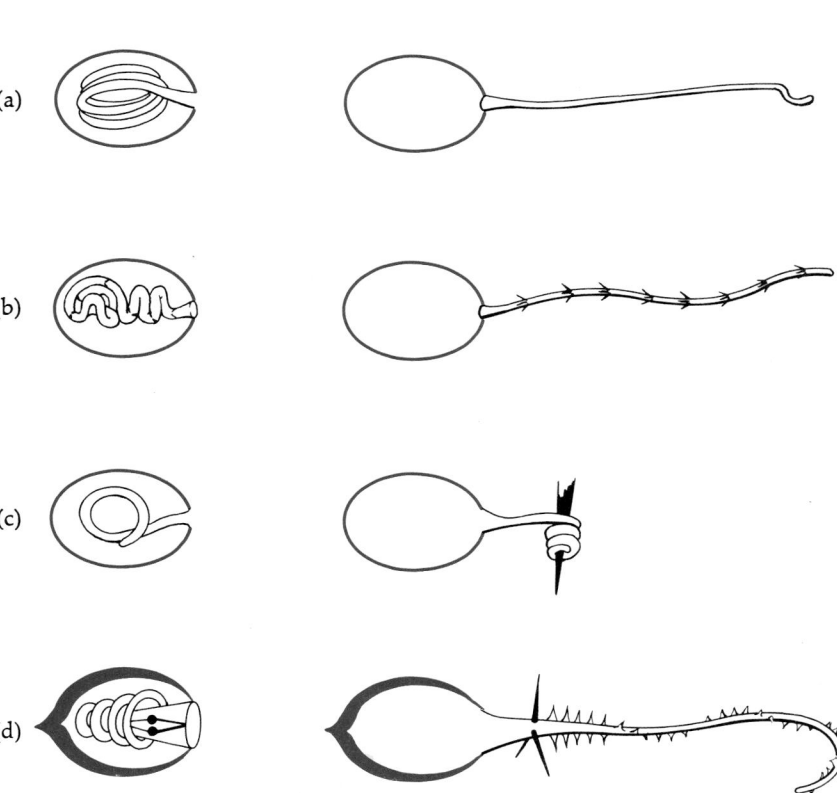

Cnidoblasts occur only in this phylum, with some interesting exceptions. Certain other invertebrates, including nudibranchs (a kind of mollusk) and flatworms, can eat coelenterates without triggering the nematocysts. The nematocysts then migrate to the surface of the predator and can be fired in their new host's defense.

Classes of Coelenterates

There are three major classes of coelenterates: Hydrozoa, in which the polyp is the dominant form; Scyphozoa, predominantly medusoid, exemplified by the common jellyfish; and Anthozoa, which include the sea anemones and the reef-building corals.

Class Hydrozoa: Hydra

One of the most thoroughly studied of the coelenterates is *Hydra*, which is a small, common freshwater form, convenient to keep in the laboratory. Figure 22–12 shows a small section of the body wall of *Hydra*. The endoderm is mostly made up of cells concerned with digestion. The ectoderm is composed largely of epitheliomuscular cells, which perform a covering, protective function and also serve as muscle tissue. Each cell has contractile fibers, myonemes, at its base and so can contract individually, like the contractile epithelial cells of the sponge. In *Hydra*, however, as distinct from the sponge, a network of nerve cells underlies the epitheliomuscular cells and coordinates their muscular contractions, making possible a wide variety of activities. In *Hydra*, as in other polyps, the muscle cells in the ectoderm are oriented at right angles to those in the endoderm so the body walls can stretch or bulge.

22–11 Hydra *has a nervous system that integrates the body into a functional whole, making possible a range of fairly complex activities. For instance,* Hydra *may swim, glide on its base, or, as shown here, it may travel by a somersaulting motion.*

22–12 *The structure of the body wall of* Hydra. *The outer layer of cells, the ectoderm, is primarily for protection, while the inner layer, the endoderm, performs the digestive function. One type of digestive cell, the gland cell, secretes the digestive enzymes that are released into the coelenteron. The other type, the nutritive cell, using its flagella, mixes the food as it is being processed, and then extends pseudopodia that collect the food particles for further digestion.*

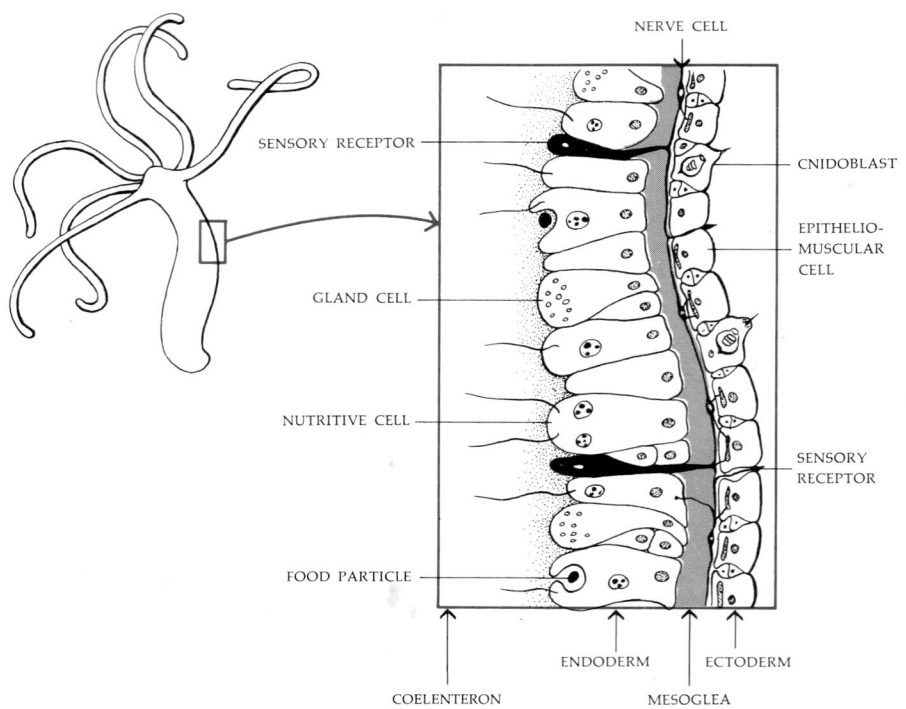

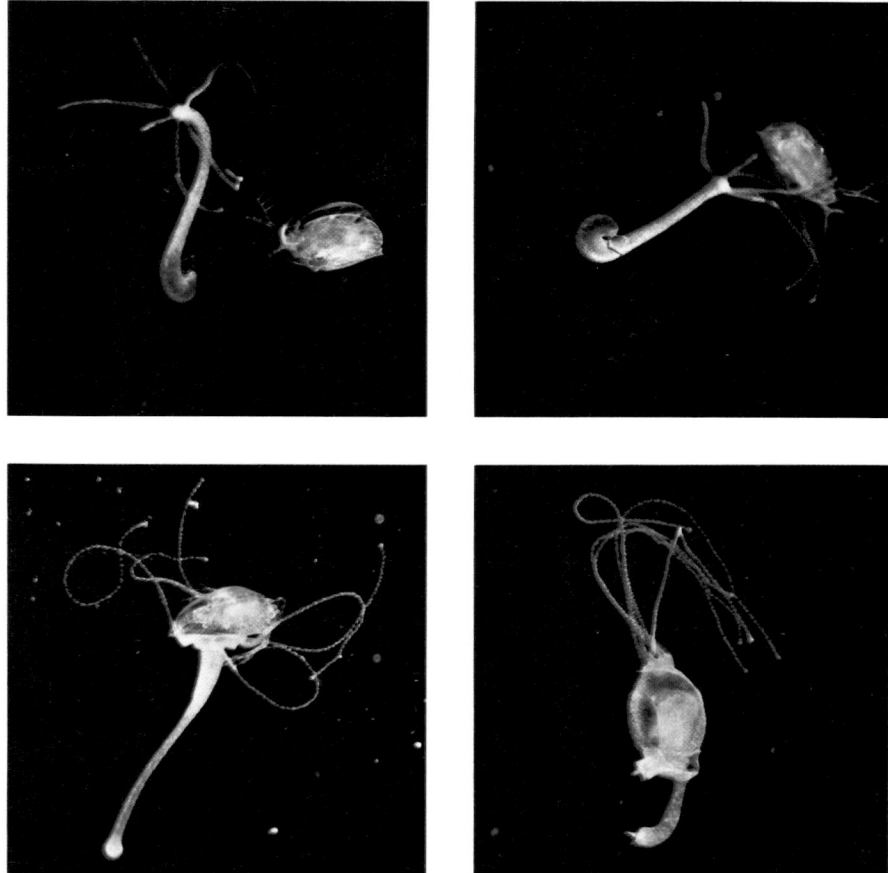

-13 *A* Hydra *encounters a small crustacean* (Daphnia), *grasps it with its tentacles, engulfs it, and digests it.*

In addition to cnidoblasts and epitheliomuscular cells, which are independent effectors—cells that both receive and respond to stimuli—the *Hydra* also contain sensory receptor cells which are more sensitive than other epithelial cells to chemical and mechanical stimuli. When stimulated, they transmit their impulses to an adjacent cell or cells. The adjacent cell may be simply an epitheliomuscular cell, an effector, which then responds. Note that this system is one step more complicated than the epitheliomuscular cell or cnidoblast, which act as both receptor and effector. *Hydra*, and some other coelenterates, have a loose network of nerve cells lying at the base of the epithelial layers. The cells of this conducting system make up the third type of specialized nerve cells (including the cnidoblast) found in *Hydra*. This nerve net is the simplest example of a nervous system that links an entire organism into a functional whole.

Although *Hydra* has only the polyp form, many hydrozoans have both hydroid (polyp) and medusoid forms at different times in their life cycles. Coelenterates of the genus *Obelia*, for example, spend most of their lives as colonial polyps. The colony arises from a single polyp, which multiplies by budding.

22–14 (a) *Coral polyps extending their tentacles from the hard exoskeleton that forms the coral reef. Reef-forming corals are colonial forms that arise from a single polyp by budding, as does* Obelia. *The exoskeleton is secreted by their ectodermal cells. In the upper left is a newly released free-swimming medusa.* (b) *Coelenterates of the order Siphonophora are large floating colonies made up of both polyps and medusas. The polyps are feeding forms and the medusas are reproductive forms; in some species, nonreproductive medusas are swimming bells. The colony produces a gas-filled float, or pontoon. In the Portuguese man-of-war shown here, the large float serves also as a sail. The blue strands are composed of reproductive and feeding individuals; the purple strands, which may grow as long as 50 feet, are made up of stinging, food-gathering polyps armed with nematocysts.*

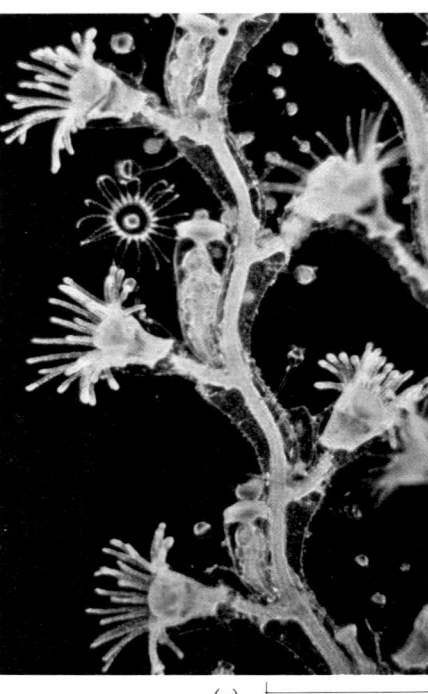

(a) |—— 1 mm ——| (b)

The new polyps do not separate but remain interconnected so that their body cavities form a continuous channel, through which food particles are circulated. Within the colony are two types of polyps: feeding polyps with tentacles and cnidoblasts, and reproductive polyps from which tiny medusas bud off. These medusas produce eggs or sperm that are released into the water and fuse to form zygotes. This sort of life cycle, in which the sexually reproductive form is distinctly different from the asexual form, superficially resembles alternation of generations in plants. There is, however, no alternation between haploid and diploid forms as there is in plants; the only haploid forms are the gametes.

Thus colonial polyps, with their division of labor between feeding and reproductive forms, are very like superorganisms. Such a high degree of specialization of function among social organisms is seen in other phyla only among the social insects.

Class Scyphozoa: Jellyfish

A second major class of coelenterates are the scyphozoans, or "cup animals," in which the medusa form is dominant. Scyphozoans, more commonly known as jellyfish, range in size from less than an inch in diameter up to animals 12 feet across and trailing 30-foot tentacles. In the adult animal, the mesoglea is so firm that a large freshly beached jellyfish can easily support the weight of a human being. The mesoglea of some jellyfish is filled with wandering, amoebalike cells, which serve to transport food from the nutritive cells of the endoderm. Unlike *Hydra*, scyphozoans have true muscle cells; these underlie the ectoderm, contracting rhythmically to propel the medusa through the water.

22–15 *The moon jellyfish (*Aurelia limbata*) has four frilly mouth lobes that gather minute organisms into the coelenteron. It has short tentacles and grows to only about a foot in diameter. This specimen was photographed in the cold waters of the Atlantic.*

-16 *The life cycle of* Aurelia. *Sperm and egg cells are released from adult medusas into the surrounding water. Fertilization takes place, and the resulting zygote develops first into a hollow sphere of cells, the blastula, and then elongates and becomes* a ciliated larva called a planula. The planula eventually settles to the bottom, attaches by one end to some object, and develops a mouth and tentacles at the other end, thus transforming into the polyp stage. The body of the polyp grows and, as it grows, begins to form medusas, stacked upside down like saucers, which bud off, one by one, and grow into full-sized jellyfish.

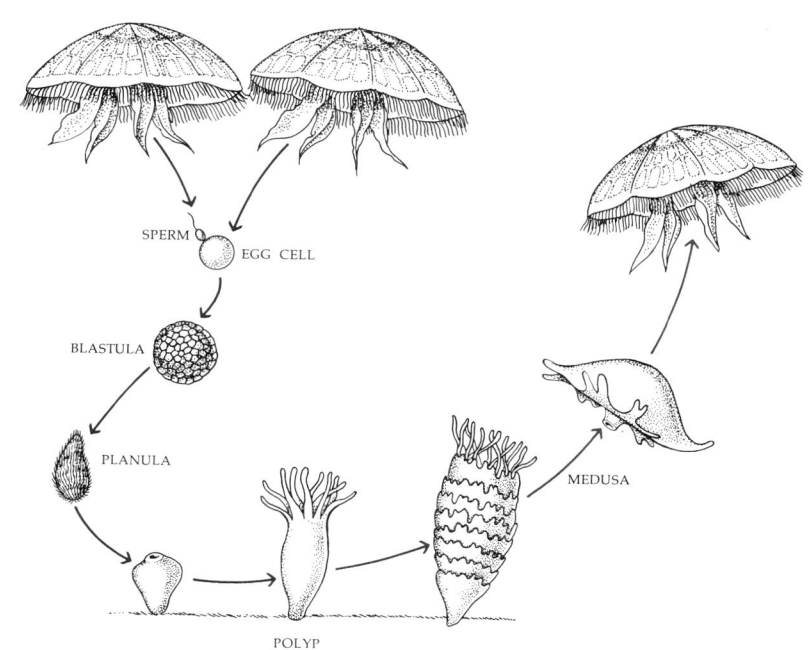

SPERM EGG CELL

BLASTULA

PLANULA

POLYP

MEDUSA

-17 (a) *The statocyst is a specialized receptor organ which orients the jellyfish with respect to gravity. When the bell tilts, gravity pulls the statolith, a grain of hardened calcium salts, down against the hair cells. This stimulates the nerve fibers and signals the animal to right itself.* (b) *Eyespot (ocellus), the simplest type of photoreceptor organ. Ocelli of this sort are found among coelenterates.*

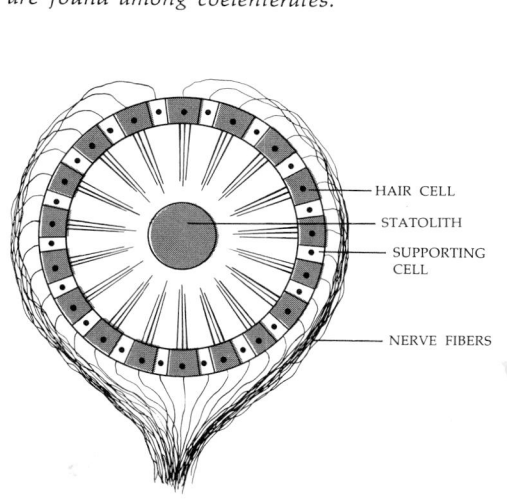

HAIR CELL
STATOLITH
SUPPORTING CELL
NERVE FIBERS

(a)

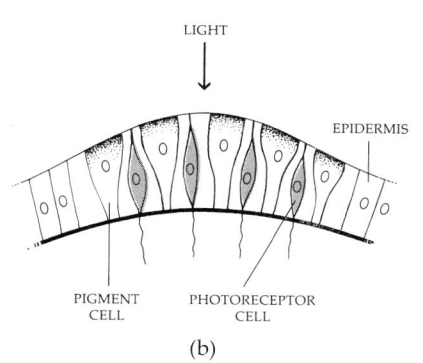

LIGHT

EPIDERMIS

PIGMENT CELL   PHOTORECEPTOR CELL

(b)

The life cycle of one of these coelenterates, the common jellyfish *Aurelia*, is shown in Figure 22–16. The ancestral coelenterates, it is believed, went through a similar polyp-to-medusa life cycle.

Nervous System of Medusa

In the medusa, there are concentrations of nerve cells in the margin of the bell. These nerve cells connect with fibers innervating (providing the nerve supply for) the tentacles, the musculature, and the sense organs.

The bell margin is liberally supplied with sensory receptor cells sensitive to mechanical and chemical stimuli. In addition, the jellyfish has two types of true sense organs: statocysts and light-sensitive ocelli. *Statocysts* are specialized receptor organs that provide information by which an animal can orient itself with respect to gravity (Figure 22–17a). The statocyst, which seems to have been one of the first special sensory organs to have appeared in the course of evolution, has persisted virtually unchanged to the present day, appearing in many animal phyla. *Ocelli*, which are probably even more primitive, are groups of pigment cells and photoreceptor cells. They are typically located on the outer bases of the tentacles.

22–18　*A sea anemone. The flowerlike appearance is deceptive; anemones are carnivorous animals belonging to a class of coelenterates which, like* Hydra, *have dropped the medusa stage. In common with other coelenterates, their tentacles are equipped with stinging nematocysts.*

Class Anthozoa

Anthozoans ("flower animals")—the corals and sea anemones—are members of a class of coelenterates that, like *Hydra*, have lost the medusa stage. They differ from *Hydra* in having a gullet lined with epidermis and a coelenteron divided by vertical partitions. In most corals, which are colonies of anthozoans, the epidermal cells secrete protective outer walls, usually of calcium carbonate (limestone), into which each delicate polyp can retreat. The coral-forming polyps are the most ecologically important of the coelenterates. The 1,200-mile-long Great Barrier Reef off the northeast shores of Australia and the Marshall Islands in the Pacific are examples of coral-created land masses. A coral reef is composed primarily of the accumulated limestone skeletons of coral coelenterates, covered by a thin crust occupied by the living colonial animals.

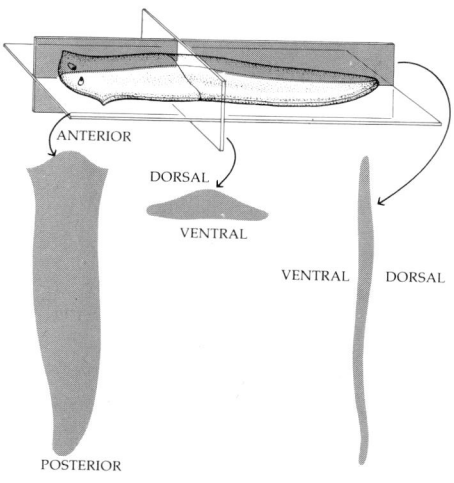

2–19 *In a bilaterally symmetrical organism, the right and left halves of the body are mirror images of one another. The upper and lower (or back and front) surfaces are known as dorsal and ventral. The end that goes first is termed anterior and the rear, posterior.*

ANTERIOR

DORSAL

VENTRAL

VENTRAL   DORSAL

POSTERIOR

PHYLUM PLATYHELMINTHES: FLATWORMS

The flatworms are the simplest animals, in terms of body plan, to show bilateral symmetry. In bilaterally symmetrical animals, the body plan is organized along a longitudinal axis, with the right half an approximate mirror image of the left half. It also has a top and a bottom, or in more precise terms (applicable even when it is turned upside down or, as in the case of humans, standing upright), a dorsal and a ventral surface. Like most bilateral organisms, the flatworm also has a distinct "headness" and "tailness," anterior and posterior. Having one end that goes first—cephalization—is characteristic of actively moving predatory animals. In such animals, for obvious reasons, many of the sensory cells are also collected into the anterior end. With the aggregation of sensory cells, there came a concomitant gathering of nerve cells; this gathering is a forerunner of the brain.

Flatworms have three distinct tissue layers—ectoderm, mesoderm, and endoderm—characteristic of all animals above the coelenterate level of organization. Moreover, not only are their tissues specialized for various functions, but two or more types of tissue cells may combine to form organs. Thus, while coelenterates are largely limited to the tissue level of organization, flatworms can be said to have gained the organ level of complexity.

The flatworms are believed by some zoologists to have evolved from the coelenterates, not by way of either adult form, however, but from the ciliated larval form.

The Planarian

Platyhelminthes form a large and varied group, and we shall single out just one for special examination, the freshwater planarian. The ectoderm of the planarian is made up of cuboid epithelial cells, many of which are ciliated, particularly those on the ventral surface. Ventral ectodermal cells secrete

-20 *A polyclad flatworm. The many species belonging to this phylum, though widely various in color and shape, are nearly all small and flat-bodied, like a planarian (Figure 22–3).*

22–21 *Example of a flatworm: the freshwater planarian. The planarian, which is carnivorous, feeds by means of its extensible pharynx. The nervous system is indicated in color in the left-hand diagram. Note that some of the fibers have been aggregated into two cords, one on each side of the body, and there is a cluster of nerve cells in the head, the beginnings of a brain. The ocelli are light-sensitive areas.*

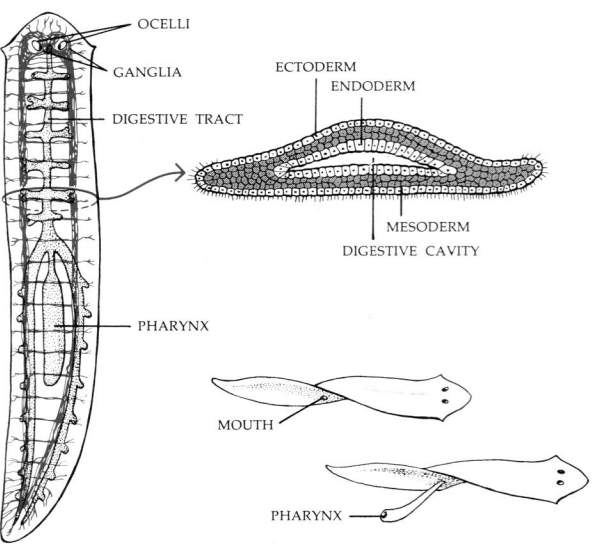

22–22 *Flatworms have a tubular excretory system. The system usually consists of two or more branching tubules running the length of the body. In the planarian and its relatives, the tubules open to the body surface through a number of tiny pores. At the ends of the side branches are small bulblike structures known as flame cells. Within each of these cells, a tuft of cilia in constant motion resembles the flickering of a flame. Water and some waste materials from the tissue fluids are moved by the cilia through the tubules to the excretory pores, where the collected liquid leaves the body.*

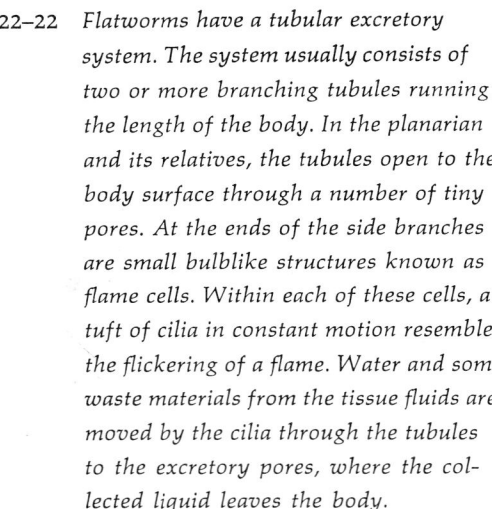
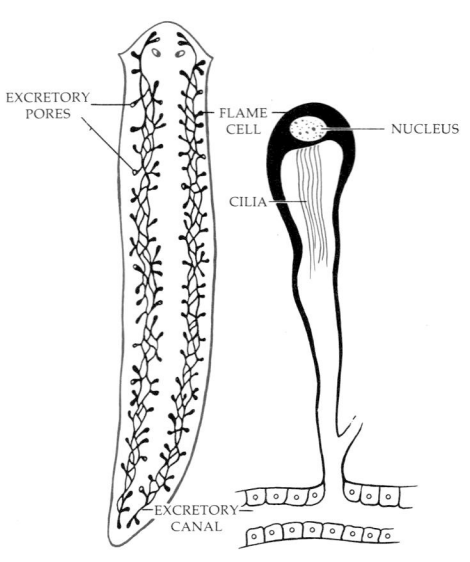

mucus, which provides traction for the planarian as it moves by means of its cilia along its own slime trail. Planarians are among the largest animals that can use cilia for locomotion; the cilia in larger species are usually employed for moving water or other substances along the surface of the animal, as in the human respiratory tract, rather than for propelling the animal through the water.

The planarian has an endoderm composed largely of amoeboid cells, which, although they are phagocytic, are not wandering cells like the amoebocytes of the jellyfish. Between the ectoderm and the phagocytic endoderm is a *mesoderm*, or middle tissue layer. In planarians, as in higher animals, the muscle cells and the principal organ systems are of mesodermal origin.

The planarian, like the coelenterates, has a digestive cavity (gut) with only one opening, located on the ventral surface. This digestive cavity has three main branches, which is why planarians are placed in the order of flatworms known as Tricladida.

Like other flatworms, the planarian is carnivorous. It eats either dead meat or slow-moving animals it can fasten itself to or subdue by sitting on, such as smaller planarians. It feeds by means of a muscular tube, the pharynx, which is free at one end. The free end can be stretched out through the mouth opening. Muscular contractions in the tube cause strong sucking movements, which tear the meat into microscopic bits and draw it, along with the juices of the food, into the internal cavity, where it is phagocytized by the endodermal cells.

Unlike the sponges or coelenterates, most flatworms have an excretory system. In the planarian, the system is a network of fine tubules that runs the length of the animal's body. Side branches of the tubules contain flame cells, each of which has a hollow center in which a tuft of cilia beats, flickering like a tongue of flame, moving water along the tubules to the exit pores between the epidermal cells. The flame-cell system appears to function largely to regulate water balance; most of the metabolic waste products probably diffuse out through the ectoderm or the endoderm.

–23 *Planarians have both male and female reproductive structures, and mating involves a mutual exchange of sperm. The erect penis of each partner is inserted into the sperm receptacle of the other. Planaria may also reproduce asexually, by fragmentation or by fission. Fragments of the tail, for example, may regenerate new heads, and sometimes the animal divides longitudinally, forming two mirror images.*

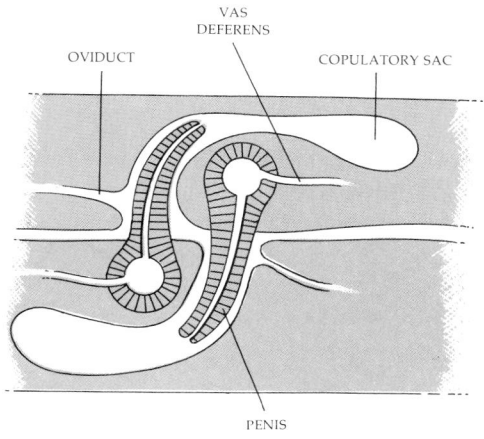

24 *Ocellus of the planarian. The light stimulus is received by the ends of the photoreceptor cells adjacent to the pigment cup. Thus the light travels first through the fibers carrying the signals to the cerebral ganglia. In this inside-out organization, the planarian ocellus resembles the vertebrate eye.*

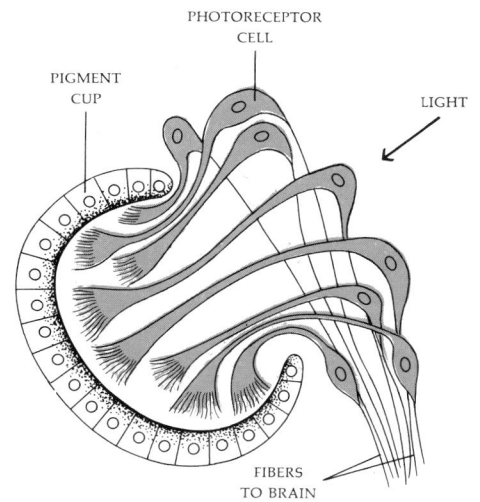

Planarians have a complicated reproductive system. The eggs are fertilized internally. At mating, each partner deposits sperm in the copulatory sac of the other partner. These sperm then travel along special tubes, the *oviducts*, to fertilize the eggs as they become ripe. Organisms in which male and female gametes are produced by different individuals are known as dioecious ("in two houses"). Organisms like planarians, in which both types of gametes are produced by one individual, are known as monoecious or, sometimes, hermaphroditic. Solitary slow-moving animals, such as earthworms and snails, are often monoecious since, under these conditions, every other adult member of the species encountered is a potential mate and every mating results in two sets rather than one set of fertilized eggs. Some types of monoecious animals can fertilize themselves, although they do not usually do so if another individual is present.

The Planarian Nervous System

The evolution of bilateral symmetry brought with it marked changes in the organization of the nervous system as well as of other systems. Even among primitive flatworms, the neurons (nerve cells) are not dispersed in a loose network, as in *Hydra*, but are instead condensed into longitudinal cords. In the planarians, this condensation is carried further and there are only two main conducting channels, one on each side of the flat, ribbonlike body, which carry impulses to and from the aggregation of nerve cells in the anterior end of the body. (See Figure 22–21.) Such aggregations of nerve cell bodies are known as *ganglia* (singular, ganglion).

The ocelli of the planarian are usually inverted pigment cups (Figure 22–24). Since they have no lenses, they cannot form an image. A planarian can probably see about as well as you can see with your eyes closed; that is, it can distinguish light from dark. Planarians are photonegative; if you shine a light on a dish of planarians from the side, they will move quickly and steadily away from the source of light. Among the epithelial cells are receptor cells sensitive to certain chemicals and to touch.

The head region in particular is rich in *chemoreceptors*. If you place a small piece of fresh liver in the culture water so that its juices diffuse through the medium, the planarians will raise their heads off the bottom and, if they have not eaten recently, will lope directly and rapidly (on a planarian scale) toward the meat, to which they then attach themselves to feed. The animal locates the food source by repeatedly turning toward the side on which it receives the stimulus more strongly until the stimulus is equal on both sides. If the chemoreceptor cells are removed from one side of the head, the animal will turn constantly toward the intact side.

Planarians, with their simple nervous systems, their capacity to react to a variety of stimuli, and their powers of regeneration, have been the subject of a number of experiments. In one group, planarians trained to avoid electric shock were fed to untrained planarians. The result, it was claimed, was that the untrained planarians that had eaten the trained ones behaved as if they had been trained. These experiments, still not verified, on the "transfer of training by cannibalism" led to a great deal of controversy in the 1960s and also, as you might expect, to a number of jokes about new possibilities for more meaningful student-teacher relationships.

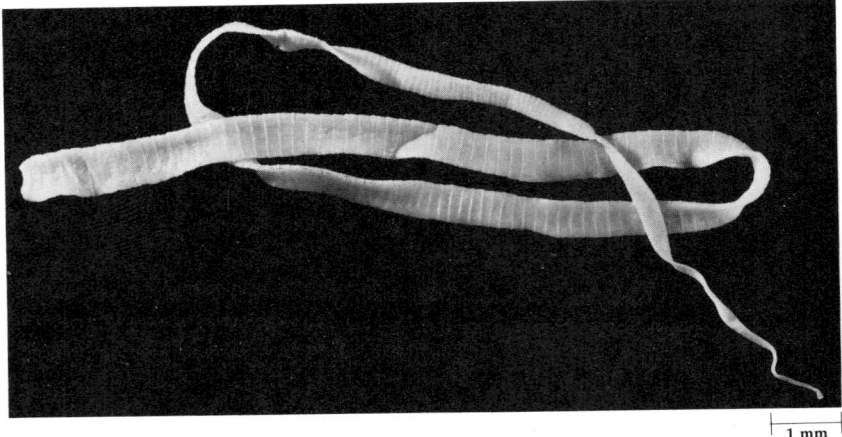

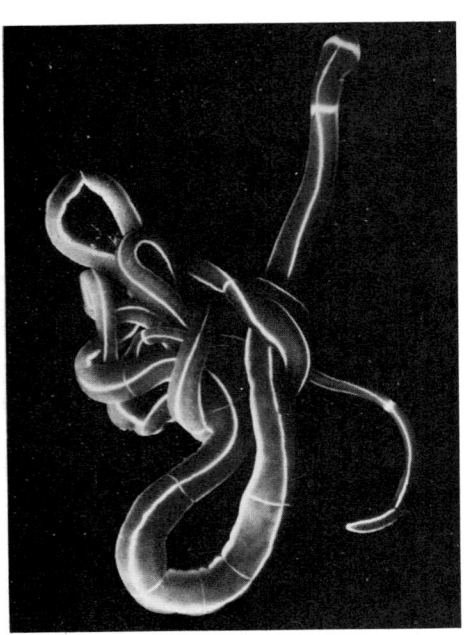

22–25 *Human tapeworm,* Taenia solium. *Tapeworms are intestinal parasites that lack any digestive system of their own. They cling by their heads, which are equipped with hooks and suckers, and absorb the food molecules digested by their hosts through their body walls. Their bodies, posterior to the heads, are divided into segments, each of which is a sexually complete hermaphroditic unit in which sperm and ova are produced. Segments break off when the ova are mature and new segments are formed. Characteristically, the ova are eaten and develop to a larval form in one species (in this case, pigs) and the adult worm parasitizes members of another species (here, man).*

22–26 *A ribbon worm in an embarrassing situation. Ribbon worms range in length from less than an inch to 100 feet in length and through virtually all the colors of the spectrum. All members of the speies, however, have flat, thin (seldom more than ¼ inch thick) bodies, a mesoderm, a mouth-to-anus digestive tract, and a long muscular tube which can be thrust out to grasp prey.*

Tapeworms and Flukes

Phylum Platyhelminthes includes also the tapeworms and flukes, parasitic forms that can cause serious and sometimes fatal diseases among the higher animals. Members of both of these parasitic classes have a tough outer layer of cells that is resistant to digestive fluids and, usually, suckers or hooks on their anterior ends by which they fasten to their victims. The flukes (or trematodes) feed through the oral sucker, but the tapeworms, which have no mouths or digestive cavities, merely hang on and absorb food molecules. Tapeworms are found in the intestines of many vertebrates, including man, and may grow as long as 15 to 20 feet. They cause illness not only by encroaching on the food supply but also by producing wastes and by obstructing the intestinal tract. The most common human tapeworm, the beef tapeworm, infects people who eat the undercooked flesh of cattle that have eaten fodder contaminated by human feces.

All parasites, including parasitic flatworms, are believed to have originated as free-living forms and to have lost certain tissues and organs (such as the digestive tract), as a secondary effect of their parasitic existence, while developing adaptations of advantage to the parasitic way of life.

PHYLUM RHYNCHOCOELA: RIBBON WORMS

The ribbon worms (sometimes called nemertines), although a small phylum, are of special interest to biologists attempting to reconstruct the evolution of the invertebrates, because they appear to be closely related to the flatworms with an important difference: They have a one-way digestive tract beginning with a mouth and ending with an anus. This is a far more efficient arrangement than the one-opening digestive system of the coelenterates and flatworms. In the one-way tract, food moves assembly-line fashion, with the consequent possibility that various segments of the tract can become specialized for different stages of digestion. The ribbon worms also have a circulatory system, usually consisting of one dorsal and two lateral blood vessels that carry the colorless blood.

This phylum is called Rhynchocoela ("beak" plus "hollow") because these worms are characterized by a long, retractile, slime-covered tube (proboscis),

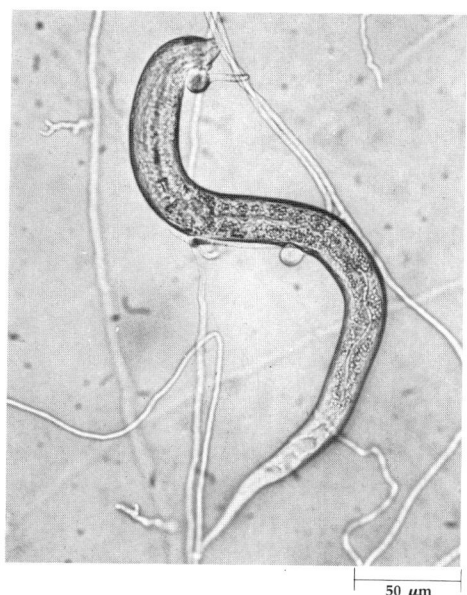

–27 *A soil-dwelling roundworm (nematode). This particular organism is being attacked by a predatory fungus, which traps the worm by means of small adhesive knobs. Once captured, fungal hyphae grow into the body of the worm and digest it.*

50 μm

which lies in a special cavity. The proboscis, sometimes armed with a barb, seizes prey and draws it to the mouth where it is engulfed. Some inject a paralyzing poison into their prey.

PHYLUM NEMATODA: ROUNDWORMS

The number of species of roundworms (nematodes) has been variously estimated as low as 10,000 and as high as 400,000 to 500,000. Most are free-living, microscopic forms. It has been estimated that a spadeful of good garden soil usually contains about a million nematodes. Some are parasites; most species of plants and animals are parasitized by at least one species of nematodes. Man is host to about fifty species, including hookworms, pinworms, and *Trichinella*. The latter causes trichinosis, which is transmitted by eating uncooked pork, a single ounce of which may contain 80,000 cysts, or resting forms, of *Trichinella*.

Nematodes have a three-layered body plan and a tubular gut with a mouth and an anus. They are unsegmented and are covered by a thick, continuous cuticle, which is molted periodically as they grow. An interesting, and unique, feature of nematode construction is the absence of circular muscles. The contraction of the longitudinal muscles acting against the tough, elastic cuticle gives the worm its characteristic whipping movement in water. The sexes are usually separate.

Nematodes are considered to be evolved from the Platyhelminthes, having, like them, a three-layered body plan without a true coelom (Figure 22–28c). They have, however, what is known as a pseudocoelom, which is between the endoderm and the mesoderm. Six other minor (in terms of species and numbers) phyla, mostly small, wormlike animals have body plans based on the pseudocoelom, but Nematoda is the only major pseudocoelomate phylum.

28 *Basic body plans of the animal phyla, as shown in cross section. (a) A body which consists fundamentally of only two layers is characteristic of coelenterates. (b) Flatworms and ribbon worms have three-layered bodies, with the layers closely packed on one another. (c) Nematodes have three-layered bodies with a pseudocoelom between the endoderm and mesoderm. (d) Annelids and most other animals, including vertebrates, have bodies which are three-layered with a cavity, the coelom, within the middle layer (mesoderm). The mesodermal mesenteries suspend the gut within the body wall.*

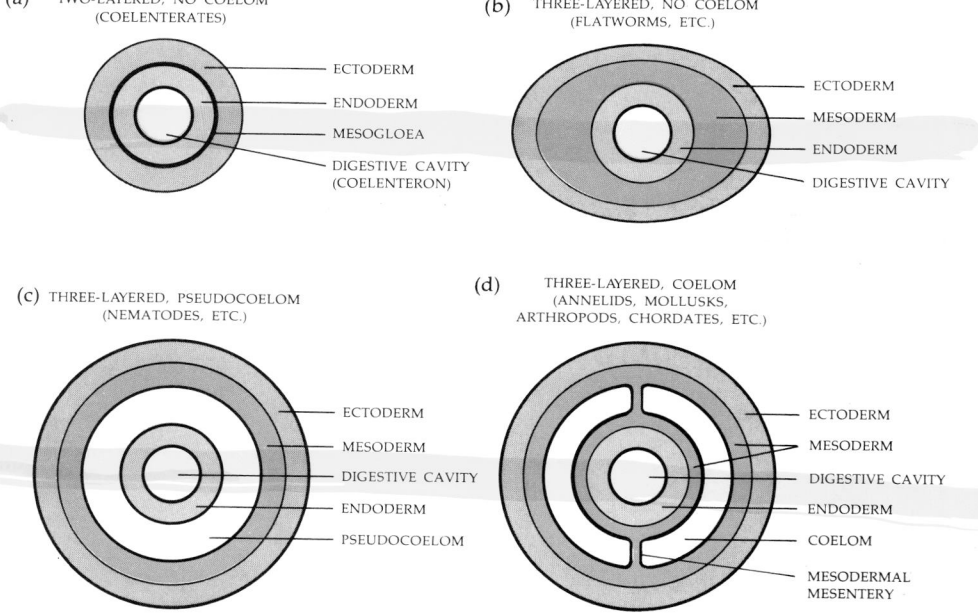

(a) TWO-LAYERED, NO COELOM (COELENTERATES)
— ECTODERM
— ENDODERM
— MESOGLOEA
— DIGESTIVE CAVITY (COELENTERON)

(b) THREE-LAYERED, NO COELOM (FLATWORMS, ETC.)
— ECTODERM
— MESODERM
— ENDODERM
— DIGESTIVE CAVITY

(c) THREE-LAYERED, PSEUDOCOELOM (NEMATODES, ETC.)
— ECTODERM
— MESODERM
— DIGESTIVE CAVITY
— ENDODERM
— PSEUDOCOELOM

(d) THREE-LAYERED, COELOM (ANNELIDS, MOLLUSKS, ARTHROPODS, CHORDATES, ETC.)
— ECTODERM
— MESODERM
— DIGESTIVE CAVITY
— ENDODERM
— COELOM
— MESODERMAL MESENTERY

PHYLUM ANNELIDA: SEGMENTED WORMS

This phylum includes almost 9,000 different species of marine, freshwater, and soil worms, including the familiar earthworm. The term annelid means "ringed" and refers to the most distinctive feature of this group, which is the division of the body into segments, not only by rings on the outside but by partitions on the inside. This segmented pattern is found in a modified form in higher animals, too, such as dragonflies, millipedes, and lobsters, which are thought to have evolved from ancestors that probably gave rise also to modern annelids.

The annelids have a three-layered body plan, a tubular gut, and a well-developed circulatory system that transports oxygen (diffused through the skin or through fleshy extensions of the skin, such as gills or parapodia; see Figure 22–33) and food molecules (from the gut) to all parts of the body. The excretory system is made up of specialized paired tubules, nephridia, which occur in each segment of the body except the head. Annelids have a nervous system and a number of special sense cells, including touch cells, taste receptors, light-sensitive cells, and cells concerned with the detection of moisture. Some also have well-developed eyes and sensory antennae.

In the flatworms and ribbon worms, the mesoderm is packed solid with muscle and other tissues, but in the annelids there is a fluid-filled cavity, the *coelom,* in this middle layer. (Note that the term coelom, although it sounds similar to coelenteron and comes from the same Greek root, meaning "cavity," refers quite specifically to a cavity *within* the mesoderm, whereas the coelenteron is a digestive cavity lined by endoderm.) The gut is suspended within this cavity by mesenteries consisting of double layers of mesoderm. The fluid in the coelom constitutes a hydraulic skeleton for the annelid, stiffening the body in somewhat the same way water pressure stiffens and distends a fire hose.

Although the development of space within the mesoderm may seem less dramatic than other evolutionary innovations, it is extremely important. Within such a space, organ systems can bend, twist, and fold back on themselves, increasing their functional surface areas and filling, emptying, and sliding past one another, surrounded by lubricating coelomic fluid. Consider the human lung, constantly expanding and contracting in the chest cavity, or the 20 or so feet of coiled human intestine; neither of these could have evolved until the coelom made room for them.

The pseudocoelom, discussed previously, is obviously an alternative solution to the same problem. The pseudocoelom has no mesenteries (tissues connecting two layers of mesoderm). The mesenteries suspend the internal organs in the body cavity and are found not only in annelids but in all higher animals. Moreover, the mesoderm layer surrounding the endoderm provides the musculature of the gut, making possible the contractions (peristalsis) that move food masses through the digestive system. Persistalsis is not possible in pseudocoelomates.

Earthworms

The earthworm is the most familiar of the annelids. Figure 22–29 shows a portion of the body of an earthworm. Note how the body is compartmented into regular segments. Most of these segments, particularly the central and posterior ones, are identical, each exactly like the one before and the one after. Each identical segment contains four pairs of bristles, or *setae;* two *nephridia,* excretory tubules that pick up waste materials from the body fluids and excrete

2–29 *Three segments of the earthworm, an annelid. On each segment are four pairs of bristles, which are extended and retracted by special muscles. These are used by the worm to anchor one part of its body while it moves another part forward. Two excretory tubes, or nephridia, are in each segment (except the first three and the last). Each nephridium really occupies two segments since it opens externally by a pore in one segment and internally by a ciliated funnel in the segment immediately in front of it. The intestine, nephridia, and other internal organs are suspended in the large coelom, which also serves as a hydraulic skeleton.*

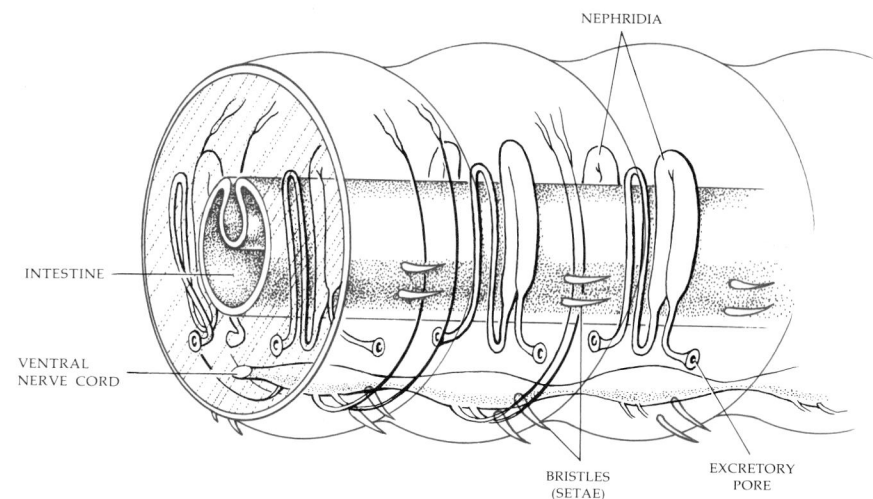

2–30 *The digestive tract and nervous system of an earthworm. The mouth leads into a muscular pharynx, which sucks in decaying vegetation and other material. These are stored in the crop and ground up in the gizzard with the help of soil particles. The rest of the tract is a long intestine (gut) in which food is digested and absorbed. The nervous system consists of a double ventral nerve cord with fibers branching from it to each segment. The nerve cord divides at the anterior of the worm, circles the pharynx, and meets again in two clusters of nerve cell bodies (ganglia) that form the primitive brain.*

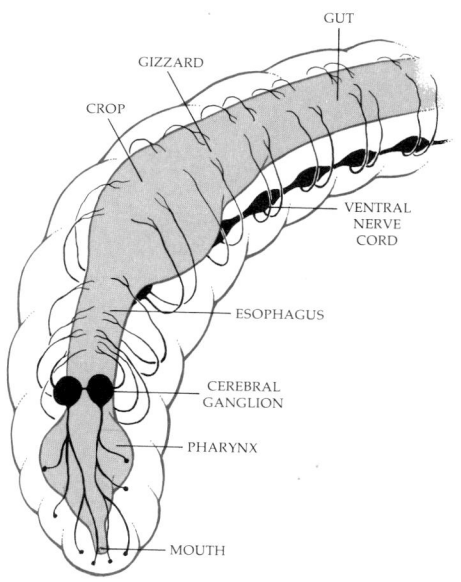

them through pores on the ventral surface of the worm; four sets of nerves branching off from the central nerve cord running along the ventral surface; and a portion of digestive tract. The chief exceptions to this rule of segmented structure are found in the most forward segments. In these, sensory cells, a cluster of nerve cells (ganglia), and specialized areas of the digestive, circulatory, and reproductive systems are found.

The tubelike body is wrapped in two sets of muscles, one set running longitudinally and the other encircling the segments. When the earthworm moves, it anchors some of its segments by its setae, and the circular muscles of the segments anterior to the anchored segments contract, thus extending its body forward. Then its forward setae take hold and the longitudinal muscles contract while the posterior anchor is released. And so the earthworm moves forward.

Digestion in Earthworms

The digestive tract of the earthworm is a long, straight tube. The mouth leads into a strong, muscular pharynx, which acts like a suction pump, drawing in decaying leaves and other organic matter, as well as dirt, from which organic materials are extracted. The earthworm makes burrows in the earth by passing such material through its digestive tract and depositing it outside in the form of castings, a ceaseless activity which serves to break up, enrich, and aerate the soil. The narrow section of tube posterior to the pharynx, the *esophagus,* leads to the crop, where food is stored. In the gizzard, which has thick muscular walls lined with protective cuticle, the food is ground up with the help of the ever-present soil particles. The rest of the digestive tract is made up of a long intestine, which has a large fold along its upper surface that increases its surface area. The intestinal epithelium consists of enzyme-secreting cells and ciliated absorptive cells.

Circulation in Earthworms

In the protists and in the smaller and simpler animals, food molecules and oxygen are supplied to cells largely by diffusion, aided by movement of external fluids and, sometimes, as we saw in coelenterates, by wandering, amoeboid

22–31 *The circulatory system of the earthworm comprises longitudinal vessels running the entire length of the animal, one dorsal and several ventral. Smaller vessels in each segment collect the blood from the tissues and feed it into the muscular dorsal vessel through which it is pumped forward. In the anterior segments are five pairs of hearts—muscular pumping areas in the blood vessels—whose irregular contractions force the blood downward to the ventral vessels from which it returns to the posterior segments. The arrows indicate the direction of blood flow.*

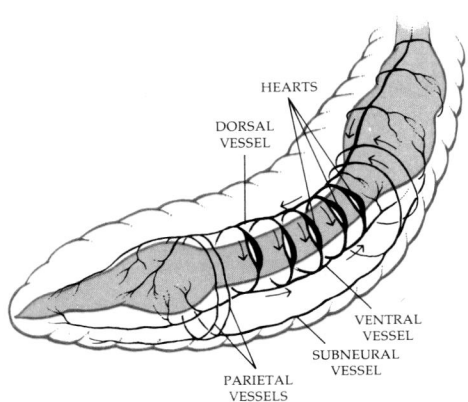

cells. A circulatory system that propels extracellular fluid around the body is a solution to the evolutionary challenge of providing each tissue cell with a more direct and rapid line of supply.

The circulatory system of the earthworm is composed of longitudinal vessels running the entire length of the worm, one dorsal and several ventral. The largest ventral vessel underlies the intestinal tract for its entire length, collecting nutrients from it and distributing them by means of many small branches to all the tissues of the body and to the three smaller ventral vessels that surround the nerve cord and nourish it. Numerous small capillaries in each segment carry blood from the ventral vessels through the tissues to the dorsal vessel. Also in each segment are larger parietal ("along the wall") vessels transporting blood from the subneural vessels to the dorsal vessel. Fluids collected in this way from all over the animal's body are fed into the muscular dorsal vessel, which propels the blood forward.

Connecting the dorsal and ventral vessels (Figure 22–31), and so completing the circuit, are five pairs of hearts, muscular pumping areas in the blood vessels. Their irregular contractions force the blood down to the ventral vessels and also forward to the vessels that supply the more anterior segments. Both the hearts and the dorsal vessel have valves that prevent backflow. Note that the blood flows entirely through vessels. Such a system is known as a closed circulatory system. Evolution of a closed circulatory system, in effect, added a new compartment to the body plan and, in so doing, made possible a degree of control not previously feasible over the content of the circulating body fluid.

Excretory System of Earthworms

The excretory system consists of pairs of tubules, the nephridia; one pair for each segment. Each nephridium consists of a long tubule that terminates in a ciliated funnel opening into the coelomic cavity. Coelomic fluid is carried into the funnel by the beating of the cilia and is excreted through an outer pore. As the fluid makes its way through the long tubule, sugar, salts, and other needed material are returned to the coelomic fluid through the walls of the tubule, while other materials are absorbed into the tubule for excretion. Thus the excretory system is concerned not only with the problem of water balance,* as are the contractile vacuoles of paramecia and the flame cells of planarians, but also with the homeostatic regulation of the chemical composition of the body fluids.

Respiration in Earthworms

The earthworm has no special respiratory organs; respiration takes place by simple diffusion through the body surface. The gases of the atmosphere dissolve in the liquid film on the surface of the earthworm's body, which is kept moist by secreted mucus and excreted water. Oxygen travels inward by diffusion since the surface film, exposed to the oxygen-rich atmosphere, contains more oxygen than the blood in the network of capillaries just underlying the body surface. The oxygen is consumed by body cells as the blood circulates. Carbon dioxide moves out to the surface film and then into the air by the same principle. All gas exchange in animals, whether the organism is land-dwelling or water-dwelling, takes place on a moist membrane.

* See Chapter 36.

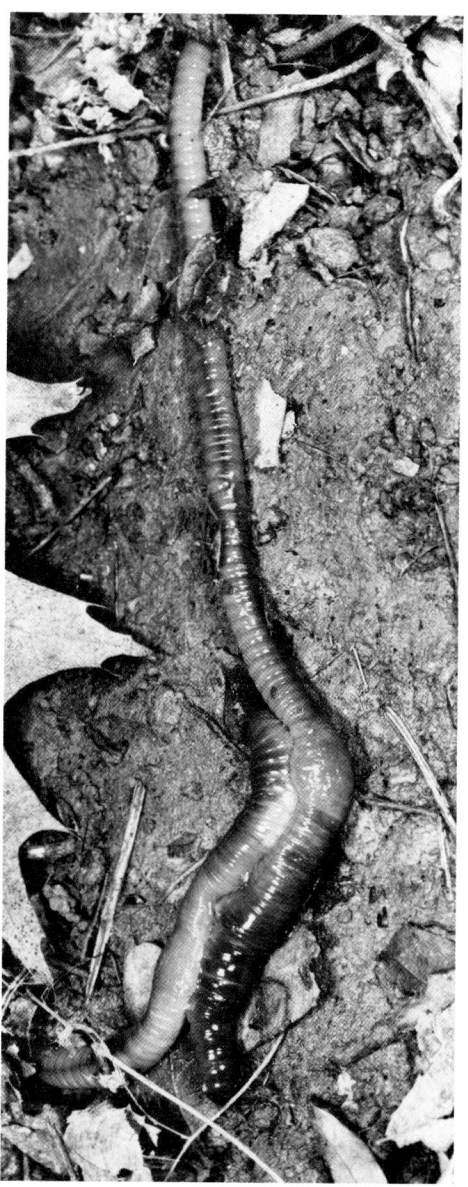

2–32 *Earthworms mating. The worms' heads are facing in opposite directions and their ventral surfaces are in contact. The clitellum, a thickened band which surrounds the body of each, secretes mucus which holds them together during copulation, which may take as long as 2 hours. Sperm cells are released through pores in specialized segments of one worm into the sperm receptacles of its partner. After the partners separate, the clitellum secretes a cocoon into which first the eggs and then the sperm are released. The eggs are fertilized within this cocoon.*

Nervous System in Earthworms

The earthworm has a variety of sensory cells. It has touch cells, or <u>mechanoreceptors</u>. These contain tactile hairs, which, when stimulated, trigger a nerve impulse. Patches of these hair cells are found on each segment of the earthworm. The hairs probably also respond to vibrations in the ground, to which the earthworm is very sensitive. The earthworm does not have ocelli—as one might expect, since it lives most of its life in complete darkness—but it does have light-sensitive cells. These cells, which are not visibly different from the other epidermal cells of the worm, are more abundant in its anterior and posterior segments, the parts of its body most likely to be outside of the burrow. These cells are not responsive to light in the red portion of the spectrum, a fact exploited by anglers who search for worms in the dark using red-lensed flashlights.

Among the earthworm's most sensitive cells are those concerned with detecting moisture. The cells are located on its first few segments. If an earthworm emerging from its burrow encounters a dry spot, it swings from side to side until it either finds dampness or, failing that, retreats. However, when the anterior segments are anesthetized, the earthworm will crawl over dry ground. The animal also appears to have taste cells. In the laboratory, worms can be shown to select, for example, celery in preference to cabbage leaves and cabbage leaves in preference to carrots.

Each segment of the worm is supplied by nerves that receive impulses from sensory cells and nerves that cause muscles to contract. The cell bodies for these nerves are grouped together in clusters (ganglia). The movements of each segment are directed by a pair of ganglia. Movement in each segment is triggered by movement in an anterior segment; thus a headless earthworm can move in a coordinated manner. However, an earthworm without its cerebral ganglia moves ceaselessly; in other words, these ganglia modulate activity.

There are also, as in planarians, conducting channels made up of nerve fibers bound together in bundles, like cables, which run lengthwise through the body. These nerve fibers are gathered together in a double nerve cord that runs along the ventral surface of the body. The nerve cords contain fast-conducting fibers that make it possible for the earthworm to escape danger by contracting its entire body very quickly and withdrawing into its burrow.

Reproduction in Earthworms

Earthworms are monoecious (hermaphroditic). Each earthworm deposits sperm in the two paired spermathecas of its partner by means of a tube formed from the mucus-producing glands of the clitellum, a special collection of glandular cells. Two or three days after the worms separate, the clitellum forms a second mucous sheath surrounded by an outer tougher protective layer of chitin. This sheath is pushed forward along the animal by muscular movements of its body. As it passes over the female gonopores, it picks up a collection of mature eggs, and then, continuing forward, it picks up the sperm deposited in the spermathecas. Once the mucous band is slipped over the head of the worm, its sides pinch together, enclosing the now fertilized eggs into a small capsule from which the infants hatch.

22–33 *Polychaetes, unlike the more familiar oligochaetes, characteristically have a well-differentiated head with sensory appendages and lateral parapodia ("side feet") with many setae.*

22–34 *Leeches are primarily blood suckers with digestive tracts specially adapted for storage of blood. They range in size from ½ inch to a foot and are found mostly in inland waters or in damp places on land, where they parasitize fish, turtles, and other vertebrates. A few species are predatory, feeding on worms and insects which they swallow whole.*

Other Annelids

Other annelids resemble the earthworm in that they are cylindrical worms divided into a series of similar segments and they have a complex circulatory system of blood vessels, a main ventral nerve trunk, a complete digestive tube, and a coelom. The phylum is usually divided into three principal classes: Oligochaeta, Polychaeta, and Hirudinea. Oligochaeta is the group that includes the earthworms and related freshwater species. The polychaetes, which are marine animals, differ from the earthworms and other oligochaetes in a number of ways. The most obvious difference is that they typically have a variety of appendages, including tentacles, antennae, and specialized mouthparts. Each segment contains two fleshy extensions, parapodia, which function in locomotion and also, because they contain many blood vessels, are important in gas exchange. Many polychaetes live in elaborately fashioned tubes constructed in the mud or sand of the ocean bottom. The sexes are separate and fertilization is usually external. Hirudinea consists of the leeches, which have flattened, often tapered, bodies with suckers at each end. The bloodsucking leeches are parasitic. They attach themselves to their hosts by their posterior sucker, then, with their anterior sucker, either slit the host's skin with their sharp jaws or digest an opening through the skin by means of enzymes, and finally secrete a special chemical (hirudin) into the host's blood to prevent it from coagulating.

PHYLUM ECHINODERMATA: STARFISH

The starfish and its relatives are known as echinoderms, or "spiny skins." Adult echinoderms are radially symmetrical, like most coelenterates, although the symmetry is imperfect with vestiges of bilaterality in the adults and with bilaterally symmetrical larvae.

Starfish

The most familiar of the echinoderms is the starfish, whose body consists of a central disk from which radiate a number of arms. Most starfish have five arms, but some have multiples of five. A starfish has no head, and any arm may lead in its sluggish creeping movements along the sea bottom. The central disk contains a mouth on the ventral surface, above which is the stomach. Like all echinoderms, the starfish has an interior skeleton that typically bears projecting spines, the characteristic from which the phylum derives its name. The skeleton is made up of separate calcium-containing plates held together by the skin tissues and by muscles. Each arm contains a pair of digestive glands and also a nerve cord, with an eyespot at the end. These latter are the only sensory organs, strictly speaking, of the starfish, but the epidermis contains thousands of neurosensory cells (as many as 70,000 per square millimeter) concerned with touch and chemoreception. Each arm also has its own pair of gonads, which open directly to the exterior through small pores.

The circulatory system consists of a series of channels within the coelomic cavity. Respiration is accomplished by many small fingerlike projections, the skin gills, which are protected by spines. Amoeboid cells circulate in the coelomic fluid, picking up the wastes and then escaping through the thin walls of the skin gills, where they are pinched off and ejected.

The water vascular, or hydraulic, system is a unique feature of the phylum. Each arm of a starfish contains two or more rows of water-filled tube feet (see Figure 22–36). These tube feet are interconnected by a central ring and radial canals. Water filling the soft hollow tubes makes them rigid enough to walk on. Each tube foot connects with a rounded muscular sac, the ampulla. When the ampulla contracts, the water is forced under pressure through a valve into the tube foot; this extends the foot, which attaches to the substrate by its sucker. When the muscles contract, the animal is pulled forward.

–35 *Some echinoderms: (a) starfish, (b) sea urchins, (c) a sea cucumber.*

(a)

(b)

(c)

22–36 *The water vascular system of the starfish is its means of locomotion. Water enters through minute openings in the sieve plate and is drawn, by ciliary action, down a tube to the ring canal. Five radial canals, one for each arm, connect the ring canal with many pairs of tube feet, which are hollow, thin-walled cylinders ending in suckers. Each tube foot connects with a rounded muscular sac, the ampulla. When the ampulla contracts, the water in it, prevented by a valve from flowing back into the radial canal, is forced under pressure into the tube foot. This stiffens the tube, making it rigid enough to walk on and extends the foot until it attaches to the substrate by its sucker. The longitudinal muscles of the foot then contract, forcing the water back into the ampulla. If the tube feet are planted on a hard surface, such as a rock or a clamshell, the collection of tubes will exert enough suction to pull the starfish forward or to pull open the clam.*

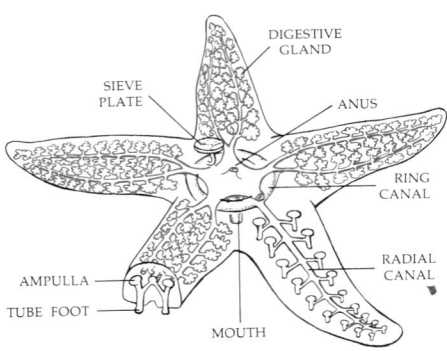

DIGESTIVE GLAND

SIEVE PLATE

ANUS

RING CANAL

RADIAL CANAL

AMPULLA

TUBE FOOT

MOUTH

If the tube feet are planted on a hard surface, such as a rock or a clam shell, the collection of tubes will exert enough suction to pull the starfish forward or to pull apart a bivalve mollusk, a feat that will be appreciated by anyone who has ever tried to open an oyster or a clam. When attacking bivalves, which are its staple diet, the starfish everts its stomach through its mouth opening and then squeezes the stomach tissue through the minute space between the bivalve shells. It is reported that the stomach tissues can insinuate themselves through a space as little as 0.1 millimeter and begin to digest the soft tissue of the resident.

Evolutionary History

The echinoderms are believed to have evolved from an ancestral, bilateral, mobile form that settled down to a sessile life, and then became radially symmetrical. The sea lilies represent this second hypothetical stage. In the third evolutionary stage, some of the animals, as represented by the starfish and sea urchins, became mobile again. Following this line of reasoning, one might expect an eventual return to bilateral symmetry in this group, and, in fact, this is seen to some extent in the soft, elongated bodies of sea cucumbers.

PHYLUM MOLLUSCA: MOLLUSKS

The mollusks constitute one of the largest phyla of animals, both in numbers of species and in numbers of individuals. They are characterized by soft bodies within a hard, calcium-containing shell, although in some forms the shell has been lost in the course of evolution, as in slugs and octopuses, or greatly reduced in size and internalized, as in squids. There are three major classes of mollusks: (1) the gastropods, such as the snails, whose shells are generally in one piece; (2) the bivalves, including the clams, oysters, and mussels, which have two shells joined by a hinge ligament; and (3) the cephalopods, the most active and most intelligent of the mollusks, including the cuttlefish, squids, and octopuses.

Although the mollusks are diverse in size and shape, they all have the same fundamental body plan. There are three distinct body zones: a head-foot, which contains both the sensory and the motor organs; a visceral mass, which contains the organs of digestion, excretion, and reproduction; and a mantle, which hangs over and enfolds the visceral mass and which secretes the shell. The mantle cavity, a space between the mantle and the visceral mass, houses the gills, and the digestive, excretory, and reproductive systems discharge into it. Water sweeps into the mantle cavity propelled by cilia on the gills, passing through the gills and aerating them. It then passes by the nephridia, gonopores, and rectum, which are always downstream from the gills. Water leaving the mantle cavity carries excreta and, in season, gametes.

The digestive tract is far more convoluted and so provides more working surface than that of the annelids. In all mollusks, the digestive tract is extensively ciliated, with many different working areas. Food particles are absorbed by the cells lining the stomach and the anterior intestine, and, from them, it is passed into the bloodstream.

-37 *Mollusks are characterized by soft bodies composed of a head-foot, a visceral mass, and a mantle, which can secrete a shell. They breathe by gills except for the land snails in which the mantle cavity has been modified for air breathing. The hypothetical primitive mollusk is shown in (a). The three major modern groups are the bivalves, such as the clam (b), which are generally sedentary and feed by filtering water currents, created by beating cilia, through large gills; the gastropods, exemplified by the snail (c), in which the visceral mass has become coiled upward and the gut turned back so that mouth, anus, and gills all share the same small aperture in the mantle; and the cephalopods, such as the squid (d), in which the head is modified into a circle of tentacles and part of the head-foot forms a tubelike siphon through which water can be forcibly expelled, providing for locomotion by jet propulsion. The arrows indicate the direction of water movement.*

A characteristic organ of the mollusk, found only in this phylum, and in all mollusks except the bivalves, is the radula, a tooth-bearing strap of movable pieces of cartilaginous tissue covering the tongue. The radula apparatus, which operates with a rhythmic back-and-forth movement, serves both to scrape off algae and other food material and also to convey them backward to the digestive tract.

Mollusks, as we noted previously, have gills. To understand the basic plan of gill structure and function, it is necessary only to recall the earthworm with its moist epidermis, through which oxygen diffuses, and the blood vessel lying close beneath it, which transports the oxygen to other parts of the body. A gill is a structure with an increased amount of surface area, through which gases can diffuse, and a rich blood supply for transport of these gases. Oxygen diffuses inward, along the gradient, because the cells of the animal have removed oxygen from the bloodstream by cellular respiration. Carbon dioxide, produced by cellular respiration, diffuses outward.

Mollusks have three-chambered hearts; two of the chambers (atria) collect oxygenated blood from the gills, and the third (the ventricle) pumps it to the oxygen-depleted tissue. Cephalopods, which are extremely fast and active animals, have accessory hearts that propel blood into the gills.

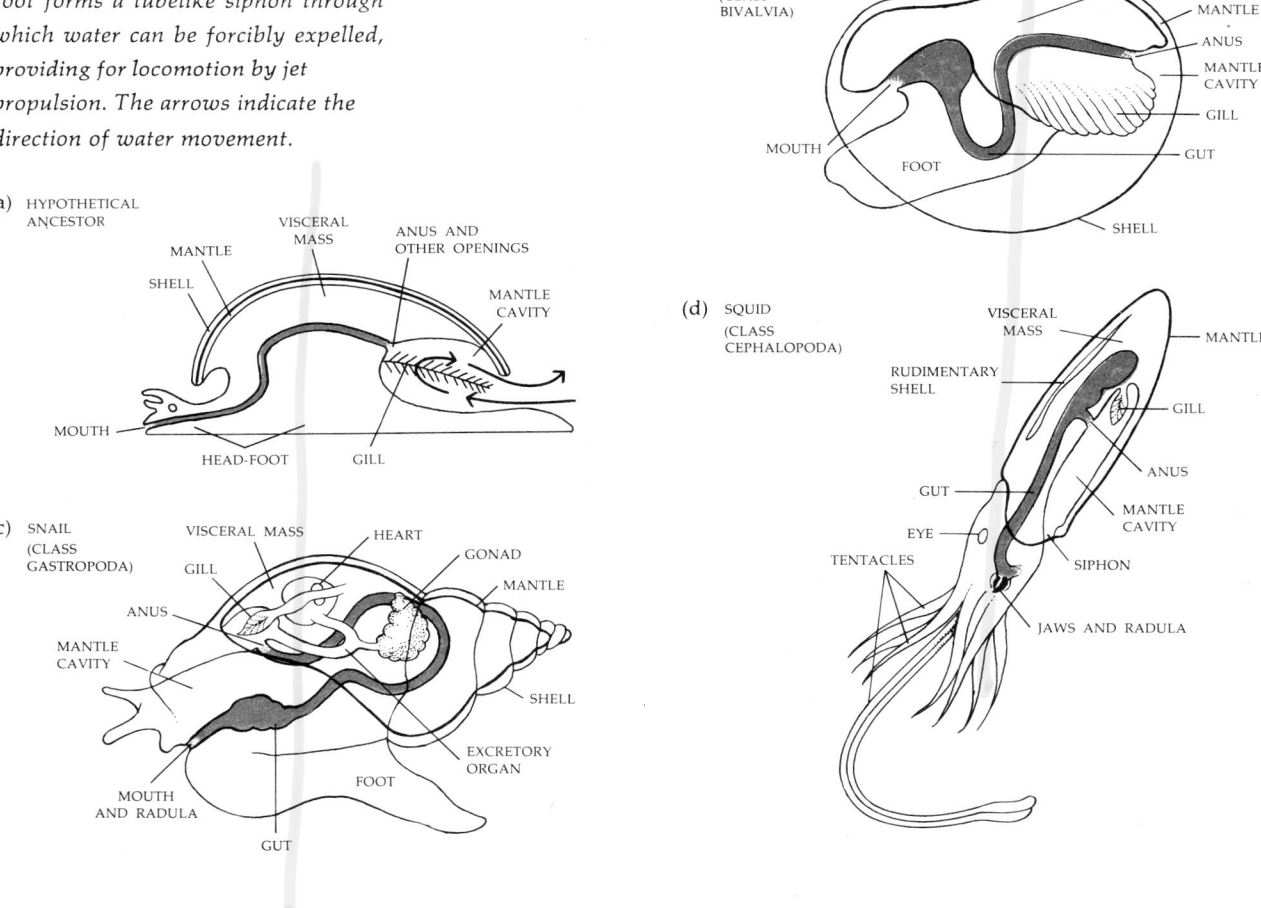

22–38 *A vertical section through the head of a snail to show the radula. The radula rubs back and forth against the hardened roof of the mouth, rasping and tearing plant materials.*

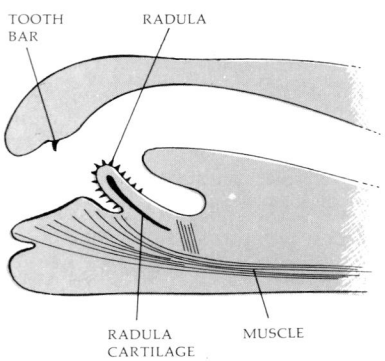

TOOTH BAR   RADULA

RADULA CARTILAGE   MUSCLE

The basic primitive molluscan body plan is shown in Figure 22–37. As you can see, this hypothetical animal was bilaterally symmetrical. Among modern mollusks, only the chitons, a relatively small group, resemble the archetypal model. Each of the three major groups represent variations on this plan.

Class Bivalvia

In the bivalves, the two-shelled mollusks, the body has become flattened between the two shells, and all "headness" has disappeared. The bivalves are sometimes called Pelecypoda—"spade foot"—because the muscular foot is often highly developed in this group; a clam, using its "spade foot," can dig itself into sand or mud with remarkable speed. However, the bivalves are generally sessile forms, and many of them secrete strong, collagenlike strands by which they anchor themselves to rocks or to the bottom.

Bivalves are filter-feeding herbivores; they live largely on microscopic algae. The gills of this group, which are large and elaborate, collect food particles. Water is circulated through the sievelike gills by the beating of gill cilia. Small organisms and particles of food are trapped in mucus on the gill surface and swept toward the mouth by the cilia; the gills also sort particles by size, rejecting sand and other larger particles. Bivalves and also many snails have a peculiar and characteristic digestive organ, the crystalline style, which is a mucoprotein rod extending into the stomach. The style contains digestive enzymes, which it releases slowly into the stomach. Continually rotated by the cilia, it serves as a stirring rod for the stomach contents, and in many species, it acts as a windlass, pulling in a continuous mucous rope of food material. The shells are held together at the hinge by a strong ligament and are drawn closed by one or two large muscles connecting the two shells near the hinge.

22–39 *Like other bivalves, scallops are filter feeders, drawing water currents bearing food particles into their gills by ciliary action. Unlike other bivalves, they have bright blue eyes. Although their eyes do not form images, they are capable of discriminating light from dark and so of detecting the warning shadow of a predator, such as a starfish.*

Bivalves (and all other mollusks except cephalopods) have what is known as an open circulation; that is, the blood does not circulate entirely within vessels —as it does in the earthworm, for example—but is oxygenated and released directly into spaces in the tissues from which it returns, deoxygenated, to the veins or the heart. In the bivalves, the coelom has been reduced to a single space, the pericardial cavity, surrounding the heart.

Throughout the molluscan phylum, there is a wide range of development of the nervous system. The bivalves have three pairs of ganglia of approximately equal size—cerebral, visceral, and pedal (supplying the foot)—and two long pairs of nerve cords interconnecting them. They have a statocyst, usually located near the pedal ganglia, and sensory cells for discrimination of touch, chemical changes, and light. The scallop has quite complex eyes; a single individual may have a hundred or more eyes located among the tentacles on the fringe of the mantle. The lens of this eye cannot focus on images, however, so it does not appear to serve for more than the detection of light and dark and movement.

Class Gastropoda

The gastropods, which include the snails, whelks, periwinkles, abalones, and slugs, are the largest group of mollusks. They have either a single shell or, as a secondary evolutionary development, no shell. Another feature common to all members of this group, as compared with the ancestral mollusk, is that all of them have undergone torsion. In other words, the internal organs, the shell, and the mantle have been twisted 180° so that in the modern animal, the mouth and anus and also the gills share the same comparatively small mantle cavity. Second, the stomach and digestive gland have become twisted upward into a spirally coiled visceral mass. As a result of this displacement and consequent crowding of the internal organs, the gill and nephridium of the right side have been lost in many species. In some close relatives of the snails, such as the slugs, the digestive tract has become straightened out again by another course of evolutionary events, in which the shell was lost but the missing organs were not regained.

Land-dwelling snails do not have gills, which are especially adapted for extracting oxygen from water. The area in their mantle cavities once occupied by gills is extremely rich in blood vessels very near the surface, and the snail's blood is oxygenated in this area. Some snails which were once land dwellers have returned to the water, but they have not regained gills. Instead, they must bob up to the surface at regular intervals to entrap a fresh bubble of air in their mantle cavities. Thus the gill has, in effect, become a lung. Moreover, as with all lungs, the opening is reduced to retard evaporation.

Gastropods, which lead a more mobile, active existence than bivalves, have a ganglionated nervous system with as many as six pairs of ganglia connected by nerve cords. There is a concentration of nerve cells at the anterior end of the animal, where the tentacles, which have chemoreceptors and touch receptors, and the eyes are located. In some of the animals, the eyes are quite highly developed in structure; they appear, however, to function largely in the detection of light, like the eyes of the scallop.

2–40   A land-dwelling gastropod. The shell, which is secreted by the mantle and grows as the soft body grows, covers and protects the visceral mass. The head contains sensory organs, including two eyes at the tips of the longer tentacles. The animal secretes mucus, which lubricates its movements and leaves a slime trail behind it.

–41   The ganglionated nervous system of a gastropod. Each dot represents a pair of ganglia. The cerebral ganglia supply the tentacles and eyes; the pleural ganglia, the mantle; the pedal ganglia innervate the foot muscles; and the visceral ganglia supply the visceral mass.

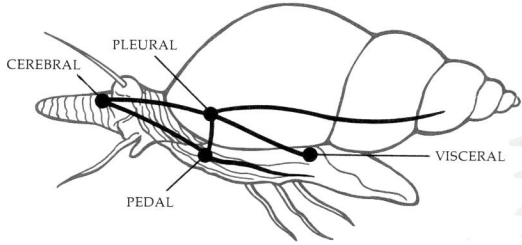

PLEURAL
CEREBRAL
VISCERAL
PEDAL

Class Cephalopoda

In the cephalopods (the "head-foots"), the "foot" has come to be wrapped around the head and divided up into "arms," some 70 or 80 in the chambered nautilus, 10 in the squid, and 8 in the octopus. The nautilus, as the only modern shelled cephalopod, offers an indication of some of the steps by which this order disposed of the shell entirely. The animal occupies only the outermost portion of its elaborate and beautiful shell, which serves primarily as a flotation chamber. In the squid and its relative, the cuttlefish, the shell has become an internal, stiffening support and flotation device, and in the octopus, it is lacking entirely.

The octopus body seldom reaches more than a foot in diameter (except on the late late show), but giant squids sometimes attain true sea-monster proportions. One caught in the Atlantic some hundred years ago was 50 feet long, not counting the tentacles, and was estimated to weigh 2 tons.

Freedom from the external shell has given the mantle more flexibility. The most obvious effects of this is the jet propulsion by which cephalopods dart through the water. Usually, water taken into the mantle cavity bathes the gills and is then expelled slowly through a tube-shaped structure, the siphon; but when the cephalopod is hunting or being hunted, it can contract the mantle cavity forcibly and suddenly, thereby squirting out a sudden jet of water. Contraction of the mantle-cavity muscles usually shoots the animal backward, head last, but the squid and the octopus can turn the siphon in almost any direction they choose. In addition to the siphon, cephalopods have sacs from which they can release a dark fluid that forms a cloud, camouflaging their retreat and confusing their enemies. These colored fluids were at one time a chief source of commercial inks. *Sepia* is the name of the genus of cuttlefish from which a brown ink used to be obtained.

The cephalopods have well-developed brains, composed of many groups of ganglia, in keeping with their highly developed sensory systems and their lively, predatory behavior. These large brains are covered with cartilaginous cases.

22–42 *Three cephalopods: (a) a cuttlefish, Sepia, (b) a squid, Loligo, and (c) a chambered nautilus. The shell of the nautilus contains 30 or more air-filled chambers. The squid and the cuttlefish have only small internal shells, which serve as stiffeners, and the octopus has none.*

(a)

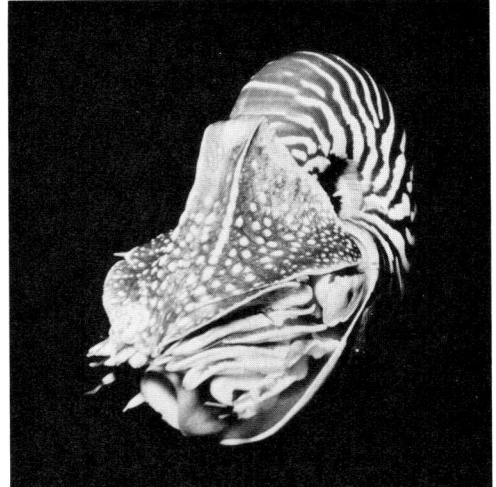

(b)

(c)

2–43 Octopus vulgaris. *Notice the numerous suckers on the underside of the tentacles. Each tentacle carries numerous sensory receptors and can make fine discriminations concerning both texture and taste.*

The rapid responses of the cephalopods are made possible by a bundle of giant nerve fibers which control the muscles of the mantle. Many of the studies on conduction of the nerve impulse are made with the giant axon of the squid, which is large enough to permit the insertion of an electrode.

The Octopus

Behaviorally, the octopus is the best studied of the cephalopods. The octopus is a sea dweller that creeps about actively on its "arms" or swims rapidly through the water by strong rhythmic muscular contractions that expel water from the mantle cavity. The octopus, like the other cephalopods, is carnivorous, living on smaller sea animals, usually crabs. It bites the crab, or other prey, with its parrotlike beak, injecting a toxin from its salivary glands. It then bundles up the paralyzed animal in its web and carries it home to eat. When it is not actively in pursuit of food, the octopus, lacking any protective shell, lives in small caves behind rocks or in reefs or wreckage.

Curious, perpetually hungry, and able to use its arms with great dexterity, the octopus makes an extremely apt experimental subject. In a seawater tank, it will gather together bricks, shells, and any movable debris into a crude sort of house, and there it sits and watches, often bobbing its head up and down. Although its eye is remarkably similar to our own and equally acute, apparently it does not have stereoscopic vision, and this head bobbing seems to be the way it estimates distance, fixing on an object from two points, just as surveyors triangulate a distant landmark.

The sight of a crab can so excite an octopus that its arms weave about and changes appear in its skin color. Unlike the response to threat of danger, in which the animal becomes lighter, the octopus darkens on seeing a crab, breaking out in patches of bright blue, pink, or purple, depending on the species.

-44 *A simplified view of the brain and nervous system of the octopus. The relatively huge lobes behind the eyes are the optic lobes, concerned with collecting and analyzing data from the retina. The octopus often finds its prey by reaching its tentacles into a crevice into which it cannot see. The suckers on these tentacles are softer and more flexible than our fingertips and so can make fine distinctions among textures. They also contain chemoreceptors which can detect sugars, salts, and other chemicals in dilutions well below the range of discrimination of the human tongue.*

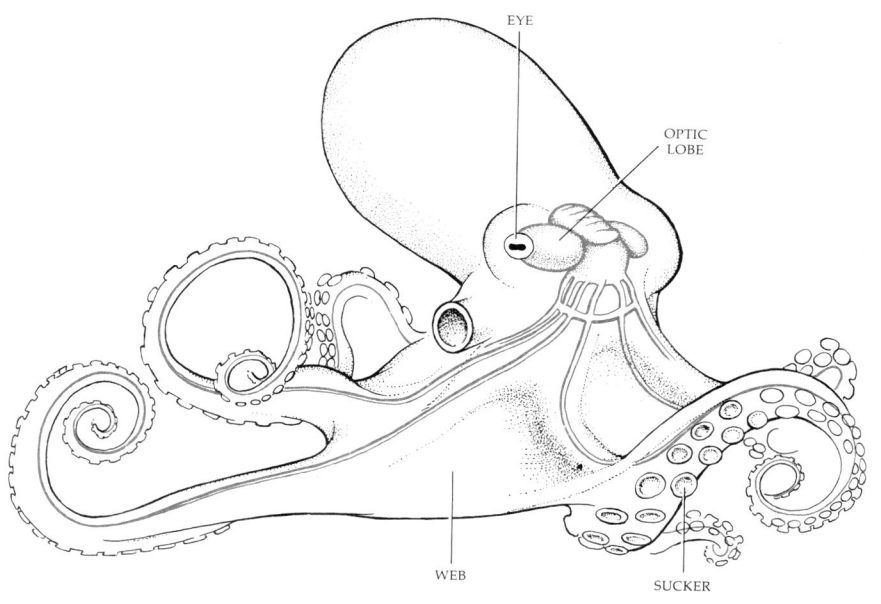

EYE

OPTIC LOBE

WEB

SUCKER

22-45 *Octopus vulgaris literally pales when confronted by an object larger than itself. It flattens and turns almost white except for a dark area around the eyes and a violet trim along the edges of its web. This makes the octopus seem larger than it is and probably serves to deter a would-be predator. The same response can be obtained in the octopus by stimulating its brain with electric currents.*

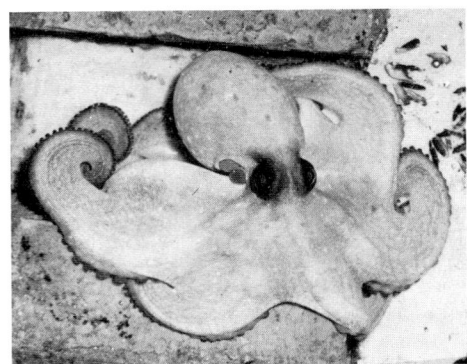

These skin changes are brought about by the contraction of muscles which draw out small sacs of pigment, the chromatophores, to form a flat plate.

By using a system of rewards (crabs, for example) and punishments (mild electric shock), the investigator can readily teach the octopus to seize certain objects and not to seize others. Such experiments approximate what an octopus must learn in nature—that some objects are edible, some are not, and some may bite or sting. A small crab, for instance, is a meal, but if it is carrying a sea anemone on its shell, it had better be left alone. Because of the comparative ease with which such tests can be performed, owing to the octopus's natural curiosity and appetite, the animal has been the subject of a great many experiments on vision, touch, and learning.

The manipulatory powers of the octopus are great, and the tentacles are very sensitive to texture and rich in chemoreceptors. However, the animal seems to have difficulty processing and coordinating sensory data. For example, if an octopus sees a crab behind a glass partition, it will rush directly toward it, flushed with excitement, and when it reaches the glass, will press itself against the pane, writhing its tentacles. One of the tentacles may chance over the top of the glass and reach the crab. In this case, the tentacle will close about the crab. The octopus, however, will continue to respond to the visual stimulation of the crab and press excitedly against the glass as if it were still in pursuit of the prey. Apparently the impulses received from the tactile stimulation of the arm are not integrated with those received from the eye, nor is the movement of the arm detectably influenced by the fact that the octopus can see its arm and the prey. For similar reasons, the octopus is unable to discriminate between objects on the basis of their weight. The observer can tell which of two objects is the heavier by watching the strain in the animal's muscles, but the octopus lifting the objects cannot.

As Martin Wells, of Churchill College, Cambridge, who has carried out many of the studies, points out, the problems of handling data received from eight extensible arms, each of which has several hundred suckers and can move separately and in any direction, would probably be insuperable. As we shall see in the following chapter, the severe limitations of movement provided by an articulated skeleton prove to be a great advantage for many types of specialized activities.

SUMMARY

Organisms must solve a number of basic problems in order to exist. A primary need is to supply each cell with oxygen, water, and nutrients, and to dispose of wastes. In one-celled organisms, these materials are exchanged across cell membranes largely by diffusion, but larger and more complex organisms have developed special systems for circulation, respiration, excretion, and other essential functions. Also, as organisms become more complex, special systems for organizing and integrating the various functions are required.

Animals are multicelled heterotrophs. More than 90 percent of the known species are invertebrates, animals without backbones. There are about 30 different phyla.

Table 22–1 *Characterisitcs of Some Major Invertebrate Phyla*

Phylum	No. of species	Body plan					Characteristic features	Distribution
		Cell layers	Symmetry	Body cavities	Segmentation	Anus		
Porifera (Sponges)	4,200	2, 3, or aggregate	None or radial	None	—	—	Unique body plan, choanocytes	Aquatic (mostly marine)
Coelenterata (Jellyfish)	11,000	2	Radial	None	—	—	Cnidoblasts, with nematocysts, mesoglea	Aquatic (mostly marine)
Platyhelminthes (Flatworms)	15,000	3	Bilateral	None	—	—	Flame cells	Aquatic, parasitic, some terrestrial
Rhynchocoela (Ribbon worms)	600	3	Bilateral	Circulatory, rhynchocoel	—	+	Retractile proboscis	Aquatic (mostly marine) and in moist soil
Nematoda (Roundworms)	80,000	3	Bilateral	Pseudocoelom	—	+	Only longitudinal muscle, periodically molted cuticle	Aquatic, in soil
Annelida (Segmented worms)	8,800	3	Bilateral	Circulatory, coelom	+	+	Setae, segmentation	Aquatic, in soil
Echinodermata (Starfish)	6,000	3	Larvae bilateral; adults secondarily pentaradial	Compartmentalized coelom, including water vascular system	—	+	Tube feet, internal dermal skeleton, penta-radial symmetry	Aquatic (entirely marine)
Mollusca (Snails, clams, octopuses)	110,000	3	Bilateral	Hemocoel, coelom	In some classes, perhaps	+	Body divided into head-foot; visceral mass; mantle-secreting shell; radula	Aquatic, some terrestrial forms

Sponges, phylum Porifera, are composed of a number of different cell types, including choanocytes, or collar cells, which are the feeding cells of the sponge; epithelial cells, some of which are contractile; and archeocytes, largely undifferentiated cells from which other cell types, such as eggs and sperm, may arise. In the sponge, there is little coordination among the various cells. The animal, although it may grow very large, is limited nutritionally to very small particles of a size that can be ingested by the choanocytes.

The distinctive features of the coelenterates, phylum Coelenterata, are (1) a two-layered body, in which the two layers, the ectoderm and the endoderm, are divided by a jellylike substance, the mesoglea, (2) the coelenteron, a cavity in which the food can be partially digested extracellularly, and (3) the cnidoblasts, special stinging cells found only rarely outside this phylum. The coelenterates may take the form of either the polyp or the medusa; in many species, both forms are seen in the course of each life cycle. *Hydra*, a common freshwater coelenterate, in which only the polyp form is seen, has a variety of sensory cells and a nerve net that coordinates the movements of the animal. *Obelia* is an example of a coelenterate in which the polyp form predominates. The polyps, which are joined, live in colonies in which some of the individuals feed and others are reproductive forms, giving rise to medusas which produce gametes. In *Aurelia*, the medusa form predominates. In addition to special sensory cells, two specialized sense organs, statocysts and ocelli, are found in this phylum.

The flatworms, phylum Platyhelminthes, are bilaterally symmetrical and elongated, with a distinct "headness" and "tailness" and a concomitant clustering of nerve cells in the anterior region. The flatworms have three distinct tissue layers—ectoderm, mesoderm, and endoderm—and also an excretory system, involving flame cells, that serves largely to maintain water balance.

The ribbon worms, phylum Rhynchocoela, are characterized by a retractile, prey-seizing proboscis, a one-way (mouth-to-anus) digestive tract, and a simple circulatory system.

The roundworms, or nematodes, phylum Nematoda, of which there are, conservatively, 80,000 species, are pseudocoelomates. A pseudocoelom is a cavity between the endoderm and the mesoderm.

Members of phylum Annelida—the earthworm is an example—have segmented bodies, coeloms, which are cavities within the mesoderm and within which internal organs are supported by mesenteries, one-way digesive tracts, and closed circulatory systems, often with contractile vessels. Excretion is accomplished by special organs, the nephridia, which are convoluted tubules. The nephridia collect fluids from the coelom and exchange salts and other substances with the body fluids as the urine passes along the tubules for excretion. The earthworm has a relatively complex nervous system, consisting of a ventral nerve cord, which divides to encircle the pharynx at the anterior end of the animal, and pairs of ganglia (clusters of nerve cells), one pair to each segment, which receive sensory impulses and trigger motor activities in each segment.

Phylum Echinodermata includes the starfish, the sea urchins, and the sea lilies, all of which are radially symmetrical in their adult forms. An unusual characteristic of this phylum is the water vascular system, which provides a hydraulic skeleton for the animal and also provides suction for the clinging and pulling activities associated with moving and predatory behavior.

There are three main classes of phylum Mollusca: the mollusks, represented by the bivalves (oysters and clams), the gastropods (snails), and the cephalopods (octopuses and their relatives). The body is always divided into a head-foot, visceral mass, and mantle. In each of these classes, the basic body plan is the same, but it has been modified in the course of adaptation to a particular environment. In most mollusks, respiration is carried out by means of a gill, a thin-walled structure that is an extension of the epidermis. It is richly endowed with blood vessels that serve as an area of gaseous exchange (respiration). Mollusks are also characterized by an efficient three-chambered heart and a toothed tongue, the radula. Nervous systems and behavior vary among the species, reaching a zenith of complexity in the brainy octopus.

QUESTIONS

1. Define the following terms: coelom, pseudocoelom, choanocyte, monoecious, cnidoblast, radula, flame cell.
2. Compare the nervous systems of a hydra, a planarian, a clam, an annelid, an octopus.
3. Describe respiration in a coelenterate, an earthworm, an aquatic snail.
4. What is a closed circulatory system? An open one?
5. A larval or medusoid stage is common in the life histories of many marine invertebrates. Yet in group after group of freshwater and terrestrial invertebrates, each supposedly separately evolved from marine ancestors, the free-living larva or medusa has been lost. What might be the selective factor underlying the repeated loss?
6. Nematodes have only longitudinal muscles in their body walls but have very high internal fluid pressures; earthworms, with both longitudinal and circular muscles, have only low pressures. Similarly, spiders have only joint-bending muscles but have high blood pressures, whereas insects, with joint-extending muscles as well, usually have low blood pressure. Can you suggest a mechanism by which internal fluid at high pressure can circumvent the need for certain muscles?
7. We commonly discuss the various "inventions" in animal phylogeny—coelom, anus, segmentation, head, hearts, etc.—as advances, a kind of progress toward better and better adaptation. We also assert that evolution selects the better adapted variant. How, then, can we account for the continued presence on Earth of simple creatures lacking various of these splendid accouterments?
8. Label this drawing.

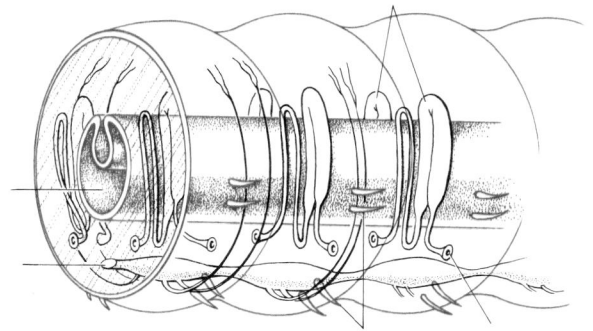

Chapter 23

The Animal Kingdom II: Arthropods and Chordates

This chapter will be concerned chiefly with the arthropods, the largest by far of all the phyla. Almost 1 million species of insects and other arthropods have been classified to date, and estimates of the total number are as high as 10 million. As to the number of individuals, it has been calculated that, of insects alone, as many as 10^{18}—a billion billion—are alive at any one time. Over every square mile in the temperate zone, someone has estimated, there are some 25 million, layered in the atmosphere, like plankton.

At the end of this chapter, we shall discuss the evolution of the Chordata, a much smaller phylum. This latter discussion will serve as an introduction to Section 5, which deals principally with man and other chordates.

PHYLUM ARTHROPODA

The arthropods, "joint-footed" animals, include the crustaceans (crabs and lobsters), arachnids (spiders), and insects; the latter constitute by far the largest group. Despite the huge size of the arthropod phylum and the diversity of its members, there are a number of features shared by all the members of this group.

The Exoskeleton

First, all arthropods have an articulated (jointed) exoskeleton. This exoskeleton is secreted by the underlying epidermis and is attached to it; it is made up of an outer waxy layer composed of lipoprotein, and a middle horny layer and inner flexible one, both of chitin. The exoskeleton not only covers the surface of the animal but also extends inward at both ends of the digestive tract and, in insects, lines the tracheae (breathing tubes) as well. Muscles are attached to the various segments of the exoskeleton, just as they are attached to the various bones of the endoskeleton in vertebrates, and when the muscles contract, the exoskeleton moves. Chitin is a remarkable material. It may form a veritable coat of armor, as it does in some of the crustaceans (in which it is often infiltrated with calcium salts), but at the joints it is flexible and thin, permitting free movements. It serves as protection against predators. It is water-

23–1 *Arthropods are characterized by a variety of specialized appendages, such as the long sucking tongue of this butterfly,* painted lady *(Cynthia annabella).*

Molting is another arthropod characteristic. An exoskeleton, once formed, does not grow and so must be periodically discarded. The old shell is at the top of the photograph and below is the newly emerged "soft-shelled" crab. The new exoskeleton, already formed, will expand and harden.

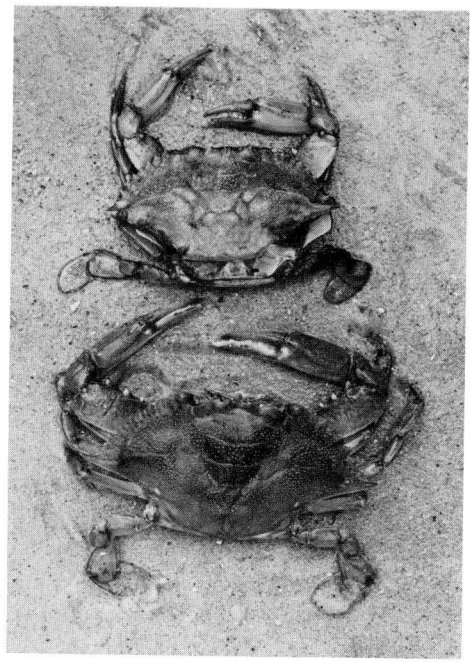

proof, keeping exterior water out and interior water in. It can be used for biting jaws, for grinders in the stomach, for wings, and for parts of various sensory organs; it even forms the lens of the arthropod eye.

The exoskeleton has certain disadvantages. It does not grow (as the bony vertebrate endoskeleton does) and so it must be discarded and re-formed many times as the animal grows and develops.

Molting is dangerous since the newly molted animal is particularly vulnerable to predators and in the case of terrestrial forms, subject to water loss. Many arthropods go into hiding until their new cuticle has hardened. Molting is also costly in terms of metabolic expenditures, although a number of insects and some freshwater crustaceans limit their losses by thriftily eating the old exoskeleton.

Other Arthropod Characteristics

All arthropods are segmented, a characteristic that suggests that they share a common ancestry with the annelids. In some of the arthropods the segments have become fused, forming a head, a thorax (sometimes fused to form a cephalothorax), and an abdomen. But the basic segmented pattern is still clearly evident in the immature stages (witness the caterpillar) and can be discerned in the adult by examination of the appendages, the musculature, and the nervous system.

Arthropods have a large number of segmented appendages. Especially among the insects and crustaceans, these appendages constitute a veritable tool kit of highly specialized appliances, including jaws, gills, poison fangs, tongs, egg depositors, sucking tubes, claws, antennae, paddles, and pincers.

The insects and some other terrestrial forms have unusual respiratory machinery consisting of a system of chitin-lined air ducts (tracheae) that pipe air directly into various parts of the body. Air flow is regulated by the opening and closing of special pores (spiracles) on the exoskeleton. Terrestrial arthropods that do not have tracheae have book lungs, structures that are also unique to this phylum. Excretion in terrestrial forms is by means of tubes (called Mal-

23–3 *Arthropods are segmented, like the annelids from which they presumably arose. However, unlike annelids, arthropods, such as the millipede shown here, have exoskeletons and segmented, jointed appendages.*

23–4   (a) *Respiration by means of a system of internal tubes (called tracheae) is found only among arthropods. Tracheae are usually branched, like those shown here, and open to the outside by spiracles that may be closed to conserve water. The tubes are lined with spiral rings and cuticle, which keeps them open. A tracheal system is one of the most efficient respiratory systems in the animal kingdom because it delivers gaseous oxygen directly to the tissues. (b) Malpighian tubules represent another exclusively arthropod characteristic, although like tracheae, they are not found in all classes. These tubules collect water and nitrogenous wastes from the coelom and empty them into the gut. The wastes (in the form of uric acid) are excreted with the feces.*

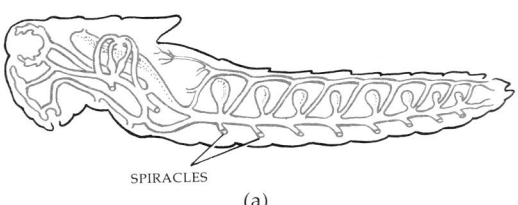

SPIRACLES

(a)

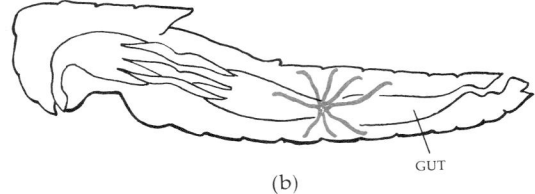

GUT

(b)

pighian tubules) attached to and emptying into the hindgut. Respiration by tracheae or book lungs or book gills and excretion by Malpighian tubules are exclusive arthropod characteristics.

Arthropods also share a number of characteristics, besides segmentation, with the annelids. Like them, they are bilaterally symmetrical, three-layered, and have a tubular, mouth-to-anus gut. They have a coelom, as do the annelids, but the arthropod coelom is markedly reduced, consisting only of the cavities that surround the gonads and the nephridium. (As you will recall, the coelom serves as a hydrostatic skeleton in annelids, but the arthropods with their exoskeletons do not require such internal stiffening.)

The arthropods, like the bivalve mollusks, have an open circulatory system in which blood flows through free spaces among the tissues as well as through vessels. The large space occupied by the blood is called the hemocoel. Blood returns from the hemocoel to the tubular heart through special valved openings.

The Arthropod Nervous System

Arthropods have a ladderlike nervous system, with a double chain of segmental ganglia running along the ventral surface, splitting in two to encircle the esophagus and ending in a pair of dorsal ganglia. The dorsal ganglia are enough larger than the other ganglia to be called a brain. However, many arthropod activities are controlled at the segmental level, as in annelids. For example, members of a number of species can move, eat, and carry on other functions normally, after the brain is removed. In fact, in arthropods generally, the brain appears to act not so much as a stimulator of the action of the animal but as an inhibitor, as in the earthworm. The grasshopper, for example, with its brain removed, can walk, jump, or fly; in fact, the brainless grasshopper responds to the slightest stimulus by jumping or flying. The extreme consequences of this releasing of inhibition can be seen in the praying mantis. Mantises are carnivorous and cannibalistic, and the female, being larger than the male, frequently

23–5   *A female mantis will often decapitate a courting male. Fortunately for the species, as with other arthropods, loss of the brain removes inhibitions to certain activities, and a headless male is more likely to copulate than an intact one. After mating, she may leisurely complete her meal.*

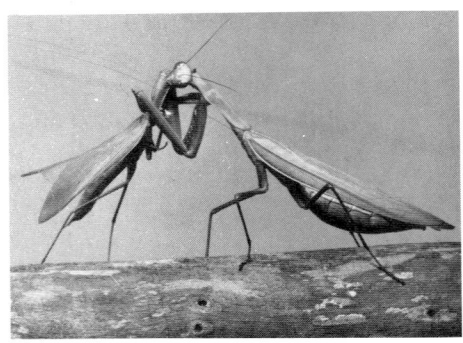

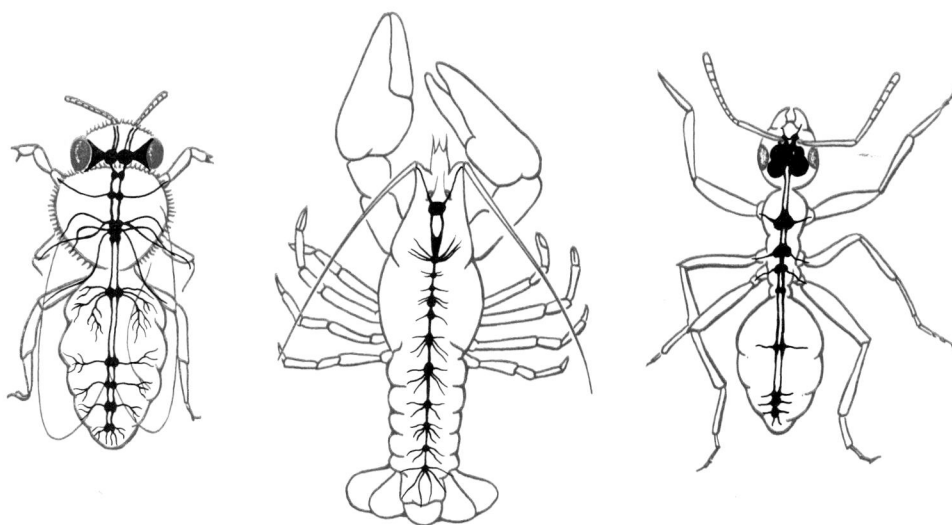

3–6 *Arthropod nervous systems: a bee, a crayfish, and an ant. The brain is a pair of ganglia at the end of a chain of ganglia which are interconnected by two bundles of nerve fibers running along the ventral surface. Because of the arthropod's "ladder" type of nervous system, many arthropod activities are controlled at a local level, and a number of species can carry on a few of their normal functions after the brain has been removed.*

3–7 *Because the body of the horseshoe crab is largely covered with a heavy shield, or carapace, its segmented body plan is only evident from the underside. The operculum is a flat, movable plate that covers and protects the book gills dorsal to it. Horseshoe crabs are often called living fossils because their remarkably similar ancestors date back to the Cambrian period. There are five living species.*

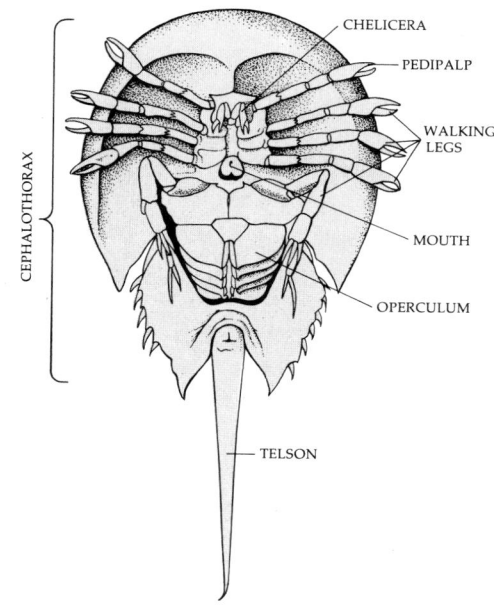

CHELICERA

PEDIPALP

WALKING LEGS

MOUTH

OPERCULUM

CEPHALOTHORAX

TELSON

captures her approaching mate and, grasping him in her forearms, begins to eat him, head first. This decapitation results in the release of strong motor activities by which the headless male struggles loose from the grasp of the female, mounts her, and mates with her. The headless male is more likely to copulate than the intact male; investigators seeking to breed mantises have found that males of strains that do not mate readily in captivity will often do so after decapitation.

SUBDIVISIONS OF THE PHYLUM

The arthropods are often classified in two subphyla, the Mandibulata and the Chelicerata, which can be clearly distinguished by even an inexperienced eye. In the chelicerates, which include horseshoe crabs, sea spiders, and arachnids (spiders, scorpions, and their relatives), the head and thorax segments are fused into a cephalothorax, whereas in the mandibulates, the head and thorax are separate. There are also conspicuous differences in the appendages. The most anterior appendages in the mandibulates are *antennae*, one or two pairs, and the next are *mandibles* (jaws). The chelicerates have no antennae and no mandibles. Their first pair of appendages consist of *chelicerae* (singular, chelicera), which usually take the form of pincers or fangs. In spiders, ducts from a pair of poison glands lead through the chelicerae, which are sharp and pointed and are used for capturing and paralyzing prey. Chelicerates may have book lungs or book gills; these structures, which are not present in mandibulates, derive their name from their resemblance to the leaves of a partially opened book.

Each of these subphyla are, in turn, divided into several classes. The three largest groups are the crustaceans, the arachnids, and the insects. Minor chelicerate classes include the trilobites, a now extinct class; the xiphosurans, which include the horseshoe crabs; and the sea spiders. Minor mandibulate classes include the centipedes and millipedes (see Figure 23–3).

Class Crustacea

The crustaceans include crabs, crayfish, lobsters, barnacles, shrimp, prawns, *Daphnia* (water fleas), and a number of smaller forms found largely in freshwater and marine plankton, as well as some terrestrial forms such as the familiar pillbugs or sowbugs. Crustaceans are mandibulates. They differ from the insects, which are also mandibulates, in that they have legs or leglike appendages on the abdomen as well as the thorax, and have no wings.

Lobster

Figure 23–8 shows the structure of a lobster, a representative crustacean. A crayfish differs morphologically from the lobster only in minor respects, the most obvious of which is the size of the claws.

The lobster has 20 segments, the first 13 of which are united on the dorsal side in a combined head and thorax, or cephalothorax. A heavy shield, or carapace, covers its back. Carapaces are common among crustaceans. The abdomen consists of six distinct segments. (To lobster eaters, the abdomen is the "tail," but to biologists, the designation "tail" is usually reserved for areas of the body posterior to the anus.) The various appendages have special functions. The antennae, of which there are two pairs, are sensory. The mandibles, or jaws, are used for crushing; like all arthropod jaws, they move laterally, open-

23–8   *A representative crustacean, the American lobster,* Homarus americanus.

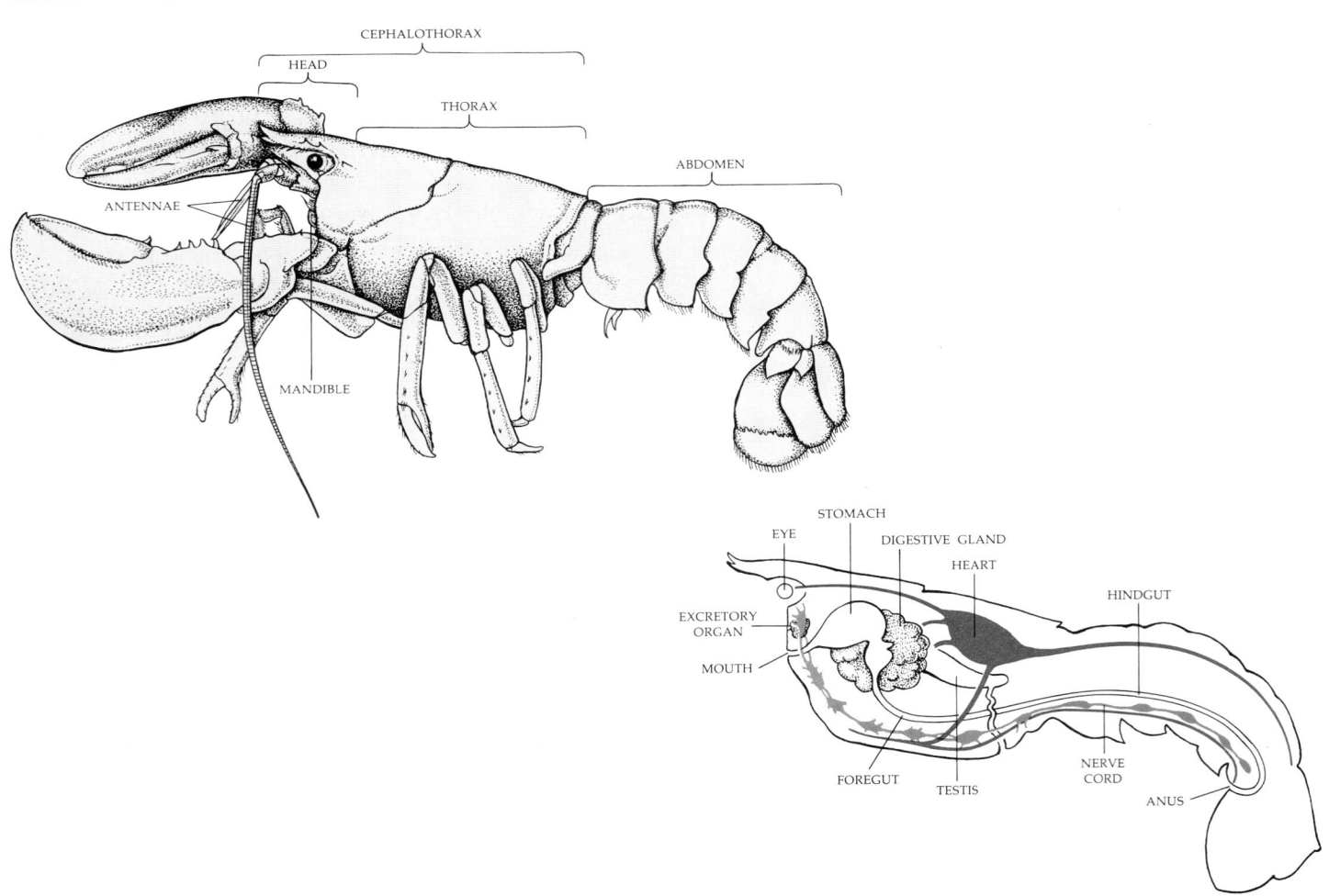

Some crustaceans. (a) Hermit crab, which houses itself in the abandoned shells of mollusks or other crustaceans; (b) pillbugs (also known as sowbugs or wood lice), terrestrial arthropods found in damp places; (c) crab (Gecarcinus quadratus) on a beach in Mexico.

(a)

(b)

(c)

ing and closing from side to side like a pair of ice tongs rather than up and down like vertebrate jaws. The maxillae and the maxillipeds serve chiefly to collect food, mince it, and pass it on to the mouth. The claws are unequal in size in the full-grown lobster; the larger claw is used for crushing, and the smaller claw, which has the sharper teeth, seizes and tears the prey. The first two pairs of walking legs have small pincers, which can seize prey. The last pair of walking legs also serves to clean the abdominal appendages. The flattened posterior appendages are used, like flippers, for swimming. The claws and abdomen are filled almost entirely with large and edible striated muscles. These muscles are extremely powerful. A lobster can snap its "tail" ventrally with enough force to shoot backward through the water, and a large lobster can shatter the shell of a clam or oyster with its crushing claw. Any one of its appendages can be regenerated at the next molt if lost, and a lobster will often break off a claw or a leg that is held by a predator, in order to escape.

The anterior and posterior regions of the digestive tract, the foregut and hindgut, are lined with cuticle. As a consequence, most of the food is absorbed through the midgut and through the cells of the large digestive gland, or "liver." The anterior segment of the stomach is used for storage. The midsection, which has sharp chitinous teeth, functions as a gastric mill, grinding the food by muscular contractions. The third section sorts the food, routing the smallest particles to the digestive glands, the larger ones to the intestinal tract, and the coarsest ones to the gizzards.

Respiration is accomplished by 20 pairs of feathery gills, which lie in the thorax between the body wall and the exoskeleton, where water can flow freely over them. The nephridium is in the head; wastes extracted from the blood are collected into a bladder and excreted from a pore at the base of each of the second antennae. Fertilization is external; the sperm, which are deposited near the female gonopores, fertilize the eggs as they are laid, and the fertilized eggs cling by means of a sticky secretion to the swimmerets of the female until they hatch.

Terrestrial Crustaceans

Unlike other arthropod groups, almost all the crustaceans are aquatic, but some, crabs in particular, are amphibious or land dwellers. Amphibious crabs continue to breathe with gills, carrying around water in their thoracic cavities with which to keep the gills wet and aerating the water through holes in their exoskeletons. The true land crab has lost the gill structures and instead has an area of highly vascularized epithelial tissue through which oxygen is exchanged. The land snail, you will recall, solved the respiratory problems involved in the transition from water to land in an analogous way.

Class Arachnida

The arachnids, which include spiders, ticks, mites, scorpions, and daddy longlegs, are chelicerates. The first pair of appendages are chelicerae, which are sharp and pointed and are used for capturing and then paralyzing prey by the injection of a poison. The second pair are the pedipalps, which are used for handling and tearing food. Like most other arachnids, spiders have four pairs of walking legs.

23–10 *Two spiders (class Arachnida). (a) A female green lynx spider on a zinnia, lying in wait for prey. She uses her silk as a drag line for jumping. (b) A wolf spider. These common spiders are ground dwellers. They have eight eyes, four small ones in a row beneath four large ones.*

(a)

(b)

Spiders, like other arachnids, live on a completely liquid diet. The prey is masticated by the mouthparts, and then enzymes from the midgut are poured out over the torn tissues to produce a partially digested broth. The liquid is pumped into the stomach by the muscular pharynx, where digestion is completed and the juices absorbed. Spiders and scorpions respire by means of *book lungs*, which are a series of leaflike plates within a chitin-lined chamber. Air is drawn into the chamber and expelled by muscular action.

On the posterior portion of the spider's abdominal surface is a cluster of spinnerets, fingerlike organs from which a fluid protein exudes that hardens into silk as it comes into contact with the air. Silk is used not only for the variety of webs made by the different species but for a number of other purposes as well, such as for a drop line, on which the spider goes sky diving, for a cocoon, for lining a burrow, for the shroud of a victim, or for wrapping an edible offering presented to the female of certain species by the courting male. Most spiders can spin several kinds and thicknesses of silk.

The weaving of a web is not learned behavior, nor does it require any practice. A spider confined from egg to adult can weave a perfect web on its first try.

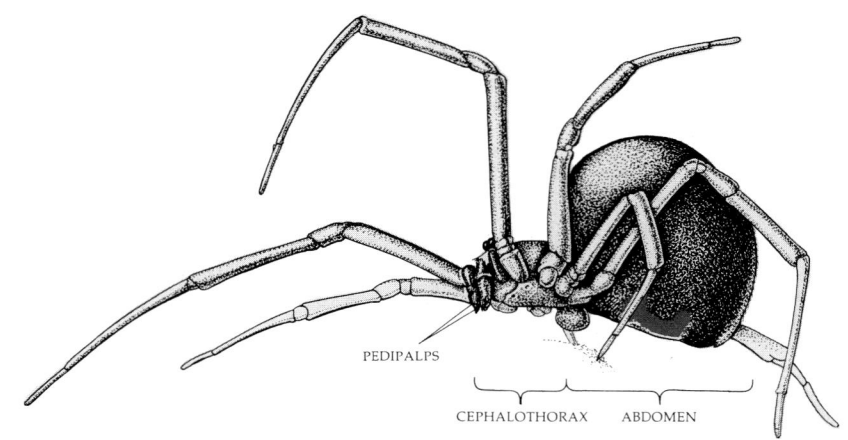

PEDIPALPS

CEPHALOTHORAX   ABDOMEN

23–11 *A spider, an arachnid. Ducts from the poison gland open at or near the tips of the chelicerae. The flow of poison is voluntarily controlled by the spider. Only a few spiders are dangerous to man; perhaps the most dangerous is the species shown here, the black widow.*

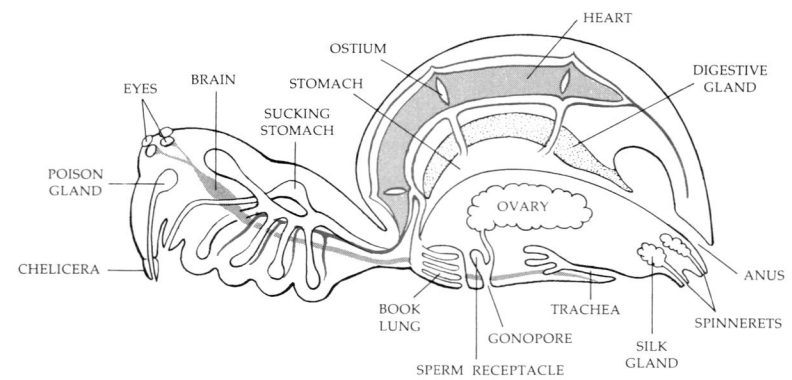

EYES   BRAIN   STOMACH   HEART

SUCKING   DIGESTIVE
STOMACH   GLAND

POISON
GLAND

OVARY

CHELICERA

ANUS

BOOK   TRACHEA   SPINNERETS
LUNG

GONOPORE   SILK
GLAND

SPERM RECEPTACLE

23–12 *A grass spider* (Agelenopsis) *has caught an insect, a lacewing, in its web. Webs of this species are sometimes 3 feet wide.*

23–13 *The late J. B. S. Haldane, noted for his agnosticism and crusty disposition, as well as for his scientific achievements, was once asked what his study of biology had revealed to him about the mind of God. "Madame," he replied, "only that He had an inordinate fondness for beetles." Shown here is everybody's favorite beetle, a ladybug* (a), *accompanied by a stag beetle* (b), *so called because of the antlerlike mandibles, characteristic of the pugnacious males.*

(a)

(b)

Class Insecta

The insects constitute the largest class, by far, of the arthropods. In fact, there are more species of insects than of all other animals combined. About 800,000 are known, of which about 275,000 are beetles.

Figure 23–14 shows a grasshopper. Here you can see many of the characteristic features of insects: three main divisions—the head, the thorax, and the abdomen—a single pair of antennae, and a set of mouthparts somewhat similar to those of the lobster. In the more primitive insects, such as the grasshopper, the mouthparts are used for handling and masticating food, but in the more highly evolved and specialized groups, the mouthparts are often molded into sucking, piercing, slicing, or sponging organs, many of which are exquisitely adapted to the nectaries (nectar-holding organs) of special flowers.

Most adult insects have two pairs of wings made up of light strong sheets of chitin; the veins in the wings are chitinous tubules that serve primarily as braces. The wings of the various orders of insects have evolved separately from one another. In some, such as the fleas and lice, they have been partially or totally lost, returning the insect to the condition of its wingless ancestors.

23–14 *In the grasshopper, an insect, the head consists of six fused segments which have appendages specialized for biting and chewing. Each of the three segments of the thorax carries a pair of legs (three pairs in all), and two of them carry wings (in the grasshopper, the forewings are hardened as protective covers). The spiracles in the abdomen open into a network of chitin-lined tubules through which air circulates to various tissues of the body. This sort of tubular breathing system is found only among insects and other land-dwelling mandibulates (the centipedes and millipedes). Excretion takes place by Malpighian tubules that empty into the hindgut.*

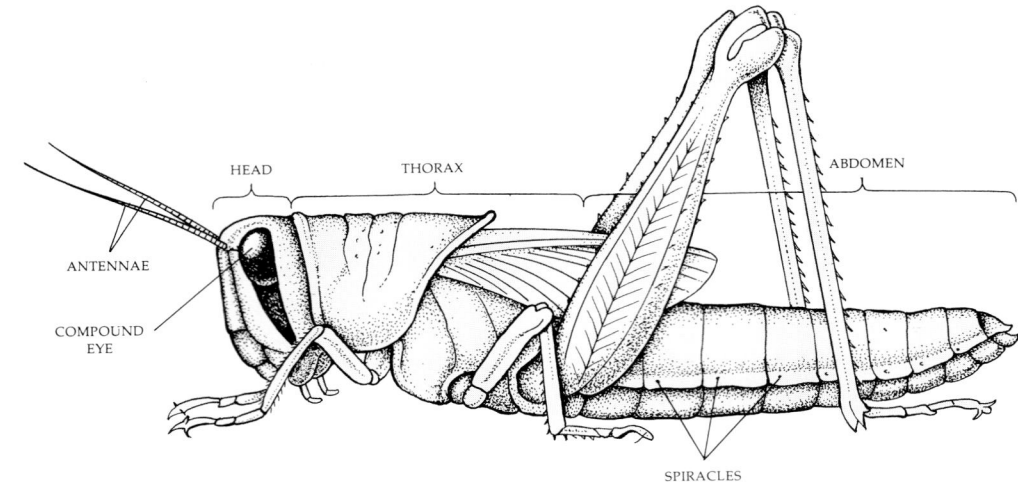

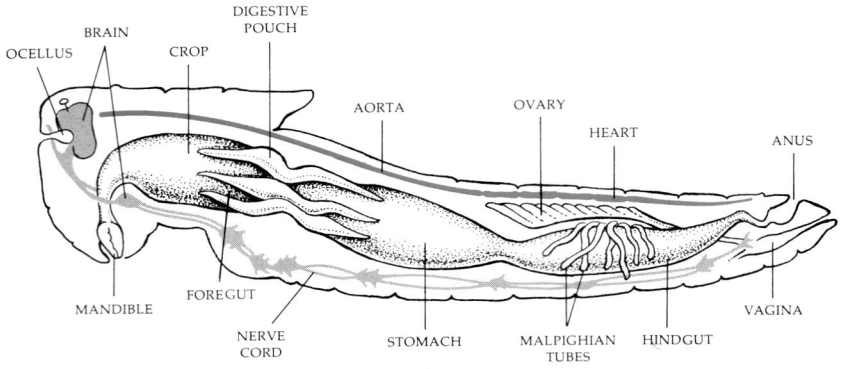

23–15 *Mouthparts of a grasshopper. The mandibles are crushing jaws. The labium and the labrum are the lower and upper lip. The maxillae move food into the mouth.*

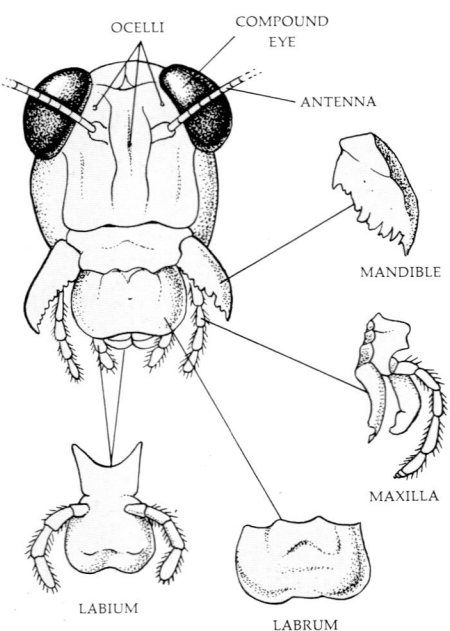

Digestive and Respiratory Systems

The foregut and hindgut of the insect digestive tract are lined with chitin. Salivary gland fluids are carried with food into the crop, where digestion begins. The stomach, which lies mainly in the abdomen, is the chief organ of absorption. In the grasshopper, digestive pouches open into the stomach, releasing digestive juices that are especially adapted to the grasshopper's diet of leaves and grasses. Insects have digestive enzymes as specialized as their mouthparts; the structure of the enzyme depends on whether the insect dines on blood, seeds, other insects, eggs, flour, cereal, glue, wood, paper, or your woolen clothes. Excretion is carried out through Malpighian tubules. In the grasshopper and many other insects, the nitrogenous and other wastes are eliminated in the form of nearly dry crystals, an adaptation that promotes water conservation.

The respiratory system consists of a network of chitin-lined tubules through which air circulates to the various tissues of the body, supplying each cell directly. Muscular movements of the animal's body improve the circulation of air. The amount of incoming air and also the degree of water loss can be regulated by the opening and closing of small openings, or spiracles.

23–16 *Some immature forms.* (a) *Scarab beetle larva.* (b) *Black swallowtail larva (caterpillar).* (c) *Mosquito larvae and a pupa (on right). Mosquito larvae are aquatic, hanging by their tails on the undersurface of the water with respiratory tubes extended above the surface.* (d) *Tent caterpillars.*

(a)

(b)

Metamorphosis

Most insects go through definite developmental stages. In some species, the infant, although sexually immature, looks like a small copy of the adult; it grows larger by a series of molts until it reaches full size. In others, like the grasshopper, the newly hatched young is wingless with somewhat different proportions from the adult, but it is otherwise similar. These immature, nonreproductive forms are known as *nymphs*. Almost 90 percent of the insects, however, undergo a complete metamorphosis, so that the adult is completely different from the immature form. The immature eating and feeding forms may all correctly be referred to as *larvae*, although they are also commonly known as caterpillars, grubs, or maggots, depending on the different species. Following the larval period, the insect undergoing complete metamorphosis enters an outwardly quiescent pupal stage in which extensive remodeling of the organism occurs. The adult insect emerges from the pupa.

The insect that undergoes complete metamorphosis exists in four different forms in the course of its life history. The first form is the egg and the embryo. The second form is the larva, the animal that hatches from the egg; larvae eat and grow. In many larvae, such as those of flies, growth takes place not by an increase in the number of cells, as in most animals, but by an increase in the size of the cells, in somewhat the same way that it takes place in certain plant tissues. During the course of its growth, the larva molts a characteristic number of times —twice in the fruit fly, for example. The stages between molts are known as *instars*. Then, when the larva is full-grown, it molts to form the pupa. During this outwardly lifeless pupal stage, many of the larval cells break down and entirely new groups of cells, set aside in the embryo, begin to proliferate, using the degenerating larval tissue as a culture medium. These groups of cells are known as imaginal disks since they form the imago, the adult insect, which, according to Aristotle, is the perfect form or ideal image that the immature

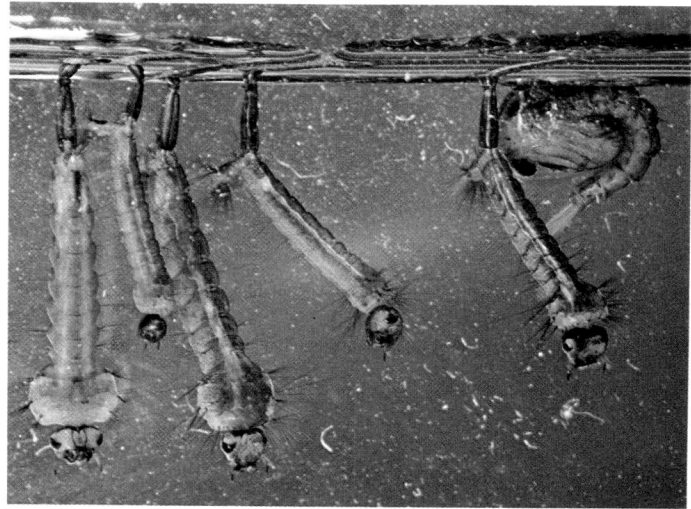

(c)

(d)

(a)

(b)

(c)

(d)

(e)

(f)

(g)

3–17 *Metamorphosis in a blue swallowtail* (Battus philenor). (a) *The beginning, mating,* (b) *eggs,* (c) *hatching,* (d) *several larval stages,* (e) *a full-grown larva, after several molts,* (f) *pupa,* (g) *adult emerging. The larvae feed on dutchman's pipe, a poisonous plant. The poison accumulates in the larva and is retained by the adult butterfly, making it poisonous to predators.*

18 *Molting is under hormonal control. In insects, a hormone (brain hormone) produced by neurosecretory cells in the brain and released from corpora cardiaca stimulates the prothoracic gland, which, in turn, produces molting hormone (ecdysone). Although all molts require ecdysone, whether or not metamorphosis occurs depends on a third hormone, juvenile hormone, produced by the corpora allata. Continued presence of juvenile hormone at high concentrations ensures that larval moths occur during the first portion of the life history. In later larval life, production of juvenile hormone declines, permitting metamorphosis to take place.*

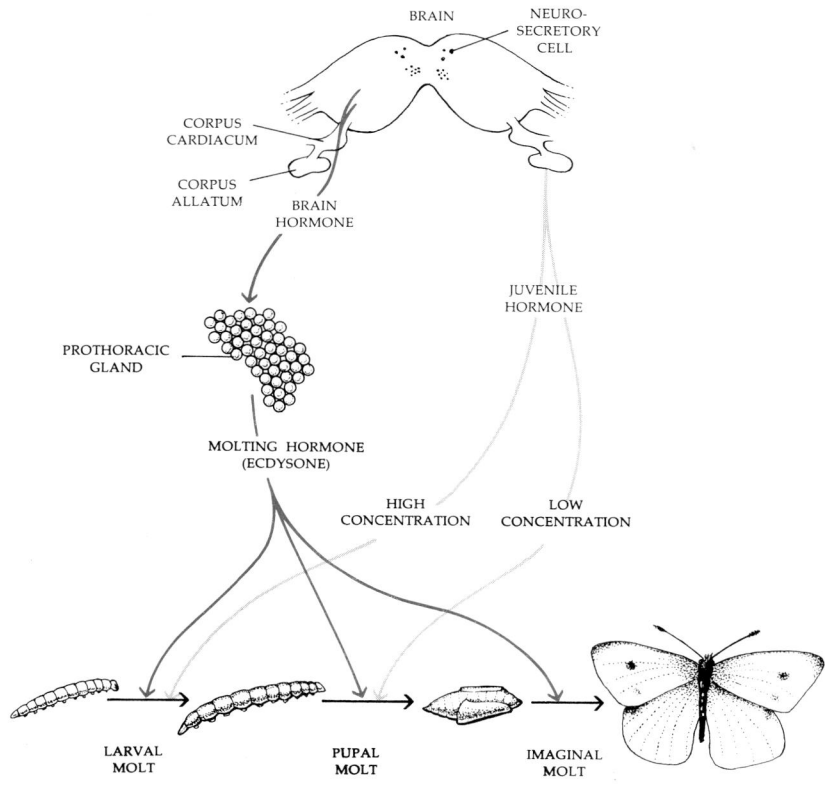

form is "seeking to express." These imaginal disks develop into the complicated structures of the adult.

Molting and metamorphosis are under hormonal control. Like many hormone-controlled processes, they are the end result of an interplay of several hormones: brain hormone, molting hormone (ecdysone), and juvenile hormone. At intervals during larval growth, brain hormone, produced by neurosecretory cells in the brain, is released into the blood. It stimulates the release, in turn, of molting hormone from a gland in the thorax. The molting hormone stimulates not only molting but also the formation of a pupa and the development of adult structures. The latter are held in check, however, by a third hormone, the juvenile hormone. Only when the production of juvenile hormone declines, in later larval life, can metamorphosis take place.

REASONS FOR ARTHROPOD SUCCESS

Among all the invertebrates, why are the arthropods, in general, and the insects, in particular, so spectacularly successful? One important reason is undoubtedly the nature of chitin, which waterproofs and provides protection and, above all, makes possible the evolution of the many finely articulated appendages characteristic of this phylum. A second reason, which applies

particularly to the insects, is the high specificity of diet and other requirements of each species. As a consequence, many different species can live in a single small area—in a few cubic inches of soil, on a small plant, or beneath the fur of a single animal—without competing with one another. Among the insects with complete metamorphosis, even the larval forms do not compete with the adult forms for the same food and the same territory. A third reason for success is undoubtedly the arthropod nervous system, with its fine control over the various appendages and the many extraordinarily sensitive sensory organs found in great diversity throughout the phylum.

The rest of this discussion deals with sensory perception among arthropods and some examples of arthropod behavior, especially in the areas of communication among members of the same species.

Arthropod Senses

Vision: The Compound Eye

The most conspicuous sensory organ of the arthropods is the compound eye (Figure 23–19), which is an evolutionary development characteristic of this one phylum. The basic structural unit of this eye is the *ommatidium*. A dragonfly has some 30,000 ommatidia. Each ommatidium is covered by a cornea, usually with a square or hexagonal surface; these are visible under low-power magnification as individual facets of the eye. Underlying the cornea is a group of eight retinal cells surrounded by pigment cells. The light-sensitive portion

23–19 *Structure of the compound eye. The eye is composed of a large number of structural and functional units called ommatidia. Each ommatidium has its own cornea, which forms one of the facets of the compound eye, and its own light-focusing lens. The light-sensitive part is the rhabdom, which is surrounded by the retinal cells, which transmit the stimulus. The ommatidium is surrounded by pigment cells that prevent light from traveling from one ommatidium to another.*

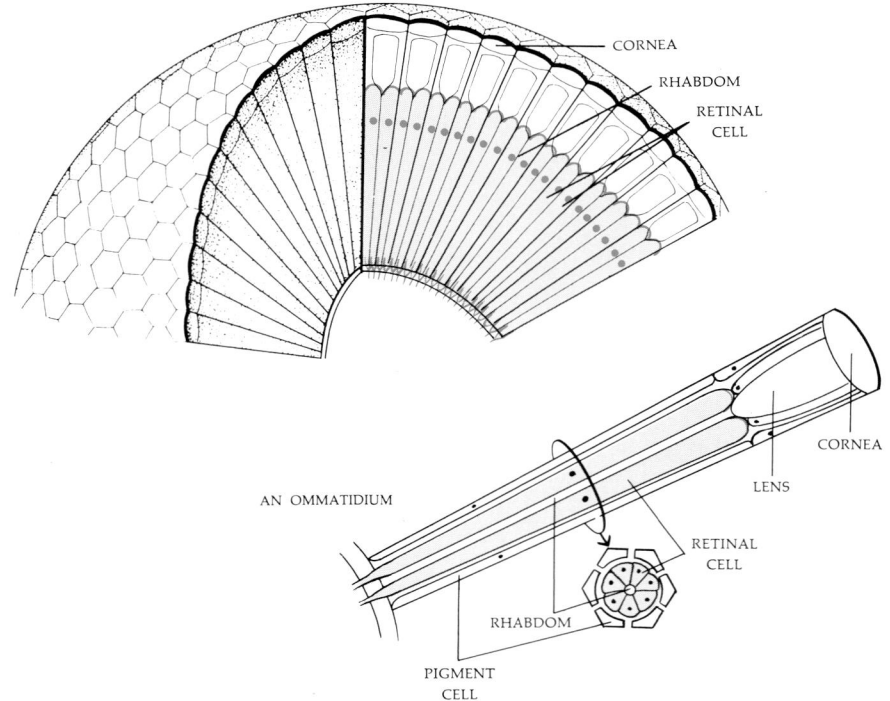

3-20 *Compound eye of* Drosophila, *as shown by the scanning electron microscope. Although insect eyes cannot change focus, they can observe objects only a millimeter from the lens, a useful adaptation for an insect.*

of the ommatidium is the *rhabdom*, which is the central core of the ommatidium. Nerve fibers carry the stimulus from each ommatidium to the brain. The pigment cells prevent light from traveling from one ommatidium to another. An ommatidium is much larger than a vertebrate photoreceptor, and so there are far fewer in an equivalent space. Hence, the image has less resolution, like a newspaper picture under high magnification.

Although the compound eye is deficient in acuity, offering less detail than the vertebrate eye, it is better for detecting motion because each ommatidium is stimulated separately and so has a separate visual field. Also, each ommatidium responds to stimuli more rapidly than does a vertebrate photoreceptor. Ability to detect motion can be measured accurately in the laboratory by testing a phenomenon known as flicker fusion. In this test, a light is flicked on and off with increasing rapidity until the observer sees the flicker as a continuous beam. The beam is perceived as continuous because stimulation of any retinal cell persists for a brief period even after the stimulus disappears. So, in effect, the flicker-fusion test is a measurement of how quickly the photoreceptor cell recovers from one stimulus and becomes sensitive to another. It is possible to test flicker-fusion rates in animals by training experiments in which the animal

23–21 *The hairs on the legs of this tiger beetle are touch receptors. At the base of the hairs are sensory cells. When a hair is touched or bent, nerve impulses are initiated.*

23–22 *Campaniform cell, a type of propriocep-tor common in arthropods.*

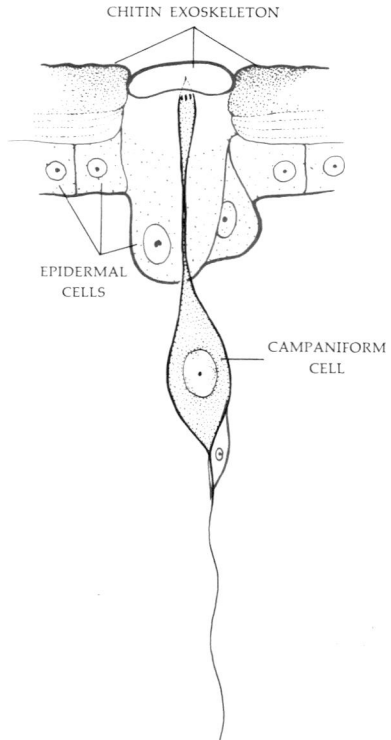

CHITIN EXOSKELETON

EPIDERMAL
CELLS

CAMPANIFORM
CELL

learns to associate a flickering light with a reward (usually food) and a steady beam with no reward, or vice versa. Such tests have proved that the compound eye greatly exceeds the camera eye in this respect. A bee would see in clear outline a moving figure that we would see as blurred, and if the bee went to the movies, the film seen by us as a continuous picture would jerk along from frame to frame in the bee's eye. The ability to perceive motion is extremely important for an insect since it must be able to make out objects when it is flying at high speed (which, as far as the visual apparatus is concerned, presents the same problems as following a moving object).

In addition to, or instead of, compound eyes, many of the arthropods possess simple eyes, or ocelli, which seem generally to serve only for light detection. Most insects have two or three ocelli, and spiders, which do not have compound eyes, may have as many as eight ocelli, depending on species.

Touch Receptors

The body surfaces of terrestrial arthropods are often covered with sensory-receptor units known as sensilla, or "little sense organs." Most of the sensilla take the form of fine spines, or setae, composed of hollow shafts of chitin. At the bases of these shafts are sensory cells. In their simplest form, the sensilla are touch receptors. In these, when the hair is touched or bent, the sensory cell responds and initiates nerve impulses. Such receptors are found, in particular, on the antennae and the legs. In addition to being stimulated by direct contact, they can also be stimulated by vibrations. A spider keeps in touch with what is going on in its web by sensing through its legs vibrations that are transmitted through the threads when the web is touched. Soldier termites of certain species strike the ground or the walls of their nest with their heads when threatened or disturbed; the vibrations they produce warn their colony mates. A fly perceives the air currents from the movement of a hand or fly swatter and so escapes; a fly in a glass jar is much less likely to be disturbed by such movements.

Proprioceptors

Proprioceptors are sensory receptors that provide information about the position of various parts of the body and the stresses and strains on them. A type of propriocepter common in the arthropods is the campaniform sensillum (Figure 23–22). They are located in thin, stretchable areas of the cuticle. When the cells are twisted or stretched, a nerve fiber signals the central nervous system. Touch receptors can also serve as proprioceptors. The praying mantis, for example, is capable of firing a lightning-swift strike at a moving object. When it sights a potential victim, the insect moves its entire head to bring it into binocular range, since the eyes themselves do not move. (See Figure 23–23.) Movement of the head results in the stimulation of proprioceptive hairs on the head and thorax of the insect. On the basis of the impulses received from these hairs, the position of the prey and the movement of its own legs are automatically coordinated by the mantis. If these hairs are removed, the mantis can strike a moving object only if the object is directly in front of it.

23 *Since its eyes do not move, the praying mantis must move its entire head to bring its victim into binocular range. The movement of its head sends impulses through proprioceptive hairs on its head and thorax to its legs. The position of its prey and the movement of its legs are thus automatically coordinated, giving the mantis the ability to strike swiftly at a moving object.*

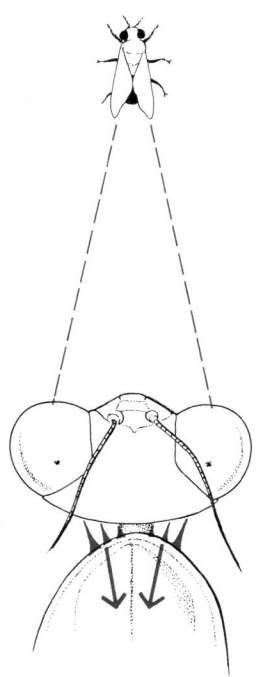
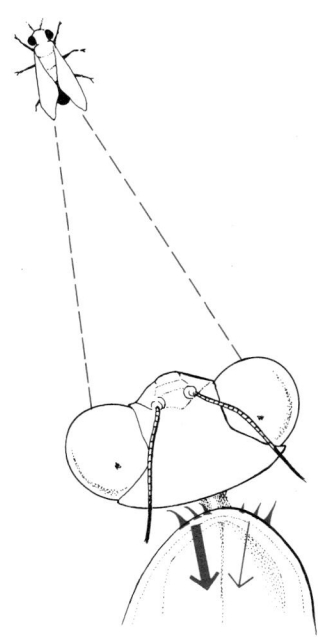

24 *The most elaborate of the insect sound receptors is the tympanic organ. The tympanic air sacs are covered by a membranous drum, and the sensory cells are so arranged in the organ that they are stimulated by movements of the drum or the air-sac walls. Tympanic organs respond to vibrations in the air or in water. This is a tympanic organ on the inside front tibia of a katydid.*

Sound Receptors

Arthropods have a variety of sound receptors. The simplest is a sensillum with a tactile hair that vibrates as a result of being "touched" by sound waves. The antennae of the male mosquito contain thousands of such hairs, which are responsive to the vibrations made by the wings of the female mosquito in flight and so serve to bring the sexes together. When the male mosquito first emerges from its pupal shell, it is sexually immature and also deaf, with its antennal hairs lying flat along the shafts. When the male matures sexually, some 24 hours later, the hairs almost simultaneously become erect.

Other insects have developed special groups of cells for hearing; these are known as tympanic organs. (See Figure 23–24.) In these organs, a fine membrane, the tympanum (or eardrum), is stretched across an air-filled cavity. The tympanic membrane vibrates in response to sounds of certain frequencies, and this vibration is transmitted to underlying receptor cells.

Communication

Communication by Sound

Arthropods, particularly insects, have developed complex forms of sensory communication. A number of species, such as the locusts, grasshoppers, and crickets, call to one another by sounds made by rubbing their legs or wings together or against their bodies. Five distinct types of calls are known: (1) calling by males and (2) calling by females, both of which are long-range sounds; (3) courtship sounds by males and (4) aggressive sounds by males, both of which are short-range; and (5) alarm sounds, which may be given either by

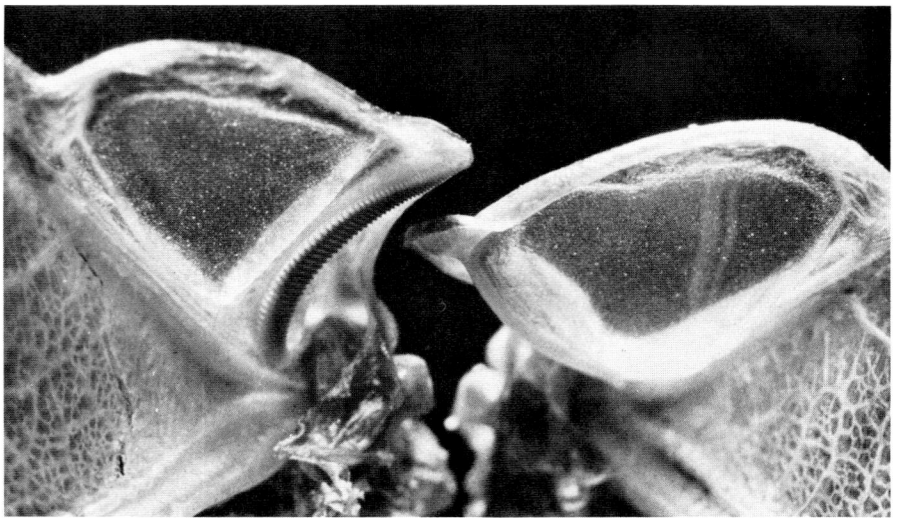

23–25 *The katydid produces its characteristic loud, shrill sounds by rubbing the scraper at the base of its right wing against the file at the base of its left wing.*

males or by females. Recognition of and response to the sound seem to be based on pattern and on rhythm because insects apparently are not able to distinguish frequencies, or the differences between high and low notes, and so are essentially "tone deaf." The effectiveness of calling songs is often increased by group singing, such as the famous chorus of male seventeen-year cicadas, which can attract females from distances far greater than an individual "voice" would reach. Insects produce songs and respond to appropriate songs without ever having heard a song before.

Communication by Pheromones

The use of chemicals for communication is common among animals, and the substances employed range from the sex attractants of the little algal cell *Chlamydomonas* to Chanel No. 5. Many insects communicate by chemicals; such chemicals are known as *pheromones*. Pheromones are chemical messengers. They are usually produced in special glands and are discharged into the environment, where they act on other members of the same species.

Among the best studied of the pheromones are the mating substances of moths. One female gypsy moth, by the emission of minute amounts of a pheromone commonly known as gyplure, can call male moths from several miles away. In fact, a single female contains enough gyplure, somewhere around a millionth of a gram, to attract more than a billion males, supposing that it were distributed with the maximum of efficiency. Since the male can detect as little as a few hundred molecules per cubic centimeter of the attractant, gyplure is still potent even when it has become widely diffused.

The male moth characteristically flies upwind, and the pheromone, of course, disperses downwind. Therefore, when a male moth detects the odor of a female of the species, he will fly toward the source. If he loses the scent, he flies about at random until he either picks it up again or abandons the search. It is not until he is quite close to the female that he can fly "up the gradient" and use the intensity

23–26 *A cricket. The large appendages extending from either side of the head segment contain chemoreceptors (taste organs).*

The plumed antennae of the male cecropia moth are adapted for detecting the pheromone emitted by the female.

The waggle dance of the honeybee communicates to her fellow workers the location of a food source she has found. The bee locates the source with reference to the position of the sun. (a) If she performs her dance outside the hive on a horizontal platform, the straight run of the dance will point directly toward the food source. (b) If she performs her dance inside the hive on the vertical wall of the comb, she orients herself by gravity and the point directly overhead takes the place of the sun. The angle X is the same for both dances. The distance of the food source from the hive is indicated by how fast she performs the dance pattern.

of the odor as a locating device. Figure 23–27 shows the lavishly plumed antennae of the male cecropia moth by which he detects the pheromone emitted by the female.

The Dancing Bees

Probably the most remarkable example of communication among insects is the "language of the bees" discovered by Karl von Frisch. The honeybee returning to the hive performs a dance upon the comb, which signals quite precisely to her hivemates the location of the food she has found.

When the nectar source is near the home, the bee does a round dance, which arouses other workers to seek the food source. From the scent on the bee's body, the workers are informed at the same time about the type of flower she has found. When the food source is some distance from the hive (more than 275 feet, in the case of an Austrian strain of honeybee, *Apis mellifera carnica*, with which Von Frisch did his original studies), the bee does the waggle dance, in which she runs a short distance in a straight line, waggling her abdomen from side to side, and returns in a semicircle to the starting point. Then she repeats the straight run and comes back in a semicircle on the opposite side. This cycle, which is shown in Figure 23–28, is repeated many times. The time taken by the straight run indicates the distance of the source from the hive. If the source is 1,000 feet away from the hive, the worker bee performs a straight run lasting ½ second; if the source is 2,000 feet away, the straight run takes about twice as long, or 1 second.

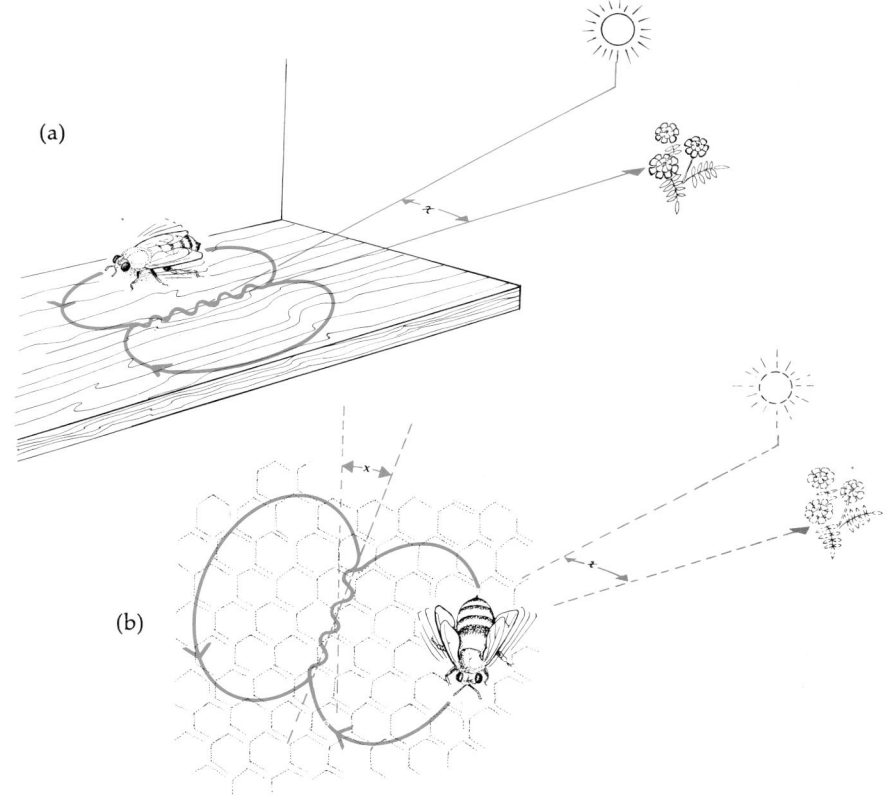

(a)

(b)

More remarkable, the direction of the straight run of the waggle dance indicates the direction of the food source. The bee locates the food source for herself by reference to the position of the sun, and this is how she finds the food again. Sometimes when she returns, she dances first on a horizontal platform outside the hive. In this case, the straight run of the waggle dance will point directly toward the food source. When she reenters the hive, however, to share and store the nectar, her only dancing spot is on the vertical surface of the comb in the dark interior of the hive. Here she orients herself, not by the sun, but by gravity, and the point directly overhead takes the place of the sun in the bee's map, just as our maps are generally oriented so that the North Pole is on the top. If the food source lies at an angle of 40° to the left of the sun, for example, the waggling run points 40° to the left of the vertical. The bees that follow the dancer translate the information back into a direction for orientation to the sun and start off toward the food source. A young bee that has been raised away from other bees outside the hive is able to respond to the dance the first time she is exposed to it in the hive; no learning is required.

Complex patterns of unlearned, genetically transmitted behavior—such as web-building among spiders and the waggle dance of the honeybee—are another arthropod characteristic. This rigid programming of behavior is a necessary correlate to the shortness of the life span of the smaller arthropods and is an interesting contrast to the more flexible behavior patterns of the higher mammals, with their comparatively long life spans and long periods of learning.

Table 23–1   *Major Classes of the Phylum Arthropoda*

	No. of species	Body parts	Head appendages	Legs	Wings	Eyes	Respiration	Distribution
Subphylum Chelicerata								
Class Arachnida (Scorpions, spiders, daddy longlegs, mites, ticks)	35,000	Cephalo-thorax, abdomen	Chelicerae, pedipalps	4 pairs	None	Simple	Book lungs, tracheae, or both	Terrestrial, some aquatic (secondarily)
Subphylum Mandibulata								
Class Crustacea (Lobsters, crabs, barnacles, *Daphnia*, copepods)	30,000	Head (6 segments), thorax (8 segments), abdomen (6 segments)	Antennae (2 pairs), mandibles	Numerous (on both thorax and abdomen)	None	Usually compound	Gills	Marine and freshwater, some terrestrial
Class Insecta	800,000	Head (6 segments), thorax (3 segments usually), abdomen	Antennae (1 pair), mandibles	3 pairs (on thorax only)	2 pairs, 1 pair, or none	Compound and simple	Tracheae	Terrestrial, some aquatic (secondarily)

The phylum Chordata comprises three subphyla: the Cephalochordata, or lancelets, which includes *Amphioxus;* the Tunicata, or tunicates, of which the most familiar are the sea squirts; and the Vertebrata, or vertebrates.

Amphioxus (Figure 23–29), is a small, sliver-shaped, semitransparent animal found in shallow marine waters all over the warmer parts of the world. Although it can swim very efficiently, it spends most of its time buried in the sandy bottom, with only its mouth protruding above the surface. This animal exemplifies all four of the salient features of the chordates. The first is the notochord, a gelatinous rod that extends the length of the body and serves as a firm but flexible axis. The notochord is a structural support. Because of it, the *Amphioxus* can swim with strong undulatory motions that move it through the water with a speed unattainable by the flatworms or aquatic annelids.

The second chordate characteristic is the nerve cord, a hollow tube that runs beneath the dorsal surface of the animal above the notochord. (The principal nerve cord in the invertebrates, as you will recall, is always double and ventral.)

The third characteristic is a pharynx with gill slits. The pharyngeal gill slits become highly developed in fishes, in which they serve a respiratory function, and traces of them remain even in the human embryo. In *Amphioxus,* they serve primarily for collecting food. The cilia around the mouth and at the opening of the pharynx pull in a steady current of water, which passes through the pharyngeal slits into a chamber known as the atrium and then exits through the atrial pore. Food particles are collected in the sievelike pharynx, mixed with mucus, and channeled along ciliated grooves, the endostyles, to the intestine.

The fourth characteristic is a tail, blocks of muscle around an axial skeleton posterior to the anus. Man is one of a small minority of chordates that do not have a tail.

Most of the body tissue of *Amphioxus* is made up of blocks of muscles, myotomes, which extend the length of the body.

23–29 Amphioxus, *a lancelet, exemplifies four distinct chordate characteristics: (1) a notochord, the dorsal rod that extends the length of the body, (2) a dorsal, tubular* nerve cord, (3) pharyngeal gill slits, and (4) a tail. The fleshy tentacles, which are just anterior to the mouth, are covered with cilia.

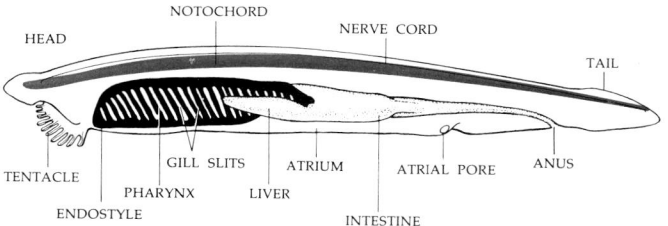

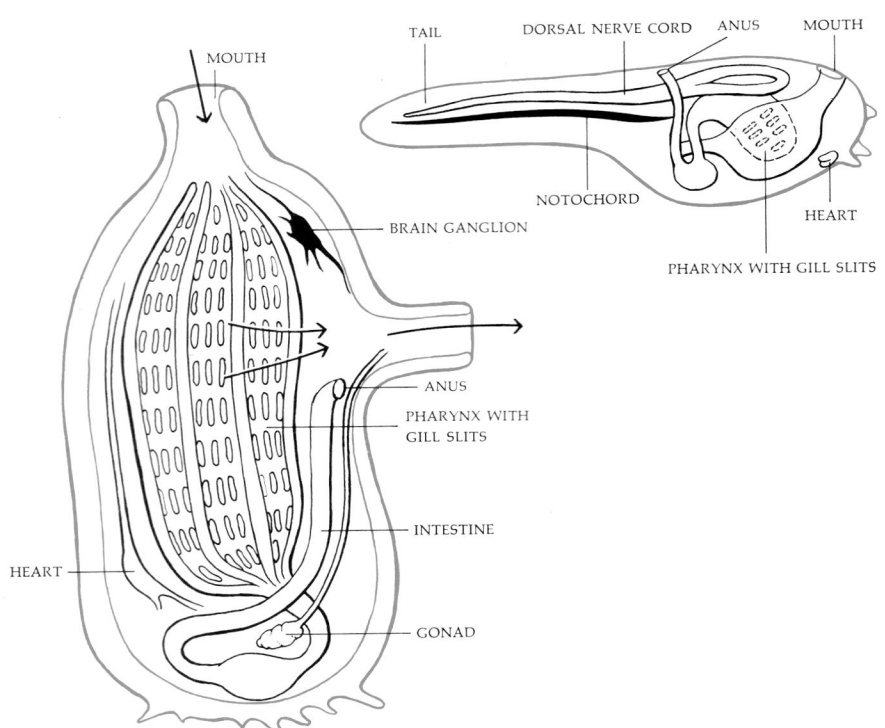

23–30 *Two stages in the life of a tunicate. At the right is the "tadpole" larva; the adult form is shown on the left. After a brief free-swimming existence, the larva settles to the bottom and attaches at the anterior end. Metamorphosis then begins. The larval tail, with the notochord and dorsal nerve cord, disappears, and the animal's entire body is turned 180°. The mouth is carried backward to open at the end opposite that of attachment, and all the other internal organs are also rotated back. As some biologists reconstruct the past, tunicate larvae wriggled up the rivers, where they gave rise, after many generations and millions of years, to the ancestral vertebrates.*

23–31 *X-ray of an elk fetus. The bones have been stained to show them more clearly, so that you can see the extent to which the skeleton is still cartilaginous. Notice the legs, for example. Only the dark areas are bone; these will gradually grow and replace the cartilage as the animal matures.*

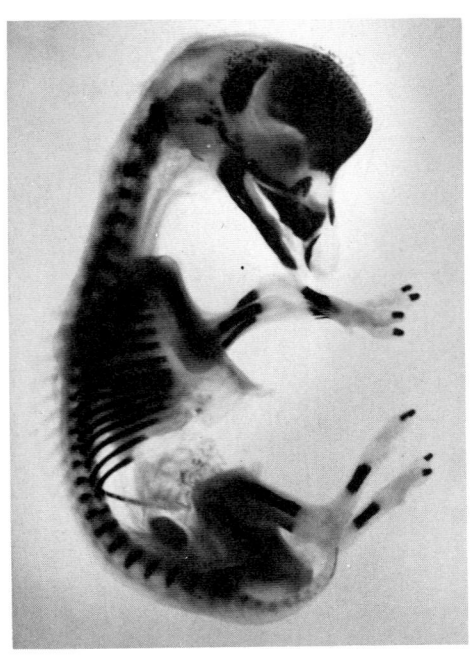

Although *Amphioxus* usefully exemplifies the chordates, many biologists believe it is more likely to be a degenerate form of a primitive fish rather than a truly primitive member of the phylum. A more probable candidate is the tunicate. Although the adult form does not have all of the typical chordate features, the larva, which resembles a tadpole, is clearly a chordate possessing the four chordate characteristics. (See Figure 23–30.)

SUBPHYLUM VERTEBRATA

The vertebrates are a large (about 43,000 species) and familiar subphylum of the chordates. All vertebrates have a backbone, or vertebral column, as their structural axis—a flexible bony support that develops around the notochord, supplanting it entirely in most species. Dorsal projections of the vertebrae encircle the nerve cord along the length of the spine. The brain, the culmination of the nerve cord, is similarly enclosed and protected by bony skull plates. Between the vertebrae are cartilaginous disks, which give the vertebral column its flexibility. Associated with the vertebrae are a series of muscle segments, myotomes, by which sections of the vertebral column can be moved separately.

One of the great advantages of this bony endoskeleton, as compared with the exoskeletons of the invertebrates, is that it is composed of living tissue that can grow with the animal. In the developing embryo, the skeleton is largely cartilaginous, with bone gradually replacing cartilage in the course of maturation. Figure 23–31 shows the developing bone tissue of a vertebrate embryo. In the vertebrates, the growing parts of the bones remain cartilaginous until the animal reaches its full adulthood.

Table 23–2 *Evolution of Animals: Major Physical and Biological Events in Geologic Time*

Millions of years ago	Era	Period	Epoch	Life forms	Climates and major physical events
	CENOZOIC	Quaternary	Recent Pleistocene	Age of man. Planetary spread of *Homo sapiens;* extinction of many large mammals, including woolly mammoths. Deserts on large scale.	Fluctuating cold to mild. Four glacial advances and retreats (Ice Age); uplift of Sierra Nevada.
1½–7		Tertiary	Pliocene	Large carnivores. First known appearance of man-apes.	Cooler. Continued uplift and mountain building, with widespread extinction of many species.
7–26			Miocene	Whales, apes, grazing animals. Spread of grasslands as forests contract.	Moderate uplift of Rockies.
26–38			Oligocene	Large, browsing mammals. Apes appear.	Rise of Alps and Himalayas. Lands generally low. Volcanoes in Rockies area.
38–53			Eocene	Primitive horses, tiny camels, modern and giant types of birds.	Mild to very tropical. Many lakes in western North America.
53–65			Paleocene	First known primitive primates and carnivores.	Mild to cool. Wide, shallow continental seas largely disappear.
65–136	MESOZOIC	Cretaceous		Age of reptiles, extinction of dinosaurs. Marsupials, insectivores. Angiosperms become abundant.	Lands low and extensive. Last widespread oceans. Elevation of Rockies cuts off rain.
136–195		Jurassic		Dinosaurs' zenith. Flying reptiles, small mammals. Birds appear.	Mild. Continents low. Large areas in Europe covered by seas. Mountains rise from Alaska to Mexico.
195–230		Triassic		First dinosaurs. Primitive mammals appear. Forests of gymnosperms and ferns.	Continents mountainous. Large areas arid. Eruptions in eastern North America. Appalachians uplifted and broken into basins.
230–280	PALEOZOIC	Permian		Reptiles evolve. Orgin of conifers and possible origin of angiosperms; earlier forest types wane.	Extensive glaciation in Southern Hemisphere. Appalachians formed by end of Paleozoic; most of seas drain from continent.
280–345		Carboniferous Pennsylvanian Mississippian		Age of amphibians. First reptiles. Variety of insects. Sharks abundant. Forests, ferns, gymnosperms, and horsetails.	Warm. Lands low, covered by shallow seas or great coal swamps. Mountain building in eastern U.S., Texas, Colorado. Moist, equable climate, conditions like those in temperate or subtropical zones, little seasonal variation, root patterns indicate water plentiful.
345–395		Devonian		Age of fish. Amphibians appear. Shellfish abundant. Lungfish. Rise of land plants. Extinction of primitive vascular plants. Origin of modern subclasses of vascular plants.	Europe mountainous with arid basins. Mountains and volcanoes in eastern U.S. and Canada. Rest of North America low and flat. Sea covers most of land.
395–440		Silurian		First terrestrial plants. Rise of fish and reef-building corals. Shell-forming sea animals abundant. Modern groups of algae and fungi.	Mild. Continents generally flat. Mountain building in Europe. Again flooded.
440–500		Ordovician		First fish. Invertebrates dominant.	Mild. Shallow seas, continents low; sea covers U.S. Limestone deposits; microscopic plant life thriving.
500–600		Cambrian		Age of marine invertebrates. Shell animals.	Mild. Extensive seas. Seas spill over continents.
	PRECAMBRIAN			Earliest known fossils.	Dry and cold to warm and moist. Planet cools. Formation of Earth's crust. Extensive mountain building. Shallow seas.

23–32 *Evolutionary relationships in the animal kingdom, according to some hypotheses. Relationships are based on structural similarities, such as the segmentation in annelids and arthropods, and also on resemblances among larval forms and in developmental patterns. The first fossils available are from the Cambrian period and, by this time, the major groups of invertebrates had already diverged.*

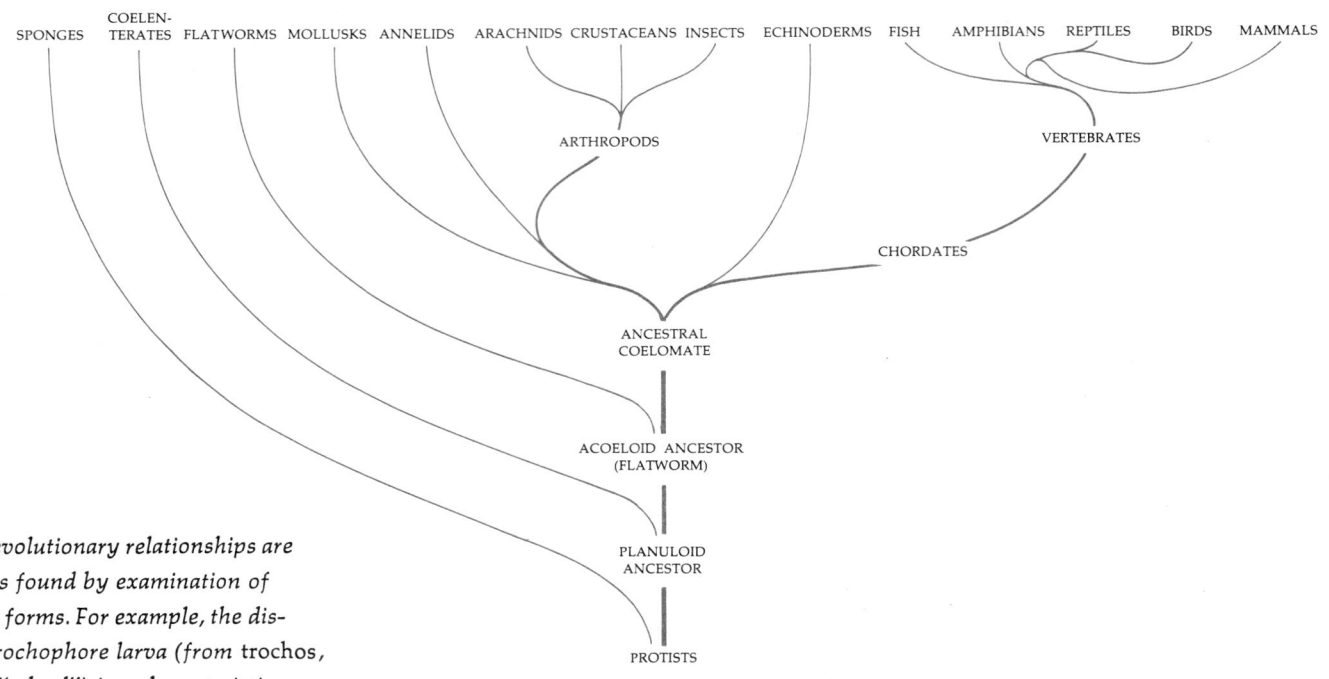

23–33 *Clues to evolutionary relationships are sometimes found by examination of immature forms. For example, the distinctive trochophore larva (from* trochos, *meaning "wheel") is a characteristic of both annelids and mollusks. Although their adult forms are very different, they both have larvae of this type. This particular larva will develop into a polychaete worm.*

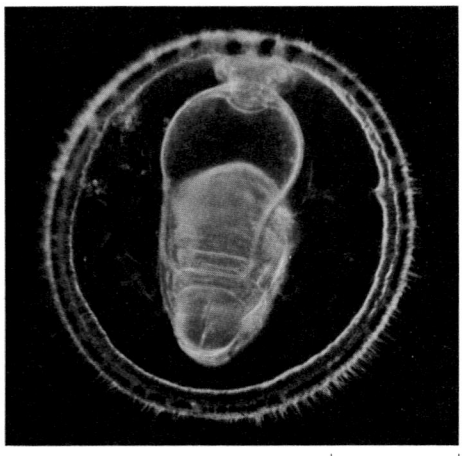

0.25 mm

Vertebrate Evolution

Representatives of the major invertebrate phyla coexisted in the warm pre-Cambrian seas (see Table 23–2) from which our most ancient invertebrate fossils derive. The earlier history of these groups and ancestral forms common to two or more of them can only be established indirectly. In the case of the vertebrates, however, which evolved more recently, the fossil record is much more complete.

There are seven living classes of vertebrates: the fish (comprising three classes), the amphibians, the reptiles, the birds, and the mammals.

Classes Agnatha, Chondrichthyes, and Osteichthyes: Fish

The first fish were jawless and had a strong notochord running the length of their bodies. Today these jawless fish (class Agnatha), once a large and diverse group, are represented only by the hagfish and the lampreys, which are the most primitive living vertebrates. They have a notochord throughout their lives, like *Amphioxus*. Although their ancestors had bony skeletons, modern agnaths have a cartilaginous skeleton. Lacking true bones, they are very flexible; a hagfish can tie itself in a knot. These cyclostomes ("round mouths"), as they are called, are highly predatory, attaching to other fish by their sucker-

34 *The heavily armored placoderm is the ancestor of two major classes of present-day fish—the Chondrichthyes (cartilaginous fish) and the Osteichthyes (bony fish).*

like mouths and rasping through the skin into the viscera of their hosts. The juvenile lamprey, however, feeds by sucking up mud containing microorganisms and organic debris—as, most probably, did the primitive Agnatha.

The sharks (including the dogfish) and skates, the Chondrichthyes, the second major class of fish, also have a completely cartilaginous skeleton. Their ancestors were also bony animals. Their skin is covered with small pointed teeth (denticles), which resemble vertebrate teeth structurally and give the skin the texture and abrasive capacity of the coarsest grade of sandpaper.

The third major class of fish includes those with bony skeletons, the Osteichthyes. This group includes the trout, bass, salmon, perch, and many others—most of the familiar fresh- and saltwater forms.

According to present evidence, fish evolved in fresh water. The chondrichthyans returned to the sea early in their development, while the bony fish went through most of their evolution in fresh water and spread to the seas at a much later period. Some still recapitulate in each lifetime this difficult physiological transition. Salmon, for example, return to fresh water to spawn, while eels leave the fresh waters of the continents to return to the Sargasso Sea at breeding time, from which distant point the young begin the long, difficult journey, often lasting many years, back to the rivers and lakes.

35 *A modern osteichthyan, an Atlantic salmon, swimming up the rapids of the Metamek River. When salmon are sexually mature, they return to fast-moving fresh water, apparently always to the place where they themselves were spawned. The eggs are laid there and fertilized, and the young remain in fresh water for several years until they are mature. Salmon do not feed on their journey upstream. Pacific salmon die, but Atlantic salmon can sometimes repeat the migration several times.*

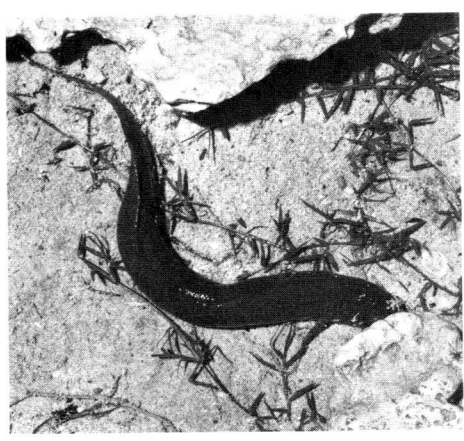

23–36 *A modern lungfish. When the dry seasons come, members of this African species wriggle downward into the mud, which eventually hardens around them. Mucus glands under the skin secrete a watertight film around the body, preventing evaporation. Only the mouth is left exposed. During this period, they take a breath only about once every two hours.*

23–37 *Carpenter frog. Frogs, like other amphibians, are clearly fishlike in their larval (tadpole) stages. As adults, they require water to reproduce and their moist skins are an important accessory breathing organ.*

The Transition to Land

Some primitive osteichthyans had swim bladders that functioned in some species as lungs. These lungs, however, were not efficient enough to serve as more than accessory structures to the gills. These lungs were a special adaptation to fresh water, which, unlike ocean water, may stagnate and, because of decay or of algal bloom, become depleted of oxygen.* A few modern lungfish exist. They can live in water that does not have sufficient oxygen to support other fish life. Lungfish surface and gulp air into their lungs in much the same way that certain aquatic but air-breathing snails bob to the surface to fill their mantle cavities.

Using their supplementary oxygen supply, the primitive osteichthyans could waddle, dragging their bellies on the ground, up the muddy bottom of a drying stream bed to seek deeper water or perhaps even make their way from one water source to another one nearby. Lungfish were the most common fish in the later Devonian seas. In most of them, the lung evolved into an air bladder, or swim bladder; many modern osteichthyans have air-filled swim bladders that serve as flotation chambers or organs of sound production. A fish raises or lowers itself in the water by adding gases to the air bladder or removing oxygen from it. The oxygen is supplied and removed by the bloodstream. Other lungfish retained the lung and increased its efficiency, making way for the first amphibians.

Class Amphibia

By the end of the Devonian period, some of the lungfish had begun to develop skeletal supports in their fleshy fins. Three modern genera of lungfish have fins with internal bones corresponding to the limb bones of terrestrial vertebrates. They resembled four-legged fish more than the modern amphibians; they were heavy and clumsy, with short sprawling legs, big flattened heads, and stubby tails. Some of them were large, as much as 4 to 5 feet long. Although they could walk on land, they probably spent most of their lives in the water; nor have their descendants, the modern amphibians, freed themselves entirely from the water.

Amphibians lay their eggs in water, and the males fertilize them externally after they are laid. Consequently, the amphibians, in order to reproduce, must live near water or return to it at regular intervals. Also, if they are not in water or moist air, water evaporates rapidly through their thin, scaleless skin, which serves as an important accessory breathing organ.

In their early stages of development, all amphibians are clearly fishlike. The larvae (tadpoles) live entirely in the water and breathe through gills, later changing into lung-breathing forms with limbs. Modern amphibians include toads and frogs (which are tail-less as adults) and newts and salamanders (which have tails throughout their lives).

Some of the salamanderlike amphibians, such as the mud puppy and the axolotl, never complete their metamorphosis, remaining essentially aquatic larval forms. In some species, these larvalike forms can be induced to metamorphose into adult forms by administration of hormones, indicating that the genetic capacity for this later developmental stage has not yet been lost completely.

* Lungs and gills, and their comparative functions, will be described in more detail in Chapter 33.

38 *Alligators and crocodiles, the largest modern members of the class Reptilia, lay their eggs on land, and their skins are reinforced with epidermal horny scales. Crocodiles have more slender snouts than alligators, and the fourth pair of teeth on the lower jaw remains exposed when the mouth is closed. Alligator jaws are broader and rounded anteriorly, and none of the teeth are exposed when the mouth is closed; they are also reported to be less aggressive. The animal shown here, as you can perhaps tell from the bland smile, is an American alligator.*

39 *Black snakes hatching. With the evolution of an egg in which the embryo could develop on land, some groups of vertebrates were able to become fully terrestrial.*

Class Reptilia

As you will recall, the vascular plants freed themselves from the water by the development of the seed. Analogously, the vertebrates became truly terrestrial with the evolution in the reptiles of the amniote egg, an egg that carries its own water supply and so can survive on land. The reptilian egg, which is much like the familiar hen's egg in basic design, contains a large yolk, the food supply for the developing embryo, abundant albumin, and a water supply. A membrane, the amnion, surrounds the developing embryo in a liquid-filled space that substitutes for the ancestral pond. In mammals also, although their eggs typically develop internally, the embryos are enclosed in water within an enveloping membrane.

A major advantage to reptiles of this evolutionary development, it is believed, is that at the time the reptiles emerged onto the land, few land predators existed, in contrast to the numerous aquatic predators. In keeping with their terrestrial existence, the reptiles also evolved a dry skin covered with protective scales (the forerunners of feathers and hair).

From one branch of primitive reptiles all the birds evolved, another branch gave rise to the mammals, and several other branches retained mostly reptilian characteristics and are represented today by lizards, snakes, turtles, and crocodiles.

(a)

(b)

23–40   (a) *The oldest known fossil bird, Arch-aeopteryx, dates from the middle Jurassic period, about 150 million years ago. It still has many reptilian characteristics. The teeth and the long, jointed tail are not found in modern birds.* (b) *A modern bird, the saw-whet owl, in pursuit of a white-footed mouse.*

Class Aves: Birds

In their skeletal structure, birds are essentially reptiles highly specialized for flight. They have feathers, however, one of the distinctions of this class, and, unlike the reptiles, they maintain a high and constant body temperature, which may in part permit the high energy output required for flight. Their bodies are lightened by air sacs and also by hollow bones. The frigate bird, an oceanic species with a 7-foot wingspread, has a skeleton that weighs only 4 ounces. The most massive bone in the bird skeleton is the breastbone, or sternum, which provides the keel for the attachment of the huge muscles that operate the wings. Flying birds have jettisoned all extra weight; the female's reproductive system has been trimmed down to a single ovary, and even this becomes large enough to be functional only in the mating season. Most of the common nonflying birds, such as the penguin and the ostrich, are believed to have evolved secondarily from flying types.

Class Mammalia

Mammals also descended from the reptiles. Characteristics which distinguish mammals from other vertebrates are (1) mammals are hairy rather than scaly, (2) mammals nurse their young, and (3) mammals (like birds but unlike other vertebrates) maintain a constant body temperature. In nearly all mammalian species, the young are born alive, as they are in some fish and reptiles, which retain the eggs in their bodies until they hatch. Some very primitive mammals, however, such as the duckbill platypus, lay shelled eggs but nurse their young after hatching. The *marsupials*, which include the opossum and the kangaroo, also bear their young alive, but they differ from the major group of mammals

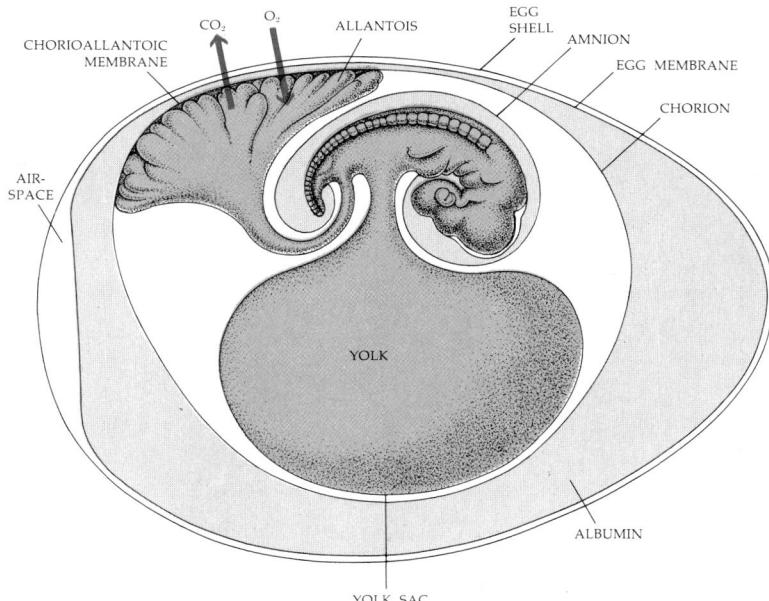

1 *Amniote egg. The membranes, which are produced as outgrowths from the embryo as it develops, surround and protect the embryo and the yolk (its food supply). The egg shell and egg membrane, which are waterproof but permeable to gases, are added as the early embryo passes down the maternal reproductive tract.*

2 *Marsupial infants are born at an immature stage and continue their development attached to a nipple in a special protective pouch of the mother. This tiny kangaroo accidentally became dislodged from its mother's pouch. As you can see, it is still attached to the nipple. After the picture was taken, the baby was restored to the pouch, with no apparent ill effects from its premature introduction to the outside world.*

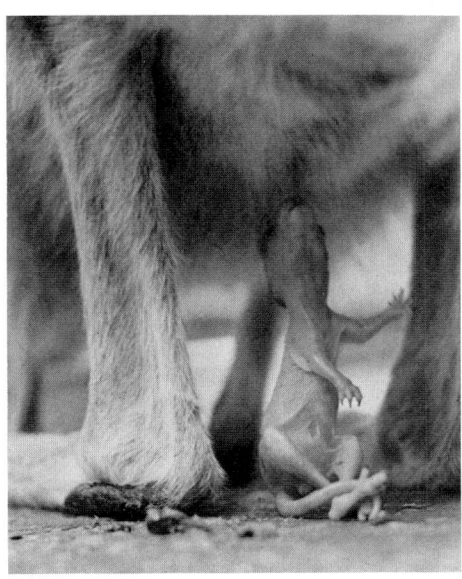

in that the infants are born at a tiny and immature stage and are kept in a special protective pouch in which they suckle and continue their development. Most of the familiar mammals are <u>placentals</u>, so called because they have an efficient nutritive connection, the placenta, between the uterus and the embryo. As a result, the young develop to a much more advanced stage before birth. Thus the young are afforded protection during their most vulnerable period without seriously interfering with the mobility of the mother. The earliest placentals were small, shy, and probably nocturnal; they undoubtedly lived mostly on insects, grubs, worms, and eggs, devoting much of their energies to avoiding the carnivorous dinosaurs. Shrews, which closely resemble these primitive mammals, have retained their elusive habits.

Mammals have fewer, but larger, skull bones than the fish and reptiles, an example of the fact that "simpler" and "more primitive" may have quite opposite meanings. In the mammals, a bony platform or partition has developed that separates nasal and food passages far back in the throat, making it possible for the animal to breathe while eating. The jawbones have been fused to one, which is far larger and more powerful than the reptilian jaw, although mammals have lost the capacity of the reptile to unhinge its jaw—an ability which makes it possible for an anaconda, for example, to swallow a pig whole.

Among the placental mammals, there are four major evolutionary lines. One group comprises the rabbits and rodents. A second is made up of the marine mammals, such as whales and seals. The third includes the carnivores, such as dogs and cats, and the hoofed animals—horses, cattle, sheep, pigs, and the like. The fourth group consists of the insect eaters (shrews, moles, and hedgehogs), bats (the only flying mammals), and primates (lemurs, monkeys, apes, and man).

23–43 *From left to right, the leg of a horse, the arm and hand of a man, the flipper of a whale, the foreleg of a crocodile, the wing of a bird. The crocodile is the closest to the ancestral type; the others represent adaptations to a particular way of life. Note that in the horse, the two bones of the forearms are fused. Colors indicate homologous structures—that is, structures with a common ancestral background.*

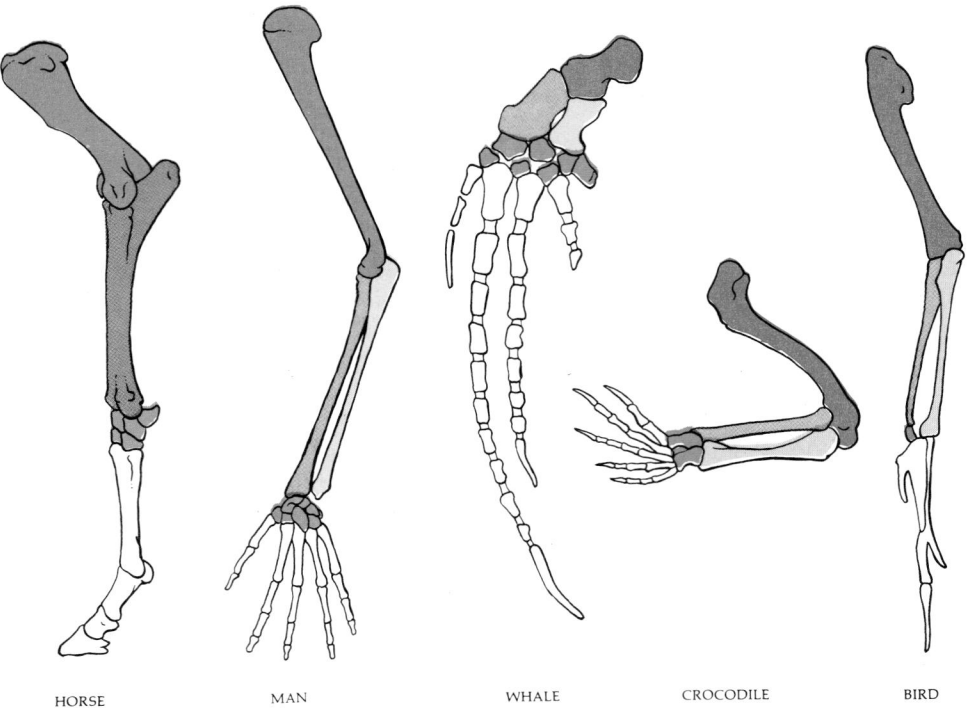

HORSE   MAN   WHALE   CROCODILE   BIRD

The primates are characterized by possession of (1) a placenta, (2) three kinds of teeth (canines, incisors, molars), (3) opposable first digits (thumbs), (4) two pectoral mammae, (5) expanded cerebral cortex, and (6) a tendency toward single births. Man is distinguished from the other primates by his upright posture and his lack of body hair. Among all the mammals, he is one of the least specialized. He is an omnivore, eating a wide variety of fruits, vegetables, and other animals. His hands closely resemble those of a primitive insectivore, in contrast to the highly specialized forelimbs developed by, for example, the whales, bats, gibbons, and horses. His sensory organs are crude compared with those of insects or of many other mammals. Man has, however, one area of extreme specialization: the brain. Because of his brain, man is unique among all the other animals in his capacity to reason, to speak, to plan, and to learn.

SUMMARY

The arthropods constitute the largest animal phylum in both number of species and of organisms. Arthropods are segmented animals with jointed chitin exoskeletons and a variety of highly specialized appendages and sensory organs. In the higher forms, these segments form a head, a thorax (sometimes combined with the head as a cephalothorax), and an abdomen. The arthropods are also characterized by an open circulatory system and a ladder-type nervous system, consisting of a series of ganglia, a pair per segment, interconnected by a double

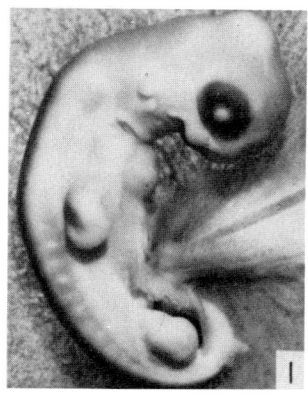
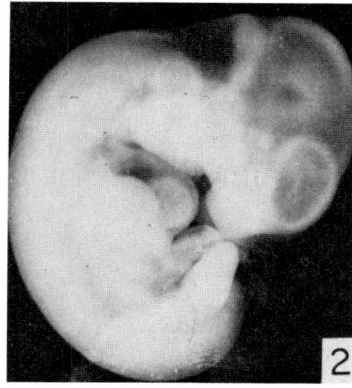
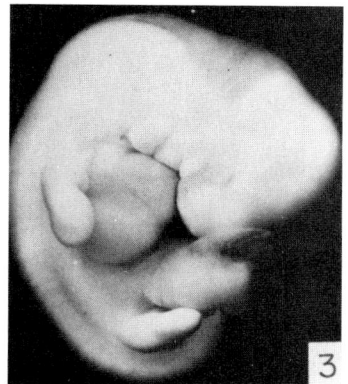
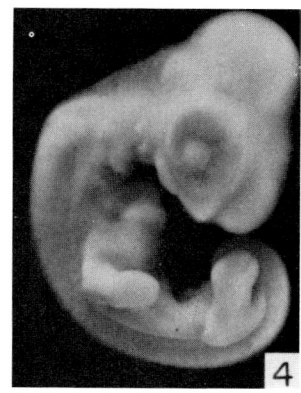

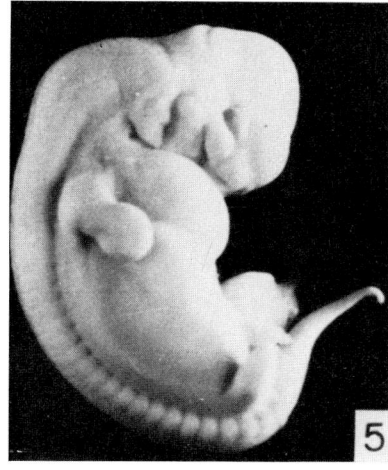

44 *Which of these is human? As this comparison of the early fetuses of various vertebrates illustrates, the conservative nature of evolution is very evident in embryological development. The human embryo in its earlier stages greatly resembles the embryos of other vertebrates; for instance, at one stage we have gill slits and a tail. At one time, in fact, some students of evolution maintained that every embryo repeats the entire history of its evolution. Actually, what are retained are certain developmental pathways—ways of getting "from here to there." The third embryo is human. The first is a turtle; the second, a mouse; the fourth, a chick; and the fifth, a pig.*

ventral nerve cord. Tracheae (chitin-lined breathing tubes) and Malpighian tubules (excretory ducts leading into the hindgut) are found exclusively among arthropods.

There are two subphyla of arthropods, the Mandibulata, characterized by mandibles (jaws) and antennae, and the Chelicerata with chelicerae (pincers) and cephalothorax but with no mandibles or antennae. The class Arachnida of the subphylum Chelicerata includes spiders, scorpions, mites, and ticks. Crustaceans, which are mandibulates, include lobsters, crabs, crayfish, barnacles, and shrimp. The class Insecta, largest of the arthropod classes, and also mandibulates, with about 800,000 species, are terrestrial animals with a single pair of antennae (crustaceans have two, arachnids have none) and usually two pairs of wings.

In the course of their development, most insects pass through a complete metamorphosis. The stages are: egg, larva (caterpillar, maggot, grub, etc.), pupa (often with a cocoon or some other protective covering), and adult. In a minority of species, including the grasshopper, the changes are more gradual; in these, the immature form is known as a nymph. Metamorphosis is under hormonal control.

Arthropods, especially insects, have a variety of sensory receptors, one of which is the compound eye. The basic structural unit is the ommatidium, which consists of photoreceptor cells surrounding a light-conducting rhabdom and covered by a cornea. It is less acute than the vertebrate eye but more sensitive to motion.

Sensory receptors (sensilla) are found on the body surface of terrestrial arthropods. Most take the form of setae, hollow shafts of chitin leading to sensory cells that respond when the shafts are stimulated by touch, vibrations, or air currents.

Proprioceptors are sensory receptors that provide information about the position of various parts of the body and the stresses and strains on them. Arthropod proprioceptors include touch receptors and campaniform cells.

Arthropods communicate with members of the same species by sound and also by pheromones, chemical messengers released in the air. Pheromones serve as mating lures and also as trail-marking and alarm substances. Honeybees

have been found to communicate by means of a waggle dance that directs other members of the hive to a food source.

Arthropod behavior is notable not only for its complexity and diversity but also for the extent to which it is programmed in the nervous system, that is, not learned.

The phylum Chordata comprises three subphyla: the Cephalochordata (lancelets), the Tunicata (tunicates), and the Vertebrata.

The primary characteristics of the chordates are the notochord, a flexible, longitudinal rod running just ventral to the nerve cord and serving as the structural axis of the body (present only in embryonic life in the vertebrates); the nerve cord, which is a hollow tube located dorsally; a pharynx with gill slits; and a tail. *Amphioxus* illustrates the basic chordate body plan. It is hypothesized that the tunicate larva, which also shows these characteristics, resembles the primitive chordate from which the vertebrates evolved. The vertebrates, the largest subphylum of the chordates, all have a vertebral column, a flexible bony support that develops around and supplants the notochord and encloses the nerve cord. The vertebrates include the fish (three living classes), the amphibians, the reptiles, the birds, and the mammals.

QUESTIONS

1. Define the following terms: chelicera, tympanic organ, ommatidium, Osteichthyes, pheromone, ganglion, vertebra.
2. How would you distinguish an insect from an arachnid? An insect from a crustacean?
3. Name the identifying characteristics of the phylum Chordata and indicate the functional significance of each.
4. The arthropods, which include some of the most active animals, have open circulatory systems, often considered primitive or inefficient. Annelids, from which arthropods may have evolved, have closed circulatory systems; presumably, half of the annelid system must have been lost. How could the acquisition of a relatively rigid exoskeleton make superfluous the vessels returning blood to the heart?
5. Consider the graceful swimming of a fish or squid. What do you think is the role of the semirigid beam running the length of the animal, whether the "pen" of a squid, the notochord of an *Amphioxus*, or the vertebral column of a fish?
6. Cite an example to illustrate the idea that evolution may proceed through the loss of adult specializations and the retention of larval characteristics.

SUGGESTIONS FOR FURTHER READING

AHMADJIAN, V.: *The Lichen Symbiosis*, Blaisdell Publishing Company, Waltham, Mass., 1967.

In this small but fascinating book, Ahmadjian presents recent work on the nature of the relationship between the fungal and algal components of a lichen.

ALEXOPOULOS, C. J.: *Introductory Mycology*, 2d ed., John Wiley & Sons, Inc., New York, 1962.

A general, brief introduction to the fungi.

BARNES, ROBERT D.: *Invertebrate Zoology*, 2d ed., W. B. Saunders Company, Philadelphia, 1968.

One of the best general introductions to protozoans and invertebrates.

BONNER, J. T.: *The Cellular Slime Molds*, 2d ed., Princeton University Press, Princeton, N.J., 1968.

A record of experimental work with a small but fascinating group of organisms.

BUCHSBAUM, RALPH, and LORUS J. MILNE: *The Lower Animals: Living Invertebrates of the World*, Doubleday & Company, Inc., Garden City, N.Y., 1962.

A collection of handsome photos of the invertebrates accompanied by a text prepared by two noted zoologists but directed toward the general reader.

CORNER, E. J. H.: *The Life of Plants*, Mentor Books, New American Library, Inc., New York, 1968.*

A renowned botanist with a flair for poetic prose describes the evolution of plant life, telling how plants modify their structures and functions to meet the challenge of a new environment as they invade the shore and spread across the land.

CURTIS, HELENA: *The Marvellous Animals*, Natural History Press, Garden City, N.Y., 1968.

An introduction to protozoans.

DAWSON, E. Y.: *Marine Botany: An Introduction*, Holt, Rinehart and Winston, Inc., New York, 1966.

A short, lively text that covers seaweeds, marine bacteria, fungi, phytoplankton, and sea grasses.

EVANS, HOWARD E.: *Life on a Little-Known Planet*, E. P. Dutton & Co., Inc., New York, 1968.

Professor Evans is Curator in the Department of Entomology of Harvard's Museum of Comparative Zoology and also the author of many popular articles and books on insects. This book profits from his wide knowledge, clarity, and humor.

JURAND, A., and G. C. SELMAN: *The Anatomy of* Paramecium aurelia, The Macmillan Company, New York, 1964.

An exploration, mainly by electron microscopy, of the astonishing complexity of a single-celled organism.

KLOTS, ALEXANDER B., and ELSIE B. KLOTS: *Living Insects of the World*, Doubleday & Company, Inc., Garden City, N.Y., 1962.

A spectacular gallery of insect photos. As with The Lower Animals, *which belongs to the same series, the text is informal but informative, written by experts for laymen.*

LARGE, E. C.: *The Advance of the Fungi*, Dover Publications, Inc., New York, 1962.*

A fascinating popular account of the closely interwoven histories of fungi and man, first published in 1940.

ROMER, ALFRED: *The Vertebrate Story*, 4th ed., The University of Chicago Press, Chicago, 1959.

The history of vertebrate evolution, written by an expert but as readable as a novel.

ROSEBURY, THEODOR: *Life on Man*, The Viking Press, Inc., New York, 1969.

An account of our resident microbes and our attitudes toward them—"from Aristophanes to Lenny Bruce."

RUSSELL-HUNTER, W. D.: *A Biology of Lower Invertebrates*, The Macmillan Company, New York, 1968.*

*Available in paperback.

RUSSELL-HUNTER, W. D.: *A Biology of Higher Invertebrates*, The Macmillan Company, New York, 1968.*

Short, authoritative accounts, with emphasis on function.

SMITH, A. H.: *The Mushroom Hunter's Field Guide*, 2d ed., The University of Michigan Press, Ann Arbor, Mich., 1966.

A clear, concise, well-illustrated guide to edible mushrooms, enlivened with good advice and pertinent anecdotes.

STANIER, R. Y., et al.: *The Microbial World*, 3d ed., Prentice-Hall, Inc., Englewood Cliffs, N.J., 1970.

An introduction to the biology of microorganisms, with special emphasis on the properties of bacteria. It is widely considered one of the most authoritative accounts.

WELLS, M. J.: *Brain and Behavior in Cephalopods*, Stanford University Press, Stanford, Calif., 1962.

Experimental analyses of behavior in the octopus and squid.

* Available in paperback.

SECTION 4

BIOLOGY OF PLANTS

Chapter 24

Introducing the Land Plants

In the preceding two chapters, we have attempted to present, in broad perspective, a view of the diversity of animal life. In this chapter, we begin to examine in detail the very different evolutionary solutions to existence that characterize the higher plants, particularly the land plants. The land plants, in general, and the flowering plants, in particular, which are the subjects of this section, are best understood in terms of their long evolutionary history and, in particular, the transition of the plants from water to the land.*

As we noted in Chapter 21, the ancestor of the land plants was a single-celled, eukaryotic, photosynthetic organism that floated on the water's surface or just below it. Like the modern plants, its photosynthetic pigments were chlorophylls *a* and *b*, it contained beta-carotene as an accessory pigment, and it had chloroplasts, mitochondria, and other cellular organelles, as well as an external cell wall of cellulose. Its energy source was sunlight, and it obtained oxygen, carbon dioxide, and the few minerals it required from the waters in which it lived.

Today's plants have the same few and relatively simple requirements: light, water, oxygen, carbon dioxide, and certain minerals. From these simple materials they, like their ancestors, make the sugars, amino acids, and all other organic substances on which plant and animal life depends. But there is an important difference. In the simplest photosynthesizing organism—the single green cell or filament of cells—each of the needed materials is immediately available to every cell. In the land plant, however, the single cells can no longer function autonomously but can survive only through cooperation and division of labor among the many cells and tissues forming the plant body. The reasons for this division of labor are easy to understand. The water and minerals needed by land plants are found mostly below the surface of the earth; the development of a complex and extensive root system can be seen as a response to this selective pressure. Sunlight cannot reach these belowground structures, however, so photosynthesis is relegated to another part of the plant body. As the land plants began to crowd each other and compete for light, selection pressures favored those with more and more extensive and efficient light-collecting surfaces, leaves.

The stem raises the photosynthesizing areas into the sunlight. Through its specialized vascular tissues, water and minerals from the ground travel to the leaves, and the products of photosynthesis formed in the leaves are transported to flowers, roots, and other nonphotosynthetic parts of the plant.

*If you have not read Chapter 21, we recommend you do so now, particularly pages 340, and 349 to 353.

–1   *Flowers of a dogwood tree.*

24–2 *The body plan of a complex land plant. The aboveground structures constitute the shoot, consisting of the stem, the leaves, whose primary function is photosynthesis, and the flowers, the reproductive organs. Leaves appear at regions on the stem known as nodes. The portions of the stem between nodes are called internodes. The belowground structures, the roots, supply water and minerals to the plant body.*

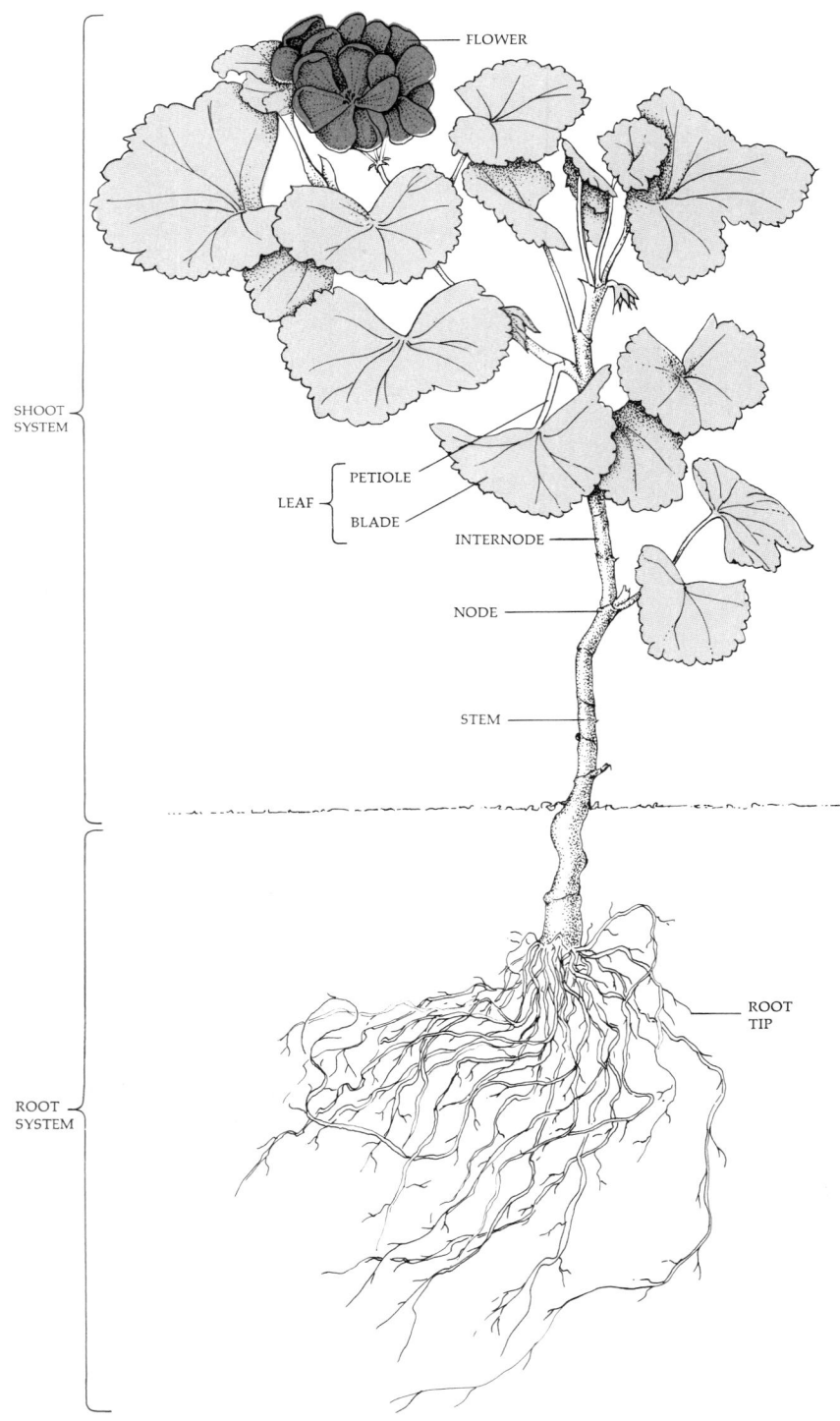

SHOOT SYSTEM

FLOWER

LEAF — PETIOLE
BLADE

INTERNODE

NODE

STEM

ROOT SYSTEM

ROOT TIP

–3 *Some representative leaf shapes.*

GINKGO

SASSAFRAS

MAPLE

BEECH

WILLOW

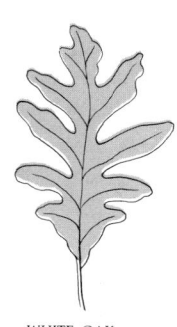
WHITE OAK

–4 *Leaves may be opposite, as in a zinnia, or alternate, as in* Ficus. *The leaves of the zinnia are sessile, that is, they lack leaf stalks (petioles).*

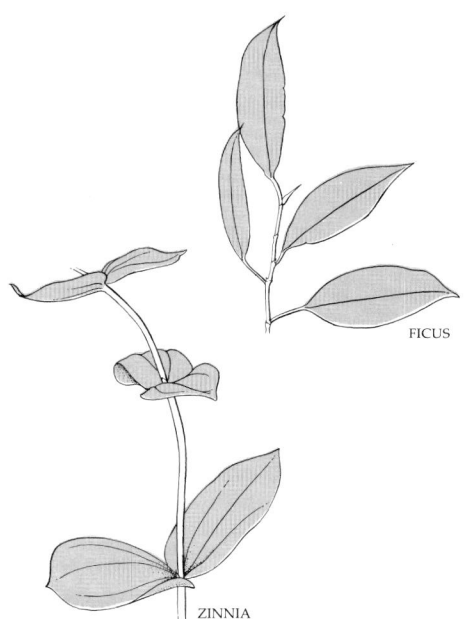
FICUS

ZINNIA

LEAVES

The primary challenge faced by plants in the transition to land was that of dehydration. Photosynthesis, as we saw in Chapter 9, requires light and carbon dioxide. The problem, evolutionarily speaking, is to find a way to collect light and carbon dioxide from the air without losing so much water from the plant body that the cell membranes and other structures collapse. The leaf represents a set of solutions to this problem—the best solutions at the present stage of plant evolution. (It would seem that a better alternative would be a specialized cell membrane impermeable to water but freely permeable to oxygen and carbon dioxide, which reminds us that evolution—chance and necessity—works within very strict limitations.)

Leaves come in all shapes and sizes, ranging from broad fronds to tiny scales. Some of these variations can be correlated with the environments in which the plants live. Small, leathery leaves are often associated with harsh, dry climates, where the light-capturing surface has to be sacrificed for water conservation. This trend reaches its extreme in the desert cacti, which have no leaves at all. In these plants, photosynthesis takes place in the fleshy stems. Conversely, large leaves with broad surfaces are often found in plants that grow under the canopy in a tropical rain forest, where water is plentiful but where there is intense competition for light. Leaves of such plants often have "drip tips" that facilitate the runoff of rainwater.

Because leaf shape is genetically determined and relatively invariant, differences among leaves are useful in plant identification (Figure 24–3). There

–5 *Compound leaves, three examples of which are shown here , are made up of a number of separate blades (leaflets) attached by a single petiole. Simple leaves, such as those shown in Figures 24–3 and 24–4, have just one blade.*

POISON SUMAC

HONEY LOCUST

HORSE CHESTNUT

24–6 *The water buttercup,* Ranunculus peni- *cillatus,* grows with much of the plant body submerged in water. Although the leaves are identical genotypically, the broad floating leaves differ markedly in both form and physiology from the finely divided leaves that develop on the same stem underwater. Such plants illustrate dramatically that the final form of an organism is the result of an interaction between heredity and environment.

may be some variations, however, among leaves of plants of the same species or even among leaves of the same plant, especially when the plants or leaves are living under different conditions. In some semiaquatic plants, the submerged leaves are very different in shape from the aerial leaves (Figure 24–6). Also, some plants, for example, *Eucalyptus,* have one type of leaf on rapidly growing shoots and another on the slowly growing shoots more often found in older trees and shrubs (Figure 24–7).

Veins, or vascular bundles, which consist of conducting tissues (xylem and phloem), are visible on most leaves. In one large group of angiosperms, the dicots, the venation (vein pattern) is netted, either palmate (fanlike) or pinnate (featherlike). In the other large group, the monocots, the venation is usually parallel. (See Figure 24–8.) Other differences between dicots and monocots are discussed elsewhere in the chapter and summarized in the boxed essay on page 440.

24–7 *Juvenile* (a) *and mature* (b) *leaves of* Eucalyptus globulus. *The juvenile leaves are softer, opposite, sessile, and held parallel to the ground. Their only layer of palisade parenchyma is just below the upper epidermis. The mature leaves are* hard, spirally arranged, and hang vertically. In the mature leaves, both surfaces are equally exposed to the light, and there is a layer of palisade parenchyma below each surface.

(a)

(b)

4–8 (a) *Netted venation of a leaf of a dicot (an avocado). (b) Parallel venation of a leaf of a monocot (a corn plant). Veins are bundles of conducting tissues, both xylem and phloem (pages 436 and 437). Continuous with the vascular tissue of the stem and root, they branch and divide into finer and finer bundles, reaching a short distance from every photosynthetic cell. Water and dissolved minerals are carried to the leaf through the xylem; sucrose and other organic molecules produced in the leaf are carried away from their sites of synthesis through the phloem.*

(a)   (b)

Leaf Structure and Function

Despite their different shapes and sizes, the internal structure of the leaves of angiosperms tends to follow the same pattern, looking much like the geranium leaf whose cross section is shown in Figure 24–9. The upper and lower surfaces of the leaf blade are covered with epidermal cells, which are transparent (permitting the passage of light) and coated with a waxy layer, the *cuticle*, which retards water loss. Sandwiched between the epidermal layers is the *mesophyll* ("middle leaf"), which includes the *palisade cells* and the *spongy mesophyll*. The palisade cells, the long columnar cells lying just beneath the upper epidermis, are the chief sites of photosynthesis. The spongy mesophyll bears that name because the cells are loosely packed with many large gas spaces between them. These cells also take part in photosynthesis.

Air laden with water vapor fills the intercellular spaces surrounding the spongy mesophyll cells. From these spaces oxygen and carbon dioxide diffuse in and out of the individual cells of the leaf, where they are used in or produced by respiration and photosynthesis. Thus, every photosynthetic cell of the leaf is in direct contact with an air space, providing an efficient gas exchange.

The palisade layer and the spongy mesophyll are both composed of a cell type known as *parenchyma*. Parenchyma cells, which are typically many-sided and thin-walled, are the most abundant type of plant cell. Some parenchyma cells, such as those in the leaf, have chloroplasts and are photosynthetic; others, which lack chloroplasts, serve a storage function in stems and roots.

The thin-walled parenchyma cells of the palisade layer and spongy mesophyll are enclosed within layers of epidermal cells and cuticle, the waxy layer secreted by them. Cuticle inhibits not only the passage of water vapor in and out of the leaf but also the exchange of gases. Gas exchange is able to take place because of the presence of *stomata* (singular, stoma, from the Greek word for "mouth"), which are pores in the epidermis. When the stomata are closed, water loss from the leaf is greatly reduced, but carbon dioxide cannot enter the leaf and so, once the carbon dioxide in the air spaces is used up, photosynthesis cannot take place. When the stomata are open, gas exchange (by diffusion) takes place between the air spaces within the leaf and the air surrounding it. Most of the water that escapes from the plant body is lost as water vapor through open stomata.

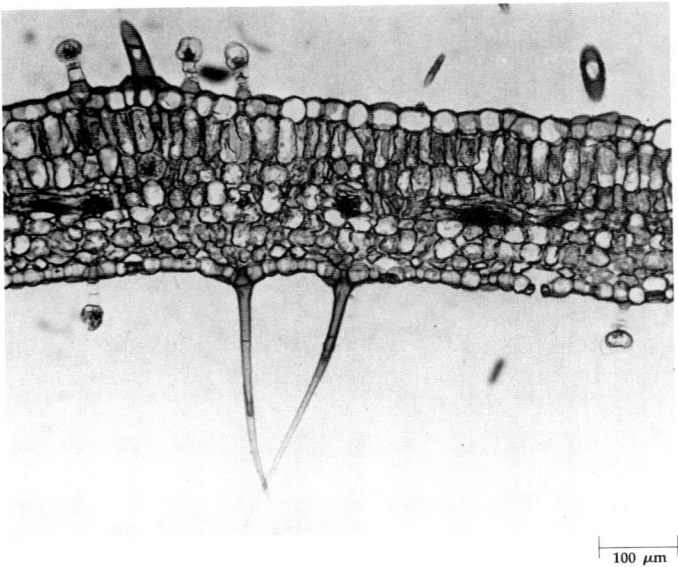

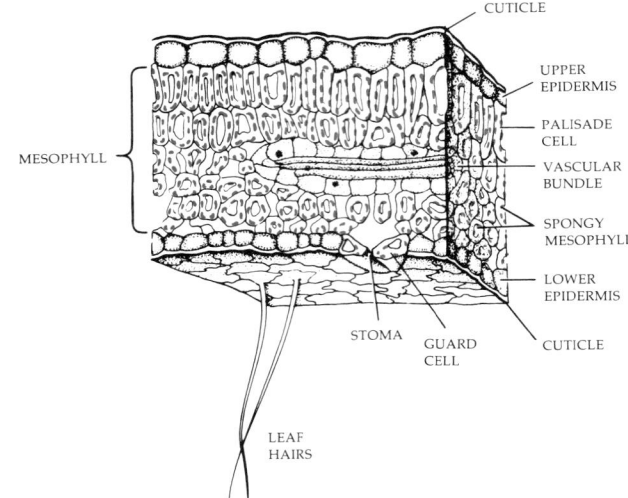

CUTICLE

UPPER
EPIDERMIS

PALISADE
CELL

VASCULAR
BUNDLE

SPONGY
MESOPHYLL

LOWER
EPIDERMIS

CUTICLE

MESOPHYLL

STOMA   GUARD
CELL

LEAF
HAIRS

100 μm

24–9   *Micrograph and diagram of cross section
of a geranium leaf.*

24–10   *Portion of the leaf of a spiderwort
(Tradescantia). Two stomata are visible.
The one at the top is partly open and the
other is closed.*

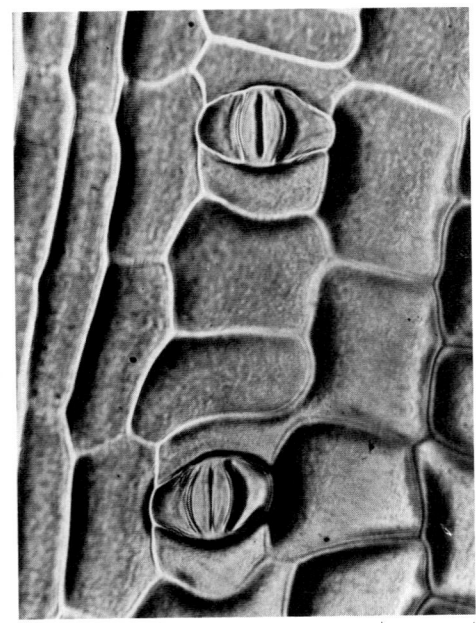

50 μm

Each stoma is enclosed by a pair of specialized epidermal cells called *guard
cells*. These cells control the opening and closing of the stoma by changing shape
in response to environmental cues (see pages 447 and 448). Thus the wax-covered
epidermal cells and the movable stomata with underlying air spaces provide a
system of checks and balances, a compromise solution to conflicting evolutionary
pressures. (Note that this solution is remarkably similar to that of insects with
their chitinous exoskeletons and tracheal systems with movable spiracles.)

Veins run through the mesophyll tissues of the leaves, bringing water and
minerals into the leaf blades and removing sugars. The veins pass through the
petiole (stalk) of the leaf and are continuous with the conducting tissues of the
stem.

Leaves of plants growing in intense sunlight tend to have a thick, shiny cuticle
and two or more layers of palisade cells, whereas leaves of the same species
growing in the shade have thinner leaves, with a single layer of palisade cells
and a thin cuticle.

Form and Function

Victor Greulach,* in the introduction to his recent book on plant physiology,
estimates that in order for a human being to be photosynthetic (like those green
creatures from outer space that sometimes appear in science fiction), he would
have to increase the surface area of his body at least twentyfold, assuming that
he lived in a warm climate, wore no clothes, and did not use energy by moving
around. He probably could not move around anyway without damage to his
photosynthetic appendages. It would be useful if he could give up certain
metabolic functions—such as digestion and circulation by a pumping system—

* Victor A. Greulach, *Plant Function and Structure*, The Macmillan Company, New York,
1973.

resulting in conservation of energy. Without mobility he would have little use for a nervous system. In fact, with increasing adaptation to the photosynthetic way of life, he might turn out to be very much like a plant.

STEMS

In an aquatic environment, the photosynthetic portions of the plant are held in the light by the buoyancy of the surrounding water. In the land plants, the stem supports the light-collecting plant surfaces and also provides for the movement of water up to these surfaces and for the transport of organic molecules from them to other parts of the plant body.

The stem is a complex organ, consisting of a number of different types of tissues. On the surface of a green stem is a wax-coated layer of epidermal cells, often punctuated by stomata, as occurs on the outer surfaces of the leaf. The interior of the stem contains ground tissue and conducting tissue, each containing a number of different cell types.

Ground Tissue

The *ground tissue*, like the mesophyll of the leaf, is composed mostly of parenchyma cells. The turgor* of these cells provides the chief support for young green stems; withhold water and, as you know, young stems wilt.

Parenchyma cells also accumulate and store food, usually in the form of starch, and in some plants, such as cacti and other succulents, they are specialized for storing water.

In green stems, the main supporting tissues are formed from *collenchyma* and *sclerenchyma* cells. Collenchyma cells (Figure 24–11) differ from the thin-walled parenchyma cells of the stem in having primary cellulose walls that are thickened at the corners or in some other uneven fashion. Their name derives from the Greek word *colla*, meaning "glue," which refers to their characteristic thick, glistening walls. Collenchyma cells are often located just inside the epidermis, forming either a continuous cylinder or distinct vertical strips.

* See page 114 for a discussion of turgor.

24–11 *Some types of cells found in the ground tissue of the stems. (a) Collenchyma cells. Their irregularly thickened cellulose walls contain pectin and large amounts of water and they are plastic, so they permit growth while providing support. (b) Fibers. Their specialization is thickened, often lignified cell walls that give them strength and rigidity. Many, but not all, are dead at maturity, like those shown here. (c) Sclereids, or stone cells. These cells, which have very thick lignified walls, are often found in seeds and fruit, as well as in stems. These are from a pear; they are what gives the fruit its characteristic gritty texture. This cell is a living cell with slender branches of cytoplasm extending through the cell walls. These terminate in plasmodesmata.*

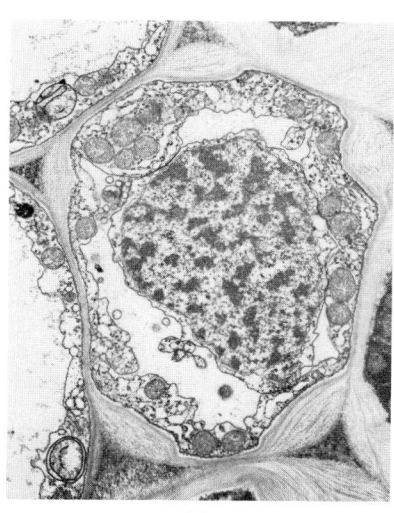

(a)

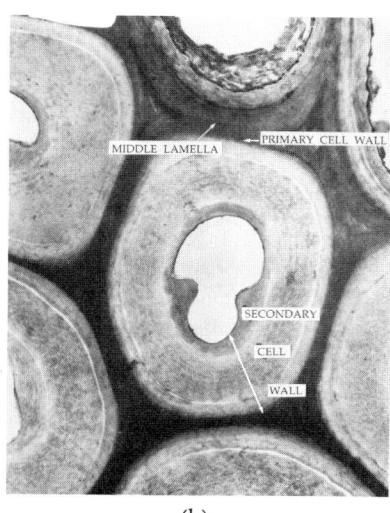

(b)

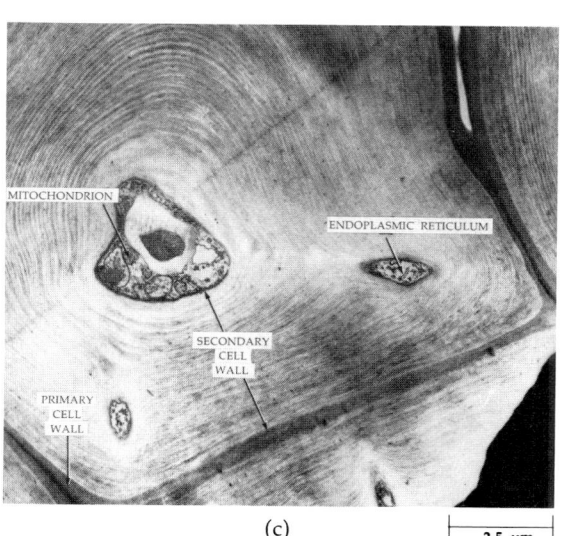

(c)

2.5 μm

Sclerenchyma cells are of two types: fibers and sclereids. The name is derived from the Greek *skleros*, meaning "hard." Fibers, which are elongated, somewhat elastic cells, typically occur in strands or bundles arranged in patterns characteristic of the plant. They are often associated with the vascular tissues. Plant fibers have long been useful to man in such forms as flax, hemp, jute, and raffia. Sclereids, or stone cells, which are variable in form, are less common in stems than fibers. Layers of sclereids are found in seeds, nuts, and fruit stones, however, where they form the hard outer coverings. Sclerenchyma cells differ from collenchyma in three respects: (1) the walls contain lignin, a complex macromolecule that impregnates the cellulose and toughens and hardens it; (2) the cells are sometimes dead at maturity—that is, without living protoplasm; (3) the secondary walls are strongly developed.

Conducting Tissues

Phloem

Phloem (pronounced flow-em) conducts the products of photosynthesis, chiefly in the form of sucrose, from the leaves to the stem, roots, reproductive structures, and other nonphotosynthetic tissues of the plant. In angiosperms, the conducting cells of the phloem are *sieve-tube elements*. A *sieve tube* is a vertical column of sieve-tube elements joined by their end walls. These end walls, called *sieve plates*, have openings leading from one sieve-tube element to the next.

The sieve-tube elements, which are alive at maturity, are filled largely with a watery substance often referred to as slime. The nucleus of a sieve-tube element disintegrates as the cell matures, as do many of the organelles. The remaining cytoplasm forms a film along the longitudinal surface of the cell. Sieve-tube elements are always associated with *companion cells*, which are thought to provide nuclear functions and energy for the sieve tubes. A sieve tube can function only if its cell membrane is intact, and it is likely that the companion cell helps to maintain the membrane. Phloem also contains parenchyma cells, which store food and water, and supporting fibers.

24–12 *In angiosperms, the conducting elements of the phloem are sieve tubes, made up of individual cells, the sieve-tube elements. These cells, which lack nuclei at maturity, are always found in close association with companion cells, which do have nuclei. Although the movement of solutes through the sieve tubes is usually downward, it may be horizontal or upward, as when sugars are moving into a flower or fruit. Two sieve-tube elements are shown in the electron micrograph and diagram. They are separated by a sieve plate. The light-colored (electron-transparent) material near the pore is callose, a polysaccharide characteristically associated with sieve plates, especially in old or injured cells. Callose plugs the pores when a sieve cell is injured, thus preventing leakage. Immediately to the left of the upper sieve-tube element, identifiable by its dense cytoplasm, is a portion of a companion cell. A parenchyma cell is to the right.*

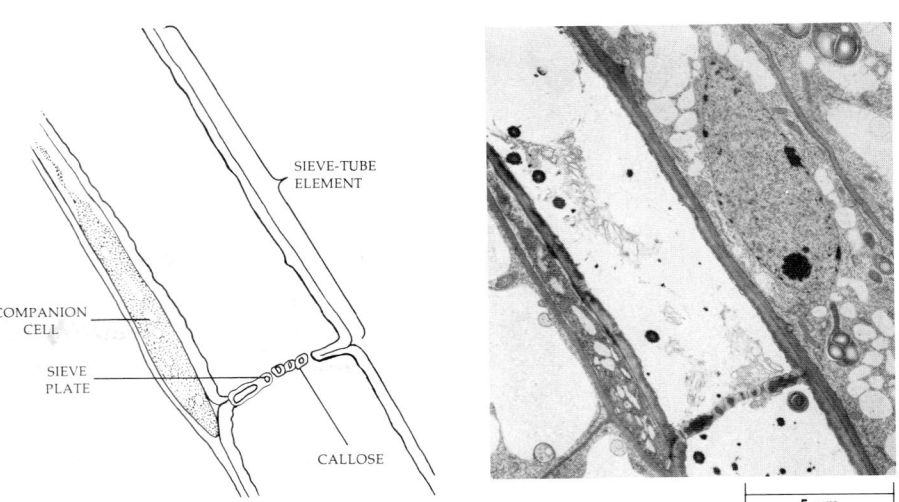

SIEVE-TUBE ELEMENT

COMPANION CELL

SIEVE PLATE

CALLOSE

5 μm

-13 *Tracheids and vessel elements are the conducting cells of the xylem in angiosperms. (a) Tracheids are a more primitive and less efficient type of conducting tissue. Water passing from one tracheid to another passes through pits. Pits are not open pores but areas in which there is no secondary cell wall. Water moving from one tracheid to another passes through two primary cell walls and the middle lamella (see Figure 6–4). Vessel elements differ from tracheids in that the primary walls of vessel elements are perforated where they are joined with other vessel elements. (b) There may be numerous small perforations on adjoining walls of vessel elements, or (c) the adjoining walls may break down completely as the cells mature, forming an open, continuous vessel. Vessel elements are also characteristically shorter and wider than tracheids and their adjoining walls are less oblique.*

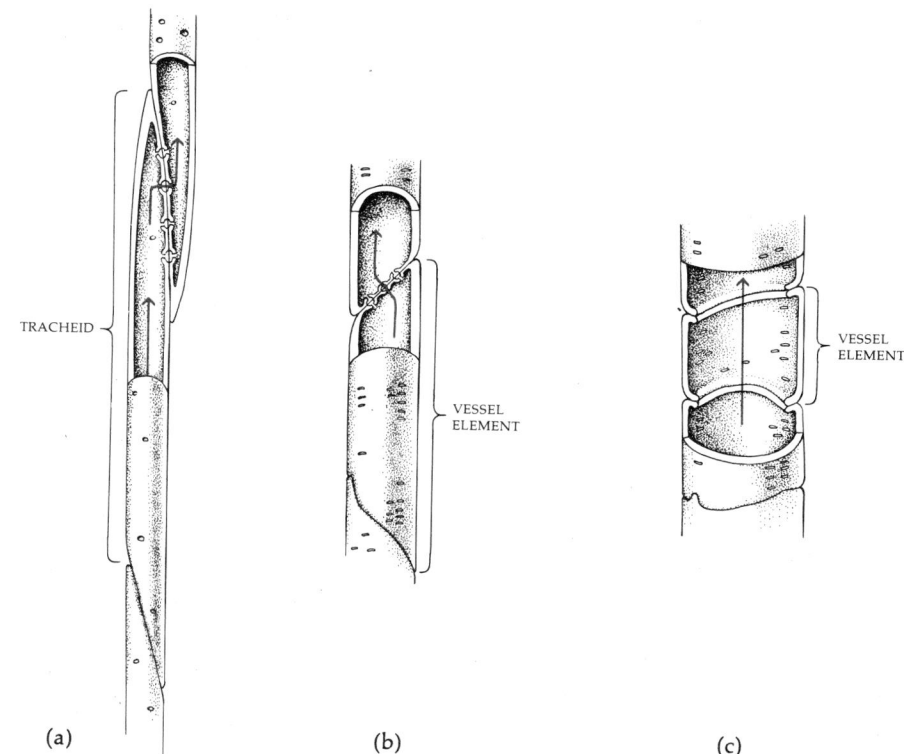

TRACHEID

VESSEL ELEMENT

VESSEL ELEMENT

(a)   (b)   (c)

Xylem

Xylem conducts water and minerals from the roots up through the plant body. The conducting cells of angiosperms are *tracheids* and *vessel elements*. Both of these cell types have heavy secondary walls impregnated with lignin, and both are dead at maturity. Tracheids are long, thin cells that overlap one another on their tapered ends (Figure 24–13a). These overlapping surfaces contain thin areas, pits, where no secondary wall has been deposited, and water passes from one tracheid to the next through these pits. Vessels differ from tracheids in that their end walls contain multiple perforations or are broken down entirely. Thus, the cells form a continuous *vessel*, which is a much more effective conduit than a series of tracheids. Gymnosperms and lower vascular plants have only tracheids.

Xylem also contains parenchyma cells, which store food and water, and fibers.

Stem Patterns

In green stems, the xylem and the phloem are usually arranged in longitudinal parallel groups, the *vascular bundles*, which are embedded in the ground tissue. In young dicot stems, the vascular bundles form a ring around a central area called the *pith*. The cylinder of tissue outside the vascular bundles is called the *cortex*. The xylem characteristically is on the inside of each bundle, adjacent to the pith, and the phloem is on the outside, adjacent to the cortex. (Pith and cortex are both ground tissue.) Among the monocots the vascular bundles are scattered through the ground tissue. (See Figure 24–14.)

24–14 (a) *In a young dicot stem, discrete vascular bundles form a ring near the outer perimeter of the stem. Phloem is on the outside of each bundle, and xylem is on the inside. The ground tissue outside the ring of vascular bundles is the cortex;* *the tissue inside the ring is the pith.* (b) *In a monocot stem, the vascular bundles, containing both xylem and phloem, are scattered throughout the ground tissue.*

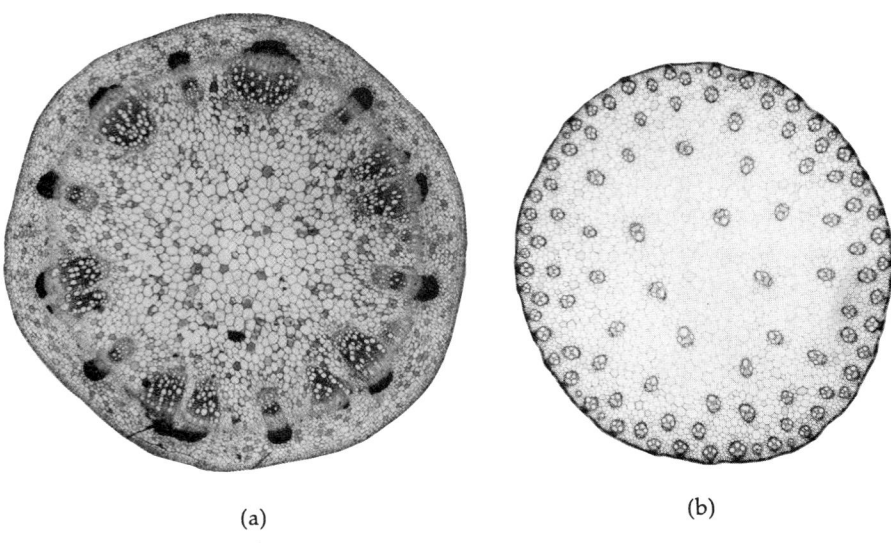

(a)   (b)

24–15 *Root hairs of radish seedling. Most of the uptake of water and solutes takes place through the root hairs, which form just behind the growing tip of the root.*

ROOTS

Roots collect water and minerals from the soil and transport them to the other parts of the plant; they also serve as an anchor.

Like the stem, the root has an outer layer of epidermis, although this layer does not have an outer surface of cuticle, which would interfere with its capacity to absorb water. The younger cells have root hairs, which are not actually hairs but are slender extensions of the epidermal cells. The nucleus of an epidermal cell is often within the root hair. Much root absorption of water and minerals takes place through root hairs.

Within the epidermal layer of the dicot root is a cylindrical zone of cortex, as there is in the stem. It is composed chiefly of parenchyma cells and is relatively larger than the cortex of the stem. The innermost layer of cortex is a cylinder of cells one cell thick, the *endodermis*. It is composed of specialized parenchyma whose cell walls contain a waxy zone, the Casparian strip. Because of the Casparian strip, the only route from the cortex to the xylem is through the living protoplasts of the endodermal cells. Just inside the endodermis is another cylinder of cells, the pericycle, which is the site of origin of lateral roots. The pericycle surrounds the vascular tissues, composed of xylem and phloem, which are continuous with the vascular tissues of the stem and contain the same cell types. In most species the central core is occupied by the xylem tissue and the phloem is arranged in strands outside the core of xylem.

There are two primary patterns of root growth. Some plants, mostly dicots, have long taproots from which lateral roots grow. Others, mostly monocots,

have no central taproot but a mass of slender roots, all about the same size, known as fibrous roots. Fibrous roots often develop as outgrowths from the stem above the point of origin of the first root which, in monocots, usually dies early in plant growth. Roots that grow from a stem or from any organ other than the primary roots or its branches are known as adventitious roots. Like leaf patterns, the pattern of growth is determined genetically, but environmental factors, such as the type of soil and the availability of water, can modify root growth.

Roots, especially taproots, may serve as storage organs for the products of photosynthesis, and many of our garden vegetables, such as beets and carrots, are plants that have been selectively bred for their fleshy, energy-rich roots. Roots provide an astonishingly extensive absorbing surface. In the 1930s, a biologist at Iowa State University grew a single winter rye plant in a small box (12" × 12" × 22") and, at the end of four months, washed the soil away from the root system and measured it. The combined length of the roots, including root hairs, totaled 7,000 miles, and the surface area was nearly 7,000 square feet.

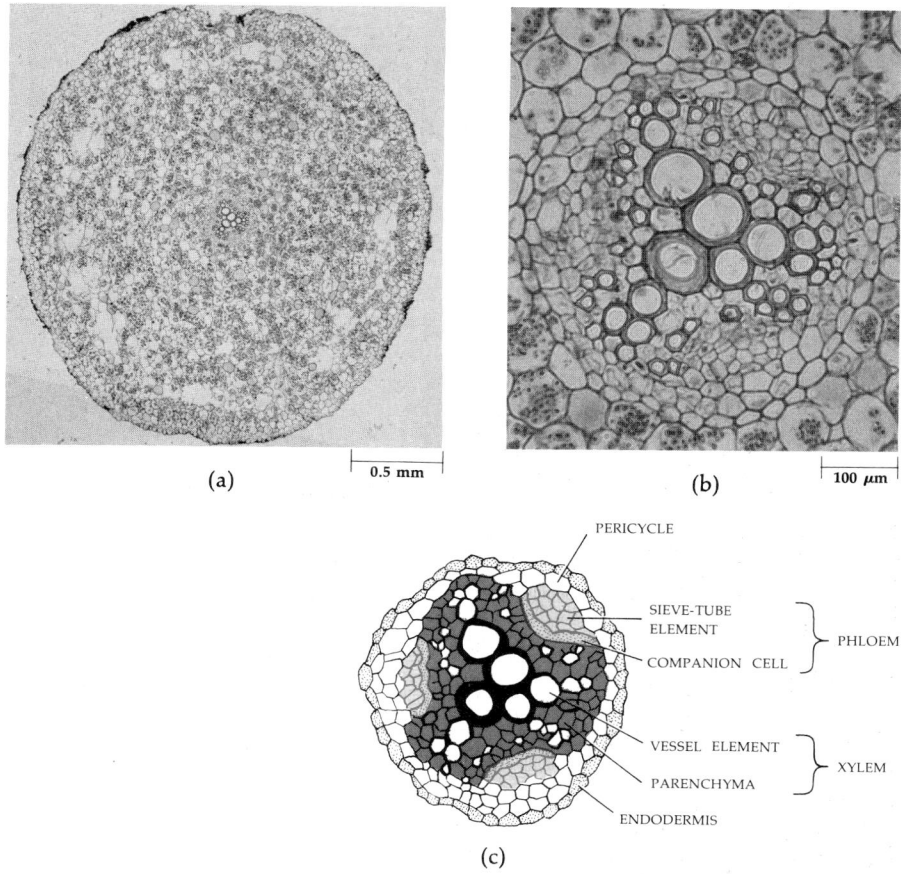

-16   (a) Cross section through the root of a buttercup (Ranunculus), a dicot. At the center of the root, within the cortex and surrounded by the endodermis and the pericycle, are the vascular tissues. This portion of the root is enlarged in (b) and diagrammed in (c).

(a)   0.5 mm

(b)   100 μm

PERICYCLE

SIEVE-TUBE ELEMENT

COMPANION CELL

} PHLOEM

VESSEL ELEMENT

} XYLEM

PARENCHYMA

ENDODERMIS

(c)

MONOCOTS AND DICOTS

The angiosperms are divided into two broad groups: the dicots (dicotyledons), with about 200,000 species, and the monocots, with about 50,000 species. The names refer to the fact that the plant embryo in the dicots has two cotyledons ("seed leaves") and in the monocots has one. In the dicots, the vascular tissues are arranged around a central core in the stem; in the monocots, they are scattered through the stem. Dicots characteristically have taproots, and monocot roots are often fibrous. Few monocots have secondary growth. The veins of dicot leaves are usually fanlike or featherlike; those of monocot leaves are usually parallel.

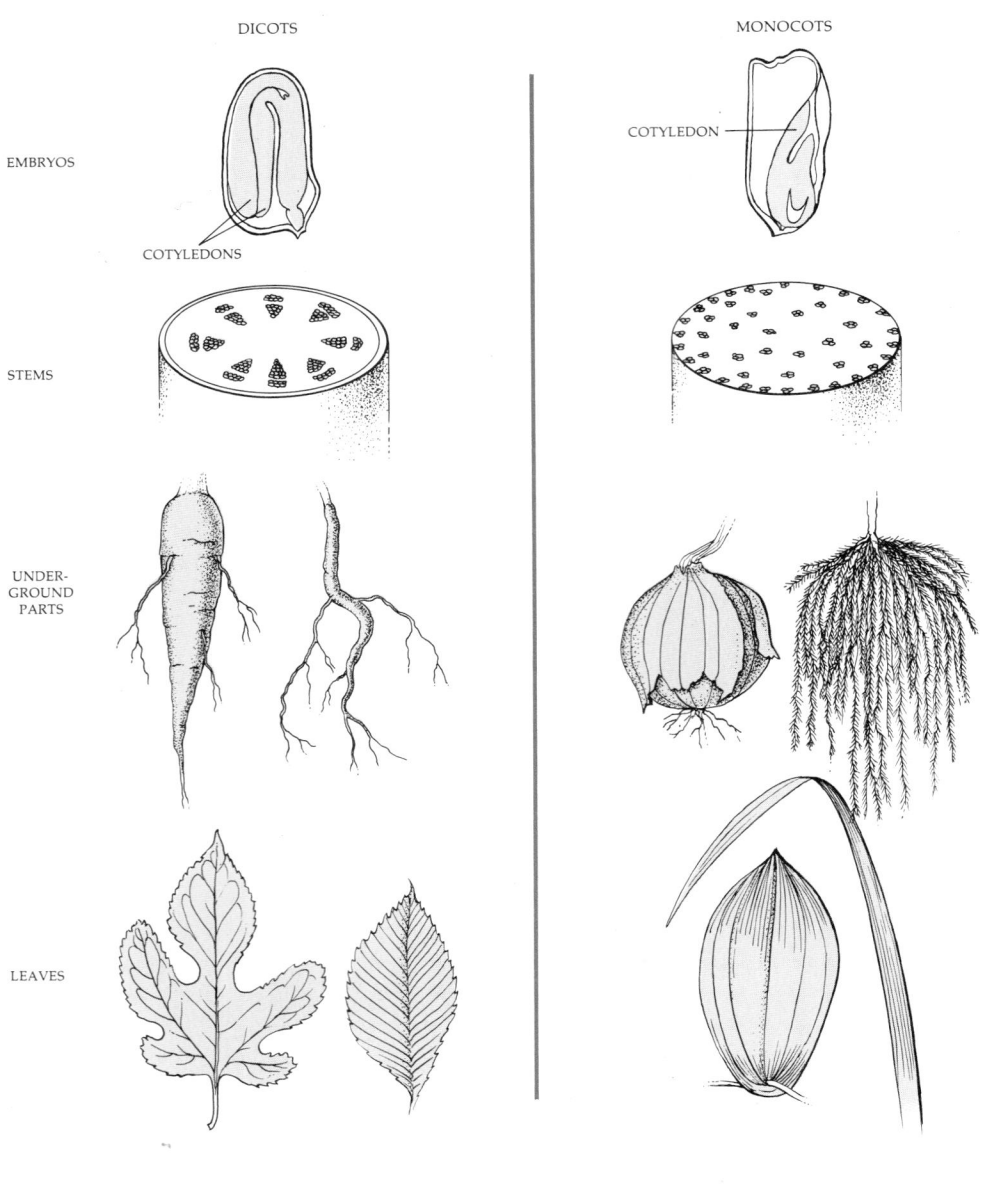

DICOTS

MONOCOTS

EMBRYOS

COTYLEDON

COTYLEDONS

STEMS

UNDER-
GROUND
PARTS

LEAVES

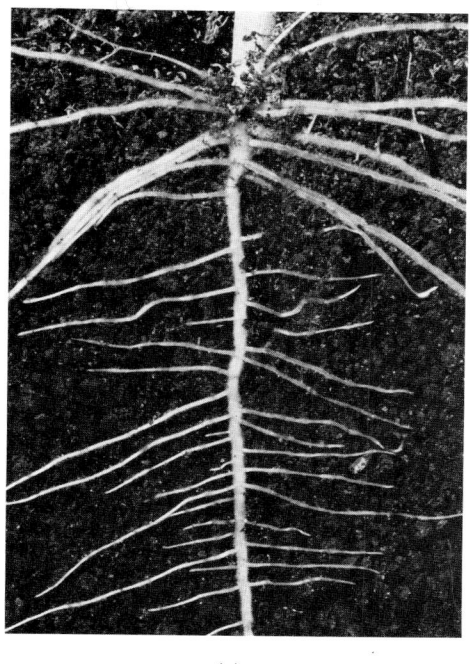

| (a) | (b) | (c) |

-17 *Types of roots. (a) Taproot of a bean, with branch roots. (b) Fibrous roots of a grass. (c) Adventitious roots, the prop roots of a corn plant, growing from its stem.*

ADAPTATIONS TO CLIMATE CHANGE

As we noted in Chapter 21, the angiosperms evolved during a relatively mild period in the Earth's history. As the climate became colder and, as a consequence, water became locked in snow and ice for part of the year, the angiosperms, which already possessed some adaptations to drought (perhaps because of highland origins), were placed under new environmental stress. Some did not survive, and some were pushed southward. Those that did survive in colder, drier areas did so because of selection for characteristics that offered advantages in these relatively unfavorable environments. Chief among such characteristics is the capacity to remain dormant during periods when water is in short supply and when climatic conditions are unfavorable for delicate growing buds, shoots, new leaves, and root tips. Modern plants are classified as annuals, biennials, and perennials, depending on their characteristic patterns of active growth and dormancy.

Annuals, Biennials, and Perennials

Among annual plants, which include many of our weeds, wild flowers, garden flowers, vegetables, and grasses and most other monocots, the entire cycle from seed to vegetative plant to flower to seed again takes place within a single growing season. All vegetative organs (roots, stems, and leaves) die, and only the dormant seed bridges the gap between one generation and the next. Plants with nonwoody stems, such as most annuals, are known as herbs.

In biennial plants, the period from seed germination to seed formation spans two growing seasons. The first season of growth ends with the formation of a

root, a short stem, and often a rosette of leaves near the soil surface. In the second growing season, extensive stem elongation, flowering, fruiting, and seed formation occur. This completes the life cycle and the vegetative organs die. In perennial plants the vegetative structures persist from year to year. Biennials may be herbaceous or woody.

The ancestral angiosperm is believed to have been a woody plant, although probably a low-growing one. In a woody plant, the stem increases in diameter and becomes hardened, with a protective layer of dead tissue, the cork (phellem), on the outside. (The common term "bark," a word not favored by botanists, includes not only the cork but also the underlying living tissues, including the phloem.) The process by which stems increase in girth and cork is formed, known as secondary growth, is described in Chapter 26. Secondary growth is also found among gymnosperms, but is rare among monocots, although some monocots, such as palms and bamboos, become woody as they grow older.

In favorable climates, such as a tropical rain forest, plants may live year after year with little change during the annual cycle. Perennials that live in areas where part of the year is unfavorable to growth show a variety of adaptations. Some, such as cacti and most conifers, undergo little visible change, although their rates of metabolism, and therefore of growth, change with the seasons.

CORK, WHICH IS A DEAD TISSUE, PROTECTS THE INNER TISSUES FROM DRYING OUT, FROM MECHANICAL INJURY, AND FROM INSECTS AND OTHER HERBIVORES. CORK AND PHLOEM TOGETHER MAKE UP THE BARK.

THE PHLOEM CONDUCTS THE SUGARS PRODUCED BY PHOTOSYNTHESIS TO THE ROOTS AND OTHER LIVING, NONPHOTOSYN-THETIC PARTS OF THE PLANT

THE CAMBIUM LAYER, ONLY ONE CELL THICK, PRODUCES NEW XYLEM AND NEW PHLOEM

SAPWOOD IS MADE UP OF XYLEM TISSUE WHICH CONTAINS THE TRACHEIDS AND VESSELS THROUGH WHICH WATER AND MINERALS MOVE FROM THE SOIL TO THE LEAVES AND OTHER LIVING PARTS OF THE TREE. AS THE PARENCHYMA AND SUPPORTING CELLS OF THE XYLEM DIE, THEY BECOME HEARTWOOD.

HEARTWOOD, COMPOSED ENTIRELY OF DEAD CELLS, IS THE CENTRAL SUPPORTING COLUMN OF THE MATURE TREE.

24–19 *A tree trunk showing the relationships of the successive concentric layers. The heartwood is composed largely of dead xylem cells. The cortex, which is outside the phloem in a green stem, is sloughed off in the process of secondary growth.*

Most dicots living in or north of the temperate zones undergo both structural and functional changes that allow them to take maximum advantage of the growing season and still survive the rest of the year. In some, all of the above-ground structures die, and the plants winter over as underground roots, underground stems (rhizomes or tubers), or underground buds (bulbs). Others, which include many of the common vines, shrubs, and trees, are deciduous—that is, all their photosynthetic leaves drop seasonally, an adaptation which reduces the rate of water loss. The dropping of the leaf comes about as a result of the formation of an abscission layer, a special growth of weak, thin-walled parenchyma cells across the petiole. As the leaf ages, the cells separate. Before the leaf falls, a layer of cork develops and seals the leaf scar.

As the leaf ages, new chlorophyll is no longer formed and the green color gradually fades, unmasking the yellows and oranges of the carotenoids and producing the blaze of colors characteristic of the deciduous forest in the fall.

SUMMARY

The bodies of land plants are specialized into photosynthetic areas (leaves), conducting and supporting structures (stems), and organs for the absorption of water and minerals (roots).

The blade of the leaf is composed principally of mesophyll cells enclosed by epidermis. The mesophyll cells include palisade cells, in which most of the photosynthesis takes place, and spongy mesophyll, which forms an additional photosynthetic layer beneath the palisade cells. Both layers of mesophyll are composed of parenchyma cells. Veins, containing both xylem and phloem, conduct water and minerals to the leaf cells and remove the products of photosynthesis. The upper and lower surfaces of the leaf are covered with a single layer of transparent epidermal cells, in which there are stomata. The outer surface of the epidermis is coated with a waxy cuticle. The blade is attached to the stem by the petiole; the conducting tissues of the leaf are continuous with those of the stem.

Green stems, like leaves, have an outer layer of waxy epidermal cells. Within the epidermis is the ground tissue of the stem, which may be divided into cortex and pith. The ground tissue is made up of turgid parenchyma cells. Also present may be collenchyma cells, which are living cells with irregularly thickened cellulose walls, and sclerenchyma cells, which have thick cell walls, often containing lignin, and may be dead at maturity. Sclerenchyma cells include fibers, which are elongated and somewhat elastic and are frequently found in various types of stems, and sclereids, irregularly shaped cells sometimes found scattered in stems and commonly found in dense layers in seed coats, nuts, and fruit stones.

The vascular tissues of the stem are the phloem and xylem. In angiosperms, the conducting cells of the phloem are the sieve-tube elements, living cells with their end walls perforated (sievelike), which form continuous tubes. Associated closely with each sieve-tube element is a companion cell. Phloem also contains parenchyma and fibers. The conducting tissues of the xylem are made up of a series of vessel elements and tracheids. Vessel elements and tracheids are

heavily walled cells, dead at maturity. Xylem also contains parenchyma cells and fibers.

In monocots, strands of xylem and phloem, paired in vascular bundles, are scattered through the ground tissue. In dicots, parallel strands of xylem and phloem form a ring around a central zone of ground tissue, the pith. The ground tissue between the vascular bundles and the epidermis (or the cork) is the cortex. Other differences between dicots and monocots include: two seed leaves (cotyledons) in dicots, one in monocots; netted (palmate or pinnate) venation in dicots, parallel venation in monocots; commonly a main taproot in dicots, numerous fibrous roots in monocots.

Roots, like stems, have an outer layer of epidermis but no cuticle. Extensions of the epidermal cells form root hairs, which greatly increase the absorptive surface of the root. Beneath the epidermis is a cortex composed mostly of parenchyma cells, often with large storage capacities. The innermost layer of the cortex is the endodermis, a single layer of specialized cells whose walls contain a waterproof zone which makes up the Casparian strip. Just inside the endodermis is another layer of cells, the pericycle, from which root branches form, and within the pericycle, the xylem and phloem. The xylem and phloem of the root resemble those of the stem in composition but not in their arrangement. In the root, xylem characteristically fills the central core and the phloem is arranged in strands outside the core of xylem.

Among the most useful adaptations of angiosperms, and probably one of the chief reasons for their evolutionary success, is the capacity, found in many species, to remain dormant during climatically unfavorable periods. Angiosperms are classified as annuals, biennials, or perennials, depending on whether the plant body dies at the end of one growing season (annual) or two years (biennial), leaving only the seeds, or whether the plant body persists (perennial). In harsh climates, perennials are often deciduous, losing all of their leaves during periods when water is in short supply.

QUESTIONS

1. Define the following terms: parenchyma, xylem, collenchyma, pericycle, phloem, fiber, endodermis, cuticle, herbaceous, sclerenchyma.
2. Sketch the layers of a leaf and identify each type of tissue. Check your sketch with the drawing on page 433. What is the primary function of each?
3. Sketch a cross section of a dicot stem, identifying epidermis, cortex, phloem, xylem, and pith. Compare your sketch with the micrograph on page 439.
4. Sketch a cross section of a root, identifying the tissue layers. What two tissues are present in roots but not in stems?
5. Herbaceous stems have a core of liquid-filled parenchyma and a layer beneath the epidermis of thickened collenchyma. How can these two tissues work in concert to maintain the rigidity of the stem? (Consider the properties of an inflated cylindrical balloon.) An earthworm can make itself stiff enough to penetrate soil. What structures in the worm are analogous to the parenchyma and collenchyma of herbaceous stems?

6. Consider a mature tree. Which parts of the tree are composed of living cells? Why might it be advantageous for living cells to make up such a small fraction of the mass of the tree? Similarly, why might it be advantageous for a deciduous tree to invest as little material as possible in making leaves? Estimate (based on leaf-raking experience) the fraction of the dry weight of a tree represented by the leaves.

Chapter 25

Plant Nutrition: Sugar, Water, and Minerals and Their Transport

As we saw in Chapter 9, the energy requirements of plants are met by the conversion, in the chloroplasts, of carbon dioxide and water into carbohydrate and oxygen. This reaction, which runs on solar energy, provides plants with their source of chemical energy and also with the carbon skeletons on which plant cells build their other organic molecules. It is also the source of chemical energy and organic carbon compounds for all of the animal world.

In this chapter, we are going to look at the process of photosynthesis in relation to the plant as a whole and examine how the plant obtains the materials required for this process and how the products of photosynthesis are transported to other parts of the plant body.

25–1 *Cross section of a broad bean leaf. In this scanning electron micrograph, you can see a one-cell-thick layer of cuboidal epidermal cells on the upper and lower surfaces. They enclose a layer of columnar palisade cells, and beneath those, the spongy mesophyll. Notice the abundant intercellular spaces where exchanges of gases (including water vapor) take place.*

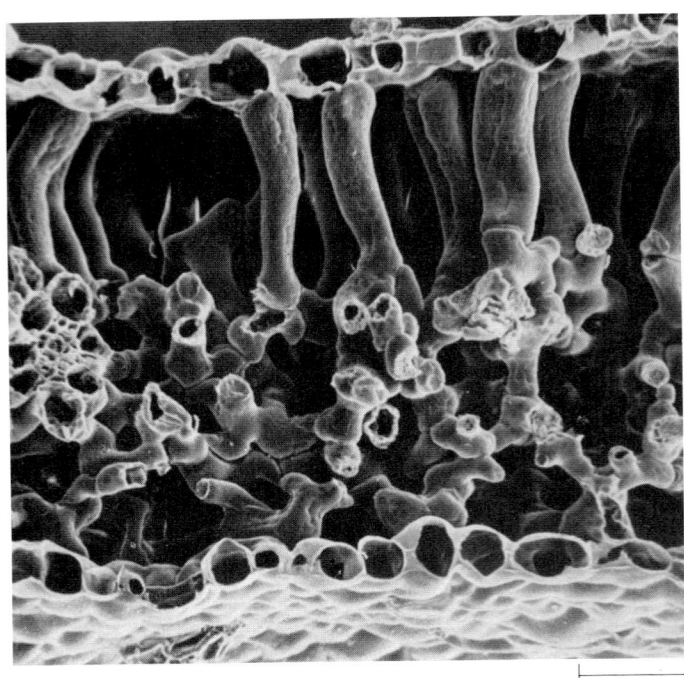

100 μm

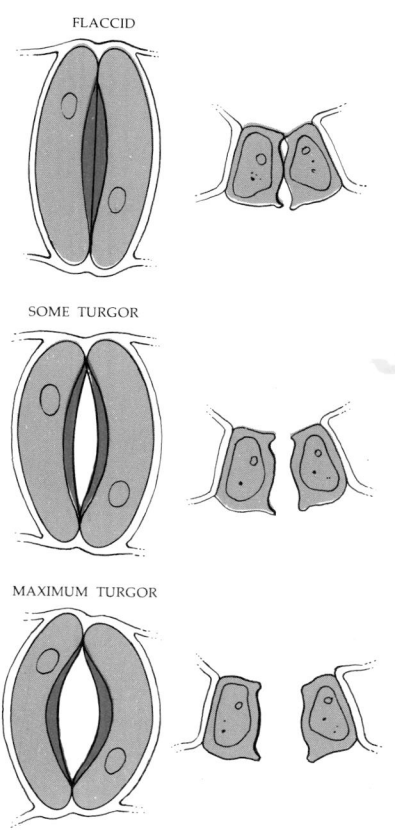

25–2 (a) *Mechanism of stomatal movements. Each stoma, shown here both in surface view and in cross section, is flanked by two guard cells which open the stoma when they are turgid and close it when they lose turgor. In many species, the guard cells have thickened walls adjacent to the stomatal opening. As turgor pressure increases, the thinner parts of the cell wall are stretched more than the thicker parts, causing the cells to bow out longitudinally and forcing the stoma to open. The micrograph (b) shows a closed stoma.*

FLACCID

SOME TURGOR

MAXIMUM TURGOR

(a)

As we noted in the previous chapter, carbon dioxide and oxygen enter the photosynthesizing cells of the leaf from intercellular air spaces in the mesophyll. When the stomata of the leaf are open, carbon dioxide diffuses into the intercellular air spaces. Because the cells use up the carbon dioxide in photosynthesis, the carbon dioxide concentration in these spaces is lower than that in the air and so CO_2 moves inward, down the gradient. (For a review of the principles of diffusion, see page 109.) Similarly, oxygen is given off in the course of photosynthesis. Some of this oxygen is used up in respiration, but when the cells are engaged in photosynthesis, they produce more oxygen than they use in respiration so, also moving down the gradient, oxygen diffuses out into the atmosphere. In most plants, this process is reversed at night; in the dark, photosynthesis stops and, as a result of cellular respiration, oxygen diffuses into the cells and carbon dioxide diffuses out. Because atmospheric gases are in direct contact with the wet cell walls of the photosynthetic cells, plants are generally sensitive to atmospheric pollution.

Water vapor is also present in intercellular air spaces. Except under conditions of unusually high atmospheric humidity, the concentration of water vapor is considerably higher within the air spaces than in the surrounding air, and so water vapor is constantly diffusing from the plant.

REGULATION OF GAS EXCHANGE

Loss of water vapor from the leaf and the exchange of carbon dioxide and oxygen between the air spaces and the air surrounding the leaves are regulated primarily by the stomata. In the daytime stomata remain open unless the roots cannot supply enough water to meet the plant's needs; then they close. The opening and closing of stomata can be shown to depend on turgor pressure in the guard cells, the crescent-shaped epidermal cells that surround the openings. In many species, the wall of the guard cells adjacent to the stomatal opening is slightly thickened and therefore less elastic than the other wall. Therefore when the guard cells absorb water they tend to bow outward, opening the air pore. As they lose turgor pressure, they return to their original position, closing the pore. (See Figure 25–2.)

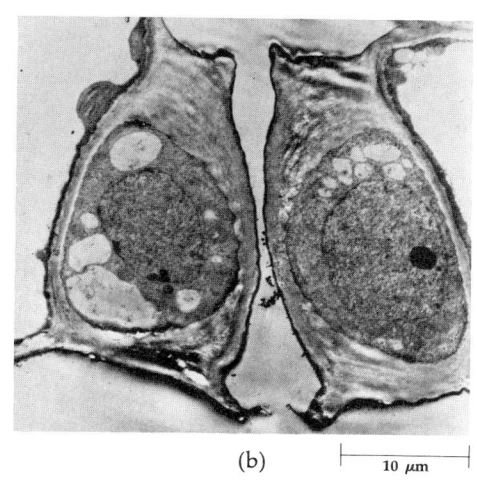

(b) |———————| 10 μm

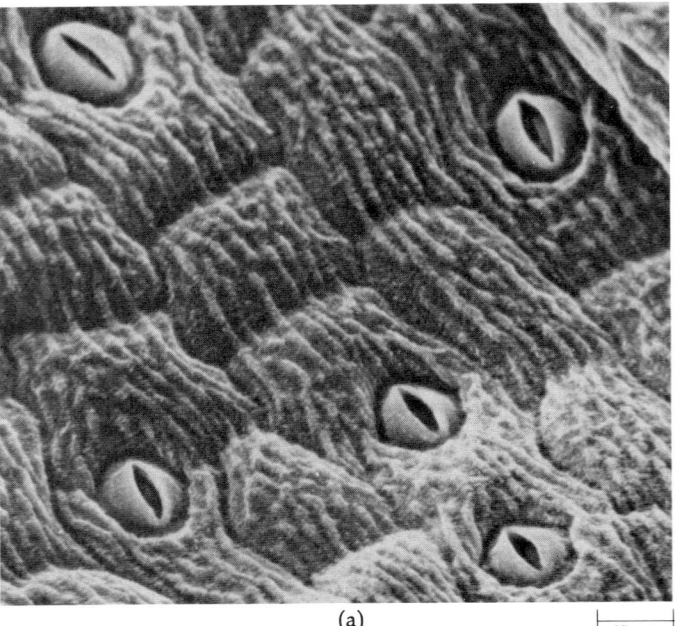

(a)

$\vdash$ 25 μm $\dashv$

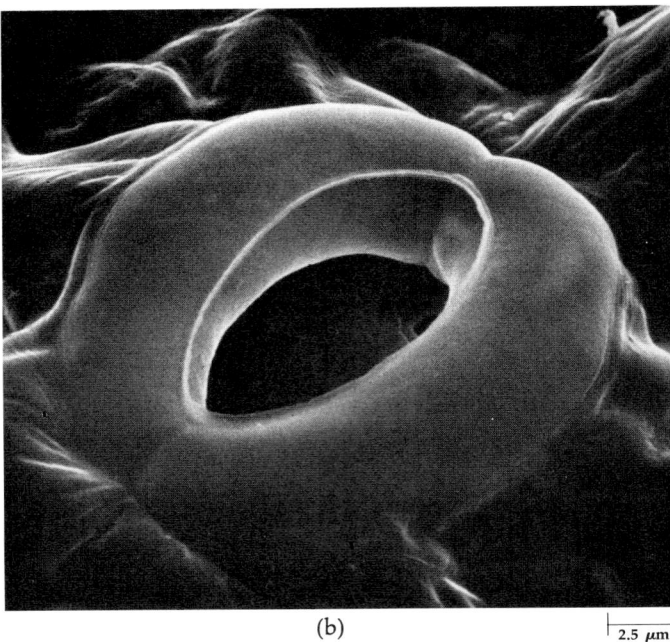

(b)

$\vdash$ 2.5 μm $\dashv$

25–3 (a) *Stomata on the lower surface of a dicot leaf, as recorded by the scanning electron microscope.* (b) *View of an opened stoma of cucumber.*

In many species, stomata also close at night, even when the plant has sufficient water, and open in the light. What causes the turgor changes under these conditions? Experimental evidence indicates that the opening in response to light is a result of increased solute concentration within the guard cells. This increased solute concentration leads to an increased uptake of water by osmosis (see page 112). There is, however, no general agreement on what causes the increases in solute concentration. One possible clue is that guard cells, unlike other epidermal cells, have chloroplasts, and these chloroplasts are believed to play a role in guard cell function. For instance, opening does not occur in plants that are albinos (mutants devoid of chloroplasts). One explanation is that sugar is produced by photosynthesis within the guard cells in the light, increasing the solute concentration within the cells; however, the opening occurs too rapidly to be explained fully by this mechanism.

Another possible clue is that low concentrations of carbon dioxide within the leaf induce stomatal opening, although how this occurs is not clear. Carbon dioxide lowers the pH by combining with water to produce carbonic acid (H_2CO_3) (see page 55), and this decrease in pH may be a factor. A long-held and still popular hypothesis is that removal of CO_2 during photosynthesis (with an increase in pH) increases the activity of the enzyme that breaks down starches to sugars within the guard cells and so increases the solute concentration. Remember that the concentration depends on the number of separate molecules or atoms in solution, not their size. However, opening occurs in guards cells that apparently contain no starch.

A third explanation is that the guard cells take up potassium and other ions by active transport (see pages 115 and 116 for a review of the mechanisms involved). The resulting increase in solute concentration then leads to uptake of water by osmosis. This hypothesis is well supported by experimental data, but it

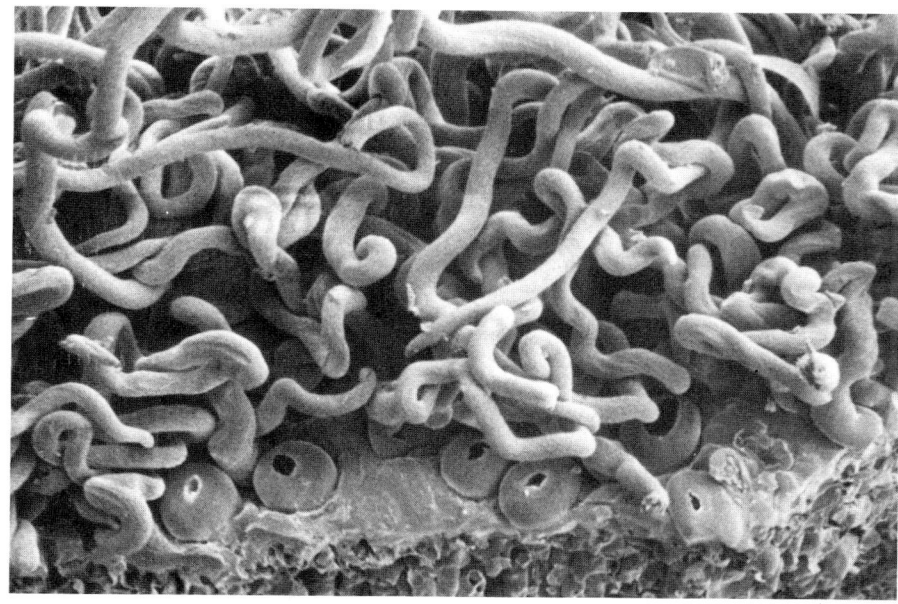

25 µm

does not explain the known effects of light and carbon dioxide. So this area of research in plant physiology remains controversial.

TRANSPORT OF SUGARS

Translocation is the process by which the products of photosynthesis are transported from the leaves and other photosynthetic areas and from storage organs to the nonphotosynthetic cells of the plant. These cells, like the cells of heterotrophic organisms, are dependent on the green, chloroplast-containing cells.

As we mentioned in the previous chapter, the tissues involved in translocation are collectively known as the phloem. If a tree is ringed—that is, if a cylinder of bark (which, you will recall, includes the phloem) is torn off—the roots and other tissues below the injury will die, thus indicating that the phloem does indeed transport the products of photosynthesis. Studies with radioactive tracers indicate that most if not all of the sugar transport in the phloem takes place in the sieve tubes. Another generally agreed upon fact is that the sieve-tube elements, and perhaps parenchyma and companion cells as well, must be alive for translocation to take place. It has been established that substances can move in opposite directions at the same time through the phloem, although it is not clear whether they move both ways at the same time in the same sieve tube.

The way in which sugars and other solutes move in the sieve tubes is not fully resolved. Translocation does not take place by simple diffusion; too much sugar is moved too fast. Rates of up to 300 centimeters per hour are common, whereas diffusion from a 10 percent solution would take place at less than 2 centimeters an hour.

(a)

1 mm

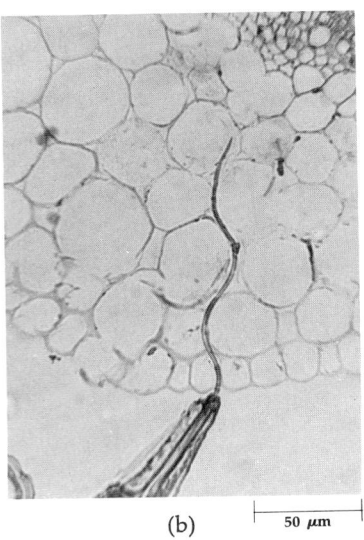

(b)

50 μm

25–5 *Assistance by aphids. (a) Aphids are very small insects that feed on plants, sucking out their juices; you probably have seen them on rosebushes. The aphid drives its sharp mouthpart, or stylet, like a hypodermic needle between the epidermal cells and then, (b) as the micrograph reveals, taps the contents of a single sieve-tube element. If the aphid is anesthetized, it is possible to sever the stylet and leave it undisturbed in the cell. The fluid will continue to exude through the stylet from the sieve tube for several days, and pure samples of the fluid flowing through the sieve tube can be collected for analysis without damaging the sieve tube or interfering with its function.*

Some investigators hypothesize that translocation is enhanced by cyclosis, or cell streaming, the cytoplasmic movements so clearly visible in many plant cells. Streaming movements can be observed in many, though not all, sieve-tube elements, and it would certainly appear to play at least a minor role in the movements of materials through cells. One attractive feature of this hypothesis is that it explains why substances can move in two directions. A principal objection is that the observed rate of cyclosis, though faster than diffusion, is slower than that observed in translocation.

Mass-flow Hypothesis

The most popular current explanation is the mass-flow hypothesis, which proposes that the solute moves as a result of differences in water potential. The principle underlying this theory can be illustrated by a simple physical model consisting of cells permeable only to water and connected by glass tubes. The first cell contains a solution in which there is a dissolved material, such as sucrose, and the second cell, to make the example as simple as possible, contains only water. When these interconnected cells are placed in distilled water, water will enter the first cell by osmosis. The entry of water will increase hydrostatic pressure within this cell and cause the solution in it to move along the tube to the second cell, where the pressure again builds up. If the second cell is connected with a third cell containing water or a sucrose concentration lower than that which is now in the second cell, the solution will flow from the second to the third by the same process, and so on indefinitely down a line of cells. This hypothesis is supported by the fact that it can account for the rate of movement. Moreover, it has been demonstrated that there are distinct gradients in the concentrations of sucrose and other sugars along the phloem tissues during the summer months.

Proponents of the mass-flow hypothesis agree that other processes must also play a role in translocation. For instance, under some conditions sugar is transported from cells of lesser hydrostatic pressure, or turgor, to cells of greater

Model of mass-flow hypothesis. Cells A and B, which are interconnected and are permeable to water, are placed in a bath of distilled water. Cell A contains a higher concentration of sucrose than cell B. Water enters cell A from the medium, increasing the hydrostatic pressure and pushing the solution to cell B. If B were connected to a third cell, C, with a still lower concentration of sucrose, as sieve-tube elements are connected in a series, hydrostatic pressure building up in B would push the solution on to C, and so on.

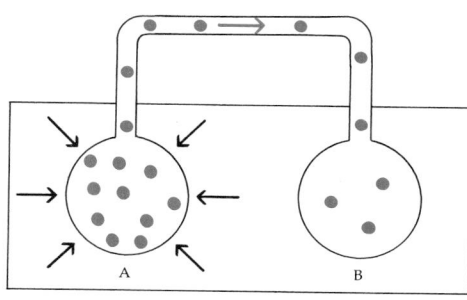

●   SUCROSE MOLECULE

→   MOVEMENT OF WATER

→   MOVEMENT OF SUGAR SOLUTION

turgor. Wilted leaves of the sugar beet, for example, continue to export sugar to the roots. Nutrients move from cotyledons and senescing petals into the plant body, presumably by active transport. In a photosynthesizing leaf, a parenchyma cell may contain about 3 percent sugar while the phloem contains 30 percent. Thus, it is clear that some form of active transport is also involved, at least in moving substances into and out of the phloem. The requirement for active transport would explain why phloem cells must be alive to be functional (whereas for mass flow they would not). Moreover, the energy requirements of active transport would account for the comparatively high rate of respiration often found in phloem.

Uses of Sugars

When the organic molecules reach the plant cells, they are metabolized in a variety of ways. Some are oxidized for energy. Others are converted to starch, or sometimes oils, and stored. Others are built up into cellulose or other structural polysaccharides. Still others are modified into amino acids and become the building blocks of enzymes and other proteins. Some may turn up in DNA, RNA, or as part of the chlorophyll molecule. These biochemical processes take place within the individual cells of the complex land plant, just as they take place within individual free-living plant cells and photosynthetic protists.

PLANTS, WATER, AND MINERALS

Plants require oxygen and carbon dioxide, which they obtain from the atmosphere. Using the energy of the sun, they produce sugars which are transported through the phloem to nonphotosynthetic cells. Plants also require water and minerals. Because minerals generally move into the plant body with the water and along the same passageways, we shall discuss the movement of water first.

As we noted previously, land plants lose water vapor during gas exchange. Largely as a consequence of this process (which is the major route of water loss), the quantity of water passing through a plant is enormous—far greater than that used by an animal of comparable weight. An animal requires less water because a great deal of its water recirculates through its body over and over again, taking the form, in vertebrates, of blood plasma and other fluids. In plants, more than 90 percent of the water that enters the roots is given off into the air as water vapor. A single corn plant needs 300 to 400 pounds of water, or 40 to 50 gallons, from bud to harvest, and an acre of corn requires 4 to 5 million pounds, or more than half a million gallons, of water a season. The loss of water vapor from the plant body is known as *transpiration.*

Water Movement

As we noted in the previous chapter, water enters the body of land plants almost entirely through the roots. During periods of rapid transpiration, water may be removed from around the roots so quickly that the soil in the vicinity of the roots becomes depleted; water will then move to some extent toward the roots through the soil. Also, the roots extend into moist soil by growing

25–7 *Guttation droplets on the edge of a wild strawberry leaf. Guttation, the loss of liquid water, is a result of root pressure. The water escapes through specialized pores located near the ends of the principal veins of the leaf. Guttation usually occurs at night when the air is moist.*

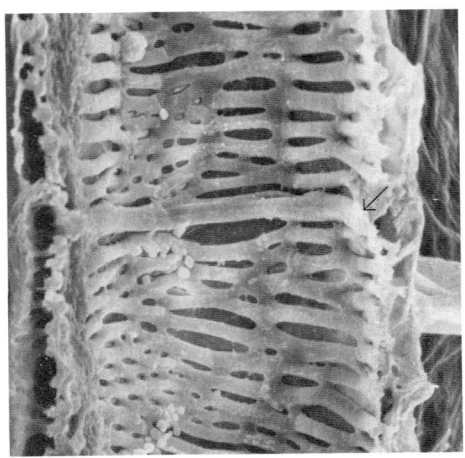

25–8 *Vessel from the stem of squash plant. The vessel is cut lengthwise so that the inner side is seen. There are numerous pits in the secondary walls. The boundary between the two elements, indicated by the arrow, is actually the collar left after the end walls are perforated. The small spheres are starch grains.*

into it. Under normal conditions, roots of apple trees elongate an average of ⅛ to ⅜ inch a day; roots of prairie grasses may grow more than ½ inch a day; and the main roots of corn plants average 2 to 2½ inches a day. Root cells, like other living parts of the plant, contain a higher concentration of salts and minerals than does soil water, and so water from the soil enters the roots by osmosis. Osmotic pressure is sufficient to move water a short distance up the stem; for example, the phenomenon known as guttation (Figure 25–7) is a consequence of osmosis. But how can water reach 60 feet high to the top of an oak tree, travel three stories up the stem of a vine, or move 300 feet up in a tall redwood? Active transport cannot be the answer, because vessel elements are dead cells.

One important clue is the observation that during times when the most rapid transpiration is taking place—which is, of course, when the flow of water up the stem must be the greatest—xylem pressures are characteristically negative (less than atmospheric pressure). The existence of negative pressure can be demonstrated readily. If you peel a piece of bark from a transpiring tree and make a cut in the xylem, no sap runs out. In fact, if you place a drop of water on the cut, the drop will be drawn in.

What is the pulling force? It is not simple suction, as the negative pressure might indicate. Suction simply removes air from a system so that the water (or other liquid) is pushed up by atmospheric pressure. But 1 atmosphere of pressure is only enough to raise water (against no resistance) about 32 feet at sea level, and many trees are much taller than 32 feet.

According to the now generally accepted theory, the explanation is to be found not in the properties of the plant but in the remarkable properties of water, to which the plant has become exquisitely adapted. As we pointed out in Chapter 7, in every water molecule, the pair of hydrogen atoms is linked to a single oxygen atom. The hydrogen atoms are also held to the oxygen atoms of the nearest water molecules by hydrogen bonds. This secondary attraction can produce a tensile strength of as much as 2,000 pounds per square inch in a thin column of water. In the leaf, water evaporates, molecule by molecule, from the cell walls as a consequence of the lower water potential of the intercellular spaces in the leaves. The water potential of the leaf cell falls, and water from the vessels or tracheids moves, molecule by molecule, into the leaf cell. But each molecule in the xylem vessel is linked to other molecules in the vessel, and they, in turn, are linked to others, forming one long, narrow, continuous strand of water reaching right down to a root tip. As the molecule of water moves into the leaf cell, it tugs the next molecule along behind it.

Because the diameter of the vessels is very small and because the water molecules adhere to the walls, even as they are cohering to one another, gas bubbles, which could rupture the column, do not usually form. The pulling action, molecule by molecule, causes the negative pressures observed in the xylem. The technical term for a negative pressure is *tension*, and this theory of water movement is known as the *cohesion-tension theory*.

The energy for the evaporation of water molecules—and thus for the movement of water and minerals through the plant body—is supplied not by the plant but directly by the sun.

-9 *Demonstration of root pressure in the cut stump of a plant. Uptake of water by the roots pushes the mercury upward in the column. Pressures of 3 to 5 atmospheres have been demonstrated by this method.*

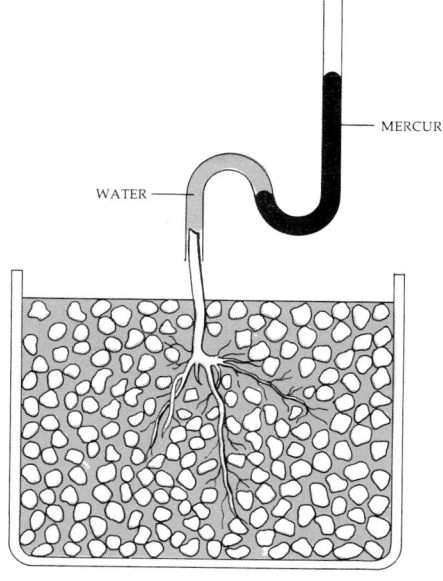

11 *Measurements in ash trees show that a rise in water uptake follows a rise in transpiration. These data suggest that the loss of water generates forces for its uptake.*

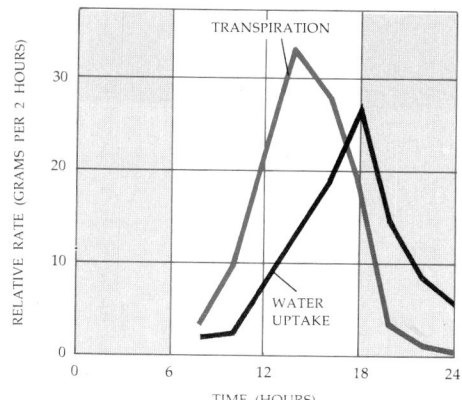

25–10 *(a) A simple model that illustrates the cohesion-tension theory. A piece of porous clay pipe, closed at both ends, is filled with water and attached to the end of a long, narrow glass tube also filled with water. The water-filled tube is placed with its lower end below the surface of a volume of mercury contained in a beaker.*

As water molecules evaporate from the pores in the pot, they are replaced by water "pulled up" through the narrow glass tube in a continuous column. As the water evaporates, mercury rises in the tube to replace it. (b) Transpiration from plant leaves results in sufficient water loss to create a similar negative pressure.

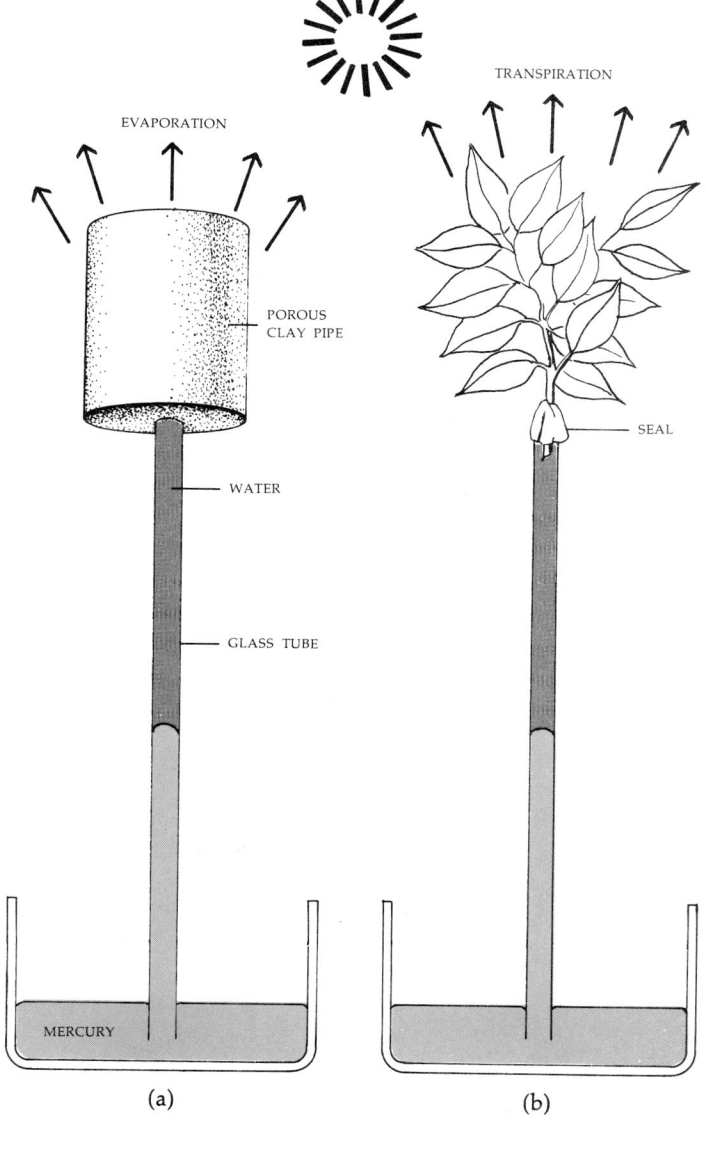

Minerals

In addition to carbon, hydrogen, and oxygen, which are obtained from water or from air, plants require a number of other elements. These elements, which are listed in Table 25–1, are obtained from minerals in the soil. A mineral is, by definition, a naturally occurring inorganic substance with a definite chemical composition. Some minerals, such as diamond, sulfur, and copper, consist of single elements. Others, such as quartz (SiO_2) and calcite ($CaCO_3$), are compounds. Rocks consist of one or more minerals, and soil characteristically contains mixtures of minerals.

Nitrogen

Although nitrogen is the most abundant element in air, most plants cannot use gaseous nitrogen and are dependent upon ammonium (NH_4^+) and nitrogen oxides—mostly nitrates (NO_3^-)—from the soil. Plant cells reduce the nitrates to ammonium, and the ammonium is then combined with carbon-containing compounds to form amino acids, nucleotides, chlorophyll, and other nitrogen-containing compounds.

Animals can make some amino acids from ammonium and organic carbon compounds, but they lack the biosynthetic pathways to make others—the so-called "essential amino acids." These they must obtain—as they obtain their carbohydrates—either directly or secondarily from green plants.

Determination of Mineral Requirements

The need of plants for elements such as nitrogen or magnesium can be deduced from an analysis of the molecules of which plant cells are composed. Other requirements are determined by studying the capacity of plants to grow in distilled water to which small amounts of various minerals are added. This sounds easier than it is. Sometimes, it has been found, a substance—chlorine, for example—is needed in such small amounts that it is almost impossible to set up experimental conditions that exclude it, and so it is difficult to prove that its absence is lethal.

25–12 (a) Plants incorporate inorganic nitrogen into organic compounds. This process is known as amination. A major pathway for the incorporation of nitrogen is the combination of ketoglutaric acid with ammonium (NH_4^+) to form the amino acid glutamic acid. NAD_{red} provides energy and hydrogen atoms for the second step of this reaction. Ketoglutaric acid is one of the intermediates in the Krebs cycle. (b) Transamination is the process by which an amino group, such as that in glutamic acid, is transferred to another carbon skeleton to form a new nitrogen-containing compound, such as the amino acid alanine.

Table 25–1 *A Summary of Minerals Essential to Plants*

Element	Form in which absorbed	Approximate concentration in whole plant (dry weight)	Some functions*
MACRONUTRIENTS			
Nitrogen	NO_3^- (or NH_4^+)	1–3%	Component of amino acids, proteins, nucleotides, nucleic acids, chlorophyll, and coenzymes
Potassium	K^+	0.3–6%	Involved in amino acid and protein synthesis; involved in turgor pressure changes
Calcium	Ca^{2+}	0.1–3.5%	Combines with pectin in cell walls; decreases cell permeability
Phosphorus	$H_2PO_4^-$ or HPO_4^{2-}	0.05–1.0%	Formation of "high-energy" phosphate compounds (ATP and ADP); component of nucleic acids and phospholipids
Magnesium	Mg^{2+}	0.05–0.7%	Component of the chlorophyll molecule; activator of many enzymes
Sulfur	SO_4^{2-}	0.05–1.5%	Component of 2 of the 20 amino acids and of coenzyme A
MICRONUTRIENTS			
Iron	Fe^{2+}, Fe^{3+}	10–1,500 ppm†	Required for chlorophyll synthesis; component of cytochromes and ferredoxin; activator of some enzymes
Chlorine	Cl^-	100–300 ppm	Probably essential in photosynthesis in the reactions in which oxygen is produced
Copper	Cu^{2+}	2–75 ppm	Activator of some enzymes
Manganese	Mn^{2+}	5–1,500 ppm	Activator of some enzymes; needed for oxygen production in photosynthesis
Zinc	Zn^{2+}	3–150 ppm	Activator of some enzymes
Molybdenum	MoO_4^{2-}	Trace	Needed for nitrogen metabolism
Boron	BO^{3+} or $B_4O_7^{2-}$ (borate or tetraborate)	2–75 ppm	Influences Ca^{2+} uptake and utilization; needed for RNA synthesis in roots

* All major ions play a role in osmosis and in the distribution of positive and negative charges. All also can affect the configuration of enzymes and other proteins since this is determined, to a large extent, by attractions between differently charged areas of the molecule or molecules.
† Parts per million.

As a result of such studies, six elements that plants require in relatively large amounts (macronutrients) and seven that are needed in smaller quantities (micronutrients, or trace elements) have been identified. As you can see in Table 25–1, nitrogen, which is absorbed by higher plants in the form of nitrogen salts, is used in the largest quantities. It is inaccurate, however, to say that nitrogen is "more essential" than the other elements, since the complete deprivation of *any* material that a plant requires, even in minute quantities, is fatal to the plant.

You might anticipate that organisms make use of what is most readily available, as indeed they seem to have done when life originated from elements in the gases of the primitive atmosphere. But the table reveals some findings that

25–13 *Most plants obtain their nitrogen in the form of inorganic salts taken up with the soil water, but for some unusual species, such as the sundew shown here, insect bodies are the chief nitrogen source. The plant has a rosette of 8 to 15 club-shaped leaves, each about 1 inch long and not quite ½ inch wide at the tip. The upper exposed surface of the leaf is covered with hairlike structures, the tentacles, each about ⅛ inch long, enlarged at the tip. A crystal droplet of sticky secretion surrounds each enlargement. These droplets trap insects or other comparably small animals and then the tentacles secrete enzymes that digest the prey.*

25–14 *Horsetails (Equisetum) incorporate silicon into their cell walls, making them indigestible to most herbivores. One species in colonial America was used for scouring pots and pans.*

you might not expect. Sodium, for instance, which is one of the most abundant of the elements, is not required at all by plants except for some salt-marsh species. The fact that plants did not find a use for sodium appears even stranger when you consider that sodium is vital to the function of animals. In the seas, where both plants and animals seem to have had their origins, sodium is the most abundant element and is far more readily available than potassium, which it closely resembles in its essential properties. Similarly, although silicon and aluminum are almost always present in large amounts in soils, few plants require silicon and none requires aluminum. On the other hand, all plants need molybdenum, which is relatively rare.

Comparisons of the mineral content of a plant cell with the mineral content of the water in which it grows show that there can be marked differences in the concentrations of various components. Thus, we can conclude that substances from the soil do not diffuse passively into the root cells but are carried across cell membranes, often against the diffusion gradient, by energy-requiring processes, facilitated by diffusion and active transport.

Diagrammatic cross section of a root, showing the two pathways of uptake of water and minerals. Along pathway A, water moves by osmosis and salts by active transport through the cellular membranes of a series of living cells. Along pathway B, the water flows through the cell walls and intercellular spaces and the solutes flow with the water or by diffusion. Notice how the Casparian strip blocks off pathway B all around the vascular cylinder of the root. In order to pass the Casparian strip, the solutes must be transported through the cell membranes of the endodermal cells.

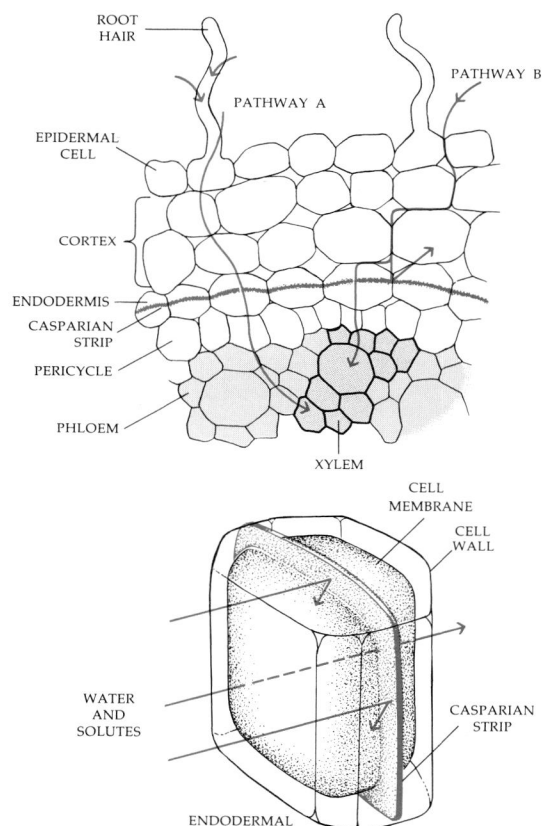

ROOT HAIR
PATHWAY B
PATHWAY A
EPIDERMAL CELL
CORTEX
ENDODERMIS
CASPARIAN STRIP
PERICYCLE
PHLOEM
XYLEM

CELL MEMBRANE
CELL WALL
WATER AND SOLUTES
CASPARIAN STRIP
ENDODERMAL CELL

Figure 25–15 shows the pathways of water and nutrients through the roots to the xylem. Along pathway A, water moves by osmosis and nutrients absorbed by epidermal cells and their root hairs move from cell to cell through cell membranes and through plasmodesmata. Water and solutes can flow freely along pathway B, moving between cells and through cell walls, but they are blocked at the endodermis, as shown, by the Casparian strip. Only those solutes taken up by the cells of the endodermis can reach the xylem. In this way, the plant regulates its uptake of mineral elements. The Casparian strip also prevents the leakage from the conducting tissues of the minerals pumped into the xylem.

By growing plants in solutions that contain the required minerals in the form of radioactive isotopes, it has been shown that, once they enter the root cells, the mineral ions are carried up the plant by the flow of water in the xylem. The various cells of the plant then select from the water in the xylem the elements they require for their maintenance and growth.

SUMMARY

Photosynthesis, which takes place mostly in the leaf, provides the plant with organic carbon compounds that supply energy to the plant's nonphotosynthetic cells and also the carbon skeletons on which all other organic molecules are built. In the course of photosynthesis, carbon dioxide moves into the photosynthetic cells of the leaf from intercellular air spaces, present particularly in the spongy mesophyll layer, and oxygen moves out of the cells.

Diffusion of gases in and out of the leaf is regulated by openings called stomata. The stomata are opened and closed by guard cells as a consequence of changes in turgor pressure.

The movement of organic carbon compounds (mostly sucrose and other sugars) from the photosynthetic parts of the plant is known as translocation. Translocation over long distance takes place in the phloem. According to the mass-flow hypothesis, sugars are carried along by the flow of water from an area of higher solute concentration in the sieve tubes of the leaves to areas of lower solute concentration in the sieve tubes of the roots. Some authorities believe that cyclosis also plays an important or, perhaps, the major role in translocation. In either case, active, energy-requiring processes are also involved. These processes, which take place in the phloem cells, move sugar into or out of the phloem system against the concentration gradient.

Plants require large amounts of water, most of which is lost to the atmosphere by transpiration. Water enters the plant through the roots and makes its way to the leaves by means of the conducting cells of the xylem. The current and widely accepted theory of how water moves through vessels and tracheids is the cohesion-tension theory. According to this theory, water within the vessels is under negative pressure because the water molecules cling together in continuous columns as they are pulled by evaporation from above.

A total of 16 elements are known to be required by higher plants for normal growth. Of these, carbon, hydrogen, and oxygen are derived from air and water. The rest are taken up in the form of minerals from the soil and selectively transported through root cells to the xylem.

From these 16 elements, plants obtain all of their nutrient requirements.

QUESTIONS

1. Define the following: photosynthesis, translocation, transpiration, macro-nutrient, tension, cohesion-tension theory, mass-flow hypothesis, Casparian strip, trace element.

2. Which of the properties of water discussed in Chapter 3 are important to phenomena described in this chapter? How have plants adapted to these properties?

3. Most plants cannot live in areas in which there is a high salt concentration, such as salt marshes. Explain.

4. How is it physically possible for an *increase* in turgor of the guard cells to *open* the stomata? Consider what would happen if you partly inflated a cylindrical balloon, applied a strip of adhesive tape along its length, and then inflated it further. What does the experiment suggest about the role of wall thickenings of stomata?

5. In Figure 25–10, why doesn't air enter the tube through the top of the enclosed porous pipe or the leaf, even though water vapor can easily escape? How can the porous pipe (and, by analogy, the leaf) be permeable to air or water but effectively impermeable to an air-water interface? Think about the consequences of the high surface tension of water (Chapter 3).

6. Consider a tree transpiring most rapidly at midday and an investigator with a sensitive instrument for measuring changes in the diameter of the trunk. If water is pulled up from the top (cohesion-tension theory), what changes in diameter should be observed from night to day? (The change was, in fact, one of the early bits of evidence for the theory.)

Chapter 26

Reproduction, Development, and Growth

Unlike the higher animals, whose reproduction is almost exclusively sexual, many higher plants reproduce both sexually and asexually. (Asexual reproduction is often referred to as vegetative reproduction, even when it occurs among animals.) Organisms produced by asexual reproduction are genetically identical to their single parent, whereas organisms produced by sexual reproduction, which involves meiosis and fertilization, are different from both parents.

There are many forms of asexual reproduction among plants. One of the most familiar occurs by means of horizontal stems growing either above ground (runners) or below ground (rhizomes). Strawberries are a common example of plants that propagate by runners, as are spider plants (walking anthericum), commonly grown as hanging plants. Plants that reproduce by rhizomes include potatoes, many flowering garden perennials, such as lilies-of-the-valley, irises, and dahlias, and the sod-forming grasses of lawns and pastures. Both runners and rhizomes develop adventitious roots.

Many members of the lily family, which includes onions and tulips as well as lilies, reproduce asexually, by bulbs. Some species of plants with arching stems, such as raspberries, develop new roots from stem tips that touch the soil, and new plants may form if the stem is subsequently broken, separating it from the parent plant. The leaves of some plants, such as African violets, develop adventitious roots when they are detached from the parent plant and may give rise to new individuals in that way. Some species of *Kalanchoe* produce plantlets in the margins of leaves which later drop to the ground and develop into separate plants. If a dandelion is injured or broken near its upper portion, a callus forms that plugs the wound and eventually two to five new plants grow from this callus tissue. Thus both suburban householders and grazing animals have highly beneficial effects on the dandelion population.

The capacity of many species of plants to reproduce asexually has been exploited by man in developing domestic varieties of plants for food or ornamental use. Such plants are, of course, genetically identical to the parent stock, and so vegetative reproduction is a way of preserving uniformity. Many plants are reproduced by stem cuttings, which simply involves sticking young stems in the ground and protecting them from drying air until adventitious roots appear.

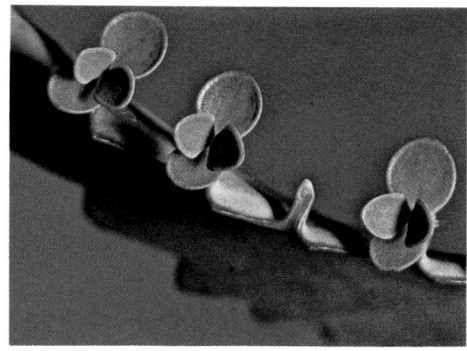

5–1 *Tiny plantlets grow along the leaf margins of* Kalanchoe. *When mature, they drop to the soil and take root.*

26–2   (a) *Wild strawberry plants reproduce asexually by means of aboveground stems, or runners. Roots and leaves develop at every second node along the modified stems. These plants also form flowers and reproduce sexually. (b) Violets also reproduce both sexually and* asexually. *The larger flowers are cross-pollinated by insects, and the windborne seeds are carried some distance from the parent plant. The smaller flowers, closer to the ground, are self-pollinated and never open. Seeds from these flowers drop close to the parent plant and produce* plants *that are genetically similar to the parent and so, presumably, apt to grow successfully near the parent. Creeping underground stems (rhizomes) also eventually produce a new series of genetically identical plants right next to the parent.*

RUNNER

(a)

SELF-POLLINATING FLOWER

RHIZOME

(b)

26–3   *Among the common harbingers of spring is the white-flowered bloodroot (Sanguinaria canadensis), a member of the poppy family, which derives its name from the orange-colored juice that flows from a broken stem.*

Rooting can often be facilitated by hormone treatment (see page 482). Another artificial form of plant propagation is grafting, in which a stem cutting is attached to the main stem of a rooted, woody plant. Most fruit trees and roses are propagated in this way.

Many economically important plants are sterile and can only be propagated vegetatively; these include pineapples, bananas, seedless grapes, navel oranges, and numerous ornamental plants.

SEXUAL REPRODUCTION: THE FLOWER

The flower is the organ of sexual reproduction of the angiosperms. Unlike the reproductive organs of animals, which are permanent structures that develop in the embryo, flowers are transitory, developing seasonally from modified shoot tips. Each consists of four sets of floral appendages, which grow in spirals or whorls; each floral part, evolutionarily speaking, is a modified leaf. The outermost parts of the flower are the *sepals,* which are commonly green and obviously leaflike in structure. The sepals, collectively known as the calyx, enclose and protect the flower bud. Next are the *petals,* collectively called the *corolla;* these are also usually leaf-shaped but are often brightly colored. They advertise the presence of the flower among the green leaves, attracting insects or other animals that visit flowers for their nectar (a sugary liquid) or for other edible substances and so carry pollen from flower to flower.

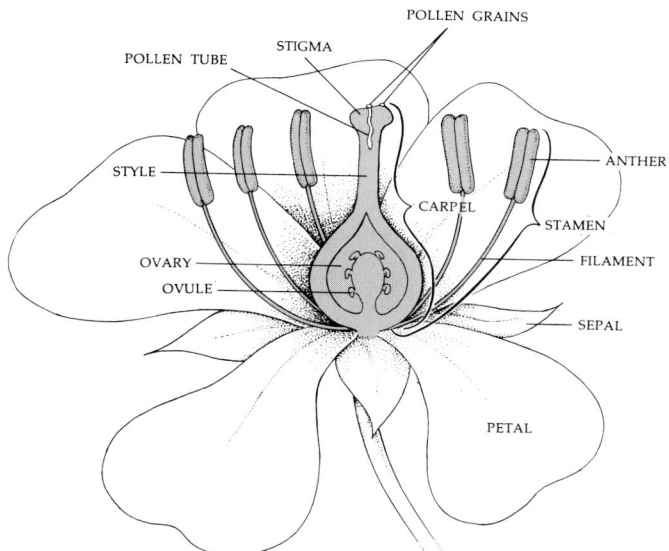

6–4 *Structure of a flower. The sex organs are shown in color. A flower that possesses both stamens and carpel, such as this flower, is known as a perfect flower.*

6–5 *Corn (Zea mays) is monoecious; that is, it has separate male and female flowers but both are borne on the same plant. The tassels are male (pollen-producing) flowers. Each thread of silk is the combined stigma and style of a female flower.*

Within the corolla are the *stamens*. Each stamen consists of a single elongated stalk, the *filament*, and at the end of the filament, the *anther*. The pollen grains, which are immature male gametophytes,* are released from the anther when ripe—often in large numbers—usually through narrow slits or pores.

The centermost appendages of the flower are the *carpels*, which contain the female gametophytes. Typically a carpel consists of a *stigma*, which is a sticky surface specialized to receive the pollen; a slender stalk, the *style*, down which the pollen tube grows; and a hollow base, the *ovary*. Within the ovary are the *ovules*, each of which encloses a female gametophyte with a single egg cell. When the egg cell is fertilized, the ovule develops into a seed.

In some species, flowers are either male (staminate) or female (pistillate). Male and female flowers may be present on the same plant, as in corn, squash, oaks, and birches, or on different plants, such as the tree of heaven (*Ailanthus*), the date palm, and the American mistletoe. Species in which both male and female flowers are borne on the same plant are known as monoecious ("in one house"); species in which the male and female flowers are on separate plants are known as dioecious ("in two houses").

Evolution of the Flower

As we noted in Section 2, sexual reproduction confers certain advantages on organisms. It also presents problems, one of the most crucial of which is finding a mate. Unlike animals, land plants cannot move about. The flower evolved as a device by which plants induced animals to transport their sperm to the egg cells.

The early gymnosperms from which the angiosperms evolved were probably wind-pollinated, as are modern gymnosperms. And like the modern gymno-

* For a review of the concept of alternation of generations, see page 338.

26–6 *Primitive flowers are believed to have been radially symmetrical, with numerous separate floral parts, such as this magnolia.*

sperms, the female gametophytes probably exuded droplets of sticky sap in which pollen grains were caught and drawn into the ovule. Insects, probably beetles, feeding on plants must have come across the protein-rich pollen grains and the sticky, sugary droplet. As they began returning regularly to these newfound food supplies, they inadvertently carried pollen from plant to plant.

Beetle pollination must have been more efficient than wind pollination for some species because, clearly, selection began to favor plants with insect pollinators. The more attractive the plants were to the beetles, the more frequently they would be visited and the more seeds they would produce. Any chance variations that made the visits more frequent or that made pollination more efficient thus offered immediate advantages; more seeds would be formed, and more offspring would survive. Nectaries evolved which lured the pollinators. Plants developed white or brightly colored flowers that called attention to the nectar and other food supplies. The carpel, originally a leaf-shaped structure, became folded on itself (page 354), enclosing and protecting the ovule from hungry pollinators. By the beginning of the Cenozoic era, some 65 million years ago, the first bees, wasps, butterflies, and moths had appeared. These are long-tongued insects for which flowers are often the only source of nutrition. From this time onward, flowers and insects had a profound influence on one another's history, each shaping the other as they evolved together.

A flower that attracts only a few kinds of animal visitors and attracts them regularly has an advantage over flowers visited by more promiscuous pollinators: Much less of its pollen is liable to be lost on a plant of another species. Similarly, it is advantageous to the insect to have a private food supply, relatively inaccessible to others. Most of the distinctive features of modern

26–7 *Examples of two large families of flowers. (a) Ox-eye daisies (Chrysanthemum leucanthemum), representative composites. The ray flower is often sterile. (b) Epipactus gigantea, stream orchids. Composites (Asteraceae) are the second largest family of flowering plants, with some 13,000 species. The Orchidacea, with about 20,000 species, is the largest family. The composites are dicots; the orchids, monocots.*

(a)

(b)

26–8 (a) Diagram of the organization of the head of a composite (family Asteraceae). The individual flowers are subordinated to the overall effect of the head, which acts as a large single flower in attracting insects. The internal structure of the disk flower is indicated in color. (b) The parts of an orchid flower. The lip is a modified petal that serves as a landing platform for insects.

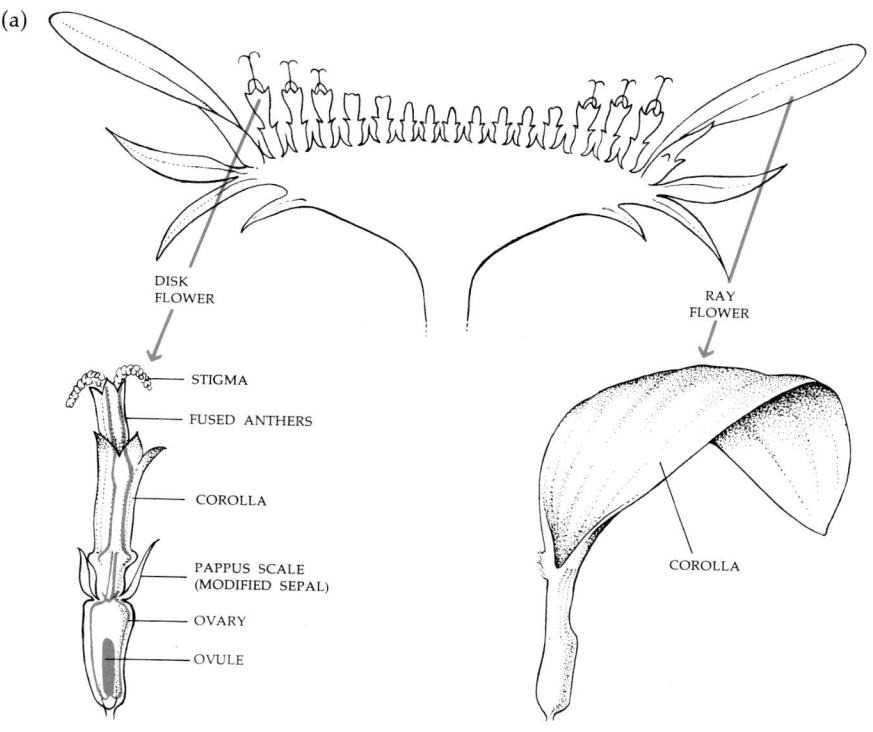

(a)

DISK FLOWER

RAY FLOWER

STIGMA

FUSED ANTHERS

COROLLA

PAPPUS SCALE (MODIFIED SEPAL)

OVARY

OVULE

COROLLA

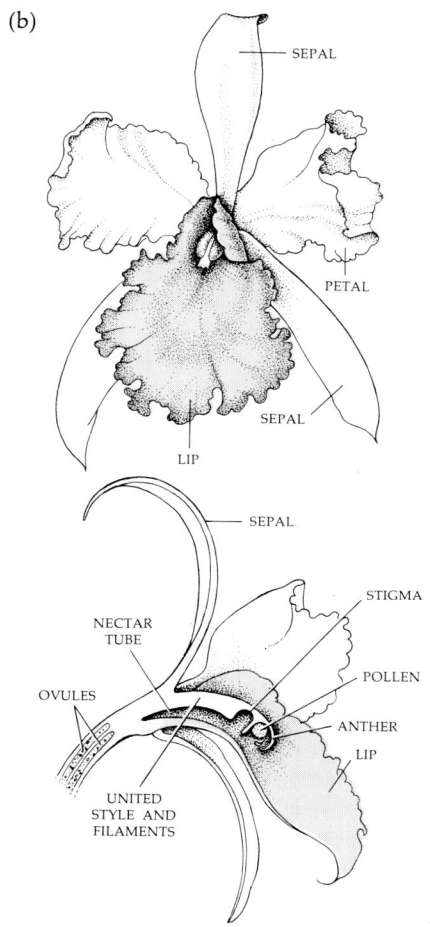

(b)

SEPAL

PETAL

SEPAL

LIP

SEPAL

NECTAR TUBE

STIGMA

OVULES

POLLEN

ANTHER

LIP

UNITED STYLE AND FILAMENTS

flowers are special adaptations that encourage constancy of particular pollinators. The varied colors and odors are "brand names" for guiding pollinators. The diverse shapes such as deep nectaries and complex landing platforms that are found, for example, in orchids, snapdragons, and irises, represent ways of excluding indiscriminate pollinators.

The most primitive flowers are believed to have resembled the modern magnolia, shown in Figure 26–6, or the hepatica, which have numerous floral parts, each clearly separate from the other. By comparing this type of flower with some of the more specialized ones, such as a composite or an orchid, it is possible to see four main trends in flower evolution:

1. The number of floral parts has been reduced. Most specialized flowers have few stamens and few carpels, and these are present in a definite number, depending on the species.

2. The floral parts, particularly carpels and petals, have become fused, sometimes elaborately so.

3. In the primitive flower, the floral parts arise at the base of the ovary (see Figure 26–4). In the more advanced flowers, the free portions of the floral parts are above the ovary; this is an important adaptation by which the ovules are protected from foraging insects.

4. The radial symmetry of the primitive flower has given way, in more advanced flowers, to bilaterally symmetrical forms that precisely accommodate particular insects and other pollinators and exclude intruders.

26–9 (a) *Longhorn beetle eating thistle pollen.*
(b) *Brazilian bronze hummingbird visiting a fuschia.* (c) *Masarid wasp in a penstemon flower. These wasps often spend the night in these tubular flowers.*
(d) *Honeybee rummaging in a cosmos.*
(e) *Acrocerid ("small-headed") fly on an iris. Note the pollen on its back.* (f) *Inornate ringlet butterfly.* (g) *Syrphid fly receiving pollen from an* Epipactus gigantea *orchid. This species is the only pollinator for this flower.*

5–10 *Unlike most angiosperms, grasses and most common species of trees are wind-pollinated. The staminate flowers of the turkey oak (Quercus laevis) hang in catkins, which are flexible, thin tassels several inches long. These catkins are whipped by passing breezes, and the pollen, when ripe, is thrown out and caught by the wind.*

Pollination

For most flowering plants, a new cycle of life begins when a grain of pollen—brushed from the body of a foraging insect—comes into contact with the stigma of a flower of the same species. By the time this pollen grain is released from its parent flower, it consists in many angiosperms of three haploid nuclei (two sperm nuclei and a so-called "vegetative" nucleus), a small amount of dense cytoplasm, and a tough wall.

Pollen is commonly produced in great quantities; the probability of any particular pollen grain reaching the stigma of an appropriate flower is very small. The pollen grain contains its own nutrients and has so tough an outer coating that intact grains have been found in peat bogs thousands of years old.

In lower plants, you will recall, there is a distinct cycle of alternation of generations in which the sporophyte produces spores that produce gametophytes that produce gametes, with gametophyte and sporophyte having separate existences. In the course of plant evolution, the gametophyte stage has been steadily reduced and, in the angiosperms, all that remains of the male gametophyte is the tough, tiny pollen grain and the pollen tube. The sperm cells are the gametes.

Once on the stigma, the pollen grain germinates, and a pollen tube grows down through the style into an ovule. The ovule holds the female gameto-

–11 *Pollen grains. The walls of the pollen grain protect the male gametophyte on the journey between the anther and the stigma. These outer surfaces, which are remarkably tough and resistant, are often elaborately sculptured. As you can see, the pollen grains of different species are distinctly different: (a) a chrysanthemum (spiny pollen grains such as these are common among composites), (b) a morning glory, (c) a lily, and (d) a horse chestnut (each grain contains three slit-like pores, a characteristic of dicots).*

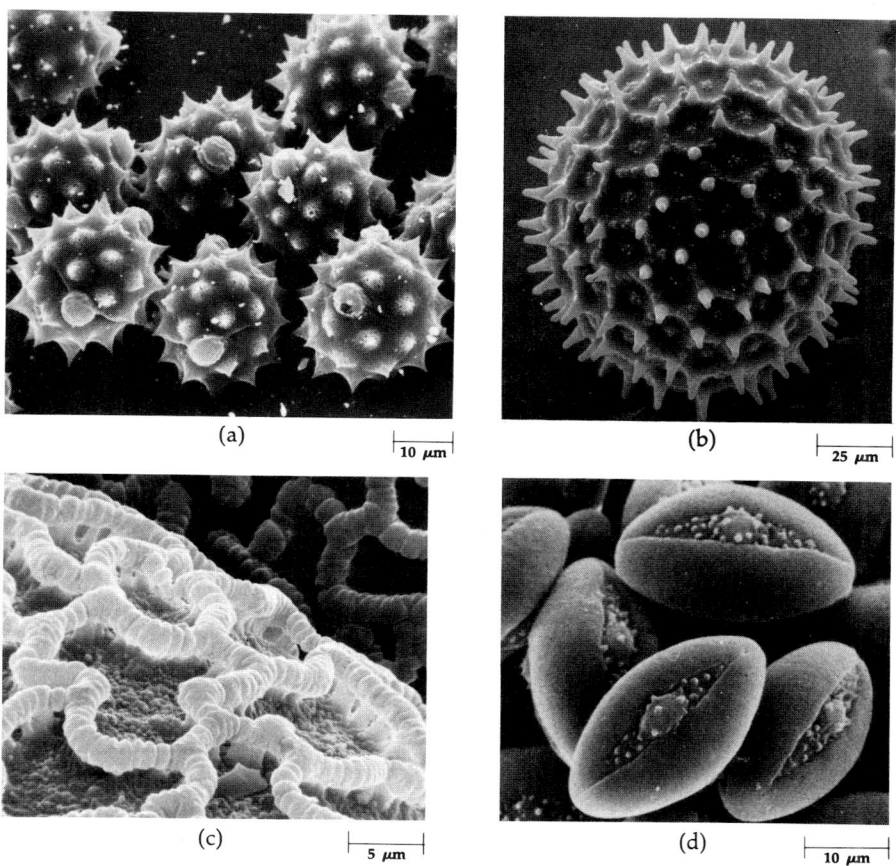

(a)   10 μm

(b)   25 μm

(c)   5 μm

(d)   10 μm

26–12 *Cross section of a pollen tube. Two lobes of the vegetative nucleus are visible and at the top is the sperm nucleus, surrounded by a small amount of cytoplasm and a cell membrane. Numerous mitochondria are visible, as are several plastids. The reproductive (sperm) cell is actually a cell within a cell.*

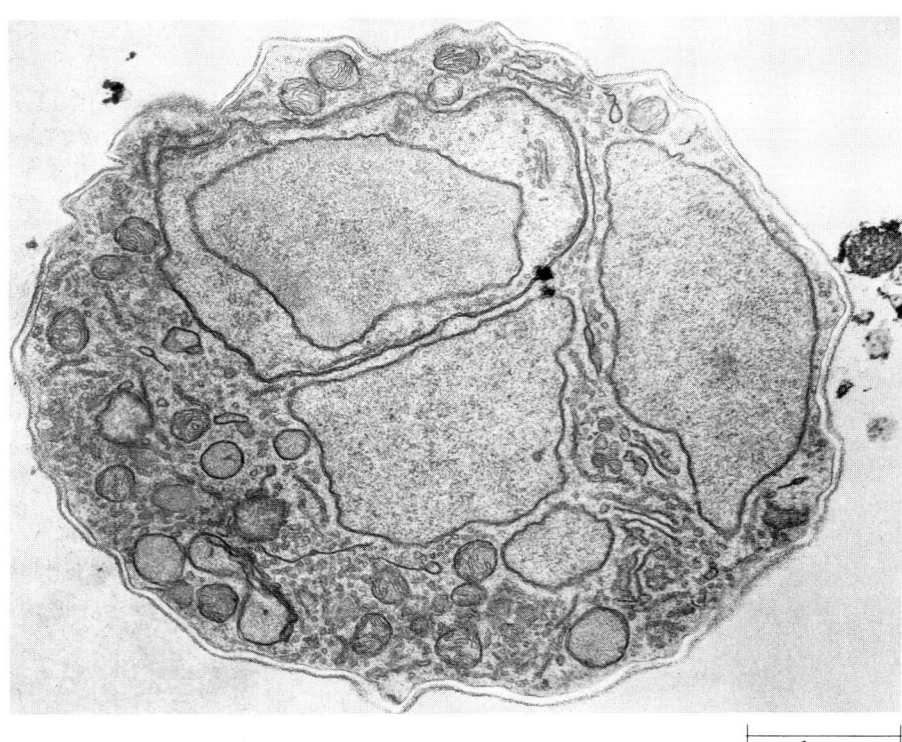

1 μm

26–13 *Fertilization in angiosperms. The pollen tube of the male gametophyte, or pollen grain, grows down through the style and enters the ovule, in which the female gametophyte has developed to a seven-cell stage. One of the sperm nuclei unites with the egg cell, forming the zygote. The other sperm nucleus fuses with the two polar nuclei that are present in a single large cell (which in the drawing fills most of the ovule). From the resulting triploid (3n) cell, the endosperm will develop. The carpel shown here contains a single ovule.*

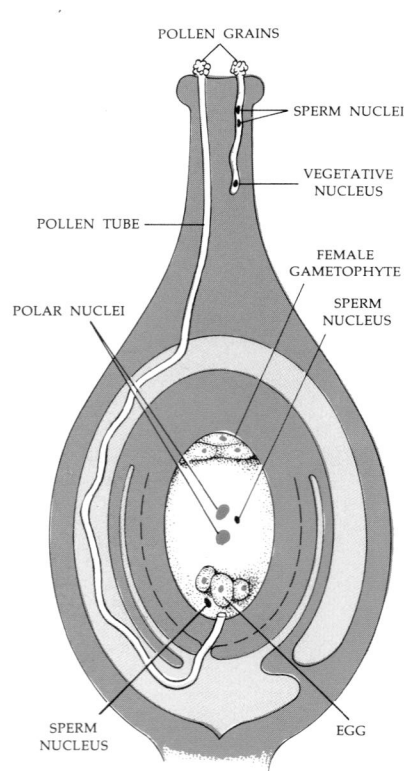

POLLEN GRAINS

SPERM NUCLEI

VEGETATIVE NUCLEUS

POLLEN TUBE

FEMALE GAMETOPHYTE

POLAR NUCLEI

SPERM NUCLEUS

SPERM NUCLEUS

EGG

phyte, which has become similarly reduced in size in the course of evolution. In many species it consists of seven cells with a total of eight haploid nuclei (a large central cell has two haploid nuclei). One of the smaller cells, containing a single haploid nucleus, is the egg.

One sperm nucleus moves down the pollen tube and unites with the egg. This fertilized cell, the zygote, develops into the embryo plant. The second sperm nucleus unites with the two polar nuclei (so called because they move to the center from each end, or pole, of the gametophyte). From the 3n cell, a specialized tissue called the *endosperm* develops, which completely surrounds and nourishes the embryo. These extraordinary phenomena of "double fertilization" and triple fusion take place, in all the natural world, only among the flowering plants.

THE EMBRYO

The zygote (the fertilized egg cell) divides mitotically, and as the embryo grows, its cells begin to *differentiate*—become different from one another—and the embryo begins to take on a characteristic shape, a process known as *morphogenesis.*

Growth Areas

In the earliest stages of embryonic growth, cell division takes place throughout the body of the infant plant. As the embryo grows older, however, the addition of new cells becomes gradually restricted to certain parts of the plant body: the _apical meristems_ of the root and the shoot. During the rest of the life of the plant, all the primary growth—which chiefly involves the elongation of the plant body—originates in these meristems.

The existence of such meristematic areas, which add to the plant body throughout the life of the plant, is one of the principal differences between plants and animals. Higher animals stop growing when they reach maturity, although the cells of certain "turnover" tissues, such as nails, hair, skin, or the lining of the intestine, continue to divide.

Plants, however, continue to grow during their entire life span. Growth in plants is the counterpart, to some extent, of mobility in animals. Plants "move" by extending their roots and shoots, both of which involve changes in size and form. By growth, a plant modifies its relationship with the environment, turning toward the light and extending its roots. The sequence of growth stages in plants thus corresponds to a whole series of motor acts in animals, especially those concerned with the search for food and water. In fact, growth in plants serves many of the functions that we group under the term "behavior" in animals.

The Seed and the Fruit

As the embryo develops, the petals and stamens of the parent flower fall away, and the wall of the ovary develops into the fruit. The seed is the ovule (which comes from the maternal sporophyte), with the $3n$ endosperm and the young sporophyte growing within it. In a peach, for instance, which contains only one ovule per ovary, the skin, the succulent edible portion of the fruit, and the stone are three distinctive layers of the wall of the matured ovary (the base of the carpel). The almond-shaped structure within the stone is the seed. In a pea, the pod is the mature ovary wall and the peas are the seeds (the ovules and their

-14   *Seeds. (a) In dicots such as the common bean* (Phaseolus vulgaris), *the endosperm is digested as the embryo grows and the food reserve is stored in the fleshy cotyledons. (b) In corn and other monocots, the single cotyledon, known as the scutellum, absorbs food reserves from the endosperm.*

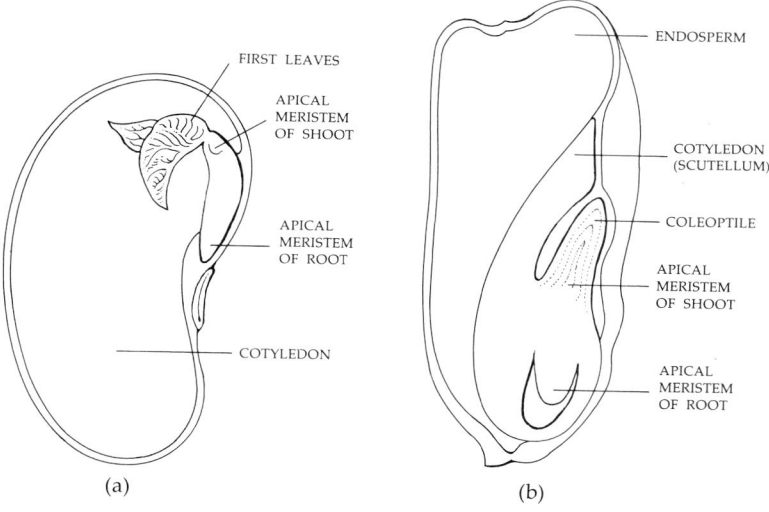

(a)   (b)

contents). A raspberry is an aggregate of many fruits from a single flower, each fruit containing a single seed and each formed from a separate carpel.

When the seed matures, the embryo enters a period of dormancy. The seed coat, which develops from the outer layers of the ovule, thickens and hardens, and the seed, which includes the embryo and its stored food, falls from the parent plant. Some of the many mechanisms for seed dispersal are shown on page 356.

SEED GERMINATION

The seeds of most wild plants require a period of dormancy before they will germinate. This genetic requirement ensures that the seed will "wait" at least until the next favorable growth period. Seeds can remain dormant and yet viable—with the embryo in a state of suspended animation—for hundreds of years. The record for dormancy, as far as is known, has been set by some seeds of Arctic tundra lupine recently found in the Yukon in a lemming burrow. Deeply buried in the permanently frozen silt, they are estimated to be at least 10,000 years old. But when a sample was planted, the seeds germinated within 48 hours.

The seed coat apparently plays a major role in maintaining dormancy. In some species, the seed coat seems to act primarily as a mechanical barrier, preventing the entry of water and gases, without which growth is not possible. In these cases, growth is promoted by the seed coat being worn away in various ways—such as being washed by rainfall, abraded by sand or soil, burned away by a forest fire, or partially digested as it passes through the digestive tract of a bird or other animal. In other species, dormancy seems to be maintained chiefly by the presence of chemical inhibitors in the seed coat. These inhibitors undergo chemical changes in response to various environmental factors, such as light or prolonged cold or a sudden rise in temperature, which neutralize their effects, or they may be washed or eroded away. Eventually, the embryo is released.

The dormancy requirement in seeds apparently evolved only recently, geologically speaking, among groups of plants subjected to the environmental stress of increasing winter cold characteristic of the most recent Ice Age. By this time—only 1½ to 2 million years ago—the angiosperms were already a highly diversified group, and different populations responded to these pressures in different ways, which explains why even closely related plants have different mechanisms for maintaining and breaking dormancy.

PRIMARY GROWTH

During dormancy, the seed contains very little moisture (only about 5 to 10 percent of its total weight). Dormancy ends with a massive entry of water (imbibition) into the seed. The seed coat ruptures and the embryo sporophyte emerges.

Primary growth, which begins immediately, involves the formation and elongation of stems, roots, and branches and the differentiation of the conduction tissues and other specialized tissues of the young shoot and root. All primary growth originates in the apical meristems of the shoot and root.

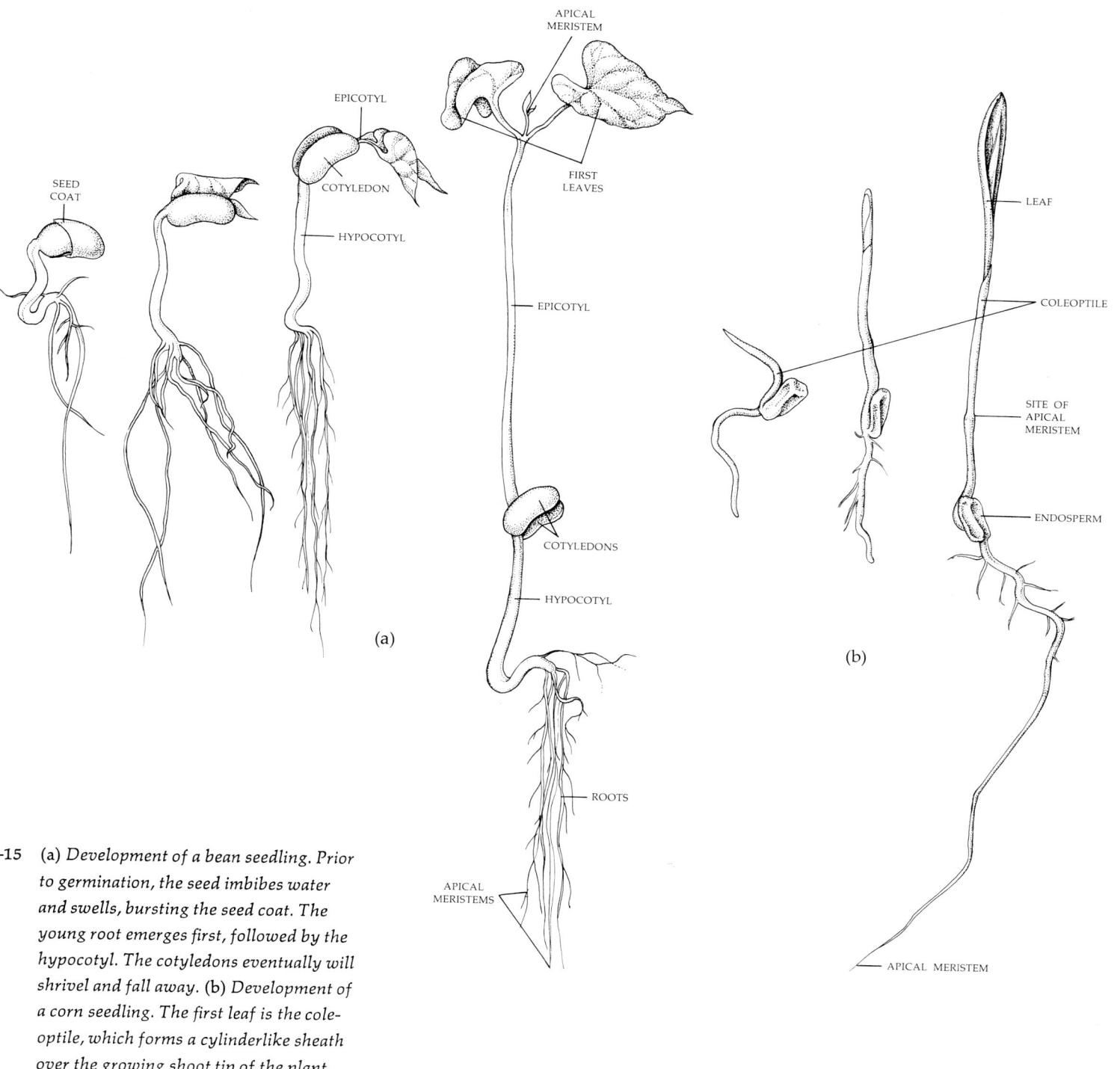

SEED
COAT

EPICOTYL

COTYLEDON

HYPOCOTYL

APICAL
MERISTEM

FIRST
LEAVES

EPICOTYL

COTYLEDONS

HYPOCOTYL

ROOTS

APICAL
MERISTEMS

(a)

LEAF

COLEOPTILE

SITE OF
APICAL
MERISTEM

ENDOSPERM

APICAL MERISTEM

(b)

–15   (a) *Development of a bean seedling. Prior
to germination, the seed imbibes water
and swells, bursting the seed coat. The
young root emerges first, followed by the
hypocotyl. The cotyledons eventually will
shrivel and fall away.* (b) *Development of
a corn seedling. The first leaf is the coleoptile, which forms a cylinderlike sheath
over the growing shoot tip of the plant.
Typically, the shriveled endosperm, with
the scutellum buried in it, is still present
in the young seedling.*

The Root

The first part of the embryo to break through the seed coat, in nearly all seed plants, is the embryonic root. Figure 26–16 diagrams the growing zone of the root of a dicot. At the very tip is the root cap, which protects the apical meristem as the root tip is pushed through the soil. The cells of the root cap wear away and are constantly replaced by new cells from the meristem. The cells in the meristem designated as *apical initials* are those that produce new cells. All the other cells in the root are the progeny of these relatively few meristematic cells. The apical initials divide continuously. Some of the daughter cells remain in the tip of the meristem. Others differentiate as they divide, some becoming cells of the root cap and others forming the complex tissues

26–16 *The developmental regions of a dicot root. New cells are produced by the division of apical initials, cells within the meristem. The cells above the meristem undergo a characteristic series of changes as the distance increases between them and the root tip. First, there is a maximum rate of cell division, followed by cell elongation, which accounts for most of the lengthening of the root. As the cells elongate, they differentiate into various specialized tissues of the root. The protoderm becomes the epidermis, the ground meristem becomes the cortex, and the procambium becomes the xylem and phloem. Some of the cells produced in the apical meristem form the protective root cap.*

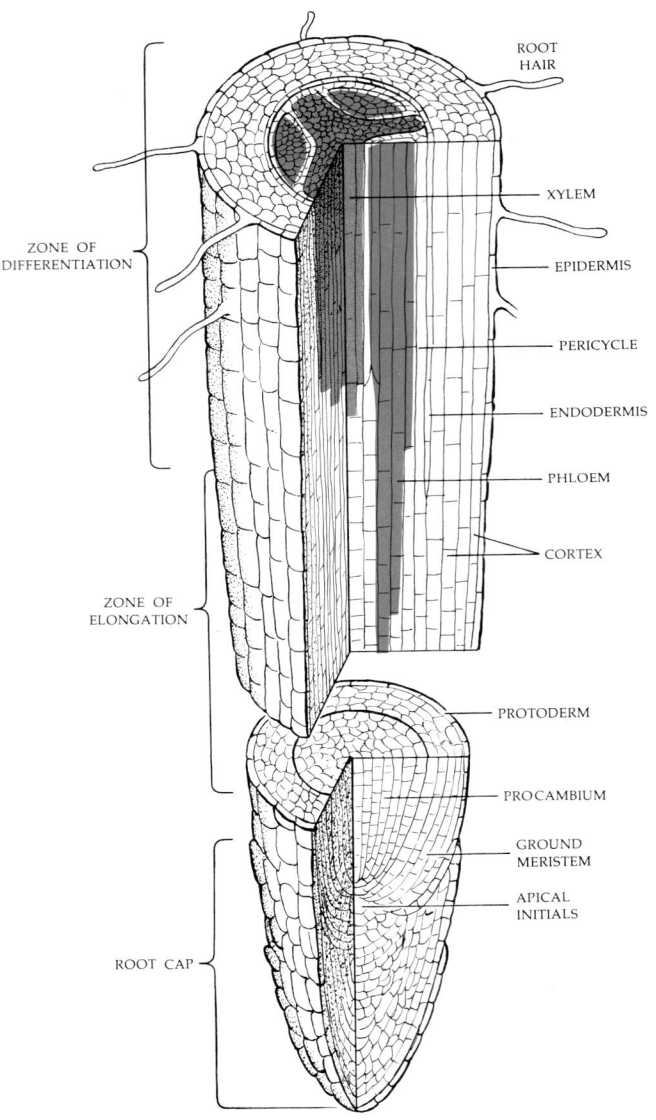

ROOT HAIR

XYLEM

EPIDERMIS

PERICYCLE

ENDODERMIS

PHLOEM

CORTEX

ZONE OF DIFFERENTIATION

ZONE OF ELONGATION

PROTODERM

PROCAMBIUM

GROUND MERISTEM

APICAL INITIALS

ROOT CAP

–17 *Two stages in the development of a branch root in a willow. Such roots originate in the pericycle and grow out laterally through the endodermis, cortex, and epidermis. They destroy the tissues in their path, partly by crushing, partly by digestion by multiple enzymes.*

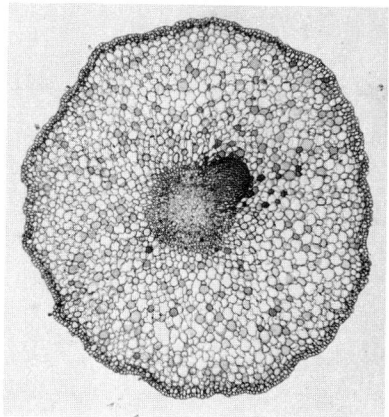 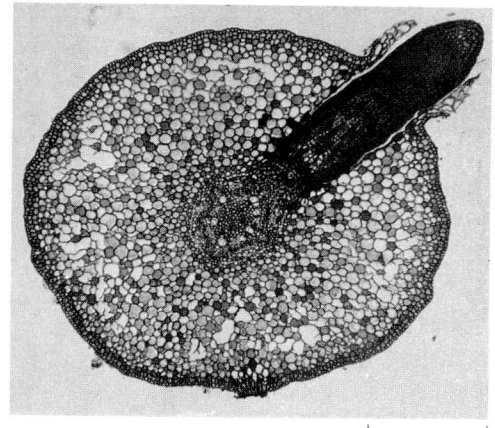

0.5 mm

-18 *Plant cell walls consist mainly of cellulose fibers embedded in a matrix (the same construction principle used in fiberglass boats). As plant cells elongate, the original fibers become reoriented longitudinally, but new fibers are laid down in the opposite direction, maintaining the cross-ply construction.*

of the root. The maximum rate of cell division occurs at a point well above the tip of the meristem. Then, just above the point where cell division ceases, the cells gradually elongate, growing to 10 or more times their previous size, often within the span of a few hours. This elongation process is the immediate cause of root growth, although, of course, growth is ultimately dependent on the production of the new cells which become part of the elongation zone.

As the cells stop dividing, they begin to differentiate, forming first the sieve-tube elements, the conducting cells of the phloem, and vessel elements, the conducting cells of xylem. Farther up the root, the endodermis takes shape. Within the endodermis, the pericycle forms. This tissue is the site of branch roots. At about this same stage, the epidermal cells differentiate and begin to extend root hairs into the crevices between the soil grains.

This same basic pattern of growth is seen in the first root of a seedling and in the growing root tips of a tree 100 feet tall.

The Shoot

The organization of the developing shoot tip is somewhat similar to that seen in the root: first, a zone in which most of the cell division takes place; next, a zone of cell elongation; and finally, a zone of differentiation. These zones are not as distinct in the shoot as they are in the root, however, because of the regular occurrence of nodes and their appendages.

As in the root, the outermost layer of cells develops into the epidermis. In the shoot, these cells have an outer surface, the cuticle. Other cells differentiate to form ground tissue, which includes the cortex and, in dicots, the pith, and the primary vascular tissues—the primary xylem and the primary phloem. The pattern of development is more complicated, however, than in the root tip since the apical meristem of the shoot is the source of tissues that give rise to new leaves, branches, and flowers. (At the time of flowering the apical meristem forms the floral parts and ceases to exist.)

26–19 *Longitudinal section of the shoot tip of a lilac* (Syringa vulgaris). *At the tip of the stem is the apical meristem, the central zone of cell division. The leaf primordia, from which new leaves will form, originate along the sides of the shoot apex. In this picture, two leaf primordia can be seen emerging from the apical meristem. Successively older leaf primordia have formed on both sides. The developing stem below the apical meristem and the lateral buds on either side are also regions of active cell division.*

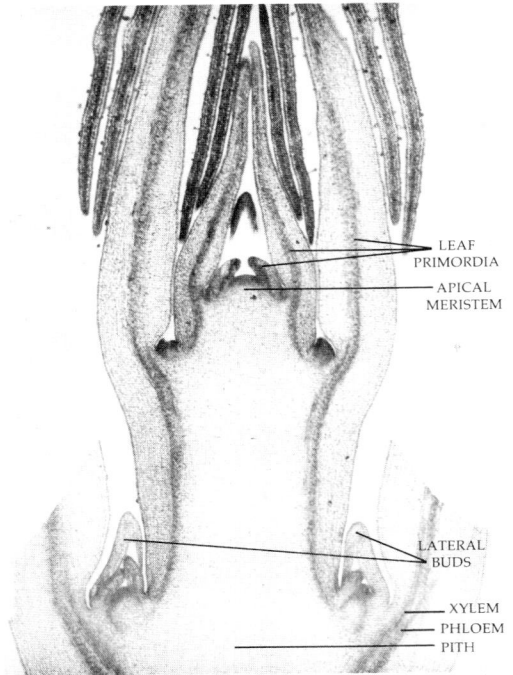

LEAF PRIMORDIA

APICAL MERISTEM

LATERAL BUDS

XYLEM
PHLOEM
PITH

Figure 26–19 shows the shoot tip of a lilac. Here you can see the apical meristem, which is very small, and the beginnings—primordia—of leaves. As you can see, leaves are formed in an orderly sequence at the shoot tip. The leaf originates by the division of cells in a localized area along the side of the shoot apex. The vascular tissue begins to differentiate in the leaf primordia, eventually becoming part of the general vascular system that connects the plant from root to leaf tip. In some species, leaves arise simultaneously in pairs opposite one another, as in our figure. In other species, the leaves occur spirally or in circles (whorls). As the internodes elongate, the young leaves become separated so that the leaf pairs clustered so tightly together around the apex in Figure 26–19 will eventually be spaced out along the stem of the plant.

Buds, Flowers, and Branches

As the growing tip of the shoot elongates, small portions of meristematic tissue are left just above the point at which the leaf joins the stem (the leaf axil). These new meristems, or buds, remain dormant until after the growth of the adjacent leaf and internode is complete. In many species, these buds do not develop at all unless the apical meristem of the shoot is removed. (This phenomenon, known as apical dominance, will be discussed further in the next chapter.) In some species, some buds are destined to become lateral branches or specialized shoots (such as rhizomes or tubers) or flowers. (A flower is regarded, evolutionarily speaking, as a modified, shortened shoot.) In other species, whether or not a particular bud will become a flower or a branch is determined by environmental conditions, particularly day length.

A sunflower and bud. Note the spiral arrangement of the florets—such arrangements are typical of composites.

In woody perennials, buds often form during one growing season and become dormant, and develop during the next growing season. Such buds in species of temperate regions are characteristically covered with bud scales that protect them from desiccation.

SECONDARY GROWTH

Secondary growth is the process by which woody plants, after primary growth has ceased, increase the thickness of trunks, stems, branches, and roots. The so-called "secondary tissues" are not derived from the apical meristems; they are the result of the production of new cells by the vascular cambium and cork cambium. These tissues are called *lateral meristems*.

The vascular cambium is a thin, cylindrical sheath of tissue completely surrounding the xylem and completely surrounded by the phloem. In plants with secondary growth, the cambium cells divide continually during the growing season, adding new xylem cells—that is, secondary xylem—on the outside of the primary xylem, and secondary phloem on the inside of the primary phloem. Some daughter cells remain as a cylinder of undifferentiated cambium. As the tree grows older, the living cells of the xylem in the center of the trunk die, and the vessels cease to function. This nonliving wood is called heartwood, as distinct from sapwood, which consists of living cells and functional vessels.

As the girth of stems and roots increases by secondary growth, the epidermis becomes stretched and torn. In response to this tearing process, a new type of cambium forms from the cortex, and from this cork cambium, cork, which is a dead tissue, is produced.

26–21 (a) *Stem of a dicot before the onset of secondary growth.* (b) *Beginnings of secondary growth. Secondary xylem and secondary phloem are produced by the vascular cambium, a meristematic tissue formed late in primary growth. As the trunk increases in diameter, the epidermis is stretched and torn, which apparently triggers the formation of the cork cambium, from which cork is formed, replacing the epidermis.* (c) *Cross section of a three-year-old stem, showing annual growth rings. Rays are strands of living cells that transport nutrients and water laterally (across the trunk). On the perimeter of the outermost growth rings of xylem is the vascular cambium, encircled by a band of secondary phloem. The primary phloem and also the cortex will eventually disappear. In an older stem, the thin cylinder of active secondary phloem is immediately adjacent to the cork cambium.*

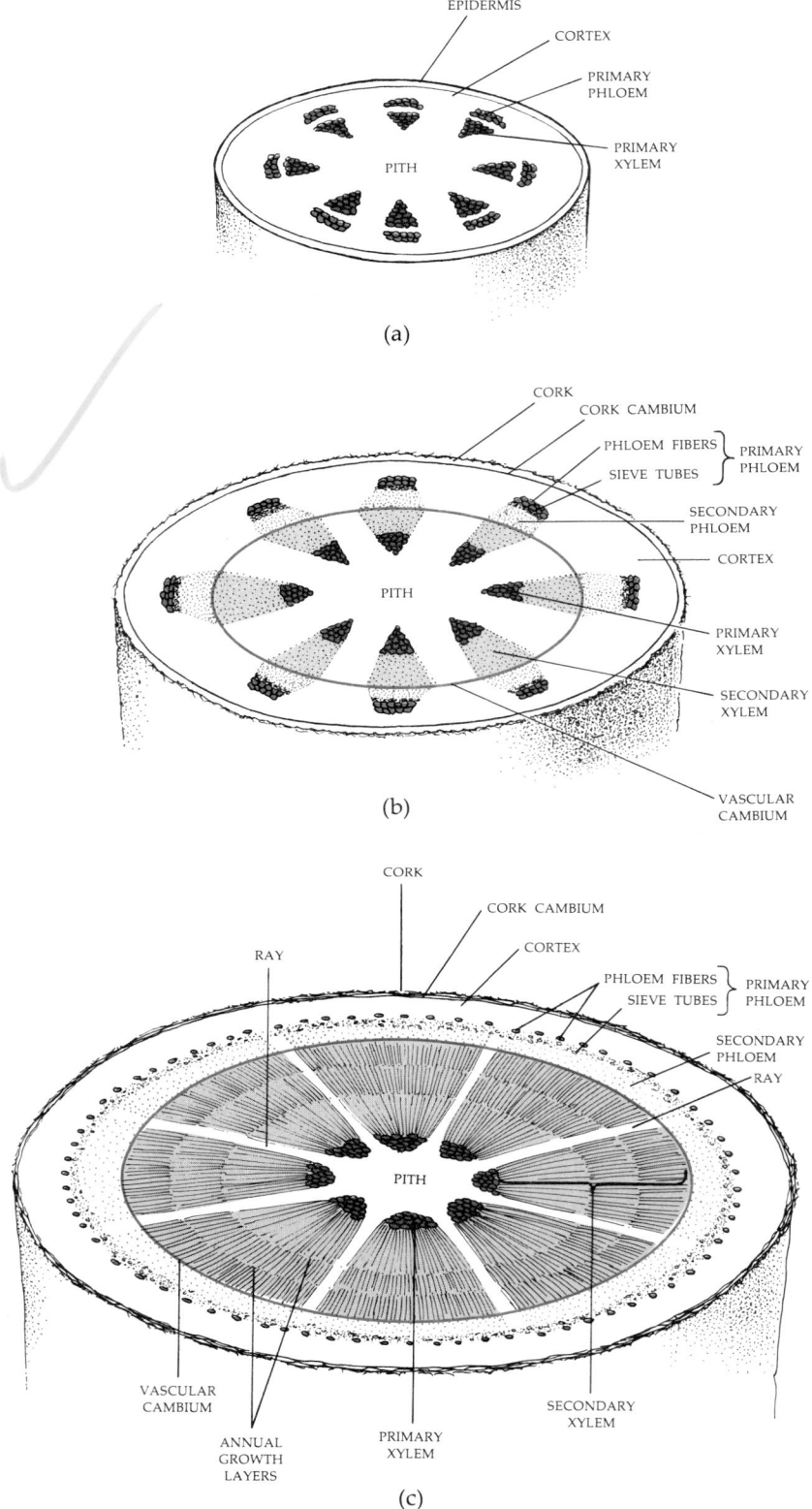

Figure 26–21 shows the cross section of the stem of a young dicot, in which some secondary growth has taken place. In the center of the stem are loosely packed parenchyma cells, the pith. The first cylinders of tissue around the pith are layers of xylem, composed of vessels and other cell types. Around the outermost layer of xylem is a layer of meristematic tissue, the vascular cambium. Each year, during the growing season, the cambium undergoes a number of mitotic divisions, forming new xylem (secondary xylem) on its inner surface and new phloem (secondary phloem) on its outer surface. By this continuous formation of layers of xylem and phloem, tree trunks increase their diameter as they increase their height. As they grow, season after season, the new xylem may form visible growth layers, or rings, each growth season leaving its trace, so that the age of a tree can be estimated by counting the number of growth rings in a section near its base. Since the rate of growth of the tree depends on climatic conditions, it is possible to determine from the width of the annual growth layers of ancient trees fluctuations in temperature and rainfall that occurred hundreds of years ago.

AGING AND DEATH IN PLANTS

Because of its meristems, a plant body is constantly renewed. Why, then, do plants die?

In the case of annual plants, flowering and fruiting apparently bring on death. Many annuals can be kept alive indefinitely if their flower buds are constantly removed. Some botanists have suggested that production of seeds and fruits so exhausts a plant's resources that it dies. However, in dioecious species—Cannabis, for example—not only does the female plant die after flowering and fruiting but the male plant perishes similarly, suggesting that the production of the flower causes death.

Another possibility is that death is a consequence of the complete pattern of development genetically programmed into the plant. According to this hypothesis, flowering and fruiting bring about profound physiological changes in the plant that result in its death. This idea is substantiated by observations of plants that may grow vegetatively for many years and then suddenly flower and die. There are bamboos, for example, which may remain vegetative for more than 50 years before a large number of plants, scattered over a wide geographic area, suddenly bloom and die. The survival curve for plants which die after blooming is shown in Figure 26–22, line A. The survival curve of human populations in the United States and other economically favored countries, where the majority of individuals live out a normal life span, is about the same shape. Line C shows the survival curve for a stand of oaks. The same curve resulted from recording the survival of glasses in a cafeteria; in other words, the mortality rate is constant, so that the number of "deaths" in a given time interval depends simply on the number of individuals "alive" at any given time. If we start with a population of 1,000 individuals and the mortality rate is such that one-half of these, or 500, die in a given time, then in the next equal interval of time only 250 will die, and in the next, 125, and so on. This curve suggests either that some perennials are essentially immortal or that processes of aging and death are masked by ecological events that kill individuals at random. Some natural populations of plants, however, show a survival curve represented by line B. This curve, which is characteristic for many animal populations under natural conditions, shows

26–22 *Three types of survival curves common in natural populations, showing the percent of individuals surviving, plotted against time. Curve A represents a common survival curve for annual plants, such as the plants in a field of corn. Curve B represents the survival curve often seen among animals, with death highest among the very young and the old. Curve C represents the constant mortality rate in a stand of oaks, in which the number dying at any particular time depends simply on the number of living plants.*

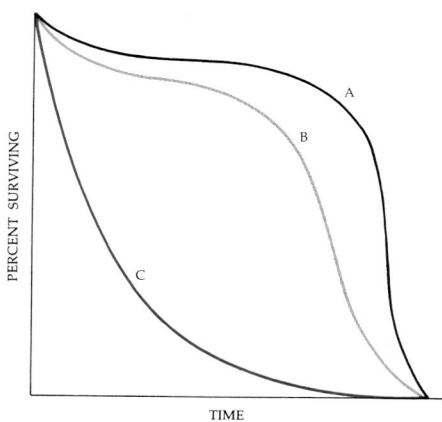

PERCENT SURVIVING

TIME

26–23 *Bristlecone pine (Pinus aristata) from the White Mountains of east-central California. These pines, which grow near the timberline of the mountains, are the oldest living trees; one reached an age of 4,900 years. In the trunk of an ancient tree, the only living tissue is a thin cylinder of cells near the periphery.*

high mortality among the very young, then a period of decreasing mortality, followed by a period in which deaths occur more rapidly. Such a curve suggests that aging itself does bring about changes in perennial plants that affect their mortality. One possible explanation is that as plants grow older, more and more nonphotosynthetic tissue is produced until it may eventually exceed the capacity of the leaves to produce carbohydrates or of the vascular tissues to provide for such an extensive leaf canopy.

Death is, of course, an essential partner of reproduction. If old organisms did not make way for new ones, there would soon be no living space for the new generations.

SUMMARY

The flower is the reproductive organ of the angiosperms. The anthers of the flower produce the pollen grain, the male gametophyte; each anther is attached to a slender filament. The entire structure, anther plus filament, is known as the stamen. The carpel typically consists of the stigma, an area on which the pollen grains germinate, a slender style, and at its base, the ovary. The ovary contains one or more ovules. Within each ovule, the female gametophyte containing the egg cell develops. The ovule later becomes the seed, and the ovary becomes the fruit.

The various shapes and colors of flowers evolved under selection pressures for more efficient pollinating mechanisms. Major trends in flower evolution include reduction and fusion of floral parts, a change in the position of the ovary relative to the other flower parts to a more protected position, and a shift from radial to bilateral symmetry. The composites and the orchids represent the most successful, evolutionarily speaking, of the dicots and monocots.

In flowering plants, a new life cycle begins when a pollen grain germinates upon the stigma of a flower of the same species, sending a pollen tube down the style and into an ovule. Within the ovule is the female gametophyte, which usually consists of seven cells with a total of eight haploid nuclei, including the egg nucleus. One of the three haploid nuclei in the pollen grain (a sperm nucleus) unites with the egg cell nucleus, and the other sperm nucleus unites with the two polar nuclei of the female gametophyte. The plant embryo develops from the first union, and the nutritive endosperm, which is a triploid (3*n*) tissue, from the second. These phenomena of "double fertilization" and triple fusion are found only among the flowering plants.

The embryo has two apical meristems—the apical meristem of the shoot and the apical meristem of the root or radicle—and either one or two cotyledons ("seed leaves"). From the endosperm, the cotyledons absorb nutrients which nourish the growing embryo. Early in embryonic life, cell division becomes confined to limited areas in the plant body: the apical meristems of the root and the shoot.

The seed consists of the embryo and the seed coat—that is, the ovule and its contents. It also often contains a nutrient tissue, the endosperm. The petals, stamens, and other floral parts of the parent plant may fall away as the ovary ripens into fruit and the seed forms. The embryo commonly enters a dormant period.

When the seed germinates, growth of the root and shoot proceeds from the

Table 26-1   *Summary of Main Cell Types in Angiosperms*

Cell type	Origin	Location	Characteristics	Function
Apical meristem	Embryonic cells	Apices of shoots and flowers; in root, beneath root cap	Many-sided, small, thin-walled cells; nucleus large; vacuoles usually small	Origin of primary meristematic tissues and leaf primordia
Epidermis	Protoderm	Surface of entire plant	Flattened, variable in shape, overlaid by cuticle in shoot; guard cells specialized	Boundary between plant body and external environment
Parenchyma	All dividing tissues except protoderm	Everywhere, usually dominant in pith, cortex, mesophyll, phelloderm	Many-sided, usually thin-walled; abundant air spaces between cells	Photosynthesis, storage, differentiation into other cell types
Collenchyma	Ground meristem of leaf and stem	Strands in cortex of stem and in leaves	Elongate, with irregularly thickened primary walls	Support for young stems and leaves
Sclereid	Ground meristem or parenchyma; protoderm	Scattered in parenchyma tissue; seed coats	Irregular; massive secondary wall; dead at maturity	Produces hard texture
Fiber	Procambium or vascular cambium	Primary and secondary xylem and phloem	Very long, narrow cell, with thick cell walls; dead at maturity	Support
Tracheid	Procambium or vascular cambium	Xylem	Similar to fiber, but shorter and thicker, and with lateral walls perforated; dead at maturity	Conduction of water and solutes
Vessel element	Procambium or vascular cambium	Interconnected series (= vessels) in xylem	Similar to tracheid, but broader, with end walls perforated; dead at maturity	Conduction of water and solutes
Sieve-tube element	Procambium or vascular cambium	Phloem, usually connected with companion cells (parenchyma), forming interconnected series (= sieve tube)	Elongate, cylindrical, with specialized sieve plates; nucleus lacking at maturity	Conduction of organic solutes
Vascular cambium	Procambium, reactivated parenchyma	Lateral, between young phloem and xylem tissues	Elongate fusiform (spindle-shaped) and short-ray initials	Produces secondary xylem and secondary phloem
Cork (phellem)	Phellogen	Surface of stems and roots with secondary growth	Flattened cells, compactly arranged; protoplast absent at maturity, the cells often air-filled	Restricts gas exchange and water loss

apical meristems of the embryo. Certain cells within the meristems divide continuously. Some of these cells remain meristematic, continuing to divide, while others elongate and then differentiate, forming, according to their position, various specialized cells, including those of the xylem and the phloem. In the root, the apical meristem forms a root cap, which protects the root tip as it is pushed through the soil. Root hairs, which are the principal pathways of absorption by the roots, are outgrowths of the epidermal cells.

Leaf primordia arise from the apical meristem of the shoot. As the nodes are separated by elongation of the internodes, small apical meristems (buds)

are left in the axils of the leaves. These buds may remain dormant or they may give rise to branches, specialized shoots (such as runners, rhizomes, or tubers), or flowers.

Secondary growth is the process by which the woody perennials increase their girth. Such growth arises primarily from the vascular cambium, a sheath of meristematic tissue completely surrounding the xylem and completely surrounded by phloem. The cambium cells divide during the growing season, adding new xylem cells (secondary xylem) outside the primary xylem and new phloem cells (secondary phloem) on the inside of the primary phloem. As the trunk increases in girth, the epidermis is eventually ruptured and destroyed and is replaced by cork.

QUESTIONS

1. Define the following terms: secondary growth, rhizome, pollen grain, endosperm, apical initial, bud, primary growth, lateral meristem.
2. Sketch a flower. Give the function of each floral part.
3. Sketch a dicot embryo at the time of seed release. Identify each part in terms of the future development of the plant body.
4. Sketch a tree trunk with secondary growth. Compare your sketch with Figure 26–21c.
5. Suppose you carve your initials 5 feet above the ground on the trunk of a mature tree that is growing vertically at an average rate of 6 inches a year. How high will your initials be at the end of 2 years? At the end of 20 years?
6. As a result of so-called secondary growth, most of our common trees increase in girth as they increase in height. What limitations would a lack of secondary growth impose on the form and mechanical support of a tree? To what extent might the shapes of palms and bamboos (both monocots) reflect their restricted secondary growth?
7. Plant development is a very orderly affair, with none of the wholesale cell migrations and changes of shape seen in animals. Describe how, from a single slice of a growing root tip, you can reconstruct the developmental history of an apical meristematic cell.

Chapter 27

Integration of Growth: The Plant Hormones

We saw in the previous chapter that as a plant grows it does far more than simply increase its mass and volume. It differentiates, forming a variety of cells, tissues, and organs, and undergoes morphogenesis, taking on the shape characteristic of the adult sporophyte. How can one single cell, the fertilized egg, be the source of the myriad tissues—shoot, root, flower, fruit, seed—that make up that extraordinary individual, the "normal plant." As we noted in Chapter 18, many of the details explaining how these processes are regulated are not known, but it has become clear that normal development depends on the interplay of a number of internal and external factors.

Chief among the internal factors are the plant hormones. Hormones, by definition, are substances that are produced in one tissue and transported to another, where they exert highly specific effects. Typically they are active in very small quantities. In the shoot of a pineapple plant, for example, only 6 micrograms of auxin, a common growth hormone, are found per kilogram of plant material. One enterprising plant physiologist calculated that the weight of the hormone in relation to that of the shoot is comparable to the weight of a needle in a 22-ton haystack. The term "hormone" comes from the Greek word meaning "to excite." It is now clear, however, that many hormones have inhibitory influences. So, rather than thinking of hormones as stimulators, it is perhaps more useful to consider them as chemical messengers. But this term, too, needs qualification. As we shall see, the response to the particular "message" depends not only on its content but upon how it is "read" by its recipient.

Of the many plant hormones that have been identified, the best studied are the auxins, cytokinins, and gibberellins. In this chapter we shall introduce these and a few other hormones by considering some of what is known about their effects in plants. Although we shall discuss each group separately, it should be remembered that the functions of these hormones overlap and that their effects depend not only on the target tissue but also on the presence or absence of other hormones. Cytokinins and auxins, for example, interact to

produce cell division and cell elongation. In cultures of cells growing in a test tube, an auxin plus a low concentration of cytokinin results in rapid cell expansion, with the production of a relatively few giant cells. On the other hand, a cytokinin plus a low concentration of auxin results in rapid cell division, with the production of large numbers of relatively tiny cells. Thus, normal growth depends upon a balance of the two, not upon either one alone.

THE AUXINS

The effects of the auxins were first observed by Charles Darwin and his son Francis and reported in *The Power of Movement in Plants*, published in 1881. The Darwins were studying the bending toward light (*phototropism*) of grass seedlings. They noted that the bending takes place below the tip, in the lower part of the shoot. Then they showed that if they covered just the terminal portion of the shoot of the seedling with a cylinder of metal foil or a hollow tube of glass blackened with India ink and exposed the plant to a light coming from the side, the characteristic bending of the shoot did not occur. If, however, the tip was enclosed in a transparent glass tube, bending occurred normally. Bending also occurred normally when the lightproof cylinder was placed below the tip (see Figure 27–2).

"We must therefore conclude," they stated, "that when seedlings are freely exposed to a lateral light some influence is transmitted from the upper to the lower part, causing the material to bend."

In 1926, the Dutch plant physiologist Frits W. Went succeeded in separating this "influence" from the plants that produced it. Went cut off the leaf (coleoptile) tips from a number of oat seedlings. He placed the tips on a slice of agar (a gelatinlike substance), with their cut surfaces in contact with the agar, and left them there for about an hour. He then cut the agar into small blocks and placed a block off-center on each stump of the decapitated plants,

27–2 *The Darwins' experiment. (a) Light striking a growing shoot tip (such as the tip of this oat seedling) causes it to bend toward the light. (b) Placing an opaque cover over the tip of the seedling inhibits this bending response, but (c) an opaque collar placed below the tip does not. These experiments indicate that something produced in the tip of the seedling and transmitted down the stem causes the bending.*

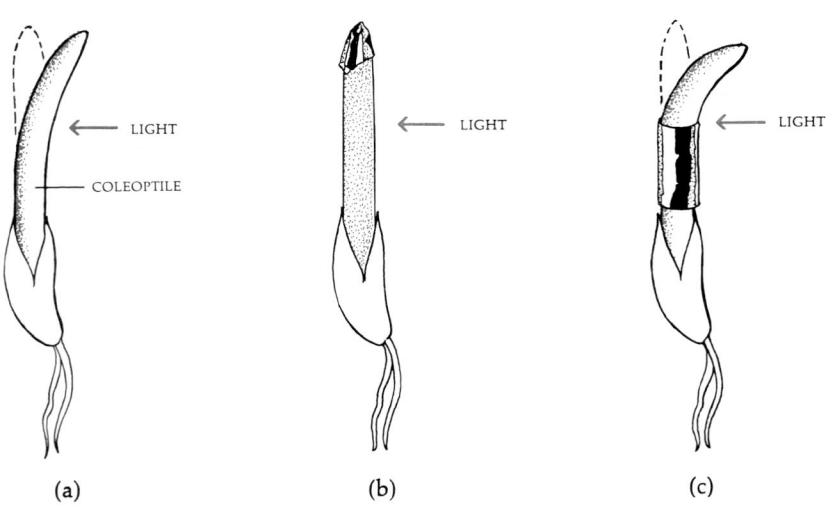

(a)   (b)   (c)

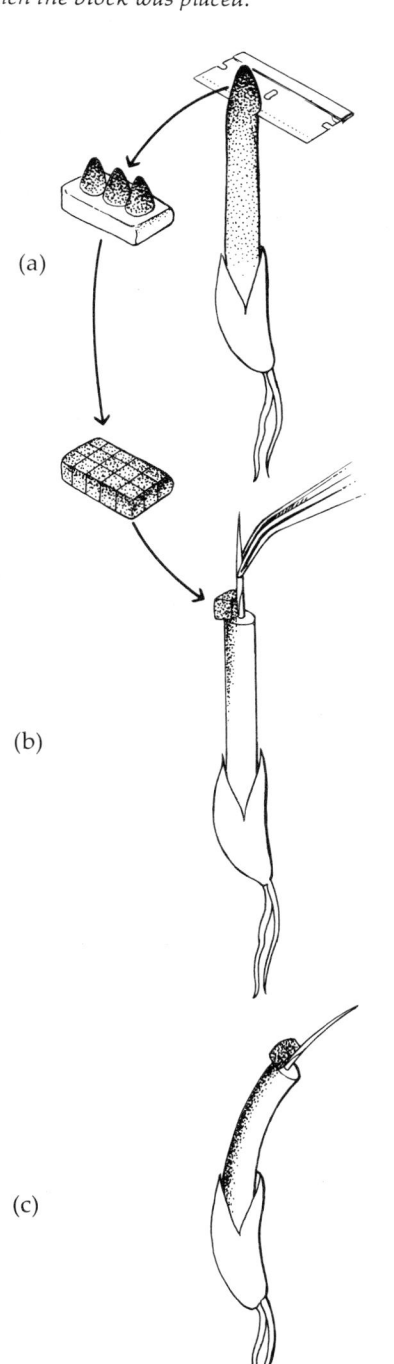

27–3 *Went's experiment. (a) He cut the tips from oat seedlings and placed them on a slice of agar. After about an hour, he cut the agar in small blocks and (b) placed each block, off-center, on a decapitated seedling (the leaf is pulled up). (c) The seedlings bent away from the side on which the block was placed.*

(a)

(b)

(c)

which were kept in the dark during the entire experiment. Within one hour, he observed a distinct bending *away* from the side on which the agar block was placed. (See Figure 27–3.)

Agar blocks that had not been exposed to a coleoptile tip produced either no bending or only a slight bending toward the side on which the block had been placed. Agar blocks that had been exposed to a section of coleoptile lower on the shoot produced no physiological effect.

Went interpreted these experiments as showing that the coleoptile tip exerted its effects by means of a chemical stimulus (in short, a hormone) rather than a physical stimulus, such as an electrical impulse. This chemical stimulus came to be known as auxin, a term coined by Went from the Greek word *auxein*, "to increase."

Several different substances with auxin activity have now been isolated from plant tissues and others have been synthesized in the laboratory; these are all known as auxins. The most common of the natural auxins is indoleacetic acid, abbreviated IAA. One of the synthetic auxins, known as 2,4-D, is commonly used as an herbicide (like many other physiologically active compounds, auxins are toxic in high doses). For reasons not known, 2,4-D and related compounds are effective on broad-leaved plants at concentrations not harmful to grasses and so are commonly used on lawns to control weeds. Auxins, in massive doses, are among the herbicides that were used in Southeast Asia during U.S. participation in the Vietnam War. The synthetic auxins, unlike IAA, are not readily broken down by natural plant enzymes or by bacteria. Their longer effective life makes them better suited for commercial purposes.

Auxins and Phototropism

The phototropism observed by the Darwins results from the fact that under the influence of light, auxins migrate from the light side to the dark side of the tip. The cells on the dark side, having more auxin, elongate more rapidly than those on the light side, causing the plant to bend toward the light, a response with high survival value for young plants. It is the unequal distribution of auxin, rather than a change in the production of it or in the response of cells to it, that accounts for the difference in growth rate of the two sides of the coleoptile tip.

27–4 *Auxins. IAA (indoleacetic acid) is a natural auxin, isolated from plant tissues. 2,4-D is a synthetic auxin used as a weed killer (herbicide). Naphthalenacetic acid, also a synthetic auxin, is commonly employed to induce the formation of adventitious roots in cuttings and to reduce fruit drop in orchard crops.*

IAA

2,4-D

NAPHTHALENACETIC ACID

27–5 *The upper row of holly cuttings was treated with a synthetic auxin 21 days before the picture was taken. The cuttings in the lower row were not. Note the growth of adventitious roots on the plants in the upper row.*

27–6 *Why shoots grow upward and roots downward. (a) When a seedling is perpendicular the auxin produced in the tip is distributed evenly and the plant grows erect. (b) However, when a seedling is placed on its side, auxin is transported laterally and accumulates in cells on the lower side of the shoot; (c) these cells then grow more rapidly and so the shoot grows upward. It has been hypothesized that auxin similarly accumulates in the cells on the lower side of the root, causing the cells to grow more slowly so that the root tip turns down.*

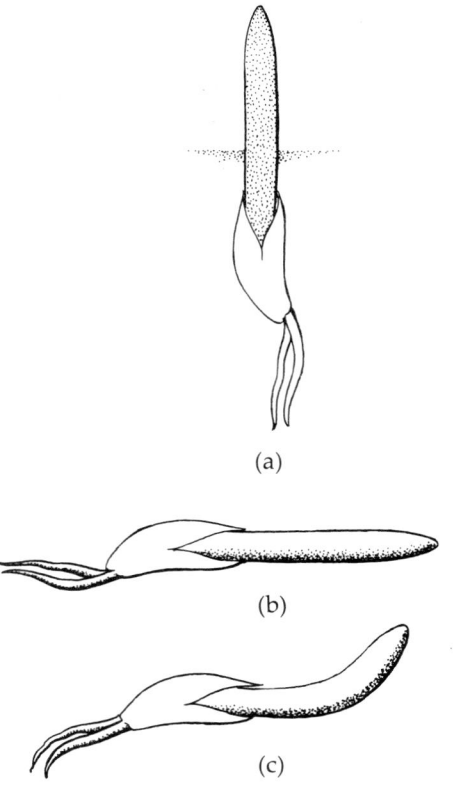

(a)

(b)

(c)

Auxins and Growth Regulation

In the intact plant, auxins promote growth of the shoot. The hormone is produced in the rapidly dividing cells of the meristem and migrates downward. If you cut off the shoot tip, growth stops. If an auxin—either in an agar block or in a paste—is applied to the cut surface, growth resumes. If the hormone is applied to the apex of an intact plant, there is usually no growth effect. Apparently, under normal circumstances, the tip produces all the auxins that the cells of the stem can respond to. Experimental evidence indicates that the concentrations normally present near growing shoot tips stimulate cell division. The lesser concentrations in the zone of elongation cause the cells to elongate.

Auxins, in very small amounts, sometimes stimulate the growth of roots. In somewhat larger amounts, they inhibit the growth of the main roots, although they may promote the initiation of new branch roots or adventitious roots. This inhibitory effect of auxins on root elongation may be involved in the downward orientation of seedling roots (Figure 27–6).

Auxins and Bud Growth

Although auxins stimulate growth in apical buds, they may inhibit growth in lateral buds. If you nip off the terminal bud of *Coleus*, a common houseplant, for example, the lateral buds begin to grow vigorously, producing a plant with

a bushier, more compact body. If you apply an auxin to the cut surface of the tip, growth of lateral buds will stop. In a potato plant, the edible portion, as we noted previously, is actually a thickened, modified stem (a tuber) and the "eyes" are lateral buds. Treating the tuber with a synthetic auxin inhibits bud growth, permitting the potatoes to be stored for longer periods. The effect of auxin, apparently, is to maintain dormancy; once the buds begin to grow, auxin stimulates them, and their growth is regulated by auxins they produce.

In other words, the effects of auxin—whether it stimulates or inhibits—depend on the hormone itself, on the target tissue, and on other growth factors, including other hormones.

Auxins and Abscission

Auxins are also produced in the blades of young leaves, but they do not appear to have any direct effects on the rate of leaf growth. They do, however, affect leaf drop. Leaves fall from trees as a result of formation of the *abscission layer*, a special layer of weak, thin-walled parenchyma cells, across the base of the petiole. As the leaf grows older, reusable substances, such as sugars, amino acids, and Mg^{2+}, are returned to the stem. Next, enzymes break down the middle lamella at the abscission zone, and corky walls are deposited. The cells in the abscission layer become drier and disintegrate until the leaf is held to the tree by only a few strands of vascular tissue. Before the leaf falls, a layer of cork develops below the abscission layer and seals the leaf scar. Abscission has been correlated with a diminished production of auxin in the leaf, among other factors.

7–7 *Abscission layer forming across the petiole of a leaf. (Carolina Biological Supply Company.)*

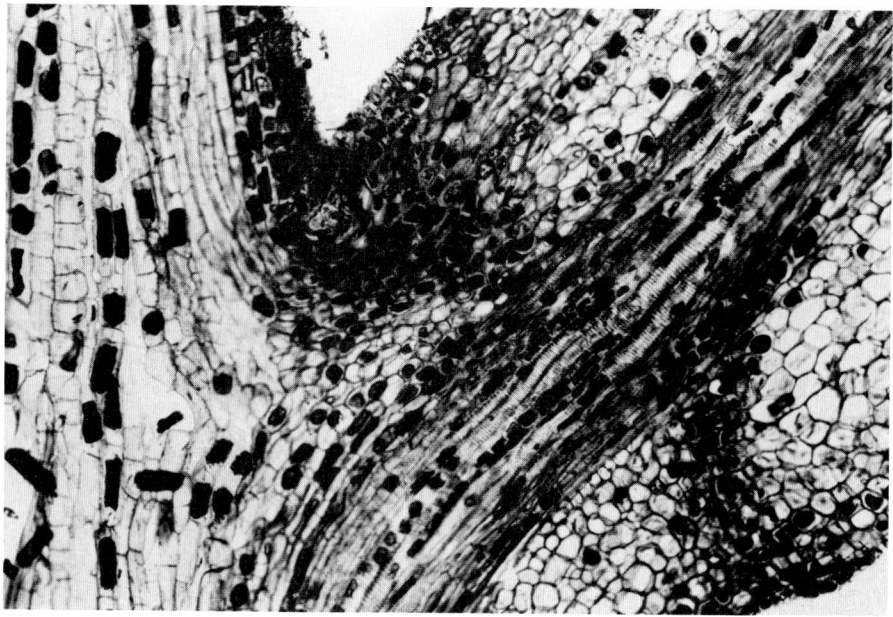

Auxin concentrations also decrease as fruits ripen, and abscission layers form in the stem, permitting the fruit to drop. Commercial fruit growers often spray their orchards with synthetic auxins in order to keep the ripe fruit on trees until it can be harvested.

Auxins and Fruit Growth

Ripening in fruit involves a number of changes. In fleshy fruits, the chlorophyll content drops, and other pigments may form, changing the fruit color. Simultaneously, the fleshy part of the fruit softens. This is a result of the enzymatic digestion of pectin, the principal component of the middle lamella. When the middle lamella is weakened, cells are able to slip past one another. During this same period, starches or—as in the case of some fruits, like the avocado—oils are broken down into sugars. As a consequence of these changes, the smells and tastes of the fruit change. They become conspicuous and so they attract animals that eat the fruit and so scatter the seed.

Ordinarily, if the flower is not pollinated, the fruit will not develop. In some plants, fertilization of one egg cell is sufficient for normal fruit development, but in others, such as apples or melons, which have many seeds, unless several egg cells (in as many ovules) are fertilized, the ovary wall will not mature and become fleshy. By treating the stigma with auxin, it is possible to induce seedless fruit, such as seedless tomatoes or melons.

Apparently, the developing embryo is the source of auxin for the development of the fruit. In the strawberry, for example, if the seeds are removed during the fruit's development, the strawberry stops growing altogether. If auxin is applied to a fruit from which seeds have been removed, growth proceeds normally. If a narrow ring of seeds is left, a bulging girdle of growth forms in the area of the seeds (see Figure 27–8).

The natural auxins move from the tips of the shoots toward the base of the plant. The reason for this unidirectional flow is not known. It is not caused by gravity; auxin is in solution and so does not settle in response to gravity. In fact, if a stem is inverted, auxin moves against gravity. Its movement is presum-

27–8 (a) *Normal strawberry,* (b) *strawberry from which all seeds have been removed, and* (c) *strawberry in which three horizontal rows of seeds were left.*

(a)   (b)   (c)

ably a result of some interaction between the chemical and plant cells. Plants have distinct polarity—an inbuilt "upness" and "downness." This polarity in the transport of auxins is clearly a basic factor in the overall phenomenon of plant polarity.

How Do the Auxins Work?

Auxins achieve their cell-elongating effect by increasing the plasticity of the cell wall. When the cell wall softens, the cell enlarges, owing to the pressure of water from within its vacuole. As the water pressure is reduced by expansion of the cell, the plant cell takes up more water and the cell thus continues to enlarge until it encounters sufficient resistance from the wall. It has not been determined whether or not the effect on the cell wall is a consequence of direct interaction between auxin and some cell wall constituent or the result of the production of a new enzyme that breaks down pectin or some other component.

One proposal is that auxin stimulates the movement of H^+ ions out of the cell, thereby lowering the pH of the cell wall. Such a decrease in pH is known to stimulate growth, perhaps by breaking some of the chemical bonds that cross-link cellulose or other molecules and give the wall its rigidity. It is not known whether or not auxin's action on the cell wall, whatever it may be, is involved in other auxin effects, such as maintaining dormancy in lateral buds.

It has been hypothesized that auxin regulates gene activity in a way analogous to the operon model proposed by Jacob and Monod for bacterial cells (page 261). Auxin, according to this concept, inactivates the repressor substance produced by the regulator gene, permitting mRNA synthesis to proceed along the operon and new enzymes to be synthesized. However, some growth responses from auxin can be detected in less than 3 minutes, which is too short a time to be consistent with this hypothesis.

THE GIBBERELLINS

Gibberellins, like auxins, are involved in growth, stimulating both cell division and cell elongation in plants.

The most remarkable results are seen when gibberellins are applied to certain plants that, because of a single mutation, have lost the capacity to synthesize gibberellins. Such plants are genetic dwarfs. Under gibberellin treatment, these dwarfs become indistinguishable from normal tall plants. The effect on dwarf plants of gibberellins cannot be duplicated by auxins or any of the other known hormones.

Gibberellins also cause stem elongation in normal (nondwarf) plants. Too much gibberellin typically causes the stems to become long and thin, with few branches and pale stems. In fact, gibberellin was first discovered by a Japanese scientist who was studying a disease of rice plants, "foolish seedling disease." The diseased plants grew rapidly but were spindly and tended to fall over. The cause of the symptoms, the scientist found, was a chemical produced by a fungus, *Gibberella fujikuroi*, which parasitized the seedlings. The substance, which he named gibberellin, was subsequently isolated not only from the fungus but from many species of plants. More than 20 gibberellins have now been identified which vary slightly in structure and considerably in activity.

7–9 *Three of the more than 20 gibberellins that have been isolated from natural sources. Gibberellic acid (GA₃) is the most abundant in fungi and the most biologically active in many tests. The minor structural differences that distinguish the other two gibberellins are indicated by arrows.*

GIBBERELLIC ACID (GA₃)

GA₇

GA₄

Bolting and flowering in cabbage following gibberellin treatment. The plant on the left was not treated.

Gibberellins and Flowering

Some plants, such as mustard and cabbage, form leaf rosettes before flowering. (In a rosette, leaves develop but the stem does not elongate between the developing leaves and separate them.) These plants can be induced to flower by exposing them to long days, to a period of cold (as in the biennials), or to both. Following the appropriate exposures, the stems elongate—a phenomenon known as bolting—and the plants flower. Application of gibberellin causes bolting and flowering without these exposures.

Gibberellins and Seed Germination

In grass seeds, there is a specialized layer of cells, the aleurone layer, just inside the seed coat. These cells are rich in protein. When the seeds imbibe water, just prior to germination, the embryo produces gibberellin, which diffuses to the aleurone layer. In response to the gibberellin, the aleurone cells produce several enzymes, including alpha-amylase, which hydrolyzes the starch in the stored food reserves of the endosperm, converting it to soluble sugars which the embryo can use. In this way, the embryo itself calls forth the substances needed for its survival and growth at the time it requires them.

When the endosperm breaks down under the action of gibberellin, it is believed to release other growth-promoting hormones.

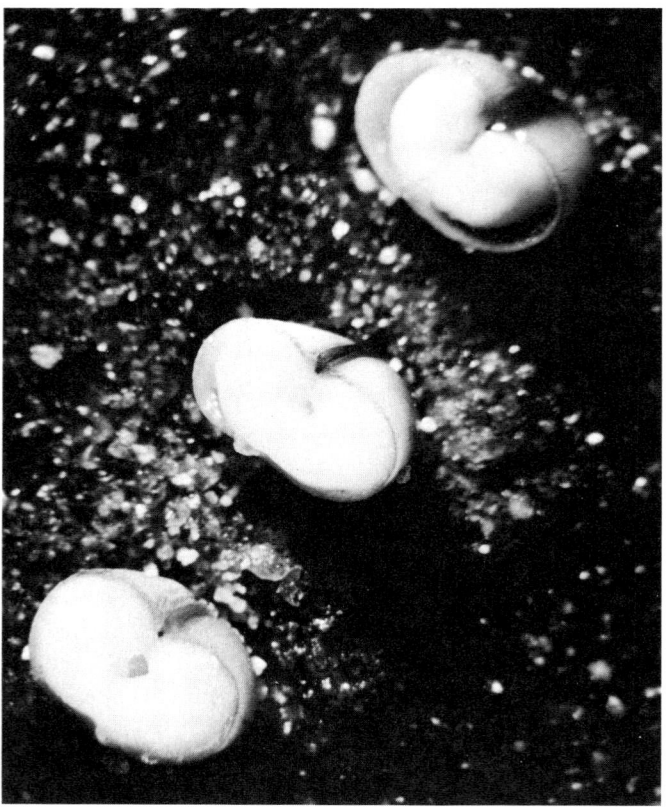

27–11 *An experiment investigating the action of gibberellin in barley seeds. Forty-eight hours before the picture was taken, each of these three seeds was cut in half and the embryo removed. The seed at the bottom was treated with plain water, the seed in the center was treated with a solution of 1 part per billion of gibberellin, and the seed at the top was treated with 100 parts per billion of gibberellin. As you can see, digestion of the starchy storage tissue has begun to take place in the seeds treated with gibberellin.*

7–12 *Mechanism of action of gibberellin in a barley seed. The embryo, shown in color, releases gibberellic acid (GA), which diffuses to the aleurone layer. The gibberellin induces the aleurone cells to synthesize enzymes that digest starch stored in the endosperm. The starch is broken down into sugars that the embryo then uses for growth.*

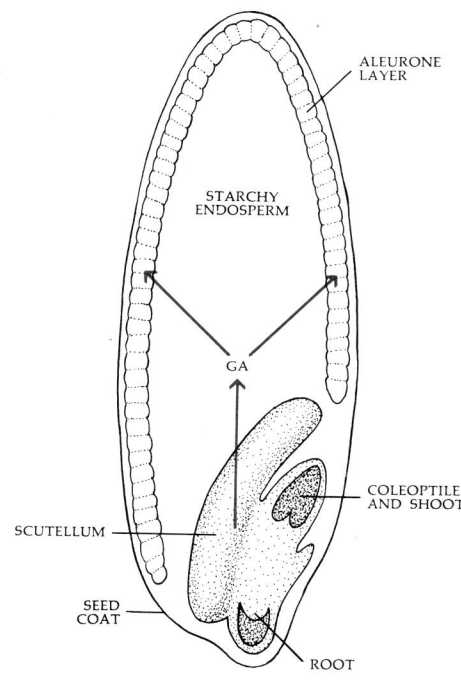

The enzymes in the aleurone cells are not merely activated by gibberellin, they are synthesized from scratch—that is, from their component amino acids. So it is hypothesized that gibberellin may also act by facilitating the activity of certain genes, causing the synthesis of specific messenger RNA molecules, which, in turn, direct the synthesis of the enzymes. It has not been proved that gibberellin acts directly on the gene, although investigators have shown that both RNA and protein syntheses are necessary for the appearance of the enzyme. Furthermore it is not known whether the way in which gibberellin works in these seeds is related to its effects on other plant organs.

THE CYTOKININS

In 1941, Johannes van Overbeek, a Dutch plant physiologist, found that in unripe coconuts, the "milk" (which is a liquid endosperm) contains a potent growth factor—one different from anything known at the time. This factor, or factors, it was found, greatly accelerated the enlargement of plant embryos and also promoted the division of cells isolated in test tubes. Because of this latter effect, this group of growth regulators became known as the cytokinins, from *cytokinesis*, a term for cell division.

Cytokinins have now been found in about 40 different species of higher plants, largely in actively dividing tissues, including seeds, fruits, roots, and bleeding sap (the sap that drips out of cracks and pruning cuts).

Cytokinins have been shown in some experiments to have effects that oppose those of the auxins. For example, application of cytokinins to lateral buds will sometimes cause them to develop even in the presence of auxin. And unlike auxin (but like gibberellin), cytokinin stimulates growth in leaves.

The seeds of certain varieties of lettuce that ordinarily require light for germination will germinate even in the dark if treated with cytokinin solutions. Cytokinins are not naturally present in the dry seed, however; they do not appear until the seed starts to grow, indicating that they are a result rather than a cause of germination. According to present theory, cytokinins (and probably auxins also) are released by the endosperm as a consequence of its breakdown by enzymes under the influence of gibberellin.

7–13 *Cytokinins. (a) Kinetin, the first of this group to be discovered, and 6-benzyl-amino purine (BAP) are commonly used synthetic cytokinins. (b) Zeatin and 2iP have been isolated from plant material.*

Note the resemblances between the purine adenine and a portion of the molecule of each of these cytokinins. The significance of the resemblance between adenine and the hormones of this group is not known.

It may merely be another example of biological economy: the use of a single major biosynthetic pathway to produce a number of functionally different products.

KINETIN

BAP

(a)

ZEATIN

2iP

ADENINE

(b)

27–14   *Two buds forming on a callus from a geranium following treatment with both an auxin and a cytokinin. Callus from some types of plants will produce undifferentiated tissue, roots, or buds, depending on the relative proportions of auxins and cytokinins.*

27–15   *Auxin-cytokinin interactions as shown on tobacco cells in tissue culture. Kinetin alone has little effect on the growth of tobacco callus (a mass of undifferentiated cells). IAA alone (not shown), regardless of the concentration used, causes the culture to grow to a weight of about 10 grams. When both hormones are present, growth is greatly increased. Notice, however, that when optimum concentrations are exceeded, the growth rate declines.*

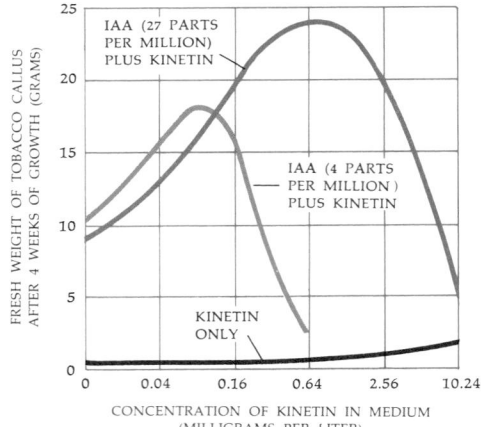

CONCENTRATION OF KINETIN IN MEDIUM
(MILLIGRAMS PER LITER)

Cytokinins also prevent senescence in leaves, both in dying plants and in leaves at the time of leaf drop.

Auxin-Cytokinin Interactions

Studies of interactions involving auxins and cytokinins are helping physiologists understand how plant hormones work to produce the total growth pattern of the plant. In general, the undifferentiated plant cell has two courses open to it: either it can enlarge, divide, enlarge, and divide again, or it can elongate without cell division. The cell that divides repeatedly remains essentially undifferentiated, or embryonic, while the elongating cell tends to differentiate, or become specialized. In studies of tobacco stem tissues, addition of IAA to the tissue culture produced rapid cell expansion, so that giant cells were formed. Kinetin alone had little or no effect. IAA plus kinetin resulted in rapid cell division, so that large numbers of relatively tiny cells were formed. In other words, addition of the kinetin (although, you will notice, not kinetin alone) switched the cells to a meristematic course.

FLOWERING HORMONE?

In many species of plants, flowering can be induced by exposing the leaves of the plant to an appropriate period of light. (Such exposure is known as photoinduction; we shall have more to say about this phenomenon in the next chapter.) Apparently something is transmitted from leaf to bud that causes the apical meristem of the bud to form floral primordia rather than leaf primordia. This "something" has been called the flowering hormone, although its existence has not yet been proved.

The earliest experiments on this hypothetical flowering hormone were carried out by a Russian scientist, M. H. Chailakhyan, in the 1930s. Working with a species of chrysanthemum, Chailakhyan found that if the upper portion of the stem was stripped of its leaves and the leaves on the lower stem were exposed to light for the appropriate induction period, the plant would flower. If, how-

ever, only the upper, leafless stem and its buds were exposed, no flowering occurred. He interpreted these results as indicating that the leaves form a hormone that moves to the apex and initiates flowering. He named this hypothetical hormone florigen, the "flower maker."

Subsequent experiments showed that the flowering response does not take place if the leaf is removed immediately after photoinduction. But if the leaf is left on the plant for a few hours after the induction cycle is complete, it can then be removed without stopping flowering. The flowering hormone can pass through a graft from a photoinduced plant to a noninduced plant. If a branch is girdled, that is, if the "bark" (the cortex and phloem) is removed, florigen movement ceases. This led to the conclusion that florigen moves by way of the phloem system, the means by which most organic substances are transported. However, the speed of movement of the flowering stimulus is much slower than sieve-tube transport.

–16 Experiments that indicate the existence of a flowering hormone. (a) When certain plants, such as the cocklebur shown here, are exposed to an appropriate light cycle, those with leaves flower and those without leaves do not. When even one-eighth of a leaf remains on a plant, flowering occurs, and the illumination of a single leaf—not necessarily the whole plant—suffices. These experiments indicate that a chemical originating in the leaves causes the plant to flower. (b) This conclusion is supported by experiments on branched plants. Exposure of one branch to the light induces flowering on the other branch as well, even when only a portion of a leaf is present on the lighted branch. (c) When two plants are grafted together, exposure of one of the plants to the light cycle induces flowering in both the lighted plant and the grafted one. This effect occurs even when a piece of paper is inserted across the graft.

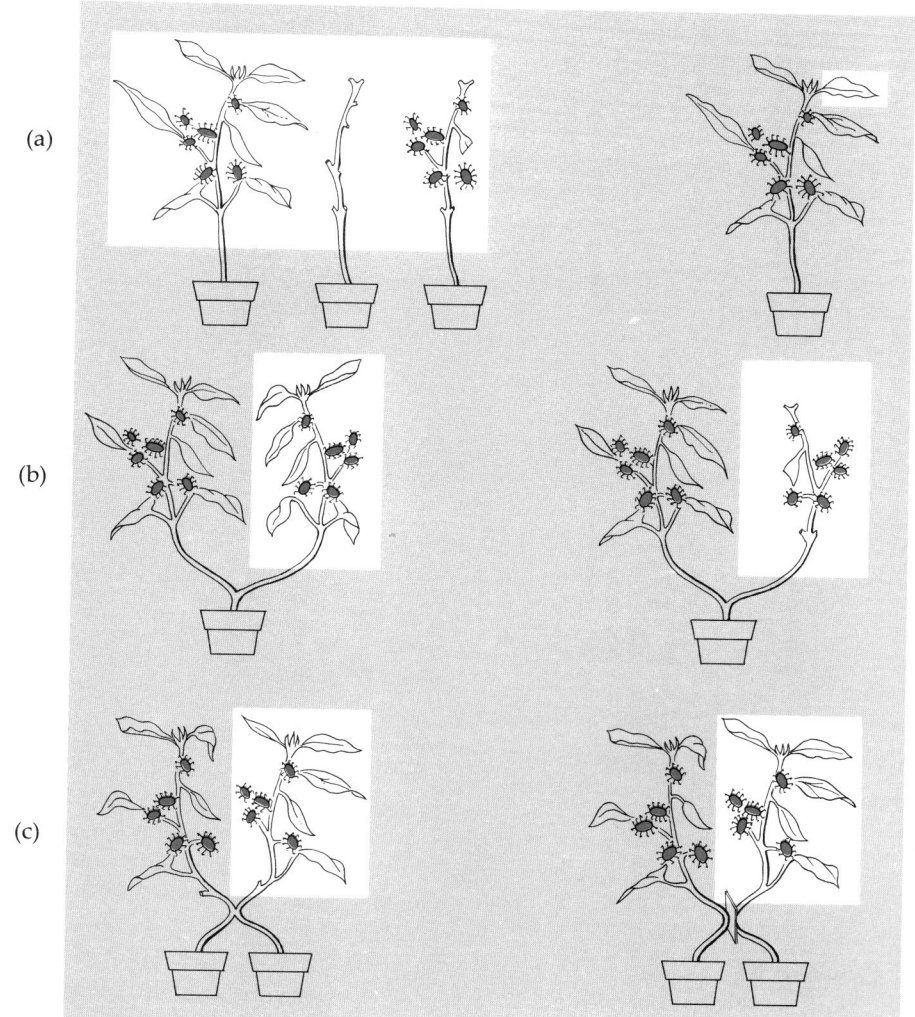

(a)

(b)

(c)

27–17 *Abscisic acid (dormin), an inhibitor that blocks the action of the growth-promoting hormones.*

27–18 *Winter bud of a pignut hickory. These buds form in one growing season and develop in the next. The delicate tissues inside the bud are protected from mechanical injury and desiccation by bud scales, which are small, tough modified leaves.*

In some plants—the Biloxi soybean is an example—leaves must be removed from the grafted receptor plant or it will not flower. This observation suggests that in noninduced plants, the leaves may produce an inhibitor. In fact, some investigators have concluded on the basis of such evidence that there is no substance that initiates flowering but rather a substance that inhibits flowering unless it is removed by the proper conditions. Strong evidence now suggests that both inhibitors and promoters are involved in the control of flowering.

ABSCISIC ACID

At certain times, the survival of the plant depends on its ability to restrain its growth or its reproductive activities. Following the early discovery of the growth-promoting hormones, plant physiologists began to speculate that regulatory hormones with inhibitory actions would also be found. Finally, in 1949, it was discovered that the dormant buds of ash and potatoes contained large amounts of inhibitors. These inhibitors blocked the effects induced by IAA in the *Avena* coleoptile. When dormancy in the buds was broken, the inhibitor contents declined. These inhibitors became known as dormins.

During the 1960s, several investigators reported the discovery in leaves and fruits of a substance capable of accelerating abscission. One of these, called abscisin, was chemically identified. In 1965, one of the dormins was also identified chemically, and the two, the abscisin and the dormin, were found to be identical. The compound is now known as *abscisic acid*.

Abscisic acid is collected largely from the ovaries of fruits. The fruit of the cotton plant has proved to be a particularly rich source. The largest amounts of abscisic acids are found at the time of fruit drop.

Application of abscisic acid to vegetative buds changes them to winter buds by converting the leaf primordia into bud scales. Its inhibitory effects on buds can be overcome by gibberellin. The appearance of the enzyme alpha-amylase, induced by gibberellin in the barley seed, is inhibited by abscisic acid. Abscisic acid has little effect on dwarf plants, but it reduces the growth of normal plants. This inhibition also can be counteracted by gibberellin.

In short, abscisic acid, in its many and varied effects, serves as a direct opponent to the action of growth-promoting hormones. Abscisic acid's role emphasizes the concept that growth is the result of a balance of different factors.

ETHYLENE

During the time of ripening of many fruits, there is a large increase in cellular respiration, evidenced by a general uptake of oxygen. This phase is known as the climacteric. The relationship between the climacteric and the other events of food ripening is not known, but the ripening of fruits can be suppressed by suppressing the intensity of the climacteric. A decrease in available oxygen suppresses respiration, which is why fruits and vegetables "keep" longer when stored in plastic bags. Cold also suppresses it, and in some fruits, cold stops the climacteric permanently. After the climacteric, senescence sets in, and the fruit becomes susceptible to invasions by fungi and other microorganisms.

In the early 1900s, many fruit growers made a practice of improving the color of citrus fruits by "curing" them in a room with a kerosene stove. (Long before this, the Chinese used to ripen fruits in rooms where incense was being burned.) It was long believed that it was the heat that ripened the fruits. Ambitious fruit growers, who went to the expense of installing more modern heating equipment, found to their sorrow that this was not the case. As experiments showed, it is actually the incomplete combustion products of the kerosene that are responsible. The most active gas was identified as ethylene. As little as 1 part per million of ethylene in the air will speed the onset of the climacteric.

As early as 1910, it was reported that gases emanating from oranges hastened the ripening of bananas, but it was not until almost 25 years later that ethylene was identified as a natural product of numerous fruits and plant tissues. The amounts produced by plants are very small, and new and extremely sensitive assay methods had to be developed before it could be proved that ethylene production began *before* the climacteric, even though the largest amounts coincide with the climacteric. When this was established, ethylene became generally accepted as a natural plant-growth regulator. It has now been found in fruits (in all the types tested), flowers, leaves, leafy stems, and roots of many different species, and also in certain types of fungi.

Auxin, in some plants, causes a burst of ethylene production, and it is believed that some of the effects on fruits and flowers once attributed to auxin are related to auxin's effects on ethylene production.

In addition to its effects on fruit ripening, ethylene causes leaves to abscise, chlorophyll to bleach, flowers to fade, and the upper surface of petioles to overgrow so that leaves bend downward. It interferes with the normal geotropic responses of seedlings, so that if they are placed on their sides, both roots and shoots grow horizontally. Some of these effects may or may not have anything to do with the growth of the plant under normal conditions.

The way in which ethylene affects the ripening of fruit is not known, but a variety of hypotheses are currently under investigation.

CONCLUSIONS

Exposing a plant tissue to a hormone has been compared to putting a coin in a vending machine. You may get your morning newspaper, a candy bar, or four minutes of country and western music. It depends not so much on the

-19   *Ethylene, a simple hydrocarbon involved in the ripening of fruit. Unlike any other known hormone, it is a gas.*

$$CH_2 = CH_2$$

-20   *The timing of hormone production for an apple throughout the growing season.*

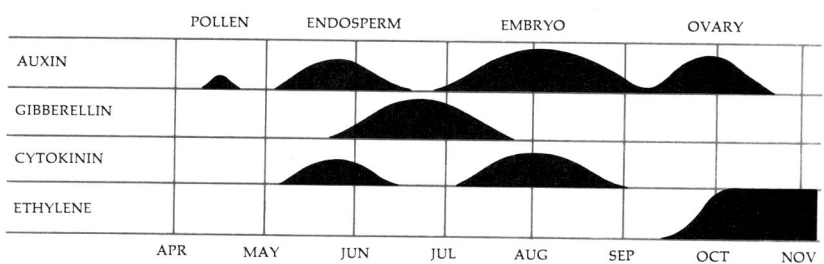

coin as on the machine in which you put it. Similarly, the effects of plant hormones depend largely on the target tissues and the chemical environment in which these tissues find themselves.

SUMMARY

Hormones are important regulators of growth in organisms. A hormone is a chemical produced in particular tissues of an organism and carried to other tissues, where it exerts specific effects. Characteristically, it is active in extremely small amounts.

Auxins are hormones that are produced principally in rapidly dividing tissues, such as coleoptile tips and apical meristems. They generally travel toward the base of the plant. They cause lengthening of the shoot and the coleoptile, chiefly by promoting cell elongation. Studies indicate that their effect on cell elongation is achieved in some indirect way by increasing the plasticity of the cell wall, permitting the cell to expand. Auxins inhibit growth in lateral buds, thus restricting growth principally to the apex of the plant. The same quantity of auxin that promotes growth in the stem inhibits growth in the main root system. Auxins promote the initiation of branch roots and adventitious roots, however. They retard abscission in leaves and fruits. In fruits, auxin produced by seeds or the pollen tube stimulates growth of the ovary wall, perhaps as a result of increased ethylene production. The capacity of auxins to produce such varied effects is believed to result from the different responses of the various target tissues.

Gibberellins were first isolated from a parasitic fungus that causes abnormal growth in rice seedlings. They were subsequently found to be natural growth hormones present in higher plants. The most dramatic effects of gibberellins are seen in dwarf plants, in which application of gibberellins restores normal growth, and in plants with a rosette form of growth, in which gibberellins cause bolting. Gibberellins are active at the time of seed germination in grasses. One of the mechanisms of action of gibberellin in the barley seed is known: The embryo releases gibberellins, which cause the aleurone layer of the endosperm to produce several enzymes including alpha-amylase. Alpha-amylase breaks down the stored starch in the endosperm, releasing sugar. The sugar nourishes the embryo and promotes the germination of the seed. It is believed that gibberellin derepresses the specific genes in the aleurone cells that code for the enzyme.

The cytokinins, a third class of growth hormone, were first discovered as a consequence of their capacity to promote cell division in plant tissue cultures. They are chemically related to adenine. Cytokinins plus auxin cause cell division in plant tissue cultures. In cultures of tobacco callus, a high ratio of auxin to cytokinin promotes root formation while a higher proportion of cytokinin promotes bud formation. In intact plants of some species, cytokinin promotes the growth of lateral buds, opposing the effects of auxins. In leaves, cytokinins stimulate protein synthesis and delay senescence.

Abscisic acid is a growth-inhibiting hormone that has been found in dormant buds and in fruits, with a maximum amount present just before fruit drop. Abscisic acid induces dormancy in vegetative buds, accelerates abscission, and thus opposes the effects of growth hormones.

Ethylene is a gas produced by the incomplete combustion of hydrocarbons. It also emanates from fruits during the ripening process. Ethylene is now considered a natural growth regulator. It produces a number of physiological effects, including ripening in fruit.

QUESTIONS

1. Name the five types of hormones that have been identified chemically and briefly describe the effects of each.
2. The distinctions we make among the plant hormones refer to general classes of their effects rather than to the precise chemical names of the substances actually produced by particular plants. Why might it be relatively easy to test chemicals for hormonal activity but be particularly difficult to determine just which chemical a plant actually produces and uses?
3. Describe Went's experiments and the conclusions that can be drawn from them.
4. What is the mechanism of action of gibberellin in barley seeds?
5. If you prune many shrubs, they grow bushier. Devise a simple experiment that can show that the change in overall shape of the plant is not simply a result of reducing apical elongation by cutting. Suggest a hormonal mechanism for the phenomenon of bushiness induced by pruning.
6. Describe the experiments that indicate the existence of a flowering hormone.

Chapter 28

Plant Responses to Stimuli

Although the term "behavior," as we noted earlier, is generally restricted to phenomena involving organisms with nervous systems—i.e., animals—plants are capable of some responses that, if they occurred in animals, would clearly be called behavior. These responses have survival value because they enable plants to modify their activities in keeping with changes in their environment.

PHOTOPERIODISM

In many regions of the biosphere, the most important environmental changes affecting land plants (and indeed, land organisms, in general) are those that result from the changing seasons. Plants are able to accommodate themselves to these changes because of their capacity to sense and even to anticipate the yearly calendar of events—the first frost, the spring rains, long dry periods, long growing spells, and even the time that nearby plants of the same species will be in flower. For many plants, all of these determinations are made in the same way: by measuring the relative periods of light and darkness. This phenomenon is known as *photoperiodism.*

The discovery of photoperiodism depended on two factors. The first was the continuing attempt to cultivate and improve plants useful to man, an effort that began almost 3,000 years ago. The second factor was the appearance of an unusual plant, one that by not following the "rules," called attention to what the other organisms had been doing quietly and efficiently for millenia. Many scientific discoveries have come about either as a result of man's attempts to harness or control natural phenomena and many also have been the result of the observance of unusual forms.

The story of the discovery of photoperiodism in plants began fifty years ago when a mutant appeared in a field of tobacco plants growing near Washington, D.C. The new variety had unusually large leaves and stood over 10 feet tall. As the season progressed, the regular plants flowered, but Maryland Mammoth, as it came to be called, merely grew bigger and bigger. Scientists from the Department of Agriculture took cuttings from the Mammoth and put them in the greenhouse, where they would be safe from frost. These cuttings flowered in December, although by then they were only 5 feet tall, half the size of their parent. New Maryland Mammoths grew from their seed, and these, too, did not flower until December.

Coincidentally, these same researchers were carrying out experiments with

−1 *Photoperiodism is the phenomenon that brings the plants of many species into flower simultaneously, year after year. This is a cherry tree, blooming in Brooklyn.*

the Biloxi variety of soybean. Agriculturalists were interested in spacing out the soybean harvest by making successive sowings of seeds at two-week intervals from early May through June. But spacing out the planting had no effect; all the plants, no matter when the seeds were sown, came into flower at the same time —in September.

The investigators started growing these two kinds of plants—Maryland Mammoth tobacco and Biloxi soybeans—under a wide variety of controlled conditions of temperature, moisture, nutrition, and light. They eventually found that the critical factor in both species was the length of daylight during the 24-hour cycle. Neither plant would flower unless the day length was shorter than a critical number of hours. Consequently, Biloxi soybeans, no matter when they were planted, all flowered after the days became short enough, which was in September, and the Maryland Mammoth, no matter how tall it grew, would not flower until December, after the days became even shorter. (Note that these mutant strains, like many mutant forms, could not have survived in nature, at least not at the latitudes at which they appeared. They would have died without reproducing.)

28-2   *The cocklebur, a short-day plant, was an*
important tool for experimental studies
on photoperiodism.

Long-day Plants and Short-day Plants

The botanists went on to test and confirm this discovery with many other species of plants. Following this single lead, they were able to answer a host of questions that had long troubled both professional botanists and amateur gardeners. Why, for example, is there no ragweed in northern Maine? The answer, they found, is that ragweed starts producing flowers when the day is less than 14½ hours long. The long summer days do not shorten to 14½ hours in northern Maine until August, and then there is not enough time for ragweed seed to mature before the frost. Why doesn't spinach reseed itself in the tropics? Because spinach needs 14 hours of light a day for a period of at least two weeks in order to flower and such long days never occur in the tropics. As you can see, the discovery of the photoperiodic control of flowering not only provided an explanation of plant distribution but was of great practical importance.

The investigators found that plants are of three general types, which they called day-neutral, short-day, and long-day. Day-neutral plants flower without regard to day length. Short-day plants flower in early spring or fall; they must have a light period *shorter* than a critical length. For instance, the cocklebur flowers when exposed to 15½ hours or less of light. In addition to the Biloxi soybean and the Maryland Mammoth already mentioned, other short-day plants are poinsettias, strawberries, primroses, and some chrysanthemums.

Long-day plants, which flower chiefly in the summer, will flower only if the light periods are *longer* than a critical length. Spinach, potatoes, some wheat varieties, clover, henbane, and lettuce are examples of long-day plants.

Note that cocklebur and spinach will both bloom if exposed to 14 hours of daylight, yet one is designated as short-day and one as long-day. The important factor is not the absolute length of the photoperiod but rather whether it is longer or shorter than a particular critical interval for that variety. And in some varieties, 5 or 10 minutes difference in exposure can determine whether or not a plant will flower.

Photoperiodism has now been demonstrated in many species of insects, fish, birds, and mammals, influencing such diverse phenomena as the metamorphosis from caterpillar to butterfly, sexual behavior, migration, molting, and seasonal changes in coat or plumage.

28-3   *The relative length of day and night*
determines when plants flower. The four
curves depict the annual change in day
length in four North American cities at
four different latitudes. The black lines
indicate the effective photoperiod of three
different short-day plants. The cocklebur,
for instance, requires 15½ hours or less
of light. In Miami, it can flower as soon as
it matures, but in Winnipeg the buds do
not appear until early in August, so late
that the frost usually kills plants before
the seed is mature.

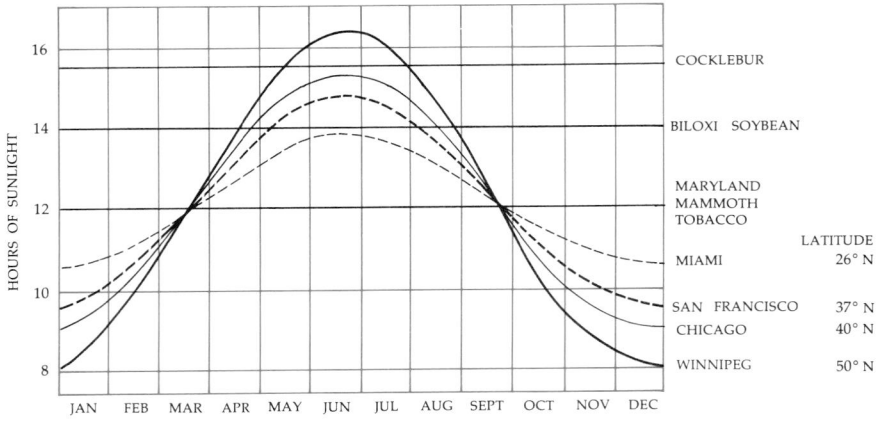

Measuring the Dark

In 1938, two other investigators, Karl C. Hamner and James Bonner, began a study of photoperiodism, using the cocklebur as their experimental tool. As we mentioned previously, the cocklebur is a short-day plant requiring 15½ hours or *less* of light per 24-hour cycle to flower. It is particularly useful for experimental purposes because a single exposure under laboratory conditions to a short-day cycle will induce flowering two weeks later, even if the plant is immediately returned to long-day conditions. The cocklebur can withstand a great deal of rough treatment, surviving even if its leaves are removed. Hamner and Bonner showed that it is the leaf blade of the cocklebur that responds to the photoperiod. A plant stripped of all its leaves cannot be induced to flower, as we saw in Figure 27–16. But if as little as one-eighth of a fully expanded leaf is left on the stem, the single short-day exposure induces flowering.

In the course of these studies, in which they tested a variety of experimental conditions, Hamner and Bonner made a crucial and totally unexpected discovery. If the period of darkness was interrupted by as little as a 1-minute exposure to a 25-watt bulb, flowering did not occur. Interruption of the light period by darkness had no effect on flowering whatsoever. Subsequent experiments with other short-day plants showed that they, too, required periods not of uninterrupted light but of uninterrupted darkness.

What about long-day plants? It was found that they also measure darkness. A long-day plant that will flower if it is kept in a laboratory in which there is light for 16 hours and dark for 8 hours will also flower on 8 hours of light and 16 hours of dark if the dark is interrupted by even a brief exposure to light.

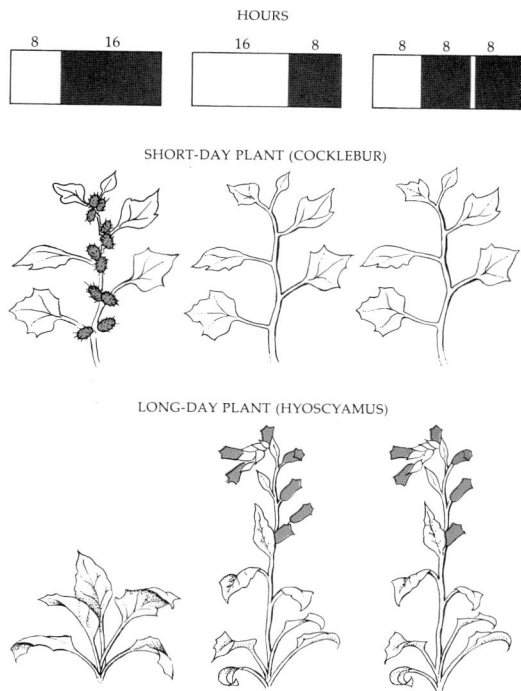

HOURS

SHORT-DAY PLANT (COCKLEBUR)

LONG-DAY PLANT (HYOSCYAMUS)

8–4 *Experiments on photoperiodism showed that plants measured the darkness rather than the light. Short-day plants flower only when the darkness exceeds some critical value. Thus, the cocklebur, for instance, will flower on 8 hours of light and 16 hours of darkness. If the 16-hour period of darkness is interrupted even very briefly, as shown on the right, the plant will not flower. The long-day plant, on the other hand, which will not flower on 16 hours of darkness, will flower if the darkness is interrupted. Long-day plants flower only when the darkness is less than some critical value.*

28–5 *Lettuce seeds of certain varieties will ger-*
minate only upon exposure to red light,
present in sunlight. Such a response is
found among small, light seeds whose
delicate seedlings would have little chance
of survival if the seeds were deeply buried
in the soil.

Thus, the investigation of photoperiodism revealed, among other insights, that the phenomenon had been called by the wrong name, but the term photoperiodism has nevertheless persisted.

PHYTOCHROME: THE CHEMICAL BASIS OF PHOTOPERIODISM

Hamner and Bonner had shown that if the dark period is interrupted by a single exposure to light from an ordinary bulb, the cocklebur will not flower. Following this lead, a team of research workers at the U.S. Department of Agriculture Research Station in Beltsville, Maryland, began to experiment with light of different wavelengths, varying the intensity and duration of the flash. They found that red light of about a 660-nanometer wavelength was the most effective in preventing flowering in the cocklebur and other short-day plants. Light of this wavelength is present in white light, that is, in daylight and artificial incandescent and fluorescent light (page 142). This wavelength (660 nanometers) was also the most effective, they found, in promoting flowering in long-day plants. (Effectiveness is measured in terms of percentage of plants that respond.)

The Beltsville group found their next clue in the report of an earlier study performed with lettuce seeds. Lettuce seeds of the Grand Rapids variety germinate only if they are exposed to light. (This is true of many small seeds, which need to germinate in loose soil and near the surface in order for the seedlings to be able to break through.) The earlier workers, in studying the light requirement of lettuce seeds, had shown that red light stimulated germination and that red light of a slightly longer wavelength (far-red) inhibited germination. In fact, in lettuce seeds exposed to far-red light, the percentage that germinated was even lower than that of the control seeds (those not illuminated at all). The Beltsville group found that when red light was followed by far-red light before the seeds were returned to darkness, the seeds did not germinate. The red light most effective in inducing germination in seeds was light of the same wavelength as that involved in the flowering response—about 660 nanometers. Furthermore, they found that the light most effective in inhibiting the response in seeds to red light was the far-red light, of a wavelength of 730 nanometers. The series of exposures could be repeated over and over; the number of exposures did not matter, but the nature of the final one did. If the series ended with an exposure of red (660) light, the great majority of the seeds germinated. If it ended with a far-red (730) exposure, the great majority did not.

Far-red light was then tried on short-day and long-day plants, with the same on-off effect. Far-red light alone, when given during the dark period, had no effect. But an exposure of far-red light immediately following an exposure of red light canceled the effects of the red light.

Discovery of Phytochrome

Here is how the Beltsville group and others have interpreted these results. The plant contains a pigment called *phytochrome* that exists in two different forms. These two forms of phytochrome came to be known as P_{660} and P_{730}. P_{660} absorbs red light and is converted to P_{730}, which is the active form. This conversion takes place in daylight or in incandescent light; in both of these lights, red wavelengths predominate over far-red. When P_{730} absorbs far-red light, it is converted back to P_{660}. The P_{730}-to-P_{660} conversion can also take place, although

28–6 *Absorption spectra of the two forms of phytochrome, P_{660} and P_{730}. This shift in absorption spectra made it possible to isolate the pigment.*

ULTRA-VIOLET VIOLET BLUE YELLOW RED FAR-RED

P_{660}

P_{730}

ABSORBENCY

WAVELENGTH OF LIGHT (NANOMETERS)

28–7 *P_{660} changes to P_{730} when exposed to red light. P_{730} reverts to P_{660} when exposed to far-red light. In darkness, P_{730} reverts to P_{660}.*

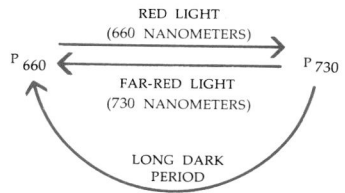

RED LIGHT
(660 NANOMETERS)

P_{660} P_{730}

FAR-RED LIGHT
(730 NANOMETERS)

LONG DARK
PERIOD

much more slowly, in the dark, which is how it usually occurs in nature. In short-day plants, P_{730} inhibits flowering under conditions in which flowering would otherwise occur. In long-day plants, P_{730} promotes flowering under appropriate conditions. Far-red (730 nanometers) light is present in white light in only very small amounts. When plants are in white light, the phytochrome is cycled between the two forms, with P_{730} predominating.

In light-requiring seedlings, red light (of 660 nanometers) promotes germination; far-red inhibits it. In this way, seeds inform themselves of light conditions. On a forest floor when the trees are in full leaf, the chlorophyll and accessory pigments of the leaves absorb light in the blue and red regions of the spectrum, shifting the balance toward far-red and so toward the presence of P_{660} phytochrome. When a forest area is cleared, many seedlings appear rapidly; their rapid appearance is apparently the result of the breakdown of the inhibitory P_{730} phytochrome.

Isolation of Phytochrome

Phytochrome is present in plants in very small amounts. To detect it, a spectrophotometer was needed which was sensitive to extremely small changes in patterns of light absorbency. (Large changes can be detected by the eye, of course, as changes in color.) Such a spectrophotometer was developed some seven years after the existence of phytochrome was proposed, and it was used to first detect and subsequently, in 1959, to isolate the pigment. The pigment proved to be blue in color (why might you expect this color?) and to show the characteristic red–far-red conversion in the test tube by reversibly changing color slightly in response to red or far-red light.

There are two portions to the phytochrome molecule, it was found: a light-absorbing portion and a large protein portion. The light-absorbing portion is a phycobilin, a type of pigment found in blue-green and red algae. It functions in these organisms as an accessory pigment in photosynthesis.

Phytochrome is found in small amounts throughout the plant body, with the largest concentration in the apical meristems of shoots and roots. How phytochrome works is not known. There are three current hypotheses. One is that the molecule acts as an enzyme. According to this theory, when red light is absorbed by the pigment portion of the molecule, the energy of the light changes the structure of the pigment and this, in turn, changes the structure of the

499 *Plant Responses to Stimuli*

28–8 *Chloroplast development. (a) The immature plastid contains small crystalline structures (top). (b) In the presence of light, this structure begins to break up into elongated vesicles. (c) The vesicles flatten into stacks of thylakoids. (d) A mature chloroplast.*

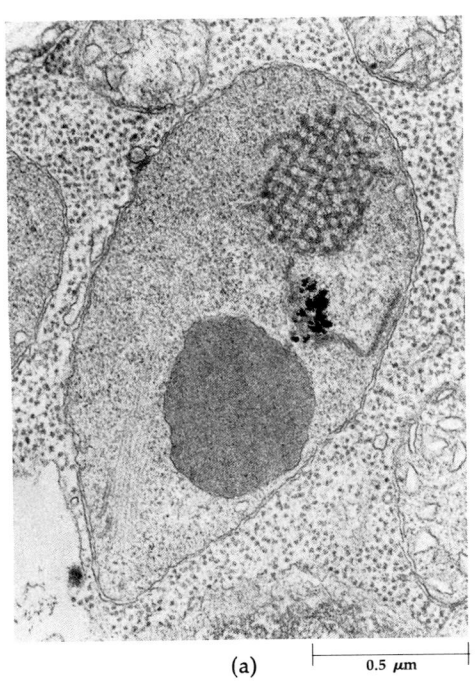

(a) 0.5 μm

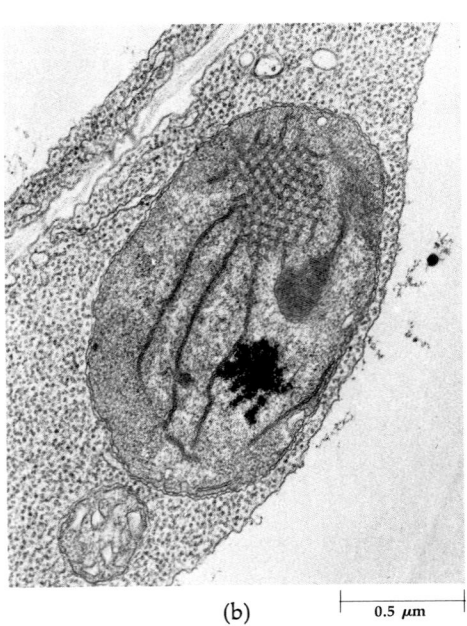

(b) 0.5 μm

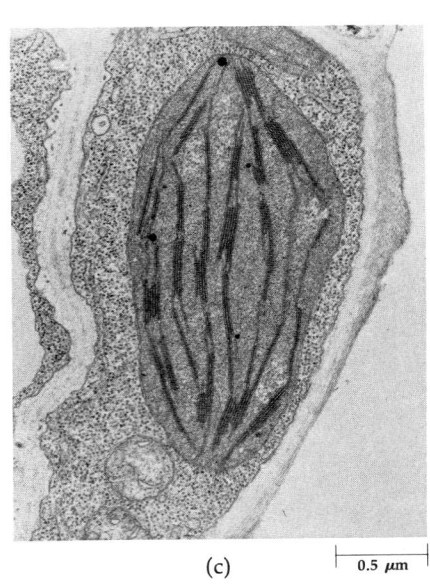

(c) 0.5 μm

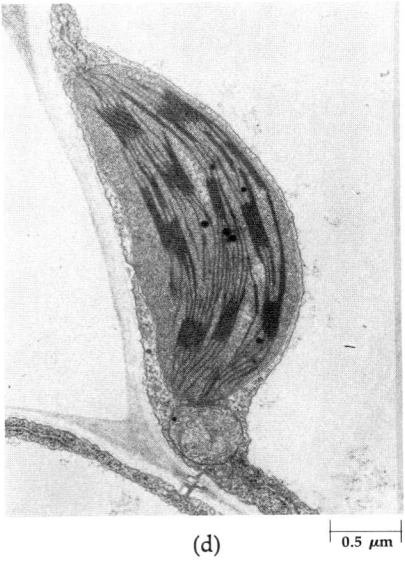

(d) 0.5 μm

protein portion, rendering the enzyme active. The enzyme becomes inactive in the dark or can be inactivated by a flash of far-red light. A second hypothesis is that the primary action of phytochrome is to alter membrane permeability, thus permitting or prohibiting the passage of hormones or other substances. The third hypothesis is that phytochrome releases gene activity by inactivating a repressor substance.

You will note that the question of the relationship between the phytochrome shift and the critical time periods involved in flower induction has not yet been answered. We shall return to it again in the course of this chapter.

Photomorphogenesis

Light plays a particularly important role in the morphogenesis of seedlings. Dark-grown seedlings are usually almost colorless. The chloroplasts develop from small proplastids under the stimulus of light. Development depends on a complex interplay of genetic and environmental factors. Proplastids are present in all the cells of a green plant, but whether or not they mature and green depends on the genetic makeup of the chloroplast itself (which, as you recall, has DNA in it), the genetic makeup of the host cell, and the differentiation of the host cell—chloroplasts develop only in certain cell types—as well as on light.

In addition to being colorless, seedlings growing in the dark are elongated, have small leaves, and, in the case of some dicots, have stems that curve, forming a hook at the top (all of which are associated with the necessity for the seedling to break its way through the soil if it is to survive). This group of symptoms is known as etiolation. On exposure to light, the elongation process slows down, the leaves expand, and the hook straightens out. In this way, all of a plant's energy resources are concentrated on reaching the light. These responses also involve phytochrome; the unfolding of the leaf is promoted by red light and inhibited by far-red.

28–9 *Leaves of the wood sorrel, day (a) and night (b). One hypothesis concerning the function of such "sleep" movements is that they protect the leaves from absorbing moonlight on bright nights, thus protecting photoperiodic phenomena. Another, proposed by Darwin almost 100 years ago, is that the folding protects against heat loss from the leaves by night. (Richard F. Trump, OMIKRON.)*

(a)

(b)

28–10 *Dark-grown seedlings, such as the ones on the right, are thin and pale with longer internodes and smaller leaves than normal seedlings, such as those on the left. This group of characteristics, known as etiolation, has survival value for the seedling because it increases its chances of reaching light before its stored energy supplies are used up. (Richard F. Trump, OMIKRON.)*

Other Phytochrome Responses

Many plants raise and fold their leaves at night, open and spread them in the daytime. Similarly, the flowers of many plants open or close at specific times during the day. In some species, these movements, like those of the stomata, appear to involve changes in turgor pressure. The light-sensing pigment involved in these responses is also phytochrome. Other phytochrome-mediated light responses include abscission of leaves and induction of dormancy, both of which involve photoperiodism, and also the stem elongation (bolting) of rosette plants.

These responses also involve plant hormones. Abscisic acid is involved in both abscission and dormancy. Gibberellins can also produce bolting and may be involved in lengthening of the internodes seen in etiolation. And flowering certainly involves flowering hormone, if one exists, or some other biochemical process apart from the red–far-red interconversion.

In short it now appears that the phytochrome system—which is a plant's way of detecting whether or not an adequate amount of white light is present for its purposes—can be coupled with the on-off switches of a variety of physiological processes of the plant.

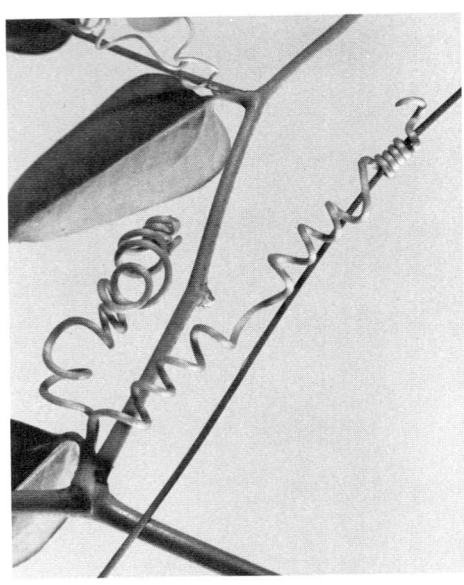

28–11 *Tendrils of smilax (greenbrier). Twisting is caused by varying growth rates on different sides of the tendril.*

TOUCH RESPONSES IN PLANTS

Plants also respond to touch. One of the most common examples is seen in tendrils (which are modified leaves) in some species of plants and in stems of others, such as twining vines. The tendrils or stems wrap around any object with which they come in contact (see Figure 28–11) and so enable the plant to cling and climb. The response can be rapid; a tendril may wrap around a support in less than a minute. Cells touching the support shrink slightly and those on the other side elongate. There is some evidence that auxin plays a role in this response.

A more spectacular response is seen in the sensitive plant, *Mimosa pudica*, in which the leaflets and sometimes entire leaves droop suddenly when touched. This response is a result of a sudden change in turgor pressure at the base of leaflets and leaf, but how the change is brought about is not known. There is also some controversy about its survival value to the plant. *Mimosa pudica* often grows in dry, exposed areas where it may be subjected to drying winds; strong winds may shake the leaves enough to make them fold up, so conserving water. Another suggestion is that the wilting response makes the plant unattractive to grazing animals.

The triggering of turgor changes by touch is also involved in the capture of prey by the carnivorous Venus flytrap. The leaves of the Venus flytrap are hinged in the middle, and each leaf half is equipped with three sensitive hairs. When an insect walks on one of these leaves, attracted by the nectar on the leaf surface, it brushes against the hairs, triggering the traplike closing of the leaf. The toothed edges mesh, the leaf halves gradually squeeze closed, and the insect is pressed against digestive glands on the inner surface of the trap.

The trapping mechanism is so specialized that it can distinguish between living prey and inanimate objects, such as pebbles and small sticks, that fall on its leaves by chance: the leaf will not close unless two of its hairs are touched in succession or one hair is touched twice.

28–12 *Sensitive plant* (Mimosa pudica). (a) *Normal position of leaves and leaflets.* (b, c) *Successive responses to touch.*

(a)

(b)

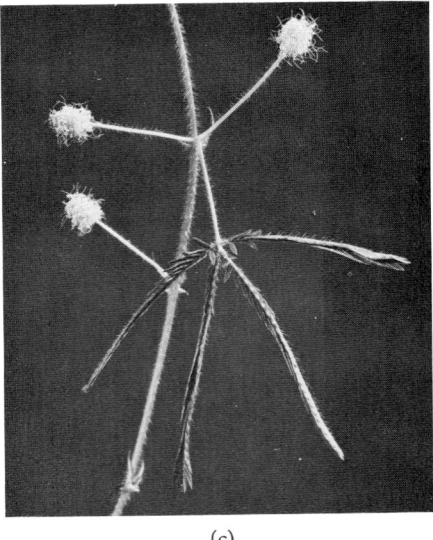

(c)

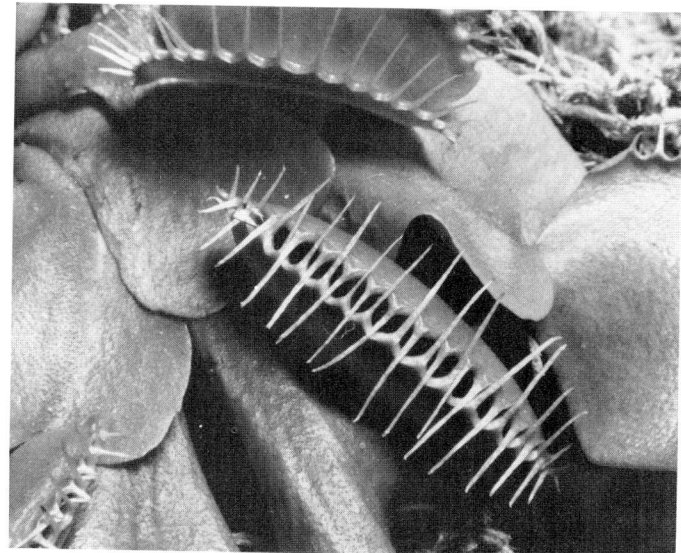

-13   *Touch responses in the Venus flytrap.*

CIRCADIAN RHYTHMS

Some plants, as we noted previously, have flowers that open in the morning and close at dusk or they spread their leaves in the sunlight and fold them toward the stem at night. As long ago as 1729, the French scientist Jean-Jacques de Mairan noticed that these diurnal (daily) movements continue even when the plants are kept in dim light. More recent studies have shown that less evident activities, such as photosynthesis, auxin production, and the rate of cell division, also have regular daily rhythms, which continue even when all environmental conditions are kept constant. These regular day-night cycles have come to be called circadian rhythms, from the Latin words *circa,* meaning "about," and *dies,* "day." Circadian rhythms now have been found throughout the plant and animal kingdoms.

Biological Clocks

Are these rhythms internal—that is, caused by factors within the plant or animal itself—or is the organism keeping itself in tune with some external factor? For a number of years, biologists debated whether it might not be some environmental force, such as cosmic rays, the magnetic field of the Earth, or the Earth's rotation, that was setting the rhythms. Attempts to settle this recurrent controversy have led to countless experiments under an extraordinary variety of conditions. Organisms have been taken down in salt mines, shipped to the South Pole, flown halfway around the world in airplanes, and, most recently, orbited in satellites. Although there is still a vocal minority that believes that circadian rhythms are under the influence of a subtle geophysical factor, most workers now agree that the rhythms are endogenous—that is, they originate within the organism. The strongest evidence in support of this belief is that the rhythms are not exact. Different species and different individuals of the same species often have slightly different, but consistent, rhythms, often as much as

28–14 *"Sleep movement" rhythms in the bean plant. Many legumes, such as the bean, orient their leaves perpendicular to the rays of the sun during the day and fold them up at night. These "sleep movements" can easily be transcribed on a rotating chart by a delicately balanced pen-and-lever system attached to the leaf by a fine thread (a). The rhythm will persist for several days in continuous dim illumination. A representative recording is seen in (b).*

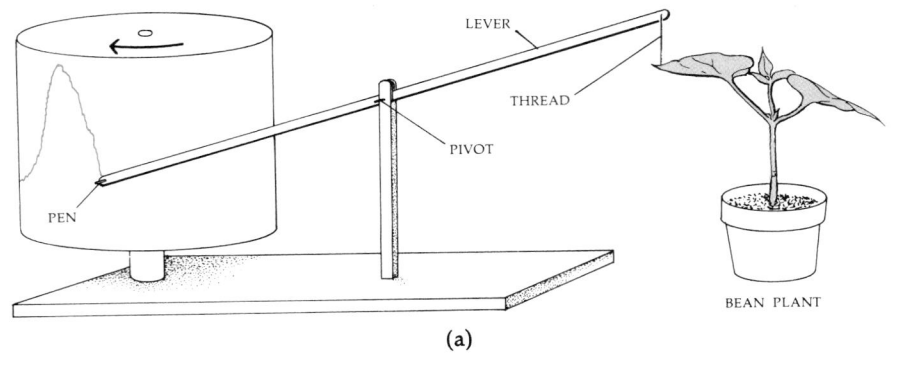

(a)

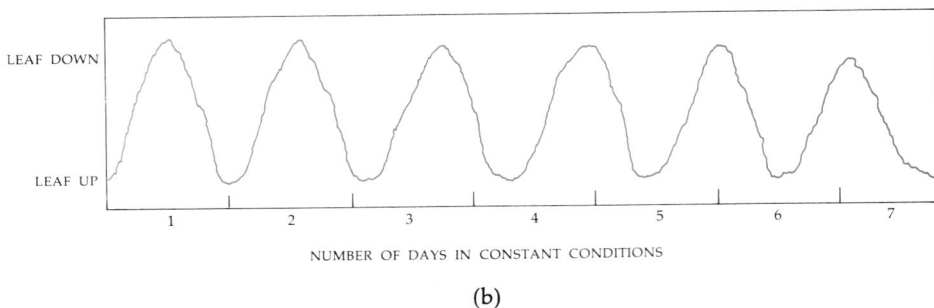

(b)

an hour or two longer or shorter than 24 hours. Nothing, however, is known about the physical or chemical nature of this internal timing device, which is often referred to as a *biological clock*.

Changing the Rhythms

Although circadian rhythms probably originate within the organisms themselves, they can be modified by external conditions—a fact that is, of course, important to the survival of both individuals and species. For instance, a plant whose natural daily rhythm shows a peak every 26 hours when grown under continuous dim light can adjust its rhythm to 14 hours of light and 10 of darkness. It can also adjust to 11 hours of light and 11 of dark (or 22 hours). Such adjustment to an externally imposed rhythm is known as entrainment. If the new rhythm is too far removed from the original one, however, the organism will "escape" the entrained rhythm and revert to its natural one. A plant that has been kept on an artificial or forced rhythm, even for a long period of time, will revert to its normal internal period when returned to continuous dim light.

Clock Functions

Biological clocks are believed to play an essential role in many aspects of plant and animal physiology. For instance, insects are more active in the early evening hours. Bats that feed on insects begin to fly each evening just when the insects are most available. Moreover, caged bats under controlled and constant laboratory conditions continue to show this sort of activity about every 24

hours, indicating that they are following an internal rhythm, not merely responding to environmental cues.

Some plants secrete nectar or perfume at certain specific times of the day. As a result, insects—which have their own biological clocks—become accustomed to visiting these flowers at these times, thereby ensuring maximum rewards for both the insects and the flowers.

The ability to tell time also appears to be involved in a number of complex and fascinating phenomena such as the extraordinary ability of migrating birds and turtles to navigate.

Biological clocks enable organisms to recognize the changing seasons of the year by "comparing" external rhythms of the environment, such as changes in day length, to their own relatively constant internal rhythms. This capacity, as we shall see, is an important factor in regulating the growth cycle of many plants.

Finally, biological clocks are clearly involved in photoperiodism, the topic with which this chapter began. In order for organisms to detect changes in day length—and remember that in some cases, they are accurate to within 15 minutes—they must have some constant, a clock, with which to compare them. The chemical nature of the biological clock—or indeed if there is just one kind of clock or many—is still not known.

SUMMARY

Plants respond to a number of environmental stimuli. Photoperiodism is the response of organisms to changing periods of light and darkness in the 24-hour day. Such a response controls the onset of flowering in many plants. Some plants will flower only when the periods of light exceed a critical length. Such plants are known as long-day plants. Other plants, short-day plants, flower only when the periods of light are less than some critical period. Day-neutral plants flower regardless of photoperiods. Interruption of the dark phase of the photoperiod, even by a brief exposure to light, can serve to reverse the photoperiodic effects, indicating that the dark period rather than the light period is critical.

Phytochrome, a pigment commonly present in small amounts in the tissues of higher plants, is the receptor molecule for transitions between light and darkness. The pigment can exist in two forms, P_{660} and P_{730}. P_{660} absorbs red light (which is also ordinarily present in white, or visible, light) of a wavelength of 660 nanometers and is thereby converted to P_{730}, which absorbs far-red light of a 730-nanometer wavelength. With ordinary day-night sequences, P_{730} is converted back to P_{660} over a period of hours in the dark; it can also be converted to P_{660} by exposure to far-red light. P_{730} is the active form of the pigment; it promotes flowering in long-day plants and inhibits flowering in short-day plants. It promotes germination in lettuce seeds, which is inhibited by P_{660}, produced by exposure to far-red light. Phytochrome appears to serve as an on-off switch for a number of physiological activities in plants. Its mechanism of action is unknown.

Other light responses of plants include phototropism and turgor movements. The latter include stomatal opening and closing, sleep movements of leaves, and the closing and opening of flowers.

The morphogenesis of seedlings is also influenced by light. Dark-grown seedlings are colorless, have long internodes, small folded leaves, and are often hooked at the top. Such seedlings, which are known as etiolated, develop normally when exposed to light. These responses are also mediated by phytochrome.

Some species of plants respond to touch. Examples include the winding of tendrils, the collapse of the leaves of the sensitive plant, and the triggering of the carnivorous Venus flytrap. The latter two responses are a result of rapid turgor changes.

Circadian rhythms are cycles of activity in an organism that recur at intervals of approximately 24 hours. They are probably endogenous, caused not by an external factor such as alternating light and darkness or the Earth's rotation but by some internal timing mechanism within the organism. Such a timing mechanism, whose chemical and physical nature is unknown, is called a biological clock. Having a biological clock makes it possible for an organism to perceive changes in daily cycles, such as the lengthening and shortening of days as the seasons progress. In this way, activities such as seed germination, flowering, and dormancy can be synchronized with changes in the external environment.

QUESTIONS

1. Define the following terms: photoperiodism, circadian rhythm, biological clock. What do they have in common? What are the important differences among these phenomena?
2. Plants must, in some manner, synchronize their activities with the seasons. Of the clues which they might use, day (or night) length seems to have been selected. What might be the advantage of using photoperiod rather than, say, temperature as a season detector?
3. Traveling from north to south, one can find varieties of the same species with different photoperiodic requirements. How would you expect them to differ?
4. How would you define "behavior"? According to this definition, do plants exhibit behavior?
5. Photoperiodic systems are not nearly as sensitive to low levels of illumination as are many visual systems. Why might extreme sensitivity be a positive disadvantage in a photoperiodic system? What havoc might be wrought by the widespread use of bright lights for street illumination or by a scheme to orbit a large reflector to lend luster to our nights?

SUGGESTIONS FOR FURTHER READING

GALSTON, A. W.: *The Green Plant*, Prentice-Hall, Inc., Englewood Cliffs, N.J., 1968.*

A convenient and concise summary of plant growth and development.

GREULACH, VICTOR: *Plant Structure and Function*, The Macmillan Company, New York, 1973.
A short, lucid introduction to botany, with emphasis on physiology.

* Available in paperback.

LEDBETTER, M. C., and KEITH PORTER: *Introduction to the Fine Structure of Plant Cells,* Springer Publishing Co., Inc., New York, 1970.

An excellent atlas of electron micrographs of plant cells, with a detailed explanation of each.

RAVEN, PETER, and HELENA CURTIS: *Biology of Plants,* Worth Publishers, Inc., New York, 1970.

An up-to-date and handsomely illustrated general botany text, particularly strong in the fields of ecology and evolution.

RAY, PETER M.: *The Living Plant,* 2d ed., Holt, Rinehart and Winston, Inc., New York, 1971.*

An outstanding short text.

RICHARDSON, MICHAEL: *Translocation in Plants,* St. Martin's Press, Inc., New York, 1968.*

An extremely useful, brief review of experimental work on the movement of water in plants.

SALISBURY, FRANK B., and CLEON ROSS: *Plant Physiology,* Wadsworth Publishing Co., Inc., Belmont, Calif., 1969.

A good, modern plant physiology text for more advanced students.

TORREY, JOHN G.: *Development in Flowering Plants,* The Macmillan Company, New York, 1967.*

How flowering plants develop, with emphasis on the underlying physiological processes.

* Available in paperback.

SECTION 5

BIOLOGY OF ANIMALS

29–1   *Genus,* Homo; *species,* sapiens. *An*
immature form.

Chapter 29

Man Among the Vertebrates

Even those of us who have most marveled at the exquisite architecture of an orchid flower or who are as content as Leeuwenhoek to watch the intricate and varied movements of a paramecium and its neighbors, approach the subject of the biology of our own systems with a quickened interest. Indeed for some, student and scientist alike, the principal and perhaps the only reasons for studying "lower forms" is the extent to which such studies bear directly on the welfare of mankind. But as *Escherichia coli*, T2 and T4 bacteriophages, and *Drosophila* remind us, there is no way to study only man, any more than it is possible to study only a fruit fly.

In this section of the book, which deals with vertebrate physiology, we shall examine man (a term which includes, equally, woman) as an example of an animal organism. This one species, however, will not command our full attention. First, we shall rely heavily on your understanding of the physiological principles governing such matters as surface and volume, diffusion and osmosis, and gas exchange and water balance, which were discussed and exemplified in earlier chapters. Second, in the chapters that follow, we shall introduce numerous examples from other organisms that illuminate our central theme—the physiology of man. Finally, we shall attempt to continue to emphasize broad principles of animal physiology rather than lose our way in a mass of details and intricate terminology. Because man is part of a continuum of nature, it is logical to study other animals in order to understand the human animal; it is equally logical to study the human animal (now one of the best understood) in order to gain a deeper understanding of animal life in general.

In 1966, George Gaylord Simpson, one of the leading modern students of evolution, wrote:

The question "What is man?" is probably the most profound that can be asked by man. It has always been central to any system of the philosophy or of theology. We know that it was being asked by the most learned humans 2,000 years ago, and it is just possible that it was being asked by the most brilliant australopithecines 2 million years ago. The point I want to make now is that all attempts to answer that question before 1859 are worthless and that we will be better off if we ignore them completely.*

* G. G. Simpson: "The Biological Nature of Man," *Science*, **152**, 472–478, 1966.

29–2 *Two vertebrate skeletons.*

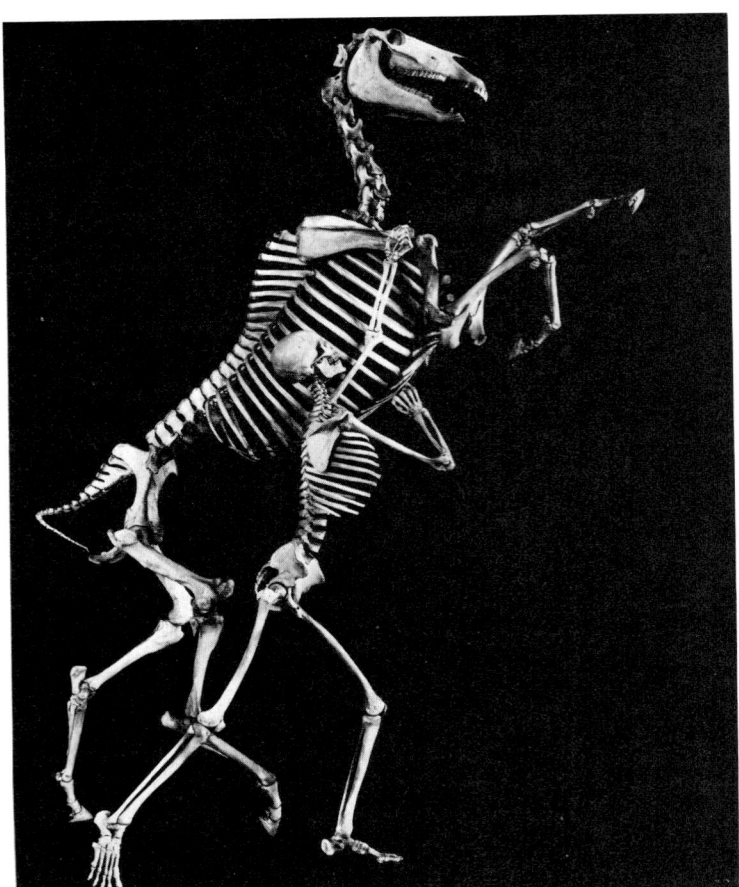

29–3 *In mammals, the coelom is divided into two cavities, the thoracic cavity and the abdominal cavity, by the diaphragm, which is a dome-shaped respiratory muscle. The upper (thoracic) cavity contains the heart, the lungs, and the upper portion of the digestive tract, as well as major blood vessels and nerves. The lower (abdominal) cavity contains digestive organs and also organs concerned with excretion and reproduction. The lower portion of the abdominal cavity is sometimes called the pelvic cavity.*

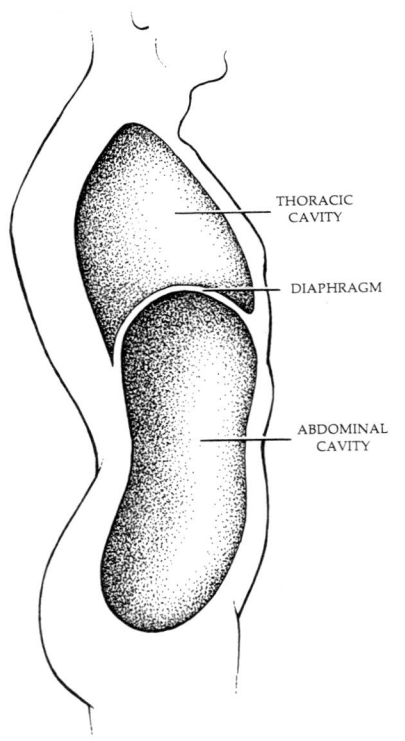

THORACIC CAVITY

DIAPHRAGM

ABDOMINAL CAVITY

Some of us may not accept Simpson's dismissal of pre-Darwinian philosophy, but there are few who would not agree about the relevance of his statement to the biology of man.

MAN AS VERTEBRATE

Let us begin by reviewing man's relationship to the rest of the animal world. Man is a vertebrate, and as such he has a bony, articulated (jointed) endoskeleton that supports his body and grows as he grows. His spinal cord is surrounded by bony segments, the vertebrae, and his brain is enclosed in a protective casing, the skull. He moves his skeleton by means of muscles, which are either attached directly to the bones or to tendons attached to the bones.

As in other vertebrates, and most invertebrates as well, man's body contains a cavity, or coelom. In man, the coelom is divided into two parts, the thoracic cavity and the abdominal cavity, which are separated by a dome-shaped diaphragm. The thoracic cavity contains the heart, the lungs, and the upper portion of the digestive tract; the abdominal cavity contains the stomach, intestines, liver, and other organs concerned with digestion, excretion, and reproduction.

9–4 *The three types of epithelial cells that cover the inner and outer surfaces of the body. Squamous cells, which usually perform a protective function, make up the outer layers of the skin and the lining of the mouth and other mucous membranes. There are usually several layers of these flat cells piled on top of one another. Cuboidal and columnar cells perform much of the chemical work of the body.*

SQUAMOUS

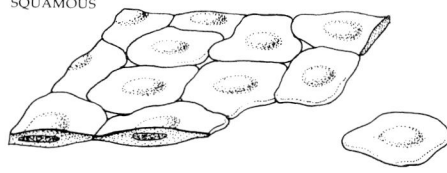

CUBOIDAL

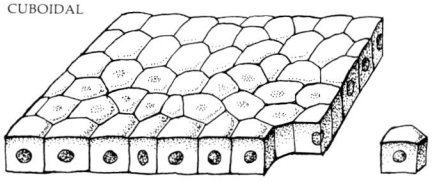

COLUMNAR

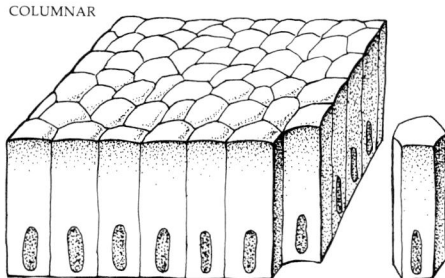

Man's body, like that of all other complex animals, is made up of a great variety of different cells. Each of these cells is, to some extent, an independent entity. Each needs oxygen, glucose (or some other source of carbon and energy), essential amino acids, and small amounts of other substances (such as iron). Each needs to dispose of carbon dioxide and other wastes. Each of man's physiological processes, like those of any other organism, is concerned with providing second-by-second 24-hour-a-day service to each of these billions of cells.

Although they greatly resemble one-celled organisms in their requirements, the cells of an animal body differ from one-celled organisms in that they develop and function as part of an organized whole. Cells are organized into *tissues*, groups of cells similar in structure and function. Different kinds of tissues united structurally and coordinated in their activities form *organs*. Organs that function together in an integrated and organized way make up *organ systems*. The digestive system, for example, is composed of a number of different organs, each of which carries out a specific activity that contributes toward the overall process. Because the focus of this section will be on the various functions of the body, we shall be examining it principally in terms of organ systems.

Although experts can distinguish more than 100 different types of cells in the human body, they are customarily classified in terms of only four tissue types: (1) epithelial tissue, (2) connective tissue, (3) muscle tissue, and (4) nervous tissue.

Epithelial tissue, or epithelium, consists of sheets of cells overlying a noncellular fibrous layer, the basement membrane. Epithelium is classified according to the shape of the individual cells as squamous, cuboidal, or columnar. (See Figure 29–4.) Epithelium forms the covering and lining membranes of the whole body and of internal organs, cavities, and passageways. Hence, everything that goes into and out of the body must pass through epithelial cells. In other regions, as in the skin, it provides a protective covering and also contains various sensory nerve endings. The epithelium of the body cavities frequently contains special glands that secrete mucus, which lubricates surfaces. Glands are special types of epithelial tissue whose cells produce specific chemicals, such as hormones or enzymes.

Connective tissue binds together and supports all other types of tissues. Unlike epithelial tissues, connective tissue has large amounts of intercellular substances and relatively few cells, widely separated from one another. The intercellular substances include connecting and supporting fibers, such as collagen (page 85) and cartilage; elastic fibers, which are often found in the walls of hollow, distensible organs, such as the stomach and the uterus; and reticular fibers, which form networks inside solid organs. These fibers are embedded in a matrix, the ground substance. Bone, like other connective tissues, consists of cells, fibers, and ground substance. It is unlike the others in that its extracellular components are calcified. Blood and lymph are connective tissues in which the ground substance is a fluid.

The other major tissue types of the body are muscle and nervous tissues. There are two general types of muscle tissue, striated muscle, so called because of its striped appearance, and smooth muscle, which is not striped. Striated muscle is the type of muscle that moves the skeleton; skeletal striated muscle is also sometimes called voluntary muscle, since we can move it at will. Cardiac

29–5 Electron micrograph and diagram of a bone cell (osteocyte). Young bone cells (osteoblasts) produce the intercellular bone matrix, an organic material consisting of collagen fibers and an unstructured ground substance. This matrix gradually mineralizes (calcifies), becoming hard and dense.

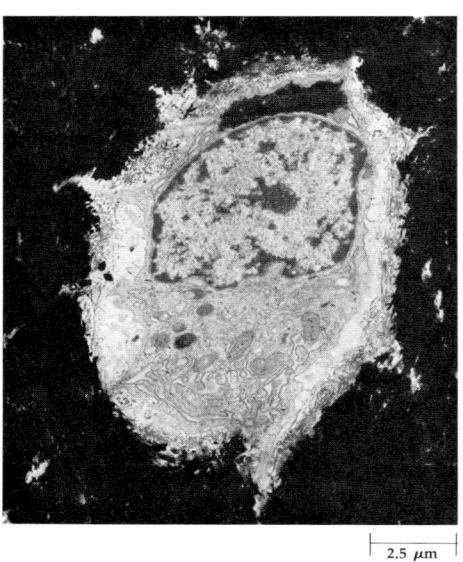

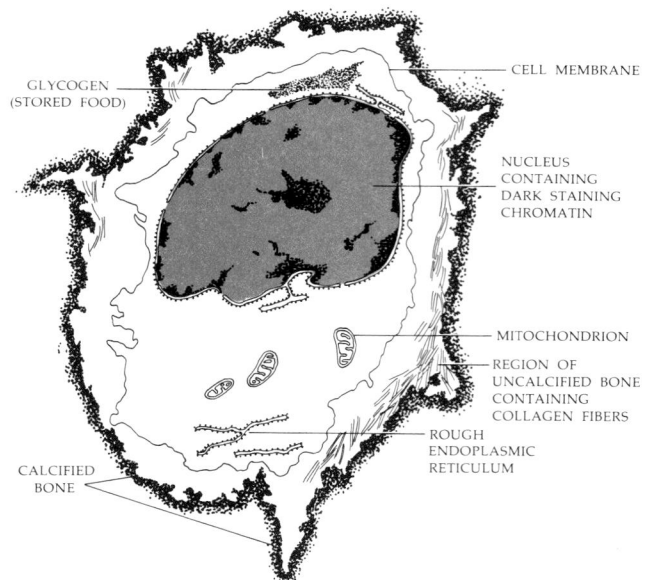

29–6 Electron micrograph and diagram of a cartilage cell (chondrocyte). The cell is embedded in a dense, nonliving material, cartilage, which is produced by the chondrocyte. The large fat droplet is the cell's supply of stored food.

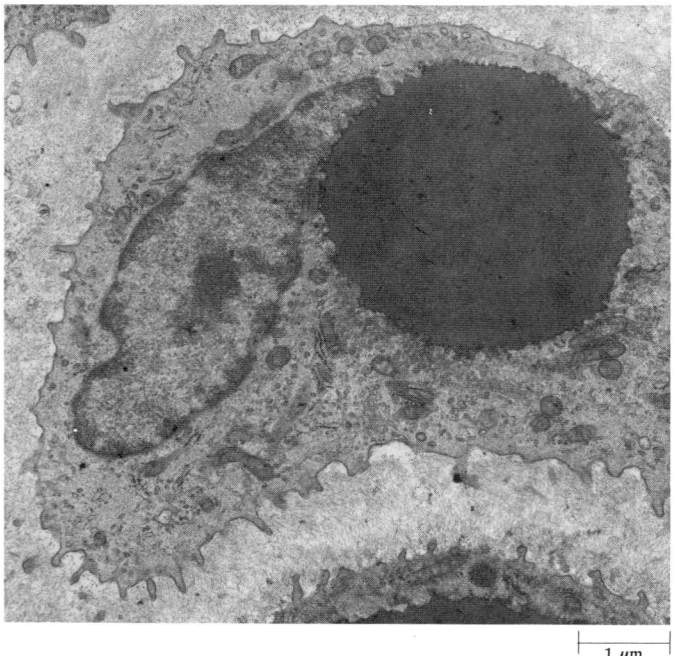

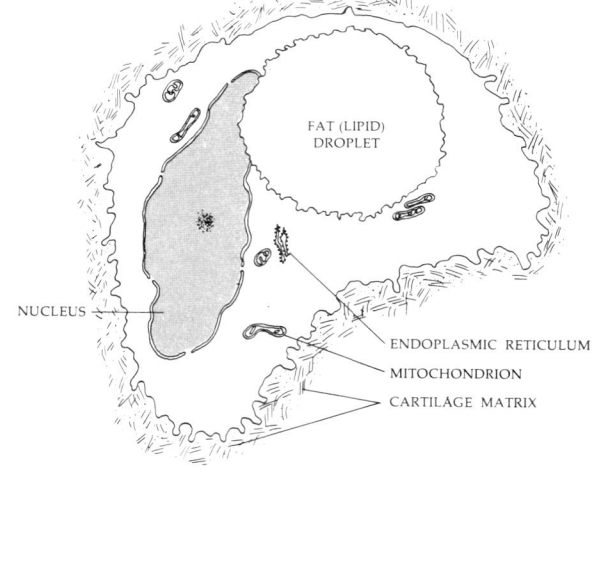

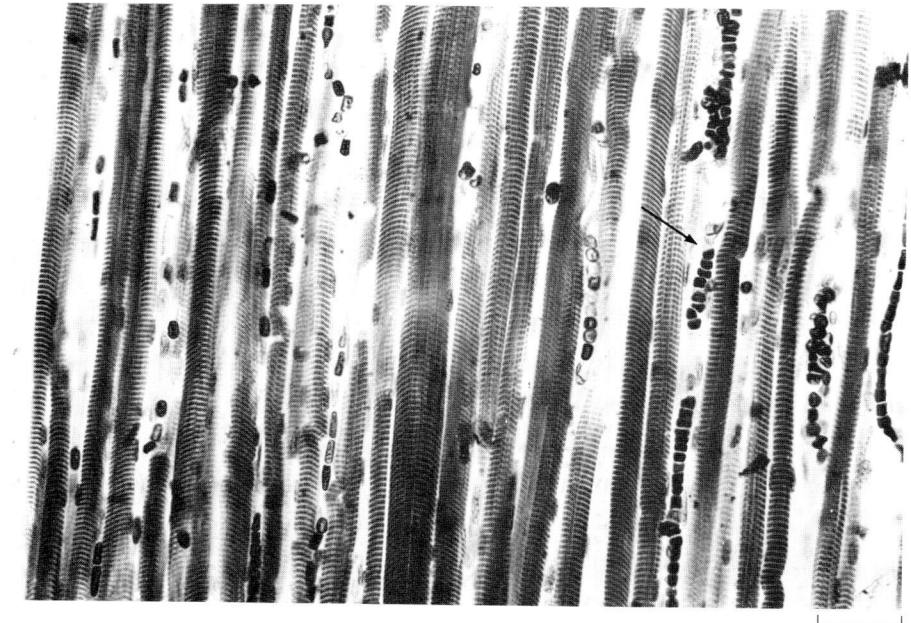

9–7 *Photomicrograph of skeletal muscle fibers, showing striated pattern. The arrow indicates a small blood vessel (capillary).*

9–8 *Muscles attached to bone move the vertebrate endoskeleton. They characteristically work in antagonistic pairs, with one slowly relaxing as the other contracts. For example, when you move your hand toward your shoulder, as shown here, the biceps contracts and the triceps relaxes. When you move your hand down again, the triceps contracts, while the biceps relaxes. The muscles that move the skeleton, such as those diagrammed here, are known as striated, or voluntary, muscles.*

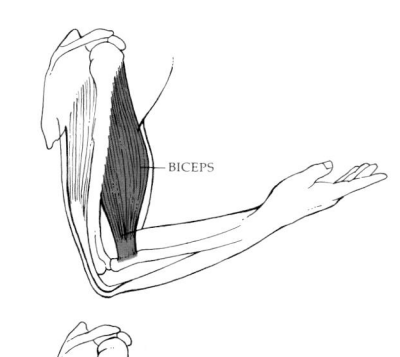

BICEPS

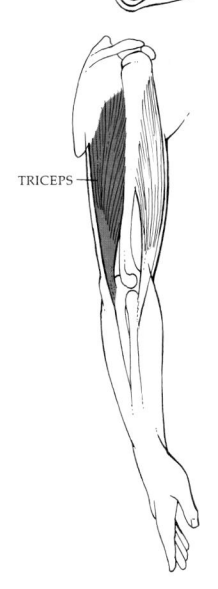

TRICEPS

150 μm

muscle is also striated muscle. Smooth muscle surrounds the walls of the internal organs, such as the digestive tract and the uterus; it is sometimes called involuntary muscle. Nerve cells, or neurons, are cells specialized to carry nerve impulses from one part of the body to another. These are the most morphologically spectacular of the body's cells. Nerve cells may reach astonishing lengths. For example, the fiber of a single motor neuron (a nerve cell that activates muscle) may extend from the spinal cord down the whole length of the leg right to the toe. Or a sensory neuron (one that transmits sensations to the brain) based near the spinal cord may send one fiber down to the toe and another fiber up the entire length of the spinal cord to terminate in the lower part of the brain. In a man, such a cell might be more than 5 feet long (15 feet long in a giraffe).

The structure of nerve and muscle cells will be examined in more detail in Chapter 32, when we discuss their functions in integrating and coordinating the body's activities.

MAN AS MAMMAL

Man is a mammal. One of the most important characteristics of mammals is that they are warm-blooded. More precisely, they are homeotherms; that is, they maintain a high and relatively constant body temperature. As a consequence, mammals (and birds, which are also homeotherms) are able to achieve and sustain levels of physical and mental activity generally far greater than animals whose temperatures rise and fall with those of their external environment. A second consequence of homeothermy is a high metabolic rate, which requires relatively large and constant supplies of food (fuel) molecules and oxygen.

515 *Man Among the Vertebrates*

MIRACLE WRAP

Human skin is composed of two principal layers: the epidermis and the dermis. Below the dermis is fatty subcutaneous tissue. Intermingled with the cells of the dermis are fine blood vessels, tactile and other nerves, the smooth muscles that raise the hairs, and a variety of specialized glands, including sweat glands.

The epidermis, the upper layer of the skin, is a turnover system. Cells in the basal layer, the layer of cells overlying the basement membrane (which separates the epidermis from the dermis), divide constantly. When a cell divides in the basal layer, a cell somewhere in the vicinity begins to migrate toward the skin surface. As it migrates, it begins to differentiate. It changes in shape from a columnar or cuboidal basal cell to a flattened, tile-shaped squamous cell, and finally dies. The upper layer of skin is composed entirely of dead cells.

As cells move from the basal layer toward the surface, they begin to synthesize keratin, which is a mixture of proteins. Keratin is also the substance from which nails, scales, feathers, claws, horns, hooves, and antlers are formed; biosynthesis of keratin is another distinguishing characteristic of the vertebrates. The differentiating cell continues to produce keratin until it dies, and the keratin deposited in dead cells makes up about half of the total substance of the dead epidermal layer. This outer layer of dead cells, known as the horny layer, or stratum corneum, is a waterproof, pliable, semitransparent film, thin and tough, like plastic wrap. Fragments of the stratum corneum are shed constantly from the human skin; they are a major component of ordinary house dust.

As William Montagna of the University of Oregon reminds us: "Skin is a remarkable organ—the largest and by far the most versatile of the body. It is an effective shield against many forms of physical and chemical attack. It holds in the body's fluids and maintains its integrity by keeping out foreign substances and microorganisms. It acts to ward off the harsh ultraviolet rays of the sun. It incorporates mechanisms that cool the body when it is warm and retard the loss of heat when it is cold. It plays a major role in regulating blood pressure and directing the flow of blood. It embodies the sense of touch. It is the principal organ of sexual attraction. It identifies each individual by shaping the facial and bodily contours as well as by distinctive marking such as fingerprints."

A section of human skin. The epidermis is a turnover system of epithelial cells: cells produced in the basal layer overlying the dermis, migrate toward the surface and die. The dermis, consisting mostly of connective tissue, contains sensory receptors, hair follicles, and sweat and sebaceous glands. The latter produce a fatty substance that lubricates the skin surface. The epidermis and dermis are separated by the basement membrane.

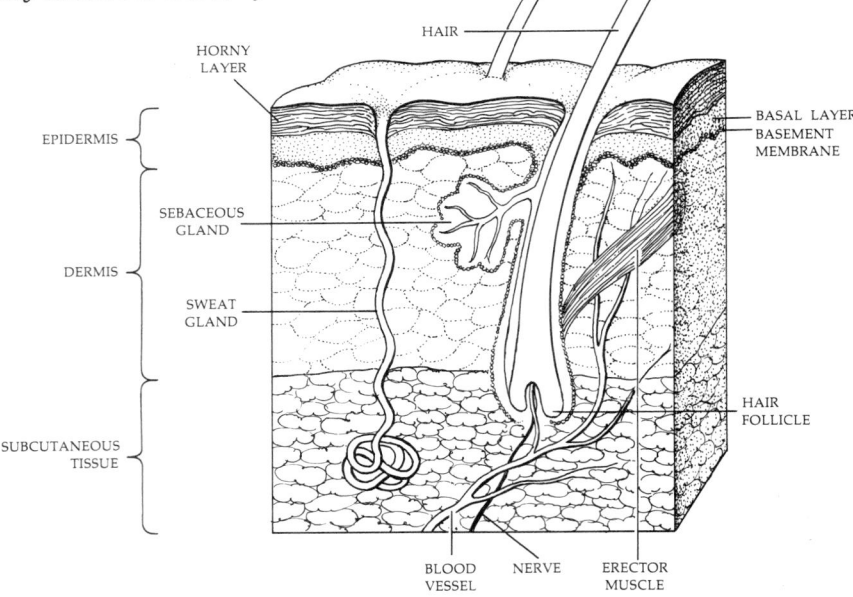

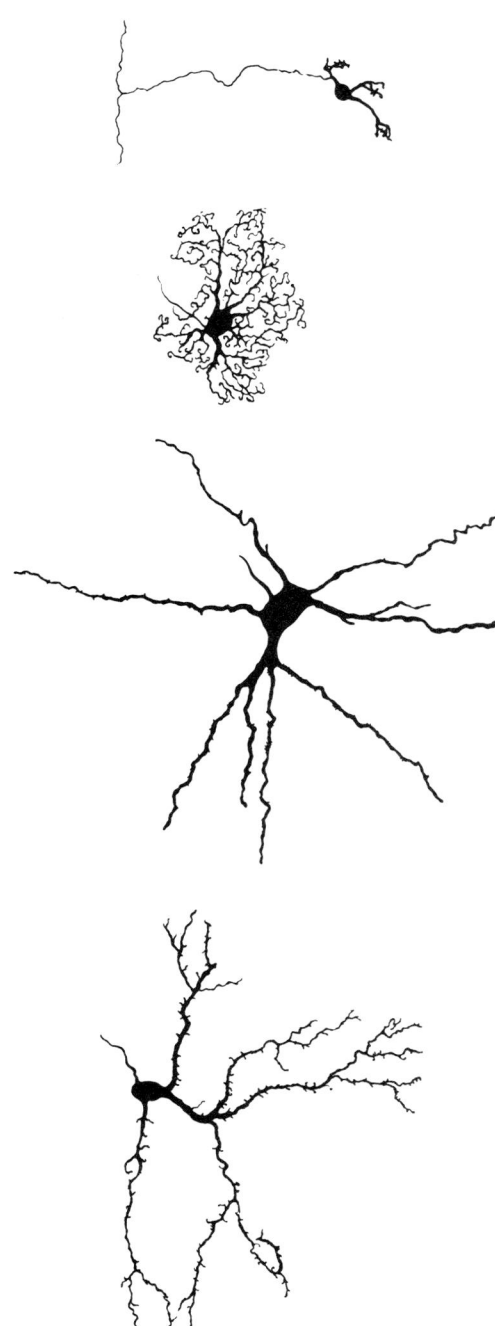

29–9 *Four neurons that carry messages within the central nervous system. These drawings were made from tissues treated with a silver stain, which selectively stains the nerve cell membrane. The stain travels along the many intricately branching nerve fibers, tracing out their delicate patterns.*

Mammals have other important characteristics. They have hair or fur rather than scales or feathers. Almost all mammals (except the monotremes) give birth to live young, as distinct from laying eggs, which all birds and most fish and reptiles do. They nurse their offspring, which involves a relatively long period of parental care. This degree of parental care is correlated with a relatively long learning period (as contrasted, for example, with insects, with most species of fish, and with reptiles, all of which are independent from the moment they hatch from the egg). There has been a tendency among the large mammals, in particular, toward fewer young and an even longer period of maternal or parental care. Humans provide a relatively extreme example, with rarely more than two surviving young per birth, only two mammae with which to nurse them, and an extraordinarily long period of infancy and childhood, with dependency on parents often lasting well past physical maturity.

As a consequence, at least in part, of homeothermy and the long learning period, mammals are, in general, by far the most intelligent of all groups of organisms. Also, they have the most highly developed systems for receiving, processing, and correlating information from the environment (although, as we noted in Chapter 23, in complexity and variety of sense organs, they find close rivals among some of the invertebrates).

The class Mammalia comprises a large number of different species, ranging from whales and dolphins, bats, moles and hedgehogs, hippopotamuses, zebras and rhinoceroses, to gorillas and marmosets. Among this class, man is distinguished first by being versatile rather than specialized. He cannot run as fast as a deer, swim as gracefully as a seal, or swing from branch to branch with the agility of a gibbon. But he can, if adequately motivated, run a mile, swim a river, and climb a tree—and few other mammals can do all three. Finally, for better or worse, he is by far the most intelligent of all mammalian species.

SUMMARY

Man, a vertebrate, has a bony, supporting endoskeleton, a vertebral column that encloses his spinal cord, and a skull. His body is made up of single cells which make up four main tissue types: epithelium, connective tissue, muscle, and nerve. Cells are organized into tissues, tissues into organs, and organs into organ systems, which include a number of different organs that work together to carry out a specific function, such as digestion.

A mammal, man is a homeotherm, which necessitates a high rate of energy consumption and makes sustained physical activity and mental alertness possible. Mammals are also characterized by a long period of dependence on parental care, and hence of learning, and by their intelligence. Man is among the least highly specialized of mammals except in his capacity to learn and his intelligence.

QUESTIONS

1. Define the following terms: coelom, organ, intercellular substances, homeotherm.
2. Name the four principal tissue types and give examples of each.
3. What are the distinguishing characteristics of the vertebrates? Of the mammals?

Chapter 30

The Continuity of Life: Reproduction

When does a human life begin? When sperm encounters egg? When the fetus "quickens" in the womb? When the infant becomes viable as an independent entity? In the past, these matters were discussed by philosophers and theologians who were concerned with the question, When does the soul enter the body? These issues have been revived in the recent ethical and legal controversies concerning abortion. In the biological sense, however, none of these events marks the beginning of life. Life began more than 3 billion years ago and has been passed on since that time from organism to organism, generation after generation, to the present, and stretches on into the future, farther than the mind's eye can see. Each new organism is thus merely a temporary participant in this continuum of life. So is each sperm, each egg, indeed, in a sense, each living cell. From the viewpoint of the species, the continuum is shorter, of course —perhaps as brief as 10,000 years. And as for the individual, he or she is a totally unique blend of heredity and experience, never to be duplicated and therefore irreplaceable, but from the perspective of the biological continuum, an individual life lasts no longer than the blink of an eye.

In this and the following chapter, we are going to discuss the process by which parent organisms produce new individuals, maintaining the continuity of life and species, and how these new individuals develop to become reasonable facsimiles of their parent organisms. In the remaining chapters, we shall discuss how the individual organism maintains itself, solving, as an organism, the problems of survival by utilizing the solutions bequeathed it by its parents and other ancestors, even to its most distant forebears.

PATTERNS OF ANIMAL REPRODUCTION

Asexual reproduction, as we have seen, is the only form of reproduction among prokaryotes and some protists. Vertebrates, however, usually reproduce sexually. As you will recall,* sexual reproduction involves two events: meiosis and syngamy (fertilization). In vertebrates, which are almost always diploid, meiosis produces gametes, the only haploid forms in the life cycle. The gametes are specialized for motility (sperm) or for storage of food (eggs) and are produced by separate individuals. In many lower organisms (insects, in particular), the generations are nonoverlapping, as is the case in annual plants, but in vertebrates, parents not only survive after their young are produced but often

* Or turn back to page 196 for a reminder.

30–1   *Asexual reproduction takes place in a variety of ways including budding, as in* Hydra. *The offspring that result are genetically identical to the single parent.*

0.5 mm

are essential to the rearing of the young. This evolutionary trend toward increasing parental care becomes pronounced among birds and reaches its fullest expression among the mammals, as we noted in the previous chapter.

Vertebrates are uniformly dioecious; that is, organisms are either male or female, so designated by the types of gametes they produce. In most species of fish and in amphibians, as in most invertebrates, fertilization is external. Among organisms that produce amniote eggs (reptiles, birds, and monotreme mammals), fertilization has to be internal because the outer protective shell, produced as the egg moves down through the female reproductive tract, is laid down after the egg cell is fertilized, enclosing the embryo and its membranes. Fertilization is also internal among marsupial and placental mammals, in which the embryo develops within the mother and is nourished by her.

In the following pages, we shall describe sexual reproduction in man. First, we shall trace the development of the male gametes, then of the female gametes, and we shall describe the special structures and activities that make fertilization and implantation of the fertilized egg possible.

-2 *Sexual reproduction always involves meiosis and fertilization. Fertilization may be external, as in frogs (a), or internal, as in reptiles (b), birds (c), and mammals (d).*

(a)

(b)

(c)

(d)

THE MALE REPRODUCTIVE SYSTEM

The primary sexual organs are the gonads, the organs in which the gametes are formed. The male gonads are the testes (singular, testis). The testes develop in the abdominal cavity of the male embryo and, in the human male, descend into an external sac, the scrotum. This descent often occurs before birth. The function of the scrotum, apparently, is to keep the testes cool. A temperature 3°C lower than that of the body is necessary for the sperm to develop. Sperm are not produced in an undescended testis, and even temporary immersion of the testes in warm water—as in a hot bath—has been known to produce temporary sterility. (This effect is not reliable enough, however, to recommend the use of this technique as a birth control measure.) When the temperature outside the scrotal sac is warm, the sac is thin and hangs loose in multiple folds. When the outside temperature is cold, the muscles under the skin of the scrotum contract, drawing the testes close to the body. In this way, a fairly constant testicular temperature is maintained.

Spermatogenesis

Each testis is subdivided into about 250 compartments (lobes), and each of these is packed with tightly coiled seminiferous ("seed-bearing") tubules. These are the sperm-producing regions of the testes. Between the tubules are the

30–3 *The human male reproductive tract. Sperm cells formed in the seminiferous ("seed-bearing") tubules of the testes enter the vas deferens, which empties into a seminal vesicle, and then pass through the prostate gland. The sperm cells are mixed with fluids, mostly from the seminal vesicles and prostate gland, and this mixture, the semen, is released through the urethra of the penis. The urethra is also the passageway for urine, which is stored in the bladder.*

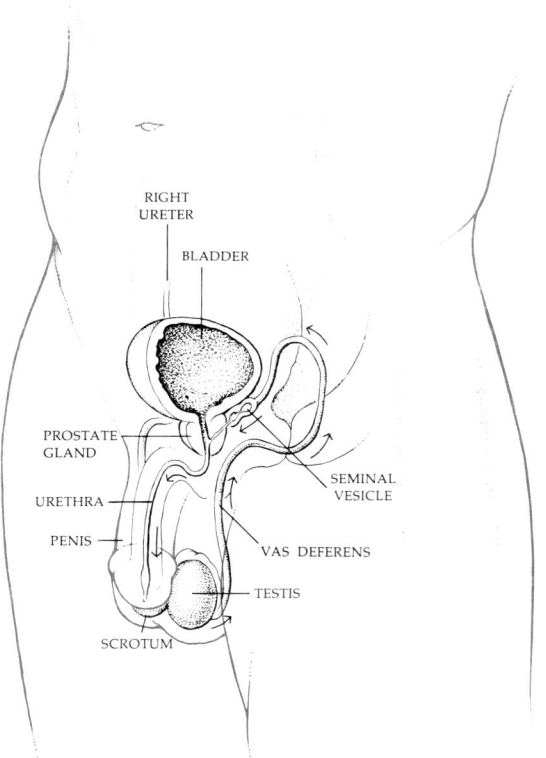

interstitial cells that produce androgens, the male hormones, of which testosterone is the most important. Each seminiferous tubule is 1 to 3 feet long, and the two testes contain a total of about 275 yards of tubules. The sperm are produced within the tubules by a turnover system analogous to the turnover system previously described in the skin.

The tubules contain two types of cells: spermatogenic (sperm-producing) cells and Sertoli cells. The spermatogenic cells pass through several stages of differentiation. Once production of sperm begins at puberty in the human male, it goes on continuously and so, in a single seminiferous tubule, it is possible to find cells in all the different stages of spermatogenesis. It is unusual, however, to find all of the stages in a single cross section, because spermatogenesis characteristically occurs in waves which travel down the tubules.

The earliest (first-stage) cells are spermatogonia, which line the basement membrane of the tubules (Figure 30–4c). Cells in the bottom layer divide continuously; some of the daughter cells remain undifferentiated in this first layer, while others, in the course of successive divisions, move away from the basement membrane and begin to differentiate, giving rise to primary spermatocytes. Spermatogonia are diploid and have, in the human, 44 autosomes and 2 sex chromosomes, an X and a Y; they divide by mitosis. Primary spermatocytes undergo the first meiotic division* to produce two secondary spermatocytes, each of which contains 22 autosomes and either an X chromosome or a Y chromosome; each of these chromosomes consists of two chromatids. The secondary spermatocytes undergo the second meiotic division to produce spermatids, each of which contains the haploid number of single chromosomes. It takes 8 to 9 weeks for a spermatogonium to differentiate into four sperm cells (spermatozoa). During this time the developing cells receive nutrient material from adjacent Sertoli cells.

* For a review of meiosis, see pages 186 to 190.

-4 *The testis (a) is made of tightly packed coils of seminiferous tubules (b) containing sperm cells in various stages of development. As shown in the idealized cross section (c), spermatogonia develop into cells known as primary spermatocytes. These divide (first meiotic division) into two equal-sized cells, the secondary spermatocytes. In the second meiotic division, four equal-sized spermatids are formed. These differentiate into functional sperm. (It is unusual to find all these stages in a single cross section.) The Sertoli cells support and nourish the developing sperm. The interstitial cells, which are found in connective tissues between the tubules, are the source of the male hormone testosterone. The sperm cells leave the testis through the epididymis and the vas deferens.*

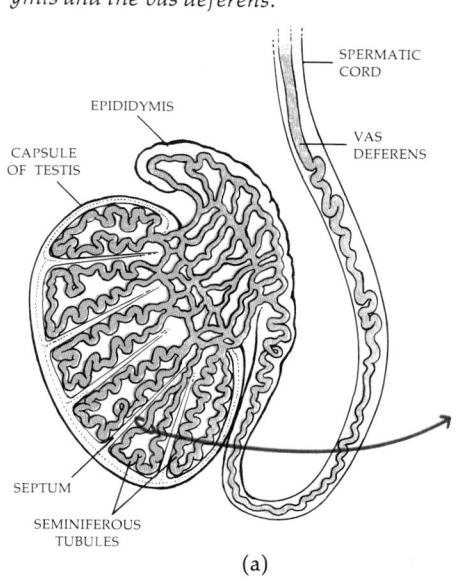

SPERMATIC CORD
EPIDIDYMIS
VAS DEFERENS
CAPSULE OF TESTIS
SEPTUM
SEMINIFEROUS TUBULES

(a)

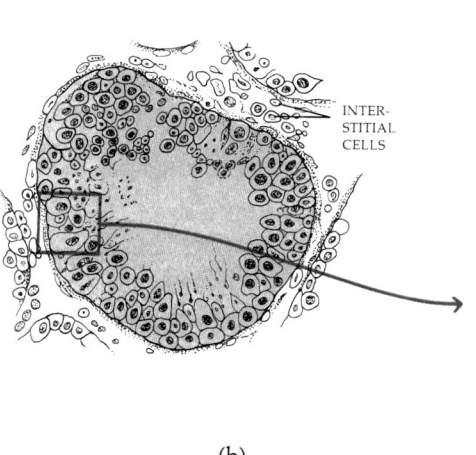

INTERSTITIAL CELLS

(b)

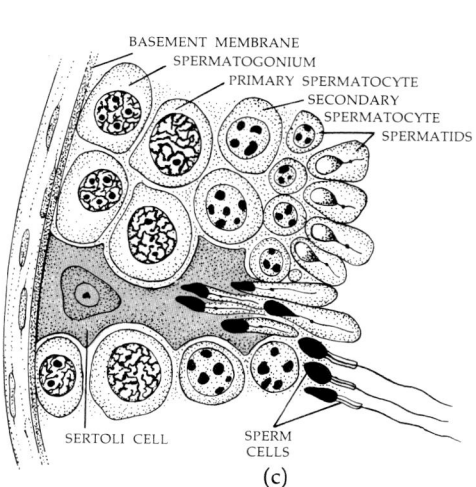

BASEMENT MEMBRANE
SPERMATOGONIUM
PRIMARY SPERMATOCYTE
SECONDARY SPERMATOCYTE
SPERMATIDS
SERTOLI CELL
SPERM CELLS

(c)

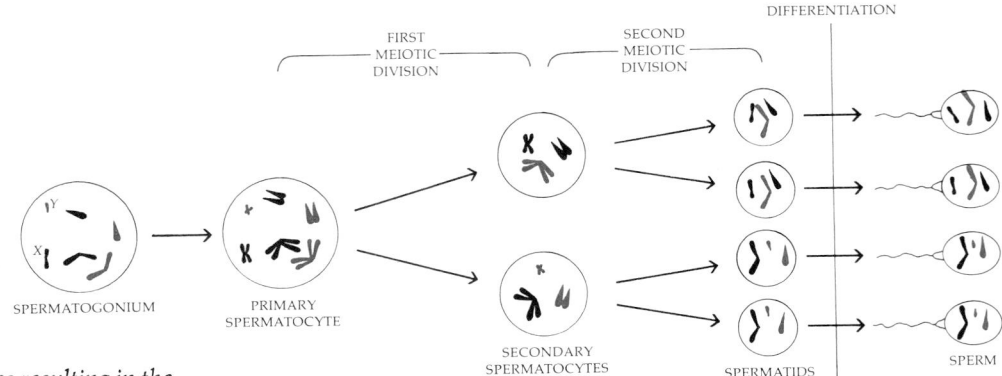

FIRST
MEIOTIC
DIVISION

SECOND
MEIOTIC
DIVISION

SPERMATOGONIUM   PRIMARY
SPERMATOCYTE

SECONDARY
SPERMATOCYTES

SPERMATIDS

SPERM

30–5   *The series of changes resulting in the formation of sperm cells begins with the growth of spermatogonia into large cells known as primary spermatocytes. At the first meiotic division, each primary spermatocyte divides into two haploid secondary spermatocytes. The second meiotic division results in the formation of four haploid spermatids. The spermatids differentiate into functional sperm.*

Differentiation of Spermatids

A spermatid is a small spherical or polygonal ("many-sided") cell that develops, without further divisions, into a sperm cell. The sequence of changes by which the quite unremarkable spermatid becomes the highly specialized, very extraordinary sperm cell is an excellent example of cell differentiation, a process which, as we shall see, plays an important role in many stages of embryonic development.

The first visible sign of differentiation of a spermatid is the appearance of vesicles containing small, dark granules within the Golgi body, which, as in most cells, lies close to the nuclear envelope. These vesicles enlarge and coalesce into a single vesicle. As the vesicle enlarges, the membrane surrounding it adheres to the nuclear envelope and folds over it until one-half to two-thirds of the nucleus is enveloped by it. This vesicle full of opaque material is known as the acrosome. Its contents help to lyse the membrane of the egg cell. The position of the acrosome determines the polarity of the sperm; that is, it establishes where the anterior end, or "head," is going to be.

During the early stages of acrosome formation, the cell's pair of centrioles moves to the outer cell membrane at the end of the cell opposite the acrosome, and from one of them a fine thread grows out, the beginning of the sperm "tail." The centrioles then move back to the nucleus, carrying the cell membrane inward with them.

30–6   *A mammalian spermatid in the process of differentiation. The nucleus almost fills the top half of the electron micrograph. The two centrioles have moved to a position just behind the nucleus and the tail has begun to form from one of them. To the left, below the nucleus, are several mitochondria.*

One centriole eventually lodges in a notch formed at the posterior end of the nucleus, while the other, the one associated with the tail, lies at right angles to the first, as shown in Figure 30–6.

As the tail grows, its axial filament shows the characteristic 9 + 2 structure found in cilia and flagella generally. Mitochondria aggregate about its basal end, forming a continuous spiral, providing a ready energy source (ATP) for the flagellar movement. The rest of the axial filament, almost to its tip, is surrounded by nine additional protein fibers tightly coiled in a helix. These fibers, which are somewhat thicker than the microtubules within the flagellum, presumably play some role in sperm motility.

During this period, the nucleus condenses, apparently by eliminating water. Once the tail has formed, the cell rapidly elongates. Longitudinal bundles of

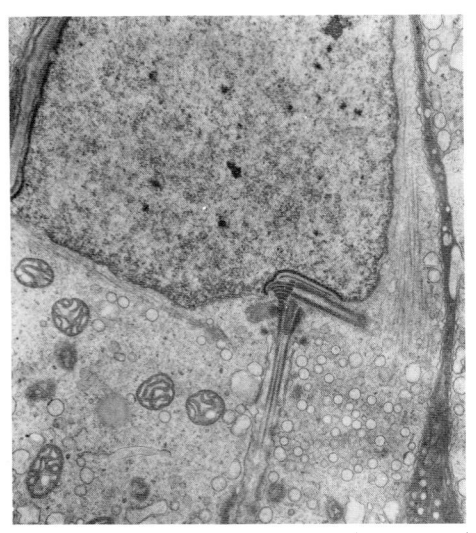

├─────────┤
2 μm

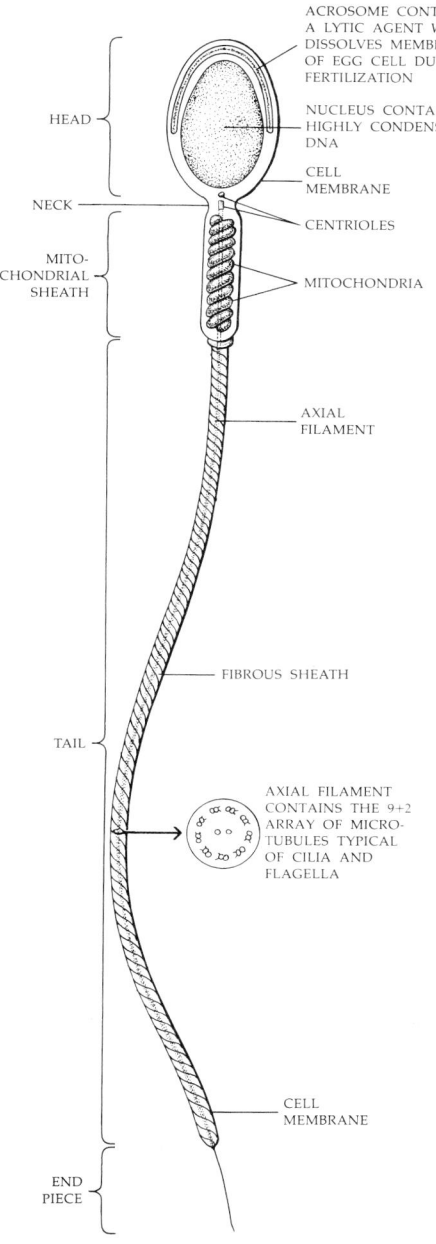

30-7 *Diagram of a human sperm cell.*

HEAD

ACROSOME CONTAINS A LYTIC AGENT WHICH DISSOLVES MEMBRANE OF EGG CELL DURING FERTILIZATION

NUCLEUS CONTAINS HIGHLY CONDENSED DNA

CELL MEMBRANE

NECK

CENTRIOLES

MITO-CHONDRIAL SHEATH

MITOCHONDRIA

AXIAL FILAMENT

FIBROUS SHEATH

TAIL

AXIAL FILAMENT CONTAINS THE 9+2 ARRAY OF MICRO-TUBULES TYPICAL OF CILIA AND FLAGELLA

CELL MEMBRANE

END PIECE

microtubules can be seen in the cell at this time, and they may play a role in changing cell shape. As the cell lengthens, the bulk of the cytoplasm, together with the Golgi body, is sloughed away.

In its final form, the sperm cell consists of the acrosome, the tightly condensed nucleus, a pair of centrioles, one of which is now serving as the basal body for the tail, and the long, powerful tail itself, all bounded, as in all cells, by a cell membrane. In the fully differentiated sperm, all other functions have been subordinated to effective motility for the delivery of the "payload," the DNA and its associated protein, condensed and coiled in the sperm head. A young adult human male may produce several hundred million sperm per day; a ram may produce several billion.

Pathway of the Sperm

From the testis, the sperm travel to the epididymis, which consists of a coiled tube 20 feet long overlying the testis. It is surrounded by a thin, circular layer of smooth muscle fibers. (Testis and epididymis together constitute the testicle.) The sperm are immotile when they reach the epididymis and gain motility only after some 18 hours there. From the epididymis, the sperm pass to the vas deferens, where most of them are stored. The vas deferens, an extension of the tightly coiled tubules of the epididymis, leads from the testicles into the abdominal cavity. Together with its accompanying nerves, arteries, and wrapping tissues, the vas deferens constitutes the spermatic cord. The spermatic cords travel upward along the same path the testes took during their descent in the embryo.

Within the abdominal cavity, the vas deferens leads around the bladder where it merges with the ducts of the seminal vesicles and then enters the urethra. The vas deferens is covered with a heavy, three-layered coat of smooth muscle whose contractions propel the sperm along it. The part nearest to the urethra is called the ejaculatory duct.

The urethra, which runs through the prostate gland and terminates in the penis, serves both for the excretion of urine and the ejaculation of sperm.

30-8 *During a vasectomy, the vas deferens on each side is severed and tied off, preventing the release of sperm from the testes. The sperm cells are reabsorbed by the body and the semen is normal except for the absence of sperm. This is a safe and almost painless procedure that does not require general anesthetic or hospitalization. Its chief drawback is that it is generally not reversible.*

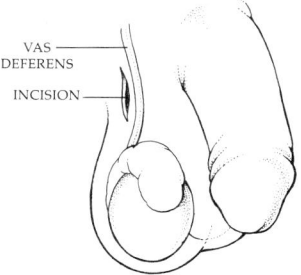
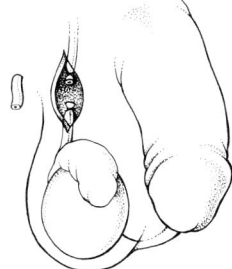
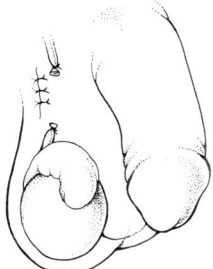

VAS DEFERENS

INCISION

The Penis

The function of the penis (plural, penes) is to deposit sperm cells within the reproductive tract of the female. The penis, in various forms, has evolved apparently independently in a number of species of insects and in other invertebrates. It is found among some reptiles and birds; all flightless birds have penes and so do all ducks, flightless or not. In most reptiles and birds, however, one opening, the cloaca, serves as the passage for eggs and sperm and also for the elimination of wastes, and these animals mate by juxtaposition of their cloacae. Only among mammals is the penis found in all species.

The penis is formed of three cylindrical masses of spongy erectile tissue, each of which contains a large number of small spaces, each about the size of a pinhead. Two of these erectile masses form the upper two-thirds of the penis; the third lies beneath them, surrounding the urethra. (See Figure 30–9.) This latter enlarges at the far end to form the glans penis, which forms a smooth protective cap over the spongy tissues. At the other end, the mass enlarges to form the bulb of the penis, which is embedded in the pelvic cavity and is surrounded by muscles that participate in orgasm. The exterior portion of the penis is covered by a loose, thin layer of skin, which at the end forms an encircling fold over the glans. This fold, the foreskin, is sometimes surgically removed (circumcision). The urethra terminates in a slitlike opening in the glans.

Erection of the penis, which can be elicited by a variety of stimuli, is caused by dilation of the blood vessels carrying blood to the spongy tissues, resulting in the collection of blood within these spaces. As the tissues become distended, they compress the veins and so inhibit the flow of blood out of the tissues. With continued stimulation, the penis and the underlying bulb become hard and enlarged.

Erection is accompanied by discharge from the bulbourethral glands (pea-shaped organs at the base of the penis) of a small amount of fluid which serves as a lubricant to facilitate the penetration of the penis into the female. Continued stimulation of the penis, such as may be produced by repeated thrusting in the

30–9 *A cross section of a human penis. The spongy tissues (corpus spongiosum and corpora cavernosa) become engorged with blood during erection. The urethra is the passageway for the urine as well as for the sperm and seminal fluid.*

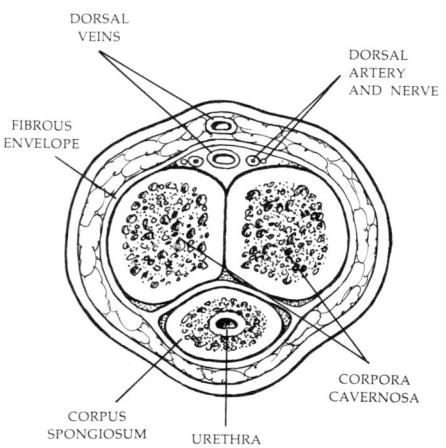

DORSAL VEINS

DORSAL ARTERY AND NERVE

FIBROUS ENVELOPE

CORPORA CAVERNOSA

CORPUS SPONGIOSUM

URETHRA

The male reproductive system, showing the penis and scrotum before (dotted lines) and during erection.

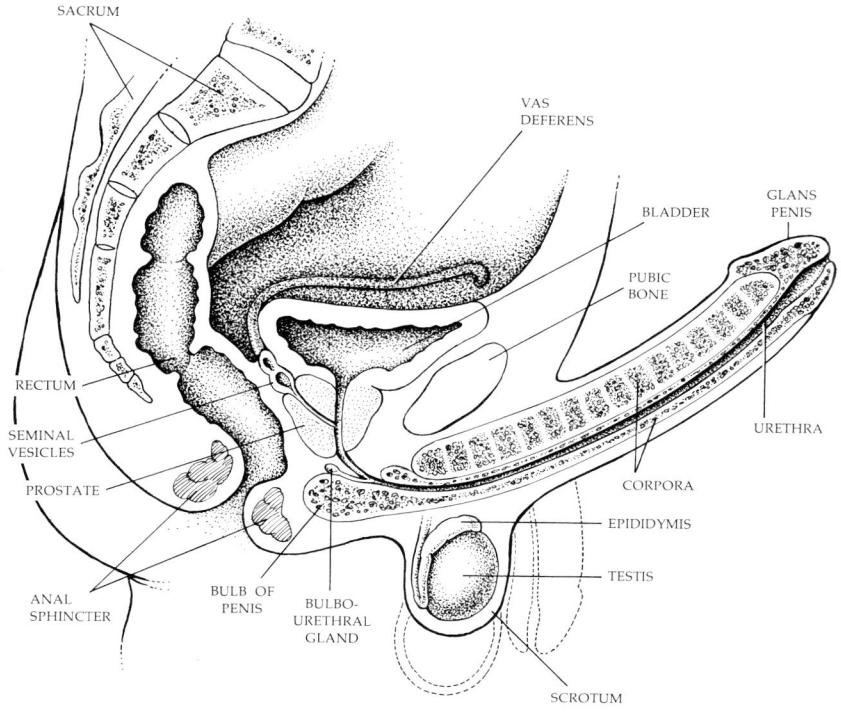

vagina, characteristically leads to contraction of the muscles in the scrotum (raising the testes close to the body) and of the muscles encircling the epididymis and vas deferens. These contractions move the semen into the urethra. Finally, the muscles surrounding the bulb are stimulated. These muscle contractions propel the semen out through the urethra and produce some of the sensations associated with orgasm. As the sperm cells are propelled along to the urethra, they are combined with fluid from the seminal vesicles, from the prostate gland, which surrounds the urethra and has numerous small ducts emptying into it, and from the bulbourethral glands. The fluids, which average about 3 milliliters per ejaculation, contain fructose, which nourishes the sperm cells. These fluids suspend the sperm, mobilize them, and neutralize the normally acid environment of the female reproductive tract to a pH more suitable for sperm survival. They also contain newly discovered hormones, known as prostaglandins, which cause contractions in the musculature of the uterus and oviducts and so may assist the sperm in reaching the egg. The fluids plus the sperm constitute the semen.

Although sperm constitute less than 10 percent of the semen, about 300 to 400 million of them are present in each ejaculate of a normal young adult male. Of those, only one can fertilize each egg cell. The rest die within three days, retaining their fertilizing ability for only about 24 hours. The other sperm cells apparently play an accessory role, however, perhaps by bringing about chemical changes necessary for fertilization. Males that produce less than 35 million sperm per milliliter of fluid are generally sterile.

30–11   *Secondary sex characteristics among
vertebrate males.*

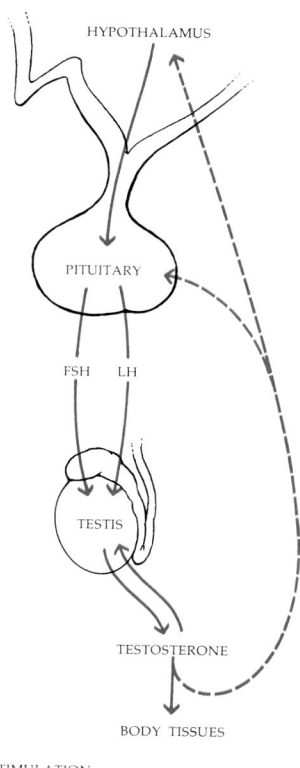

12 *The production of male sex hormone (testosterone) is controlled by a negative feedback system. The master gland is the hypothalamus, which regulates the pituitary's production of gonadotropins. The gonadotropic hormone LH stimulates production of testosterone. Under normal circumstances, the organism never produces more than is needed of either of these hormones because an increase of testosterone to a certain level of concentration in the blood acts on the pituitary gland or on the hypothalamus, a brain center. The hypothalamus ceases stimulating the pituitary, which, in turn, stops secretion of LH. FSH stimulates sperm production.*

The Role of Hormones

The principal male sex hormone is testosterone, produced principally by the interstitial cells of the testes. Testosterone and other chemicals with testosterone-like effects are known collectively as androgens. Androgens produced in the male fetus are important in the development of the external genitalia. When the human male is about 10 years old, renewed androgen production is associated with the enlargement of the penis and testes and also the prostate, the seminal vesicles, and other accessory organs. Androgens also affect other parts of the body not directly involved in the production and deposition of sperm. In the human male, these effects include growth of the larynx and an accompanying deepening of the voice, muscle development, skeletal size, and distribution of body hair. They stimulate the biosynthesis of proteins and so of muscle tissue, and also the formation of red blood cells. They stimulate the apocrine sweat glands, whose secretion attracts bacteria and so produces body odors associated with sweat after puberty. And they may cause the sebaceous glands of the skin to become overactive, resulting in acne. Such characteristics, associated with sex but not directly involved with reproduction, are known as secondary sex characteristics.

In animals other than man, the androgens are responsible for the lion's mane, the powerful musculature and fiery disposition of the stallion, the cock's comb and spurs, the bright plumage of many adult male birds, and a variety of behavior patterns, such as the scent marking of dogs, the courting behavior of sage grouse, and various forms of aggression toward other males found in a great many vertebrate species.

Since almost the beginnings of agriculture, domestic animals have been castrated in order to make them fatter or more manageable or both. Eunuchs (male castrates) were traditionally, and for obvious reasons, used as harem guards, and as recently as the eighteenth century, selected boys were castrated before puberty in order to retain the purity of their soprano voices for church and opera choirs. As a consequence of these early practices, the effects of testosterone—or more precisely, of its absence—were the first hormonal influences to be recognized and studied.

Regulation of Hormone Production

The production of testosterone (see Table 30–1) is regulated by a gonadotropic (gonad-stimulating) hormone called luteinizing hormone (LH), produced by the pituitary gland. The pituitary, which is about the size and shape of a kidney bean, is located at the geometric center of the human skull. It is lodged just beneath an area of the brain known as the hypothalamus, and the pituitary's production of gonadotropic hormones is controlled by this brain center. LH acts on the interstitial cells to stimulate their output of testosterone. The testosterone, in turn, inhibits the release of LH. This type of regulatory system, in which the system is shut off when the products of its operation reach a certain level, is known as negative feedback. The most familiar example of a negative feedback system is a thermostat that turns off a furnace when the temperature rises. Many examples of negative feedback are known, both in biological systems and in those engineered by man.

Another pituitary hormone, also under the control of the hypothalamus, is follicle-stimulating hormone (FSH), which acts on the seminiferous tubules, stimulating the development of sperm cells.

Production of sperm is a continuous process in the human male. In many animals, though, hormone production is presumably triggered by changes in temperature or daylight or other environmental cues that act upon some receptive center in the nervous system. The nervous system in turn relays the data to the hypothalamus, which sets in motion the chain of hormonal events just described. Recent studies on bulls, for example, showed that after the bull sees a cow, his blood level of LH rises as much as seventeenfold; within about half an hour, the blood level of testosterone reaches its peak. Human testosterone production has also been shown to fluctuate measurably in response to environmental stimuli.

30–13 *Rutting is the annual period of intense sexual activity characteristic of some animals, such as the red deer stags shown here. It is triggered by male hormones. During the period from spring to midsummer, while new antlers are growing, testosterone is almost undetectable. As soon as it begins to rise, the antlers die, losing their nerve and blood supply and their "velvet" cover. In the hard horn stages that follow, stags have pushing contests in which they lock antlers. Hierarchy within a group is determined by antler size. In early fall, testosterone levels begin to soar, the stag's voice "changes," his neck and shoulder muscles thicken, and he sets out to corral a harem, which he defends until rutting season is over. Then he returns to the "stag line," his antlers drop off, and the testosterone in his blood becomes once more undetectable.*

30–14 *Increasing daylight in springtime results in increased levels of testosterone, stimulating the male stickleback to seek a breeding site. Location of a suitable site—determined, apparently, by local vegetation for the most part—sets off an additional series of changes, including color changes, nest building, and the courting activities shown here.*
(a) A female ready to lay her eggs responds to a male displaying his red underbelly by turning her swollen abdomen toward him and then following him to the nest he has built. (b) At the nest, the male lies flat on his side, with his head in the nest, and the female swims past him into the nest. (c) The male then prods the base of the female's tail with his snout, and the female spawns. (d) Finally, the male enters the nest to fertilize the eggs. The male may lead as many as three or four females to his nest to deposit eggs, and he fertilizes each batch. Courtship completed, he cares for the eggs, aerating them by fanning them, and protects the young when they are hatched. Each of the activities in this chain of events depends upon the successful completion of the previous one, and the entire sequence depends upon the establishment of the appropriate internal hormonal environment.

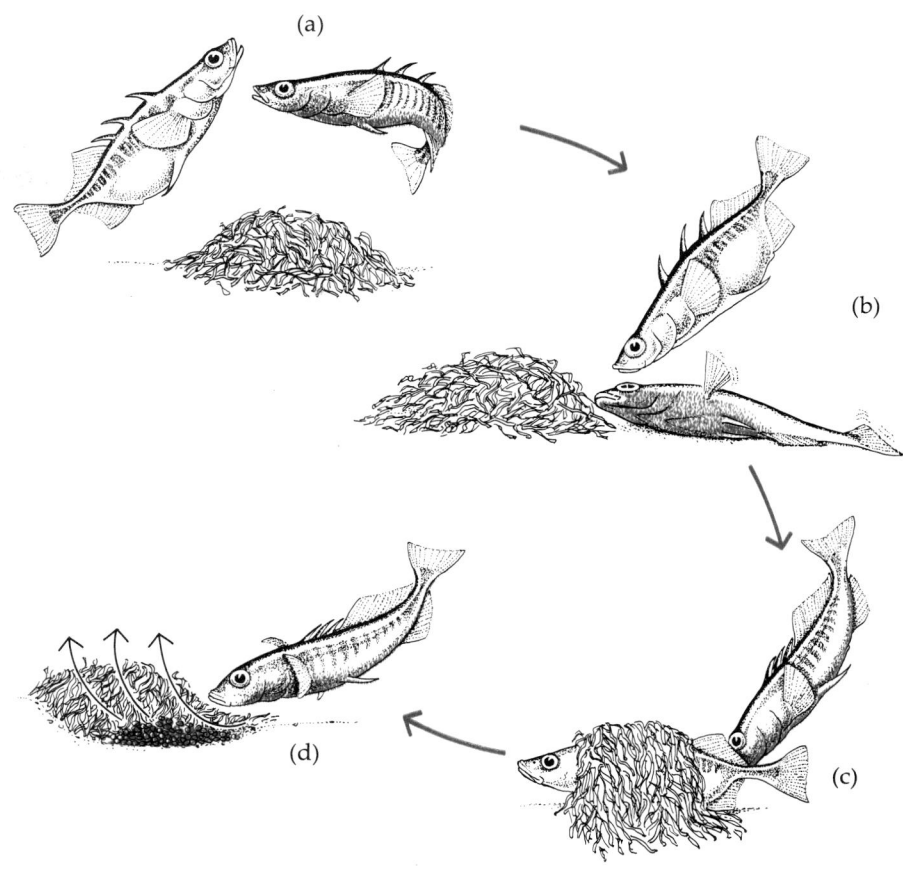

Table 30–1   *Major Mammalian Gonadotropic and Sex Hormones*

Hormone	Principal source	Principal effects	Control
FSH	Pituitary	Stimulates sperm production; stimulates growth of ovarian follicle; stimulates estrogen production	Hypothalamus
LH	Pituitary	Stimulates testosterone production; stimulates release of egg cell; stimulates progesterone production	Hypothalamus
Estrogen	Ovary (follicle, corpus luteum)	Produces and maintains female sex characteristics; thickens lining of uterus	FSH
Progesterone	Ovary (corpus luteum)	Thickens lining of uterus; inhibits uterine movement	LH
Testosterone	Testes	Produces and maintains male sex characteristics; stimulates sperm production	LH

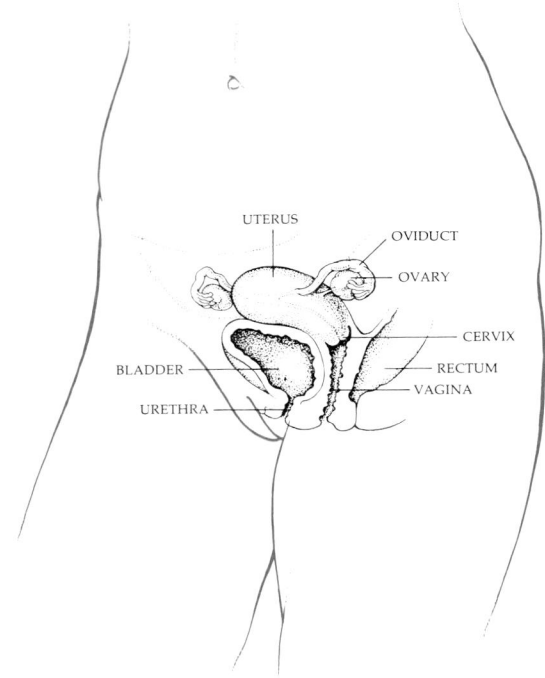

THE FEMALE REPRODUCTIVE SYSTEM

The gamete-producing organs in the female are the ovaries, each a solid mass of cells about 3 centimeters long, suspended in the abdominal cavity by ligaments (bands of connective tissue) and mesenteries. The oocytes, from which the eggs develop, are in the outer layer of the ovary. In human females, the primary oocytes begin to form about the third month of fetal development, and by the time of birth, the two ovaries contain some 400,000 primary oocytes, which have reached prophase of the first meiotic division. These primary oocytes remain in prophase until the female matures sexually. Then, under the influence of hormones, the first meiotic division resumes and is completed at about the time of the release of the oocyte from the ovary (ovulation). Of these 400,000 primary oocytes, about 300 to 400 mature, usually one at a time, to become secondary oocytes, which develop into mature egg cells (ova). In the human female after puberty, the primary oocytes usually mature one at a time about every 28 days. Therefore, as many as 50 years may elapse between the beginning and the end of the first meiotic division in a particular oocyte.

Maturation of the primary oocyte involves both meiosis and a great increase in size. At meiosis, an oocyte does not divide to form four ova. Instead, a single ovum and one to three polar bodies are formed. When the oocyte is ready to complete meiosis, the nuclear membrane fragments and the chromosomes move to the surface of the cell. As the nucleus divides, the cytoplasm of the oocyte bulges out. One set of chromosomes moves into the bulge, which then pinches off into a small cell, the first polar body. The rest of the cellular material forms the large secondary oocyte. The first meiotic division is completed a few hours before ovulation.

The second meiotic division does not take place until after fertilization. This division produces the ovum and another polar body.

As a consequence of these unequal cell divisions, all the accumulated food reserves of the oocyte are passed on to a single ovum. The first polar body may also divide, although there is no functional reason for it to do so. All the polar bodies eventually die.

30–16 *Oogenesis. A primary oocyte undergoes a meiotic division to produce a secondary oocyte and a polar body. This first meiotic division begins within the human fetus during the third month of fetal development and ends at ovulation, which may take place 50 years later. The second meiotic division, which produces the egg cell and a second polar body, does not take place until after fertilization. The first polar body may also divide.*

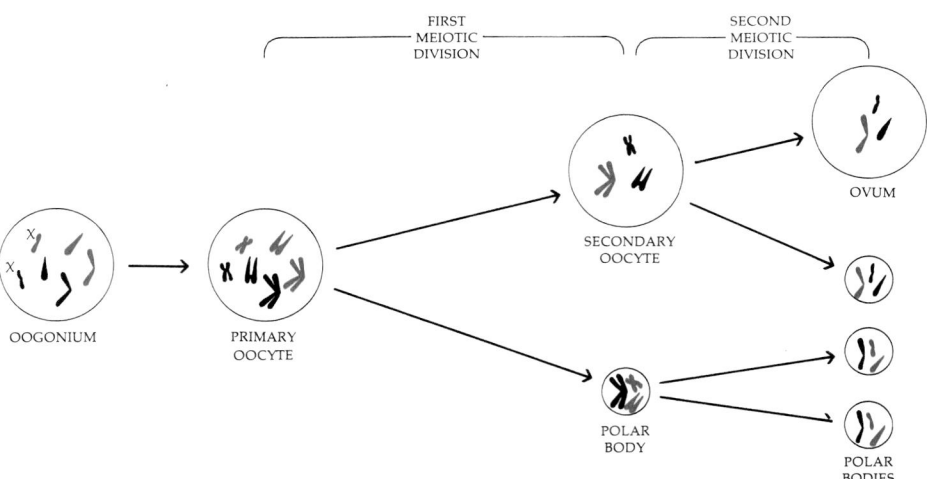

-17 *Oocytes develop near the surface of the ovary within follicles. After an oocyte is discharged from the follicle, cells of the ruptured follicle give rise to the corpus luteum ("yellow body"). If the oocyte is not fertilized, the corpus luteum is reabsorbed in two to three weeks. If the oocyte is fertilized, the corpus luteum persists, producing progesterone, which prepares the uterus for the embryo.*

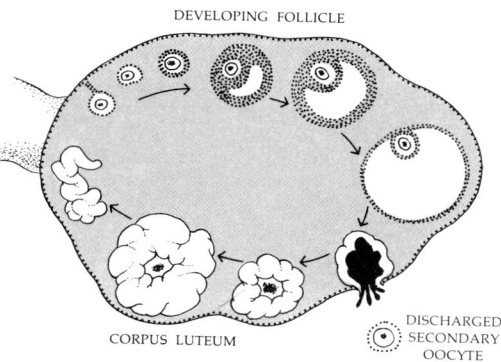

DEVELOPING FOLLICLE

CORPUS LUTEUM

DISCHARGED SECONDARY OOCYTE

18 *The female reproductive organs. Notice that the uterus lies at right angles to the vagina. This is one of the consequences of the bipedalism and upright position of* Homo sapiens *and one of the reasons that childbirth is more difficult for the human female than for other mammals.*

Oocytes develop near the surface of the ovary; the oocyte and the specialized cells surrounding it are known as an ovarian follicle. The cells of the follicle supply food to the growing oocyte. During the final stages of its growth, the follicle moves to the surface and produces a thin, blisterlike elevation, which eventually bursts, releasing the oocyte (ovulation). From the follicle, it is swept into one of the oviducts (sometimes called the Fallopian tubes) as a result of the movement of the funnel-shaped opening of the oviduct over the surface of the ovary and the beating of cilia, which line the fingerlike projections surrounding this opening.

A human oocyte is about 0.1 millimeter, or 100 micrometers, in diameter, which is very large for a cell. It contains an unusually large supply of ribosomes, enzymes, amino acids, and all the other cellular machinery that will be used in the early, rapid stages of biosynthesis characteristic of embryonic cells.

Uterus, Vagina, and Vulva

The uterus is a hollow, muscular, pear-shaped organ about 7.5 centimeters long and 5 centimeters wide. It lies almost horizontally in the abdominal cavity and is on top of the bladder (Figure 30–18). It is lined by the endometrium, which has two principal layers, one of which is shed at menstruation and another layer from which the shed layer is regenerated. The opening of the uterus is the cervix, through which the sperm pass on their way toward the oocyte and through which the fetus emerges at the time of birth. The vagina is a muscular tube about 7.5 centimeters long that leads from the cervix of the uterus to the outside of the body. It is the receptive organ for the penis and also the birth canal. Its opening is between the urethra, the tube leading from the bladder, and the anus.

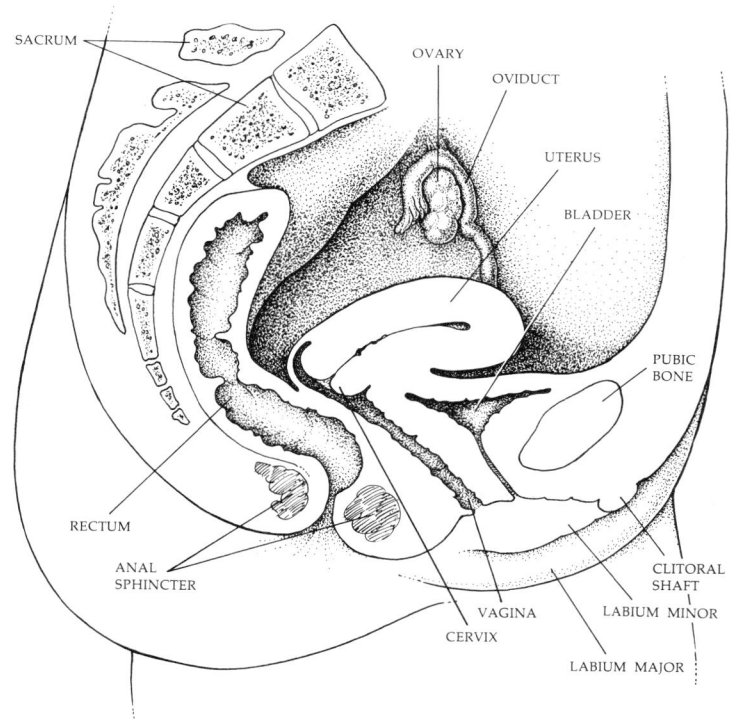

SACRUM

OVARY

OVIDUCT

UTERUS

BLADDER

PUBIC BONE

RECTUM

ANAL SPHINCTER

VAGINA

CERVIX

CLITORAL SHAFT

LABIUM MINOR

LABIUM MAJOR

The lining of the vagina is rich in glycogen which bacteria normally present in the vagina convert to lactic acid; as a consequence the vaginal tract is acid, with a pH between 4 and 5.

The external genital organs of the female are collectively known as the vulva (see Figure 30–19). The clitoris, which corresponds to the penis in the male,* is about 2 centimeters long and, like the penis, is composed chiefly of erectile tissue. The clitoris has two bulbs (analogous to the penile bulb of the male), which lie on either side of the lower third of the vagina. The labia (singular, labium) are folds of skin. The labia majora are fleshy and, in the adult, covered with pubic hair; they enclose and protect the underlying, more delicate structures. (Embryologically, they are homologous with the scrotum in the male.) The labia minora are thin and membranous.

Orgasm in the Female

Under the influence of a variety of stimuli, the clitoris, the labia, and other tissues in the pelvic region become engorged and distended with blood, as does the penis of the male. This process is somewhat slower in women than in men, largely because the valves trapping the blood in the sexual structures are not as efficient as in men. The distension of the tissues is accompanied by the secretion into the vagina of a fluid that both lubricates the walls of the vagina and neutralizes its highly acid, and therefore spermicidal, secretion.

* In the early embryo, the structures are identical.

30–19 *The external genitalia of the human female. The labia, clitoris, vaginal opening, and the mons pubis constitute the vulva. The mons pubis is a pad of fatty tissue overlying the pubic symphysis, where the two pubic bones join. It is also sometimes known as the mons veneris, the mount of Venus.*

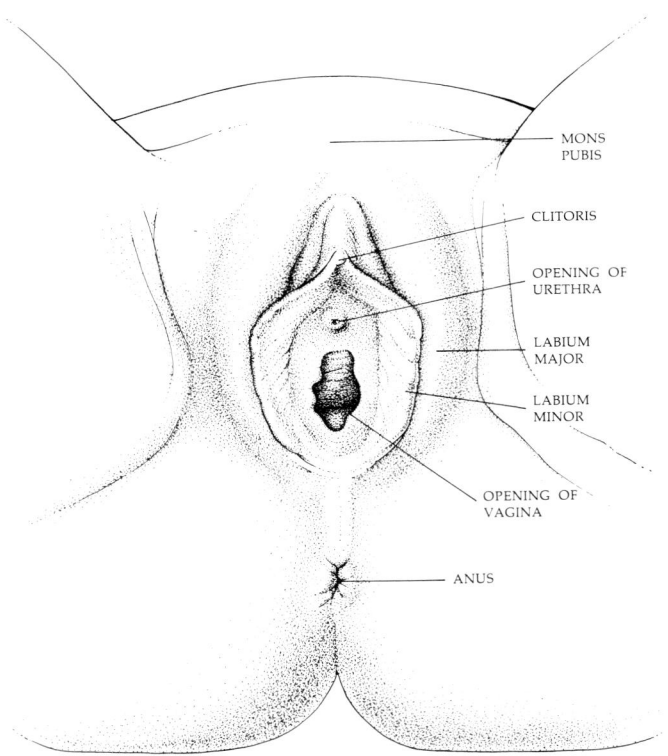

MONS PUBIS

CLITORIS

OPENING OF URETHRA

LABIUM MAJOR

LABIUM MINOR

OPENING OF VAGINA

ANUS

30–20 *Fertilization of the egg by sperm. Once a month in the nonpregnant female of reproductive age, an oocyte breaks loose from the ovary and is swept into one of the oviducts. Fertilization, when it occurs, generally takes place within the tube, after which the fertilized egg becomes implanted in the lining of the uterus. Muscular movements of the oviduct, plus the beating of the cilia that line it, propel the egg cell down the tube toward the uterus. If the egg cell is not fertilized, it dies, usually within 12 to 24 hours. A sperm cell has an average life of 48 hours within the female reproductive tract.*

30–21 *Implantation. The tiny embryo invades the lining of the uterus within a week after fertilization. Subsequently the placenta begins to form; this organ is the source of hormones that help to maintain pregnancy. The uterine glands secrete a glycogen-rich material that nourishes the embryo.*

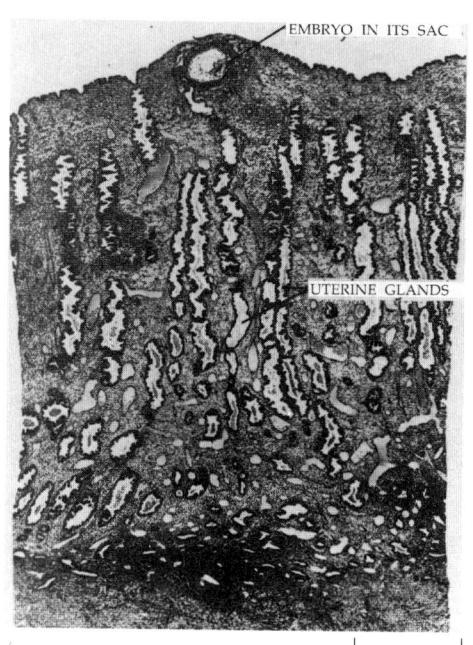

EMBRYO IN ITS SAC

UTERINE GLANDS

1 mm

OVIDUCT   FERTILIZATION

OVARY

IMPLANTATION

UTERUS

OOCYTE

CERVIX OF UTERUS

↑ PATHWAY OF SPERM

↑ PATHWAY OF EGG

The thrusting of the penis into the vagina stimulates the lower third of the vagina and also causes the prepuce of the female (which covers the clitoris just as the male foreskin covers the glans of the uncircumcised penis) to move back and forth over the clitoris, stimulating it. Orgasm in the female, as in the male, is the consequence of rhythmic muscular contractions, followed by expulsion of the blood trapped in the engorged tissues. Homologous muscles produce orgasm in the two sexes, but in the female there is no ejaculation of fluid through the urethra, as occurs in the male.

At orgasm, the cervix drops down into the upper portion of the vagina, where the semen tends to form a pool. The female orgasm also may produce contractions in the oviducts which propel the sperm upward. It has been calculated that it would take a sperm cell at least two hours to make its way up the oviducts under its own power, and sperm have been found in the oviducts as soon as five minutes after intercourse. Orgasm in the female, however, is not necessary for conception.

Fertilization and Implantation

About three days is required for an oocyte to travel down the oviducts to the uterus. An oocyte lives, however, only about 24 hours after it is ejected from the follicle. So, as you can see, fertilization, if it occurs, occurs in an oviduct. If the egg cell is fertilized, it becomes implanted in the endometrium; implantation usually occurs 24 to 48 hours after the young embryo reaches the uterus, six or seven days after the egg was fertilized. If the egg cell is not fertilized, it dies, and the endometrial lining is shed at menstruation. Fertilized eggs implanted in the endometrium are sometimes lost in abnormal menstrual flow, but it is difficult to estimate the number of such very short lived pregnancies.

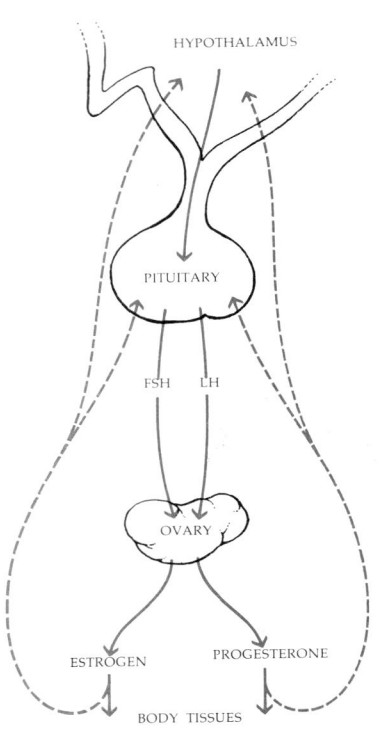

HYPOTHALAMUS

PITUITARY

FSH   LH

OVARY

ESTROGEN   PROGESTERONE

BODY TISSUES

——— STIMULATION
– – – INHIBITION

Effects of Female Sex Hormones

Like spermatogenesis, oogenesis is under hormonal control. Unlike spermatogenesis in the human male, however, oogenesis in all vertebrate females is cyclic, involving an interplay of hormones, including estrogen, progesterone, and the two gonadotropic (gonad-stimulating) hormones, follicle-stimulating hormone (FSH) and luteinizing hormone (LH). The timing and control of the cycle rests in the hypothalamus.

Estrogens are the female sex hormones. A variety of them, of which estradiol is probably the most important, are produced by the ovarian follicles under the stimulation of FSH, whose production they inhibit in a feedback system similar to that which controls the production of male sex hormones. Estrogens stimulate the development of the breasts, the external genitalia, pubic and axillary (underarm) hair, and the distribution of body fat. Both estrogens and progesterone are required to prepare the endometrium for the implantation of the embryo; neither can do the job alone. This cooperation, or synergism, in which two or more agents act together to produce an effect that is greater in magnitude than the sum of the effects produced separately, is characteristic of many hormonal responses. DES, diethylstilbestrol, which is used in the "morning-after pill," is a potent synthetic estrogen. For a time it was used commercially to fatten cattle, but this use is now forbidden by government regulation since the hormone was found to cause cancer in experimental animals when administered in high doses.

The Menstrual Cycle

The beginning of the first menstrual cycle marks the onset of puberty in human females. The average age of onset is thirteen and a half, but the normal range is very wide. Puberty in the female begins on the average about a year and a half before puberty in the male, which is generally considered as the time of onset of sperm production. Puberty in the female is usually accompanied by the appearance of the secondary sex characteristics, such as pubic and axillary hair and enlargement of the breasts.

Although menstruation does not require an environmental cue, as do reproductive cycles in many other vertebrates, such as the stickleback, it is clearly under the influence of the nervous system to some extent. For example, some women find that emotional upset delays a menstrual period or eliminates it completely. This mechanism may be linked to the sterility that occurs among animals as a consequence of population pressures or other forms of stress.

Human females living in groups—as in boarding schools—are also familiar with the tendency of the menstrual cycles of the group to become synchronized. The mechanism of this is unknown, but it has been suggested that it may be a result of the exchange of a pheromone* among the individuals involved.

The menstrual cycle (see Figure 30–23) begins with the casting off of a layer of endometrium. Under the influence of FSH, an egg cell and its follicle mature, and the follicle secretes increased amounts of estrogen. (Usually a number of follicles begin to enlarge simultaneously, but only one becomes mature enough to release its ovum, and the others regress.) The estrogen stimulates the regrowth of the endometrium. The rapid rise in estrogen triggers the production of LH (an

* For another discussion of pheromones, see page 410.

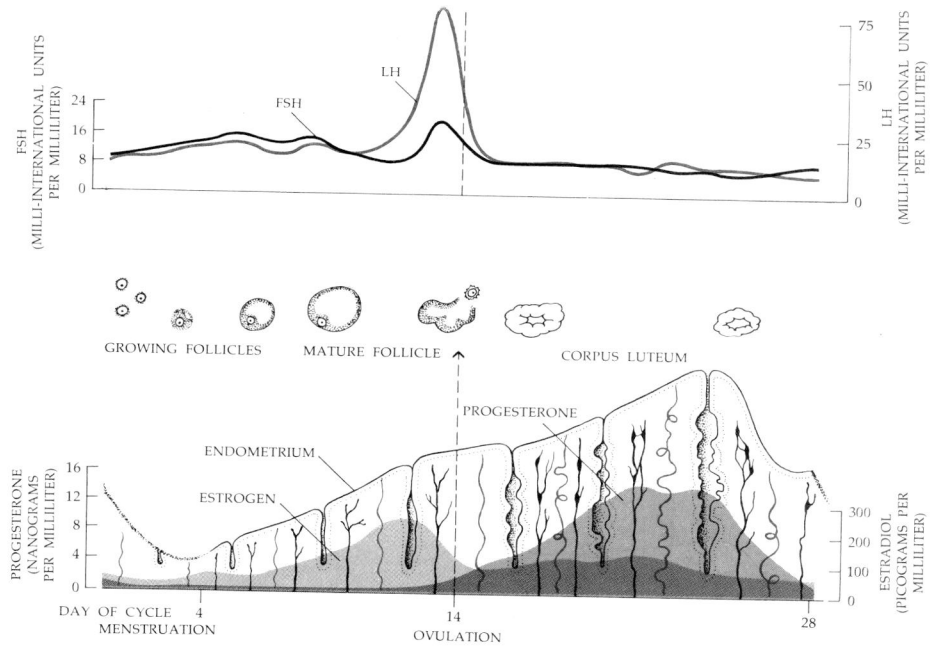

-23 Diagram of the menstrual cycle, showing blood levels of the hormones involved. An increase in the concentration of FSH from the pituitary promotes the growth of the ovarian follicles and the secretion of estrogen. The estrogen stimulates the lining of the uterus (the endometrium) to thicken. At about midcycle, there is a sharp increase in LH production by the pituitary, accompanied by a smaller increase in FSH. Under the influence of LH, ovulation occurs. The follicle is converted into the corpus luteum, which begins to secrete progesterone and estrogen, which further stimulate the growth of the endometrium. By the end of the menstrual cycle, if pregnancy has not occurred, progesterone production falls, the endometrium begins to slough off, and the cycle begins again.

24 The chemical structures of testosterone, estradiol, and progesterone. Note that the hormones differ only very slightly chemically, in contrast to the great differences in their physiological effects—another example of the extreme specificity of biochemical actions. All are steroids.

TESTOSTERONE

ESTRADIOL

PROGESTERONE

example of positive feedback). Paradoxically, toward the end of the cycle, LH and FSH production decline as a result of the increased concentration of progesterone and estrogen.

The cycle usually lasts about 28 days, but individual variation is common. Moreover, even in women with cycles of average length, ovulation does not always occur at the same time in the cycle.

All these events take place as the result of a shifting balance of hormones. LH stimulates the follicle to release the egg cell, which begins its passage to the uterus. Under the continued stimulus of LH, the cells of the emptied follicle grow larger and fill the cavity, producing the corpus luteum ("yellow body"). The cells of the corpus luteum, as they increase in size, begin to synthesize progesterone as well as estrogen. As the estrogen and progesterone levels increase, they inhibit the production of the gonadotropic hormones from the pituitary. Production of ovarian hormones then drops. The lining of the uterus can no longer sustain itself without hormonal support, and a portion of it is sloughed off in the menstrual fluid. Also, in response to the low level of ovarian hormones, the level of pituitary gonadotropic hormones begins to rise again, followed by development of a new follicle and a rise in estrogen as the next monthly cycle begins.

Fertility pills for women contain either gonadotropins or a synthetic compound with antiestrogen activities. The latter, by decreasing estrogen concentrations in the blood, stimulates the production of FSH by the pituitary. Multiple births may occur when several ova are released simultaneously as the result of such treatments.

535 *The Continuity of Life: Reproduction*

"The Pill"

"The pill" consists of a combination of estrogen and progesterone. When taken daily, it keeps the level of ovarian hormones in the blood high enough to shut off production of the pituitary hormones. In the absence of LH, no ovulation occurs. The lining of the uterus thickens under the artificial hormone influence. Then, when a woman stops taking the pill toward the end of the cycle, the endometrium sloughs off, producing a menstrual period, although ovulation has not occurred. Other contraceptive methods are summarized in Table 30–2.

Despite the availability of a number of contraceptive methods, there are many unwanted pregnancies each year, especially among teenagers, as conservatively estimated by the numbers seeking abortions. In a recent survey among 4,611 unmarried girls 15 to 19 years old, it was found that more than half (53 percent) had failed to use any form of contraception the last time they had had sexual intercourse. Of those who did use contraception, the majority used one of the less reliable methods. Those most frequently used were the condom (27 percent) and withdrawal (24 percent). Twenty-one percent used the pill. Among the reasons that teenagers failed to use contraception were the beliefs that pregnancy could not occur because they were too young, or because they had sexual intercourse too infrequently, or because it was the "wrong time of the month." (Only about two-fifths of the teenagers surveyed knew when in the menstrual cycle a woman is most likely to become pregnant.) Only 19 percent reported that they always used contraceptives; however, the percentage using medically prescribed contraceptive measures was considerably higher among those who had previously been pregnant.

Hormones and Pregnancy

If the egg cell is fertilized, it begins to develop in the oviduct, becomes implanted in the lining of the uterus, and the placenta begins to form. Almost immediately, the placenta begins to produce gonadotropic hormones of its own and in large quantities. These gonadotropic hormones act upon the corpus luteum so that it continues production of estrogen and progesterone when pituitary gonadotropic hormone production ceases under the negative feedback system. Pregnancy tests often involve testing for these gonadotropic hormones from the placenta in either blood or urine. As the placenta matures, it also begins its own production of estrogen and progesterone; both hormones are produced in relatively high quantities throughout pregnancy.

FERTILITY AND MATING

Females of almost all species except man will mate only during their fertile period, which is known as *estrus*, or heat. Estrus may occur only once a year, usually in the early spring (as in wolves and deer), about once a month (as in cows and horses), or every few days (as in rats and mice).

The periods of estrus may last from only a few hours to three or four weeks. In animals, such as dogs, that produce eggs continuously during estrus, the eggs may be fertilized at different times and by different males, which explains in part why a mixed-breed litter can contain such an astonishing variety of siblings and also why there may be a great size variation even among purebred pups, many of which are actually of different ages at birth. In some mammals, such as

Table 30–2 *Methods of Contraception Currently Available (In decreasing order of effectiveness)*

Method	Mode of action	Effectiveness if used correctly	Action needed at time of intercourse	Requires resupply of materials used	Requires instruction in use	Requires services of physician	Suitable for menstrually irregular women	Possible undesirable effects
Vasectomy	Prevents release of sperm	Very high	None	No	No	Yes, operation	Yes	May produce irreversible sterility
Tubal ligation	Prevents passage of egg cell to uterus	Very high	None	No	No	Yes, operation	Yes	May produce irreversible sterility
Oral pill, 21-day administration	Prevents follicle maturation and ovulation	Very high	None	Yes	Yes, timing	Yes, prescription	Yes	Early—some water retention, breast tenderness, nausea. Late—possible blood clots, hypertension
"Morning-after pill"	Arrests pregnancy, probably by preventing implantation, 50× normal dose of estrogen	By currently available data, high	None	Yes	Yes, timing	Yes, prescription	Yes	Breast swelling, nausea, water retention, cancer(?)
Intrauterine device (coil, loop, IUD)	Prevents implantation	High	None	No	No	Yes, to insert	Yes	Menstrual discomfort, possible displacement or loss of device, possible uterine infection
Diaphragm with spermicidal jelly	Prevents sperm from entering uterus, jelly kills sperm	High	Insertion before intercourse	Yes	Yes, must be inserted correctly each time	Yes, for sizing and instruction on use	Yes	None known
Condom (worn by male)	Prevents sperm entry into vagina	High	Yes, male must put on after erection	Yes	Not usually	No	Yes	Some loss of sensation in male
Temperature rhythm	Determines ovulation time by noting body temperature	Medium	None	No	Definitely, must learn to interpret chart	No, but physician should advise	Yes, if skilled at reading graph	Requires abstinence during part of cycle
Calendar rhythm	Abstinence during part of cycle	Medium to low	None	No	Definitely, must know when to abstain	No, but physician should advise	No	Requires abstinence during part of cycle
Vaginal foams	Spermicidal	Medium to low	Yes, requires application before intercourse	Yes	No	No	Yes	None usually, may irritate
Withdrawal	Remove penis from vagina before ejaculation	Low	Yes, withdrawal	No	No	No	Yes	Frustration in some
Douche	Washes out sperm	Lowest	Yes, immediately after	No	No	No	Yes	None

cats, rabbits, and minks, although the egg is mature and the female receptive during estrus, ovulation occurs only under the stimulus of copulation, obviously a very efficient system, ensuring maximum economy in the utilization of gametes. There is suggestive evidence that in some women, also, ovulation may be triggered by sexual intercourse.

The human female appears to be one of the few female animals receptive to mating during unfertile periods. Anthropologists speculate that this receptivity coevolved with the establishment of strong pair-bond relationships between human or prehuman males and females. A consequence of this pair-bond relationship is a society based on a family unit, in contrast to many other primate groups, in which the social and breeding unit is a troop or band. The establishment of such family units is seen, in turn, as the basis for the division of labor among the sexes, with the female concentrating on childbearing and the home and the male on hunting, defending his territory, and waging war against his enemies. If the anthropologists are right, this behavioral and hormonal adaptation on the part of the human female has profoundly influenced the shape of human civilization.

SUMMARY

In vertebrates, reproduction is always sexual and involves two parents, one of which produces sperm and the other eggs. Sperm and eggs are formed by meiosis in the gonads (the testes and the ovaries). The male gametes are produced by meiosis in the seminiferous tubules of the testes. The spermatogonia become primary spermatocytes; then, after the first meiotic division, secondary spermatocytes; and, following the second meiotic division, spermatids, which then differentiate into sperm cells. These sperm cells enter the epididymis, a tightly coiled tubule overlying the testis, where they are mobilized. The epididymis is continuous with the vas deferens, which carries sperm through the spermatic cord into the abdominal cavity and around the bladder. There the two vas deferentia merge with the urethra which leads through the penis.

The penis is composed largely of spongy erectile tissue which can become engorged with blood, enlarging and hardening it. At the time of ejaculation, sperm are propelled along the vas deferens by contractions of a surrounding coat of smooth muscle. Secretions from the seminal vesicles, the prostate, and the bulbourethral glands are added to the sperm as they pass to the urethra. They are expelled from the urethra by muscular contractions involving, among other structures, the base of the penis. These muscular contractions also pump blood out of the engorged structures, producing the sensation known as orgasm.

Production of sperm and the development of characteristics associated with masculinity are under the control of hormones, including testosterone (an androgen) and two gonadotropins, luteinizing hormone (LH) and follicle-stimulating hormone (FSH). LH acts on the interstitial cells (located between the seminiferous tubules) to stimulate the production of testosterone. FSH and testosterone stimulate the production of sperm. The gonadotropins are produced by the pituitary gland under the regulation of the hypothalamus, a brain center. Production of LH is inhibited by the presence of sex hormones through a negative feedback system.

The female gamete-producing organs are the ovaries. The primary oocytes develop within nests of cells called follicles. The first meiotic division begins in the female fetus and is completed at ovulation. The second is completed at fertilization. On the average, one oocyte is produced every 28 days; it travels down the oviduct to the uterus. If it is fertilized (which usually takes place in an oviduct), it becomes implanted in the lining of the uterus (the endometrium). If it is not fertilized, it degenerates and the endometrial lining is shed at menstruation.

The production of ova, the menstrual cycle, and development at puberty of the uterus, vulva, breasts, and other secondary sex characteristics of the female are controlled by estrogens and progesterone, both steroid hormones, and the gonadotropins LH and FSH. Estrogen is produced by the ovarian follicles before ovulation. After ovulation, the corpus luteum, which forms from the emptied follicle, produces both estrogen and progesterone. Progesterone and estrogen combined stimulate the growth of the endometrium.

In many vertebrates, gamete production and mating are cyclic. The part of the cycle during which mating occurs is known as estrus, or heat, in females. In the human male, spermatogenesis is continuous. Although ovulation is cyclic in most human females, they, unlike most other female animals, are continuously receptive to sexual intercourse, a behavior pattern which may have evolved along with pair bonding.

QUESTIONS

1. Define the following: secondary spermatocyte, vulva, Sertoli cell, corpus luteum, FSH, clitoris, vagina, primary oocyte.
2. What are the advantages of asexual reproduction to the organism?
3. What would constitute the semen of a man who had had a vasectomy? Would the vasectomy affect the structures associated with orgasm?
4. During which days in the menstrual cycle is a woman most likely to become pregnant? (Include data on longevity of eggs and sperm in making this calculation.) Why, in your opinion, is the use of calendar rhythm method of birth control so ineffective?
5. Diagram the feedback regulation of house temperature by a thermostat.
6. Do you believe the use of contraceptive methods among teenagers should be encouraged?
7. Under what circumstances would you consider involuntary sterilization acceptable?
8. John and Jane Gamble, practicing people prevention, use a contraceptive scheme estimated to be 95 percent effective; that is, the chance of a pregnancy occurring in a given month is a mere 5 percent. The chance of a pregnancy in two months is, of course, 0.95^2 and in three months, 0.95^3. How many months can they play this game before they have an even (50 percent) chance of incurring parenthood? Would you regard their method of contraception as effective? Assuming the Gambles know what you now know, should a resulting pregnancy be considered planned or accidental?

Chapter 31

The Continuity of Life: Development

In this chapter, we are going to be concerned with the changes by which a single cell, the fertilized egg, becomes a complex organism, containing many billions of cells with specialized functions and behavior, precisely organized in relationship to one another. Development is the process by which these changes are brought about. It has three principal components: (1) growth, an irreversible increase in mass, (2) differentiation, the changes by which cells become different from each other and from their parent cells in both structure and function, and (3) morphogenesis, the appearance of characteristic form (from the Greek word *morphe*, meaning "shape") in a single cell, a population of cells, such as a hand or a leaf, or an entire organism.

Although this chapter will be limited in its scope to the development of an embryo, it is important to keep in mind that embryology is only one manifestation of a more general phenomenon. Wound-healing is another everyday example of development, as is the regeneration of body parts that occurs among certain organisms. The continuous replacement of skin and other epithelial tissues is another manifestation of development, as is spermatogenesis, described in the previous chapter.

All of these processes are both commonplace and mysterious. They have been closely observed and analyzed for at least 100 years in a variety of animals. However, relatively little is known about how the remarkable changes associated with the formation of a single skin cell, or a sperm cell, or the tail of a lizard, or a human infant take place. Development has been and continues to be one of biology's most formidable challenges.

FERTILIZATION

Our story begins with the fertilization of the egg, which, in the human female, takes place high in one of the oviducts. The details of this process of fertilization have been studied most extensively in aquatic invertebrates and amphibians in which fertilization is external, but the process seems to be very similar throughout the animal phyla.

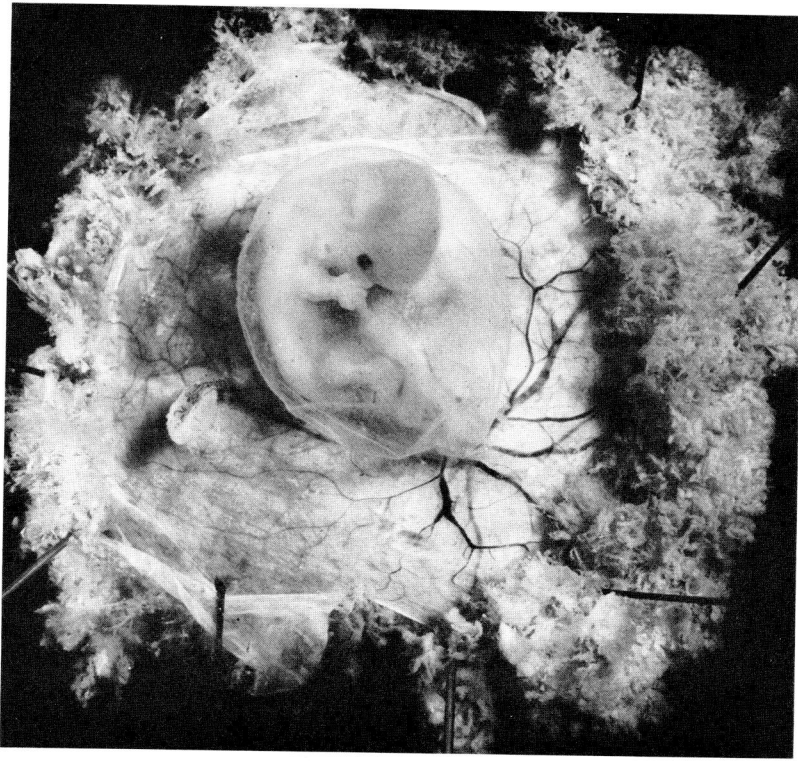

–2   *Sperm cell penetrating an egg. The zona pellucida, which appears as a dark border, and numerous follicle cells are visible outside the egg cell membrane. Note the difference in size between an egg cell, a sperm cell, and a follicle cell.*

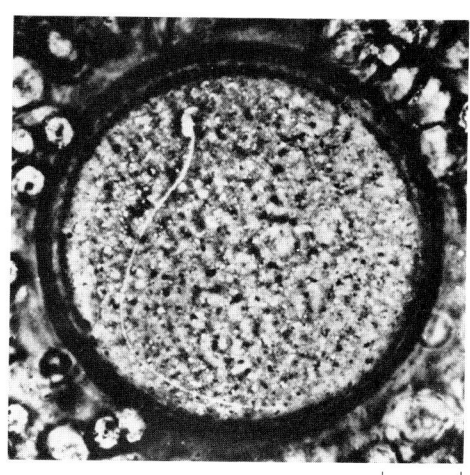

$\vdash$ 30 μm $\dashv$

In mammals, by the time the egg cell arrives in the oviduct, it is surrounded by an outer membrane, the *zona pellucida,* and by those follicle cells that immediately surrounded the oocyte within the ovary. The first polar body lies beneath the zona pellucida. The egg cell, as you will recall, has just completed its first meiotic division. The second division has begun but will only be completed if the egg is fertilized.

The sperm, which is much smaller than the egg cell, penetrates the zona pellucida and the cloud of cells around it. When the sperm touches the egg, the sperm cell membrane fuses with the egg cell membrane, and the egg cell cytoplasm bulges out at this point. The sperm head penetrates the egg through the bulge, and the sperm nucleus enters the egg cell. Changes in the egg cell membranes prevent the entry of other sperm cells. The egg cell nucleus completes its second meiotic division; then it and the sperm nucleus move to the center of the cell, where they fuse to form the zygote nucleus. The genetic identity of the new individual is now determined.

ACTIVATION OF THE EGG

The fusion of egg and sperm not only introduces new genetic material into the egg, but, in the normal course of events, penetration of the egg membrane by the sperm activates the egg, initiating its development. These two events are not necessarily related, however. Eggs can be activated artificially in a variety of

An egg shortly after fertilization. The egg has shrunk slightly, and a space has formed between the egg cell membrane and the zona pellucida. Within this space (on the right), the polar bodies appear. The nucleus of the egg is at the center, and to the right of it the nucleus of the sperm. The nuclei are migrating toward the center of the cell where they will fuse.

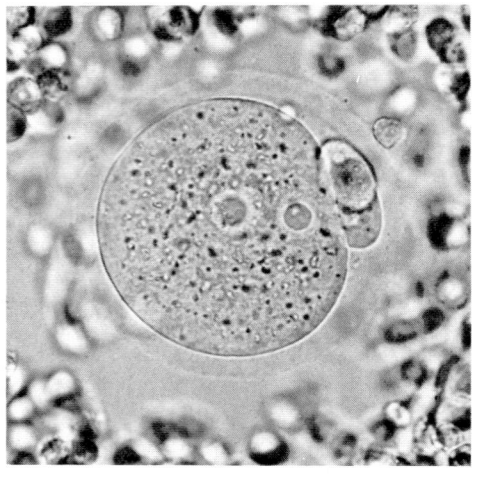

ways. Unfertilized sea urchin eggs will begin to develop if they are put in a hypotonic solution. Frog eggs can be activated by pricking them with a glass or metallic needle or by electric shock. Development of unfertilized eggs, which is called parthenogenesis, takes place naturally in a number of invertebrate species and is relatively common in some groups of fish and lizards and also among domestic turkeys. In some organisms, individuals that develop parthenogenetically have a different biological role from those that develop from fertilized eggs. Queen bees, for example, mate only once (on their nuptial flight) and store a lifetime's supply of sperm in a special organ, the spermatheca. Most eggs are fertilized as they are laid; such eggs develop into females, either workers or queens. When, in the life of the hive, new queens are being developed, the present queen lays unfertilized eggs. These unfertilized eggs develop into males, which play out their brief role in the life of the hive and then perish.

Among bees, the males are haploid (having 16 chromosomes, compared to the females' 32), but in many parthenogenetic eggs, the chromosomes divide once before the egg cell does (mitosis without cytokinesis is followed by mitosis with cytokinesis), and, as a consequence, the progeny are diploid.

The signs that activation has occurred are many and varied, often differing from species to species. One sign common to all vertebrates is the emission of the second polar body, which takes place as meiosis is completed. In many animals, activation is immediately followed by a drastic rearrangement of the cytoplasm of the egg cell. For example, in frogs of the genus *Rana*, the yolk is massed in the lower hemisphere of the unfertilized egg, and the upper two-thirds of the egg is covered by a heavily pigmented layer. (See Figure 31–4.) When the sperm penetrates the egg, the pigment cap rotates toward the point of sperm penetration and a gray crescent appears on the side of the egg, opposite the point of sperm entry. The cytoplasm associated with the crescent is of critical importance in the later development of the embryo. (See Figure 31–5.)

The appearance of the gray crescent and similar phenomena indicate that an egg cell is not simply a food reservoir or a blob of homogeneous jelly, but that

31–4 *The formation of the gray crescent in a frog's egg. (a) Before fertilization, the upper two-thirds of the egg is covered by a heavily pigmented layer and the greater part of the yolk is massed at the lower hemisphere. The nucleus of the egg is near the pole of the upper hemisphere. (b) The egg has been fertilized; the sperm has entered at the right and is moving toward the center of the egg. The trail it leaves behind it is caused by the disruption of pigment granules clustered near the egg surface. The whole pigment cap has rotated toward the point of sperm penetration. The cytoplasmic region on the other side of the egg, from which the pigment layer has moved away, becomes the gray crescent.*

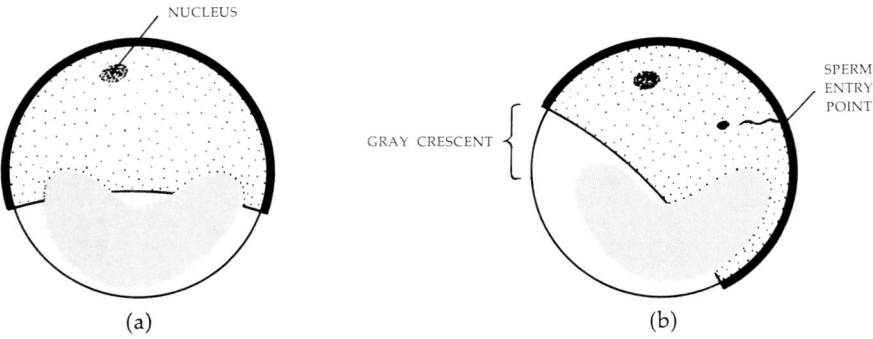

(a) (b)

1–5 *The importance of the gray crescent in development was demonstrated by separating the two cells formed by the first division of the egg. When the egg on the left divided, half of the gray crescent passed into each of the two new cells. When these cells were separated from each other, each formed a complete embryo. The first division of the egg on the right resulted in all the gray crescent going to one cell and none to the other. When these cells were separated, the one without the crescent did not develop.*

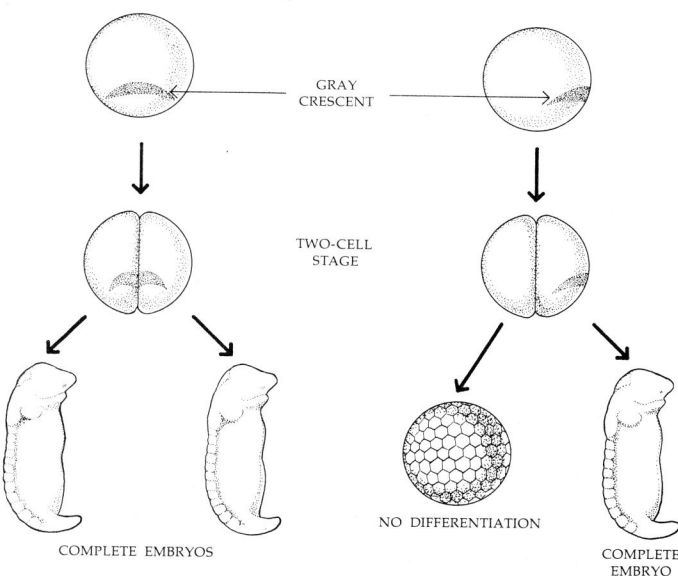

GRAY CRESCENT

TWO-CELL STAGE

COMPLETE EMBRYOS

NO DIFFERENTIATION

COMPLETE EMBRYO

its cytoplasm is heterogeneous and complex. Much evidence supports this conclusion.

The events that take place following activation are fundamentally the same in all animal species. They take place in three stages:

1. Cleavage: the fertilized egg cell (zygote) divides into smaller daughter cells
2. Gastrulation: the cells arrange themselves in distinct layers
3. Organogenesis: organs and organ systems are formed

In the pages that follow we shall trace each of these events through a series of different organisms and then in man. As you will see, the seeming complexity of human development becomes relatively simple in its basic outlines, viewed from this more general, distant perspective.

CLEAVAGE

During cleavage, the original zygote undergoes a series of divisions with no increase in volume. The first divisions produce a solid sphere of cells. Subsequently, the dividing cells pull away from the center so that a hollow (the blastocoel*) is produced. The embryo at this stage is known as a _blastula._ The mass of the blastula is less than that of the zygote.

In cell divisions not associated with cleavage, as we saw in Chapter 10, the processes of mitosis and cytokinesis alternate with periods of cell growth. The cellular divisions of cleavage probably represent a mechanism for bringing the cytoplasm of the cells involved under nuclear control. This concept is supported by evidence that these precleavage cells are not under nuclear control: If the nucleus of a fertilized egg is destroyed, it is still able, in many cases to cleave,

* The word root *coel* means "cavity," as it did in coelenteron and coelom (see page 378). However, aside from being hollow structures, they have no anatomical relationship to one another.

although further development stops short at this point. The fact that cleavage can proceed normally in the absence of a nucleus implies not only that the nucleus is not necessary but also that the materials needed for spindle formation, new cell membranes, DNA synthesis, ribosomes, and a host of other cellular necessities are either present in the cytoplasm of the fertilized egg or, at least, that the messenger RNA is there, already formed.

The patterns of cleavage are clearly influenced by at least one cytoplasmic factor, the yolk (Figure 31–6). When the yolk is absent or present in very small amounts, as in *Amphioxus*,* cleavage involves the whole egg. When larger amounts of yolk are present, the whole egg divides, but unevenly. In the presence of still larger amounts of yolk, cleavage is confined to a small yolk-free region.

Thus cleavage increases nucleus-to-cytoplasm ratios (and also surface-to-volume ratios) to approximately the same as those existing in the cells of adult individuals of the species. It stabilizes the internal organization of the egg; for

* *Amphioxus* and its position in vertebrate phylogeny were discussed in Chapter 23, page 413.

31–6 *Comparative cleavage stages of three types of egg. The differences among these early cleavage patterns are determined primarily by the distribution of the yolk in the egg. (a) In Amphioxus, the egg contains a small amount of yolk almost evenly distributed throughout the cell, and the cleavages result in daughter cells that are all about the same size. (b) Eggs that contain a large amount of yolk concentrated in one hemisphere, such as those of the frog, cleave unequally. The first two cleavages split the frog's egg longitudinally to produce four cells shaped like the segments of an orange. The third cleavage separates the yolkier (vegetal) part from the less yolky (animal) part. As you can see, the four cells in the animal hemisphere are much smaller than the four in the yolky vegetal hemisphere. Subsequently, the yolkier cells cleave much more slowly than the less yolky ones. (c) In fish, reptiles, and birds, the cytoplasm and the nucleus of the egg are located on a huge ball of yolk. The yolky area does not cleave at all, and only the small cap on top divides into cells.*

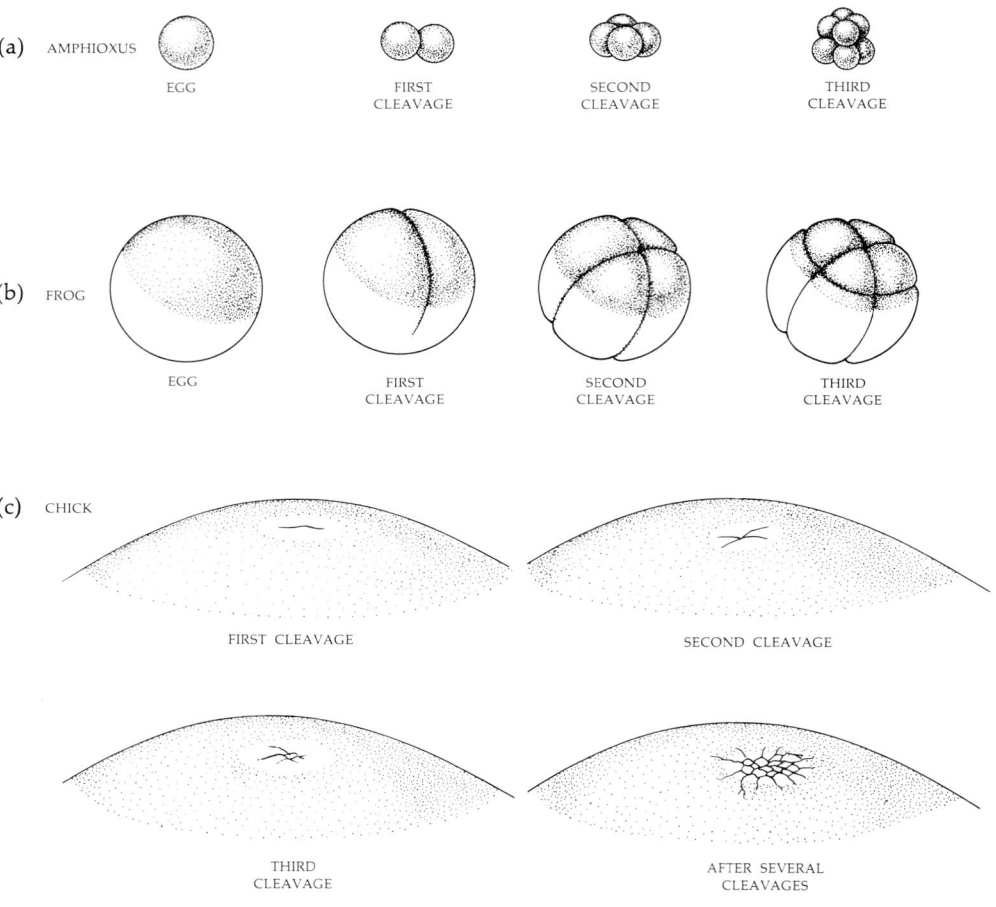

31–7 *Cleavage in a sea urchin. The sea urchin egg, like that of* Amphioxus, *contains little yolk. Notice that as the egg divides, the cells become progressively smaller, so that in the blastula stage, at the time of hatching from the egg membrane, they are barely distinguishable, although the magnification of the micrograph has not changed.*

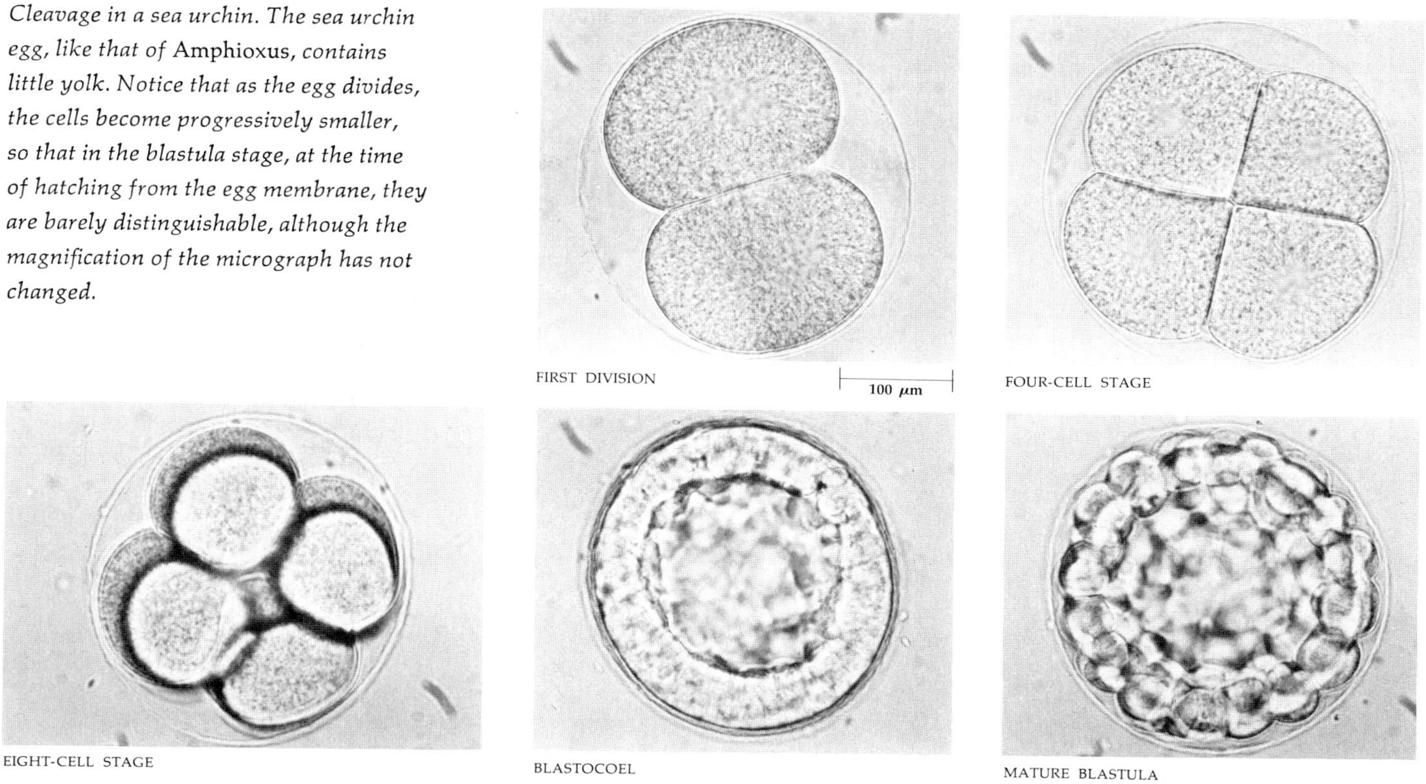

FIRST DIVISION   100 μm

FOUR-CELL STAGE

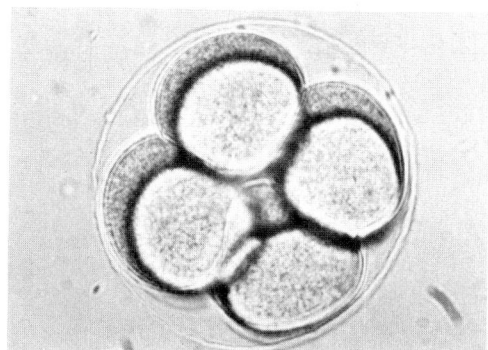

EIGHT-CELL STAGE

BLASTOCOEL

MATURE BLASTULA

1–8 (a) *An unfertilized egg of the American leopard frog,* Rana pipiens. *(b) The fertilized egg has divided into four cells. (c) Eight-cell and (d) sixteen-cell stages of frog development.*

(a)   0.5 mm

(b)   0.5 mm

(c)   0.5 mm

(d)   0.5 mm

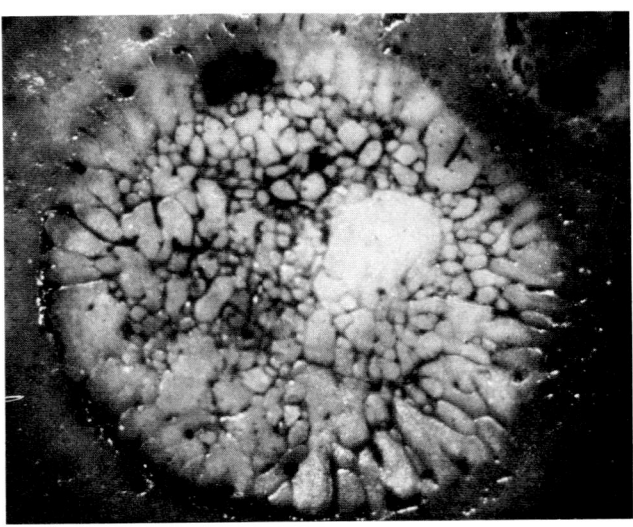

instance, before cleavage, the yolk mass can be shifted in a frog's egg simply by rolling the egg over. After cleavage this is no longer possible. Finally, cleavage prepares the embryo for the next developmental stage.

GASTRULATION

Gastrulation is complicated in detail, yet simple in principle. In all vertebrate species, the cellular mass of the blastula is transformed into an embryo composed of three layers: the ectoderm, mesoderm, and endoderm. Each of these three gives rise, as the embryo matures, to fundamentally different types of tissues.

In gastrulation, about half of the cells on the surface of the blastula migrate into its interior, moving along very specific pathways. As in cleavage, differences in the amount of egg yolk present in the different types of blastulas play a large role in determining the details of the cellular migrations.

In those species in which the egg contains little or no yolk (the mammals are an exception, as we shall see later), the cell movements are relatively simple. Figure 31–10 shows gastrulation in the primitive chordate *Amphioxus.* The cells move from the outside to the inside by a massive invagination of all cells at once through a collapse of one half of the blastula into the opposite half. The circular opening thus formed is the *blastopore,* and the new cavity formed by gastrulation, the *archenteron,* is a forerunner of the lumen (the passageway) of the gut. Although all species with yolkless eggs do not gastrulate exactly in the manner of *Amphioxus,* the general pattern is the same.

For species in which the eggs contain a fair quantity of yolk, gastrulation again takes place by material invaginating through a blastopore. Figure 31–11 shows gastrulation in an amphibian. The uneven distribution of yolk prior to cleavage has resulted in a blastula made up of small cells at one pole of the egg (the pole where the polar bodies were emitted) grading through the blastula to large cells at the opposite pole. The slower pace of cell movement in the larger

−10 *Gastrulation in* **Amphioxus**. *The blastula (a), like that of the sea urchin, consists of a single layer of cells enclosing a fluid-filled cavity. The layer of cells at the vegetal pole flattens and pushes inward, forming the beginnings of the archenteron, or primitive gut. Continued inturning results in the double-walled cup shape of the gastrula (c). In these drawings and the ones below, the ball of cells has been shown as if cut in half so as to reveal the interior. Externally, the embryo at all of these stages appears to be a solid ball of cells.*

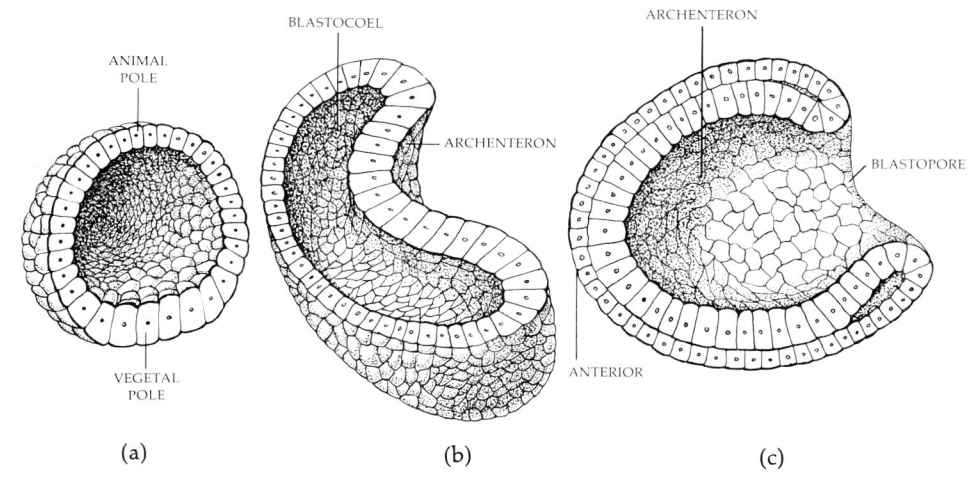

(a) (b) (c)

−11 *Development of the frog gastrula. (a) The blastopore forms in the blastula. (b) Cells from the outer surface of the blastula move into the blastopore, obliterating the blastocoel and creating the archenteron. (c) Three layers of tissues have been produced: ectoderm, endoderm, and mesoderm. (d) The ectoderm on the dorsal surface, overlying the notochord, thickens and flattens. (e) The neural plate and neural folds begin to form.*

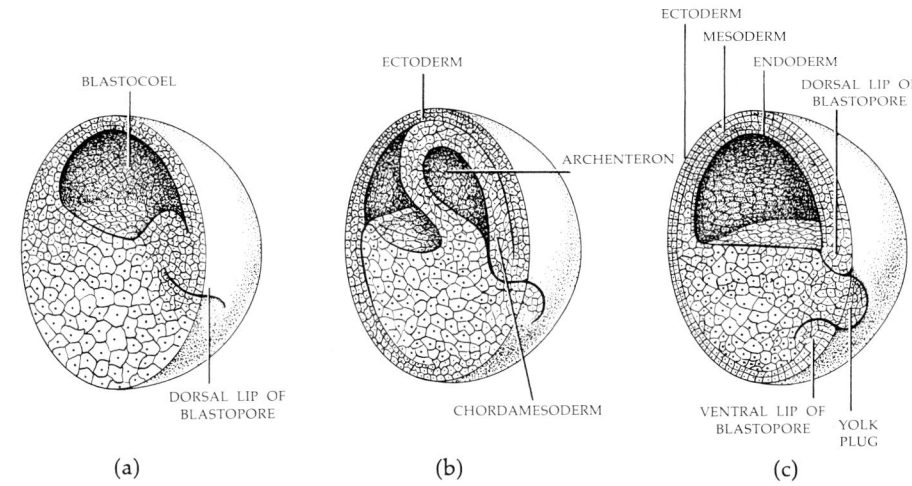

(a) (b) (c)

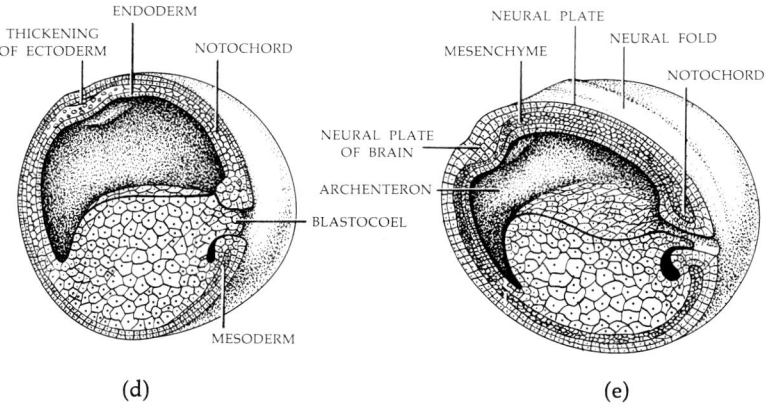

(d) (e)

547 *The Continuity of Life: Development*

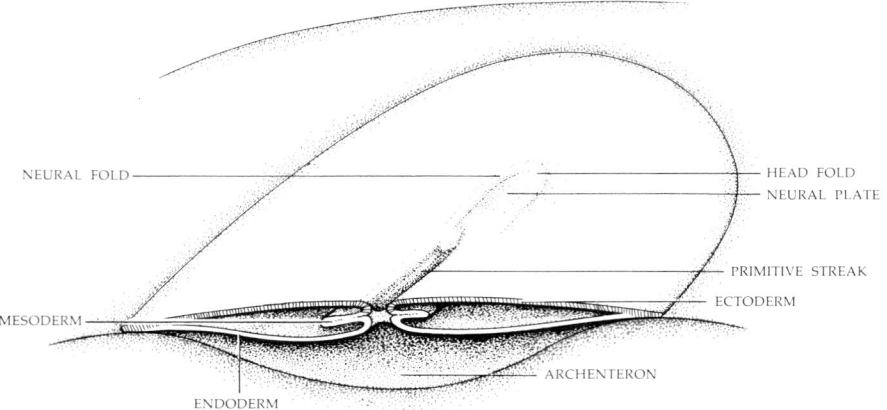

NEURAL FOLD — HEAD FOLD — NEURAL PLATE — PRIMITIVE STREAK — ECTODERM — MESODERM — ENDODERM — ARCHENTERON

cells causes the blastopore to be less open than in *Amphioxus,* and it often adopts the shape of a crescent-shaped slit, over the rims of which the cells migrate to the interior. The blastopore always forms at the boundary between the gray crescent and the lower, yolk-filled hemisphere of the egg.

A third category of egg is that in which there is so much yolk that cleavage is confined to a cap of cytoplasm on one side of the yolky sphere. This cleavage is typical of fish, reptiles, and birds and results in a blastodisc, the equivalent of the blastula. The external cells can no longer move easily through a blastopore (although some reptiles do make a very small version of one). Instead, the cells drop through from the upper layer of cells to form a second, lower layer; the site of their immigration is known as the *primitive streak.* The streak is the homologue of the blastopore (Figure 31–12).

NEURAL TUBE FORMATION

At the close of gastrulation the dorsal surface of the gastrula is transformed into a flat neural plate. The lateral edges of the plate proceed to curve dorsally and medially to form a tube, the neural tube—the primary organ rudiment of the brain and spinal cord. The neural tube, you will note, is formed from ectodermal cells. (See Figure 31–15.)

As the neural tube forms, changes take place in the underlying mesoderm, from which three longitudinal structures are formed. The central one is the notochord. The ones on either side are segmental blocks (somites) from which muscles and connective tissue ultimately form. In *Amphioxus,* as in other chordates, the number of somites increases as the embryo elongates, providing a clear index of the developmental stage. A mesodermal sheet will eventually spread to surround the gut entirely, so that mesoderm will form an inner layer against the ectoderm and an outer layer against the endoderm throughout the body with a space between the two mesodermal sheets; that space is the coelom. A continuous strip of mesoderm lateral to the somites gives rise to the kidneys, adrenals, gonads, and reproductive and excretory ducts. The floor and sides of the archenteron are composed of endoderm cells which rapidly form the gut, or gastrointestinal tract.

31–13 *A salamander embryo, showing the neural folds and neural groove.*

4–14 *Following gastrulation in Amphioxus, the neural plate forms, and the layer of cells beneath it differentiates into mesoderm, from which the notochord and the somites form. The neural plate, of ectodermal origin, folds into the neural tube. The endodermal cells form the gut.*

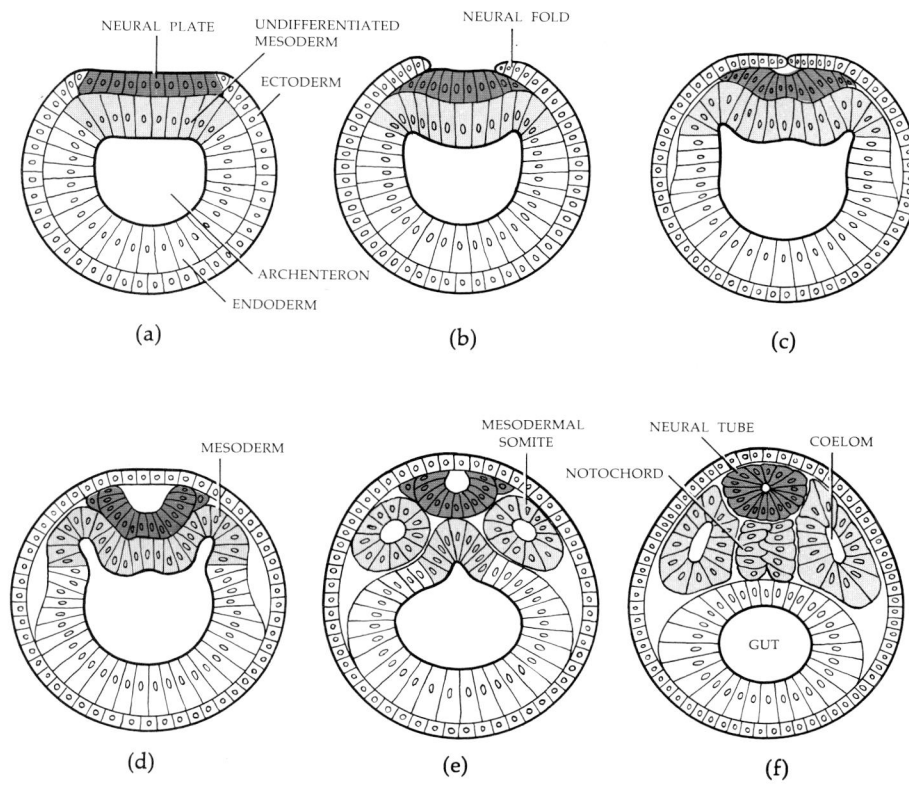

15 *The formation of the neural tube from the neural plate in the frog. The thickened elevations of ectoderm on the right and left sides of the neural plate curve inward, forming the neural groove. The ridges bordering the neural groove then meet and fuse. Finally, the resulting neural tube pinches off from the rest of the ectodermal tissue.*

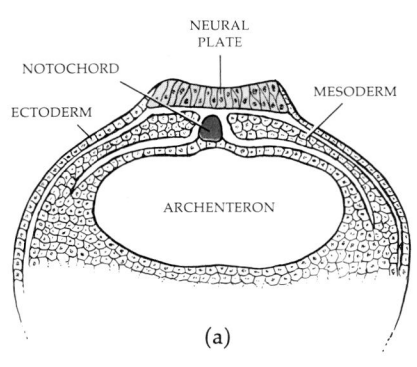

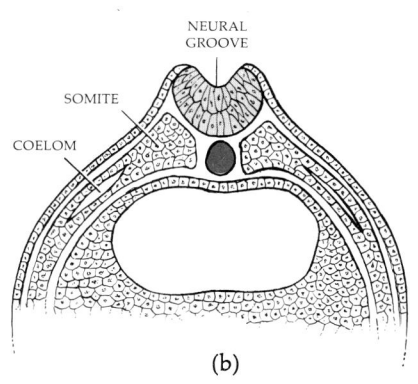

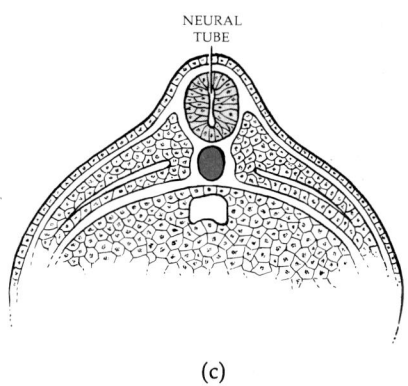

31–16 *Chick embryo at 22 hours. The head fold is at the top. Posterior to it are the neural folds and then the remnants of the primitive streak.*

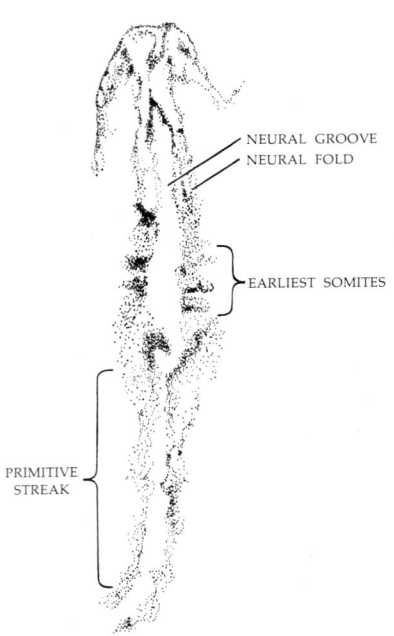

NEURAL GROOVE
NEURAL FOLD

EARLIEST SOMITES

PRIMITIVE STREAK

THE MOLECULAR PROGRAM OF THE EGG

What are the mechanisms involved in these activities? How are they regulated and controlled? As we noted previously, some of the controlling factors are clearly in the cytoplasm. The most striking evidence for this is the fact that cleavage can take place in eggs from which the nucleus has been removed. Also, early development is regulated by external events; the point at which the sperm cell penetrates the egg in *Rana* dictates the position of the gray crescent, which in turn establishes where the blastopore will form, and the formation of the blastopore determines the structural axis of the embryo.

Does the nucleus also have a role in early development? Analyses of the molecular contents of cleavage cells have shown that cleavage nuclei are very active in the production of messenger RNA and its translation to proteins important in gastrulation and succeeding stages. Among amphibians, a group that embryologists have studied particularly thoroughly, some messenger RNA and proteins are made during oogenesis and stored in an inactive state prior to fertilization. At least one of these proteins made during oogenesis does not begin to function until gastrulation. Ribosomal RNA is also made in great abundance during oogenesis, but not during cleavage and early gastrulation. Thus early development depends entirely on ribosomes made very much earlier in time. In general, the chemical events of early development precede the morphological events they are most closely associated with, and are not contemporaneous with them. The chemistry of oogenesis is a prerequisite for cleavage, while the chemistry of cleavage is the preparation for gastrulation. Whether or not this pattern is equally true for mammals, however, is debatable. The amphibian egg develops external to the body, and stockpiling and precocious synthesis may be associated only with a "closed" system—that is, one in which no externally derived materials enter the egg between fertilization and morphogenesis. (Indeed, experiments involving the incorporation of radioactive molecules into cleaving amphibian eggs are not usually successful.) By contrast, the mammalian egg is necessarily "open": Because it lacks nutritive reserves, it has to receive many substances from its environment in order to survive and develop. Therefore much has to be learned from studies in mammals before we can acquire any real understanding of genetic involvement in the earliest stages of development.

CONTROL OF CELL DIFFERENTIATION

The end product of all this genetic activity is, of course, cellular differentiation. We recognize cells as being different from one another by three distinct criteria: They may differ in their structure, they may specialize in their synthesis of particular proteins (such as hemoglobin in red blood cells), or they may come to behave differently. Such differences are genetically controlled and, in a particular cell line, genetically transmitted. (In other words, skin cells give rise to skin cells, spermatogonia to sperm.)

But how can an organism's cells be different from the original egg cell and from one another? There are two logical possibilities: The first is that the different cells contain different genetic material, which would mean that cells lost genetic material during early stages of development and that this loss took place

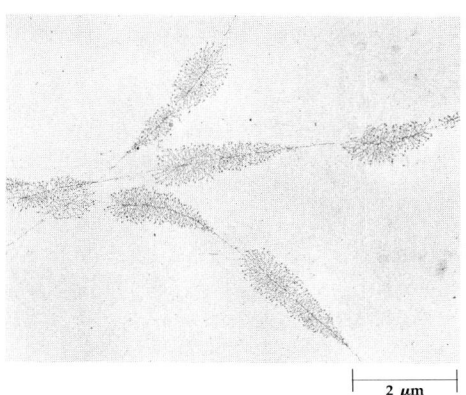

−17　*Electron micrograph of RNA being formed on an amphibian egg nucleolus.*

|← 2 μm →|

−18　*In experiments by J. B. Gurdon, the nucleus was removed from an intestinal cell of a tadpole and implanted into an egg cell in which the nucleus had been destroyed. In many cases, the egg developed normally, indicating that the intestinal cell nucleus contained all the information required for the whole organism.*

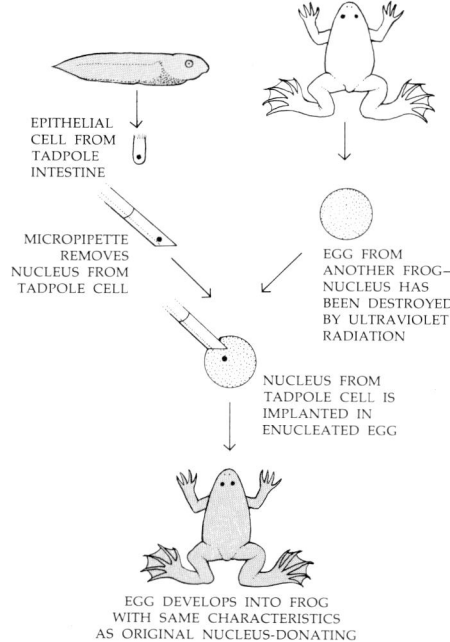

EPITHELIAL CELL FROM TADPOLE INTESTINE

MICROPIPETTE REMOVES NUCLEUS FROM TADPOLE CELL

EGG FROM ANOTHER FROG— NUCLEUS HAS BEEN DESTROYED BY ULTRAVIOLET RADIATION

NUCLEUS FROM TADPOLE CELL IS IMPLANTED IN ENUCLEATED EGG

EGG DEVELOPS INTO FROG WITH SAME CHARACTERISTICS AS ORIGINAL NUCLEUS-DONATING TADPOLE

in a very organized way. The second is that the genetic information is the same in all cells, but that control of its expression is different in different cell types.

The evidence points to the second possibility. First, the amount of DNA in all the diploid cells of any given organism is the same. Second, the precision of the chromosomal movements during mitosis would appear to ensure that each daughter cell receives the same genetic material. Finally, there are the results of a number of ingenious experiments designed to test the hypothesis.

Just over twenty years ago, R. W. Briggs and T. J. King, for example, were able to show that if nuclei were removed from frog blastula cells and injected singly into a series of unfertilized eggs whose nuclei had been physically removed, some of the eggs would begin development and grow into normal adult frogs. Thus cleavage does not result in any permanent inactivation of genes or the loss of important genes. Subsequently, J. B. Gurdon of the University of Oxford was able to do the same kind of experiment with nuclei from the intestinal cells of early tadpoles (Figure 31–18). Although very few of the recipient eggs developed normally, his results showed that even nuclei of differentiated cells retain the capacity to function as egg nuclei.

At about the same time, F. C. Steward demonstrated that, under certain conditions, a single differentiated root cell of a carrot can be persuaded to reconstitute a whole carrot plant.

Thus we are led to the conclusion that differentiation between cells depends on inactivation of certain groups of genes and activation of others.

In Chapter 18, we discussed some of the hypotheses on the subject of how large groups of genes are inactivated in differentiated cells, but we did not go into the question of how or when it is decided which sets of genes should be active or inactive. As a preliminary approach to this question, let us examine a little further the implications of the Briggs and King and Gurdon experiments. When the nucleus is in the cytoplasm of the intestinal cell of a frog, it functions one way; transplanted to egg cytoplasm, it functions another way. In other words, not only does the nucleus control the cytoplasm, but the cytoplasm regulates the nucleus.

Let us develop this consideration further. If the pattern of gene activation of a nucleus is related to the composition of the cytoplasm that surrounds the nucleus, then there are two ways in which cells are caused to differentiate. The first is the composition of the cytoplasm a cell is endowed with as a result of cleavage. The second is the composition of a cell's cytoplasm at some developmental stage later than cleavage, when the original cytoplasm has been modified by materials acquired from the extracellular environment. This extracellular environment, of course, can also consist of other cells. We will take a look at examples of each mechanism.

A number of invertebrate eggs can be seen to contain regions of cytoplasm qualitatively different from other regions. The differences may involve optical density (many invertebrate eggs are small and translucent), affinities for particular stains, and also natural colors. The egg of the sea squirt *Styela* is an example of the third type. One region of the cytoplasm is yellow, and cells that contain this yellow substance later develop into mesoderm. When fertilized eggs of *Styela* are centrifuged before cleavage, the yellow cytoplasm becomes distributed abnormally. Nonetheless, the egg begins to develop, but as a chaotically

organized monster. Within the abnormal embryo, however, all the cells with yellow cytoplasm differentiate into mesodermal cells. Thus, *Styela* is plainly an example of the way in which the direction of cellular differentiation is closely linked to the kinds of cytoplasm present in the egg.

THE ORGANIZER

An example of the effect of cellular environment involves what is sometimes called the *organizer.* Some time ago it was discovered that it is usually possible to divide a developing frog blastula in two and obtain *two* normal embryos. Once blastulation is completed and gastrulation is under way, however, only one normal embryo can be obtained. The embryo that does develop always forms from the half that contains the dorsal lip of the blastopore.

If you take two blastulas, just as gastrulation begins, and excise the tissue passing over the dorsal lip from one of them and implant it into the other, a second embryo forms within the tissues of the first (see Figure 31–19). The transplanted dorsal lip material develops into the notochord, as it would have had it not been removed in the first place. The rest of the second embryo is made up of tissues of the host. (In the original experiments, animals of two different colors were used, so that the two tissues could readily be distinguished.) Thus the transplanted tissue interacted with the host tissues and "organized" them into neural plate and neural tube. Similarly, in normal development, differentiation into different cell types begins when the organizer is "transplanted" to a new place on the embryo as a result of the cellular migrations of gastrulation.

These early experiments were carried out in the laboratories of the German embryologist Hans Spemann, and it was he who first called the dorsal lip tissue the organizer.

31–19 *Experiment showing the activity of the organizer. (a) The dorsal lip of the blastopore from one amphibian embryo is grafted onto another embryo. At gastrulation the grafted tissue moves into the interior of the embryo (b, c) through the blastopore of the host. Two neural plates are formed (d), and a double embryo develops (e). In (f) you see the structure of the secondary embryo under the yolk of the host embryo. The colored cells are derived from the graft, and the light cells are host material that has been induced to undergo these differentiations.*

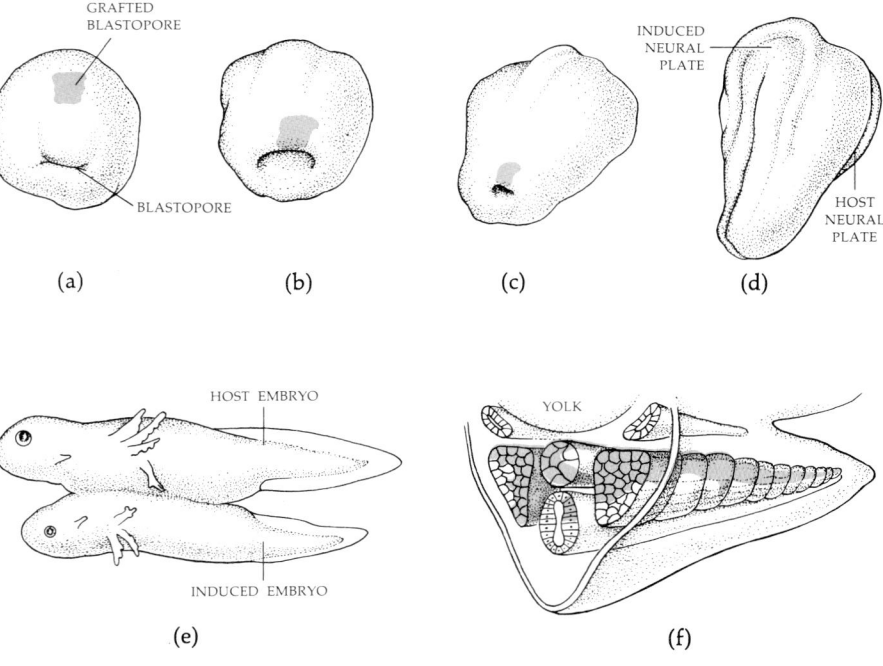

1–20 *The differentiation and proliferation of epidermal cells depend on a continued interaction between epidermis and dermis. However, the two tissues do not have to be in direct contact for the neces-* *sary interactions to occur. This has been demonstrated by experiments using the filter apparatus shown here. Epidermis and dermis were grown in culture separated by a thin, porous membrane* *that permitted the passage of molecules but not cells. The epidermal cells directly opposite the dermis continued to divide and differentiate.*

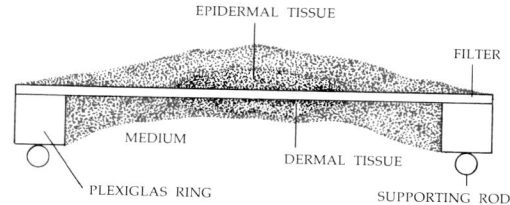

Induction and Inducers

It is now known that the action of Spemann's organizer is just one of many tissue interactions that take place in development. This general process is known as _embryonic induction._ Embryonic induction occurs when two different types of tissues come in contact with one another, as the result either of morphogenetic movement, as in the induction of the neural plate, or of growth. The induction process may continue throughout the life of the organism, as in the turnover of cells in the skin.

Numerous experiments have shown that induction is mediated through physical contact and that some chemical material passes from the inducing tissue to the tissue that is induced. Induction takes place even when direct contact is prevented between the two tissues, as by placing a porous filter between them, but not if chemical exchanges are prevented, as by a piece of metal foil. The substance or substances must be of some very general nature. For example, the capacity of the organizer to induce neural plate formation is not species specific; thus one can provoke the formation of a secondary neural tube in a chick embryo by means of a graft of primitive streak tissue from a rabbit. Moreover, experiments have shown that a number of chemicals, some of which never occur in embryos, will induce neural plate formation in embryonic ectoderm. It is not, then, that the inducing substance _endows_ the ectoderm with the capacity to form nervous tissue; rather, it merely _evokes_ a potentiality that is already present in the responding tissue. The organizing substances, in entering the responding cell, change the nature of its cytoplasm in such a way as to provoke a new pattern of gene activity, a pattern leading to formation of a neural tube.

Induction and the Egg

Where did the organizer derive its powers? Remember that the position in the embryo of the dorsal lip of the blastopore is established at the moment the sperm enters the egg and that what establishes the position of the dorsal lip is the gray crescent. Gastrulation brings the cells formerly at the gray crescent, and so containing gray crescent cytoplasm, into a position directly underlying the dorsal ectoderm, where they become notochord and organize the overlying tissues to form a neural plate. Thus, in effect, Spemann's organizer has its origins in activa-

tion of the egg by the sperm. Moreover, since the architecture of the uncleaved egg is a result of its growth in the ovary, at least some developmental events which occur late in an embryonic history are causally linked to the genetic and metabolic activities of oogenesis. The same argument can be made for those differentiations that result directly from the presence in some cells of special types of cytoplasm, as in *Styela*. Thus we may conclude that many really important chemical decisions of embryonic development are not made in the embryo at all, but before development, in its usual sense, really begins.

ORGANOGENESIS

The third major stage of embryonic development is organogenesis. In every vertebrate embryo ever studied, organogenesis begins with the interaction between ectoderm and underlying chordal mesoderm. Each of the three primary tissues formed during gastrulation now proceeds to undergo growth, differentiation, and morphogenesis. The process is essentially the same in all vertebrates.

Ectoderm

The Central Nervous System

The neural tube stretches out along the dorsal surface, growing and becoming longer and thinner as the embryo increases in size. The cells in the neural tube at first appear to be all alike, but some of them, the future motor neurons (nerve cells), begin to extend long fibers that grow out beyond the neural tube and invade the peripheral organs and tissues. When the neural tube formed, some of the ectodermal cells at the crests of the neural folds were left behind in the surrounding tissue (see Figure 31–21). Some of these cells from the neural crest now migrate into positions around the neural tube and form connections between the dorsal part of the tube and the surrounding tissues. Meanwhile, mesodermal cells have begun to migrate toward the neural tube and underlying

31–21   *During the formation of the neural tube, some of the ectodermal cells at the crests of the neural folds are pinched out as the folds come together. Eventually, some of these neural crest cells will migrate down toward the notochord and aggregate into ganglia. These cells become sensory neurons that send extensions up to connect with the dorsal part of the spinal cord and axons into the surrounding tissues. Others become Schwann cells, which surround and insulate the nerve fibers, some become the pigment cells (melanocytes) found at the base of the epidermis, and some form the adrenal medulla.*

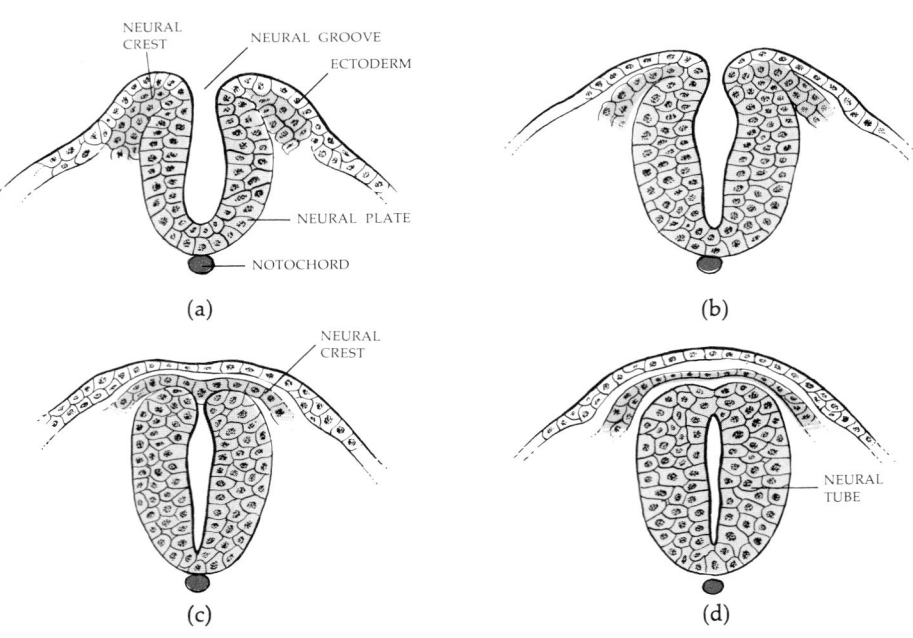

-22 *The development of the brain and associated sense organs. At the front end of the neural tube, local swellings produce three distinct bulges—the forebrain, the midbrain, and the hindbrain. The forebrain then bulges laterally, and two optic vesicles appear and develop a cuplike shape. At the same time, the surface epidermis folds inward to meet the optic cups. The ears and nostrils also appear at first as epidermal infoldings.*

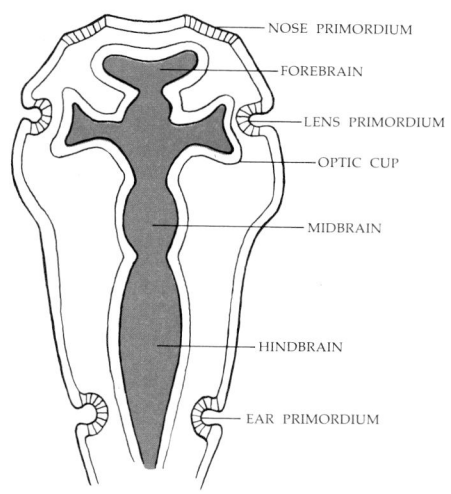

notochord and aggregate around them, differentiating into cartilage and giving rise to the hollow vertebral column. If a portion of neural tube is transplanted to another region of the mesoderm, mesodermal cells will aggregate similarly around it as a result of tissue interaction.

Brain and Eye

The brain begins as three bulges in the foremost part of the neural tube (see Figure 31–22). Almost immediately, two saclike protrusions, the optic vesicles, appear on both sides of the forming brain. These spherical vesicles enlarge and come into contact with the epidermis, while still remaining connected to the neural tube by the narrowing optic stalk. Within this stalk, the optic nerve will develop.

When the optic vesicle comes into contact with the inner surface of the epidermis, the external surface of the vesicle flattens out and pushes inward. The vesicle thus becomes a double-walled cup. The rim of the cup becomes the edge of the pupil. The opening of the cup is large at first, but the rims bend inward and converge, so that the opening of the pupil becomes smaller. At the point where the optic vesicle touches the epidermis, the epidermis begins to differentiate into a lens, becoming transparent.

The differentiation of the eye lens is an example of a *secondary induction*. This is because the inducing tissue (the optic vesicle) is itself the result of the primary induction of dorsal ectoderm by dorsal mesoderm. In fact, the formation of the complete eye involves at least six separate inductions, occurring in an orderly sequence, each one linked to the one before, with primary induction by the organizer as the starting point.

Other infoldings and differentiations of head ectoderm, associated with the presence of the embryonic brain, lead to the elaboration of the organs of smell and hearing.

3 *The eye develops as a result of interactions between the epidermis and a lateral outgrowth of the forebrain, the optic vesicle. When the optic vesicle reaches the epidermis, the overlying epidermis begins to thicken and differentiate. The optic vesicle flattens out and invaginates, becoming the double-walled optic cup. The invaginated wall will become the retina of the eye; the outer, thinner wall develops into the pigment coat. The rim of the optic cup will later form the edge of the pupil. The thickened epidermal layer pinches off to become the transparent lens, and the overlying epidermis, which also becomes transparent, forms the cornea. The connection with the brain remains as the optic stalk, within which are the optic nerves. Note that the eye is thus a differentiated extension of the brain.*

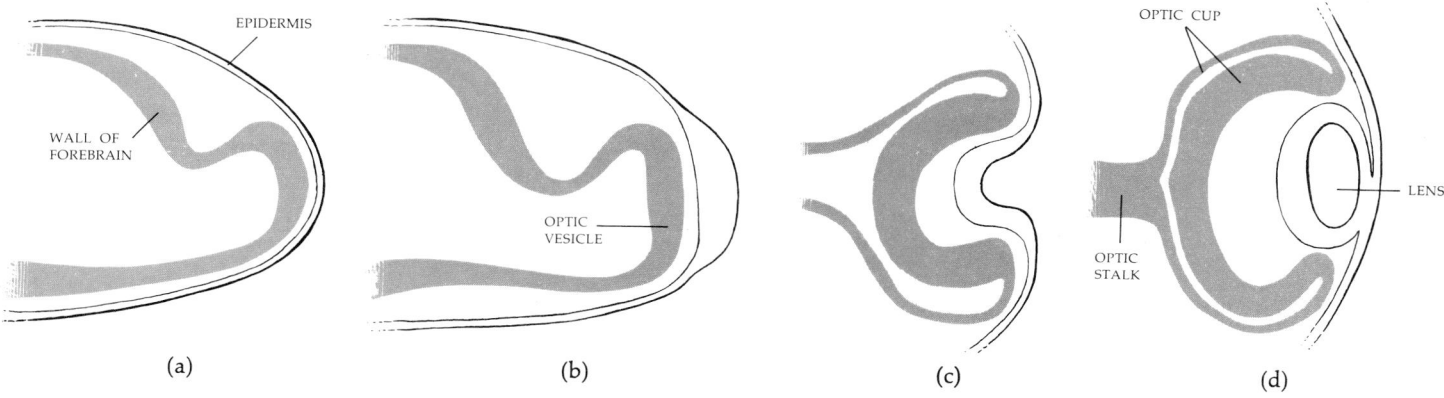

(a)   (b)   (c)   (d)

555 *The Continuity of Life: Development*

31–24　*Developmental processes characteristically involve changes in the shapes of
cells and groups of cells. One way in
which cells can change shape is by
changes in contacts between cells and
their supporting membranes. (a) Moderate contact between cells and supporting membrane. (b) An increase in
contact between cells and a reduction in
contact with the supporting membrane
results in columnar cells. (c) If contact
with the supporting membrane is not
reduced when contact between cells is
increased, the cells form an arc. (d) If
there is a reduction in contact between
the cells and the ends of the cell sheet
are fixed, the cell sheet will curve in the
opposite direction. (e, f) An increase in
contact between the cells at Y and Y'
changes the shape of the blastula.*

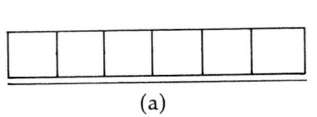

(a)

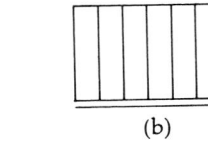

(b)

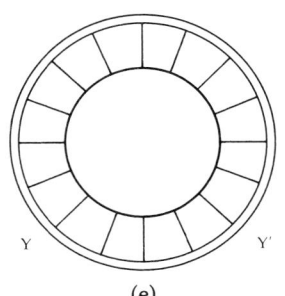

(e)

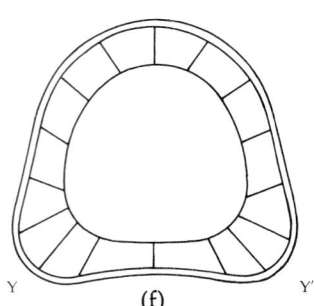

(f)

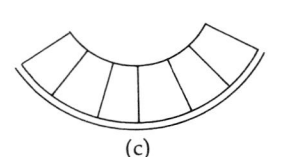

(c)

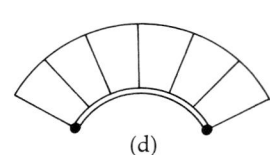

(d)

Mesoderm

The series of somites that came to lie on each side of the notochord shortly
after gastrulation now differentiate into three kinds of cells: (1) sclerotome cells,
which later form skeletal elements, (2) dermatome cells, which become part of
and contribute to the developing skin, and (3) myotome cells, which form most
of the musculature. In the lower vertebrates, the same pattern, with its simple
double series of back muscles, is retained in adulthood; it is well suited to the
side-to-side motion of the body in swimming and crawling. In the higher vertebrates, however, this initial segmented arrangement is almost obliterated by
modifications and insertions of the muscles associated with quadrupedal locomotion.

Mesoderm lateral to the somites forms the kidneys, the gonads, and the ducts
of the excretory and reproductive systems. The ventral mesoderm splits into
two layers. One sheet of mesoderm lines the thoracic and abdominal cavities, and
another becomes the outer layer of the internal organs. The discontinuity
between the two sheets is the coelom.

Endoderm

The endoderm differentiates into tissues of the respiratory and digestive tracts
and a number of related organs. Early in development, endodermal pouches
develop at the anterior end of the endodermal tube (the gut) and push laterally
until they meet the ectoderm, which folds inward to meet the pouches, producing
a series of grooves on the surface of the embryo. (See Figure 31–25.) In lower
vertebrates, endoderm and ectoderm fuse and a perforation forms, around
which gill filaments develop. In terrestrial vertebrates, including man, these
pouches develop into eustachian tubes (which connect the pharynx and the
middle ear), tonsils, parathyroid glands, and the thymus gland. Posterior to the

Schematic diagram of tissue interactions involving the endoderm. In embryonic stages of all vertebrates, pouches of endoderm that form at the upper end of the endodermal tube grow toward and meet the ectoderm of the body wall. In the fish, perforations develop, as shown on the right, around which the gills form. In mammals, the pouches develop into the eustachian tubes, the parathyroids, the thymus, and various other glands.

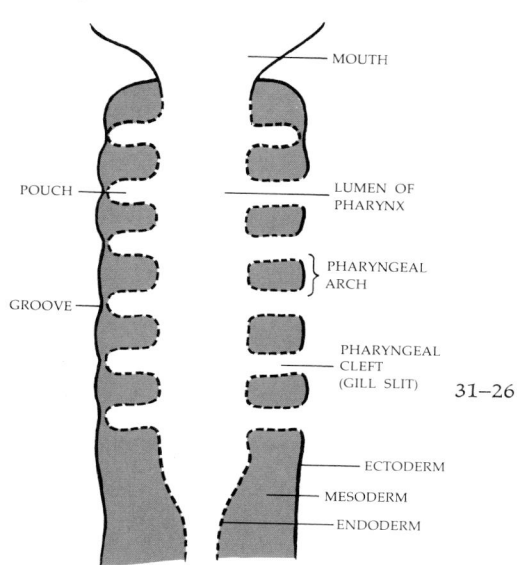

gill region, the lungs develop as similar outpocketings, branching into two sacs, and more posterior still, outpushings from the primitive gut begin to differentiate into liver, gallbladder, and pancreas.

At one time, it was popularly believed that development recapitulates phylogeny, meaning that each organism in its development goes through stages that reflect all of the various stages of its evolution. Persistence of gill pouches in land vertebrates was among the examples cited. Actually, although it does offer evidence of our evolutionary background, the persistence of gill pouches is now viewed, in modern terms, as serving a functional purpose, a way of getting from here to there. The retention of gill pouch formation is necessary for the development of the structures derived from these pouches. When one developmental event depends upon some preceding one for its own success, there is obvious selective value, from the evolutionary point of view, in keeping all parts of the system, even obsolete ones, intact.

THE ORGAN SYSTEMS

By the end of gestation, as a result of this continuing highly coordinated pattern of tissue interactions—one unfolding after another—the many organ systems of the mammalian infant are nearly complete. None of these organs in the adult animal is derived from only one type of tissue. For example, the lining of the intestine is of endodermal origin; these lining cells secrete the digestive juices

31–26 *The principal channels of vertebrate development.*

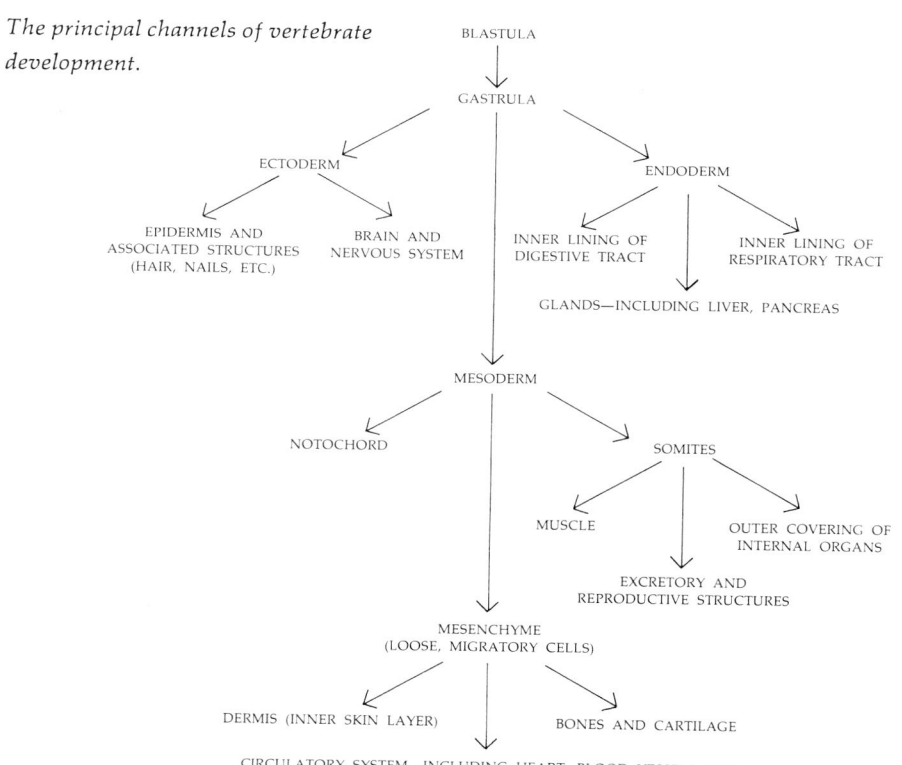

and absorb the digested materials, which are the principal functions of the intestine. However, the functional structure of the intestine also includes muscles, connective tissue, blood vessels, nerves, an outer wrapping, and so on, which are made up of tissues derived from mesoderm and ectoderm.

Moreover, as we noted at the beginning of the chapter, adulthood does not mark the end of development. In fact, apparently it does not even mark the end of what we have called embryonic induction. Again, the skin is a good example. As we noted previously, cells in the epidermal layer divide and differentiate continuously. Excised skin cells can be grown in a nutrient medium in a test tube, and under such conditions they will continue to divide and differentiate, but only if the epidermal layer (which is of ectodermal origin) remains in contact with the dermal layer (of mesodermal origin).

EXTRAEMBRYONIC MEMBRANES

The eggs of birds and reptiles, as noted before, contain so much yolk that the embryo develops from a disk of cells on top of the egg. In the course of organogenesis, the edge of the blastodisc overgrows the undivided yolk; this overgrowth involves all three of the primary tissues. The coelomic split within the mesoderm also invades this extraembryonic region. As Figure 31–27 shows, the yolk gradually becomes surrounded by a membrane of mesoderm and endoderm, the _yolk sac membrane_. The function of this membrane is nutritive; the endodermal cells digest the yolk, and the blood vessels formed in the mesodermal component carry food molecules from the yolk into the embryo proper. A yolk sac membrane is also formed in many fish eggs.

The extraembryonic ectoderm and mesoderm on the outside of the coelomic space fold upward around and over the embryonic region, and these so-called amniotic folds eventually fuse with each other over the embryo; thus the embryo comes to lie within a cavity (actually, the cavity rapidly becomes filled with a saline fluid of embryonic origin) termed the _amniotic cavity_. The fluid in the cavity serves to buffer the embryo against mechanical and thermal shocks. As Figure 31–27a shows, the fusion of the amniotic folds creates two membranes, both composed of a layer of ectoderm and mesoderm, separated by coelomic space, which is continuous with the extraembryonic coelom. The inner membrane is the _amnion_ and the outer is the _chorion_. For the genetic testing procedures described in Section 2, fluid samples are taken from the amniotic cavity. These samples contain fetal cells.

As development proceeds and the embryo becomes increasingly demarcated from the yolk mass (although always attached to it by the yolk sac stalk), a ventral and posterior pouch is formed from the primitive hindgut. This is the _allantois,_ and, as you can see in Figure 31–27b, its wall is composed of endoderm and mesoderm. At first, the function of the allantois is excretion. Since bird and reptile eggs are closed off from their environment by their shell, there is no way to dispose of the products of nitrogen metabolism, most of which are toxic. Evolution found a dual answer to this problem. First, the nitrogenous wastes are converted to uric acid, which is insoluble and of low toxicity. This uric acid is then voided from the embryo into the allantois, which essentially serves as an embryonic garbage bag. (When a chick hatches, the accumulated uric acid can be found adhering to the abandoned shell.)

1–27 Development of the extraembryonic membranes of the chick. In early development, body folds of the embryo begin to separate from the underlying yolk. One membrane, the yolk sac, grows around and almost completely encloses the yolk. A second, the allantois, arises as an outgrowth of the rear of the gut. The third and fourth are elevated over the embryo by a folding process during which the membrane is doubled. When the folds fuse, two separate membranes are formed. The inner one is the amnion and the outer one is the chorion. The chorion eventually fuses with the allantois to form the chorioallantoic membrane, which, in the later stages of development, encloses embryo, yolk, and all the other structures. Note that each membrane is composed of two types of primary tissue.

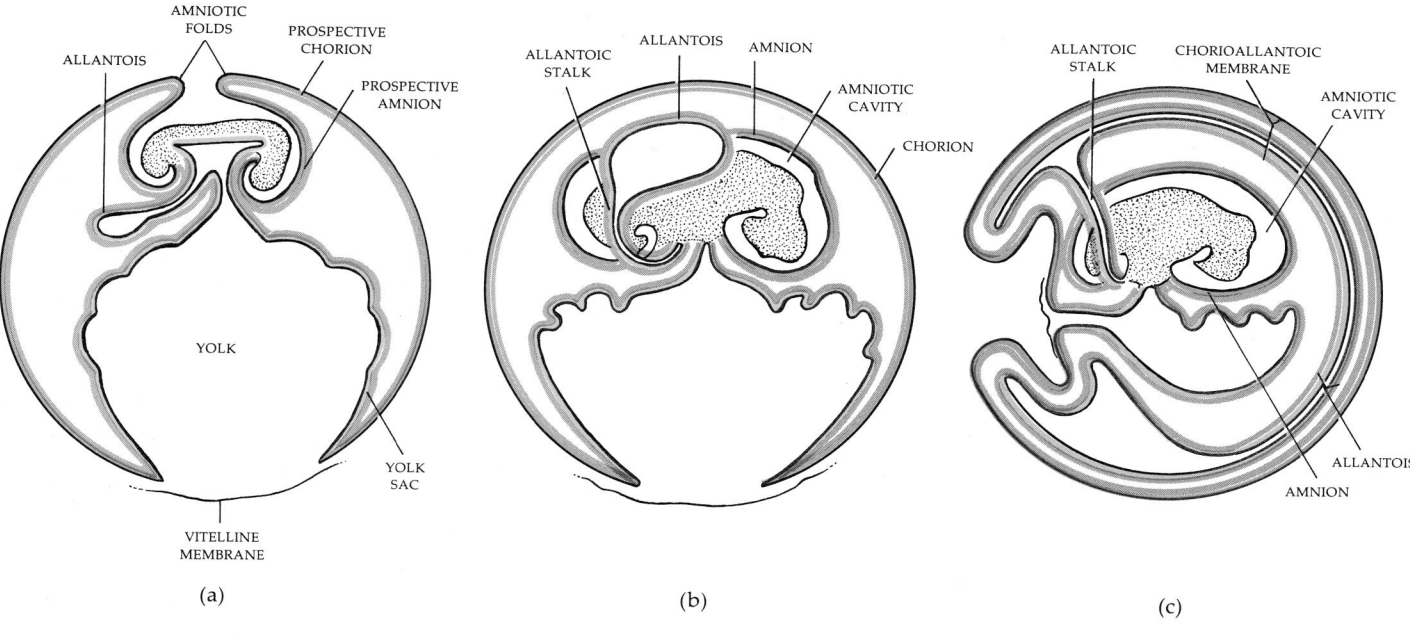

(a)

(b)

(c)

THE FOUR EXTRAEMBRYONIC MEMBRANES:
YOLK SAC (ENDODERM AND MESODERM)
ALLANTOIS (ENDODERM AND MESODERM)
AMNION (MESODERM AND ECTODERM)
CHORION (MESODERM AND ECTODERM)

As development proceeds further, the allantois expands in volume and pushes out into the extraembryonic coelom, gradually enveloping the embryonic region as it does so. Eventually, as Figure 31–27c shows, the allantois effectively obliterates the extraembryonic coelom, and its walls fuse with the chorionic membrane lying under the eggshell. Since the allantois is plentifully supplied with blood vessels, the *chorioallantoic membrane,* formed by the fusion of allantoic wall with the chorion, acts as an efficient respiratory membrane for the embryo during its later development.

31–28 (a) *A human egg. The dark strands within the nucleus are chromosomes. A polar body is at the lower right.* (b) *Human sperm.* (a, *Roberts Rugh and Landrum B. Shettles, M.D., From Conception to Birth:* The Drama of Life's Beginnings, *Harper & Row, Publishers, Inc., New York, 1971;* b, *Fritz Goro, Time-Life Picture Agency.*)

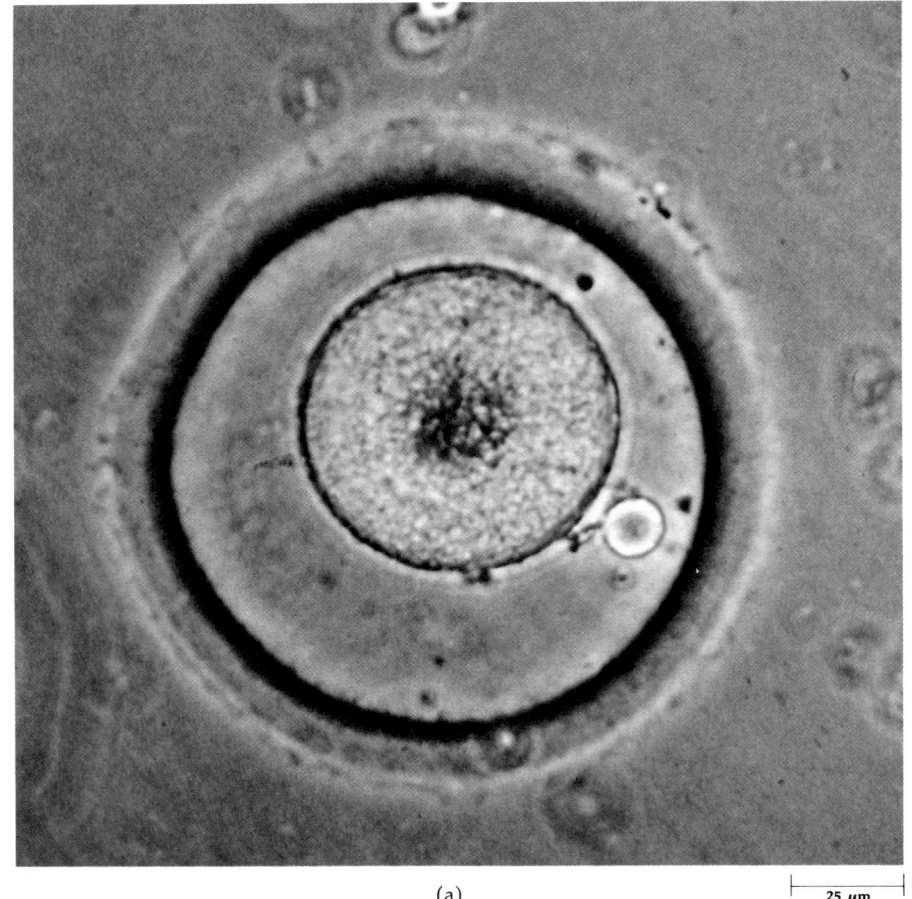

(a)

25 μm

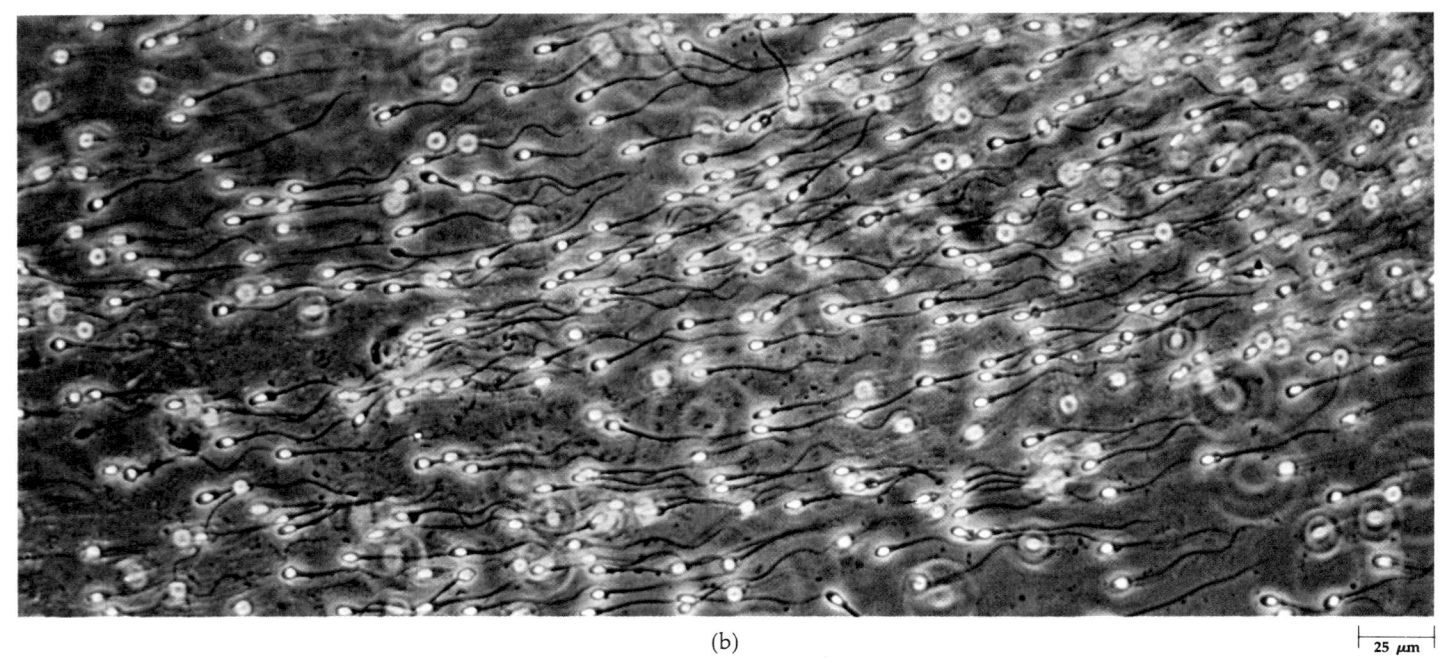

(b)

25 μm

1–29 *A human egg surrounded by sperm cells.*
At the top are two polar bodies.
Surrounding the ovum is a layer of
mucoprotein, the zona pellucida.

1–30 (a) *A human egg at the two-celled stage.*
The cells are still surrounded by an outer
membrane. Sometimes they separate at
this stage, resulting in identical twins.
(b) *The cells continue to divide, but,*
since the mass of the embryo does not
increase, they are still easily contained
within the same membrane.

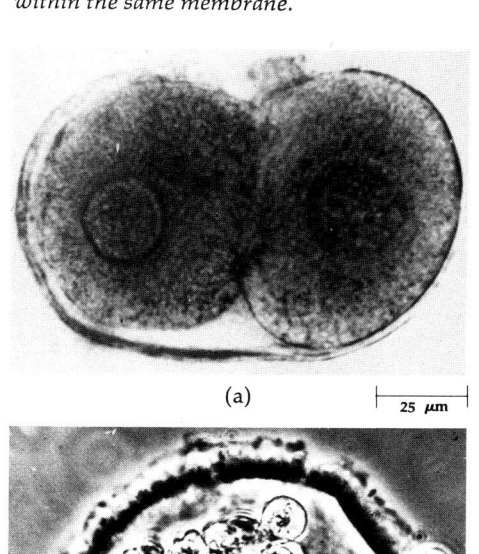

(a)   ├─ 25 μm ─┤

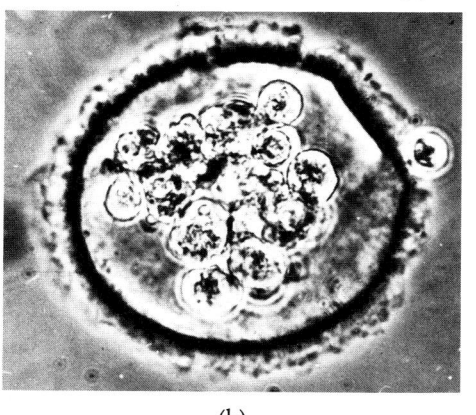

(b)

├─ 25 μm ─┤

THE HUMAN EMBRYO

Now let us look at the development of a human embryo. Figure 31–29 shows human sperm cells approaching an egg cell. The cell is surrounded by the jellylike zona pellucida, which is itself covered by a cloud of small follicle cells that accompany the egg at ovulation. The first cell division occurs about 30 hours after fertilization, when the cell is still in the oviduct. At 60 hours, the two cells divide to form four cells. At three days the four cells divide to form eight. In this early stage, all of the cells are of equal size, as they are in the sea urchin. As the blastula develops further, a large cavity forms. The embryo at this stage is a hollow sphere of cells, a blastocyst. Completely enveloping the blastocyst is a single layer of cells, the *trophoblast* (from the Greek word *trophe*, "to nourish"). The trophoblast will give rise to embryonic membranes, the chorion and amnion.

The embryo itself is a little cluster of cells within the sphere. (In the chick, the surrounding membranes, you will recall, grew out from cells at the edge of the embryo early in development. It is as if the human embryo has skipped this step, arriving, as it were, prepackaged.)

At about the sixth day, the trophoblast comes into contact with the tissues of the uterus. The trophoblast cells, multiplying rapidly by now, induce changes in the endometrium and invade it. As the embryo penetrates the endometrial tissues, it becomes surrounded by ruptured blood vessels and the nutrient-filled blood escaping from them. The trophoblast thickens and develops amoebalike projections that invade the uterine lining.

31–31 *The embryo in cross section. A cavity, the yolk sac, has formed within the sphere of cells. The cluster of cells at the bottom, within the sphere, will develop into the embryo itself. A second cavity, the amniotic cavity, is beginning to form below the embryonic cell mass. The outer cells are the trophoblast.*

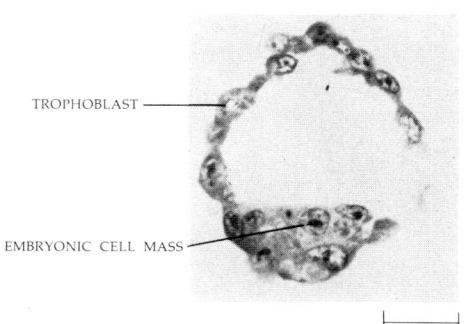

TROPHOBLAST

EMBRYONIC CELL MASS

25 μm

Human Embryonic Membranes and the Placenta

As the embryo becomes implanted, the embryonic membranes begin to develop. These have interesting similarities to and interesting differences from the avian-reptilian membranes discussed previously. In the first place, the yolk sac, which develops first, has no yolk. This lack of yolk is a secondary evolutionary development. The monotremes (egg-laying mammals) produce eggs with yolks. The marsupials produce eggs with yolks, but the yolks are discarded at the first cleavage. The placentals produce eggs with no yolks but a prominent cavity where the yolk "used to be." Cleavage and the cellular migrations at gastrulation proceed as if yolk were still there.

The allantois connects with the hindgut and eventually becomes a primitive urinary bladder, even though wastes are not segregated within the egg, as they are in the chicken, but are transported to the maternal bloodstream. In the course of development, the endodermal component of the allantois (the actual sac) becomes greatly reduced, but the allantoic mesoderm spreads out and is the source of the major blood vessels on the embryonic side of the placenta. The allantoic stalk eventually becomes the umbilical cord, the embryo's principal connection with the uterine tissues. The blood vessels of the allantois transport oxygen and nutrients that have diffused in from the mother's circulatory system and carry off carbon dioxide and other wastes.

The third membrane, the amnion, lines the amniotic cavity, which arises between the embryo and the trophoblast. The fourth is the chorion, which is a combination of ectoderm from the trophoblast and mesoderm grown out from

31–32 *The embryo at implantation. Trophoblast cells are extending through the endometrium to the rich blood supply.*

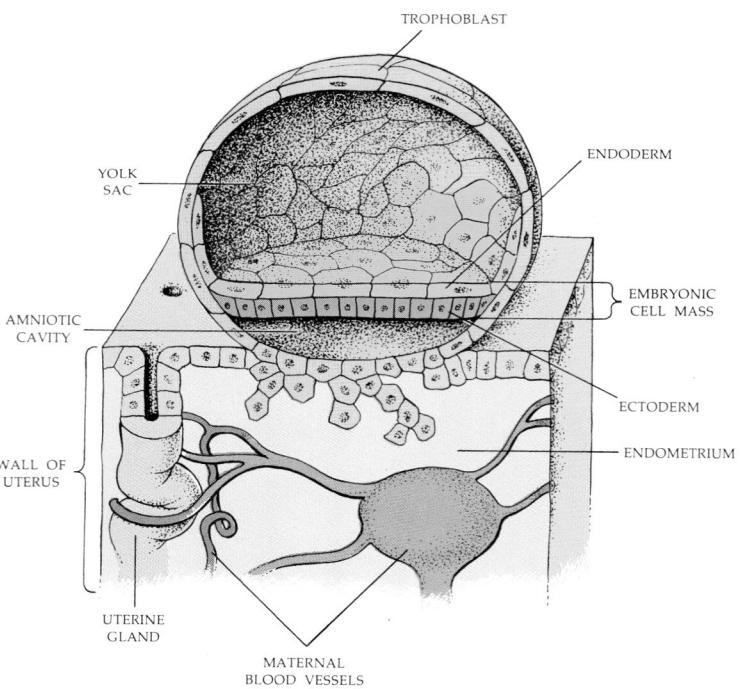

TROPHOBLAST

YOLK SAC

ENDODERM

AMNIOTIC CAVITY

EMBRYONIC CELL MASS

ECTODERM

ENDOMETRIUM

WALL OF UTERUS

UTERINE GLAND

MATERNAL BLOOD VESSELS

1–33 *Two weeks after conception, the embryo is established beneath the surface of the uterine mucosa. The chorion (the former trophoblast) surrounds the embryo and the other extraembryonic membranes. The placenta has begun to form.*

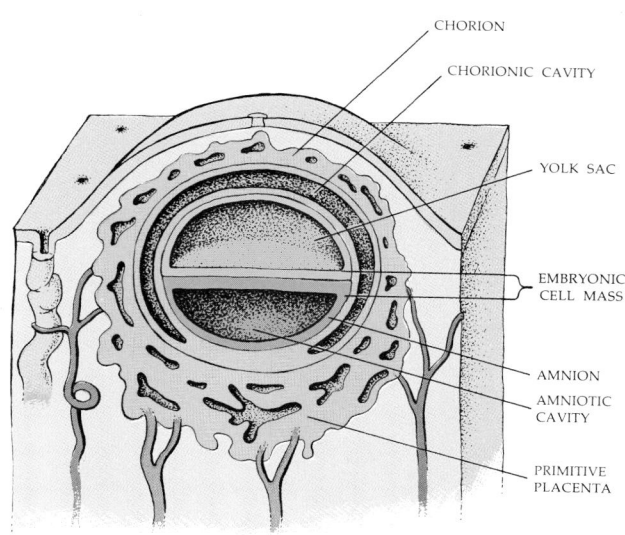

the embryo itself. By about the fourteenth day, chorionic villi begin to form; the formation of the villi represents the beginning of the mature placenta. The chorion, as we mentioned previously, is a source of gonadotropic hormones. By this time (about the time of the first missed menstrual period), a pregnancy test based on detection of these hormones will probably be positive.

By the end of the third week after conception, the placenta covers 20 percent of the uterus. It is a disk-shaped mass of tissues in which capillaries from both mother and fetus intertwine without joining. Oxygen, carbon dioxide, waste products, nutrients, and other materials are exchanged by diffusion through placental tissues. The embryo is connected to the placenta by the umbilical cord.

.–34 *From the placenta, numerous fingerlike chorionic villi project into the maternal blood space in the wall of the uterus. This space is kept charged with blood from branches of the uterine artery. Across the thin barrier separating fetal from maternal blood, exchange of materials takes place: soluble food substances, oxygen, water, and salts pass into the umbilical vein from the mother's blood; carbon dioxide and nitrogenous waste, brought to the placenta in the umbilical artery, pass into the mother's blood. The placenta is thus the excretory organ of the fetus as well as its respiratory surface and its source of nourishment.*

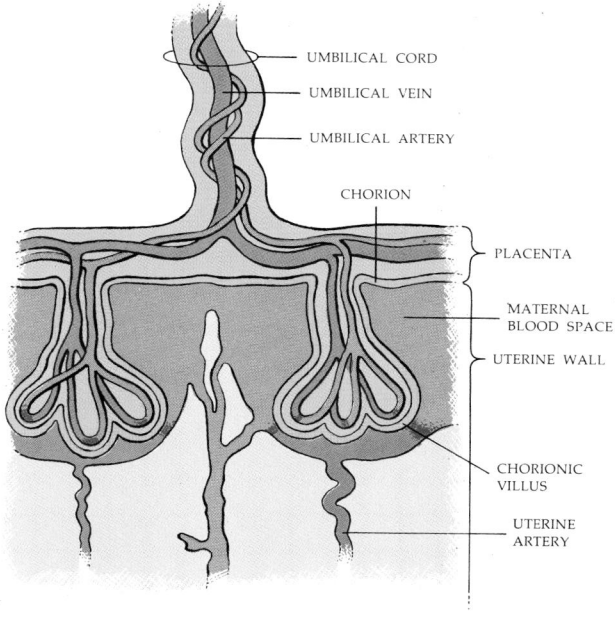

31–35　*As the embryo develops, two ridges form and enlarge, becoming the neural folds. The groove between them is called the neural groove. Within the neural groove, the notochord and the dorsal nerve cord—the two "trademarks" of the chordates—develop. These two folds soon close. This embryo is 18 days old.*

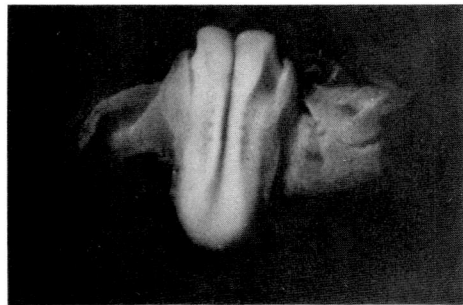

The Early Embryo

During this same period, the embryo itself has been developing. When the embryo is about two weeks old and about 1.5 millimeters in length (including the extraembryonic tissues), a primitive streak forms, very similar in appearance to the primitive streak of the chick. A neural plate appears and is soon replaced by the neural ridges, which fold to form the neural tube. At about 18 days, the earliest somites are visible. By the end of the third week, the embryo has 12 to 14 somites, and the blood vessels and the primitive gut have begun to develop. The embryo, overall, is about 2.3 millimeters long, about $\frac{1}{10}$ inch. The third week of life is the beginning of the most critical period in the physical development of a human being. Yet at this stage, most women are not even sure they are pregnant.

Fourth Week

At about 21 days, the eyes begin to form. By 24 days, the rudimentary heart, still only a tube, begins to flutter and then to pulsate; from this time on, it will not stop its 100,000 or more beats per day until the death of the individual.

By the end of the first month, the embryo is 5.2 millimeters (about $\frac{1}{5}$ inch) in length and has increased its mass 7,000 times. The neural groove has closed, and the embryo is now C-shaped, with about 30 pairs of somites. Muscles, bones, and connective tissues will develop from these somites. The arm buds are clearly visible, and the leg buds have begun to form. The heart, beating all the while, has developed from a simple contracting tube to a four-chambered vessel.

31–36　*A human embryo at 28 days. The balloonlike structure is the yolk sac. The embryo is curved toward you. At the top you can see the bulge of the rudimentary brain. The "seam" where the neural ridges closed is still visible. At the posterior of the embryo the somites are visible.*

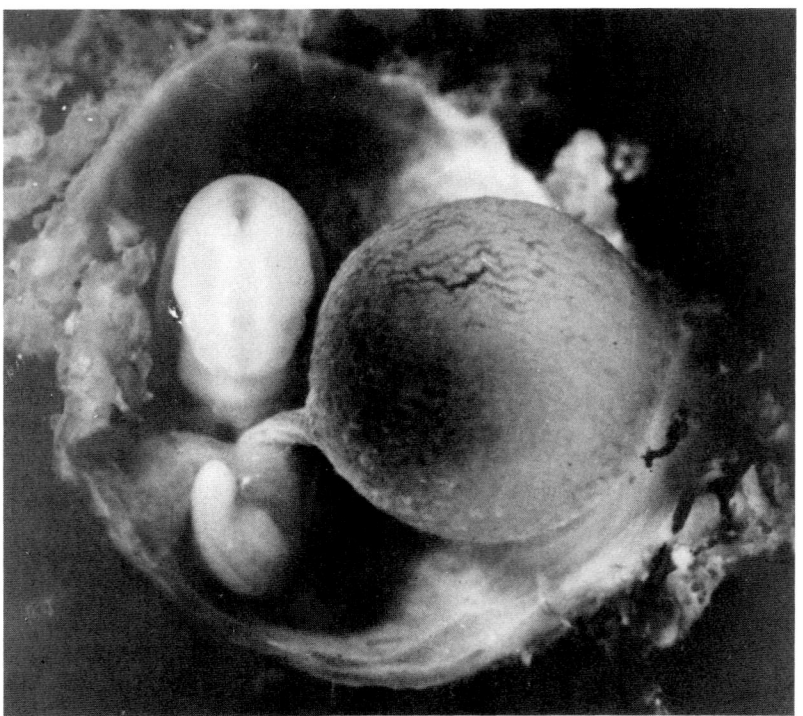

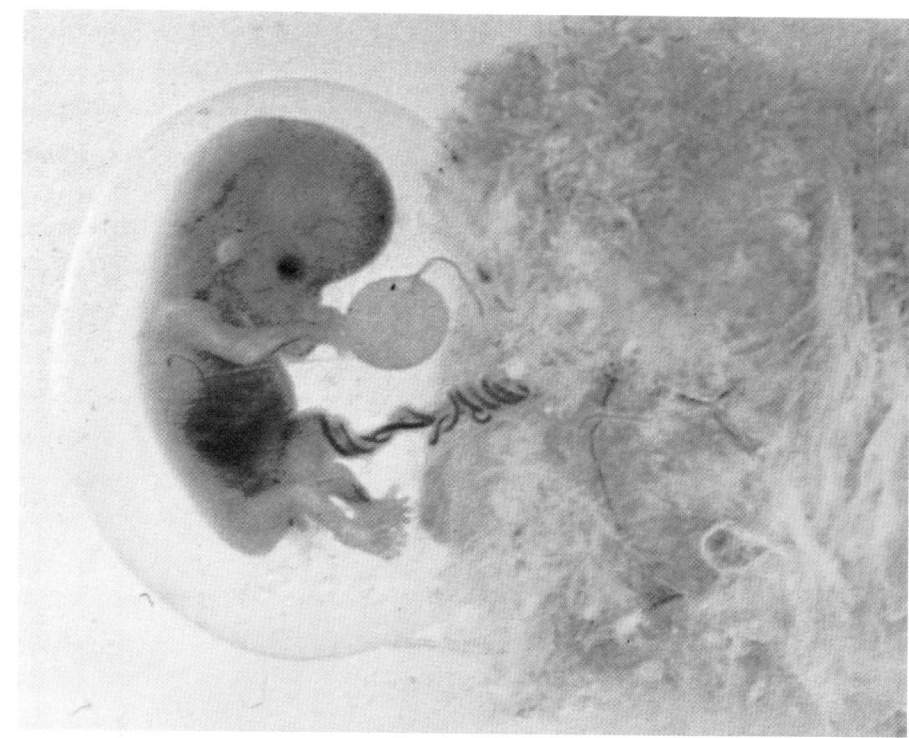

1–37 *Embryo at 2 months with persisting yolk sac. (Roberts Rugh and Landrum B. Shettles, M.D., From Conception to Birth: The Drama of Life's Beginnings, Harper & Row, Publishers, Inc., New York, 1971.)*

1–38 *Two-month-old fetus showing the skeleton. At this age, most of the body structures are soft and cartilaginous. The process of bone formation is not complete until after puberty.*

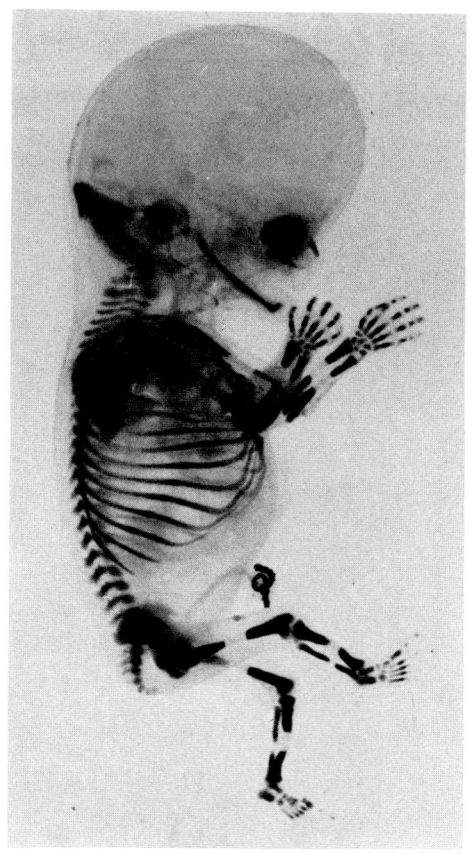

Second Month

During the second month, the embryo increases in mass about 500 times. By the end of this period, it weighs about $\frac{1}{30}$ ounce, slightly less than the weight of an aspirin tablet, and is about 1 inch long. Despite its small size, it is almost human-looking and, after two months, is generally referred to as a fetus. Its head is still relatively large because of the early and rapid development of the brain, but the head size will continue to be reduced in proportion to body size throughout gestation (and through childhood as well). Arms, legs, elbows, knees, fingers, and toes are all forming during this time, and as another reminder of our ancestry, there is a temporary tail. The tail reaches its greatest length in the second month and then gradually begins to disappear; it is entirely gone in 94 percent of all babies by the time they are born (of this 6 percent, the tail is generally detectable only by x-ray). Liver, gallbladder, and pancreas are present, and there is clear differentiation of the divisions of the intestinal tract. The liver now constitutes about 10 percent of the body of the fetus and is its main blood-forming organ.

By the end of the second month, the major steps in organogenesis are more or less complete and the embryo appears quite human in its external form. In fact, the rest of development is mostly concerned with growth and the maturation of physiological processes.

The first two months are the most sensitive period as far as the possible influence of factors of external origin is concerned. For example, when the arms and legs are mere rudiments (fourth and fifth weeks), a number of substances

31–39 (a, b) *A human embryo at 39 days. The budding limbs, the tail, and the bulge over the heart (now beating) can be clearly seen. The eyes have lenses but are covered by lids that will fuse during the third month and remain closed for the next three months. Just below the head structures is a series of branchial ("gill") pouches, another characteristic shared by all vertebrates. (c) At 42 days, the outline of the brain and the developing fingers and toes are clearly visible. The embryo is not quite ½ inch long.*

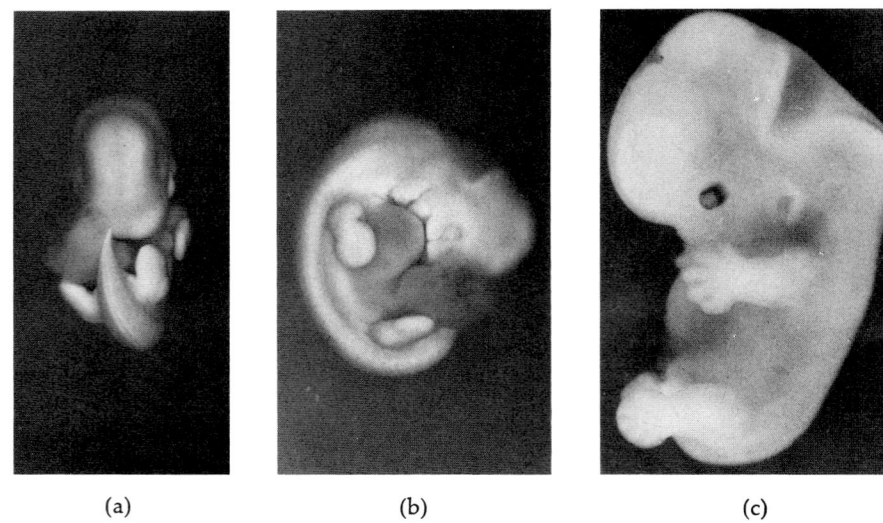

(a)　　　　　　　(b)　　　　　　　(c)

31–40 *Human fetus at 2 months, 1 week, now 1½ inches long. (Lennart Nilsson, Time-Life Picture Agency, © Time, Inc.)*

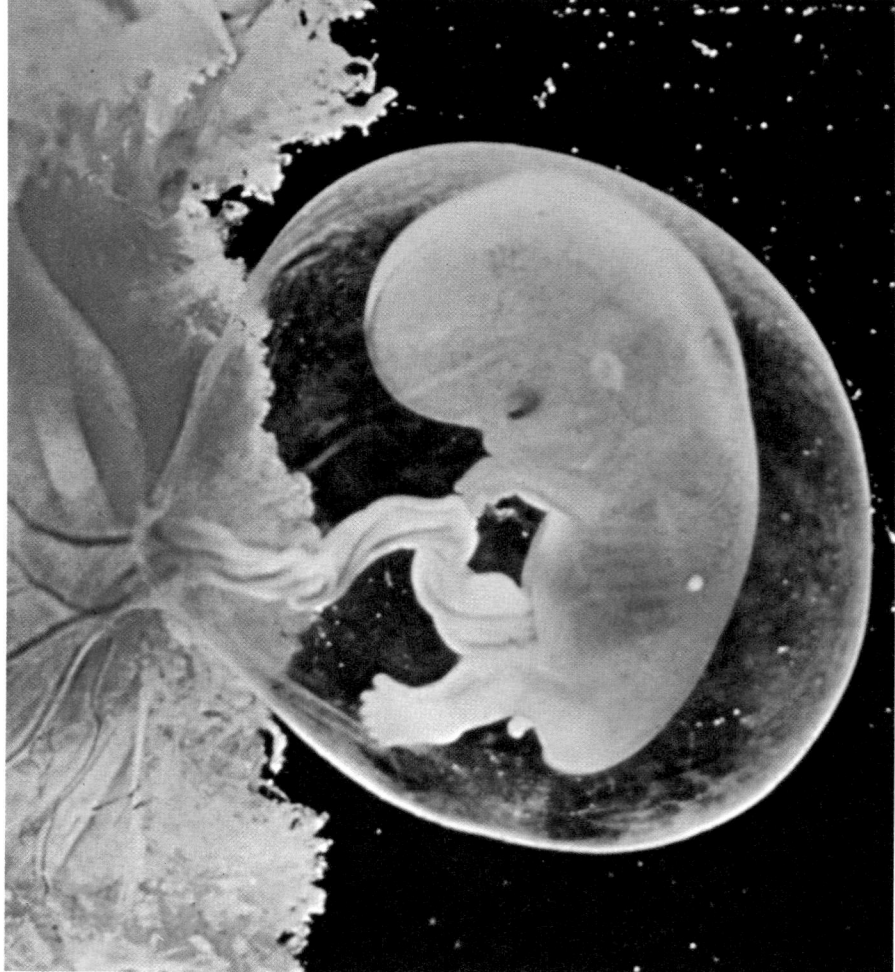

can upset the normal course of events and engender limb abnormalities. The experience with the presumably safe tranquilizer thalidomide in the 1960s is a tragic and familiar example. Because of the widespread publicity about the "thalidomide babies," some of whom are now becoming thalidomide adults, both pregnant women and their doctors have become far more cautious about the use of any kind of medication during these critical first months.

Infections may also affect the development of the embryo. Rubella (German measles) is a very mild disease in children and adults. Yet when contracted by the mother during the fourth through the twelfth weeks of pregnancy, it can have damaging effects on the formation of the heart, the lens of the eye, the inner ear, and the brain, depending on exactly when the infection occurs in relation to embryonic development. Similarly, exposure to x-rays at doses that would not affect an adult or even an older fetus may produce permanent abnormalities.

Third Month

During the third month, the fetus begins to move its arms and kick its legs, and the mother may become aware of its movements. Reflexes, such as the startle reflex and (by the end of the third month) sucking, first appear at this time. Its face becomes expressive; the fetus can squint, frown, or look surprised. Its respiratory organs are fairly well formed by this time but, of course, are not yet functional. The external sexual organs begin to develop. By the end of this month, the fetus is about 3 inches long from the top of its head to its buttocks and weighs about ½ ounce. It can suck and swallow, and sometimes does swallow some of the fluid that surrounds it in the amniotic sac. The finger, palm, and toe prints are now so well developed that they can be clearly distinguished by ordinary fingerprinting methods. The kidneys and other structures of the excretory system develop rapidly during this period, although waste products are still disposed of through the placenta. By the end of this period—the first trimester of development—all the major organ systems are laid down.

Fourth Month

During the fourth month, the baby's movements become obvious to the mother. Its bony skeleton is forming and can be visualized by x-rays. The body is becoming covered with a protective, cheesy coating. The fetus at this stage is about 5 inches long and weighs about 5 ounces.

Fifth Month

By the end of the fifth month, the fetus has grown another 2 inches and now weighs half a pound. It has acquired hair on its head, and its body is covered with a fuzzy, soft hair called the lanugo, from the Latin word for "down." Its heart, which beats between 120 and 160 times per minute, can be heard by a stethoscope. The five-month-old fetus is already discarding some of its cells and replacing them with new ones—a process which will continue throughout its lifetime. Nevertheless, a five-month-old fetus cannot yet survive outside of the uterus. The youngest fetus on record to survive was about 23 weeks old and required continuous assistance in breathing, taking food, and maintaining its body temperature. At this stage, the placenta covers about 50 percent of the uterus.

31–41 *Fetus at 16 weeks. Blood vessels are*
visible through the translucent skin. The
hands and feet are well formed; even the
fingernails are clearly visible. The fetus
is now about 5¼ inches long and fills
the uterus, which expands as the fetus
grows. (Lennart Nilsson, Time-Life
Picture Agency, © Time, Inc.)

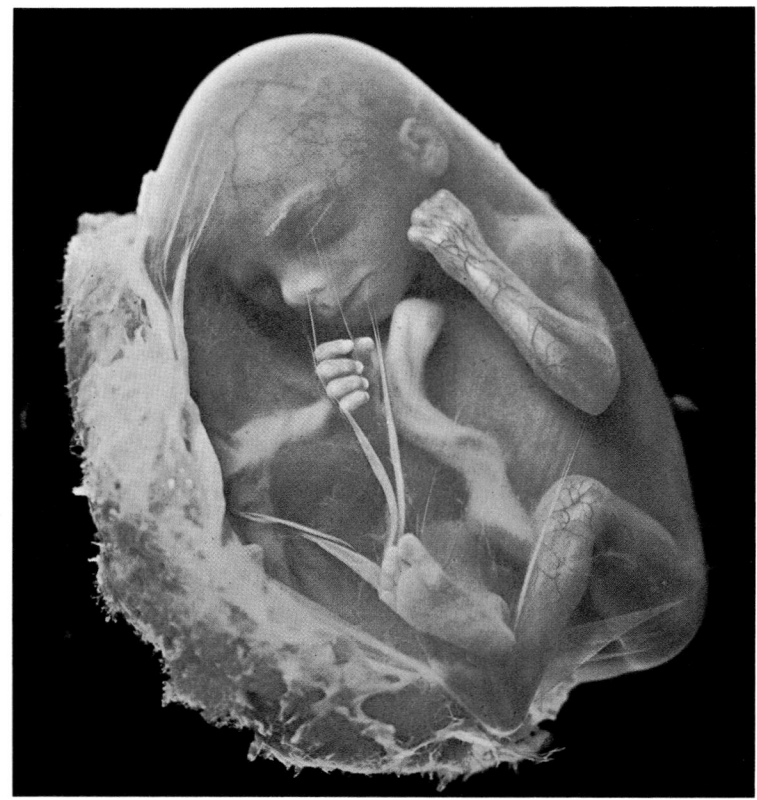

31–42 *Human fetus at 17 weeks, sucking its*
thumb. (Roberts Rugh and Landrum B.
Shettles, M.D., From Conception to
Birth: The Drama of Life's Beginnings,
Harper & Row, Publishers, Inc., New
York, 1971.)

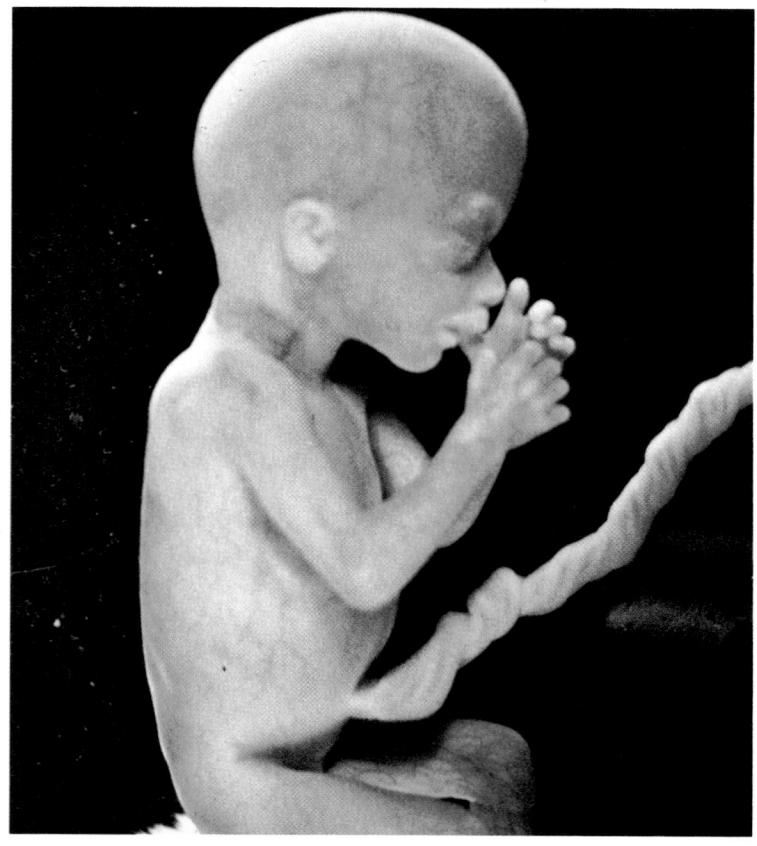

Sixth Month

Ninety percent of the fetal weight gain occurs after the fifth month. During the sixth month, the fetus has a sitting height of 12 to 14 inches and weighs about 1½ pounds. By the end of the sixth month, it could survive, although probably only with respiratory assistance in an incubator. Its skin is red and wrinkled, and although teeth are only rarely visible at birth, they are already forming dentine. The cheesy body covering, which helps protect the fetus against abrasions, is now abundant. Reflexes are more vigorous. In the intestines is a pasty green mass of dead cells and bile, known as meconium, which will remain there until birth.

Final Trimester

During the final trimester, the baby increases greatly in size and weight. In fact, it doubles in size just during the last two months. During this period, many new nerve tracts are forming and new brain cells are being produced at a very rapid rate. By the seventh month, brain waves can be recorded, through the abdomen of the mother, from the cerebral cortex of the fetus. Some recent research indicates that the protein intake of the mother is extremely important during this period if the child is to have full development of its nervous system and, hence, its intelligence potential.

-43 *Fetus at 28 weeks; it is now more than 10 inches long and weighs about 2¼ pounds. The membrane that envelopes it is the "caul," and the infant born with one still over its head, according to folklore, is destined for good fortune. (Lennart Nilsson, Time-Life Picture Agency, © Time, Inc.)*

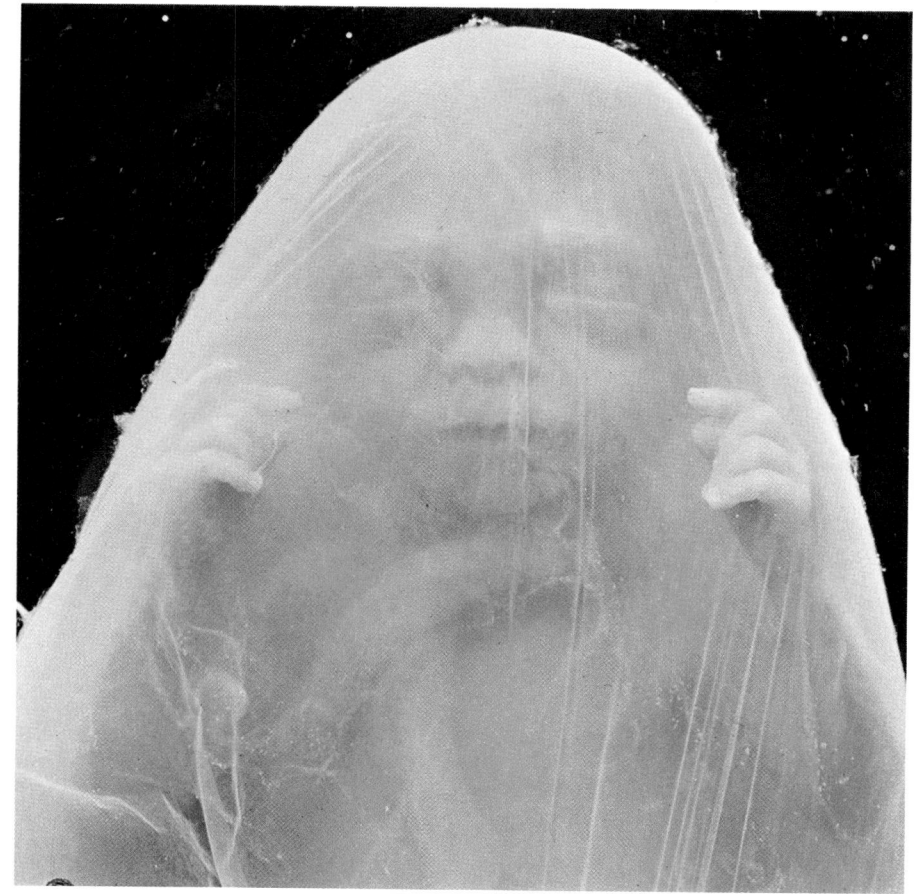

31–44 *X-ray of a human fetus in position for delivery. The pelvic bones of the mother, clearly visible here, can separate slightly to permit the baby's head and body to pass through. In 95 percent of all births, the fetus is in this head-down position.*

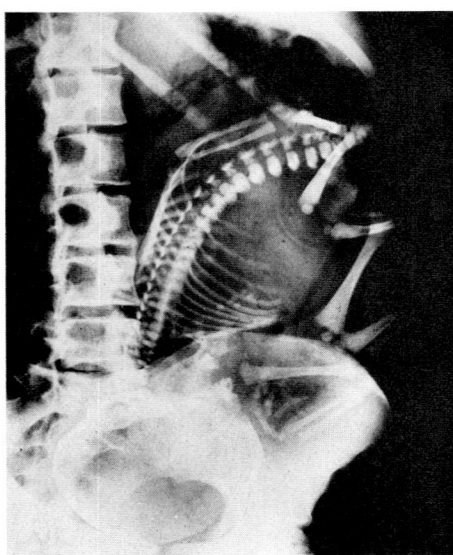

As the fetal period progresses, the physiology of the fetus becomes increasingly like the physiology of the adult, and so agents that affect the mother's physiology also threaten the late fetus. An increasingly familiar example is found in the infants being born with heroin or methadone addiction.

During the last month in the uterus, the baby usually acquires antibodies from its mother. These are globular proteins—formed against bacteria, viruses, or other foreign invaders—which defend the body against attack by such microorganisms. The baby becomes immune to whatever the mother is immune to. This immunity is only temporary. Within one to two months after birth, the maternal antibodies will be gradually replaced by antibodies manufactured by the baby's own immune system.

During the last month of pregnancy, the growth rate of the baby begins to slow down. (If it continued at the same rate, the child would weigh 200 pounds by its first birthday.) The placenta begins to regress and become tough and fibrous.

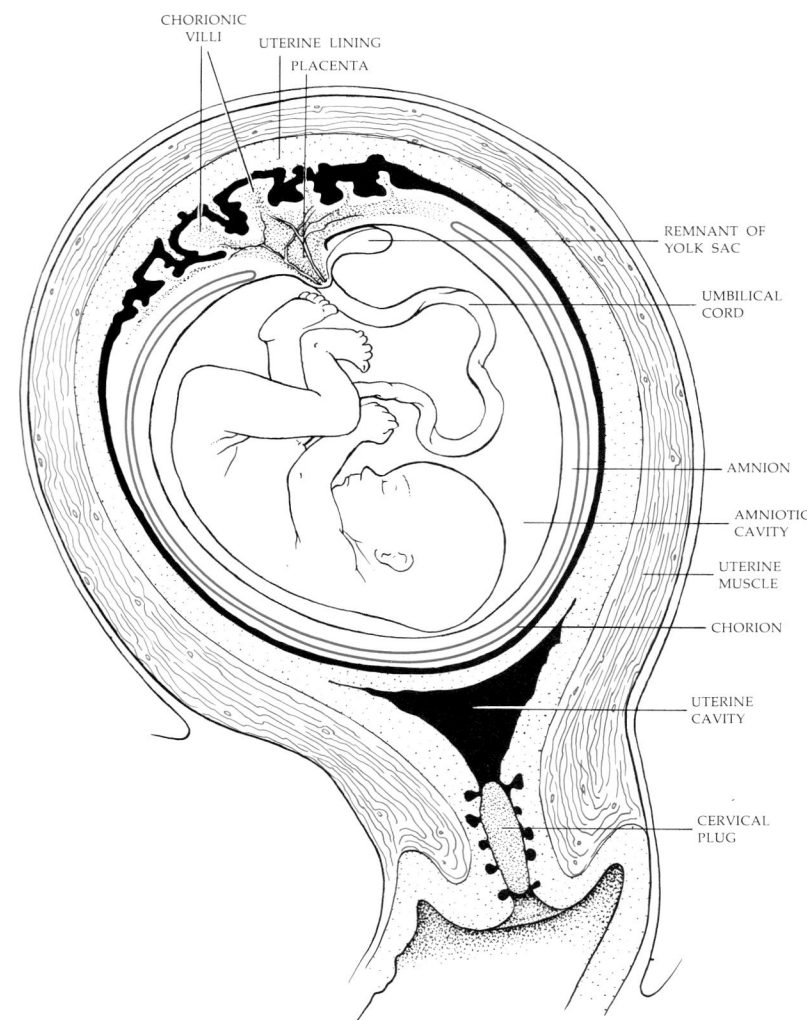

31–45 *A human fetus, shortly before birth.*

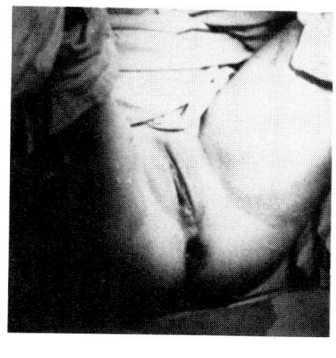
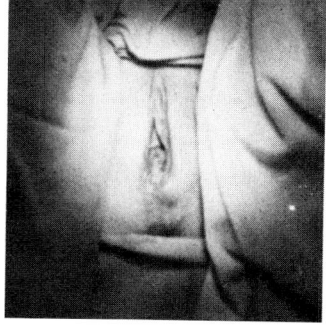
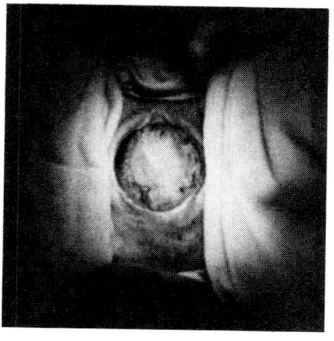
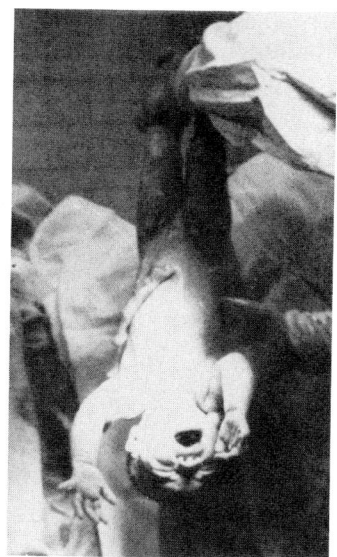
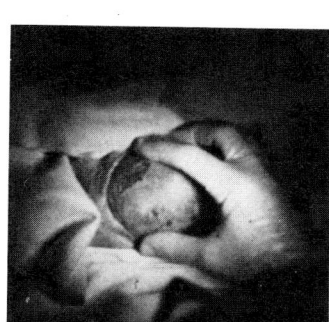
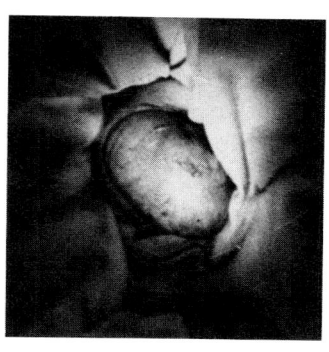
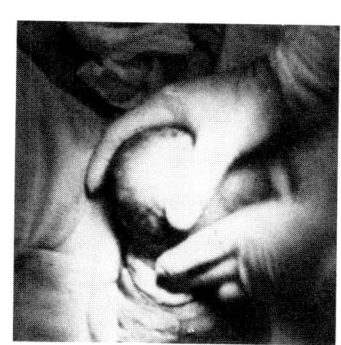

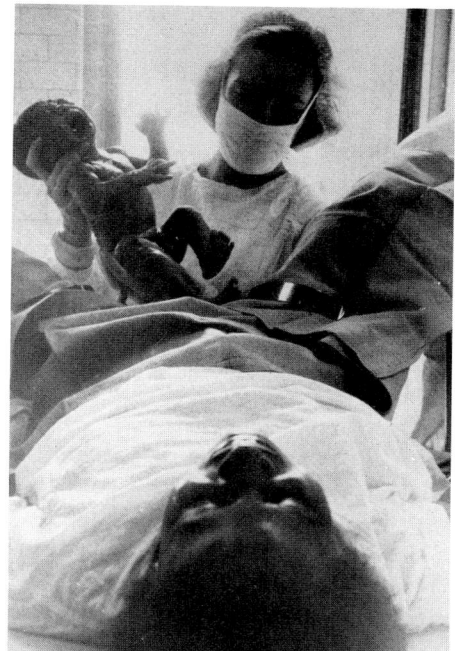

16 *Birth of a baby. Once the head and shoulders emerge, the contracting uterus expels the baby quickly.*

Birth

The date of birth is calculated as about 266 days after conception or 280 days after the beginning of the last regular menstrual period, but only some 75 percent of babies are born within two weeks of the scheduled time. Labor is divided into three stages: dilation, expulsion, and placental stages. Dilation, which lasts from 2 to 16 hours (it is longer with the first baby than with subsequent births), begins with the onset of contractions of the uterus and ends with the full dilation, or opening, of the cervix. At the beginning of the stage, uterine contractions occur at intervals of about 15 to 20 minutes and are relatively mild. By the end of the dilation stage, contractions are stronger and occur about every 1 to 2 minutes. At this point, the cervix is dilated to about 10 centimeters in diameter. Rupture of the amniotic sac, with the expulsion of fluids, usually occurs during this stage.

The second, or expulsion, stage lasts 2 to 60 minutes. It begins with the full dilation of the cervix and the appearance of the head in the cervix, called crowning. Contractions at this stage last from 50 to 90 seconds and are 1 or 2 minutes apart.

The third, or placental, stage begins immediately after the baby is born. It involves some contractions of the uterus and the expelling from the vagina of fluid, blood, and finally the placenta, with the cut umbilical cord attached. Minor uterine contractions continue; they help to stop the flow of blood and to return the uterus to its prepregnancy size and condition.

31–47 (a) *Surface view of an early chick embryo showing the original position of the germ cells, which lie in a crescent well beyond the embryo itself. If this germinal crescent is destroyed, no sex cells will be found in the gonads of the chick, although its development will be otherwise normal. (b) An early human embryo (about 4 weeks), showing the original position of the primordial germ cells.*

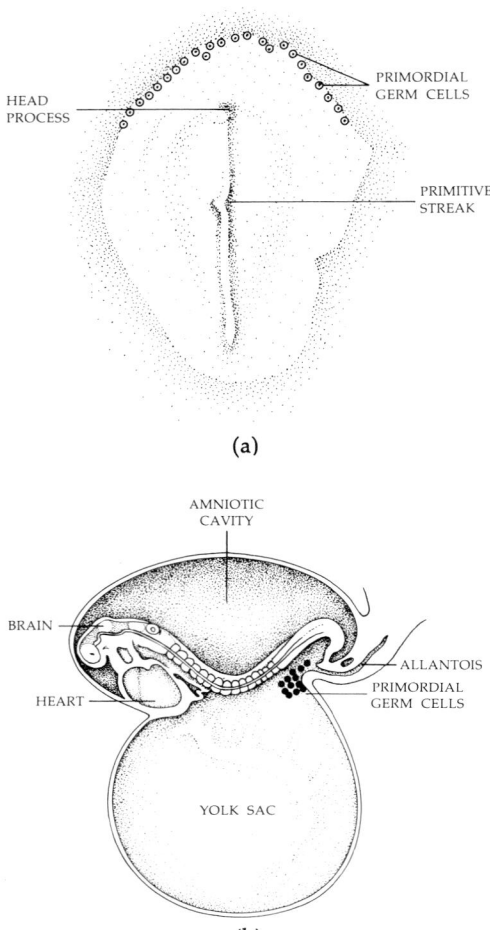

HEAD
PROCESS

PRIMORDIAL
GERM CELLS

PRIMITIVE
STREAK

(a)

AMNIOTIC
CAVITY

BRAIN

HEART

ALLANTOIS
PRIMORDIAL
GERM CELLS

YOLK SAC

(b)

The baby emerges from the warm, protective enclosure where he has been nourished and permitted to grow for nine months. The umbilical cord—until that moment, his lifeline—is severed. The baby cries as he takes his first breath (or if he does not, the obstetrician slaps him gently, causing him to cry and hence to breathe) and then starts to breathe regularly. His separate existence has begun.

Acquiring of Genetic Characteristics: Sex

As the embryo develops, its genetic potential unfolds. Although we shall examine this statement only in relation to one such characteristic, sex, it applies equally to the others. We chose this particular one because, of all the questions answered for us before we were born, "Is it a boy or a girl?" was clearly one of the most consequential. Moreover, there are some new and interesting data on this subject.

In Section 2, we saw that sex is determined in the human species by the presence or absence of a Y chromosome. In Chapter 30, we saw that sex is defined by the primary sex organs, the ovaries or the testes. What happens in the embryo, what does the presence of the Y or the absence of the Y do to start us on our long path to maleness or femaleness?

In the human embryo, germ cells (the cells that will later give rise to egg cells or sperm) can be traced to a little cluster of cells in the endoderm of the yolk sac. If these cells are destroyed (as they have been experimentally in man's useful stand-in, the mouse), no gametes are ever produced by the animal. It has been hypothesized that the germ cells are set aside so early in order to minimize the possibility of copying "mistakes," which might occur during many cellular divisions, and so help to ensure that the genetic message is passed on. Such minor mistakes might be of no consequence in somatic cells, which express only a small amount of their genetic information, but could be of major importance in a germ cell, which must carry the genetic instructions for the entire organism.

In humans, the influence of the genes determining sex is not seen in the embryo until about the fifth or sixth week of fetal life. During these first weeks, all embryos are morphologically identical.

During the fifth week of embryonic life, the germ cells migrate to the gonads. Animal experiments indicate that if the fetal gonads are removed during this critical developmental period, the infant will appear to be a normal female, whether genetically male or female, and if the appropriate hormones are administered at adolescence, he or she will be a normal-appearing (though, of course, sterile) adult female. If male hormones are administered to a genetically female fetus during early development, or even neonatally (right after birth), the reproductive structures, including the external genitalia, become masculinized. Mary Jane Sherfey* points out that the use of one hormone (androgen) for sexual differentiation in the embryo was probably an evolutionary adaptation to intrauterine life. In reptiles and birds, the fetus is essentially masculine and fetal estrogens are required to produce the female pattern. If the mammalian embryo were sensitive to estrogen, however, all male embryos would be feminized by the estrogens circulating in their mother's bloodstream during their long sojourn in the uterus.

The Nature and Evolution of Female Sexuality, Vintage Books, New York, 1973.

–48 In frogs and the higher vertebrates (including man), the sex glands are identical during the early stages of development of the embryo. They are composed of two main types of tissue: the cortex and the medulla. (a) Germ cells that do _not find their way to cortex or medulla remain sexually undifferentiated. The early, indifferent sex gland may develop into (b) a testis or (c) an ovary. In the testis, the germ cells migrate to the_ _medullary tissue, and the cordlike tissue of the medulla (shown in color) predominates in development. In the ovary, the sex cells lodge in the cortex, which becomes the dominant tissue._

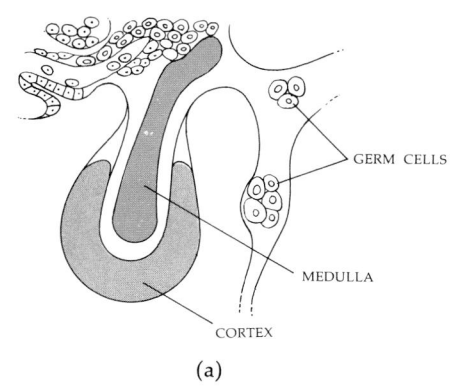

GERM CELLS

MEDULLA

CORTEX

(a)

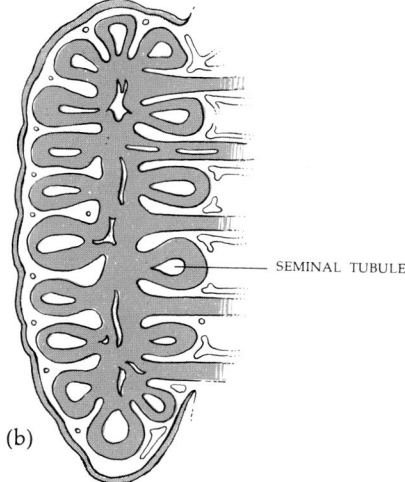

SEMINAL TUBULE

(b)

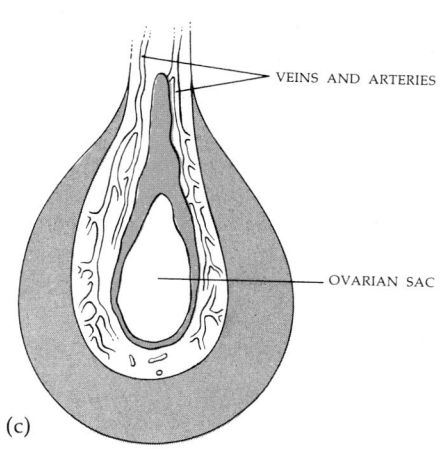

VEINS AND ARTERIES

OVARIAN SAC

(c)

Current studies in lower mammals, particularly rats, have shown that early exposure to sex hormones affects sexual behavior as well as physical development. Females injected with testosterone just before or just after birth not only show male physical characteristics, but as adults they fail to exhibit normal female sexual responses. Studies in male rats show similar results. If males are castrated a week or more after birth, their sexual behavior as adults will be normal if they are given testosterone. However, if they are castrated within the first five days after birth, they are incapable of normal mating behavior as adults, even though testosterone is administered.

SUMMARY

Development is the process by which a fertilized egg becomes a complete organism, consisting of billions of cells and closely resembling its parent organism.

In the human female, fertilization of the egg cell by a sperm cell takes place in one of the oviducts. It results in: (1) changes in the egg cell membrane that prevent fertilization by another sperm cell, (2) introduction of the genetic material of the male parent into the egg, and (3) activation of the egg cell, which then begins dividing.

Development takes place in three stages: cleavage, gastrulation, and organogenesis. During cleavage, the original egg cell is divided, with little or no change in volume. In eggs containing little or no yolk, such as those of the sea urchin or _Amphioxus_, cleavage results in cells of approximately equal size. In yolkier eggs,

PREFORMATION AND EPIGENESIS

The accompanying illustration from a medieval manuscript depicts fetuses developing in the uterus. Notice that they are represented as miniature adults. This concept is in accord with the theory of preformation held by many scientists through the seventeenth and eighteenth centuries. According to this theory, the adult organism is already present in the germ cell, somewhat as a flower is present in the bud, needing only to unfold.

The second, opposing theory was that of epigenesis. Epigeneticists believed that the egg contained only simple, essentially structureless materials—the blocks, so to speak, from which the organism might be constructed. Early epigeneticists (and the theory can be traced back to Aristotle) believed that the form of the adult was imposed in the course of development by the working of some "creative principle" or "vital force."

Modern embryologists combine the two views. Within the fertilized egg cell in the form of DNA is all the information for development—"preformed" and passed along from generation to generation. But in the course of development, new properties arise "epigenetically"—properties that were not present as such in the egg cell. This modern view is in accord with the levels-of-organization concept discussed in the Introduction.

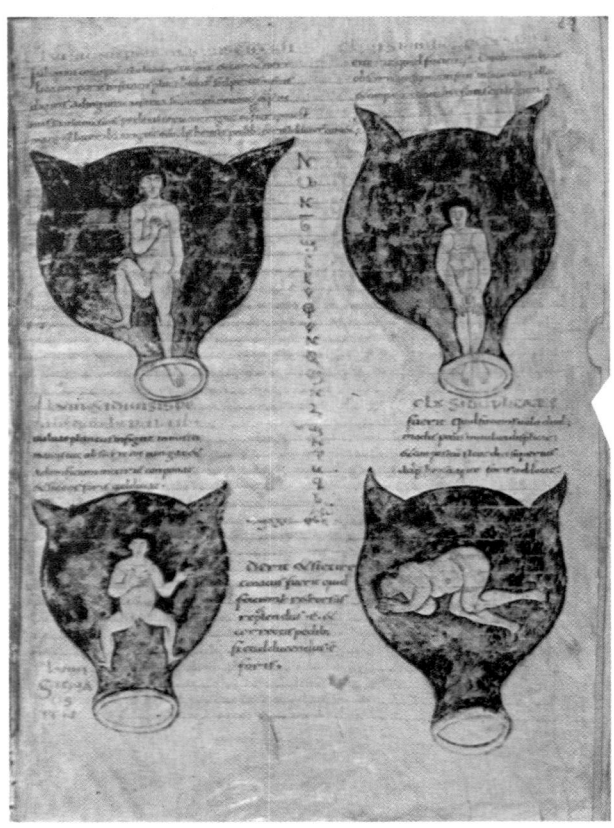

such as that of the frog, cleavage is unequal, with fewer and larger cells in the yolky (vegetal) hemisphere. In all cases, when cleavage is complete, the embryo consists of a hollow sphere of cells, the blastula, with a central cavity, the blastocoel. In eggs with a great deal of yolk, such as the hen's egg, cleavage occurs only in the nonyolky portion of the egg and results in a blastodisc.

Gastrulation involves the movement of cells into new relative positions and results in the establishment of the axes of the embryo and of the three body layers, endoderm, mesoderm, and ectoderm. In the course of gastrulation, an opening forms, the blastopore. Cells from the outer surface of the embryo move through the blastopore into the blastocoel, forming a new cavity, which will be the primitive gut. In the chick and man, the homologue of the blastopore is the primitive streak. When the movement of cells is complete, the central part of the dorsal surface of the gastrula flattens, becoming the neural plate, which folds to form the neural tube. The cells underlying the neural plate become notochord (a mesodermal tissue), and the somites become embryonic muscle tissue. The endodermal cells, on the interior of the embryo on the ventral side, become gut.

The small segment of tissue on the dorsal lip of the blastopore prior to gastrulation is known as the organizer because, as shown by experiments on amphibians, it induces the cells overlying it to form the neural tube. (It itself forms the notochord.) If it is implanted in another embryo, it will induce the formation of a second neural tube in that embryo.

Other organ systems also begin to develop by this same process of embryonic induction, in which one tissue induces changes in an adjacent tissue with which it comes in contact, whether by growth or cell migration. The nature of the inducing substances is not known.

Each of the three primary cell layers as established at gastrulation gives rise to particular tissues and organs. The epidermis (outer skin) and the central nervous system arise from ectoderm; the notochord, muscles, bones and cartilage, heart and blood vessels, excretory organs and gonads, the inner lining of the skin, and the outer linings of the digestive and respiratory tracts arise from mesoderm; the inner linings of the digestive and respiratory tracts and the digestive glands develop from endoderm. The body cavity, or coelom, is a space, or discontinuity, between two layers of mesoderm.

In amniotic eggs, such as those of reptiles, birds, and mammals, the embryo, as it develops, becomes surrounded by protective membranes. There are four: the yolk sac, which surrounds the yolk and connects the embryo to it; the amnion, which encloses the amniotic fluid in which the embryo floats and which bathes and cushions it; the allantois, which emerges from the gut and which has the primary function of collecting wastes; and the chorion, which surrounds the embryo and all its membranes.

The human embryo develops as a cell mass completely surrounded by an outer layer of cells, the trophoblast. When the embryo descends into the uterus, at about its sixth day of development, the trophoblast develops rapidly and invades the maternal tissues. The trophoblast becomes the chorion and eventually the fetal part of the placenta. Like the chick embryo, the human embryo is surrounded by an amnion filled with amniotic fluid. A yolk sac with no yolk is also present and an allantoic membrane, which develops to become the umbilical cord, linking the embryo to the placenta.

When the embryo is about 2 weeks old, a primitive streak forms, followed by the development of a neural plate and neural ridges, which fold to form a neural tube. Although the embryo is still very small (about 1/10 inch long), most of the major organs begin to form in these very early weeks, which is why damage caused to the embryo by viral infection, x-rays, or drugs during this period can be so widespread. By the end of the second month, the embryo, now called a fetus, is almost human-looking, although it only weighs 1/30 ounce. By the end of the third month, the mother becomes aware of the movements of the baby; by this time all of the organ systems have been laid down.

Laboratory experiments have shown that all of the cells of an organism contain all of the genetic information present in the fertilized egg. Thus development, at the molecular level, is visualized as a gradual, sequential expression of particular genes by particular cells with a repression of other genes. This sequential expression and repression is under the influence of the cytoplasm as formed during oogenesis and of inducing substances transmitted by other cells.

QUESTIONS

1. Define the following terms: secondary induction, blastopore, primitive streak, placenta, organizer.
2. Describe, in general terms, the end results of each of the following events: fertilization, cleavage, gastrulation, organogenesis.
3. Follow the course of a single cell from its place of origin in the fertilized egg of a frog to its position in the eye cup of an early frog embryo. List, for each stage, all of the influences on this cell that might affect its history.
4. It is illegal in the United States to keep an aborted human embryo or fetus alive for research purposes. Do you approve or disapprove of this restriction on scientific research? Why?

Chapter 32

Integration and Coordination: Hormones, Nerves, and Muscles

In the two previous chapters we examined the processes by which organisms reproduce themselves. In the chapters that follow, we shall be concerned with how they maintain themselves. As you will see, the requirements of multicellular organisms are remarkably similar to those of a single-celled organism—a paramecium, for instance, or an amoeba. Each heterotrophic cell requires energy-yielding molecules, a source for their oxidation, usually oxygen, and a means for disposing of waste products, such as carbon dioxide. One of the outstanding differences, however, between a paramecium and a multicelled animal is the problem of coordinating the activities of all the various cells and tissues and organs in order to promote the survival of the whole organism. This problem of integration and control of the body's many activities is the subject of this chapter.

The principal regulatory systems of the body are the endocrine and nervous systems. In Chapter 30, we saw some examples of the ways these two systems interact: for instance, in the regulation of estrogen concentration and in the synchrony of copulation and ovulation in cats and rabbits. In this chapter, we are going to examine these vast communication networks in order to provide a background for understanding the control of such functions as circulation, digestion, excretion, and temperature regulation, which are the subjects of the chapters that follow. The discussion of the brain and of some of the current research on brain function is reserved for the final chapter.

Glands are epithelial tissues specialized for secretion. Endocrine glands are glands that secrete their products into the bloodstream (or, more precisely, into the extracellular space from which the substances diffuse into the bloodstream). Exocrine glands secrete their products into ducts; examples are sweat glands (see page 686) and digestive glands. "Endocrine" is generally, though not quite precisely, used as a synonym for "hormone-secreting," and endocrinology means the study of hormones.

THE ENDOCRINE SYSTEM

In 1849, A. A. Berthold, a physician and professor in Göttingen, carried out the first scientific experiment in endocrinology. Using six young cockerels, he set up three experimental groups: (1) Two birds were castrated; these became typical capons in which the combs, wattles, plumage, aggressiveness, crow, and

-1 *Emotions, such as rage, have profound physiological effects upon an organism, preparing it for suitable behavioral responses. These physiological changes are brought about by dynamic interactions of the endocrine and nervous systems.*

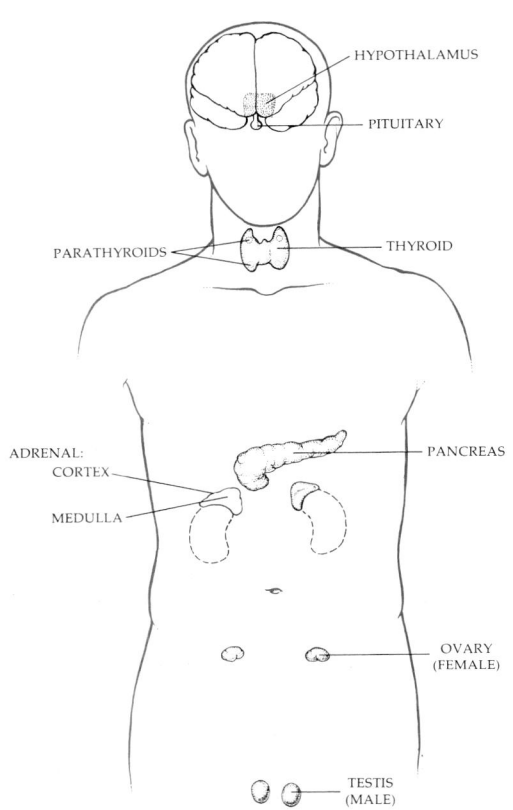

32–2 *Some of the hormone-producing organs. The pituitary releases hormones that, in turn, regulate the hormone secretions of the sex glands, the thyroid, and the adrenal cortex (the outer layer of the adrenal gland). The pituitary is itself under the regulatory control of an area in the brain known as the hypothalamus, indicated in color.*

HYPOTHALAMUS

PITUITARY

PARATHYROIDS

THYROID

ADRENAL:
CORTEX

PANCREAS

MEDULLA

OVARY
(FEMALE)

TESTIS
(MALE)

sexual urge characteristic of the mature cock failed to develop. (2) Two additional birds were castrated, and the removed testes were reimplanted in the same animal but at a new site, distant from ducts, nerves, or any possible previous connections except with the circulatory system. These birds developed into normal cocks. (3) Two birds were subjected to "sham" surgery; that is, surgical incisions were made, but no tissues were removed or transplanted. They, too, developed normally. (These last animals served as controls; without them, the experimenters could not be sure that the results obtained in the second animals were related to removal and reimplantation of the testes and not to surgical stress alone.)

This experiment initiated a wholly new concept of controlling mechanisms: that substances produced by particular tissues of the body could be carried by the bloodstream and could exert specific effects on distant tissues. Such substances came to be known as *hormones,* from the Greek word meaning "excite." It is now known, however, that many hormones act as inhibitors rather than exciters. (The juvenile hormone of insects, page 405, is an example.) The hormones are better described as chemical messengers, establishing communications between the various parts of the body. Some of the communications are concerned with homeostatic regulation, the perpetual adjustment of the internal environment. Others are involved with change—the changes that occur with sexual maturation or responses to emergencies.

Hormones act by altering the metabolism of particular cells. Because they travel through the bloodstream, all tissues in the body are equally exposed to them. Thus their specificity of action depends as much on the receptivity of the target tissue as on the chemical specificity of the hormone. Moreover, target tissues may be receptive under some circumstances and not under others. For example, prolactin causes the production of milk—but only in a mammary gland also under the influence of estrogen, progesterone, thyroxine, adrenal steroids, and growth hormone.

In the following pages, we shall discuss some of the principal endocrine glands of mammals and the hormones they secrete. The hormones involved in digestion, which are not included here, will be discussed in Chapter 34.

THE PITUITARY GLAND

The pituitary gland, or hypophysis, as it is sometimes called, is situated right at the base of the brain in the geometric center of the skull. It is about the size of a kidney bean and, like a kidney bean, has two lobes: the posterior lobe, or neurohypophysis, and the anterior lobe, or adenohypophysis.

The anterior lobe of the pituitary gland is the source of six different hormones. Four of these are tropic hormones, hormones that act upon other glands to regulate their secretions. Two of these tropic hormones, FSH and LH, which act upon the gonads, were discussed in Chapter 30. TSH, the thyroid-stimulating hormone, influences the thyroid. Like the gonadotropic hormones, TSH is regulated by negative feedback, being sensitive to the concentration of thyroid hormone in the blood. Adrenocorticotropic hormone (ACTH) has a similar regulatory relationship with certain of the hormones produced by the adrenal cortex.

Growth hormone, sometimes called somatotropin, is also produced by the anterior pituitary. It promotes the growth of bone and muscle. As is the case

CIRCADIAN RHYTHMS IN MAN

Man, like plants and other organisms, is tied to his circadian rhythms. It has long been known that he is more likely to be born between 3 and 4 A.M. and also to die in these same early morning hours. Body temperature fluctuates as much as 2°F during the course of the day, usually reaching a high about 4 P.M. and a low about 4 A.M. Alcohol tolerance is greatest at 5 P.M. Many people have an automatic internal alarm system that wakes them at the same hour every morning—whether they want to or not. Hormone secretion, heart rate, blood pressure, and urinary excretion of potassium, sodium, and calcium in man all vary according to a circadian rhythm. A recent study by the Federal Aviation Agency showed that pilots flying from one time zone to another— from New York to Europe, for instance—exhibit "jet lag," a general decrease in mental alertness and ability to concentrate, and an increase in decision time and physiological reaction time. Measurements of circadian rhythms show that the body may be "out of sync" for as much as a week after such a flight. This brings into question present policies of diplomats speeding to foreign capitals at times of international crisis or of troops being air-transported into combat.

Elaborate studies in which men have been kept isolated in constant conditions for long periods of time—in underground bunkers, for example—have indicated that these rhythms are under the control of a biological clock or clocks. Such clocks presumably are another mechanism for integrating and coordinating the body's multitude of physiological functions.

2–3 *Members of the sideshow, at one time a major circus attraction, were often persons with endocrine disorders. Giants and midgets are the result of too much or too little growth hormone.*

with most of the hormones, growth hormone is best known, both to scientists and laymen, by the effects caused by too much or too little. If there is a deficit in somatotropin production in childhood, a midget results, the so-called "pituitary dwarf." An excess of somatotropin results in a giant; most circus giants are the result of an excess of growth hormone. Excessive growth hormone in the adult does not lead to giantism, since not all adult tissues respond to growth hormone, but to acromegaly, an increase in the size of the jaw and the hands and feet, adult tissues that are still sensitive to the effects of growth hormone. Recently, it has been shown that growth hormone has effects on glucose metabolism, inhibiting the uptake and oxidation of glucose by many types of cells. It also stimulates the breakdown of fatty acids, thus conserving glucose.

Prolactin, also produced by the anterior pituitary, stimulates secretion of milk in mammals. Prolactin secretion begins in the mother shortly after delivery, as a result of the hormonal changes that take place following the expulsion of the fetus from the uterus. As long as the infant continues to nurse, the impulses produced by the suckling of the breast are transmitted by way of the central nervous system to the pituitary and cause it to produce prolactin, and the prolactin, in turn, acts upon the breast to maintain the production of milk. Once suckling ceases, the synthesis and release of prolactin stops and so does milk production. Thus supply is regulated by demand. In some birds, prolactin stimulates the production of crop milk. (See the boxed essay on page 587.)

The pituitary gland is the source of melanocyte-stimulating hormone. In reptiles and amphibians, this hormone stimulates color changes associated with camouflage or with behavior patterns such as aggression or courtship. It has no known function in man.

Relationships between the pituitary and the hypothalamus. Nerve fibers (in color) connecting the hypothalamus and posterior lobe of the pituitary transmit oxytocin and antidiuretic hormone (ADH), which are stored in and released from the posterior lobe. The anterior lobe is connected with the hypothalamus by a network of capillaries. Neurosecretory cells of the hypothalamus secrete hormones designed for the anterior pituitary directly into the capillaries. They are carried by the blood to the anterior pituitary where they affect the production of tropic hormones (hormones that affect other glands), growth hormone, and prolactin.

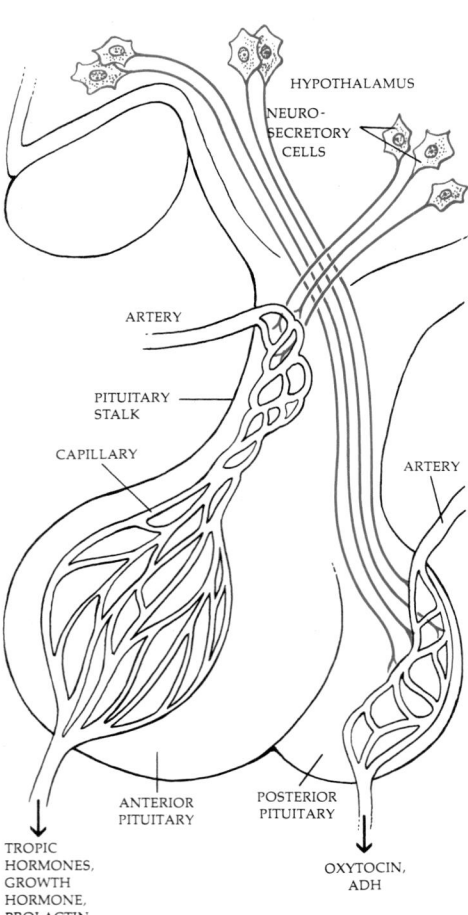

The Pituitary and the Hypothalamus

The pituitary gland lies beneath an area of the brain known as the hypothalamus and is directly under its influence and also under the influence, by way of the hypothalamus, of other brain centers. Some nine hormones have now been isolated from the hypothalamus that act either to stimulate or inhibit the secretion of hormones by the anterior pituitary. Two have been analyzed and both are small peptides, one only three amino acids in length. Released into the blood supply of the pituitary gland as it flows through the pituitary stalk, these hormones (sometimes called releasing factors) travel through the blood to their target tissue, thus technically meeting the criterion of hormones (Figure 32–4).

The hypothalamus is also the source of two hormones stored in and released from the posterior pituitary, or neurohypophysis: oxytocin and antidiuretic hormone, or ADH. (ADH is sometimes called vasopressin because it increases blood pressure in many vertebrates; it has no such effect in man, however, except in very high doses.) Oxytocin accelerates childbirth by increasing uterine contractions. Its release is under the control of the nervous system and may be triggered by increasing pressure within the uterine wall or by movements of the fetus. In experimental animals, labor can be induced by mechanical stimulation of the uterus or electrical stimulation of the hypothalamus. Oxytocin also is responsible for the "letting down" of milk that occurs when the infant begins to suckle. The hormone promotes contraction of the muscle fibrils around the milk-secreting cells of the mammary glands in much the same way that it promotes contractions of the smooth muscles of the uterus. The hormone is also present in males, but its function in the male, if any, is unknown.

ADH decreases the excretion of water by the kidneys. It achieves this effect, apparently, by increasing the permeability of the membranes of cells in the collecting ducts of the kidneys so that more water passes through them back into the blood from the urine. (We shall examine this mechanism in more detail in Chapter 36.)

Oxytocin has some ADH effect and ADH some oxytocin effect. This cross action is not surprising because each one of these hormones consists of only nine amino acids, and differences in the two hormones involve differences of only two amino acids among the nine (see Figure 32–5).

32–5 *Primary structures of antidiuretic hormone and oxytocin, hormones produced in the hypothalamus and released from the posterior pituitary. Note that each consists of only nine amino acids and that they differ by only two.*

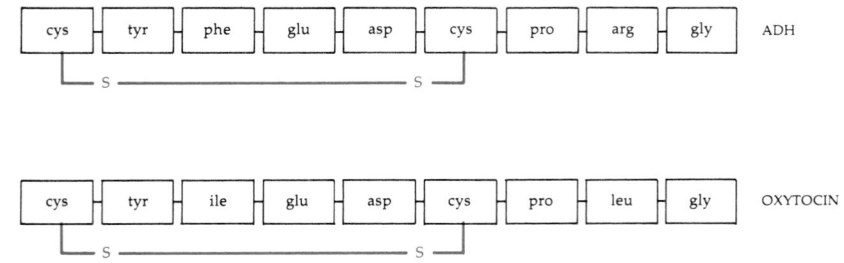

2–6 *Thyroxine, the principal hormone produced by the thyroid gland. Note the four iodine atoms in its structure. Where iodine is present in the soil, it is available in minute quantities in drinking water and in plants. Sea salts are rich in iodine, and, in the United States, table salt is ordinarily iodized or must be specifically labeled as being uniodized.*

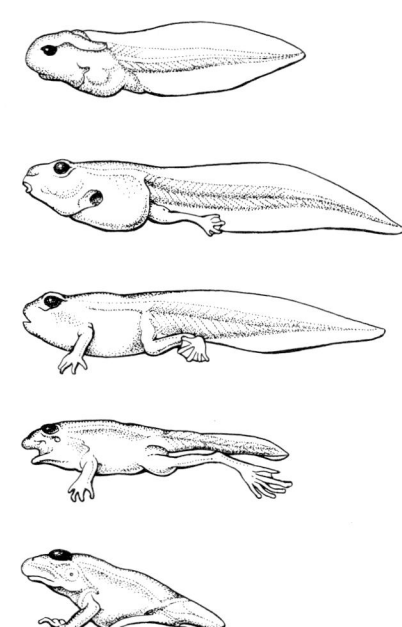

–7 *Frogs and toads change from aquatic gill-breathing vegetarian larvae to terrestrial lung-breathing carnivores. These changes, which involve almost every organ system in the body, are regulated principally by thyroid hormones. Injection of thyroxine into a tadpole causes premature metamorphosis; conversely, surgical removal of the thyroid or pituitary gland prevents metamorphosis.*

The close relationship between the emotions and the endocrine system in man is indicated by observation of institutionalized infants. Such infants, characteristically, are listless and withdrawn and appear profoundly sad. They often fail to grow normally; of 91 infants studied in one survey made in the 1940s of foundling homes in the eastern United States and Canada, 34 died in the first year of life in spite of good food and careful medical care. Infants who survived the first year were all physically retarded. A number of studies suggest that deprivation dwarfism, as it is called, is the result of underactivity of the pituitary gland, probably both in its production of growth hormone and of tropic hormones as well.

THE THYROID GLAND

The thyroid, under the influence of the thyroid-stimulating hormone from the pituitary, produces thyroxine, which is an amino acid combined with four atoms of iodine. Thyroxine accelerates the rate of cellular respiration; the rate at which oxygen is consumed at rest (basal metabolic rate) is used as an index of how much thyroid hormone the tissues are receiving. Hyperthyroidism, the overproduction of thyroxine, produces excessive nervousness and excitability, increased heart rate and blood pressure, and weight loss. Hypothyroidism in infancy affects development, particularly of the brain cells, and leads to permanent mental deficiency and dwarfism. In adults, hypothyroidism is associated with dry skin, intolerance to cold, and lack of energy.

Hypothyroidism may be caused by insufficient iodine, which is needed to make thyroid hormone, and in these cases is often associated with goiter, an enlargement of the thyroid gland. Before iodinization of salt became a common practice, goiters were so common in some geographical areas that persons without them were considered strange looking and therefore unattractive.

The thyroid gland is also the source of the hormone calcitonin, whose major action is to inhibit the release of calcium from bone. Its secretion is controlled directly by the calcium concentration of the fluid surrounding the thyroid cells.

ADRENAL CORTEX

The adrenal glands are on top of the kidneys. The outer layer, or *adrenal cortex*, is the source of a large number of steroids (see page 83) with hormonal activity. About 50 different steroids, all very similar in structure, have so far been isolated from the adrenal cortex of various mammals. Some of these undoubtedly represent steps in the synthesis of various hormones, though most of them have some hormonal activity.

One general group, the glucocorticoids, is concerned primarily with the formation of glucose from protein and fat. Only the glucose present in the bloodstream immediately following a meal is derived directly from the diet. After this immediate supply is used up, the glycogen stored in the liver is converted to glucose and the glucose is released into the bloodstream. The liver's supply of glucose is sufficient for a few hours. Subsequently, protein and fat from the body tissues are converted into glucose. In this way, a fairly stable level of glucose is maintained in the bloodstream despite an irregular intake of food.

The glucocorticoids also act to suppress inflammation and are sometimes used medically as anti-inflammatory agents in the treatment of such diseases as

32–8 *The chemical structures of two steroid hormones secreted by the adrenal cortex. Cortisol is a glucocorticoid, and aldosterone is a mineralocorticoid. Note, however, the very minor structural differences between the two.*

CORTISOL

ALDOSTERONE

arthritis. However, their severe side effects limit their usefulness. The most familiar of these glucocorticoids is cortisone, which is not itself secreted by the human adrenal gland. Instead, cortisone is produced in the liver by the breakdown of cortisol (the most important glucocorticoid in man) and is used to form other compounds. Little, if any, cortisone is actually released into the bloodstream.

Cortisol and the cortisol-like hormones are secreted in response to the tropic hormone ACTH, and they inhibit secretion of ACTH by negative feedback.

A second group, the mineralocorticoids, of which aldosterone is the primary example, is concerned with the regulation of ions, particularly sodium and potassium ions. The mineralocorticoids affect the transport of ions across the cell membranes of the kidneys and, as a consequence, have major effects both on ion concentrations in the blood and on water retention and water loss. An increase in aldosterone secretion results in greater reabsorption of sodium and chloride ions and increases the secretion of potassium. A deficiency in mineralocorticoids precipitates a critical loss of sodium from the body in the urine and, with it, a loss of water by osmosis and a reduction in blood pressure, which is dependent upon the water levels of the body.

The adrenal cortex is also a source of male sex hormones, which is why an adrenal tumor may result in increased production of these hormones and the production of facial hair and other masculine characteristics in a woman. Bearded ladies in the circus are often victims of such tumors.

32–9 *The production of many major hormones is regulated by a complex negative feedback system involving the pituitary and the hypothalamus. The hypothalamus stimulates the pituitary to secrete hormones, and these in turn stimulate the secretion of hormones from the thyroid, adrenal cortex, and gonads (the testes or the ovaries). When the hormones produced by these target glands reach a certain concentration in the blood, the hypothalamus stops stimulating the pituitary, the pituitary stops producing tropic hormones, and production of hormones by the target glands also stops. By way of the hypothalamus, which exchanges information with many other parts of the brain, hormone production can also be regulated in response to changes in the external and internal environments.*

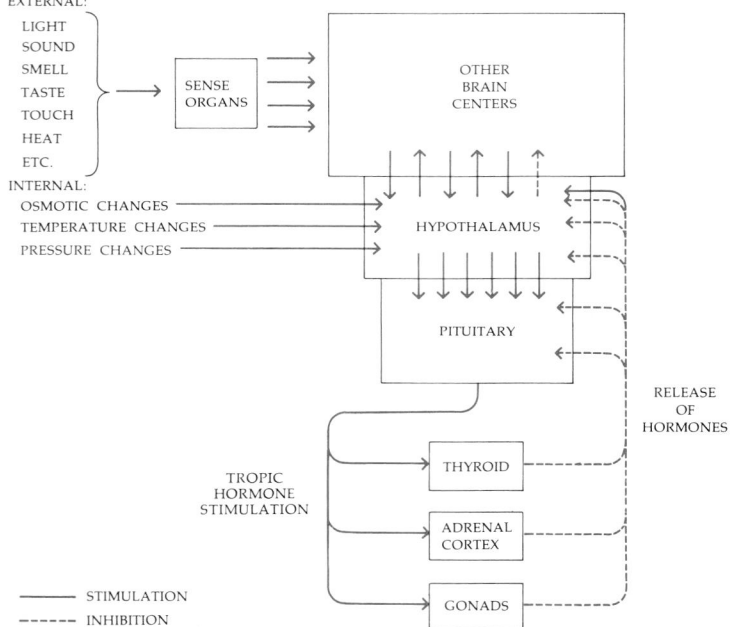

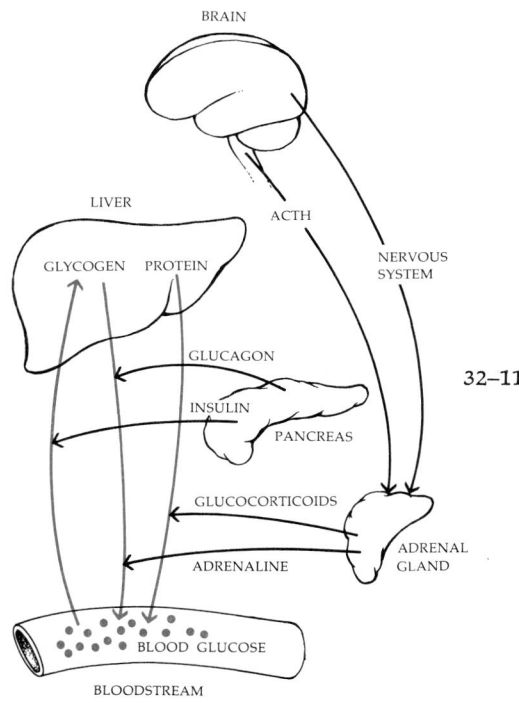

32–10 *Adrenaline (or epinephrine) is secreted by cells of the adrenal medulla. The ring structure to the left is called a catechol, and adrenaline is characterized chemically as a catecholamine, a nitrogen-containing catechol.*

ADRENAL MEDULLA

In embryonic life, the adrenal cortex arises from mesodermal tissue, in fact, from the same general mesodermal tissue that gives rise to the gonads, which also produce steroids, and to the kidneys. However, the central portion of the adrenal gland, the medulla, is of ectodermal origin, like the nerve cells, and the principal product of that gland, adrenaline (sometimes called epinephrine), is, as we shall see, also a product of nerve cells. Adrenaline raises blood pressure, stimulates respiration, dilates the respiratory passages, and stimulates the general metabolic activity of cells. It promotes the activity of the enzyme that breaks down glycogen to glucose 6-phosphate and so increases concentrations of glucose in the bloodstream. In general, adrenaline assists the body in coping with emergencies. The effects of adrenaline are rapid and shortlived; the hormone is inactivated by an enzyme in the liver within three minutes of its release.

PANCREAS

The islet cells of the pancreas are the source of two hormones concerned with the metabolism of glucose: insulin and glucagon. Insulin is secreted in response to a rise in blood sugar or amino acid concentration (as after a meal). It lowers the blood sugar by stimulating the conversion of glucose to glycogen and by stimulating cellular uptake of glucose.

When there is an insulin deficiency, as in persons with diabetes, the concentration of blood sugar rises so high that glucose entering the kidney cannot be reabsorbed; the presence of glucose in the urine forms the basis of simple tests for diabetes. The loss of glucose is accompanied by loss of water by osmotic pressure. Dehydration, which can lead to collapse of circulation, is one of the causes of death in an untreated diabetic.

Mild diabetes can be controlled by diet. Severe diabetes is treated by insulin injections. (Because insulin is a protein, it cannot be given by mouth. It would simply be digested. Certain other oral drugs, which affect sugar metabolism in different ways, can be used in diabetes therapy.)

Glucagon, produced by different islet cells of the pancreas, increases blood sugar by stimulating the breakdown of glycogen to glucose in the liver and by stimulating the breakdown of fats and protein, which decreases glucose utilization.

32–11 *Hormonal regulation of blood glucose. When blood sugar concentrations are low (hypoglycemia), the pancreas releases glucagon, which stimulates the breakdown of glycogen and the release of glucose from the liver. When blood sugar concentrations are high (hyperglycemia), the pancreas releases insulin, which removes glucose from the bloodstream by stimulating its passage into cells and promoting its conversion to glycogen, the storage form. ACTH, produced by the pituitary, stimulates the adrenal cortex to produce glucocorticoids, which cause the liver to convert stored proteins to glucose. Under conditions of stress, the adrenal medulla releases adrenaline, which also raises blood sugar, and the production of ACTH is increased.*

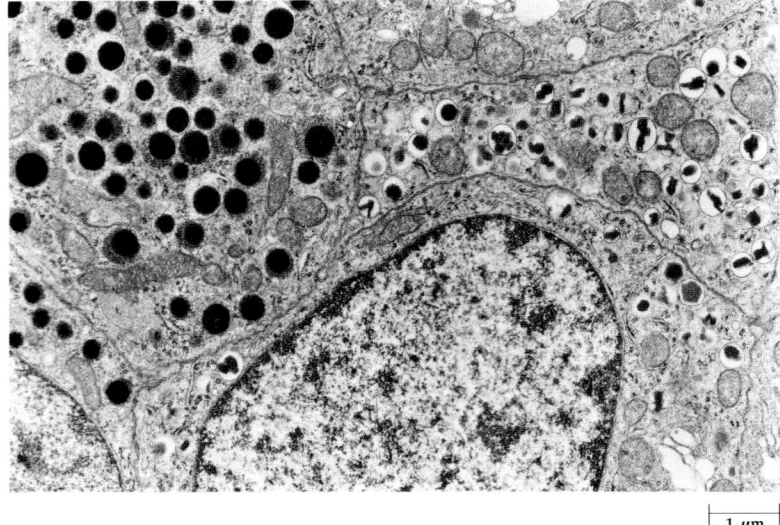

32–12 *Cells from an islet of Langerhans. The cell at the left with numerous black granules is an alpha cell. The cells at the right and bottom are beta cells. The beta cells are the source of insulin, and the alpha cells produce glucagon, which counteracts insulin in many of its effects.*

┌ 1 μm ┐

32–13 *The pea-sized parathyroid glands, the smallest of the known endocrine glands, are located behind or within the thyroid gland. They produce parathyroid hormone (parathormone), which increases concentrations of blood calcium. Calcitonin, which decreases blood calcium, was also thought to be produced by the parathyroids. It has recently been found to be a thyroid hormone and, in order to stress its new status, is sometimes called thyrocalcitonin.*

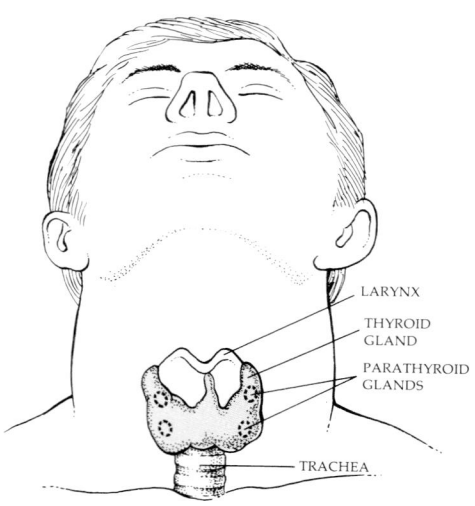

LARYNX

THYROID GLAND

PARATHYROID GLANDS

TRACHEA

Thus, as we have seen, at least five different hormones are involved in regulating blood sugar: growth hormone, cortisol, adrenaline, insulin, and glucagon. This tight control over blood glucose is particularly important for brain cells. Unlike other cells in the body, which can derive energy from the breakdown of amino acids, brain cells can utilize only glucose and so are immediately affected by low blood sugar. Fainting from hunger is an indication of the sensitivity of the brain cells to glucose deprivation, as in insulin shock, which like electric shock is sometimes used in the treatment of depression.

THE PARATHYROID GLANDS

The parathyroid glands were first discovered in 1850 by a British anatomist dissecting an Indian rhinoceros. They play an essential role in mineral metabolism, specifically in the regulation of calcium and phosphate in the blood. Calcium is normally present in the blood of man and other mammals in concentrations of about 6 milligrams per 100 milliliters of whole blood. A rise or fall of more than 2 or 3 milligrams per milliliter can lead to such severe disturbances in many functions that death may follow within a few hours. (These drastic effects occur because calcium concentrations profoundly affect the permeability of cell membranes.) Parathyroid hormone maintains blood calcium levels by increasing its absorption from foods in the intestine and by reducing its excretion by the kidneys. It also stimulates the release into the bloodstream of calcium from bone, which contains 99 percent of the body's total calcium. This release occurs as the result of bone-destroying cells known as osteoclasts. Other cells are simultaneously engaged in producing new bone. By slightly shifting the balance between these two activities, the parathyroids exercise a very tight control over blood calcium levels (another example of homeostasis).

Hyperparathyroidism, caused by tumors of the parathyroids, occasionally occurs in man. When there is too much parathyroid hormone, the bones lose too much calcium and become soft and fragile. Removal of the parathyroid gland results in violent muscular contractions and spasms, leading to death.

Table 32–1 *Some of the Principal Endocrine Glands of Vertebrates and the Hormones They Produce*

Gland	Hormone	Principal action	Mechanism controlling secretion	Chemical composition
Pituitary, anterior lobe	Thyrotropic hormone (TSH)	Stimulates thyroid	Thyroid hormone in blood; hypothalamic releasing hormone	Glycoprotein
	Follicle-stimulating hormone (FSH)	Stimulates ovarian follicle, spermatogenesis	Estrogen in blood; hypothalamic releasing hormone	Glycoprotein
	Luteinizing hormone (LH)	Stimulates interstitial cells in male, corpus luteum in female	Testosterone or progesterone in blood; hypothalamic releasing hormone	Glycoprotein
	Adrenocorticotropic hormone (ACTH)	Stimulates adrenal cortex	Adrenal cortical hormone in blood; hypothalamic releasing hormone	Protein
	Growth hormone	Stimulates bone and muscle growth, inhibits oxidation of glucose, promotes breakdown of fatty acids	Hypothalamic releasing hormone	Protein
	Prolactin	Stimulates milk production and secretion in "prepared" gland	Hypothalamic inhibiting hormone	Protein
Thyroid	Thyroxine, other thyroxinelike hormones	Stimulate and maintain metabolic activities	TSH	Iodinated amino acids
	Calcitonin (thyrocalcitonin)	Inhibits release of calcium from bone	Concentration of calcium in blood	Peptide (32 amino acids)
Parathyroid	Parathyroid hormone (parathormone)	Maintains normal calcium level and bone growth	Concentration of calcium in blood	Protein
Ovary, follicle	Estrogens	Develop and maintain sex characteristics in females	FSH	Steroids
Ovary, corpus luteum	Progesterone and estrogens	Promote growth of uterine tissue	LH	Steroids
Testis	Testosterone	Develops and maintains sex characteristics of males	LH	Steroid
Hypothalamus	Oxytocin	Stimulates uterine contractions, milk ejection	Nervous system	Peptide (9 amino acids)
	Antidiuretic hormone (vasopressin)	Controls water excretion	Osmotic concentration of blood; nervous systems	Peptide (9 amino acids)
Adrenal cortex	Cortisol, other cortisol-like hormones, aldosterone	Control carbohydrate, protein, and lipid metabolism, salt and water balance	ACTH	Steroids
Adrenal medulla	Adrenaline	Increases blood sugar, dilates some blood vessels, increases heartbeat	Nervous system	Catecholamine
Pancreas	Insulin	Lowers blood sugar, increases storage of glycogen	Concentration of glucose and amino acids in blood	Protein
	Glucagon	Stimulates breakdown of glycogen to glucose in the liver	Concentration of glucose in blood	Protein

PROSTAGLANDINS

Prostaglandins are the most recently discovered hormones. These hormones were given this name because they were first detected in seminal fluid and were thought to be produced by the prostate gland. Actually, most of the prostaglandins in semen are synthesized in the seminal vesicles. These prostaglandins, which are found in the female reproductive tract after sexual intercourse, induce rhythmic contractions in the muscular wall of the uterus. The semen of infertile males has been found to be poor in prostaglandins, and the uterus of infertile females is often unresponsive to these hormones. For these reasons, prostaglandins are believed to play a role in fertilization.

Since their initial discovery, 16 natural prostaglandins have been found. One major source is menstrual fluid. The prostaglandins in menstrual fluid also induce contractions in the uterus, which presumably aid in dispelling the endometrial lining. Because of their effects on the uterus, they are being studied for their possible use in birth control (by inducing menstruation in women in very early stages of pregnancy and so producing early abortions) and in facilitating labor. More recently they have been implicated in inflammatory and immune reactions (see Chapter 33) and, by extension, in diseases such as rheumatoid arthritis and asthma. Some have antagonistic effects; for example, one causes smooth muscle to relax, another causes the same muscle to contract.

Among the known hormones, prostaglandins are unusual for a number of reasons. (1) They are fatty acids, the only fatty acids known with hormonal activity. (2) They are the only mammalian hormones known whose target tissues are those of another individual (if their role in fertilization has been correctly interpreted). (3) They are among the most potent of all known biological materials, producing marked effects in very small doses. A concomitant of their extraordinary potency is the fact that they are produced in very small amounts and are rapidly broken down by enzyme systems in the body. If it were not for their presence in unusually large amounts in semen, they might never have been discovered. (4) They appear to be produced by cell membranes. According to a current hypothesis, they exert their effects by changing the composition of cell membranes, which would help to explain the extraordinary variety of their effects and their potency.

MECHANISM OF ACTION OF HORMONES

Although the general effects on the organism of the important hormones are now fairly well known, the way hormones act at the cellular level still remains to be clarified. The specificity of hormone action seems to depend on the presence of specific receptor sites on particular cells for particular hormones; the interaction of hormone and receptor is probably analogous to that of enzyme and substrate (page 75). Hormones appear to affect their target cells in a variety of ways. Some hormones, as we have noted, affect the entry of substances into cells. Other hormones, some investigators hypothesize, may act by affecting enzyme systems, either by stimulating synthesis of the enzymes or by promoting their activity. Some, either directly or indirectly, may act on the genes to stimulate messenger RNA synthesis.

Recently, cellular physiologists have discovered that a chemical known as cyclic AMP is involved in many hormone actions. The hormone itself, "the first messenger," apparently produces its effects by stimulating the action of adenyl

MATING BEHAVIOR IN RINGDOVES

The ringdove is a small relative of the domestic pigeon. The late Daniel S. Lehrman at Rutgers University carried out a series of studies with these birds on the relationships between behavior and endocrine activity.

As is the case with many species of birds, the reproductive cycle in ringdoves is initiated by seasonal changes, such as temperature and day length. Although it is always springtime in the temperature- and light-controlled laboratory, the solitary male or female dove will show no signs of mating or nesting behavior. If a male and female are placed together, however, they will carry out repeated cycles of reproductive behavior, each lasting six to seven weeks and each accompanied by profound physiological changes.

The first day of the cycle is spent in courtship, with the male bowing, cooing, and strutting. A nest site is selected—an empty bowl provided by the experimenters—and during the following week both birds cooperate in gathering material for the nest. During this period, they mate. Between 7 and 11 days after the beginning of courtship, the female lays the eggs, which the male and female sit on in turn. In about 14 days, the eggs hatch, and the parents begin to feed their young crop milk, a liquid secreted by the lining of the adult's crop (a pouch in the bird's gullet) under the influence of prolactin from the pituitary. The young birds are fed for about two weeks, after which the parents lose interest in them. The young birds by then are able to peck grain on the floor of the cage. When the young are 15 to 25 days old, the adult male begins once again to bow and coo and the cycle starts over.

During the period from the beginning of courtship to the egg-laying, the oviducts of the female bird increase in size some 500 percent, from 800 to 4,000 milligrams. This increase in size can also be seen in a female bird which, although in the same cage as a male bird, is separated from the male by a glass partition. If the male bird is castrated, however, and does not behave like a male, the female's oviducts do not enlarge.

Physiological changes and behavior changes develop in parallel. For instance, if a nest containing eggs is put into the cage at the same time as the male and female are placed in it, the nest is absolutely ignored. When the time comes for the birds to build their nest, they usually build it right on top of the already present eggs. If, however, the birds are allowed to court for 7 days, they will sit on a newly introduced nest and eggs quite promptly. Presumably, the courtship prepares them for nesting by stimulating hormone production. To test this hypothesis, isolated birds were injected with the hormone progesterone for one week. They were then put into cages together and immediately given nests and eggs; the great majority sat promptly on the nest.

During the 14 days when the doves are sitting on their eggs, their crops increase enormously in weight, from 900 to as much as 3,000 milligrams. This increase in crop weight, which contributes to the parents' production of "milk," is stimulated by production of the hormone prolactin. If male birds are removed from the cage during the incubation period, their crops do not increase in weight. If they are merely separated from the female by a glass partition, however, and are permitted to watch her incubate the eggs, the visual stimulus alone is sufficient to induce the secretion of prolactin and the consequent enlargement of the crop.

Thus, as Lehrman indicated, there is a constant interplay between the higher brain centers and the endocrine system. The result of this interplay is a sequence of profound physiological changes accompanied by the appropriate behavioral changes necessary to complete the reproductive cycle.

(a)

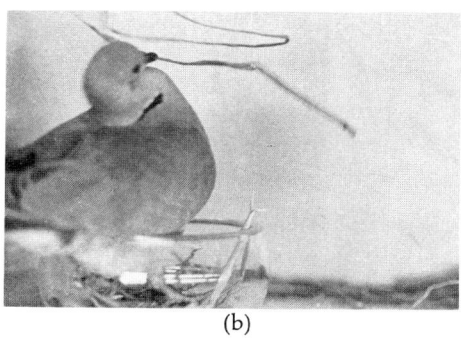

(b)

(a) Male ringdove bowing and cooing in courtship. (b) Female ringdove building nest.

32–14 *Cyclic AMP (adenosine monophosphate) acts as a "second messenger" within the cells of vertebrates. Following stimulation by various hormones—the "first messengers"—cyclic AMP is formed from ATP. "Cyclic" refers to the fact that the atoms of the phosphate group form a ring.*

cyclase, an enzyme bound to the target cell membrane, through interaction with a receptor for the hormone that is coupled to the enzyme. Adenyl cyclase, activated by this "first messenger," produces cyclic AMP from ATP. The cyclic AMP then acts as a "second messenger," activating another enzyme system—such as the enzyme that breaks down glycogen—and so initiating a sequence of metabolic activities within the cell. Hormones that trigger the production of cyclic AMP include adrenaline, glucagon, ACTH, ADH, TSH, LH, and prostaglandins. The differences in their effects are presumably due to the presence of different enzyme systems that respond to cyclic AMP within the different target cells.

At almost the same time that cyclic AMP was identified by these research workers in mammalian physiology, biologists studying a peculiar group of organisms known as the cellular slime molds isolated a chemical of great importance in this biological system. The cells of the cellular slime mold begin as individual amoebas and then come together to form a single organism (page 319). The chemical that calls them together was named acrasin, after Acrasia, the mythological siren who lured seamen in Homeric legend. Acrasin has now been identified as cyclic AMP, another example of evolution finding a new use for an ancient chemical.

THE NERVOUS SYSTEM

The nervous system, like the endocrine system, is a means of communication between the different parts of the organism and between the organism and the outside world. Functionally, it differs from the endocrine system chiefly in its capacity for rapid response. A nerve impulse can travel through the body in a matter of milliseconds. It may take seconds or even minutes, however, for the stimulation of hormone release, the transport of the needed hormone, and the initiation of its effects on the target tissue in the same animal. Moreover, the nervous system can transmit stop signals as well as initiate activities. Hormones, however, are a much more economical means of communicating with millions of cells at a time—such as the cells of a gonad—in situations where speed is not essential. Finally, whereas hormones act on many different kinds of cells, the nervous system acts primarily on muscles, thus stimulating movement of the animal or its parts, or on glands, stimulating the release of hormones or other secretions.

THE NEURON

The functional unit of the nervous system is the neuron, or nerve cell. A neuron consists of a cell body, which contains the nucleus, and thin cell extensions, which protrude from the cell body. The *dendrites* are one type of extension. Usually highly branched, the dendrites serve as reception areas for signals coming to the neuron from other cells; they usually carry impulses toward the cell body. Another type of extension, the *axon*, usually carries impulses away from the cell body. A single neuron may have many dendrites, but it usually has only one axon, although the axon may be branched.

Vertebrate axons are often enveloped in a myelin sheath formed by Schwann cells. The sheath provides electrical insulation and thus speeds up the transmission of impulses along the axon. In a large animal, an axon may be many feet in length.

-15 *A motor neuron, a nerve cell that trans-mits signals to muscles. The stimulus is received at any point on the naked nerve surface, usually by the dendrites, which conduct the nerve impulse to the cell body and to the axon. The signal travels along the axon, which is insulated by a myelin sheath composed of Schwann cells. The nodes of Ranvier are gaps in the axon sheath that occur at the junctions of adjacent Schwann cells. The nerve endings shown here are tiny un-sheathed filaments terminating in synap-tic knobs. These knobs form synapses with other cells. (See Figure 32–19.)*

32–16 *Cross section of a nerve showing myelin-ated and unmyelinated fibers. The myelinated fibers are those surrounded by the dark borders, myelin sheaths. These sheaths are formed from Schwann cells by the process illustrated in the diagram. As the Schwann cell*

grows, it extends itself around and around the axon and gradually excludes its cytoplasm from between the layers. In the electron micrograph, the cyto-plasms and nuclei of Schwann cells can be seen on the outer edges of some of the myelin sheaths.

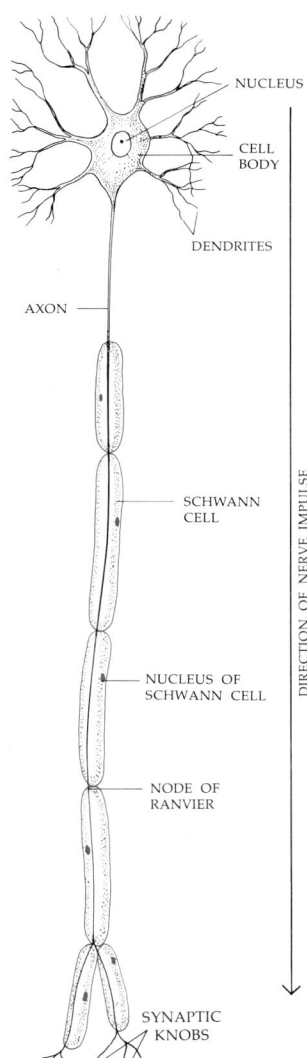

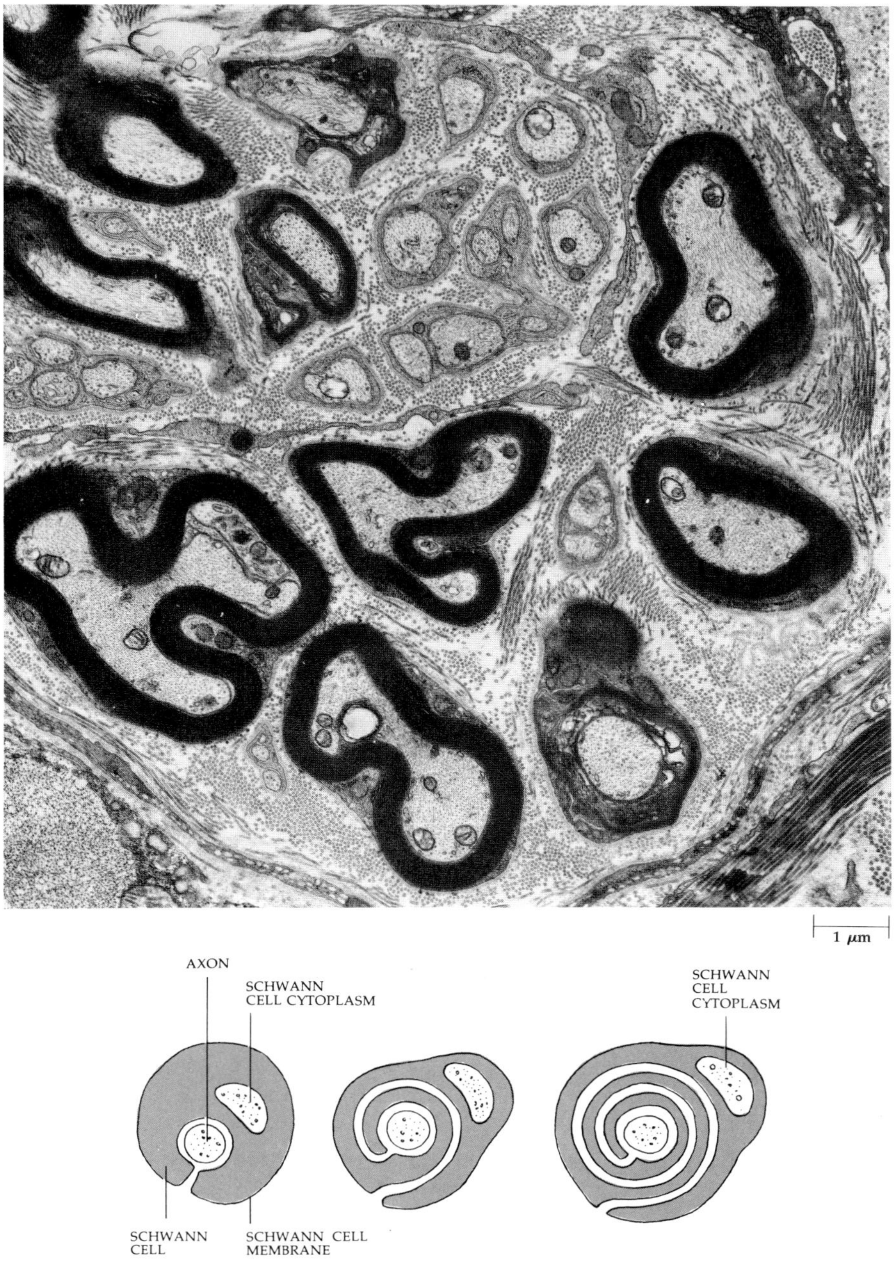

NUCLEUS

CELL BODY

DENDRITES

AXON

DIRECTION OF NERVE IMPULSE

SCHWANN CELL

NUCLEUS OF SCHWANN CELL

NODE OF RANVIER

SYNAPTIC KNOBS

1 μm

AXON

SCHWANN CELL CYTOPLASM

SCHWANN CELL CYTOPLASM

SCHWANN CELL

SCHWANN CELL MEMBRANE

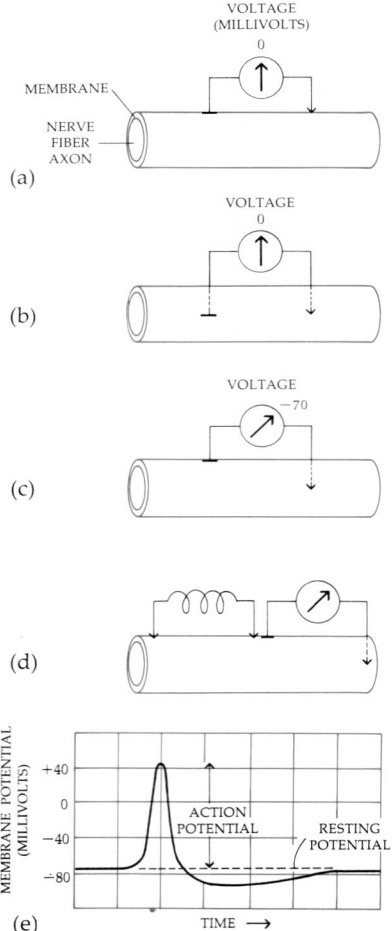

32–17 *The axon and the action potential. (a) If the recording electrodes are both placed on the outer surface of the axon, no electrical potential is recorded. (b) Similarly, no electrical potential is recorded if both of the electrodes are inserted within the axon. (c) If one electrode is on the outside and one on the inside, however, an electrical potential of −70 millivolts is registered. This is the resting potential. (d) An action potential is produced experimentally by stimulating the membrane electrically; this action potential travels down the axon. (e) An action potential occurring at any time on a particular axon or at any place will be the same; that is, it will have the same duration and the same voltage. The early experimental work on the action potential was carried out with the giant axon of the squid.*

THE NERVE IMPULSE

Axons are the communication lines of the nervous system. In order to understand how the nerve impulse travels along an axon, we need to look more closely at the axon itself, particularly at the characteristic properties of its cell membrane. First, the membrane is relatively impermeable to sodium ions (Na^+). Second, by the processes of active transport described in Chapter 7 (page 116), the membrane pumps out Na^+ ions and pumps in potassium ions (K^+). Third, the membrane of the nerve cell has the unusual capacity to reversibly reverse the distribution of these ions.

The Resting Potential

As a consequence of the sodium barrier and the active transport of sodium out of the axon, the concentration of K^+ ions is about 30 times higher inside the axon than in the fluid outside, and the concentration of Na^+ ions is about 10 times higher outside of the axon than inside. In its resting state, the membrane is relatively impermeable to Na^+ ions. Thus these ions are not able to move down the steep outside-inside concentration gradient. The resting membrane is slightly permeable to K^+ ions, however, and these therefore tend to leak out of the cell and down the steep K^+ inside-outside concentration gradient. The diffusion of positive (K^+) ions out of the cell creates a deficit of positive charge, or a net negative charge in the interior of the axon. The cell membrane thus becomes electrically polarized; that is, the inside of the axon is negative in relation to the outside.

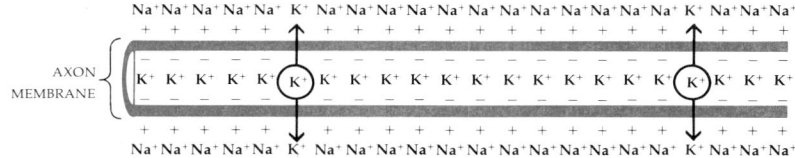

As a consequence of this ionic imbalance, an electrical potential exists across the cell membrane. This electrical potential is analogous to the electrical potential in the charged battery of a flashlight or the water potential of water behind a dam at the top of a hill. It can be looked upon as a supply of electrons (as in a copper wire) or ions (as in a battery) that can be released by flipping a switch. When they are released, they move in a current; an electric current, as you probably know, is simply the movement of electrons or ions. The force with which they move, analogous to the force of water running downhill, is measured in volts. The electrical potential across the membrane at rest is about 70 millivolts (a millivolt is 1/1,000 volt). This is called the *resting potential.*

The Action Potential

According to current hypotheses, a stimulus affects the axon by causing a small section of its membrane to become permeable to Na^+ ions. The Na^+ ions then rush in, moving down the concentration gradient, attracted, in addition, by the negative charge inside the membrane. Because of the movement of Na^+, the

electrical potential across the membrane drops. In fact, the inside of the membrane may even, for a fraction of a second, become positively charged in relation to the outside; this reduction in electrical potential across the membrane is called depolarization. This change in permeability lasts for only about 1.5 milliseconds.

Within this fraction of a second, as measured by recording electrodes (Figure 32–17), the electrical potential between the interior and exterior of the axon switches from a negative potential of about 70 millivolts to a positive potential of about 50 millivolts. This abrupt transitory change is called the *action potential*. The total action potential is therefore about 120 millivolts. Then the membrane regains its previous impermeability to Na$^+$, the Na$^+$ ions are pumped out, the K$^+$ ions are pumped in, and the original electrical potential, the resting potential, is once more restored:

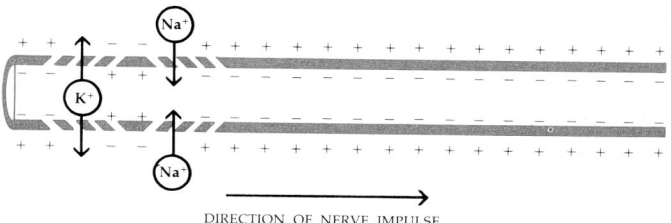

DIRECTION OF NERVE IMPULSE

Propagation of the Impulse

We are now ready to consider how the action potential impulse travels along the axon. First, as a consequence of a stimulus received by the dendrites, a small segment of the cell membrane at the base of the axon bordering the cell body becomes depolarized and gives rise to an action potential. At the height of this action potential, this segment of the axon becomes electrically positive, not only with respect to the fluid outside the cell but also with respect to the adjacent segment of the axon. As a result of this difference in electrical potential within the cell, there is a net flow of positive charges from the active zone down the axon. This flow depolarizes the neighboring, previously inactive zone sufficiently to set off another action potential in an adjacent segment, and so on:

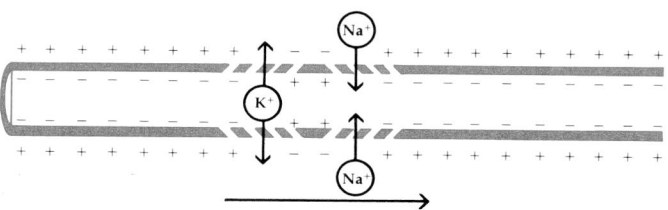

In this way, the impulse travels the length of the axon as a self-regenerating chain reaction in which each action potential sets off another action potential. The critical factor that controls the propagation of the nerve impulse is the sequential change in the membrane potential, permitting influx of Na$^+$ ions and so initiating a change in the next segment of the axon.

As a consequence of this chain reaction, an axon, which would be a very poor conductor of an ordinary electric current, is capable of transmitting a nerve impulse with absolutely undiminished intensity, often over a considerable distance:

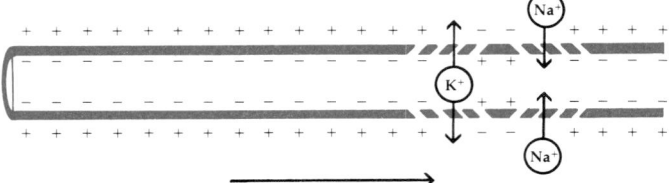

The transmission of the nerve impulse in a single neuron is an all-or-nothing reaction. The transient change in voltage produced by the action potential is determined by the concentration of ions on either side of the membrane, which does not vary to any appreciable extent. As a result, every time the nerve cell is stimulated, the size of the resulting action potential is the same.

Because the strength and duration of each impulse from a given neuron are the same, the signal carried along an axon can vary only in the frequency of transmission of the impulse (see Figure 32–18).

Following the passage of each impulse, a brief refractory period occurs when the membrane cannot produce another action potential. Thus there is an upper limit to the frequency of impulses to which an axon can give rise: this limit is about 1,000 impulses per second.

Many local anesthetics, such as novocaine, decrease the permeability of the membrane to sodium and so inhibit transmission of impulses from pain receptors to the brain.

32–18 *Nerve impulses can be monitored by electronic recording instruments. The action potentials from any one neuron are all the same; that is, each action potential has the same duration and voltage as any other. However, the frequency—the number of action potentials per second—and the pattern of action potentials vary, depending on the intensity of the stimulus. This recording from a single neuron of a cat shows the change in the frequency action potentials when the nerve is stimulated.*

Electron micrograph and diagram of a synaptic area in the spinal cord of a bat. In the micrograph, part of the cell body of one motor neuron occupies the left-hand corner of the picture. Two of the synaptic knobs of another motor neuron fill most of the rest of the micrograph. These contain numerous synaptic vesicles, which appear in the picture as small, grayish granules. At the synaptic junction, the membranes of the two cells are intact and clearly separated from one another by an intercellular space, the synaptic gap. Apparently, the vesicles, triggered by an electric impulse in the axon, release their chemical contents across the gap. The chemicals stimulate the adjacent neuron and thus relay the impulse from one nerve cell to another, as indicated by the arrows. The outlines of the axons above the synaptic knobs were not included in the extremely thin slice of tissue prepared for this electron micrograph.

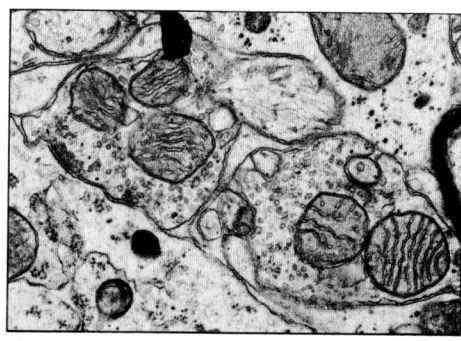

0.5 μm

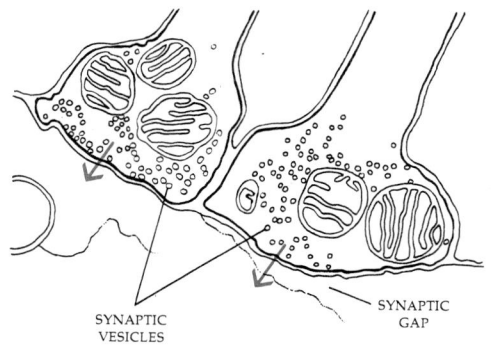

SYNAPTIC
VESICLES

SYNAPTIC
GAP

THE SYNAPSE

A nerve impulse travels from one cell to another across a junction known as a *synapse.* Transmission across most synapses is by chemical means. In synaptic knobs at the end of the axon are numerous small vesicles, visible in the electron microscope (Figure 32–19), which contain a transmitter substance. Arrival of the nerve impulse causes these vesicles to empty their contents into the synaptic gap. The transmitter substance crosses the gap and combines with receptor molecules on the membrane of the postsynaptic cell, changing the permeability of the membrane. Synapses are one-way junctions; thus synapses confer directionality on nerve impulses.

A number of different chemical transmitters have been tentatively identified, including serotonin, acetylcholine, adrenaline, and noradrenaline (an adrenaline-like compound). The adrenal medulla is actually a collection of adrenaline-secreting nerve cells, releasing adrenaline and noradrenaline into the bloodstream rather than into a synaptic junction.

The transmitters that carry signals across nerve junctions are rapidly destroyed by specific enzymes after their release. Such destruction is, of course, an essential feature in the tight control of the activities of the nervous system.

For example, acetylcholine, which is a transmitter substance in invertebrates as well as vertebrates, is rapidly destroyed by an enzyme called cholinesterase. Many insecticides, as well as certain military "nerve gases," act by inhibiting cholinesterase. The animal's nervous system then runs wild, causing spasms and eventually death. Recently it has been found that noradrenaline and related compounds are recaptured by the nerves that release them, and it is suggested that this may be a more important route of transmitter inactivation than enzymatic destruction in mammals.

Unlike the nerve impulse along the axon—which is an all-or-nothing proposition—signals transmitted by chemicals across a synaptic junction can modulate one another. A single neuron may receive molecular signals from hundreds of synapses.

Each transmitter-receptor reaction can cause a very small decrease in membrane voltage. If many such reactions occur at about the same time, they can initiate the nerve impulse. The chemicals released at some synapses, however, have a reverse effect, increasing the membrane voltage and thereby decreasing the likelihood of an impulse being generated. Synapses are, therefore, relay and integration points that are extremely important in the functioning of the nervous system.

TYPES OF NEURONS

There are three types of neurons in the nervous system: (1) *motor neurons* (sometimes called efferent neurons), which carry impulses outward to the effectors (muscles or glands), (2) *sensory neurons* (sometimes called afferent neurons), which transmit impulses received by sensory receptors to the central nervous system, and (3) *interneurons* (sometimes called association neurons), which receive the messages arriving from the sensory neurons and send out appropriate impulses to other interneurons or to motor neurons. There are about 12 billion nerve cells in the human body, the great majority of which are interneurons.

32–20 *The spinal column and skull form a bony outer covering for the central nervous system of the vertebrate. The central and peripheral nervous systems are connected by fibers carried in the nerves, which enter and emerge through the spaces between the vertebrae. This x-ray shows the lower (lumbar) portion of a human vertebral column.*

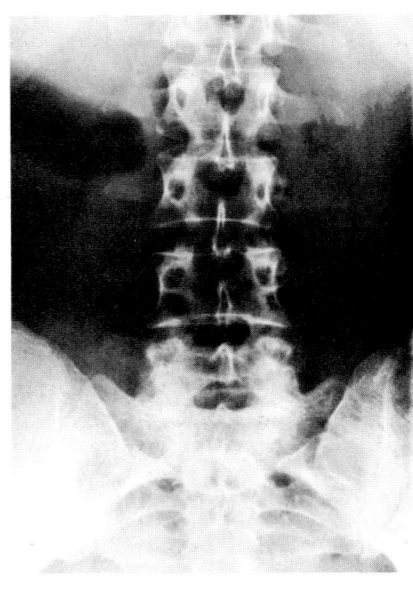

THE PERIPHERAL AND CENTRAL NERVOUS SYSTEMS

A nerve usually consists of a bundle of hundreds or even thousands of separate nerve fibers (axons), each capable of transmitting separate signals, like the wires in a telephone cable. The nerve has no cell bodies in it, except the cell bodies of the Schwann cells that form the myelin sheaths; the cell bodies of the neurons themselves are located in the brain and spinal cord or are aggregated in other parts of the body in clusters known as *ganglia* (singular, *ganglion*). The nervous system lying outside of the brain and spinal cord constitutes the *peripheral nervous system.*

Most nerves contain both motor and sensory fibers, which are sorted out when the nerve makes its connection with the spinal cord. Sensory fibers feed into the spinal column on the dorsal side and ascend toward the brain. Fibers descending from the brain make synapses with motor neurons, the fibers of which emerge from the spinal column on the ventral side.

The brain and the spinal cord constitute the *central nervous system.* The central nervous system of the vertebrates is encased in bony supporting and protecting structures—the vertebral column and the skull. The vertebrate spinal column, the "trademark" of this subphylum, is much the same from fish to man.

Figure 32–21 shows a segment of the human spinal cord, with pairs of spinal nerves entering and emerging from the cord through spaces between the vertebrae. Each of these pairs innervates the skeletal muscles of a different and distinct area of the body. In mammals, there are 31 such pairs.

Within the spinal cord, sensory and motor neurons synapse with interneurons clustered in *nuclei,* which is the term used to describe groups of cell bodies in the central nervous system, and fibers carrying information from various parts of the body (ascending fibers) and those relaying instructions from the brain (descending fibers) are clustered in groups called *fiber tracts.*

32–21 *A segment of the human spinal cord. Each spinal nerve divides into two fiber bundles, the sensory root and the motor root, at the vertebral column. The sensory bundle connects with the cord dorsally (toward the back); the cell bodies of the sensory neurons are in the dorsal root ganglia. The motor bundle connects ventrally (toward the front) with the spinal cord. The sympathetic ganglia are part of the autonomic nervous system (see page 596). The butterfly-shaped gray matter within the spinal cord is composed of groups of cell bodies, and the surrounding white matter consists of ascending and descending tracts of fibers.*

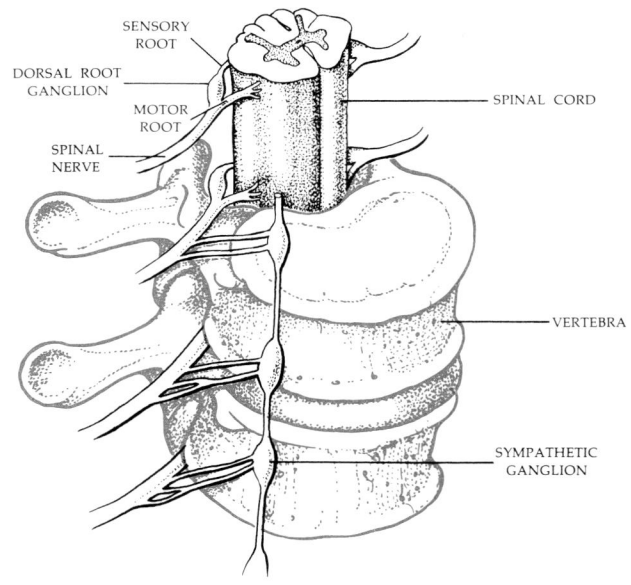

Scientific monitoring of autonomic functions has shown that yogis are able to control activities such as oxygen consumption and the rate of heartbeat. In confirmation of the fact that such functions may be brought under voluntary control, Neal Miller of the Rockefeller University has trained rats and, more recently, human volunteers to reduce blood pressure, for example. These techniques hold possibilities for medical treatment which are now being explored.

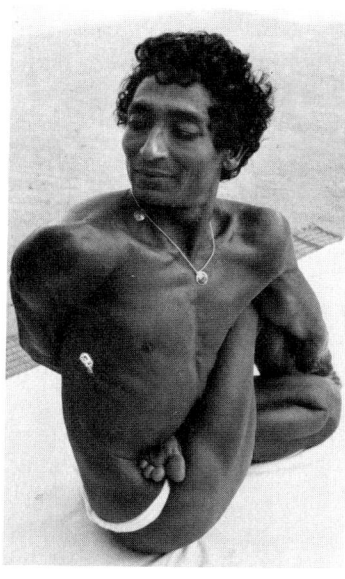

23 *Diagram of the reflex arc. Impulses from a receptor cell travel along the sensory fiber to the spinal cord. The cell body of the sensory neuron is located in a ganglion lying just outside the spinal cord. The sensory axon enters the cord and synapses with an interneuron in the gray matter of the cord. The interneuron relays the impulse to a motor neuron, which stimulates an effector muscle. The response is automatic and does not involve the brain, although it is informed about what is taking place.*

THE SOMATIC AND AUTONOMIC NERVOUS SYSTEMS

There are two important subdivisions of the peripheral nervous system: the *somatic* and the *autonomic*. The somatic nervous system includes both motor and sensory neurons. The autonomic nervous system is entirely a motor system, consisting of the nerves that control heart muscle, glands, and smooth muscle (the type of muscle found in the walls of blood vessels and in the digestive, respiratory, and reproductive tracts). The autonomic nervous system is thus generally categorized as an "involuntary" system, in contrast to the somatic system, which controls the muscles that we can move at will, that is, the skeletal muscles.

You will readily recognize that the distinction here between "voluntary" and "involuntary" is not clear-cut. Skeletal muscles often move involuntarily, as in a reflex action, and it is reported that some persons, particularly practitioners of yoga, can control their rate of heartbeat and the contractions of some smooth muscle. Anatomically, the motor neurons of the somatic system are distinct and entirely separate from those of the autonomic nervous system, although fibers of both types may be carried in the same nerve. Also, the cell bodies of the motor neurons of the somatic system are located within the central nervous system, with long nerve fibers running without interruption all the way to the skeletal muscle. The motor fibers of the autonomic nervous system also originate in cell bodies inside the central system; however, they do not travel all the way to their target organ, or effector, but instead form a synapse with a second neuron, which innervates the cell of the muscle or gland. This two-neuron pathway with postganglionic fibers, as they are called, constitutes a characteristic difference between the autonomic and somatic systems.

The Somatic Nervous System: The Reflex Arc

Figure 32–23 is a diagram of the neural connections in a reflex arc. In a simple reflex action—the withdrawal of your finger from a hot surface, for example—the stimulus is received by a receptor cell (a pain-sensitive cell, in this case)

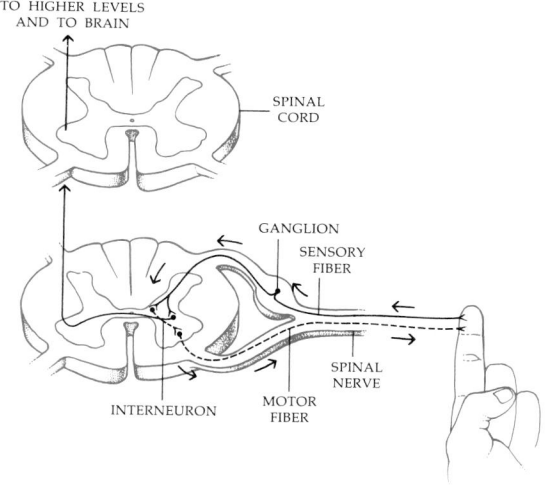

and is transmitted to the spinal cord. Here this sensory neuron synapses with an interneuron, which, in turn, relays the signal to a motor neuron. The motor neuron causes the appropriate muscle to contract, which moves your arm.

Thus you move your hand away "automatically" before your brain has had a chance to process the information. However, your brain has been informed and can call forth the next appropriate activity—such as muttering or turning off the stove.

The Autonomic Nervous System

The autonomic nervous system has two divisions, the sympathetic and the parasympathetic, which are anatomically and functionally distinct. The parasympathetic system consists of the vagus nerve, which originates in the medulla of the brain with branches to the heart and digestive organs and certain other cranial nerves and nerves originating in the sacral region (the lower region of the spinal cord). The sympathetic nervous system originates in the thoracic and lumbar areas of the spinal cord.

In the parasympathetic system, the points of synapse of the neurons are near or in the target organ, whereas (as shown in Figure 32–24) in the sympathetic system, the synapses are at some distance from the target organ, either in a regular chain of ganglia running parallel to the spinal cord (the sympathetic chain) or in other ganglia near the spinal cord. Most postsynaptic sympathetic nerve endings release adrenaline or noradrenaline, whereas postsynaptic parasympathetic endings release acetylcholine.

As you can see in Figure 32–24, most of the major internal organs of the body are innervated by neurons from both the sympathetic system and the parasympathetic system. The effects of the sympathetic system are often antagonistic to those of the parasympathetic system, and vice versa. As a consequence, whereas the somatic nervous system can only excite or not excite a particular effector, the autonomic nervous system can have both excitatory and inhibitory effects. Such antagonistic relationships are not invariable, however, and there is no completely applicable simple rule that predicts whether sympathetic or parasympathetic stimulation will excite or inhibit a particular organ.

Table 32–2 *Autonomic Nervous System*

Divisions	CNS connections	Point of synapse	Chemical transmitter	Effects
Sympathetic	Thoracic and lumbar regions of spinal cord	Ganglia near spinal cord	Adrenaline or noradrenaline	Triggers stress responses, inhibits activities
Parasympathetic	Brain stem and sacral (tail) area of spinal cord	Ganglia near target organ	Acetylcholine	Triggers activity, inhibits stress reactions

-24 The autonomic nervous system, consisting of the sympathetic and the parasympathetic systems. The presynaptic fibers of the parasympathetic system exit from the base of the brain and from the sacral region of the spinal cord and synapse at or near the target organs. The sympathetic system originates in the thoracic and lumbar regions; presynaptic fibers of the sympathetic system synapse in the sympathetic chain or in other ganglia, such as the celiac ganglia, part of the solar plexus. Most, but not all, internal organs are innervated by both systems, which usually function in opposition to each other. In general, the sympathetic system produces the effect of exciting organs involved in fight or flight reactions, and the parasympathetic system stimulates only those organs involved with more tranquil functions such as digestion. Each also inhibits most organs excited by the other.

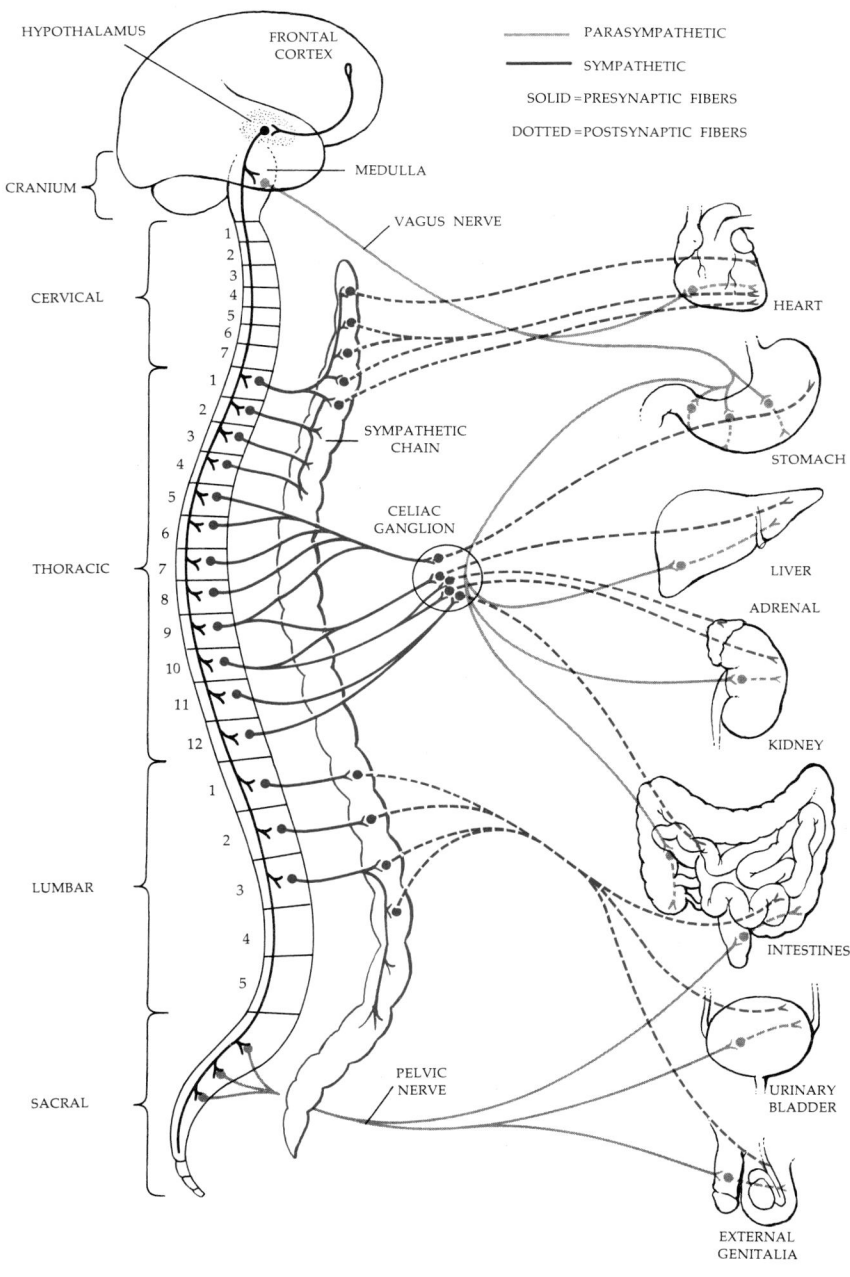

Table 32–3 *Functions of the Autonomic Nervous System*

Tissue	Location	Effect of parasympathetic stimulation	Effect of sympathetic stimulation
Smooth muscle	Iris	Contraction of circular fibers, constricting pupil	Contraction of radial fibers, dilating pupil
	Ciliary muscles (eye)	Contraction, making lens thinner	Relaxation, making lens thicker
	Stomach wall	Contraction and increased motility	Inhibition of contraction
	Pyloric sphincter	Inhibition of contraction	Contraction
	Intestinal wall	Increase in tone and motility	Decrease of motility
	Anal sphincter	Inhibition of contraction	Contraction
	Bladder wall	Contraction	Inhibition of contraction
	Bladder sphincter	Inhibition of contraction	Contraction
	Uterus (pregnant)	None known	Contraction
	Uterus (nonpregnant)	None known	Inhibition of contraction
	Bronchioles	Constriction	Dilatation
	Hair follicle	None known	Contraction
Gland	Eye (lacrimal)	Secretion	None known
	Mouth (salivary)	Secretion of copious thin saliva	Secretion of scanty thick saliva
	Stomach (gastric)	Secretion	Inhibition of secretion
	Liver	Inhibition of glycogen breakdown	Increase in glycogen breakdown
	Pancreas	Increase in enzyme secretion	No effect
	Adrenal medulla	None known	Secretion
	Skin (sweat)	None known	Secretion
Blood vessel	Cerebrum (arteries)	Dilatation	Constriction
	Heart (arteries)	Constriction	Dilatation
	External genitalia	Dilatation (erection)	Constriction (ejaculation)
	Skin	None known	Constriction
Heart muscle	Heart	Deceleration of rate	Acceleration of rate

32–25 *Muscle contraction may be isometric or isotonic. In isometric contractions, tension increases, but the muscles do not shorten. In isotonic contractions, the muscle shortens, but there is little increase in tension. If you analyze your own muscle movements as you carry out a few normal tasks, you will observe both isometric and isotonic contractions. When muscles are developed, as in a trained athlete, muscle protein and the diameter of the muscle cells increase, but no new cells are added.*

MUSCLES

As we noted earlier in this chapter, the chief targets of nerve stimulation are muscles. There are three principal types of muscles: skeletal, smooth, and cardiac. We shall begin with the discussion of skeletal muscle because it has been the subject of intensive biological research, and much of what is known about the other types is inferred from these studies.

Skeletal muscles, like all muscles, act by contracting. A skeletal muscle is typically attached to two or more bones, either directly or by means of tendons (tough strands of connective tissue). Some of these tendons, such as those that connect the fingers and their muscles, may be very long. When the muscle contracts, the bones move around a joint, which is held together by ligaments and which contains a lubricating fluid. As we noted in Chapter 29, most of the skeletal muscles of the body work in antagonistic pairs in this way.

The function of muscles is the conversion of chemical energy to kinetic, or mechanical, energy. Studies of how muscle carries out this function are of particular interest, because in these investigations, involving as they do both electron microscopy and biochemistry, scientists are tantalizingly close to being able to visualize the precise role of individual molecules.

A muscle, such as the biceps, consists of bundles of muscle fibers—often hundreds of thousands of fibers—held together by connective tissue. Each fiber is a single cell with many nuclei; the fibers arise from the fusion of a large number of embryonic muscle precursor cells. These fibers are very large cells—50 to 100 micrometers in diameter and, often, several centimeters long. They are surrounded by an outer cell membrane that has been given the special name of *sarcolemma*.

32–26 Muscle tissue is composed of individual muscle cells, the muscle fibers. These are cylindrical cells, often many centimeters long, with numerous nuclei. Each muscle fiber is made up of many cylindrical sub-units, the myofibrils. These are rods of contractile proteins that run from one end of the fiber to the other. The fibril is divided into segments, sarcomeres, by thin, dark partitions called Z lines, which run from myofibril to myofibril across the fiber, giving the muscle cell its striated appearance. Each sarcomere is composed of an array of thick and thin protein filaments arranged longitudinally. The H zone is the part of the A band from which thick filaments are absent. According to microscope studies, muscle contraction involves the sliding of the thin filaments between the thick ones. Chemical analysis shows that the thick filaments consist of bundles of a protein called myosin; each individual myosin molecule is composed of two protein chains wound in a helix. The end of each chain is folded into a globular structure. Each thin filament consists of a molecule of F-actin. F-actin is composed of two helical chains each made up of globular subunits, G-actin molecules. According to the widely held current hypothesis, the globular myosin heads protruding from the thick filaments serve as hooks or levers, attaching to the F-actin molecules (the thin filaments), and pulling them toward each other, shortening the sarcomere and contracting the myofibril. When the myofibrils contract, the fiber contracts, and when enough fibers contract, the entire muscle shortens, producing skeletal movements.

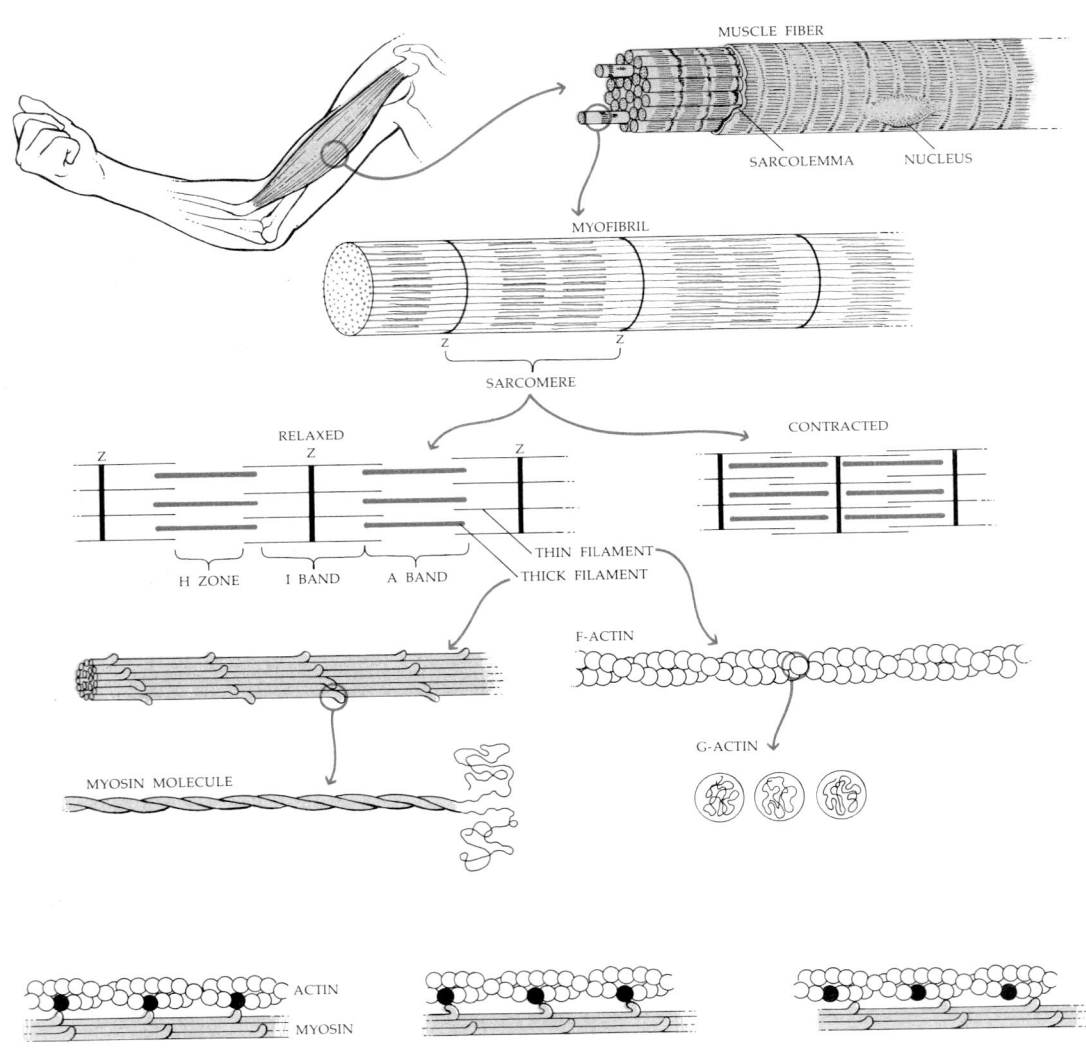

MUSCLE FIBER

SARCOLEMMA   NUCLEUS

MYOFIBRIL

Z   Z

SARCOMERE

RELAXED
Z   Z   Z

CONTRACTED

THIN FILAMENT
THICK FILAMENT

H ZONE   I BAND   A BAND

F-ACTIN

MYOSIN MOLECULE

G-ACTIN

ACTIN
MYOSIN

Embedded in the cytoplasm of each muscle cell (fiber) are some 1,000 to 2,000 smaller structural units; they appear ribbonlike in electron micrographs but they are actually cylindrical strands. These strands, which are called *myofibrils* (from *myo*, the prefix for "muscle") run parallel the length of the cell. The nuclei are crowded to the periphery of the cytoplasm by the myofibrils and can typically be found at the cell surface, just beneath the sarcolemma.

Each myofibril is, in turn, composed of units called <u>sarcomeres.</u> The repetition of these units gives the muscle a characteristic pattern. Because of this pattern, skeletal muscle is also known as striated (striped) muscle.

Figure 32–27a shows a sarcomere as seen in a longitudinal section of muscle. Each sarcomere is about 2 or 3 micrometers in length. The Z line is the dense black line seen in the electron micrograph; the I band is the relatively clear, broad stripe that the Z line bisects; and the A band is the large, dense stripe in the center of the sarcomere bisected by the central H zone. As the diagram shows, each sarcomere is composed of two types of filaments running parallel to one another. The thicker filaments in the central portion of the sarcomere are composed of a protein known as myosin; the thinner filaments are made of actin, also a protein. The Z line is where the actin filaments from adjacent sarcomeres interweave. In Figure 32–27b, you can see the actin and myosin filaments arranged in a regular pattern.

In addition to the nuclei and the myofibrils, each muscle cell, or fiber, contains a network of channels and vesicles, somewhat similar to the endoplasmic reticulum of other cells, which is called the sarcoplasmic reticulum (Figure 32–28). The sarcoplasmic reticulum runs parallel to the myofibrils. Running perpendicular to both is a system of transverse tubules, known as the T system, leading to the outside of the fiber. Muscle cells also contain numerous mitochondria. In very active muscles, such as the flight muscles of the hummingbird and of certain insects, the mitochondria are unusually large and are located exactly opposite the junction of A and I bands.

-27 (a) *Electron micrograph of sarcomeres, contracted. You should be able to identify the Z lines, the A bands, the H zones, the I bands, and the actin and myosin filaments. (b) Transverse section of myofibrils, showing the hexagonal array of filaments. Each thick (myosin) filament is surrounded by six thin (actin) filaments.*

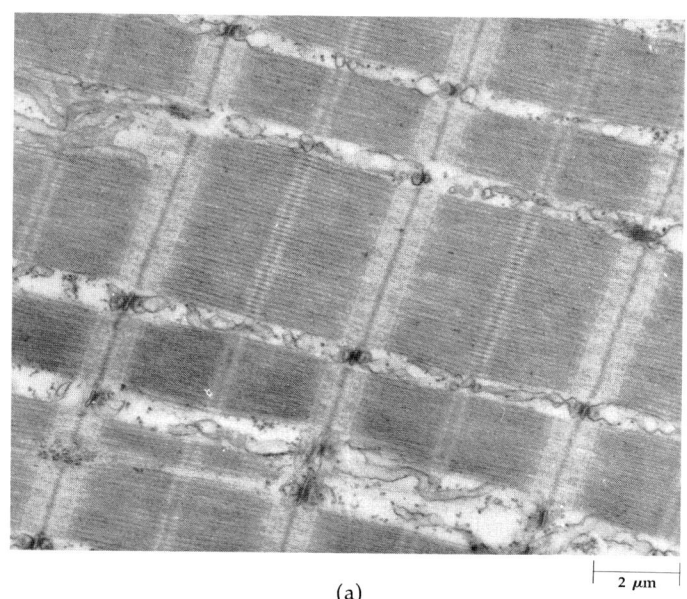

(a)   2 µm

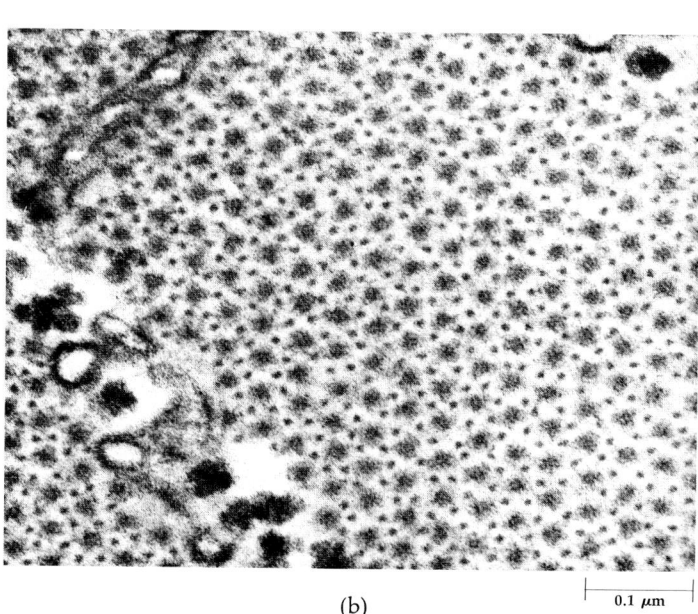

(b)   0.1 µm

32–28 *Six myofibrils, each of which contains numerous thick and thin filaments. The specialized cytoplasm of muscle cells is often referred to as sarcoplasm, and the unusual endoplasmic reticulum of muscle cells as sarcoplasmic reticulum. The sarcoplasmic reticulum is a longitudinal system that occupies the sarcoplasm between myofibrils. As shown here, the regular, longitudinal pattern of the sarcoplasmic reticulum is interrupted at the Z lines of the sarcomeres. Traversing the myofibrils, perpendicular to the sarcoplasmic reticulum, is the T system. This system of tubules can be visualized as a grid with perforations through which the myofibrils pass. Both the sarcoplasmic reticulum and the T system are involved in the movement of calcium into and out of the myofibrils.*

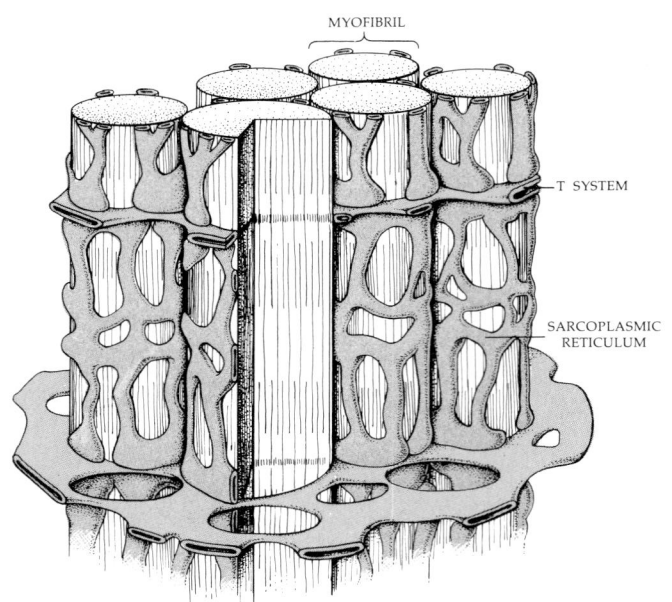

MYOFIBRIL

T SYSTEM

SARCOPLASMIC RETICULUM

Muscles function by contracting. What is the relationship between the structure of the sarcomere and the contraction of muscles? According to the most widely accepted hypothesis, the actin filaments of the sarcomere slide past the myosin filaments. Since the actin filaments are anchored into each Z line, this causes each sarcomere to shorten, and thus the myofibril as a whole contracts. There are crossbridges between the thick and thin filaments that may be formed rapidly and broken as one filament "walks" along the other.

The actin molecules in muscle are composed, it has been found, of many smaller globular subunits assembled in a long chain. Each actin filament in muscle appears to be composed of two such chains wound around each other. The myosin molecule has the longest protein chains known, with each one consisting of some 1,800 amino acid units. Each long protein chain has a globular head attached. The myosin molecule consists of two of these chains wound around each other, with the globular heads free. The thick filaments, in turn, are composed of bundles of myosin molecules. These globular heads apparently have two crucial functions: They are the binding sites which link the actin and myosin molecules, and they also act as enzymes to split ATP to ADP, thus providing the energy for muscle contraction.

According to a current model of muscle contraction, each crossbridge consists of the head of one myosin molecule linked to one globular actin unit. Breaking the link requires hydrolysis of one molecule of ATP,* also carried out by the myosin head. A new link forms between the myosin head and another actin subunit.

---

* Very little ATP is stored in muscle fibers. In vertebrates, the immediate energy supply is in the form of creatine phosphate: ATP + creatine ⇌ ADP + creatine phosphate. When the muscles are at rest, the reaction goes from left to right; when they are active, it goes from right to left.

(a) *The axon of each motor nerve divides into branches, each forming a neuro-muscular junction with a different muscle fiber. The motor nerve and the numerous muscle fibers that it innervates are known as a motor unit. Stimulation of a motor nerve stimulates all of the fibers in that motor unit. Within a given muscle, fibers of different motor units are inter-mingled. (b) A neuromuscular junction. An action potential conducted along the axon of a motor neuron releases acetyl-choline from synaptic vesicles. This transmitter agent combines with reactive sites on the motor end plate, the portion of the sarcolemma underlying the synap-tic knob. The combination alters the membrane permeability and triggers an action potential in the muscle fiber.*

MUSCLE CONTRACTION AND THE NERVE IMPULSE

The contraction of the sarcomeres is triggered by impulses arriving at the axon terminal of the motor neuron that makes synaptic contact with the muscle fiber. The nerve impulse causes release of the transmitter chemical from the synaptic knobs, as at a neuron-to-neuron synapse. The transmitter diffuses across the synaptic gap to the muscle fiber membrane. There it effects a temporary change in the permeability of the sarcolemma to ions; this change in permeability sets off an action potential that travels throughout the muscle fiber along the mem-branes of the transverse tubules. The reversal in the electrical potential of the membranes of the sarcoplasmic reticulum causes the sarcoplasmic reticulum to release stored calcium ions into the cytoplasm of the fiber. The released calcium ions then, in turn, facilitate the enzymatic splitting of ATP to ADP and the subsequent interaction between actin and myosin. The individual sarcomeres shorten and, with that shortening, the muscle fibers contract.

Individual muscle fibers follow the all-or-none law; that is, in response to a single stimulus, they either contract all the way or not at all. The contraction of the muscle as a whole depends on how frequently nerve impulses are received by the muscle fibers and how many fibers are stimulated. Also, the strength of a muscle contraction depends upon the number of individual fibers contracting. Many more fibers contract when you are lifting a heavy book, such as this one, than when you are lifting a piece of paper. Smoothly coordinated activities re-quire the asynchronous contractions of different groups of fibers in antagonistic muscles.

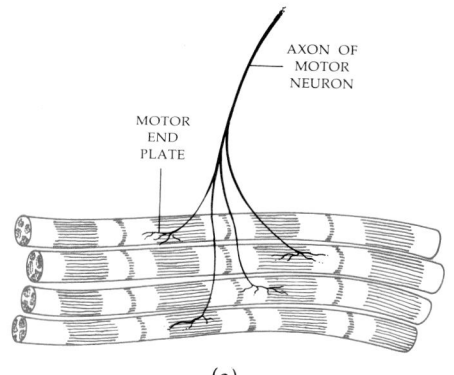

AXON OF MOTOR NEURON

MOTOR END PLATE

(a)

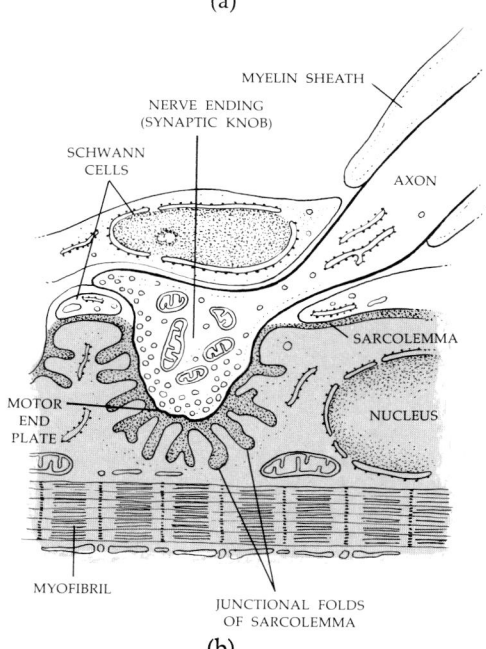

MYELIN SHEATH

NERVE ENDING (SYNAPTIC KNOB)

SCHWANN CELLS

AXON

SARCOLEMMA

MOTOR END PLATE

NUCLEUS

MYOFIBRIL

JUNCTIONAL FOLDS OF SARCOLEMMA

(b)

32–30 *Electron micrograph of a neuromuscular junction. The bottom half of the micro-graph is filled with the sarcomeres of the muscle. Above those are the junctional folds of the sarcolemma, and above the* *folds, the bulb of the axon in which nu-merous synaptic vesicles are visible. Above the synaptic bulb is a portion of a Schwann cell. At the very top of the micrograph are some collagen molecules.*

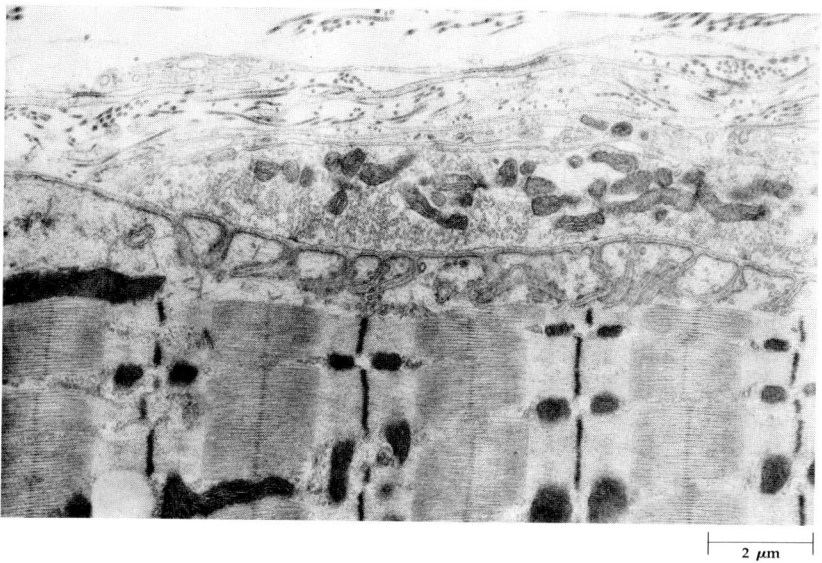

2 μm

32–31 *Low-power electron micrograph of heart muscle. The dark areas are special areas of attachment—splices—that bind the fibers (cells) together longitudinally. Characteristic of heart muscle, they are known as intercalated disks. A number of mitochondria are also visible.*

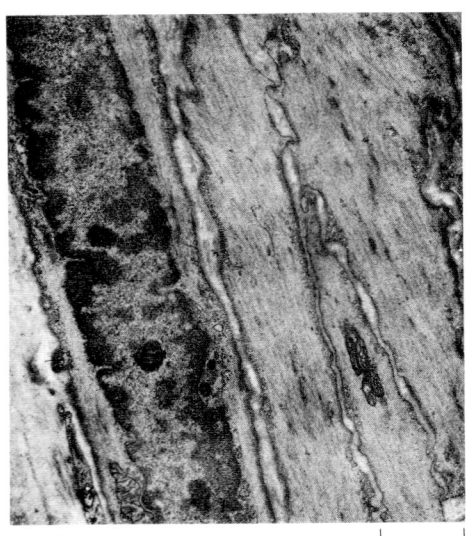

1 μm

32–32 *Electron micrograph of smooth muscle from the lining of a trachea (windpipe) of a bat. Smooth muscle is made up of long, spindle-shaped cells containing contractile proteins. Portions of a number of these can be seen in the picture. Part of the nucleus of one cell, distinguishable by its dark color and granular texture, can be seen in the left half of the micrograph. Very thin filaments run lengthwise through the cells and make up the bulk of the cytoplasm. The cell membranes are indicated by the fine lines which appear to be encrusted with tiny vesicles. These vesicles are actually invaginations, or pits, in the membranes. They are found in all smooth muscle cells, but what role they play in the cell is not known.*

CARDIAC MUSCLE

Cardiac muscle, the striated muscle of heart tissue, is very similar in appearance to striated skeletal muscle, except that its fibers are branched and they are connected with one another in such a way that impulses spread through all the fibers so that their contractions are coordinated.

Cardiac muscle differs from skeletal muscle in one very important way. Over a period of time, skeletal muscle may become fatigued. Cardiac muscle, however, has a long refractory period; after one contraction it is not able to contract again as quickly as striated muscle. This refractory period enables cardiac muscle to recover completely from the first contraction, helping to prevent fatigue. Cardiac muscle is also more generously supplied with blood vessels and with mitochondria than skeletal muscle.

SMOOTH MUSCLE

Smooth muscle is the type of muscle that lines the walls of many body organs, including the uterus, the vas deferens, the intestinal tract, and the blood vessels. Smooth muscle, unlike striated muscle, consists of separate cells with single nuclei. It lacks the characteristic striped pattern of skeletal and cardiac muscle. Smooth muscle cells contain numerous fibrils running through the cell longitudinally, but these fibrils are not arranged in any discernible pattern. However, these cells contain large amounts of actin and myosin (presumed but not proved to be in these fibrils). They also require ATP for contraction. Thus the contractile fibrils of smooth muscles also seem to be related to the contractile filaments found both in amoebas (page 100) and in dividing cells (page 160), and it seems possible that biologists are coming close to the formulation of a unified concept of contractile movements in a variety of different cells.

In vertebrates, skeletal muscle is innervated only by motor nerves of the somatic system; therefore, it is subject only to excitation. Cardiac and smooth muscles are innervated by the autonomic nervous system and so are under the dual control of sympathetic and parasympathetic neurons, one acting to excite, the other to inhibit contraction.

2.5 μm

INTEGRATION OF THE ORGANISM

Do you recall what it feels like to be in an absolute rage? The physical characteristics of rage result from a simultaneous discharge of sympathetic neurons. Certain of these act on the smooth muscles of the blood vessels of the skin, causing them to contract; this contraction increases the return of blood to the heart, raising the blood pressure and sending more blood to skeletal muscles. Some neurons stimulate cardiac muscle, making the heart beat faster, and others dilate the arteries. The respiratory rate increases, increasing the oxygen supply. Other muscles dilate the pupils. The muscles underlying the hair follicles in the skin contract; this is probably a legacy from our furry forebears, who looked larger and more ferocious with their hair standing on end. The rhythmic movement of the intestines stops, and sphincters, the muscles that close the intestines and the bladder, contract; these reactions inhibit digestive operations, but the constriction of the sphincters may also have the decidedly unuseful consequence of causing involuntary defecation or urination. Adrenaline pours out, which causes the release of large quantities of glucose from the liver into the bloodstream; this glucose provides an extra energy source for the muscles. The body is prepared for "fight or flight."

The parasympathetic system, on the other hand, is more concerned with the restorative activities of the body, that is, with rest and rumination. It cannot be totally mobilized except, perhaps, after a heavy meal or following orgasm. Parasympathetic stimulation slows down the heartbeat, increases the movements of the wall of smooth muscles surrounding the intestine, and stimulates secretions of the salivary gland. As mentioned earlier, most large organs, such as the heart, are under the control of both sympathetic and parasympathetic nerves, and these work in close cooperation for the ultimate homeostatic regulation of the body's functions.

Usually an animal, such as the human, is somewhere in between these two extremes. However, there is always, at every moment, a delicate interplay between hormonal system and nervous system acting in concert on the glands, muscles, and other target tissues of the body and so enabling the body to integrate and coordinate its responses to changes in the external and internal environments.

SUMMARY

The endocrine and nervous systems provide the precise and rapid communication necessary to coordinate the numerous internal functions that enable an animal to regulate its internal environment. Hormones act principally on metabolic systems of target cells. Nerves act principally on muscles, and also on endocrine organs. The two systems are interlocking.

Hormones are the agents of the endocrine system. These are specialized chemicals that are synthesized and secreted by specialized epithelial (gland) cells and are released into the bloodstream in response to specific stimuli. They, in turn, produce specific effects on certain organs and tissues.

The pituitary gland is a principal endocrine gland in vertebrates. The anterior pituitary is the source of at least six hormones: two gonadotropic hormones, follicle-stimulating hormone (FSH) and luteinizing hormone (LH), both of which act upon the ovaries and testes; thyroid-stimulating hormone (TSH); adreno-

corticotropic hormone (ACTH), which stimulates the adrenal cortex; prolactin, which stimulates diverse phenomena, including milk production; and growth hormone. All of these hormones are proteins or glycoproteins. Hormone release by the anterior pituitary is regulated by the hypothalamus, which produces a number of hormones that stimulate or inhibit the biosynthesis and release of specific pituitary hormones. Many of the pituitary and hypothalamic hormones are further regulated by feedback control of hormones whose production they stimulate.

The hypothalamus is also the source of two hormones, oxytocin and antidiuretic hormone (ADH), which are transferred to and stored in the posterior lobe of the pituitary gland. Both are peptides (nine amino acids each). Oxytocin stimulates the wall of the uterus during labor. ADH regulates the permeability to water of ions through the collecting ducts of the kidney.

The thyroid gland is the source of thyroxine, an iodinated amino acid derivative. Thyroxine accelerates cellular respiration and is essential for normal growth, particularly of the brain in infants, and for a normal rate of metabolic activity in adulthood. Thyroxine production is regulated by a feedback system involving the hypothalamus and the pituitary. Thyroxine deficiency may be caused by iodine shortage, in which case it often results in an enlargement of the thyroid gland known as a goiter.

Calcium concentration is regulated by two hormones: parathormone, a hormone of the parathyroid that increases blood calcium, and calcitonin, a thyroid hormone that decreases blood calcium.

The adrenal cortex is the site of biosynthesis of three types of steroid hormones: (1) glucocorticoids, such as cortisol and cortisol-like hormones, which are primarily involved with the synthesis of glucose from amino acids and fat, (2) the mineralocorticoids, such as aldosterone, which are concerned with the regulation of ions and of water balance, and (3) sex hormones, particularly androgens.

The adrenal medulla produces adrenaline, which raises blood pressure, stimulates respiration, dilates the respiratory passages, and stimulates the breakdown of glycogen to glucose. By these activities, it assists the body in emergency situations.

The pancreas is the source of two hormones concerned with the metabolism of sugar: insulin, which promotes the cellular uptake of glucose and the biosynthesis of glycogen and so decreases the glucose concentration in the blood, and glucagon, which stimulates the breakdown of glycogen and so increases the concentration of glucose in the blood. Both are proteins.

The prostaglandins, a group of newly discovered hormones, are fatty acids. Their most outstanding effects are on smooth muscle. They are apparently produced by cell membranes and perhaps exert their effects by affecting its composition.

Cyclic AMP, a "second messenger," is an intermediary in the activities of a number of hormones, including adrenaline, glucagon, ACTH, LH, and TSH. These hormones bind with a receptor on the cell membrane of target cells. This activates the enzyme which then produces cyclic AMP from ATP. The cyclic AMP then activates another enzyme and sets in motion a series of enzymatic reactions, which produce the effects associated with the original hormone.

In general, the nervous system elicits more rapid response than the endocrine

system. The nervous system also integrates and coordinates a variety of body functions.

The unit of the nervous system is the neuron, or nerve cell, which consists of (1) the dendrites, which receive impulses, (2) the cell body, and (3) the axon, which relays the impulses to other cells. Nerve impulses are transmitted between cells across a junction called the synapse. The signal crosses the synaptic gap by triggering the release of a chemical transmitter, which stimulates adjacent cells. Nerves are bundles of motor and sensory fibers.

The central nervous system consists of the brain and the spinal cord, which are encased, in vertebrates, in the skull and vertebral column. From here, spinal nerves emerge in pairs, each pair innervating the muscles of a different area of the body. Motor neurons convey impulses to the effectors—the muscles and glands. Sensory neurons transmit to the central nervous system impulses resulting from sensory stimuli. The nervous system outside the central nervous system constitutes the peripheral nervous system.

In vertebrates, the nervous system has two major subdivisions: the somatic nervous system, which consists of motor and sensory neurons, and the autonomic nervous system, which controls the smooth muscles and glands involved in the digestive, circulatory, respiratory, and reproductive functions. The autonomic system has its own network of effector neurons, separate from that of the somatic system, which are linked indirectly from the central nervous system to the target organs by secondary, or postganglionic, neurons. The system has two subdivisions: (1) the sympathetic system, which is largely responsible for the excitatory reactions of the body, and (2) the parasympathetic system, which controls the restorative activities, such as digestion and rest.

Muscles are the principal effectors of the nervous system. There are three major types in the vertebrate body: skeletal muscle, cardiac muscle, and smooth muscle. The skeletal muscles move the bones of the skeleton by contracting and tugging on the bones to which they are attached. Each muscle is made up of muscle fibers—long, multinucleated cells bound together by connective tissues. Every fiber is bounded by an outer membrane, the sarcolemma, and traversed by a sarcoplasmic reticulum and by transverse tubules.

Each muscle cell contains 1,000 to 2,000 smaller strands, the myofibrils, running parallel to the length of the cell. Myofibrils are made up of units called sarcomeres. Sarcomeres are made up of alternating thin and thick filaments, actin and myosin, respectively. Actin and myosin are protein molecules. Contraction comes about as a result of the actin and myosin filaments sliding between one another. The globular heads of the myosin molecules serve as enzymes for the splitting of ATP, which provides the power for the muscle contraction.

Muscle fibers contract in response to stimulation by a motor neuron, whose axon terminal releases a transmitter and causes depolarization of the fiber membrane. The resulting action potential travels along and into the muscle fiber by way of the membrane of the transverse tubules. The reversal in electric potential causes the sarcoplasmic reticulum to release stored calcium ions into the fiber cytoplasm. The sudden increase in the intracellular Ca^{2+} concentration permits the interaction between actin and myosin that brings about contraction of the sarcomeres.

Cardiac muscle resembles skeletal muscle in its striated pattern. However, its

cells are connected in such a way that the action potential spreads from fiber to fiber. As a consequence, excited fibers excite adjacent fibers so that contraction of the various parts of the heart is coordinated. Because cardiac muscle has a refractory period following each contraction, it is not as subject to fatigue as skeletal muscle is.

Smooth muscle is composed of long, uninucleate cells, which contain actin and myosin and fibrils. These elements are not arranged in a discernible pattern, and the mechanism of contraction of these muscles is unknown at this time.

Unlike skeletal muscle, which responds only to the nerves of the somatic system and thus, in vertebrates, receives only excitatory stimuli, cardiac and smooth muscles respond to autonomic nerves which can either inhibit or stimulate contraction.

Integration and control of the body's functions depend upon an interplay between the endocrine and nervous systems and their effects on glands, muscles, and other target tissues.

QUESTIONS

1. Define the following terms: endocrine gland, neuron, nerve, resting potential, action potential, muscle fiber, myofibril.
2. Draw a diagram of the sarcomere relaxed and contracted.
3. What are the classical experimental steps by which an unknown organ or tissue can be diagnosed as endocrine in function? How might this procedure prove deceptive?
4. Both nervous and endocrine systems transmit information and thereby integrate and coordinate the activities of the body. Compare their operation with regard to speed of transmission, specificity of target, and the nature of the messages handled.
5. Assume that three presynaptic neurons, A, B, and C, make adjacent synapses on the same postsynaptic neuron, D. No postsynaptic impulse is initiated as a result of single presynaptic impulses in A, B, or C, nor are they initiated if impulses arrive simultaneously in all three, in A and B, or in A and C. Only if impulses arrive together at the synapses of B and C will D fire an impulse. Explain these results in terms of excitatory and inhibitory transmitter substances and their effects on the postsynaptic membrane voltage.

Chapter 33

Circulatory Systems

As we saw in Section 1, every cell in the body of a complex organism builds up its own membranes and organelles, makes its own ATP, and assembles its own enzymes and other proteins. In addition to these processes common to all cells, it is likely to carry out a number of specialized functions. To engage in these activities, cells usually need oxygen and nutrients and must dispose of carbon dioxide and other wastes. In single-celled organisms and very small multicellular ones, the needs of each cell are supplied directly by the medium the cell lives in. In larger, more complex animals, these needs are supplied by a transport system. In man and other vertebrates, the transport system is a closed circulatory system consisting of the blood, which is the actual transport medium; the vessels, or tubes, the blood flows through; and the heart, which provides the force that moves the blood through the vessels.

The blood is pumped from the heart into the aorta, the artery leading from the heart, then into other arteries and on to the capillaries. The latter are very

33–1 *A primary function of the circulatory system is to deliver oxygen to all the cells of the body. In vertebrates, oxygen is transported in red blood cells, shown here in a scanning electron micrograph. Because of this characteristic biconcave ("doughnut") shape, red blood cells are flexible and so are able to twist and turn in their passage through the capillaries. They are only about 7 or 8 micrometers in diameter.*

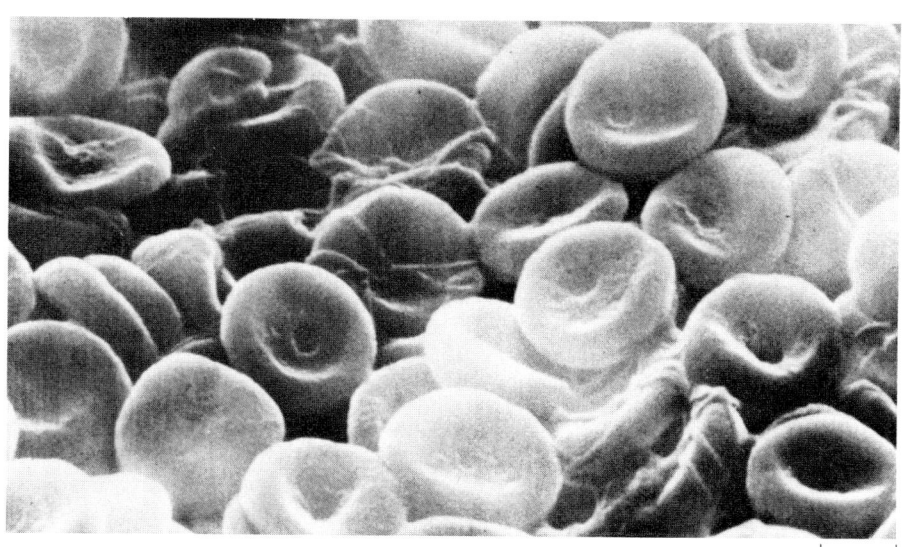

3 μm

small blood vessels; through the thin walls of the capillaries, nutrients, oxygen, carbon dioxide, and other molecules are exchanged between the blood and the fluids surrounding the body cells (the interstitial fluids). The blood then returns to the heart through the veins. Thus the heart, arteries, and veins are, in essence, the means for getting the blood to and from the capillaries, where the actual function of the circulatory system is carried out.

THE HEART

Evolution of the Heart

In its simplest form—as in the earthworm—the heart is a muscular contractile part of a circulatory vessel. In the course of vertebrate evolution, the heart has undergone some structural adaptations, as shown in Figure 33–2.

Fish have a single heart divided into the _atrium,_ which is the receiving area for the blood, and the _ventricle,_ which is the pumping area from which the blood is expelled into the vessels. The ventricle of the fish heart pumps blood directly into the capillaries of the gills, where it picks up oxygen and releases carbon dioxide. From the gills, oxygenated blood is carried to the tissues. By this time, however, most of the propulsive force of the heartbeat has been dissipated by the resistance of the capillaries in the gill, so that the blood flow to and through the rest of the tissues (the systemic circulation) is relatively sluggish.

In amphibians, there are two atria; one receives oxygenated blood from the lungs and the other receives deoxygenated blood from the systemic circulation. Both atria empty into the single ventricle, which pumps the mixed blood simultaneously through the lungs and the systemic circulation. By this arrangement, the blood enters the systemic circulation under high pressure.

33–2 _Vertebrate circulatory systems. Oxygenated blood is indicated in color. In the fish, the heart has only one atrium (A) and one ventricle (V). Blood oxygenated in the gill capillaries goes straight to the systemic capillaries without first returning to the heart. In amphibians, the single primitive atrium has been divided into two separate chambers. Oxygenated blood from the lungs enters the left atrium, where it is mixed somewhat with oxygen-poor blood. The blood is then pumped by the ventricle through the body tissues. In birds and mammals, the ventricle is divided into two separate chambers, so that there are, in effect, two hearts—one for pumping oxygen-poor blood through the lungs and one for pumping oxygen-rich blood through the body tissues._

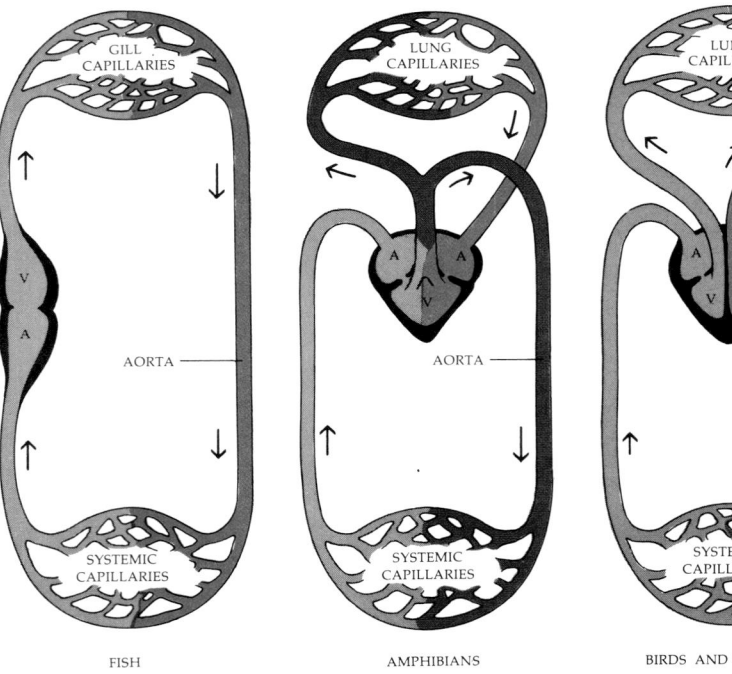

33–3 *The human heart. Blood returning from the systemic circulation through the superior and inferior venae cavae enters the right atrium and is pumped to the right ventricle, which propels it throughout the pulmonary arteries to the lungs, where it is oxygenated. Blood from the lungs enters the left atrium through the pulmonary veins and is pumped to the left ventricle and then through the aorta to the body tissues.*

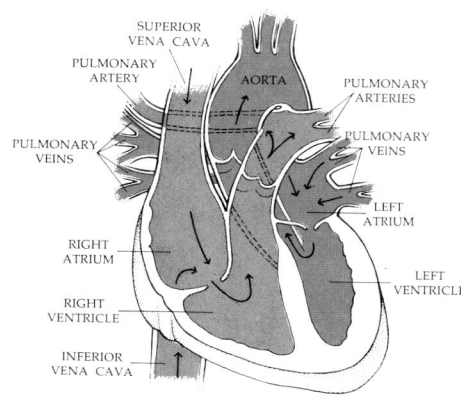

In the birds and mammals, the heart is completely divided longitudinally, and thus is functionally separated into two organs. In mammals, the right heart (your right heart is the atrium and ventricle on the right side of your body) receives blood from the tissues and pumps it into the lungs, where it becomes oxygenated. From the lungs it returns to the left heart, from which it is pumped into the body tissues. In this way, fully oxygenated blood is pumped into the systemic circulation under full pressure. This high-pressure double circulation system is apparently correlated with the high metabolic rate of both birds and mammals, with their constant body temperature, and with their generally high level of physical and mental activity.

The Heartbeat

Figure 33–3 shows a diagram of the human heart. Blood returning from the body tissues enters the right atrium through two large veins, the superior and inferior venae cavae. Blood returning from the lungs enters the left atrium through the pulmonary veins. The atria, which are thin-walled compared with the ventricles, expand as they receive the blood. Both atria then contract simultaneously, assisting the flow of blood through the open valves into the ventricles. Then the ventricles contract simultaneously; the valves are closed by the pressure of the blood in the ventricles. The right ventricle propels the blood into the lungs through the pulmonary arteries, and the left ventricle propels it into the aorta, from which it travels to the other body tissues. Valves in the openings of the ventricles into the pulmonary artery and the aorta close after the ventricles contract, thus preventing backflow of blood.

If you listen to a heartbeat, you hear "lubb-dup, lubb-dup." The deeper, first sound ("lubb") is the closing of the valves between the atria and the ventricles; the second sound ("dup") is the closing of the valves leading from the ventricles to the arteries. If any one of the four valves is damaged, as from rheumatic fever, blood may leak back through one of the valves, producing the noise characterized as a "heart murmur" (a "ph-f-f-t" sound).

33–4 *When blood enters the left atrium, the mitral valve opens, permitting blood to flow into the left ventricle. Ventricular contraction pushes the blood hard against the valve, closing it and thereby ensuring the movement of blood through the aortic semilunar valve into the aorta. A similar valve in the right heart, called the tricuspid, prevents the blood from re-entering the atrium.*

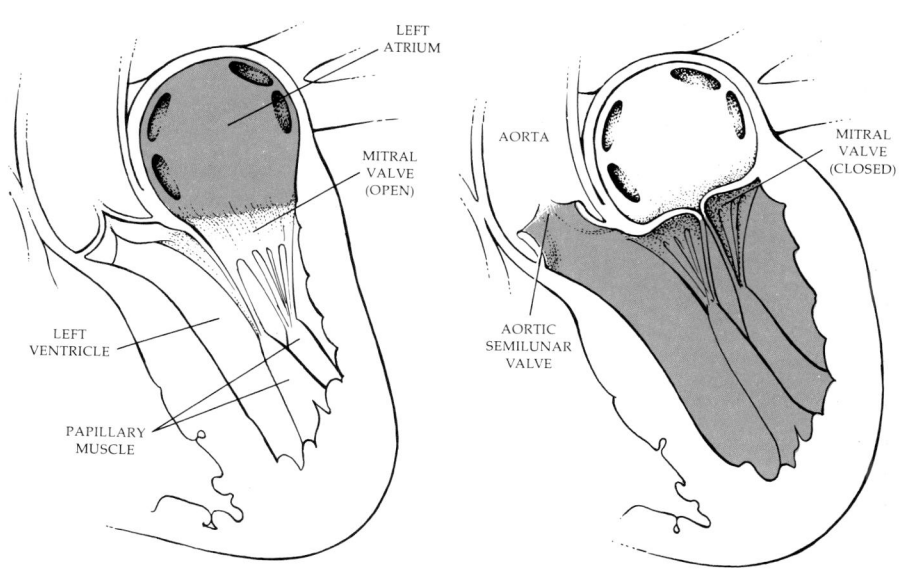

33–5 *The beat of the mammalian heart is controlled by a region of specialized muscle tissue in the right atrium, the sinoatrial node, which functions as the heart's pacemaker. Some of the nerves regulating the heart have their endings in this region. Excitation spreads from the pacemaker through the atrial muscles, causing both atria to contract simultaneously. When the wave of excitation reaches the atrioventricular node, its conducting fibers pass the stimulation to the bundle of His, which triggers simultaneous contraction of the ventricles. Because the fibers of the atrioventricular node conduct relatively slowly, the ventricles do not contract until the atrial beat has been completed.*

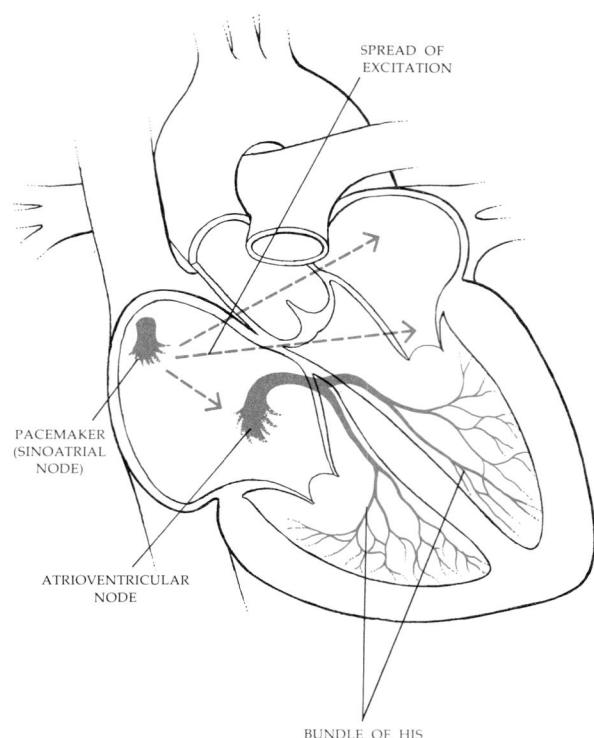

SPREAD OF
EXCITATION

PACEMAKER
(SINOATRIAL
NODE)

ATRIOVENTRICULAR
NODE

BUNDLE OF HIS

In a healthy adult at rest, the heart beats about 70 times a minute. Under strenuous exercise, the rate more than doubles. The heartbeat also quickens, as you know, at times of emotional stress.

Most muscle contracts only when stimulated by a motor nerve, but the contractions of cardiac muscle originate in the muscle itself. A vertebrate heart will continue to contract even after it is removed from the body if it is kept in a nutrient solution. In vertebrate embryos, as we saw, the heart begins to beat very early in development, before the appearance of any nerve supply. In fact, embryonic heart cells isolated in the test tube will beat.

The beat of the cardiac muscle is initiated by a special area of the heart, the *sinoatrial node*, which is located in the right atrium and functions as the pacemaker. The pacemaker is composed of nodal tissue. Nodal tissue is unique in that it can depolarize spontaneously, initiating its own action potential, contract like a muscle, and also transmit impulses like a nerve.

From the sinoatrial node, a wave of excitation spreads over the right atrium to the left atrium. About 100 milliseconds after the pacemaker fires, impulses from both atria stimulate a second area of nodal tissue, the *atrioventricular node*. The atrioventricular node is the only electrical bridge between the atria and the ventricles. It consists of slow-conducting fibers, which impose a delay between the atrial and ventricular contractions, so that the atrial beat is completed before the beat of the ventricles begins. From the atrioventricular node, stimulation passes to the bundle of His (named after its discoverer), which sets up a simultaneous contraction of the ventricles. Thus, though the heart is functionally two organs—the left and right heart—the two beat in unison.

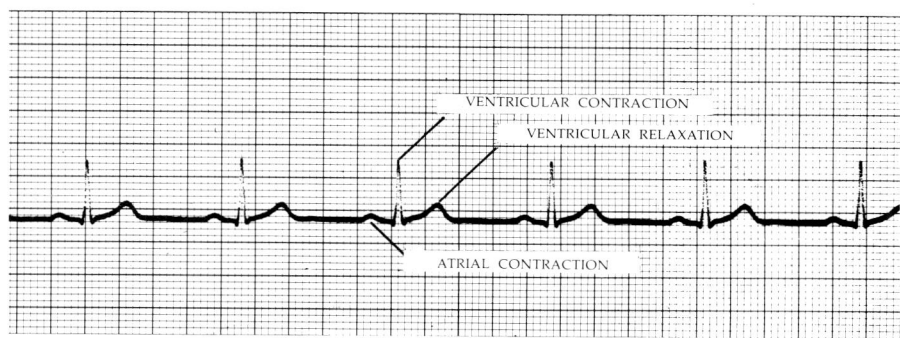

3–6 An electrocardiogram showing six normal heartbeats. Each beat is denoted by a series of waves that record the electrical activity of the heart during contraction. Analysis of the rate and character of these waves can reveal abnormalities in heart function.

Cardiac Output

The total volume of blood pumped by the heart depends both on the heart rate (beats per minute) and on the amount of blood ejected at each beat, the stroke volume. This total volume is called the cardiac output, and it is defined as the amount (usually expressed in liters) of blood pumped per minute: cardiac output (liters per minute) = heart rate (beats per minute) × stroke volume (liters per beat). Thus if the heart beats 72 times per minute and ejects 70 milliliters into the aorta with each beat, the cardiac output is about 5 liters per minute (72 times × 0.07).

Control of Cardiac Output

Cardiac output is affected by both the endocrine and nervous systems. Adrenaline both increases the rate of heartbeat and increases the stroke volume of the heart, by increasing the force of contraction of the ventricles. Both sympathetic and parasympathetic nerves act on the sinoatrial (pacemaker) and atrioventricular nodes. Sympathetic stimulation increases the discharge rate of the pacemaker. Parasympathetic stimulation (by way of the vagus nerve) decreases the discharge rate of the pacemaker and increases the refractory period for conduction along the bundle of His.

–7 Autonomic regulation of the rate of heartbeat. Sympathetic fibers stimulate the sinoatrial node, whereas parasympathetic fibers, which are contained in the vagus nerve, inhibit it.

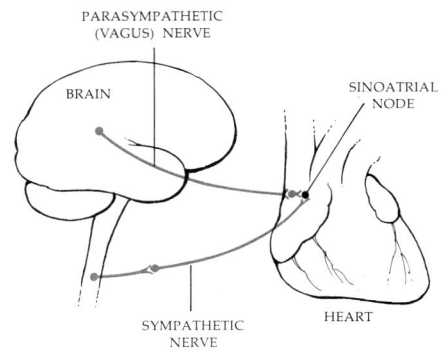

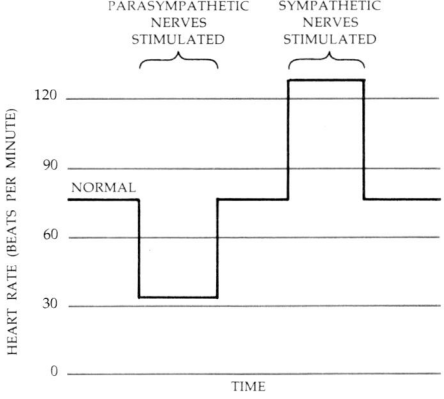

33–8 *Structure of blood vessels. Arteries have thick, tough, elastic walls that can withstand the pressure of the blood as it leaves the heart. Capillaries have walls only one cell thick. Exchange of gases, nutrients, and wastes between the blood and the interstitial fluid takes place through these thin capillary walls. Veins have larger lumens and thinner, more readily extensible walls that minimize resistance to the flow of blood on its return to the heart. They also contain valves that prevent backflow. (See Figure 33–15.)*

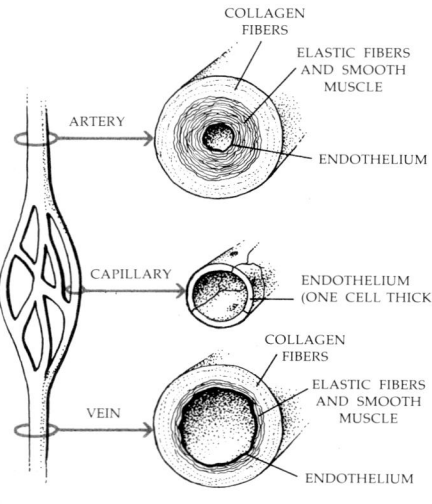

33–9 *Photomicrograph of a vein and an artery. The artery can be identified by its thick, muscular wall.*

ARTERIES, CAPILLARIES, AND VEINS

There are characteristic structural differences among the different types of blood vessels: the arteries, capillaries, and veins. Of the three types of vessels, the arteries have the thickest, strongest walls, made up of three layers. The inner layer, or endothelium (a type of epithelial tissue), forms the lining of the vessels; the middle layer contains smooth muscles and elastic tissues; the outer layer, also elastic, is made of collagen and other supporting tissues. Because of their elasticity, the arteries expand when the blood is pumped into them, and then relax slowly; as a consequence, by the time the blood leaves the arteries it is flowing smoothly through the vessels, rather than in spurts, as it does when it leaves the heart. The pulsation felt when the fingertips are placed over an artery close to the body surface—as in the wrist—represents the alternating expansion and recoil of an elastic arterial wall.

The blood is pumped out of the left ventricle of the heart to the largest artery in the body, the aorta, which is 1 to 1½ inches in diameter. Slightly smaller arteries branch off from the aorta, and these in turn give rise to still smaller branches. The smallest are the arterioles, from which blood flows into the capillaries.

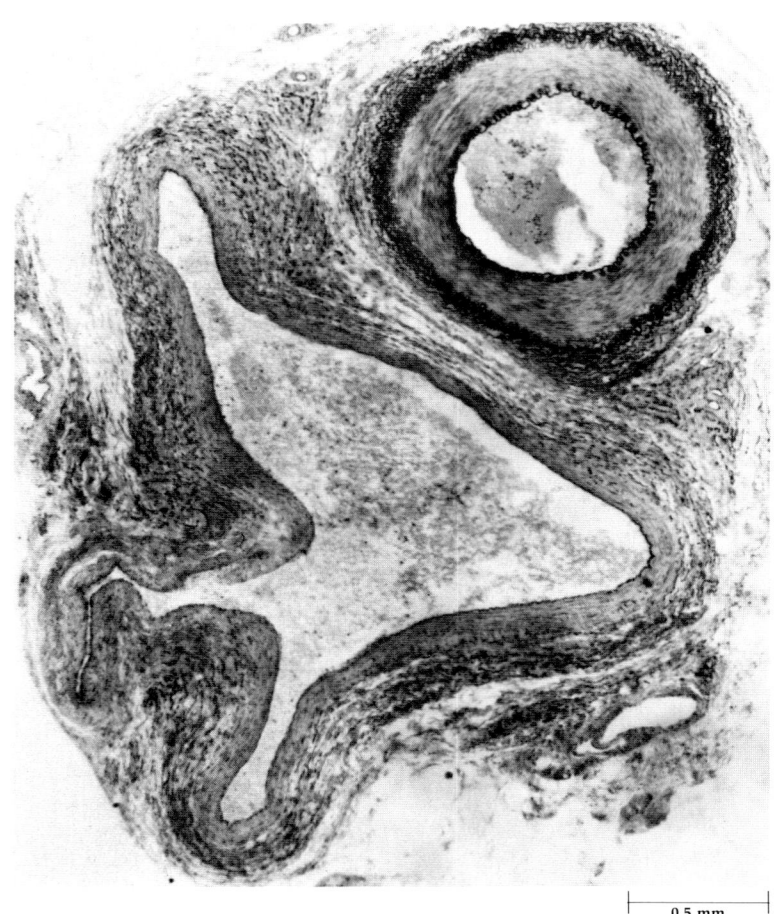

0.5 mm

3–10 *Electron micrograph and diagram of a capillary from cardiac muscle. The small pinocytosis vesicles are believed to be involved with the transport of dissolved substances across the capillary wall.*

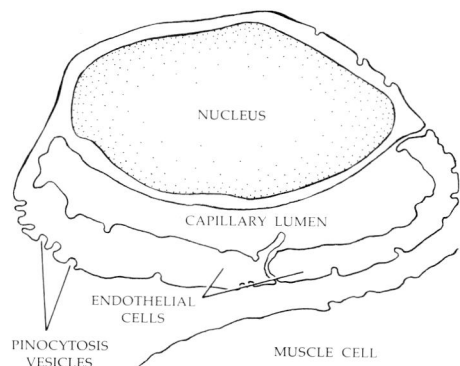

NUCLEUS

CAPILLARY LUMEN

ENDOTHELIAL
CELLS

PINOCYTOSIS
VESICLES   MUSCLE CELL

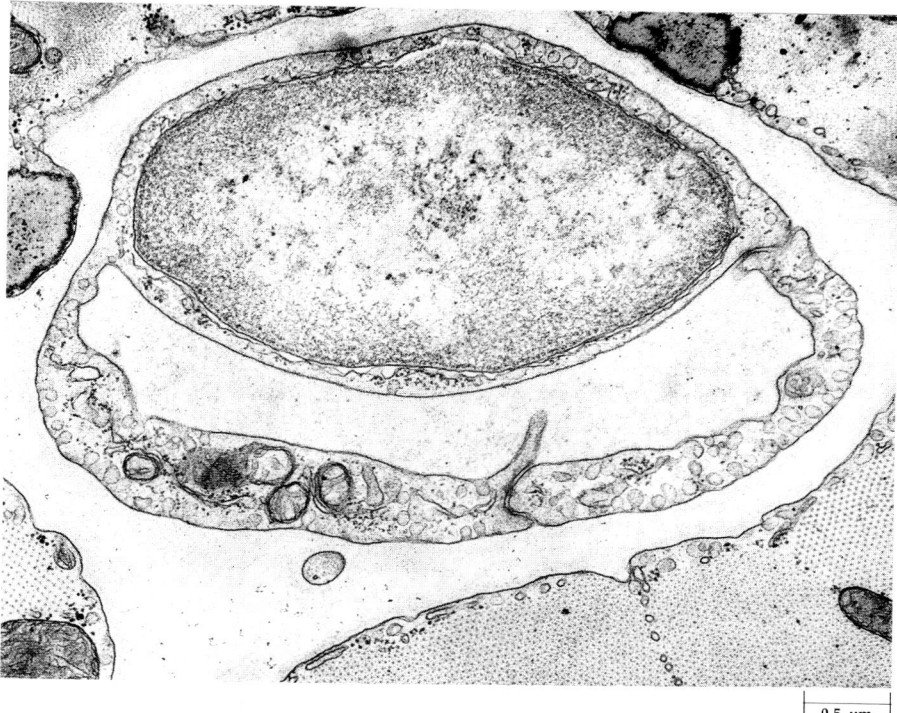

0.5 μm

BLOOD PRESSURE

Blood pressure is the force with which blood pushes against the walls of the blood vessels. It is usually measured in terms of how high it can push a column of mercury. As we shall see, it varies at different points in the circulatory system. For medical purposes, it is usually measured at the artery of the left upper arm. Normal blood pressure in a young man is 120 millimeters of mercury (120 mm Hg) when the ventricles are contracting (the systolic blood pressure) and 80 mm Hg when the ventricles relax (diastolic pressure). The blood pressures of a young woman are each 8 to 10 mm Hg less.

Constriction of the arteries by loss of elasticity or by the formation of fatty deposits within their walls increases the blood pressure and also increases the work load on the heart.

Gravity affects blood pressure, as you know if you have experienced dizziness on standing suddenly, decreasing the flow of blood to the brain. Fainting and falling down are protective responses, which prevent serious damage to brain cells as a result of inadequate blood supply, responses often thwarted by well-meaning bystanders anxious to get the affected individual "back on his feet." In fact, holding a fainting person upright can lead to severe shock and death.

Peripheral Resistance

The flow of any fluid is directly proportional to the difference in pressure between the two ends of the tube through which it flows and inversely proportional to the resistance. Resistance is essentially a measure of friction, the friction between the tube wall and the fluid flowing through it. We know from

Blood pressures in different parts of mammalian circulation. The pressure fluctuations produced by three heartbeats are shown in each section. Note the fall in pressure as blood traverses the arterioles of the systemic circulation.

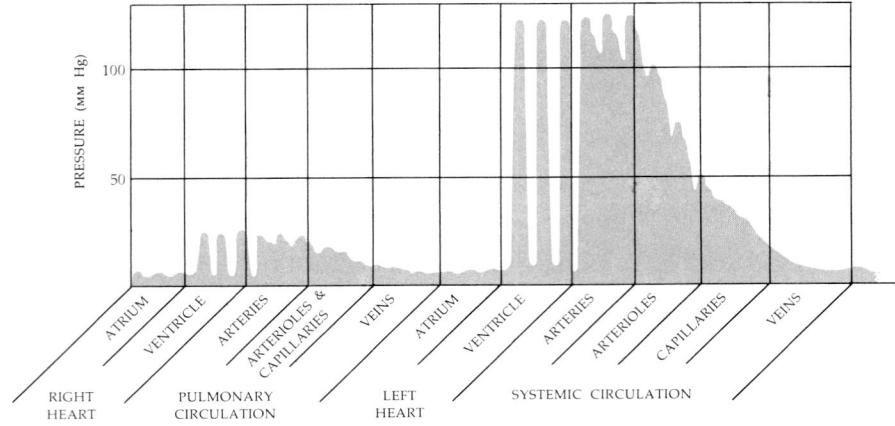

common experience that the smaller a tube, the more resistance it offers to the flow of a given volume of fluid. The exact relationship is given by a simple geometrical formula that states that resistance to flow of a given fluid is inversely proportional to the fourth power of the radius (r^4) of the vessel. Thus, if you double the radius of a tube, the flow through it increases sixteenfold. Conversely, sixteen times less fluid will flow through a tube of half the radius. As the blood reaches the branching arterioles, the radius of the vessels decreases, resistance increases greatly, and the flow of blood decreases (see Figure 33–11). Peripheral resistance is the term applied to the resistance that the many small arterioles offer to the flow of blood out of the arterial system into the capillaries and veins.

Regulation of Blood Flow

The flow of blood through the arterioles is regulated by contraction and relaxation of the arteriolar muscles. These muscles are always in a state of partial contraction, producing what is called arterial tonus, or tone. Relaxation and further constrictions of the arterioles are partly under the control of the vasomotor nerves, which are nerves of the autonomic nervous system that act on the smooth muscle of the blood vessels. Vasoconstrictor nerves, which are a part of the sympathetic nervous system, cause the blood vessels to constrict, decreasing their diameter. They are dilated either by a decrease in the output of the vasoconstrictor nerves or by the inhibiting influences of vasodilator nerves. In either case, the result is an increase in the diameter of the arterioles in a particular area with an increase in blood flow through them. Thus the oxygen supply can be distributed according to the requirements of different parts of the body at different times, with muscles getting more during exercise, for example, or the intestinal tract more during digestion.

If cardiac output is constant, generalized vasoconstriction causes a rise in blood pressure, such as accompanies anger and fear and muscular exercise. Psychological events influence vasomotor activities in other ways also; familiar examples are blushing, turning pale with fear, angina pectoris (pain in the chest and left arm caused by a shortage of oxygen to the heart muscle) precipitated by emotion, and erection of the penis or clitoris following erotic stimulation.

Capillary Exchange

The walls of the capillaries consist of only one layer of cells, the endothelium, and the cavity (lumen) of the smallest capillary is only 8 micrometers in diameter, just wide enough to permit the passage of red blood cells in single file. The total length of the capillaries in a human adult is more than 62,000 miles. Because of their large total volume, blood moves slowly through the capillary system. As it moves, gases (oxygen and carbon dioxide) and other materials are exchanged with the surrounding tissues by diffusion through the cytoplasm of the endothelial cells. Organic molecules, such as glucose, are probably moved by transport systems of the cells. Pinocytosis vesicles (see Figure 33–10) have been observed in the capillary endothelial cells; they presumably serve to ferry dissolved materials through the cytoplasm and so in or out of the capillaries.

At the arteriolar end of the capillaries, the hydrostatic pressure of the blood is greater than its osmotic pressure (see Figure 7–5, page 112) and blood (minus cells and large molecules) leaves the capillaries along with nutrient molecules and oxygen. At the venous end of the capillaries, the hydrostatic pressure is reduced. Thus the osmotic pressure of the blood exceeds its hydrostatic pressure, and fluids move back into the capillaries. As a consequence, there is little total fluid loss from the blood as it moves through the capillaries (the relatively small amount that is not returned by osmosis is channeled back into the bloodstream by the lymphatic system, as we shall see later in this chapter). When loss of water into tissues does occur, as, for example, when the endothelium is damaged by a blow, swelling (edema) is produced. Severe fluid loss from the blood vessels is the cause of the condition known as shock, which is severe hypotension (low blood pressure).

No cell in the human body is farther than 130 micrometers—a distance short enough for rapid diffusion—from a capillary. Even the cells in the walls of the large veins and arteries depend on this capillary system for their blood supply, as does the heart itself, like all the other organs of the body.

At the ends of the capillary beds, blood passes into the venules, smallest of the veins, then into larger veins, and finally back into the heart through either the superior (anterior) or inferior (posterior) vena cava. Like the arteries, the walls of the veins are three-layered, but they are less elastic and much weaker. An empty vein collapses, whereas an empty artery remains open. The veins, with their thin walls and relatively large diameters, offer little resistance to flow, facilitating the return of the blood to the heart.

12 *Diagram of a capillary bed. The sphincter muscle at the end of the arteriole controls the blood flow through the capillaries. These muscles, which are innervated by the autonomic nervous system, make it possible to regulate the blood supply to different regions of the body depending on the physiological state of the organism.*

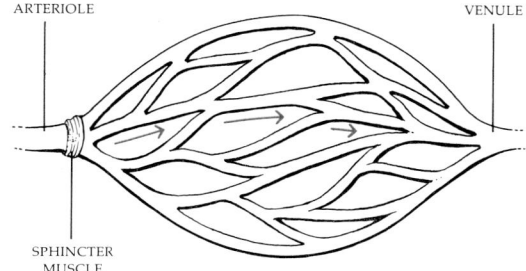

ARTERIOLE

VENULE

SPHINCTER
MUSCLE

Vascular Circuitry

As we mentioned previously, there are two principal circuits of the vascular system: the pulmonary circuit and the systemic circuit. In the pulmonary circuit, blood leaves the right ventricle of the heart through the pulmonary artery. This artery divides into right and left branches, which carry the blood to the right and left lungs, respectively. Within the lungs, the arteries divide into smaller and smaller vessels that finally open into capillaries through which oxygen and carbon dioxide are exchanged. Blood flows from the capillaries into small venules and then into larger and larger veins, finally draining into the four pulmonary veins that carry the blood, now oxygenated, to the left atrium of the heart. The pulmonary arteries are the only arteries that carry deoxygenated blood; and the pulmonary veins are the only veins that carry fully oxygenated blood.

The systemic circuit is much larger. Many arteries branch off the aorta after it leaves the left ventricle. The first two branches are the right and left coronary arteries, which bring oxygenated blood to the heart muscle itself. Another major subdivision of the systemic circulation supplies the brain. If the circulation of freshly oxygenated blood to the brain is cut off for even five seconds, unconsciousness results, and after four to six minutes, brain cells are damaged irreversibly.

The hepatic portal system is another special subdivision of the systemic circulation. Venous blood collected from the digestive organs is shunted via the hepatic portal vein through the liver, where it goes through a second capillary system, before it is emptied into the inferior vena cava. (This passage from capillaries to veins to capillaries is called a portal system.) In this way, the products of digestion can be directly processed and stored in the liver or detoxified and excreted. The liver also receives freshly oxygenated blood directly from a major artery, the hepatic artery.

33–13　*A heart attack is caused by stoppage of blood flow in a branch of the coronary artery, the vessel that supplies oxygen and nutrients to the heart tissue. When such a stoppage occurs, the heart cells normally serviced by this vessel die. Recovery from a heart attack depends on how much of the heart tissue is damaged and whether other blood vessels in the heart can enlarge their capacity and supply these tissues, which may then recover to some extent. Formation of a clot, as shown here, is one probable cause of a heart attack, or "coronary."*

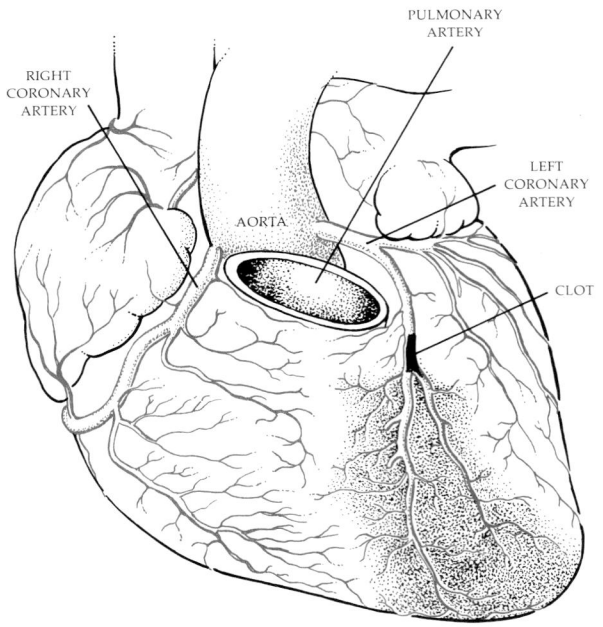

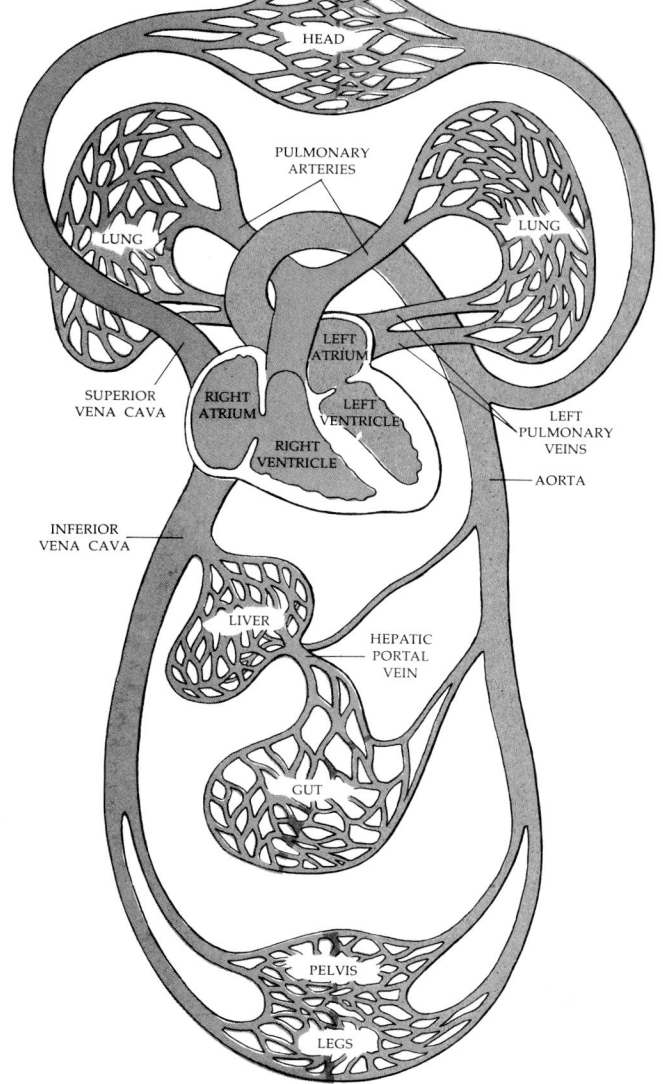

3–14 *Diagram of the human circulatory system. Oxygenated blood is shown in color. Every tissue of the body contains capillaries that supply oxygen and nutrients to every cell of these tissues and carry off carbon dioxide and other wastes.*

HEAD

PULMONARY
ARTERIES

LUNG

LUNG

LEFT
ATRIUM

SUPERIOR
VENA CAVA

RIGHT
ATRIUM

LEFT
VENTRICLE

LEFT
PULMONARY
VEINS

RIGHT
VENTRICLE

AORTA

INFERIOR
VENA CAVA

LIVER

HEPATIC
PORTAL
VEIN

GUT

PELVIS

LEGS

3–15 *Valves in the veins open to permit movement of blood toward the heart but close to prevent backflow.*

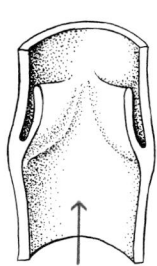 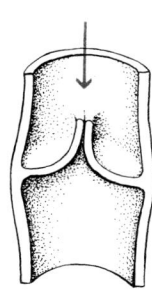

There is a large drop in pressure as the blood goes through the capillaries. The return of blood to the heart through the veins depends in large part on body movements, which squeeze the veins between contracting muscles and force the blood upward. (If you have to stand still for long periods of time, try contracting your leg muscles periodically to move blood back toward the heart and prevent blood pooling.) Valves in the veins prevent backflow. Also, the thoracic cavity is under negative pressure (less than atmospheric pressure), and as the thorax expands in respiration, the elastic walls of the veins in the thorax dilate, so that blood is both pulled and pushed upward.

HARVEY AND THE CIRCULATION OF THE BLOOD

William Harvey's discovery of the circulation of the blood is often cited as a model of biological research, not only because of its great importance to medicine but also as an example of how reason can push its way against the general current of scientific opinion. The stubbornness of Harvey's logic is even more impressive when one realizes that the microscopic observations that offered the final proof of Harvey's theory were not made until after his death.

Harvey lived from 1578 to 1657. By this time, the function of the heart as a pump had been recognized, but it was thought that the blood ebbed and flowed in the veins, like tides in the ocean, being gradually used up and replaced. The arteries were thought to carry only air, because the arteries of dissected cadavers were empty.

Harvey, working with animals, observed that a beating heart expelled the blood within it. By calculating the amount of blood expelled from the heart at each beat as compared with the total amount in the human body, he convinced himself that all of the blood in the body must return to the heart every few minutes. (His figures were not entirely accurate, but accurate enough for the purpose. Actually, the amount of blood pumped per minute by the heart—5 liters—is equal to the entire amount of blood in the human body.)

One of the simple demonstrations Harvey used to support his theory is shown in the drawings below. If you tie a ligature around the upper arm, the valves in the veins show up as small swellings. If blood is pushed up toward the heart to a point above one of these swellings, it fails to flow back again, even if pushed. Concerning these observations, Harvey wrote "so provident a cause as nature had not so plac'd many valves without design."

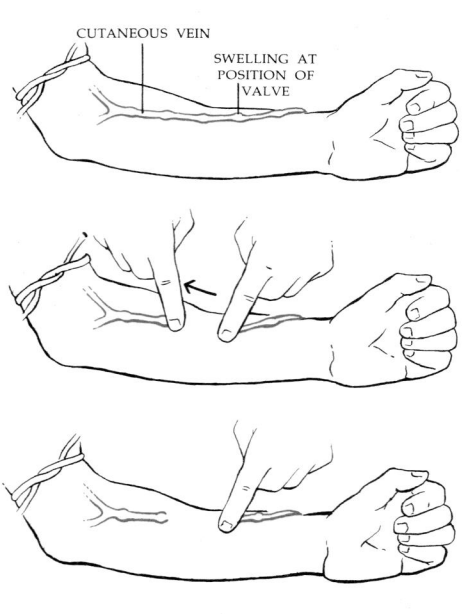

THE BLOOD

An individual weighing 165 pounds has about 5 liters of blood. About 55 percent of it is a straw-colored liquid called *plasma*. The plasma, which is more than 90 percent water, carries a large number of different kinds of ions and molecules. They include fibrinogen, the protein from which clots are formed; nutrients, such as glucose, fats, and amino acids; various ions; antibodies, hormones, and enzymes; and waste materials such as urea and uric acid.

The other 45 percent of the blood is made up of *erythrocytes,* or red blood cells, *leukocytes,* or white blood cells, and *platelets,* which are cytoplasmic fragments of cells. The relative amount of plasma varies among species and among individuals who live under different conditions. For example, persons living at high altitudes (low oxygen pressure) have relatively less plasma and more red blood cells.

Red Blood Cells

Red blood cells transport oxygen. There are about 5 million of them per cubic millimeter of blood—some 25 trillion (25×10^{12}) in the whole body. In man, an individual red blood cell has a life span of about 120 to 130 days. New ones are produced in the bone marrow of adults at the rate of about 1½ million per second.

Red blood cells are among the most highly specialized of all cells. As a mammalian red blood cell matures, it extrudes its nucleus and its mitochondria, and its other cellular structures dissolve. Almost the entire volume of a mature red blood cell is filled with hemoglobin, about 300 million molecules per cell. Anemia is a deficiency either in the number of red blood cells or in total hemoglobin.

Invertebrates also have oxygen-carrying molecules, some of which resemble hemoglobin. However, they are carried in solution in the bloodstream and not in specialized cells. Why is the hemoglobin packed in red blood cells in vertebrates? One answer to this question seems to lie in the fact that vertebrates in general, and mammals in particular, require transport of comparatively large amounts of oxygen and concomitantly have large amounts of hemoglobin. If the hemoglobin were dissolved in the bloodstream, the blood would be so hypertonic in relation to the interstitial fluid that the fine balance between the amount of plasma pushed out into the tissues and that returned to the capillaries by osmosis would be completely disrupted. By carrying the hemoglobin in red blood cells, the blood is maintained at a concentration of solutes that renders it only slightly hypertonic in relation to the surrounding fluids.

We shall discuss the process of the exchange of oxygen and carbon dioxide in more detail in Chapter 35.

White Blood Cells

There are about 6,000 to 9,000 white blood cells per cubic millimeter of blood— 1 or 2 white blood cells for every 1,000 red blood cells. These cells are nearly colorless, are larger than red blood cells, contain no hemoglobin, and have a nucleus. Although they appear spherical in some micrographs, they are all flattened cells like amoebas, and, also like amoebas, they move by means of pseudopodia and many are phagocytic.

A white blood cell phagocytizing a chain of streptococci. Numerous red blood cells are also visible.

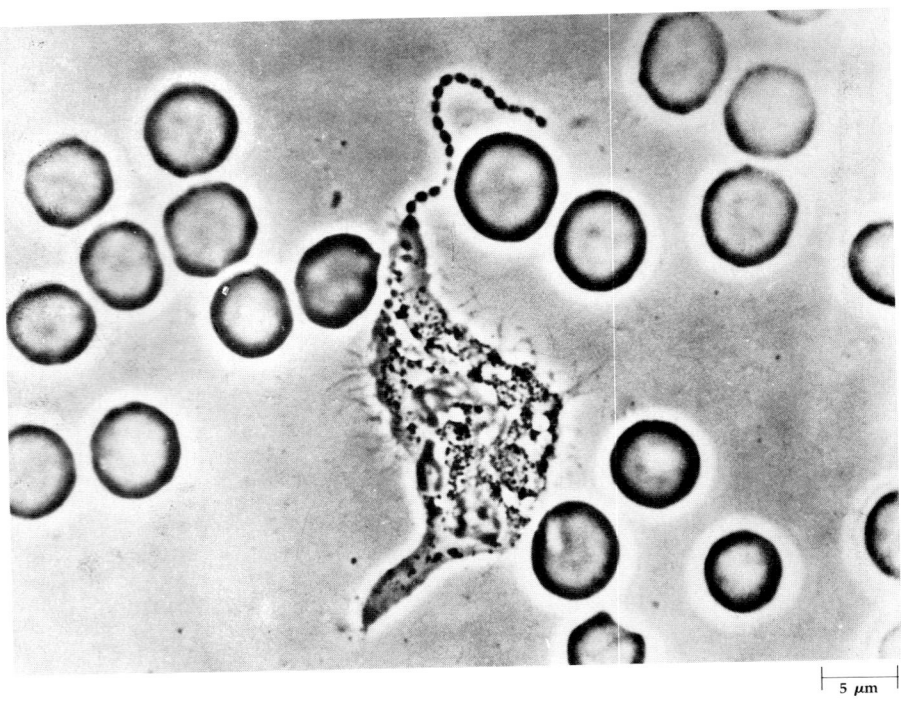

5 µm

White blood cells are involved in defending the body against viruses, bacteria, and other foreign intruders; they do so by phagocytosis and by the production of antibodies.

There are two principal types of white blood cells: *neutrophils,* which are produced in the bone marrow, and *lymphocytes,* which are produced in the lymphatic tissue. Neutrophils, which make up 60 to 70 percent of all leukocytes, are the first white blood cells to arrive at the scene of an injury. They are very active in phagocytosis and may play a role in promoting wound healing. Lymphocytes are of two types: B-cells, which are produced in the lymph cells and are involved in the synthesis of antibodies, and T-cells, which are produced by cells that originate in the thymus gland and are involved in graft rejection and other immune responses (see Figure 33–22).

White blood cells are unusual among somatic cells because of their active locomotion. They are capable of moving against the bloodstream and of migrating through the walls of the capillaries and making their way into damaged or infected tissues. They are attracted to sites of infection, probably by the chemical substances diffused from such sites. Whereas red blood cells are part of the blood transport system, white blood cells are passengers in it.

Platelets

Platelets, so called because they look like little plates, are colorless, round or biconcave disks smaller than erythrocytes (about 3 micrometers in diameter). They are cytoplasmic fragments of unusually large cells, megakaryocytes, found in the bone marrow. Platelets play an important role in plugging breaks in blood vessels and in the formation of clots.

Blood Clotting

Blood clotting is a complex phenomenon; at least 15 factors involved in the process have been identified. It is not initiated when blood encounters air, as was reasonably assumed for a long period of time. The sequence of events begins when platelets encounter a rough surface, such as a torn tissue. This stimulates the release of substances called thromboplastins. These substances can also be released by the injured tissues themselves. (The same effect occurs when platelets touch a glass surface, such as that of a test tube, which is one of the reasons the natural process was so difficult to analyze.) Thromboplastin acts to convert the enzyme prothrombin, a plasma protein, which is produced in the liver, to its active form, thrombin:

$$\text{Prothrombin} \xrightarrow{\text{thromboplastin}} \text{thrombin}$$

Several other factors normally present in the bloodstream are also required for this reaction. Thrombin converts fibrinogen, a soluble plasma protein, to fibrin:

$$\text{Fibrinogen} \xrightarrow{\text{thrombin}} \text{fibrin}$$

The fibrin molecules aggregate together, forming an insoluble network, which enmeshes red blood cells and platelets and forms a clot. The clot contracts, pulling together the edges of the wound. When blood is removed from the body and placed in a test tube, it congeals to form a clot. The clot eventually contracts, leaving a clear fluid, the serum.

17 *Beginning of the formation of a clot. Red blood cell enmeshed in fibrin.*

1 μm

Clot formation, although essential to the survival of the organism, also poses a threat because clots can block the circulatory system. The clot-forming process is very sensitive; just a few molecules of thromboplastin can set it off. On the other hand, its many steps provide numerous opportunities for interrupting the process. A number of natural inhibitors of clotting are known. For example, heparin, produced by cells of the liver, lungs, and other tissues, inhibits the conversion of prothrombin to thrombin. Heparin is used medically to limit clot formation after surgery or following a heart attack.

Hemophilia is an inherited deficiency in blood clotting (see page 218). The most common type of hemophilia involves a defective protein (known as Factor VIII) involved in the production of prothrombin. Hemophiliacs can be treated by administration of plasma concentrates of Factor VIII from normal blood.

LYMPHATIC SYSTEM

As we noted previously, not quite all of the fluid forced out of the capillaries by the pressure of the circulating blood is returned to the capillaries by osmosis. In higher vertebrates, this fluid, amounting to 1 or 2 percent of the total fluid, is collected by another circulatory system, the lymphatic system, which routes it back to the blood. The lymphatic system is like the venous system in that it consists of an interconnecting network of progressively larger vessels. The larger vessels are, in fact, similar to veins in their structure. The small vessels are much like the blood capillaries; the most important difference is that, rather than forming part of a continuous circuit, the lymph capillaries end blindly in the tissues. Interstitial fluid seeps into the lymph capillaries, from which it travels in the form of lymph to large ducts from which it is emptied into the vena cava. Some vertebrates have lymph "hearts," which help to move the fluid. In mammals, lymph is moved largely by contractions of the body muscles with valves preventing backflow, as in the venous return system of the blood. The lymph also serves to transport some food, particularly fats, absorbed from the digestive tract and picked up by the lymph capillaries.

Lymph Nodes

As the fluid travels through the lymphatic system, it passes through lymph nodes (sometimes incorrectly referred to as lymph glands). As you can see in Figure 33–18, single lymph nodes are distributed throughout the body, but most are found in clusters in particular areas, such as the neck, the armpits, and the groin. These nodes range from the size of a pea to that of a lima bean. A lymph node is a mass of spongy tissue separated into compartments by connective tissue. Lymph nodes have two functions; they manufacture the type of lymphocyte (the B-cells) involved in antibody production, and they remove foreign particles from the lymph before it enters the blood. This removal system is in part mechanical filtration; the nodes catch dead cells, dust, and other particles that may have entered the body fluids. Lymph nodes near the respiratory system, for example, are often filled with carbon particles from soot or tobacco smoke. The lymph nodes also contain leukocytes that attack bacteria and other invaders. Cancer cells that have broken loose from the principal site of cancer growth are often caught in the lymph nodes, and hence in major cancer operations, surgeons often routinely remove lymph nodes adjacent to the cancer

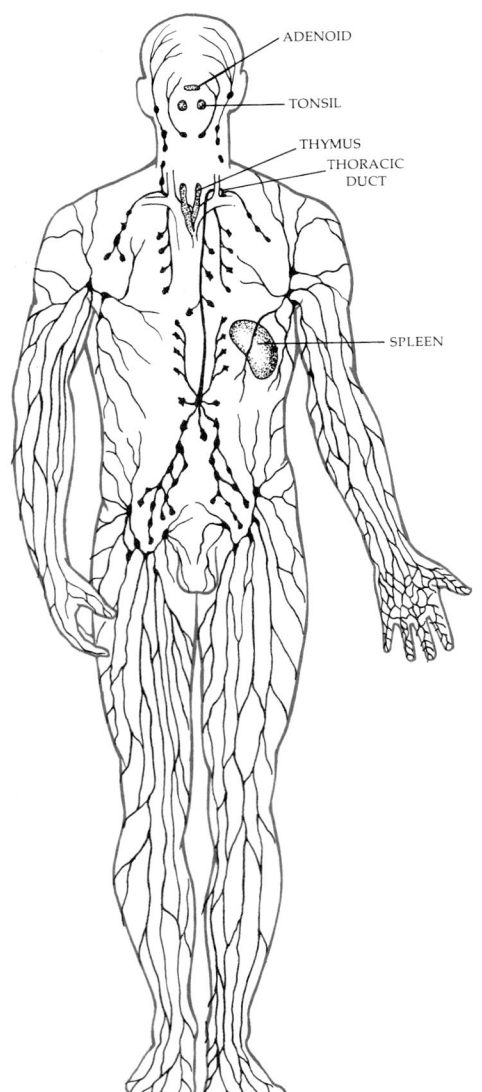

33–18 *Human lymphatic system. Lymphatic fluid reenters the circulatory system of the blood through the thoracic duct.*

ADENOID

TONSIL

THYMUS
THORACIC
DUCT

SPLEEN

growth. If lymphatic vessels and nodes are damaged by disease or injury or are removed by surgery, fluids that would have been drained off may collect in the tissues, producing the condition known as lymphedema. Elephantiasis, which causes severe, often grotesque, swelling of the appendages, is a form of lymphedema caused by a parasitic worm that infects and blocks lymph nodes.

Other Lymphatic Organs

Spleen

The spleen, which is largely lymphoid tissue, also manufactures lymphocytes. Another and principal function of the spleen is the culling of red blood cells from the circulation. As aged, damaged, and otherwise defective erythrocytes are broken down in the spleen, the iron is removed from the hemoglobin and recycled by the bone marrow for use in the production of new hemoglobin.

In many vertebrates, the spleen also serves as a reservoir of red blood cells that can be squeezed out into the circulation following hemorrhage. Whether the human spleen serves this same function is a matter of controversy.

Thymus

The human thymus is a spongy, pinkish gray, two-lobed organ that lies high in the chest, in front of the aorta and behind the breastbone. It is a relatively large organ in infancy, continues to grow until puberty, and then becomes smaller. Recent research suggests that the thymus is the original site of production of the type of lymphocyte known as the T-cell and that all of the body's T-cells are derived from cells that originate in the thymus and then leave the thymus to colonize the blood, lymph nodes, and bone marrow. After puberty, these thymus cells produce new T-cells. The thymus may also be the source of a hormone that stimulates maturation of the lymphocytes.

Tonsils and adenoids are lymphoid tissues concerned primarily with lymphocyte production.

DEFENSE MECHANISMS

The circulating blood also carries a variety of agents that serve as defenses against bacteria and other microscopic invaders that are potential causes of disease. These may enter the body through a break in the skin or through the lining of the respiratory or digestive tracts. Suppose, for example, you nick your skin, a not uncommon occurrence. The injured cells immediately release histamine and other chemicals that lead to the distension of the nearby capillaries, thus increasing their permeability. Circulating white blood cells make their way through the distended capillary walls, crowding into the site of the injury. They appear to move by chemotaxis, the same sort of sensory mechanism that directs bacteria and protists toward food sources (page 297). The neutrophils, which are the first to arrive, engulf any foreign invaders by phagocytosis. The next on the scene are the lymphocytes. Blood clots begin to form, walling off the injured area. Pus may accumulate; it is made up chiefly of white blood cells, dead and alive, combined with tissue debris. The local temperature in the area often rises, creating an environment unfavorable to the multiplication of microorganisms while accelerating the motion of the white blood cells.

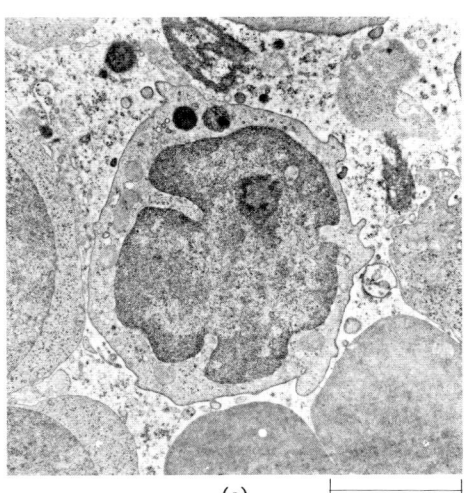

33–19 *As a lymphocyte* (a) *differentiates into a plasma cell* (b), *it grows larger, its nucleus becomes relatively smaller, and there is a large increase in endoplasmic reticulum and ribosomes.*

(a) 5 μm

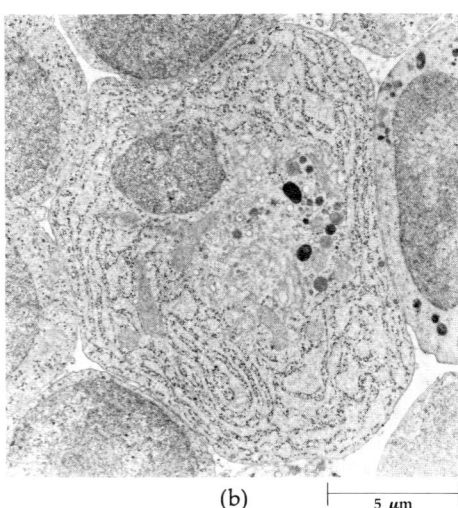

(b) 5 μm

Development of Immunity

If this inflammatory response, as it is called, is insufficient, a second defense mechanism comes into play, which is the immune system. The immune response differs from the other defenses of the body because it is highly specific, involving recognition of a particular invader and the tailoring of an attack against it.

We noted in Section 3 (page 299) that an organism that has been infected with a particular bacterium or virus is often immune to reinfection with that pathogen. Common examples of diseases that confer immunity are measles, mumps, and chicken pox. This sort of immunity is caused by the formation in the bloodstream of complex globular proteins known as *antibodies*. Antibodies, like enzymes, are highly specific, and their specificity, like that of enzymes, is based on their combining in a very precise way with another molecule. The molecule with which an antibody combines is known as an *antigen*. Virtually all proteins and some polysaccharides can act as antigens. A single cell, such as a bacterial cell, may carry a number of antigens, each of which can combine with a specific antibody.

Antibodies act against invaders in one of three ways: (1) they may coat the foreign particle in such a way that it can be taken up by the phagocytic cells, (2) they may combine with it in such a way that they interfere with some vital activity—such as, for example, covering the protein coat of a virus at the site where the virus attaches to the cell membrane, or (3) they may themselves, in combination with another blood component known as *complement*, actually lyse and destroy the foreign cell.

How Antibodies Are Produced

According to current hypotheses, antibody synthesis begins with an encounter between an antigen and an immature lymphocyte. The exact nature of this encounter is not known, but as a result of it, the immature lymphocyte differentiates into a mature cell, called a plasma cell, with the capacity to make large numbers of antibodies against that particular antigen (and probably against no other). The antigen also stimulates the cell to divide, thereby greatly increasing antibody production. This process takes four or five days, at the end of which time antibody production often catches up with the multiplication of the infectious organism.

After the first bout of infection, memory cells sensitized to that particular antigen and ready to begin immediate antibody production against it persist indefinitely in the circulation.

The Structure of Antibodies

In 1972, Gerald Edelman of Rockefeller University and R. R. Porter of Oxford were awarded the Nobel prize for their work on antibody structures. These researchers elucidated the primary structure (the amino acid sequence) of an antibody for the first time. Moreover, their work has shown that each antibody molecule consists of four subunits, two "light" and two "heavy." (The heavy ones are heavier because they have more amino acids.) Each of the light units is made up half of a sequence of amino acids apparently common to all

3–20 *A hypothetical mechanism for the immune response. (a) An immature lymphocyte with antibodies on its surface encounters the antigen molecules for which the antibodies are specific and (b) antibodies and antigens bind together. (c) The lymphocyte then begins to divide and differentiate (d), forming plasma cells and memory cells. Plasma cells secrete circulating antibodies. Memory cells persist in the circulation, secreting antibodies only following an encounter with the specific antigen.*

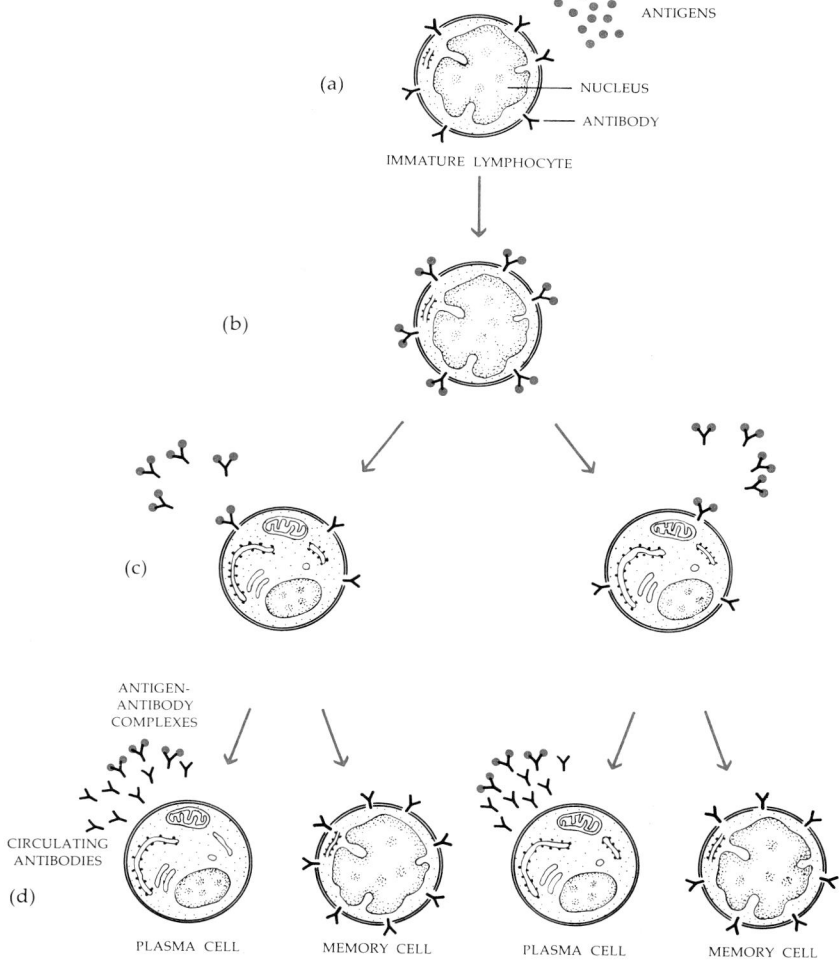

3–21 *Each antibody molecule is made up of four polypeptide chains, two light (short) and two heavy (long), connected by disulfide bonds. Part of the sequence of each chain is the same in all antibody molecules and part (indicated in color) is variable. The variable parts determine the specificity of the antibody—that is, what antigen it will combine with.*

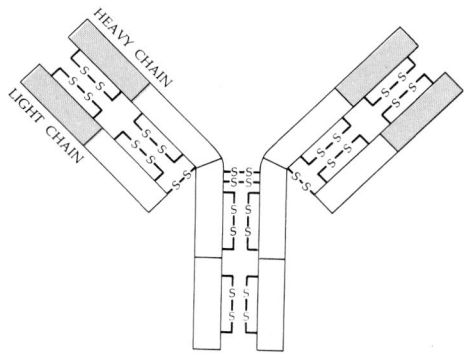

antibodies, and the other half of a sequence unique to a particular antibody. Similarly, each heavy unit is composed partly of a common sequence and partly of a unique one. The antibody combining site, it is not surprising to find out, involves the unique sequences. Each antibody molecule has two such sites, at the points at which the heavy and light chains join. According to current hypotheses, the general sequence of each protein subunit is coded by one gene and the specific sequence by another. The molecule assembles itself from its component units.

Disorders Related to the Immune System

The *immune response* is a powerful bulwark against disease, but it sometimes goes awry. Hayfever and other allergies are the result of interactions of pollen dust, or other substances that are weak antigens (antigens to which most people do not react), and a special kind of antibody which fixes onto the cells of the skin or mucous membranes where the antigen makes contact.

Another medical problem caused by the immune system is hemolytic anemia of the newborn. During the last month in the uterus, the human baby usually acquires antibodies from its mother. Most of these antibodies are beneficial. An important exception, however, is found in the antibodies formed against a blood factor, the Rh factor (named after the rhesus monkeys in which the research leading to its discovery was carried out). The Rh factor is a genetically determined substance which is found on the surface of red blood cells. If a woman who lacks the Rh factor (that is, Rh negative) has children fathered by a man homozygous for the Rh factor, all the children will be Rh positive; if he is a heterozygote, about half of the children will be Rh positive. During the course of her first pregnancy carrying an Rh-positive fetus, the mother is likely to produce antibodies against Rh antigens that enter her bloodstream from the fetel blood due to leakage across the placenta. The antibodies are produced slowly and the leakage occurs late enough in the pregnancy so that the Rh-positive child normally is not threatened. In subsequent pregnancies, these antibodies may be transferred to the fetus, causing erythroblastosis—a destruction of the red blood cells—which can be fatal, either before or just after birth. One method of treatment, which is quite an intricate procedure, is to transfuse the infant completely, replacing all its blood and washing out the mother's antibodies shortly after birth. Such transfusions save about 35 percent of the babies who would have died of erythroblastosis. Recently, two medical scientists at Columbia have developed a substance, which they have called RhoGAM, that contains antibodies against the antibodies that form in the mother against the Rh factor. RhoGAM, if injected into an Rh-negative woman at the birth of her first Rh-positive child, destroys her anti-Rh antibodies.

Self and Not Self

The immune system can ordinarily distinguish between "self" and "not self." Substances which are present during embryonic life, when the immune system is developing, will not be antigenic in later life. This recognition system occasionally breaks down, however, and the immune system attacks the body. One type of anemia and certain other rare disorders have been identified as autoimmune diseases—that is, diseases in which an individual makes antibodies against his own proteins—and it is possible that other disorders, such as rheumatoid arthritis, whose causes are not yet known, may prove to have the same basis.

The Immune Response and Tissue Transplant

Because it is programmed by evolution to act against foreign materials of all kinds, the immune system works vigorously against tissues—such as skin, kidney, or heart—transplanted from another individual (except an identical twin). The blood transfusion reactions described in Chapter 20 are examples of this same phenomenon. Surgeons and medical research workers concerned with extending the use of tissue transplants are seeking ways to suppress or paralyze the immune response in such a way that these foreign but potentially lifesaving tissues can survive. X-ray treatment and certain chemicals that inhibit white blood cell production suppress immune reactions. Such treatments, of course, also render the patient more susceptible to disease. Immunity against

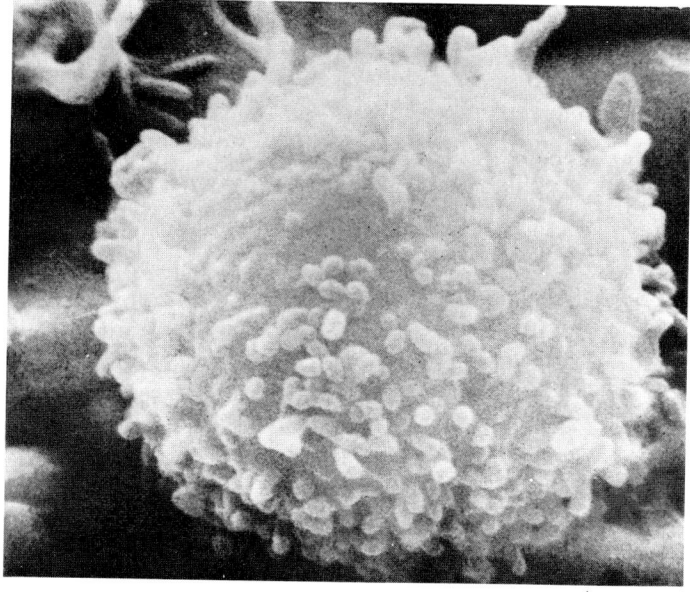 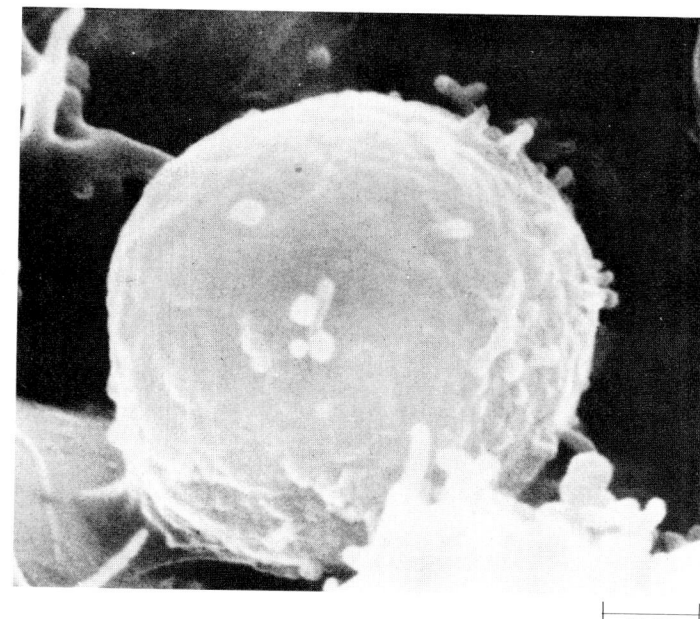

−22 *Lymphocytes, as shown in scanning electron micrographs. The B-cells (left) have about 150 fingerlike protrusions on their surfaces. Found mostly in the spleen and lymph nodes, they differentiate to become the plasma cells that produce circulating antibodies. T-cells (right), which are smooth surfaced, congregate and proliferate at the site of an infection and participate in killing the invading cell. These are the cells that trigger graft rejection. They may also be involved in immune responses to cancer.*

transplanted tissue appears to involve white blood cells—again lymphocytes derived from the thymus—rather than circulating antibodies. These white blood cells can be seen aggregating around a tissue transplant before and during its rejection.

Immunity and Cancer

Cancer cells resemble the host's own cells in many ways. Yet, within the host, they act like foreign organisms, invading and destroying normal tissues. Moreover, they can be shown to have antigens on their cell surfaces that can be distinguished from the antigens of the normal host cells. Does this mean that people can form antibodies against their own cancers? A growing number of cancer researchers believe that not only can cancer induce an immune response but that it usually does so, and in fact it usually does so successfully, overwhelming the cancer before it is ever detected by the patient or his physician. The cancers that are discovered represent occasional failures of the immune system. This conclusion suggests, of course, that bolstering the patient's immune defenses may provide a means for cancer prevention or control. This suggestion is now being vigorously pursued in a number of research laboratories.

SUMMARY

Transport of oxygen, nutrients, hormones, and wastes is accomplished in vertebrates by closed circulation systems in which the blood is pumped by the muscular contractions of the heart into a vast circuit of arteries, arterioles, capillaries, veins, and venules. This network ultimately services every cell in the body. The essential function of the circulatory system is performed by the capillaries, through which substances are exchanged with the interstitial fluid surrounding the individual cells.

Heart structure in the vertebrates varies with the demands of their differing metabolic rates. The fish has a two-chambered heart; the atrium receives deoxygenated blood and the ventricle expels blood through the oxygenating gill capillaries into the systemic circulation. In amphibians, both oxygenated and deoxygenated blood are received in a two-chambered atrium, are mingled somewhat in the ventricle, and are then pumped simultaneously through the lungs and through the systemic circulation. Birds and mammals have a double circulation system, made possible by a four-chambered heart that functions as two separate pumping organs, one pumping deoxygenated blood to the lungs and the other pumping oxygenated blood into the body tissues.

Synchronization of the heartbeat is controlled by the sinoatrial node (the pacemaker) located in the right atrium, and secondarily by the atrioventricular node, which delays the stimulation of ventrical contraction until the atrial contraction is completed. The rate of heartbeat is under neural control. Acetylcholine inhibits the pacemaker; adrenaline stimulates it.

The blood is composed of plasma, white blood cells (leukocytes), red blood cells (erythrocytes), and platelets. The fluid part of the blood is plasma, chiefly composed of water, in which are dissolved or suspended the nutrients, antibodies, enzymes, waste substances, and other specialized compounds necessary to the life functions. Leukocytes defend the body against invaders by the manufacture of antibodies and by phagocytosis. Erythrocytes are the vehicles of oxygen-bearing and carbon dioxide–bearing hemoglobin. Platelets serve in the formation of clots.

Fluid that seeps out of the capillaries as a consequence of hydrostatic pressure is returned to the blood by the lymphatic system. The lymph also picks up bacteria, cellular debris, and other foreign particles that may become trapped in the lymph nodes. Lymph nodes are also the primary sites for the production of lymphocytes, the special white blood cells involved in the immune reaction. The spleen, thymus, tonsils, and adenoids are also components of the lymphatic system. All produce lymphocytes. The spleen also breaks down damaged or defective red blood cells.

The immune system is a defensive system of vertebrates which involves recognizing a substance entering the circulatory system as foreign and developing a specific defense against that substance. One important part of the system involves the production of antibodies by plasma cells, which are lymphocytes that have differentiated from more generalized lymphocytes after contact with an antigen. Antibodies are complex proteins that combine with specific antigens and inactivate them. The immune system is also involved in reactions against blood transfusions and tissue transplants. Disorders involving the immune system include allergies, Rh disease of infants, and possibly cancer.

QUESTIONS

1. Define the following terms: plasma, peripheral resistance, systemic circulation.
2. Blood serum is the portion of the plasma remaining after a clot is formed. Name some of the components of the plasma that would not be present or would be present in a lesser amount in serum.

3. Trace the course of a single red blood cell from the right ventricle to the right atrium in a mammal.
4. What is probably the immediate cause of death by crucifixion? (If you need a clue, see page 615.)
5. The valves of the heart are not directly controlled by nerves. Yet, in most individuals, they open and shut at precisely the right points in the cardiac cycle for efficient heart operation. How is this precise timing possible? What does determine just when the valves will open and shut?
6. Label the diagram below.

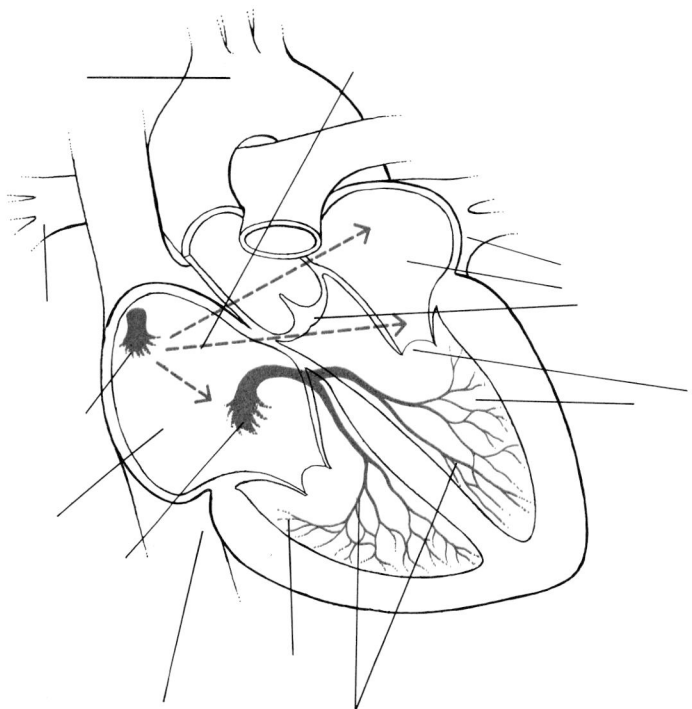

Chapter 34

Energy and Nutrients I: Digestion and Absorption

Digestion, the subject of this chapter, and respiration are the means by which each cell of the animal body obtains (1) energy, (2) the various atoms and molecules with which it builds new protoplasm and carries out the constant replacement of the various parts of the cellular machinery, and (3) ions, coenzymes (vitamins), and other substances necessary for the various chemical processes of the cell. Digestion provides the atoms and molecules needed by the individual cells and also the cellular fuel for their various activities. Respiration, which will be discussed in the next chapter, provides the oxygen for burning this fuel.

DIGESTION

Digestion is the process by which heterotrophs break down the tissues of other organisms to molecules that can be used by their own cells. Sometimes digestion takes place within the cells themselves, as in sponges, but intracellular digestion places strict limits on the size of particles that can be used for food. Some organisms digest their food outside of the body by secreting enzymes into the food mass and then absorb the food molecules from this digested mass. Fungi feed in this way and so do spiders and other arachnids. Most higher organisms digest their food in internal cavities; the simplest form of such a cavity is the coelenteron of jellyfish and other coelenterates. These types of digestion are described in Section 3.

DIGESTIVE TRACT IN VERTEBRATES

In the course of digestion, food substances of various kinds are broken down into molecules small enough to travel in the bloodstream and to enter cells. This process takes place in a long tube, running from mouth to anus, known as the digestive tract or, rather less elegantly but just as accurately, as the gut. The surface of the gut is continuous with the surface of the body, and, from this point of view, the cavity of the gut can be seen to be exterior to the animal body.

Although the requirements of animals are remarkably similar on a cellular level, the ways these requirements are met vary widely. An animal's diet determines not only its dentition and digestive enzymes but dictates, to a large degree, its way of life (and, of course, vice versa). Vegetarianism, however, does not ensure a peaceful disposition: the rhinoceros is among the most short-tempered of mammals.

The food molecules actually enter the body only when they pass through the walls of the digestive tract. Therefore, the process of digestion actually involves two components: breakdown of food molecules and their absorption into the body.

The digestive tube begins with the oral cavity and includes the pharynx, esophagus, stomach, small intestine, large intestine, and anus. Each of these areas of the gut is specialized for a particular phase in the overall process of digestion, but the fundamental structure of each is similar. The tube, from beginning to end, has four layers: (1) the *mucosa,* which is made up of glandular epithelial tissue, an underlying basement membrane, and connective tissue, with a thin outer coating of smooth muscle in some places, (2) the *submucosa,* which is made up of connective tissue and contains nerves, blood, and lymph vessels, (3) the *muscularis externa,* muscle tissue, and (4) the *serosa,* an outer fibrous coating. Along most of the digestive tract, the muscularis is made up of two layers of smooth muscle, the inner layer, in which the orientation is circular, and the outer layer, in which the cells are longitudinally arranged. Coordinated

34–2 *The layers of the digestive tract include (1) the mucosa, (2) the submucosa, which contains nerves and blood and lymph vessels, (3) the muscularis externa, and (4) the serosa, an outer coating. All of the outer surfaces are covered by peritoneum, a tissue of mesodermal origin that makes up the mesenteries, which support the various organs, and also lines all of the abdominal cavity. Glands outside the digestive tract, principally the pancreas and liver, discharge digestive enzymes and bile into the tract through various ducts.*

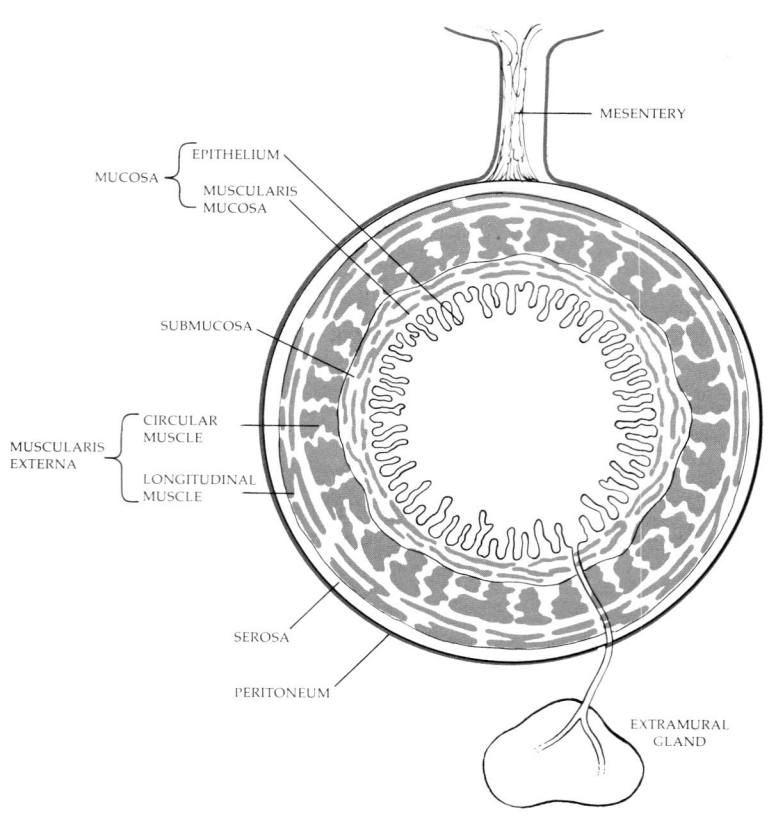

34–3 *The human digestive tract.*

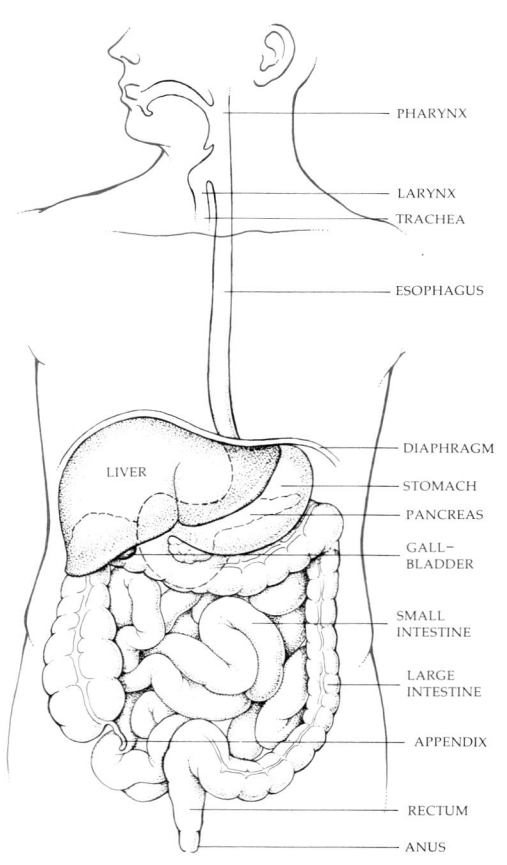

movements of these muscles produce the wavelike motions known as *peristalsis,* which move food along the digestive tract. At several points the circular layer of muscle thickens into heavy bands, called *sphincters.* These sphincters, by relaxing or contracting, control passage of food from one area of the digestive tract to another. They are located at the junctions of the esophagus and stomach, of the stomach and small intestine, of the small intestine and large intestine, and at the anus.

The Oral Cavity

The mechanical breakdown of food begins in the mouth. In man, as in most other mammals, the first tearing and grinding of food is done by the teeth. (The gizzards of birds and earthworms serve the same function.) Children have 20 teeth, which are gradually lost and replaced by a second set of 32 teeth as the jaw grows larger. Of the 16 adult teeth in each jaw, four are incisors, flat chisel-like structures specialized for cutting; two are canines; four are premolars ("in front of the molars"), each of which has two cusps, or protuberances, and so are also called bicuspids; and six are molars, each of which has four or five cusps. The premolars and molars are for grinding. Molars do not replace temporary teeth but are added as the jaw grows. The crown of the tooth—the visible part—is covered with enamel, the hardest substance in the body (mostly

Teeth of various mammals. Rodents (whose name comes from the Latin rodere, *to gnaw) are characterized by their sharp chisel-shaped incisors. Most rodents have no canines. Large predatory carnivores, such as the lion, have canines adapted for stabbing and slicing and large molars which can easily crush the bones of large herbivores. In elephants, tusks are modified incisors; they are used for attack and defense and for rooting food from the ground or breaking branches. (The tusks of other mammals, such as walruses, are modified canines.) The modern horse is a grazing animal; its incisors clip off grasses, which are then ground by the large, flat molars. Man, with a relatively unspecialized diet, has a correspondingly unspecialized dentition.*

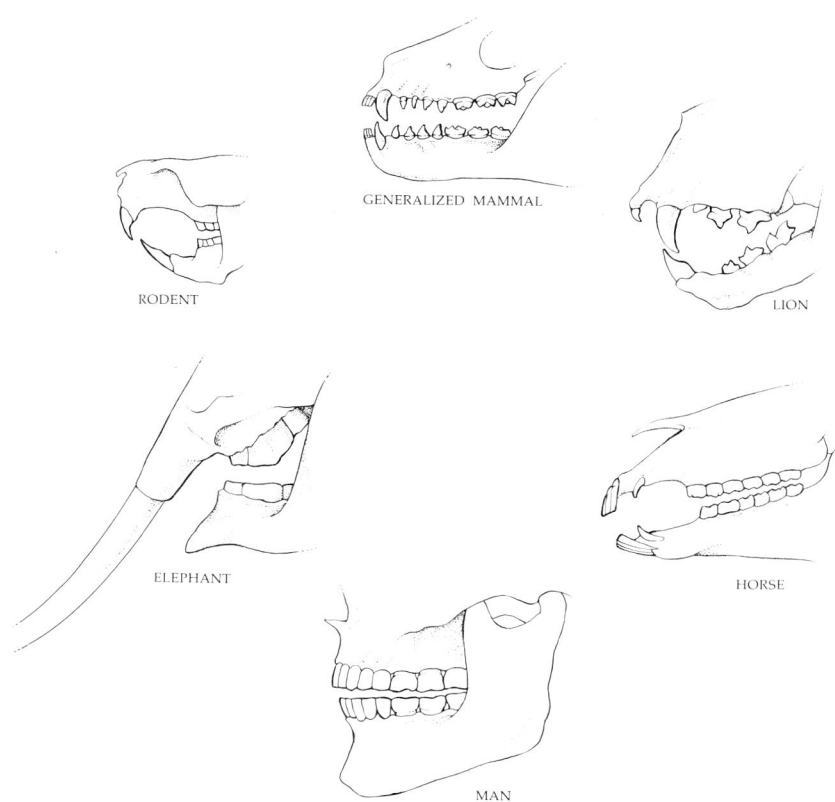

RODENT
GENERALIZED MAMMAL
LION
ELEPHANT
HORSE
MAN

In the human mouth, taste buds at the tip of the tongue are more sensitive to sweet substances, those at the sides to salt (toward the front) and to sour (further back), and those at the back to bitter—which is why saccharin, for instance, changes taste from sweet to bitter as it moves from front to rear.

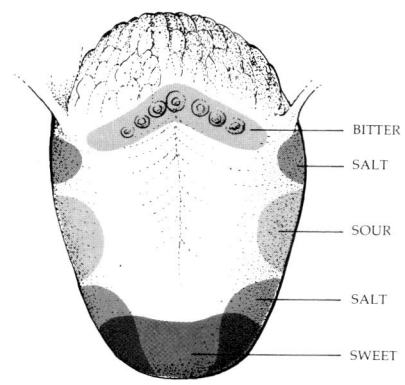

BITTER
SALT
SOUR
SALT
SWEET

calcium phosphate); the root, within the gum, is covered with cement, a substance similar to bone. The bulk of the tooth is composed of dentin, another bonelike material, which forms slowly during the life of the tooth. The pulp cavity contains the cells that produce dentin and also the nerves and blood vessels.

In mammals, the tongue serves largely to move and manipulate food. Some vertebrates, though, such as the hagfish and lampreys, have tongues equipped with horny "teeth," and the tongues of frogs and toads flip out (they are attached at the front, not the back) to catch insects. Mammalian tongues carry the taste buds, and in man the tongue has developed a secondary function of formulating sounds for communication.

While the food is being chewed in the mouth, it is moistened by saliva, a watery secretion produced by three pairs of large salivary glands plus numerous minute glands, the buccal glands, in the mucosa lining the mouth. The saliva, which is 99.5 percent water, also contains ions and mucus. It lubricates the food so that it moves down easily. On the average, 1,000 to 1,500 milliliters—1 to 1½ quarts—of saliva are produced a day. The saliva is slightly alkaline, owing to the presence of sodium bicarbonate. In man, saliva also contains a digestive enzyme, amylase, which begins the breakdown of starches. (Carnivores, such as dogs, which characteristically tear and gulp their food have no digestive enzymes

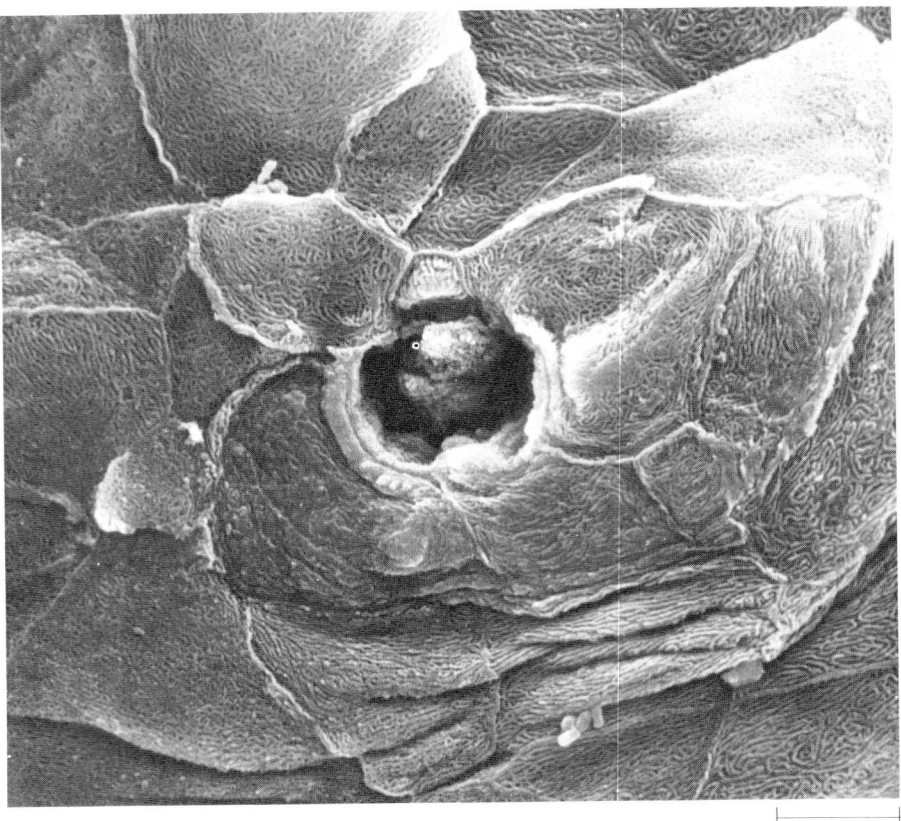

34–6 *Surface of human tongue, showing an outer taste pore. The sensory cells are in the structure just visible within the pore. The epithelial cells surrounding the pore die and are shed like the cells on the surface of the skin.*

|———| 2 μm

34–7 *The bulk of the saliva is produced by three pairs of salivary glands. Additional amounts are supplied by minute glands in the mucous membrane lining the mouth. The parotid glands are the sites of infection of the mumps virus.*

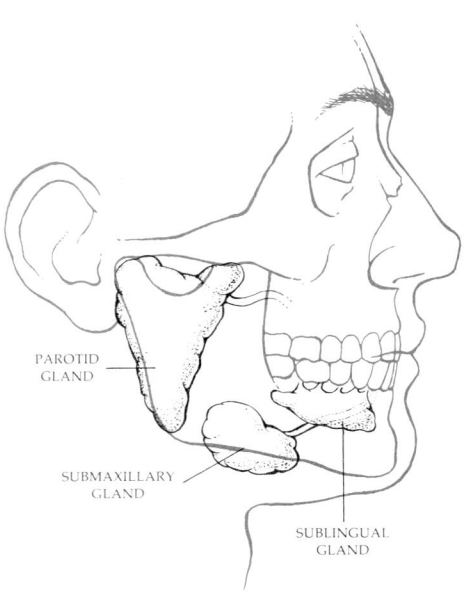

PAROTID
GLAND

SUBMAXILLARY
GLAND

SUBLINGUAL
GLAND

in the saliva.) Like all digestive enzymes, amylase works by hydrolysis (see page 61); that is, the breaking of each bond involves addition of a molecule of water.

The secretion of saliva is controlled by the autonomic nervous system. It can be initiated by reflexes originating in taste buds and in the walls of the mouth, and also by the mere smell or anticipation of food. (Think hard, for a moment, about a lemon.) Fear inhibits salivation; at times of great danger or stress, the mouth may get so dry that speech is difficult.

The Pharynx and Esophagus: Swallowing

From the mouth, food is propelled backwards toward the pharynx, an aperture at the back of the mouth which connects with the larynx and the esophagus. The pharynx also serves as a resonating cavity in sound production. Swallowing is the passing of food to the esophagus, a muscular tube (about 10 inches long in man), and through the esophagus to the stomach. Swallowing begins as a voluntary action and once under way in man, continues involuntarily. In man, the upper part of the tube is striated muscle but the lower part is smooth muscle. (Dogs, cats, and other animals that gulp their food have striated muscle along the whole length of the esophagus.) Both liquids and solids are propelled along the esophagus by rhythmic muscular contractions (peristalsis).

34-8 *Separation of the digestive and respiratory systems in mammals makes it possible for mammals to breathe while* *eating. The pharynx is the common cavity of the two systems. Notice how an inward bulge of the muscles of the pos-* *terior pharyngeal walls moves downward, pushing the food mass ahead of it.*

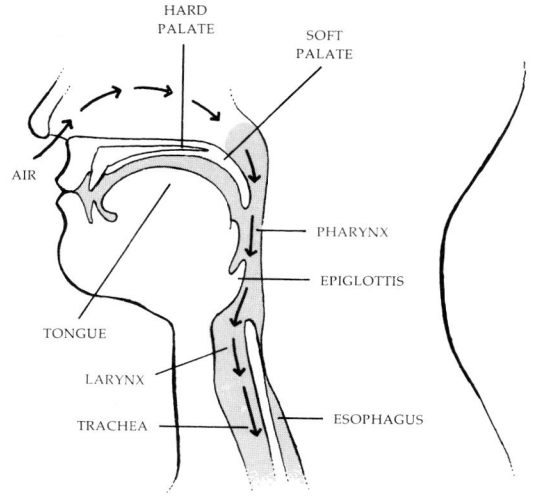

FACE AND NECK, SHOWING PARTS OF
RESPIRATORY AND DIGESTIVE SYSTEMS

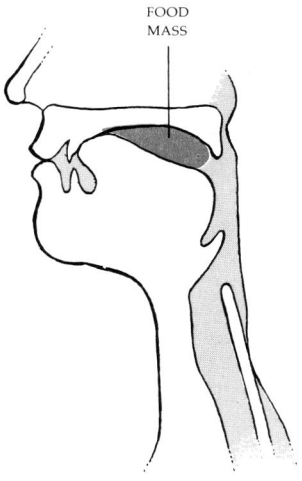

FOOD MASS IS IN MOUTH,
SOFT PALATE IS DRAWN UPWARD

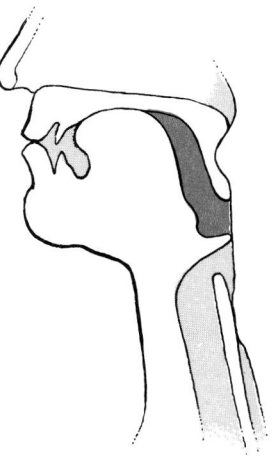

TONGUE PUSHES FOOD FURTHER BACK,
OPENING BETWEEN SOFT PALATE AND
PHARYNGEAL WALL CLOSES

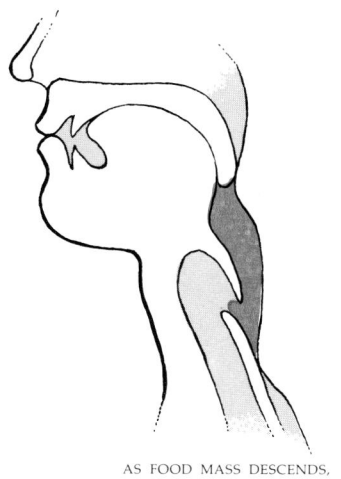

AS FOOD MASS DESCENDS,
EPIGLOTTIS TIPS DOWNWARD

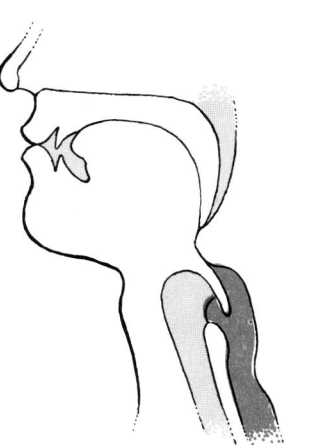

FOOD MASS PASSES INTO ESOPHAGUS

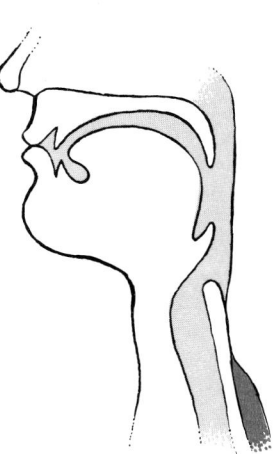

AS FOOD PASSES THE TRACHEA,
EPIGLOTTIS TURNS UP AGAIN AND
PHARYNX IS OPENED ONCE AGAIN

The Stomach: Holding and Churning

The esophagus passes through the diaphragm separating the thoracic and abdominal cavities and opens into the stomach, which with the rest of the digestive organs lies in the abdomen. All the viscera are suspended in the abdominal cavity, which is completely lined by a thin layer of connective tissue covered by moist epithelium known as the _peritoneum_. The intestines and other organs are held in place by folds of peritoneum known as _mesenteries_ (Figure 34–2). Mesenteries consist of double layers of tissue, with blood vessels, lymphatic vessels, and nerves lying between the two layers.

The stomach is essentially a collapsible, elastic bag, which, unless it is fully distended, lies in folds. Distended, it holds two to four quarts of food. The mucosal layer is very thick and contains numerous gastric pits (see Figure 34–9). Mucus-secreting epithelial cells cover the surface of the stomach and also line the gastric pits. Within the pits lie the parietal cells, which produce hydrochloric acid, and the chief cells, which produce pepsinogen.

As a consequence of the HCl secretion, the pH of gastric juice is normally between 1.5 and 2.5, far more acid than any other body fluid. The burning sensation you feel if you vomit or regurgitate food is caused by the acidity of gastric juice acting on unprotected membranes. Normally, the mucus forms a barrier between the stomach epithelium and the gastric juices and so prevents the stomach from digesting itself. The HCl kills most bacteria and other living cells in the ingested food and also loosens the tough, fibrous components of tissues and erodes the cementing substances between cells. It initiates the conversion of pepsinogen to its active form, pepsin, by splitting off a small portion of the molecule. Once pepsin is formed, it acts on other molecules of pepsinogen to form more pepsin. Pepsin, which breaks proteins down into peptides, is active only at the low pH of the normal stomach.

34–9 *Surface of the stomach, as shown in a scanning electron micrograph. The numerous indentations are gastric pits.*

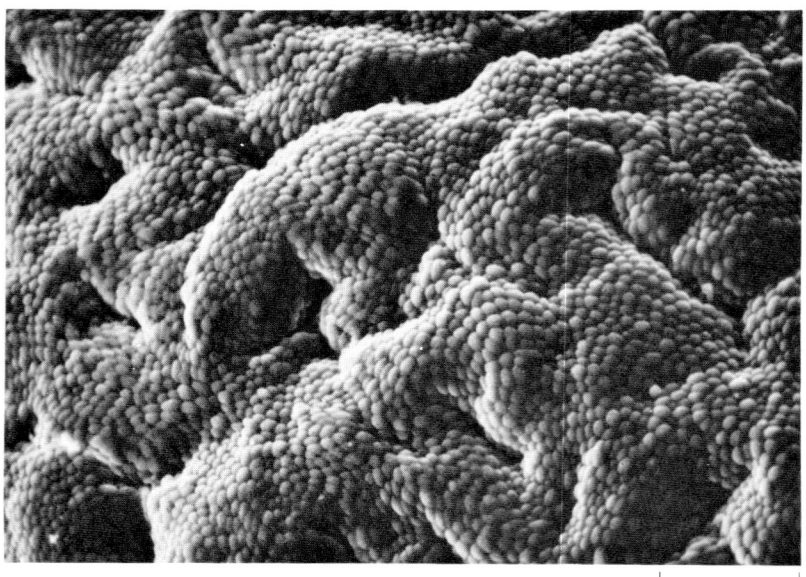

100 μm

4–10 *A cross section of stomach mucosa. The parietal cells secrete hydrochloric acid and the chief cells produce pepsinogen. Mucus, secreted by cells of the surface epithelium, coats the surface of the stomach and lines the gastric pits, protecting the stomach surface.*

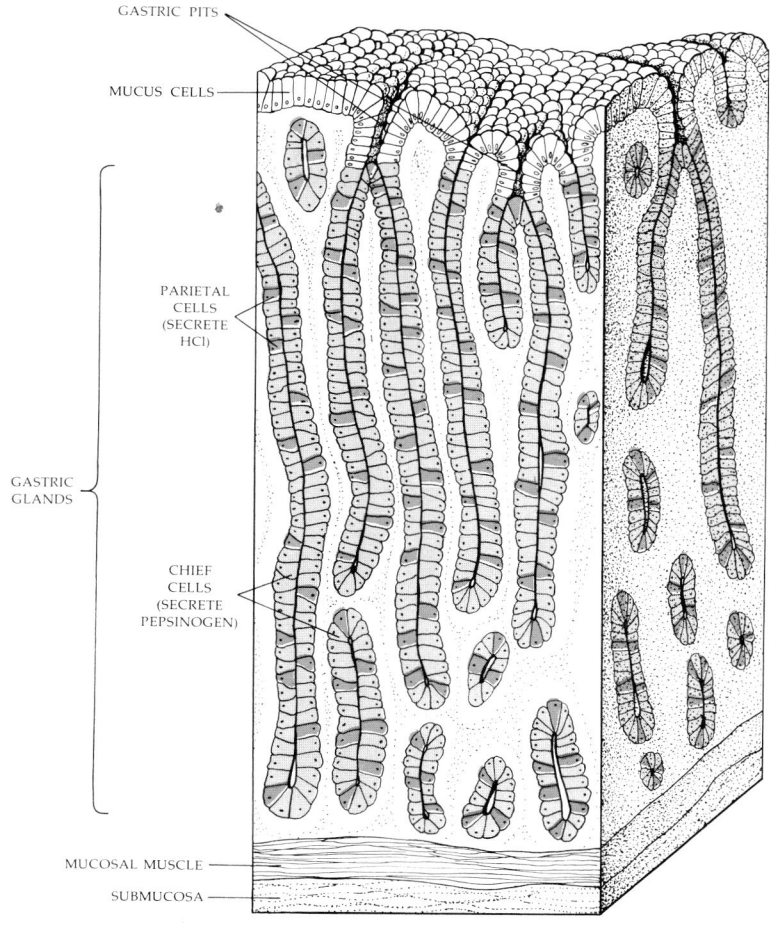

GASTRIC PITS

MUCUS CELLS

PARIETAL CELLS (SECRETE HCl)

GASTRIC GLANDS

CHIEF CELLS (SECRETE PEPSINOGEN)

MUCOSAL MUSCLE

SUBMUCOSA

With the exception of small molecules such as glucose and alcohol, little food is absorbed through the stomach. (Because alcohol is absorbed through the stomach, its effects are much more rapid when the stomach is empty.)

The stomach is under both neural and hormonal control. Anticipation of food and the presence of food in the mouth stimulate churning movements of the stomach and the production of gastric juices. Fear and anger decrease motility. When food reaches the stomach, its presence causes the release of a hormone, gastrin, from gastric cells into the bloodstream. This hormone acts on the cells of the stomach to increase their secretion of gastric juices.

In the stomach, as a result of the action of the mucus, the pepsin, the hydrochloric acid, and the churning motions, food is converted from a solid into a semiliquid mass. The stomach empties gradually through the pyloric sphincter, which separates the stomach and small intestine; the food mass is forced out, little by little, by peristalsis. The stomach is usually empty by the end of 4 hours after ingestion of a meal.

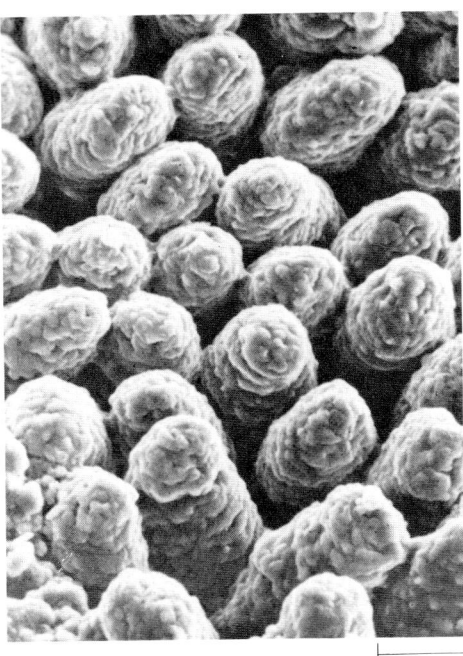

34–11   *Scanning electron micrograph of intestinal villi. The villi are associated with the areas of the digestive tract principally involved with absorption of food molecules.*

100 μm

The Small Intestine: Digestion

In the small intestine, the breakdown of food molecules begun in the mouth and stomach is completed, and the food molecules are then absorbed from the digestive tract into the circulatory system of the body, from which they are delivered to the individual cells.

Anatomically, the small intestine is characterized by circular folds in the submucosa, numerous fingerlike projections, *villi*, on the mucosa, and tiny cytoplasmic projections, *microvilli*, on the surface of the individual epithelial cells. All of these structural features increase the surface area of the small intestine. The epithelial cells of the mucosa secrete mucus. The small intestine is 10 to 13 feet long in the living adult; the total surface area of the human small intestine is about 2,000 square feet, the area of a singles tennis court. The upper 8 inches, known as the duodenum, is the most active in the digestive process; the rest is principally concerned with absorption of food nutrients.

The small intestine contains a variety of digestive enzymes, some of which are produced by the intestinal cells and some by the pancreas. (See Table 34–1.) The pancreatic enzymes enter the small intestine through the pancreatic duct, about 10 centimeters below the pyloric sphincter.

In the small intestine, amylases continue the breakdown of starch and glycogen begun in the mouth. Lipases hydrolyze fats into glycerol and fatty acids. Three types of enzymes break down proteins. One group breaks apart the long protein chains. Each enzyme in this group acts only on the bonds linking par-

34–12   *Longitudinal section of intestinal villi. Food molecules are absorbed through the walls of the villi and, with the exception of fat molecules, enter the bloodstream by means of the capillaries. Fats, hydrolyzed to fatty acids and glycerol, are taken up into the lymphatic system. The villi can move independently of one another; their motion increases after a meal.*

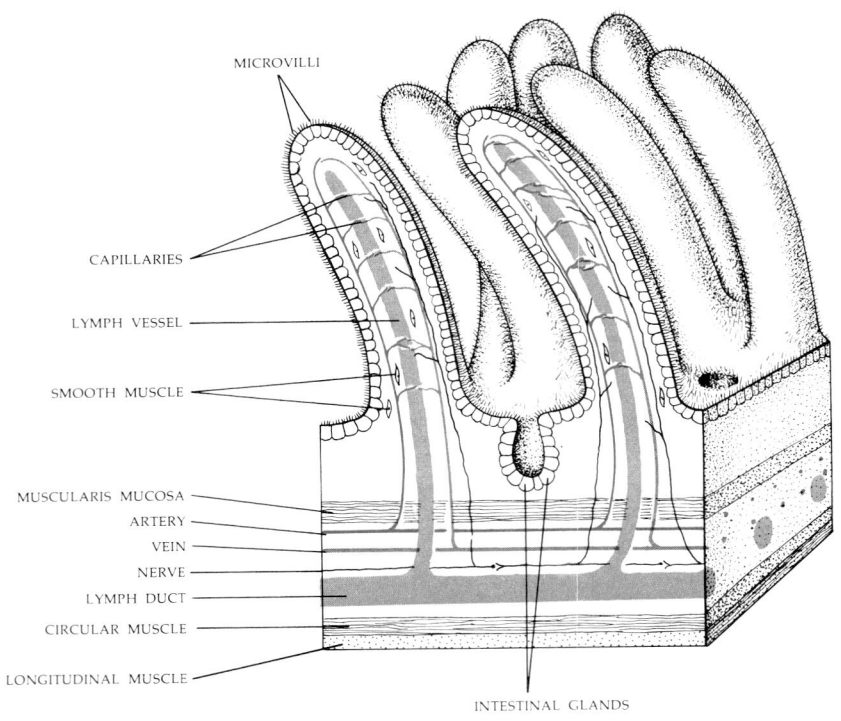

MICROVILLI

CAPILLARIES

LYMPH VESSEL

SMOOTH MUSCLE

MUSCULARIS MUCOSA
ARTERY
VEIN
NERVE
LYMPH DUCT
CIRCULAR MUSCLE
LONGITUDINAL MUSCLE

INTESTINAL GLANDS

-13 *Microvilli on two cells of intestinal epithelium. Near the outer surface of the cells, the membranes of the two cells are fused in what is known as a tight junction. Such tight junctions prevent extracellular materials from entering the space between the cells. In addition, desmosomes, the dense areas of cytoplasm below the tight junction, also appear to hold the cells tightly together.*

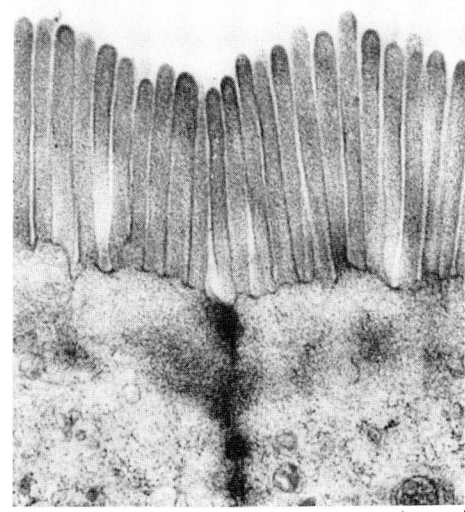

0.25 µm

Table 34-1 *Principal Digestive Enzymes*

Enzyme	Source	Substrate	Site of action
Salivary amylase	Salivary glands	Starches	Mouth
Pepsin	Stomach mucosa	Proteins, pepsinogen	Stomach
Pancreatic amylase	Pancreas	Starches	Small intestine
Lipase	Pancreas	Fats	Small intestine
Trypsin	Pancreas	Proteins, chymotrypsinogen	Small intestine
Chymotrypsin	Pancreas	Proteins	Small intestine
Carboxypeptidase	Pancreas	Peptides	Small intestine
Deoxyribonuclease	Pancreas	DNA	Small intestine
Aminopeptidase	Small intestine	Peptides	Small intestine
Dipeptidase	Small intestine	Dipeptides	Small intestine
Maltase	Small intestine	Maltose	Small intestine
Lactase*	Small intestine	Lactose	Small intestine
Sucrase	Small intestine	Sucrose	Small intestine
Enterokinase	Small intestine	Trypsinogen	Small intestine
Phosphatases	Small intestine	Nucleotides	Small intestine

* Not present in most non-Caucasian adults.

ticular amino acids, so that several are required to break a single large protein into shorter peptide chains. A second type of enzyme acts only on the end of a chain splitting off dipeptides, some on the amino end and some on the carboxyl end. A third group of enzymes comes into action breaking the remaining dipeptides (pairs of amino acids) into single amino acids. The amino acids and dipeptides are absorbed through the epithelial cells of the intestine and finally enter the bloodstream.

In addition to these many different enzymes, the small intestine receives an alkaline fluid from the pancreas, which neutralizes the stomach acid, and bile, which is produced in the liver and stored in the gallbladder. Bile contains a mixture of salts that, like laundry detergents, emulsify fats, breaking them apart into droplets. In this form they can be attacked by the lipases. Bile is rich in sodium bicarbonate, which neutralizes the stomach acid. Neutralizing the acidity is essential because the intestinal enzymes are optimally active at a pH of 7 to 8 and would be denatured by the acid pH of the gastric juices.

Bile also contains pigments that are derived from a breakdown of hemoglobin; the chief pigment is known as bilirubin. If the liver is not functioning properly, owing to damage, infection, or some other disease process, the bile pigments may be removed from the blood. Abnormal amounts of bilirubin in the blood cause the eyes and skin to become yellow, a condition known as jaundice.

RUMINANTS

The ruminants—a group that includes cattle, sheep, goats, camels, and giraffes—are able to utilize cellulose as a source of energy. Ruminants do not themselves digest this tough polysaccharide; it is broken down by the action of resident microorganisms in their digestive tracts. In the ruminant digestive system is a series of four stomachs, the first two of which (the rumen) contain a rich concentration of bacteria and protozoans. These microorganisms secrete enzymes that break cellulose and fats down into simple fatty acids and gases (carbon dioxide and methane). The fatty acids pass through the walls of the rumen into the bloodstream of the animal and are utilized as energy sources in various parts of the body. The gases are belched forth by the animal. The last two stomachs of the ruminants secrete enzymes which break down the proteins of the bacteria and protozoans that continuously arrive from the first two stomachs. Thus, in effect, the ruminant obtains all its dietary essentials from this rich culture of bacteria and protozoans. The microorganisms, in turn, are provided with an environment rich in carbohydrates and maintained at a constant temperature favorable for growth.

THE REGULATION OF DIGESTION

The digestive activities of the small intestine are coordinated and regulated by hormones (Table 34–2). In the presence of food, the duodenum releases secretin, a hormone that stimulates the pancreas to secrete its alkaline fluid and the liver to make bile, and cholecystokinin (also called pancreozymin), a hormone that acts on the pancreas to stimulate the secretion of enzymes and that triggers the emptying of the gallbladder. A fourth hormone, enterogastrone, is stimulated by the presence of fatty acids and acts on the stomach to inhibit its release of HCl. It is important, however, to appreciate the complex interplay of stimuli and checks and balances that serves to activate and inactivate digestive enzymes and to adjust the chemical environment. In addition to hormonal influence, the intestinal tract is also regulated by the autonomic nervous system. Stimulating the vagus nerve, which carries parasympathetic nerve fibers, causes the pancreas to secrete digestive enzymes.

The Pancreas and Liver

The pancreas and the liver are specialized digestive organs, developing, both in the embryo and in the course of evolutionary history, from the digestive tract. The bulk of the tissues of the pancreas resemble salivary gland tissue. In addition, little clusters of cells in the pancreas, known as the islets of Langerhans (named after their discoverer, not some romantic Pacific atoll), secrete insulin and glucagon, which are released into the bloodstream rather than into the gut, as are the digestive enzymes.

The liver carries out an extraordinary variety of functions. It is the source of bile and also of prothrombin, plasminogen, and other plasma proteins involved in blood clotting. It breaks down a variety of toxic substances, including alcohol and other drugs, and inactivates a number of hormones, thus playing an important role in hormone regulation. Liver cells break down excess amino acids, converting the nitrogen waste to urea (see page 668), which is excreted by the kidney. They are the source of angiotensinogen (page 679), one of the regulators of kidney function, and are storage sites for many vitamins, especially B_{12} and A, and minerals. Moreover (see page 645), the liver plays the central role in the regulation of blood glucose.

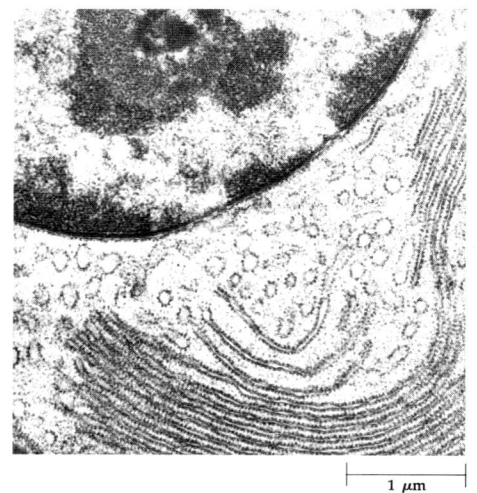

34–14   *Portion of pancreatic cell involved in the production of digestive enzymes. Notice the rough endoplasmic reticulum characteristic of cells that synthesize proteins for export. The cell nucleus can be seen in the upper half of the micrograph.*

34–15 *Electron micrograph of a liver cell. The dark granules are glycogen. Despite the variety of activities carried out by the liver, the cells of the liver all resemble one another in appearance and, apparently, in function. There seems to be no division of labor, as there is in the pancreas, for example. Liver cells, as you would expect, are rich in mitochondria and in rough endoplasmic reticulum.*

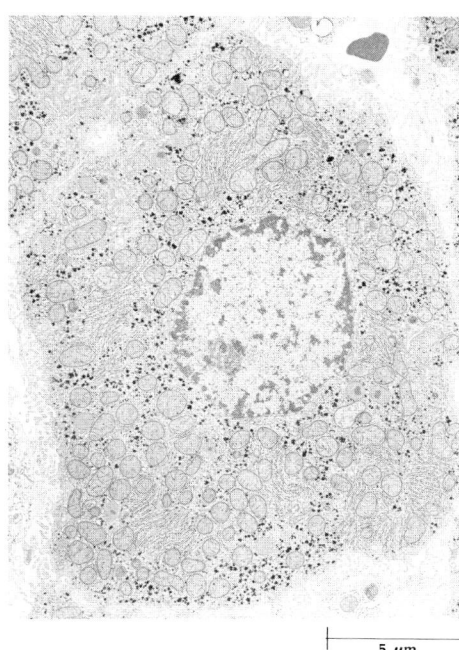

5 μm

34–16 *Gastric hormone interactions in the human digestive system.*

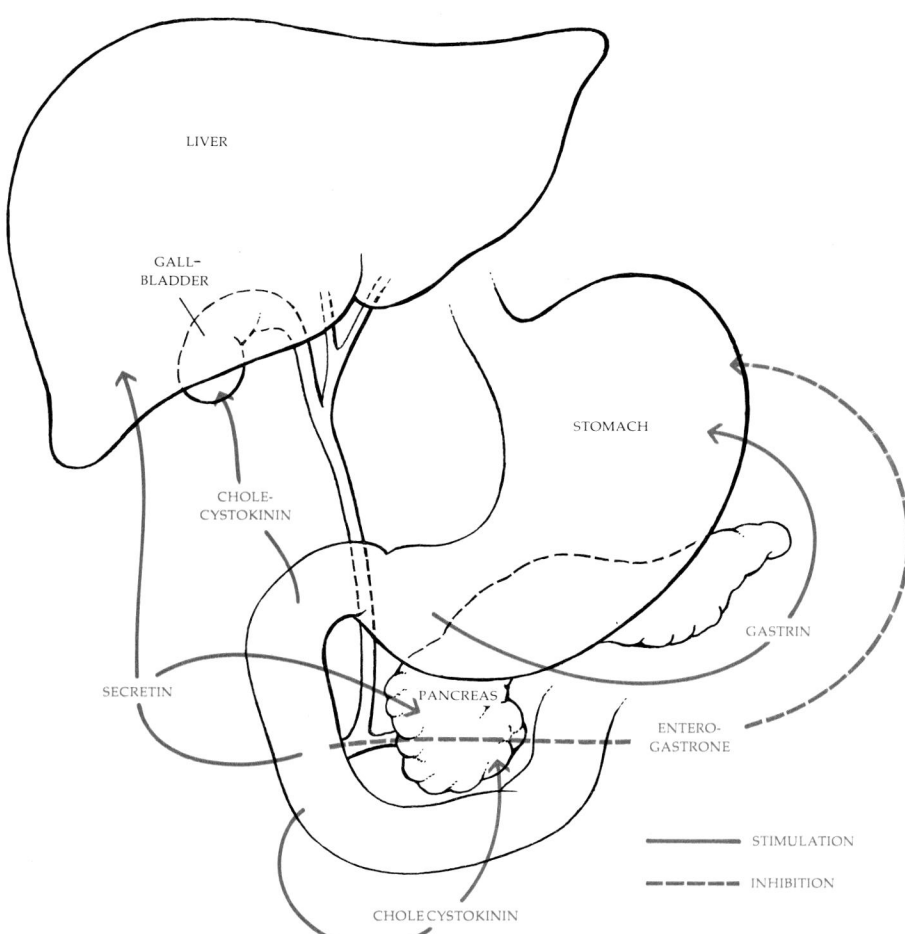

LIVER

GALL-
BLADDER

CHOLE-
CYSTOKININ

SECRETIN

PANCREAS

STOMACH

GASTRIN

ENTERO-
GASTRONE

——— STIMULATION

- - - - - INHIBITION

CHOLECYSTOKININ

Table 34–2   *Gastrointestinal Hormones*

Hormone	Source	Stimulus for production	Action
Gastrin	Stomach	Food in stomach	Stimulates secretion of gastric juices
Enterogastrone	Small intestine	Fatty acids in intestine	Inhibits secretion of HCl and gastric motility
Secretin	Small intestine	HCl in duodenum	Stimulates secretion of pancreatic fluids containing bicarbonate; stimulates production of bile by liver
Cholecystokinin (pancreozymin)	Small intestine	Food in duodenum	Stimulates release of pancreatic enzymes; stimulates release of bile from gallbladder

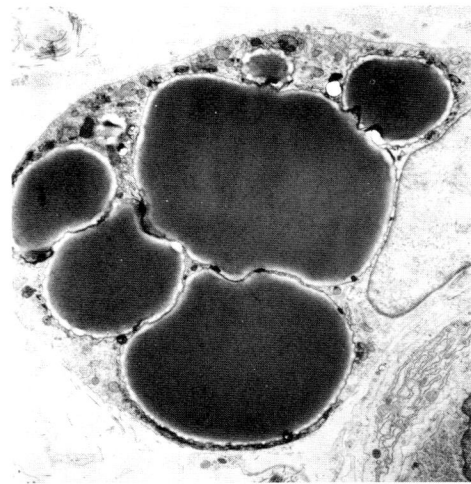

34–17 *A fat cell from the skin of a newborn rat. The nucleus is to the right. The large stored droplets of fat that nearly fill the cell have been stained with osmium for electron microscopy. When excess calories are taken in in the diet, fat accumulates in these specialized cells, and when caloric intake is less than sufficient, fat is mobilized, broken down to glucose, and released into the bloodstream.*

2.5 μm

ABSORPTION

The Small Intestine: Absorption of Food Molecules

The food moves by peristalsis along the small intestine. Here most of the absorption of food molecules takes place. Monosaccharides are rapidly absorbed from the small intestine into the blood vessels of the villi, as are amino acids and dipeptides, the breakdown products of proteins. Fats are hydrolyzed to glycerol and fatty acids; these are taken up by the epithelial cells of the villi, resynthesized to fats, and in this form enter the lymphatic vessels. Proteins present in the lymphatics coat the fat molecules and so prevent their coalescing. These lipoprotein droplets are transferred to the venous circulation via the thoracic duct. After you consume a meal high in fats, your blood plasma actually turns a milky color owing to the presence of these fat droplets. While they are in the bloodstream, the fat molecules are hydrolyzed once again, this time by a plasma enzyme. The fatty acids and glycerol are then broken down by cellular respiration (see page 125) or are stored as fat in subcutaneous tissue, mesenteries, muscle, or liver.

The Large Intestine: Absorption of Water and Elimination

The chief function of the large intestine is the resorption of water. Its epithelial cells secrete mucus, which lubricates the drying mass of the food residue. In the course of digestion, large amounts of water—some two or three quarts—enter the stomach and small intestines by osmosis from the body fluids or as secretions of the glands lining the digestive tract. When the resorption process is interfered with, as in diarrhea, severe dehydration can result. Infants dying from infectious diarrhea, until recently a chief cause of infant deaths in this country, died principally as a consequence of water loss.

The appendix is a blind pouch off the large intestine, which may become irritated, inflamed, and then infected. If it ruptures, as a consequence of inflammation and swelling, it spills its bacterial contents into the abdominal cavity. The appendix has no known function.

The large intestine also harbors a considerable population of symbiotic bacteria (including the familiar *Escherichia coli*), which break down additional food substances. Living on these food substances, largely materials we lack the enzymes to digest, they synthesize amino acids and also vitamins, which are absorbed into the bloodstream. These bacteria are the chief source of vitamin K.

The pigments that result from hemoglobin breakdown in the liver pass into the intestine, giving the feces their characteristic color. Also, excess calcium and iron are secreted as salts into the large intestine. However, the bulk of the fecal matter consists of bacteria (mostly dead cells) and cellulose fibers, along with other indigestible substances. It is stored briefly in the rectum and eliminated through the anus as the feces.

REGULATION OF BLOOD GLUCOSE

As we noted at the beginning of the chapter, the function of digestion is to provide an energy source for each of the individual cells of which the body is composed. Although most animals ingest food only at more or less regular intervals, the cellular energy supply remains extraordinarily constant. The liver

In 1536, explorer Jacques Cartier, following the example of some American Indians, saved some of his party from scurvy with a cedar-needle brew. About two hundred years later it was observed that fresh fruits, especially citrus fruits, prevented scurvy. Scottish physician James Lind, the central figure in the drawing shown here, was among the first to urge that ships on long voyages take along a supply of citrus fruits, a practice which led to British seamen being known as limeys. Ascorbic acid was subsequently discovered to be the vitamin supplied by the plant material, but despite extensive research, little is known about its exact function at the cellular level.

19 *Vitamin D is a steroid that is produced in the skin by the action of ultraviolet rays from the sun on a cholesterol-like compound (the absorbed energy opens the second ring of the molecule). According to one current hypothesis, the ancestors of modern* Homo sapiens *originated in the tropics and were all dark-skinned. In those populations that moved northwards, however, the screening effects of the darker pigment, which inhibits the production of vitamin D, caused selection for lighter skin color, whereas no such selection pressures occurred in the sun-drenched areas nearer the equator.*

VITAMIN D

plays the central role in this critical process. Glucose and other monosaccharides are absorbed into the blood from the intestinal tract and are shunted directly to the liver by way of the portal vein (see page 618). The liver converts most of these monosaccharides to glycogen (the liver stores enough glycogen to satisfy the body's needs for about 4 hours) and to fat. Some of the fat is stored in the liver and some in fat cells. The liver similarly breaks down amino acids (excess amino acids are not stored) and converts them to glucose, excreting the nitrogen in the form of urea and storing the glucose as glycogen. Whether the liver takes up or releases glucose and the amount it takes up or releases is determined primarily by the concentration of glucose in the blood, is regulated by a number of hormones (page 583), and is influenced by the autonomic nervous system.

SOME NUTRITIONAL REQUIREMENTS

Because of the liver's activities in converting various types of food molecules into glucose, the energy requirements of the body can be met by either carbohydrates, proteins, or fats—the three principal types of food molecules. They are ordinarily met by a combination of the three. Carbohydrates and proteins supply about the same number of calories per gram, and fats about twice as much as either of them.

In addition to the need for calories, the cells of the body need the twenty amino acids required for assembling proteins. Of these twenty, vertebrates can synthesize twelve, either from a simple carbon skeleton or from another amino acid (amination or transamination, page 454). The other eight, which must be obtained in the diet, are known as essential amino acids,* which, as you can see, is something of a misnomer. Plants are the ultimate source of these essential amino acids, but it is difficult (although by no means impossible) to obtain them by eating a completely vegetable diet, largely because plant proteins are relatively deficient in lysine and tryptophan.

Mammals also require but cannot synthesize certain polyunsaturated fats that provide fatty acids needed for fat synthesis. These can be obtained by eating plants or insects (or eating animals that have eaten plants or insects).

Vitamins are an additional group of molecules required by living cells, which cannot be synthesized by animal cells. They are characteristically required only in small amounts. Many of them function as coenzymes. Table 34–3 indicates some of the principal vitamins required in the human diet. There is no clear evidence that the ingestion of amounts of any particular vitamin in excess of the amounts available in well-balanced diets has any beneficial effects on a normally healthy individual.

The body also has a dietary requirement for a number of inorganic substances, minerals. These include calcium and phosphorus for bone formation, iodine for thyroid hormone, iron for hemoglobin and cytochromes, and sodium and chloride and other ions essential for ionic balance. Most of these are present in the ordinary diet or in drinking water. Like the vitamins, however, they must be given in a supplementary form when the dietary intake is inadequate or when the individual is not able to assimilate them normally.

---

* The essential amino acids are isoleucine, leucine, lysine, methionine, phenylalanine, threonine, tryptophan, and valine.

Table 34–3 *Vitamins*

Designation letter and name	Major sources	Function	Deficiency symptoms
A, carotene	Egg yolk, green or yellow vegetables, fruits, liver, butter	Formation of visual pigments, maintenance of normal epithelial structure	Night blindness; dry, flaky skin
B-complex vitamins: B_1, thiamine	Brain, liver, kidney, heart, whole grains	Involved in decarboxylations, formation of enzyme involved in Krebs cycle	Beri-beri, neuritis, heart failure, mental disturbance
B_2, riboflavin	Milk, eggs, liver, whole grains	Flavoproteins in electron transport	Photophobia, fissuring of skin
B_6, pyridoxine	Whole grains, liver, kidney, fish	Coenzyme for amino acid metabolism and fatty acid metabolism	Dermatitis, nervous disorders
B_{12}, cyanocobalamin	Liver, kidney, brain	Nucleoprotein synthesis (RNA), prevents pernicious anemia	Pernicious anemia, malformed erythrocytes
Biotin	Egg white, synthesis by intestinal bacteria	Concerned with protein synthesis, CO_2 fixation, and transamination	Scaly dermatitis, muscle pains, weakness
Folic acid	Liver, leafy vegetables	Nucleoprotein synthesis, formation of erythrocytes	Failure of erythrocytes to mature, anemia
B_3, niacin (nicotinic acid)	Whole grains, liver and other meats	Part of coenzymes NAD, NADP, and CoA	Pellagra, skin lesions, digestive disturbances
Pantothenic acid	Present in most foods	Forms part of coenzyme A (CoA)	Neuromotor disorders, cardiovascular disorders, GI distress
C, ascorbic acid	Citrus fruits, tomatoes, green leafy vegetables	Vital to collagen and ground (intercellular) substance	Scurvy, failure to form connective tissue fibers
D_3, calciferol	Fish oils, liver, milk and other dairy products, action of sunlight on lipids in the skin	Increases Ca absorption from gut, important in bone and tooth formation	Rickets (defective bone formation)
E, tocopherol	Green leafy vegetables	Maintains resistance of red cells to hemolysis	Increased red blood cell fragility
K, naphthoquinone	Synthesis by intestinal bacteria, leafy vegetables	Enables prothrombin synthesis by liver	Failure of blood coagulation

SUMMARY

Digestion is the process by which food is broken down into molecules that can be taken up by the bloodstream and so be distributed to the individual cells of the body. It occurs in successive stages, regulated by an interplay of hormones and nervous stimuli. In man, food is processed initially in the mouth, where the breakdown of starch begins, and then in the stomach, where highly acid gastric juices destroy bacteria and begin to break down proteins. Most of the digestion occurs in the duodenum. Digestive hormones (enzymes) are secreted by the intestinal cells, and these enzymes, in turn, stimulate the functions of the pancreas and the liver. The pancreas synthesizes and secretes a number of digestive enzymes and also an alkalizing fluid; the liver produces bile. The pancreas is also the source of insulin and glucagon, hormones involved in a variety of other metabolic activities.

Absorption of food molecules takes place largely in the small intestine. Monosaccharides, amino acids, and dipeptides are absorbed into the blood vessels of the villi, and fats into the lymphatic vessels.

The chief energy source for cells in the mammalian body is glucose. The organ principally responsible for maintaining a steady supply of glucose is the liver, which stores glucose (in the form of glycogen or fat) when glucose levels in the blood are high and breaks down glycogen or fats, releasing glucose, when the levels drop. These activities of the liver are regulated by at least five different hormones.

Water is resorbed from the residue of the food mass as it passes through the large intestine. The large intestine contains symbiotic bacteria which are the source of certain vitamins. Dietary requirements include calories, essential amino acids, certain unsaturated fats, vitamins, and minerals. These can be obtained in amounts adequate for normal individuals from the diet.

QUESTIONS

1. Define the following terms: mucosa, sphincter, incisor, peristalsis, mesentery, jaundice, villi, essential amino acids.
2. Diagram the human digestive tract, including all of the major organs. Check your drawing against Figure 34–3.
3. Trace the chemical progress of a hamburger on a bun through your digestive tract.
4. Humans are omnivores rather than strict herbivores or carnivores or specialists on a single organism as a food source. Humans have an unusually large number of substances specifically required in their diets. How might these two observations be related? Can you make an educated guess as to which way cause-and-effect operates in this case?
5. If you oxidize a pound of fat, whether butter or human adipose tissue, 3,500 kilocalories are released. Susan Avoirdupois is on one of her diets, eating only 1,000 kilocalories per day. Sedentary and well-insulated, she will probably expend an estimated 2,000 kilocalories per day. How many pounds can she expect to lose in each week of dieting?

Chapter 35

Energy and Nutrients II: Respiration

Respiration has two meanings in biology. At the intracellular level, "respiration" refers to the process by which cells break down (oxidize) molecules and so provide energy for the cell. For multicellular organisms, respiration also means taking in oxygen and getting rid of carbon dioxide. This latter process—which is, of course, essential for the former—is the subject of this chapter.

The means by which organisms extract energy without oxygen are not very efficient, as we saw in Chapter 8. Before oxygen accumulated in the atmosphere, the only forms of life were one-celled organisms, and the only present-day forms that can carry on life processes without oxygen are very simple organisms, including a few types of bacteria and yeasts. The transition to an atmosphere containing free oxygen made possible one of the giant steps in evolution.

A second revolution in terms of energy came about with the transition to land. Air is a far better source of oxygen than water; one-fifth of the air of the

35–1 *In animals, chemical energy is converted to kinetic energy by the oxidation of food molecules. Oxygen consumption of a running animal increases (linearly) with its speed. These speedy animals are Thompson's gazelles, photographed on an African savanna.*

Table 35–1	Composition of Dry Air	
Nitrogen	77	%
Oxygen	21	
Argon	1	
Other gases*	0.97	
Carbon dioxide	0.03	

* Includes hydrogen, neon, krypton, helium, ozone, and xenon.

modern atmosphere is made up of free oxygen, whereas at ordinary temperatures, even when water is saturated with air, only about 1 part in 250 (by weight) is free oxygen. Not only must more water be processed for oxygen, but water weighs a great deal more. A fish spends up to 20 percent of its energy in the muscular work associated with respiration, whereas an air breather expends only 1 or 2 percent of its energy in respiration. Also, oxygen diffuses much more rapidly through air than through water, about 300,000 times more rapidly, and so can be replenished much more quickly as it is used up by respiring organisms. All the higher vertebrates—the birds and the mammals—are air breathers, even those that live in the water. On the other hand, all oxygen absorption by animals occurs across wet membranes, and respiratory surfaces must therefore be built in such a way as to remain moist.

AIR UNDER PRESSURE

At sea level, the air around us exerts a pressure on our skin of about 15 pounds per square inch. This pressure is enough to raise a column of water 32 feet high or a column of mercury 760 millimeters (29.91 inches) in the air. (Atmospheric pressure is generally measured in terms of mercury simply because mercury is relatively heavy—so the column will not be inconveniently tall—and because air does not dissolve in it.) Of this 760 millimeters, 155 millimeters result from the pressure of the oxygen in the air. This is known as the partial pressure of oxygen (P_{O_2}). The total pressure of a mixture of gases is the sum of the partial pressures of the gases in the mixture. It is as a result of the atmospheric pressure that oxygen enters living organisms.

We are so accustomed to the pressure of the air around us that we are unaware of its presence or of its effects on us. However, if you visit a place—such as Mexico City—which is at a comparatively high altitude and therefore

−2 *Atmospheric pressure is usually measured by means of a mercury barometer.* (a) *To make a simple mercury barometer, fill a long glass tube, open at one end only, with mercury. Closing the tube with your finger,* (b) *invert it into a dish of mercury. Remove your finger. The mercury level will drop until the pressure of its weight inside the tube is equal to the atmospheric pressure outside. At sea level, the height of the column will be about 76 centimeters (29.91 inches).* (c) *If the tube and dish are enclosed in an airtight covering and the air pressure is lowered by pumping air out, the mercury level will fall.*

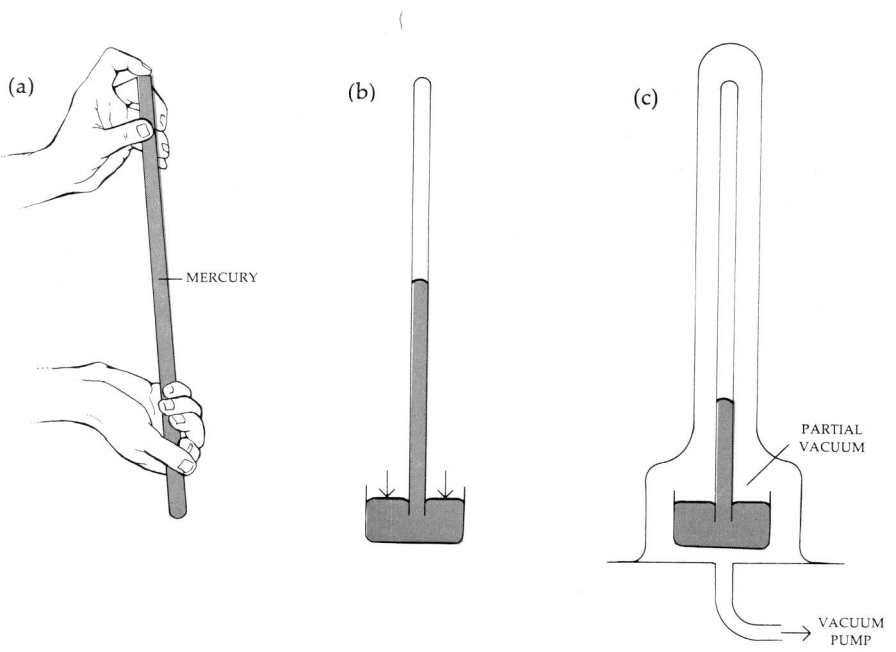

(a) MERCURY

(b)

(c) PARTIAL VACUUM

VACUUM PUMP

35–3 (a) *Gaseous exchange across the entire surface of the body is found in a wide range of small animals from protozoans to earthworms.* (b) *Gaseous exchange across the surface of a flattened body. Flattening increases the surface-to-volume ratio and also decreases the distance over which diffusion has to occur within the body. Flatworms are examples.*

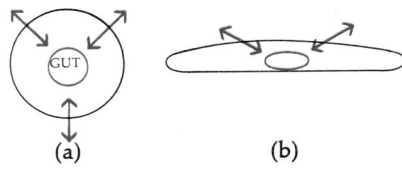

(a)   (b)

(c) *External gills. These increase the surface area, but note that they are unprotected and therefore easily damaged. Gaseous exchange usually takes place across the rest of the body surface as well as the gills. External gills are found in polychaete worms and some amphibians.* (d) *Highly vascularized internal gills. A ventilation mechanism draws water over the gill surfaces, as in fish.*

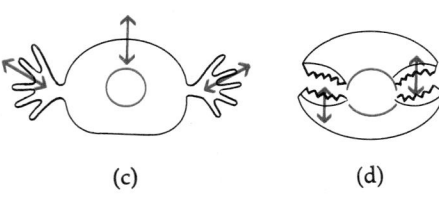

(c)   (d)

(e) *Gaseous exchange at the terminal ends of fine tracheal tubes that branch through the body and penetrate into all the tissues, found in insects and other arthropods.* (f) *Highly vascularized lungs consisting of sacs connected to a pharynx. Air is drawn into them by a ventilation mechanism. Lungs are found in all air-breathing vertebrates.*

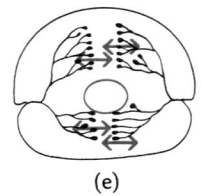
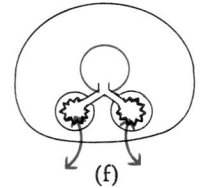

(e)   (f)

has a lower atmospheric pressure, you will feel lightheaded at first and will tire easily. Regular inhabitants of such places breathe more deeply, have enlarged hearts that circulate oxygen-carrying blood more rapidly, and have more red blood cells for oxygen transport.

The consequences of increased atmospheric pressures are seen in deep-sea divers. Early in the history of deep-sea diving, it was found that if divers come up from the bottom too quickly, they get the "bends," which is always painful and sometimes fatal. The bends result from breathing air under pressures higher than atmospheric pressure. Nitrogen is relatively insoluble in water, but high pressure forces nitrogen from the air contained in the lungs into solution in the blood and tissues. If the body is rapidly decompressed, the nitrogen bubbles out of the blood, like the bubbles that appear in a bottle of soda water when you first take off the top. These bubbles lodge in the capillaries, stopping blood flow, or invade nerves or other tissues.

HOW OXYGEN GETS INTO CELLS

Oxygen enters and moves within cells by diffusion. Cell membranes are freely permeable to oxygen, even those that are not permeable to water. This is true of all cells, whether an amoeba, a paramecium, a liver cell, or a brain cell. However, substances can move effectively by diffusion only for very short distances —less than 1 millimeter. Also, the respiratory surface must be exposed to a medium, such as air or water, over which the diffusion process can take place.

You will recall from Chapter 7 that diffusion occurs only *down* a concentration gradient. The diffusion of oxygen into a cell or into an organism takes place because there is a higher concentration of oxygen outside than inside the cell or organism (which, remember, is using up oxygen). Similarly, the passage of carbon dioxide out of the cell or organism takes place because the concentration is higher inside than outside the respiratory surface.

HOW OXYGEN GETS INTO ORGANISMS

The larger the volume of an animal, the smaller, in proportion, is its exposed surface area (as was demonstrated in Figure 6–1). Some small aquatic or subterranean animals respire through their outer surfaces, just as single-celled organisms do, and the cells receive their oxygen by simple diffusion. This very simple means of respiration is usually found only in organisms with each cell quite close to the surface. Many eggs and embryos also respire in this simple way, particularly in their early stages of development.

A second major means of respiration also involves diffusion through the general surface area but with the assistance of circulation. Many of the worms respire in this way. Earthworms, for example, have a network of capillaries just one cell layer below the surface (see page 380). The blood picks up oxygen by diffusion as it travels near the surface of the animal and releases oxygen by diffusion as it circulates past the oxygen-poor cells in the interior of the earthworm's body. The tube shape of a worm exposes a proportionally large surface area. Some worms can adjust their surface area in relation to oxygen supply. If you have an aquarium at home, you may be familiar with tubifex worms, which

are frequently sold as fish food. If these worms are placed in water low in oxygen, such as a poorly aerated aquarium, they will stretch out as much as 10 times their normal length and thus increase the surface area through which oxygen diffusion occurs. Frogs (see Figure 35–7) also respire through their skin, which is richly supplied with capillaries.

Gills and lungs are other ways of increasing the respiratory surface. Gills are outgrowths, whereas lungs are ingrowths, or cavities. The respiratory surface of the gill, like that of the earthworm, is a layer of cells, one cell thick, exposed to the environment on one side and to circulatory vessels on the other. The layers of gill tissue may be spread out flat, stacked in layers, or convoluted in various ways. The gill of a clam, for instance, is shaped like a steam-heat radiator, which is also designed to provide a maximum surface-area-to-volume ratio.

Lungs are internal cavities into which the oxygen-laden air is taken. They have certain disadvantages as compared with gills; from the point of view of diffusion, the continuous flow of oxygen across the respiratory surface provided by the gill is more efficient than the intermittent flow of oxygen in the lung. However, they have one overwhelming advantage in air; the respiratory surface can be kept moist without a large loss of water by evaporation. Lungs are also more efficient; 20 percent of a fish's energy expenditure goes into extracting oxygen from water, whereas only 1 to 2 percent is required by air breathers. Lungs are a principal characteristic of terrestrial vertebrates, but they are found in some lower animals. Air-breathing (pulmonate) snails, for example, have independently evolved lungs that are remarkably similar to the lungs of some amphibians.

–4 *Insects and some other terrestrial arthropods breathe by means of tracheae. These are inward tubular extensions of the exoskeleton, which are thickened in places, forming spiral supports that hold the tubes open. The lining of the trachea is molted when a new exoskeleton is formed. This micrograph shows part of the tracheal system of a cockroach.*

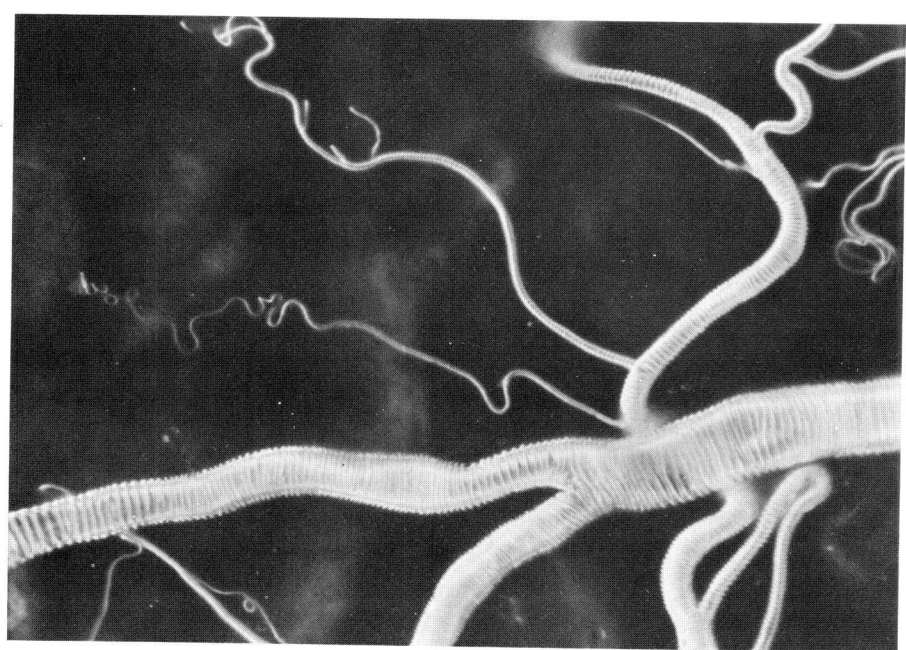

20 μm

EVOLUTION OF THE GILL

The vertebrate gill probably originated primarily as a feeding device. Primitive vertebrates breathed mostly through their skin (as do modern frogs). They filtered water into their mouths and out of what we now call their gill slits, extracting bits of organic matter from the water as it went through. (*Amphioxus*, which is believed to resemble closely the ancestral vertebrate, functions in this same way.)

In the course of time, numerous selection pressures, chiefly involved with predation, came into operation. As one consequence, there was a trend toward an increasingly thick skin, even one armored or covered with scales. Such a skin was not, of course, useful for respiratory purposes. At the same time, similar forces were operating to produce animals that were larger and swifter and so more efficient at capturing prey and escaping predators. Such animals also had larger energy requirements—and consequently larger oxygen requirements. These problems were solved by the "capture" of the gill for a new purpose: respiratory exchange. The surface area of the epithelium beneath the gill slowly increased in size and in blood supply over the millennia. The modern gill is the result of this evolutionary process.

35–5   *Rates of diffusion are proportional not only to the surface areas exposed but to differences in concentration. The greater the difference in concentration of a gas or liquid, the more rapid its diffusion. In the gill of the fish, the circulatory vessels are arranged so that the blood is pumped through them in a direction opposite to that of the oxygen-bearing water. This arrangement, which is an example of what is known as countercurrent exchange (see facing page) results in a far more efficient transfer of oxygen to the blood than if the blood flowed in the same direction as the water.*

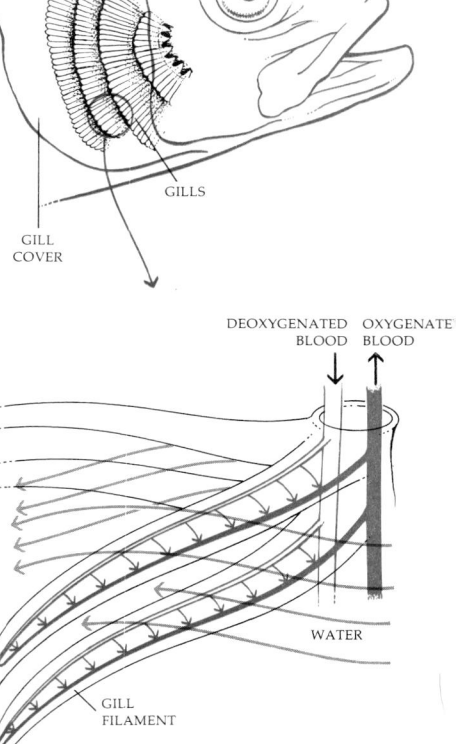

GILLS

GILL COVER

DEOXYGENATED BLOOD   OXYGENATE BLOOD

WATER

GILL FILAMENT

COUNTERCURRENT EXCHANGE

The principle of countercurrent exchange is illustrated by a hot-water pipe and a cold-water pipe placed side by side to achieve maximum temperature transfer under ideal conditions. In (a), the hot water and cold water flow in the same direction. Heat from the hot-water pipe warms the cold water until both temperatures equalize, at 5. Thereafter, no further exchange takes place and the water in both pipes remains lukewarm. In (b), the flow is in opposite directions, so that heat transfer continues for the length of the pipes. The result is that the hot water transfers most of its heat as it travels through the pipe and the cold water is warmed to almost the initial temperature of the hot water. The principle of countercurrent exchange applies equally well to oxygen exchange (Figure 35–5) or, as we shall see in Chapter 36, to exchange of solutes such as sodium ion (Na^+).

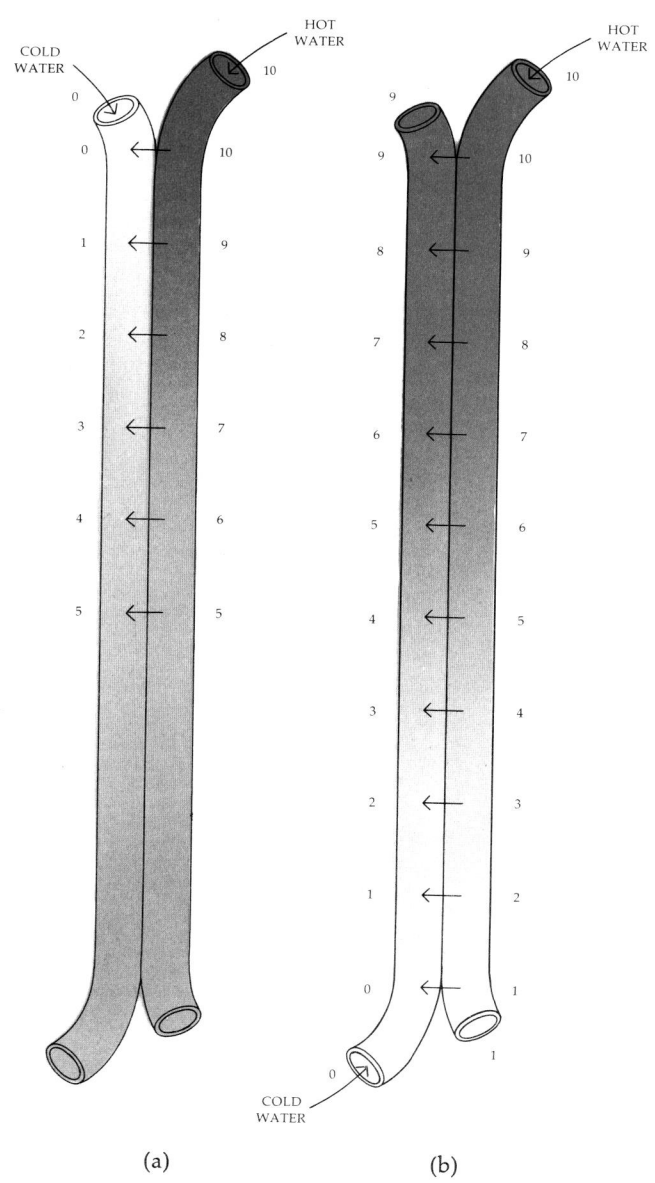

(a)   (b)

35–6 *Swimming is far less expensive, in terms of energy requirements, than is running —that is, if you are an aquatic animal, such as the porpoises shown here. However, a man, whose body is basically unfit for moving through the water, uses five times as much energy—as measured by his oxygen consumption—in swimming as in running.*

35–7 *Frogs have relatively small and simple lungs, and a major part of their external respiration takes place through the skin. The chart shows the results of a study of a frog in which oxygen and carbon dioxide exchanges through the skin and lungs were measured simultaneously for one year. Can you explain why the respiratory activity peaks in April and May and is lowest in December and January?*

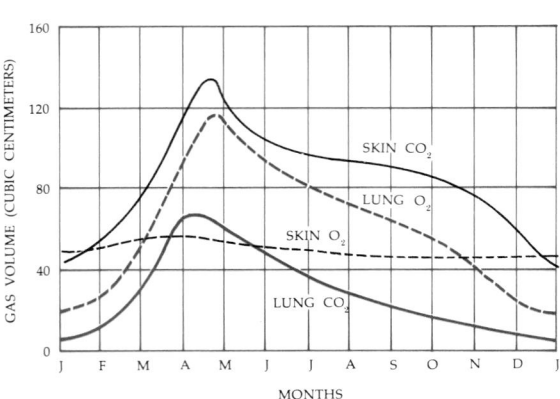

In most fish, the water (in which oxygen is dissolved) is pumped in at the mouth by oscillations of the bony gill cover and flows out across the gills. The fish can regulate the rate of flow, and sometimes assist it, by opening and closing its mouth. Fast swimmers, such as mackerel, obtain enough oxygen only by keeping perpetually on the move, driving water through their gills. Such fish cannot be kept in an aquarium or any other space where their motion is limited because they will die of a lack of oxygen.

EVOLUTION OF THE LUNG

Some, perhaps all, primitive fish had lungs as well as gills, although the lungs were not efficient enough to serve as more than accessory structures. These lungs were a special adaptation to fresh water (in which the vertebrates probably arose), which, unlike ocean water, may stagnate and, because of decay or algal bloom, become depleted of oxygen. A few species of lungfish still exist. These surface and gulp air into their lungs and so can live in water that does not have sufficient oxygen to support other fish life.

Amphibians and reptiles have relatively simple lungs, with small internal surface areas, although their lungs are far larger and more complex than those of the lungfish. The lungs of lungfishes developed directly from the pharynx, the posterior portion of the mouth cavity which leads to the digestive tract. In amphibians, reptiles, and other air-breathing vertebrates, we see the evolution of the windpipe, or trachea, guarded by a valve mechanism, the glottis, and nostrils, which make it possible for the animal to breathe with its mouth closed. In moist-skinned amphibians, the skin is still an important factor in exchange of oxygen and carbon dioxide, but reptiles and dry-skinned amphibians (toads) breathe almost entirely through their lungs.

An important feature of all vertebrate lungs is that the exchange of air with the atmosphere takes place as a result of changes in lung volume. Such lungs are known as ventilation lungs. Frogs gulp air and force it into their lungs in a swallowing motion; then they open the glottis and let it out again. In reptiles, birds, and mammals, air is sucked into the lungs as a consequence of changes in the size of the lung cavity, brought about by activity of the chest muscles.

5–8   *In terms of energy consumption, flying is less expensive than walking and running. Many birds can fly more than 1,000 kilometers nonstop, whereas an animal of similar weight could not travel on the ground for that distance without stopping for food. However, flying is more expensive, in energy-consumption terms, than swimming is for a fish.*

–9   *The lungs of birds are extraordinarily efficient. They are small and are expanded and compressed by movements of the body wall. Each lung has several air sacs attached to it which empty and fill like balloons at each breath. No gas exchange takes place in the sacs. They appear rather to act as bellows. Thus, the lung is almost completely flushed with fresh air at every breath; there is little residual "dead" air left in the lungs, as there is in mammals.*

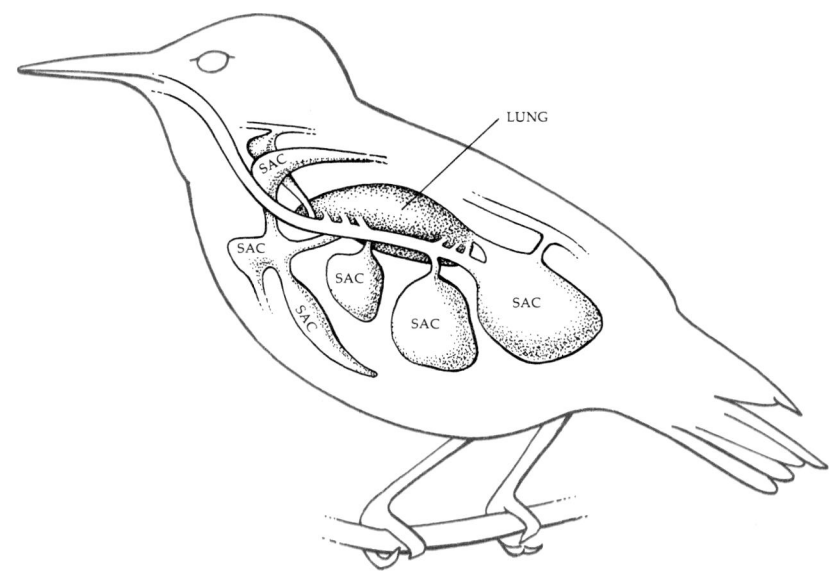

35–10 *The human respiratory system. (a) Air enters through the nasopharynx and passes down the trachea, bronchi, and bronchioles to the alveoli (b) in the lungs. Within each alveolus, of which there are some 300 million in a pair of lungs, oxygen and carbon dioxide diffuse into and out of the bloodstream, through the capillary walls. The electron micrograph (c) shows an alveolar capillary in cross section, with several irregular thin sections of red blood cells.*

In man, inspiration (breathing in) and expiration (breathing out) usually take place through the nose. The nasal cavities are lined with hairs and also cilia, both of which trap dust and other foreign particles. The epithelial cells that line the cavities also secrete mucus, which humidifies the air and collects debris that can be removed by swallowing, sneezing, or spitting. The cavities have a rich blood supply, which keeps their temperature high, warming the air before it reaches the lungs.

From the nostrils, the air goes to the pharynx and from there to the *larynx*, which is located in the upper front part of the neck. An adult human larynx is shaped somewhat like a triangular box, with its broad base above it. Across it are stretched the vocal cords, which are two ligaments stretched across the lumen of the respiratory tract. Vibrations of these cords, by expired air, cause the sounds made in speech. These cords are influenced by the male hormone. At puberty, the cords become longer and thicker, sometimes so rapidly that the adolescent male temporarily loses control over them, occasionally emitting embarrassing squeaks. Laryngitis, which is simply an inflammation of the vocal cords, interferes with their vibration, so you "lose your voice."

From the larynx, inspired air travels through the *trachea*, which is a long membranous tube, also lined with ciliated, mucus-producing epithelial cells, and strengthened by rings of cartilage that prevent it from collapsing during inspiration. The trachea leads into the *bronchi* (singular, *bronchus*), which subdivide into smaller and smaller passageways, the *bronchioles*.

The bronchi and bronchioles are surrounded by a thin layer of smooth muscle. Adrenaline causes the bronchioles to dilate; histamine causes them to contract. Asthma is a spasm of these muscles resulting in labored breathing.

Cilia along the trachea, bronchi, and bronchioles beat continuously, pushing mucus and foreign particles embedded in mucus up toward the pharynx, from which it is generally swallowed. We are usually aware of this production of

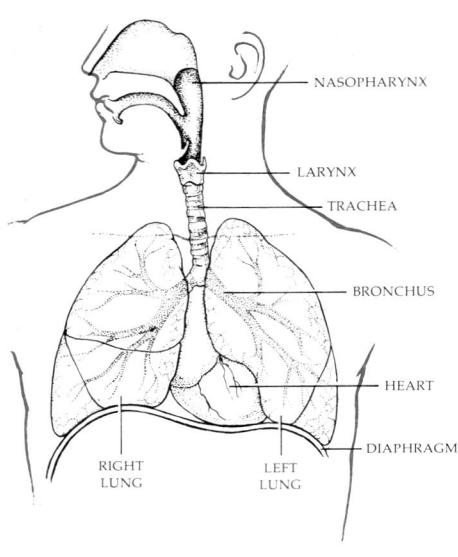

NASOPHARYNX

LARYNX

TRACHEA

BRONCHUS

HEART

DIAPHRAGM

RIGHT LUNG   LEFT LUNG

(a)

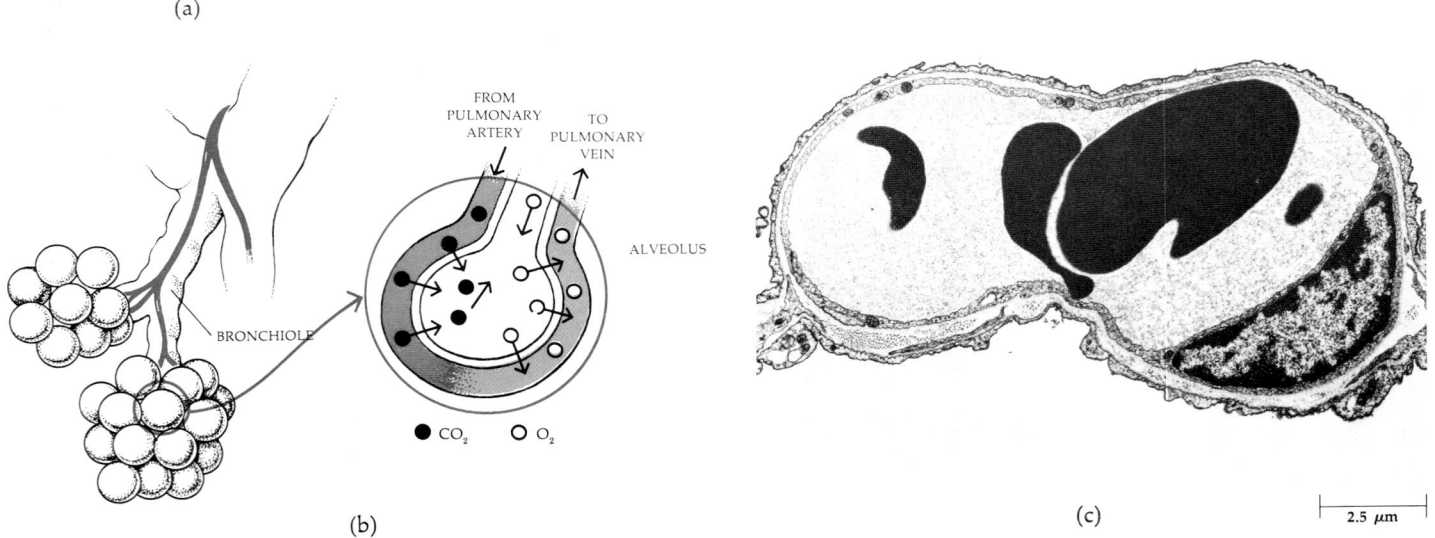

FROM PULMONARY ARTERY

TO PULMONARY VEIN

ALVEOLUS

BRONCHIOLE

● CO₂   ○ O₂

(b)

(c)

2.5 μm

35–11 *Electron micrograph and diagram of tracheal epithelial cells. As you can see, the epithelium contains both ciliated cells (marked C in the diagram) and mucus-secreting cells (M), sometimes called goblet cells. The beating of the cilia removes foreign particles from the trachea and distributes a protective coating of mucus. In the micrograph, the white circles in the cytoplasm of the middle goblet cell are secretion droplets of mucus.*

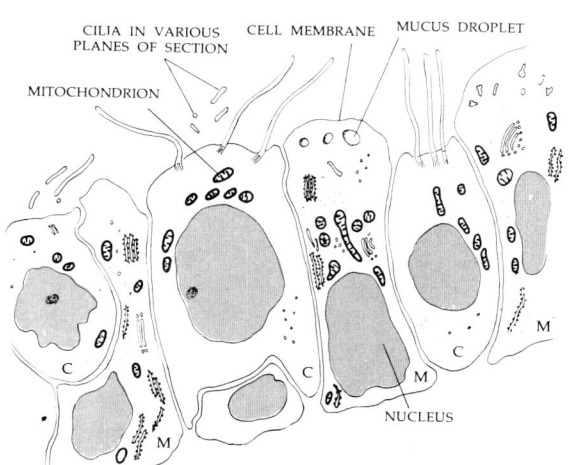

CILIA IN VARIOUS PLANES OF SECTION   CELL MEMBRANE   MUCUS DROPLET

MITOCHONDRION

NUCLEUS

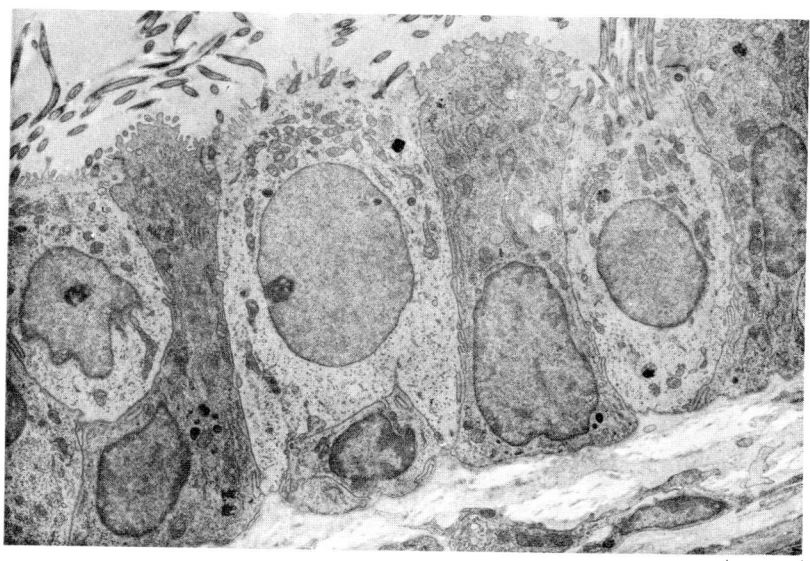

2.5 μm

mucus only when it is increased above normal as a result of an irritation of the membrane.

The actual exchange of gases takes place in small air sacs, the *alveoli*, which are clustered in bunches like grapes around the ends of the bronchioles. Each is about 1 or 2 millimeters in diameter, and each is surrounded by capillaries. The walls of the capillaries and of the alveoli each consist of only a single layer of flattened epithelial cells firmly attached to one another; thus the barrier between an alveolus and its underlying capillaries is only about 0.3 millimeter. Gases are exchanged between the air in the alveoli and the blood in the capillaries by diffusion.

A pair of human lungs has about 300 million alveoli, providing a respiratory surface area of some 70 square meters, or about 750 square feet—approximately 40 times the surface area of the entire human body.

The lungs are surrounded by a thin membrane and the thoracic cavity (part of the coelom) is lined by a similar membrane. These are known as the pleura. They secrete a small amount of fluid which lubricates them so they slide past one another as the lungs expand and contract. Pleurisy is the inflammation of these membranes which sometimes, when they are irritated, secrete excess fluid that collects in the thoracic cavity.

LUNG CANCER

(a) *Normal epithelial tissue, showing ciliated columnar cells similar to those seen in Figure 35–11 and an underlying layer of basal cells. In healthy tissue, the basal cells divide continuously, replacing the ciliated epithelial cells as these are worn away and slough off. A basement membrane separates the epithelial cells from the underlying connective tissues.*

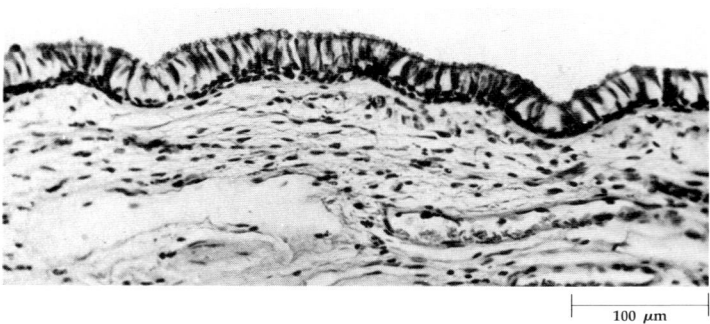
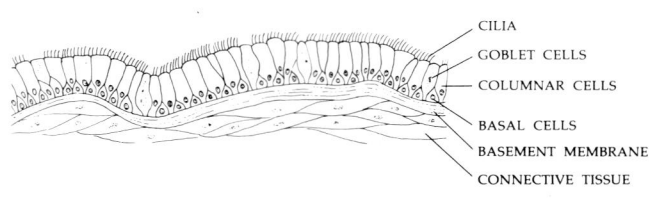

100 μm

(b) *Tissue showing hyperplasia, an overgrowth of the basal cells. This is typically the first change in the epithelium seen among cigarette smokers.*

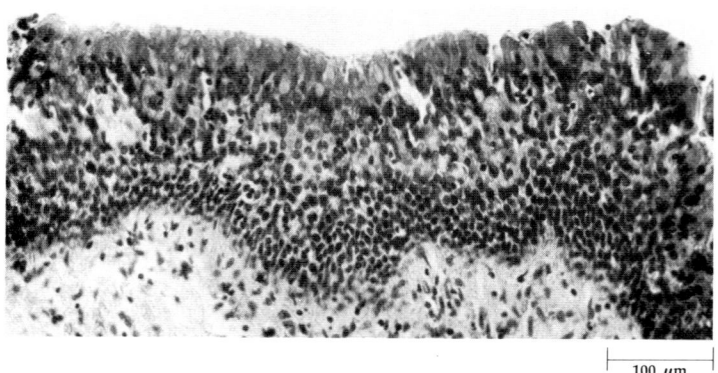
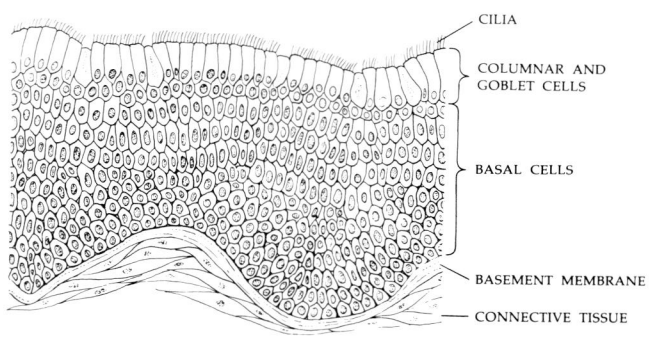

100 μm

(c) *A later stage. The ciliated cells have been replaced by squamous (flattened) cells. With the disappearance of the cilia, the epithelium has lost a primary defense against tiny particles of soot, tobacco, tar, dust, etc.*

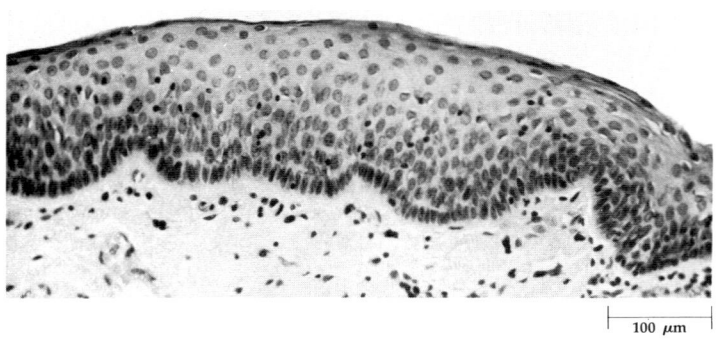
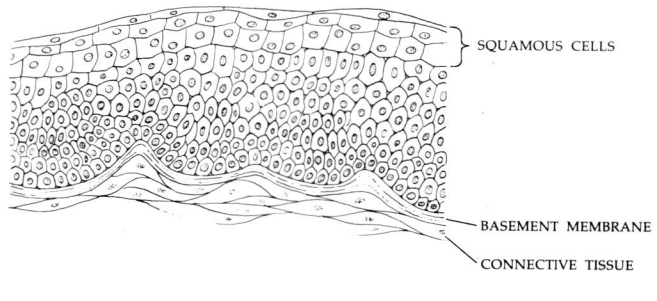

100 μm

(d) *The cells have developed atypical nuclei and become disordered. This condition is called carcinoma in situ. There are usually no symptoms of disease at this stage.*

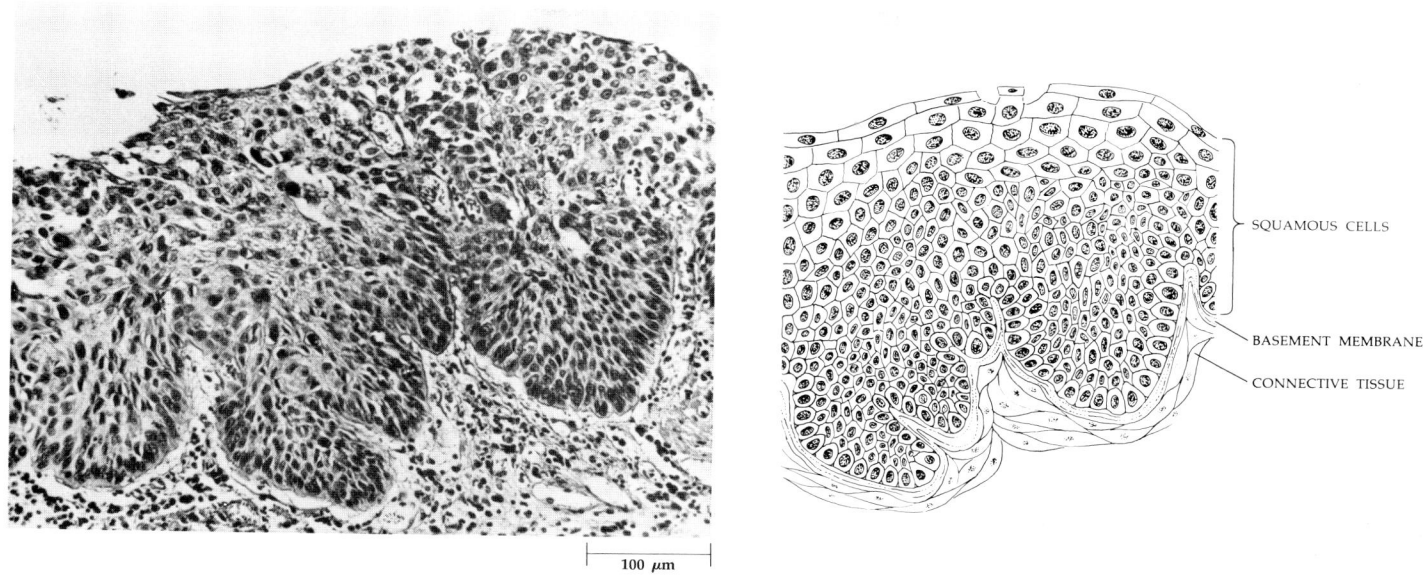

SQUAMOUS CELLS

BASEMENT MEMBRANE

CONNECTIVE TISSUE

100 μm

(e) *When these cells break through the basement membrane, as they have here, they may spread throughout the lungs and to the rest of the body. Lung cancer has become the most common form of cancer among males in the United States and is rapidly increasing in incidence.*

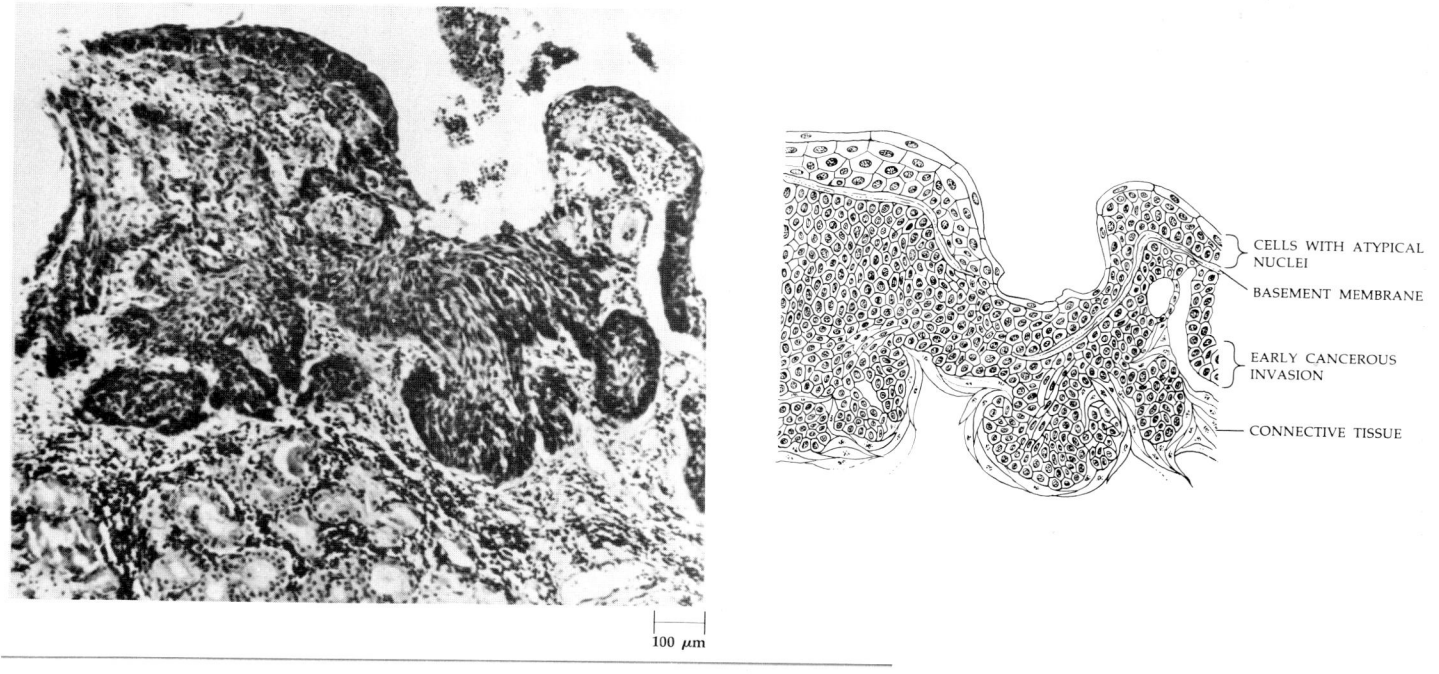

CELLS WITH ATYPICAL NUCLEI

BASEMENT MEMBRANE

EARLY CANCEROUS INVASION

CONNECTIVE TISSUE

100 μm

35–12   *Oxygen-hemoglobin dissociation curve. This curve represents figures for normal adult hemoglobin at 38°C and at a normal pH. As the partial pressure of oxygen drops, the oxygen and the hemoglobin dissociate.*

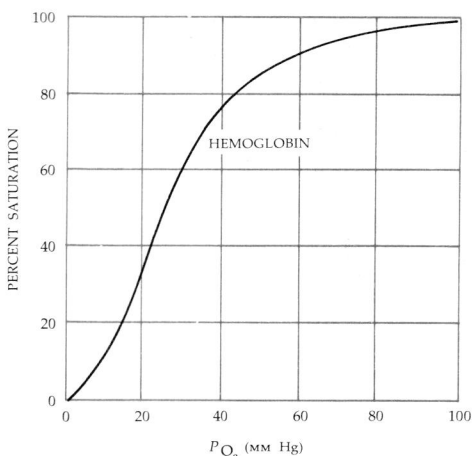

Gas Exchange

Oxygen

Gases are exchanged between the alveoli and their underlying capillaries by diffusion. Oxygen, which is only moderately soluble in water, is carried by hemoglobin molecules packed within red blood cells, the erythrocytes. As we noted in Chapter 5, hemoglobin is composed of a protein (globin) plus an iron-containing porphyrin (heme). Four such globin-heme units are found in each hemoglobin molecule. Each hemoglobin molecule combines with four atoms of oxygen, enabling whole blood to carry about 60 times as much oxygen as could be transported by an equal volume of water or plasma. The amount of oxygen carried by the hemoglobin molecules in the blood is related to the partial pressure of oxygen (P_{O_2}) in the blood. In adult humans, the partial pressure of oxygen of the blood as it leaves the lungs is 100 millimeters of mercury (100 mm Hg); at this pressure, the hemoglobin is saturated with oxygen. As the hemoglobin molecules travel through the bloodstream, the P_{O_2} drops, and as it drops, the oxygen in the hemoglobin molecules is given up. (See Figure 35–12.) Little oxygen is yielded as the P_{O_2} drops from 100 mm Hg to 60 mm Hg. This is a built-in safety factor that protects individuals at high altitudes or those who have heart or lung diseases that decrease blood P_{O_2}. However, as the pressure drops below 60 mm Hg, oxygen is given up much more readily.

35–13   *These curves show how the amount of oxygen carried by the hemoglobin is related to oxygen pressure. When oxygen pressure reaches 100 mm Hg—the pressure usually present in the human lung, for instance—the hemoglobin becomes totally saturated with oxygen. As the pressure drops, the oxygen in the hemoglobin molecule is given up. Therefore, when blood carrying oxygen reaches the capillaries, where pressure is only about 40 mm Hg or less, it gives up its oxygen to the tissues. A curve located to the right signifies that the oxygen is given up more readily, that is, at a higher pressure.*
(a) Small animals have higher metabolic rates and so need more oxygen per gram of tissue than larger animals. Therefore, they have blood that gives up oxygen more readily. (b) The llama, which lives in the high Andes of South America, has a hemoglobin which enables its blood to take up oxygen more readily at the low atmospheric pressures. (c) The fetus must take up all its oxygen from the maternal blood. The hemoglobin of mammalian fetuses has a greater affinity for oxygen than does the hemoglobin of adult mammals, and so the oxygen tends to leave the maternal blood and enter the fetal blood.

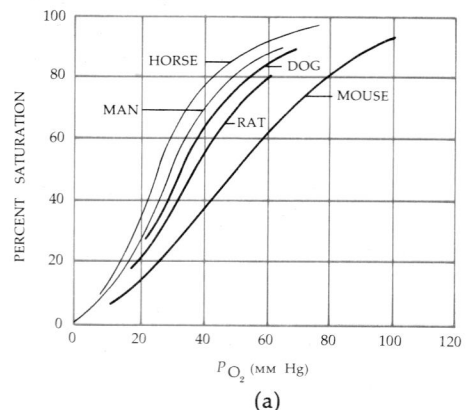

(a)

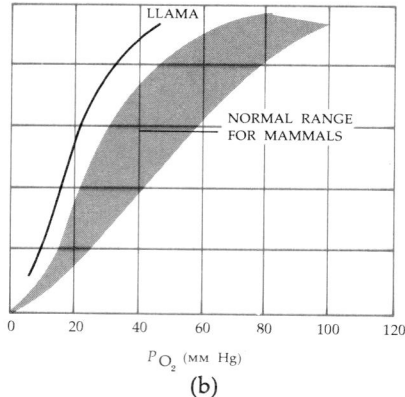

(b)

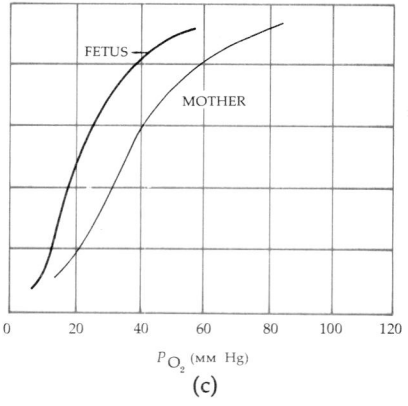

(c)

5–14 *Tertiary structure of myoglobin, as deduced from x-ray diffraction analyses. Myoglobin closely resembles a single chain of the four-chain hemoglobin molecule. The heme portion is shown in color.*

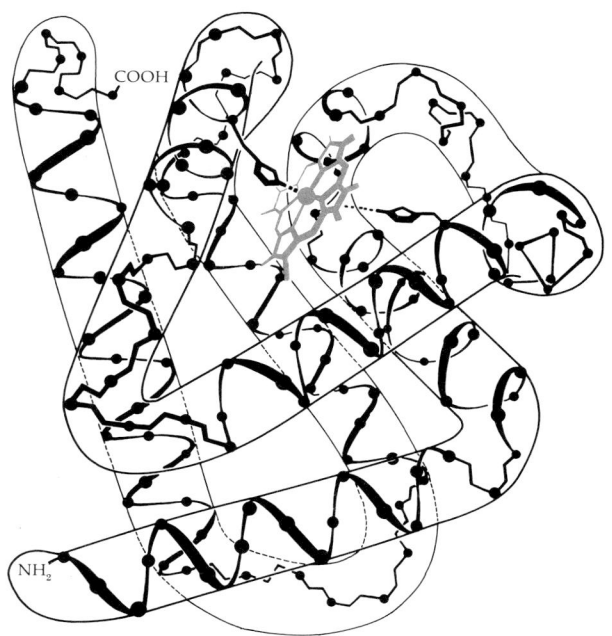

15 *Comparison of the oxygen dissociation curves of myoglobin and hemoglobin. Note that myoglobin remains 80 percent saturated with oxygen until the partial pressure of oxygen falls below 20 mm Hg. Therefore, myoglobin retains its oxygen in the resting cell and relinquishes it only when strenuous muscle activity uses up the available oxygen provided by hemoglobin.*

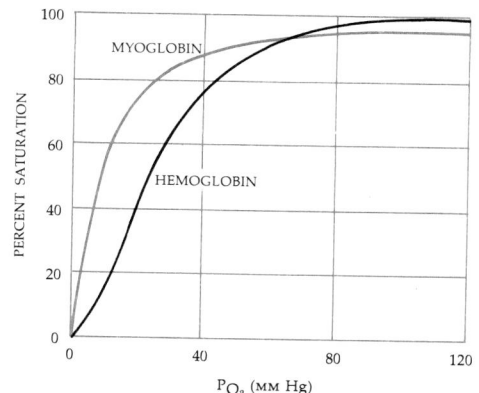

The P_{O_2} of the blood in the capillaries is normally about 40 mm Hg. As a consequence, when the blood leaves the capillaries, its hemoglobin is still usually 70 percent saturated. This extra O_2 represents an emergency supply of oxygen should the demand increase—as a result, for example, of exercise.

Myoglobin and Its Function

Myoglobin is a protein molecule with an iron-containing (heme) group (Figure 35–14); in its structure, it resembles a single unit of the hemoglobin molecule. Myoglobin is found in skeletal muscle. It has a greater affinity for oxygen than hemoglobin does and begins to release significant amounts of oxygen only when the P_{O_2} falls below 20 mm Hg. Thus, when the muscle is at rest or engaged in only moderate activity, the myoglobin holds on to its oxygen. During strenuous exercise, however, when muscle cells are using oxygen rapidly and when the partial pressure of oxygen in the muscle cells drops toward zero, myoglobin gives up its oxygen. Thus myoglobin provides an additional reserve of oxygen for active muscles.

Carbon Dioxide

A small amount of carbon dioxide is carried in the blood in the form of dissolved CO_2. Some (about 25 percent) is bound to the hemoglobin molecule. Carbon dioxide does not combine with the heme units of the hemoglobin molecule, as oxygen does, but rather with the amino groups of the hemoglobin molecule. However, most of the carbon dioxide (about 65 percent) is carried in the blood as bicarbonate. Bicarbonate is produced in a two-stage reaction. First, carbon dioxide combines with water to form carbonic acid. This reaction is catalyzed by

the enzyme carbonic anhydrase found in red blood cells. Carbonic acid, a weak acid, dissociates to yield bicarbonate and hydrogen ions:

$$CO_2 + H_2O \underset{\text{anhydrase}}{\overset{\text{carbonic}}{\rightleftharpoons}} H_2CO_3 \rightleftharpoons HCO_3^- + H^+$$

As more carbon dioxide is taken up by the blood, the blood becomes increasingly acid. As the acidity increases, hemoglobin gives up its oxygen more readily. Thus, as carbon dioxide enters the capillaries, the acidity of the blood increases and the yield of oxygen increases.

Mechanics of Respiration

Inspiration and expiration are the results of changes in the volume of the thoracic cavity, brought about by the contraction and relaxation of the intercostal ("between-the-ribs") muscles and the muscular diaphragm separating the thoracic and abdominal cavities. We inhale by contracting the intercostal muscles, pulling the rib cage up and out, and by contracting the dome-shaped diaphragm, which flattens it and increases its diameter. These movements enlarge the thoracic cavity, increasing its volume. This puts the cavity under negative pressure; that is, the pressure within it becomes less than atmospheric pressure. Air—which is, of course, at atmospheric pressure—enters the lungs and expands them. The lungs empty as the muscles relax. Usually, only about 10 percent of the air in the lung cavity is exchanged at every breath, but as much as 80 percent can be exchanged by deliberate deep breathing. Whales and other large aquatic mammals suffocate on land because of the inability of their intercostals to expand their massive chests when the expansion is counteracted by gravity.

35–16   *A model illustrating the way air is taken into and expelled from the lungs.*

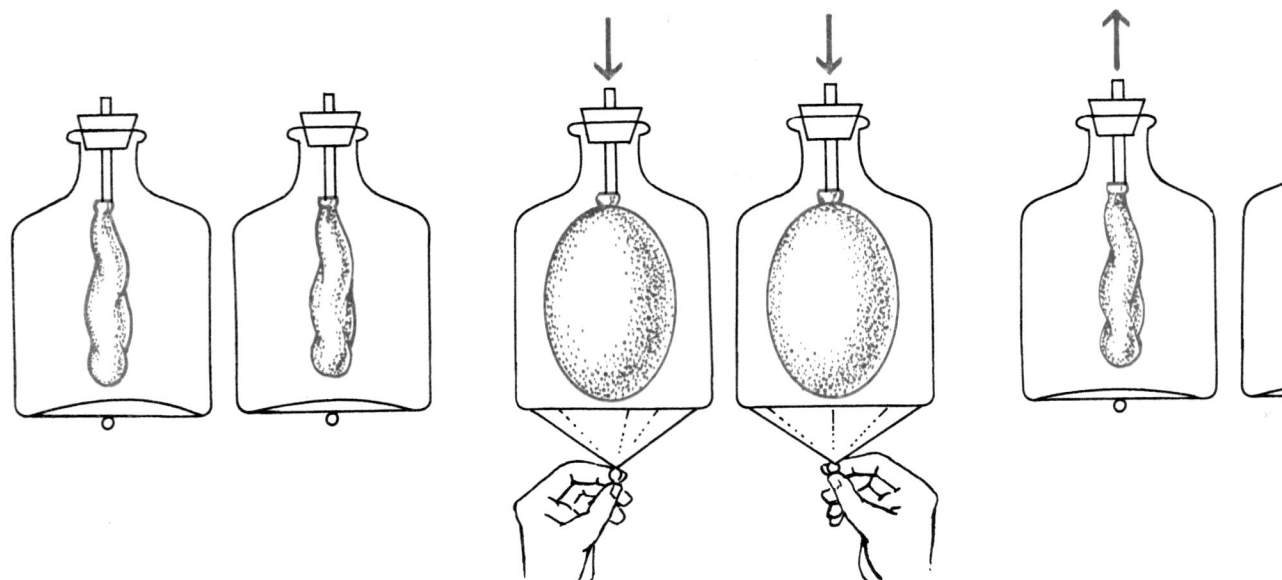

DIVING MAMMALS

Most mammals have about the same oxygen requirements as man. Structurally, their respiratory organs are very similar; yet marine mammals can dive to great depths and stay submerged for long periods of time. A sperm whale, for example, has been found at a depth of more than 1,000 meters and has been known to stay submerged for 75 minutes. A Weddell seal, a much smaller animal, can dive to 600 meters and stay submerged for 70 minutes, and can swim submerged for between 2,000 and 4,400 meters—1 to 2 miles. Are their lungs different from man's? Or their blood? Do they have special oxygen reserves, like a submarine?

Studies in a variety of diving mammals have shown that none have lungs significantly larger, in proportion, than man's. In fact, seals (and perhaps other diving mammals also) exhale before diving or early in the dive. Blood volume is increased, however; in man, blood is about 7 percent of the body weight, whereas in diving marine mammals, it is 10 to 15 percent. The blood vessels are proportionately enlarged, and these appear to serve as a reservoir of oxygenated blood. The proportion of red blood cells is higher and myoglobin is more concentrated, giving the muscles a very dark color. Still, physiologists calculate, these adaptations would not provide enough oxygen for long dives.

Other adjustments are made. The principal one is apparently the slowing down of the heart rate and the reduction of blood supply to tissues that are tolerant of oxygen lack, such as the digestive organs and the muscles. The muscles obtain energy by anaerobic glycolysis, producing large amounts of lactic acid and building up an oxygen "debt." The heart and brain, which are more sensitive to oxygen lack, are kept better supplied.

Hydrostatic pressure increases at the rate of about 1 atmosphere for every 10 meters of seawater. Within diving mammals, air spaces, such as lungs and tracheae and the middle ear spaces, collapse; otherwise the air within them would be under such pressure that the tissues would rupture. Compression of air spaces also protects the animals from the "bends."

One further difference between diving mammals and other mammals, including man, is that their respiratory centers do not function automatically to produce breathing at a certain concentration of carbon dioxide. Experiments in which diving mammals are anesthetized—as for the implantation of recording instruments—have to be carried out with extreme precaution to ensure that the animal receives sufficient oxygen during the procedure.

With these questions answered, only a few hundred remain. How do diving mammals "know" when to surface? How do they find their way from one breathing hole to the next beneath the ice? What tells them when to turn back? Is there a relationship between the solutions to these problems and the enormous brains of the most highly evolved of the marine mammals, the whales and porpoises?

35–17 *Location of the carotid body, one of the receptors that monitors the concentration of dissolved oxygen (P_{O_2}) in the blood and also, to a lesser extent, of P_{CO_2}. Carbon dioxide concentrations are also measured directly by centers in the brain.*

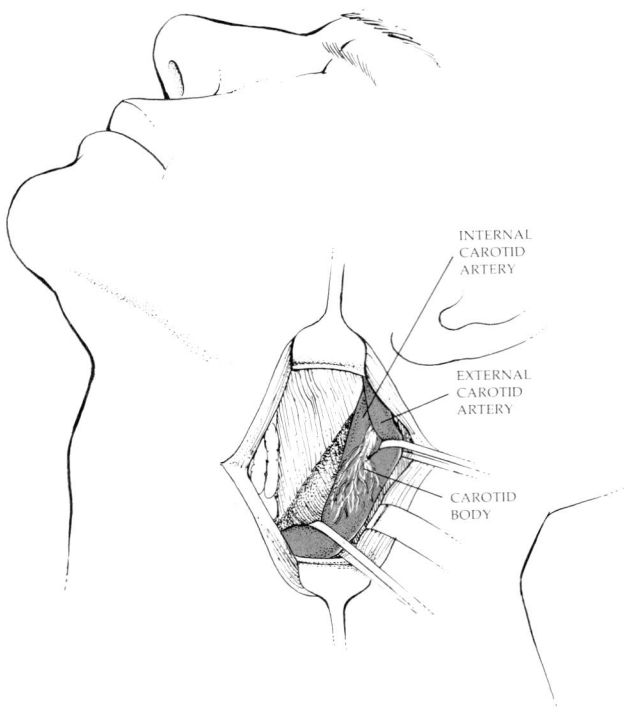

INTERNAL CAROTID ARTERY

EXTERNAL CAROTID ARTERY

CAROTID BODY

Control of Respiration

The rate and depth of respiration are controlled by a respirator center at the base of the brain. This center, which is in a portion of the brain known as the medulla oblongata, is responsible for normal breathing, which is rhythmic and involuntary, like the beating of the heart. Unlike the beating of the heart, however, which few of us can control voluntarily, breathing may be brought under voluntary control within certain limits.

The brain center receives and coordinates information about the carbon dioxide concentration of the blood. Signals are relayed from this center to nerves originating in the spinal cord that control the intercostal muscles and the muscles of the diaphragm. It is also responsive to signals from other parts of the body. Chemoreceptor cells located in the aorta and in the carotid arteries (which supply oxygen to the brain) signal the respiratory center when the concentration of oxygen in the blood decreases. These also monitor the concentration of dissolved carbon dioxide (carbonic acid), which is simultaneously monitored by centers in the brain. Thus, a number of regulatory mechanisms are at work. Control of P_{CO_2} is of overriding importance. If the concentration of CO_2 increases only slightly, breathing immediately becomes deeper and faster, permitting more carbon dioxide to leave the blood until the carbon dioxide level has returned to normal. If you deliberately hyperventilate (breathe deeply and rapidly) for a few moments, you will feel faint and dizzy because of the blood's increased alkalinity. You can, as we noted, deliberately increase your breathing rate by contracting and relaxing your chest muscles, but breathing is normally under involuntary control. It is impossible to commit suicide by deliberately holding

your breath; as soon as you lose consciousness, the involuntary controls take over once more. Deaths due to barbiturates and other drugs with a depressant activity are usually the result of a damping of the activity of this vital brain center.

As a consequence of all these various tissues and molecules and control mechanisms, highly specialized for efficient gas exchange, each cell in the body is assured of a steady supply of blood-borne oxygen under conditions of normal functioning. This oxygen is thus available for use by the individual cells in the oxidation of glucose and other energy-yielding organic molecules. This process, cellular respiration, is the means by which cells, and so organisms, derive their energy for chemical and mechanical work and for heat.

SUMMARY

Heterotrophic organisms obtain most of their ATP energy from enzymatic processes that take place in the mitochondria. These processes require oxygen and release carbon dioxide. Respiration is the means by which an organism obtains the oxygen required by its cells and rids itself of carbon dioxide.

Oxygen is available in both water and air. It enters into cells and body tissues by diffusion. However, movement of oxygen by diffusion requires a relatively large, moist surface area exposed to the source of oxygen and a short distance over which the oxygen has to diffuse. Selection pressures for increasingly efficient means of obtaining oxygen and getting rid of carbon dioxide led to the evolution in vertebrates of gills and lungs. Both gills and lungs present enormously increased surface areas for the exchange of gases and have a rich blood supply for transporting these gases to and from other parts of the animal's body.

In humans, air usually enters the body through the nostrils, where it is warmed, humidified, and filtered. It passes from the nostrils to the pharynx, then through the larynx to the trachea. From the trachea, it passes to the bronchi, which divide to form the bronchioles. Trachea, bronchi, and bronchioles are lined, like the nostrils, with ciliated mucus-producing epithelial cells.

Gas exchange takes place in the alveoli, thin membranous sacs surrounded by capillaries. Hemoglobin in the red blood cells picks up oxygen from the alveoli and releases it as it passes through the tissue capillaries. Carbon dioxide, which is carried principally in the form of bicarbonate (HCO_3^-), is picked up in the tissues and released in the lungs.

The lungs empty and fill as a result of changes in the volume of the thoracic cavity produced by intercostal muscles and muscles in the diaphragm. The muscles that control respiration, although they may be moved voluntarily, are under the involuntary control of nerves from a respiratory center in the medulla oblongata that responds to very slight changes in the H^+ and HCO_3^- concentrations of the blood.

QUESTIONS

1. Define the following terms: bronchioles, ciliated epithelium, gill, alveoli, pleura.
2. Sketch the human respiratory system. When you have finished, compare your drawing to Figure 35–10.

3. Trace the path of an oxygen atom from the air into a body cell.
4. What are the advantages and disadvantages of obtaining oxygen from air rather than from water? You might be able to think of several of each besides those mentioned in the text.
5. Carbon monoxide (CO) attaches to hemoglobin; the resulting compound does not readily disassociate, can no longer combine with oxygen, and is a brighter red than normal hemoglobin. From these facts, suggest how you might recognize and give assistance to a victim of carbon monoxide inhalation.
6. Frogs breathe air with lungs, as do we. But our lungs are in a separate thoracic cavity separated by a muscular wall from the abdominal cavity; the frog's lungs and viscera are in the same cavity. What, as a result of this difference, can we do that a frog can't?

Chapter 36

Homeostasis I: Excretion and Water Balance

One of the identifying properties of living systems is that they are homeostatic, meaning simply that they are able to maintain a relatively constant internal environment despite constant exchanges with and changes in the external environment. This capacity exists even in the simplest living organisms—a bacterial cell, for example, has a chemical composition quite distinct from the medium in which it lives.

In many invertebrates and all vertebrates, the internal chemical environment is to a large extent under the regulation of a special excretory system. Among terrestrial vertebrates, the most important component of this system is the kidney. In fact, it is possible to relate major advances in vertebrate evolution—such as the transition to land and the development of homeothermy—to increasing efficiency of kidney function. Indeed, one can trace vertebrate evolution in terms of the evolution of the kidney, as Homer Smith did in his classic book, *From Fish to Philosopher.*

As we noted in Chapter 33, the blood plays a major role in determining the chemical environment of the body. It is the supply line for chemicals taken up by the individual cells, and it carries away the wastes released by these cells. The blood can function as an efficient supply and sanitation system only because cellular wastes are constantly removed from the bloodstream. Notice that this sort of excretory mechanism is quite different in principle from the elimination of feces from the intestinal tract. In the latter case, the bulk of what is eliminated is material, such as cellulose, which was never actually in the body in the first place, in that it never passed through the walls of the digestive tract. Removal of substances from the blood, by contrast, is a very selective process of monitoring, analysis, selection, and rejection.

We shall look first at the problems involved in the control of the internal chemical environment of an organism. Second, we shall examine the evolution, the structure, and the function of the kidney—the chief organ involved in solving these problems in land-dwelling vertebrates. Finally, we shall see what is known of the control mechanisms that regulate kidney function.

36–1 *Rocky islands off the west coast of South America are the nesting places for several species of seabirds such as boobies (shown here), cormorants, and pelicans. Their nitrogen-rich excreta, known as guano, which collects on these islands, is probably the richest natural fertilizer known.*

36-2 *"I'm just glad that cows can't fly."*

—Anon.

$$NH_2-\overset{\overset{\displaystyle O}{\|}}{C}-NH_2$$

36-3 *Urea, the principal form in which nitrogen is excreted in mammals. It is formed in the liver by the combination of two molecules of ammonia with one of carbon dioxide through a complex series of energy-requiring reactions. What would be the other end product of this reaction?*

FUNCTIONS OF THE EXCRETORY SYSTEM

Regulation of the chemical environment involves four distinct processes: (1) excreting metabolic wastes, (2) adjusting particular ion concentrations, (3) maintaining acid-base balance, and (4) regulating water content.

Metabolic Wastes

Cells function extremely efficiently biochemically. In an industrial chemical laboratory, the proportion of waste to end product is characteristically very high. In the cell, by contrast, the waste products of one system are often the starting materials for another, and the amount of true waste is proportionately small. The two chief metabolic wastes that cells release into the bloodstream are carbon dioxide and the nitrogen compounds, such as ammonia, produced by the breakdown of amino acids. Carbon dioxide is eliminated from the lungs or it diffuses out into water through the skin or the respiratory organs. Animals that have unlimited access to water most often excrete nitrogenous wastes in the form of ammonia through the body surface, the gills, or the kidney. In land-dwelling organisms, which do not have this unlimited water supply, it must be converted to some other form since ammonia is highly toxic, even in low concentrations.

Birds, reptiles, and insects eliminate waste nitrogen in the form of uric acid, which can be excreted as crystals. In birds, the uric acid is mixed with the undigested wastes in the cloaca (the common exit chamber for the digestive and urinary tracts) and the combination is dropped as a semisolid paste, familiar to frequenters of public parks and admirers of outdoor statuary. This nitrogen-laden substance forms a rich natural fertilizer; guano, the excreta of seabirds, accumulates in such quantities on the small islands where these birds gather in great numbers that at one time it was harvested commercially. (One of the major industries at Woods Hole, Massachusetts, now the home of a famous marine biological laboratory, was a guano-processing plant.)

Mammals excrete nitrogenous waste products largely as urea, which comes principally from amino acids. Amino acids are used to make proteins for the growth and repair of cells. Like most of the other molecules in the body, amino acids are constantly renewed. In the course of this turnover, they may undergo deamination, the removal of the amino group. (The remainder of the molecule, which consists of carbon, hydrogen, and oxygen, is broken down by cellular respiration or converted to glycogen or fat for storage.) Deamination, which takes place principally in the liver, results in the formation of ammonia (NH_3). For example:

$$\underset{\text{alanine}}{NH_2-\overset{\overset{\displaystyle CH_3}{|}}{\underset{\underset{\displaystyle COOH}{|}}{C}}-H} + \tfrac{1}{2}O_2 \longrightarrow \underset{\text{pyruvic acid}}{\overset{\overset{\displaystyle CH_3}{|}}{\underset{\underset{\displaystyle COOH}{|}}{C}}=O} + NH_3$$

In the liver, the ammonia is quickly converted to urea, which is relatively nontoxic, and is then released into the bloodstream. Urea, unlike uric acid, must be dissolved in water for excretion.

Ions and Osmotic Pressure

Chemical regulation also involves maintaining closely controlled concentrations in the blood and other body fluids of ions such as Na^+, K^+, H^+, Mg^{2+}, Ca^{2+}, Cl^-, and HCO_3^-. These ions are important for their specific roles in various chemical processes, such as maintenance of protein structure, membrane permeability, action potentials, and muscle contraction. They are also important as determinants of osmotic pressure in the blood and body fluids.

Acid-Base Balance

A third element in the regulation of the chemical environment of the body involves adjustment of the acid-base balance. As we noted previously (page 662), the lungs are one of the principal regulators of acid-base balance of body fluids because they can quickly alter the carbon dioxide–bicarbonate equilibrium. In addition, in higher organisms, excretion of an acid or alkaline fluid by the kidneys can adjust the acid-base balance toward neutral. The pH of urine may range from 4.5 to 8, depending on the acid-base balance of the body fluids.

Regulation of Water Content

The fourth function of the excretory system is the regulation of water content. The concentration of a particular substance in the body depends not only on the absolute amount of the substance but on the amount of water in which it is dissolved. Thus, although the fundamental problem is always the same—the chemical regulation of internal environment—the solution to the problem varies widely. The most important influence is the availability of water in the animal's habitat.

The problem of water balance is such a universal one, biologically speaking, and is so important to the survival of the organism that we shall digress for a moment to discuss it in more detail before examining the anatomy and physiology of the kidney.

—4 *One of the functions of the kidney is maintaining control over the pH of body fluids. Note the very narrow range of pH within which the body functions normally. The boxes at the bottom show some of the mechanisms by which the pH is adjusted to normal.*

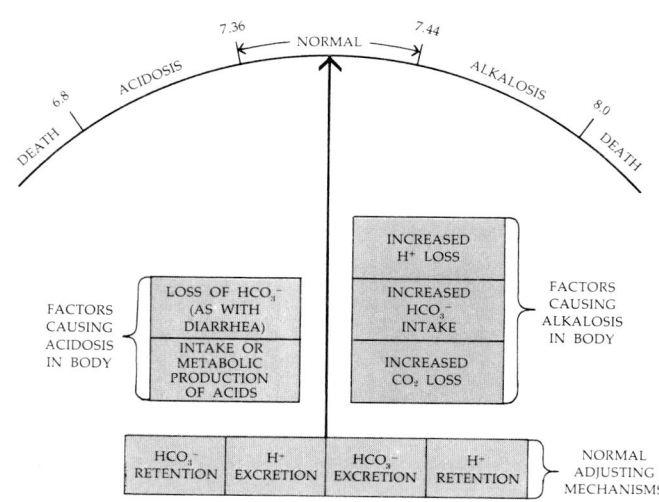

Water Balance—An Evolutionary Perspective

The earliest organisms probably had a salt and mineral composition much like that of the environment in which they lived. The early organism and its surroundings were probably also isotonic; that is, each had the same total effective concentration of dissolved substances, so water did not tend to move either in or out of the cell. When organisms moved to fresh water (a hypotonic—less concentrated—environment), they had to develop systems for "bailing themselves out," since fresh water tended to move into their bodies by osmosis; the contractile vacuole of the paramecium is an example of such a bailing device.

If the vertebrates evolved in fresh water, as is generally believed, the first function of the kidneys, phylogenetically speaking, was probably to pump water out and to keep salt and other desirable solutes, such as glucose, in. In freshwater fish, the kidney works primarily as a filter and reabsorber and the urine is hypotonic—that is, it has a concentration of solutes lower than that of body fluids.

Saltwater fish have a different problem. Their body fluids are generally less concentrated than their environment and so they tend to lose water by osmosis. Their need is to conserve water and thereby keep their body fluids from becoming too concentrated. This problem has been solved in different ways by different groups of fish. In hagfish, for example, body fluids are about as salty as the salt waters of the surrounding ocean, and so are isotonic with them. Cartilaginous fish such as the shark adapted in a unique way. They developed an unusual tolerance for urea, so instead of constantly pumping it out, as do all other fish, they retain a high concentration of it in the blood. This high concentration of urea makes their body fluids almost isotonic in relation to seawater; hence, they do not tend to lose water by osmosis. (This is a striking example of what has been called the "opportunism of evolution.")

36–5   *Because of their high concentration of blood urea, cartilaginous fish, such as this sand shark, have body fluids isotonic with seawater.*

36–6 *Some marine animals, such as the turtle, have special glands in their heads that can excrete sodium chloride at a concentration about twice that of seawater. Since ancient times, turtle watchers have reported that these great armored reptiles come ashore to lay their eggs with tears in their eyes, but it is only recently that biologists have learned that this is not caused by an excess of sentiment—as is the case of Lewis Carroll's mock turtle—but is, rather, a useful solution to the problem of excess salt. Marine birds have similar salt-secreting glands.*

The bony fish spread to the sea at a much later period than did the cartilaginous ones, and their body fluids are hypotonic in relation to the environment, having an osmotic pressure only about one-third that of seawater. Thus, like terrestrial animals, they have the problem of losing so much water to their environment that the solutes in the body fluids become too concentrated and the cells die. In bony marine fish, this problem has been solved by the evolution of special gland cells in the gills that excrete excess salt. Hence these fish can take in salt water and still remain hypotonic. (Freshwater fish, conversely, have salt-absorbing cells in their gills.)

Since terrestrial animals do not always have automatic access to either fresh or salt water, they must regulate water content in other ways, balancing off gains and expenditures.

Sources of Water Gain and Loss

Animals gain water by drinking fluids, by eating water-containing foods, and as an end product of the oxidative processes that take place in the mitochondria, as we saw in Chapter 8. When 1 gram of glucose is oxidized, 0.6 gram of water is formed. When 1 gram of protein is oxidized, only about 0.3 gram of water is produced. Oxidation of 1 gram of fat, however, produces 1.1 grams of water because of the high hydrogen concentration in fat (the extra oxygen comes from the air).

Some animals can derive all their water from food and do not require fluids. The kangaroo rat of the American desert, for example, can live its entire existence without drinking water if it eats the right type of food. It is not surprising that it prefers a diet of fatty seeds. If it is fed high-protein seeds, such as soybeans—the oxidation of which produces a large amount of nitrogen waste and a relatively small amount of water—it will die of thirst unless some other source of water is available.

On the average, a human takes in about 2,300 milliliters of water (about 2½ quarts) a day in food and drink, and gains an additional 200 milliliters a day by oxidation of food molecules. Water is removed from the blood and excreted as urine, is lost from the lungs in the form of moist exhaled air, is eliminated in the feces, and is lost by evaporation from the skin. (See Table 36–1.)

Table 36–1 *Major Subdivisions of the Body Water*

Compartment	Approximate percent of body weight	Approximate percent of body fluid	Amount in a 155-pound man (liters)
Extracellular fluid:	25	35	17
Plasma	5	7	3
Interstitial fluid	20	28	14
Intracellular fluid	40	65	29
Total body water	65	100	46

36–7 *The kangaroo rat, a common inhabitant of the American desert, may spend its entire life without liquid water. It lives on seeds and other dry plant materials, and in the laboratory, it can be kept for months on a diet of only fatty seeds such as barley or rolled oats. Analysis shows the kangaroo rat is highly conservative in its water expenditures. It has no sweat glands, and being nocturnal, it searches for food only when the external temperature is relatively cool. Its feces have a very small water content and its urine is highly concentrated. Its major water loss is through respiration, and even this loss is reduced by the animal's long nose in which some cooling of the expired air takes place, with condensation of water from it.*

36–8 *The fluid compartments of the body, showing the routes of exchange among them.*

Water Compartments

There are three principal water compartments in the body: (1) the plasma (7 percent of body fluid), (2) the interstitial fluid and the lymph (28 percent of body fluid), and (3) the intracellular fluid, the fluid within the cells (65 percent of body fluid). Water absorbed from the digestive tract, the major source of water gain, passes largely into the capillaries of the tract, entering the plasma. (See Figure 36–8.) This passage is largely a result of osmosis. Because of active transport of food molecules into the capillaries from the intestinal tract, the plasma becomes hypertonic in relation to intestinal fluid, and so water tends to follow the dissolved molecules. Hydrostatic pressure forces fluid through the capillary walls into the interstitial tissues, as we mentioned on page 617. However, most of this fluid reenters the plasma osmotically. (Because protein molecules are not forced through the capillary walls, the plasma remains hypertonic in relation to the interstitial fluid.) Other fluid is returned to the plasma via the lymphatic vessels. The loss or gain of water from the plasma is of extreme importance to the blood pressure and hence to cardiac function.

Cell membranes are freely permeable to water. Thus water is constantly moving from compartment to compartment, and the figures in Table 36–1 represent a dynamic rather than a static condition. [Rates of exchange are measured by administering a traceable substance—such as heavy water (^2H_2O) or a harmless dye molecule that passes readily through cell membranes—and analyzing the amount of the tracer material that turns up in each compartment.]

A number of factors affect the movement of water from one compartment to another. Dehydration (water loss greater than water intake) increases the solute concentration of the extracellular fluid, and so water moves out of the cells,

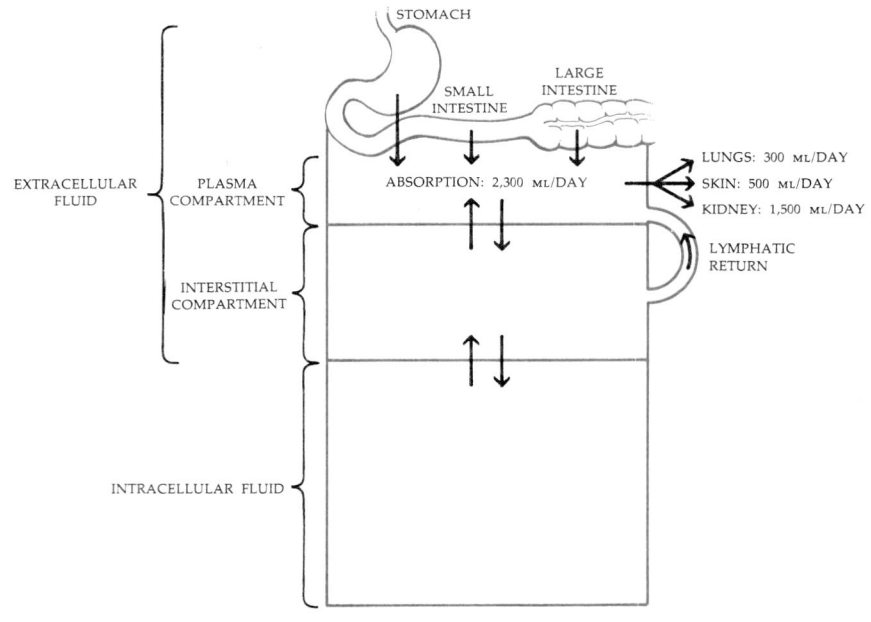

Seals and some whales get their water by eating fish, most of which have a high water content and an osmotic concentration only about one-third that of seawater. A man on a life raft cannot survive by eating fish, however, rumor to the contrary. Man cannot form urine with a salt concentration higher than about 2.2 percent. Although a fish's fluids are less concentrated than this, a fish contains a large amount of protein; so our hypothetical shipwreck survivor is placed in the same position as that of the kangaroo rat eating soybeans. He would, of course, be much worse off if he drank seawater, since the salt concentration of seawater is about 3.5 percent and he would have to expend some of his precious body water to dispose of the excess salt. (A kangaroo rat, however, which excretes a highly concentrated urine, can maintain water balance by drinking seawater.)

including the cells of the mucous membrane of the oral cavity, producing the sensation of dryness we recognize as thirst.

Human sweat, unlike that of most other mammals, contains salt. The reason for salt excretion by the skin is not known, but it has been suggested that the little crumbs of salt so produced clung to the fur of our primate ancestors and that these delicacies served as rewards for their companions who groomed them.

In cases of profuse sweating, if the water is replaced without the salt, water will move into body cells, diluting their contents. The effects of such dilution are particularly severe in the central nervous system, and water intoxication may produce disoriented behavior, convulsions, coma, and even death before the excess water can be excreted.

A number of physiological malfunctions, such as a loss of plasma proteins or the retention of salts, which may occur as a consequence of kidney disease, or the loss of plasma proteins from starvation, lead to the accumulation of interstitial fluids (edema).

—9 *A mammal is in water balance when the total amount of water lost in expired air, evaporation from skin, and in urine and feces equals the total amount of water gained by the intake of food and drinking water and by the oxidation of food molecules.*

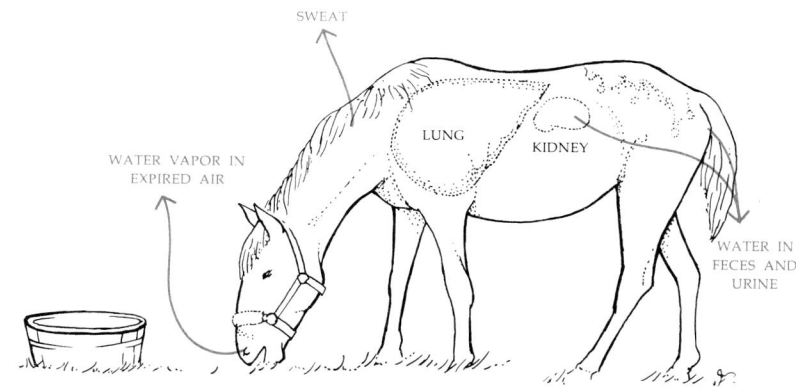

THE KIDNEY

In land-dwelling vertebrates, the organ chiefly responsible for the complex functions involved in regulating the chemical environment of the body is the kidney. The human kidneys are dark red, bean-shaped structures that lie just behind the stomach and liver.

The functional unit of the kidney is the *nephron*, which consists of a long tubule with a closed end called Bowman's capsule (see Figure 36–11b). The other end of the tubule, called the collecting duct, drains into the ureter through a funnel, the *renal pelvis*. In man, each of the two kidneys contains about a million nephrons.

Each Bowman's capsule contains a tight, twisted cluster of capillaries known as the *glomerulus*. The blood passes through the glomerulus under high pressure; the afferent vessel, the one leading into the glomerulus, is a branch of the renal artery. Moreover, the efferent vessel—the blood vessel leading out of the glomerulus—has a smaller lumen than the afferent vessel, and therefore more pressure builds up in the glomerulus. The pressure can be varied by means of a contractile muscular cuff at the entrance to the efferent vessel. As a consequence of this pressure, the liquid part of the blood (water and solutes) filters out. Blood cells and large blood proteins, both of which are too large to pass through the capillary walls (unless the capillaries are damaged or irritated by infection), are left behind. Caffeine increases the glomerular filtration rate by dilating the afferent arterioles and so increases urinary output.

Except for the absence of large molecules, the glomerular filtrate resembles the blood plasma in its chemical composition. About 180 liters (approximately 45 gallons) of fluid are filtered from the plasma into Bowman's capsule from the glomerular vessels every 24 hours. (In other words, the entire plasma volume is filtered about 60 times a day.) However, only about 1 to 1½ liters of urine are produced every day. As you can see, a tremendous amount of liquid—about 99

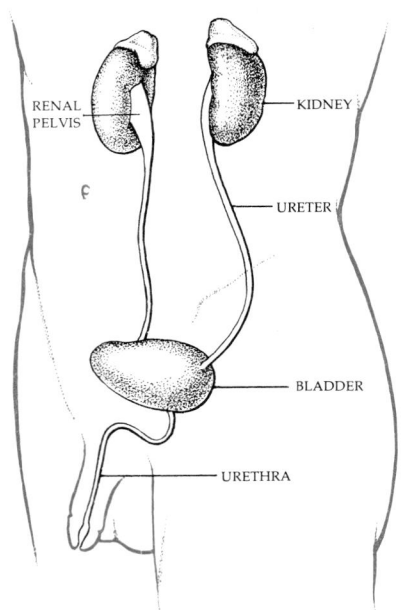

36–10 *Urinary system in a human male. Urine formed in the kidneys travels through the ureters to the bladder where it is stored until it is released through the urethra.*

RENAL PELVIS

KIDNEY

URETER

BLADDER

URETHRA

-11 (a) *In longitudinal section, the human kidney is seen to be made up of an outer region, the cortex, which contains the blood-filtering mechanisms, and an inner region, the medulla, through which collecting ducts carrying the urine merge and empty into the funnel-shaped renal pelvis, which enters into the ureter. (b) The nephron is the functional unit of the kidney. Each human kidney contains about a million nephrons. Blood enters the nephron through the afferent arteriole leading into the glomerulus. Unlike other capillaries, a glomerular capillary lies between two arterioles. The pressure in these capillaries is 50 to 70 mm Hg (about twice that of other capillaries), and the capillary walls are thin. As a consequence, fluid is forced out through the capillary walls into Bowman's capsule. The capsule connects with the long tubule that twists down into the medulla and up again. The fluid entering the nephron contains small molecules from the blood but not the large elements, such as the blood cells and large blood proteins. As the fluid travels through the tubule, almost all the water and other useful substances are reabsorbed into the bloodstream through the capillaries surrounding the tubule. Other substances are secreted from the capillaries into the tubules. Waste materials and some water pass into the ureter and are excreted from the body.*

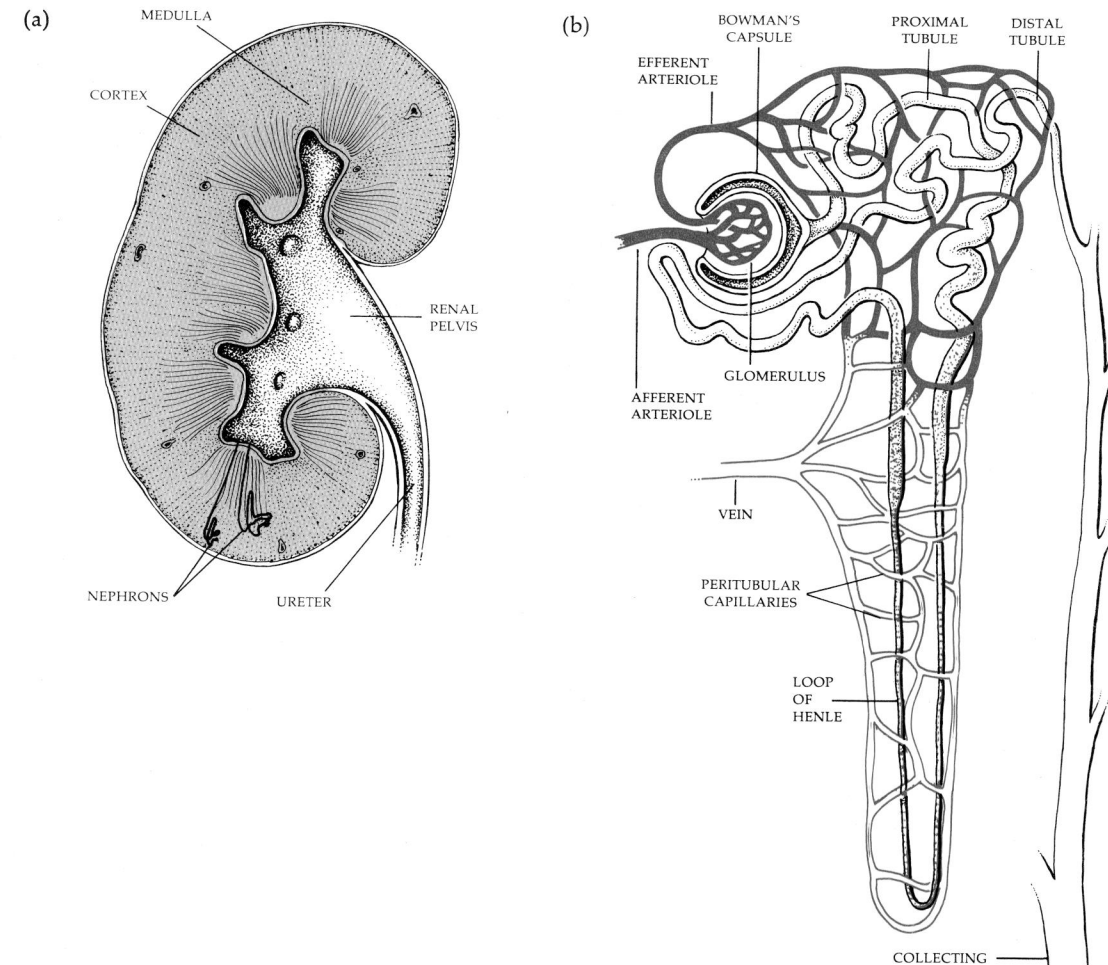

(a)

MEDULLA

CORTEX

RENAL PELVIS

NEPHRONS

URETER

(b)

EFFERENT ARTERIOLE

BOWMAN'S CAPSULE

PROXIMAL TUBULE

DISTAL TUBULE

AFFERENT ARTERIOLE

GLOMERULUS

VEIN

PERITUBULAR CAPILLARIES

LOOP OF HENLE

COLLECTING DUCT

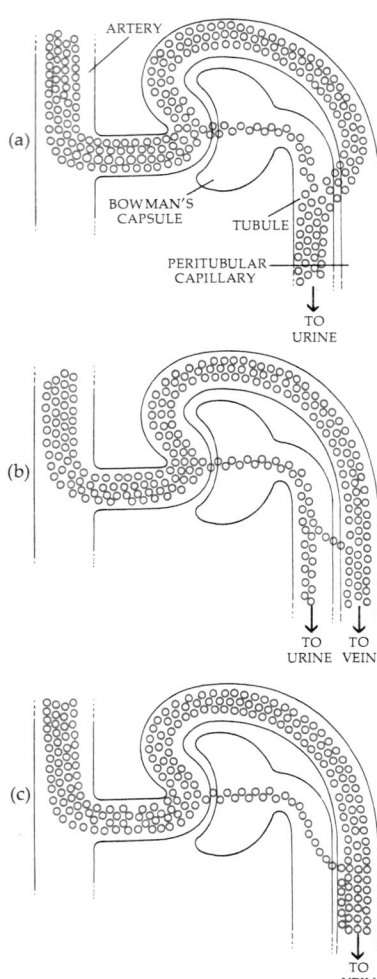

36-12 *Three ways in which a substance present in the bloodstream can be processed by the kidney. (a) Some of the substance is filtered through the glomerular capillaries and some is secreted from the peritubular capillaries into the tubule. None is reabsorbed. (b) Some of the substance is filtered. Most is reabsorbed from the tubule into the peritubular capillaries. (c) Some of the substance is filtered and all that is filtered is reabsorbed.*

percent—is reabsorbed. Some solutes, such as glucose, are normally 100 percent reabsorbed.

The fluid forced out through the capillary walls passes down the long tubule, which is lined with epithelial cells specially adapted for active transport. As the filtrate passes through the tubule, most of the water is reabsorbed as are the nutrient molecules, such as glucose, amino acids, and vitamins. (However, molecules present in excess amounts are not fully reabsorbed; for example, if the glucose concentration of the plasma is greater than 0.17 percent, as in diabetes mellitus, glucose is excreted.) About 40 percent of the urea is reabsorbed. H^+, Na^+, and other ions are generally reabsorbed, but may not be, depending on the physiological status of the body.

In addition to glomerular filtration, which moves substances from the bloodstream into the renal tubules, and reabsorption, which moves substances from the renal tubules back into the bloodstream, substances are also secreted from the peritubular capillaries into the tubule (see Figure 36–12).

The Loop of Henle

As a water-conservation measure, mammals have developed the ability to excrete a hypertonic urine, that is, one more concentrated than their body fluids. This ability is associated with a hairpin-shaped section of the nephron known as the *loop of Henle* (Figure 36–13). By sampling the fluids in and around the nephron and analyzing the amount of sodium present in each sample, physiologists have been able to determine how this deceptively simple-looking structure makes this function possible.

As shown in the diagram, the fluid entering the renal capsule first enters the proximal tubule, descends the loop of Henle, ascends it, and then passes through the distal tubule into the collecting duct.

The fluid entering the proximal tubule is isotonic with the blood plasma; that is, it has the same solute concentration as that of the blood plasma. As the urine descends the loop of Henle, it becomes increasingly concentrated. In the ascending branch, it becomes less and less concentrated, and as it enters the distal tubule, it is hypotonic. By the end of the distal tubule, it is once more isotonic or still hypotonic in relation to the blood. In animals needing to conserve water, the fluid descending the collecting duct becomes increasingly concentrated and the urine that is excreted is distinctly hypertonic in relation to the circulating blood.

This is how these findings are explained: Water diffuses freely through the cells lining the walls of the nephron, with the exception of the ascending branch of the loop of Henle and, in the absence of ADH (antidiuretic hormone), the collecting duct. Sodium* is actively pumped from the loop of Henle back to the interstitial fluid. The sodium pumped out of the ascending loop passes back into the descending loop by simple diffusion since the fluids surrounding the descending loop are more concentrated than those within it. Thus the salt recirculates to the ascending loop, where it is pumped out again. This recirculation of the sodium has two consequences: (1) As the urine passes through the loop of Henle, much salt but little water is removed, and (2) the lower part of the loop

---

* Recent evidence suggests Cl^-, rather than Na^+, is actively pumped out of the ascending branch and that Na^+ follows passively. Stay tuned for the next edition.

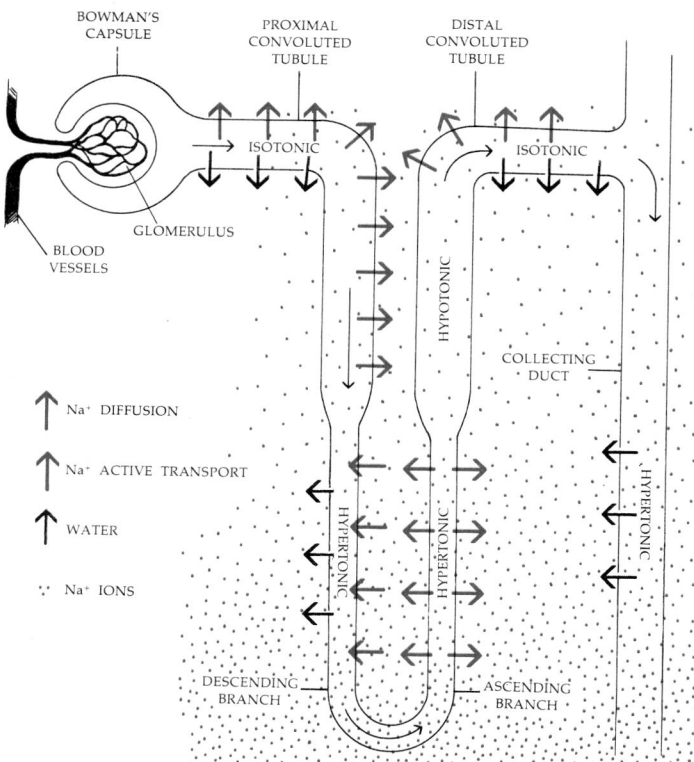

6–13 *The formation of the hypertonic urine in the human nephron. As the urine passes up the ascending branch of the loop of Henle, the cells lining the tubule pump out sodium into the surrounding fluid. The sodium ions diffuse passively into the descending branch and are recirculated in the urine to the ascending branch, where they are pumped out again. As a consequence of this recirculation of sodium, the loop of Henle and the lower section of the collecting duct are constantly bathed in a salty fluid. The cells of the ascending loop are impermeable to water, so it does not diffuse out when the sodium is pumped out. The membrane of the collecting duct, however, is permeable to water if ADH is present, so water diffuses out into the surrounding salty fluid, and a hypertonic urine is passed down the duct to the renal pelvis, the ureter, the bladder, and finally out the urethra.*

of Henle and also the lower part of the collecting duct are bathed in a fluid containing many times the level of salt normally found in tissue or blood. The urine —which has become isotonic by the time it leaves the distal tubule—then descends the collecting duct, which traverses the zone of high salt concentration. Water pours out of the collecting duct by osmosis if ADH is present, leaving within the duct a urine that is isotonic with the surrounding briny fluid but hypertonic in relation to the body fluids as a whole. In this way, the body is able to excrete a fluid, the urine, far more concentrated than the plasma from which it is derived. The loops of Henle, in short, maintain a briny (hypertonic) region through which the fluid to be excreted from the body passes just before it leaves the kidney.

Not all nephrons have long loops of Henle; in man, only about 20 percent of the loops are long enough to dip deep into the medulla of the kidney. Animals with longer loops and more of them, such as the kangaroo rat, are able to produce a more concentrated urine.

Regulation of Kidney Function

In a normal adult, the rate of water excretion in urine averages 1,500 milliliters a day, as we noted in Table 36–1. The actual amount of urine produced may vary from a few hundred milliliters—500 is about an essential minimum, since this much water is needed to remove waste products—to several thousand a day. However, despite this wide variation in water output, there will be a variation of less than 1 percent in the fluid content of the body.

Countercurrent exchange, as we saw in Chapter 35 (page 653), takes place by passive transfer (conduction or diffusion) between one element of the system and the other. It can help to maintain a difference in temperature or in solute concentration, but it cannot create one. If the transfer between the two elements of the system involves active transport, differences between the two can be created as well as maintained. Such a system is known as a countercurrent multiplier system. The loop of Henle is an example of a countercurrent multiplier system. The figures refer to comparative concentrations of Na^+. As you can see, the longer the loop, the greater the differences that can be created between the Na^+ concentration at the extremity of the loop and those on entry and departure.

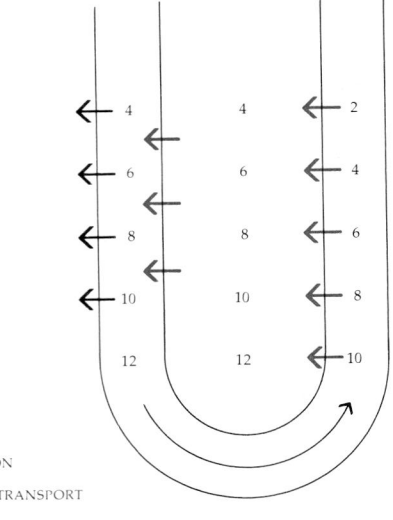

← WATER
← Na^+ DIFFUSION
← Na^+ ACTIVE TRANSPORT

36–14 *Regulation of ADH production. Increased blood pressure stimulates baroreceptors in the left atrium. These receptors then transmit impulses to the hypothalamus that inhibit ADH (antidiuretic hormone) production. ADH production is stimulated by increased concentration of solutes in the blood, detected by osmoreceptors located in the hypothalamus. As its name suggests, ADH inhibits the excretion of water.*

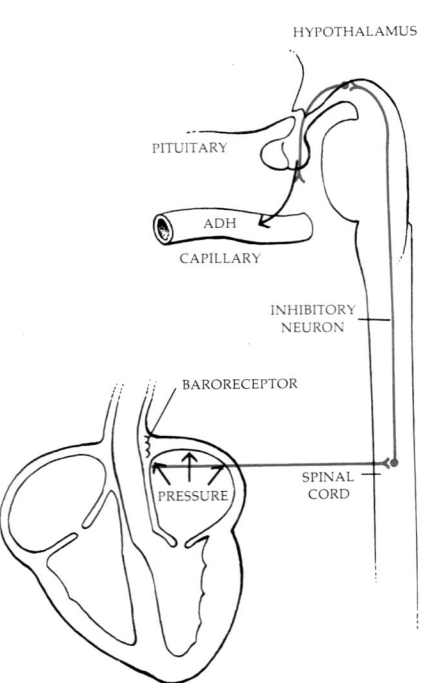

HYPOTHALAMUS
PITUITARY
ADH
CAPILLARY
INHIBITORY NEURON
BARORECEPTOR
PRESSURE
SPINAL CORD

The physiological adjustments that regulate water balance are principally under the control of the hormone ADH. ADH acts on the membranes of the collecting ducts of the nephrons and apparently increases their permeability to water. If no ADH is present, the collecting ducts are relatively impermeable to water and so less water passes back into the blood from the urine by osmosis before it enters the ureters. In short, ADH allows the kidney to conserve water. In amphibians, ADH acts on the skin to increase the uptake of water from the surrounding medium.

ADH, as we noted in Chapter 32, is a neurohormone secreted by nerve endings of the hypothalamus. It is stored in the posterior pituitary gland. Whether or not ADH is released depends primarily on the osmotic receptors located in the hypothalamus, which monitor the solute content of the blood. Pressure receptors, which detect changes in the blood volume, are found in the walls of the heart, in the aorta, in the carotid arteries, and in extreme situations, such as hemorrhage, they may also stimulate secretion of ADH. Stimuli received by these receptors are transmitted to the hypothalamus. Thus, a high concentration of solute in the blood or low blood pressure stimulates the production of ADH and the conservation of body water. Factors that decrease blood concentration, such as the ingestion of large amounts of water, or that increase blood pressure—high doses of adrenaline, for instance—signal the posterior pituitary to stop secreting ADH, and so water is excreted. Alcohol suppresses ADH secretion and thus increases urinary flow; pain and emotional stress trigger ADH secretion, decreasing urinary flow. In the disease known as diabetes insipidus, there is a deficiency of ADH and a person with this disease may excrete up to 15 liters of urine a day.

The adrenal cortical hormone aldosterone stimulates reabsorption of sodium from the kidney. When the adrenal glands are removed or when they function poorly (as in a disease called Addison's disease), sodium chloride and water are lost in the urine and the tissues of the body become depleted of them. Generalized weakness results, and if a patient with Addison's disease is untreated, the

6-15 *Reduction of sodium transport across the tubules of the kidney stimulates the kidney to produce renin. Renin, an enzyme, splits angiotensinogen, a large protein produced by the liver, to form angiotensin. Angiotensin stimulates the adrenal cortex to produce the steroid hormone aldosterone. Aldosterone increases the reabsorption of sodium and chloride ions by the renal tubules.*

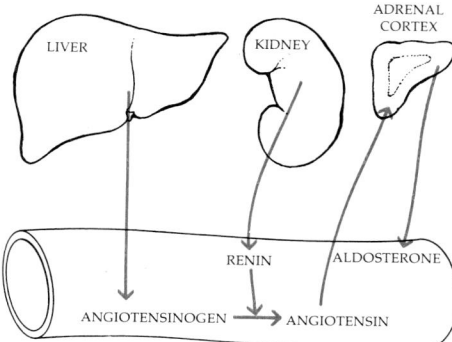

fluid loss can eventually become fatal. Most of the ill effects, however, can be overcome by giving the patient enough sodium, chloride, and water to make up for those lost by excretion.

The kidney itself influences aldosterone production by releasing an enzyme, renin, when the sodium concentration within the renal tubule decreases. According to present evidence, renin acts on a blood protein, angiotensinogen, made by the liver, and converts it to angiotensin. Angiotensin stimulates the adrenal cortex to secrete more aldosterone. This is an interesting example of a hormonal feedback mechanism that avoids the hypothalamus and pituitary.

The way in which the cells of the kidney act to select and reject particular molecules and the way in which the hormones alter this selection and rejection process are a part of the larger and largely unsolved problem of how substances are transported across cell membranes.

SUMMARY

The excretory system enables an animal to regulate its internal chemical environment by selectively removing substances from the bloodstream. This regulatory process involves (1) eliminating the by-products of metabolism, especially the nitrogen compounds produced in the breakdown of amino acids, (2) controlling the ionic content of the body fluids, (3) regulating acid-base balance, and (4) maintaining water balance.

In mammals, nitrogen is excreted chiefly in the form of urea, which is formed in the liver from ammonia (produced by deamination of amino acids) and carbon dioxide.

Water regulation involves equalizing water gain and loss as well as maintaining a balance between the various water compartments of the body. The principal source of water gain in most mammals is in the diet; water is also formed as a result of oxidation of foodstuffs. Water is lost in the feces and the urine, by respiration, and from the skin. Although the amount of water taken in and given off may vary widely from animal to animal and also from time to time in the same animal, depending largely on environmental circumstances, the volume of water in the body is very nearly constant. The principal water compartments of the body are the plasma, the interstitial fluids, including the lymph, and intracellular water. The major factor determining exchange of water among compartments of the body is osmotic pressure.

The kidney is the organ chiefly responsible for the regulation of the chemical composition of the blood. The functional unit of the kidney is the nephron, which consists of a closed end, Bowman's capsule, and a long tubule. Each Bowman's capsule contains a cluster of capillaries, the glomerulus. Blood is pumped into the glomerulus under sufficient pressure to force plasma (less the larger proteins) through the capillary walls. Other molecules are secreted from the peritubular capillaries into the renal tubules. Most of the water and solutes are reabsorbed from the filtrate as it passes through the renal tubule and returned to the blood. Some of the water, along with some of the urea and other substances that are not reabsorbed, is excreted as urine.

An important means of water conservation in mammals is the capacity for excreting a urine that is hypertonic in relation to the blood. The loop of Henle is the

portion of the mammalian nephron that makes possible the production of hypertonic urine.

Epithelial cells of the renal tubules reabsorb molecules from the glomerular filtrate and so control the chemical composition of the body fluids. These cells are influenced by hormones, chiefly antidiuretic hormone (ADH), secreted by the posterior lobe of the pituitary gland, and aldosterone, an adrenal cortical steroid hormone. ADH increases the return of water to the blood and so decreases water excretion. Aldosterone increases the reabsorption of sodium, which is followed by chloride and water.

QUESTIONS

1. Define the following terms: homeostasis, Bowman's capsule, nephron, water compartment, urine, loop of Henle.
2. Sketch a glomerulus, indicating the pathways of the blood and the glomerular filtrate.
3. Trace the path of urea through the human kidney.
4. Viewed as a whole, the kidney selectively removes substances from the bloodstream. A closer look, however, reveals that it does so by filtering out all small molecules through the glomerulus and then reabsorbing those which are needed in the tubules. (We'll neglect tubular secretion for the moment.) Thus, the system need not identify wastes as such; rather, it must identify useful substances. What advantage can you see in this arrangement in a world in which some 500 new chemicals, most of them toxic, are introduced each year?
5. Why does a high-protein diet require an increased intake of water? (You should be able to think of two different reasons.) Why does a person lose some weight after shifting to a low-salt diet, even without reducing his caloric intake? Why isn't this a useful way to lower the fat content of the body?
6. Compare the rate of urine production in a marine and freshwater teleost (bony) fish; in a marine teleost fish and a shark.
7. Why do birds lose less water in eliminating nitrogenous wastes than do mammals? Compare their relative urinary losses under conditions of excess salt.

Chapter 37

Homeostasis II: Temperature Regulation

For life forms, the seeming wastelands of the deserts and the Poles represent great extremes of hot and cold. Yet, in fact, measured on a cosmic scale, the temperature differences between them are very slight. Life exists within only a very narrow temperature range. The upper and lower limits of this range are imposed by the nature of biochemical reactions, all of which are extremely sensitive to temperature change. Biochemical reactions take place almost entirely in water, the principal constituent of living things, and the slightly salty water characteristic of living tissues freezes at $-1°$ or $-2°C$. Molecules that are not immobilized in the ice crystals are left in such a highly concentrated form that their normal interactions are completely disrupted. In Section 1, we saw that as temperature rises, the movement of molecules increases and the rate of enzymatic reactions goes up rapidly. In fact, it is a convenient generalization that the rates of most biochemical reactions about double for every $10°C$ increase in temperature. The upper temperature limit for life is apparently set by the point at which proteins begin to lose their functional three-dimensional shapes (the process known as denaturation). Once denaturation occurs, enzymes

37–1 *Reptiles and other "cold-blooded" animals cannot maintain the high, constant body temperature characteristic of birds and mammals. However, they can regulate their internal temperature to some extent. For example, by absorbing solar heat, the zebra-tailed lizard shown here is raising its temperature well above that of the air around it.*

37-2 *Life processes can take place only within a very narrow range of temperature. The temperature scale shown here is the Kelvin, or absolute, scale. Absolute zero ($0°K$) is equivalent to $-273.1°C$, or $-459°F$, and is the temperature at which all molecular motion ceases.*

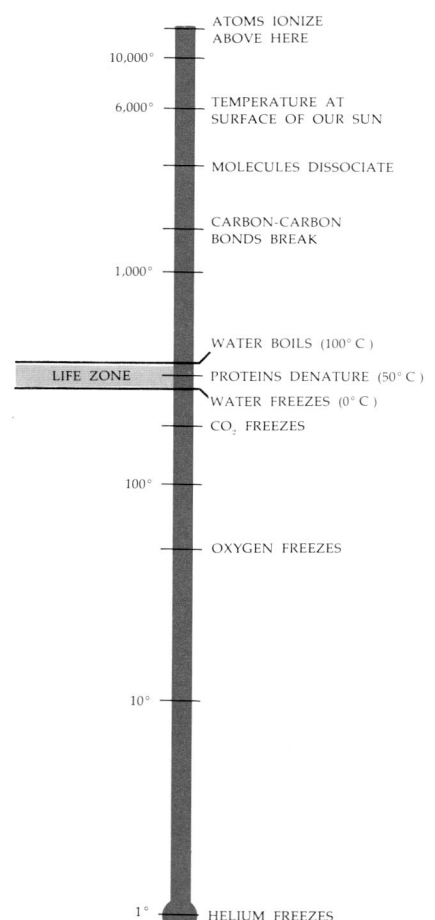

ATOMS IONIZE ABOVE HERE
10,000°

TEMPERATURE AT SURFACE OF OUR SUN
6,000°

MOLECULES DISSOCIATE

CARBON-CARBON BONDS BREAK

1,000°

WATER BOILS (100°C)

LIFE ZONE — PROTEINS DENATURE (50°C)
WATER FREEZES (0°C)
CO_2 FREEZES

100°

OXYGEN FREEZES

10°

1° — HELIUM FREEZES

and other proteins whose activity depends on a specific shape are no longer able to carry out their functions. As a consequence of this restriction to a very narrow temperature zone, animals must either find external environments that range from just below freezing to between 45° and 50°C or must create suitable internal environments. (However, there are exceptions: Blue-green algae have been found in hot springs at 85°C and living bacteria at 80°C.)

The ways in which temperature requirements are met—and, in particular, the internal regulation of temperature—are the subject of this chapter. Before going into the details of these processes, however, it is worth taking a moment to consider what excellent temperature regulators most mammals are. One of the simplest and most dramatic demonstrations of this capacity was given some 200 years ago by Dr. Charles Blagden, then secretary of the Royal Society of London. Dr. Blagden, taking with him a few friends, a small dog in a basket, and a steak, went into a room that had a temperature of 126°C (260°F).* The entire group remained there for 45 minutes. Dr. Blagden and his friends emerged unaffected. So did the dog. (The basket had kept its feet from being burned by the floor.) But the steak was cooked. This regulation of internal temperature is another example of homeostasis.

"COLD-BLOODED" VERTEBRATES

Most animals are what we call "cold-blooded." A more precise term for cold-blooded animals is *poikilotherms*, from *poikilo*, meaning "changeable." Mammals and birds are the only homeotherms.

Fish are poikilotherms. Their metabolic processes produce heat, but, in most fish, this heat is rapidly carried from the core of the body by the bloodstream and lost by conduction into the water. A large proportion of this body heat is lost from the gills. Exposure of a large, well-vascularized surface to the water is necessary in order to acquire enough oxygen, as you will recall from Chapter 35. This same process rapidly dissipates heat. However, a few very large, fast-swimming fish, such as tuna, can maintain a temperature in their trunk muscles considerably above that of the surrounding water (Figure 37–3). They are able to do this largely because of heat-conserving countercurrent mechanisms such as that described on page 653. Fish also cannot maintain a temperature lower than that of the water. In contrast, land animals cope with temperatures higher than those of their bodies by unloading heat largely by evaporation, a process not available to water dwellers.

In general, large bodies of water, for the reasons we discussed in Chapter 3, maintain a very stable temperature. At no place in the open ocean is the annual range of temperature more than 10°C. (By contrast, temperatures on land may vary annually in a given area by as much as 60° to 70°C.) Because water expands as it freezes, ice floats on the surface of the water, and life continues beneath this surface. In shallower water, where greater temperature changes occur, fish seek an optimal temperature, presumably the one to which their metabolic processes are adapted. However, because they can do almost nothing to make their own temperature different from that of the surrounding water, they are quickly victimized by any rapid, drastic changes in water temperature.

* A centigrade to Fahrenheit temperature conversion scale can be found in Appendix A at the back of the book.

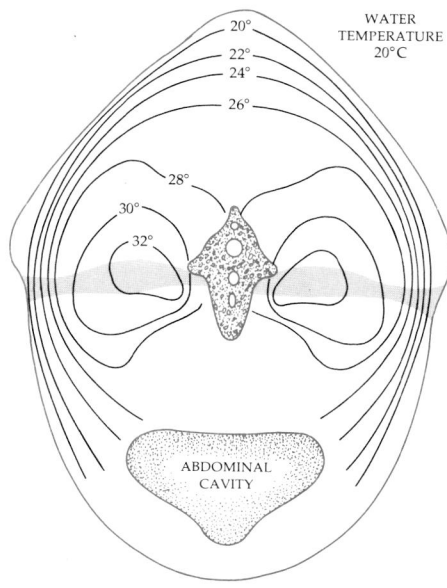

7–3 *Distribution of heat in a cross section of a 155-pound big-eye tuna. The heat is produced by metabolism, particularly in the hard-working, red, myoglobin-containing muscle tissue, indicated in light color.*

WATER
TEMPERATURE
20°C

20°
22°
24°
26°
28°
30°
32°

ABDOMINAL
CAVITY

With the transition to land, vertebrates gained an opportunity to maintain a high and constant temperature. Amphibians and reptiles are also poikilotherms, but behavioral adaptations, particularly in the case of the reptiles, leave them less at the mercy of their environment than are the fishes. Moreover, because of their low metabolic rate, they can, when they seek suitable conditions of temperature and humidity during periods of low temperature or drought, go for long intervals without eating. Toads that live in arid parts of Australia have been known to remain in a burrow for as long as two years and emerge alive with the advent of a rainy season. Many species of salamander, although they are amphibians and hence air breathers, spend most of their lives in water, as do aquatic reptiles such as alligators and turtles. These animals, maintaining a body temperature essentially the same as that of the surrounding water, seek water of the preferred temperature.

Fully terrestrial reptiles, the snakes and the lizards, often maintain remarkably stable body temperatures during their active hours. Unlike mammals, which are warmed from the inside out, reptiles often warm themselves from the outside in, depending to a large extent on solar radiation as a heat source. By careful selection of suitable sites, such as the slope of a hill facing the sun, and by orienting their body with a maximum surface exposed to sunlight, they can heat themselves rapidly (as rapidly as 1°C per minute) even on mornings when the air temperature, as on deserts or in the high mountains, may be close to 0°C. As soon as their body temperature goes above the preferred level, lizards move to face the sun, presenting less exposed surface, or seek shade. By staying in motion, lizards are able to keep their temperatures oscillating within a quite narrow range. At night, when temperatures drop, the animals seek the safety of their shelters where they will not become immobilized in an exposed position and thus vulnerable to attack.

-4 *Spadefoot toad. These animals, although they are thin-skinned, are able to survive in hot, dry climates without excessive loss of water largely because of behavioral adaptations. Using their large hindlegs as digging tools, they are able to burrow several feet under ground and may remain there for weeks or even months during dry periods. They emerge only at night.*

37–5 (a) *In laboratory studies, the internal temperature of reptiles was shown to be almost the same as the temperature of the surrounding air. It was not until observations were made of these animals in their own environment that it was found that they have behavioral means for temperature regulation that enable them to maintain a surprisingly constant temperature. This chart is based on observations of the behavior of a horned lizard in response to temperature fluctuations. Changes in albedo (whiteness) are produced by pigment in the epithelial cells dispersing or collecting. (b) Horned lizard (often but not accurately known as a horned toad).*

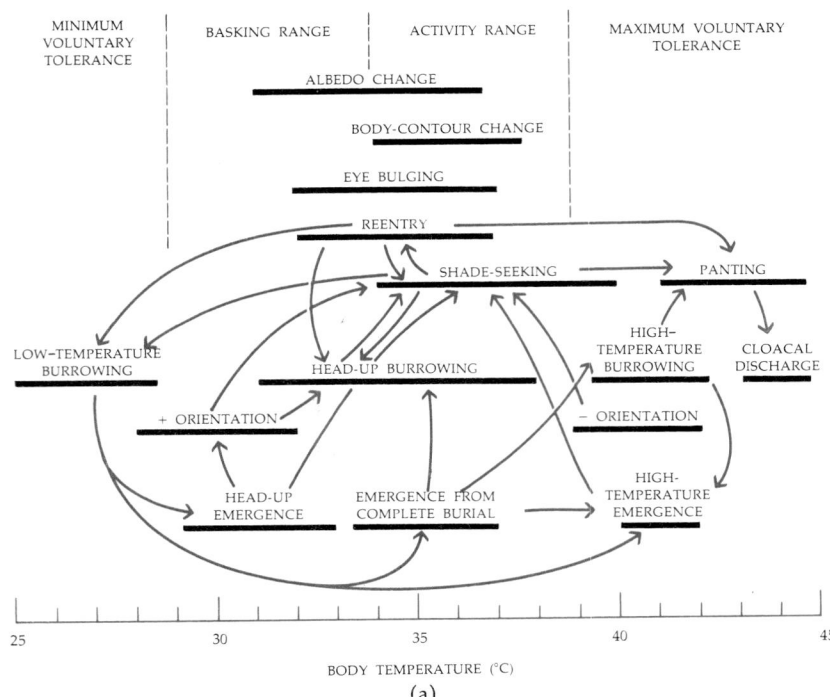

(a)

(b)

Some large reptiles retain some of the heat produced by their oxidative mechanisms. In fact, certain insects, such as honeybees, during periods of activity, maintain body temperatures substantially above those of their environment. Even some plants, such as *Philodendron*, may temporarily elevate their temperatures 10°C or more by heat produced during flowering.

HOMEOTHERMS

Homeotherms are organisms that hold their body temperature at some closely regulated level (usually between 35° and 42°C, depending on the species) that is independent of environmental temperatures. As we noted previously, the only true homeotherms are the birds and the mammals. Early mammals were small, about the size of a rat or mouse. They had sharp teeth, but since they were too small to attack other vertebrates, it is assumed that they lived mostly on insects and worms. They first appeared in the middle of the Mesozoic era, at about the time of the first dinosaurs. For some 80 million years, they coexisted with the huge carnivorous reptiles. Then suddenly, as geologic time is measured, most of the great reptiles disappeared and the mammals began to increase in number and variety, soon becoming by far the most prominent and successful group of large land animals.

Paleontologists do not agree about why the dinosaurs became extinct. One reason that the primitive mammals survived may have been their ability to maintain a high and constant body temperature, which enabled them to be both fast and alert regardless of external conditions. Also, and perhaps most important at the beginning, their warm-bloodedness would have enabled them to be

active at twilight and throughout the night, when the reptiles were rendered inactive by the cold. In fact, most modern species of mammals are still largely nocturnal, and even those we regard as the most primitive are intelligent and energetic to a degree seldom attained by any poikilotherm.

Like the mammals, the birds evolved from reptiles. Even today, they share many characteristics with the reptiles, such as certain skeletal and muscular structures and similar eggs. Paleontologists believe that the first birds began to appear some 150 million years ago, shortly after the first mammals, but from a different reptilian stock. The principal distinguishing feature of these early birds was the fact that they had feathers, which are important as insulators. This is still a unique avian characteristic; no animals except the birds have feathers, and all birds possess them.

TEMPERATURE REGULATION IN HOMEOTHERMS

In man and other homeotherms, heat is provided for the organism as one of the products of the breakdown of glucose and other energy-containing molecules within cells. The cost of living in terms of energy productivity is high for homeotherms. Approximately 80 to 90 percent of their oxidative energy is expended in maintaining thermal homeostasis. Thus they pay a high price for their independence.

A warm-blooded mammal is warmer at the core than at the periphery. (Man's temperature usually does not reach 37°C or 98.6°F until some distance below the skin surface.) Heat is transported from the core to the periphery by the circulatory system. At the surface of the body, the heat is transferred to the air. Temperature regulation involves increasing or decreasing heat production and increasing or decreasing heat loss at the body surface.

7–6 Heat exchanges between a mammal and its environment. The core body temperature of the man is 37°C. The air temperature is 30°C and there is no wind movement.

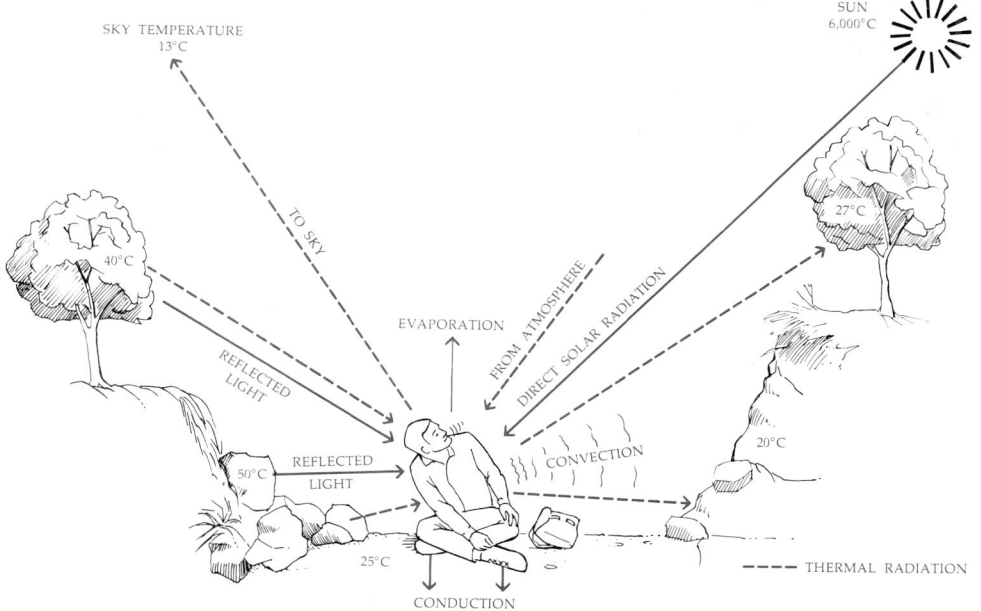

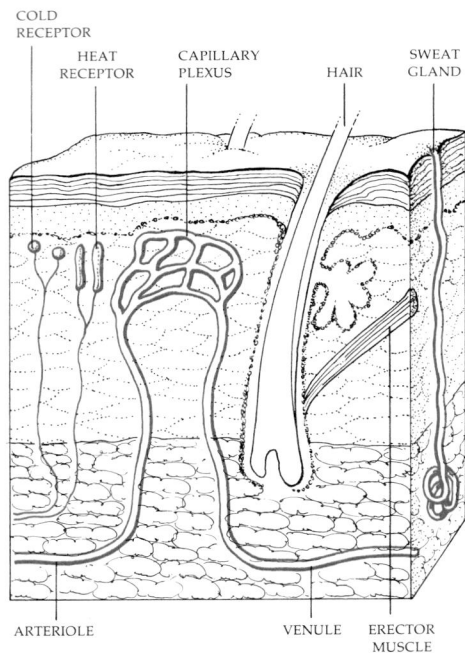

COLD
RECEPTOR

HEAT
RECEPTOR   CAPILLARY
            PLEXUS   HAIR   SWEAT
                                  GLAND

ARTERIOLE   VENULE   ERECTOR
                                MUSCLE

37–7   *Cross section of human skin showing
structures involved in temperature regulation. In cold, the arterioles constrict,
reducing the flow of blood through the
capillaries, and the hairs, each of which
has a small erector muscle under nervous control, stand upright. In animals
with body hair (*Homo sapiens *is one of
the few terrestrial mammals that is
virtually hairless), air trapped between
the hairs insulates the skin surface, conserving heat. In heat, the arterioles dilate
and the sweat glands secrete a salty
liquid. Evaporation of this liquid cools
the skin surface, dissipating heat (approximately 540 calories for every gram
of* H_2O*). Beneath the skin is a layer of
subcutaneous fat that serves as insulation, retaining the heat in the underlying
body tissues.*

The remarkable constancy of temperature is maintained by an automatic system—a thermostat—in the hypothalamus, which precisely measures the body temperature and triggers the appropriate control mechanisms. Receptor cells in this center monitor the temperature of the blood flowing through the hypothalamus. Although the surface of the skin is covered with receptors for hot and for cold, these are not the principal agents in the regulation of internal temperature, as has been demonstrated by some interesting experiments with human subjects. For example, in a room in which the air is warmer than body temperature, if the blood circulating through a person's hypothalamus is cooled, he will stop perspiring, even though his skin temperature continues to rise.

The elevation of temperature known as fever is due not to a malfunction of the thermostat but to a resetting. Thus, at the onset of fever, an individual typically feels cold and often has chills; although his body temperature is rising, it is still lower than his thermostat setting. The substance primarily responsible for the resetting of the thermostat is a protein released by the neutrophils as part of the inflammatory response (see page 625).

Regulation at High Temperatures

As the external temperature rises, the blood vessels near the skin surface dilate and the supply of blood to the skin surface increases. In man and many other large animals, when the external temperature rises above body temperature, perspiration begins. Evaporation from the body surface requires heat energy, which comes from the surface, and the blood just below the skin surface cools as this heat energy is used. Men and horses sweat from all over their body surfaces. Dogs and some other animals pant, so that air passes rapidly over the moist nasal passages and tongue, evaporating water. Cats lick themselves all over as the temperature rises, and evaporation of this water cools their body surface. For animals that sweat, temperature regulation at high temperatures is necessarily a principal route of water loss. (Dr. Blagden and his friends were probably very thirsty, illustrating that the solution of one evolutionary problem often creates another.)

Regulation at Low Temperatures

As external temperatures go down, cutaneous blood vessels constrict, limiting heat loss across the skin surface. Another early and important response is shivering. A naked man begins shivering when the air temperature reaches 20°C. Shivering is another means of increasing heat production; it mobilizes glucose and ATP stored in muscle tissues and releases their energy in the form of motion and heat. Metabolic processes may increase, although metabolic responses to cold are less common in man than in other animals. In response to cold, the thyroid glands increase their production of thyroid hormone. Thyroid hormone increases the metabolic rate; it apparently exerts its effects directly on the mitochondria, although the exact way in which it works is not known. The skin of a person whose thyroid glands are not functioning at a normal level often feels cold, and such people often complain about never being warm enough. The thyroid, like the sex glands, is under the control of the pituitary,

37–8 *Temperature-regulating mechanisms in mammals. The dotted lines represent hormonal pathways; these are apparently of minor importance in man.*

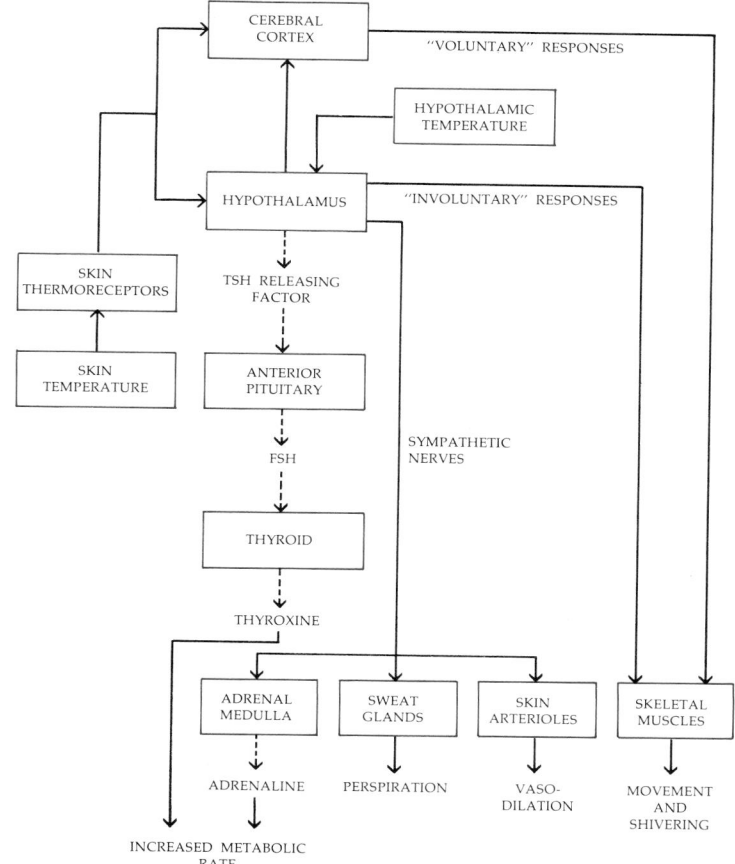

37–9 *Dogs unload heat by panting, which involves short, shallow breaths. When it is hot, a dog breathes at a rate of about 300 to 400 respirations a minute, compared with a respiration rate of from 10 to 40 a minute in cool surroundings. Most of the air is inhaled by the nose and exhaled by the mouth. Thus the nose is a primary site of water evaporation.*

which, in turn, is regulated by the hypothalamus. With a further drop in temperature, the adrenal gland releases adrenaline, which also raises the organism's metabolic rate, increasing the supply of heat.

TEMPERATURE, SIZE, AND ENERGY

Heat radiates out from the body surface, and radiation of heat, like diffusion of gases, is proportional to the surface area that is exposed. Animals minimize heat loss by curling up at night with their legs underneath them, reducing the exposed surface area.

It is more difficult for a small animal to maintain a high body temperature than it is for a large one, since the large surface-to-volume ratio of small animals makes them such efficient radiators. Mice nest, making a communal heat. Rabbits burrow.

Most birds and mammals maintain an almost constant body temperature. Birds require it for their special activities, including sustained flight. (Insects fly but not at colder temperatures.) In the mammals, high constant temperature is undoubtedly associated with activity and alertness and, as a consequence, the

37–10 *Observed metabolic rates of mammals. Each division on the ordinate represents a tenfold increase in weight. The metabolic rate of the very small mammals is much higher than that of larger mammals, owing principally to their greater surface-to-volume ratios.*

RELATIVE METABOLIC RATE

SHREW

HARVEST MOUSE

KANGAROO MOUSE

CACTUS MOUSE

MOUSE

FLYING SQUIRREL

RABBIT

RAT

CAT

DOG

SHEEP

MAN

HORSE

ELEPHANT

0.01   0.1   1   10   100   1,000

BODY WEIGHT (KILOGRAMS)

37–11 *Hibernating animals store calories in a special kind of fatty tissue known as brown fat. This electron micrograph shows the nucleus of a brown fat cell from a bat recently aroused from hibernation. Note the large spherical mitochondria surrounding the nucleus; these are typical of brown fat cells. Brown fat cells differ from those of ordinary fat in that the lipid is distributed within the cells in multiple droplets rather than in one large drop. These droplets are metabolized in the course of hibernation, as they have been in this animal, leaving only a compact mass of cells. The color of the cells is due largely to the cytochromes of the many mitochondria.*

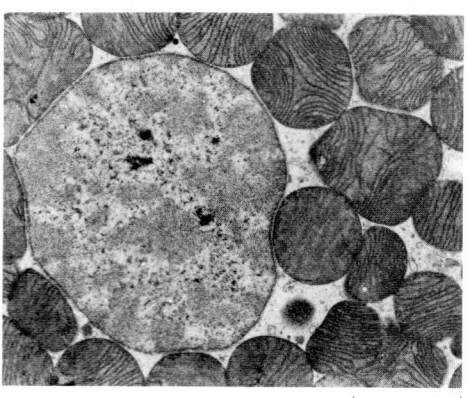

2 μm

evolution of intelligence. Maintaining constant temperature adds significantly to an animal's energy requirements, however. Birds maintain a comparatively high body temperature of 40° to 42°C. Unlike other small animals, they spend most of their time exposed, and flying birds have high energy requirements for flight. Flying birds have a particularly difficult problem since they must keep their weight down and therefore cannot store much food for energy; thus they need to refuel constantly to keep warm and to keep flying. A bird eating high-protein foods, such as seeds and insects, commonly consumes as much as 30 percent of its body weight per day. Fall migrations are necessary not so much because of the colder weather, since a well-fed bird can maintain its temperature, but rather because of the reduced food supply and the shortened day, which gives the diurnal bird less time to feed. Similarly, birds return to the temperate zones to breed partly because the longer days give them more time to seek food to meet the high-energy demands of their young. A hummingbird consumes about four times as much oxygen in hovering as it does when awake but not flying. Its reserves are insufficient for this rate of metabolism at night; in cool weather, its body temperature drops every evening, and so it conserves fuel and also water (which would be lost by respiration).

HIBERNATION

Hibernation, which comes from *hiber*, the Latin word for "winter," is another means of adjusting energy expenditures to food supplies. Hibernating animals do not stop regulating their temperatures altogether—they turn down their thermostats. Life at a lower temperature conserves energy. Only a few animals truly hibernate; some others, such as the bear, sleep for long periods but do not

hibernate in the strict sense of the word. True hibernators include a few insectivores, hamsters, ground squirrels, some South American opossums, some bats, and a few birds, such as the poor-will of the southwestern United States. In these animals, the body temperature is lowered, often to that of the surrounding air (if the air is above 0°C). The heartbeat slows; the heart of an active ground squirrel, for instance, beats 200 to 400 times a minute whereas that of a hibernating one beats 7 to 10 times a minute. The metabolism as measured by oxygen consumption is reduced 20 to 100 times. Apparently even aging stops; hibernators have much longer life spans than similar nonhibernators. Many of the monitoring processes continue during hibernation, however. If the animals are exposed to carbon dioxide, for example, they breathe more rapidly and those of certain species wake up. Similarly, animals of many species wake up if the temperature in their burrows drops to a life-threatening low of 0°C. Hibernators of some species can be awakened by sound or by touch.

Arousal from hibernation can be rapid. In one experiment, bats kept in a refrigerator for 144 days without food were capable of sustained flight after 15 minutes at room temperature. As indicated, however, by the arousals at 0°C, the arousal is a process of self-warming rather than of collecting environmental heat. Breathing becomes more regular and then more rapid, increasing the amount of oxygen available for consumption, and then the animal "burns" its stored food supplies as it returns to its normal temperature and the fast pace of a homeothermic existence.

12 (a) *Prior to hibernating, mammals, such as the dormouse shown here, store up food reserves in body fat.* (b) *Note the heat-conserving reduction in surface-to-volume ratio in this hibernating chip-* *munk.* (c) *Bats have a number of special adaptations which enable them to conserve heat despite their small body size. Some hibernate. In others, the temper-* *ature drops when they are at rest (as it does in hummingbirds). Members of the species shown here also roll up their ears when they get cold.*

(a)

(b)

(c)

37–13 *By facing the sun, a camel exposes as small an area of body surface as possible to the sun's radiation. Its body is insulated by fat on top, which minimizes heat gain by radiation. The under part of its body, which has much less insulation, radiates heat out to the ground cooled by the animal's shadow. Note also the loose-fitting garments worn by the camel driver. When such garments are worn, sweat evaporates off the skin surface. With tight-fitting garments, the sweat evaporates from the surface of the garment instead, and much of the cooling effect is lost. The loose robes, which trap air, also serve to keep desert dwellers warm during the cold nights.*

ADAPTATIONS TO EXTREME TEMPERATURES

Man relies mainly on his technology, rather than his physiology, to permit him to live in extreme climates, but many animals are comfortable in climates that we consider inhospitable or, indeed, uninhabitable.

Adaptations to Extreme Heat

Man is a good regulator of his own temperature at high external temperatures, as Dr. Blagden's experiment showed. His chief limitation in this regard is that he must evaporate a great deal of water in order to unload body heat and so must keep his water consumption high. The camel, the philosophical-looking "ship of the desert," has several advantages over man in the desert. For one thing, the camel excretes a much more concentrated urine; in other words, it does not need to use so much water to dissolve its waste products. In fact, man is very uneconomical with his water supply; even dogs and cats excrete a urine twice as concentrated as man's.

Also, the camel can lose more water proportionally than man and still continue to function. If a man loses 10 percent of his body weight in water, he becomes delirious, deaf, and insensitive to pain. If he loses as much as 12 percent, he is unable to swallow and so cannot recover without assistance. Laboratory rats and many other common animals can tolerate dehydration of up to 12 to 14 percent of body weight. Camels can tolerate the loss of more than 25 percent of their body weight! They can go without drinking for one week in the summer months, three weeks in the winter.

Finally, and probably most importantly, the camel can tolerate a fluctuation in internal temperature of 5° to 6°C. This tolerance means that it can let its temperature rise during the daytime (which the human thermostat would never permit) and cool during the night. The camel begins the next day at below its normal temperature—storing up coolness, in effect. It is estimated that the camel saves as much as 5 liters of water a day as a result of these internal temperature fluctuations.

Camels' humps were once thought to be water-storage tanks, but actually they are localized fat deposits. Physiologists have suggested that the camel carries its fat in a hump instead of distributed all over the body because fat acts as an insulator and impedes the heat flow to the skin. An all-over distribution is thus decidedly useful in Arctic animals but not in inhabitants of hot climates.

Small desert animals usually do not unload heat by sweating or panting; because of their relatively large surface area, such mechanisms would be extravagant in terms of water loss. Rather, they regulate their temperature by selecting a diet with a high fat content and therefore a high water yield, as well as a low urea yield (as we noted in the previous chapter), by excreting a highly concentrated urine, and by avoiding direct heat. Most small desert animals are nocturnal, like the earliest mammals from which they are descended.

Adaptations to Extreme Cold

When you step out of bed barefoot on a cool morning, you probably prefer to step on a wool rug rather than on the bare floor. Although both are at the same temperature, the rug feels warmer. If you touch something metal, such as the door handle, it will feel even cooler than the floor. These apparent differences in

temperature are actually differences in the speed at which these different types of materials conduct heat away from your body. The doorknob, like all metals, is an excellent conductor, and wood is a better conductor than wool.

Water is a better conductor than air. A man is quite comfortable in air at 70°F but may be uncomfortable in water at the same temperature. Asbestos is such a poor conductor that it acts as an insulator, interrupting the flow of heat. Fat is also a poor conductor.

Poor conductors, such as asbestos, wool, air, and fat, serve as insulators, materials that interrupt the flow of heat from an object. Animals adapt to extreme cold largely by increased amounts of insulation.

Fur and feathers, both of which trap air, provide insulation for Arctic land animals. Fur and feathers are usually shed to some extent in the spring and regrow in the fall.

Whales, walruses, seals, and penguins are insulated with fat (neither fur nor feathers serve as effective insulators when wet). In general, the rate of heat loss depends both on the amount of surface area and on temperature differences between the body surface and the surroundings. These marine animals, which survive in extremely cold water, can tolerate a very great drop in their skin surface temperature; measurements of skin temperature have shown that it is only a degree or so above that of the surrounding water. By permitting the skin temperature to drop, these animals, which maintain an internal temperature as warm as that of man, expend very little heat outside of their fat layer and so keep warm—like a diver in a wet suit.

Many animals similarly permit temperatures in the extremities to drop. By so doing, they conserve heat. Also, for many animals, particularly those that stand on the ice, this capacity is essential for another reason. If, for example, the feet of a seabird were as warm as its body, they would melt the ice, which might then freeze over them, trapping the animal until the spring thaws set it free. The extremities of such animals are actually adapted to live at a different internal temperature from that of the rest of the animal. For example, the fat in the foot

37–14 *Aquatic mammals, such as this bull fur seal, have a heavy layer of fat beneath the skin that acts, in effect, like a diver's wet suit, insulating them against heat loss. On land, they have a problem unloading heat. One way they do it is by keeping their fur moist; another is by sleeping, which, in some species, reduces heat production by almost 25 percent.*

37–15 *An example of countercurrent mechanism. The legs of these Canada geese are much cooler than their bodies—a useful adaptation that prevents heat loss from their bodies and also circumvents the problems that would occur if ice melted (and then refroze) around their feet.*

of an Arctic fox has a different thermal behavior from that of the fat in the rest of its body, so that its footpads are soft and resilient even at temperatures of −50°C.

Countercurrent Exchange

In many Arctic animals, the veins and arteries leading to and from the extremities are juxtaposed in such a way that the chilled blood returning from the legs (or fins or tail) through the veins picks up heat from the blood entering the extremities through the arteries. The veins and arteries are closely apposed to give maximum surface for heat transfer. This arrangement, which serves to keep heat in the body and away from the extremities, is accomplished, largely, by countercurrent exchange (see page 653).

SUMMARY

Life can exist only within a very narrow temperature range, from about 0° to about 50°C, with few exceptions. Animals must either seek out environments with suitable temperatures or create suitable internal environments. The warmer the environment, the higher the rate of biochemical reactions within the life zone. Poikilotherms are animals whose temperature is variable. Homeotherms are animals that maintain a constant and comparatively high temperature.

Most poikilothermic vertebrates live in the ocean or other large bodies of water where the temperature is relatively constant.

With the emergence to land, animals entered a life zone in which temperatures are much more variable and also in which temperature regulation is possible. Amphibians and, in particular, reptiles modify their temperature by a number of behavioral adaptations.

The first homeotherms were the primitive mammals, small nocturnal insectivores. In homeotherms—the birds and the mammals—temperature is monitored by centers in the hypothalamus. The regulation of temperature requires a means for altering heat production and dissipation. Heat is produced by the breakdown of glucose and other energy-yielding compounds; homeothermy is costly in terms of energy expenditure. The rates of heat production can be increased to some degree by thyroid hormone and also by adrenaline.

Heat is carried by the bloodstream from the core of the body to the surface, where it is dissipated. In hot weather, the cutaneous blood vessels dilate, increasing the flow of blood to the skin. As external temperatures rise above body temperatures, man and most other large animals begin to sweat. Evaporation of sweat from body surfaces requires heat and cools the surface. In the cold, cutaneous blood vessels constrict. As temperatures drop, shivering begins; the muscular activity involved in shivering produces heat. Temperature is regulated by an automatic system, or thermostat, in the hypothalamus that measures the body temperature and sets in motion the response mechanisms.

QUESTIONS

1. Define the following terms: temperature, heat, poikilotherm, homeotherm, insulation.
2. Compare the surface-to-volume ratio of an Eskimo igloo with that of a California ranch house. In what way is the igloo well suited to the environment in which it is found?
3. Compare hibernation in animals with dormancy in plants. In what ways are they alike? In what ways are they different?
4. Diagram a heat-conserving countercurrent mechanism in the leg of an Arctic animal.
5. Plants, as well as animals, are warmed above air temperature by sunlight, and plants cannot actively seek shade. Why might the leaves in the sunlight on the top of an oak tree be smaller and more extensively lobed (more fingerlike) than the leaves in the shady areas lower on the tree?
6. The primary temperature-measuring device in our bodies is located in the hypothalamus, at the base of the brain. Why is this a better location than some more peripheral point, such as the skin? The latter would permit the body to measure the temperature around it and take action when the environment got too hot or too cold. (And why do we mount the thermostats of our homes on an inside wall rather than on the doorposts?)

Chapter 38

The Senses

Our sensory equipment is the means by which we keep in touch with the world around us. It provides the mechanisms by which we and other animals are informed of predator and prey, friend and foe, danger, whether or not something is good to eat, or changes in the weather or the seasons. It identifies infant to mother, mother to infant, and mate to mate. It is also our source of pleasure and of esthetic enjoyment and provides our tools for learning. Our sense organs, like the other organs of the body of any animal, are the products of adaptation and evolution, and by these long processes, they have been tailored to our specific requirements. For example, we think we see what is visible and hear what is audible, but visible and audible are actually indefinite and relative. What we see is very different from what an insect sees, and a bat or a fish, acoustically speaking, might as well be living on another planet. Moreover, it is likely that our insensitivity to a great number of the potential stimuli that surround us is almost as useful as our sensitivity to others.

Our sensory receptors are part of the nervous system. In Chapter 32, we discussed some aspects of nervous system function with particular attention to the autonomic ("involuntary") nervous system, which along with the endocrine system, regulates the circulation of the blood, digestion, and other vital functions. In this chapter and the one that follows, we shall be more concerned with the reception and processing of information received from the environment and with higher functions of the brain, including dreams, learning, and memory.

The somatic component of the peripheral nervous system, as we noted previously, is made up of both sensory and motor neurons. The motor neurons exert their effects by stimulating muscles. The sensory neurons receive information from various kinds of sensory receptors, including receptors for light, sound, pain, touch, pressure, heat, cold, odor, and taste. They also include stretch receptors, gravity receptors, and other internal receptors which coordinate the movements of the various parts of our bodies. The particular agency to which the sensory receptor responds is called the stimulus.

In this chapter, we shall describe some of these sensory receptors, in particular the special sense organs of man.

SENSORY RECEPTORS

Sensory receptors signal change in the environment or in the animal in relation to the environment, and relay this information to the central nervous system. One kind of internal receptor (interoceptor) monitors the internal organs. Examples are the receptors that measure blood pressure and CO_2 concentration in the carotid arteries. We are usually not conscious of signals from these receptors, although sometimes these signals are perceived as pain, hunger, thirst, nausea, or the sensations, produced by stretch receptors, of having a full bladder or bowel.

Another type of internal receptor, the proprioceptor (from the Latin *proprius*, meaning "self"), keeps us informed about the relative positions of various parts of our body, enabling us to coordinate our muscular activities.

The external environment is monitored by a class of receptors called exteroceptors. The skin contains a number of such exteroceptors, each specialized for touch, or pressure, or heat, or cold. These receptors are not uniformly distributed in the skin of various parts of the body. For instance, a square centimeter of skin on the tip of your finger contains approximately 60 pain and 100 touch spots, whereas a square centimeter of skin on the back of your finger has only about 9 touch spots but about 100 pain spots. (These spots are mapped by marking off the skin in a fine grid and testing the various points on the grid with a metal stylus, either hot, cold, or at skin temperature.)

Other exteroceptors are located in specialized organs such as eye, ear, and olfactory tissues. These organs are clusters of sensory receptor cells with accessory structures, such as the lens of the eye or the tympanic membrane of the outer ear, which focus or amplify the stimulus. The important difference between different types of receptors is the kind of stimulus to which each responds; the response, however, is always the same, the generation of a nerve impulse.

8–1 *Receptors found in skin. Free nerve endings are pain receptors, end-bulbs of Krause are primarily receptors for cold, Meissner's corpuscles are receptors for superficial touch, and the onion-shaped* *Pacinian corpuscles are for pressure. Nerve fibers are shown in color. In the Pacinian corpuscle, the best studied of the four, a single myelinated nerve fiber (axon) enters the corpuscle; it loses its* *myelin sheath in the central core. The corpuscle itself is composed of many concentric layers of connective tissue. Pressure on these outer layers produces an action potential in the nerve fiber.*

FREE
NERVE
ENDING

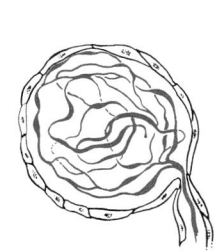

END-BULB
OF KRAUSE

MEISSNER'S
CORPUSCLE

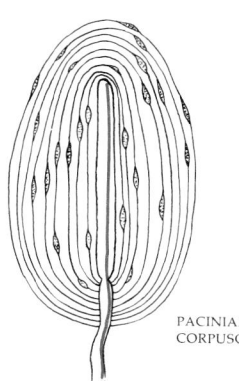

PACINIAN
CORPUSCLE

LIGHT RECEPTION: VISION

Eyes have evolved independently several times in the course of evolutionary history. Among the most highly developed of modern light-receptor systems are the compound eye of the arthropods (see page 406), the eye of the octopus (see page 389), and the vertebrate eye, of which the human eye, shown in Figure 38–2, is an example.

The vertebrate type of eye is often called a camera eye. In fact, it has a number of features in common with the ordinary camera equipped with several expensive accessories, such as a built-in cleaning and lubricating system, an exposure meter, and an automatic field finder. As indicated in Figure 38–3, light from the object being viewed passes through the transparent cornea and lens, which focus an inverted image of the object on the light-sensitive retina in the back of the eyeball.

In mammals, the eye is focused by contracting or relaxing the ciliary muscles, thus changing the curvature, and hence the focal length, of the lens. (Fish and amphibians, which lack ciliary muscles, focus their eyes the same way one focuses a camera, with a lens of fixed focal length. Muscles within the eye change the position of the lens, drawing it back toward the retina in order to focus more distant objects.) The iris controls the amount of light entering the eye by regulating the size of the pupil.

Depth vision, or stereoscopic vision, depends on viewing the same visual field with both eyes simultaneously. When, under these conditions, both eyes are trained on a distant object, then that object can be seen as a single image.

38–2 *The eyeball is a fluid-filled sphere composed of three layers of tissue. The outer layer, the sclera, is white fibrous connective tissue that serves a protective function. The anterior portion of the sclera, the cornea, is transparent. The middle layer, the choroid, contains blood vessels. Its anterior portion is modified into the ciliary body, the suspensory ligament, and the iris. The ciliary body is a circle of smooth muscle from which the suspensory ligaments extend that hold the lens in place. The colored part of the eye, the iris, is a circular structure attached to the ciliary body. The pupil is a hole in the center of the iris. The innermost layer, the retina, has no anterior portion. It consists of the rods and cones and a pigmented layer of epithelium that adheres to the choroid layer. Only one-sixth of the eye is exposed; the rest of the eyeball is recessed in the orbit, protected by the bony socket of the skull.*

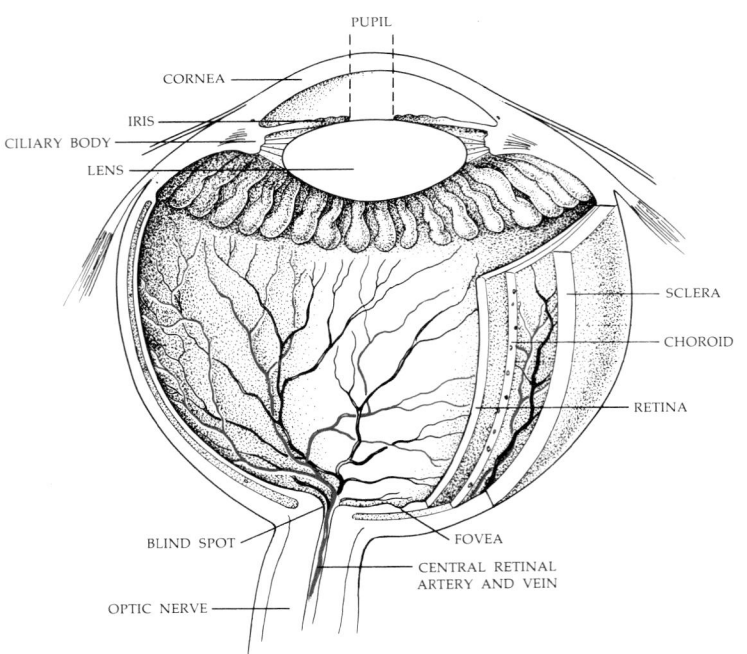

38–3 *Focusing the eye. The lens, which is like a jelly-filled transparent rubber balloon, is held in place by suspensory ligaments attached to the ciliary body, which encircles the lens. The ciliary body contains two sets of smooth muscles, radial and circular, controlled by the autonomic nervous system. When the radial muscles contract, the lens is pulled outward and flattened; when the circular muscles con-tract, the lens is released and, as it returns to its more nearly spherical shape, its curvature is increased. (a) The normal eye views distant objects with the lens stretched and flattened; (b) for viewing near objects, the lens is relaxed and more convex, bending the light rays more sharply. (c) Nearsightedness occurs when the eyeball is too long for the lens to focus a distant image on the retina. (d) It is corrected by a concave lens that bends the light rays out enough so that the image is focused. (e) Farsightedness is the result of an eyeball too short for the lens to focus a near object. (f) It is corrected by a convex lens that bends the light rays in before they reach the lens of the eye.*

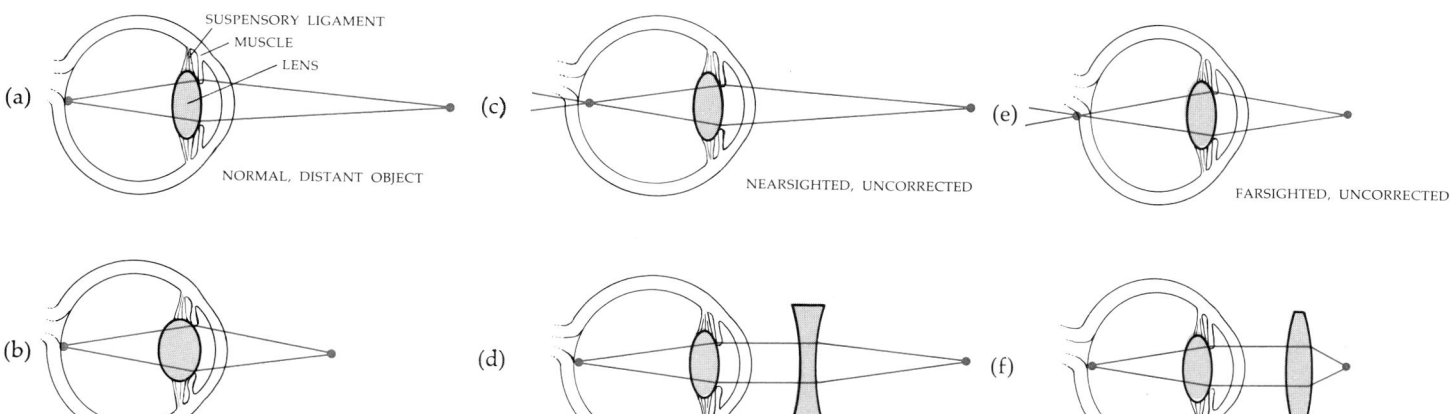

(a) NORMAL, DISTANT OBJECT
SUSPENSORY LIGAMENT
MUSCLE
LENS

(b) NORMAL, NEAR OBJECT

(c) NEARSIGHTED, UNCORRECTED

(d) NEARSIGHTED, CORRECTED

(e) FARSIGHTED, UNCORRECTED

(f) FARSIGHTED, CORRECTED

38–4 *Some vertebrate adaptations. (a) Mud skipper eyes are directed upwards like periscopes. (b) Eagles and other predatory birds often have eyes as large as man's (although the skull and brain are much smaller), two foveas (see page 699) in each retina, and strong powers of near and far accommodation. (c) Horses have eyes set high in their skulls, an adaptation that allowed their ancestors to watch for predators while they were grazing. (d) This primitive primate, a lemur, is nocturnal. Unlike many other nocturnal animals, it has relatively large eyes. Nocturnal animals characteristically have only rod cells, which are more light-sensitive than cone cells, and so they lack any color vision.*

(a)

(b)

(c)

(d)

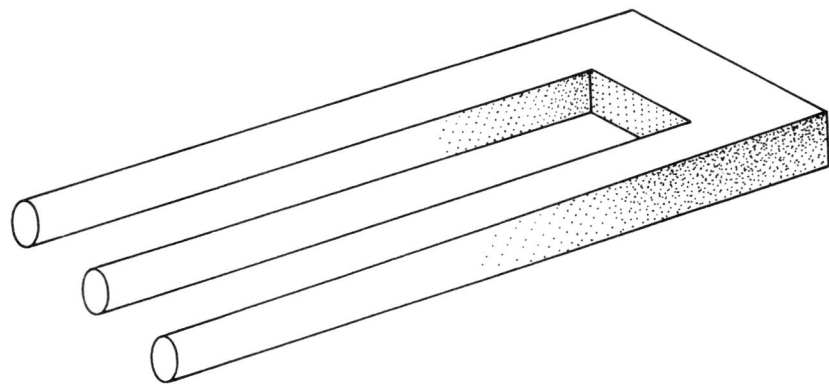

38–5 *Believing is seeing. Because we cannot interpret this picture in terms of visual conventions, we cannot make up our minds what it is we are seeing.*

However, a nearby object will present a slightly different image to each eye (parallax). On the basis of the disparity of the two images of the nearby object, we automatically compute its distance from the distant object. When the eyes are trained on a very distant object, however, the disparity is too small; hence, we are not able to make such judgments for faraway objects. Instead, we estimate their distance on the basis of known size of the objects or using such visual clues, such as houses, people, or cars. A young child viewing a distant object—an airplane, for example—may see it as little rather than distant.

Tree-dwelling animals, such as the probable forerunners of *Homo sapiens*, usually have overlapping visual fields in the front that provide them with the stereoscopic vision essential for the distance judgments involved in moving from branch to branch. Predatory animals also tend to have stereoscopic vision, but animals which are more likely to be hunted than to hunt usually have eyes on the side of the head, giving a wide total visual field. Some birds with large, laterally placed eyes—the woodcock, the cuckoo, and certain species of crow—have binocular fields of vision both in front of them and behind them.

The Retina

The retina of the vertebrate eye is anatomically inside out—that is, the photoreceptors of the eye are pointed toward the back of the eyeball—and light must reach the light-sensitive area by passing through a layer of nerve cell bodies (see Figure 38–6). Light that is not captured by the photoreceptors is absorbed by pigmented epithelium that lines the back of the eyeball. Some nocturnal vertebrates have a reflecting layer behind the photoreceptors that increases the likelihood of dim light stimulating the photoreceptors. This reflecting layer makes the animal's eyes seem to shine at night.

The photoreceptor cells transmit their electrical signals to neurons known as bipolar cells, which pass them on to the ganglion cells whose axons form the optic nerve. As you can see in the figure, these pathways of transmission are not direct, but varied and elaborate interconnections are made among the cells even before nerve impulses leave the retina, indicating that some preliminary analysis of visual information takes place in the retina itself. The human retina contains about 125 million photoreceptors and the optic nerve about 1 million fibers.

38–6 *The retina of the vertebrate eye. Light (shown here as entering from the top) must pass through a layer of cells to reach the photoreceptors (the rods and cones) at the back of the eye. Signals from the photoreceptor cells are then transmitted through the bipolar cells to the ganglion cells, whose axons converge to become the optic nerve. The transmission paths are not direct but involve elaborate interconnections.*

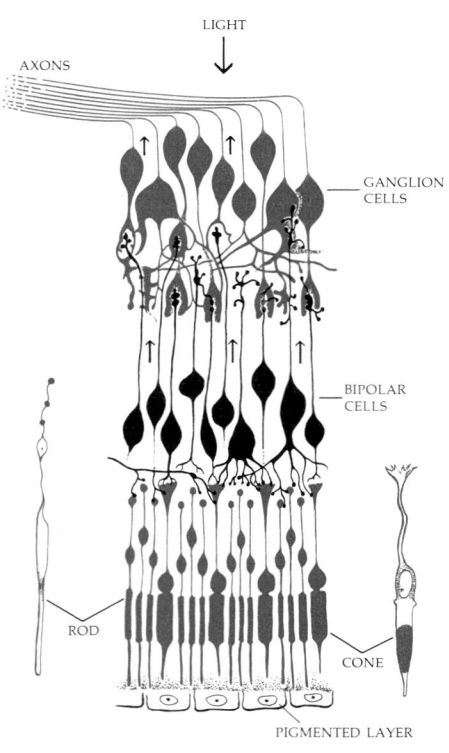

LIGHT

AXONS

GANGLION CELLS

BIPOLAR CELLS

ROD

CONE

PIGMENTED LAYER

38–7 *With this diagram, you can prove for yourself that you have a blind spot in each eye. Hold the book about 12 inches from your face, cover your left eye, and gaze steadily at the X while slowly moving the book toward your face. Note that at a certain distance, the image of the dot becomes invisible. Then cover your right eye and gaze steadily at the dot while you move the book toward your face. What happens to the X?*

 X

 ● / /

Ganglion cell axons from all over the retina converge at the rear of the eyeball and, bundled together like a cable to form the optic nerve, pass through the retina and out the back. The point at which the axons pass out of the eyeball is a blind spot in the eye since photoreceptor cells are absent. We generally are not aware of the existence of the blind spot since we usually see the same object with both eyes and the "missing piece" is always supplied by the other eye. If you are not aware of your own blind spot, you can demonstrate its existence with the help of Figure 38–7.

As shown in Figure 38–2, there is an area in the human retina just beside the blind spot known as the *fovea*. This is the only area in the retina at which the imperfect lens of our eye forms a really sharp image. In this area, the photoreceptor cells are packed much more closely together and many have a one-to-one connection with the bipolar cells. Birds, which rely on vision above all the other senses, may have two or three foveas. Most vertebrates, however, have no fovea, and their vision is comparable in acuity to what we see out of the corners of our eyes.

Rods and Cones

The human retina contains two types of photoreceptors, named, because of their shapes, rods and cones. The cones, of which the human eye contains about 5 million, provide greater resolution of vision, giving a "crisper" picture. They are generally concentrated in the center of the retina; the fovea consists entirely of cones. Man has about 160,000 cones per square millimeter of eye. A hawk, with an eye of about the same size, has some 1 million cones per square millimeter and therefore a visual acuity about eight times that of man.

Rods do not provide as great a degree of resolution as cones do, but they are more light-sensitive; in dim light, our vision depends entirely on rods. In man, rods are generally concentrated around the periphery of the retina. Some nocturnal animals, such as toads, mice, rats, and bats, have retinas made up entirely of rods, and some diurnal animals, such as some reptiles, have only cones.

Cones are responsible for color vision, which is why the world becomes gray to us at twilight, when the light is too dim to stimulate the cones and we must depend on the rods.

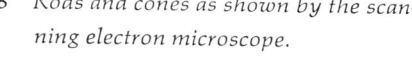 **38–8** *Rods and cones as shown by the scanning electron microscope.*

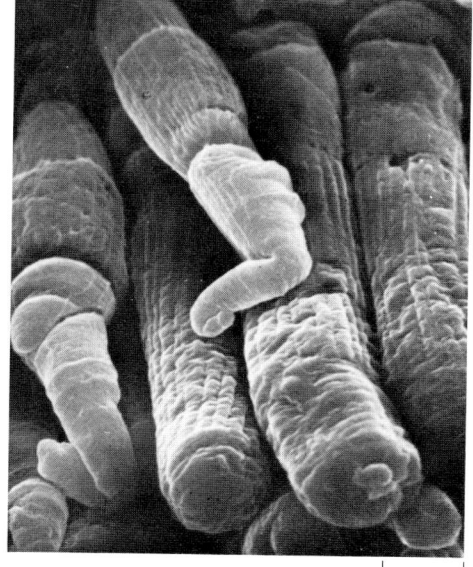

5 μm

38–9 (a) *Diagram of a rod from the human retina. The molecules of light-sensitive pigment, rhodopsin, are thought to be located on the lamellae. The foot makes synaptic contact with a bipolar cell that connects with a ganglion cell.* (b) *Photoreceptor cells in an electron micrograph. The ends pointing toward the back of the eyeball consist of a stack of membranes, piled up one on top of the other like poker chips. The light-sensitive pigment is built into these membranes. In the lower part of the cell, new pigment molecules and other substances required by the light receptor area are synthesized and transported through a narrow stalk (indicated by the arrow), separating the two portions of the cell, to the membrane-packed portion, where the actual work of this highly specialized cell is done. A cross section of this stalk reveals, surprisingly, that it has the internal structure of a cilium, lacking only the two central fibers. The cells shown here are rods, but cones are built along the same general principles.*

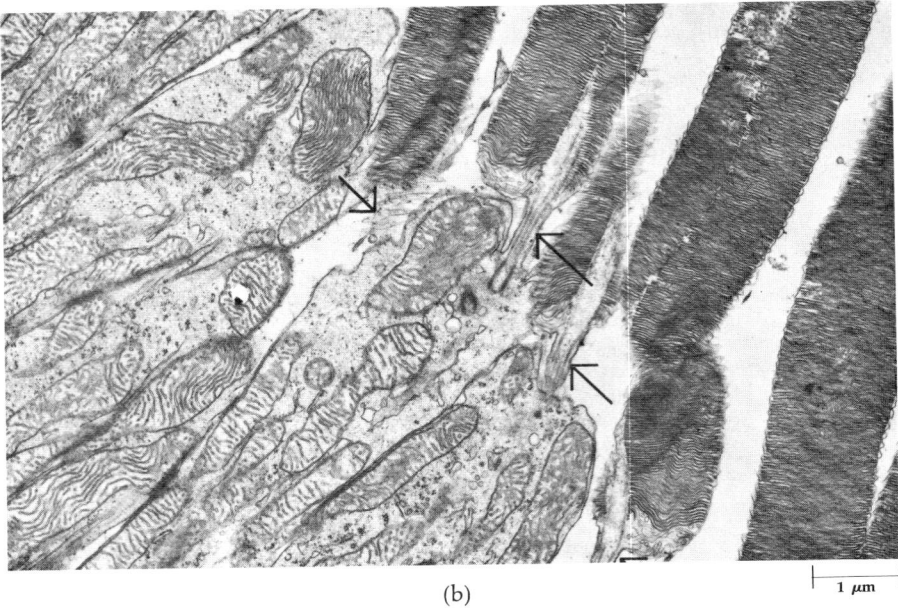

(b)

| | 1 μm |

(a)

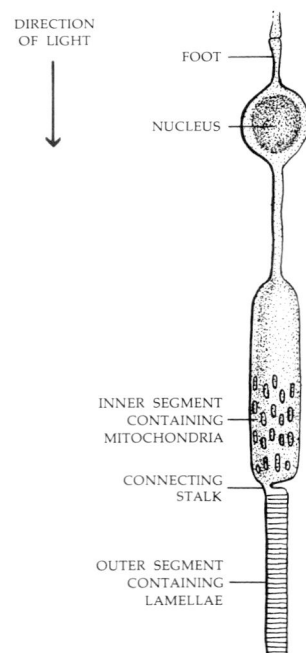

Both rods and cones contain light-sensitive compounds, each consisting of a protein, called an opsin, and a carotenoid, which, as you may recall, is a pigment derived from vitamin A. As we noted previously, the eyes of arthropods, mollusks, and vertebrates all evolved separately, and yet each of the three types contains almost identical visual pigments. Moreover, since no animals can synthesize carotenoids, all vision depends on substances derived in the diet directly or indirectly from plants.

The light-sensitive compound in the rods of the vertebrates is a reddish material named rhodopsin, sometimes called visual purple. The carotenoid portion of the rhodopsin molecule is retinal (formerly called retinene). When rhodopsin absorbs light, retinal changes shape (see Figure 38–10). This change in shape causes retinal to dissociate from its opsin. The opsin then also changes shape. In some way, as yet unknown, this breakdown of the rhodopsin results in a change in the membrane of the rod cell that causes a nerve impulse to be transmitted to the bipolar cell and then along pathways to the brain.

An enzymatic reaction changes retinal back to its original shape and recombines it with opsin, restoring rhodopsin and so completing the cycle.

Eyes become adapted to the dark by slow buildup of rhodopsin from vitamin A stored in the pigmented epithelium of the eye. Persons who have a vitamin A deficiency cannot build up sufficient rhodopsin and therefore suffer "night blindness." A fully dark adapted rod can normally respond to the absorption of a single photon of visible light.

There are three different types of cones, each containing one of three different visual pigments (each of these pigments is made up of retinal and a slightly different opsin). As long ago as 1802, a three-receptor hypothesis for color vision was proposed, and this hypothesis, supported by a number of psychological studies, has been greatly bolstered by the identification of these three separate cone types and their pigments. Each pigment is most sensitive to the wavelength

–10 *When the retinal portion of the rhodopsin molecule is excited by light, it changes shape, setting in motion a series of events that result in the sensation of light. Rhodopsin is changed back to its original conformation by an enzyme, completing the visual cycle. The arrow indicates the bond around which the side chain of the "resting" retinal rotates to form the "excited" retinal.*

RESTING   EXCITED

of one of the three "primary" colors—blue, green, or red. Different shades of color stimulate different combinations of these cones, and these signals, after being processed in the retina, are further processed by the brain and so become what we perceive as color.

Vision and Behavior

Vision is a direct stimulus to action in many of the vertebrates. Predatory fish are vision-stimulated, and many such fish literally "cannot help" striking at a moving object of appropriate size. The frog can learn not to strike at a small moving object, but only with difficulty. If it comes to associate an unpleasant experience with a particular shape, it can restrain itself, although apparently with effort, from striking at that shape. On the other hand, it can never learn to take an object that is not moving. Even a frog that has eaten thousands of live flies will starve to death surrounded by motionless ones.

Recently, it has been found that certain ganglion cell axons leaving the retina of the frog produce impulses only when the frog's eye sees a small moving object, rounded on the top. In the laboratory, the visual stimulus for these ganglion cells can be simulated by a variety of objects—such as paper clips and pencil erasers—provided they are in motion. In a frog's natural environment, however, the objects likely to elicit this response most frequently are flying insects. A large, moving object or a small, still one provoke no response. These

11 *Frog with bug.*

38–12 *Colors elicit behavioral responses in many animal species. In experiments by Niko Tinbergen with sticklebacks, the crude models, colored red, caused much more positive reactions in both male and female sticklebacks (aggressiveness in the males and attraction in the females) than did the exact replica of a male, which was not colored. Tinbergen shared a Nobel prize in 1973 for his work in animal behavior.*

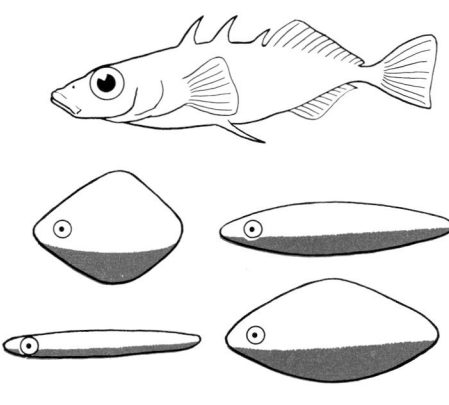

38–13 *Longitudinal section of a taste receptor.*

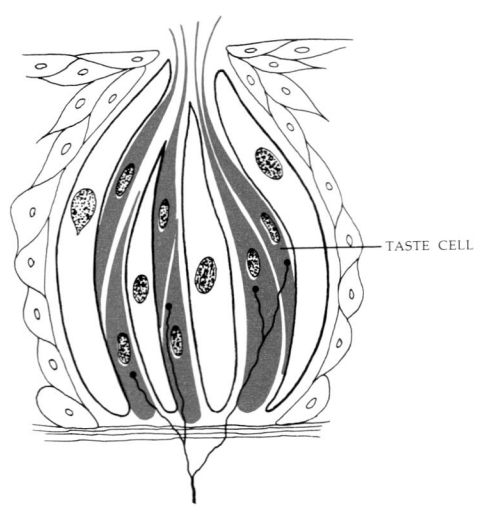

TASTE CELL

ganglion cell axons terminate in a special layer of cells in the frog's midbrain, and although no further connections have been traced, it is tempting to speculate that signals are relayed from this part of the midbrain to motor nerves that control the muscles involved in striking.

Colors often play an important role in eliciting behavior in vertebrates. Studies of the three-spined stickleback, a small freshwater fish, have shown that fighting behavior among the males is elicited by the red color of their underbellies, which develops during the mating season. Niko Tinbergen, who kept a laboratory in England full of aquariums containing sticklebacks, found that the male sticklebacks would rush to the sides of their tanks and assume threatening postures every time a red mailtruck passed on the street outside. Similarly, it has been found that fighting behavior among English robins (which are a different species of bird from our American robins) can be evoked by a small tuft of red feathers. Among birds, bright plumage, especially of the males, plays an important part in attracting the female and in courtship ceremonies.

In mammals, in general, vision is less important than it is in many of the other vertebrates. Many of the smaller mammals are nocturnal, and most of them live close to the ground in habitats in which their highly developed sense of smell provides the greatest amount of information about their environment. Presumably, it was the tree-dwelling habits of our immediate ancestors that made the sense of smell less useful for the primates. Most mammals do not have color vision, and, among the mammals, only primates have a central fovea for sharp vision. In man, the highly developed vision is undoubtedly closely correlated with the development of the use of the hands for fine manipulative movements. This reasonable hypothesis is supported by the fact that the human eye is so constructed that the sharpest images—images within the fovea—can be made of objects which are within the reach of the hand.

CHEMORECEPTION: TASTE AND SMELL

If we were to lose our sense of taste or smell, our life would lose many of its pleasures but we would not be greatly handicapped in our essential activities. Blindness or deafness is a far more serious threat. For many animals, however, the sense of smell is their window on the world, much as vision is ours. Watch a dog the next time it is let out of the house in the morning: It does not look around, human fashion; it sniffs the air and then explores the ground with its nose in order to bring itself up to date on the events of the previous night.

Although the organs of chemoreception seem much simpler than those of vision or hearing, we know less about how they work. We do not know (1) how a chemical can give rise to an electrical signal in a nerve cell or (2) how the various different chemicals that we can distinguish by taste or smell manage to identify themselves to the brain.

Taste

Fish, particularly the bottom feeders, have taste cells scattered over the surfaces of their bodies, and these cells play a major role in determining their behavior. In the carp, for example, the part of the brain that receives information from these taste cells is larger than all the other sensory centers combined. The catfish, also a bottom feeder, has taste cells on its body and, in addition, trails

−14 *Amphibians and reptiles have, in addition to nostrils, a special olfactory organ, the vomeronasal organ, in the roof of the mouth. Food substances are apparently tested in this organ. When a snake lashes its forked tongue in the air, like the Eastern hognose snake shown in the photograph, it is collecting samples for testing in its vomeronasal cavities.*

barbels, or "whiskers," along the ground. These are richly supplied with gustatory nerve endings and when they are stimulated, the fish turns and snaps at the source of the stimulus.

In the terrestrial animals, taste cells are located inside the mouth, where they act as sentinels, providing a final judgment on what is and what is not to be swallowed. The taste receptors and the supporting cells around them form the goblet-shaped clusters known as taste buds. As noted in Chapter 34, man is able to distinguish four primary tastes: sweet, sour, salty, and bitter. While each primary taste appears to stimulate a different type of taste bud, there are indications that a single taste bud may respond to more than one category of substance.

Most animals seem to have the same general range of taste discrimination, although there is some difference in what animals "like" and "don't like." Birds, for example, will readily eat seeds and insects with a bitter taste, whereas for most other vertebrates, bitterness serves as a warning signal. Cats are among the few animals that do not prefer substances with a sweet taste, and very few taste receptors that respond to sugar can be found in the cat's tongue. Most animals will work for a sweet reward even though they are not hungry, and the cat's indifference to sweetness is one of the reasons it is difficult to train—although its general social independence is probably a more significant one.

Smell

In fish, taste and smell are operationally very similar since both involve the detection of substances dissolved in the surrounding water. Taste receptors and smell receptors differ anatomically in fish as in higher vertebrates, however, and the centers of taste and smell within the brain are entirely distinct.

In terrestrial animals, smell can be defined as the chemoreception of airborne substances. Even these substances, however, must be dissolved in the watery layer of mucus overlying the olfactory epithelium. In man, this epithelium, which is located high within the nasal passages, is comparatively small, the part within each nostril being only about as large as a postage stamp. Each of these areas contains some 600,000 receptor cells. Even with our relatively insensitive olfactory equipment, we are able to discriminate some 10,000 different odors.

38–15　(a) *A patch of special tissue, the olfactory epithelium, arching over the roof of each nasal cavity, is responsible for our sense of smell.* (b) *The olfactory epithelium is composed of three types of cells: supporting cells, basal cells, and olfactory cells. Supporting cells are tall and columnar, wider near the surface than they are deep within the tissue, and their outer surfaces are covered with micro-* *villi, similar to the microvilli found on the surface of intestinal cells. The basal cells, triangular in shape, are found along the deepest layer of the epithelium. Their function is unknown. They may give rise to new supporting cells when these are needed. The olfactory cells are the sensory receptors.* (c) *The olfactory cell is long and narrow. The cilia protruding from its upper surface are believed to be* *the odor receptors, although the way in which they function is not known. This outermost part of the cell is connected with the cell body by a long stalk containing microtubules arising from deep inside the cell. From the deepest portion of the cell body, a nerve fiber extends through the underlying tissue, transporting the sensory message to the brain.*

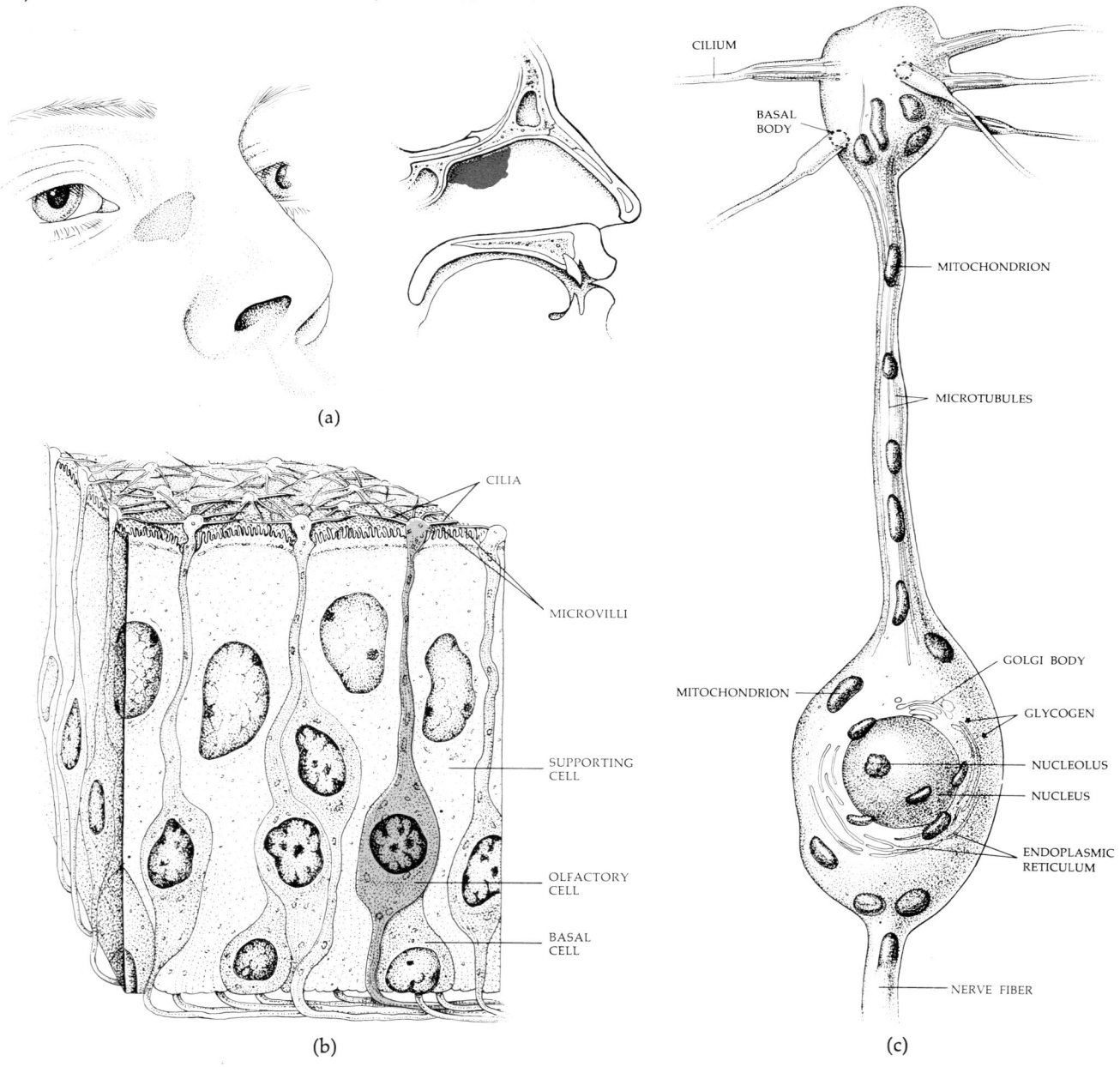

(a)

(b)

(c)

CILIA

MICROVILLI

SUPPORTING
CELL

OLFACTORY
CELL

BASAL
CELL

CILIUM

BASAL
BODY

MITOCHONDRION

MICROTUBULES

GOLGI BODY

GLYCOGEN

NUCLEOLUS

NUCLEUS

ENDOPLASMIC
RETICULUM

MITOCHONDRION

NERVE FIBER

Most of what we call flavor in food is actually a result of volatile substances reaching the olfactory epithelium.

One of the current theories of odor discrimination was developed in the 1960s by John Amoore while he was still an undergraduate at Oxford University in England. According to Amoore's theory, all scents are made up of combinations of seven "primary" odors: camphoric, musky, floral, pepperminty, etherlike, pungent, and putrid. There are also, he hypothesized, seven different types of olfactory receptors, each one of which has its own special shape. Molecules of a certain shape fit into correspondingly shaped receptors, just as a piece of a jigsaw puzzle fits into its own particular space. All substances that have a pungent odor, for example, have the same general shape and fit into the same type of receptor. Some molecules, depending on which way they are oriented, can fit into more than one receptor and so can evoke in the brain two different types of signals. From the signals received from the various different types of cells, the brain constructs a "picture" of an odor.

Smell plays a large role in the behavior of most mammals. The males of many species—including domestic dogs and cats—scent-mark their territories with urine as warning signals to other males. Males are attracted to females by special odors associated with estrus. Such chemical messages among members of the same species are known as *pheromones*. (See page 410 for a discussion of pheromones in insects.)

Among mice, juvenile females reach sexual maturity earlier when exposed to the odor of an adult male. Pregnant females may resorb their fetuses in the presence of the odor of a strange male. Both of these effects appear to be mediated by pheromones in the urine.

Recently a sex attractant has been isolated from female rhesus monkeys and identified as a fatty acid. Whether or not sex attractants and other chemical signals are exchanged among members of the human species is not known. However, the fact that the apocrine sweat glands, the sources of "body odor," begin to function at puberty strongly suggests that the chemicals they produce (also fatty acids) originally played such a role. The pubic and axillary hair that develops at about the same time would appear to have the function of retaining and amplifying these odors. It may be that society's current obsession with soaps, deodorants, and feminine hygiene products represents merely another device by which civilization attempts to shape and curb our natural proclivities.

SOUND RECEPTION: HEARING

Hearing (and also balance) in the vertebrates depends on the movement of tiny hairs that stimulate a response in underlying sensory cells. Figure 38–17 shows such cells as they appear in the organ of Corti, the hearing receptor of the mammal, but the sensory cell from any hearing or balancing organ in any of the vertebrates would appear much the same. The hairs on the surface of this cell have the characteristic structure of cilia and are undoubtedly derived from them.

In some fish, hair cells of this sort appear along the surface of the body, but in most species, they are protected beneath the surface in canals that are open to the exterior. The network of canals, known as the lateral-line system, runs along the side of the fish, branching out around the head. The movement of

the water causes the hairs in the lateral-line canals to bend, and this bending stimulates the dendrites of the underlying neurons. These movements keep the fish informed about other animals in the vicinity—whether prey, predator, or potential mate. They also make it possible for the fish to hear sounds produced by other fish. Fish are able to make a variety of noises, but there is a lack of general agreement on the extent to which such sounds are actually used in communication. Undoubtedly, they serve at the very least to advise fish of the presence of other fish.

Within the head of the fish is another organ, the labyrinth, which contains hair cells. The primary function of the labyrinth is balance. It is from this organ that the hearing mechanisms of all the terrestrial vertebrates have evolved. The uppermost part of this structure contains three fluid-filled semicircular canals. These canals are oriented in three directions, representing the three-dimensional space in which we live, and movements of the head cause the fluids within these canals to press upon the hair cells. Impulses from the hair cells report these movements to the brain. Below the semicircular canals and connected with them are three hollow chambers, each containing statocysts (page 371), which respond to gravity. This part of the labyrinth—the semicircular canals and the chambers containing the statocysts—is found in all vertebrates and has changed little in the course of evolution.

In some fish, the lower chambers of the labyrinth are connected either directly through the skull or by a small series of bones to the swim bladder, a gas-filled pouch which gives the fish buoyancy. Sound waves set up vibrations in the swim bladder, which are transmitted to the labyrinth. The statoliths in one of the three chambers, the lagena, respond to these sound vibrations. In fish which do not have a swim bladder, sound impulses are transmitted directly through the head, just as we can feel vibrations through our fingertips.

The Evolution of the Ear

In the course of evolution, the lagena became larger and more elaborate. The outer structures of the ear also underwent changes. In the fish, as we have seen, there is no opening connecting the labyrinth with the outer surface of the animal. In amphibians, there is a tympanic membrane on the surface of the animal's head, and underlying it is a small bone, the stapes, which conducts sound from the membrane to the labyrinth. (This bone is "borrowed" from one of the jawbones of the fish.) In reptiles, a ridge or fold of skin may guard the tympanum, and in birds and mammals, the eardrum is sunk far below the surface, with a deep tube (the outer ear) leading to it. Many mammals have a highly developed external ear, which serves as an adjustable scoop for catching sound waves and funneling them into the ear tube. Some few human beings retain the ability to wiggle their ears, using the same muscles other animals use to cock these special appendages toward the source of a sound.

The Structure of the Human Ear

Figure 38–16 diagrams the structure of the human ear. Sound travels through the outer ear to the tympanic membrane (the eardrum), in which it sets up a vibration. This vibration is transferred to a series of three very small and delicate bones, which are called, because of their shapes, the hammer, the anvil,

-16 *The structure of the human ear. Sound waves enter the outer ear and pass to the tympanic membrane which they cause to vibrate. These vibrations are transmitted through the hammer and anvil to the stirrup. The space containing these three small bones (and which connects with the eustachian tube) is the middle ear. The stirrup is attached to another membrane (the oval window), and as the stirrup vibrates, it pushes against the oval window, leading to the cochlea, which is in the inner ear. Through the inner ear, the signal is transmitted to the brain as an action potential. The semicircular canals are concerned with balance. They are fluid-filled ducts embedded in a bony labryinth in the skull. The canals contain hair cells that are stimulated by movements of the fluid and convey information about the movements and position of the head.*

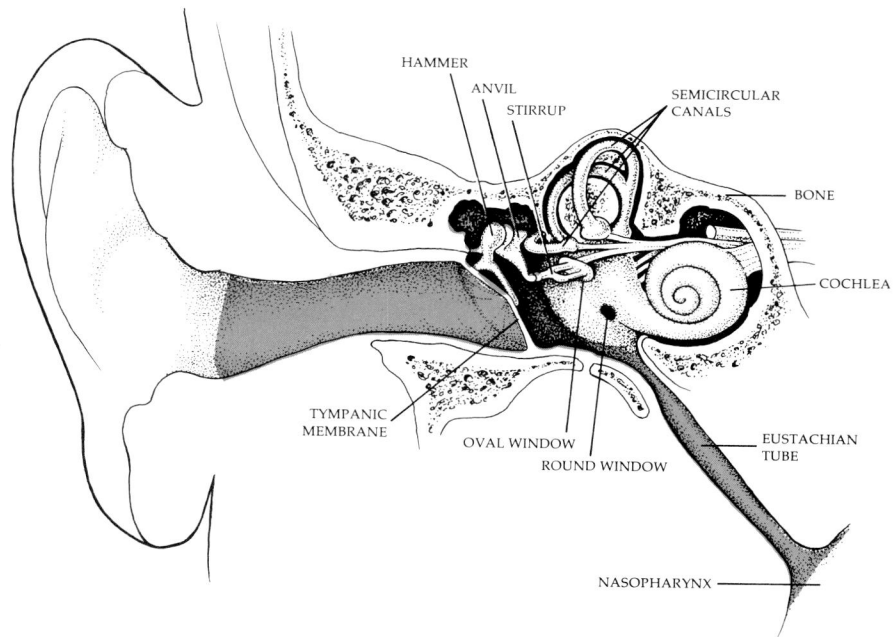

and the stirrup. Movements in the tympanic membrane cause the stirrup to tap gently and rapidly against a membrane, the oval window, leading to the inner ear, and from the inner ear, the signal is transmitted by nerve fibers to the brain.

The middle ear, as the area between the eardrum and the inner ear is called, is connected with the upper pharynx by the eustachian tube. This connection makes it possible to equalize the air pressure in the middle ear; it also unfortunately makes the middle ear a fertile breeding ground for infectious microorganisms which enter the body through the nose or throat.

The sensory receptors lie in the inner ear, in a bony spiral shell, the *cochlea,* which had its evolutionary origins in the lagena. Figure 38–17 shows a diagram of the cochlea uncoiled. It consists essentially of three fluid-filled canals separated by membranes. The upper and lower canals are connected with one another at the far end of the spiral. At the base of each of these two canals are the movable membranes known as the oval and round windows. The little stirrup vibrates against the oval window membrane at the base of the upper canal, and the sound waves travel the length of the cochlea, around the far end, and back again to the membrane at the base of the lower canal, the round window. As the oval window moves in, the round window moves out, keeping the pressure equalized. You will notice that, although hearing is generally considered to be the detection of airborne sounds, our hearing cells are ultimately stimulated by movements in fluid, just as are the cells in the lateral-line system of the fish.

The central canal contains the organ of Corti which rests on the basilar membrane of the canal. The movement of the fluid along the outer surface of the central canal causes vibrations in the basilar membrane which, in turn, stimulate individual sensory cells, the hair cells, within the organ of Corti.

The basilar membrane does not vibrate uniformly along its length; instead different areas of the membrane respond to different frequencies of sound. In this way, different sensory hairs are stimulated by different frequencies. The fine discriminations of sound human beings are capable of are made by the brain on the basis of the signals received from the various hair cells.

In general, the human ear can detect sounds ranging from 16 to 20,000 cycles per second, although some children can hear up to 40,000 cycles per second. From middle age onward, there is a progressive loss in ability to hear the higher frequencies, owing to loss in the elasticity of the tympanic membranes.

Our inability to hear very low-pitched sounds (below 16 cycles) is a useful adaptation since if we could hear at low pitch, we would be barraged with sounds caused by the movements of our own bodies and conducted through the bones of our own skeletons. We do hear some sounds conducted directly through the skull—for example, the noise we make when we chew celery or the grating noise we hear when a tooth is pulled. We also hear our own voices, mainly through the skull. This is why a recording of one's own voice sounds so startlingly unfamiliar.

38–17 *The part of the inner ear concerned with hearing is the cochlea, a coiled tube of 2½ turns—shown here (a) as if it were uncoiled. Vibrations transmitted from the tympanic membrane to the stirrup cause the stirrup to push against the oval window, resulting in pressure waves in the fluid that fills the cochlear canals. The short pressure wave in the fluid sets up a vibration in the basilar membrane, stimulating the individual sensory cells in the organ of Corti, which rests on the basilar membrane. Sounds at different frequencies (or pitch) have their maximum effect on different areas of the membrane. The round window permits the pressure waves to leave the cochlea. A cross section of the cochlea is shown in (b) and a close-up of the organ of Corti in (c).*

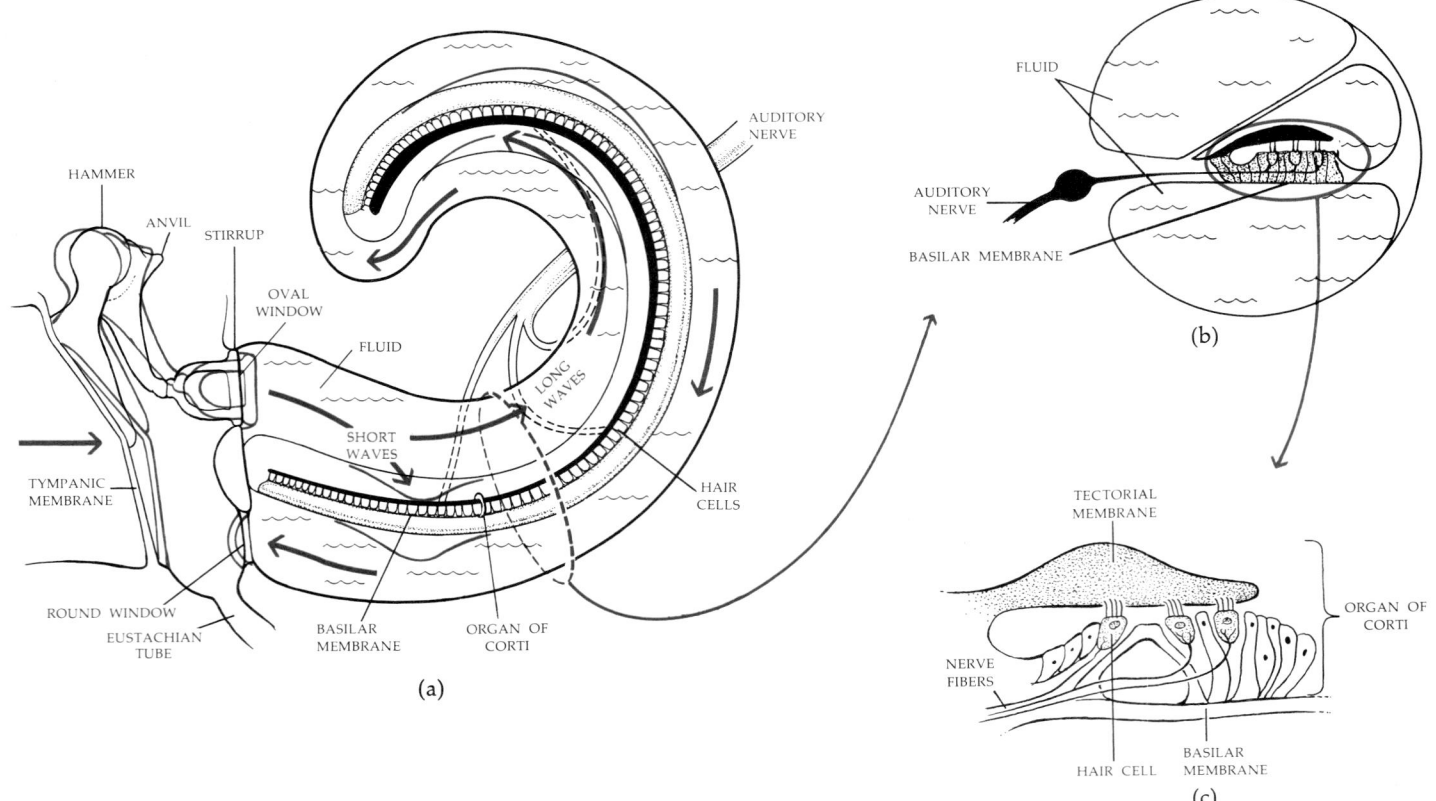

SENSORY RECEPTORS AND THE BRAIN

Despite the elaborate construction of the special sense organs, sensory perception ultimately takes place, as we shall see, in the brain. The difference between what we see and what a frog sees, for example, has much less to do with differences among the receptor organs than with differences in the way the signals received are analyzed and processed.

SUMMARY

Sensory receptors are cells or groups of cells particularly adapted to recording change in the environment and so initiating appropriate responses by the animal. Receptors include interoceptors that monitor internal conditions, proprioceptors that respond to muscle tension or stretch within the body and provide our positional sense, and exteroceptors in the skin that respond to conditions of the external environment, such as touch, pressure, heat, and cold.

The major sense organ in man is the eye, which provides most of our environmental information. The eye comprises an outer, transparent layer, the cornea, below which lies the lens. The eye is focused by changes in the shape of the lens produced by the ciliary muscles. The amount of light entering the lens is governed by the iris, which controls the size of the light aperture (the pupil). Light passes through the eyeball to the retina where it is received by specialized cells, the rods and the cones. These point toward the back of the retina, toward a pigmented layer that lines the back of the eyeball and captures light not absorbed by the photoreceptor cells. Cones provide greater resolution and are responsible for color vision. Rods, which are more light-sensitive, are involved in night vision. Some vertebrates, notably birds and man, have areas specialized for acute vision; such an area is known as a fovea. The photoreceptor cells transmit signals to overlying bipolar cells from which they are relayed to a network of ganglion cells whose axons form the optic nerve. The optic nerve passes through the retina, creating a blind spot, and to the brain.

Chemoreception involves the detection of specific molecules. In man, detection of dissolved molecules takes place in the oral cavity and is the function of the taste buds. Taste buds are specialized for the discrimination of four primary tastes: sweet, sour, salty, and bitter. Detection of airborne molecules takes place in the olfactory epithelium. The mechanism or mechanisms of chemoreception are unknown; according to current hypotheses, chemoreceptors respond to the three-dimensional, hand-in-glove fit of particular molecules.

Hearing is the function of specialized hair cells. These "hairs" vibrate in response to sound waves, and the vibrations trigger nerve impulses that are transmitted to the brain. The function of the outer and middle ear is to focus and transmit sound waves to the hair cells which, in man, are located in the inner ear in the organ of Corti. The outer ear, which leads to the tympanic membrane, functions, in most vertebrates, as a funnel for capturing and routing sound waves. Such sound waves cause vibrations in the tympanic membrane. These vibrations are transmitted to the middle ear where they move three delicate bones: anvil, hammer, and stirrup. The stirrup pushes against a membrane (the oval window) leading to the cochlea. Coiled within the cochlea is the organ of Corti, surrounded by fluid in which the motion of the stirrup causes vibrations.

These vibrations are transmitted to the basilar membrane underlying the organ of Corti, and from there to the individual hair cells.

Although sensory receptors are specialized as to the type of stimuli they respond to, the stimuli are all transmitted in the same way: in the form of a nerve impulse traveling along an axon to a particular area of the brain where the stimuli are processed.

QUESTIONS

1. Define the following terms: exteroceptor, proprioceptor, chemoreceptor, cochlea, fovea, tympanic membrane.
2. Describe the structure and function of the organ of Corti.
3. Label this drawing:

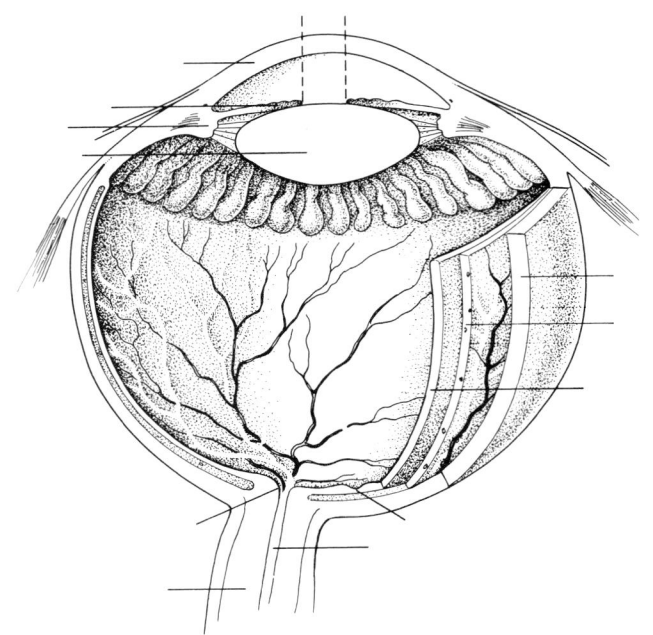

Chapter 39

The Brain

The human brain, which weighs about 3 pounds, has the consistency of semi-soft cheese. Like the spinal cord, the brain is made up of both "white matter"—the fiber tracts—and gray matter. The gray matter consists of nerve cell bodies and glial ("glue") cells, which apparently support and nourish the neurons and may also play a role in ionic balance and thus in electrical potentials. In some areas of the brain, neurons and glial cells are so densely packed that a single cubic inch of gray matter contains some 100 million cell bodies, with each connected to as many as 60,000 others.

For the past 2,000 years, men have wondered about the relationship between the brain—this 3 pounds of semisoft substance—and the mind, the center of consciousness, thought, and emotion. Even René Descartes—"I think; therefore, I am"—who believed that the body is essentially a complex machine, conceived of mind and body as two separate entities, whose meeting place was a small gland, the pineal, located within the skull. The persistence of this dualistic concept is reflected in such phrases, still in use today, as "mind over matter." Many years after the cell doctrine was accepted for all other parts of the body, the brain was considered to be an exception. One reason was that it was not possible to visualize neurons as readily as other types of cells because special staining techniques are required to reveal them. Another and undoubtedly more important reason is the feeling that mental processes are somehow special, different from other physiological functions.

There is, of course, no longer any scientific doubt that mind and matter are one. Even today, however, many of us who accept without difficulty the fact that the tissues and cells of the digestive tract are responsible for digestion or that respiratory gases are exchanged across the surface of the lungs find it less easy to comprehend—to *really* believe—that the ideas, ideals, dreams, fantasies, thoughts, hypotheses, loves, hates, fears, and aspirations which make up the content of the mind somehow can be explained by the interactions of the cells within our heads. Nor, indeed, can scientists yet fully explain "mind" in terms of brain, despite the many intriguing discoveries that have been made in this field. This is one of the reasons why the brain is one of the most exciting of modern biological frontiers.

–1 *A brain, viewed from above. The corpus callosum is visible within the deep groove separating the two cerebral hemispheres, and the cerebellum can be seen below the hemispheres at the base of the brain. Note the many convolutions of the cerebral cortex. By these, you can immediately distinguish this brain as human.*

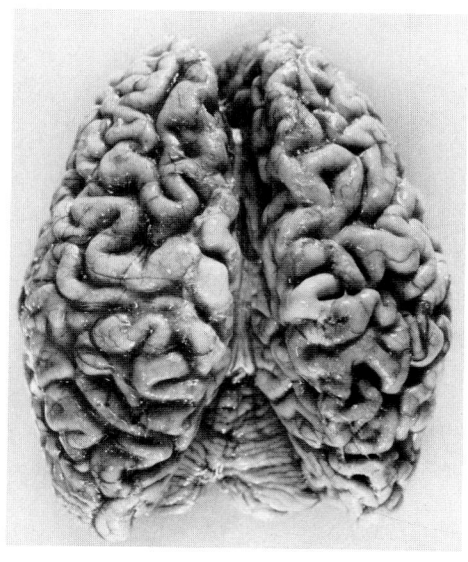

39–2 *Paddy, a chimpanzee, is one of a group of experimental animals used by José Delgado of Yale to establish effects of electrical stimulation on various areas of the brain. A hundred fine electrodes planted in various areas of Paddy's brain send signals to a remote computer, which analyzes them and sends back responses to other parts of Paddy's brain. For example, when the computer receives signals from Paddy associated with aggressiveness, it sends back counteracting signals, which have the effect of making Paddy docile.*

39–3 *A continuous discharge of alpha waves seems to be associated with tranquility and well-being. By concentrating on a musical tone that sounds when alpha waves are produced, it is possible to learn to turn on those waves at will. This technique is known as biofeedback training, or BFT.*

HOW THE BRAIN IS STUDIED

The Electroencephalogram

One of the ways in which the brain is studied is by the electroencephalogram (EEG). The electroencephalogram is a record of continuous electrical activity from the brain as measured by the difference in electrical potential between electrodes placed on specific areas of the scalp and a "neutral" electrode placed elsewhere on the body or the difference between pairs of electrodes on the head. The voltages that arise are very weak—about 300 microvolts (a microvolt is a millionth of a volt) is the maximum in a normal adult—and extremely sensitive recording equipment is required.

Such recordings obviously reflect only gross electrical activity. Donald Kennedy of Stanford University, one of the leading investigators in the field of nerve cell function, compares the electroencephalogram to measurements that an observer on Mars might make of the roars emanating from the Houston Astrodrome during a baseball game. They would certainly indicate that something was going on in Houston, but it would be very difficult to infer the rules or determine the progress or the nature of the game from these recordings.

Nevertheless, there are characteristic EEG patterns, and these patterns can be correlated with the level and type of brain activity.

One such pattern is the alpha wave, a slow, fairly irregular variation in electrical potential, with a frequency of about 8 to 12 cycles per second, recorded at the back of the head. Alpha waves are usually produced during periods of relaxation; in most persons they are conspicuous only when the eyes are shut.

39–4 (a) *Brain wave recordings of a student with conspicuous alpha rhythm. Electrodes were placed at four positions on the head. When the student was asked to multiply 18 × 14, the alpha rhythm was suppressed and then resumed after the problem was solved (10 seconds of the recording are omitted in order to keep it on the page). (b) Brain waves from a student with a complete absence of alpha rhythms. According to a recent study, people who have almost no alpha rhythms under any conditions think almost exclusively by visual imagery, whereas people with a persistent alpha rhythm tend to be abstract thinkers.*

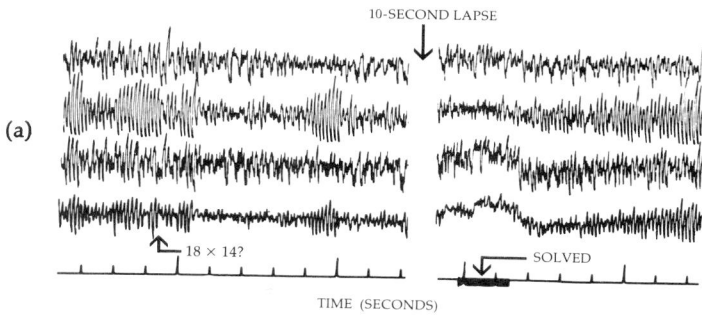

(a)

10-SECOND LAPSE

18 × 14?   SOLVED

TIME (SECONDS)

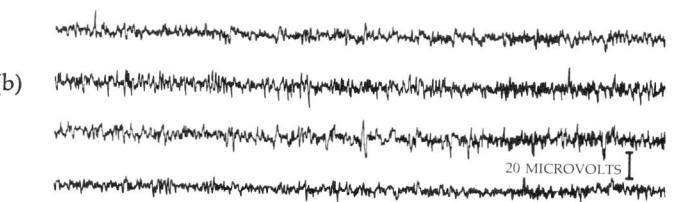

(b)

20 MICROVOLTS

About two-thirds of the general population have alpha waves that are disrupted by attention. Of the remaining third, about one-half have almost no alpha rhythm at all and about one-half have persistent alpha rhythms that are not easily disrupted by attention.

Another EEG pattern is represented by the beta wave. The beta-wave variations in electrical potential are of lesser amplitude than those of the alpha wave, but their frequency is greater, being about 18 to 32 cycles per second. They occur in bursts in the anterior part of the brain and are associated with mental activity and excitement.

Waves of lower frequencies (delta waves) are seen in infants or in adults during sleep, but in waking adults they are signs of mental disturbance or brain injury.

Other Methods for Studying the Brain

Electroencephalograms have been in use for monitoring brain activity since 1929. More recently, it has become possible to insert microelectrodes into the brain tissues and so to record the activity of smaller groups of nerve cells or even of single cells. To return to Kennedy's analogy, this is somewhat like isolating a single spectator from the group at the Astrodome and recording his comments. The problem here is that, as far as the observer can tell, this particular spectator may not be watching the ballgame at all, but may just be ordering a hot dog or another can of beer.

In other types of studies, particular areas of the brain are stimulated by pulses of electric current or by chemicals, and the effects are recorded and analyzed. Also, portions of the brain can be removed or connections severed either by accidental injury, in the course of treatment for epilepsy or other diseases, or experimentally in animals, and the results observed.

713   *The Brain*

The experimental animals most frequently used in brain studies are primates, because of their resemblance to man, and rats and cats, because they are readily available to research laboratories. Fortunately for both experimenter and subject, there are no pain receptors within brain tissue, so that procedures can be carried out painlessly and with a minimal disturbance of function using only local anesthesia.

Perhaps as a consequence of the techniques presently available, researchers have been more successful in "mapping" the brain—that is, correlating certain functions with certain areas—than in discovering the cellular basis of thought, memory, or learning.

THE ANATOMY OF THE BRAIN

The human brain has three subdivisions:

1. The *brainstem,* which is an enlarged, knobby extension of the spinal cord. It includes the *medulla, pons,* and *midbrain.*
2. The *cerebellum* ("little brain"), a bulbous mass of tissue which, in man, lies at the back of the head.
3. The *cerebrum,* and its subdivisions, including the *cerebral cortex,* the *thalamus, hypothalamus,* and associated structures. The cerebrum is divided into two hemispheres by a clearly visible deep furrow running from front to back. In man, the cerebral hemispheres and the cortex overlie the brainstem and the cerebellum.

Embryologically, the medulla, pons, and cerebellum are derived from the hindbrain, the more posterior of the three bulges in the neural tube from which the brain develops. The midbrain arises from the middle bulge, and the cerebrum and assorted structures from the forebrain (see Figure 31–22.)

39–5   *Photograph of a human brain, in sagittal section, with a diagram indicating some of the principal structures. The gray matter, characteristic of the cortex, consists of nerve cell bodies, glial cells, and unmyelinated fibers (fibers without myelin sheaths). The "white matter" is made up of myelinated nerve fibers.*

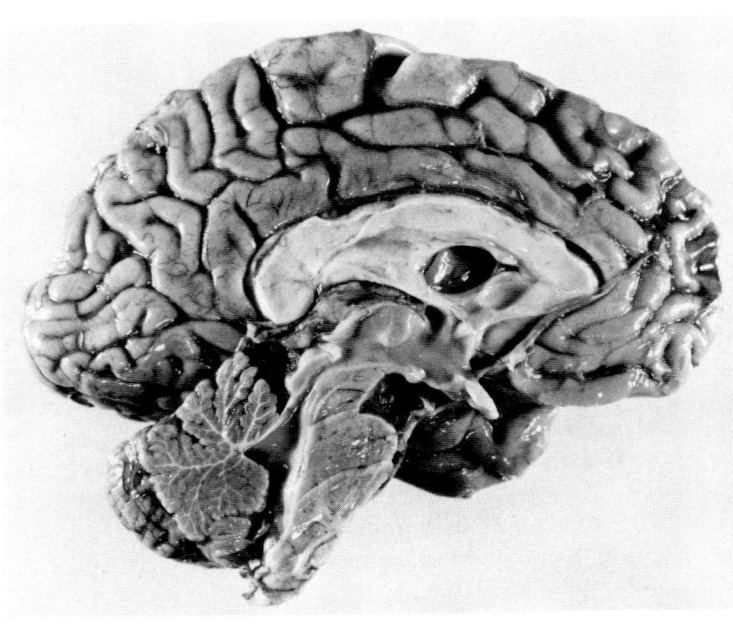
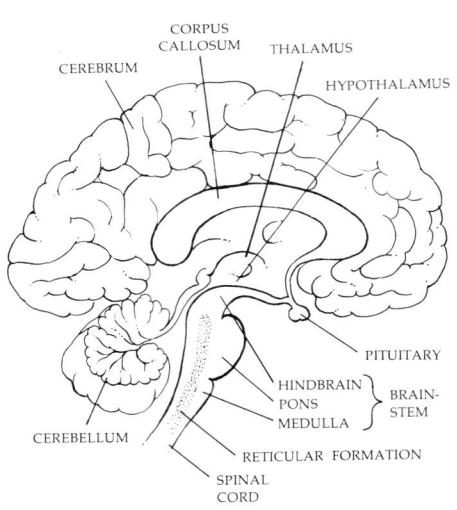

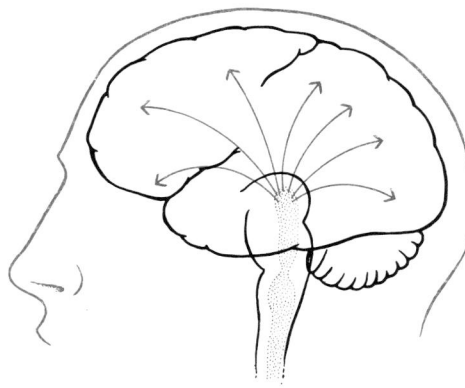

39–6 *The reticular activating system is a core of tissue running through the brainstem. It monitors incoming stimuli, analyzes them, and sends appropriate arousal signals to other areas of the brain.*

The Brainstem

The brainstem contains all the fiber tracts connecting the higher brain structures with the spinal cord. Sensory fibers run dorsally and laterally along the brainstem, and motor fibers ventrally. Within the brainstem, many of these fibers cross from left to right, and vice versa, so that the right side of the brain controls the motor functions of, and receives the sensory reports from, the left side of the body. The brainstem also contains nuclei (nerve cell bodies) of most of the sensory and motor neurons serving the muscles and sensory organs of the head. Nuclei concerned with heart action, respiration, and gastrointestinal function are located at the base of the brainstem, in the medulla. The pons ("bridge") contains, in addition to ascending and descending tracts and nuclei, a very large bundle of transverse fibers that interconnect the brainstem and cerebellum.

The midbrain was the chief control center for the early vertebrates. In the course of evolution, this function of integration and control has shifted somewhat to the thalamus and, eventually, to the cerebral cortex. In mammals, the midbrain functions as a relay center; it also carries out reflex functions, such as those associated with respiration and circulation of the blood.

Reticular System

As we mentioned previously, the dorsal part of the brainstem is concerned mainly with sensory input and the ventral part with motor and response processes. Between these two zones is a core of tissue that runs centrally through the entire brainstem. It is made up of a weblike network of fibers and neurons; hence, its name, *reticular* ("netlike") *system.*

The reticular system is of particular interest to physiological psychologists because it is involved with arousal and attention—that hard-to-define state we know as consciousness. All the sensory systems have fibers that feed into the reticular system, which apparently filters incoming stimuli and "decides" whether or not they may be important. Electrical stimulation of this system, either artificially or by these incoming impulses, results in increased electrical activity in other areas of the brain. The existence of such a filtering system is well verified by ordinary experience. We are not generally conscious of the pressure of our clothing on our body, for instance, or of background noise. A person may sleep through the familiar blare of a subway train or a loud radio or TV program but wake instantly at the cry of the baby or the stealthy turn of a doorknob. Similarly, we may be unaware of the contents of a dimly overheard conversation until something important—our own name, for instance—is mentioned, and then our degree of attention increases.

The most recent research indicates that the reticular system is not, as was first thought, *the* arousal zone but rather works in conjunction with other areas in the brain to produce wakefulness and consciousness. In fact, modern research is tending to undermine previous concepts of particular "centers" for the control of specific functions and to put more stress on complex interrelationships of various parts of the brain. Nevertheless, the reticular system continues to be recognized as having important functions concerned with consciousness and sensory filtering. It has been suggested that acupuncture works by jamming the signals going into the reticular system and so interfering with the perception of pain.

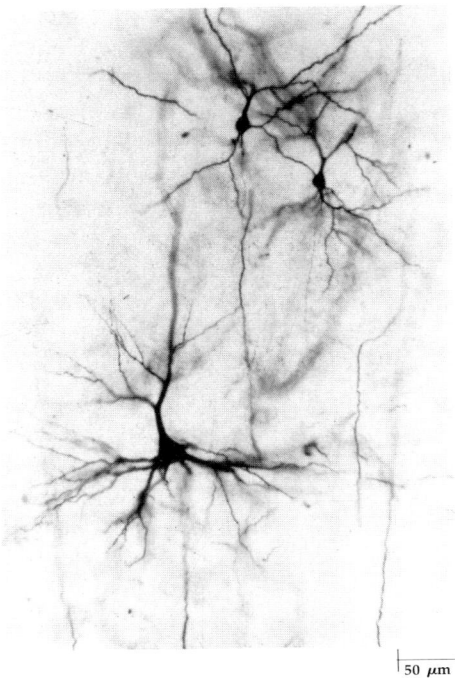

39–7 *Photomicrograph of several brain cells from the cerebral cortex of the cat, showing their interconnections in what Sir Charles Sherrington described as "the enchanted loom."*

|50 μm|

The Cerebellum

The cerebellum is primarily concerned with the regulation and coordination of muscular activities. There are connections between the cerebellum and nearly all the other parts of the brain, including spinal sensory fibers and areas concerned with hearing and vision as well as the thalamus and other parts of the brainstem. Because of the coordinating capacities of the cerebellum, we are able to carry out complex functions such as walking, riding a bicycle, driving a car, swallowing (Figure 34–8), and playing a piano. Like the brainstem, the cerebellum is an old structure, evolutionarily speaking. It reaches its largest size, proportionately, in the brains of birds, where it is concerned with the activities involved in flying.

The Thalamus

The thalamus consists of a pair of egg-shaped masses of gray matter located below the cerebral hemispheres. In man, it functions primarily as a relay station. All sensory pathways, except olfactory pathways, lead through the thalamus to the overlying cerebral cortex. It also has interconnections with the reticular formation and with certain areas of the cerebral cortex.

The Hypothalamus

The hypothalamus is a small (about 4 grams or $\frac{1}{10}$ ounce) but functionally mighty area of the brain. We noted in Chapter 32 that the hypothalamus produces oxytocin and ADH (antidiuretic hormone), released from the posterior lobe of the pituitary. Moreover, it regulates the secretion of tropic hormones from the anterior pituitary and by so doing modulates and integrates much of the endocrine function of the body. The hypothalamus is also a control center for sexual drive, anger, hunger, thirst, and pleasure; stimulation of particular areas in the hypothalamus can result in sexual arousal, anger, food-seeking behavior, overeating, overdrinking, and so on in experimental animals. Rats with electrodes implanted in hypothalamic "pleasure centers" will spend much of their time pressing levers that stimulate these electrodes; some observers of the human condition have noted that similar implantation in the brains of humans would enable a political leader to exert an effective, strictly nonviolent control over a suitably stimulated population. Factors affecting the internal environment, such as blood pressure and heart rate, are also modified by particular hypothalamic nuclei.

39–8 *(a) Rat with an electrode implanted in its hypothalamus. (b) By pressing a lever, it is able to stimulate its "pleasure center" with a weak electric current. Some animals press the lever as often as 8,000 times an hour.*

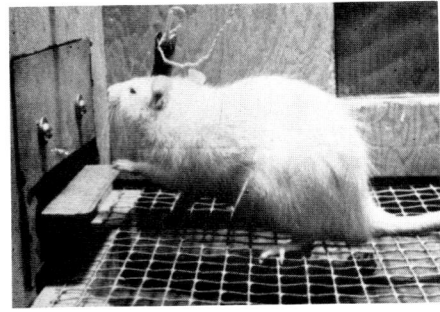

(a)   (b)

The brains of various vertebrates. Note that the brainstems (indicated in color) are approximately the same in the various species but that the cerebrum is much larger in the homeotherms than in any of the "cold-blooded" animals. The cerebral cortex, the upper surface of the cerebral hemispheres, which is relatively smooth and small in lower homeotherms, is larger in more intelligent ones, such as dogs and cats, and reaches its greatest development in the primates, particularly man.

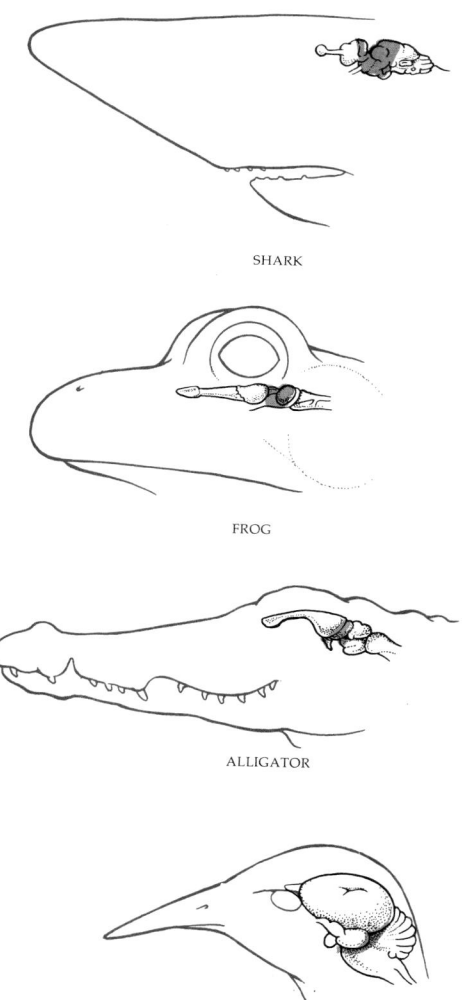

SHARK

FROG

ALLIGATOR

BIRD

The Cerebrum and the Cerebral Cortex

The cerebrum makes up the bulk of the human brain—about 80 percent, in fact. It consists of two large hemispheres, the interiors of which are composed of fiber tracts ("white matter") and relay nuclei. The two hemispheres are connected by three bundles of fibers, the corpus callosum, and the anterior and posterior commissures. The surface of the cerebral hemispheres is a 2-millimeter-thick layer of gray matter, the cerebral cortex. The cortex of the human brain is wrinkled and folded, which greatly increases its surface area.

Evolution of the Cerebrum

The cerebrum has undergone striking evolutionary changes in the vertebrates. In fish, amphibians, and reptiles, the most imposing parts of the cerebrum are the olfactory bulbs, which directly underlie the special organs for smell and which are attached to the rest of the brain by the olfactory stalks. Behind the olfactory bulbs in these primitive vertebrates are two round swellings that make up the cerebrum. In fish and amphibians, the cerebrum is largely concerned with olfactory sensations, but in reptiles we can find the trace of new evolutionary developments that have proved to be of major importance.

In the reptiles, there is a small patch of nervous tissue, the neocortex, at the forward upper part of the cerebral hemisphere. In mammals, this neocortex is greatly expanded, folding back to cover the entire surface of the cerebral hemispheres. In some mammals, such as rats, the neocortex is relatively small and smooth, but in primates and man the neocortex is large and convoluted. Generally speaking, there is a correlation between the extent of neocortical development in a species and the complexity and modifiability of its behavior.

As the neocortex increased in size during mammalian evolution, it took over more and more functions from the older, underlying parts of the brain. As a result of these progressive changes, in which the patterns of control shifted as the brain evolved, many different areas in the brain may now be concerned with the same function, such as vision or hearing, and among different mammals the relative importance of the cerebral cortex differs. For instance, if the area of the cortex concerned with vision is removed from a rat's brain, the animal retains some vision but cannot discriminate patterns. If the visual cortex is removed from a monkey, the animal can only distinguish light from dark. A man whose visual cortex has been damaged is totally blind.

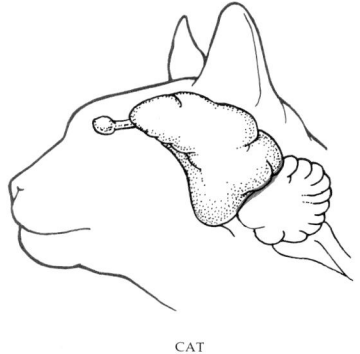

CAT

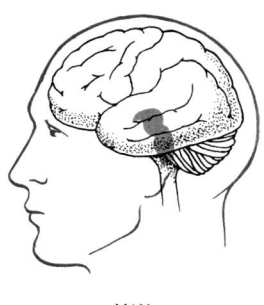

MAN

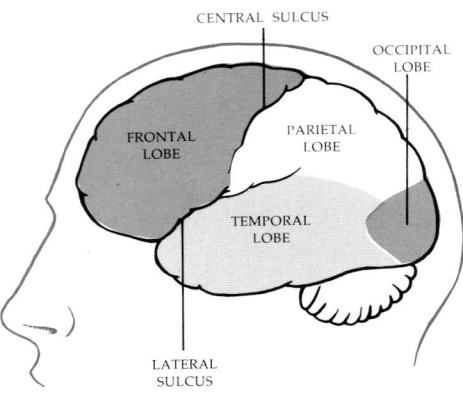

39–10   *The principal fissures and lobes of the human cerebral cortex.*

CENTRAL SULCUS

OCCIPITAL LOBE

FRONTAL LOBE

PARIETAL LOBE

TEMPORAL LOBE

LATERAL SULCUS

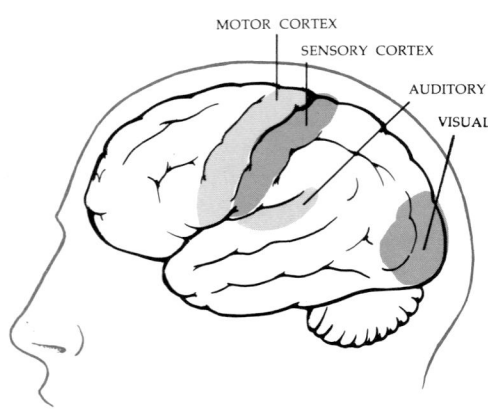

39–11   *The human cerebral cortex, showing the location of the motor and sensory areas, on either side of the central sulcus, and the auditory and visual zones.*

MOTOR CORTEX

SENSORY CORTEX

AUDITORY

VISUAL

An important difference between the cortices of the higher and lower mammals is that in the latter almost the whole surface of the cortex is concerned with definite sensory or motor activities, whereas in man, about three-fourths of the cortex is made up of so-called association cortex, which does not have definite sensory or motor functions. Scientists agree that learning ability and intelligence in man and the other primates are generally correlated with the size and degree of development of these large areas of association cortex, although at present we know very little about the anatomical organization of these abilities.

Mapping the Cortex

Because of its relative accessibility, just below the surface of the skull, the cerebral cortex is the most thoroughly studied area of the human brain, and some parts of it have been mapped in exquisite detail. In primates, each of the hemispheres is divided into lobes by two deep fissures, or grooves, in the surface: the central sulcus, which runs laterally—from left to right—across the top of each hemisphere, and the lateral sulcus (Figure 39–10).

The area just anterior to the central sulcus, in the frontal lobe, is the somatic motor cortex, and the area posterior to it, in the parietal lobe, is the somatic sensory cortex. These are involved in the initiation of muscular actions and the receipt of tactile stimuli, respectively. Because of the crossing of fibers in the brainstem, the motor and sensory areas for the left side of the body are in the right hemisphere, and vice versa.

The temporal lobe of the brain lies below the frontal and parietal lobes, separated by the other deep fissure, the lateral sulcus. In the temporal lobe, buried within this fissure, is the auditory cortex. By measuring electrical discharges in this area of the brain in dogs and cats exposed to sounds of varying frequencies, investigators have been able to show that different regions of the auditory cortex respond to different frequencies of sound.

The area behind the central sulcus is divided into two main lobes, the most posterior of which is concerned with vision. Stimulation of an animal's retina with light results in electrical discharges by cells in this area of the cortex. By using a tiny point of light to stimulate very small regions of the retina, one after the other, investigators have been able to show that each region of the retina is represented by a corresponding but larger region on the visual cortex. This cortical region contains a variety of cells, different groups of which respond to different types of visual stimuli.

The somatic sensory, auditory, and visual zones of the cortex appear to function as terminals for messages from somatic sensory neurons and from the special organs of the eye and the ear. The motor cortex is apparently one of the areas of origin for outgoing signals to the somatic muscles. The function of the rest of the cortex is largely unknown.

Association Cortex

The unmapped area of the cortex is sometimes known as the association, or "silent," cortex. The term "association" was introduced when this part of the cortex was thought to function as a sort of giant switchboard, interconnecting the motor and sensory zones. More recent studies suggest, however, that or-

–12 *A combined transverse section of one hemisphere of the human cerebrum, indicating the functional areas of the motor and sensory cortices. The motor cortex is indicated in black; stimulation of these areas causes responses in corresponding parts of the body. The sensory cortex is in color; stimulation of various parts of the body produces electrical activity in corresponding parts of this cortex. Notice the relatively huge motor and sensory areas associated with the hand and the mouth. This map is based largely on studies done by neurosurgeon Wilder Penfield on patients undergoing surgical treatment for epilepsy. The motor and sensory cortices are located on either side of the central sulcus. Because nerve fibers cross in the brainstem, the motor and sensory cortices of the right hemisphere are associated with the left side of the body, and vice versa.*

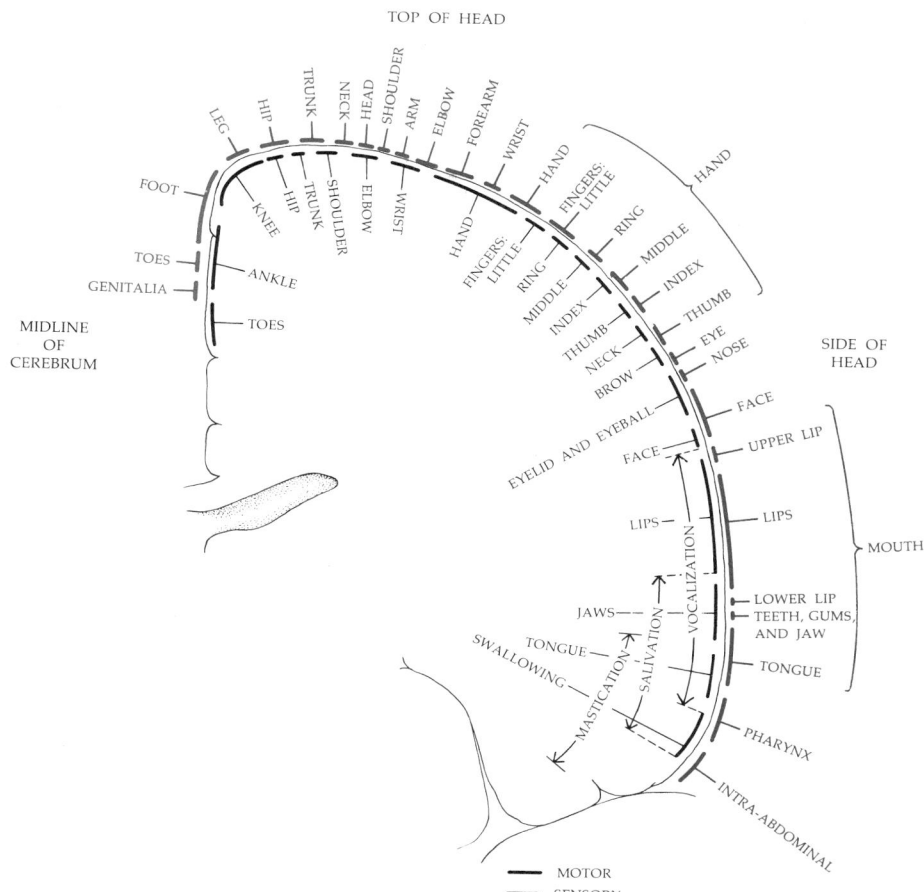

ganization of the brain is more vertical than horizontal; for example, severing the motor cortex from the sensory areas by deep vertical cuts appears to have no effect at all on an animal's behavior. Anatomical studies support the interpretation that communication between the sensory areas and motor areas of the cortex takes place primarily in lower brain centers, particularly the thalamus.

Nor are the association areas actually "silent." Although sensory stimuli do not evoke local responses in these areas, and direct electrical stimulation of these areas does not elicit muscle movements (as in the somatic sensory and motor areas, respectively), the association cortex shows steady spontaneous activity, apparent (but nonspecific) responses to stimuli, and increased activity following electrical stimulation of the thalamus or the reticular system. Clearly, something is taking place in these "silent" areas. Since the proportion of association area to motor and sensory area is much larger in primates than in other mammals, even higher mammals such as cats, and is very large in man, we assume that these areas have something to do with what is special about the mind of man.

Frontal Lobes

About half of the association area of the cortex is in the frontal lobes, the part of the brain anterior to the motor cortex. This part of the brain developed most rapidly during the recent evolution of man. It is responsible for the high forehead of modern man, as compared with the beetle brow of our most immediate ancestors, and public appraisal of its function is reflected in the terms "high brow" and "low brow."

During a particularly bleak period in the recent history of treatment of the mentally disturbed, a large number of mental patients had their frontal lobes removed or the connections severed between these lobes and the rest of the brain in order to make the patients more docile and hence less inconvenient to their families and society (a goal now attained by the long-term use of massive doses of tranquilizers). Despite the large number of persons "treated" in this way, there is little agreement on the results of frontal lobotomy—or, more precisely, on the normal role of the frontal lobes. Most, but not all, psychologists agree that the frontal lobes are somehow involved with intention, will, planning, and the marshaling of efforts to carry out long-range programs—all of which are particularly human characteristics.

Posterior Lobes

The posterior lobes, consisting of the parietal and occipital lobes, have largely been studied by experiments on brains in which parts of the cortex have been damaged, removed, or otherwise interfered with. Their functions are not quite so elusive. In both experimental animals and humans, damage to specific areas of the posterior lobes seems to result in problems of discriminating specific stimuli. For example, monkeys can no longer tell the difference between patterns, objects, or sounds that they were previously able to distinguish if particular cortical areas of the posterior lobes are damaged. In humans, damage to specific areas of the cortex of the left posterior lobe (in most left-handed as well as all right-handed people) results in impairment of speech. The corresponding area in the right brain is associated with music-making. It is probable, although far from proved, that the functions of memory and learning and the organization of ideas take place in the "silent" areas of the posterior lobes.

39–13 *The human cerebral cortex, showing the areas associated with speech. The more anterior area (Brocas' area) is adjacent to the region of the motor cortex that controls the movements of the muscles of the lips, tongue, jaw, and vocal cords. Damage to this area results in slow and labored speech but does not affect comprehension. When the more posterior area (Wernicke's area) is damaged, speech is fluent but often meaningless, and comprehension of both spoken and written words is impaired. The two speech areas are joined by a nerve bundle.*

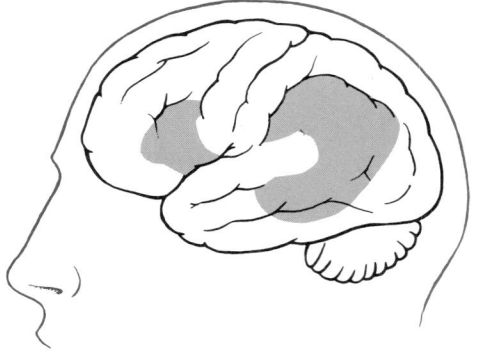

THE EYE AND THE BRAIN

When we look at an object, as we noted in Chapter 39, the lens of our eye, like the lens of a camera, projects the object onto the retina, stimulating the photoreceptor cells. Impulses are carried from the retina by way of fibers of the optic nerve. These fibers lead to a visual relay center in the thalamus, where the fibers from the two eyes come together on each side of the brain and synapse with fibers leading to the visual cortex. Various experiments have shown that the spatial arrangement of neurons on the cortex corresponds topographically with the spatial arrangement of the image as it is received on the retina, except for "over-representation" of the fovea. (The fovea, which occupies only about 1 percent of the area of the retina, projects to nearly 50 percent of the visual cortex.)

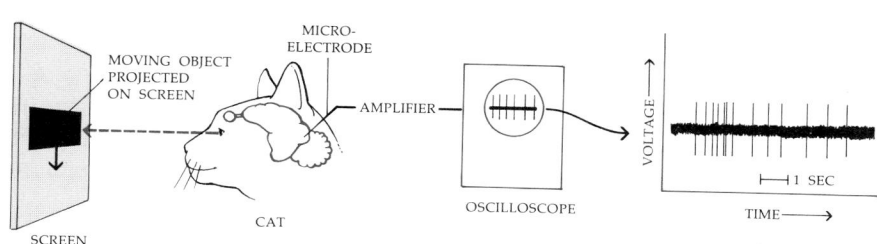

The experimental method of Hubel and Wiesel. A cat with a small electrode implanted in a single cell in its visual cortex observes a screen on which a moving shape is projected. (a) and (b) represent tracings from two different neurons. The first neuron (a) responds to a horizontal bar; as the bar is turned toward the vertical, the action potential ceases. The second neuron (b) responds to a right angle. Note the reduction in response as the angle increases.

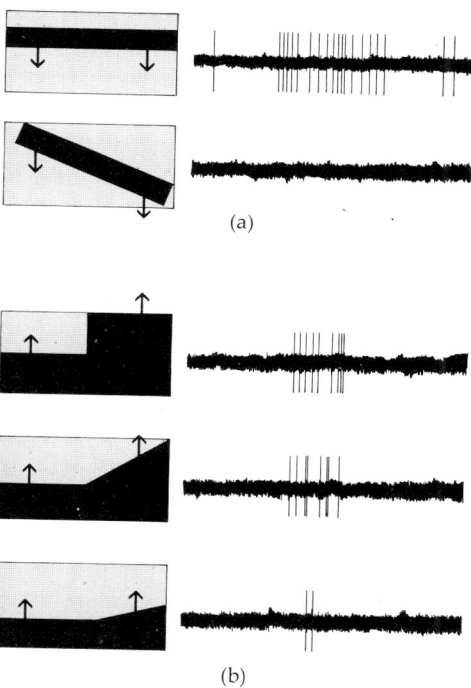

How do we perceive form? The logical answer is that we perceive form simply because of the way the visual image is arranged on the visual cortex, allowing for the foveal distortion. However, recent studies on cats by David H. Hubel and Torsten N. Wiesel of Harvard University reveal that quite a different mechanism is involved. They implanted microelectrodes into single cells of the visual cortex, recorded the responses of individual cells to visual stimuli, and came up with some very surprising results. One cortical neuron, for example, was found to respond only to a horizontal, bar; as the bar was tipped away from the horizontal, discharges from the cell slowed and ceased. Another responded to a vertical bar, ceasing to fire as the bar was oriented toward the horizontal. A third neuron fired only when the bar was moved from left to right; another only to a right-to-left movement. One cell turned out to be a right-angle detector; it responded only to an angle moving across the visual field and was most excited when the angle was a right angle. Thus an important component of the visual information-processing system is a set of very specifically tuned neurons, each responsive to one aspect of the shape of the object perceived. The information-processing system synthesizes the final perception by integrating the information received from its individual elements.

You will remember that neurons in the frog's retina carried out much this same kind of data processing. These very fundamental discoveries would seem to be leading to a new and far deeper level of understanding of the ancient dilemma about the relationship between the mind of man and the outside world.

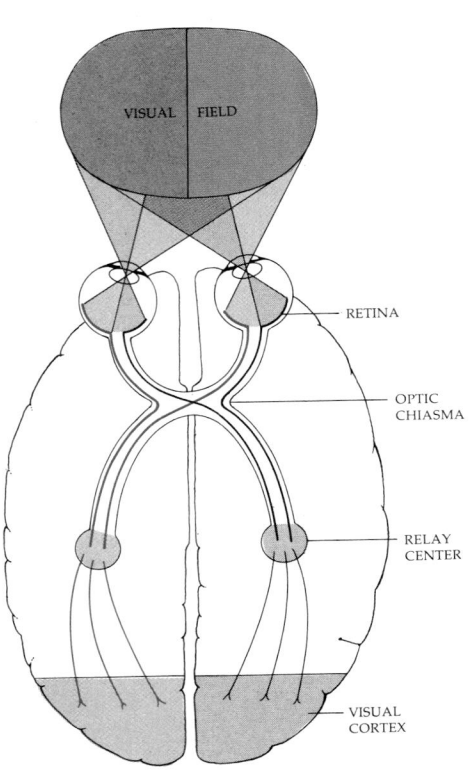

The optic chiasma viewed from below. The chiasma is the structure formed by the crossing-over of fibers traveling from the retina to the visual cortex. Because of this crossing-over, the right and left halves of the center each "see" only the opposite half of the visual field.

VISUAL FIELD

RETINA

OPTIC CHIASMA

RELAY CENTER

VISUAL CORTEX

SPLIT-BRAIN STUDIES

Additional interesting implications concerning processing of information in the cortex has come from split-brain studies carried out over the past 10 years by Roger Sperry and coworkers at the California Institute of Technology. Sperry severed the connections between the two cerebral hemispheres in cats and monkeys by cutting through the corpus callosum and the optic chiasma (see Figure 39–14). The animals' everyday behavior was normal, but by using specially contrived testing techniques, the investigators were able to show some remarkable and unexpected effects. The animals responded to test procedures as if they had two brains. When the right eye of an animal was covered, for instance, it could learn tasks and discriminations using only the left eye. But then if the left eye were covered and the same tests were presented to the right eye, the animal had to learn all over again. In fact, it was possible, without creating any confusion at all, to have both sides of the brain learn opposite solutions to the same problem.

More recently, similar observations have been made in patients in whom the corpus callosum has been severed for the treatment of epilepsy. The patients are able to carry out their normal activities, but specially devised tests indicate the presence of what Sperry calls "two realms of consciousness." If such a patient is asked to identify by touch objects that he cannot see, he can name those he can feel with his right hand but not those he can feel only with his left hand; apparently, because the speech centers of the brain are located in the left hemisphere, the right brain is mute. If, for instance, the patient's left hand is given a plastic object shaped like the number 2, which the patient cannot see, he is unable to identify the object verbally but he can readily tell the experimenter what it is by extending two fingers. If a picture is flashed before his eyes as his head is held in a fixed position (in human patients with the optic chiasma intact, the right brain sees only the left side of a picture briefly presented and the left brain sees only the right side) and he is asked what he saw, he reports only what he saw on the right side of the picture; however, the left side of the picture can have an emotional effect on him. The patient laughs, for instance, but cannot explain why.

One conclusion that can be drawn from these experiments is that there is a great deal of redundancy in the human cortex. This redundancy has long been suspected by surgeons who have observed patients functioning normally after removal of or damage to extensive areas of the right temporal lobes. In a number of cases, children have, after an interval, learned to speak again after left temporal lobe damage, as a result of the right lobe taking over the speech functions. (Around puberty, apparently some of this plasticity is lost and the right speech centers cannot take over.)

Sperry's experiments also demonstrate quite clearly that the site of complex learned behaviors in humans *is* the cerebral cortex, rather than the brainstem, since this structure was still intact in all the subjects tested.

There is some indication that there are differences in mental capacity between the two brains, with the left brain excelling in verbal skills, obviously, and the right one in discriminations involving shape, form, tune, and texture. The implication of this finding for the normal person is not clear, however.

9–15　*A split face (a) used in tests of patients whose brains have been surgically divided is made up from two of the faces shown in the selection (b). The patient, wearing a headgear that restrains eye movements, sees the picture projected briefly on a screen. The left side of the brain recognizes the child; the right side sees the woman with glasses.*

(b)

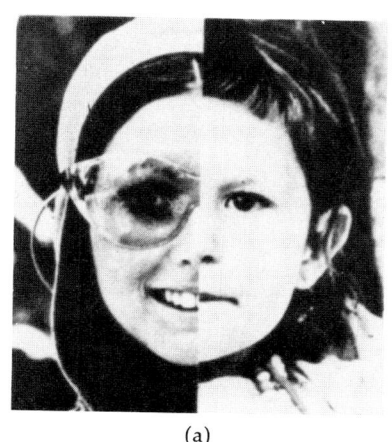

(a)

SLEEP AND DREAMS

Considering the prominence of sleep in our daily existence, it is remarkable how little we know about it. Why does it occupy one-third of our lives? What happens during sleep to restore our capacity to function? Why does sleep, or the lack of it, have such profound effects on our alertness, judgment, good temper, coordination, and emotional stability? Guinea pigs have survived, never sleeping, after destruction of areas in the midbrain. There are also two well-documented cases of persons, one in Italy and one in Australia, who never sleep, presumably because of some brain dysfunction, such as that inflicted on the guinea pigs. Thus, sleep is certainly not necessary for survival in the same way as oxygen, food, or water.

One of the oldest hypotheses concerning sleep is that it results from the buildup of some chemical in the brain that induces sleep and is then altered or destroyed during the sleep process. Certain chemicals, including serotonin and acetylcholine, have been discovered to induce sleep, but whether any of these is the natural sleep promoter—or if, indeed, there is a sleep-promoting chemical—is not known. In fact, some studies seem to refute this attractive hypothesis; for example, Siamese twins sharing a common circulation may sleep at different times.

During sleep, heart rate and arterial pressure decrease, breathing becomes more shallow, the CO_2 concentration of the blood increases, and rectal temperature drops as much as 2°F. There are also obvious changes in the electroencephalogram, which can be divided into five stages.

723　*The Brain*

Stage 1. Drowsiness, in which alpha waves are slowed slightly. This stage is not classified as true sleep but rather as a transition period between sleep and wakefulness. Sustained Stage 1 EEG patterns do not fulfill sleep requirements in experiments with human volunteers.

Stage 2. Light sleep, which is a mixture of slow (3 to 6 per second) low-amplitude waves interrupted by short bursts of faster (14 to 15 per second) waves known as sleep spindles.

Stage 3. Intermediate sleep with high-amplitude waves at about 2 per second (delta) and with occasional sleep spindles.

Stage 4. Deep sleep with very slow waves of comparatively high voltage and without spindles.

REM sleep. Every 80 to 120 minutes, there occurs a period of rapid, low-voltage activity similar to that seen in alert persons. This is also called paradoxical sleep and is associated with rapid eye movements (REMs), in which the eyeballs move jerkily as if the sleeper were watching some scene of intense activity.

As shown in Figure 39–16, sleepers pass through a regular sleep cycle, moving from one stage of sleep to another. The average young adult spends 50 percent of his sleeping time in Stage 2 and about 20 percent in REM sleep, divided into three to five episodes.

39–16 *Stages of sleep as shown by an electro-encephalogram. Five tracings are shown for each stage; the upper two are from electrodes attached beside the eye and the lower three from electrodes attached on the scalp. Electrodes were placed according to an international system (the 10–20 system); such an exact placement permits investigators from all over the world to compare their results. The record length for each sleep stage represents 20 seconds. The electrodes placed beside the eye record eye movement as well as EEG activity.*

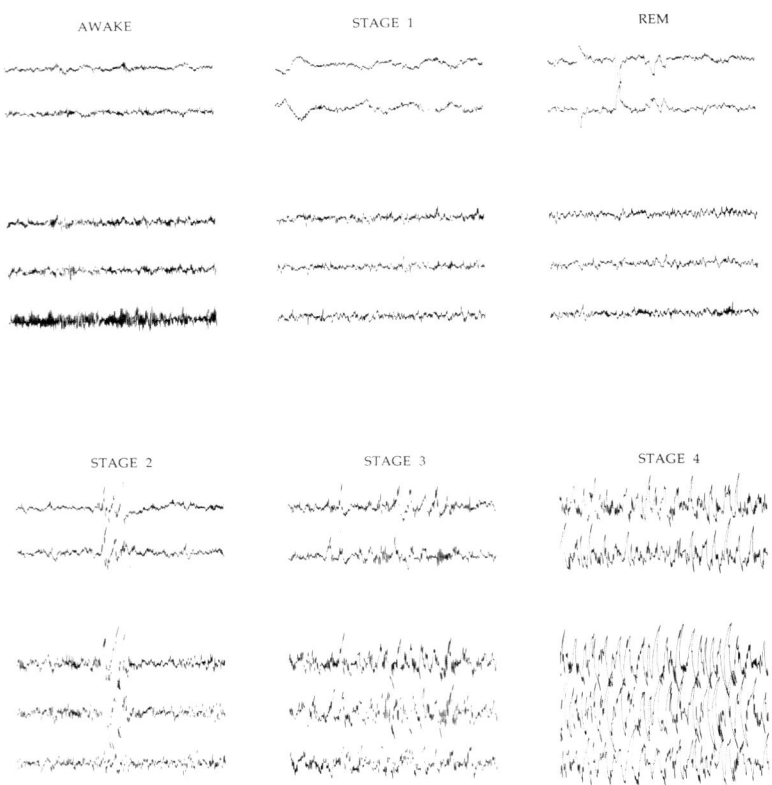

AWAKE   STAGE 1   REM

STAGE 2   STAGE 3   STAGE 4

100 MICRO-VOLTS

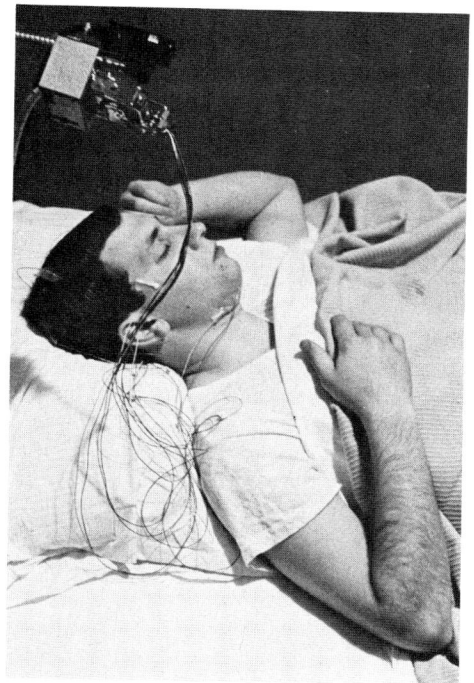

REM sleep differs from the other sleep stages. For example, although there is electroencephalographic evidence of alertness, the muscles are more relaxed and the sleeper is harder to awaken than in ordinary Stage 4 sleep, although, once awakened, he is more alert. Fluctuations in heart rate, blood pressure, and respiration are also seen during REM sleep. Many male subjects experience erections of the penis during REM sleep, and women have similar erections of the clitoris with secretions of vaginal fluid. A person awakened during a period of REM sleep nearly always reports that he or she has been dreaming, and at one time, REM sleep was thought to be the period when all dreaming occurred. It is now generally believed, however, that dreaming can occur at any time during sleep but that conditions for recalling dreams are most favorable when subjects are awakened during REM sleep.

Some of the most interesting experiments on sleep, first reported in 1960, were performed by William Dement of Stanford University School of Medicine. Dement undertook to deprive volunteers of REM sleep by awakening them as soon as rapid eye movements occurred. After a while, he found they began to display anxiety, became irritable, and had difficulty concentrating, although their total sleep periods had been of normal duration. Those deprived of REM sleep for as long as 15 days underwent temporary personality changes, becoming hostile and suspicious. Dement also noted that a person deprived of REM sleep on one night will have increasing periods of REM sleep on the subsequent night, as if to make up for the loss.

Based on these results and on the now-disputed theory that dreaming occurs only during periods of REM sleep, Dement and others formulated the hypothesis that one of the reasons sleep is necessary is that dreaming is necessary and that REM sleep occurs in order to permit dreaming.

More recent work, however, including some studies by Dement himself, have cast doubt on this interesting hypothesis, and some alternative ones have been suggested. One proposal is that REM sleep serves as an information-processing period during which data from the previous waking cycle are sorted, processed, and stored, and so this stage of sleep is essential to the memory process. According to other hypotheses, Stage 4 sleep or the total sleep cycle is necessary for the restorative functions of sleep, and the principal function of REM sleep is a partial arousal—somewhat similar to the occasional arousals seen in hibernating animals and presumably serving the same sort of sentinel function. These studies and the controversies they engender are continuing.

LEARNING AND MEMORY

For scientists interested in brain research, the biggest present challenge is to understand the mechanisms of memory and learning. If we define learning as a change in behavior based on experience, and if the functions of the brain—the mind—are to be explained in terms of the atoms and molecules and the structures composed of them, then learning must involve changes in these atoms or molecules or structures. But what is this change, and where and how does it take place?

Almost 50 years ago, the late Karl Lashley of Harvard set out to locate this physical change, the "scar" of memory, which he called the engram. He taught rats and other animals to solve particular problems and then performed opera-

tions to see whether he could remove the portion of the brain containing that particular engram. But he never found the engram. As long as he left enough brain tissue to enable the animal to respond to the test procedures at all, he left memory as well. And the amount of memory that remained was generally proportional to the amount of remaining brain tissue. Lashley concluded that memory is "nowhere and everywhere present."

Although much more sophisticated techniques are now available for brain research, modern neurologists have had little more success than Lashley in finding memory's hiding place. Moreover, the problem has become, if anything, more complex. It is now clear that there are two types of memory, short-term and long-term. A simple example of the former is looking up an unfamiliar number in a phone book; you can usually remember it just long enough to dial it. A less common but also well-known phenomenon is a loss of memory for the events immediately preceding a blow to the head; only rarely (except in soap operas) is there any loss of the records of earlier experiences. The existence of long-term and short-term memory has also been convincingly demonstrated in memory tests in human volunteers who are called upon to memorize, for instance, lists of unrelated words. Not surprisingly, it was found that the memorizing of new lists hastens the forgetting of the previous ones, suggesting that the short span of short-term memory may serve the useful function of clearing the mind for new information. In experimental animals, it has been shown that a physical insult to the brain, such as electric shock, given soon after the learning of a simple task, can apparently erase the memory just acquired without interfering with earlier ones.

The working hypothesis held by most neurobiologists concerning memory and learning is that these processes involve changes in the connections among cells. The functioning of the brain clearly depends on the physical organization of its billions of nerve cells—its wiring plan. And thus it seems plausible that learning and memory are changes in the wiring plan induced by experience. But no direct evidence of this hypothesis has yet been produced. One of the chief difficulties in finding such evidence is, of course, that there is no way of knowing which of the multitude of synapses that may converge on a single neuron is actually operative. As a way around this difficulty, a number of neurobiologists have turned to working with simpler nervous systems, such as that of the sea slug, *Aplysia*, a mollusk, which has huge neurons with relatively few synaptic junctions. The extent to which such invertebrate nervous systems will provide a useful model for exploring memory and learning in vertebrate animals is not yet clear.

SOME SPECULATIONS

Earlier in this century of remarkable biological progress, the work of a number of brilliant scientists—including physicists, physicians, and biochemists as well as biologists—focused with great success on the problem of the molecular basis of heredity. In the last several years, a number of outstanding research workers, including some prominent molecular geneticists, have abandoned their previous lines of research and turned their attention to the tantalizing subject of the molecular basis of learning and memory. It will be interesting to see if their ventures in neurobiology meet with the same spectacular success as the earlier efforts to discover the nature of the gene.

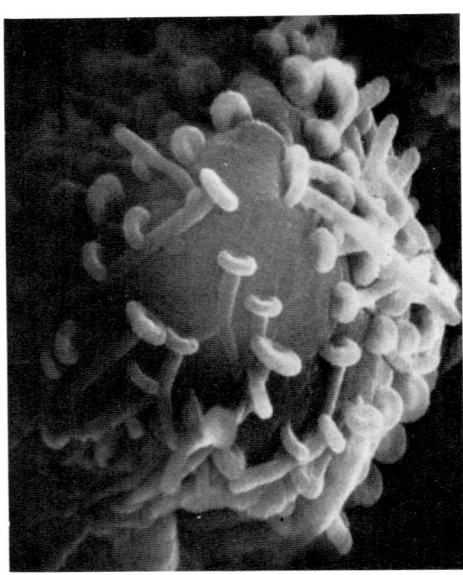

39–18 *Synaptic knobs of the sea slug* Aplysia *as shown by the scanning electron microscope. Much current brain research is being carried out in invertebrate systems.*

2.5 μm

SUMMARY

The function of the brain is to process information received from the environment and to initiate, by means of motor neuron impulses, appropriate actions. As, in the course of evolution, behaviors have become more complicated, this processing done by the brain has become more and more complex, and more and more subject to modification on the basis of experience—i.e., learning. All of the many functions of the brain, including information analysis, storage and retrieval, and modification of these functions by learning are, according to modern neurobiological thought, explicable (though, clearly, far from explained) in terms of the physical-chemical structure of the brain—that is, the organization of its atoms and molecules.

Brain function can be studied by recording wavelike variations in electrical potential from the brain surface (electroencephalography). Three principal types of brain waves can be recorded in normal subjects: beta waves, of relatively high frequency, associated with alertness and attention; alpha waves, of lower frequency, associated with relaxation; and delta waves, of still lower frequency, which normally appear only during sleep. Electrical impulses also can be recorded from groups of cells within the brain, and from single cells. Other methods of brain study in live subjects include removal of portions of the brain and the severing of connections between areas of the brain.

Anatomically, the brain is made up of "white matter" (myelinated nerve fiber tracts) and gray matter (unmyelinated fibers, nerve cell bodies, and glial cells). The brain has three principal divisions: (1) the brainstem, which includes the medulla, pons, and midbrain, (2) the cerebellum, and (3) the cerebrum, which is divided into two cerebral hemispheres and, in the higher mammals, has a greatly convoluted outer layer of cells, the cerebral cortex—and the subdivisions of the cerebrum, the thalamus and hypothalamus.

The brainstem, which connects the spinal tract with the brain centers, contains many ascending and descending fibers and relay nuclei. Motor and sensory nerve fibers cross over in the brainstem, which is why the right side of the brain controls the left side of the body, and vice versa. Nuclei concerned with heart action, respiration, and gastrointestinal function are located in the brainstem. The optic and auditory lobes, which in mammals serve complex sensory reflex functions, are located in the midbrain. The reticular system of the brainstem is involved in arousal and sensory filtering. The thalamus relays auditory, visual, and tactile information to the cerebral cortex and to the hypothalamus. The hypothalamus regulates the release of hormones from the pituitary and so mediates much of the endocrine control of the body. It is also the seat of basic drives and emotions such as sex, hunger, and pleasure.

The cerebellum integrates and coordinates motor and sensory information, making it possible to carry out complex muscular activities.

The cerebral cortex, which is the outer layer of the cerebral hemispheres, reaches its greatest degree of development in man. Mapping studies of the human cortex have succeeded in identifying a motor cortex, in which stimulation of particular areas results in motor response in particular parts of the body; a sensory cortex, which appears to be a terminal for incoming somatic sensory impulses; and auditory and visual cortices, where impulses from the ear and eye, respectively, are received and analyzed. On the left cerebral hemisphere are also the so-called speech areas; damage to these areas interferes with the

complex processes involved in the formulation and organization of speech. The rest of the cortex—which constitutes about three-fourths of the total area—is known as the association cortex. The relative amount of association cortex is much larger in man than it is in other mammals, and these areas are believed to be involved in planning, learning, memory, and other abstract mental processes.

Split-brain studies have been carried out on experimental animals and human patients in which connections have been severed between the cerebral hemispheres. Results of such studies indicate that the two cerebral hemispheres of the same individual can function independently and that they differ somewhat in their talents.

Sleep involves metabolic changes, such as decrease in heart rate, in ventilation, and in body temperature and also changes in the electroencephalogram. The slow, rhythmic waves characteristic of deep sleep are interrupted by periods of rapid, high-frequency waves (REM, or paradoxical, sleep) typically accompanied by rapid eye movements. The function of sleep is not understood.

Studies on the mechanisms of memory and learning involve the search for changes in the physical-chemical structure of the brain cells or the connections among them. No convincing evidence regarding the nature of these changes in the organizational relationship among brain cells is as yet available.

QUESTIONS

1. Define the following terms: gray matter, white matter, association cortex, reticular system.
2. If a split-brain patient, working under the experimental conditions specified, were asked to identify the face in Figure 39–15, what answer would you expect?
3. On the drawing below, label or note the approximate position of the following structures: brainstem, midbrain, medulla, cerebellum, thalamus, hypothalamus, pituitary, cerebral cortex.

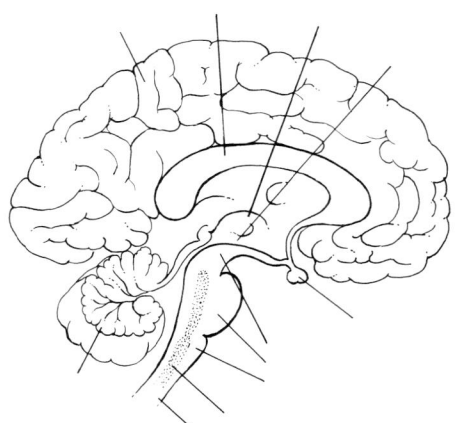

SUGGESTIONS FOR FURTHER READING

BECK, WILLIAM S.: *Human Design*, Harcourt Brace Jovanovich, Inc., New York, 1971.

A comprehensive, modern treatise, designed for the undergraduate student, covering "molecular, cellular, and systematic physiology" of man.

BODEMER, CHARLES W.: *Modern Embryology*, Holt, Rinehart and Winston, Inc., New York, 1968.

Designed for a college course in embryology, this book strikes a good comprehensive balance between classical embryology and the emerging field of molecular embryology.

CALDER, NIGEL: *The Mind of Man*, The Viking Press, Inc., New York, 1970.

A well-written, fast-moving, journalistic report on the "drama of brain research."

CROUCH, JAMES E., and ROBERT MCCLINTIC: *Human Anatomy and Physiology*, John Wiley & Sons, Inc., New York, 1971.

A good basic text, clearly written, with excellent illustrations.

FREEDMAN, RUSSELL, and JAMES E. MORRIS: *The Brains of Animals and Man*, Holiday House, Inc., New York, 1972.

A simple, well-illustrated account of the brain and modern research on its function.

GORDON, MALCOLM S., et al.: *Animal Functions: Principles and Adaptations*, The Macmillan Company, New York, 1972.

A good animal physiology textbook, intended for an advanced course. The authors emphasize function as it relates to the survival of organisms in their natural environments.

GREGORY, R. L.: *Eye and Brain: The Psychology of Seeing*, 2d ed., McGraw-Hill Book Company, New York, 1973.*

A vivid introduction to the science of vision.

HANDLER, PHILIP (ed.): *Biology and the Future of Man*, Oxford University Press, New York, 1970.*

A widely acclaimed survey of the state of biology today, written by a committee of experts in the various fields. Its purpose is to inform the general public both of present knowledge in major fields of biology and of current problems and probable areas of future investigation. It deals with a wide variety of subjects, including most of the ones covered in this book. There are outstanding sections on animal physiology, development, and behavior.

NOSSAL, G. J. V.: *Antibodies and Immunity*, Basic Books, Inc., Publishers, New York, 1969.

An account for the general public of research in this important field by a young scientist who has himself made many significant contributions.

OATLEY, KEITH: *Brain Mechanisms and Mind*, E. P. Dutton, New York, 1972.

A readable, up-to-date, well-illustrated survey written by an experimental psychologist for the general reader.

ROMER, ALFRED: *The Vertebrate Story*, 4th ed., The University of Chicago Press, Chicago, 1959.

The history of vertebrate evolution, written by an expert but as readable as a novel.

RUGH, ROBERTS, and LANDRUM B. SHETTLES: *From Conception to Birth: The Drama of Life's Beginnings*, Harper & Row, Publishers, Inc., New York, 1971.

This is an account of the history of life before birth. The book describes in detail the development of the unborn child from the moment of fertilization and also the changes in the mother during pregnancy. It discusses such related topics as birth control, con-

* Available in paperback.

genital malformation, labor, and delivery of the baby. There are a large number of illustrations, including a group of magnificent color photographs of the developing fetus.

SCHMIDT-NIELSEN, KNUT: *Animal Physiology*, 2d ed., Prentice-Hall, Inc., Englewood Cliffs, N.J., 1964.*

A useful, short paperback text.

SCHMIDT-NIELSEN, KNUT: *Desert Animals*, Oxford University Press, New York, 1964.

Although considered the definitive work on the physiological problems relating to heat and water, this readable book also contains numerous anecdotes—such as that about Dr. Blagden—and many fascinating personal observations.

SCHMIDT-NIELSEN, KNUT: *How Animals Work*, Cambridge University Press, New York, 1972.*

Schmidt-Nielsen is always concerned with underlying principles of animal physiology. In this short book, he analyzes the relationships of body size, energy supply, and environment on animal form and function.

SHERFEY, MARY JANE: *The Nature and Evolution of Female Sexuality*, Vintage Books, New York, 1973.

Psychiatrist Sherfey presents the argument, carefully based on biological principles, that the human female possesses a virtually insatiable sexual drive that, through the centuries, has been suppressed to permit the evolution of modern civilization.

SMITH, HOMER W.: *From Fish to Philosopher*, Doubleday & Company, Inc., Garden City, N.Y., 1959.*

Smith is an eminent specialist in the physiology of the kidney. Writing for the general public, he explains the role of this remarkable organ in the story of how, in the course of evolution, organisms have increasingly freed themselves from their environments.

VANDER, A. J., J. H. SHERMAN, and DOROTHY S. LUCIANO: *Human Physiology*, McGraw-Hill Book Company, New York, 1969.

Most highly recommended. The text is a model of clarity, and the diagrams, many of which we have borrowed, are splendid.

WILENTZ, JOAN STEEN: *The Senses of Man*, Thomas Y. Crowell Company, New York, 1968.

This lively and provocative book, although written for laymen, is sufficiently profound and thorough to instruct as well as intrigue the serious student.

WILKIE, D. R.: *Muscle*, Studies in Biology, No. 11, Edward Arnold, Publishers, Ltd., London, 1968.

A useful, concise (63 pages) description of the composition and function of muscle.

* Available in paperback.

PART III

Populations

SECTION 6

ECOLOGY

Chapter 40

-1 *The redwood forest along the coast of northern California and southern Oregon is, like every biological community, the result of the interaction of numerous factors, including temperature, altitude, rainfall, seasonal changes, and a particular evolutionary history.*

-2 *The Earth's atmosphere is seen as a sliver-thin, bright shell in this photograph taken from an orbiting spacecraft.*

The Biosphere

Ecology is the study of the interactions of organisms with their physical environment and with each other. As a science, it seeks to discover how an organism affects, and is affected by, its environment and to define how these interactions determine the kinds and the numbers of organisms found in a particular place and time.

Life, on this planet at least, requires water and energy from the sun. In this chapter we are going to set the stage for the study of organisms and their environments by looking at Earth as if from outer space. We shall see how water and solar energy are distributed on the surface of the planet and how, in turn, their nonuniform distribution determines the broad patterns of life forms.

That part of the Earth where life exists is called the biosphere. It forms a thin, patchy film at the surface of the planet, extending about 5 or 6 miles above the surface of the Earth and about equally far into the depths of the sea. This film of life is not uniform in either its depth or its density. In fact, in some areas, such as portions of the deserts or the ice sheets of the Antarctic and Greenland, the only evidence of life to be found is in the form of dormant spores. The key

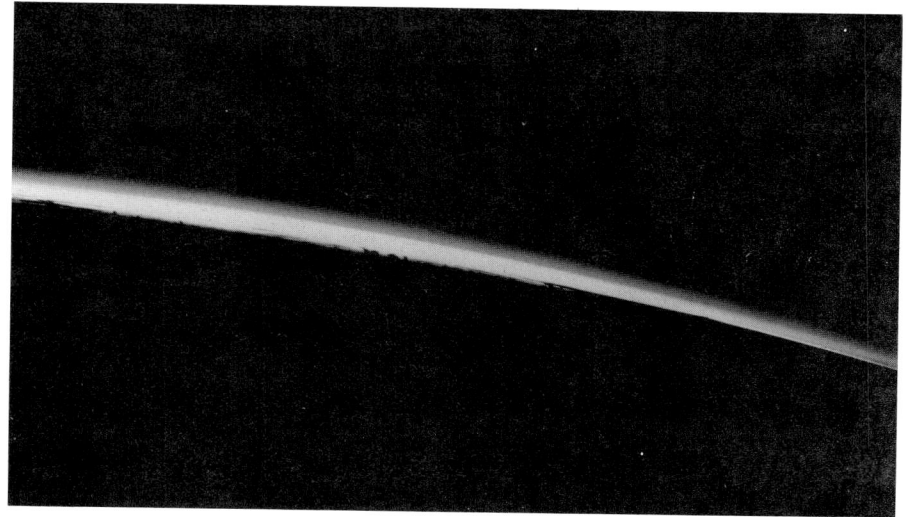

40–3 (a) *A beam of solar energy striking the Earth near one of the poles is spread over a wider area of the Earth's surface than is a similar beam striking the Earth near the equator. (b) In the northern and southern hemispheres, temperatures change in an annual cycle because the Earth is slightly tilted on its axis in relation to its pathway around the sun. In winter, the northern hemisphere tilts away from the sun, which decreases the angle at which the sun's rays strike the surface and also decreases the duration of daylight, both of which result in lower temperatures. In the summer, the northern hemisphere tilts toward the sun. Note that the polar region of the northern hemisphere is continuously dark during the winter and continuously light during the summer.*

requirements for the presence of living organisms on our planet are an energy source (the sun), water, and a temperature which permits the water to be in a liquid form. Presumably there are many other biospheres in the universe, and it is likely that they also are characterized by the presence of a similar energy source—a sun—and liquid water.

A number of factors are involved in producing the changing patterns of life on the surface of the Earth. The major one is the planet's relationship to the sun. The Earth is about 93 million miles from this star, which is the source of its heat and light energy and, indirectly, of its chemical energy as well. At the equator, the sun's rays are almost perpendicular to the Earth's surface, and this sector receives more energy from the sun than the areas to the north and south, with the polar regions receiving the least. Moreover, because the Earth, which is tilted on its axis, rotates once every 24 hours and completes an orbit around the sun once about every 365 days, the amount of energy reaching different parts of the surface varies hour by hour and season by season. (See Figure 40–3.)

Another major influence on the biosphere is the mixture of gas molecules and, usually, dust and other solid particles held in a blanket over the Earth's surface by gravity. Of the radiant energy directed toward the Earth from the sun, about 29 percent (averaged over the surface of the globe) is reflected back into space. About 20 percent is absorbed by the atmosphere (see page 737). The remaining 51 percent travels through the atmosphere and is absorbed at the Earth's surface. Much of the light energy absorbed by the Earth is reradiated as heat energy. The atmosphere impedes the loss of this heat energy from the Earth's surface, just as a blanket impedes the loss of heat from the body of a person. The major contributors to this blanketing effect are carbon dioxide and water vapor. A somewhat similar effect is produced inside a greenhouse. Light energy passes through the glass, is absorbed by the plants and soil in the greenhouse, and is reradiated as heat energy. Because heat energy passes through the glass much less readily than the higher-energy light waves, the

(a)

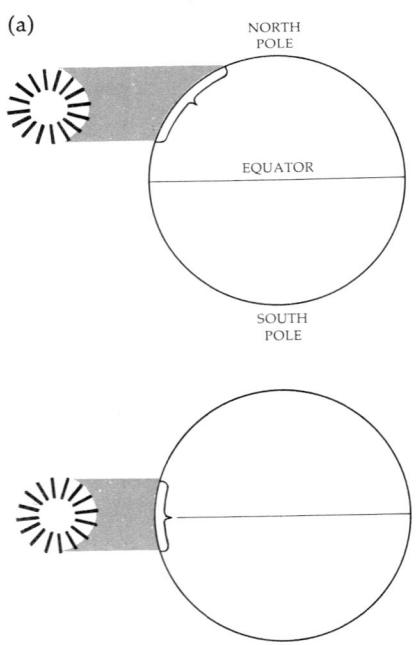

(b)

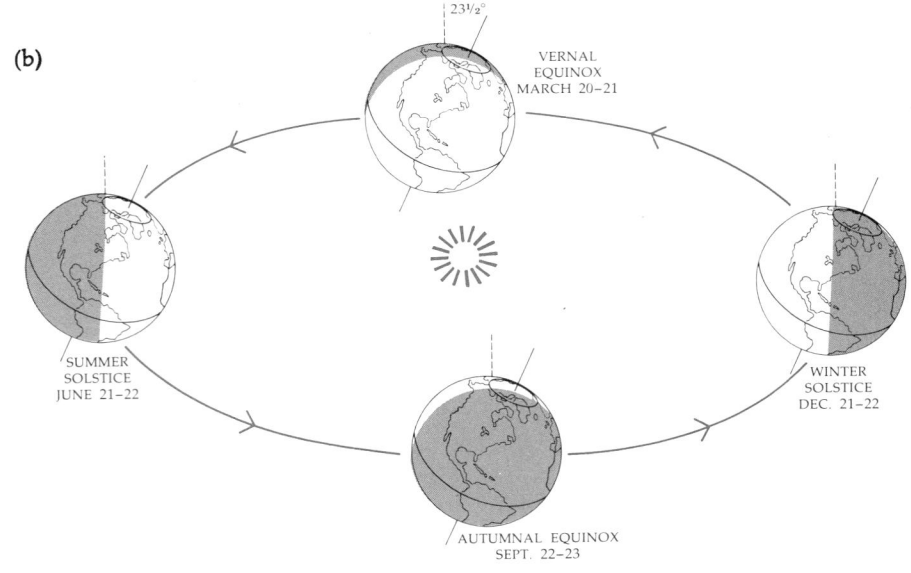

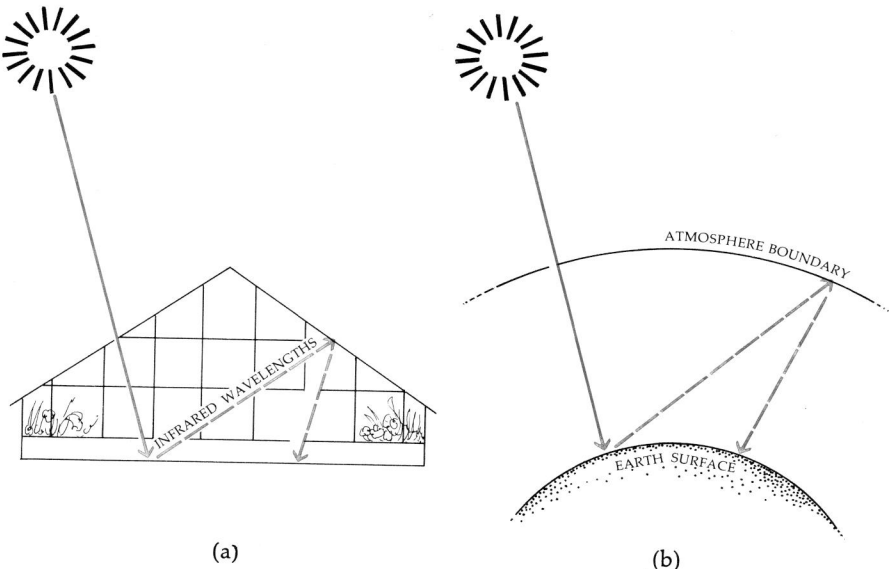

4 *The greenhouse effect. (a) Light rays penetrate the glass of the greenhouse, are absorbed by the plants and soil, and are then reradiated as longer-waved heat rays. The glass does not permit these rays to escape, and so the heat remains within the greenhouse. (b) Carbon dioxide, like glass, is transparent to light but absorbs the infrared (heat) rays, preventing their escape. An increase in atmospheric carbon dioxide, such as may be caused by the burning of fossil fuels, could lead to an increase in the temperature of the biosphere as a whole.*

(a) (b)

-5 *The Earth's surface is covered by many belts of air currents, which determine the major patterns of rainfall and wind. Air rising at the equator loses moisture in the form of rain, and falling air at latitudes of 30° north and south is responsible for the great deserts found at these latitudes.*

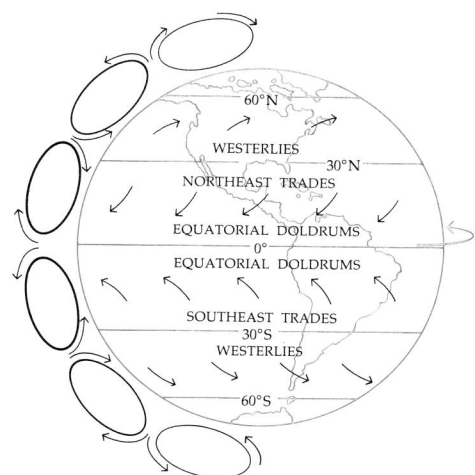

temperature inside the greenhouse remains warmer than that outside. By analogy, the atmosphere's retention of heat at the Earth's surface is often called the "greenhouse effect." (Note that the air in the greenhouse and the atmosphere of the Earth are both heated from below not from above.)

The differences in solar energy received at the various latitudes of the Earth's surface and also the Earth's rotation together establish the major patterns of planetary air circulation and rainfall. These patterns depend, to a large extent, on the fact that cold air is denser than warm air. As a consequence, hot air rises and cold air falls. As air rises, it encounters lower pressure and consequently expands, and as a gas expands, it cools. Cooler air holds less moisture than warmer air, so rising, cooling air tends to lose its moisture in the form of rain or snow.

The air is hottest along the equator, the region heated most by the sun. This air rises, creating a low-pressure area (the doldrums), moves away from the equator, cools, loses its moisture through rain, and falls again at latitudes of about 30° north and south, the regions where most of the great deserts of the world are found. This air warms, picks up moisture, and rises again at about 60° latitude (north and south); this is the polar front, another low-pressure area. A third, weaker convection cell rising at the polar front descends again at the pole, producing a region in which again, as in other areas of descending air, there is relatively little rainfall. These cells of air circulation, combined with rotational movements of the Earth, are responsible for the major wind currents across the Earth's surface.

In the spring and fall, the cells of circulation are symmetrical, as shown in Figure 40–5. In winter and summer, the cell between the equator and 30° latitude in the hemisphere where summer is occurring has a much stronger circulation than the other cells.

The worldwide patterns are modified locally by a variety of factors. For example, along the West Coast of the United States, where the winds are prevailing westerlies (coming from the west), the western slopes of the Sierra

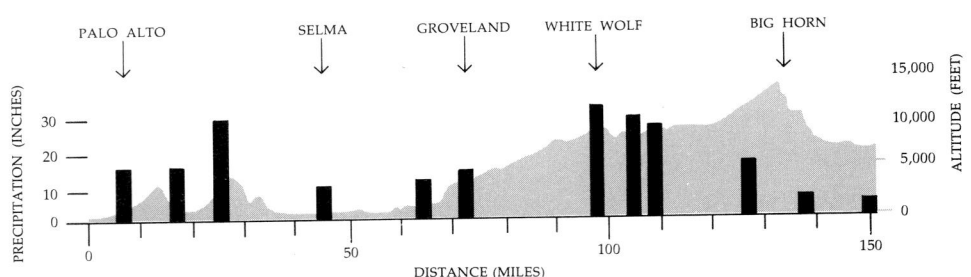

40–6 The mean annual rainfall (vertical columns) in relation to altitude at a series of stations from Palo Alto on the Pacific Coast across the Coastal Range and the Sierra Nevada mountains. The prevailing winds are from the west and there are "rain shadows" on the eastern slopes of the two mountain ranges.

Nevadas have abundant rainfall while the eastern slopes are dry and desertlike. As the air from the ocean hits the western slope, it rises, is cooled, and releases its water. Then, after it passes the crest of the mountain range, the air descends again, becoming warmer, and its water-holding capacity increases.

EFFECTS OF ALTITUDE

The mean temperature of the atmosphere decreases about 0.5° for each degree of latitude. Increases in elevation produce a similar effect, and, in general, a change in mean atmospheric temperature corresponding to 1°C in latitude occurs with a rise of about 100 meters (about 100 yards) in elevation. As a consequence, plants and animals of almost arctic characteristics are sometimes found near the equator at high altitudes.

LIFE ON LAND: THE BIOMES

The major different life formations on land are called _biomes._ These biomes, the big landscapes, are generally characterized and identified by their most abundant, or dominant, plants. The dominant plants, which are often, but not always, the tallest plants in the area, provide both protection and food for much of the animal life, and the way in which these plants grow strongly determines what can grow around their roots and in their shadows.

40–7 Because of similarities between environmental conditions at different altitudes and at different latitudes, as one ascends a mountain, one may move from forest to taiga to tundra as one approaches the summit. In this photograph taken in the Colorado Rockies, quaking aspens in the foreground give way to conifers higher up the slopes. The bare peak in the background rises above the timberline.

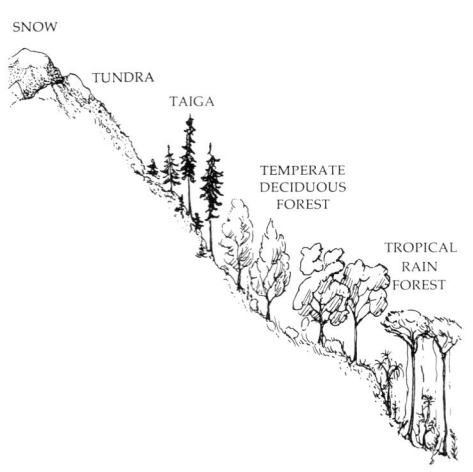

THE ATMOSPHERE

The atmosphere that lies between the Earth and outer space forms four concentric shells, or layers, distinguished by temperature differences. The lowest layer, closest to the Earth, is called the troposphere. It extends out about 6 miles and so covers all points on the Earth's surface, although the tallest mountain peaks extend almost to its limits. About 75 percent of all the molecules in the atmosphere are contained in this layer. At its outer boundary, the temperature is below −50°C. It derives its name from tropos, "to turn"; it is the layer that turns over (see Figure 40–5). Almost all of the phenomena included under the broad general heading of weather take place within the troposphere.

The layer above the troposphere is the stratosphere, which extends to an altitude of about 30 miles. In the stratosphere, the temperature increases with altitude; the temperature at the outer boundary of the stratosphere is only slightly cooler than the temperature at the Earth's surface. The major reason for the warming of the stratosphere is its layer of ozone, which is proportionately densest near the outer boundary. Ozone (O_3) is formed when molecules of oxygen gas (O_2) are broken apart by radiant energy and recombine. Ozone molecules absorb most of the ultraviolet rays of the sun, rays that would be lethal to most forms of life should they reach the Earth's surface. The solar energy absorbed by the ozone is changed to heat and so the temperature rises. About 99 percent of the atmosphere lies below the outer boundary of the stratosphere.

In the mesosphere, the third layer, there is a gradual decrease in temperature as the concentration of ozone decreases in the steadily thinning atmosphere. Then, above the mesosphere, we encounter another of those surprising temperature reversals: the temperature increases after we enter the thermosphere. This increase is, in fact, easy to understand if you recall that temperature and heat are not the same; temperature is the average kinetic energy of molecules, whereas heat also involves volume (page 51). The individual molecules of the mesosphere, unshielded from the sun's rays, move at great speed, but there are few molecules, so that there is little heat. At the outer boundary of the thermosphere, however, the thin atmosphere blends with the hydrogen and helium atoms in the space between the stars.

There are four main divisions of the atmosphere: troposphere, stratosphere, mesosphere, and thermosphere. The boundaries between them are determined by abrupt changes in average temperature.

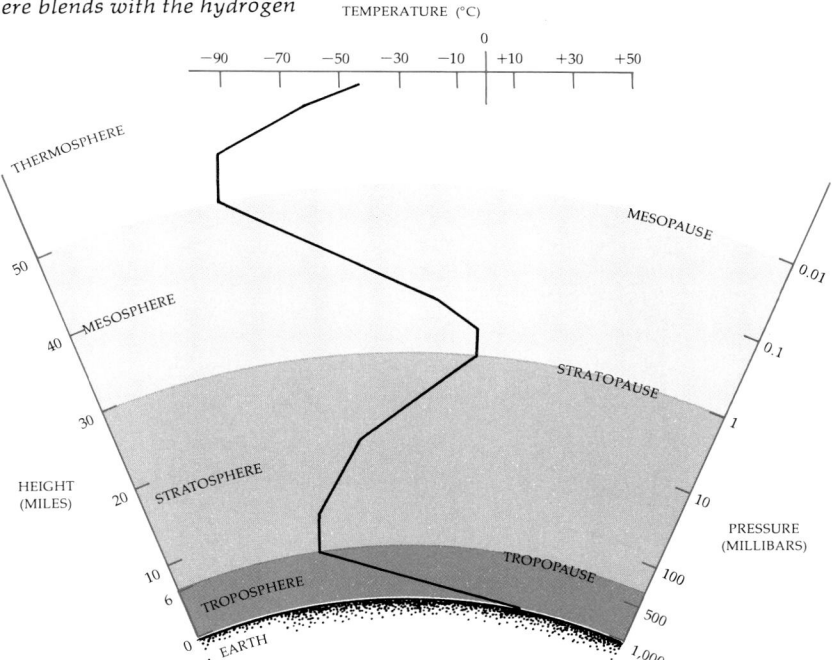

40–8　*A tropical rain forest in Uganda. Dense undergrowth, such as that shown here, occurs only where there are clearings in the forest, as by this river bed, allowing direct sunlight to reach all the levels of vegetation. Deep within the forest, the tree tops form such a thick multilayered canopy that little light reaches the forest floor, which is often almost bare.*

40–9　*Epiphytes, such as this bromeliad, grow in the canopy of the tropical rain forest, obtaining water and minerals from the moist air. Bromeliads are members of the pineapple family that are especially adapted to the tropical forest biome.*

Tropical Rain Forest

The richest of all the biomes is the tropical rain forest, where neither shortage of water nor extreme temperatures limit growth during any part of the year.

The rain forest is far richer in species than most other forests. No plant or tree is likely to be the same as its neighbors. In North Queensland, Australia, 141 different species were found among the 1,261 trees counted on 1¼ acres. (The count for a similar area in the temperate zone would be 10 to 15 species.) Naturalist Marston Bates tells of finding 18 different species of swallowtail butterflies within a 10-minute walk from his home in tropical South America; in all the United States, there are only 20 species of swallowtails, and some of these are sporadic migrants. Food is plentiful, termites and fruits providing the richest diet. Insects abound, eating the plants and one another and providing food for birds, amphibians, reptiles, and small mammals. In a 6-square-mile area in the Panama Canal Zone, some 20,000 different species of insects were cataloged.

The most abundant plants are trees. They are tall—the average height is about 150 feet—and they branch only near the crown, so the canopy is high and dense. The tropical rain forest has been described as a great green cathedral. The leaves of the trees are often large, leathery, and dark green with shiny upper surfaces. Their smooth surfaces and the drip tips found on many species

help them to shed water. The roots are shallow and so the trunks are often buttressed near the base to provide a firm, broad anchorage. There are numerous woody vines, lianas, which hang from the trees like cables, and epiphytes (plants which use the trunks and limbs of trees as their sole physical support). Epiphytes obtain water and minerals directly from the humid air of the canopy. In addition, they often have spongy roots or cup-shaped leaves that capture moisture and organic debris, and many of them can take up nutrients from decaying organisms in these storage tanks.

Many kinds of plants, including ferns, orchids, mosses, and lichens, have exploited this life style. An extraordinary variety of insects, birds, and other animals have moved into the treetops along with the vines and epiphytes to make it the area of the tropical rain forest that is most abundantly and diversely populated. Mammals are largely nocturnal or arboreal.

A great variety of palms, cycads, and ferns, some of which reach 60 feet or more in height, grow beneath the canopy. Little light reaches the forest floor. There is almost no accumulation of leaf litter, such as we find in our northern forests; decomposition is too rapid. Everything that touches the ground disappears almost immediately—carried off, consumed, or decomposed. In many places the ground is bare.

The soils of tropical rain forests are relatively infertile. Many are chiefly composed of a red clay; these red soils are known as laterites, from the Latin word for "brick." When laterite soils are cleared, in many cases they either erode rapidly or form thick, impenetrable crusts which cannot be cultivated after a season or two. Tropical soils are generally deficient in minerals; most of the nitrogen, phosphorus, calcium, and other nutrients are in the plants rather than the soil. Also, those minerals that are present are leached out by the heavy rainfall. As a consequence, soils where tropical forests have stood are very poor agriculturally.

In past geologic eras, the tropical rain forests may have extended over great areas of the world. It is believed that they represent the point of origin for many groups of plants and animals.

0–10 *Some inhabitants of New World (Western Hemisphere) rain forests:* (a) *a toucan,* (b) *a squirrel monkey,* (c) *an emerald tree boa.*

(a)

(b)

(c)

A deciduous forest of oak and hickory with a prominent field layer of ferns.

Temperate Deciduous Forest

Deciduous forests occupy areas where the year contains a warm growing season and a colder season that is less suited to plant growth and where there are moderate precipitation and mild temperatures during the growing season. These forests once covered most of eastern North America, as well as most of Europe, part of Japan and Australia, and the tip of South America. In the United States, only scattered patches of the original forest still remain. There are also deciduous forests in the tropics in areas where rainfall is limited in certain seasons.

A deciduous forest, as the name implies, is made up primarily of trees that shed their leaves seasonally. The leaves may be lost during the winter or dry season (an adaptation that conserves water during the period when most of it is locked in the soil in the form of ice or is otherwise unavailable) or, as in the foothills of the Sierra Nevadas, before the rainy season, when light intensities are low.

The dominant trees of the deciduous forests vary from region to region, depending largely on the local rainfall. In the northern and upland regions, oak, birch, beech, and maple are the most prominent trees. Before the chestnut blight struck North America, an oak-chestnut forest ran from Cape Ann, Massachusetts, and the Mohawk River valley of New York to the southern end of the Appalachian highland. Maple and basswood predominate in Wisconsin and Minnesota, and maple and beech in southern Michigan, becoming mixed with hemlock and white pine as the forest moves northward. The southern and lowland regions have forests of oak and hickory. Along the southeastern coast, the wet warm climate supports an evergreen forest of live oak and magnolia.

The Great Smoky Mountains of North Carolina and Tennessee (named for

the wisps of fog and haze that hover over them constantly) are the home of at least 130 species of trees. Each time the glacier (see Figure 40–32) moved down the face of the continent, trees were forced southward. The climate just in advance of the glacier was probably near-arctic. The regions just to the south—in Kentucky, for example—perhaps had a climate much like that of most of modern Canada. Only farther south—in what is now Tennessee, for instance—was it temperate enough for most of the deciduous trees to survive. As a consequence, new species were added to the ancient forest of the Great Smokies, and their representatives can still be found there. In western Europe, where there was no haven for the trees retreating before the glacier, there are far fewer tree species than in the United States.

In deciduous woodlands there are up to four layers of plant growth.

1. The tree layer, in which the crowns form a continuous canopy. The canopy is usually between 25 and 100 feet high.
2. The shrub layer, which grows to a height of about 15 feet. Shrubs and bushes resemble trees in that they are woody and deciduous, but they branch at or close to the ground.
3. The field layer, made up of grasses and other herbaceous (nonwoody) plants, including the wild flowers, which typically bloom in the spring before the trees regain their leaves. Bracken and other ferns, whose large leaf areas make them efficient interceptors of light, are also often conspicuous members of the field layer.
4. The ground layer, which consists of mosses and liverworts. It is also often covered with a layer of leaf litter.

Beneath the ground layer is the soil of the forest, often a rich, gray-brown topsoil. Such soil is composed largely of organic material—decomposing leaves and other plant parts and decaying insects and other animals—and the bacteria, protozoans, fungi, worms, and arthropods that live on this organic matter. The roots of plants penetrate the soil to depths measurable in feet and add organic matter to the soil when they die. Carnivorous arthropods carry fragments of their prey to considerable depths in the soil. The myriad passageways left by the dead roots and the fungi and by the earthworms and other small animals that

2 *Floor of the deciduous forest with sphagnum moss, snowberry, and mushrooms.*

A birch forest, showing layers of plant growth. The shrub layer is made up of thimbleberry and mountain maple, a favorite browse of deer.

inhabit the forest, make the soil into a sponge that holds the water and minerals. The land where deciduous forests have stood makes good farmland.

The deciduous forest supports an abundant animal life. Smaller mammals, such as the chipmunk, vole, squirrel, raccoon, opossum, shrew, and white-footed mouse, live mainly on the nuts and fruits of the forest. The wolves, bobcats, gray foxes, and mountain lions, in the areas where they have not been driven out by the encroachments of civilization, feed on these smaller mammals. Deer live mainly on the forest borders, where shrubs and seedlings grow.

Except in the winter, the forest is a noisy place. Here vision is often less important than hearing since visibility is limited. In the spring, birds proclaim their territories, chipmunks and squirrels chatter and call. Frogs and insects chorus through the summer nights. When autumn comes, a hush begins to fall over the forest. Often, more than three-quarters of the birds leave for the winter. Those that remain abandon their territories and gather in flocks, seeking shelter on the lee side of the forest or in the river valleys. The mammals den up in hollow logs or trees during the coldest periods. Woodchucks and bats hibernate, their body temperatures dropping almost to freezing and all their metabolic processes slowing down. Bears, larded with fat, maintain their body temperature but stay in a state of winter torpor, during which their cubs are born. Chipmunks nap and doze, waking up from time to time to eat their stores of nuts. The deer move into the forest from its borders to seek shelter, grazing on the buds and bark of forest trees and on cedar foliage and twigs. Most insects and other invertebrates move out of the trees and shrubs and migrate down into the soil. Many species survive only in the egg or other immature stages or as dormant or semidormant forms, emerging in the spring when the forests come to life once more.

(a)

(b)

(c)

0–14 *Some animals of the deciduous forest:*
(a) raccoon, (b) female cardinal nesting
in a multiflora rose, (c) whitetail deer
fawn, (d) red fox, (e) screech owl, and
(f) white-footed deer mouse.

(d)

(e)

(f)

Southern Forests

South of the oak-hickory deciduous forests of the eastern United States grow the pine forests of the southern states, beginning in New Jersey and spreading south and west until they merge with the grasslands beyond the Mississippi Delta. In moist locations, southern hardwoods, which are evergreen trees—tupelo, live oak, magnolia, bay, holly, and cypress—grow through the pinewoods. The magnolia forest extends along the Gulf and Atlantic coasts from east Texas to central South Carolina, including most of the Florida peninsula. Wolves no longer inhabit the forest, but bears and mountain lions are found there occasionally, and deer still abound. As one travels southward, chameleons and spadefoot toads are encountered. Along the coasts, the forest merges with salt and brackish water marshes and, in the lower Mississippi valley and near other sluggish or still fresh water, with cypress swamps.

Coniferous Forests

Conifers are gymnosperms, "naked-seed" plants. With the exception of a few groups, such as the larch and the tamarack, conifers are evergreen. They

40–15 (a) *A coniferous evergreen forest of white pine and balsam fir in northern Minnesota.* (b) *Floor of a northern coniferous forest, with moss and lichens. Decay is slower than on the warmer, wetter floors of the deciduous forest.*

(a)

(b)

Bull moose feeding on the edge of a coniferous forest.

possess several adaptations that enable them to live with a persistent cover of snow in regions with long, cold winters and a short growing season. One of these adaptations is their needle-shaped leaves which have thick cuticles; both the needle shape and the cuticle serve to reduce the rate of transpiration. Conifers also have a thick bark, which protects the inner woody tissues, and a root system that enables them to grow in shallow soils.

Taiga

The northern coniferous forest is characterized by severe winters and a constant cover of winter snow. It is composed chiefly of evergreen needle-leaf trees such as pine, fir, spruce, and hemlock. A thick layer of needles and dead twigs, matted together by fungus mycelium, covers the ground. Along the stream banks grow tamarack, willow, birch, alder, and poplar. There are virtually no annual plants.

The principal large animals of the North American coniferous forest are elk, moose, and mule deer. Among the smaller animals are porcupines, red-backed mice, snowshoe hares, shrews, wolverines, lynxes, warblers, and grouse. The small animals use the dense growths of the evergreens for breeding and for shelter. Wolves feed upon these other mammals, particularly the larger ones. The black bear and grizzly bear eat everything—leaves, buds, fruits, nuts, berries, fish, the supplies of campers, and occasionally the flesh of other mammals. Porcupines are bark eaters, and many seriously damage trees by girdling them. Moose and mule deer are largely browsers. The ground layer of the coniferous forest is less richly populated by invertebrates than that of the deciduous forest because the accumulated litter is slower to decompose.

POLAR ICE CAP
TUNDRA
NORTHERN CONIFER FOREST (TAIGA)
TEMPERATE DECIDUOUS AND RAIN FOREST
TEMPERATE GRASSLAND
CHAPARRAL
DESERT
TROPICAL RAIN FOREST
TROPICAL DECIDUOUS FOREST
TROPICAL SCRUB FOREST
TROPICAL GRASSLAND AND SAVANNA
MOUNTAIN

40–17 *Biomes of the world.*

A lofty coniferous forest, more than 2,000 miles long, lines the Pacific coast from western Alaska to central California. (a) The Douglas fir grows throughout the Sierras and the Rockies but usually reaches great heights only in the Coastal Range of Washington and Oregon. Here the trees often soar to 200 feet or more, with the lowest branch usually at least 60 feet above the ground. (b) Another giant, the Sitka spruce, rarely grows more than 50 miles inland from the moist air of the coast. Those shown here are in the rain forest of the Olympic Peninsula, where as much as 12 feet of rain fall annually. Mosses lie inches thick on the ground, cling to the tree trunks, and hang from the branches.

The Pacific Coast

An unusual and magnificent coniferous forest grows along the mountain range of the Pacific Coast of North America. In Alaska, it is composed largely of hemlock and Sitka spruce. Between British Columbia and Oregon, the Douglas fir is the dominant tree (Figure 40–18a). In southern Oregon and northern California are found the coastal redwoods. *Sequoia sempervirens,* the "always green" redwood, has a life span of more than 2,000 years (Figure 40–19a). Nearly all these redwoods grow within 20 miles of the ocean. The trees of the Pacific forests are all very tall, taller than those of the deciduous forests. The tallest living specimen of redwood measures 359 feet, 4 inches. Douglas firs range up to 260 feet tall and 50 feet in circumference.

The giant sequoia, *Sequoiadendron giganteum,* occurs only on the western slopes of the Sierra Nevadas. (See Figure 40–19b.) Some of these trees are known to be more than 2,000 years old. Recent studies have shown that these forests may owe their existence to fires that cleared the ground of the seedlings of other trees, while leaving the thick-barked sequoias virtually unharmed. Most of the ancient trees have extensive fire scars. In one of the groves that was studied in detail, 18 large fires, started either by lightning or by man, occurred between

(a)

(b)

(a)

(b)

40–19   (a) *Redwoods* (Sequoia sempervirens) *grow only along the coasts of California and southwestern Oregon. Their distribution is determined by the moisture-laden coastal fog. The tallest trees in the world, some reach heights of up to 385 feet and diameters in excess of 35 feet.*

(b) Sequoiadendron giganteum, *the giant sequoia, is found only in widely scattered groves on the western slopes of the Sierra Nevadas. Giant sequoias are not as tall as the coastal redwoods (the tallest known sequoia is not quite 300 feet), but they are heavier. In fact, the giant sequoia is the largest living organism. These two are among the last surviving species of a redwood empire that, 25 million years ago, stretched across North America.*

1760 and 1889. Borings into the trees, which reveal the annual rings, show that each of the fires was followed by a growth spurt. The forest has been fire-free since that time, and now trees such as the fir are moving in, making the shade so dense that the young sequoias cannot grow. For, despite the enormous size of the parent tree, sequoia seeds are small. Small seeds have only a slight chance of getting started in dense shade since the amount of stored food material in them permits only a short growth of seed roots.

The Tundra

Where the climate is too cold and dry even for the hardy conifers, the taiga grades into the tundra. The tundra is a form of grassland that occupies one-tenth of the Earth's land surface, forming a continuous belt across northern North America and Europe. Its most characteristic feature is permafrost—a layer of permanently frozen subsoil. During the summer the ground thaws to a depth of a few inches and becomes wet and soggy; in winter it freezes again. This freeze-thaw process, which tears and crushes the roots, keeps the plants small and stunted. Drying winter winds and abrasive driven snow further reduce the growth of the arctic tundra.

40–20 (a) The ptarmigan and (b) the varying hare, also known as the snowshoe hare because of its large feet which enable it to walk on snow. Both are year-round residents of the tundra that change their coats with the changing seasons. The ptarmigan is unusual in having three changes of costume: brown for summer, gray for autumn, and white for winter. The hare, which has two seasonal coats, was caught in the process of changing for winter. The arctic tern (c) represents one of a number of bird species that breed in the tundra, taking advantage of the long summer days to gather food for their nestlings. The terns winter in the Antarctic, following migration routes of 8,000 to 10,000 miles. Within three months after hatching, the young must be ready to migrate.

The virtually treeless vegetation of the tundra is dominated by herbaceous plants, such as grasses, sedges, rushes, and heather, beneath which is a well-developed ground layer of mosses and lichens, particularly the lichen known as reindeer moss. The flowering plants are mainly perennials.

The largest animals of the arctic tundra are the caribou of North America and the reindeer of the Old World. Lemmings, small furry rodents with short tails, and ptarmigans, which are pigeon-sized grouse, feed off the tundra. The ptarmigan is one of the few birds that lives year around on the tundra. It is also notable because it has not just two, but three changes of wardrobe to suit the season—drab brown for summer, gray for autumn, and white for winter. The white fox and the snowy owl of the Arctic live largely on the lemmings.

During the brief arctic summer, insects emerge in great numbers; the most numerous, or at least the most memorable from the point of view of human visitors, are the devastating hordes of mosquitoes, blackflies, and deerflies. Migratory birds visit in the summer, taking advantage of the insect hordes and the long periods of daylight. With the approach of winter, however, the land becomes almost deserted again. The insects become dormant, mainly in immature stages. The voles and other small mammals retreat to nests or runways, the ptarmigans tunnel into the snowbanks, and the caribou move slowly southward to the edges of the coniferous forest.

(a)

(b)

(c)

40–21 A caribou on the tundra of interior Alaska.

Grasslands

Grasslands, areas in which grasses are the dominant plants, are found in the interior areas of continents where there is not enough rainfall to support the growth of trees. They are characterized also by periodic droughts and rolling-to-flat terrain. The great grasslands of the world include the plains and prairies of North America, the steppes of Russia, the veld of South Africa, and the pampas of Argentina.

The vegetation is largely bunch or sod-forming grasses, often mixed with legumes (clover and alfalfa) and other annuals. In North America, there is a transition from the more desertlike, western, short-grass prairie (the Great Plains), through the moister, richer, tall-grass prairie (the corn belt), to the eastern temperate deciduous woodland. Grasslands become drier and drier at increasing distances from the Atlantic Ocean and the Gulf of Mexico, which are the major sources of moisture-bearing winds in the eastern half of the continent.

The grasslands of the world support large herbivores such as the bison of early America, the gazelles and zebras of the African veld, the wild horses of the Asiatic steppes, and now the domestic herbivores raised by man. These large grass-eating mammals, in turn, support the carnivores, such as the lions, tigers, and wolves, as well as omnivorous man.

The large animals graze in herds to afford some protection on the open plains, while the smaller mammals, such as the ground squirrels, prairie dogs, and gophers, burrow underground. The herds of grazing animals serve, as do the ground fires, to maintain the nature of the landscape. Shrubs and trees that venture into the prairies are eaten as seedlings. In nature, the carnivores' constant cropping of the grass eaters protects the land from overgrazing, but man has sometimes permitted sheep and cattle to destroy the roots and trample the ground to the point where the grasses can no longer recover.

40–22 *Short-grass prairie with two familiar inhabitants, bison and a prairie dog. Bison immigrated from Asia about a million years ago. In 1600, American bison numbered an estimated 45 million. They were almost exterminated during the nineteenth century, but now, under the protection of the federal government, are making a comeback. Prairie dogs are actually burrowing squirrels that live in vast underground cities, one of which was estimated to number 400 million animals.*

3 *A June day on a tall-grass prairie in North Dakota. The cottonwood grove by the prairie creek is characteristic of this biome. A thunderstorm is gathering on the horizon.*

4 *South American grasslands (pampas). The ostrichlike bird is a rhea. Like ostriches, rheas are flightless, but they are able to outrun the fastest horse.*

40–25 *The bushy vegetation that characterizes the chaparral grows as dense as a mat on the foothills of southern California. It is the result of long, dry summers, during which much of the plant life is semi-dormant, followed by a brief, cool rainy season. The name comes from* chapparo, *the Indian word for the scrub oak that is one of the prominent components of the chaparral. Chaps, the leather leggings worn by the cowboys making their way through this dense, dry growth, has the same derivation.*

Chaparral

Regions with mild, rainy winters and long, hot, dry summers, such as the southern coast of California, are dominated by small trees or often by spiny shrubs with hard, thick evergreen leaves. In the United States, such areas are known as chaparral. Similar communities are found in areas of the Mediterranean (where they are called the *maquis,* which became the name of the French underground in World War II), in Chile (where they are the *matorral*), in southern Africa, and along portions of the coast of Australia. Although the plants of these various areas are essentially unrelated, they closely resemble one another in their characteristics, as a consequence of adaptations to similar environments.

Mule deer live in the North American chaparral during the spring growing season, moving out to cooler regions during the summer. The resident vertebrates—brush rabbits, lizards, wrentits, and brown towhees—are generally small and dull-colored, matching the dull-colored vegetation. As the vegetation dries in the late summer, fires may sweep the slopes. Following a fire, many chaparral shrubs sprout vigorously, while others seed abundantly on the burned ground.

The Desert

The great deserts of the world are located at latitudes of about 30°, both north and south. These are areas of falling, warming air and, consequently, little rainfall. The Sahara Desert, which extends all the way from the Atlantic coast

of Africa to Arabia, is the largest in the world. Australia's desert, covering 44 percent of that continent, is next in size. About 5 percent of North America is desert.

Desert regions are characterized by less than 10 inches of rain a year. Because there is little water vapor in the air to prevent the escape of heat, the nights are often extremely cold. The temperature may drop as much as 30°C (54°F) at night, in comparison with the humid tropics, where day and night temperatures vary by only a few degrees.

Many of the plants in the desert are annuals that race from seed to flower to seed during periods when water is available. Because the perennials must compete for the very low supply of water, there is no dense covering to inhibit the growth of annuals—as there is in the grasslands, for example—and during the brief growing seasons, the desert may be carpeted with flowers. The perennials are either succulents, such as the North American cacti, which store water, or are plants with small leaves that are very leathery or are shed during the long droughts.

The animals that live in the desert have special adaptations to this extreme climate. Reptiles and insects, for example, have waterproof outer coverings and dry excretions. The few mammals of the desert are small and nocturnal and get what little water they require from the plants they eat. One of the most remarkable adaptations is seen in the desert species of the spadefoot toad. Most amphibians require two or three months for their metamorphosis, but the desert spadefoot must complete transformation from egg to young toad before its birthplace, a rain puddle, becomes a mudhole in the searing sun. The tiny spadefoot has been observed to progress from egg to tadpole to small hopping toadlet in only 9½ days.

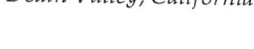

26 *Death Valley, California.*

40-27   *The North American deserts extend for nearly 500,000 square miles. The Sonoran stretches from southern California to western Arizona and down into Mexico. A dominant plant, the giant saguaro cactus (a), is often as much as 50 feet high, with a widespreading network of shallow roots. Water is stored in a thickened stem which expands, accordionlike, after a rainfall.*

To the southeast is the Chihuahuan desert, one of whose principal plants is the agave (b), or century plant, a monocot.

North of the Sonoran is the Mojave, whose characteristic plant is the grotesque Joshua tree (c), named by early Mormon colonists who thought that its strange, awkward form resembled a bearded patriarch gesticulating in prayer. The Mojave contains Death Valley, the lowest point on the continent (280 feet below sea level), only 80 miles from Mt. Whitney, whose elevation is more than 14,000 feet.

The Mojave blends into the Great Basin, a cold desert, bounded by the Sierra Nevada to the west and the Rockies to the east. It is the largest and bleakest of the American deserts. Here the dominant plant form is sagebrush (d).

(a)

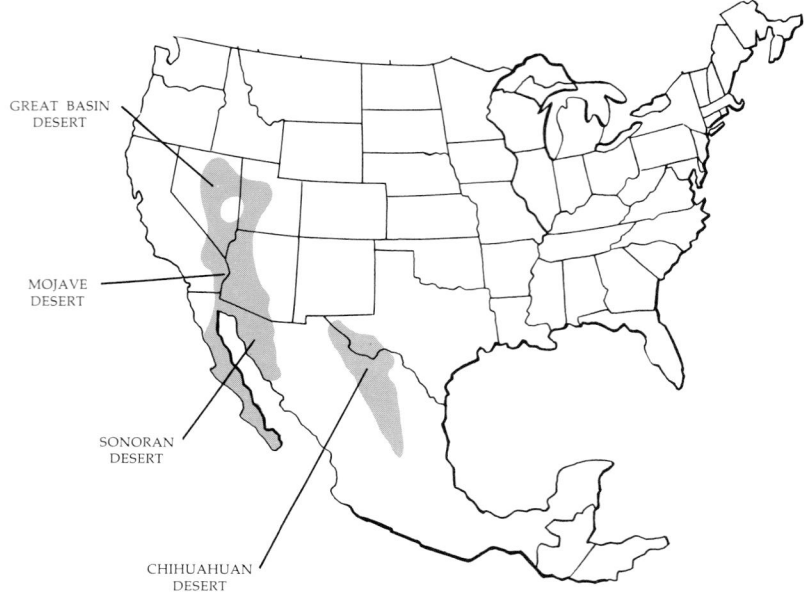

GREAT BASIN DESERT

MOJAVE DESERT

SONORAN DESERT

CHIHUAHUAN DESERT

(b)

(c)

(d)

-28 *Inhabitants of the North American deserts:* (a) *Gamble quail,* (b) *giant hairy scorpion,* (c) *golden carpet, found only (and rarely) in Death Valley,* (d) *horned lizard with a grasshopper,* (e) *roadrunner with a whiptail lizard,* (f) *bighorn sheep of the southwestern desert mountains, and* (g) *gila monster.*

(a)

(b)

(c)

(d)

(e)

(f)

(g)

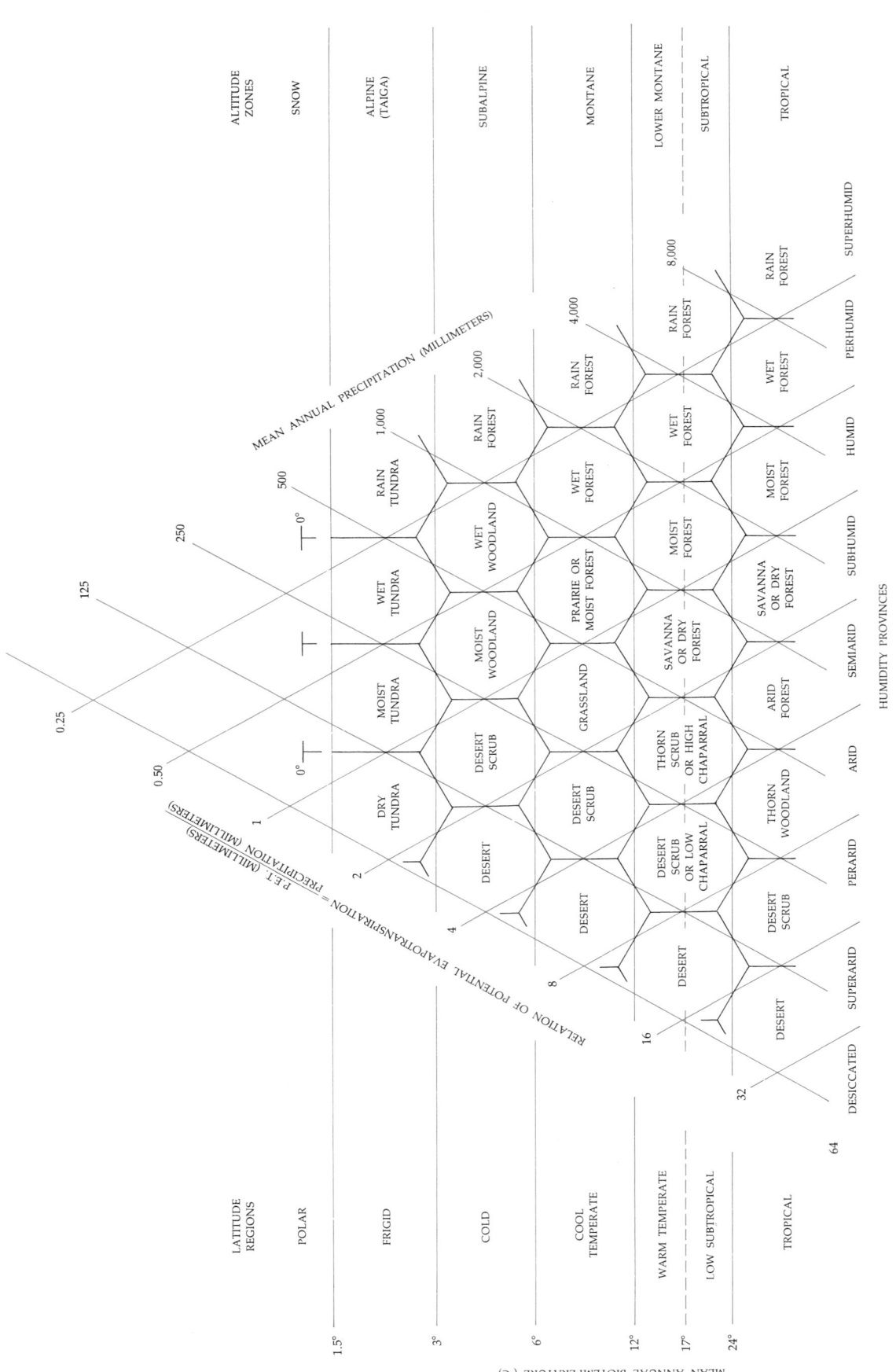

40–29 This chart showing the Holdridge system of classification differentiates the vegetation of the dry land areas of the world on the basis of temperature, precipitation, and evaporation. Latitude regions indicated on the left are equivalent to the altitude zones indicated at the right (see also Figure 40–7). Mean temperature values are those of the growing season. The dashed lines separating the warm temperate region from the low subtropical and the lower montane zone from the subtropical indicate the frost line. Potential evapotranspiration (P.E.T.) refers to evaporation from soil plus transpiration from vegetation. Local factors, such as soil conditions or the annual distribution of rainfall, may cause variations in these patterns, but in general this system permits not only description of known biomes but also enables prediction, on the basis of simple climatic data or types of plant formations.

Savannas and Tropical Deciduous Forests

Savannas and tropical deciduous forests occupy transitional areas between the evergreen tropical rain forest and the deserts. They usually have much lower annual rainfall than the tropical forest—frequently in the range of 90 to 150 centimeters a year—and the seasonal distribution of this precipitation is much more varied than in the rain forests. There is also a wider fluctuation in average monthly temperatures, owing to the seasonal drought, and a sparse covering of vegetation.

Savannas are tropical grasslands that differ from their temperate counterparts in having wet-dry annual cycles rather than a warm-cold cycle. Savannas are dominated by grasses and other herbaceous vegetation. Where the average annual rainfall is higher, trees begin to replace the grassland and the savannas merge with the tropical forest. If the rainfall is still highly seasonal, however, the forests are deciduous, rather than evergreen like the rain forest.

Trees of the tropical deciduous forests lose their leaves at the start of the dry season and flower during their leafless season. Their leaves are smaller than those characteristic of the evergreen trees of the rain forest and so lose less water by transpiration. Many of the shrubs and trees of the tropical deciduous forests are spiny or thorny and therefore discouraging to would-be herbivores.

The best-known savannas are those of Africa, which are inhabited by the most abundant and diverse group of large herbivores in the world, including the gazelle, the impala, the eland, the buffalo, the giraffe, the zebra, and the wildebeest.

-30 *A water hole on a savanna in Kenya, with zebras and gnus. The trees in the background are acacias.*

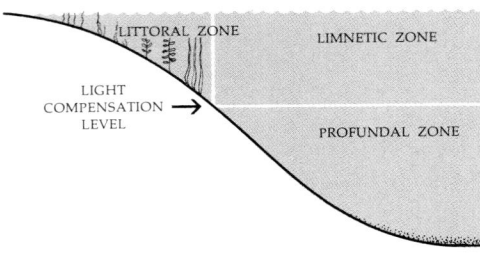

40-31 *Zones of the lake. The light compensation level is the depth at which photosynthesis (food production) just balances plant respiration (food utilization).*

LITTORAL ZONE

LIMNETIC ZONE

LIGHT
COMPENSATION →
LEVEL

PROFUNDAL ZONE

40-32 *Ice sheets, or glaciers, form when summer heat is not sufficient to melt as much water as freezes in the winter. In the most recent Ice Age—which is just now ending (or perhaps only subsiding)— four successive sheets of ice pushed down over the face of North America. At their farthest advance, shown on the map, glaciers covered most of Canada and much of the northern United States and the western mountains. At times, the wall of ice pushed forward as much as a foot a day. Most of the major lakes of the Northern Hemisphere were scraped into the continents by these glaciations. The photograph shows Found Lake in Glacier Peak Wilderness, Washington. Composed of glacial meltwater, the lake contains white pulverized rock from the glacier and therefore reflects a chalky green color; the lake is not frozen.*

Lakes

Lakes are depressions in the surface of the continents that contain standing water. They vary in size from large seas, covering thousands of square miles, to small ponds. Lake Baikal in southern Siberia has a surface area of about 13,200 square miles and is almost a mile deep at its greatest depth. The majority of the Earth's lakes are in the northern parts of North America and Eurasia. Many of these were formed in the scratches and recesses scraped on the face of the continent by the glaciers. In a geologic sense, therefore, they are relatively young. Prominent lake regions also occur in the Andes mountains of South America and in Africa, notably the East African Rift lakes.

Lakes contain three distinct environments, or zones: (1) the *littoral zone* along the shore, in which rooted vegetation lives, (2) the *limnetic zone* of open water, dominated by plankton, and (3) the deep-water *profundal zone*. (See Figure 40-31.)

The littoral zone, at the edge of the lake, is the most richly inhabited zone. Here the most conspicuous plants are angiosperms, which are rooted to the bottom, such as cattails and rushes. Water lilies grow farther out from shore; their roots are fastened to the bottom, but their leaves spread out on the surface, shading the water. There is often a green blanket of duckweed, a small free-floating angiosperm. Other pond weeds grow entirely beneath the water. Such plants lack cuticle and so can absorb minerals through their epidermis as well as through their roots. Their submerged plant surfaces harbor large numbers of small organisms. Snails, small arthropods, and mosquito larvae feed upon the plants. Other insects, such as the larvae of the dragonfly and damselfly, and the

A great blue heron at the edge of a lake. The water is covered with duckweed.

-34 *Seasonal cycle of temperature changes in a temperate-zone lake. Water, like air, increases in density as it cools, reaching a maximum density at 4°C. In the summer, the top layer of water, called the epilimnion, becomes warmer than the lower layers and therefore remains on the surface. Only the water in this warm, oxygen-rich layer circulates. In the middle layer, which is called the thermocline, there is an abrupt drop in temperature. Since the thermocline water does not mix, it cuts off oxygen from the third layer, the hypolimnion, producing summer stagnation (a). In the fall, the temperature of the epilimnion drops until it is the same as that of the hypolimnion. The warmer water of the thermocline then rises to the surface, producing the fall overturn (b). Aided by the fall winds, all the water of the lake begins to circulate (c), and oxygen is returned to the depths. As the surface water cools below 4°C, it expands, becoming lighter, and remains on the surface; in many areas, it freezes. The result is winter stratification (d). In spring, as ice melts and the water on the surface warms to 4°C, it sinks to the bottom, producing the spring overturn (e), after which the water again circulates freely (f).*

water scorpion, are carnivorous. Clams, worms, snails, and still other insect larvae burrow in the mud. Frogs, salamanders, water turtles, and water snakes are found almost exclusively in these littoral zones, where they feed primarily upon the insects and other small animals. Fish, too, are more abundant along the lake margins. Ducks, geese, and herons feed on plants, insects, mollusks, fish, and amphibians abundant in this zone. The shallow margins of some lakes and ponds are marshy. Among the inhabitants of marshes are such invertebrates and vertebrates as snails, frogs, ducks, herons, bitterns, muskrats, otters, and beavers.

In the limnetic zone, the zone of open water, small floating algae—phytoplankton—are usually the only photosynthetic organisms. This zone, which extends down to the limits of light penetration, is the habitat, for example, of smallmouth bass, bluegills, and, in colder waters, of trout.

The deep-water profundal zone has no plant life. Its principal occupants are bacteria and fungi that decompose the organic debris that filters down from the overlying water.

Ice

Liquid water, like other liquids, contracts as it gets colder. Then, when it reaches 4°C, it expands, and because it expands, its density decreases and it becomes lighter. This, of course, is quite within the range of common experience. We know that ice cubes float at the top of a water glass and that icebergs cruise the surface of the oceans. But have you thought about what would happen if water behaved like an "ordinary" liquid and just got denser and denser (as ammonia does) as it froze? Not only would ice cubes and icebergs plummet to the bottom, but ponds and lakes and perhaps even oceans would begin to freeze from the bottom up. Spring and summer might stop the freezing process, but laboratory experiments have shown that if ice is held to the bottom of even a relatively shallow tank, water can be boiled on the top without melting the

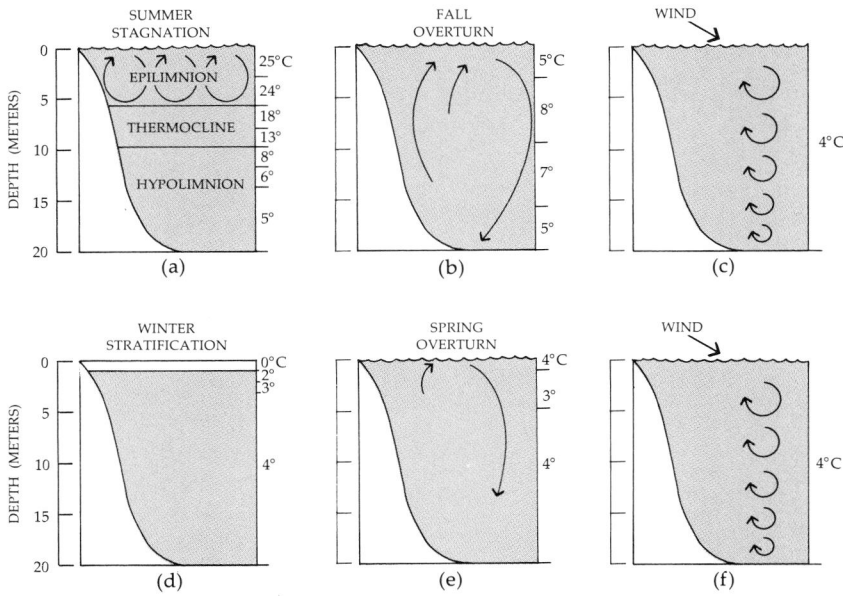

759 *The Biosphere*

THE WATER CYCLE

The Earth's supply of water is constant and is used over and over again. Most of the water (98 percent) is in liquid form—in the oceans, lakes, and streams. Of the remaining 2 percent, some is frozen in polar ice and glaciers, some is in the soil, some is in the atmosphere in the form of vapor, and some is in the bodies of living organisms.

The water cycle is driven by solar energy. Energy from the sun evaporates water from the oceans, lakes, and streams, from the moist soil surfaces, from the leaves of plants, and from the bodies of other living organisms. The water molecules are carried up into the atmosphere by air currents and eventually fall again as rain.

Some of the water that falls on the land percolates down through the soil until it reaches a zone of saturation. In the zone of saturation, all pores and cracks in the rock are filled with water. The upper surface of this zone of saturation is known as the water table. Below the zone of saturation is solid rock, through which the water cannot penetrate. The deep ground water, moving extremely slowly, eventually emerges in lakes and streams and even in the ocean floor as seepages and springs, thereby completing the water cycle.

The Earth's large bodies of water have moderating effects on climate. As we noted in Chapter 3, water has a high heat of vaporization; it requires more than 500 calories to change a gram of liquid water into water vapor. Water also has a high latent heat of melting. The latent heat of melting is, by definition, the number of calories required to convert 1 gram of solid at the freezing point to 1 gram of liquid at the same temperature. In the case of water the same quantity of heat must be employed to melt 1 gram of ice as to raise the temperature of the resulting ice water to 80°C. Conversely, when water turns back to ice, 80 calories of heat are given up to the surroundings. This giving up and absorbing of heat as water freezes and ice melts acts as a sort of thermal buffer system. When a body of water becomes cooled to its freezing point, so long as water and ice exist in contact, additional heat will tend only to melt the ice and cooling will tend to freeze the water. As a consequence, a body of water at the freezing point will remain at about the same temperature despite a fairly wide range of changes in air temperature. Thus, because of these unusual properties, water serves to stabilize the temperature of the surrounding air and land. Great extremes of temperature, either hot or cold, occur only far from large bodies of liquid water.

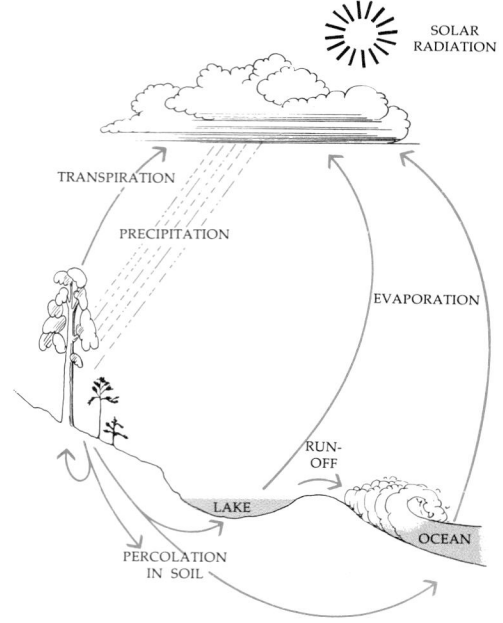

ice. Thus if water did not expand when it froze, it would continue to freeze from the bottom up, year after year, and never melt again, leaving half of our planet locked in ice.

Because water increases in density as its temperature drops (until just above freezing), the upper part of the lake during the summer will consist of a layer of water that has been heated by the sun, and the denser, cold water will form a bottom layer. In winter, the coldest water (below 4°C) will remain on top, and the slightly warmer water will sink to the bottom. This stratification effectively seals off the depths from any contact with the air. Also, the nutrients derived from the organic debris that collects on the lake bottom cannot rise to the surface. Consequently, during the summer and winter, the supply of oxygen in the profundal zone and of nutrients in the limnetic zone may run short, imposing special stresses on the inhabitants of these zones. In the spring and fall, when the entire body of water approaches the same temperature, mixing again occurs, often resulting in "blooms" of phytoplankton.

Ponds

The term pond is used generally for small bodies of water so shallow that rooted plants grow over most of the bottom. They range in life span from a few weeks or months—in the case of small seasonal ponds—to several hundred years.

Rivers and Streams

Rivers and streams are characterized by continuously moving water. They may begin as outlets of ponds or lakes, as runoffs from melting ice or snow, or they may arise from springs, flows of groundwater emerging from bedrock.

The character of life in a stream is determined to a large extent by the swiftness of the current, which characteristically changes as a stream moves downward from its source and, fed by tributaries, increases in volume and decreases in speed. In swift streams, most organisms live in the riffles, or shallows, where small photosynthetic algae and mosses cling to rock surfaces. Many insects, both adult and immature forms, live on the underside of rocks and gravel in the riffles. These often have extremely flattened bodies, which fit in narrow crevices,

–35 *Pond with water lilies.*

40–36 *A fast-moving stream in Glacier Park, Montana.*

40–37 *Special adaptations are necessary for fish living at great depths in the ocean. The deep-water oceanic angler fish shown here is a female, and the structure above her eye is a parasitic male. This arrangement presumably solves the problem of finding partners in the dark.*

and special adaptations that enable them to cling to the rocks. For those small organisms that can survive the swiftly moving current, there is an abundance of oxygen and of nutrients swept along by the flowing waters. As the stream travels along its course, the riffles are often interrupted by quieter pools where organic material may collect and be decomposed. Few plants can gain footholds on the shifting bottoms of stream pools, but some invertebrates, such as dragonflies and water striders, are typically found in or about the pools. Some organisms, notably trout, move back and forth between the riffles and the pools.

As streams broaden and become slower, they begin to take on the characteristics of lakes and ponds.

THE OCEANS

The oceans cover almost three-fourths of the Earth's surface and reach down to an average depth of almost 3 miles. At its deepest point, the Marianas trench in the Pacific, the ocean is about 7½ miles deep.

Oceans, like lakes, are characterized by a euphotic ("well lit") zone, corresponding to the limnetic zone of lakes, and the abyssal, lightless zone. The boundary between them depends on local conditions, such as the intensity of the sunlight, the turbidity of the water, and the density of plankton or algae. In the clearest parts of the ocean, the euphotic zone may extend as far down as 650 feet. As one goes deeper into the ocean, the long-waved red and yellow rays of light are absorbed before the shorter rays, which is why the sea looks blue. As you reach the boundary of the euphotic zone, there is only a dim blue twilight, and below 1,600 to 1,900 feet, there is complete darkness.

There are two divisions of life in the open ocean: pelagic and benthic. One major component of the pelagic division is plankton. It is composed of photosynthetic algal cells (phytoplankton) intermingled with small shrimp and other crustaceans and the eggs and larval forms of many fish and invertebrates (zoo-

40–38 *There are two divisions of life in the ocean: pelagic and benthic. The benthic division is subdivided into tidal, littoral, and deep-sea zones. Only the first 200 meters or less of open ocean are euphotic (well-lit); the lower pelagic and deep-sea benthic zones are in perpetual darkness.*

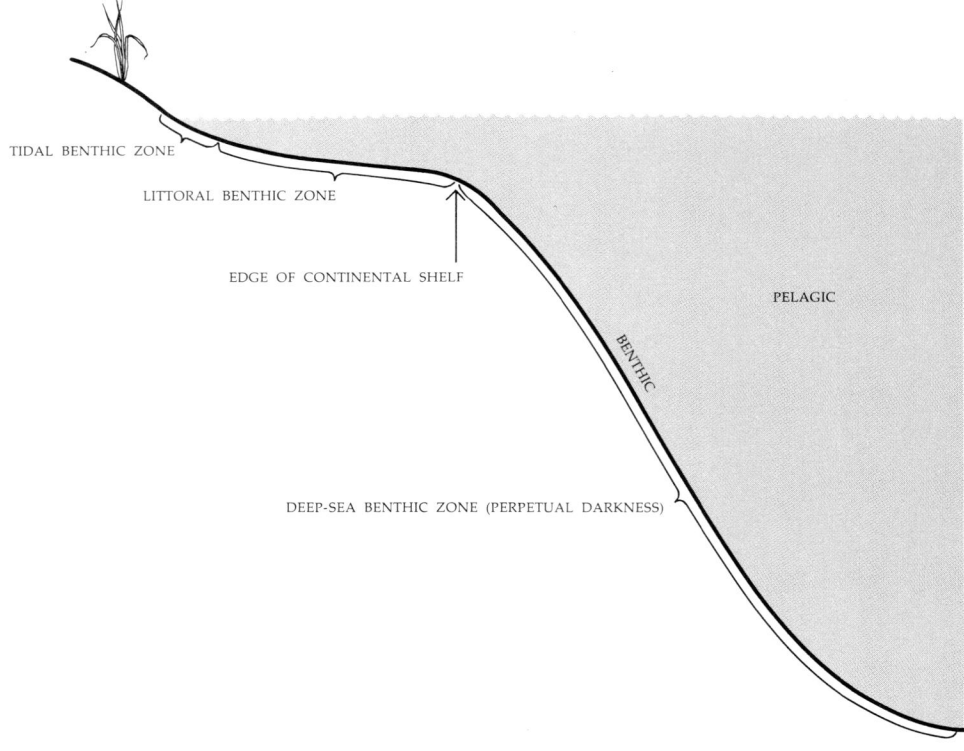

TIDAL BENTHIC ZONE

LITTORAL BENTHIC ZONE

EDGE OF CONTINENTAL SHELF

PELAGIC

BENTHIC

DEEP-SEA BENTHIC ZONE (PERPETUAL DARKNESS)

The major currents of the ocean have profound effects on climate. Because of the warming effects of the Gulf Stream, *Europe is milder in temperature than is North America at similar latitudes. The eastern coast of South America is warmed* *by water from the equator, and the Humboldt Current brings cooler weather to the western coast of South America.*

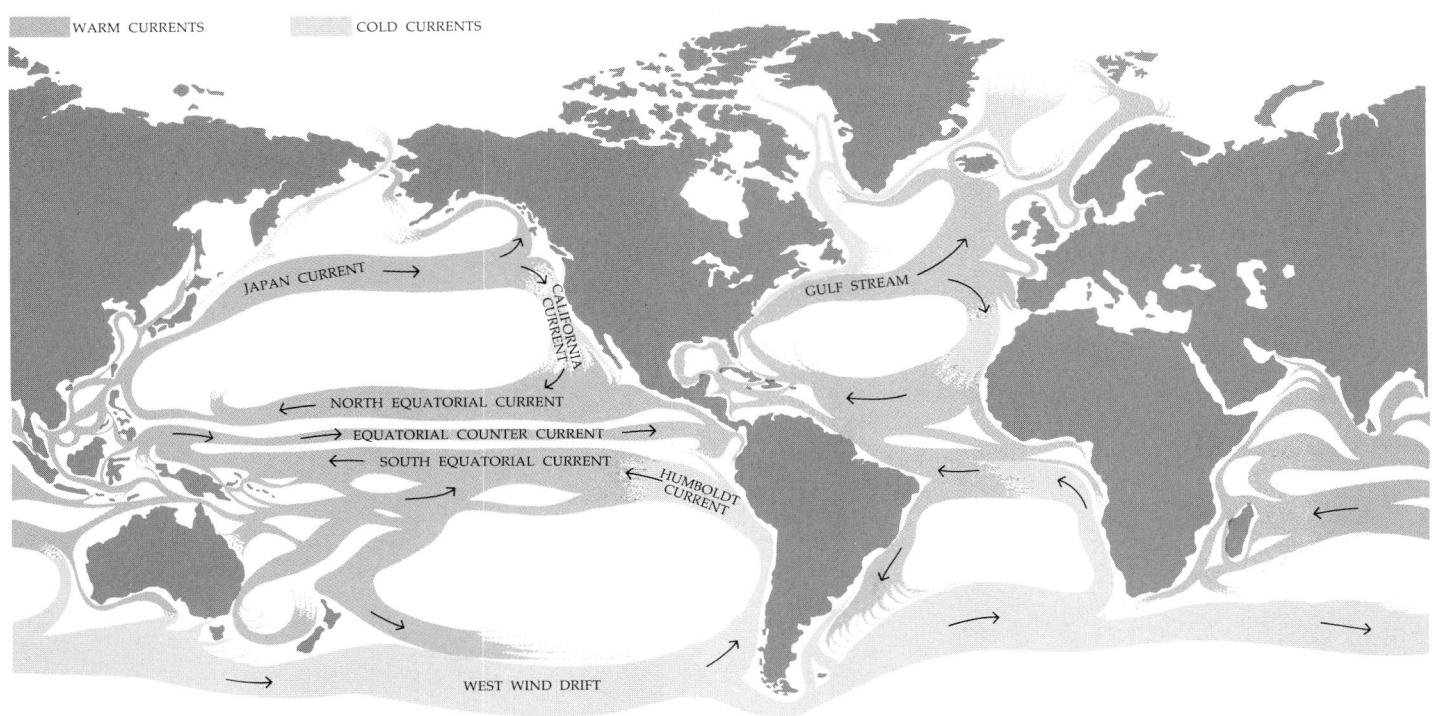

WARM CURRENTS   COLD CURRENTS

JAPAN CURRENT →

CALIFORNIA CURRENT

GULF STREAM

← NORTH EQUATORIAL CURRENT

→ EQUATORIAL COUNTER CURRENT →

← SOUTH EQUATORIAL CURRENT

HUMBOLDT CURRENT

WEST WIND DRIFT

A sample of living phytoplankton showing several species of diatoms.

plankton). Twelve million single-celled algae have been found in a cubic foot of seawater from the sunlit surface of the pelagic division. These planktonic forms provide food for the fish and other animals that are also part of the pelagic division. The benthic division contains the bottom-dwellers, the sessile animals, such as sponges and sea anemones, and a variety of fungi and bacteria that live on the sparse but steady accumulation of debris drifting from the more populated levels of the ocean.

The major currents of the ocean, which are produced by a combination of winds and the Earth's rotation, profoundly affect life in the oceans and alter the climate along the ocean coasts. These patterns of water circulation—clockwise in the Northern Hemisphere and counterclockwise in the Southern Hemisphere —move currents of warm water north and south from the equator. One such current, the Gulf Stream, warms a portion of the eastern coast of North America and the western shores of Europe, and another warms the eastern coast of South America. The same patterns of circulation bring cold waters to the western coasts of North and South America. Where the winds move the water continuously away from the shores, as off the coasts of Portugal and Peru, cold water rich in nutrients is brought to the surface in upwellings. Such areas traditionally have highly profitable fishing industries. However, in recent years a periodic change in water temperatures coupled with overfishing have enormously depleted the fishing industry of Peru. At this writing, it is not clear whether or not it will recover.

A reef in the New Hebrides, islands in the Pacific, east of Australia. Staghorn coral is in the foreground.

The Coral Reef

The coral reef is the most diverse of all marine communities. The coelenterates that form the reef can grow only in warm, well-lighted surface water, where the temperature seldom falls below 21°C (70°F). The oceans are rising slowly, 4 to 6 inches every 100 years as the glaciers melt, but the upward growth of the coral keeps the reef tops in the well-lighted zone. The longest reef in the world is the Great Barrier Reef of Australia, which extends some 1,200 miles. Other reefs are found throughout the tropical waters and as far north as Bermuda, which is warmed by the Gulf Stream. The upper Keys of Florida, which begin just south of Miami, were built by coral animals during a warm period before the last glaciation. When the glacial ice built up in the north, the water level fell, exposing the upper surface of the Keys and killing the coral coelenterates.

Each polyp in the colony secretes its own calcium-containing skeleton, and when it dies, the next generation builds on top of it. The reef is not the work of the animals alone, however; symbiotic algae occur on and in the inorganic skeleton, as well as inside the bodies of the polyps. In fact, the entire colony contains about three times as much plant as animal tissue. Laboratory studies have shown that while the coral animals can live independent of the algae, algae are necessary for the rate of reef formation seen under natural conditions. The algae photosynthesize by day, contributing oxygen and organic compounds to the coral animals, and at night the carnivorous polyps extend their tentacles and capture zooplankton from the surrounding waters.

The reef furnishes both food and shelter for other sea animals, including invertebrates and numerous species of reef fishes.

0–42 (a) *Sea otter in a kelp bed off the coast of California.* (b) *Tidal pools at low tide on Alaska's rocky southeast coast.* (c) *Dowitchers feeding on a sandy beach.* (d) *A ghost crab.*

(a)

(b)

(c)

The Seashore

The edges of the continents extend out into the sea from 10 to 150 miles. Along these edges, the continental shelves, nutrients are washed out from the land, and life is much denser than in the open seas. Sessile animals, such as sponges and corals, are found all over the ocean bottom, but they are abundant only in areas that are close to the shores. They are especially numerous in warmer waters. Predators, such as mollusks, crustaceans, and many kinds of fish, roam over the bottoms of the continental shelves. Eel grasses and turtle grasses, which are angiosperms, and seaweeds provide shelter for myriad animals and increase the supply of oxygen. Sea urchins and some species of fish actually eat the grasses and seaweeds, and microscopic algae and small animals fasten themselves to these plants. Snails, slugs, and worms crawl over the plant surfaces, eating off the encrusted growth, and fish nibble the smaller animals that cling to them.

Seashores—where the sea and the land join—are of three general types: rocky, sandy, and muddy. The organisms that live on rocky coasts, like those that live in the riffles of fast-moving streams, have special adaptations for clinging to the rocks. The algae have strong holdfasts. The starfish of the rocky coasts lies spread-eagled on the rocks, clinging with its suction cups. The abalone holds tight with its well-developed muscular foot. Mussels, scallops, and oysters secrete coarse ropelike strands that anchor them to the bottom of rocky surfaces.

(d)

(a)
(b)
(c)
(d)

40–43 *Mud flats and marshes.* (a) *A salt marsh.* *cranes probing for marine worms.* (c) *The grasses are* Spartina. (b) *Whooping* *Yellow rail.* (d) *Prothonotary warbler.*

The organisms of the rocky coast face the additional problem of the rising and falling tides that threaten them alternately with inundation and desiccation. Largely as a consequence of the effects of the tidal ebb and flow and the adaptations of organisms to it, life along the rocky coasts is highly stratified. The supralittoral zone, which is flooded only occasionally by high tides, is a zone of dark algal and lichen growth. The littoral zone, submerged and exposed daily, is characterized by rockweeds (*Fucus*), often intermixed with Irish moss (a red alga) and other seaweeds. Animal life includes barnacles, oysters, blue mussels, limpets, and periwinkles. The sublittoral zone, exposed only occasionally, contains forests of the large brown alga, *Laminaria,* sea squirts, starfish, and various other invertebrates. When the tides go out, pools of water are left behind in rocky basins and crevices. Despite the fact that these pools are subjected to drastic changes in temperature, pH, and salinity as water evaporates, they contain such a diversity of life that they have been called the "microcosms of the seas." In one such microcosm—a small tidal pool—investigators counted 5,185 organisms (not including microscopic forms), representing 45 species.

Sandy Shores

Sandy beaches have fewer bottom dwellers because of the constantly shifting sands. Clams, numerous ghost crabs, sandfleas, lugworms, and other small invertebrates live below the surface of the sand, sometimes emerging at low tide to feed on the debris washed in and out by the tide. Along the sandy beaches, beach grasses are important for stabilizing the shifting dunes. These grasses spread by means of underground stems, which eventually form a network that knits the dunes together and stabilizes them enough so that other types of plant life can take hold.

Mud Flats and Estuaries

The mud flat, while not so rich or diverse in species as the rocky coast, is the most suitable of all three types of seacoast for animal growth. It is a constant bustle of changing activities. At low tide on the mud flats, the bottom forms retreat into their burrows and the mobile forms scurry seaward. The snails continue to browse on the algae, and insects and birds come into the exposed area for prey and debris. As the tide comes in, insects and birds retreat toward the land, but the clams extend their siphons once more, the annelids rise from their burrows, and shrimps, crabs, and fish move about over the surface again. A mud flat can support tens of thousands of individuals per square meter.

All streams and rivers eventually drain into the seas; the area where fresh water joins salt water is an *estuary*. The principal vegetation here is grass, such as *Spartina,* which is specially adapted to survive in salt water. These salt marshes are rich in nutrients brought down from the land, and so are the spawning places and nurseries for many forms of marine life. Estuaries and saltwater swamps, like many of the natural biomes of the world, are threatened with extinction as wetlands are filled and paved to make room for man's steady expansion of his domain. Their protection is of special importance because of their role in nurturing the life of the oceans.

SUMMARY

The biosphere is the part of the Earth that contains living organisms. It is a thin film on the surface of the planet, irregular in its thickness and its density. The Earth's atmosphere is a blanket of gas and dust molecules held close to the surface by gravity. The atmosphere filters out harmful ultraviolet rays from the sun and impedes the escape of heat from the Earth's surface. The biosphere is affected by the position and movements of the Earth in relation to the sun and the movements of air and water. These conditions cause wide differences in temperature and rainfall from place to place and season to season on the Earth's surface. These differences are reflected in differences in the kinds of plant and animal life.

The major life formations on land are called biomes. The richest terrestrial biome is the tropical rain forest, where neither water shortage nor extreme temperatures limit plant growth. Here the trees are broad-leaved evergreens, which are characteristically covered with vines and epiphytes. There is almost no collection of decomposing material or humus. Tropical soils are often red clays (called laterites) which erode or solidify when the forest is cleared.

The temperate deciduous forest is an important biome of eastern North America and Eurasia, where temperatures are warm in the summer and cold in the winter. In the deciduous forest, the trees lose their leaves in the fall, which reduces water loss during the winter months, when the water is locked in ice.

North of the deciduous forest is the taiga, the worldwide subarctic coniferous forest. The trees of this forest, the conifers, have a number of special adaptations for conservation of water and protection against extreme cold. Other needle-leafed evergreen forests are found along the Pacific Coast and in the Southeast in the United States and also in Southeast Asia.

Between the taiga and the northern polar region is the tundra, on which there are few trees and the dominant vegetation consists of low-growing perennials.

The grasslands lie between the deciduous forests and the deserts and have a rainfall intermediate between the two. The growth of trees is prevented not only by a shortage of rain but also by grazing animals, which eat the young shoots, and by recurring prairie fires. Savannas are tropical grasslands intermediate between tropical rain forests and deserts. Tropical deciduous forests occur where rainfall is highly seasonal.

Chaparral, a shrubland biome found on the southern California coast and in the Mediterranean area, is characterized by dry summers but abundant winter rainfall.

Desert regions have very little rainfall and high daytime temperatures. The chief vegetation is made up of annual plants with extremely short growing seasons; cacti and other perennial desert plants are highly adapted for the conservation of water.

Freshwater biomes include lakes and ponds, which are standing water, and rivers and streams, which are flowing water.

Lakes are conveniently divided into three ecological zones: the littoral zone, along the shore, characterized by rooted vegetation; the limnetic zone of open water extending as far down as the light penetrates; and the profundal zone. The littoral zone contains a rich variety of plant and animal life, including snails, many insects, and other small invertebrates. The photosynthetic or-

ganisms of the limnetic zone are chiefly phytoplankton; the larger fish are found principally in this zone. There are no plants in the profundal zone; the principal organisms here are decomposers, including bacteria and fungi. Lakes in temperate zones characteristically are thermally stratified and turn over in spring and fall, redistributing oxygen and nutrients. A pond is a body of water so shallow that rooted plants can grow over much of the bottom.

The character of rivers and streams is determined in large part by the swiftness of the current. In fast-moving fresh water, living organisms are found principally in the riffles, clinging to rocks and wedged into crevices.

The open ocean has two distinct life divisions: the pelagic (planktonic and swimming) division and the benthic (bottom) division. The photosynthetic vegetation of the open ocean is almost entirely plankton—floating single-celled algae. The plant plankton (phytoplankton), mixed with animal plankton (zooplankton), provides the basic food supply for fish and deepwater mammals and also for the sparse but diverse animal, fungal, and bacterial life of the ocean bottoms.

Patterns of air circulation combined with the Earth's rotation produce the major currents of the oceans. These modify the temperatures of the coastal waters and the continents.

The richest zone of ocean life is the coral reef, composed of the bodies and skeletons of coelenterates combined with symbiotic algae.

The seashores include the rocky coasts, the sandy beaches, and the mudflats and estuaries, each with its characteristic variety of life. The salt marshes of the estuaries are the nursery for many immature forms of marine life.

QUESTIONS

1. Define the following terms: biome, troposphere, littoral, limnetic, tundra, taiga, greenhouse effect.
2. What are the major influences on plant life in the desert? The tropical rain forest? The taiga? The tundra? Describe some special adaptations of plants in each of these biomes.
3. What is the position of the Earth in relation to the sun at the vernal equinox? At the winter solstice? What is the day-night cycle at the South Pole on these dates?
4. What are the dominant plants in the area in which you now live? How far do you have to go to find examples of this vegetation? What kinds of animals are characteristically associated with it? What other areas of the world are part of this same biome?

Chapter 41

Ecosystems

In the preceding chapter, we described the major biomes and aquatic communities of the world, some of the forces that determine their characteristics, and some of the organisms that inhabit each of them. If we look more carefully at any one part of any of these communities—at a patch of woodland, a pasture, a pond, or one coral reef—we begin to see that none of the organisms living in this particular area exists in isolation; rather, each is involved in a number of relationships, both with the other organisms and with factors in the nonliving environment. The details of these relationships vary from place to place. In all cases, however, the interactions of the various organisms in a natural setting have two consequences: (1) a flow of energy through photosynthetic autotrophs —green plants or algae—to heterotrophs, which eat either the green plants or other heterotrophs, and (2) a cycling of inorganic materials, which move from the nonliving environment through the bodies of living organisms and back to the environment.

41–1   *A pride of African lions with a zebra kill.*

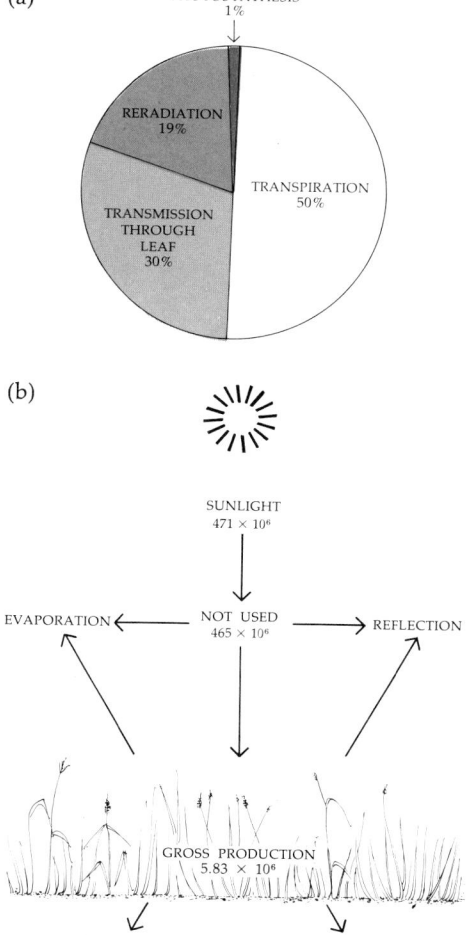

4-2 (a) *What happens to the light energy that falls on a green leaf? Note that only about 1 percent is used in photosynthesis.* (b) *Calculation of the productivity of a field in Michigan in which the vegetation was mostly perennial grasses and herbs. Measurements are in terms of calories per square meter per year. In this field, the net production—the amount of chemical energy stored in plant material—was 4,950,000 calories per square meter per year. Thus, slightly more than 1 percent of the 471,000,000 calories per square meter per year of sunlight reaching the field was converted to chemical energy and stored in the plant bodies.*

Such a combination of living (biotic) and nonliving (abiotic) elements through which energy flows and minerals recycle is known as an *ecosystem*. Taking a large, astronautical view, the entire biosphere can be seen as a single ecosystem. This view is useful when studying materials that are circulated on a worldwide basis, such as carbon dioxide, oxygen, and water. A suitably stocked aquarium or terrarium is also an ecosystem, and such man-made models may be useful in studying certain ecological problems, such as the details of transfer of a particular mineral element. Most studies of ecosystems have been made, however, on more or less self-contained natural units—on a pond, for example, a mud flat, or, as in a recent, interesting study, a watershed area of a deciduous forest.

The portion of an ecosystem in which a particular organism lives is known as the organism's *habitat*. Thus, within a pond ecosystem, an amoeba's habitat would be a square millimeter of bottom ooze, the surface of the pond the habitat of a water strider, and the open water the habitat of a largemouth bass.

THE FLOW OF ENERGY

The Source of Energy

The ultimate source of energy for all natural ecosystems is the sun. The Earth receives from the sun an average of 2 calories of radiant energy per minute for every square centimeter of the Earth's surface—a total of 13×10^{23} calories every year. About half of this energy never reaches the surface, but is reflected back into space by the clouds and dust of the atmosphere. (Because of these reflections, Earth, from outer space, is a shining planet, almost as bright as Venus.)

Of the solar energy that reaches the ground, some is dissipated in the evaporation of water. Most of it is absorbed into the ground and radiated back in the form of heat. Only a small fraction of the total available energy from the sun enters the food chain of living organisms. Even if the light falls where vegetation is abundant, as in a forest, a cornfield, or a marsh, only 1 or 2 percent of that light (calculated on an annual basis) is used in photosynthesis. Yet this fraction, as small as it is, may result in the production—from carbon, oxygen, water, and a few minerals—of several thousand pounds (dry weight) of organic matter per year in a single square meter of field or forest, adding up to a worldwide total of 150 to 200 billion tons of such organic matter per year.

Trophic Levels

The passage of energy from one organism to another takes place along a particular food chain, which is made up of trophic (feeding) levels.

The first step in the chain is always a primary producer, which on land is a green plant and in an aquatic ecosystem is usually a photosynthetic alga. These photosynthetic organisms use light energy to make carbohydrates and other compounds, which then become sources of chemical energy. Ecologists speak of the *productivity* of ecosystems; productivity is measured by the amount of energy fixed in chemical compounds or by increase in biomass in a particular length of time. (*Biomass* is a convenient shorthand term meaning organic matter; it includes woody parts of trees, stored food, arthropod shells, bones,

etc.) The biomass in a particular area at a particular time is known as the standing crop. _Net productivity_ represents the amount of light energy converted to chemical energy less the amount dissipated (that is, converted to heat energy) by the plants in the course of respiration. Both productivity and net productivity are measurements of rate and so must be expressed in terms of a time unit.

The Consumers

Energy enters the animal world through the activities of the herbivores, animals that eat plants (including fruits and other plant parts). An herbivore may be a blue whale, a caterpillar, an elephant, or a field mouse; each type of ecosystem has its characteristic complement of herbivores. Of the plant material consumed by herbivores, much is excreted undigested. Some of the chemical energy is transformed to other types of energy—heat energy or kinetic energy—or used in the digestive process itself. A fraction of the material is converted to animal biomass.

The next level in the food chain, the second consumer level, involves a carnivore, a meat-eating animal that devours the herbivore. The carnivore may be a lion, a minnow, a robin, or a spider, but in each of these cases, only a small part of the organic substance present in the body of the herbivore becomes incorporated into the body of the consumer. Some chains have third and fourth consumer levels, but five links are usually the absolute limit, largely because of the loss involved in the passage of energy from one trophic level to another. Other important consumers in the food chain are decomposers and parasites. The roles of these two groups will be discussed later in this section.

In a stable ecosystem, a large proportion of the chemical energy fixed by the primary producers is stored in plant or animal biomass or is used in respiration. (In certain systems, such as flowing streams, some may be exported.)

41–3 (a) *Harvest mouse (primary consumer) eating the winter seed capsules of a trumpet creeper. (b) Leopard (secondary consumer) with kill. (c) Female bear and cub (at top of food chain) feeding on a salmon.*

(a)

(b)

(c)

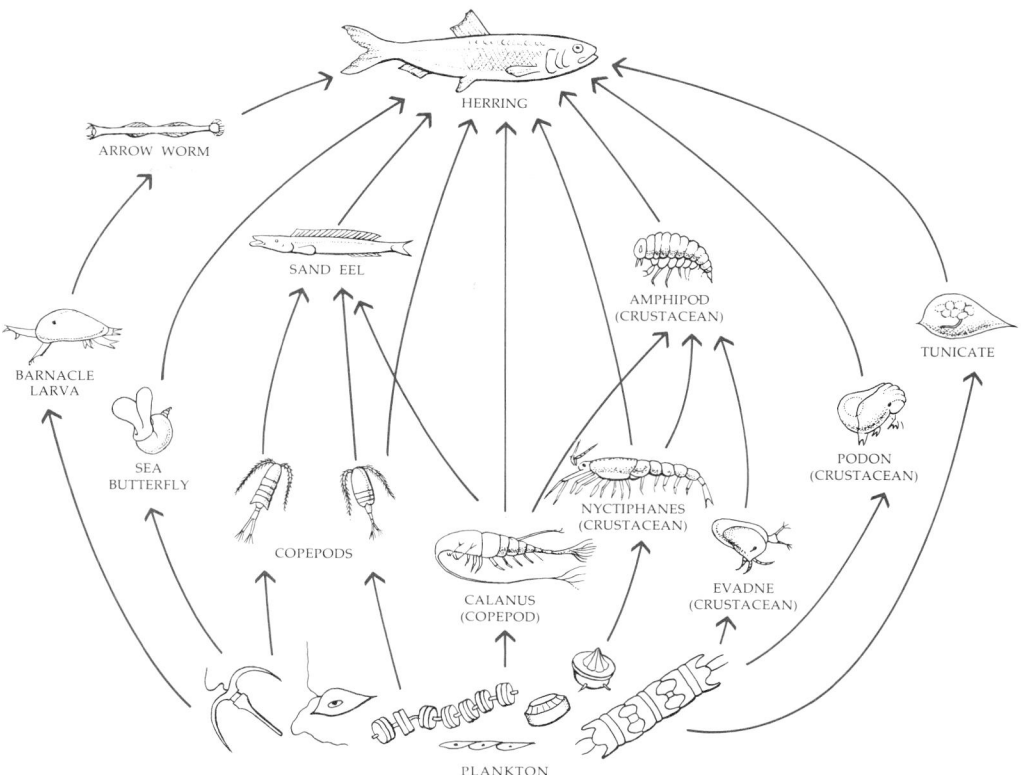

41–4 *The feeding relationships of the adult herring in the North Atlantic illustrate the complex interconnections of a food web.*

Ecological Pyramids

The flow of energy through a food chain is often represented by a graph of quantitative relationships among the various trophic levels. Because large amounts of energy and biomass are dissipated at every trophic level, so that each level retains a much smaller amount than the preceding level, these diagrams nearly always take the form of pyramids. An *ecological pyramid*, as such a diagram is called, may be a pyramid of numbers, showing the numbers of individual organisms at each level; a pyramid of biomass, based either on the total dry weight of the organisms at each level or on the number of calories at each level; or a pyramid of energy flow, showing the productivity of the different trophic levels.

The shape of any particular pyramid tells a great deal about the ecosystem it represents. Figure 41–5a, for example, shows a pyramid of numbers for a grassland ecosystem. In this type of food chain, the primary producers (in this case, the grasses) are small, and so a large quantity of them is required to support the primary consumers (the herbivores).

In a food chain in which the primary producers are large (for instance, trees), one primary producer may support many herbivores, as indicated in Figure 41–5b.

41–5 *Pyramids of numbers for (a) a grassland ecosystem in which the numbers of primary producers (grass plants) is large and (b) a temperate forest in which a single primary producer, a tree, can support a large number of herbivores.*

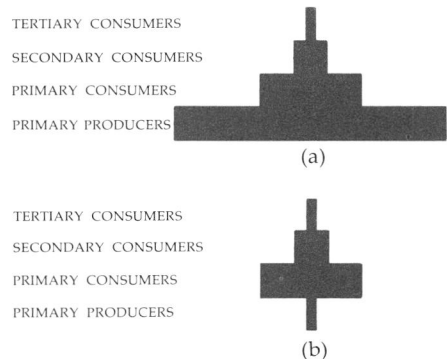

TERTIARY CONSUMERS
SECONDARY CONSUMERS
PRIMARY CONSUMERS
PRIMARY PRODUCERS

(a)

TERTIARY CONSUMERS
SECONDARY CONSUMERS
PRIMARY CONSUMERS
PRIMARY PRODUCERS

(b)

773 *Ecosystems*

Table 41–1 *Trophic Levels in Various Biomes, Indicating Some Major Producers and Consumers*

	Arctic tundra	Antarctic ocean	Taiga	Deciduous forest
Super-carnivores		Killer whales, leopard seals, skuas	Man	Man
Secondary carnivores	Man	Penguins, petrels, Ross seals, Weddell seals, fish	Wolves, lynx, great horned owls, martins, minks	Foxes, wolves, owls, hawks, skunks, raccoons, opossums
Primary carnivores	Arctic wolves, polar bears, snowy owls, ravens, wolf spiders, warble flies	Crabeater seals, blue whales, some seabirds, squids, small fish	Black bears, woodpeckers, nuthatches, thrushes, predatory insects	Tufted titmice, jays, squirrels, chipmunks, turtles, snakes, woodland mice, spiders, salamanders, toads
Herbivores	Ptarmigans, muskoxen, caribou, lemmings, arctic hares, springtails, bumblebees	Krill	Moose, crossbills, spruce grouse, mice, hares, squirrels, porcupines, budworms, other plant-eating insects	Woodchucks, woodland deer, slugs, snails, earthworms, mites, millipedes, insects
Producers	Lichens, mosses, arctic willows	Phytoplankton	Spruces, firs, birches, aspens, algae, mosses, lichens	Maples, beeches, oaks, hickories, algae, mosses, ferns, wildflowers, shrubs

41–6 *Pyramids of biomass for (a) a field in Georgia and (b) the English Channel. Such pyramids reflect the mass present at any one time, hence the seemingly paradoxical relationship between phytoplankton and zooplankton.*

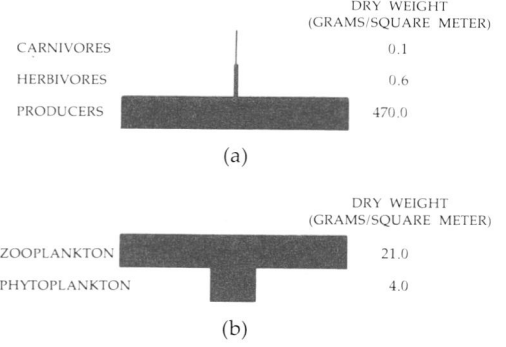

(a)

	DRY WEIGHT (GRAMS/SQUARE METER)
CARNIVORES	0.1
HERBIVORES	0.6
PRODUCERS	470.0

(b)

	DRY WEIGHT (GRAMS/SQUARE METER)
ZOOPLANKTON	21.0
PHYTOPLANKTON	4.0

A pyramid of biomass for a grasslands ecosystem, like the pyramid of numbers for that system, takes the form of an upright pyramid, as shown in Figure 41–6a.

Inverted pyramids of biomass occur only when the producers have very high reproduction rates. For example, in the ocean, the standing crop of phytoplankton may be smaller than the biomass of the animal plankton (zooplankton) that feeds upon it (Figure 41–6b). Because the growth rate of the phytoplankton is much more rapid than that of the zooplankton (animal plankton), a small biomass of phytoplankton can supply food for a larger biomass of zooplankton. Like pyramids of numbers, pyramids of biomass indicate only the quantity of organic material present at any one time; they do not give the total amount of material produced or, as do pyramids of energy, the rate at which it is produced.

Productivity and biomass, although related, are not the same. As we saw, a small biomass of phytoplankton can support a larger biomass of zooplankton because of the phytoplankton's greater productivity. Energy-flow pyramids, which show the productivity relationships of the trophic levels, have therefore the same general shape for every ecosystem. (Remember, as we noted in Chapter

Southern swamp	Prairie	American desert	African forest	Savanna
Man	Man	Man		Man
Raccoons, egrets, herons, water moccasins, alligators, snapping turtles, soft-shelled turtles	Coyotes, wolves, grizzly bears, ferrets, snakes, badgers, owls, hawks, kit foxes	Desert foxes, roadrunners, owls	Man	Lions, leopards, cheetahs, hyenas, vultures, maribou, storks
Waterfowl, warblers, predatory insects, fish, frogs	Ground squirrels, some birds, praying mantises, prairie dogs	Poorwills, cactus wrens, toads, lizards, snakes, woodpeckers	Leopards, crowned eagles, snakes, frogs, lizards	Herons, flamingos, baboons
Insects, insect larvae, snails, shrimps	Pronghorns, bison, mice, horned larks, prairie chickens, grasshoppers, jack rabbits, pocket gophers	Peccaries, quail, kangaroo rats, snails, grasshoppers, moths, ants	Monkeys, lemurs, bongos, okapis, mice, hornbills, parrots, ants, bees, termites, butterflies, moths	Ostriches, elephants, hippos, rhinos, giraffes, zebras, impalas, gnus, topis, fish, grasshoppers, other insects, crustaceans
Phytoplankton, water grasses	Grasses	Joshua trees, ocotillos, saguaros, bunch grasses, creosote bushes, chollas, wildflowers	Woody vines, broad-leaved trees, shrubs	Palms, acacias, euphorbias, phytoplankton, grasses, shrubs

-7 *Pyramid of energy flow for a river system in Florida. This type of pyramid, regardless of the ecosystem under analysis, is never inverted.*

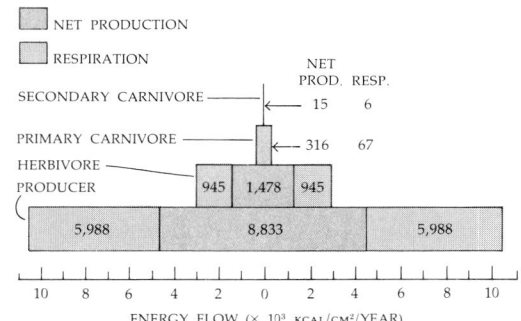

8, that less than half of the total energy of a food molecule is "recaptured" as usable energy when the molecule is broken down. In addition, when speaking of net energy transfer, one has to count the energy costs of other activities, such as obtaining the energy source in the first place.)

In general, of the total amount of energy in the biomass that passes from one trophic level to another in a food chain, only about 10 percent is stored in body tissue—whether the plants are single-celled algae and the consumers are tiny crustaceans or the plants are grasses and the consumers are cows. About 90 percent of the total calories is either unassimilated or "burned" in respiration by the eating animals at that level. Hence the tertiary consumers, the carnivores that eat the carnivores that eat the herbivores, are reduced to approximately $1/10 \times 1/10 \times 1/10 = 1/1,000$ the energy stored in the plants that are eaten. Supercarnivores, which eat these tertiary carnivores, are reduced to one-tenth of this, or 1/10,000 the energy in the plant material. Individuals at the top of the ecological pyramid, such as man, may have to be quite large, as individuals, to capture and eat other animals, but even if they are larger as individuals, they are usually smaller in total number and total biomass. And they are always lower in total captured energy than the organisms they eat.

41–8 *Some common soil animals of the temperate deciduous forest. They play a role in breaking down organic matter and, in turn, as they molt and die, become a part of the biogeochemical cycle.*

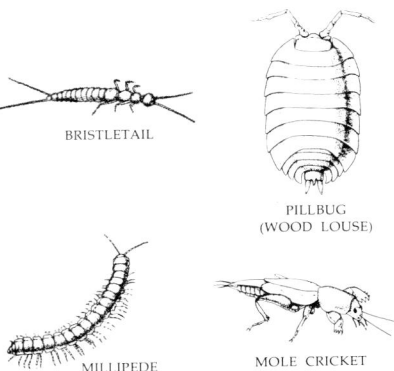

BRISTLETAIL

PILLBUG
(WOOD LOUSE)

MILLIPEDE

MOLE CRICKET

THE CYCLING OF MINERALS

Energy takes a one-way course through an ecosystem, but many inorganic substances cycle through the system. Such substances include water, nitrogen, carbon, phosphorus, potassium, sulfur, magnesium, sodium, and calcium and also a number of minerals, such as cobalt, which are required by some living systems in only very small amounts. These movements of inorganic substances are referred to as _biogeochemical cycles_, because they involve geological as well as biological components of the ecosystem. The geological components are the atmosphere, which is made up largely of gases, including water vapor, but also contains dust and other particles; the lithosphere, the solid crust of the Earth; and the hydrosphere, comprising the oceans, lakes, and rivers, which cover three-fourths of the Earth's surface. The biological components include the producers and consumers, which we have already discussed, and a third and extremely important group, the decomposers. The decomposers, which are primarily bacteria and fungi, break down dead and discarded organic matter, completing the oxidation of the energy-rich compounds formed by photosynthesis. Without their activities, all the living world would soon be inundated with its own waste products. As a result of the metabolic work of the decomposers, these waste products—dead leaves and branches, the roots of annual plants, feces, carcasses, even the discarded exoskeletons of insects—are broken down, and the chemicals of which they are composed are returned to the soil or the water. From the soil or water, they once more enter the bodies of plants and begin their cycle again.

41–9 (a) *Fungal hyphae, entwining what appears to be the remnants of a* Pinus *pollen grain.* (b) *Soil bacteria on a sand grain from a forest floor. In soil composition, the kind of rock from which the soil forms is less important than temperature, moisture, and the accumulated organic matter (humus) broken down by decomposers such as these.*

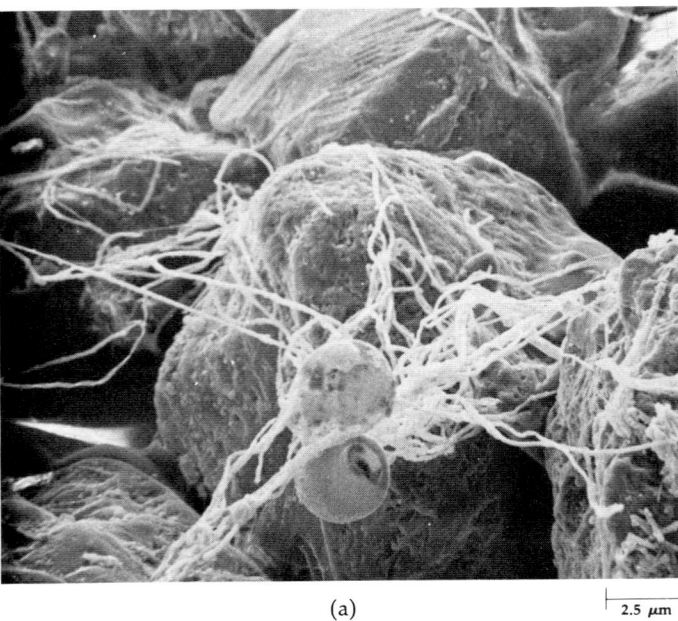

(a)

2.5 μm

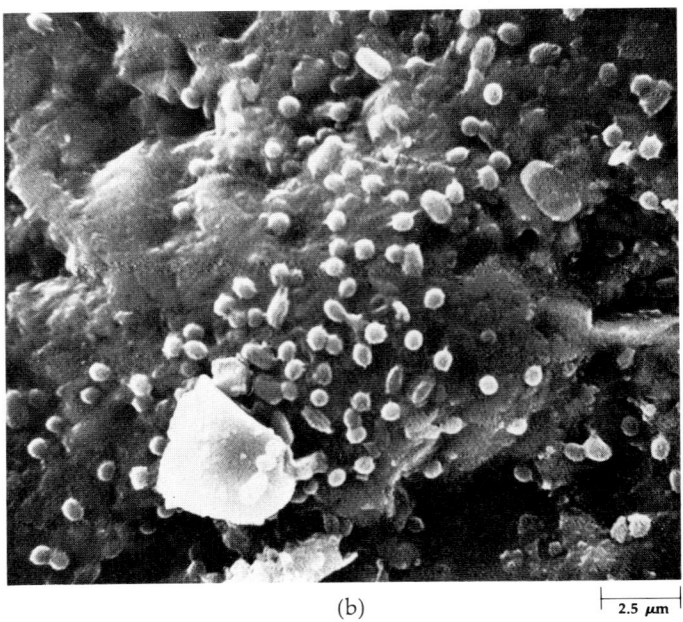

(b)

2.5 μm

CARBON AND OXYGEN CYCLES

Carbon and oxygen have closely interrelated worldwide cycles. The reservoir of both is in the atmosphere; about 21 percent of the atmosphere at sea level is oxygen and about 0.03 percent (or about 320 parts per million) is carbon dioxide. By the process of photosynthesis, land plants incorporate an estimated 20 to 30 billion metric tons of carbon (a metric ton equals 1,000 kilograms or 2,200 pounds) into organic compounds each year. In the oceans, phytoplankton incorporate about 40 billion more. A single carbon atom may move through a great variety of organic compounds and trophic levels before it is returned to the atmosphere by the process of respiration. A small proportion of the circulating carbon may be trapped in organic deposits. Such carbon deposits were laid down in unusual abundance some 300 million years ago when large amounts of plant and other organic materials were buried before they had a chance to decompose; this organic carbon, transformed by time and geological processes into coal, oil, and natural gas, is now being burned as fuel and returned to the atmosphere as carbon dioxide.

Because the oceans take up or give off carbon dioxide depending upon the relative quantities of the compound in the atmosphere and the water, the carbon dioxide content over the surface of the globe tends to remain relatively constant, despite diurnal, seasonal, and geographic variations in carbon dioxide production and the great recent increase in the burning of fossil fuels.

The oxygen of the atmosphere is constantly being renewed by the photosynthetic processes of green plants and phytoplankton. Free oxygen first appeared in the atmosphere about 2 billion years ago as a consequence of the activities of oceanic phytoplankton and began to approach its present concentration only about 75 million years ago.

Free oxygen is provided by decomposition of water molecules by the process of photosynthesis; the 1.5 billion cubic kilometers of water on the Earth are split by photosynthesis about once every 2 million years and reconstituted by respiration. Thus, oxygen gas and carbon dioxide are part of a continuous global cycle, involving atmosphere, hydrosphere, and biosphere.

41–10 *Diagrams of three major soil types. The litter of the northern coniferous forest (center) is acid and slow to decay, and the soil has little accumulation of humus, is very acid, and is leached of minerals. In the cool, temperate deciduous forest, decay is somewhat more rapid, leaching less extensive, and the soil more fertile. Such soils have been used extensively for agriculture, but they need to be prepared by lime (to reduce acidity) and fertilizer. In the grasslands, almost all of the plant material above the ground dies each year as do many of the roots, and so much organic matter is constantly returned to the soil. In addition, the finely divided roots penetrate the soil extensively. The result is highly fertile soil, often black in color, with a topsoil sometimes 3 to 4 feet deep.*

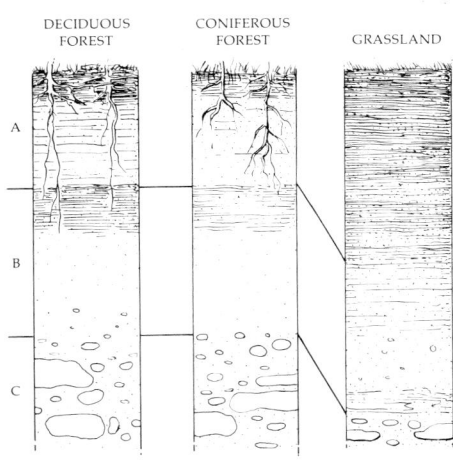

DECIDUOUS FOREST CONIFEROUS FOREST GRASSLAND

Table 41–2 *Soil Classification*

	Diameter of fragments (micrometers)
Coarse sand	200–2,000 (0.2–2 millimeters)
Sand	20–200
Silt	2–20
Clay	Less than 2

Soils and Mineral Cycles

Soil, the uppermost layer of the Earth's crust, is composed of weathered rock associated with organic material, both living and in various stages of decomposition. It typically has three layers: The A horizon, the B horizon, and the C horizon. (See Figure 41–10.)

The A horizon, or topsoil, is the zone of maximum organic accumulation (humus). The B horizon, or subsoil, consists of inorganic particles in combination with mineral nutrients that have leached down from the A horizon. The C horizon is made up of loose rock that extends down to the bedrock beneath it.

The mineral content of the soil depends in part on the parent material from which the soil is formed. These soil differences may be extremely localized—as compared to factors such as temperature and precipitation—with sharp lines of demarcation. Plants are sometimes used by experienced field geologists as indicators of mineral deposits. Copper moss (*Merceya latifolia*) is abundant only in soils with a high copper concentration. Iron deposits often cause stunting, cobalt may cause white spots on leaves, and unusually large amounts of molybdenum can make leaves turn yellow-orange.

On most soils, the mineral content is more dependent on the biotic component of the ecosystem. In an undisturbed environment, as we noted earlier, most of the mineral nutrients stay within the ecosystem. If, however, the vegetation is repeatedly removed, as when crops are harvested or grasslands are overgrazed, or the top, humus-rich layer of the A horizon is eroded, the soil rapidly becomes depleted and can be used for agricultural purposes only if it is heavily fertilized.

A third factor influencing the mineral content of soils is the soil composition. The smaller fragments of rock are classified as sand, silt, or clay, according to size (see Table 41–2). Water and minerals drain rapidly through soil composed of large particles (sandy). Soil composed of small particles (clay) holds the water against gravity. Moreover, the small clay particles are negatively charged and so hold positively charged ions such as calcium (Ca^{2+}), potassium (K^+), and magnesium (Mg^{2+}). However, a pure clay soil is not suitable for plant growth because it is usually too tightly packed to let in enough oxygen for the respiration of plant roots, soil animals, and most soil microorganisms. Clay soils that contain enough large particles to keep the soil from packing are known as loams, and these are generally the best soils for plant growth.

The pH of the soil also affects its capacity to retain minerals. In an acid soil, hydrogen ions replace other positively charged ions clinging to clay particles, and these nutrient ions easily leach out of the soil. Soil pH also affects the solubility of certain nutrient elements. Calcium, for example, increases in solubility (and therefore in its availability to plant roots) as pH increases. Crops such as alfalfa, sweet clover, and other legumes, have high calcium requirements and so grow well only in an alkaline soil. Rhododendrons and azaleas, on the other hand, have high requirements for iron, which is abundant only when the soil is acidic.

Soils and plant life interact. Plants constantly add to the humus, thereby changing not only the content of the soil but also its texture and its capacity to hold minerals and water. In turn, the plants are dependent upon the mineral content of the soil and its holding capacity. As these improve, plants increase in biomass and also often change in kind, thereby producing further changes in the soil. Thus, under natural conditions, the soil is constantly changing its composition.

1–11 *Prairie soil. Note how the roots of the grasses bind the topsoil.*

The Nitrogen Cycle

Nitrogen makes up 78 percent of our atmosphere. Since most living things, however, cannot use elemental atmospheric nitrogen to make amino acids and other nitrogen-containing compounds, they must depend on nitrogen contained in soil minerals. So despite the abundance of nitrogen in the biosphere, a shortage of nitrogen in the soil is often the major limiting factor in plant growth. The process by which this limited amount of nitrogen is circulated and recirculated throughout the world of living organisms is known as the *nitrogen cycle.* The three principal stages of this cycle are (1) ammonification, (2) nitrification, and (3) assimilation.

Much of the nitrogen found in the soil reaches it as a result of the decomposition of organic materials and is in the form of complex organic compounds, such as proteins, amino acids, nucleic acids, and nucleotides. However, these nitrogenous compounds are usually rapidly decomposed into simple compounds by soil-dwelling organisms. Certain soil bacteria and fungi are mainly responsible for the decomposition of dead organic materials. These microorganisms use the protein and amino acids as a source of their own needed proteins and release the excess nitrogen in the form of ammonia (NH_3) or ammonium (NH_4^+). This process is known as *ammonification.*

41–12 *The nitrogen cycle. Nitrogen is returned to the soil in the form of dead tissues, feces, and urine.*

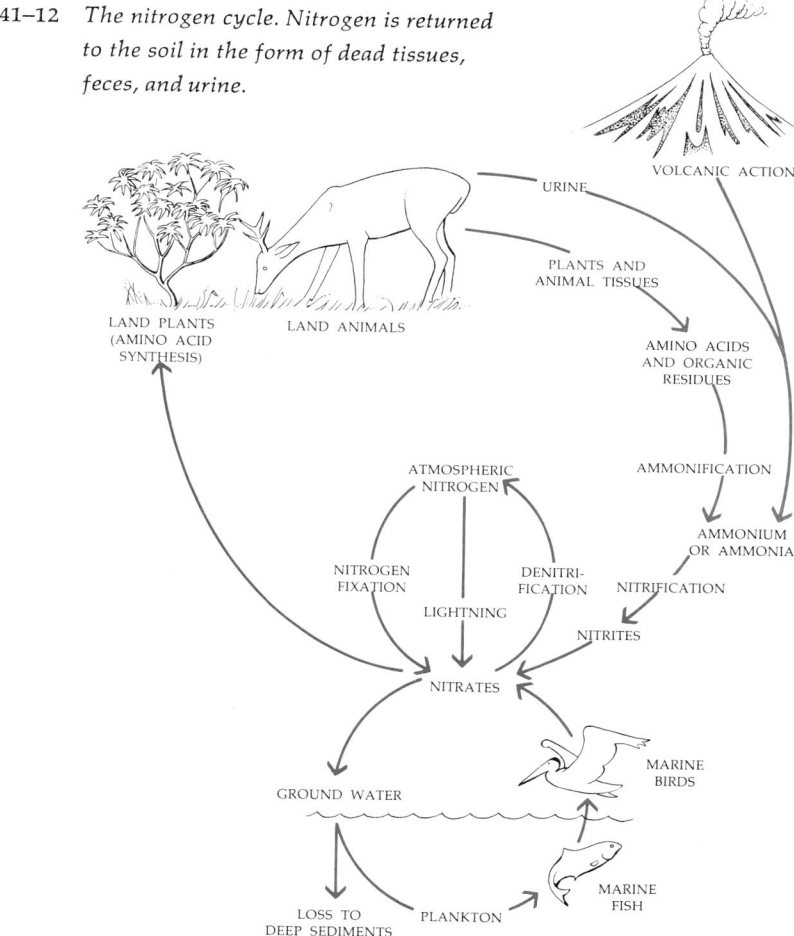

Several species of bacteria common in soils are able to oxidize ammonia or ammonium. The oxidation of ammonia or ammonium, known as *nitrification*, is an energy-yielding process, and the energy released in the process is used by these bacteria as their primary energy source. One group of bacteria oxidize ammonia (or ammonium) to nitrite (NO_2^-):

$$2NH_3 + 3O_2 \longrightarrow 2NO_2^- + 2H^+ + 2H_2O$$

Nitrite is toxic to higher plants, but it rarely accumulates. Members of another genus of bacteria oxidize the nitrite to nitrate, again with a release of energy:

$$2NO_2^- + O_2 \longrightarrow 2NO_3^-$$

Although plants can utilize ammonium directly, nitrate is the form in which most nitrogen moves from the soil into the roots.

Once the nitrate is within the cell, it is reduced back to ammonium. In contrast to nitrification, this assimilation process requires energy. The ammonium ions thus formed are transferred to carbon-containing compounds to produce amino acids and other nitrogen-containing organic compounds.

Nitrogen Fixation

The nitrogen-containing compounds of green plants are returned to the soil with the death of the plants (or of the animals which have eaten the plants) and are reprocessed by soil organisms and microorganisms, taken up by the plant roots in the form of nitrate dissolved in the soil water, and reconverted to organic compounds. In the course of this cycle, a certain amount of nitrogen is always "lost," in the sense that it becomes unavailable to the land plants.

41–13 (a) *Nitrogen-fixing nodules on the roots of a bird's-foot trefoil, a legume. These nodules are the result of a symbiotic relationship between a soil bacterium* (Rhizobium) *and root cells.* (b) *Tip of a root hair of a clover seedling, with several rhizobia and some soil particles.*

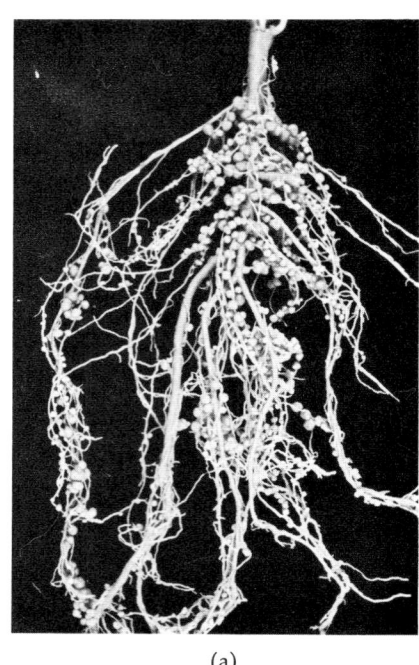

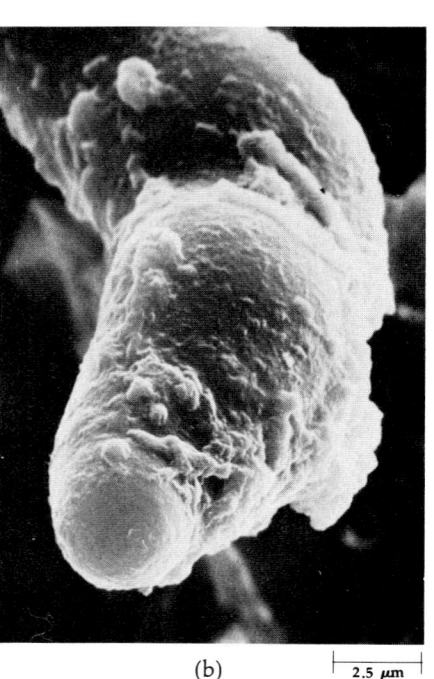

(a) (b) 2.5 μm

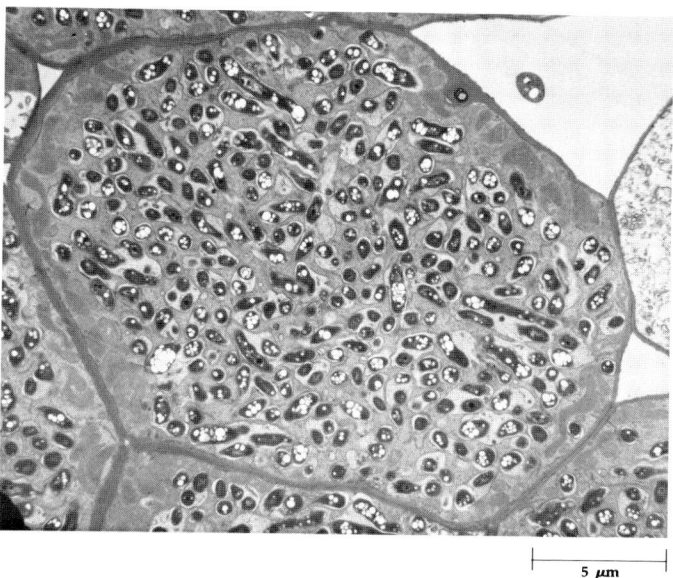

5 μm

The main source of nitrogen loss is the removal of plants from the soil. Soils under cultivation often show a steady decline of nitrogen content. Nitrogen may also be lost when topsoil is carried off by soil erosion or when ground cover is destroyed by fire. Nitrogen is also leached away by water percolating down through the soil to the groundwater. In addition, numerous types of bacteria are present in the soil that, when oxygen is not present, can break down nitrates, releasing nitrogen into the air and using the oxygen for the oxidation of carbon compounds (respiration). This process, known as denitrification, takes place in poorly drained (hence, poorly aerated) soils.

As you can see, if the nitrogen lost from the soil were not steadily replaced, virtually all life on this planet would finally flicker out. The "lost" nitrogen is returned to the soil by *nitrogen fixation.*

Nitrogen fixation is the process by which gaseous nitrogen from the air is incorporated into organic nitrogen-containing compounds and thereby brought into the nitrogen cycle. It is carried out to a small extent by abiotic processes, such as the production of nitrogen oxides by interactions between nitrogen and other atmospheric gases; lightning is a possible energy source. Most nitrogen fixation, however, is carried out by a few types of free-living microorganisms—chief among which are the blue-green algae and some free-living bacteria—and by symbiotic combinations of bacteria and higher plants. And just as all organisms are ultimately dependent on photosynthesis for energy, they all depend on nitrogen fixation for their nitrogen.

One hundred million tons of nitrogen are added to the Earth's surface each year, of which 90 million tons are biological in origin. (The other 10 percent is largely in the form of chemical fertilizers.) Of the various classes of nitrogen-fixing organisms, the symbiotic bacteria are among the most important in terms of total amounts of nitrogen fixed. The most common of the nitrogen-fixing symbiotic bacteria is *Rhizobium*, a type of bacterium that invades the roots of leguminous plants such as clover, peas, beans, vetches, and alfalfa.

Phosphorus and Other Minerals

The chief reserve supply of nitrogen is in the air. The other elements that cycle through ecosystems are found in the largest amounts in the soil. Such elements are said to undergo sedimentary cycles, as distinguished from the atmospheric cycles of gases. An example of a sedimentary cycle is the phosphorus cycle, diagrammed in Figure 41–15. As we saw in Section 1, phosphorus is directly involved in the transfer of energy in biological systems. It is also a component of nucleic acids, both DNA and RNA. In these capacities, it is required by every living system. Many kinds of rocks contain phosphorus, and when such rocks are eroded by water, minute amounts of phosphorus dissolve and become available to plants and so enter the biogeochemical cycle. A certain amount of phosphate runs off each year, eventually finding its way to the oceans, where most of it is lost to terrestrial ecosystems. This "lost" phosphorus, in natural systems, is replaced by the further release of phosphorus from the rocks beneath the soil.

41–15 *The phosphorus cycle.*

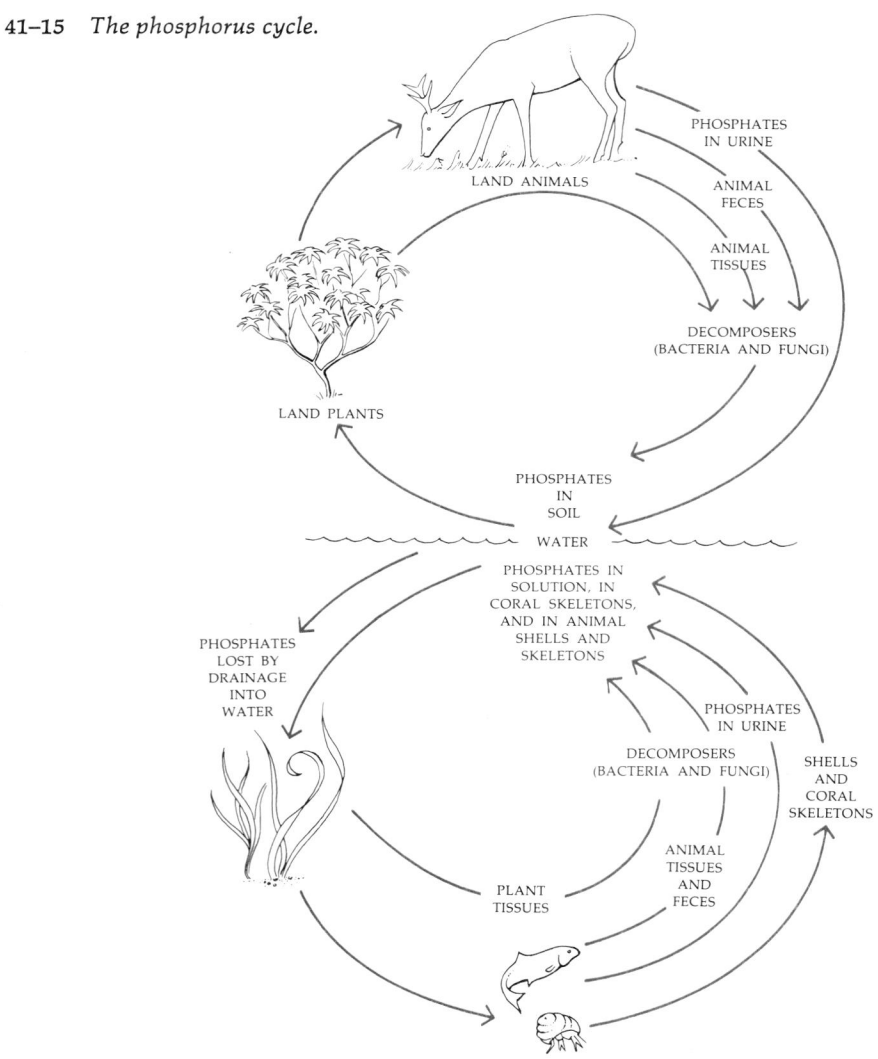

Concentration of Elements

Many of the elements needed by living systems are present in the tissues of these systems in far higher concentrations than in the surrounding air, soil, or water. This concentration of elements comes about as a result of selective uptake of such compounds by living cells, amplified by the channeling effects of the ecological pyramids of biomass and energy. Under natural circumstances, this concentration effect is usually valuable; animals usually have a greater requirement for minerals than do plants because so much of the biomass of plants is cellulose. In man-made environments, however, the results of concentration can be surprising and even catastrophic. We have an example in strontium-90, a by-product of the testing of atomic weapons, which will be discussed in Section 8.

LIMITING FACTORS

Of the 92 elements naturally present on Earth, living organisms require six in relatively large amounts (CHNOPS) and a number of others in small amounts. An organism can grow and multiply only if all of these requirements are met; a living thing is no stronger than the weakest link in its ecological chain of requirements. This principle was first expressed for plants in 1840 by Justus Liebig. His statement that "growth of a plant is dependent on the amount of foodstuff which is presented to it in minimum quantity" has come to be known as Liebig's *law of the minimum.* This law has become somewhat modified by studies that show that minimum requirements can be modified by interactions of available factors. For example, when calcium is deficient and strontium abundant, organisms can substitute one for the other.

Growth may be limited not only by too little of a substance but also by too much. In other words, living organisms have a range of tolerance and can survive only within the limits of this range (see Figure 41–16). Particular organisms typically have different ranges of tolerance for different elements in their environment; thus an organism that can tolerate wide fluctuations in the availability of calcium may have a very narrow range of tolerance for sodium chloride. The particular nutrient element that determines whether an organism can grow in a given environment is known as a *limiting factor.*

The ways in which limiting factors affect growth of organisms has been well illustrated by some recent pollution problems. For instance, phosphorus is required by plants as a macronutrient and is often a limiting factor in plant growth. When phosphorus is the limiting factor in the growth of freshwater algae in a lake or slow-moving stream, phosphate-containing detergents, added to the water in the outfall of sewage systems, will produce a spectacular bloom of algae. This period of rapid growth continues until the available supply of another essential element—perhaps calcium—is consumed. Then the algae begin to die, the decomposing bacteria take over, and the respiration of the bacteria begins to use up the oxygen in the water. Eventually the amount of oxygen present drops below the tolerance of fish and other organisms, and these too begin to die. If the process is not interrupted, the lake becomes completely stagnant, and only bacteria and other microorganisms can survive in it.

–16 *The principle of limiting factors. Every species has a characteristic limiting factor curve for each factor in its environment. The three critical points on the curve are the lower tolerance limit, the optimum concentration, and the upper tolerance limit.*

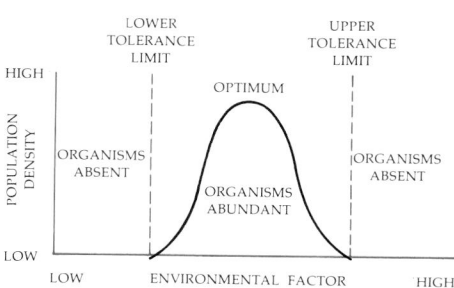

Ducks vs. Oysters

Shellfishing is one of the important industries of Long Island, New York. In the 1950s, fishermen began to notice a decline in the number of oysters available in Long Island's Great South Bay. The demise of the oysters was traced to the operation of duck farms along the island's south shore. Fertilization of the waters by duck manure was causing a great increase in phytoplankton. Because the oysters live on phytoplankton, this increase might appear beneficial at first glance. However, because the nitrogen-phosphorus ratio of duck manure is low, nitrogen became a limiting factor. As an unexpected consequence, the normal phytoplankton of the area became almost completely replaced by very small green flagellates, so rare that they were completely unfamiliar to most marine biologists. Whereas the diatoms and other members of the normal phytoplankton community use nitrogen in the form of nitrates, the new green flagellates could use it in the form of urea, uric acid, or ammonia, and so they were able to short-circuit the nitrogen cycle, thereby cutting off the other organisms' supply of this critical factor. Oysters, which cannot utilize the unusual green flagellates for food, starved to death in the midst of this abundance.

Now strict antipollution ordinances have forced most of Long Island's duck farms to close down. The land they once occupied is being taken over by real estate interests, which have a wide range of tolerance for most environmental factors and which may be expected, in turn, to create their own pollution problems.

Although the concept of limiting factors, as originally formulated, applied solely to nutrient elements, it is now used by many biologists to describe any environmental factor, such as temperature, light, available water, or nesting spaces, that determines whether a particular group of organisms can survive in a particular region.

Recycling in a Forest Ecosystem

As we noted in the preceding chapter, in our discussion of the tropical rain forest biome, most of the nutrients of the tropical forest are present in the bodies of the plants, animals, and other organisms that make up the rain forest community, and little is present in the soil. Recent studies of an ecosystem in a deciduous forest have shown that the plant life of this community also plays a major role in retaining the nutrient elements. The studies were made in the Hubbard Brook Experimental Forest in the White Mountain National Forest in New Hampshire. The investigators first established a procedure for determining the mineral budget—input and output, "profit" and "loss"—of areas in the forest. By analyzing the content of rain and snow, they were able to estimate input, and by constructing concrete weirs that channeled the water flowing out of selected areas, they were able to calculate output. (A particular advantage of the site is that bedrock is present just below the soil surface so that little material leaches downward.) They discovered, first, that the natural forest was extremely efficient in conserving its mineral elements. For example, the annual net loss of calcium from the ecosystem was 9.2 kilograms per hectare (a hectare is about 2½ acres). This represents only about 0.3 percent of the calcium in the system. In the case of nitrogen, the ecosystem was actually accumulating this

41–17 *Weir at Hubbard Brook Experimental Forest in New Hampshire. Water from each of six experimental ecosystems was channeled through a weir, such as this one built where the water leaves the watershed, and was analyzed for chemical elements. The watershed behind the weir in this photograph has been stripped of vegetation. The experiments showed that such deforestation greatly increased the loss of elements from the system.*

element at a rate of about 2 kilograms per hectare per year. There was a similar, though somewhat smaller, net gain of potassium.

In the winter of 1965–1966, all of the trees, saplings, and shrubs in one 15.6 hectare area of the forest were cut down completely. No organic materials were removed, however, and the soil was undisturbed. During the following summer, the area was sprayed with a herbicide to inhibit regrowth. During the four months from June through September, 1966, the runoff of water from the area was 4 times higher than in previous years. Net losses of calcium were 10 times higher than in the undisturbed forest, for example, and potassium 21 times higher. The most severe disturbance was seen in the nitrogen cycle. Plant and animal tissues continued to be decomposed to ammonia or ammonium, which then were acted upon by nitrifying bacteria to produce nitrates, the form in which nitrogen is usually assimilated by higher plants. However, no higher plants were present, and the nitrate, a negatively charged ion, was not held in the soil. Net losses of nitrate nitrogen averaged 120 kilograms per hectare per year from 1966 to 1968. As a side effect, the stream draining the area has become polluted with algal blooms, and its increase in nitrate concentrations now exceeds the levels established by the U.S. Public Health Service for drinking water.

ECOLOGICAL SUCCESSION

If land is laid bare as the result of a landslide, erosion, excavation, the eruption of a volcano, the retreat of a glacier, the rising of a new island from the sea, or some other such phenomenon, it will, if the environment is not too harsh, slowly become covered with vegetation and its accompanying animal life. The vegetation that initially colonizes the bare land is usually replaced in the course of time by a second type, which gradually crowds out the first and which itself may eventually be replaced. In the southeastern United States, where these stages of replacement have been studied by ecologists over a number of years, it has been found that such communities of living organisms replace one another in a predictable and orderly sequence. This process of community change is known as _ecological succession_.

The process of ecological succession is carried out by the living organisms themselves. Each temporary community changes the local conditions of temperature, light, humidity, soil content, and other abiotic factors, and so sets up favorable conditions for the next temporary community. When the site has been modified as much as possible (here the limits are set by the environment), succession ceases, or at least slows down considerably. The final community is known as the _climax community_. In a sense, the climax community, too, is in a state of constant change; individual organisms die, and their places are taken by new individuals. However, these individuals will probably be of the same sort and of the same species as the ones being replaced. Some ecologists maintain that further changes in the character of the community will take place only if the community is disturbed or if the climate undergoes a drastic change (as in the Ice Ages). Others, pointing out that there is no proof of this hypothesis, hold that succession continues in a mature community, but at a much slower rate.

(a) (b)

41–18 (a) *Yapoah Crater, a volcanic cinder cone east of the Cascade Mountains in central Oregon. Succession leading to the establishment of climax forest on such a cone may take centuries and may often be interrupted by further volcanic activity long before it is complete.* (b) *An early stage of succession on a rocky slope. Lichens have begun to accumulate soil, and a bladder fern has sprung up in a small crevice.*

The physical characteristics of the environment determine the nature of the mature community. In regions where conditions are particularly unfavorable, such as the tundra, the process of succession involves relatively few stages and the climax community is correspondingly simple. Where physical conditions are less limiting, the mature community is rich and diverse.

Primary Succession

The occupation by plants of an area not previously covered by vegetation is known as *primary succession*. Rocks and cliffs are common sites of primary succession. The first stage in such areas is the formation of soil. The solid rock is broken down by weathering processes, such as freezing and thawing or heating and cooling, which cause substances in the rocks to expand and contract, thus splitting the rocks apart. Water and wind exert a scouring action that breaks the fragmented rock into smaller particles, often carrying the fragments great distances. Water enters between the particles, and soluble materials such as rock salt dissolve in the water. Water in combination with carbon dioxide from the air forms a mild acid which dissolves substances that will not dissolve in water alone. Chemical reactions that contribute to the disintegration of the rock begin to take place. Soon, if other conditions such as light and temperature permit, bacteria, fungi, and then small plants begin to gain a foothold. Growing roots split rock particles, and the disintegrating bodies of the plants and those of the animals associated with them add to the accumulating material. Finally, the larger plants move in, anchoring the soil in place with their root systems, and a new community has begun.

PRIMARY SUCCESSION

An example of primary succession: a pond gradually becomes woodland. The first forms of life to appear are plankton (stage 1). As plankton multiply and die, a layer of ooze develops on which bottom plants can gain a hold (stage 2). These plants bind the bottom soil together and add to the deposition of organic matter so that the pond becomes increasingly shallow. Bottom plants are replaced by plants whose leaves and flowers float on the pond's surface (stage 3). These overshadow the bottom dwellers, which are slowly eliminated. As the pond grows increasingly shallow and the bottom increasingly stable, cattails and bullrushes replace the water plants, after which the bottom fills in even more rapidly. The pond is now a marsh. As the bottom continues to rise, the pond exists only temporarily (stage 4), drying in the summers and freezing completely in the winters. As the land builds higher, drainage improves, and meadow grasses gain a foothold. Finally, woody plants move in and the pond becomes indistinguishable from the area that surrounds it (stage 5).

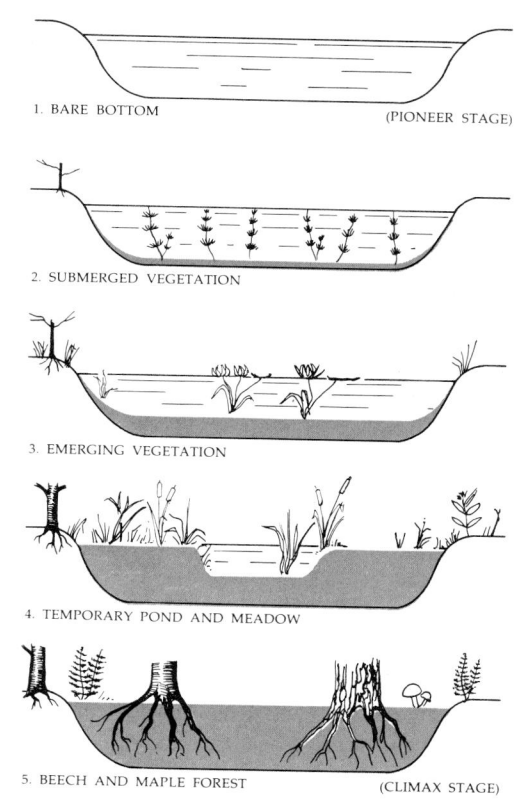

1. BARE BOTTOM   (PIONEER STAGE)

2. SUBMERGED VEGETATION

3. EMERGING VEGETATION

4. TEMPORARY POND AND MEADOW

5. BEECH AND MAPLE FOREST   (CLIMAX STAGE)

Emerging vegetation grows along the edge of a pond.

Plants with surface-floating leaves grow across the surface of a pond, choking out bottom-dwelling plants.

Marsh grasses, sedges, and cattails growing on an old pond bed.

787 *Ecosystems*

41–19 *Sand dune near Lake Michigan with beach grass in the foreground. Later successional stages can be seen on the more remote dunes.*

Lake Michigan Sand Dune Succession

Primary succession has been particularly well studied on the sand dunes of Lake Michigan. The dunes studied are part of beach systems running parallel to the shoreline and about 25, 40, and 55 feet above the level of the lake, which has been falling with the retreat of the glaciers. Because the soil, the climate, and the plant and animal life of the surrounding area are basically the same, these dunes have offered an unusual opportunity to study changes associated with time alone.

Bare sand surface is colonized at first by grasses, such as Marram grass, which propagates almost exclusively by rhizomes. It can stabilize a bare area in as little as six years. Before a sand dune is so stabilized, it shifts continually and no other plant life can gain a foothold. Once the sand dune is stabilized, the Marram grass dies out for some unknown reason.

The first tree to appear is usually the cottonwood, after which the dune is likely to be invaded quickly—in about 50 to 100 years—by pines. Black oak replaces the pines, appearing in about 100 to 150 years. As this occurs, the understory changes, with shade-tolerant shrubs replacing the light-requiring shrubs associated with the pine forests. The oldest dunes (about 12,000 years old) are still populated with black oaks and associated vegetation, and many ecologists doubt that they will ever reach the beech-maple stage, which represents the climax community for that area.

During the early stages of succession, the originally very alkaline pH drops rapidly as carbonates are leached from the soil very quickly. Nitrogen and organic carbon also increase markedly in these early stages. Most of the soil changes in the original barren dune sand occur within about 1,000 years.

Secondary Succession

When succession occurs on a site previously occupied by vegetation, it is known as *secondary succession.* Secondary succession commonly occurs on areas laid bare by man, such as abandoned farmland and strip mines, roadsides, and landfills. (As you can see, the distinctions between primary and secondary succession are not biologically clear cut, and some ecologists recommend that the terms be dropped.) Such an open area is bombarded by the seeds of numerous plants and is captured by those that can germinate most quickly. In an open field, these are the plants that can survive the sunlight and drying winds—weeds and grasses and such trees as cedars, white pines, poplars, and birch. For a while, these plants are dominant, but eventually they eliminate themselves because their seedlings cannot compete in the shade cast by the parent trees.

41–20 *Secondary succession. A cleared field is first invaded by weeds, grasses, and seedling trees. The first trees to overtop the other vegetation are species that require full sunlight. Once they are mature, they die out—their own seedlings cannot grow in the shade they produce. They are gradually replaced by trees whose seedlings thrive in shade. These trees are generally taller, longer-lived, and more varied in species than the pioneer types. (After "The Tropical Rain Forest" by Paul W. Richards. Copyright © December 1973 by Scientific American, Inc. All rights reserved.)*

-21 *Secondary succession. A young stand of loblolly pine is taking over an abandoned field in Wake County, North Carolina. The plow furrows are still clearly visible. The field was abandoned perhaps 20 years ago.*

41–22 (a) *The seedling of a red maple rising above needles of white pine. Mature white pines filter the light so that their own seedlings cannot survive and only those tolerant of shade, such as maple and oak seedlings, can gain a foothold.* (b) *Seedling trees of balsam fir growing up under and replacing quaking aspen in northern Minnesota—a stage in forest succession leading to a climax community of white spruce and balsam fir.*

(b)

(a)

Some seedlings can thrive in partial shade, however—for instance, oaks, red maples, white ash, and tulip trees. As the forest matures, these trees grow tall, and finally they shut out so much light that even *their* seedlings cannot grow. Eventually, the only young trees that can grow in the forest are those that can survive in the dimmest light, such as the hemlock, beech, and sugar maple, and these ultimately take over the forest. Nothing else can compete with them in the conditions that have been established. This is the climax forest.

Fire

In some areas, fire plays an important part in determining the final stage in forest succession. Young seedlings of deciduous trees are very susceptible to fire, while pines are resistant to it. Jack pines, in fact, open their cones and release their seeds only after they have been heated, so they tend to spread most readily after a fire. In the southern pine forests and the northern lake area of the United States, recurrent ground fires maintain the pinewoods, keeping the forest perpetually young. Similarly, the forests of redwoods and giant sequoias on the western coast of the United States are maintained by fires, which destroy competitive trees that are faster growing and less fire resistant than the redwoods and lay bare the ground so the small seeds can germinate. These light ground fires, characteristic of the fire-type ecosystem, are distinctly different from the uncontrolled crown fires of northern forests, which spread through the treetops, destroying entire communities of plants and animals and leaving the ground barren. Also, there is now good evidence that, in prehistory, grass fires played a significant role in determining the distribution and age characteristics of various grass species in the central prairies or plains of the United States and southern Canada.

-23 *When fire sweeps through a forest, secondary succession—with regeneration from nearby unburned stands of vegetation—is initiated. Some plants produce sprouts from the stumps; others seed abundantly on the burn. In one group of pines, the closed-cone pines, the cones do not open to release their seeds until they have been heated by fire.*

Immature and Mature Ecosystems

As ecosystems pass through the various stages of succession, one type of plant and animal community is replaced by another, distinctly different type. Although the changes that take place as the ecosystem matures differ in detail as to the exact species involved and the rates of change, many ecologists believe that they all have certain results in common.

First, there is an increase in total biomass. Compare, for example, a recently abandoned field, which is an immature ecosystem, with a deciduous forest, which is a more mature one.

Second, there is a decrease in net productivity in relation to biomass (standing crop) or to respiration. In other words, the biomass of a mature system does not tend to increase with time, as does that of a system in earlier stages of succession.

Third, as we would expect from the Hubbard Brooks experiments, mature systems have a greater capacity to entrap and hold nutrients within the system.

Fourth, during the early stages of succession, the diversity of species increases. (There is some indication that, later in succession, the number of species in the area evens off or may even decrease slightly. Complete species counts are difficult to obtain.) There is also a general increase in the size of organisms, the lengths of their lives, and the complexity of their life cycles.

Fifth, it is hypothesized that the ecosystem becomes more stable as it matures. Mature ecosystems show a greater capacity for homeostasis and are less likely to be drastically affected by changes in the environment.

SUMMARY

An ecosystem, or ecological system, is a unit of biological organization made up of all the organisms in a given area and the environment in which they live. It is characterized by interactions between the living (biotic) and nonliving (abiotic) components that result in (1) a flow of energy from the sun through autotrophs to heterotrophs and (2) a cycling of minerals and other inorganic materials.

Within an ecosystem, there are trophic (feeding) levels. All ecosystems have at least two such levels: autotrophs, which are plants or photosynthetic algae, and herbivores, which are usually animals. The autotrophs, the primary producers in the ecosystem, convert a small proportion (above 1 percent) of the sun's energy into chemical energy. The herbivores, which eat the autotrophs, are the primary consumers. A carnivore which eats the herbivore is a secondary consumer, and so on. About 10 percent of the energy transferred at each trophic level is stored in body tissue. There are seldom more than five links in a food chain.

The movements of water, carbon, nitrogen, and minerals through ecosystems are known as biogeochemical cycles. In such cycles, inorganic materials from the air, water, or soil are taken up by primary producers, passed on to consumers, and eventually transferred to decomposers, chiefly bacteria and fungi. The decomposers break down dead and discarded organic material and return it to the soil or water in a form in which it can be used again by the primary producers. Some characteristics of soil affect the presence, retention, and recycling of minerals. These characteristics include the rock from which the soil was formed, the presence of humus on the soil surface, and soil composition and pH. Positively charged ions are held in soil by negatively charged clay particles. Soils and plant life interact. Plants increase the availability of minerals in the soil and affect soil texture; these changes, in turn, improve the soil's capacity to maintain plant life, leading to a further increase in the humus content.

The cycling of nitrogen from the soil, through the bodies of plants and animals, and back to the soil again is known as the nitrogen cycle. It involves several stages. Nitrogen reaches the soil in the form of organic material of plant and animal origin. This material is decomposed by soil organisms. Ammonification, the breakdown of nitrogen-containing molecules to ammonia (NH_3) or ammonium (NH_4^+), is carried out by certain soil bacteria and fungi. Nitrification is the oxidation of ammonia or ammonium to form nitrites and nitrates; these steps are carried out by two different types of bacteria. Nitrogen enters plants almost entirely in the form of nitrates. Within the plant, nitrates are reduced to ammonium with a carbon-containing group. Nitrogen-containing organic compounds are eventually returned to the soil, principally through death and decay, completing the nitrogen cycle.

Nitrogen is lost from the soil by harvest, erosion, fire, leaching, and denitrification. Nitrogen is increased in the soil by nitrogen fixation, which is the incorporation of elemental nitrogen into organic components. Biological nitrogen fixation is carried out entirely by microorganisms, including blue-green algae, or by a symbiotic association of bacteria (*Rhizobium*) and leguminous plants.

An organism has certain requirements for inorganic nutrients, and growth and reproduction cease if any one of these is absent. Also, any such substance

present in excessive amounts may be inhibitory or toxic. Each organism has a range of tolerance within which it can grow. This range differs for different nutritional elements and also for other environmental factors. The element that determines whether a particular organism can grow in a particular area is known as a limiting factor.

Ecological succession is an orderly sequence of changes in the type of vegetation and other organisms on a particular site. Primary succession occurs in areas previously unoccupied by living organisms, whereas secondary succession occurs in disturbed areas where vegetation has been removed. Both types of successions result from modifications produced by the living organisms themselves and culminate in the establishment of a climax community, which is the characteristic community for that area.

Maturation in ecosystems is accompanied by increases in biomass and in number of species. The mature system is characterized by greater stability than is present in the immature system.

QUESTIONS

1. Define the following terms: trophic level, ecological pyramid, biomass, B horizon, nitrification, ecological succession.
2. Is it possible to have an inverted pyramid of energy transfer? Explain.
3. Describe what happens to the energy in light striking a temperate forest ecosystem. What happens when it strikes a cornfield? A pond? A field on which cattle are grazing?
4. Describe what happens to a nutrient mineral in each of those areas.
5. What are the characteristics of a mature ecosystem? Using these characteristics as criteria, how would you classify land under agriculture in terms of ecosystem maturity? On the basis of this answer, what would you expect to be some of the consequences of intensive agriculture?

Chapter 42

Communities and Populations

At the beginning of this section, we defined ecology as the scientific study of the interactions that determine the distribution and abundance of organisms. In Chapter 40, we examined broad factors, such as light, temperature, and precipitation that dictate, in very general terms, the kinds and numbers of plants and other living things found in various parts of the biosphere—the tropical rain forest versus the tundra, the grasslands versus the coniferous forest, and so on. In Chapter 41, we subdivided these broad areas, the biomes, into functioning units, ecosystems, and saw how the flow of energy and recycling of minerals that define an ecosystem call for the presence of organisms of various categories and in certain numbers. In other words, an ecosystem will probably contain herbivores and first-level carnivores, and the biomass of the former will usually be at least ten times that of the latter.

In this chapter we are going to move in still more closely and look at communities and populations. A community consists simply of all the plants,

42–1 *Some members of a seashore community: starfish, barnacles, whelks, kelp, and mussels.*

animals, and other organisms that live in a particular area. A population is a group of organisms of the same species occupying a particular area at the same time. Thus, communities are made up of populations, and, as we shall see, interactions among populations within a community are major factors in determining the numbers and kinds of organisms found there.

KINDS OF ORGANISMS: THE ECOLOGICAL NICHE

As we saw in the previous chapter, organisms play different roles in ecosystem function, and these roles are interdependent. Herbivores depend on producers, first-level carnivores depend on herbivores, decomposers on producers and consumers, and so on. Thus, each type of organism has a particular role that it performs in the community, and this role, in anthropomorphic terms, is its job, the way it makes its living. An organism's position in the ecosystem is called its _ecological niche._ A buffalo, for example, has the ecological niche of herbivore. A broad ecological niche, such as herbivore, is almost always divided into smaller niches, such as browser, which eats the leaves and tender shoots of trees and bushes, or grazer (grass eater). These smaller niches are often further subdivided as to just what kind of shrub or grass the browser or grazer selects. Probably the finest division of ecological niches occurs among the decomposers, where every small single step in the decomposition process is carried out by a particular microbial expert. We saw examples of this high degree of specialization in the bacteria involved in the cycling of nitrogen.

A concept coupled with that of the ecological niche is the principle of competitive exclusion, sometimes referred to as Gause's principle, after the Russian biologist G. F. Gause. According to Gause's principle, in any given community, only one species (that is, one population) can occupy any given ecological niche for an extended period of time.

Gause bolstered his hypothesis by a number of laboratory experiments. His simplest, now classic experiment involved laboratory cultures of two species of paramecia, _Paramecium aurelia_ and _Paramecium caudatum_. When the two species were grown under identical conditions in separate containers, _P. aurelia_ grew much more rapidly than _P. caudatum_, indicating that the former used the available food supply more efficiently than the latter. When the two were grown together, the former rapidly outmultiplied the latter, which soon died out.

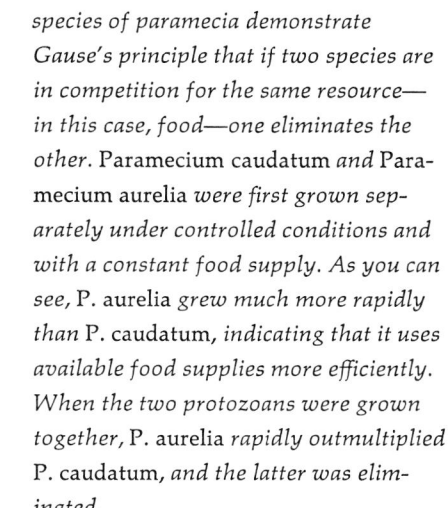

2–2 _Results of Gause's experiment with two species of paramecia demonstrate Gause's principle that if two species are in competition for the same resource—in this case, food—one eliminates the other._ Paramecium caudatum _and_ Paramecium aurelia _were first grown separately under controlled conditions and with a constant food supply. As you can see,_ P. aurelia _grew much more rapidly than_ P. caudatum, _indicating that it uses available food supplies more efficiently. When the two protozoans were grown together,_ P. aurelia _rapidly outmultiplied_ P. caudatum, _and the latter was eliminated._

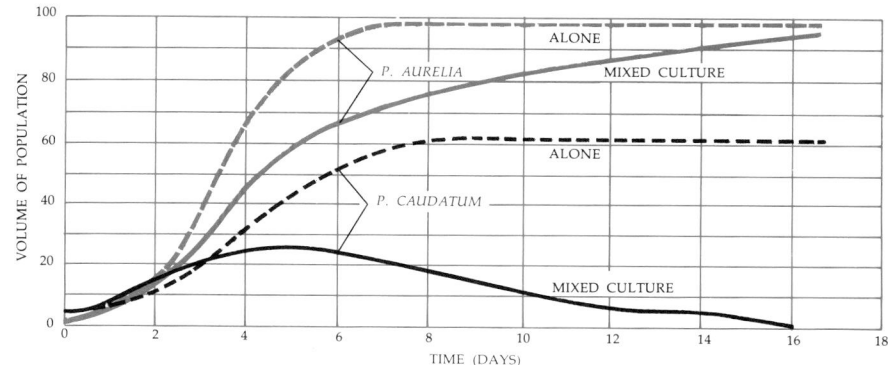

A demonstration of Gause's principle: the feeding zones in a spruce tree of five species of North American warblers. The colored areas in the tree indicate where each species spends at least half its feeding time. In this way, all five species feed in the same trees with diminished competition.

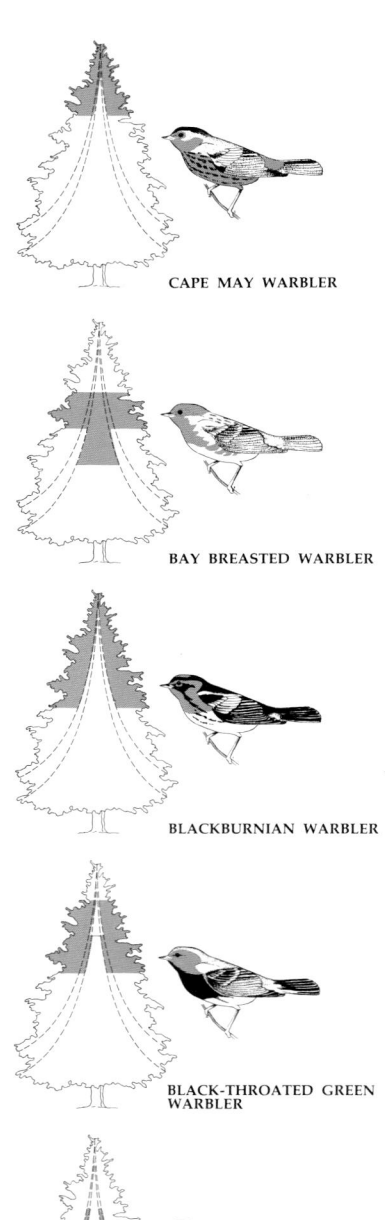

CAPE MAY WARBLER

BAY BREASTED WARBLER

BLACKBURNIAN WARBLER

BLACK-THROATED GREEN WARBLER

MYRTLE WARBLER

A similar simple experiment was made with two very similar species of clover. Either could grow well alone, it was found, but when the two were mixed together, one soon eliminated the other. The reason was simple: the successful species holds its leaves slightly higher than those of its competitor and so, once the cultures grow dense enough, completely overshadows it.

The Niche in Nature

Gause's principle leads one to predict that in a situation in which two very similar species coexist in nature, it will be possible to demonstrate differences in the niches they inhabit. The prediction has been fulfilled by a number of observations. For example, some New England forests are inhabited by five closely related species of warbler, all about the same size and all insect eaters. Why do these birds not eliminate one another by competitive exclusions? An analysis by Robert MacArthur showed that these warblers have different feeding zones in the canopy (Figure 42–3), and because of these different zones they are able to coexist without direct competition.

Niches, by definition, involve not only the type of food consumed and where it is found, as shown in the foregoing example, but also other factors, such as time. Organisms that feed only in the spring, such as certain insects, are not in competition with similar species whose feeding forms emerge in the summer, and, similarly, organisms that feed in the early morning are not considered to compete directly with those that feed at twilight. Organisms that eat mostly large seeds are not directly competitive with those that eat small seeds, for example. Moreover, in some definitions of ecological niche, additional factors, such as shelters or nesting habits, are included. In the broadest sense, an ecological niche is the entire spectrum of environmental resources used by a particular species.

In natural communities environmental heterogeneity is so great that competitive exclusion is hard to demonstrate. At present, some ecologists are inclined to abandon or ignore the competitive exclusion principle, whereas others consider it as the most important concept of modern ecology. The latter group points out that recognizing the existence of distinct and exclusive ecological positions for different organisms in a biological community makes it possible not only to understand but also to predict the overall structure of the community. As we shall see in Section 7, the concept of the niche is valuable in terms of evolutionary theory.

We shall return to this question of the kinds of organisms later on in the chapter, but let us first look at the closely related question of the numbers of organisms.

NUMBERS OF ORGANISMS

The size of a population is the end result of a number of different forces. Whether or not a population decreases or increases in size is determined by three basic factors: (1) natality (birth rate), (2) mortality (death rate), and (3) immigration and emigration, movement of organisms in and out of a particular community.

The Reproductive Potential

Natality—the number of *possible* offspring—varies enormously from one species to another. An oyster lays more than a million eggs a season; a large

2–4 Parental care is a major component in determining survivorship. (a) Within this long egg case, composed of a parchmentlike material, a whelk is depositing hundreds of eggs. When the baby mollusks hatch, they will break through thin areas in the individual capsules and emerge as miniature copies of their parents. Once the egg case is produced and the eggs laid, the mother whelk does not participate further in raising the young. (b) Most amphibians lay their eggs in the water and leave them there to develop. An exception is the midwife toad, Alytes obstetricians; male toads of this genus carry their eggs with them in special epidermal pits on their back. Immediately after fertilizing the eggs, the male shifts his position forward and then backward, sliding his legs under the newly laid egg mass which has collected on the outstretched hind legs of the female. He spends his days in his burrow; his nocturnal foragings through dewy grass keep the eggs from drying out. (c) Baby Canada geese remain with their parents through the summer after they are hatched and migrate with them to the winter feeding grounds. In proportion to life span, the time allotted to parental care by birds is longer than for any other animal group. (d) Elephant calves are suckled by their mothers for at least two years. Care of the young is primarily the responsibility of the female, but young and females are protected by the bull elephant that is the leader of the group.

(a)

(b)

(c)

(d)

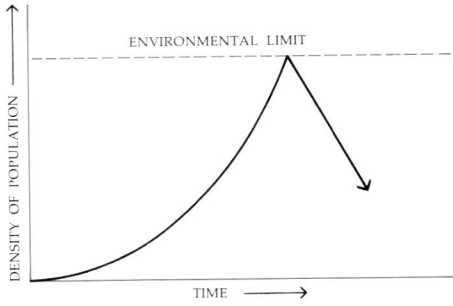

42–5 *Exponential growth curve. After an initial establishment phase, the population increases compound-interest fashion until an environmental limit causes a population crash.*

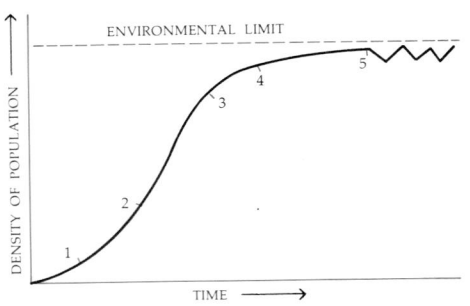

42–6 *A more usual growth curve is sigmoid, or S-shaped. As with exponential growth, there is an establishment phase (1), and a phase of rapid acceleration (2). Then, as the population approaches environmental limits, the growth rate slows down (3 and 4), and finally stabilizes (5).*

mammal can produce and care for only one infant every two or three years. It has been calculated, for instance, that a single female housefly producing an average of 120 eggs per laying, with half developing into females and with seven generations per year per female, could produce 6,182,442,727,320 houseflies in the space of one year. At the other end of the reproduction scale is a pair of elephants, which have a long gestation time and usually give birth to only one offspring at a time. If all offspring and their offspring, in turn, survived and reproduced, this one pair of elephants would have 19 million descendants at the end of 750 years, as noted by Darwin.

These hypothetical figures of the housefly and the pair of elephants represent examples of exponential growth, which operates on the same principle as compound interest. The more money in a compound savings account, the more dollars are added to the account per unit time. In short, the interest earns interest. Similarly, in a population increasing exponentially, the more individuals that are added to the population, the faster it increases. In natural populations, exponential growth curves, such as that seen in Figure 42–5, are sometimes seen when an alien species is introduced into a new environment, in the seasonal blooms of algae, or in an insect invasion. Often, as indicated in the figure, the growth is followed by a crash, which occurs when the population uses up one of its natural resources (remember the principle of limiting factors) or, as happens more typically among insect populations, is overcome by predators, also increasing exponentially, or by disease microorganisms.

The more common natural growth curve is the sigmoid (S-shaped) curve shown in Figure 42–6. Increase in the number of individuals is slow at first, because there are a relatively small number of individuals reproducing. Then, as the number increases, the curve comes to resemble the exponential curve. Typically, it then slows down and finally levels off, reaching a point at which there is no net change in the population. This leveling off occurs when the population reaches the limit at which the environment can support it. This limit, also known as the carrying capacity, is determined by any one of a number of factors, such as the availability of food, water, or oxygen, or the accumulation of wastes. Altering the environment may shift the point or the time at which the population reaches equilibrium. In most well-established communities, most of the populations have reached the leveling-off point of the sigmoid curve. In other words, for every two houseflies now present, there were two houseflies ten years ago, and there will probably be two ten years from now. And the same is true of elephants.

Death Rates

The rate of increase or decrease of a population is the difference between the birthrate and the death rate. Death rates are high in natural populations. For example, in a study made of the saltmarsh song sparrows of San Francisco Bay, it was estimated that of every 100 eggs laid, 26 are lost before hatching. Of the 74 live nestlings, only 52 leave the nest, and of these 52, 80 percent die the first year. The remaining 10 breed the following season, but during the next year, 43 percent of these die, leaving only 6 out of the original 100. Each subsequent year, mortality among the survivors amounts to 43 percent.

Once a bird survives its first risk-laden year, the mortality rate for the years before old age remains more or less constant. In nearly all species, the probabil-

–7 *Representative survivorship curves, plotted from life-table data on the basis of survivors per 1,000 log scale (vertical coordinate) and age in relative units of mean life span (horizontal coordinate). As you can see, the oyster's chances of survival greatly improve with age, while mortality among* Hydra *remains constant throughout the life span. The curve at the top of the graph is for a hypothetical population in which all individuals live out the average life span of the species—a population, in other words, in which all individuals die at about the same age. The fact that man's curve approaches this hypothetical curve indicates that the human population as a whole is reaching a uniform age of mortality.*

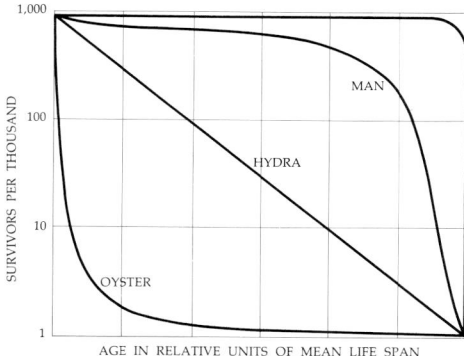

ity of survival for one more year or one more period of time seems to be highest during the middle of life and lowest for both very young and very old individuals. Patterns of mortality differ from species to species, however. Figure 42–7 shows the basic types of survivorship curves. In the oyster, for example, mortality is extremely high during the free-swimming larval stage, but once the individual attaches itself to a favorable substrate, life expectancy levels off. Among *Hydra*, the mortality rate is the same at all ages.

If all individuals in a population lived out the average life span of their species, their survivorship curve would resemble the hypothetical curve shown at the top of the graph. The mortality curve for man, as you can see, approaches this curve, indicating that the human population as a whole is reaching a uniform age of mortality.

Emigration

Emigration is the departure of individuals from a population. Arctic lemmings, among the most famous emigrants, increase periodically in number and then emigrate in large numbers, apparently in response to crowding. In northern Europe, they have been known to march by the millions across the countryside and, on reaching the coasts, reportedly drown in the sea. Similar spectacular emigrations occur among locusts, resulting in the plagues of locusts recorded from Biblical times. When a locust population increases in number, hormonal changes begin to take place among members of the new generations. These changes result in longer wings, more slender bodies, and darker colors. The locust swarms are made up of these individuals. Similarly, swarming occurs in honeybees, also apparently in response to crowding. In the case of bees, the old queen leaves the hive, taking a group of workers with her. Emigration may also occur continuously but less spectacularly, involving only small numbers of individuals at a time.

The stimuli that cause emigration undoubtedly differ from species to species, but the result is always the same: a drastic reduction in the size of the local population. (Note that emigration and migration are two quite different phenomena. Emigration is a one-way trip, whereas migration refers to periodic, seasonal movements.)

–8 *Lemmings are inhabitants of the North American and Eurasian tundra. Lemming populations, like those of many other small rodents, undergo drastic cyclic fluctuations in numbers. Ecologists are divided as to whether these cyclic changes are the results of extrinsic factors, such as food supply, predators, or disease, or intrinsic factors, such as changes in genetic composition or increase in hostility among members of the population.*

42–9 *The number of lynx and snowshoe hare pelts received yearly by the Hudson Bay Company over a period of almost 100 years, indicating a pattern of 10-year oscillations in population density. The lynx reaches a population peak every 9 or 10 years, and these peaks are followed in each case by several years of sharp decline. The snowshoe hare follows the same cycle, with a peak abundance generally preceding that of the lynx by a year or more. Presumably a surge in the population of snowshoe hares, the prey, results in a corresponding increase in the predator population, with a resultant decrease in prey followed by a crash in the predator population. This type of pattern is characteristic of very simple or immature ecosystems.*

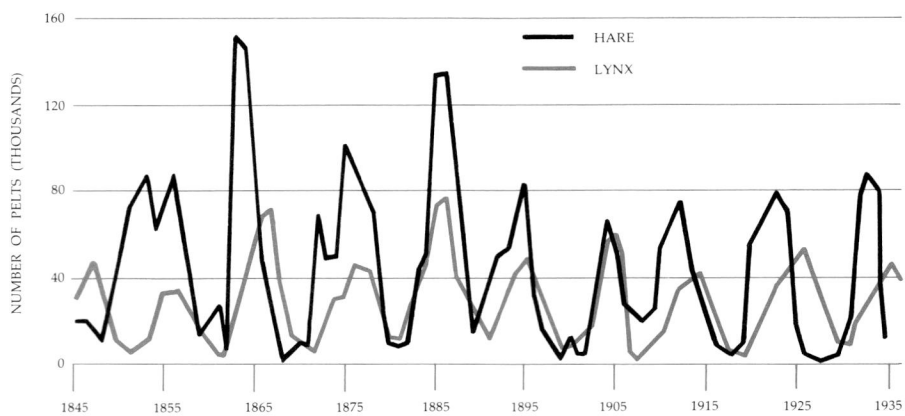

42–10 *The snowy owl, which preys on lemmings. In peak lemming years, snowy owls lay 11 to 12 eggs; in years of moderate lemming populations, 4 to 5 eggs; and in years of low lemming numbers, the owls do not reproduce. Thus the population oscillations of owls closely follow those of the lemmings.*

The Size of Populations

Many factors may limit the growth—decrease the natality or increase the mortality or emigration rate—of populations. These factors are sometimes grouped under the term environmental resistance and are frequently divided into density-independent and density-dependent factors. The density of the population is simply the number of individuals in a given unit of area at a particular time.

Temperature is, for example, a density-independent factor. Extreme cold or extreme heat may destroy large numbers of organisms in a population; the number destroyed (with perhaps a few exceptions) is not dependent upon the size of the population in the area. Forest fires, floods, severe storms, and other catastrophes are also grouped with density-independent factors.

The most obvious density-dependent factor is food supply. The per capita food supply decreases as the number of organisms in the population increases. Others include shelter or nesting places and communicable diseases. Predation is another density-dependent factor; when an organism is a prey organism for another species, it is likely to be particularly heavily preyed upon when it is in abundance. When it comes to be in short supply, the predator also tends to decrease in number or to turn to another, more abundant food source. The snowy owl and the lemming provide an example.

INTERACTIONS AMONG POPULATIONS

Let us now ask ourselves the question that Charles Darwin asked himself some 150 years ago; if only two elephants, or houseflies, or oak trees, or paramecia, survive out of the millions that are mathematically possible, why do these two survive and not the others? We have touched on this question many times previously in this book, and we shall deal with it again in greater depth in Section 7. At this moment, let us focus the question in terms of communities and populations and see how interactions between populations within communities determine the numbers of organisms, where they are found, and, more particularly, the kinds of organisms, not just in terms of broad categories but specific characteristics. In the rest of this chapter, we shall identify the different kinds of relationships that may exist among organisms in a community and give some examples of the evolutionary effects of such relationships.

Interspecific Competition

As we noted earlier in this chapter, competition appears to limit the number of similar species in a particular community. Which of two competing species proves the victor may be determined by any one of a variety of factors.

A situation similar to the competition between clover species that we saw previously occurs between two species of duckweed, *Lemna gibba* and *Lemna polyrrhiza*. Each grows equally well alone, but *L. gibba* always replaces *L. polyrrhiza* when they are grown together. Again, evolution has provided one with an advantage. The plant bodies of *L. gibba* have air-filled sacs that elevate them like little pontoons, so that they form a mass over the other species, cutting off the light. As with the clover, the species in the shade dies out. (Figure 42–11.)

These are, of course, laboratory examples. Many ecologists believe that, in nature, competition results not so much in exclusion or extinction as in adaptations that reduce the intensity of the competition, as in the example of the New England warblers.

Competition between two species of barnacles in Scotland is a well-studied example. Barnacles are a kind of crustacean. Before they change from their immature, larval forms, in which they are free-swimming, into their adult, feeding forms, they cement themselves to a rock and secrete a shell. One species of barnacle, *Chthamalus stellatus*, occurs in the high part of the intertidal seashore, and another barnacle, *Balanus balanoides*, occurs lower down. Although young members of the species *Chthamalus*, after their short period of drifting in the plankton, often attach to the rock in the lower, *Balanus*-occupied zone, no adults are ever found there.

The history of a population of barnacles can be recorded very accurately by holding a pane of glass over a patch of barnacles and noting with glass-marking ink each spot where a barnacle is. Once attached, barnacles remain fixed, so that by returning later, one can check exactly which barnacles have died and which new ones have arrived.

By doing this, the investigator was able to show that in the lower zone, *Balanus*, which grows faster, ousts *Chthamalus* by crowding it off the rocks or growing over it. When *Chthamalus* was isolated from contact with *Balanus*, it lived with no difficulty in the lower zone, showing that competition with *Balanus* restricted *Chthamalus* to the higher, less favorable zone. However, when

An experiment with two species of floating duckweed, tiny angiosperms found in ponds and lakes. One species, Lemna polyrrhiza, *grows more rapidly in pure culture than the other species,* Lemna gibba. *But* L. gibba *has tiny air-filled sacs which, like pontoons, float it on the surface, and so it shades the other species, making it a victor in the competition for light.*

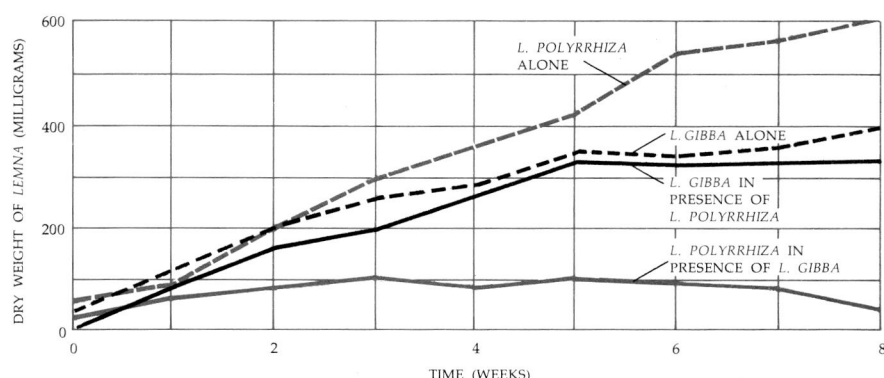

Chthamalus was removed, members of the *Balanus* species were unable to live in its range. *Chthamalus* survives in the intertidal community by special physiological adaptations that make it possible for it to inhabit a marginal area.

A somewhat similar competitive relationship is seen between two species of blackbird, redwings and tricolors. Both nest in marshes and feed on insects. Redwings are territorial; that is, each male establishes a territory and defends it against members of his own species. Tricolors live in colonies. Redwings nest in late winter, earlier than tricolors, and usually have filled the marsh by the time the tricolors arrive, but the tricolors, which appear in flocks of hundreds and sometimes thousands, easily drive them out. However, tricolors, unlike the redwings, are highly nomadic and so do not nest regularly in the same areas. As a consequence, the redwings, like the *Chthamalus* barnacles, do not leave the community completely but are relegated to a less favorable position in it.

Some species of plants also have forms of aggression against competing species, though not ordinarily as raucous as those of blackbirds. The leaves of several species of shrubs in the chaparral near Santa Barbara, California, produce toxic substances that are carried to the ground by rainfall. There they prevent the germination of seeds or the encroachment of roots of other plants. The characteristic aromatic odor of some chaparral communities is the result of the release into the air of volatile chemicals, which settle to the ground and attach to dry soil particles. These chemicals also are toxic to plant seedlings (Figure 42–12). This phenomenon, which is known as allelopathy, has also been observed in various forms among other species.

Environmental disruptions may prolong competition between species. In England, the grasses on chalk soils grow more vigorously than the annual flowering plants, but the annuals are able to grow in the area because rabbits crop the grasslands. However, when a virus epidemic severely reduced the rabbit population, the grasses took over and the wildflowers disappeared. Similarly, it has been hypothesized that members of the phytoplankton community, which are all photosynthetic and which utilize a common pool of nutrients, do not eliminate one another because of environmental instability. Before competitive displacement has time to occur, drastic seasonal changes take place in the lake or the ocean and the conditions of competition are greatly changed.

SYMBIOSIS

Symbiosis ("living together") is a close and permanent association between organisms of different species. There are several types of symbiotic relationships. If the relationship is beneficial to both, it is called mutualism. If one species benefits from the association while one is neither harmed nor benefited, it is called commensalism. If one species benefits and the other is harmed, the relationship is known as parasitism. However, since all the details of symbiotic relationships between species are not fully understood, these distinctions are not always useful.

The examples of symbioses in nature are many and varied (see Figure 42–18). One, which we have mentioned previously, is the association between a ruminant animal, such as a cow, and the bacteria that inhabit its stomach and break down cellulose. *Trichonympha*, a protist which is the source of the spectacular cross section of cilia shown in Figure 6–15, has a similar association with wood termites. Similarly, symbiotic algae, such as those found in the coral reef, provide

42–12 *Shrubs of purple sage,* Salvia leucophylla, *produce volatile terpenes that inhibit the growth of other plants in their vicinity. Here, near Santa Barbara, California, they are surrounded first by a completely bare zone and then by a zone of inhibited grassland inhabited by stunted annual herbs. Some plants are restricted to similar zones of inhibited grassland, having the ability to tolerate the toxins and being protected from their competitors.*

-13 *Lichen on the trunk of a tree. Lichens, which are symbiotic combinations of an alga and a fungus, are able to survive in a state of suspended animation even though their water content is reduced to less than 10 percent. When water is restored, photosynthesis resumes. The symbiotic association is unusual in that it originates, apparently, with the capture of a free-living alga by a fungus especially adapted to this purpose.*

14 *One of the most ecologically important symbioses is the mycorrhiza ("fungus-root"), a relationship between certain fungi and the roots of plants. In the experiment whose results are shown in this picture, both groups of white pine seedlings were raised in a sterile nutrient solution for two months. Those on the left were transplanted directly to prairie soil. Those on the right were placed for two weeks in forest soil, where mycorrhizae developed, before they were planted in the prairie. The fungi apparently convert minerals in the soil to a form more readily usable by plants, and the plant roots seem to secrete organic compounds the fungi can utilize.*

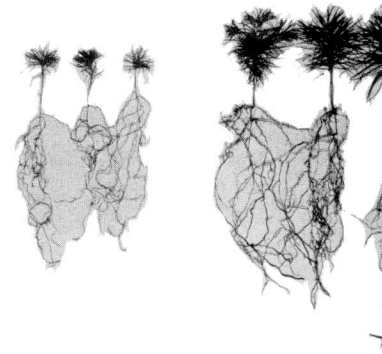

their host with the products of photosynthesis and receive protection from this relationship.

A classic example of symbiosis is provided by the lichens, which are part alga, part fungus. The body of the lichen is composed largely of fungal mycelium, and held within this mycelium are numerous photosynthetic algal cells. The two organisms together form a closely integrated unit that can grow under conditions where neither the fungus nor the alga alone could survive. Lichens occur from arid desert regions to the Arctic; they grow on bare soil, tree trunks, sunbaked rocks, and windswept alpine peaks all over the world. They are often the first colonists of bare rocky areas.

One of the most important symbioses from the ecological point of view is the one that exists between certain plants, particularly legumes, and the bacteria that infect their roots. As we saw in Chapter 41, this relationship is crucial to the essential process of nitrogen fixation. Many plants have symbiotic relationships with fungi that form a feltlike coating (mycorrhizae) on their roots. Many plants grow less well if they are cultivated in soil that does not contain their characteristic root fungi, although it may have all of the same nutrients.

The Fungus-gardening Ants

One of the most elaborate symbiotic relationships is that which exists between a particular group of ants and their fungus gardens. The ants involved in this relationship are known as the Attini, a group that comprises several hundred species (the principal genus is *Atta*). They live in North, South, and Central America largely in tropical regions, although one species is found along the Atlantic Coast as far north as Long Island. The center of their activities is their garden, which is cultivated by the workers and on which the queen and her brood live. In most species, the workers supply the garden with cuttings of fresh green leaves or other plant material. A species that lives in eastern Texas and southern

Louisiana does considerable damage to citrus groves and gardens by these leaf-cutting activities. The Attini inhabit extensive underground metropolises that may be as much as 50 feet in diameter and 20 feet deep, with numerous chambers containing fungus gardens.

The ants carry the leaf cuttings into their nests, usually traveling along well-marked trails, chew them into a pulpy mass, deposit anal droplets on them, and then plant them, usually around the base of the garden. They pick up bits of fungi and dot the surface of the chewed leaf with them. When a queen leaves a colony to found a new one, she carries a speck of fungus with her and cultivates it herself on her excreta until the first brood of workers matures.

The ants eat nothing but fungus. The fungus, in turn, is completely dependent on the ants. The particular types of fungi in the ant gardens are found nowhere else. If the ants are removed, the fungus garden is soon overrun by other types of fungi or by bacteria.

Parasites and Hosts

Parasitism, as we noted earlier, is a symbiotic relationship between two organisms of different species in which one lives at the expense of the other. Most diseases in organisms (except some in higher animals, such as heart disease, which is linked to a prolonged life span) are caused by parasites, chiefly microorganisms. Such diseases are special cases of predator-prey relationships. In these instances, the predator is always considerably smaller than the prey, and the prey is known as the host, which implies gracious acceptance and is therefore somewhat misleading.

Large animals may sometimes be parasites. The hagfish, which sucks out the flesh of other fish, might be classified either as a predator or as a parasite. The tapeworm, often yards long, is indisputably a parasite. (See Figure 22–25.)

Most plants and animals in a natural community support hundreds of parasites of many species—in fact, perhaps millions, if one were to count viruses. Parasites do not usually kill their host, and they almost never wipe out entire populations. We know this from our own experience. Bacterial disease is fatal only when the bacteria find themselves a particularly favorable place to multiply, such as in the blind pouch of an infected appendix, in an open wound, or when the host's resistance is reduced. Most bacteria are usually harmless, and some are helpful, such as those in the stomachs of the herbivores and the vitamin-supplying bacteria in our own intestinal tracts. Modern antibiotics often cause nausea and diarrhea because they destroy the useful and protective bacteria of the intestinal tract. And of all the viral diseases of man, only one, rabies, is regularly fatal if unchecked.

In animal and plant communities, diseases are most likely to wipe out the very young, the very old, and the disabled—either directly or, often, indirectly by making them more susceptible to other predators or to the effects of climate. Most infectious diseases are density-dependent; that is, they tend to spread more rapidly when populations are crowded and to subside as the population thins out. They thus serve as a check on population density in much the same way that food shortage does.

It is logical that a parasite-caused disease should not be too virulent or too efficient. If a parasite were to kill all the hosts for which it is adapted, it, too,

16 *Rabbits crowd a water hole in Australia. Imported from Europe as potentially valuable herbivores, they soon overran the countryside. The myxoma virus was introduced to control them, but now host and parasite are coexisting.*

would perish. This principle is particularly well illustrated by the changes that have taken place in the myxoma virus of rabbits in Australia. The myxoma virus causes only a mild disease in the South American rabbit, its normal host, but a usually fatal one, laboratory studies have shown, in the European rabbit, which has had no previous evolutionary experience with this disease. To control the invasion of introduced rabbits in Australia, myxoma-infected rabbits were set free on the continent. At first the effects were spectacular and the rabbit population steadily declined, yielding a share of pastureland once more to the sheep herds on which much of the economy of the country depends. But then occasional rabbits began to survive, and their litters were also resistant. Now some 95 percent of the rabbits survive an attack of myxoma virus. A double process of selection took place. The virus, as originally introduced, was so rapidly fatal that often a rabbit died before it could be bitten by a mosquito and thereby infect another rabbit; the virus strain then died with the rabbit. Strains less drastic in their effects, on the other hand, had a better chance of survival since they had a greater opportunity to spread to a new host. (After the initial infection, the rabbit is immune to the virus, just as human beings usually become immune to mumps or measles after one infection.) So, first, selection began to work in favor of a less virulent strain, as proved by tests of the Australian virus on European rabbits in the laboratory. Almost simultaneously, rabbits that were resistant to the original virus began to appear, as proved by tests of the South American virus on Australian rabbits. Now the two are reaching a peaceful equilibrium, like most hosts and parasites.

Malaria

Many parasites have two or more hosts. Malaria, for example, is caused by protozoans of the genus *Plasmodium*. The plasmodia are passed back and forth between man and the *Anopheles* mosquito (see page 312). The protozoan is distinctly different in form in the two stages of its life cycle in its two hosts. It is not harmful to the mosquito, indicating that its relationship with the mosquito probably long predated its relationship with man.

(a)   (b)

42–17   (a) *A baby cuckoo ejecting a newly*
hatched hedge sparrow from its nest. As
you can see on the right, another young
hedge sparrow and an egg have already
been pushed out and are now lodged in
the twigs beside the nest. (b) A bedrag-
gled hedge sparrow feeding a young
cuckoo. Only the English species of
cuckoo lays its eggs in the nests of other
birds. In North America similar behavior
is seen in the cowbird.

Social Parasites

European cuckoos and North American cowbirds lay their eggs in the nests of other birds. A single female may lay as many as twelve eggs, one at a time, each in a nest containing other eggs. The European cuckoo, the more accomplished of the two parasitic species, ejects an egg from the nest as she deposits her own, thus maintaining the "egg count." Immediately upon hatching, the young cuckoo burrows under and pushes out legitimate eggs and young members of the brood. Even when other fledglings remain, the interloper is treated preferentially. Parent birds, on returning to their nest with food, push it down the throat of whichever fledgling is gaping the widest. Since the cuckoo or cowbird usually has a larger head (and so a wider gape) than the young of the songbird it parasitizes, it regularly receives the most food, and soon outgrows its nestmates and, often, its duped, hard-working foster parents. (See Figure 42–17.)

TROPHIC RELATIONSHIPS

As we saw in the previous chapter, the various trophic levels of an ecosystem are linked by predator-prey relationships. Animals prey on plants, and other animals prey on the herbivores. Predator-prey relationships tend to regulate numbers of organisms but, unlike competition, seldom result in the elimination of a species. In mature (complex) ecosystems, most predators prey on a wide variety of other organisms. When one species begins to decrease in number, its members are preyed upon less frequently and the predators turn their attention to one increasing in number. In very simple ecosystems, such as the desert, the tundra, or areas of human agriculture, predator and prey numbers may fluctuate widely. When prey is abundant, the population of the predator increases. The population of the prey is then reduced drastically in number, and

−18 *Symbioses. (a) Sea anemones on the back of a hermit crab. The anemone protects and camouflages the crab and, in turn, gains mobility—and so a wider feeding range—from its association with the crab. Hermit crabs, which periodically move into new, larger shells, will coax their anemones to move with them. (b) Cleaner fish are permitted to approach larger fish with impunity because they feed off the algae, fungi, and other microorganisms on the fish's body. The fish recognize the cleaners by their distinctive markings. Other species of fish, by closely resembling the cleaners, are able to get close enough to the large fish to remove large bites of flesh. What would probably happen if cleaner mimics began to outnumber cleaners? (c) Oxpeckers live on ticks which they remove from their hosts. An oxpecker forms an association with one particular animal, such as the young impala shown here, conducting most of its activities, including courtship and mating, on its back. (d) Aphids suck phloem, removing certain amino acids, sugars, and other nutrients from it and excreting most of it as "honeydew," or "sugar-lerp," as it is called in Australia where it is harvested as food by the aborigines. Some species of aphids have been domesticated by some species of ants. These aphids do not excrete their honeydew at random, but only in response to caressing movements of the ant's antennae and forelimbs. The aphids involved in this symbiotic association have lost all their own natural defenses, including even their hard outer skeletons, relying upon their hosts for protection.*

(a)

(b)

(c)

(d)

42–19
Upsetting the balance. Schistosomiasis is a debilitating disease caused by a snail-carried parasite that attacks the human liver. In South Africa, a campaign was waged against hippopotamuses. Un-lovely and unloved, they cluttered up rivers, were "useless," and so were shot on sight. Without the hippos, which served as natural bulldozers, the rivers silted up, the snails proliferated, and schistosomiasis has become as great a hazard in some parts of Africa as malaria was 50 years ago.

42–20
Prickly pear cactus on a homestead in Australia. Such rapid and environmentally destructive spread is often seen among alien organisms introduced into a region where they have no natural enemies.

many of the predators, lacking other prey, starve to death. The prey then increases in number, and the cycle begins again.

The introduction of a new species in a community where there are no natural predators for that species may sometimes have disastrous results. When prickly pear cactus was brought to Australia from South America, it soon escaped from the garden of the gentleman who imported it and spread into fields and pastureland until more than 30 million acres were so densely covered with prickly pears that they could support almost no other vegetation. The cactus then began to take over the rest of Australia at the rate of about a million acres a year. It was not brought under control until a natural predator was imported. This was a South American moth (*Cactoblastis cactorum*), whose caterpillars live only on the cactus. Now only an occasional cactus and a few moths can be found.

Predation is not necessarily detrimental to the population as a whole, particularly when there is competition within the population for a limited resource such as food. Wolves, for instance, have great difficulty overtaking healthy adult caribou or even healthy calves. A study of Isle Royale, an island on Lake Superior, showed that in some seasons more than 50 percent of the animals the wolves killed had lung disease, although the incidence of such individuals in the population was less than 2 percent. (Human hunters, however, with their superior weapons and their desire for a "prize" specimen, are more likely to injure or destroy strong, well-adapted animals.)

Bats and Moths

If food is a limiting factor, natural selection strongly favors the most efficient predator. Thus, over the course of the generations, the big cats become swifter and more cunning, the necks of the giraffes become progressively longer, the grosbeaks' bills grow thicker and stronger, and eagles become more eagle-eyed.

−21 *A bat emitting ultrasonic cries. The megaphone-shaped mouth serves to beam the echolocation pulses while the funnel-shaped ears receive the echoes. The auditory system is so sharply tuned that the echoes can be distinguished from the cries even when both are received simultaneously.*

In addition to strength, swiftness, cunning, and acuity, evolution has produced some unusual predatory stratagems. Bats, for example, hunt by sonar, emitting shrill cries well above the range of human hearing. (These cries were not detected until about 20 years ago when Donald Griffin, while still a graduate student at Harvard, carried a cageful of bats into a physics laboratory that contained an electronic device tuned to detect high frequencies.) When these sound waves hit a solid object, they echo back to the ears of the bat. On the basis of these echoes, bats can navigate skillfully through a dark room strung with wires little thicker than a human hair or can catch an insect as small as a fly.

Most animal species that have hearing organs have organs tuned to receive sounds from members of their own species. A notable exception is the noctuid moth. These moths are among the principal prey of bats and their tympanic organs have been precisely tuned in the course of evolution to receive, you will not be surprised to hear, the shrill cries of bats. On hearing these cries, they take appropriate evasive action.

NATURAL DEFENSES

Just as natural selection favors the most efficient predator, it also favors the prey best able to avoid predation. Under strong evolutionary pressures from predators, plants and animals have developed some interesting and, to the human eye, extremely ingenious defenses.

Natural Defenses in Plants

Some natural defenses in plants are structural, such as the sharp-toothed edges of the holly leaf, the thorn of the rosebush, the spine of the cactus, and the sting of the nettle. Plants also produce a number of chemical substances for which the plant itself has no physiological use and which appear to function as defenses against leaf-eating insects and other predators. Eating foxgloves (*Digitalis purpurea*) can cause convulsive heart attacks in vertebrates. Other plants containing digitalislike toxins include oleander (*Nerium oleander*), of which a single leaf may be fatal to man, and the members of a large family of plants known as the milkweeds.

Some plants contain chemicals with insect-hormone activities. A number of gymnosperms have been found to contain chemicals with molting-hormone (ecdysone) activity that fatally accelerate insect metamorphosis. The balsam fir has been found to produce a chemical that resembles juvenile hormone (page 405) and arrests the development of certain insect species. R. H. Whittaker of Cornell describes these chemicals as "a fiendishly subtle mechanism of defense."

Plant defense chemicals include many substances presently useful to man, among them digitalis, quinine, castor oil, and peppercorns and other spices; some substances of more questionable value, such as nicotine, caffeine, and morphine; and others, such as the active principles in marijuana, mescaline, and peyote, whose desirability for the animal world is a matter of opinion. Plants appear to have been waging such chemical warfare long before the coming of man, and it is possible that some small leafhopper was the first of all animals to have its mind expanded in a psychedelic experience.

(a)

(b)

(c)

42–22 (a) *Like many other plants of arid regions, this species of acacia has thorns that protect it against herbivores. The graceful gerenuk, however, which has coevolved in the savanna, is in turn specially adapted to circumvent these defenses. Acacias and other trees and shrubs provide gerenuks with water as well as food; these giraffe-necked gazelles never need to drink. (b) The biting flavor of wild mustard discourages grazers. (c) The dose of digitalis in a foxglove may be fatal to a small insect.*

In cultivating plants for agricultural purposes, man has carefully bred out their toxic and bitter-tasting chemicals and so, having destroyed their natural defenses, must protect them with insecticide. (Another, more important reason for the vulnerability of cultivated plants as compared to wild ones is that the former are often grown in large pure stands that attract and support large populations of insects and other parasites and predators.)

Natural Defenses in Animals

Natural defenses in animals include structural adaptations and the capacity to produce noxious chemicals, both of which are seen in plants, and behavioral strategies as well. Some animals are formidably armored—for example, the armadillo, the porcupine, and the sea urchin. Some have ingenious behavioral devices. The armadillo and the pillbug roll themselves up in tight armor-plated balls. Some species of cephalopods vanish, jet-propelled, leaving behind only an ink cloud. Many lizards have brightly colored tails that break off when they are attacked and, conspicuous and wriggling, divert the predator from the prey—which is escaping, tailless but with all its vital organs intact. Hermit crabs often carry sea anemones as passengers on their shells. The anemone protects the crab from predators while enjoying the benefits of mobility (and perhaps some stray morsels of food). When the crab changes shells, it often coaxes its sea anemone from its old home to the new one.

23   *Some well-defended animals:* (a) *a*
      skunk, (b) *an armadillo,* (c) *a sea urchin,*
      (d) *an 8-week-old horned owl, bluffing.*

(a) *Seventeen-year cicada emerging after a long wait.* (b) *A killdeer, injury feigning. By flopping helplessly on the ground, dragging a wing, it lures predators away from its young. Once a safe distance is reached, the bird flies away.*

(a)

(b)

Grazing animals feed in herds, and birds, when threatened, tend to tighten their flock. It is almost impossible for a hawk to catch a bird within the flock, since the hawk will be subjected to a constant bombardment of other bird bodies moving at high speed.

The periodical cicadas protect one another in a quite different way. By emerging all together from their well-hidden pupal casings only after a long interval (17 years in one species), they greatly reduce the likelihood that a natural predator dependent on cicadas will be present; most predators simply cannot wait that long. (A possible exception is a species of fungus parasitic on the cicada that does wait that long, lying dormant between the emergences of the cicadas.)

Some birds emit alarm cries that warn the flock of an approaching predator. This seeming altruism has been interpreted by some scientists as running counter to theories of natural selection within species, since the bird that calls attention to itself in this way is more likely to be lost—and its genes with it—to the predator than the bird that is not so altruistic. Other scientists point out, however, that since the bird is nearly always closely related genetically to the other members of its own flock, the genotype it saves, while perhaps not its own, is very similar. Therefore, genotypes dictating warning behavior would be more likely to be preserved than those that do not. A somewhat similar but simpler situation is seen in the worker bee that dies on stinging an intruder; in so doing, she may save her hive of sisters, and, most important, her queen.

Birds also take risks to save their young. "Injury feigning" is a common device among many types of ground-nesting birds. The parent bird flutters on the ground as though crippled, uttering piteous cries but always managing to stay just beyond reach of the intruder. Some scientists prefer to call this behavior "distraction display" to emphasize that the bird does not know what it is doing, although this latter term implies that the bird is being distracting, which it is equally unlikely to know. Regardless of the bird's intention, anyone who has seen this display cannot fail to marvel at the perfection of the performance, shaped as stringently and delicately by the forces of evolution as a bone is shaped, or a beak, or a feather.

Concealment and Camouflage

Hiding is one of the chief means of escape from predators. The young of many birds and of some other vertebrates respond to the warning cries of their parents by "freezing" or by running to cover. Small mammals often have nests or burrows or makeshift residences in hollow logs or beneath tree roots in which they conceal themselves from their enemies.

Protective coloration is common. Mice, lizards, and arthropods that live on the sand are often light-colored, and such light-colored species, if removed from their home territories, will immediately try to return to them. Snails that live on mottled backgrounds are often banded. Grass snakes are grass-colored, as are many of the insects that live among the grasses.

Some animals are countershaded for camouflage. The next time you pass a fish market, look at the specimens laid out on view. Fish are nearly always darker on the top than on the bottom. Countershading reduces the contrast

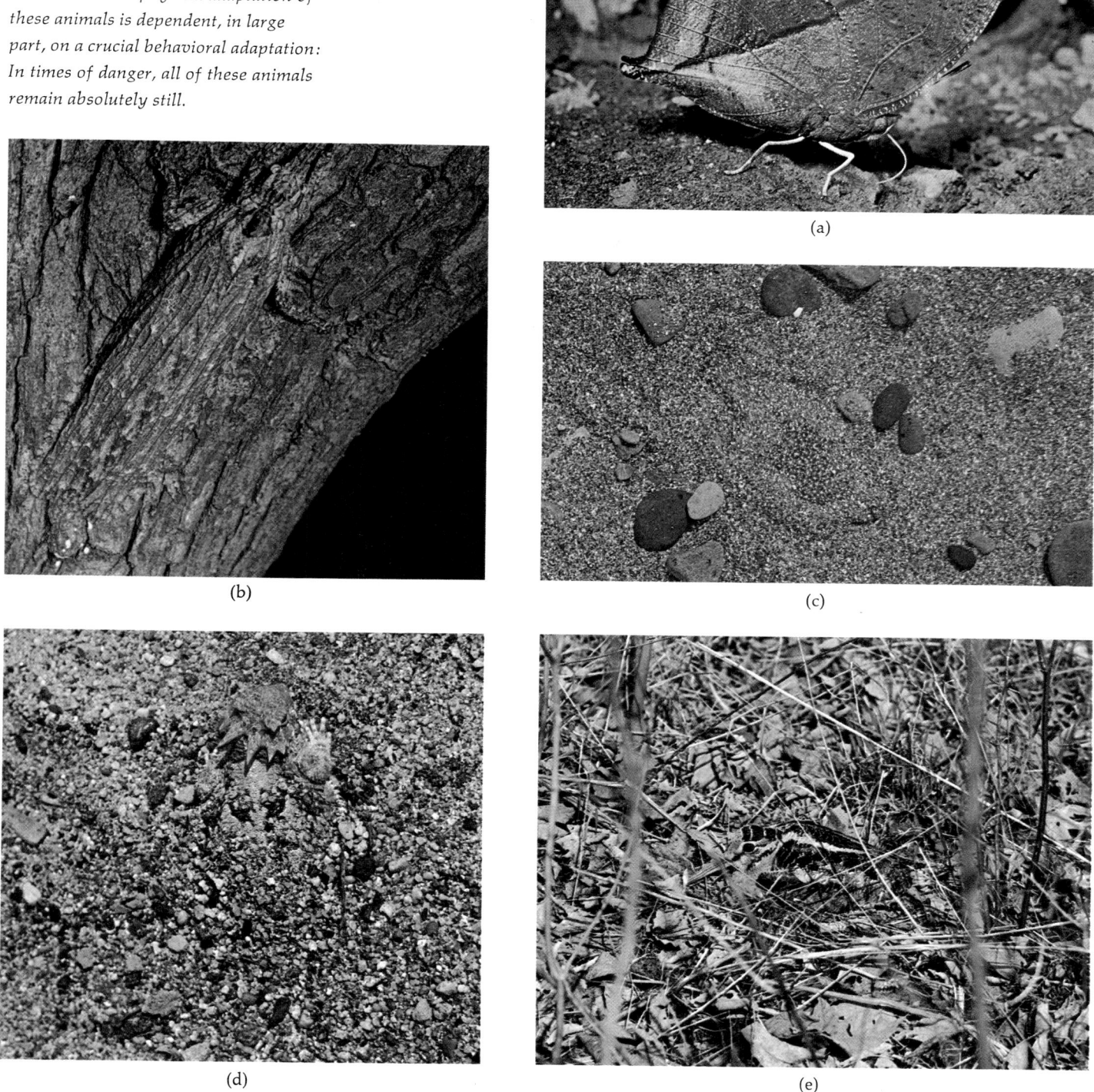

–25 *Concealment and camouflage. (a) A leaflike insect (Anaea), (b) bark katydid, (c) flounder on sandy bottom, (d) horned lizard in desert, (e) American woodcock on her nest. The physical adaptation of these animals is dependent, in large part, on a crucial behavioral adaptation: In times of danger, all of these animals remain absolutely still.*

(a)

(b)

(c)

(d)

(e)

42–26 *Deception and disguise. (a) Frog fish. (b) Predatory flower mantis, waiting for an insect in search of nectar. (c) Owl butterfly. With wings folded, the butterfly is camouflaged. The startling eyespots are a second line of defense.*

(a)

(b)

(c)

between the shaded and unshaded areas of the body when the sun is shining on the organism from overhead. A fish was once found—the Nile catfish—that was reverse-countershaded; that is, its dorsal surface was light and its ventral surface dark. The selective theory of camouflage was momentarily threatened, but scientific order was restored when it was discovered that the Nile catfish characteristically swims upside down.

Other organisms hide by looking like something else. One type of insect, the treehopper, looks like a thorn; another, the walkingstick, like a twig. Young larvae of some swallowtail butterflies look like bird droppings. In order for such disguises to work, animals must behave appropriately. *Biston betularia*, the peppered moth, lies very flat and motionless on its tree trunk, so that its colors blend with those of the tree. These and other moths that sit exposed on the bark of trees even habitually orient themselves so that the dark markings on their wings lie parallel to the dark cracks in the bark. Some desert succulent plants look like smooth stones, revealing their vegetable nature only once a year when they flower.

Some insects manage through warning coloration to frighten off their would-be predators. Large spots that look like eyes are commonly found on the backs of butterflies or the bodies of caterpillars, where they will suddenly appear when the insect spreads its wings or arches its body. Small birds reported to flee at the sight are probably birds that themselves are likely to be the prey of larger, large-eyed birds, such as owls or hawks. Smaller eyespots, while probably not

frightening, seem to have the effect of deflecting the point of attack away from the head. Examination of wounded butterflies has shown that if a part of the wings bears beak marks or is missing, it is most often the part that contains the eyespots. One investigator has tested and confirmed this conclusion by painting eyespots on the wings of living insects, releasing them, and later recapturing them for examination.

On Being Obnoxious

Some animals defend themselves, or their species at least, by having a disagreeable taste, odor, or spray, often derived from distasteful chemicals in the plants they eat. Such animals are, generally, carefully avoided. Monarch and other related butterflies feed in their larval stages on milkweeds (which they, in turn, as butterflies pollinate), and the digitalislike compounds of the plants are concentrated in their tissues. The caterpillars are immune to these poisons, but the birds that eat the caterpillars become violently ill.

Obviously, tasting bad, while useful, is not an ideal defense from the point of view of the individual, since making this fact known may demand a certain amount of personal sacrifice. (Actually, birds drop the monarch butterfly after the first bite, but they often inflict fatal injury in the process.) Obnoxious sprays and odors have the advantage of warding off predators before they harm the prey. To man, the most familiar of such animals is the skunk, which advertises its malodorous threat by its distinctive coloration. Many insects and other arthropods have developed defense secretions. In some millipedes, for example, the secretion oozes out of a gland onto the surface of the animal's body. Other arthropods are able to spray their secretion over a distance, in some instances even aiming it precisely at the attacker. The caterpillar *Schizura concinna*, whose single spray gland opens ventrally just behind its head, directs the spray

–27 (a) *Monarch butterflies lay their eggs on milkweeds, such as the swamp milkweed shown here. The bitter sap of plants of the milkweed family contains a potential heart poison, to which the larvae of the monarch butterfly are immune. By feeding on milkweeds, monarch caterpillars incorporate the poison into their bodies, and they and the butterflies are protected from predators by the poison's bitter taste and toxic effects. (b) Newly hatched stinkbugs grouped next to their empty egg shells. This clumping behavior and their bright coloration, combined with a repellent smell, warn predators of the stinkbug's unpleasant taste. They harbor symbiotic bacteria in their intestines that aid digestion of plant juices. These they obtain shortly after hatching by drinking moisture adhering to the egg shells contaminated by maternal feces.*

(a)

(b)

(a)

(b)

(c)

42–28 *(a) The beetle* Eleodes longicollis *has glands in its abdomen which secrete a foul-smelling liquid. When disturbed, it stands on its head and sprays the liquid at the potential predator. (b)* Eleodes longicollis *is on the right. On the left is* Megasida obliterata, *which, as you can see, resembles* E. longicollis *and emphasizes this resemblance by also standing on its head. However,* M. obliterata *has no similar glands, no noxious secretion, and no spray. (c) A grasshopper mouse which has solved the problem of how to eat* E. longicollis. *The mouse drives the posterior end of the beetle into the ground and eats it head first. It would probably eat* M. obliterata *the same way.*

simply by aiming its front end. In the beetle *Eleodes longicollis*, the spray glands are in the rear, and the beetle, when disturbed, does a quick headstand and spreads a secretion from its abdominal tip. The soldiers of certain termite species possess a pointed cephalic nozzle from which their defensive spray is ejected. This spray not only can incapacitate a small predator but also acts as an attractant to summon more troops. In the whip scorpion, two glands open at the tip of a short knob that moves like a gun turret. Many arthropods possess a number of glands but discharge only from those closest to the point of attack, thus saving ammunition and gaining efficiency. The secretions usually act as topical irritants, especially to the mouth, nose, and eyes of the predator. Because of their relatively permeable skin, frogs and toads, common predators of arthropods, are sensitive to these irritants over their entire bodies.

Vertebrates learn quickly to recognize an obnoxious prey. Blue jays, for instance, having once been sprayed by a walkingstick, will remain aloof from it even when the insect is presented two or three weeks later. Predators also may learn to counter these defenses. Grasshopper mice feed on beetles like *Eleodes longicollis* by jamming them butt-end into the earth, so that their chemical arsenal is harmlessly expended in the soil, and then eating them head first.

A monarch butterfly (below) and one of its mimics, a viceroy butterfly.

Pseudocopulation. The Ophrys *orchid resembles a female bee and even emits an odor similar to hers. Hence males try to mate with the flowers and in the process of pseudocopulation, as it is called, carry pollen from one flower to another. The ruse of the* Ophrys *is abetted by the fact that both male bees and the orchid flowers appear earlier in the season than the true females.*

Müllerian Mimicry: Advertising

For animals that have a highly effective protective device, such as a sting, a revolting smell, or a poisonous or bad-tasting secretion, it is advantageous to advertise. The more inconspicuous or rare such an animal is, the larger the proportion of individuals that must be sacrificed before the bird or other predator learns to avoid it. Müllerian mimics, named after F. Müller, who first described the phenomenon, are groups of insects that, although not closely related phylogenetically, all have effective obnoxious defenses and all resemble one another. Bees, wasps, and hornets probably offer the most familiar example; even if we cannot tell which is which, we recognize them immediately as stinging insects and keep a respectful distance. Similarly, large numbers of bad-tasting butterflies are look-alikes. Müllerian mimicry is adaptive for all species involved because each prospers from a predator's experience with another.

Batesian Mimicry: Deception

Batesian mimicry, first described by the British naturalist H. W. Bates in 1862, is deceptive mimicry. In Batesian mimicry, the innocuous mimic fools its predator by resembling a stinging or bad-tasting "model" which the predator has learned to avoid. Some species of harmless flies resemble bees or hornets, and many species of butterflies resemble monarchs or other unpalatable butterflies or moths.

Laboratory experiments have clearly demonstrated Batesian mimicry in operation. Jane Brower, working at Oxford, made artificial models by dipping mealworms in a solution of quinine, to give them a bitter taste, and then marking each one with a band of green cellulose paint. Other mealworms, which had first been dipped in distilled water, were painted green like the models, so as to produce mimics, and still others were painted orange to indicate another species. These colors were chosen deliberately: orange is a warning color, since it is clearly distinguishable, and green is usually found in species that are not repellent and for whom, therefore, there would be no survival value in advertising.

The painted mealworms were fed to caged starlings, which ordinarily eat mealworms voraciously. Each of the nine birds tested received models and mimics in varying proportions. After initial tasting and violent rejection, the models were generally recognized by their appearance and avoided. In consequence, their mimics were protected also. Even when mimics made up as much as 60 percent of the green-banded worms, 80 percent of the mimics escaped the predators.

Batesian mimicry obviously works to the advantage only of the mimic. The model, on the other hand, suffers from attacks not only from inexperienced predators but from predators who have had their first experience with mimic rather than with model. The mimetic pattern will be at its greatest advantage if the mimic is rare, that is, less likely to be encountered than the model, and also if the mimic makes its seasonal appearance after the model, thus reducing its chances of being encountered first. However, if the model is sufficiently distasteful—as in the case of the quinine-soaked mealworm—it may protect mimics even if the latter are very common.

CONCLUSIONS

In this chapter, we have tried to show some of the ways in which interactions of organisms shape the community and affect the characteristics of its inhabitants, viewing the community as it exists not only in space but through time. In each of our examples, we have spoken about the relationships between two populations. Obviously, community interactions involve many different species, all linked together in what has aptly been called the "web of life." The sum total of forces acting upon even one population is so complex that it can probably never be understood; however, isolated examples, such as we have given here, afford at least a glimpse of the forces at work that create the great diversity of living things.

SUMMARY

A population is a group of organisms of the same species occupying a particular area at the same time. A community consists of all of the populations in a particular area. The position, or role, of a population within a community is called its ecological niche. The hypothesis of competitive exclusion predicts that only one species can occupy the same ecological niche at the same time and that when two species compete for the same niche, one will be eliminated.

The size of any given population is determined by its birthrate and its death rate. The theoretical birthrate of a population—its reproductive potential—is exponential (that is, 2, 4, 8, 16, 32); the more individuals in the population, the faster the population grows. The actual growth rate of an expanding population can usually be graphed by a sigmoid curve, beginning slowly, increasing exponentially for a time, and then leveling off as the population reaches the limits of some available resource, such as food, space, or, in the case of aquatic organisms, oxygen. In most communities, the death rate of a species approximately equals the birthrate, and the population remains relatively stable from one generation to the next.

Both biotic and abiotic factors play a role in the natural regulation of the abundance of organisms. These factors may be density-independent (such as temperature or length of daylight) or density-dependent (such as food supply or predation).

The kinds and numbers of organisms in a community are shaped not only by abiotic factors, such as those described in the previous chapter, but also by biotic factors, the individual interactions among the various populations.

Among the types of interactions are competition, which may result in the elimination of one species (as with the duckweed species) or its restriction to a noncompetitive position (barnacles and blackbirds). Plants sometimes compete with one another by producing toxic chemicals that limit the growth of neighboring species; this phenomenon is known as allelopathy. Symbiosis is the close association between organisms of different species that live together. The association may be beneficial to both (mutualism), beneficial to one and harmless to the other (commensalism), or beneficial to one and harmful to the other (parasitism). In some cases of symbiosis, such as with lichens and the fungus-gardening ants, neither species can survive alone.

The majority of diseases in organisms are caused by parasites. Most parasites do not kill their host, and they almost never wipe out entire populations. Para-

sites tend to become so completely adapted to their hosts that they are entirely dependent on them.

The trophic levels of an ecosystem are linked by predator-prey associations. These associations exert a regulating force on the size of populations and also have profound evolutionary effects on the various species involved.

Plants and animals have developed a variety of defenses against predation. These include "armor" and other forms of physical protection, as seen in cacti, armadillos, turtles, and numerous other organisms, and chemical weapons, such as the plant poisons and noxious secretions of insects. Many organisms are camouflaged.

Some insects have come to resemble organisms of other species either to advertise an effective protective device that they possess in common with the other species (Müllerian mimicry) or to "pretend" they possess such a device when they actually do not (Batesian mimicry).

All these associations help to determine the character of the community and of the organisms within it.

QUESTIONS

1. Define the following terms: population, exponential growth, sigmoid curve, ecological niche, mimicry (Müllerian and Batesian), allelopathy.
2. Introducing a new species into a community can have a number of possible effects. Name some of these possible consequences both to the community and to the species. What types of studies should be made before the importing of an "alien" organism? Some states and many countries have laws restricting such importations. Has your own state adopted any such laws? Are they, in your opinion, ecologically sound?

Chapter 43

Societies and Social Behavior

A society is a group of individuals belonging to the same species and organized in a cooperative manner. It is thus something more than an aggregation of individuals. Often similar organisms—bacteria, for instance, or paramecia, or mealworms—are found gathered in the same place because of environmental conditions such as humidity, shelter, temperature, food supply, and so on. These aggregations do not represent societies in the sense that the word is used by students of animal behavior. In a society, stimuli exchanged among members of the group serve to hold the group together. These exchanges of stimuli, which are essentially forms of communication, result in what is defined as social behavior.

43–1 *Societies of animals are held together by various forms of communication. In insect societies, which are often very large, these communications are entirely impersonal, depending primarily on exchange of chemical signals (phero-mones), and responses to these stimuli are genetically programmed. In non-human primate societies, by contrast, re-lationships are personal and are based on recognition of other members of the group as individuals. The identity of other members of the group and one's behavior toward them depends on learn-ing as well as instinct. The photograph shows two young baboons engaged in this learning process.*

Like the spines of a cactus, the plates of an armadillo, and the symbiosis between a lichen fungus and its alga, societies and patterns of social behavior are the product of natural selection acting upon individual organisms and so affecting their kinds and their numbers.

We shall begin with a description of some insect societies, since they are by far the most ancient of all societies, and also, with the single exception of those of modern man, by far the most complex. We shall follow the bees, termites, and ants with a description of some of the foundations of vertebrate social behavior and close with brief analyses of two vertebrate societies, wolves and baboons.

INSECT SOCIETIES

Of all animal organizations, the insect societies are probably the best understood, in terms both of their evolution and of the interplay of forces that keep them together. Like many other animal societies, they are based on the family unit, and the way in which the mother-egg relationship was gradually developed over the course of millennia into a tightly knit large group is indicated by observations of the behavior of existing species. In fact, the type and degree of care provided to the young insect by the wasp or bee are coming to be used as a taxonomic characteristic for determining relationships between the many species, just as slight differences in the shape of the body or the color of the wing might be used.

BEES

Solitary and Subsocial Species

Although we think of all bees as living in communal hives, most of the species are actually solitary, as are most of the wasp species, which are closely related. Among the solitary species, the female bee builds a small nest, either burrowing into soil or wood or piling up earth or plant material, lays her eggs in it, stocks it with a mixture of honey and pollen (which is the protein source for the larvae), seals it off, and leaves it permanently. In the great majority of solitary species, the mother bee dies before the offspring reach maturity.

In subsocial species, the mother returns to feed the young larvae; similar behavior is shown by many of the wasps. In one type of subsocial bee (*Allodape pringlei*), the eggs are set in a curved row, like a spiral staircase, winding up the inside wall of a hollow stem. After hatching, the larvae hold themselves in place by short, plump "arms." The mother drops food into the hollow stem for the larvae, and they hold the food in these special fleshy protuberances and feed themselves.

In another type of subsocial bee (those belonging to the genus *Halictus*), the mother forms a crude comb of 16 to 20 cells underground, packs in food continuously for some time, and finally closes off the chamber. She then guards the entrance so long as she survives, which may be until the young bees emerge. In some of these species, the emergent young extend the old comb and, in turn, remain to watch over their own offspring. This small community is usually destroyed during the winter, and in the spring, each female that has survived founds a new colony.

3–2 *Carpenter bees (Xylocopidae) are among the largest and most powerful of bees. The female (a) lays her eggs in a hole bored in hard wood and leaves the larvae (b) to develop there.*

(a)

(b)

43–3 *Bumblebees (members of the genus Bombus) are large, hairy bees, primarily adapted to colder climates. They are social bees, but their societies are smaller and simpler than those of the highest social bees, the honeybees (members of the genus Apis). The life cycle is annual, with only the fertilized queen surviving the winter.*

Bumblebees

The bumblebee is a representative of the next stage of socialization. As with the *Halictus* bees, the bumblebees (*Bombidae*) must found their colony anew each spring. Every bumblebee you see in flight in the early spring is a queen and the potential founder of a colony. When she finds a suitable nest site, she constructs two cells from the wax that exudes from the surface of her abdomen. One cell she fills with nectar and pollen from her foraging trips. In the second cell, she lays a group of eggs, usually about eight. She then caps the egg cell with wax and settles down on it like a broody hen. In some three to five days, depending on the species, the eggs hatch; the mother then feeds the larvae on the nectar and pollen. About seven days after hatching, the larvae spin cocoons. The mother continues to guard the cocoons, and she also constructs additional cells, laying eggs in each; in her spare time, she forages for nectar. The larvae within the cocoons pupate for about two weeks, and then, helped by the mother, the damp, soft-bodied, pale-colored young bees crawl out of their cocoons. These workers are all females. In two or three days, the young bees (called callows) become hardened, develop the "furry" coat and bright colors of their mother, and go to work, helping to gather nectar and pollen and to care for each successive brood of younger sisters.

After the emergence of the first group of workers, the mother bee devotes almost all her time to egg laying. The workers do not lay eggs of their own but rather enlarge the nest, which gradually assumes the form of a rough comb, and devote the rest of their time to foraging for the insatiable and evergrowing brood. In other words, the workers care for their sisters (and, eventually, their brothers) rather than reproducing. Insects such as these, in which only certain individuals are reproductive and sterile workers assist the reproductive individuals and help in the care of the young, are known as "eusocial," or "truly social," species.

The nectar gathered by a foraging worker is carried in a special honey stomach of the bee and is regurgitated into a wax container in the hive, where it evaporates into honey. The pollen is transported simultaneously in special pollen baskets formed of long stiff hairs on the bees' hind legs. The returning worker scrapes the pollen off her hind legs by means of her middle legs and into one of the pollen bins in the hive.

The colony usually numbers only a few hundred bees, but it may grow as large as a thousand or so in the course of spring and summer. Toward the end of the season, the production of young males and young queens begins, and the young queens mate. The workers, males, and old queens die at the end of the summer, and the young queens scatter and hibernate, emerging in the spring to found new colonies.

Honeybees

Winter Organization

The honeybee colony, which usually has a population of 30,000 to 40,000 workers, differs from that of the bumblebee and many other social bees or wasps in that it survives the winter. This means that the bees must stay warm despite the cold. Like other bees, the isolated honeybee cannot fly if the tem-

Honeybee workers. (a) The first segment of each of the three pairs of legs (tarsi) has a patch of bristles on its inner surface. Those of the first and second pairs are pollen brushes which gather the pollen that sticks to the bee's hairy body. On the third pair of legs, the bristles form a pollen comb that collects pollen from the brushes and the abdomen. From the comb, the pollen is forced up into the pollen basket, a concave surface fringed with hairs on the upper segment of the third pair of legs. Transfer of pollen to the pollen basket occurs in midflight. The sting is at the tip of the abdomen. (b) The mouth parts are fused into a sucking tube containing a tongue with which the bee obtains nectar. The antennae, attached to the head by a ball-and-socket joint, contain both touch receptors and olfactory organs. The large compound eyes cannot see red (which is black, or colorless, to them) but can see ultraviolet, which is colorless to human eyes.

(a)

(b)

perature falls below 10°C (50°F) and cannot walk if the temperature is below 7°C (45°F). Within the wintering hive, bees maintain their temperature by clustering together in a dense ball; the lower the temperature, the denser the cluster. The clustered bees produce heat by constant muscular movements of their wings, legs, and abdomens. In very cold weather, the bees on the outside of the cluster keep moving toward the center, while those in the core of the cluster move to the colder outside periphery. The entire cluster moves slowly about on the combs, eating the stored honey from the combs as it moves.

The Life of the Worker

Egg laying begins early in the year, in January or February, with each egg deposited in a separate wax cell. The white, grublike larvae that hatch from the eggs are fed almost continuously by the nurse workers; each larval bee eats about 1,300 meals a day. After the larva has grown until it fills the cell, a matter of about six days, the nurses cover the cell with a wax lid, sealing it in. It pupates for about 12 days, after which an adult emerges. The adult rests for a day or two and then begins successive phases of employment. She is first a nurse, bringing honey and pollen from storage cells to the queen, drones, and larvae. This occupation usually lasts about a week, but it may be extended or shortened, depending on the conditions of the community. Then she begins to produce wax, which is exuded from the abdomen, passed forward by the hind legs to the front legs, chewed thoroughly, and then used to enlarge the comb. These houseworking bees also remove sick or dead comrades from the hive, clean emptied cells for reuse, and serve as guards at the hive entrance. During this period, they make brief trips outside, seemingly to become familiar with the immediate neighborhood of the hive. In the third and final phase of their existence, the worker bee forages for honey and nectar. The life span of a worker is usually only about six weeks.

The Queen

Each hive has only one adult queen. The queen caste is genetically the same as the worker caste. All the differences between the two castes depend on the substance fed the queen-to-be in the larval stage and on the pheromonal influences she, in turn, exerts upon her subjects. For the first two days of life, all bee larvae are fed "brood food," a white paste produced in the glands of young workers and secreted from the mouth. Worker and drone larvae are then fed honey and pollen, while the larva being made ready for queenhood is fed only the glandular secretions (hence known as royal jelly) all during its larval stage. Attempts have been made to identify the substance in royal jelly that confers queenhood, but so far they have not been successful.

Queens are raised in special cells larger than the ordinary cells and shaped somewhat like a peanut shell. If a hive loses its queen, workers will notice her absence very quickly, sometimes in only 15 minutes, and will become quite agitated. Very shortly, they begin enlarging worker cells to form emergency queen cells. The larvae in the enlarged cells are then fed exclusively on royal jelly until a queen develops. Any diploid larva so treated will become a queen.

(a)

(b)

(c)

(d)

43–5 *The life of the hive. (a) Workers tending honey and pollen storage cells. The honey is made from nectar processed by special enzymes in the workers' bodies and is thickened by evaporation during repeated regurgitation and swallowing. (b) Queen inspecting a cell before laying an egg in it. (c) Royal larvae. Nourished on royal jelly throughout their early development, such larvae are the future queens. (d) Another task of the worker is air-conditioning. Here a worker beats its wings to cool the hive.*

The Queen's Pheromones

The queen exerts influences on her subjects by means of pheromones, of which there appear to be several. As shown by the British entomologist C. G. Butler and his co-workers, the influence of one of the pheromones (9-ketodecenoic acid) inhibits ovarian development in the worker bees and prevents them from becoming queens or producing rival queens. This queen substance, as it is sometimes called, passes through the workers of the hive orally. As the workers meet, they often exchange the contents of their stomachs. Studies in which queen substance has been tagged with a radioactive label have shown that, as a result of this activity, the pheromone travels through the hive with remarkable rapidity. Within only half an hour after removal of the queen, the shortage of 9-ketodecenoic acid is already noticed, and the hive begins to grow restless. It is difficult to understand how a single queen can produce enough pheromone to influence the entire hive of as many as 40,000 workers or more, as well as tend to her stupendous egg-laying chore, but it has been suggested that, after the pheromone is passed among the workers, it is fed back to the queen in a reduced form and she need simply oxidize it to reactivate it.

3–6 Homecoming worker (at center left) shares nectar with other members of the hive. The hivemates extend their tongues to receive droplets of regurgitated fluid.

-7 Spring swarm of honeybees.

New Colonies

In the spring, when the nectar supplies are at their peak, so many new broods are raised that the group separates into two colonies. The new colony is always founded by the old queen, who leaves the hive taking about half of the workers with her. This helps to ensure survival of the new colony since this queen is of proven fertility. The group stays together in a swarm for a few days, gathered around the queen, after which the swarm either will settle in some suitable hollow tree or other shelter found by its scouts or will be picked up by a beekeeper and transferred to an empty hive, which it will then furnish with wax.

In the meantime, in the old hive, new queens have begun to develop, often even before the old queen has left, and ovarian development has begun in some of the workers, some of which lay eggs which, since they are unfertilized, develop into males or drones. After the old queen leaves the hive, a new young queen emerges, and any other developing queens are destroyed. The young queen then goes on her nuptial flight, exuding a pheromone (apparently also 9-ketodecenoic acid) that entices the drones of the colony to follow. She mates only on this one occasion, although she may mate with more than one male, and then returns to the hive to settle down to a life devoted to egg production. During this single nuptial flight, she receives enough sperm to last her entire life, which may be some five to seven years. These are stored in her spermatheca and are released, one at a time, to fertilize each egg as it is being laid. The queen usually lays unfertilized eggs only in the spring, at the time males are required to inseminate the new queens, so the release of the sperm cells from the spermatheca is apparently under the control of her internal or external environment. The drones' only service to the hive is their participation in the nuptial flight. Since they are unable to feed themselves, they become an increasing liability to the hive. As nectar supplies decrease in the fall, they are stung to death or driven out to starve by their sisters.

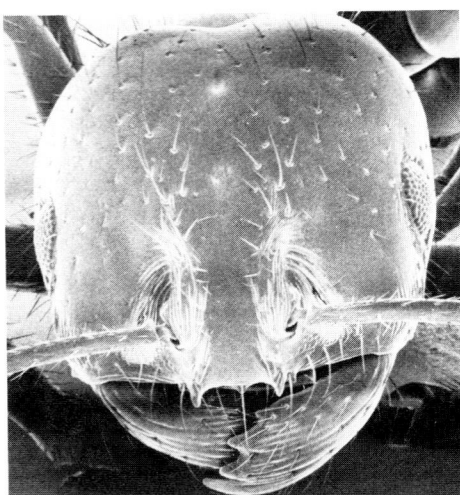

43–8 *Head of an ant* (Monomorium phara-
onis). *In the lower part of this scanning
electron micrograph, above the mandibles, are the lower segments of the
two antennae, which are covered with
sense organs of touch, taste, smell, and
perhaps water detection. The bristles on
the head also probably have a sensory
function. Compound eyes are visible on
either side; however, vision is less important in this terrestrial species than are
the other senses in maintaining communication with other members of the
colony. Upon meeting, ants will caress
one another with their antennae, exchange droplets of liquid, and groom one
another. Recognition is based on colony
odor. Although their sensory equipment
is complex, like other insects, ants have
small brains with relatively few cells
and synapses compared to the brains of
even the simplest vertebrates. This
species, which originated in Africa, has
been transported by commerce to nearly
all parts of the world and is a common household pest, living in walls of
houses and feeding omnivorously on
smaller insects and particles of grease
and sugar.*

ARMY ANTS

All ants are eusocial. Their range of social behavior is so great that it alone is
the subject of several books as large as this one.

One group of ants is known as the legionary, army, or doryline ants. Theodore Schneirla, working in Panama, studied the army ants *Eciton hamatum* and
Eciton burchelli. Like all dorylines, members of this species are raiders, living
almost entirely on arthropod prey, and they are nomadic, with no permanent
nesting place.

The temporary nest, or bivouac, as it is called, of the army ant is not made of
wood or dirt, like that of other ants, but is a seething cylindrical cluster of ants
hooked one to another. This is made possible by strong hooks on the last tarsal
segment of each leg of every worker. The clusters hang from a natural ceiling,
such as the underside of a fallen tree, a branch, or a hollow log. Within the
cluster is a labyrinth of corridors and chambers in which the eggs are laid and
the queen and brood are sheltered and fed.

Each dawn, the ants awaken and the workers unhook themselves, tumbling
into a churning throng on the ground. As more fall, the excitement increases
until a raiding column literally bursts forth. There is no leader; some of the workers are pushed forward by the surging mass behind, and the others follow the
chemical trail they lay down. An "impulse to follow" and tactile stimuli are also
important in forming the ant columns, but it is basically the odor trail secreted
from the abdomen that the ants follow. These raiding columns often advance
as quickly as 25 meters an hour and may finally extend 250 meters from the
nest in an unbroken stream.

The movements of *Eciton* follow a functional cycle with two alternating
phases, as do those of all of the army ants. During the nomadic phase the
colony carries out vigorous daily raids, which are followed by nightly emigrations of the entire colony to new nesting sites. During the second, or statary,
phase, the raids are shorter or sometimes absent and emigrations are rare. The
length of the cycle is species-specific; in *E. hamatum*, the nomadic phase lasts
16 to 18 days, and in *E. burchelli*, it ranges from 12 to 17 days. The statary
phase in both species is 20 or 21 days long. The length of both phases is determined by the developmental stage of the brood.

The nomadic phase begins when the new workers emerge from their cocoons.
This stimulates the entire colony. Daily raids increase in size and tempo and
are followed by night marches, often lasting five or six hours, to new nesting
places. These nightly emigrations are continued throughout the nomadic phase.
Midway in the phase, when the excitement generated by the new workers has
begun to wear off, the larvae hatch and, in turn, become the motivating force.
The workers constantly stroke them with their antennae, lick them with their
mouthparts, feed them, and carry them about. When these larvae pupate, the
level of excitement in the colony falls sharply, and the statary phase begins.

During this phase, the quiescent pupae seem to exert a soothing influence on
the workers, which clutch the cocoons in their mandibles. This phase lasts for
the 20 to 21 days it takes for the pupae to mature. After the first week, the
queen, stimulated by the licking and grooming of the workers, goes into a
stupendous 8- to 10-day labor. In *E. hamatum*, she may lay as many as
80,000 eggs; in *E. burchelli*, 200,000. As the pupae approach maturity and begin

-9 *African army ants, members of the genus Dorylus. These army ants live in colonies numbering up to 20 million workers and forage in columns, such as this one. Their prey is characteristically larger arthropods, including other social insects.*

to stir and twitch within their cocoons, the workers become excited and start to snatch the cocoons from one another. The level of activity in the colony then builds up, and as new workers emerge from the cocoons, the daily raids become more extensive until one night an exceptionally vigorous daily raid is followed by an emigration. And so another nomadic phase begins.

The queen is the most important factor in clustering and in keeping the colony together. Army ants in the laboratory will cluster wherever the queen has rested long enough to leave a trace of her pheromone. If a colony loses its queen, it can maintain a fairly normal functional cycle so long as the brood she has produced is still developing. After that, the colony enters a statary condition in which it will eventually perish unless fusion occurs with another colony. Normally, workers of different colonies, even though of the same species, will not mix when their raiding columns meet. But if one of the colonies has been without a queen for as little as 12 hours, its workers will mingle readily with those of a colony with a queen.

TERMITES

Termite societies are by far the most ancient communities in the world. Man has existed as a social animal for less than 1 million years, ants for about 100 million years, and termites for something like 200 million years. There are five different families and almost 2,000 species of termites. All species are eusocial; some termitaries contain as many as 3 million individuals.

10 *Army ant carrying a scorpion pincer.*

(a)

(b)

43–11   (a) *Termite queen, her body swollen with eggs, is surrounded by workers that feed her. She can lay an egg approximately every 3 seconds. (b) Cross section of wood infested with termites.*

Termite societies are organized by a rigid caste system in which differences of occupation are clearly reflected in body size and shape. The largest member of the colony is the queen, which may in some species reach several inches in length, far larger than any member of her "court." Huge, grublike, and almost shapeless, she serves as an egg factory, fed constantly by the workers and laying as many as 30,000 eggs a day. Even if she were able to move, her swollen body could not pass through the narrow chambers of the termite colony, so she spends her entire existence, often 10 or 15 years, sealed in her nuptial chamber. In many species, the king, the sole reproductive male of the colony, is sealed in the chamber with her.

The two largest termite castes are the soldiers and the workers. The soldiers, defenders of the community, possess large, strong mandibles and also, in some species, a syringelike beak from which they can eject a thick, irritating liquid. In many species, the headparts of the soldiers have become so massive that they are no longer able to feed themselves and must be taken care of by the workers, who also tend the royal couple, feed the young, and maintain the many-chambered colony.

The eggs laid by the queen develop into nymphs. All nymphs are the same through the first stages. If the colony is short of workers, the nymphs will develop, in their final molt, into workers; if soldiers are needed, a proportion of nymphs will become soldiers. (On the other hand, if there are too many soldiers, the workers will destroy some of them.) If the king or queen should die, a nymph will develop into a reproductive form and assume the burdens of the royal chamber. At specific periods of the year, depending on the species and the climate, termite colonies produce winged, reproductive males and females. These leave the colony on a brief nuptial flight, form pairs, lose their wings, and dig a burrow in which they will close themselves forever from the outside, thereby founding a new colony.

Termites lead most of their lives in hermetically sealed, warm, damp burrows, and many species perish of desiccation if exposed to the drying effects of light and air outside their burrows. This is why, incidentally, they can do so much damage when they infest a building, since they will hollow out entire beams without exposing themselves. The common destructive form prevalent in the United States has one important point of vulnerability: a need to return to the soil, apparently for moisture or nutrients. Termites are eradicated by insecticides applied to the soil.

ORGANIZATIONAL PRINCIPLES OF INSECT SOCIETIES

Insect societies, although they resemble human societies in many ways, differ from them in at least two important respects. One major point of difference is that in insect societies reproduction is carried out by only a very few individuals. For this reason all of the insects in one society are likely to be extremely closely related genetically. In fact, in some respects, an insect society is more like a supraorganism with different specialized parts than a collection of individuals.

Secondly, recognition among members of the society is based entirely on chemical or other signals; there is no personal recognition of individual members as there is in primate or other mammalian societies. This stereotyping of behavior is, of course, in keeping with (and demanded by) the relatively short life span of insects. A worker bee, for example, lives only about 50 days, which gives her little time for complex learning. It is also in keeping with the large, sometimes enormous, size of the societies themselves.

VERTEBRATE SOCIETIES

ADVANTAGES AND DISADVANTAGES OF GROUP LIFE

Group life affords the individual protection against predators. Large groups are unlikely to be attacked; even birds as savage as the hawk nearly always select a solitary flyer or a straggler. Fish that form large schools are less likely to be attacked by larger fish. It is believed that the barracuda, for instance, will mistake a large school of very small fish for one large fish and so leave it alone. Sometimes the sheer numbers of individuals in a large group are confusing to the hunter. *Daphnia,* which are small crustaceans, are common aquarium food, as fish fanciers know. If fish are fed many daphnids all at once, a smaller number are eaten than if a few daphnids are presented at one time; apparently the fish get confused by the conflicting stimuli. If one member of a group senses danger, its actions warn the rest; in some groups of birds and of mammals, certain individuals appear to act as sentinels.

Social organization is also of great value in mate selection. Within a social organization, animals can be assured of finding mates, and also, in many societies, superior females are able to choose among superior males, or superior males are able to mate with more females.

There are other advantages. Small warm-blooded animals, such as mice, provide each other with warmth, and groups of aquatic animals can sometimes survive in a toxic environment more successfully than isolated animals since their combined metabolic processes can serve to detoxify it.

43–12 *Caribou migrating. Herds sometimes number thousands of individuals.*

On the other hand, group living is not always advantageous. For example, the food resources of a herbivore or a predator may be so distributed that sharing is impossible. Thus, whether an organism is social or solitary, or some combination of the two, is determined by selective pressures acting on the individual organism.

SOME BASES FOR SOCIAL ORGANIZATION

Vertebrate societies range from small, often transient family groups, in which the social focus is on the raising of the young, to larger, permanent, quite stable groups, such as those found among the primates. Some of the societies—such as flocks of birds, schools of fish, and large migratory herds of herbivores—may number in the thousands, but they are not usually as large as the societies of most eusocial insects.

How are these vertebrate groups held together? There is no simple answer to this question, partly because the different kinds of groups are undoubtedly bound together in different ways and partly because most explanations depend heavily on concepts such as "sexual drive," "maternal instinct," or "affectional systems," which in themselves are almost impossible to define. As an indication of the complexities involved, we shall touch on two types of behavior that serve to keep animals in groups: imprinting and rituals.

3–13 *Many species of precocial birds (birds that are able to walk as soon as they're born) will follow the first moving object they see after hatching and will continue to show this following response as they mature. The phenomenon is known as imprinting. The birds shown here are goslings, and the object of their affection is animal behaviorist Konrad Lorenz. (Nina Leen, Time-Life Picture Agency.)*

Imprinting

Certain birds, such as swans, chickens, and turkeys, which are physiologically mature enough to leave the nest soon after they are hatched, follow the first moving object that they see. For ducklings, the effective objects can range from a matchbox on a string to a walking man. Once one of these newly hatched birds has become attached to a particular object, it will follow only that object. Under natural circumstances, of course, this object will be a parent, and it is this following response that keeps the young birds close behind and well within the protective range of the mother until the end of their juvenile period, when the response is lost.

The learning pattern that involves this act of recognition is called _imprinting_, and it differs from other types of learning in that it can occur only within a limited period. Studies of the newly hatched mallard duckling, using a mechanically operated decoy, showed that imprinting was most effective between 13 and 16 hours after hatching. After 24 hours, 80 percent of the ducklings failed to follow moving objects, and after 30 hours, all the ducklings avoided them. Similarly, chicks will not follow a moving object when they are only a few hours old or when they are several days old but only during the intervening period. It is believed that the end of the following period is marked by the development of fear responses, which seem to arise in normally reared birds at about the same time that the imprinting period is over.

The gosling is especially susceptible to imprinting. The only way to make an incubator-hatched gosling follow other geese is to bundle it up in a bag and rush it to the barnyard just as it is hatching, thus ensuring that the first object it sees is a goose.

Imprinting also seems to influence mate selection in the adult birds. Lorenz reports many examples of ducks and geese becoming sexually fixated on objects or on members of other species, including Lorenz himself. Controlled experiments in ducklings support these observations. Ducklings raised with females of another species courted females of that species when they reached maturity, while ducklings raised with females of their own species did not.

The term "imprinting" is usually applied only to phenomena associated with the following response in birds, but similar effects are seen in other animals. For example, if a baby lamb is taken from its mother right after birth and bottle-raised, it will not follow other sheep when it is mature, as do lambs raised normally.

Rituals

Although few vertebrates are cannibals, many of them, fish and birds in particular, have a distinct preference for maintaining discreet distances between themselves and other animals, whether of their own or another species. Such avoidance is adaptive; for many animals, being touched means being captured. Among birds lined up on a telephone wire, for instance, each will be found almost equidistant from the next, as though the spaces were measured off by an invisible yardstick. Even those animals that fly in flocks or swim in schools tend not to touch one another.

The mating ceremonies of these animals are interpreted as a means for allaying those highly adaptive feelings of hostility and suspicion so that mating can

43–14 *Sandhill cranes in a courtship dance flap their wings, hop, and leap into the air. Both sexes join in these ceremonies which start in winter and continue until eggs are laid in the spring.*

take place. As a consequence, many of these ceremonies incorporate both aggression and appeasing behavior. In the mating behavior of the stickleback, for example, described in Figure 30–14, the male's "zig" toward the female is identical to a motion of attack; and when the male "zags," the motion entices her toward the nest. Most females flee the attack motion, but the one whose need to lay her eggs is sufficiently great stands still, turns sideways (an appeasing gesture on her part since an antagonist would zig back), and then follows the male's zag.

In mating ceremonies between birds, feelings of fear and aggression brought into play by the closeness of another individual may be handled by being redirected at either a real or an imaginary antagonist. Konrad Lorenz describes the so-called "triumph ceremony" in the graylag goose, one of his favorite research subjects and companions. As a form of greeting to his partner, the gander proceeds to attack an "enemy." This attack is performed, as is a real attack, with the head and neck pointing obliquely forward and upward and is accompanied by a raucous trumpeting. After the "enemy" is routed or defeated, the gander returns to his partner. On his return, the gander holds his head lowered and pointed forward, but instead of pointing directly at the goose, as he would at an enemy, he points obliquely past her. She comes forward to meet him, her head inverted submissively, and he cackles to her triumphantly.

This ceremony, which probably had a purely sexual origin, now serves to hold entire flocks together, and even small goslings participate in elements of the triumph ceremony. When a young male performs the ceremony with a strange young female, it usually marks the beginning of a mating bond that may last for the entire lifetime of the individuals. The ceremony is typically performed between young geese the year before mating and breeding begin, and it will continue to be performed by the partners throughout their entire lives whenever they encounter one another even after a short separation. By the intensity with which the ceremony is performed, an experienced observer can judge the length and strength of the bond between the partners.

Although few animals—except perhaps humans of ambassadorial rank—have such elaborate social rituals, higher animals of many species exhibit obligate social formalities upon meeting. Think, for example, of the greeting ceremonies between domestic dogs, and how they vary with sex, rank, and familiarity.

An unusual form of social dominance is seen among the wrasses, a species of small iridescent blue fish that inhabit the waters around Australia's Great Barrier Reef. Each group of wrasses consists of a male and a number of females; the group is territorial, with territoriality exhibited largely among males. The single male and his females are organized in a hierarchy; the largest, oldest individual is the male, who dominates all the females of the group. Next in rank is the largest, oldest female, who occupies the center of the territory with the male and is dominant over the other smaller and younger females. When the male dies, either neighboring males invade the territory, or, as occurs in the majority of cases, the ranking female becomes dominant. Within 1½ to 2 hours after the male's death, she begins to show male aggressive behavior and within a few hours she is exerting dominance over the other females and patroling the borders of the territory. Within a few days, the new dominant has begun male courtship and spawning behavior, and small regions of dormant testicular tissue within her ovaries have begun to develop. By the end of about two weeks, the ovaries have become testes and "she" is fully male, able to produce sperm and fertilize the eggs of the subordinate female family members.

Wrasses photographed at a reef off New Providence Island, Bahamas.

—15 (a) *A subordinate baboon turns his buttocks toward a superior. This gesture, known as presenting and used by females to indicate their readiness to mate, is also used between males and between females to signify submission or conciliation or to beg for grooming. (b) The subordinate baboon is reassured with a pat on the back.*

(a)

(b)

Rituals serve the function both of allaying the anxieties of the individuals involved and of identifying them to one another as members of the "in" group.

SOCIAL DOMINANCE

Animal societies are often arranged in hierarchies. One type of social dominance that has been studied in some detail is what is called the *pecking order* in chickens. A pecking order is established whenever a flock of hens is kept together over any period of time. In any one flock, one hen usually dominates all the others; she can peck any other hen without being pecked in return. A second hen can peck all hens but the first one; a third, all hens but the first two; and so on through the flock, down to the unfortunate pullet that is pecked by all and can peck none in return. Hens that rank high in pecking order have privileges such as first chance at the food trough, the roost, and the nest boxes. As a consequence, they can usually be recognized at sight by their sleek appearance and confident demeanor. Low-ranking hens tend to look dowdy and unpreened and to hover timidly on the fringes of the group. During the period when a pecking order is being established, frequent and sometimes bloody battles may ensue, but once rank is fixed in the group, a mere raising or lowering of the head is sufficient to acknowledge the dominance or submission of one hen in relation to another. Life then proceeds in harmony. If a number of new members are added to a flock, the entire pecking order must be reestablished, and the subsequent disorganization results in more fighting, less eating, and less tending to the essential business, from the poultry dealer's point of view, of growth and egg laying.

Pecking orders reduce the breeding population. Cocks and hens low in the pecking order copulate much less frequently than socially superior chickens.

Thus the final outcome is the same as if the social structure did not exist: The stronger and otherwise superior animals eat better, sleep better, and leave the most offspring. However, because of the social hierarchy, this comes about with a minimum expenditure of lives and energy.

833 *Societies and Social Behavior*

TERRITORIES AND TERRITORIALITY

Most vertebrates stay close to their birthplaces, occupying a home range that is likely to be the same home range occupied by their parents. Even in the case of migratory birds that travel great distances, they are likely to return year after year to the same areas. Often these home ranges are defended, either by individuals (or more likely, mating pairs) or by groups against other individuals or groups of the same species. Areas so defended are known as territories, and the behavior of defending an area against intruders of the same species is known as territoriality.

Territoriality in some form has now been recognized in animals as widely diverse as crickets, howler monkeys, fur seals, dragonflies, red deer, beaver, prairie dogs, many types of lizards, and a large number of species of birds and fish.

A territory may be an area of plain, a corner of a small wood, or a few feet at the bottom of a pond. Sometimes the territory consists of little more than the nest itself and the immediate area around it. For the male bitterling, a small fish, the territory is an area immediately surrounding a freshwater mussel. The bitterling admits only egg-laden females into his territory, then leads them to the mussel, where they lay their eggs within its gills. The male then injects his sperm while swimming over the siphon of the mussel.

Territoriality in Birds

Territoriality was first recognized by an English amateur naturalist and bird watcher, Eliot Howard, who observed that the spring songs of male birds served not only to court the females but also to warn other males of the same species away from the terrain that the prospective father had selected for his own. In general, a territory is established by a male. Courtship of the female, nest building, raising of the young, and often feeding are carried out within this territory. Often the female also participates in territory defense.

In some species of birds, the usual sex roles are reversed. The female red-necked phalarope establishes and defends the territory while the drab male defends the nest, incubates the eggs, and takes care of the young. Some birds, such as the American robin and the European swallow, tend to return to the same territory year after year, often with the same mate. Storks and night herons, however, are attached to a nesting place rather than to a mate, and if the mate of the previous year does not return the following year, he or she is readily exchanged for a new incumbent.

By virtue of territoriality, a mating pair is assured of a monopoly of food and nesting materials in the area and of a safe place to carry on all the activities associated with reproduction and care of the young. Some pairs carry out all their domestic activities within the territory. Others perform the mating and nesting activities in the territories, which are defended vigorously by the males, but do their food gathering on a nearby communal feeding ground, where the birds congregate amicably together. A third type of territory functions only for courtship and mating, as in the bower of the bowerbird or the arena of the prairie chicken. In these territories, the male prances, struts, and postures—but very rarely fights—while the females look on and eventually indicate their choice of a mate by entering his territory. Males that have not been able to secure a territory for themselves are not able to reproduce; in fact, there is

(a)

(b)

−16 (a) *A male sage grouse courts females and intimidates other males by inflating air sacs in his chest and fluffing up his feathers. Females choose among males by entering their territories, which are simply adjacent areas on a common display ground.* (b) *Greeting ceremony of storks at their nest.*

evidence from studies of some territorial species, such as the Australian magpie, that adults that do not secure territories do not mature sexually.

Territories vary greatly in size. The golden eagle, for instance, defends a territory of 93 square kilometers, or about 37 square miles; the European robin, a territory of about 6 square kilometers; and the king penguin, a territory of only ½ square meter. Individuals of the same species often have territories of somewhat different sizes, depending partly upon the density of food and shelter in a given area, partly upon population pressure, and partly upon the aggressiveness of the individual.

Territorial Defense

Even though territorial boundaries may be invisible, they are clearly defined and recognized by the territory owner. With birds, for example, it is not the mere proximity of another bird of the same species that elicits aggression, but his presence within a part of a particular area. The territory owner patrols his territory by flying from tree to tree. He will ignore a nearby rival outside his territory, but he will fly off to attack a more distant one that has crossed the border. Animals of other species are generally ignored unless they are prey or predators.

Once an animal has taken possession of a territory, he is virtually undefeatable on it. Among territory owners, prancing, posturing, scent marking, and singing and other types of calls usually suffice to dispel intruders, which are at a great psychological disadvantage. If a male stickleback, for example, is placed in a test tube and moved into the territory of a rival male, he visibly wilts, his posture becoming less and less aggressive the farther within the territory he is transported. Similarly, a male cichlid will dart toward a rival male within his territory but as he chases the rival back into his own territory, he begins to swim more slowly, his caudal fin seemingly working harder and harder, just as if he were making his way against a current that increases in strength the farther he pushes into the other male's home ground. The fish know just where

43–17 *Territories come in many shapes and size (a) The male Uganda kob displays on his stamping ground, which is about 50 feet in diameter and is surrounded by similar stamping grounds on which similar males display. A female signifies her choice by entering one of the stamping grounds and grazing there. Only a small proportion of males possess stamping grounds, and those that do are the only ones that breed. For a fiddler crab (b), it is a burrow, from which he signals with his large pink claw, beckoning females and warning off other males. (c) Howler monkeys shift their territories as they move through the jungle canopy, but maintain spacing between groups by chorusing. (d) Territoriality is common among reef fish. For many species, a territory is a crevice in the coral, but for others, such as the skunk clown fish shown here, the territory is a sea anemone. The fish is covered by thick slime that partially protects it from the poison of the tentacles, but its acceptance by the anemone is chiefly a consequence of behavioral adaptation of the fish, which even mates and raises its brood among the tentacles.*

(a)

(b)

(c)

(d)

the boundaries are and, after chasing each other back and forth across them, will usually end up with each one trembling and victorious on his own side of the truce line.

Similarly, the expulsion from communal territories is typically accomplished by ritual rather than by force. For example, among the red grouse of Scotland, the males crow and threaten only very early in the morning, and then only when the weather is good. This ceremony may become so threatening that weaker members of the group leave the moor. Those that leave often starve or are killed by predators. Once the early-morning contest is over, the remaining birds flock together and feed side by side for the rest of the day.

Territories and Population Regulation

As we mentioned previously, animals that do not gain possession of territories or that are excluded from the "home range" or breeding area do not produce young. As older animals die, younger ones will contend for their places, keeping the breeding population stabilized. If conditions are particularly favorable, or if the range is enlarged, more animals can gain access to the breeding community.

Limitation of the reproductive rate, as by social dominance or territoriality, is sometimes referred to as "self-regulation of animal populations." Some ecologists object to this term, pointing out that, in these populations, as in all others, it is environmental resistance that is doing the limiting and not the population itself. However, within the limits imposed by the environment, dominance and territoriality, while allowing the survival and reproduction of dominant animals (which would occur anyway), minimize physical combat.

-18 *An Alaskan fur seal harem; the bull is in the upper right-hand corner of the picture. The only breeding males are those adult bulls large and strong enough to secure a beach territory and round up a harem for the summer. Females arrive at the breeding grounds in June, about a month after the males have established their territories. There is a period of almost constant conflict while each bull herds as many females as he can into his own territory. In a very short time, the females give birth to pups conceived during the preceding summer. There is then a period of mating, and for the rest of the summer the bull will continue to guard the harem, neither eating nor leaving his territory for several months. In the fall, the males and females separate, the females and young going south and the bulls moving to the Gulf of Alaska. In the spring, the bull will gather another family together again, but his harem may consist of an entirely new group of females.*

43–19 *Among animals of the same species, fighting rarely results in death or even injury to either of the combatants. (a) Two male grunts mouth-pushing. The* loser, the one who backs down first, is allowed to swim away unharmed. (b) *Two male ibex engaged in a territorial dispute for a courtship arena. (c) Kan-*garoo rats attack one another with their hind legs. The battle ends when one gives up and bounds off.

(a)

(b)

(c)

SOME NOTES ON AGGRESSION

The examples cited in the discussion of social hierarchies and territoriality serve to remind us that actual combat is rare among members of the same species, whether or not they are members of complex societies.

When fighting does occur among animals of the same species, it is often ritualized in such a way that both animals remain unharmed. For instance, male iguanas of the Galapagos Islands fight by pushing their heads against one another; the one that drops to its belly in submission is no longer attacked. Male cichlid fish of one species first display, presenting themselves head on and then side on, with their dorsal fins erected, and then beat water at each other with their tails. If this does not bring about a decision, each grasps the other by its thick strong lips; they then pull and push with great force until one lets go and, unharmed but defeated, swims away. Rattlesnakes, which could kill each other with a single bite, never bite when they fight but glide along side by side, each pushing its head against the head of the other, trying to push it to the ground in a form of Indian wrestling. Many antlered animals, such as stags of the fallow deer, which have long and vicious horns, follow an equally careful ceremony and attack only when they are facing each other, so that their horns are used only for dueling and not for goring.

These proscriptions against intraspecific killing make sense biologically. It is to the advantage of each individual, in terms of its own survival, not to waste its strength in needless bloodshed, and therefore genetic traits that channel aggressions and reduce conflict will be selected under evolutionary pressures. Except for certain fish that regularly cannibalize their young, and rats operating under severe population pressures, man is one of the few vertebrates that regularly kills its own kind.

SOCIAL BEHAVIOR AMONG WOLVES

Wolves are large members of the genus *Canis*, closely related to domestic dogs. Adult males average about 95 to 100 pounds (about the size of a good-sized German shepherd). A few individuals may weigh more; a 175-pound male was reported from Alaska. Wolves are carnivorous and predatory; their usual prey consists of large animals, such as moose, caribou, and deer, although they sometimes eat smaller herbivores, such as rabbits and mice. Their only predator is man, but they are host to a large number of parasites, such as ticks and tapeworms, and diseases, including rabies. The habitat of the wolf once included all of the biomes of the Northern Hemisphere, excluding only tropical rain forests and deserts. At present, the only substantial numbers of wolves in North America are found in Canada (the estimated number there is 17,000 to 28,000) and Alaska (5,000–10,000). The only population left in the rest of the United States, in northern Minnesota, numbers about 500 to 1,000 individuals. The state of Wisconsin placed the wolf under legal protection in 1957, but that was too late; there are no more wolves in Wisconsin. An additional 20 to 30 wolves live on Isle Royale in Lake Superior. There are also a few wolves left in upper Michigan and there may be a small population in the Rocky Mountains. Now the Endangered Species Act of 1973 provides legal protection for all the wolves in the 48 contiguous states.

20 *A pair of wolves* (Canis lupus). *The female (lying down) is characteristically smaller than the male. The species is variously known as the gray wolf, the timber wolf, and the tundra wolf.*

The Wolf Pack

All animals that prey on large mammals either weigh as much as their prey (for example, tigers) or hunt in packs; wolves belong to the latter category. Wolf packs are usually made up of two to eight members, although much larger groups have been reported occasionally, as have solitary wolves. A pack in this size range seems to represent a maximally efficient hunting unit; L. David Mech, author of what is considered the definitive study of wolves, reports that only about 6 wolves of a large pack of 15 actually hunt at any one time. Natural selection would also work against tendencies to form very large packs since a single kill would not feed all members of such a pack. However, since packs vary greatly in size—as compared to clutch size in birds, for instance—selection pressures must also be either varied or not very strong.

Although a number of field studies of wolves have been done, no one has been able to observe a single wild wolf pack over a long enough period of time to determine how a pack originates. Most packs are made up of a breeding pair, pups, and extra adults old enough to breed (but which usually do not). The social bonds among members of a wolf pack are very strong. In both strength and expression, they closely resemble the bonds between a dog and his master, and, indeed, it seems certain that the existence of a strong bonding system among the evolutionary ancestors of the domestic dog is the reason why the dog is so uniquely domesticable.

Wolf packs travel, sleep, and hunt together. They share their kills, either by regurgitating food, by carrying home part of their prey, or by leading other members of the pack to a fresh kill. This sharing of food among adult members of a group is rare among wild animals; for example, among the primates organized sharing of food occurs only in our own species.

Studies made of animals in captivity indicate a strong inhibition among wolves against taking food away from an animal already in possession of it, even by wolves that are both superior in strength and hungry.

One of the strong bonds that holds the wolf pack together is the sexual bond between the breeding pair. Wolves may begin choosing mates when they are about one year old, although a wolf is not sexually mature until the age of 22 months, and there is some evidence that these pair bonds, once formed, are maintained until one member of the pair dies.

Unlike the sexual bond between humans, the bond between a mated pair of wolves is not related to the constant sexual availability of the female. Female wolves come into heat only once a year, and although courtship (which closely resembles the play behavior of younger animals) is carried out over a period of several weeks, copulation occurs only in the one week of estrus, the time when ovulation takes place.

It has been suggested that this bonding is reinforced by the copulatory tie, a phenomenon unique to members of the dog family. In the copulatory tie, the female's vaginal sphincter muscles constrict about the swollen penis of the male after intromission and as a consequence the two animals remain attached by their genitals for 15 minutes or more following ejaculation, during which time they stand or lie rump to rump. The copulatory tie immobilizes both animals in an extremely awkward position, highly vulnerable to harassment and attack. To the casual observer it would not seem likely to increase the affection of the

43–21 *Courtship. It is the female that is making the advances. A breeding pair, such as this one, usually consists of the dominant male and the dominant female.*

3–22 *Evolution of the wolf. Other members of the genus* Canis *(dogs) include the domestic dog, the coyote, the jackals, and the dingo. Creodonts ("flesh-teeth") were early carnivorous mammals, ancestors of the modern carnivores and descendents of insectivores.*

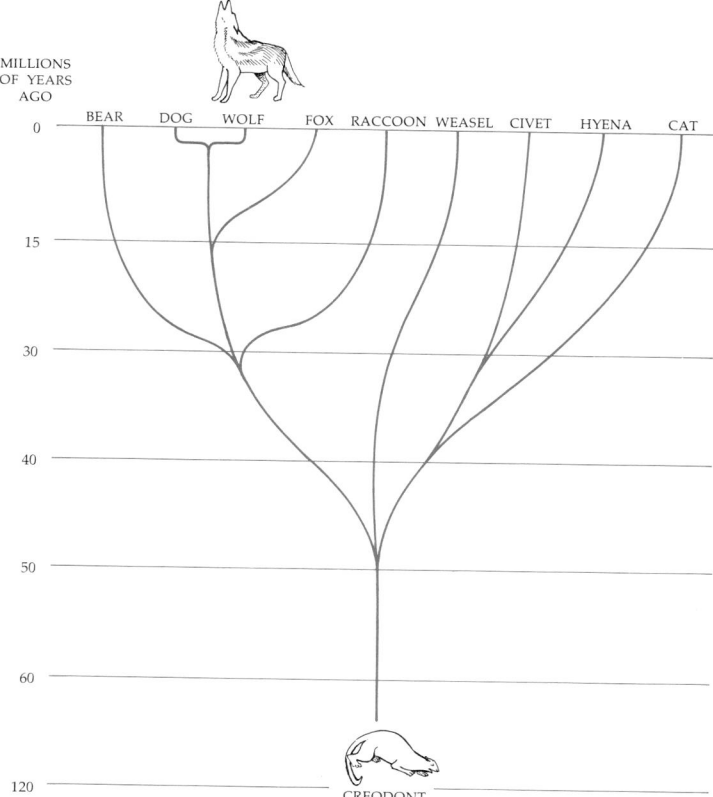

3–23 *Wolf with pups. Dens are dug a few weeks before the birth of the young and the young remain in the dens until they are about 8 weeks old. Care of the pups is shared among members of the pack. An average litter contains 5 or 6 young.*

individuals involved toward one another, but no other explanation for it has been advanced.

Strong bonds are formed between mothers and pups and between the pups and other members of the pack. During the pups' first three weeks of life, they remain in the den, in close contact with one another, and are nursed by their mother. After this time, they begin to eat predigested food that they obtain by touching their mouths to the mouths of adult wolves. This gesture stimulates the wolves to regurgitate meat for the pups. Wolf pups are fed in this manner not only by their mother but by any adult member of the pack. As the pups grow older, they are played with by all members of the pack, and numerous instances have been reported of wolf mothers going off to hunt with other members of the pack while leaving another adult behind as a baby sitter. The adult that remains behind is probably fed regurgitated food, as are the nursing mothers.

Persons who have raised wolves in captivity report that it is easy to establish friendly relationships with wolf pups less than 20 weeks old, but that after this time it becomes increasingly difficult. Wild wolf pups would not normally encounter any other animals except members of their own pack during this early socialization period. Hence this behavioral characteristic keeps members of a wolf pack closely united with one another and unfriendly or hostile toward outsiders.

Social Dominance in Wolves

There are separate dominance ("pecking") orders within each pack, one for males and one for females. Usually the dominant male and the dominant female are the mating pair, but sometimes the pack is led by an older male that does not mate. The dominant male usually leads the pack when it travels, has first choice of food, takes the initiative in hunting, and is most aggressive in defending the pups.

In this case, as with other social dominance orders, there is little fighting within established wolf packs. Among dominant males, the most frequent threat is a fixed stare; in more extreme situations, dominants may bare their teeth with the corners of their mouths pulled forward, raise their ears, wrinkle their brows, and emit rumbling growls.

Subordinates show submission toward dominants in much the same ways as domestic dogs do, by lowering their heads and tails and flattening their ears, and, under extreme threat, rolling over on their backs. Subordinate members also often express affection for the dominant wolf, nuzzling the mouth of the pack leader, using the food-begging gesture of the pups. Sometimes all the members of a pack will surround the leader and lick his face and poke his mouth with their muzzles. This ceremony often occurs when wolves first wake up, when the group has been separated from one another for some time, and when the pack has scented prey and is ready to start on a hunt.

All the other male members of the pack prefer the dominant female, and the female members prefer the dominant male. Also the leading male disrupts courtship attempts by subordinate males, and the leading female prevents other females from mating. Hence social dominance has the effect of limiting the size of the group. According to some studies, as many as 40 percent of all adult female wolves fail to breed in any given year and fewer than half—in some studies as few as 6 percent—of the pups survive for one year. When packs are

43–24 (a) *Dominant male (center) surrounded by pack in greeting ceremony. (b) A subordinant female licking the muzzle of the dominant female. This same muzzle-nuzzle gesture is used by pups begging for food.*

(a)

(b)

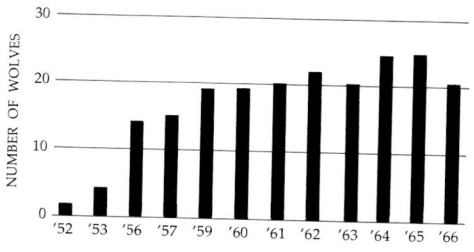

3-25 *Population of wolves on Isle Royale.*

heavily hunted, the ratio of pups to adults increases, indicating that more pups are produced or survive, or both, under these conditions. In short, wolf packs tend to remain stable in size. This concept is supported, for example, by actual population counts from Isle Royale.

When the dominant male or female dies, subordinate wolves will then mate and bear pups. It has been suggested that new packs are formed at times of social unrest after a leader has died, leaving two adult males of almost equal rank.

Territoriality

Wolf packs are territorial, each pack occupying a home range of 50 square miles or more. A wolf or wolf pack intruding on another's home range is attacked aggressively. Two adjacent ranges may overlap, but since the ranges of overlap are small and the territories very large, the chance of two packs encountering one another is not great. The boundaries of territories are scent marked with urine; scent marking may convey information about both the extent of the territory and the amount of time that has passed since the pack passed by, thus making it easy for packs to avoid one another. It also might provide information about the number of wolves in the pack, since it is likely that the urine of each wolf has a different odor.

Howling may also be related to territoriality, analogous to singing in male birds. When a pack howls, one wolf begins and the others join in, finally working up to a full chorus. A session lasts about a minute and a half and is obviously pleasurable for the wolves, as are howling sessions for domestic dogs—somewhat like a community sing. Howling can be detected by the human ear up to four miles away, and probably for a greater distance by other wolves. The howl of each wolf in the pack is different. One observer reported, in fact, that if he howled on the same note as his pet wolf, she would shift her howl by a few

-26 *Howling in wolves, which communicates information about both location and individual identity, is probably related to territoriality.*

notes. Thus a howl could provide information to a neighboring pack not only of the presence of another wolf pack but also of the number of members in it. Howling also seems to serve as a way to regroup pack members that may have been separated during a hunt.

The Wolf as Predator

Wolves prey mostly on large animals including deer, moose, caribou, elk, and bighorn sheep. Characteristically, they consume almost all of their prey, including hair, hide, and all but the largest bones. If the entire carcass cannot be eaten on the spot, the remains are either carried away or revisited.

Wolves often locate prey by scent, sometimes from a distance of more than a mile. Staying downwind, they first stalk the animal slowly. They begin to run only when the quarry runs; apparently the movement of the prey animal serves as a stimulus. The beginning of the chase is the crucial part of the hunt; this is actually a testing period. Generally, the only prey animals that are chased for more than a few minutes are those that falter because of immaturity, old age, or illness, or calves that have been deserted by their mothers. Healthy adult animals of all prey species can easily outrun wolves. Larger animals often simply stand their ground; a full-grown moose often weighs over a thousand pounds and is well able to kill a wolf. Domestic animals such as sheep and cattle have no such advantages, however. Having evolved under the protection of man, they are typically unable to defend themselves either by fight or flight and so are an easy target for a wolf pack or even for a lone animal.

There have been occasional reports of wolves using cunning and complex strategies in their hunts—such as driving the prey to a predetermined ambush—but if such tactics occur at all, they would appear to be rare.

43–27   *Results of interactions between all or part of a large wolf pack (15 to 16 wolves) and 131 moose on Isle Royale. The circled figures indicate those animals actually "tested" by the wolves.*

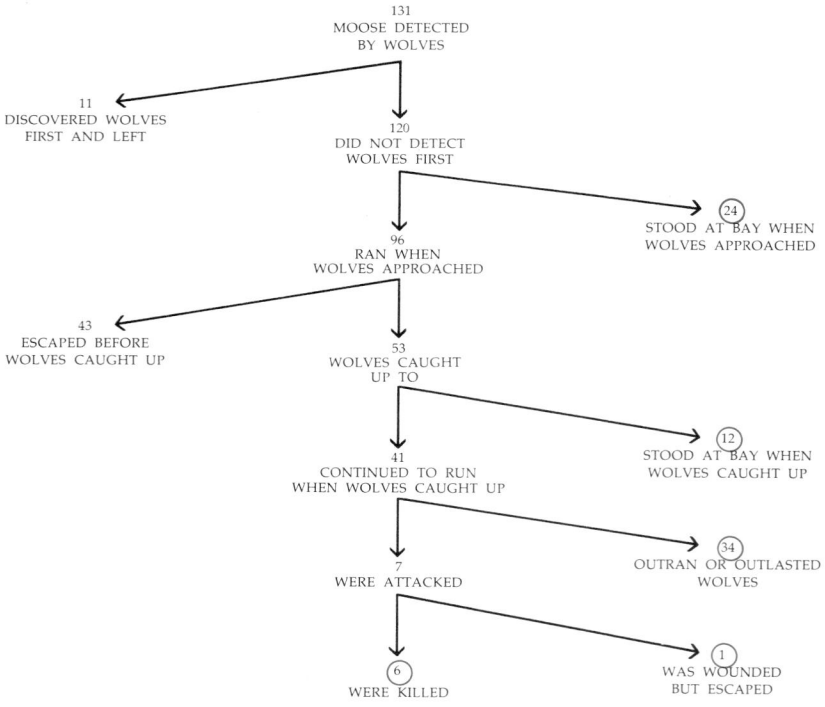

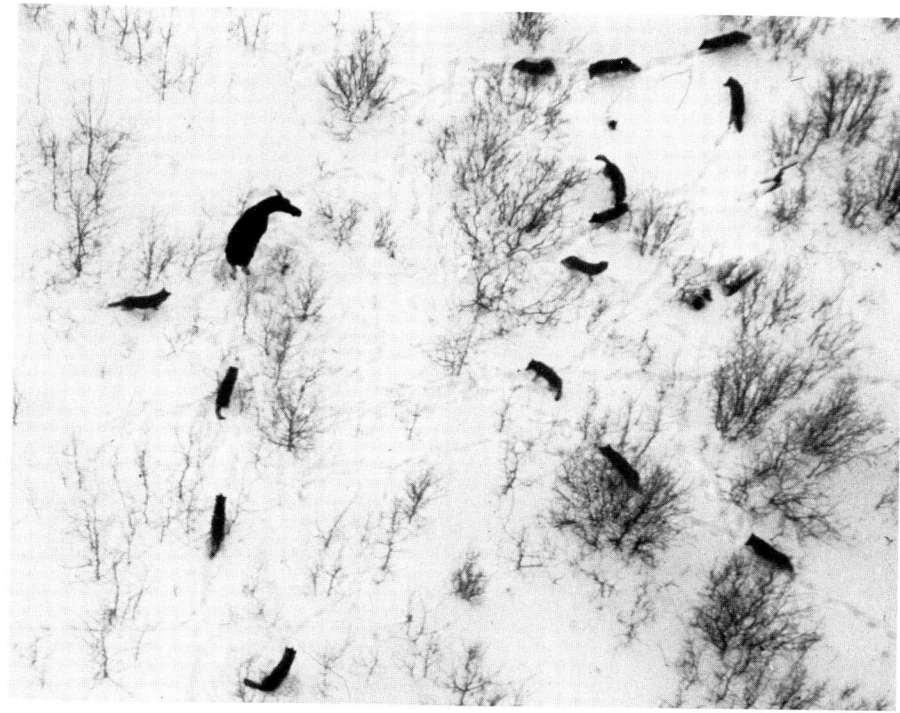

-28 *Moose under attack by a pack of wolves. The moose stood its ground for 5 minutes, after which the wolves gave up and left.*

It is difficult to state with certainty whether predation by wolves serves as a form of population control for prey species. The principal limitation on the population size of large herbivores appears to be the food supply, yet, as we noted previously, only rarely does a population under natural conditions expand to the point where large numbers of individuals die from starvation. Therefore it may be argued that the prey population remains at about the size that resources can support, whether or not wolves are present. (In this regard, it is interesting that before wolves arrived on Isle Royale, almost no twin calves were ever seen among the moose population; by 1963, the twinning rate had reached 38 percent.) Others contend that the "culling" of the population by wolves benefits the prey, not only by removing ill and unfit individuals but also by holding down the population. In Yellowstone National Park, where wolves have been exterminated, great herds of elk have ravaged the vegetation and starve periodically. Clearly wolves and their prey can coexist for long periods of time, and regardless of the presence or absence of wolves, populations of large herbivores cannot expand indefinitely. Moreover, the very young, the old, the ill, and the inept will be the least likely to survive against almost any selective force.

The wolf is in the unfortunate position of being in direct competition with man for the position of top-level carnivore. Every place the competition has been intense, as throughout the United States, the wolf has been exterminated, and, with the new hunting methods available, the problem of wiping out the remaining relatively small wolf populations in Alaska and Canada is not a difficult one should man, pursuing Gause's principle, choose to do so.

43–29   *Male baboons are much larger than the females. Other conspicuous differences are the long mane, or mantle, of the male and his bright red hindquarters (pale pink in the nonestrous female). The infants are born black and their coats become lighter as they mature.*

43–30   *Baboons sleep in groups, either in tall trees, as shown here, or on ledges or steep cliffs. The size of the group and the choice of a sleeping site appear to depend more on the terrain than on the species.*

SOCIAL ORGANIZATION AMONG BABOONS

Baboons are large, quadrupedal African monkeys with protruding, doglike muzzles. One genus (*Papio*) and four species are commonly recognized: a savanna species (*Papio anubis*), of which there are at least two and probably more distinct races; two forest species (the drill and the mandrill), both short-tailed baboons living in West Africa; and a desert species, found in North Africa and Arabia, *Papio hamadryas*, the hamadryas baboon. The geladas are large, terrestrial baboonlike monkeys that live in the mountainous grasslands of Ethiopia.

The baboons are of particular ecological interest for two reasons. First, unlike most monkeys, they live on the ground and have done so for several million years. For this reason, differences between baboons and other monkeys can be related to some degree to their environments. Second, there is a marked difference in social organization between the savanna and forest species on the one hand, and the desert species on the other. Since the groups are closely related genetically, obviously sharing a common ancestor in the not-too-distant past, it is very likely that the differences can be explained ecologically rather than genetically. Both of these statements must be tentative, however; field studies of baboons have been made only since the late 1950s, and most other monkey groups have been studied inadequately or not at all.

The Genus Papio

Baboons are large; adult males weigh about 120 pounds. Their size is clearly related to a terrestrial existence; only large primates, such as baboons, the patas and geluda monkeys, some of the great apes, and man, are terrestrial. Among all baboons, adult males are very different in appearance from adult females. This phenomenon, known as sexual dimorphism, is much less pronounced among other primates, in general, and other monkeys, in particular. The males have mantles and prominent canines and, most obviously, they are about twice the size of the females. Sexual dimorphism, which is much less pronounced among tree-dwelling primates, appears to be related to the increased hazards of life on the ground and to the male role of defender.

Like most primates, baboons are nearly exclusively vegetarian, living on leaves, fruits, flowers, and young stems, all of which they pick and consume on the spot. They also dig for roots, bulbs, and tubers. This vegetable diet is supplemented by insects and occasionally by meat, such as a snake, lizard, fledgling bird, or young gazelle, when they are encountered accidentally. There is no organized hunting for prey animals and no sharing of prey. Baboons sleep in tall trees or on cliffs, returning to a nesting site every night. Hence they must often travel to find food, sometimes for many miles across open country. Baboons have a harder time finding food than do arboreal primates, which are more likely to live in the lush tropical rain forests and which sleep where they eat. This hardship is probably related to the difference in size between the two sexes. The male is large enough to act as protector, whereas the female's reduced size reduces the food requirements of the band as a whole.

The Structure of the Band

Baboons are organized into multimale bands; those observed have ranged in size from 8 to more than 185 individuals. Adult females outnumber adult males

by about two to one. Much of this difference can be accounted for by the much slower maturation rate of the males. Although an adult male can probably produce viable sperm by the time he is about four years old, he does not reach full size nor are his canines fully erupted until he is eight or more, and therefore is not counted as an adult male either in the social organization of the band or by the field observer peering through his binoculars. Females, on the other hand, are adults by the age of two or three.

The bands are territorial, with fairly distinct home ranges, and bands rarely meet. When they do, the one farthest from the center of its customary range generally moves away, sometimes after an exchange of vocalizations, canine displays, and gestures, but more often with no visible reaction at all, except perhaps a slight display of nervousness. Bands that know each other may share large sleeping groves. Overt fighting was observed once, between two groups trying to sleep in the same clump of trees.

When the bands move in open country, females and infants characteristically are in the middle of the group, close to adult males. Mass counterattacks by adult males, which interposed themselves between the group and the attackers, have been observed to rout a leopard and disperse a large dog pack, killing or wounding several of its members.

Another function of the band is the pooling of information. Hans Kummer, one of the most experienced and imaginative observers of baboons, describes a band at the beginning of a day's march as resembling a giant amoeba, stretching out and then withdrawing one pseudopod after another, as small segments begin to move off in one direction, only to rejoin the band if the others do not follow.

1   A band of baboons moving through open country.

43–32 *A dominant male, threatening. The huge canines are found only in the males, indicating that they are related less to diet (male and female eating habits are the same) than to dominance and aggression. However, surprisingly, altercations among baboons seldom result in serious injury.*

Finally, a movement is initiated that meets with general approval, and the day's march begins.

The size of the group seems to be somewhat related to its habitat. Very small groups are correlated with small areas of vegetation, quite widely separated from one another, whereas larger groups may be found where a typical feeding site might be a clump of trees. Logically speaking, troop size must be a compromise between the availability of food resources and the need of the troop to defend itself.

Social Dominance

In all species except hamadryas baboons, the band is organized around a dominance hierarchy of adult males, which, as in other dominance hierarchies, is maintained mostly by behavioral conventions with only occasional fights, which rarely result in injury. Dominant males threaten subordinates by staring, yawning (with the ears laid back), raising the eyebrows, and, at close range, by grinding their teeth. Subordinate males demonstrate submissiveness by looking away, grinning, turning their backs, and presenting their hindquarters. The dominant male may mount the presenting male briefly. The gesture of presentation, which is also used by females in estrus as an invitation to copulate, is the most reliable index of hierarchical position. The top-ranking male presents to no other individual, the second-ranking male only to the top one, and so on down the social ladder. (Presentation as a gesture of submission or appeasement is seen among many primate species.) Adult females are subordinate to males but also have a separate dominance hierarchy, although not as clearly structured as that of the males. Adult females present to superior females.

Socially superior males mate more often than inferior males. In some groups, it was observed that a number of males copulated with females when they were not in full estrus, when the likelihood of fertilization is considerably less. But when superior females are in full estrus, as revealed by maximum swelling and coloring of the genital area, only the highest-ranking male mates with them. Thus the highest-ranking male is the one most likely to father offspring. Dominant males also have first choice of nesting sites. However, because there is no shared food supply, dominants are not allocated a larger share of food than subordinates, as they are in some other pecking orders.

Dominant males play a leading role in territorial disputes, thus maintaining the spacing among bands. They are also the most aggressive in defending the band against intraspecific aggression. If another member of the troop barks a warning or screams in fear, the dominant male steps forth and inspects the danger.

Social Bonds

As among almost all higher primates, grooming is a prominent form of social behavior. It involves a brisk parting of the fur of another individual with the fingers of both hands and the picking off of small particles, including insects; these items are often swallowed. Dominant males, females in estrus, and females with infants are groomed most frequently, but all members of the group receive some grooming attention. The function of grooming is clearly not only hygenic but also a continual reinforcement of social bonds.

-33 *As among all social primates (except man), grooming is one of the primary bonds among group members. The amount of grooming an individual receives is closely related to his (or her) position in the social hierarchy. Here an adult female grooms a dominant male.*

-34 *(a) Mother and child. (b) A female baboon presenting, in social deference, to a mother and newborn infant because she wants to approach and touch the infant.*

(a)

(b)

Infants are another clear social bond. The infant, which is all black and therefore very conspicuous for its first few weeks of life (perhaps so it can be guarded more zealously), usually clings to its mother, who may support it with her arm, but, with the mother's permission, it may be touched or even held briefly by other members of the troop. The infant or infant-mother pair are clearly attractive to other members of the group who tend to cluster around, especially when the infant is newborn. Mothers with infants are groomed frequently, especially by other females, and dominant males attend them closely as the troop moves. An infant that loses its mother will be adopted, sometimes by a childless female, but often by a young male.

The Hamadryas Baboon

The hamadryas baboon is unlike the other members of the genus *Papio* in that there are three distinct levels of social organization. The principal unit is a male with one or more females and their young, totaling up to seven or eight members. These one-male units group together into bands, similar in size and organization to the bands of the other baboon species. Finally the bands come together in troops, some of which have been counted as containing 750 members.

Students of primate behavior hypothesize a direct relationship between these social organizations and the environmental resources of the hamadryas. The habitat, which is on the edges of deserts, is typically arid grassland, interspersed with thorny acacias and other small trees and bushes. There are no tall trees, and the baboons of this area sleep on ledges on the vertical slopes of steep cliffs. (This is learned behavior. The hamadryas will sleep in tall trees if they are available, and other species will sleep on cliffs if there are no tall trees.) The baboons move across open country, often for long distances, between their sleeping cliffs and their feeding sites. They travel in bands, break up into one-male units for feeding—typically one unit to a tree—regroup into bands, return

to the cliffs, where they often sleep in troops, and then recongregate each morning in the band. Thus each social unit serves a clear and important function. The one-male unit is an optimal foraging unit, provided with a protector, the band provides for mutual defense when traveling, and the troop makes possible maximum utilization of safe sleeping sites within the habitat.

The strongest bonds are those that hold the one-male unit together. The females mate exclusively with the unit leader, and almost all social interactions, such as grooming, are carried out within the unit. Juveniles sometimes leave to play with those of other units within the band.

The structure of the band is generally similar to those of other baboon species. There is a male dominance hierarchy, and males act as defenders. About 20 percent of the adult males are not part of any family unit, but these males too are part of the band. The band moves as a whole, apparently acting on information from knowledgeable members of the group. The ties that hold the band together are less strong than those that hold the one-male unit. If a member, because of age, illness, or injury, drops out of a nonhamadryas band, he or she is left behind. If a female drops out of a hamadryas band, her male leaves the band to stay with her. The band, however, like the one-male unit, is a stable social group, always composed of the same members. The troop, on the other hand, is made up of varying groups of bands from the same territory that apparently recognize one another and are mutually tolerant.

Because the male-dominated band is the primary social structure of the other *Papio* species (and also of the macaques, arboreal monkeys who are their closest relatives), the hamadryas band appears to be a legacy from their common ancestors. Similarly, the troop seems easy to understand as a direct outgrowth of environmental pressures; baboons of other species share water holes and large sleeping groves when they have to. What, however, are the bonds that

43–35   *A hamadryas band. Note the grouping by one-male units.*

3–36 *An adult male hamadryas with a juvenile female that he has stolen from another one-male unit. New family groups originate in this way.*

-37 *Baboon eating water lilies.*

hold the one-male unit together? The cohesion of the unit turns out to depend primarily on two behavioral patterns, both those of the male. The first is a herding instinct toward the females. The females are taught to follow the male. He constantly watches them over his shoulder and, if one drops back or attempts to slip away, he stares, threatens, and sometimes nips her on the back of the neck, after which she immediately follows. If a male hamadryas is presented with a female olive baboon (*Papio anubis*), he accepts her quite readily and immediately trains her to follow him.

The second behavioral pattern is a strong inhibition among baboon males against taking another male's females. This, too, was tested experimentally. Two adult males, who knew each other, and a strange adult female were trapped. One male was enclosed with the female while the other was permitted to watch from a cage 10 yards away. The male immediately made grooming, mounting, and herding advances toward the female, as he would toward a member of his harem. Fifteen minutes later, the second male was put in the same enclosure with the pair. He not only refrained from fighting over the female, but he avoided even looking at them. The animals were separated, and two days later the experiment was repeated again, this time switching the males. Once again, the male that had watched the activities between the male-female pair was strongly inhibited in his behavior.

Another adaptive feature of this inhibition is that it applies only to adult males and females. Young males are attracted to juvenile females, which they "kidnap" from their family groups before they are sexually mature and train to follow them. Thus, the younger male has a detour around the inhibition barrier, and new family units can come into existence.

Lest you be tempted to extrapolate to human behavior from these accounts of baboon behavior, it is of some interest that, among geladas, one-male groups are formed and held together by activities of both sexes. Dominant males pair with dominant females, and wife number one is dominant over wife number two, and

851 *Societies and Social Behavior*

so on. The geladas are large, terrestrial baboonlike monkeys that live in the mountainous grasslands of Ethiopia; though they are referred to as baboons in common usage, they are different enough to be recognized as a separate genus (*Theropithecus gelada*).

The baboons remind us that social behavior in animals is ecologically adaptive—that is, that it promotes the survival of members of the social organization (and particularly of their young) within a given habitat or range of environmental conditions. Specialized behavior is undoubtedly acquired in a number of different ways—ranging from the rigid genetic programming seen among the social insects to the learned behavior seen among baboons and in human cultures.

SUMMARY

A society is a group of individuals of the same species organized in a cooperative manner and communicating with one another in some way.

Among the most complex societies are those of insects. These societies are matriarchies, centering around the care of the queen and the raising of the brood. The behavior of members of the society and even their physical characteristics, as evidenced by the very diverse members of termite society, are determined by chemical substances—pheromones—exchanged among members of the society. In eusocial insects, only two or a relatively few members of the society reproduce; the others care for the reproducing individuals and the young.

Vertebrate societies are varied both in their size and their organization and also in the behavioral factors that hold them together. Examples of two such behavioral influences are imprinting and rituals. Imprinting involves a very specific sort of learning that takes place in a very narrow developmental time limit and results, in nature, in the recognition of members of one's own species. Rituals are stylized behavioral patterns, shared by members of a group, that allay anxieties and promote recognition of group members by one another.

Some animal societies are organized in terms of social hierarchies. Socially inferior animals—those low in the pecking order—reproduce less frequently than their social superiors and are often psychological castrates. They are also the first to starve if food is limited or to be driven out if shelter is limited.

Territories are areas defended by an individual or a society against others of the same species. Territories may be "real"—that is, they may be actual areas of land containing food and nesting material to support a mating pair and young —or they may be symbolic, such as an arena. In either case, the only animals that breed are those with territories. Animals without territories provide replacements for territory owners and a reserve for population expansion if the range or food supply of the population is increased.

Acts of aggression are restricted in animal societies by social dominance, territoriality, and other forms of social behavior. Intraspecific killing is found only rarely among animals other than man.

Wolf societies (packs) are organized in a dominance hierarchy. Usually a pack contains only one breeding pair, the socially dominant male and female. Because all other females prefer the dominant male and all males prefer the dominant female, mating rarely occurs among other members of the pack, and so the size of the pack is limited. All members of the pack share in caring for the pups, and strong social bonds develop between pups and other members of

the band, which also appears to contribute to the coherence of the pack. Wolves are territorial, using scent marking, and perhaps howling, to mark their territories.

Baboons are large, quadrupedal terrestrial monkeys, with pronounced sexual dimorphism. They are organized into multimale bands. The males have a pronounced social hierarchy; superior males mate more often, have first choice of nesting sites, and play a more important role in defending the band than their social inferiors. The hamadryas baboon is unusual in that the band is subdivided into one-male family units. These units group into bands, and the bands come together in large troops, which utilize the sleeping cliffs. Thus, social organization is hypothesized to be related to the habitat. The one-male family is an optimal foraging unit, the band is the traveling unit, and the troop provides for the most efficient use of the sleeping resources.

QUESTIONS

1. Define the following terms: eusocial insects, pheromone, drone, imprinting, territoriality, social dominance.
2. Give a specific example of each of the above.
3. In what ways are human societies different from insect societies? How are they similar?
4. What behavioral conventions limit intraspecific aggression in man? Do any foster it?

SUGGESTIONS FOR FURTHER READING

"The Biosphere," *Scientific American*, September, 1970.*

A reprint of the September 1970 issue of the Scientific American. *Its eleven chapters by different authors are devoted entirely to energy flow and biogeochemical cycles and their relationship to current human problems of food consumption and pollution.*

CARSON, RACHEL: *The Sea Around Us*, New American Library, Inc., New York, 1954.*

Miss Carson was a rare combination of scientist and poet. This deservedly popular book traces the history of the formation of the oceans, describes their role in the origin of life and in its evolution, and discusses their present-day importance to human life.

EMMEL, THOMAS C.: *An Introduction to Ecology and Populations*, W. W. Norton & Company, New York, 1973.*

A short, sound text for the beginner, logically presented and sensibly organized.

FARB, PETER: *Face of North America*, Harper & Row, Publishers, Inc., New York, 1963.

An interesting and comprehensive account for a general audience of the natural history of the continent—the interplay of geological change, climate, and plant and animal life.

HUTCHINSON, G. EVELYN: *The Ecological Theater and the Evolutionary Play*, Yale University Press, New Haven, Conn., 1965.

By one of the great modern experts on freshwater ecology, this is a charming and sophisticated collection of essays on the influence of environment in evolution—and also on an astonishing variety of other subjects.

KORMONDY, EDWARD J.: *Concepts of Ecology*, Prentice-Hall, Inc., Englewood Cliffs, N.J., 1969.

This good short text differs from Emmel's largely in its more thorough coverage of biomes and ecosystems and its lesser emphasis on populations and their interactions.

* Available in paperback.

KREBS, CHARLES J.: *Ecology: The Experimental Analysis of Distribution and Abundance,* Harper & Row, Publishers, Inc., New York, 1972.

This excellent modern text focuses on populations and on interactions among them. Requires some knowledge of both biology and of mathematics.

KRUTCH, J. W.: *The Desert Year,* The Viking Press, Inc., New York, 1960.*

A description by one of the best contemporary American nature writers of the animal and plant life of the American desert.

KUMMER, HANS: *Primate Societies: Group Techniques of Ecological Adaptation,* Aldine, Atherton, Inc., Chicago, 1971.

Most of this book is concerned with behavior in baboons. This book is greatly enriched both by the author's personal experience in observing baboons and other primates and by his interpretation of primate behavior patterns in terms of evolutionary and ecological principles. Some wonderful and unusual photographs, also by the author.

MECH, L. DAVID: *The Wolf: The Ecology and Behavior of an Endangered Species,* Natural History Press, Garden City, N.Y., 1970.

The definitive book on the social history of wolves.

MARGALEF, RAMON: *Perspectives in Ecological Theory,* The University of Chicago Press, Chicago, 1968.

A short, stimulating, and often controversial book about some important ecological theories, especially those concerning succession and maturity, with some words of wisdom for conservationists.

MORAN, JOSEPH M., MICHAEL D. MORGAN, and JAMES H. WIERSMA: *An Introduction to Environmental Sciences,* Little, Brown and Company, Boston, 1973.

A clear, matter-of-fact presentation of the basic features of our planet and, in particular, of the biosphere. It provides a good background for the study of ecology in general and problems of environmental pollution and technological change in particular.

SMITH, ROBERT L.: *Ecology and Field Biology,* 2d ed., Harper & Row, Publishers, Inc., New York, 1974.

The outstanding sections of this text are those that deal with descriptions of biomes and communities and the plants and animals within them. Although meant to accompany a course in field biology, this book would be an asset and a pleasure to the amateur naturalist.

STORER, JOHN H.: *The Web of Life,* New American Library, Inc., New York, 1966.*

One of the first books ever written on ecology for the layman. In its simple presentation of the interdependence of living things, it remains a classic.

TINBERGEN, NIKO: *Curious Naturalists,* Natural History Library, Doubleday & Company, Inc., New York, 1968.*

Some charming descriptions of the activities and discoveries of scientists studying the behavior of animals in their natural environment.

WICKLER, WOLFGANG: *Mimicry in Plants and Animals,* World University Library, London, 1968.*

Many examples and illustrations of a delightful subject.

WILSON, E. O.: *The Insect Societies,* Harvard University Press, Cambridge, Mass., 1971.

A comprehensive and fascinating account of the social insects. An unusual nominee for the National Book Award.

*Available in paperback.

SECTION 7

EVOLUTION

44–1 *As Darwin observed, variations exist among the members of any given species.*

Chapter 44

The Extent and Origins of Variations

Although it is now more than a hundred years since the first publication of *The Origin of Species,* Darwin's original concept of how evolution comes about still provides the basic framework for our understanding of the process. His concept rests on four premises:

1. Like begets like—in other words, there is stability in the process of reproduction.
2. In any given population, there are variations among individual organisms, and some of these variations are inheritable.
3. In almost every species, the number of individuals that survive to reproductive age is very small compared with the number produced.
4. Which individuals will survive and reproduce and which will not are determined, in general, not by chance but by reason of these variations. Those with favorable variations will pass their characteristics on to the next and future generations in greater numbers. This is the process that Darwin termed *natural* *selection.*

Modern evolutionary theory conceives of evolution as Darwin conceived of it with the important addition of an understanding of the mechanisms of inheritance. Twentieth-century genetics answers two questions that Darwin was never able to resolve: (1) why genetic traits are not "blended out" but can disappear and reappear (like whiteness in pea flowers) and (2) how the variations appear on which natural selection acts. This combination of evolutionary theory and genetics is known as the synthetic theory of evolution. (Here "synthetic" does not mean artificial, which is the connotation it has for us in these days of man-made fabrics and artificial colors and flavors, but has its original meaning of the putting together of two or more different elements.)

The branch of genetics that emerged from this synthesis of Darwinian evolution and Mendelian principles is known as *population* *genetics.* A population, for the geneticist, is an interbreeding group of organisms. For instance, all the fish of one particular species in a pond are a population and so are all the fruit flies in one bottle. The population is defined and united by its gene pool, which is simply the total of all the alleles of all the genes of all the individuals in a population.

In this view, the individual is only a temporary vessel, holding a small portion of the gene pool for a short time, testing a particular combination. If the individual has a favorable combination of genes, its genes are more likely to be returned to the pool through his progeny and so will be present in an increased proportion in the next generation. If the combination is not favorable, its contribution to the gene pool will be reduced or perhaps eliminated. Thus the final criterion—the measurement—of an individual's fitness is the number of surviving offspring it produces, which, of course, determines its contribution to the gene pool of future generations. Evolution is the end product of such accumulated changes in the gene pool.

LIKE BEGETS LIKE

Like begets like, we know now, because of the remarkable precision with which the DNA is transmitted from cell to cell so that the DNA in the sperm or eggs of any individual is, except for occasional mutations, a true copy of the DNA that that individual received from its father or mother. In fact, this mechanism of copying serves not only to link man to his immediate ancestors but to link all living things to one another. For instance, human cells have some of the enzymes found in protists and even in prokaryotes, which indicates that certain sequences of DNA have been copied over and over for billions of years.

44–2 *Even in populations in which the individuals appear almost identical, such as the members of this group of gannets, we know that variations exist because individuals have no difficulty in identifying their own mates or offspring. To gannets, all people probably look alike, even Wilt Chamberlain and Willie Shoemaker.*

4–3 *Nature vs. nurture. A white-barked pine growing near the timberline on a mountain in California. Genotypically a tall straight tree, the constant strong winds in which it has grown have produced this phenotype.*

This constancy is, of course, essential to the survival of the individual organisms of which the population is composed. By and large, an organism, the product of tens of thousands of years of evolution, is well in tune with its environment, and unless the environment changes, wide variations in structure or function are almost always doomed to failure.

However, if evolution is to occur, variations among individuals must be present. Such variations make it possible for species to change as conditions do and provide the raw material on which natural selection acts. Population genetics thus becomes the study of this perpetual tug-of-war between the forces of constancy and the forces of change.

VARIABILITY IN NATURAL POPULATIONS

In this chapter and the one that follows, we shall be principally concerned with variations among individuals and populations—how they originate and how they are maintained. In the remaining two chapters of this section, we shall look at the processes of natural selection and some of their consequences.

All sexually reproducing organisms vary. These variations, as we noted in Section 2, may be in the genotype or in the phenotype. The genotype is simply the sum total of an individual's genes. The phenotype, the observable properties of an organism, is the result of the interactions of these genes with one another and with the environment in the course of the individual's history. Natural selection acts upon the phenotype; as a consequence, only those variations in phenotype that result from genetic variations are important in evolution. For example, a period of unusually cold weather might act as a selective force for individuals of larger size. As we noted in Chapter 13, both nature and nurture contribute to size. Only the selection of individuals with genes producing large size in the phenotype (or with other genes that might affect size in particular environments) would be of evolutionary significance.

The breeding of special strains of domestic animals offers a familiar example of the extent of variability in a single population. In Darwin's time, pigeon fanciers had produced a number of different and often very exotic breeds by means of what he termed "artificial selection." In our own time, the most clear-cut example is probably found among "purebred" dogs, strains that have been selectively bred for special traits appealing to mankind.

4–4 *Subspecies of* Potentilla glandulosa, *a relative of the strawberry. Notice the correlation between the height of the plant and the altitude at which it grows; there are other morphological differences between the plants and some physiological differences as well. When plants of the four subspecies are grown in an experimental garden under identical conditions, many of these differences persist and are passed on to the next generation, indicating that these plants are genotypically, as well as phenotypically, different.*

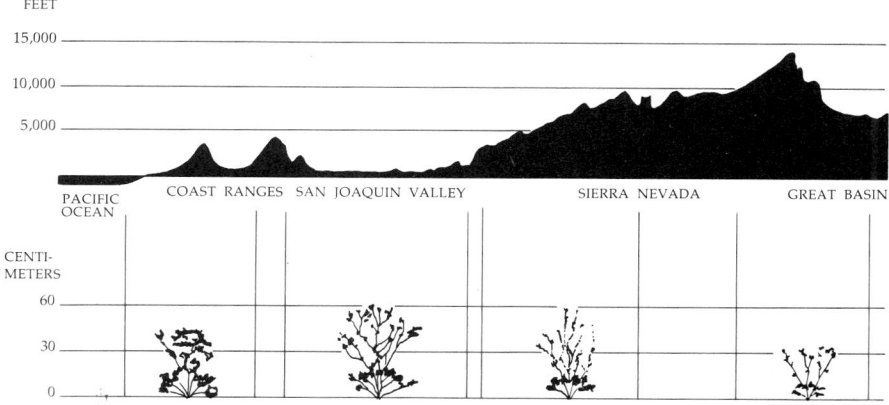

44–5 *The winners. All members of the species* Canis familiaris *belong to the same freely interbreeding population and so share a common gene pool. Artificial selection produces and maintains special "breeds" and reveals the extent of variability latent in the population.*

Similarly, groups of organisms of the same species may become separated in time and space and be placed under different selection pressures. Such groups, or populations, are still able to interbreed with members of other populations, but their gene pools are not identical. Such geographically distinct subunits of species are known as *subspecies*, or *races*. By subspeciation, differences between individual organisms become extrapolated into differences between groups of organisms.

Bristle Number in Drosophila

A number of laboratory experiments have been carried out that confirm the existence and extent of heritable variations. For example, the extent of latent variability in a natural population was demonstrated in the laboratory by experiments in which a population of *Drosophila melanogaster* was selected for an easily ascertainable hereditary trait: the number of bristles on the ventral surface of the fourth and fifth abdominal segments. In the starting stock, the average number of bristles was 36. Two selection groups were run, one for increase of bristles and one for decrease. In every generation, individuals with the fewest bristles were selected and crossbred, and so were individuals with the highest number of bristles. Selection for low bristle number resulted in a drop after 30 generations to an average of 30 bristles. In the high-bristle-number line, progress was at first rapid and steady. In 21 generations, bristle number rose steadily from 36 to an average of 56. It was apparent that, within the single

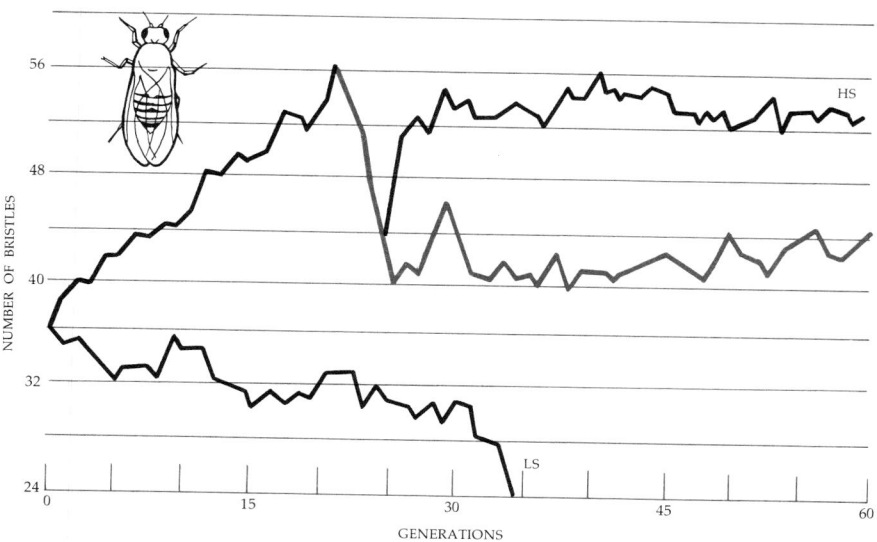

4–6 *The results of an experiment with* Drosophila melanogaster, *demonstrating the extent of latent variability in a natural population. From a single parental stock with 36 bristles, one group was selected for an increase in the number of bristles on the ventral surface (HS, high selection line) and one for a decrease in the bristle number (LS, low selection line). As you can see, the HS line rapidly reached a peak of 56, but the stock began to become sterile. Selection was abandoned at generation 21 and begun again at generation 24. This time, the previous high bristle number was regained and there was no apparent loss in reproductive capacity. Note that after generation 24, the stock interbreeding without selection was also continued, as indicated by the line of color. After 60 generations it had 45 bristles. The LS line died out owing to sterility.*

population, a very wide range of possible variation existed and that this had become manifest by selection pressures. Subsequent experiments with other traits in *Drosophila* and with other organisms have shown that the choice of bristle number was not a mere fortunate accident. Many other characteristics show a comparable range of natural variability.

There is a second part to the bristle-number story. The low-bristle-number line soon died out owing to sterility. Presumably, changes in factors affecting fertility had also taken place during selection. When sterility became severe in the high-bristle line, a mass culture was started; members of the high-bristle line were permitted to interbreed without selection. The average number of bristles fell sharply, and in five generations went down to 40. Thereafter, as this line continued to breed without selection, the bristle number fluctuated up and down, usually between 40 and 45, which still was higher than the original 36. At generation 24, selection for high bristle number was begun again for a portion of this line. The previous high bristle number was regained, and this time there was no loss in reproductive capacity. Apparently, the genotype had rearranged and reintegrated itself so that the genes controlling bristle number were present in more favorable combinations.

Mapping studies have shown that bristle number is controlled by a large number of genes, at least one on every chromosome and sometimes several at different sites on the same chromosome. Selection for bristle type, therefore, although the trait itself would appear to be neither useful nor harmful, in some way disrupted the entire genotype. Livestock breeders are well aware of this consequence of artificial selection. Loss of fertility is a major problem in virtually all circumstances in which animals have been purposely inbred for particular traits. This result emphasizes the fact that in natural selection it is the entire phenotype that is selected rather than certain isolated traits, as is often the case in artificial selection.

NATURE AND NURTURE: BEHAVIOR AND INTELLIGENCE

The "nature versus nurture" problem, which we touched upon in Chapter 13, recurs constantly in new forms. One group of current controversial questions concerns behavior.

In animals, some patterns of behavior are clearly inherited. A spider, for instance, although raised in solitary confinement, can build an extremely complex web on its first try. Among vertebrates, many species of fish and birds have elaborate courtship and territory-defending rituals whose basic patterns are part of their genetic equipment— that is, instinctive rather than learned. Mammals also show inherited behavior. For example, the first time a flying squirrel raised in a bare laboratory cage isolated from other squirrels is given nuts or nutlike objects, it will "bury" them in the bare floor, making scratching movements as if to dig out the earth, pushing the nuts down into the "hole" with its nose, covering them over with imaginary earth, and stamping on them.

Is there such a thing as inherited behavior among human beings? Do we have instinctive likes and dislikes, instinctive fears? It has recently become popular to think of mankind as instinctively aggressive and to use this fact to "explain" the persistence of war and other types of human brutality. Certainly, aggressiveness is inherited in some mammalian species. Wild rats are much more aggressive in their behavior than the white rats used in laboratories, which are generally docile and often affectionate toward their keepers. The offspring of wild rats, even though raised in a laboratory, are also aggressive. (These wild animals have adrenal glands about twice as large as those of their gentle cousins, which appears to provide a simple physiological basis for their behavior.)

There have been a number of recent attempts, by both scientists and popular authors, to elucidate human behavior by comparing it with that of other animals. Depending on which group of animals is chosen for comparison, apparently almost any point can be "proved." Of greater importance, however, is the fact that it is now clear that we never see man's "nature"—if by "nature" we mean the genotype. All the behavioral characteristics of man are the results of a unique interaction involving man's environment and personal experience and his biological makeup.

Another human trait that has become the subject of a nature-nurture controversy in recent years is intelligence. One problem is that it is difficult to generate a scientifically useful definition of intelligence. As a measurable quantity, intelligence can only be defined, essentially, as the ability to do well on intelligence tests, which necessarily reflect our society's conception of intelligence. As a consequence, the concept of intelligence, operationally speaking, has come to involve a rather limited set of socially useful skills, and standard intelligence tests seek primarily those skills alone.

Although there is evidence that intelligence has, to some extent, a genetic basis,* the effects of diet, perception, environmental stimulation, motivation, health, and numerous other factors are so massive that any possible genetic component is largely obscured. At present, geneticists are reluctant to say more than that intelligence is probably a complex mixture of factors, some inborn, some environmental in nature.

* The genetic basis of intelligence is implied by the results of studies of identical twins. Such twins, developing from the same fertilized egg, have identical genetic material. Identical twins separated at birth and growing up in different homes usually have IQ ratings within 12 points of each other.

Variability in Gene Products

Molecular biology has put forward another method for studying latent variability. Genes, as we saw in Section 2, dictate proteins. Experimenters from the University of Chicago ground up fruit flies and extracted proteins from them. From this extract they selected eighteen different enzymes that they were able to isolate on the basis of their enzymatic activity. Then they analyzed each of the eighteen samples separately to see if it was composed of a single protein or different proteins. (The method they used was electrophoresis—the same method used by Pauling to separate the proteins of normal and sickle cell hemoglobin.) Of the eighteen enzyme samples studied in this way, nine were composed of identical molecules; in other words, the genes that had produced these proteins were the same throughout the entire group of fruit flies studied. Nine of the groups, however, contained structurally different molecules. Therefore, without any direct analysis of the genes themselves, the investigators were able to conclude that, among the fruit flies studied, there were two or more alleles of the gene responsible for that enzyme. In fact, seven of the eighteen enzyme samples analyzed (39 percent of those studied) contained relatively common variants, often several different ones. In one specimen, there were as many as six slightly different structural forms; that is, six alleles for that gene were shown to exist in the species as a whole. Each population tested was heterozygous for almost a third of the genes tested. Each individual fruit fly, it was estimated, is probably heterozygous for about 12 percent of its genes.

Similar studies in human populations, using an accessible tissue such as blood or placenta, indicate that most genes are represented by at least two alleles and most individuals are heterozygous for at least 8 percent of their genes. (You will recall that only those structural changes that invoke replacement of one amino acid by another of a different charge can be detected by electrophoresis; therefore, these figures are minimums.)

THE ORIGIN OF VARIATIONS

Mutation

Hereditary variations originate as mutations, which are, by definition, sudden changes in genetic character. A mutation may involve a large part of a chromosome or a change in only a single nucleotide. As we saw in Chapter 5, the change in a single base pair is responsible for the abnormal hemoglobin associated with sickle cell anemia.

Mutations, as we noted previously, can be produced by exposure to x-rays, ultraviolet rays, and other agents. Most occur "spontaneously"—meaning simply that we do not know the reason for them.

The rate of spontaneous mutation is low; estimates of rates of mutation for human genes range from 10 to 80 per million gametes. In other words, the chance of a gamete carrying a mutation in a particular gene is less than 100 per million, or less than 1 in 10,000. However, there is an excellent possibility that any given gamete will carry a mutation in at least one gene.

Different genes have different rates of mutation. The reason for these differences is not known but may have to do with the position of the gene within the chromosome or with the chemical nature of the gene itself. Some nucleotide sequences may be less stable than others.

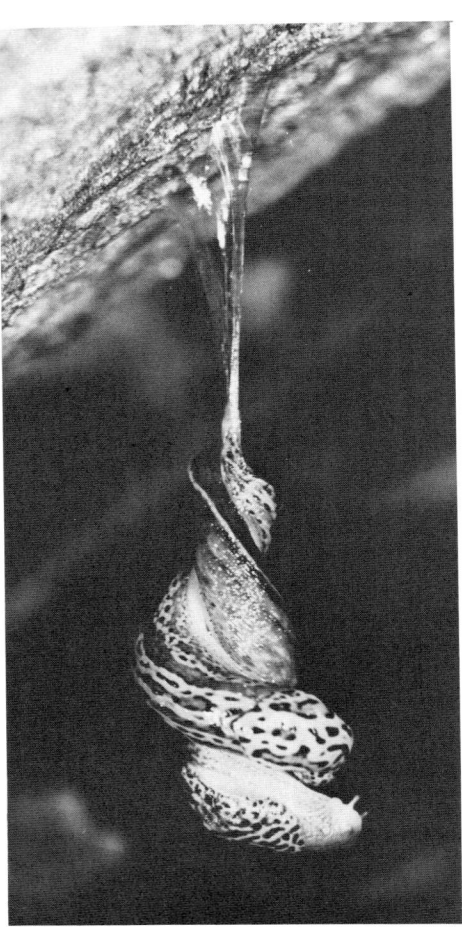

44–7 *Like many solitary slow-moving animals, slugs are hermaphrodites. However, they seldom self-fertilize. The photo shows two slugs mating, suspended from a branch by a cord of mucus.*

Many mutations appear to have no obvious effect on the phenotype, although they might make slight differences that are difficult to detect—for example, the temperature range at which a particular enzyme might function.

Recombination

Genes are recombined at meiosis, by crossing over and by the segregation of homologous chromosomes, and at fertilization, when new pairs of alleles are formed. By these processes, variations arising from mutations are reshuffled and so brought into a series of new combinations with other genes, giving rise to new phenotypes upon which natural selection can operate. The effect of a particular allele varies, depending both on the other allele and on the rest of the genotype.

Mechanisms for Promoting New Combinations

In many organisms, fertilization can occur only when the organisms are of different mating strains, as in *Chlamydomonas*, or of different sexes, as in the higher animals. Even among those invertebrates that are hermaphrodites, such as earthworms, an individual seldom fertilizes its own eggs. Some plants, such as the holly, have male flowers on one tree and female on the other, ensuring cross-fertilization. In others, such as the avocado, the pollen of a particular plant matures at a time when the stigma is not receptive. Others have genes for self-sterility. These are genes that prevent the pollen from fertilizing the ovule if it falls on the stigma of the same plant. Typically, such a gene has multiple alleles—s^1, s^2, s^3, and so on. A pollen grain carrying the allele s^1 cannot fertilize a plant with an s^1 allele, and one with an s^1/s^2 genotype cannot fertilize any plant with either of those alleles, and so forth.

In one population of about 500 evening primrose plants, 37 different self-sterility alleles were found, and it has been estimated that there are more than 200 alleles for self-sterility in red clover. As a consequence, the chances of a particular pollen grain fertilizing an egg cell are great if it has a rare self-sterility allele (which will nearly always be accompanied by a rare genotype) and slight if it has a common self-sterility allele. Such a system strongly encourages variability in a population; selection for the rare allele makes it more common, whereas more common alleles get rarer. As a consequence, variability itself is selected for.

Pin and Thrum

In many species of primroses, cross-pollination is ensured by an anatomical mechanism. In these species, two types of flowers occur in approximately unchanging proportions in every population. In one of these (known as pin), the style is long, raising the stigma high in the flower, and the stamens are short, so that the anthers are located low in the body of the flower. In the other form (thrum), the stigma is low and the anthers are high (see Figure 44–8). A pollinator that moves from primrose to primrose will receive pollen from one type on the part of its body most likely to come into contact with the stigma of the other type. In this way, cross-pollination is promoted.

In addition to the anatomical differences between the two forms, there are other physiological differences that promote cross-pollination of the two types. For instance, the cells of pin stigma react against pin pollen grains, inhibiting the formation of pollen tubes, but these cells promote the growth of tubes from thrum pollen.

Flowers have many mechanisms that prevent self-pollination. (a) In a bee-pollinated orchid, a bee with pollen on her back lands on the lip of the flower and crawls inside to gather the nectar. As she backs out, the flap separating the stigma from the pollen mass first scrapes the pollen off her back onto the stigma, and next, as she pushes the cover of the anther back, the pollen from the flower she is leaving becomes deposited on her back and is thus transported to another flower. (b) Cross sections of a pin type and a thrum type of a primrose. Notice that the anthers of the pin flower and the stigma of the thrum flower are both situated about halfway up the corolla tube and that the pin stigma is level with the thrum anthers. An insect foraging for nectar in these plants would collect pollen in different areas of his body, so that thrum pollen would be deposited on pin stigmas, and vice versa.

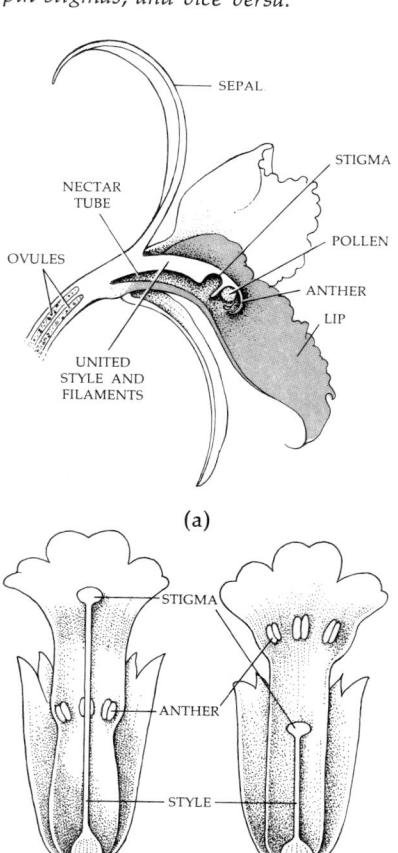

(a)

(b)

The existence of these elaborate systems is an indication that genetic recombination and the patterns of variation it helps to maintain have a survival value for the species.

Gene Flow

Another source of variability in populations is _gene flow_. Gene flow is simply the movement of genes from one population to another, as a consequence of the immigration of individuals from a different population. When these new individuals interbreed with members of the local population, the alien genes become part of the gene pool of the population.

Gene flow tends to reduce differences between populations. For example, if a bird carried seeds of *Potentilla glandulosa* from the Coastal Range to the San Joaquin Valley and if those seeds grew to mature plants that crossed with the plants previously in residence, the genetic differences between the two groups would be slightly diminished.

SUMMARY

Evolution is the product of natural selection acting on random heritable variations. Natural populations can be shown, by a variety of methods, to harbor a wide spectrum of latent variations. Subspecies, or races, are genetically different, geographically distinct subunits of species.

Variations are introduced into the gene pool in the form of mutations. They are brought into new genetic combinations by meiosis and fertilization. Both plants and animals have developed mechanisms to prevent self-fertilization. These include mating strains, sexes, self-sterility alleles (in plants), and special anatomical adaptations, such as the pin and thrum types among species of primrose.

New genes are also introduced into local populations by the immigration of individuals of the same species but with slightly different genotypes. This process is known as gene flow.

Mutation, recombination, and gene flow act to promote variations among the individuals in a population.

QUESTIONS

1. Define the following terms: genotype, phenotype, gene pool, natural selection, evolution, self-sterility, gene flow.
2. As we noted in the previous chapter, male baboons high in the dominance order are the most aggressive toward predators. Assuming that aggression also involves personal risks, how can the male baboon's behavior be reconciled with the modern theory of natural selection?
3. Even in populations growing in an environment with completely random mortality rates—that is, unrelated to genetic variations—natural selection could still occur. In this very special (highly hypothetical) case, what would be the basis for selection? (This question is answered on page 874.)
4. In the experiment on bristle number in *Drosophila*, the average number of flies at the end of the experiment had 56 bristles. No fly at the beginning of the experiment had as many as 56, however. How do you explain this fact?

Chapter 45

Hardy-Weinberg Equilibrium:
The Preservation of Variations

45–1 *Certain diseases are the result of the coming together of two recessive alleles. The Hardy-Weinberg law predicts that such alleles, even though harmful in the homozygote, will persist in the population. In this year-old infant with Tay-Sachs disease, deterioration of the brain, already begun, will progress rapidly. The child will probably die before he is 4 years old. The disease is caused by absence of an enzyme (hexosaminidase) that metabolizes a brain lipid. Among Jews, there is 1 chance in 30 of carrying an allele for this disease.*

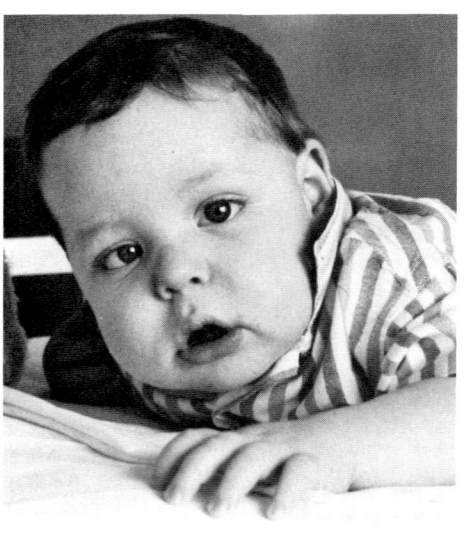

Darwin, as we saw in the preceding chapter, correctly perceived the existence of heritable variations in natural populations and their role in evolution. However, the theories of inheritance at that time were unable to explain how the small heritable variations, so important to evolutionary theory, were not blended in, like a drop of colored water in a beaker, but managed to persist, like a single blue marble in a bag of different colored ones. In fact, troubled by this very question, Darwin somewhat modified his thinking in later editions of *The Origin of Species* to move toward a more Lamarckian viewpoint, accepting Lamarck's premise that acquired characteristics may also be inherited. For similar reasons, some of Darwin's followers were prepared, at the beginning of the twentieth century, to place far too great an emphasis on the role of mutation in evolution. Their concept, which is sometimes called the "hopeful monster" theory, is that new strains of organisms appeared virtually overnight as a result of sudden large-scale changes in the hereditary material.

Mendel's great contribution was to demonstrate that traits—even when they seem to disappear—are actually preserved, "hidden" in the heterozygote. But as put forth by Mendel and other classical geneticists, the laws of heredity had been demonstrated only for ideal conditions in which simple, controlled matings were performed. For the purpose of evolutionary theory, it was necessary to establish how genes behave under the much more complex mating conditions that occur in nature. One needed to be able to detect and follow changes in the genetic makeup of a population, to analyze, for example, the fate of a new allele under natural breeding conditions. Population genetics is concerned with the origin and dynamics of variations in populations. Just as Mendel's laws form the cornerstone of classical genetics, the cornerstone of population genetics is a principle known as the Hardy-Weinberg law. Basically, the Hardy-Weinberg law* describes the circumstances in which two different alleles will both remain in the population indefinitely *and in the same proportions*. In other words, the

---

* The Hardy-Weinberg law was formulated in the early 1900s in response to a question raised by biologists: how could both dominant and recessive genes remain in populations? Why didn't dominants simply drive out recessives? If brachydactylism was a dominant gene, why didn't most or all of the population have short fat fingers? If you asked this question yourself, here is the answer.

Hardy-Weinberg law describes a situation in which there is no evolution; thus it provides a theoretical basis for measuring evolutionary change.

According to the Hardy-Weinberg law, the proportions of different alleles, and therefore of different genotypes, will remain the same in a sexually reproducing population, generation after generation, if four conditions are met: (1) the population is large, (2) no mutations occur, (3) there is no gene flow, and (4) there is no natural selection. The equilibrium is expressed mathematically in terms of the relative frequencies of the two alleles p and q by the relationship: $p^2 + 2pq + q^2 = 1$, a formula which we shall explain in the pages to follow. (The frequency of an allele is simply the proportion of that allele in a gene pool in relation to other alleles of the same gene.)

What the Hardy-Weinberg law says, in effect, is that sexual recombination does not *by itself* change the overall composition of the gene pool. A dominant allele, for instance, does not tend to increase in a population (a concept that was not understood even by many biologists in the early twentieth century), and a recessive one does not tend to decrease. A rare variant will continue to persist in the population, despite the shuffling and reshuffling of genes that takes place every generation. The simple interbreeding of organisms does not destroy variability; in fact, it preserves it.

THE HARDY-WEINBERG EQUILIBRIUM

The Hardy-Weinberg formula for equilibrium of the proportion of dominant and recessive alleles under random mating ($p^2 + 2pq + q^2 = 1$) is based both on Mendel's principles and on the mathematics of probability. (See page 181.) How do the laws of chance operate in random mating? Suppose, for instance, that in a large group of female mice, one-fourth of the females are brown. The chance of picking a brown female, at random, is therefore one in four, or $\frac{1}{4}$. In the group of male mice, one-fourth are also brown, and so the chance of picking a brown male is one in four, or $\frac{1}{4}$. But what are your chances of picking a brown female and a brown male, at random, at the same time? There are 4×4, or 16, possible combinations of male and female mice. (You can easily work this out in a diagram.) Therefore, your chances of picking a brown female and brown male together are one in sixteen. According to the laws of chance, multiplying the probability of one event (picking a brown female) by the probability of another event (picking a brown male) will give you the probability of the two events occurring simultaneously (picking a brown female and a brown male): $\frac{1}{4} \times \frac{1}{4} = \frac{1}{16}$.

Hardy-Weinberg in Mendelian Terms

To demonstrate the operation of the Hardy-Weinberg principle, let us begin by crossbreeding two populations, one of which is homozygous for a dominant allele (*AA*) and one of which is homozygous for a recessive allele (*aa*). (These are the sorts of populations with which Mendel began his studies.) All the first filial generation will be heterozygous (*Aa*). When members of this generation interbreed (*Aa* × *Aa*), the typical Mendelian ratio results; one-fourth of the population will be *AA*, one-fourth *Aa*, one-fourth *aA*, and one-fourth *aa* (Figure 45–2). Remember that in these Punnett squares, like those in Section 2, each square represents a group of individuals rather than a single individual.

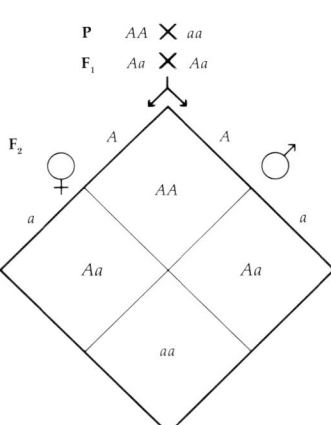

−2 *Crossbreeding of members of two homozygous populations (AA and aa) produces an F₁ generation that is all heterozygous (Aa). Random breeding among the F₁ generation produces, on the average, ¼ AA, ¼ aa, and ½ Aa.*

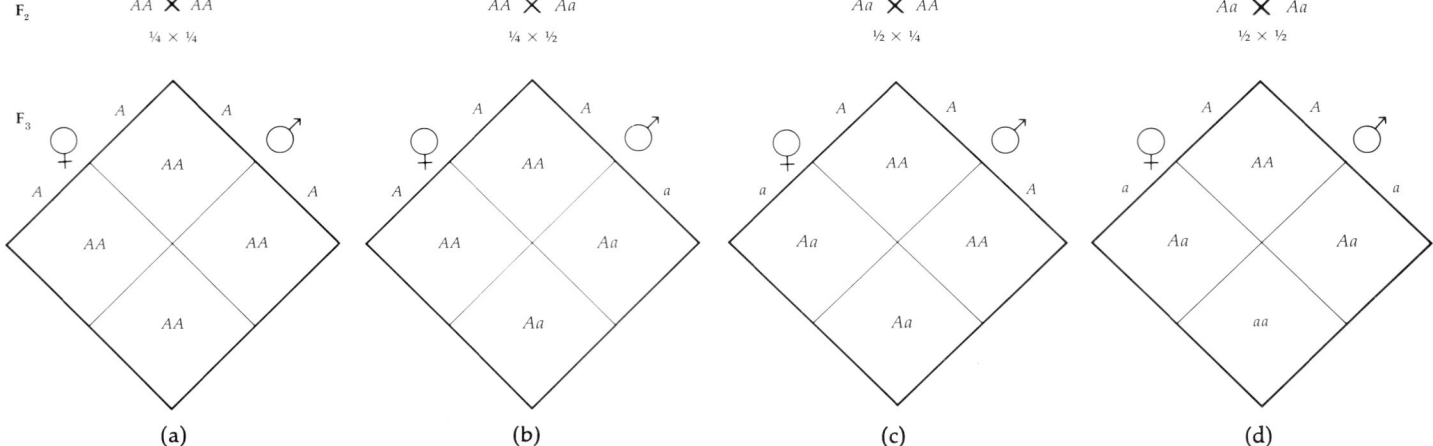

F_2 $AA \times AA$ $AA \times Aa$ $Aa \times AA$ $Aa \times Aa$
$\frac{1}{4} \times \frac{1}{4}$ $\frac{1}{4} \times \frac{1}{2}$ $\frac{1}{2} \times \frac{1}{4}$ $\frac{1}{2} \times \frac{1}{2}$

(a)   (b)   (c)   (d)

45–3 *If AAs are bred with aas, the probability of producing AAs in the third generation is the following: (a) If AAs crossbreed, the progeny will all be AA. The chances of this happening are $\frac{1}{4} \times \frac{1}{4}$ or $\frac{1}{16}$. (b) If AAs cross with Aas, the chances of producing AAs are $\frac{1}{16}$ because only half of the offspring are AAs. (c) If Aas cross with AAs, the chances of AAs are, again, $\frac{1}{16}$. (d) If Aas cross with Aas, the chances of producing AAs are $\frac{1}{16}$ because only a fourth of the offspring are AAs. Thus, in the third generation, AAs, on the average, will number $\frac{1}{16} + \frac{1}{16} + \frac{1}{16} + \frac{1}{16}$ or $\frac{1}{4}$—the same number as in the second generation, and as in the fourth, fifth, sixth, and so on. In short, sexual recombination alone does not change the proportions of different alleles.*

What will happen to these genotype proportions in the next generation? Let us consider first all the different kinds of crosses that will produce AA in the next generation. The first combination, female $AA \times$ male AA, will produce all AAs, and the chance of this occurring is $\frac{1}{4} \times \frac{1}{4}$, or $\frac{1}{16}$ (Figure 45–3a). Female $AA \times$ male Aa (or aA) will produce some AAs. The chance of these matings occurring is $\frac{1}{4} \times \frac{1}{2}$, or $\frac{1}{8}$, but only one-half of the offspring will be AAs, so the chance of getting AA from these crossings is $\frac{1}{4} \times \frac{1}{2} \times \frac{1}{2}$, or $\frac{1}{16}$ (Figure 45–3b). The third combination that will produce some AAs is female $Aa \times$ male AA. Here again, the chance of the offspring being AA is $\frac{1}{16}$ (Figure 45–3c). The fourth combination that will produce AAs is $Aa \times Aa$. There is one chance in four that such a mating will occur and one chance in four that it will produce an AA, again $\frac{1}{16}$ (Figure 45–3d). Now add these up. The chance of producing an AA in the third generation is $\frac{1}{16}$ plus $\frac{1}{16}$ plus $\frac{1}{16}$ plus $\frac{1}{16}$, or $\frac{4}{16}$ ($= \frac{1}{4}$), the same as in the F_2 generation. In other words, the probability is that the proportion of dominants in the third generation will be just the same as the proportion in the second generation. And, in fact, the proportions of Aa, aA, and aa will remain the same also, as you can readily demonstrate in this same way. In short, in a large population in which there is random mating, in the absence of forces that change the proportion of genes (conditions 2 to 4), the original frequencies of the two alleles will be retained from generation to generation.

Hardy-Weinberg in Algebraic Terms

For studies in population genetics, the Hardy-Weinberg principle is usually stated in algebraic terms, with the fractions that we used as examples before expressed as decimals. In the case of a gene for which there are two alleles in the gene pool, the frequency (represented by the symbol p) of one gene and the frequency (q) of the other must together equal the frequency of the whole, or 1; $p + q = 1$. Or, to put it another way, $1 - p = q$ and $1 - q = p$. This is equivalent to saying that if there are only two alleles, A and a, for instance, of

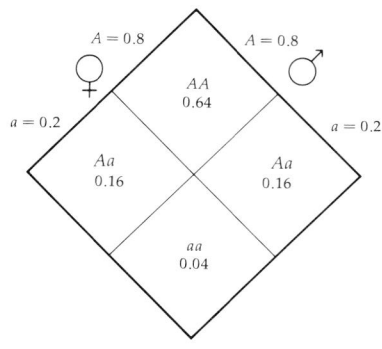

-4 *Results of random breeding in a popula-*
tion in which the frequency of A (p)
equals 0.8 and the frequency of a (q)
equals 0.2.

-5 *The relationship between the frequency*
of allele a in the population and the
frequency of the genotypes AA, Aa, and
aa. Naturally, the more AAs there are,
the lower the frequency of a. Because of
the interrelationship of AA, Aa, and aa,
a change in the frequency of either allele
results in a corresponding and symmet-
rical change in the frequencies of the
other allele and of the genotypes.

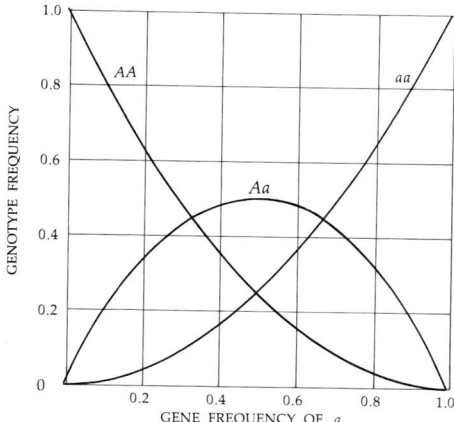

a particular gene and if half (0.50) of the alleles in the gene pool are A, the other half (0.50) have to be a. Similarly, if 99 out of 100 (0.99) are A, 1 (0.01) is a.

Then how do we find the relative proportions of individuals who are AA, Aa, and aa? As we established previously, these proportions can be calculated by multiplying the frequency of A (male) by that of A (female), A (male) times a (female), a (male) times A (female), and a (male) times a (female). We can express these multiplications in algebraic terms as $p^2 + 2pq + q^2$—which equals, as you probably know, $(p + q)^2$. Using this expression, we can see that if half (0.50) of the gene pool is A and half a, the proportion of AA will be 0.25, the proportion of Aa will be 0.50, and the proportion of aa will be 0.25, which is exactly what Mendel said—although in slightly different terms—in the first place. [$A^2 + 2Aa + a^2 = (0.50)^2 + 2(0.50)(0.50) + (0.50)^2 = (0.25) + 2(0.25) + (0.25) = 0.25 + 0.50 + 0.25 = 1$.]

Now let us look at a gene pool in which p (the frequency of A) equals 0.8 and q (the frequency of a) equals 0.2. The genotypic frequencies are AA equals 0.64, Aa equals 0.32, and aa equals 0.04, providing that mating takes place at random (Figure 45–4). The resulting population will thus be made up of 64 percent AAs, 32 percent Aas, and 4 percent aas. And, as we have just proved, the percentages will tend to remain in this equilibrium.

Multiple Alleles

So far we have talked about alleles only in terms of pairs, but actually there may be many mutant forms of the same gene. For instance, in the case of the gene for red eyes carried on the female chromosome in *Drosophila* (see Chapter 14), there are a variety of intermediate forms known by names such as apricot, buff, eosin, and honey, as well as the mutant that results in white eyes. More than 15 alleles have been recognized, and probably more could be detected if there were ways of distinguishing finer gradations of color.

The Hardy-Weinberg formula applies equally well when there are more than two alleles of the same gene, although the calculations then become more complex. The genotypic frequencies are expressed by the equation $(p + q + r)^2$, for example, with r representing the frequency in the gene pool of the third allele.

APPLICATIONS OF THE HARDY-WEINBERG PRINCIPLE

The Hardy-Weinberg equation is also useful to medical science. For instance, if we know the numbers of babies born annually with phenylketonuria (PKU) or with sickle cell anemia, we can calculate the frequency (q) of this gene and from it can calculate the number of persons heterozygous for this gene—that is, the numbers in the population who are carriers. Thus it is possible, for example, to calculate the probability of a known carrier of a harmful recessive gene marrying another individual also carrying this recessive.

For example, the incidence of PKU in the general population is estimated to be 1 in 15,000. Thus q^2, the proportion of homozygous recessives, is $\frac{1}{15,000}$, or 0.000067. The frequency of the allele for PKU is q, which equals $\sqrt{0.000067}$, or about 0.0082. If $q = 0.0082$, p equals $1.00 - q$, or 0.9918, $pq = 0.0081$, and $2pq =$ about 0.016. In other words, in the population under study, 16 in every 1,000 persons are heterozygous carriers for the PKU gene.

Protection of the Recessive Gene

Natural selection acts on the phenotype. As a consequence, a recessive gene is exposed to natural selection only in an individual homozygous for the gene. The importance of this fact becomes evident when we apply the Hardy-Weinberg formula:

Frequency of allele a in gene pool	Genotype frequencies			Percentage of a in heterozygotes
	AA	Aa	aa	
0.9	0.01	0.18	0.81	10
0.1	0.81	0.18	0.01	90
0.01	0.9801	0.0198	0.0001	99

As you can see, the lower the frequency of allele a, the smaller the proportion of it exposed in the homozygotes, and the removal of the allele by natural selection slows down accordingly. This result is of special interest to students of eugenics. For instance, in the case of a genetic disorder manifest only in the homozygous recessive (like PKU) with a frequency of about 0.01 in the human population, aa individuals make up 0.0001 of the population (1 child for every 10,000 born). You can calculate that it would take 100 generations, roughly 2,500 years, of a program of sterilization of defective homozygote individuals to halve the gene frequency and reduce the number to 1 in 40,000. The lack of effectiveness of such a program is obvious. If a dominant gene is lethal, however, it will be removed from a population in one generation unless it is replaced by new mutations. (Note that there is a difference between genotype frequency, which is the proportion of individuals in the population, and gene frequency, which refers to alleles in the gene pool.)

Importance of Heterozygote Superiority

If the allele is a rare one, as we can see in the table, it is present largely in the heterozygous state. This fact has an interesting consequence: If the heterozygote (Aa) is in some way superior to either homozygote (AA or aa), then both alleles will be maintained in the population.

One of the most dramatic instances of this phenomenon, known as _heterozygote superiority,_ is found in association with sickle cell anemia. Individuals homozygous for sickling seldom live to reproduce. Therefore, almost every time one sickling gene encounters another, two sickling genes are removed from the population. At one time, it was thought that the sickling gene was maintained in the population by a steady influx of new mutations. Yet in some African tribes, heterozygotes for sickling number as high as 45 percent, and to replace the loss of sickling genes by mutations alone would require a rate of mutation more than one thousand times greater than other mutation rates in man.

In the search for an alternative explanation, it was discovered that the heterozygote is maintained because it confers a selective advantage. In many African tribes, malaria is one of the leading causes of illness and death, especially among

young children. Studies of the incidence of malaria in this group show that susceptibility to malaria is significantly lower in individuals heterozygous for sickling. Moreover, women who carry the sickling gene are more fertile.

Among blacks in the United States the frequency of the allele is lower; only about 9 percent are heterozygous for the sickling allele. Since no more than half of this loss of the sickling allele can be explained by the black-white admixture in America, the conclusion is that, once selection pressure for the heterozygote is relaxed by the elimination of malaria, the recessive allele will tend to be slowly eliminated from the population.

Other genes that are harmful in the homozygous state appear to be maintained in the population by heterozygote superiority. For example, for reasons that are not known, women who are carriers of hemophilia (heterozygotes) have a fertility rate about 20 percent greater than that of the rest of the human female population.

Hybrid Vigor

A hybrid is a cross between two genetic types. *Hybrid vigor* (sometimes called heterosis) is the superiority of a hybrid over either of its parents in one or more measurable characteristics. Hybrid corn is a striking example. The development of hybrid corn caused a revolutionary improvement in the corn crop of the United States in the 1930s, because of the increased size and hardiness of the plants derived by crossing two different varieties to produce the seeds for each planting. (See Figure 45–6.)

Hybrid vigor is apparently the result of the fact that the plants are heterozygous at far more loci than most natural varieties. The superiority of the hybrid may be due to heterozygote superiority or to the fact that hybrids are less likely to be homozygous for deleterious alleles.

(a) *Hybrid corn is derived by first crossing strains A with B and C with D and then crossing the resulting single-cross plants to produce double-cross seed for planting. The increased size and hardiness of the hybrid are probably due to its increased heterozygosity.* (b) *Collecting pollen for hybridization.*

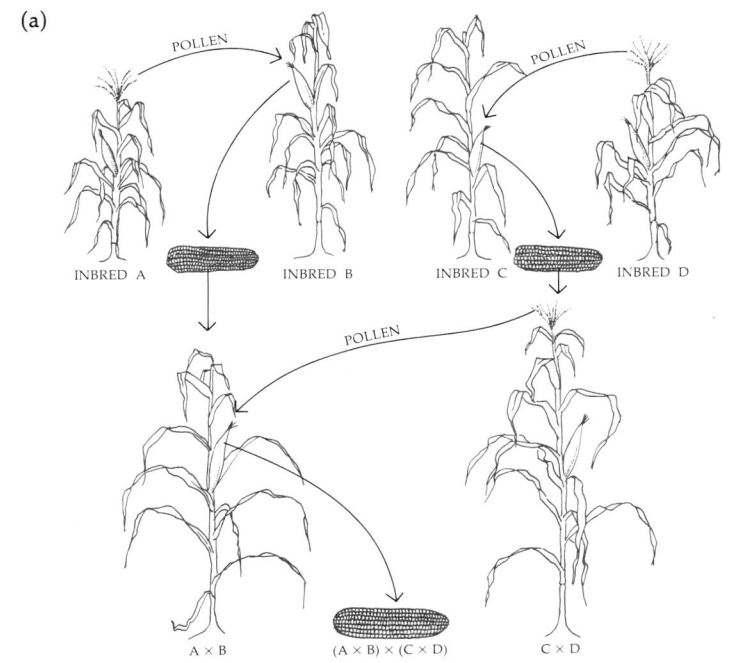

(a)

POLLEN

POLLEN

INBRED A   INBRED B   INBRED C   INBRED D

POLLEN

A × B   (A × B) × (C × D)   C × D

(b)

Incest Taboos

Most human cultures discourage inbreeding (breeding between closely related individuals). Many pretechnological societies, for example, demand that a young man choose a wife from another village rather than from his own community, and virtually all cultures have strong prohibitions against incest. Such prohibitions are particularly interesting when you consider that intermarriages of brother and sister or father and daughter would tend to keep property or power within the family and so could be socially and economically advantageous. Followers of Freud would maintain that incest is forbidden by strong psychological taboos, as reflected in the Oedipus myth. Some biologists hold that the prohibition stems from the observed ill effects of incestuous matings (through greater chances of deleterious alleles becoming homozygous) and has been merely reinforced by cultural restraints. However, no really satisfactory explanations exist.

THE IMPORTANCE OF POPULATION SIZE

As we stated previously, the Hardy-Weinberg law holds true only if the population is large. This qualification is necessary because the Hardy-Weinberg equilibrium depends on the laws of probability. These laws—the laws of chance—apply equally as well to flipping coins, rolling dice, betting at roulette, and taking a poll to see who is going to win an election. In flipping coins, it is possible for heads to show up five times in a row, but on the average, heads will show up half the time and tails half the time; the more times the coin is flipped, the more closely the expected frequencies of half (0.50) and half (0.50) are approached. Similarly, in a presidential poll, it is possible that the first five people interviewed will be Republicans, and it is not until a sufficiently large sample of the voting population is queried that accurate predictions become possible. In a small population, as in a small sample, chance plays a large role.

Genetic Drift

Consider, for example, an allele that is present in 2 percent of the individuals ($q = 0.02$). In a population of a million, 20,000 representatives of this particular allele would be present in the gene pool. But in a population of 100, there would only be two copies of this allele, and in a population of 50, only one individual would carry this allele. If this individual should fail to mate or should be destroyed by chance before leaving offspring, this allele would be completely lost from the gene pool. Similarly, if 10 of the 49 without this allele were lost, the frequency would jump from 1 in 50 to 1 in 40. This phenomenon, a change in the gene pool that takes place as a result of chance, is known as _genetic drift_, or _random drift_. The extent to which genetic drift is important in changing the Hardy-Weinberg equilibrium in natural populations is a subject of current controversy among population geneticists.

The Founder Principle

A small population that branches off from a larger one may or may not be representative of the larger population from which it was derived. Some rare alleles may be overrepresented (like the Republicans in the poll) or may be lost completely. As a consequence, even when and if the small population increases in size, it will have a different genetic composition—a different gene pool—from that of

5-7 *Among the Old Order Amish, a group founded by only three couples some 200 years ago, there is an unusually high frequency of a rare allele which, in its homozygous state, results in extra fingers and dwarfism. This Amish child is a six-fingered dwarf.*

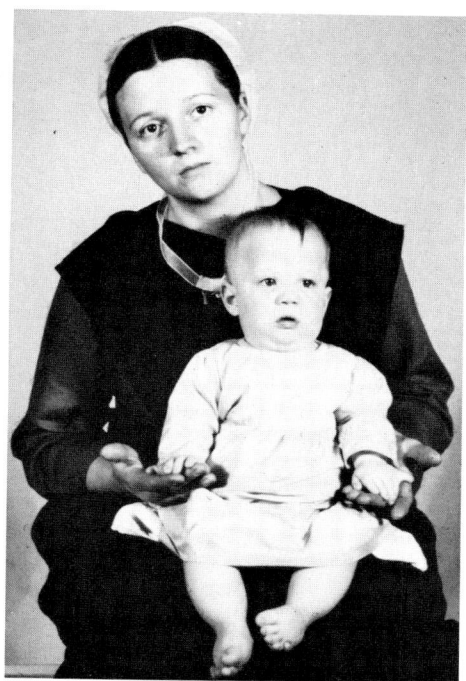

the parent group. This phenomenon is known as the *founder principle*. In a small population, there is likely to be less variation in the population, because of the loss of rare alleles either by genetic drift or by sampling error, and hence a decrease in heterozygosity. Thus, even though matings may be at random, the mates will be more closely related to each other—more similar genetically—with a consequent increase in homozygosity.

An example of the founder principle is found in the Old Order Amish of Lancaster, Pennsylvania. Among these people, there is an unprecedented frequency of a gene which, in the homozygous state, causes a combination of dwarfism and polydactylism (extra fingers). Since the group was founded in the early 1770s, some 61 cases of this rare congenital deformity have been reported, about as many as in all the rest of the world's population. Approximately 13 percent of the persons in the group, which numbers some 17,000, are estimated to carry this rare mutant gene.

The entire colony, which has kept in virtual isolation from the rest of the world, is descended from three couples. By chance, one of the six must have been a carrier of this gene. Although inbreeding does not cause disease or deformity, it may lead to its manifestation if such alleles are already present.

SUMMARY

The Hardy-Weinberg law is a basic principle of population genetics. It states that: The proportion of different alleles in a population will remain the same if (1) the population is large, (2) no mutations occur, (3) there is no gene flow, and (4) there is no natural selection. The mathematical expression of the Hardy-Weinberg equilibrium is $(p + q)^2$, which represents the proportions of genotypes in the population when p equals the frequency of one allele and q equals the frequency of an alternate allele. Thus the homozygotes for one allele are represented by q^2, and the heterozygotes by $2pq$.

Analysis of the Hardy-Weinberg equation demonstrates that in the case of a rare allele, the allele is present largely in the heterozygous state. If the allele is recessive—that is, if its effects are not manifest in the phenotype—it is thereby protected against selection pressures. Selection for the heterozygote tends to increase the proportion of a rare allele in the population even though the allele is harmful in the homozygous state.

The Hardy-Weinberg equilibrium operates only if the population is large enough. In a small population, chance events may cause the frequency of certain alleles to increase or to decrease and perhaps even disappear; this phenomenon is known as genetic drift. A small population that has branched off from a larger population may not be a representative sample of the larger population; this phenomenon is known as the founder principle.

QUESTIONS

1. Define the following terms: allele, gene frequency, heterozygote superiority, founder principle.
2. In terms of molecular genetics, what is a possible explanation for the many different eye colors in *Drosophila*?
3. What is the difference between gene flow and genetic drift? How do they each affect the Hardy-Weinberg equilibrium?
4. In a given population, why don't the dominants drive out the recessives?

Chapter 46

Natural Selection and Adaptation

In the last two chapters, we discussed the extent of heritable variations among individuals in populations and the ways in which these variations are introduced into the populations and maintained there, thus providing the raw material for natural selection. As we have seen, the forms that the variations take are entirely accidental. They become significant only as they are acted upon by a large group of environmental factors which, in combination, result in what Darwin called natural selection. Selection in no way causes the individual variations. However, by continuously eliminating particular genotypes from a population, it channels variations already in the population.

It is important to remember that when we talk about selection, we are describing events that have already taken place. Only after we find that the proportion of a certain allele or a certain genetic trait is higher in a particular generation than it was in a previous generation can we say that selection has occurred.

The result of selection is the adjustment of populations to their environment, or *adaptation*. Adaptation results in the development of differences between groups of organisms. Evolution, as we shall see in the next chapter, is the consequence of adaptation.

NATURAL SELECTION

Natural selection is "the differential change in relative frequency of genotypes due to differences in the ability of their phenotypes to obtain representation in the next generation." * This process is always in operation. Darwin's original concept of the process of evolution was based upon the observation that comparatively few organisms in most populations survive and reproduce. This principle is still true, as we noted on page 796. However, selection would continue to take place even in a population expanding without limit because some individuals would be more fertile than others and hence leave more representatives of their own genotype.

There are three requirements for the process of natural selection:

1. Organisms capable of reproduction
2. The existence of heritable differences among these organisms
3. The testing of these organisms in the environment

* E. O. Wilson and William H. Bossert, *A Primer of Population Biology*, Sinauer Associates, Inc., Stamford, Conn., 1971.

46–1 *Special adaptations of a woodpecker include two toes pointing backward with which it clings to the tree bark, a strong beak that can chisel holes in the bark, strong neck muscles that make the beak work as a hammer, and a very long tongue that can reach grubs under the bark.*

Table 46–1 *Survival in Relation to Number of Young in Swiss Starling*

Number of young in brood	Number of young marked	Number of individuals recaptured after 3 months, per 100 birds ringed
1	65	—
2	328	1.8
3	1,278	2.0
4	3,956	2.1
5	6,175	2.1
6	3,156	1.7
7	651	1.5
8	120	0.8

Three general types of selection operate within populations: stabilizing, disruptive, and directional.

Stabilizing selection, a process which goes on at all times in all populations, is the continual elimination of extreme individuals. Most mutant forms are probably immediately weeded out in this way, many of them in the zygote or the embryo. Clutch size in birds is an example of stabilizing selection. Clutch size (the number of eggs a bird lays) is determined genetically. As you can see in Table 46–1, it is a disadvantage, in terms of surviving young, for the Swiss starling to have a clutch size of more or less than five. The female whose genotype dictates this particular clutch size will have more surviving young, on the average, than a member of her species that lays more or fewer eggs.

A somewhat similar example is seen in a study that correlated birth weight and survival of 13,730 babies born in London over a ten-year period (1935–1945). As you can see in Figure 46–3, the optimum weight for this population of babies was just a little over 8 pounds. At this birth weight, 98.2 percent of the babies survived. The overall survival rate was 95.5 percent. To the extent that birth weight of offspring is determined by heredity, stabilizing selection was at work in this population to maintain this birth weight.

The second type of selection, *disruptive selection*, occurs when two extreme types in a population increase at the expense of intermediate forms. Among populations of an African swallowtail butterfly, for example, females may mimic more than one model; in such a situation, selection would, at least hypothetically, work steadily against the intermediate forms.

The third type, which is the one we shall be most concerned with, is *directional selection*. Directional selection selects for the most extreme phenotypic

46-2 *The three different types of natural selection. Stabilizing selection involves the elimination of extremes. In disruptive selection, intermediate forms are eliminated, producing two divergent populations. Directional selection, which is the gradual elimination of one phenotype in favor of another, produces adaptive change. In this graph, the y (vertical) axis denotes the proportion of individuals in a population with a particular characteristic and the x (horizontal axis), the varying dimensions of whatever characteristic is being considered, as in Figure 13–7, page 207.*

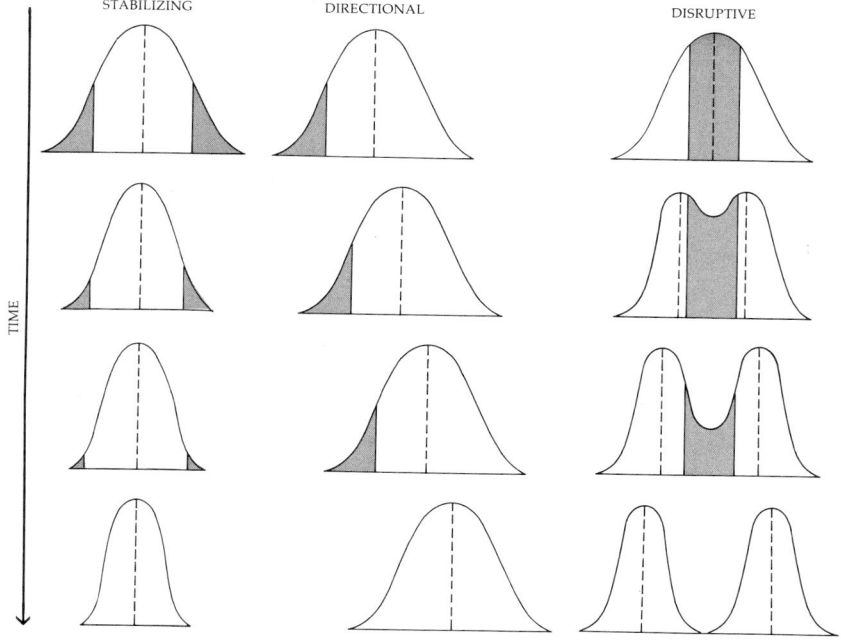

STABILIZING   DIRECTIONAL   DISRUPTIVE

TIME

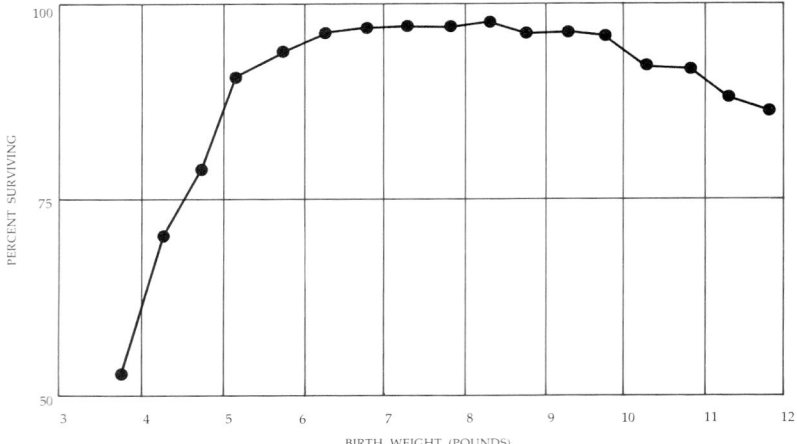

46–3 *Relationship between weight at birth and survival in a group of human babies. The death rate is higher for babies weighing either more or less than about 8 pounds. This is an example of stabilizing selection.*

characteristics and so is likely to result in the gradual replacement of one allele by another in the gene pool. Directional selection is the type of selection exercised by plant and animal breeders seeking certain traits. The experiment on bristle number in *Drosophila* was an example of directional selection. In nature, directional selection results in genotypic and phenotypic changes in population, producing adaptive changes in anatomy, physiology, and behavior. In the rest of this section, we shall be concerned with directional selection.

THE FORCES OF NATURAL SELECTION

As we have noted in previous chapters, the face of our planet and the conditions of life upon it have undergone constant change. Some 3 billion years ago, oxygen began to accumulate in the atmosphere, rendering many forms of life obsolete and opening the way for the evolution of new types of organisms. Similarly, the transition from a warm, moist climate to an arid one some 250 million years ago favored the selection of plants that did not require free water for fertilization and in which the dormant embryos were protected within seeds. The glaciations of the recent Ice Age led to the extinction of many of the large mammals and strongly influenced the evolutionary course of many types of others, including man, that have survived. The interactions of different species have also affected the course of their evolution.

The phrase "survival of the fittest" is often used in describing the Darwinian theory. In the early twentieth century, the doctrine of survival of the fittest in natural populations was used by some businessmen to defend ruthless competitive tactics in industry on the grounds that they were merely following a "law" of nature. This business philosophy was referred to by some as social Darwinism. However, in actual fact, very little in the process of adaptive change fits the concept of "nature red in tooth and claw." One fuchsia flowering a little brighter than its neighbors and better able to catch the attention of a passing hummingbird is a more pertinent model of the struggle for survival.

(a)

(b)

(c)

-4　Among most of the higher animals, including the higher invertebrates, the female of the species does the choosing. The female will usually select only a male clearly of her own species; even a slight deviation, such as a difference in eye color in a male fruit fly, is grounds for rejection. As a consequence, special anatomical and behavioral characteristics identifying males of particular species have been under heavy selection pressure. The huge antlers of the now extinct Irish elk (a) presumably served mainly for courtship and territorial displays. The spectacular tail (b) of the otherwise quite modestly endowed male lyre bird (c) serves a similar function.

Differential Fertility

One of the strongest forces in directional selection is relative fertility. The factor or factors that make a particular protozoan, plant, or animal able to produce more surviving offspring than another of the same species undoubtedly vary from individual to individual, but are clearly potent selective forces in natural populations.

Nonrandom Mating

Another strong selective pressure is nonrandom mating. As Hardy and Weinberg pointed out, alleles tend to remain in equilibrium in the population *if* mating occurs at random. However, nonrandom mating is the rule rather than the exception in most natural populations. The bright colors of flowers, the striking markings of many birds and fishes, and complex patterns of mating behavior all serve as inducement to nonrandom matings. Territoriality and dominance hierarchies, by excluding certain individuals from the breeding population, also ensure that some individuals make more contributions to the gene pool than others.

Among most insects and vertebrates, males do the courting and females do the choosing; therefore, the selection process of nonrandom mating affects males more than females. Females are usually quick to reject suitors who deviate in any way from the "ideal." For instance, female fruit flies will not mate with a white-eyed male if a red-eyed male is present. Clearly, nonrandom mating often results in stabilizing selection, but this is not necessarily the case. The selection pressures that led to the evolution of the elaborate plumage and huge tails of certain birds or the giant antlers of some deer or the long, intricate mating rituals of many birds and insects appear to be related almost entirely to the attraction of females or to the intimidation of rival males. Both of these social activities result, of course, in the most successful male leaving more offspring than his contemporaries.

Adaptation, in population biology, is the adjustment of populations to their environment as a consequence of the operation of natural selection. The resulting phenotype is a compromise, resulting from a response to numerous selection pressures, some of which may have even been antagonistic. Moreover, as R. C. Lewontin points out, the compromises may be out of date: " . . . genetic changes take place to increase the fitness of the population in *today's* environment but the result of that change is a population in the next generation living in *tomorrow's* environment."

Clines

Sometimes variations among groups—adaptations—follow a geographic distribution related to temperature, humidity, or other environmental conditions. Such a graded variation in a trait or a complex of traits is known as a <u>*cline*</u>.

Small mammals, for example, tend to have a smaller body size in the warmer parts of the range of the species and a larger body size in the cooler parts; in cooler regions, a larger body size is advantageous because it is easier for larger animals to conserve heat. Conversely, tails, ears, bills, and other extremities of animals are relatively longer in the warmer parts of a species range than in the cooler parts; such adaptations are correlated with the usefulness of the extremities in radiating heat. Plants growing in the south often have slightly different requirements for flowering or for breaking dormancy than the same plants growing in the north, although they all may belong to the same species.

Fish of cool waters tend to have a larger number of vertebrae than those living in warm waters. In the gypsy moth, a gradual increase in natural incubation time can be found as one moves from south to north, ensuring that the adults do not emerge before cold weather is over. Land snails tend to have smooth, glossy brown shells in cold climates and white or strongly sculptured shells in hot, dry climates. In these cases, although the physiological basis is not always clear, the range of natural variation is broadened and maintained by the external environment.

46–5 *The size of the extremities in a particular type of animal can often be correlated with the climate in which it lives. The fennec (a) of the North African desert has large ears which help it to dissipate body heat. The red fox (b) of the eastern United States has ears of intermediate size, and the Arctic fox (c) has relatively small ears. Similar correlations between size, weight, color, etc., and environment can also sometimes be made among animals of a single species living over an extended geographic range.*

(a)

(b)

(c)

(a) *Polymorphism in land snails.* (b) *An "anvil," where song thrushes break land snails open in order to obtain the soft, edible parts. From the evidence left by the empty shells, investigators have been able to show that in areas where the background is fairly uniform, unbanded snails have a survival advantage over the banded type. Conversely, in colonies of snails living on dark, mottled backgrounds (such as woodland floors), banded snails are preyed upon less frequently.*

(a)

(b)

POLYMORPHISM

Sometimes clear-cut, genetically determined differences may show up between large groups in the same populations. This phenomenon is known as *polymorphism.* Pin and thrum in primroses, mentioned previously, is an example of polymorphism.

Color and Banding in Snails

One of the best studied cases of polymorphism is found among land snails of the genus *Cepaea.* In one species (*Cepaea nemoralis*), for instance, the shell of the snail may be yellow, brown, or any shade from pale fawn through pink and orange to red. The lip of the shell may be black or dark brown (normally) or pink or white (rarely), and up to five black or dark-brown longitudinal bands may decorate it. (See Figure 46–6.) Fossil evidence shows that these different types of shells have coexisted since before the Neolithic period.

Studies among English colonies of *Cepaea nemoralis* have revealed selective forces at work in some of the colonies. An important enemy of the snail is the song thrush. Song thrushes select snails from the colonies and take them to nearby rocks, where they break them open, eating the soft parts and leaving the shells. By comparing the proportions of types of shells around the thrush "anvils" with the proportions in the nearby colony, the investigators have been able to show correlations between the types of snails seized by the thrushes and the habitats of the snails. For instance, of 560 individuals taken from a small bog near Oxford, 296 (52.8 percent) were unbanded; whereas of 863 broken shells collected from around the rocks, only 377 (43.7 percent) were unbanded.* In other words, in bogs, where the background is fairly uniform, unbanded snails are less likely to be preyed upon than banded ones.

Studies of a wide variety of colonies have confirmed these correlations. In uniform environments, a higher proportion of snails is unbanded, whereas in rough, tangled habitats, such as woodland floors, far more tend to be banded. Similarly, the greenest habitats have the highest proportion of yellow shells, but among snails living on dark backgrounds, the yellow shells are much more visible and are clearly disadvantageous, judging from the evidence conveniently assembled by thrushes.

Many of the snail colonies studied were at distances so great from one another that the possibility of immigration between populations could be entirely ruled out. How, then, are the differences in shell types maintained in the face of such strong selection? One would expect populations living on uniform backgrounds to be composed almost entirely of unbanded snails, and colonies on dark mottled backgrounds to lose most of their yellow-shelled individuals. The answer to this problem is not fully worked out, but it seems that there are physiological factors which are correlated with the particular shell patterns and which form a part of the same group of genes that control color and banding. Experiments have shown, for instance, that unbanded snails (especially yellow ones) are more heat-resistant and cold-resistant than banded snails. In other words, other strong selection pressures besides predation are at work, and they may maintain the genetic variations of color and banding.

* This difference (about 10 percent) may seem too small to be significant, but actually the number of shells counted was sufficiently large that there is only 1 chance in 1,000 that this difference could have occurred by chance.

(a)

(b)

(c)

(d)

46–7 *Some unusual adaptations. (a) Among a number of different species of ants that live in arid habitats, certain individuals, known as repletes, are used as storage containers for honey. A replete is shown on the left, contrasted with an ordinary worker on the right. (b) A cave-dwelling crayfish, like other animals adapted for life in the lightless interior of caves, is sightless and unpigmented. Its antennae are longer than those of an ordinary crayfish, and moving slowly and silently, it is able to detect even slight disturbances in the water. (c) A red-eyed tree frog of Central America. These animals, like many others that dwell in the canopy of the tropical rain forest, never descend to the forest floor. The frogs' eggs are laid in water held in the cupped leaves of epiphytes, and the young develop and metamorphose there. The bulges at the ends of the toes are suction pads. (d) Anteaters include marsupials and placentals, Old World and New World forms, closely related to one another only by their ecological role. They all have long noses, very long, wormlike tongues, and stout digging claws. This is the giant anteater of tropical America and, like most anteaters, it eats termites.*

Clutch Sizes

The English swift (*Apus apus*) is polymorphic for clutch size. Females lay either two or three eggs per clutch; more females lay two than lay three eggs. The parent swifts feed their young on airborne insects. In continued bad weather, such insects virtually disappear, and the young swifts often starve to death. An analysis of mortality among nestling swifts in four colonies in southern England between 1946 and 1950 showed that in broods of two young, 82 percent survived to leave the nest, whereas in broods of three young, with the food being shared among three mouths instead of two, only 45 percent left the nest. As a result, broods of two gave rise to an average of 1.6 survivors per brood, and broods of three resulted in only 1.4 survivors. In the summer of 1949, the weather was warm and unusually sunny, and the average number of young surviving per brood was higher from the broods of three than from the broods of two. This environmental change serves to maintain polymorphism in clutch size.

Blood Groups in Man

The human blood types A, B, AB, and O are familiar examples of polymorphism. Apparently, the three alleles associated with these blood types are a part of our ancestral legacy, since the same blood types are also found in other primates. Some evolutionary biologists regard the blood types as probably neutral in terms of their selective value. Others maintain that polymorphism in human blood groups is a result of selection factors. For example, among white men, the life expectancy is greatest for those with group O and least with group B; exactly the opposite is true for white women. Persons with type A blood run a relatively higher risk of cancer of the stomach and of pernicious anemia. Men with type B blood are about 2 inches taller, on the average, than those with other blood types. Persons with type O blood have a higher risk of duodenal ulcers. Furthermore, there are irregular geographic distributions of the A, B, AB, and O groups, and these differences may reflect some selective force favoring particular blood groups under particular conditions.

-8 *The blue and snow geese migrate from the Canadian Arctic, where they breed, across the Great Plains to Texas and Louisiana, where they winter along the coast. These geese were photographed en route at a National Wildlife Refuge in South Dakota.*

The Blue and White Snow Geese

At one time the lesser white snow goose (*Chen hyperborea*, "the goose from beyond the north wind") and the blue goose (*Chen caerulescens*) were believed to represent distinct species. More recently it has been found that they actually represent one species, polymorphic for color. The white is recessive; heterozygotes and homozygotes for blue are both the same dark blue color. However, birds mate preferentially with animals of their own color, probably because of imprinting (page 831). As a result, there are more homozygotes than if mating were random.

Blue and white geese differ significantly in their physiological adaptations for breeding. One major difference is that the white geese, which breed earlier, are less visible on snow, whereas the later-breeding blue geese are better camouflaged on the muddy ground that follows. Thus, polymorphism for color enables the species to utilize the short breeding time of the Canadian Arctic, where they breed.

46–9   *Populations of human beings are poly-morphic for their ability to hydrolyze milk sugar as adults. This capacity, conferred by the ability to synthesize the enzyme lactase, is highest among groups in Northern Europe where it has long been customary to keep large domesticated mammals for milk as well as for meat, work, recreation, and, legend has it, the seasonal delivery of parcels. It is lowest among groups, such as Africans, among whom unfermented milk is rarely a staple part of the adult diet.*

Lactose, Lactase, and Natural Selection

Lactase is the enzyme that breaks down lactose, the disaccharide found in milk. This enzyme, present in most human infants, is absent in adults of most animal species and in most human beings after the age of four. Lactose ingestion by individuals lacking the enzyme may cause nausea and diarrhea. (Whole milk also contains some protein and some fat so it may still have some food value for those who cannot digest lactose, its only carbohydrate.) Polymorphism for lactose digestion appears to be a result of natural selection—perhaps occurring some 10,000 years ago when certain groups began to domesticate and milk mammals. The individuals of northern European descent are much more likely to have this enzyme as adults than are those of African descent. This distribution of the allele (or of the gene that turns it on or off) is apparently correlated with the fact that northerners were more likely to drink milk (including reindeer milk) as adults, whereas people from hot climates, if they used milk at all, were likely to use fermented milk products, such as cheese or yogurt. In any case, this finding throws considerable doubt on the policy of attempting to improve the diet of black children here and in other countries by programs emphasizing milk or dried milk.

COEVOLUTION

Adaptation, as we noted at the beginning of this chapter, is the adjustment of populations to their environments. When two or more populations interact so closely that each is a strong selective force on the other, simultaneous adjustments occur that result in coadaptation or, as it is more commonly called, *coevolution.*

We have previously mentioned several examples of coevolution. One of the most important, in terms of sheer numbers of species and individuals involved, was the coevolution of flowers and their pollinators, described in Chapter 26 (pages 462 to 465). Others include the bat and the noctuid moth, ants and their fungus gardens, and monarchs and milkweeds. We shall give just one more example, also involving ants.

Ants and Acacias

Trees and shrubs of the genus *Acacia* grow throughout the tropical and subtropical regions of the world. In Africa and tropical America, acacia species are protected by thorns. (The acacias of Australia, which has few large grazing animals except recently introduced species, have no thorns.)

On one of the African species of *Acacia*, ants of the genus *Crematogaster* gnaw entrance holes in the walls of the thorns and live permanently in them. Each colony of ants inhabits the thorns on one or more trees. The ants obtain food from nectar-secreting glands on the leaves of the acacias and eat caterpillars and other herbivores that occur on the trees. Both the ants and the acacias appear to benefit from this symbiotic association.

In the lowlands of Mexico and Central America, the ant-acacia relationship has been extended to even greater lengths. The bull's-horn acacia, a common plant of that area, is found particularly frequently in cutover or disturbed areas, where the competition for light is intense. It grows extremely rapidly. This

(a)

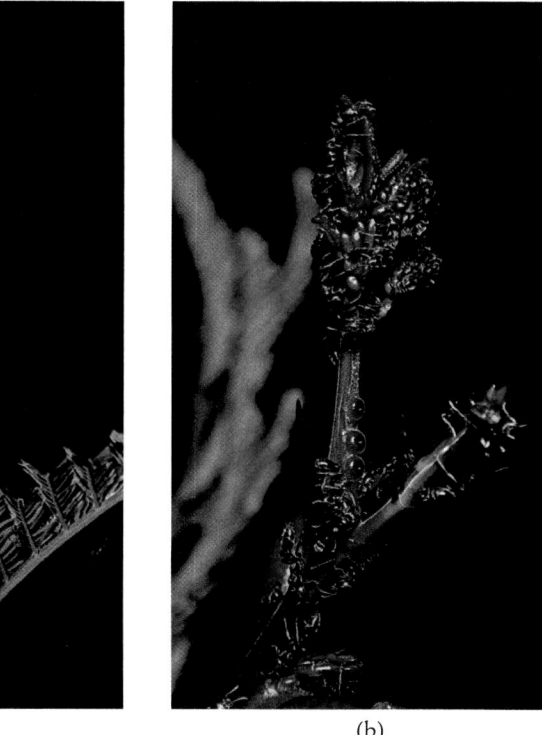

(b)

(c)

(d)

10 (a) *Growing shoot tip of an acacia with normal complement of ants on patrol.* (b) *A new leaf of an acacia, with ants on night patrol. Three petiolar nectaries are visible at the base of the new leaf.* (c) *Queen ant cutting an entrance hole into a new thorn on a seedling bull's-horn acacia. She will hollow out the thorn and rear her first brood inside it.* (d) *Seedling bull's-horn acacia, showing the thorns (at bottom of photo), the nectary with a red drop of nectar, and leaflets. Beltian bodies are conspicuous on the tips of the leaflets near the stem. These photos were taken by Daniel Janzen in the course of his investigations in Mexico and Central America.*

species of acacia has a pair of greatly swollen thorns more than an inch in length at the base of most leaves. The petioles bear nectaries, and at the very tip of each leaflet is a small structure rich in oils and proteins known as a Beltian body. Thomas Belt, the naturalist who first described these bodies, noted that their only apparent function was to feed the ants. Ants live in the thorns, obtain sugars from the nectaries, eat the Beltian bodies, and feed them to their larvae.

Worker ants, which swarm over the surface of the plant, are very aggressive toward animals of all sizes. They become alert at the mere smell of a man or a cow, and when their tree is brushed by an animal, they swarm out and attack at once, inflicting painful burning stings. The effect has been described as similar to walking into a large nettle. Moreover, and even more surprisingly, alien plants sprouting within as much as a meter of occupied acacias are chewed and mauled, and twigs and branches of other trees that touch an occupied acacia are similarly destroyed. Not so surprisingly, acacias inhabited by these ants grow very rapidly, soon overtopping other vegetation.

Daniel Janzen of the University of Michigan, who first analyzed the ant-acacia relationship in detail, removed ants from acacias artificially, by insecticides or by removing thorns or entire occupied branches. Acacias without their ants grew slowly and usually suffered severe damage from insect herbivores. Their stunted bodies were soon overshadowed by competing species of plants and vines.

As for the ants, according to Janzen, these particular species can live only on acacias.

6-11 Biston betularia, *the peppered moth,* and *its melanic form,* Biston betularia carbonaria, *at rest on a lichen-covered tree trunk in unpolluted countryside near Birmingham, England. A striking example of an evolutionary change resulting from a drastic environmental change, the black form of the moth began to appear in the latter part of the nineteenth century as the English countryside became increasingly polluted by industrial smoke.*

EVOLUTION IN A MAN-MADE ENVIRONMENT

Darwin believed that evolution is such a slow process that it can never be observed directly. Within recent years, however, the effects of human civilization have been producing new and extremely strong selection pressures, leading to what is probably an unprecedented rate of directional selection. One of the most striking and best studied examples is that of *Biston betularia*, the peppered moth. These moths fly at night and spend their days at rest. They were well known to naturalists of the nineteenth century, who remarked that they were usually found on lichen-covered trees and rocks. Against this background, their light coloring made them practically invisible. Until 1845, all reported specimens of *Biston betularia* had been light-colored, but in that year one black moth of this species was captured at the growing industrial center of Manchester.

With the increasing industrialization of England, smoke particles began to pollute the foliage in the vicinity of industrial towns, killing the lichens and leaving the tree trunks bare. In heavily polluted districts, the trunks and even the rocks and the ground became black. During this period, more and more black *Biston betularia* were found. This mutant black form spread through the population until black moths made up 99 percent of the Manchester population. Where did the black moths come from? Eventually, it was demonstrated that the black color was the result of a rare, recurring mutation.

Why were they increasing so rapidly? H. B. D. Kettlewell was among those who hypothesized that the color of the moths protected them from predators, notably birds. In the face of a number of strongly opposed entomologists—all of whom claimed they had never seen a bird eat a *Biston betularia* of any color —he set out to prove his hypothesis. He marked a sample of moths of both colors, carefully putting a spot on the underside of the wings, where it could not be seen by a predator. Then he released known numbers of marked individuals into a bird reserve near Birmingham, an industrial area where 90 percent of the local population consisted of black moths, and another sample into an unpolluted Dorset countryside, where no black moths ordinarily occurred. He returned at night with light traps to recapture his marked moths. From the area around Birmingham, he recovered 40 percent of the black moths but only 19 percent of the light ones. From the area around Dorset, 6 percent of the black moths and 12.5 percent of the light moths were retaken. Clearly, if you are a moth, it is advantageous to be black near Birmingham but light around Dorset.

To clinch the argument, Kettlewell placed moths on tree trunks in both locations, focused hidden movie cameras on them, and was able to record birds actually selecting and eating the moths, which they do with such rapidity that it is not surprising that it was not previously observed. Near Birmingham, when equal numbers of dark and light moths were available, the birds seized 43 light-colored moths and only 15 black ones; near Dorset, they took 164 black moths but only 26 light-colored forms.

Until recently, only a few of the light-colored populations could be found, and these were far from industrial centers. Because of the prevailing westerly wind in England, the moths to the east of industrial towns tended to be of the black variety right up to the east coast of England, and the few light-colored populations were concentrated in the west, where lichens still grow. A similar

As the snow geese make their long fall migration from the Arctic to the Gulf of Mexico, they are hunted intensively by man. The white-plumaged birds, which are more conspicuous and also more highly prized as trophies, are shot in much larger numbers than those with blue plumage, thus placing a new selection pressure on the species and perhaps threatening its finely tuned polymorphism.

tendency for dark-colored forms to replace light-colored forms has been found among some 70 other moth species in England and some 100 species of moths in the Pittsburgh area of Pennsylvania.

Most recently, strong controls have been instituted in Great Britain on particulate content of smoke, and the heavy soot pollution has begun to decrease. The light-colored moths are already increasing in proportion to the black forms, but it is not yet known whether a complete reversal either in pollution or in adaptive coloration will come about.

Bacterial Resistance

A familiar example of evolutionary change resulting from man-made selective pressures was first seen in the 1940s following the rapid progress in the development of antibiotic drugs, which selectively destroy many types of bacteria. It was not long before bacteriologists and physicians began to notice that bacterial cells "developed resistance to these drugs." Was this resistance a result of changes produced in the individual bacteria by the effects of the drugs—analogous to the changes produced in a heroin addict which make him tolerant of doses that would be fatal to a nonaddict—or was bacterial resistance another example of adaptation by natural selection? The question was, and still is, important medically.

It was also a serious issue for molecular biologists. As you will recall, much of the work in molecular biology has been carried out with bacteria and other microorganisms. For the purposes of these studies, scientists assumed that bacteria were representative biological systems, a supposition that was already under attack since bacteria are not eukaryotes.

If the resistance was the result of changes produced in individual bacteria, then such changes were heritable, but this Lamarckian concept violated the genetic principles governing higher organisms, giving the question broad significance.

This dilemma was resolved by a beautifully simple experiment performed by Joshua and Esther Lederberg. (See Figure 46–13.) Bacterial cells grown in a broth were spread thinly on a dish containing agar, a type of jelly that provides a firm medium for bacterial growth. Once visible colonies began to appear—a matter of about 24 hours—the cells were transferred to different dishes containing different kinds of media, including, for example, doses of penicillin and streptomycin.

The method the Lederbergs developed for transferring the cells was ingenious. As you can see in Figure 46–13, a piece of velveteen was held on a cylindrical block in the same way that a piece of material is held on an embroidery hoop. The block was marked by an arrow, and so were all the culture dishes. The velveteen was pressed gently against the original culture dish and then onto the new medium, with the arrows in careful alignment each time. As a consequence, it was possible to relate each colony that grew on the new culture to a colony in the original "mother culture."

After transfer of cells to a penicillin-treated agar, a small colony usually arose. But how did it arise? Were the cells random mutants, or did some cells "adapt themselves" to the new environment? To answer this question, the Lederbergs took more cells from the spot on the original colony corresponding

46-13　The Lederbergs' replica-plating method for detecting and isolating drug-resistant bacteria. (a) The bacteria are cultured in a liquid medium containing a diverse collection of organic molecules referred to as a "broth." (b) A sample of the cell suspension is spread over the surface of a petri dish containing broth solidified with agar. (c) The plate is incubated until colonies appear. A piece of velveteen held by a ring that fits snugly around a cylindrical block is used to transfer a sample of each colony to a replicate plate, another petri dish containing a solidified medium (d).

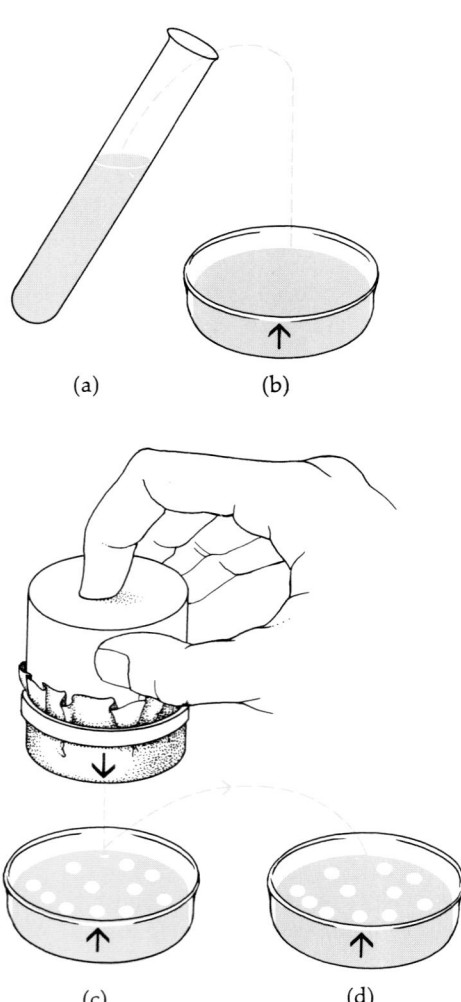

(a)　　　(b)

(c)　　　(d)

to the spot from which the penicillin-resistant cells arose and transplanted them to a new agar dish that contained no penicillin. Free from the competition of other bacterial cells, these cells grew and covered the entire dish. When members of the colony were tested, it was found that they were all penicillin-resistant—yet they had never been exposed to penicillin. The mutation had occurred independently of the exposure. Penicillin-resistant cells had arisen spontaneously in the original bacterial colony, and when placed in an environment for which they were particularly well suited, they flourished and multiplied. They were, therefore, genetic mutants, selected by the environment but not produced by it.

This finding restored the status of bacteria as experimental models for medical genetics. It also led to the search for new drugs and drug combinations.

Insecticide Resistance

The development of resistance to insecticides by flies, cockroaches, and similar unwelcome visitors is a phenomenon only too familiar to the general public. Insecticide resistance in a population develops gradually over a number of generations, but from man's point of view, this development is remarkably rapid. For instance, during World War II, DDT was highly effective in controlling body lice, but by the time of the Korean War, it was virtually useless. In fact, within two years after DDT was first used, resistant strains of many types of insects had developed independently in different parts of the world.

A particularly striking example of insecticide resistance has been found in the scale insects (*Coccidae*) that attack citrus trees in California. In the early 1900s, a concentration of hydrocyanic gas sufficient to kill nearly 100 percent of the insects was applied to orange groves at regular intervals with great success. By 1914, orange growers near Corona began to notice that the standard dose of the fumigant was no longer sufficient to destroy one type of scale insect, the red scale. A concentration of the gas that had left less than 1 in 100 survivors in the nonresistant strain left 22 survivors out of 100 in the resistant strain. By crossing resistant and nonresistant strains, it was possible to show that the two differed in a single gene. The mechanism for this resistance is not known, but one group of experiments suggests that the chief point of difference is that the resistant individual can keep its spiracles closed for 30 minutes under unfavorable conditions whereas the nonresistant insect can do so for only 60 seconds.

Flax Mimics

Another evolutionary change resulting from man-made selection pressures can be found in a weed of the mustard family. This weed (*Camelina sativa*) has developed races capable of growing in the flax fields of northern Europe, where flax is still an important commercial crop.

The parental form of the weed is a branched bushy annual with small white flowers, small seed pods, and rather small rounded seeds; this type is commonly found on roadsides and in open fields. Even if it encroaches on the flax fields, it does not survive there for more than a generation because its seeds are so different in size and shape from the large flat seeds of flax that they are automatically discarded with the chaff when the flax seeds are purified by winnowing and so are not replanted.

THE RECORD IN THE ROCKS

The Earth's long history is recorded in the rocks that lie at or near its surface, layer piled upon layer, like the chapters in a book. These layers, or strata, are composed mainly of sedimentary rock, a type of rock formed from pebbles, grains, and dust worn away from preexisting rocks and carried to another location by water, wind, or ice. First they formed part of the sand or soil, then slowly they became compacted and cemented to a solid form as new material was deposited above them. As the shape of the continents and ocean basins changed, some strata sank below the surface of an ocean or a lake, others were forced upward into a mountain range, and some were worn away in turn by water, wind, or ice, or deformed by heat or pressure.

Individual strata may be paper thin or many feet thick. They can be distinguished from one another by the types of parent material from which they were laid down, the way the material was transported, and the environmental conditions under which the strata were formed, all of which leave their traces in the rock. They can be distinguished, moreover, by the type of fossils they contain. Small marine fossils in particular can be associated with particular periods in the Earth's history. The fossil record is never continuous in any one place, but because of the identifying characteristics of the strata, it is possible to piece together the evidence from many different sources. It is somewhat like having many copies of the same book, all with chapters missing—but different chapters, so it is possible to reconstruct the whole.

The geologic eras—Precambrian, Paleozoic, Mesozoic, and Cenozoic—which were the major subdivisions of the book—were established and named in the early nineteenth century. These eras were subdivided into periods, each named, quite simply, for the areas in which the particular strata dating was first studied or studied most completely: the Devonian for Devonshire, in southern England; the Permian from the province of Perm in Russia; the Jurassic, for the Jura Mountains in northern Switzerland; and so on.

Attempts to date the various eras and periods were completely inaccurate. All that was known for sure was their relative ages; obviously, a stratum occurring regularly above another was younger than the one below it. In the mid-1800s, the famous physicist Lord Kelvin turned to the problem of estimating the age of the Earth. On the basis of his calculations of the time necessary for the Earth to have cooled from its original molten state, Kelvin maintained that the planet could not have been more than 25 million years old. This estimate greatly troubled Charles Darwin, who knew that it was too little time for evolution to have taken place.

In the last 20 years, new methods for determining the ages of strata have been developed involving measurements of the decay of radioactive isotopes. As a result of these methods the estimated age of the Earth has increased, in little more than a century, from 25 million years to more than 4.5 billion—time enough by any calculation for evolution to have brought about the changes to which the geologic strata bear witness.

Some radioactive isotopes change to other elements—uranium 238 to lead 206, for example. Because the rate of decay is constant (or so it is assumed to be), the proportion of ^{238}U to ^{206}Pb in a particular sample can be used to calculate the number of years since a particular stratum was formed. The method is not quite as precise as it sounds. One source of error is in the analysis of the indicator materials, which are present, if at all, only in very small quantities. Another source lies in the likelihood that the material has been disturbed in some way, such as by heat, which may separate the parent-daughter atoms. Because the daughter atoms are fewer and so more likely than the parent atoms to go undetected, dates set by radioactive isotopes are more likely to be too young than too old. Hence, as the methods become more precise, geologic time stretches farther and farther backward.

However, in areas where flax is commonly grown, new forms of *Camelina* have evolved. The plants are tall and unbranched and thus able to compete for sunlight with the tall, unbranched, closely growing flax plants. More important, they have much larger seeds than the races of mustard weed found along the roadside. These seeds, although they are not of the same shape as flax seeds, are not separated by the winnowing machines and so are resown each year by the flax growers. This new race, the flax mimic, produces fewer seeds in each seed capsule than do wild strains, and since the plant is unbranched, there are fewer seed capsules. On the other hand, the seeds enjoy the uncommon luxury of being sown in a carefully prepared field.

MACROEVOLUTION

Small adaptive changes produced in populations by natural selection acting upon genetic variations can be observed and verified, both in nature, as with *Biston betularia*, and in the laboratory, as with penicillin-resistant bacteria.

Let us now look at an example of large-scale change, sometimes called macroevolution to distinguish it from the microevolutionary process just described. Macroevolution can be observed only by its traces in the fossil record. One of the best known examples is the evolution of the horse, whose abundant representatives in the fossil record were studied by the great paleobiologist George G. Simpson. At the beginning of the Tertiary period, some 65 million years ago, the horse lineage was represented by *Eohippus* ("dawn horse"), a smallish (collie-sized) herbivore with three toes on its hind feet, four toes on its front feet, with doglike pads on which the weight was carried. Its eyes were halfway between the top of its head and the tip of its nose, and its teeth had small grinding surfaces and low crowns which probably could not have stood much wear. Its teeth indicate that it did not eat grass; in fact, there was probably not much grass to eat. It probably lived on succulent leaves. *Eohippus* evolved by stages into a group of larger, three-toed herbivores. Some of these lines survived, and some did not. Among those that did were those whose teeth had larger crowns that continued to grow as they were worn away (as do those of modern horses).

During the Pliocene, horses, of which there were still several distinct groups recognizable as different species, became still larger. The middle or third digit of each foot expanded and became the weight-bearing part of the foot. Some of the species remained three-toed, while in others the other digits became greatly reduced. The three-toed species did not survive, but those in whom the digits were greatly reduced—to one—did, and were the direct ancestors of the modern horse.

By hindsight, it is possible to correlate these changes with changes in the environment and so to interpret them as adaptive. In the time of little *Eohippus*, the land was marshy and the chief vegetation was leaves; the teeth of *Eohippus* were adapted for browsing. By the Miocene, the grasslands began to spread; groups of horses whose teeth became adapted to grinding grasses (coarser than leaves) survived, whereas those who remained browsers did not. The climate became drier and the ground became harder; reduction of the digits, with the development of the spring-footed gait characteristic of the modern horse, was an adaptation to harder ground and to larger size. Little *Eohippus* was probably

-14 The modern horse and some of its ancestors. Only one of the several branches represented in the fossil record is shown. Over the past 60 million years, small several-toed browsers, such as Eohippus, were replaced in gradual stages by members of the genus Equus, characterized by, among other features, a larger size, broad molars adapted to grinding coarse grass blades, a single toe surrounded by a tough, protective keratin hoof, and a leg in which the bones of the lower leg had fused, with joints becoming more pulleylike and motion restricted to a single plane.

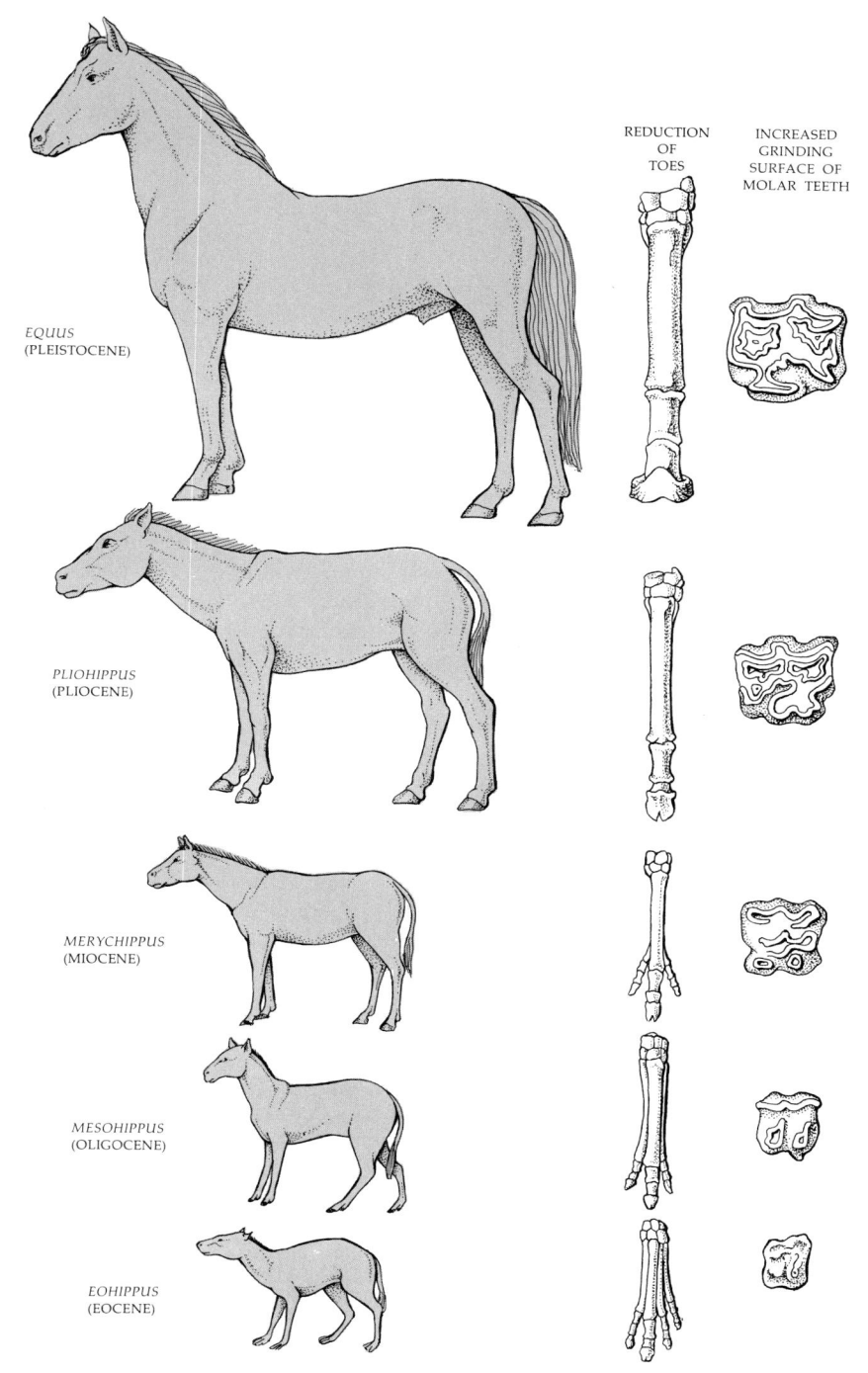

REDUCTION OF TOES

INCREASED GRINDING SURFACE OF MOLAR TEETH

EQUUS (PLEISTOCENE)

PLIOHIPPUS (PLIOCENE)

MERYCHIPPUS (MIOCENE)

MESOHIPPUS (OLIGOCENE)

EOHIPPUS (EOCENE)

as fast as the modern horse, but a large horse with the foot and leg structure of *Eohippus* would have been too slow to escape from predators, which during this same period were developing adaptations that rendered them better able to catch large herbivores, including horses.

It is not possible to *prove* that the changes described among fossil horses were produced by adaptation resulting from natural selection acting upon genetic variations, but it is certainly a likely explanation and, indeed, even opponents past and present of evolutionary theory have been unable to propose another.

Macroevolution is interpreted as representing the accumulation of microevolutionary change achieved by natural selection occurring throughout geologic time periods.

SUMMARY

Natural selection is the nonrandom reproduction of genotypes. It occurs as the result of one individual having the ability to produce more offspring than another. Natural selection does not in itself produce variations but it channels them by continuously eliminating particular genotypes. There are three general types of selection: stabilizing, disruptive, and directional. The result of selection is the adjustment of populations to their environment, a phenomenon known as adaptation. Forces that channel the direction of selection include changes in climate and other abiotic factors and in the relationships with other species in the community, leading to patterns of coevolution. Within species, differential fertility and nonrandom mating can exert powerful selective pressures.

A cline is a graded series of variations between populations or races of the same species that is correlated with geographical differences.

Polymorphism is the existence within a population of two or more discontinuous traits, such as shell color in snails, blood groups in man, or the capacity to digest lactose. The persistence of polymorphism in populations is believed to reflect the existence of different selective pressures for the particular characteristics.

Environmental changes brought about by human technology have produced strong selective pressures, resulting in rapid adaptive changes in affected populations. Examples include industrial melanism in moths, drug resistance in bacteria, insecticide resistance in insects, and mimicry among weeds.

QUESTIONS

1. What selection pressures are likely to be in operation for the human species at the present time?
2. Cave-dwelling species of fish are often blind. How would Lamarck have explained this phenomenon? How would modern evolutionary theory explain it?

Chapter 47

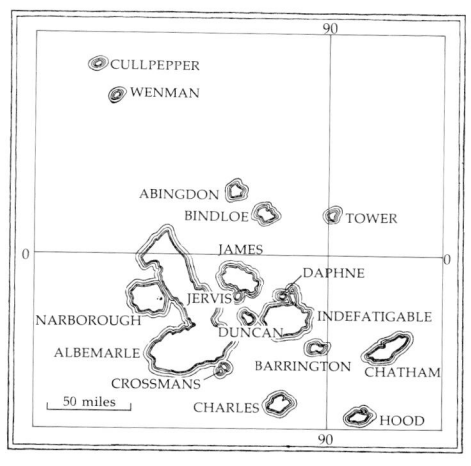

47-1 *The Galapagos Islands, some 600 miles west of the coast of Ecuador, have been called "a living laboratory of evolution." Species and subspecies of plants and animals that have been found nowhere else in the world inhabit these islands. "One is astonished," wrote Charles Darwin in 1837, "at the amount of creative force . . . displayed on these small, barren, and rocky islands. . . ."*

The Origin of Species

Evolution is the accumulation of adaptive changes in populations. As we observed in the examination of the fossil record of the horse, macroevolution appears to be the result of the continuing adaptation of species to a changing environment and the formation of new species—groups of individuals that cannot interbreed with other groups. How do new species—new kinds of organisms—come into existence? Although Darwin never dealt directly with this problem, his recognition of its central importance is reflected in the title of his major work, which is also the title of this chapter.

What is a species? In the words of Terrell H. Hamilton, a leading young evolutionary biologist: "A species may be envisioned as an isolated pool of genes flowing through space and time, constantly adapting to changes in its environment as well as to the new environments encountered by its extension into other geographic regions." *

Speciation is the formation of one or more new species. It is now generally agreed that a number of factors are involved in speciation, of which the three most important are (1) genetic isolation of a population, (2) the availability of a new ecological niche, and (3) time. Each of these factors will be discussed in greater detail later in this chapter.

How do we decide when a new species has come into being? The moment at which one species splits to become two contemporary species is difficult to pinpoint in time—although, as we shall see, it now seems to have been observed in the laboratory. There is no doubt, however, that one species has become two when the two populations can coexist in the same natural environment without interbreeding, and for the purposes of this discussion, this will be the working definition of a species.

DARWIN'S FINCHES

Admirers of Charles Darwin find it particularly appropriate that one of the best examples of speciation is provided by the finches observed by Darwin on his voyage to the Galapagos Islands. All the Galapagos finches are believed to

* Terrell H. Hamilton, *Process and Pattern in Evolution*, The Macmillan Company, New York, 1967.

47–2 *Six of the 13 different species of Darwin's finches. There are six species of ground finch, six species of tree finch, and one warblerlike species—all derived from a single ancestral species. Except for the warbler finch, which resembles a warbler more than a finch, the species look very much alike; the birds are all small and dusky-brown or blackish, with stubby tails. The differences between them lie mainly in their bills, which vary from small, thin beaks to huge, thick ones. (a) The small ground finch (Geospiza fuliginosa) and (b) the medium ground finch (Geospiza fortis) are both seed eaters. Geospiza fortis, with a somewhat larger beak than G. fuliginosa, is able to crack larger seeds. The cactus ground finches Geospiza scandens (c) and Geospiza conirostris (d) live on cactus blooms and fruit. Notice that their beaks are much larger and more pointed than those of the other two ground finch species. The tree finches Camarhynchus parvalus (e) and Camarhynchus pauper (f), both insectivorous, take prey of different sizes.*

have arisen from one common ancestral group—perhaps a single pair—transported from the South American mainland, some 600 miles away. As we saw in Chapter 45, a small population is apt to be nonrepresentative of its larger parent population and, as a consequence, lacks the genetic stability of the parent population and so is more susceptible to rapid change. The small size of the founding group of finches was probably an important factor in the remarkable changes that subsequently took place. How they got there is, of course, not known, but it is probable that they were the survivors of some particularly severe storm. (Every year, for instance, some American birds and insects appear on the coasts of Ireland and England after having been blown across the North Atlantic.)

We do have an idea of what greeted these unwilling adventurers. The Galapagos archipelago consists of 13 main volcanic islands, with many smaller islets and rocks. On some of the islands, craters rise to heights of 3 or 4 thousand feet. The islands were pushed up from the sea more than a million years ago, and most of them are still covered with black basaltic lava.

The major vegetation is a dreary grayish-brown thornbush, making up miles and miles of dense leafless thicket, and a few tall tree cactuses—"what we might imagine the cultivated parts of the Infernal regions to be," young Charles Darwin wrote in his diary. Inland and high up on the larger islands, the land is more humid, and there one can find rich black soil and tall trees covered with ferns, orchids, lichens, and mosses, kept damp by a mist that gathers around the volcanic peaks. During the rainy season, the area is dotted by sparkling shallow crater lakes. The vegetation may have differed when the finches first arrived, but clearly plants preceded the birds or the birds would have been unable to survive.

(a)   (b)   (c)

(d)   (e)   (f)

7–3 *The woodpecker finch (Camarhynchus pallidus) is a rare phenomenon in the bird world because it is a tool user. Like the true woodpecker, it feeds on grubs which it digs out of trees using its beak for a chisel. Lacking the woodpecker's long barbed tongue, it resorts to an artificial probe, a twig or cactus spine, to dislodge the grub. The woodpecker finch shown has selected a cactus spine which it inserts into the grub hole. The bird has succeeded in prying out the grub, which it eats. If the pick selected turns out to be an efficient tool, the bird will carry it from tree to tree in its search for grubs.*

The differences between the lush vegetation of the South American coast and the much harsher environment of the Galapagos undoubtedly put the new arrivals under strong selective pressures.

Thirteen different species of finches live on the archipelago, and an additional species lives on Cocos Island, several hundred miles to the north. The ancestral type is believed to have been a ground finch, and six of the Galapagos finches are ground finches. Four species of ground finches live together on most of the islands. Three of them eat seeds and differ from one another mainly in the size of their beaks, which in turn, of course, influences the size of seeds they eat. The fourth lives largely on the prickly pear and has a much longer and more pointed beak. The two other species of ground finch are usually found only on outlying islands, where some supplement their diet with cactus.

In addition to the ground finches, there are six species of tree finches, also differing from one another mainly in beak size and shape. One has a parrotlike beak, suited to its diet of buds and fruit. Four have insect-eating beaks, each adapted to a different size range of insects. The sixth and most remarkable of the insect eaters is the woodpecker finch, which has a beak like a woodpecker's that it uses as a chisel. Lacking the woodpecker's long prying tongue, it carries about a twig or cactus spine to dislodge insects from crevices in the bark. (See Figure 47–3.)

The thirteenth species of finch is hardly a finch at all. Classical taxonomists, using all ordinary standards of external appearance and behavior, would classify it as a warbler, but its internal anatomy and other characteristics clearly place it among the finches, and there is general agreement that it, too, is a descendant of common ancestors.

As students of the Galapagos finches reconstruct the story, the original founding party from the mainland probably made its landfall on one of the larger islands, gained a foothold, and hung on. From time to time, members of the group were carried to other islands. Each founder group that survived must have remained isolated long enough to construct a new gene pool. This might have been a matter of at least 10,000 years for each of the species, although since there are 13 main islands, several species were probably evolving at the same time.

Finches are not particularly good flyers; if they had been, they might not have speciated. Shore plants, for example, even when found on widely separated

islands, are likely to be of the same species since most of these plants are well able to survive long ocean voyages. To speciate, a population must remain genetically isolated; otherwise, its emerging genetic differences will be swamped, swallowed back up into the original gene pool. Once speciation has occurred, however, the different species can live together and maintain their genetic identities. Because the isolated groups of finches became adapted to different ecological niches—in particular, to the utilization of different food resources—they were able to coexist once reunited. (Remember that we do not know how many populations were lost as a result of inability either to adapt rapidly enough or to compete with established groups.) As many as 10 species of finch can now be found together on some of the larger islands.

The Galapagos, in short, offered a situation in which the three criteria for speciation could readily be met: genetic isolation, a number of unoccupied ecological niches, and, because of the geographic distances involved, time.

GENETIC ISOLATION

Islands

Because of the importance of genetic isolation in speciation, remote islands provide many other examples of evolution in progress. The tortoises of the Galapagos, for which the archipelago is named, are different from mainland tortoises and seem to offer an example of speciation in process. The races, or subspecies, of Galapagos tortoises are so different from each other that, as Darwin observed, the sailors who frequent the Galapagos can tell by the appearance of a tortoise which of the islands it came from. Perhaps in another several thousand years, they will evolve into distinct species. Unfortunately, this experiment will probably never take place; the huge, lumbering tortoises are virtually defenseless, and their numbers have been seriously reduced. For years, sailors stopped regularly at the Galapagos to provision their ships, carrying away two or three hundred tortoises each time. The tortoises, known to be able to live for as long as a year without food or water, were stacked in the hold as a welcome source of fresh meat. In recent years, the tortoises have been protected against man, but they are still threatened by introduced mammals that have gone wild. Donkeys and goats compete with them for food, and wild pigs, dogs, and rats prey upon the eggs and the young.

Most animals on remote islands, including the Galapagos tortoises, have had no experience of man and so have no fear of him. Many island visitors have reported their feelings of delight at having wild birds alight casually on their shoulders or at being able to reach out and stroke a sunning sea lion or walk unnoticed through a crèche of nesting penguins. Nature moves slowly, and by the time mechanisms of fear of man and defense against his predations have evolved, hunters will have reaped a rich harvest of meat, fur, and feathers, and many island animals will have gone forever. One of the finch species observed by Darwin has already disappeared.

Hundreds of examples of island speciation have now been studied, particularly on the Pacific islands. Here one can sometimes trace the repeated ocean voyages of a genus of land snails, for example, from New Guinea, Australia, or the Philippines eastward to New Caledonia, the Samoas, Tahiti, the Mar-

47–4 *Geographic isolation combined with local environmental conditions have favored the development of bizarre species, such as this flightless bird, the kiwi, found in the forests of New Zealand. The kiwi forages at night on the damp forest floors and spends its days sleeping in burrows or hollow logs, with its bill tucked to the side of the body where a wing would be if it had one. Presumably responding to similar selection pressures, island plants tend to have large, heavy seeds which will not be blown out to sea and washed away. Such plants and birds have, of course, burned their bridges behind them since they can no longer be dispersed by the mechanisms that brought them to the island in the first place.*

5 *Depending on the species, islands, genetically speaking, may be true islands (a), mountaintops (b), ponds, lakes, or even oceans (c), or isolated clumps of vegetation (d).*

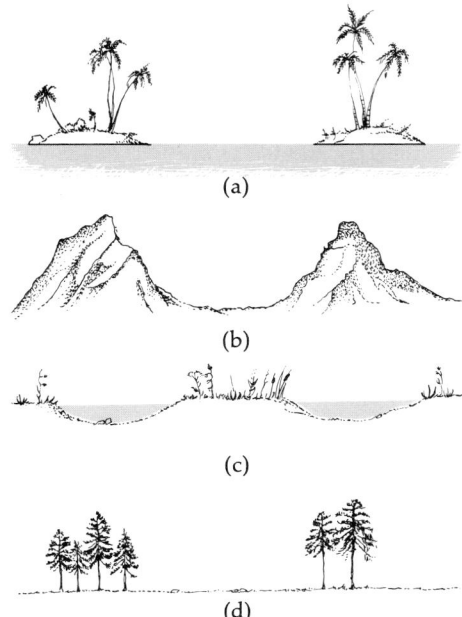

(a)

(b)

(c)

(d)

quesas, the Hawaiian Islands, or the Galapagos or westward from the coast of South America across the same island chain. Some of these organisms show only the beginnings of species formation, like the Galapagos tortoises, and some are so changed that their genealogy has become almost impossible to trace.

Other Types of Geographic Barriers

What is an island? For a plant, it may be a mist-veiled mountaintop, and for a fish, a freshwater lake. A forest grove may be an island for a small mammal. A few meters of dry ground can isolate two populations of snails. Islands of this sort usually form by the creation of barriers between formerly contiguous geographic zones. The Isthmus of Panama, for instance, has repeatedly submerged and reemerged in the course of geologic time. With each new emergence, the Atlantic and Pacific oceans became "islands," populations of marine organisms were isolated, and some new species formed. Then, when the oceans joined again (with the submergence of the Isthmus), the continents separated and became, in turn, the "islands."

There are a number of natural geographic barriers in the United States. The Grand Canyon and Upper Colorado River, for example, although not barriers for birds, are very effective at separating populations of small land animals. Several kinds of pocket gophers, wood mice, pocket mice, and field voles reach their northern limits at the south rim of the Canyon; there are related but different species that inhabit the plateau of the other rim. Two such subspecies of squirrel can easily be distinguished. The Abert squirrel, which lives south of the Grand Canyon, has a grayish tail, a white belly, and great tufted ears. The Kaibab squirrel, found only on the north rim, has a black belly and a pure white tail. These squirrels are very much the same; both kinds obviously had a common ancestor in the distant past, before the Canyon formed and became a barrier.

-6 *Two subspecies which have formed on either side of a natural geographic barrier, the Grand Canyon. The Abert squirrel (a) lives only on the south side of the Canyon, and the Kaibab squirrel (b) lives only on the north side. The most obvious differences between them are in the color of tail and belly.*

(a)   (b)

47–7 *A well-known example of subspeciation is offered by the leopard frog (Rana pipiens). The frogs are found as far north as Quebec and as far south as Mexico. The subspecies shown are from (a) Quebec, (b) western Texas, (c) Louisiana and (d) southern Florida.*

(a)

(b)

(c)

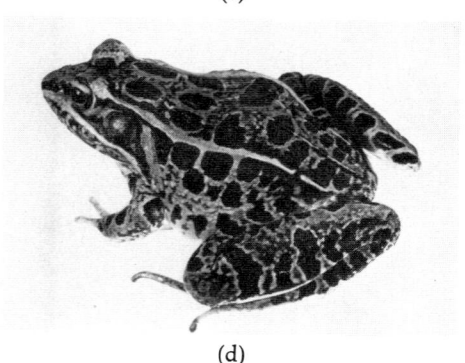

(d)

Similarly, San Francisco Bay is an important barrier for some mammals. Of 24 species of small mammals studied in the area, 11 occur both north and south of the bay, whereas 8 are found only in the north and 5 only in the south. What constitutes a barrier for some kinds of animals is not a barrier for others.

Another example of the effects of geographic barriers is provided by the *Oeneis* butterfly, which is found only on the tops of the Rocky Mountains in Colorado and the White Mountains of New England and along the cool coast of Labrador.

Recently a study was made comparing populations of wood frogs (*Rana sylvatica*) from the Colorado-Wyoming area with those from southern Canada. Presumably the more southerly populations represent a group that was displaced southward by the most recent glaciation, about 10,000 years ago, and the northern populations represent survivors of the parent group. The groups differ in several physical characteristics, including average leg length and markings, and the mating calls of the southern group tend to be lower-pitched. Most important, in cross-fertilization experiments, 99 percent of the eggs failed to develop, indicating that the southern group has become a new species.

Geographic Races

Every widespread species that has been carefully studied has been found to contain geographically representative populations that differ from each other to a greater or lesser extent. A species composed of subspecies of this sort is obviously particularly susceptible to speciation if geographic barriers arise, as with the wood frog.

Because of the work of John Moore of the University of California, the leopard frog (*Rana pipiens*) has become one of the best-known examples of subspeciation. Leopard frogs are found in North America as far north as Quebec and as far south as Mexico. Moore has studied and compared a large number of characteristics in 29 separate populations of this species. He has found that there are marked variations, as might be expected, from population to population. Crossbreeding of individuals from different populations produces normal offspring when parents are drawn from populations that are geographically adjacent, such as central and southern Florida, or that lie at roughly the same latitude, such as Texas and central Florida. However, the greater the north-south gap separating the home populations of the parents, the greater also is the proportion of defective, often inviable, offspring. In Texas-Vermont hybrids, for example, mortality among the developing embryos may reach as high as 100 percent. Yet because there is a gene flow among adjacent populations, *Rana pipiens* is still one species. One can see, however, that if the intermediate, bridging populations were eliminated, separate species would evolve quite quickly, as they have done with *Rana sylvatica*.

HOW SPECIES ARE MAINTAINED

Many species remain geographically isolated, but often species that are members of the same genus and that are obviously closely related will be found living in the same forest or on the same island. How do they maintain their genetic identities; that is, what keeps them from interbreeding? Studies of such

species have revealed the presence of a variety of isolating mechanisms, which can be divided into two general categories: (1) physiological isolating mechanisms and (2) behavioral isolating mechanisms.

Physiological Isolating Mechanisms

These barriers arise because of physiological incompatibilities between members of the two species. There are a number of such incompatibilities. For instance:

1. Differences in the shape of the genitalia may prevent insemination.
2. The sperm may not be able to survive in the reproductive tract of the female.
3. The pollen tube (in the case of plants) may not be able to form.
4. The sperm cell may not fuse with the ovum.
5. The ovum, once fertilized, may not be able to develop. Anomalies in development, as we saw, were the chief difficulties encountered in crossbreeding experiments among the races of the leopard frog.
6. The young, or some of the young, may survive but may not become reproductively mature.
7. The offspring may be hardy—the mule, for instance—but sterile. Sterility in such cases is apparently often caused by the fact that the dissimilar chromosomes cannot pair at meiosis.

Crosses between horses and zebras, like those between horses and donkeys, produce viable but sterile offspring. Zebroids, the results of these crosses, as well as horses and mules, mingle in the Kenya farm herd shown here.

THE BREAKUP OF PANGAEA

In the past ten years, the theory of continental drift—viewed at one time with as much suspicion as reports of flying saucers and extrasensory perception—has become firmly established. According to this theory, the outermost layer of the Earth is divided into a number of segments, or plates. These plates, on which the continents rest, slide across the surface of the Earth, moving in relationship to one another. The geologic expression of this relative motion occurs mainly at the boundary between the adjacent plates. Where plates collide, volcanic islands such as the Aleutians may be formed or mountain belts such as the Andes or Himalayas may be uplifted. At the boundaries where plates are separating, volcanic material wells up to fill the void. It is here that ocean basins are created. Plates may also move parallel to the boundary that joins them but in opposite directions. Such is the case of the San Andreas fault.

About 200 million years ago, all the major continents were locked together in a supercontinent, Pangaea. Several reconstructions of Pangaea have been proposed, one of which is shown here. It is generally agreed that Pangaea began to break up about 190 million years ago, about the time the dinosaurs approached their zenith and the first mammals began to appear. First the northern group of continents (Laurasia) split apart from the southern group (Gondwana). Subsequently, Gondwana broke into three parts, Africa-South America, Australia-Antarctica, and India. India drifted northward and collided with Asia about 50 million years ago. This collision initiated the uplift of the Himalayas, which continue to rise today as India still pushes northward into Asia.

By the end of the Cretaceous period, according to current reconstructions, about 65 million years ago, South America and Africa had separated sufficiently to have formed half the South Atlantic and Europe. North America and Greenland had begun to drift apart; however, final separation between Europe and North America–Greenland did not occur until the Eocene (43 million years ago). During the Cenozoic era, Australia finally split from Antarctica and moved northward to its present position, and the two Americas were joined by the Isthmus of Panama, which was created by volcanic action.

Geology now finds itself in somewhat the same position as astronomy did at the time of Galileo and Copernicus, and geology texts are being hastily rewritten in accordance with the new doctrines. Students of evolution have not yet felt the full impact of the new geology, but it is already beginning to provide answers to evolutionary problems and raise some problems of its own. Moreover, it seems certain that studies of fossil forms will help to resolve some of the configurations of Pangaea and its subsequent history.

The San Andreas fault is a boundary between two giant plates that are moving past one another. The fault, running through San Francisco and continuing down southeast of Los Angeles, is responsible for California's notorious earthquakes.

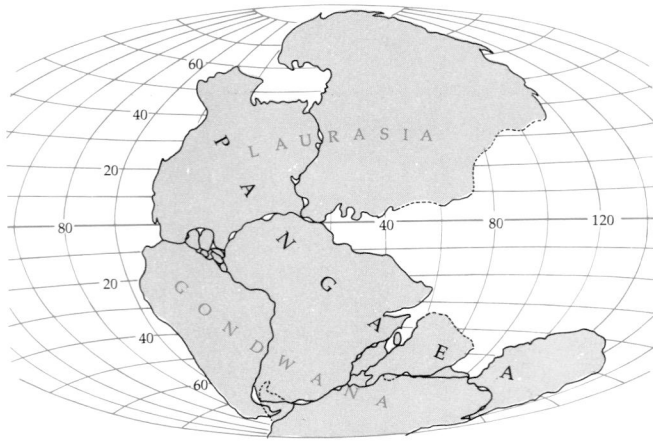

47–9 *Genetic isolation is often the result of behavioral patterns, such as the elaborate courtship rituals of the Laysan albatross. Most such behavioral patterns are both genetically determined and learned; that is, they vary in detail from population to population and, to some extent, from individual to individual, but the capacity to perform them at all is genetically determined.*

Behavioral Isolating Mechanisms

One of the most significant things about the physiological (postmating) isolating mechanisms is that, in nature, they are rarely tested. In animals, physiologically incompatible genotypes are usually prevented from mating by behavioral isolating mechanisms. These premating behavioral mechanisms can take many forms.

Some involve elaborate courtship behavior. The stickleback may make advances toward a female of another species, as described in Chapter 31, page 529, but if she does not give the appropriate social response, the affair is at an end.

Visual recognition is important, particularly among birds, which are visually oriented in all their behavior. According to observers, the Galapagos finches clearly recognize each other by their beaks. A male finch may mistakenly pursue a finch of another species—the finches often look very much alike from the rear—only to lose interest as soon as he sees the beak. He will not court a female of another species. The beak is a conspicuous feature in courtship, during which food is passed from the beak of the male to that of the female.

Among many invertebrates, pheromones are the chief isolating mechanisms, serving as signals, as in the case of the cecropia moth, to attract the male or, as with the oyster, to trigger the release of gametes by the female.

Bird songs, frog calls, the strident love notes of cicadas and crickets serve to identify members of a species to one another.

The flashings of fireflies (which are actually beetles) are sexual signals. Each species of firefly has its own particular flashing pattern, which is different from that of other species both in duration of flashes and in intervals between them. (See Figure 47–10.) For example, one common flash pattern consists of two short pulses of light separated by about two seconds, with the phrase repeated every four to seven seconds. The lights also vary among the species in intensity and color. The male flashes first, and the female answers, returning the species-specific signal. If you watch the firefly signals, you can learn to distinguish different species, and by mimicking signals with a flashlight, you can attract the males of a given species.

Among mammals, smell is probably the most important single factor. In addition to their normal distinctive odors, mammals also probably produce sexual pheromones that may serve as attractants and isolating mechanisms.

-10 *Flashing patterns of various species of fireflies found in Delaware. Each division on the horizontal lines represents one second, and the height of each curve represents the brilliance of the flash. The black curves are male flashes, and the colored ones are the female responses. The male flashes first and the female answers, returning the species-specific signal. Firefly flashes differ from species to species, not only in duration, intensity, and timing, but also in color.*

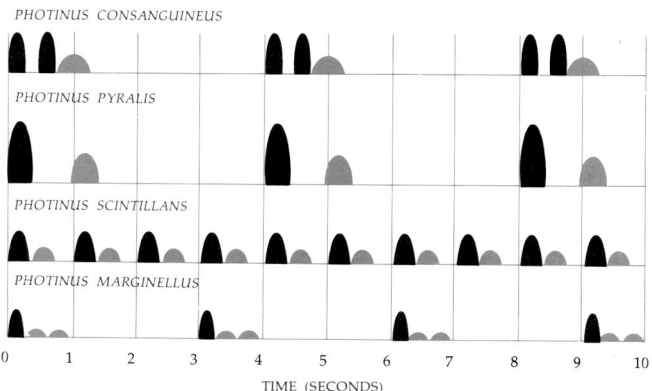

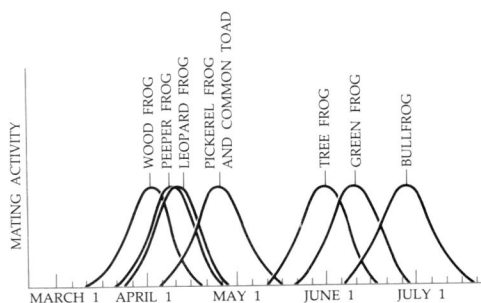

47–11 *Mating timetable for various frogs and toads that live near Ithaca, N.Y. In the two cases where two different species have mating seasons that coincide, the breeding sites differ. Peepers prefer woodland ponds and shallow water; leopard frogs breed in swamps; pickerel frogs mate in upland streams and ponds; and common toads use any ditch or puddle.*

Animal behavior serves as a genetic isolating mechanism for those plants that have evolved species-specific relationships with pollinators.

Temporal mechanisms also play an important role in sexual isolation. Species differences in flowering times are important isolating mechanisms in plants. Most mammals—man is a notable exception—have seasons for mating, often controlled by temperature or by day length. Figure 47–11 shows the mating calendar of species of frogs near Ithaca, New York. The spawning of freshwater fishes is often regulated by water temperatures. Members of the species *Drosophila pseudoobscura* and *Drosophila persimilis*, which often are found in the same areas, congregate and mate almost every day but differ in their mating hour.

The fact that hybrid progeny are often less viable serves to reinforce behavioral isolating mechanisms in the species. A female cricket that answers to the wrong song or a frog whose personal calendar is not synchronized with that of the rest of the species will contribute less to the gene pool. As a consequence, there will be a steady selection for behavioral isolating mechanisms.

POLYPLOIDY

Polyploidy is a means by which genetic isolation is produced without geographic isolation. A haploid cell, you will recall, has one set of chromosomes (n). A diploid cell has two sets ($2n$), one from each parent. A cell that has more than two sets of chromosomes (or an individual made up of such cells) is known as a polyploid.

Polyploidy originates when the chromosomes replicate and divide but the cell does not. A polyploid cell may have three, four, five, or more sets of chromosomes.

Polyploidy rarely occurs in animals but is very frequent in plants. Among the largest group of plants, the angiosperms, about half of the approximately 300,000 known species are polyploids.

A polyploid plant may reproduce vegetatively, or a flower may form from a part of the plant that is polyploid. Cells that are $4n$ or $6n$ (or any even multiple of n) can undergo meiosis because each chromosome has a suitable partner. However, they cannot produce fertile hybrids with the n parent type (such a hybrid would be $3n$). They can, however, produce fertile offspring with other polyploids. In effect, these polyploid types are genetically isolated from their parent strains.

Polyploidy also makes it possible for hybrid crosses to occur in plants. The man-made hybrid *Raphanobrassica*, formed from the radish and the cabbage, provides an example. Some years ago, an enterprising plant geneticist was struck by the commercial possibilities of such a hybrid since, in the one plant, the roots are edible and, in the other, the leaves. In each plant, the haploid (n) number is 9. A cross between the two produced a plant with nine radish chromosomes and nine cabbage chromosomes, which was, like the mule, a sturdy hybrid but infertile. But since the chromosomes could not pair at meiosis, an artificial tetraploid ($4n$) was developed in which the somatic cells had 18 radish chromosomes and 18 cabbage chromosomes, making it possible for the gamete to receive a full complement (9) of each. As their originator had hope-

(a) *A hybrid organism (such as a mule) produced from two 1n gametes can grow normally because mitosis is normal, but it cannot reproduce because the chromosomes cannot pair at meiosis.* (b) *However, a hybrid organism produced from two 2n gametes can both divide normally mitotically and undergo meiosis. Since each chromosome will have a partner, the chromosomes can pair at meiosis.*

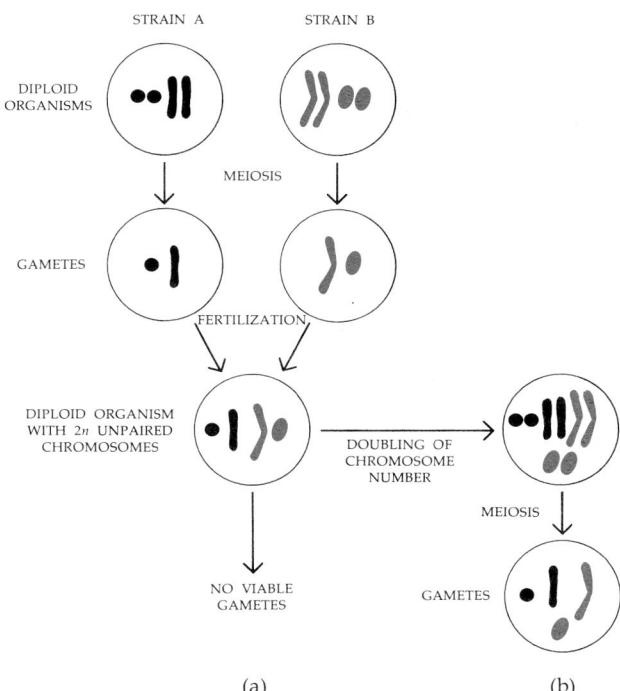

Polyploidy can be induced in plants by treating dividing cells with a chemical called colchicine which prevents the development of the spindle and so blocks the separation of the chromosomes during mitosis. Because polyploidy may increase the size of a plant or the number of its floral parts, many cultivated plants are polyploids, such as the dahlia shown here.

fully anticipated, the hybrids were both sturdy and fertile. Their only drawback: they had a root like that of a cabbage and a head like that of a radish.

Most of the more important domestic plants are polyploids, including cotton, wheat, potatoes, tobacco, oats, bananas, many apples, and some cherries, as well as such garden plants as irises, chrysanthemums, and tulips.

THE AVAILABLE NICHE

The second major factor required for speciation, as we noted earlier, is an available ecological niche. Again, one of the best illustrations is the Galapagos finches. On these deserted islands, there was not only a niche for finches but also niches for warblers and woodpeckers, and so the immigrant finches became "warblers" and "woodpeckers."

TIME

Evidence of the length of time required for a new species to develop is based largely on what is known about the ages of various islands. The British Isles, for instance, have been separated from the continent of Europe only since the last glaciers retreated, about 10,000 years ago. During this time Britain is not known to have developed any distinctly new species, so that a period of 10,000 years or so may be inadequate, in most cases, although it was apparently sufficient for *Rana sylvatica*. On the other hand, islands whose age is probably a few million years seem to have many unique species. So between 10,000 and a few million years seems necessary for most organisms under natural conditions.

ADAPTIVE RADIATION

Major evolutionary change appears to take place in sudden bursts, rather than slowly and steadily. Such bursts characteristically take the form of _adaptive radiation_, which is the diversification of a group of organisms that share a common ancestor (often, itself, newly evolved) to fill all the ecological space available. Adaptive radiation is believed by some population biologists to be the major pattern of evolution. Darwin's finches are an example of adaptive radiation on a miniscale, the marsupials of Australia (Figure 47–14) on a larger one.

The fossil record contains many examples of adaptive radiation. For example, some 425 million years ago, the first terrestrial plants and animals suddenly emerged. A large number of different populations apparently began the same move at about the same time. Most undoubtedly perished, but a few survived and then differentiated into the many different types of terrestrial plants and animals existing today. A similar, even more rapid burst of evolution gave rise to the birds. Apparently, according to the fossil record, a number of populations of reptiles must have taken to climbing and gliding at about the same time. The shift was made very rapidly from the land to the air, however, and these intermediate populations, the gliding, cold-blooded reptiles disappeared rapidly and permanently, as flying, homeothermic birds filled the ecological space. The mammals similarly burst forth on the evolutionary scene, with many different kinds appearing simultaneously in the fossil record.

What is necessary for these great evolutionary moves to take place? First, there must be an ample store of genetic varition in the gene pool. Second, there must be genetic isolation and thus opportunity for speciation.

There are important limits to what can occur. Evolution must proceed by certain "logical" steps. Water, which is 800 times heavier than air, can support an animal like a jellyfish and can provide sufficient resistance for ciliary motion, but a jellyfish could never become terrestrial. Animals needed to develop strong supporting skeletons and strong muscles before they could invade the land. A third necessary condition for evolution is, therefore, preadaptations, the existence of structures that could become modified to other uses.

For example, the lungs and lobes of the lungfish made it possible for animals to begin the invasion of land once the plants had prepared the way for them. These structures developed in the lungfish not as an adaptation to terrestrial life but as an adaptation to conditions of drought. The lungs and lobes had been in existence for millions of years. Similarly, although birds and reptiles are very different in their modes of life, very little modification of the basic plan of the reptile skeleton was necessary to produce the skeleton of the bird; for the bird, feathers were the big evolutionary breakthrough. The famous fossil _Archaeopteryx_ (Figure 23–40a) is essentially a reptile with feathers. With insects, the rapid development of a great multitude of species depended upon the existence of appendages that could be modified into a large number of specialized structures.

The fourth condition necessary for an evolutionary breakthrough to occur is that a new adaptive zone must be available—a constellation of vacant ecological niches. The land was once such an adaptive zone, and the air another. Once an adaptive zone becomes available, divergent lines of ancestral stock are tested in this new zone; probably most do not make it. For example, there were many

-14 In Australia, marsupials fill the niches occupied by placental mammals on other continents. The animals pictured here are all marsupials (the common names of their placental counterparts are in parentheses): (a) wombat (woodchuck), (b) koala bear (bear), (c) kangaroo rat (rat), and (d) Tasmanian wolf (dog).

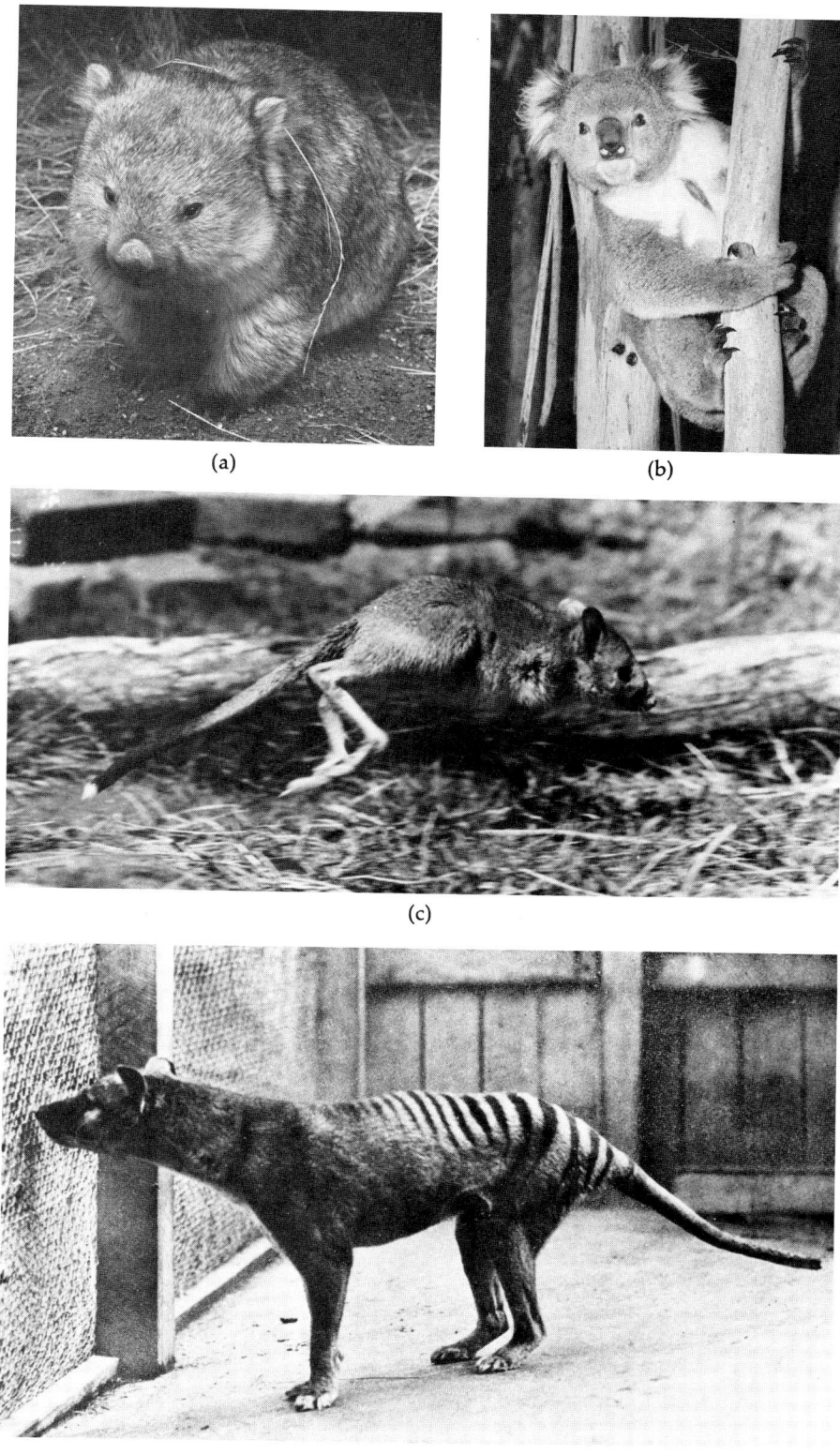

(a)

(b)

(c)

(d)

Evolutionary relationships may be established by comparisons of living species and by examination of fossil forms. Most recently a new method has become available: the study of macromolecules. Amino acids from similar proteins of different species are different, and the degree of difference between two species corresponds quite well with the degree of evolutionary divergence between them. For example, the protein chain of cytochrome c is exactly the same in men and chimpanzees, but differs from that of Neurospora *by 44 out of 100 amino acids. The composition of cytochrome c from 40 species has now been determined, and it appears that about 20 million years are required to produce a change of 1 percent in this particular protein.*

The altered protein is able to function because the changes involve only certain amino acids; most of the amino acid sequence is the same for all species. Moreover, when a substitution is made, it almost always involves an amino acid of similar chemical properties, so that the conformation of the chain is altered little if at all. Presumably mutations occur that involve other amino acids but these are weeded out as deleterious.

Other proteins change at different rates. Hemoglobin, for example, appears to change at a rate of 1 percent in 6 million years, or approximately one substitution in 3.5 million years, presumably because it is more tolerant of change.

Although the macromolecules evolve at different rates, they are in good accord with one another. Moreover, they are generally, but not always, in accord with accepted interpretations of the fossil evidence. (A notable exception, involving man and the great apes, is discussed in Chapter 48.) Perhaps most important, these ancient proteins may well hold the key to important and unsolved evolutionary problems, such as those of the relationships of the invertebrate phyla to one another and to vertebrate origins.

groups of flying reptiles at one time but the birds, as far as is known, evolved from just one ancestral stock. The shift either is rapid or fails completely. Once the shift is made, radiation takes place until the zone is filled. Other bursts of adaptive radiation appear to have taken place with the subsiding of the glaciers, when essentially new land was opened for habitation. For Darwin's finches, the underoccupied Galapagos Islands represented a new adaptive zone.

CONVERGENT EVOLUTION

Organisms that move into similar ecological niches as a result of adaptive radiation often come to resemble one another. The Australian marsupials, so strikingly similar to their placental counterparts, are a good example. The whales, a group that includes the dolphin and porpoises, are very similar in many exterior features to sharks and other large fish; the fins of whales conceal the remnants of a vertebrate hand, they are warm-blooded, like their land-dwelling ancestors, and they have lungs rather than gills. (Gills are clearly more useful than lungs in aquatic respiration—another example of the fact that evolutionary processes work only with what is already available.) Similarly, two families of plants invaded the desert, giving rise to the cacti and the euphorbs. Both evolved large fleshy stems with water-storage tissues and protective spines and appear superficially similar. However, their quite different flowers reveal their widely separate evolutionary origins.

(a)

(b)

Although porpoises (a) are evolutionary descendants of primitive mammals, their shapes closely resemble those of large, primitive ocean-dwelling fish, as exemplified by the sand shark (b). Organisms that come to occupy similar ecological niches as a result of adaptive radiation often come to resemble one another.

EXTINCTION

Only a small fraction of all the species that have ever lived are presently in existence—certainly less than $\frac{1}{10}$ of 1 percent, perhaps less than $\frac{1}{1,000}$ of 1 percent. Extinction is very much a part of the evolutionary process, with the elimination of old species making room for new ones. Certain times in the Earth's history have seen the elimination of major life forms: the trilobites that dominated the early Paleozoic seas; the dinosaurs that disappeared at the end of the Mesozoic era; and the great land mammals that became extinct during the Pleistocene, victims of its violent climatic changes or, more likely, of the predations of a new life form, man. Based on the history of evolution, it seems likely that man, too, will eventually become extinct. We do not, as yet, have any indication as to what new species, if any, will take his place.

SUMMARY

The species—an isolated gene pool moving through time and space—is the unit of evolution. Three major factors are required for speciation: genetic isolation of a population, the availability of a new ecological niche, and time. Genetic isolation is often achieved by geographic isolation, as occurs on islands. Apparently, genetic isolation can also occur by behavioral isolation. Probably between 10,000 and a few million years are required for an isolated population to become a new species. Polyploidy is another means of genetic isolation particularly common among plants. It makes possible new hybrid combinations.

Evolution often takes place by adaptive radiation, the diversification of species to fill the ecological niches in a new adaptive zone. Requirements for adaptive radiation probably include—in addition to the factors needed for species formation—the existence of a population with much latent variation, the existence

within this population of appropriate preadaptations, and the availability of a new adaptive zone.

Convergent evolution occurs when two different types of organisms with different evolutionary histories move into the same adaptive zone.

The great majority of species that have ever existed are now extinct. In the past, extinction has made way for new species. Man is now the dominant force in the extinction of species. We do not yet know what species, if any, will fill the niches now being vacated.

QUESTIONS

1. Define the following terms: polyploidy, adaptive radiation, subspecies, behavioral isolation.
2. Describe the separate steps involved in the formation of the distinct species of Galapagos finches.
3. In your opinion, why is there at present only one species in the genus *Homo*? Is a second *Homo* species likely to arise?
4. What sort of new ecological niches have been created by man's activities?

SUGGESTIONS FOR FURTHER READING

CALDER, NIGEL: *The Restless Earth*, The Viking Press, Inc., New York, 1972.

A handsomely illustrated report on the new geology, written for the general reader.

CARLQUIST, SHERWIN: *Island Life: A Natural History of the Islands of the World*, The Natural History Press, Garden City, N.Y., 1965.

An exploration of the nature of island life and of the intricate and unexpected evolutionary patterns found in island plants and animals.

DE BEER, SIR GAVIN: *Adaptation*, Oxford Biology Readers, Oxford University Press, London, 1972.*

One of a series of short (16 pages) booklets on special subjects, all written by experts and designed for beginning students and other interested readers.

DOBZHANSKY, THEODOSIUS: *Genetics and the Origin of Species*, 3d ed., Columbia University Press, New York, 1951.*

A presentation of the basic themes of population genetics by one of the foremost leaders in studies in Drosophila.

EICHER, DON L.: *Geologic Time*, Prentice-Hall, Inc., Englewood Cliffs, N.J., 1968.*

A short supplementary text that describes the fossil strata and their significance.

HAMILTON, TERRELL H.: *Process and Pattern in Evolution*, The Macmillan Company, New York, 1967.*

This short book, designed as a supplementary text, is outstanding for its clarity of definition and presentation of evolutionary concepts.

HUTCHINSON, G. EVELYN: *The Ecological Theater and the Evolutionary Play*, Yale University Press, New Haven, Conn., 1965.

By one of the great modern experts on freshwater ecology, this is a charming and sophisticated collection of essays on the influence of enviroment in evolution—and also on an astonishing variety of other subjects.

*Available in paperback.

LACK, DAVID: *Darwin's Finches*, Harper & Row, Publishers, Inc., New York, 1961.*
This short, readable book, first published in 1947, gives a marvelous account both of the Galapagos, its finches, and other inhabitants and of the general process of evolution.

LEWONTIN, R. C.: *The Genetic Basis of Evolutionary Change*, Columbia University Press, New York, 1974.
This book is intended for the advanced student or another worker in this field, but it is also highly recommended for any reader interested in the problems and perspectives of population genetics.

MAYR, E.: *Animal Species and Evolution*, Harvard University Press, Cambridge, Mass., 1963.
A masterly, authoritative, and illuminating statement of contemporary thinking about species—how they arise and their role as units of evolution.

SIMPSON, GEORGE GAYLORD: *Horses*, Oxford University Press, New York, 1951.
The "story" of the horse family in the modern world and through sixty million years of history, as recorded by one of the nation's leading paleontologists.

SMITH, JOHN MAYNARD: *The Theory of Evolution*, 2d ed., Penguin Books, Inc., Baltimore, Md., 1966.*
Written for the general public, this book is notable for its many concrete examples of evolution, past and present.

STEBBINS, G. LEDYARD: *Processes of Organic Evolution*, Prentice-Hall, Inc., Englewood Cliffs, N.J., 1966.*
A brief review of the entire field by one of its outstanding practitioners.

WILSON, E. O., and WILLIAM H. BOSSERT: *A Primer of Population Biology*, Sinauer Associates, Inc., Stamford, Conn., 1971.*
An introduction to the mathematics of population biology.

* Available in paperback.

SECTION 8

EVOLUTION OF MAN

48–1 *Our closest relatives. Recent studies on chimpanzees have revealed that they share numerous behavioral traits with man—including meat eating and tool using—and that their karyotypes are similar to ours and their hemoglobins and cytochromes are identical.*

Chapter 48

The Primates

Where does the story of human evolution begin? We might start with a chance combination of chemicals in some warm Precambrian sea. Or perhaps even with the formation of a small planet 93 million miles from a star. Or it might begin 4 billion years later, when some little tribe of man-apes found they could sharpen a digging stick or hone the flat edge of a stone. In any case, it is a very long story, measured in human terms, and many of its details are lost to us, probably forever.

For present purposes, let us start the story about 200 million years ago, in the early Mesozoic era, at about the time of the first dinosaurs. In this same period of time—give or take a few million years—the first mammals appeared, arising from a primitive reptilian stock. Our information about these mammals is very slight. The entire length of the Jurassic and Cretaceous periods has left us with only a few fragments of skulls and some occasional teeth and jaws. From these scraps of evidence, we know that the first mammals were about the size of a rat or mouse. According to these fossil remains, they resembled the modern insectivores, of which shrews (Figure 48–2) are a common example. They had sharp teeth, indicating that they were carnivores, but since they were too small to attack most other vertebrates, they are assumed to have lived mostly on insects and worms, supplementing their diet with tender buds, fruit, perhaps eggs, and probably whatever else they could find. Although they were puny and unimpressive compared with the reptiles that were their contemporaries, they had several great advantages over them. The most crucial perhaps is that they were probably warm-blooded. We surmise this because they had a bony palate separating the nasal cavity from the mouth. This is an adaptation of animals that breathe continuously and regularly, even when eating, which most warm-blooded animals do and reptiles do not. Being warm-blooded meant that these early mammals had a higher rate of metabolic activity than their reptile contemporaries and could supply their bodies with oxygen and energy more efficiently. Also, the chief characteristic of warm-bloodedness, a high and constant body temperature, would have enabled them to be active at twilight and throughout the night, as they most likely were, when the reptiles were rendered inactive by the cold. (The evidence for nocturnal existence is the large size of the eye sockets, which is a characteristic of modern nocturnal primates.)

The presence of a bony palate is one of the criteria by which paleontologists classify a transitional animal as a mammal. Other criteria are whether the jaw is a single bone and whether the two bones that form the joint between the

Shrews are active, nervous, bad-tempered nocturnal animals that eat their own weight in food daily. They have long snouts, sensitive to both smell and touch, with which they snuffle through the grass after seeds and very small animals. Fossils of the most primitive mammals indicate that they resembled modern ground shrews, such as the short-tailed shrew shown here.

skull and jaw in reptiles have moved to join the stapes and so form the delicate, three-bone mechanism of the mammalian middle ear (page 707).

For about 80 million years, these small animals led their secretive, probably nocturnal existences in a land dominated by carnivorous reptiles. Then suddenly, as geologic time is measured—that is, within several million years—the dinosaurs disappeared. By the end of the Cretaceous period the very large reptiles had disappeared forever, and about 65 million years ago an explosive radiation of the mammals began.

The early mammals immediately diverged into the two dozen or so different lines that included the monotremes, or egg-laying mammals, of which the duck-bill platypus is one of the few remaining examples; the marsupials, such as the kangaroos, wallabies, opossums, koala bears, and others whose young are born in embryonic form and continue their development in pouches; and the placentals. Among the placentals are carnivores, ranging in size from the saber-toothed tiger down to small, weasel-like creatures; herbivores, or ungulates, which include not only the many wild grazing animals but also most of our domesticated farm animals; the omnipresent rodents, so called because of their gnawing ability; and such odd groups as the whales and dolphins, the

Table 48–1 *Evolution of Primates: Major Physical and Biological Events in Geologic Time*

Millions of years ago	Era	Period	Epoch	Life forms	Climates and major physical events
	CENOZOIC	Quaternary	Recent Pleistocene	Age of man. Planetary spread of *Homo sapiens*; extinction of many large mammals, including woolly mammoths. Deserts on large scale.	Fluctuating cold to mild. Four glacial advances and retreats (Ice Age); uplift of Sierra Nevada.
1½–7		Tertiary	Pliocene	Large carnivores. First known appearance man-apes.	Cooler. Continued uplift and mountain building, with widespread extinction of many species.
7–26			Miocene	Whales, apes, grazing animals. Spread of grasslands as forests contract.	Moderate uplift of Rockies.
26–38			Oligocene	Large, browsing mammals. Apes appear.	Rise of Alps and Himalayas. Lands generally low. Volcanoes in Rockies area.
38–53			Eocene	Primitive horses, tiny camels, modern and giant types of birds.	Mild to very tropical. Many lakes in western North America.
53–65			Paleocene	First known primitive primates and carnivores.	Mild to cool. Wide, shallow continental seas largely disappear.
65–136	MESOZOIC	Cretaceous		Age of reptiles, extinction of dinosaurs. Marsupials, insectivores. Angiosperms become abundant.	Lands low and extensive. Last widespread oceans. Elevation of Rockies cuts off rain.
136–195		Jurassic		Dinosaurs' zenith. Flying reptiles, small mammals. Birds appear.	Mild. Continents low. Large areas in Europe covered by seas. Mountains rise from Alaska to Mexico.
195–225		Triassic		First dinosaurs. Primitive mammals appear. Forests of gymnosperms and ferns.	Continents mountainous. Large areas arid. Eruptions in eastern North America. Appalachians uplifted and broken into basins.

bats, the modern insectivores, which include the moles and the hedgehogs, and the primates.

SOME PRIMATE CHARACTERISTICS

The order Primata is divided into two suborders: the prosimians, or lower primates, which include tree shrews, tarsiers, lemurs, and lorises, as well as a large number of fossil forms, and the anthropoids, which include the monkeys, the apes, and man.

What distinguishes the primates from other mammalian orders? First, compared to other mammals, primates are relatively unspecialized. Their extremities resemble those of the primitive mammals—indeed, of the reptiles—more closely than do the extremities of mammals of most of the other major orders (Figure 23–43). The first four-legged mammals all had five separate digits on each hand and foot, and each digit except the thumb and the first toe had three separate segments that made it flexible and capable of independent movement. In the course of evolution, most mammals developed hooves and paws more suited for running, springing, seizing prey, tearing, and digging; the primates retained and elaborated on the primitive pattern.

Similarly, in the basic quadrupedal structure, the forelimb is supported by two bones (the radius and the ulna), a pattern that provides for flexibility, but among mammals, it is the primates that have retained the ability to twist the radius, the bone on the thumb side, over the ulna so that the hand can be rotated through a full semicircle without moving the elbow or the upper arm. Try it. You do it all the time but perhaps without appreciating how important it is for the efficient use of your hands.

Most other mammals (bats are a notable exception) have also lost the ability to move the upper arm freely in the shoulder socket. A dog or horse, for instance, usually moves its legs in only one plane, forward and backward; some South American monkeys, apes, and man are the only higher mammals that can rotate the arm widely in the socket.

Dentition

Reptiles have peglike teeth, all about the same size and shape, which are replaced as they wear out. Mammals have only two sets of teeth: baby teeth, or milk teeth, which are replaced by larger, "permanent" teeth that, among the fortunate, last the rest of their lifetime. Mammalian teeth are not uniform but are specialized for different functions. The primitive mammal has, from front to back: incisors, which are flat, sharp-edged teeth used for cutting; canines used for holding and tearing prey; and premolars and molars used for grinding. These latter teeth have more than one cusp and are progressively larger from front to back. Thus the largest teeth are closest to the joint of the jaw, where crushing power is greatest.

There have been two trends in dental evolution among the primates. First, primates, with their quite general diet, have tended to retain the generalized pattern of dentition just described, in contrast to the other mammals, whose dentition is far more specialized (see Figure 34–4). Some familiar examples of such specializations are the powerful tusks of elephants; the seizing, tearing, and bone-crushing teeth of the large carnivores; the nibbling teeth of rabbits;

48–3 *Some primate hands. One of the distinguishing features of the primate is its ability to grasp, but except in man, all primate hands are specialized in some degree for a particular function. The hand of the tarsier has enlarged skin pads for grasping branches. In the orangutan, the fingers are lengthened and the thumb reduced for more efficient brachiating. The gorilla's hand, which is used in walking as well as handling, has shortened fingers. Man's thumb is larger proportionately than that of any other primate, and opposition of thumb and fingers, on which the handling ability depends, is greatest in man.*

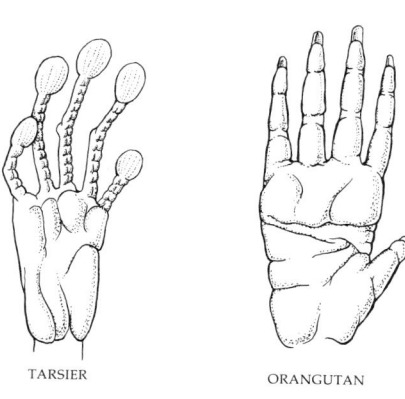

TARSIER   ORANGUTAN

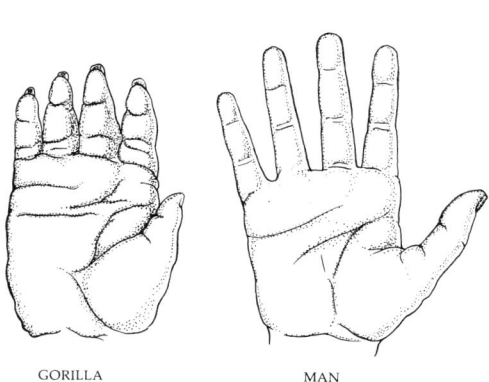

GORILLA   MAN

48–4 (a) *Chimp with palm extended.* (b) *Although the fingers of chimpanzees, like those of gorillas, are considerably shorter than man's, they are capable of a fine precision grip.*

(a)

(b)

the gnawing teeth of beavers; and the broad, flat, blade-edged incisors of herbivores. In the unlikely event that a domestic cat would permit you to inspect its teeth, you would find just four molars, one in each quadrant.

The second general trend, also seen among other mammals, is a reduction in the number of teeth. Primitive mammals seem to have had 44. Fossil primates had 34 or 36. New World monkeys have 36. Old World monkeys, apes, and man all have 32. Their dental formula is described by the following equation:

$$\left.\begin{array}{c} 2{:}1{:}2{:}3 \\ 2{:}1{:}2{:}3 \end{array}\right\} \times 2 = 32$$

Translated, this means simply that, starting in the middle and reading from front to back, there are two incisors, one canine, two premolars, and three molars. Check it out in your own jaw.

Some interesting exceptions to the general lack of specialization among primate teeth are the canines of male gorillas and male baboons, used primarily in threat displays, and the lower front teeth (both canines and incisors) of lemurs which, in keeping with their generally elegant appearance and emphasis on personal grooming, form a neat comb for their silky coats.

The only highly specialized organ of the primates is the brain. Primates are the most intelligent and also the most curious, inventive, mischievous, and destructive of the mammals.

PRIMATE ADAPTATIONS

The first distinguishing feature of the primates, as we have just noted, is their relatively unspecialized anatomy. The second distinguishing feature, closely related to the first, is their adaptation to an arboreal (tree-living) existence. The most important event in the history of the primates occurred when a mouse-sized primitive mammal took to the trees.

Prehensile Extremities

The primate hand evolved as an adaptation to movement through the trees. With a few exceptions, primates have five digits, a divergent thumb, and nails rather than claws. The divergent thumb, which can be brought into opposition to the forefinger, greatly increases gripping powers and dexterity. (The exceptions include marmosets, which climb like squirrels with clawed digits, and spider monkeys, which have lost their thumbs altogether and hang from branches with their hooklike fingers.) Nails leave the tactile surface of the digit free and so greatly increase the sensitivity of the digits for exploration and manipulation.

Among New World monkeys, the thumb characteristically moves against the forefinger with a scissorlike action. Old World monkeys, however, bring the thumb end against some part of the index finger, sometimes partially opposing the tips. In general, ground-dwelling monkeys have a more highly developed precision grip than arboreal species. Apes have very marked precision grips, in which the thumb is partly rotated and opposed to the side or tip of the index finger. Thus there is an evolutionary trend among the primates toward finer manipulative ability that reaches its culmination in man.

Sight vs. Smell

A second corollary of the move to the trees is the high premium placed on visual acuity with a decreasing emphasis on the role of olfaction, the most important of the senses among many of the other mammalian orders. (Flying produced similar evolutionary pressures among the birds, which were also evolving rapidly during this same period.) This shift from dependence on smell to dependence on sight has anatomical consequences. For instance, the terrestrial shrew (Figure 48–2), at the end of its long sensitive snout, has a comparatively large area of naked, moist skin called the rhinarium. Such "wet noses" are familiar in dogs, pigs, moles, and a number of other mammals. They are organs both of smell and of touch. All prosimians except tarsiers have naked, moist muzzles, whereas monkeys, apes, and man have dry noses.

Among active tree dwellers there is a high selection pressure for stereoscopic vision. Tree shrews, like many other animals, have eyes that are directed laterally, but among the other primates there can be traced a steady evolutionary trend toward frontally directed eyes and hence stereoscopic vision. The muzzle has become progressively reduced; in animals depending more on vision than on smell, a large muzzle can actually be a disadvantage, interfering with close vision. (An exception is the baboon. Male baboons have doglike snouts, presumably to accommodate their large canine teeth. However, their olfactory sense appears to be about the same as other monkeys'.)

Almost all primates have cones as well as rods; cones (page 700) are photoreceptor cells that are concerned with color vision and also with fine visual discrimination. All primates also have foveas, areas of closely packed photoreceptor cells, which produce sharp visual images.

Change from emphasis on olfaction to emphasis on vision has behavioral effects as well. Monkeys and apes communicate less by odors than most mammals and more by visual signals, such as gestures and facial expressions, and also by verbal communications. Observers of social groups of monkeys and apes have been impressed by the amount of noise that accompanies the troop on its daily activities, a characteristic that can be presumed to have favored the development of speech in man.

The fact that the higher primates do not use urine or feces for territorial marking, the way cats and dogs do, for example, but rather let fly indiscriminately from the treetops seems to be related to the fact, reported by many involuntary students of primate excretory habits, that domesticated nonhuman anthropoids, despite their intelligence, are difficult if not impossible to house train.

Uprightness

Another adaptation to arboreal life is an upright posture. Even quadrupedal primates, such as the monkeys, sit upright; these sitters characteristically have ischial callosities, hard skin pads on their rumps. The brachiators, primates such as gibbons and chimpanzees, that move about by swinging from branch to branch have trunks that are oriented vertically. One consequence of this vertical posture is a change in the orientation of the head, allowing the animal to look straight ahead; it is this characteristic, above all others, that makes primates look so "human" to us. For instance, a dog's eyes, which normally

–5 *Eye-hand coordination is another characteristic of higher primates, such as the female olive baboon shown here with her child.*

Life in the treetops made maternal solicitude a major factor in infant survival. Also the necessity for carrying the young for long periods resulted in strong selection pressures for reduced numbers of offspring. Higher primates, such as this black-faced vervet monkey, usually have single births.

48–7 *At the bottom of the primate ladder is the tree shrew, which the earliest prosimians probably resembled. If you look closely you will see five-digited paws. Although clawed, they can be spread out and used for grasping.*

look up and out would be directed toward the ground if its head had the same relation to the spine that our head has, and this is also true of the primitive mammal. In the course of evolution, the primate's face has rotated down about 90°.

Vertical posture was, of course, an important preadaptation for the upright stance of modern man.

Care of the Young

Another principal trend in primate evolution is toward increased care of the young. Because mammals, by definition, nurse their young, they tend to have longer, stronger mother-child relationships than other vertebrates. Arboreal life demanded even more maternal solicitude because of the constant danger of fatal falls and so increased the dependency and learning period of the young.

Among the few instincts visible in the human infant is a fear of falling. Perhaps it is this primeval fear that is also the source of the nightmares, common apparently to all mankind, of falling great distances. In any case, the danger of falling was and is a very real one for young primates and must have proved a strong selective force for judgment, alertness, and parental concern.

PROSIMIANS

The Paleocene and Eocene epochs, which ushered in the Cenozoic ("new life") era, seem to have been characterized by warm temperatures that permitted the tropical and subtropical forests to extend much farther north and south of the equator than they do today. According to the fossil record, nearly 60 genera of prosimians inhabited forests in the Northern Hemisphere during these epochs. Those characterized by elongated front teeth—nibblers and gnawers —were not successful and became extinct, apparently yielding their ecological niches to the rodents, which were becoming abundant during this same time. The more successful included forms resembling the modern prosimians. This group includes lemurs, lorises, bush babies, tarsiers, and tree shrews. Prosimians are all arboreal. Some, such as the tarsiers and bush babies, are leapers and vertical clingers. Others, including the lemurs, are quadrupedal, climbing and running along branches on all fours. In some classification systems, tree shrews are grouped with the primates, and in others, they are grouped with the insectivores, which simply indicates the closeness of the two evolutionary branches.

Nocturnal prosimians, like most nocturnal animals, live alone or in small family groups. A few species of lemurs are diurnal, and, in their social organization, they range from solitary dwellers to pair-bonded pairs to small families to social groups with up to 20 members.

Lemurs

The largest concentration of prosimians flourishing today is a highly diversified group of lemurs on the island of Madagascar, off the south coast of Africa, whose ancestors made their way to the island early in their evolutionary history. There are no other primates on Madagascar except man, who is a very recent arrival.

8–8 *A ring-tailed lemur. Lemurs have flat-tened nails on all digits of the hands and feet except the second digit of each foot, which has a "grooming claw."*

8–9 *A native of Indonesia, the little tarsier (about the size of a kitten) is the most "advanced" of the prosimians. Its upper lip, like ours, is free from the gum below it, giving it the ability (as you can see) to make faces. It has stereoscopic vision and a larger brain than the lemur's. Living entirely in trees, it has hands and feet with enlarged skin pads for grasping branches, and its hind limbs are specialized for hopping—rather like a kangaroo's except that the limb ends with a hand. As you may have guessed from the owl-like eyes, tarsiers are nocturnal.*

Lemurs are somewhat larger than shrews; many of them are about the size of a kitten and some are as large as a good-sized dog. They are also more monkeylike than tree shrews. Lemurs have claws only on the second digit of each hind foot, which they use for grooming and scratching. The rest of the claws have been modified into flattened nails similar to the nails of all the primates higher on the phylogenetic scale.

Tarsiers

The tarsier, although certainly not a direct ancestor of the more advanced primates, closely resembles them in a number of ways. One of its most striking features is the fact that its upper lip is free from the gum like that of the higher primates but not of the lemurs. This permits the tarsier to have a wide range of facial expressions. See Figure 48–9. Also, as we noted previously, it does not have a naked and moist muzzle. It has stereoscopic vision; in other words, both eyes cover most of the same visual field and permit a composite picture of the field in the brain. This is important for judging short-range distances, an essential ability for an active treetop dweller. The size of the eyes of the tarsier are a special adaptation for night vision, like the eyes of the owl. According to W. E. Le Gros Clark, the British anatomist, the little saucer-eyed tarsier has existed relatively unchanged for some 50 million years.

MONKEYS

Monkeys, along with the apes and man, make up the higher primates, the anthropoids. The monkeys are generally larger than prosimians, their skulls are more rounded, and they are more intelligent. Like tarsiers, but unlike other prosimians, they have full stereoscopic vision, and also color vision. They are all diurnal.

48–10 *New World monkeys are platyrrhine, or flat-nosed (left); Old World monkeys and apes are catarrhine, or downward-nosed (right).*

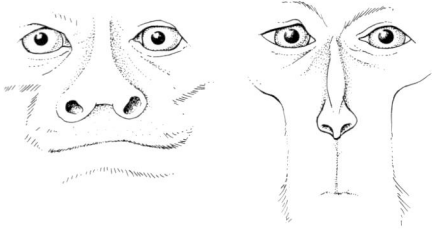

48–11 *Spider monkeys have lost their thumbs altogether and hold on with their hooklike digits 2 to 5. As you can see, this is a New World monkey.*

48–12 *The macaques are Old World terrestrial monkeys found in India, Pakistan, southeast Asia, China, and Japan. They are closely related to the baboons, which are found throughout Africa south of the Sahara and on the Arabian peninsula, and like baboons, they have a well-organized social structure with a dominance hierarchy. This monkey is a lion-tailed macaque from Asia.*

Monkeys generally move in bands composed of adult males and females, infants, and juveniles. The females protect and otherwise care for the young, and the males often serve protective functions for the group, ranging from observing and warning to attacking predators outright. In some species, males join together to mob a predator, harassing it by hooting and calling, jumping on branches until they fall on enemy heads, sometimes throwing sticks and branches, and, in some species, by group defecation and urination.

The monkeys arose from prosimian stock apparently during the Eocene period; related fossil forms are found in abundance in the Oligocene. There are two principal groups, the New World monkeys, also known as platyrrhines (meaning flat-nosed) and the Old World monkeys, the catarrhines (downward-nosed). The separation of these groups took place very early in their evolution, during the Eocene, when Gondwana broke apart to form the continents of South America and Asia.

Despite their long chronological separation, members of the two groups closely resemble each other, an example of parallel evolution.

The New World monkeys, from South and Central America, are all strictly arboreal in their habits, and many of them use their tails as a fifth prehensile limb, which none of the Old World monkeys can do. Any monkey you see hanging by its tail is definitely a New World monkey. The New World monkeys include marmosets, howler monkeys, the spider monkey, and the capuchin—the familiar monkey of the organ-grinder.

The Old World monkeys are both arboreal and terrestrial. The tree dwellers include the colobus monkeys, langurs, the mangabeys, and the guenons. Their tails are used for balancing rather than as prehensile organs. The ground dwellers, which came from tree-dwelling ancestors, include the baboons and macaques, or rhesus monkeys, all of which walk on all fours. (The social behavior of baboons was described in Chapter 43.

CULTURAL EVOLUTION AMONG MACAQUES

The little Japanese island of Koshima, a high, wooded mountain surrounded by a beach, is inhabited by macaques. The monkeys lived and fed in the forest until a decade or so ago when a group of Japanese researchers began throwing sweet potatoes on the beach for them. The group quickly got used to venturing onto the beach, brushing sand off the potatoes, and eating them. One year after the feeding started, a two-year-old female the scientists had named Imo was observed carrying a sweet potato to the water, dipping it in with one hand, and brushing off the sand with the other. Soon, other macaques began to wash their potatoes too. Only macaques that had close associations with a potato washer took up the practice themselves. Thus it spread among close companions, siblings, and their mothers, but adult males, which were rarely part of these intimate groups, did not acquire the habit. However, when the young females that learned potato washing matured and had offspring of their own, all of them learned potato washing from their mothers. Today all of the macaques of Koshima dip their potatoes in the salt water to rinse them off, and many of them, having acquired a taste for salt, dip them between bites.

This was only the beginning. Later, the scientists began scattering wheat kernels on the beach. Like the others, Imo, now four years old, had been picking the grains one by one out of the sand. One day she began carrying handfuls of sand and wheat to the shore and throwing them into the water. The sand sank, the wheat kernels floated to the top, and Imo collected the wheat and ate it. The researchers were particularly intrigued by this new behavior since it involved throwing away food once collected, much less a part of the macaques' normal behavioral repertory than holding on to food and cleaning it off. Washing wheat spread through the group in much the same way that washing potatoes had. Now the macaques, which had never even been seen on the beaches before the feeding program began, have taken up swimming. The youngsters splash in the water on hot days. Some of them dive and bring up seaweed, and at least one has left Koshima and swum to a neighboring island, perhaps as a cultural missionary.

THE APES

The apes comprise four principal genera: *Hylobates* (gibbons), *Pongo* (orang-utans), *Pan* (chimpanzees), and *Gorilla* (gorillas). Apes are, with the exception of the gibbons, larger than monkeys. Their brain case is larger in proportion to their size. Like the Old World monkeys, they are catarrhines. They are all capable of brachiation—that is, swinging from one arm and then the other with their bodies upright. Although, among modern genera, only the gibbons move primarily in this way, brachiation may have played a role in the transition from the body structures associated with the crouching position characteristic of the lower primates to the body structure that makes possible the erect posture of man. Apes have relatively long arms and short legs. As a result, even when they are on all fours, their bodies are partially erect. Their feet are excellent grasping structures.

The apes and man make up the hominoids. The earliest hominoid fossils date from the Oligocene.

Relatively early in hominoid development, the hominoid line diverged; two distinct ancestral forms are recognized. One is represented by fossils of the genus *Pliopithecus*, of which dozens of specimens have been found in both Europe and Asia in formations of the Miocene and Pliocene epochs. The *Pliopithecus* line led to the modern gibbons.

Contemporary with *Pliopithecus* was an apparently large and widespread group of apes that have been given the generic name *Dryopithecus* ("tree-ape"). This group consists of several recognized species, some as large as modern gorillas. Based on an abundance of fossil fragments, mostly jaws and teeth, *Dryopithecus* is known to have ranged through most of Africa, Asia, and Europe. The best-known and most complete specimens of *Dryopithecus* have been collected by the British anthropologists Louis and Mary Leakey in an area near Lake Victoria in East Africa. One of the first to be described was given the name of Proconsul, as an affectionate tribute to a favorite performing chimpanzee named Consul. The modern descendants of *Dryopithecus* include the orangutan, the chimpanzee, and the gorilla.

48–13 *Gibbons, which are spectacular acrobats, move through the trees by brachiating— that is, by swinging hand over hand, with their bodies upright and suspended. Other apes are less frequent brachiators; in chimpanzees and gorillas, brachiation is only seen among the young at play. Nevertheless all apes have the shoulder girdle and other structural features associated with brachiation and are believed to have evolved from brachiators. It is a matter of debate among physical anthropologists whether man's upright posture and bipedal gait also had their origins in brachiation. (Ralph Morse, Time-Life Picture Agency.)*

A gibbon family. Gibbons and some human beings are the only monogamous anthropoids.

An orangutan. Among all the anthropoids, only the orangutan pursues a solitary existence. In older males, such as this one, the cheeks become enlarged and flattened.

Gibbons

Gibbons, limited today to Southeast Asia, are by far the smallest of the apes —averaging 3 feet tall—and the lightest in build, with slim bodies and long spidery arms. They can stand erect and walk, using their arms for balancing as though walking on a tightrope. Usually, however, they move through the trees, swinging arm over arm (brachiating) with great agility. Unlike other apes, which have sparse body hair, gibbons are completely covered by dense hair.

Gibbons form permanent pairs, each couple living by itself with its younger offspring. As the offspring grow older they leave the family group, or perhaps are driven from it, find a mate, and establish a separate family unit. They are territorial and generally gentle and unaggressive except when defending a territory. As they move through the trees they vocalize—"hooting in an enchanting and haunting manner," as reported by an observer. Hooting apparently serves as a spacing signal to other gibbon family units, serving the same function as bird song.

Orangutans

The orangutan, the redheaded, orange-bearded "man of the woods," as its name means in Malayan, is shy and solitary and, although large, it is strictly arboreal. Males, which are about twice as large as females, average in the neighborhood of 160 pounds. Orangutans are long-armed like gibbons. When they stand erect, their fingertips reach their ankles. Their feet are prehensile, as those of the other apes, and quite "handlike," with very small heels and widely divergent big toes, which are also greatly reduced in size. They move on the ground only rarely, and when they do, they walk on the curled edges of their feet, rather

A mother chimpanzee with her infant. Field studies suggest that bonds between mother and offspring and perhaps also among siblings last well into adulthood, perhaps for a lifetime.

than flat on the soles, and on their fists. The rarest and the least studied of the large primates, the modern orangutan is found only in tropical rain forests in Sumatra and Borneo. They are usually encountered alone or in groups of only two or three. The groups apparently consist of a mother and offspring; males are usually alone.

Chimpanzees

Chimpanzees, like gorillas, are found in tropical Africa. They are tree dwellers and brachiators, feeding and sleeping in the trees. However, they spend a large part of their time on the ground. They walk quadrupedally, on the flats of their feet and the knuckles of their hands. A full-grown chimp weighs about 100 pounds.

They generally move in groups, feeding on wild fruits, seeds, and pods of a large variety, and occasionally killing smaller animals for meat. Chimps have been observed to use simple tools, such as a grass stem for fishing out termites (which they eat) or a sponge made of leaves to take up water. The groups are noisy, shouting, crashing through bushes, beating on logs; the noise seems to be a way of keeping the group together. The chimps of the Budongo Forest of Uganda call and beat on tree trunks when they find a tree with abundant food, attracting other chimps from as far as 2 miles away.

Within a group there is a male dominance hierarchy; females are usually subordinate to adult males, but there is also a female dominance hierarchy. Temporary (two- or three-day) bonds may form between a male and a female in estrus, but otherwise there are no pair bonds, and receptive females typically mate with a series of males. However, there are often friendships among members of the group, and friends may spend as much as two hours a day in mutual grooming. Males cooperate in capturing prey, which consists most often of other primates, such as infant baboons and colobus monkeys. Jane van Lawick–Goodall observed a party of adult male chimpanzees hunt a red colobus monkey. A male chimp occupied the foot of every tree that was connected at its crown with the one in which the colobus sat, and another male climbed the monkey's tree to catch it. Prey may be shared among the group, even when the group is large and includes females, young, and males who were not part of the hunt, and the prey animal is small. The brain is evidently the preferred portion of the prey animals. Offspring maintain strong bonds with their siblings and their mothers, apparently throughout their lifetimes.

Chimps are gregarious, curious, boisterous, and extroverted. Although groups move through home ranges, there is not a strong sense of territoriality and the bands are not well knit and exclusive as they are in some other primate groups. The composition of the group changes as newcomers are taken in and other members wander off.

Gorillas

Gorillas are the largest of living primates and, according to recent field studies, among the shyest and the gentlest. Gorillas have almost completely abandoned trees, with the exception of some of the smaller animals that sleep in nests in the lower branches. The larger gorillas sleep on the ground, and all feed on ground plants.

-17 *An adult male mountain gorilla. Gorillas spend most of their time on the ground. Usually, they walk on all fours, carrying their weight on the second joints of their fingers and on the soles of their feet. When they walk upright, they step on the side of the foot.*

A male gorilla is about as tall as a tall man but weighs several times more (some 350 to 450 pounds). Its legs are much shorter in proportion to its body than are man's, while its arms are much longer and extremely powerful. The gorilla has an arm span of about 9 feet, much longer than a man, who usually has an arm span about equal to his height. The gorilla's hind feet are flat-footed on the ground, and it ordinarily moves by walking quadrupedally on the second joints of its fingers. It can stand erect, as it does when challenging an enemy, and it can also walk erect, although not nearly as well and as easily as it can on all fours. Gorillas eat mostly soft, leafy vegetables, in prodigious quantities. Both male chimpanzees and gorillas have powerful canines and strong incisors with which they rip the tough rinds of fruits and other vegetable products that are their staple diet. However, because these large canines are not present in the females, their primary function would appear to be that of defense.

The gorilla's foot lacks the arch and the long great toe which adapt our own feet so well for walking, but unlike the other apes, it has a pronounced heel. Gorillas have a beetling brow, a strong heavy jaw, and, on the top of the skull of adult males, a bony crest to which the strong jaw muscles are attached. These are homologous to the jaw muscles you can feel moving in your own temples when you chew or grind your teeth. Unlike chimpanzees, gorillas, as far as we know, do not eat meat or use tools.

Gorillas live in groups ranging from 8 to 24 individuals, with about twice as many females as males, and a number of juveniles and infants. Each gorilla troop has a large, mature (silver-backed) male as a leader.

When threatened, males will give fearsome exhibitions of power, thrashing with broken branches, standing erect, beating their chests, barking and roaring, even hitting themselves under the chin to make their teeth rattle. Given the massive size of an adult male, the display is impressive.

Chimpanzees and gorillas, despite differences in size, are closely related to one another. Most of their obvious structural differences, such as the heavy skull of the gorillas, are simply a necessary consequence of the differences in bulk between the two groups.

Karyotypes and Proteins

Cytological and molecular studies confirm that the chimpanzees and the gorillas in particular are closely related to one another and also to man. Chimpanzees and gorillas (and also orangutans) have 48 chromosomes, as compared to man's 46. (Gibbons have 44.) The karyotypes of gorilla, chimpanzee, and man are very similar in detail. The difference in number has been interpreted as the result of a fusion of two pairs of chromosomes in the human karyotype.

As we noted previously, men and chimpanzees have the same amino acid sequence in cytochrome *c* (see page 904). They also have identical hemoglobins; the gorilla differs by two substitutions. (The hemoglobin of the rhesus monkey differs from that of the human and chimpanzee by 15 amino acids.) Thus, according to the protein time clock, man and the apes diverged only comparatively recently.

Ramapithecus

Contemporary with dryopithecines, there lived a type of hominoid that was apparently more manlike than *Dryopithecus*. This fossil primate, which was given the name *Ramapithecus*, is believed by some to represent a fork in the evolutionary pathway. One road led to the chimpanzees and the gorillas, whose routes, according to this interpretation, did not diverge for about another 12 million years. The other led to the hominids, the group to which man belongs.

48–18 *A comparison of the upper jaws of man,* Ramapithecus, *and a chimpanzee. The jaw of* Ramapithecus *is more rounded and thus more manlike than that of the ape.*

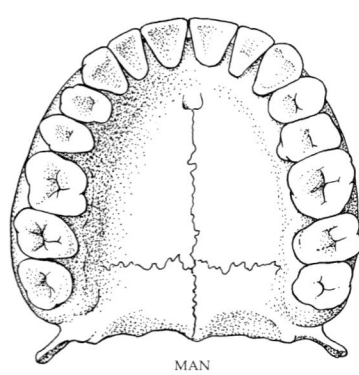

MAN

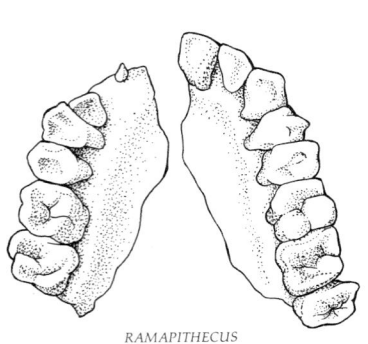

RAMAPITHECUS

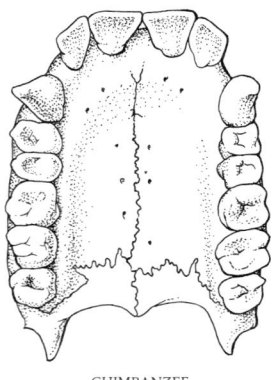

CHIMPANZEE

Ramapithecus is known only from fossil fragments of upper and lower jaws from India and Africa. The fossils are dated about 12 to 14 million years ago. Thus the evidence provided by *Ramapithecus* is at odds with that offered by the protein clocks, which suggest a much more recent evolutionary divergence between the groups. What do these jaw fragments indicate? First, when the bits of jawbone are pieced together, it is clear that *Ramapithecus* had a smaller and broader dental arch than other large contemporary primates or, indeed, than modern apes. The jaw fragments also indicate that *Ramapithecus* was comparatively small, about the size of a chimpanzee. Adjacent molars on the jaw fragments of *Ramapithecus* show marked differences in wear, in comparison to *Dryopithecus* specimens, in which almost no wear gradient is evident. This suggests, of course, that the molar eruption sequence of *Ramapithecus* was much slower than that of *Dryopithecus*, and, as a consequence, there is reason to believe that *Ramapithecus* matured much more slowly than its hominoid contemporaries—as does modern man.

Finally, the teeth and their condition indicate that tasks such as biting off and tearing up vegetation, for which apes use their front teeth, were not carried out to the same extent by the teeth of *Ramapithecus*. Physical anthropologists hypothesize that he (or it, as you prefer) used his forelimbs for these purposes, and on the basis of this hypothesis, they further hypothesize a trend toward bipedalism.

And then the curtain falls for another 11 to 12 million years, leaving many questions unanswered.

Australopithecines

The earliest hominoid universally accepted as a hominid is the form generally known as *Australopithecus* ("southern ape"). The first *Australopithecus* specimen was described in 1925 by anatomist Raymond Dart, who recognized it to be a new genus of hominid. The fossil, representing the skull of a young primate, had been recovered from a limestone fissure in South Africa. Evidence presented by Dart included the rounded appearance of the skull, the size of the fossilized brain within it, and the roundness of the jaw. Also, the point of attachment of the vertebral column to the skull (foramen magnum) indicated that the young animal was a biped. Subsequent fossil finds built up a picture

19 *The skull of a child, found in a limestone quarry in Taung, Africa, in 1924, was the first specimen discovered of the earliest known hominid,* Australopithecus africanus. *Shown here are the skull and a reconstruction drawing of the head of the child, which was about five or six years old when it died. An erect-walking creature with a brain larger than the gorilla's and with teeth that are similar to man's,* Australopithecus *lived on the ground, probably ate meat, and used tools of his own making.*

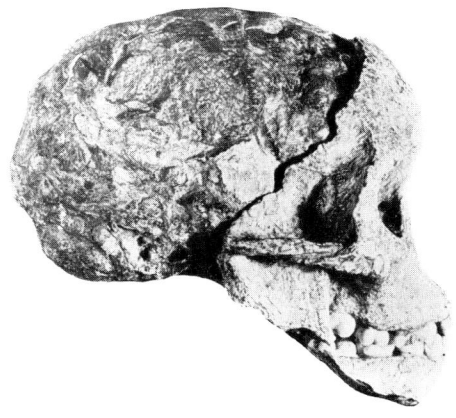 

A tentative phylogenetic tree of the higher (anthropoid) primates, based on fossil evidence.

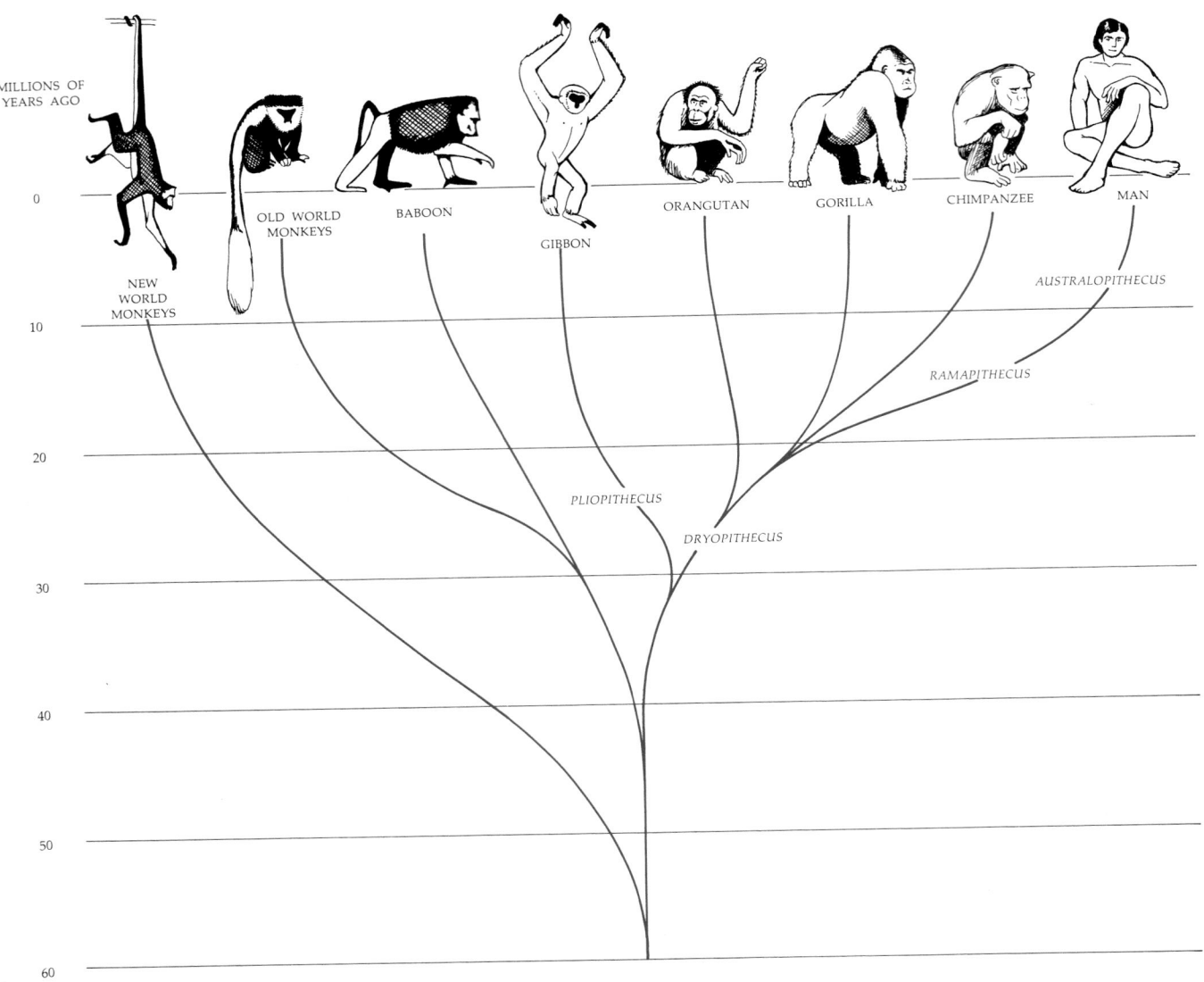

MILLIONS OF
YEARS AGO

0

10

20

30

40

50

60

NEW
WORLD
MONKEYS

OLD WORLD
MONKEYS

BABOON

GIBBON

ORANGUTAN

GORILLA

CHIMPANZEE

MAN

AUSTRALOPITHECUS

RAMAPITHECUS

PLIOPITHECUS

DRYOPITHECUS

of a small, lightly built hominid, not more than five feet tall, with cranial capacities ranging from 450 to 700 cubic centimeters, as compared to 340 to 750 for male gorillas (a much larger animal) and 1,000 to 2,000 for modern man. Their hands had a broad flat thumb and the beginnings of a human-style grip. The teeth were very much like our own, although the molars were larger. Australopithecines walked upright, probably covering the ground with quick, short steps in a sort of jog trot, rather than our far more efficient heel-and-toe striding movement. This hominid is known as *Australopithecus africanus.*

From here on, things get more complicated. Further discoveries, all from Africa, produced the remains of another form of australopithecine, much larger than the first, with much more massive molars and premolars and comparatively small incisors and canines. One of the best known of these large australopithecines is a specimen known as Zinjanthropus, the "nutcracker man," discovered by the Leakeys in 1959 in Olduvai Gorge in East Africa. The fossil, which consists only of a portion of the skull, is dated at 1.7 million years ago.

In the last decade, a number of australopithecine fragments have been found, particularly from the area east of Lake Rudolf in Kenya. Well over 300 specimens have now been uncovered. The new data indicate that these two quite different types of hominids lived together in this same area at this same time. Were they merely two quite different members of the same species? At first, some anthropologists suggested that the differences were simply sexual, male and female. Now it is generally believed that the two are different species, *A. africanus* and *A. robustus.* (Some authorities place them in two different genera: *Australopithecus* and *Paranthropus.*) There are suggestions that *A. africanus* was a maker and user of simple tools, and *A. robustus* (or *Paranthropus*) was not, and that *A. robustus,* as judged by his molars, ate a diet that was mostly tough vegetable matter, whereas *A. africanus* ate a more varied diet

Louis B. Leakey (right), the British anthropologist, with his wife and son at Olduvai Gorge in the Serengeti Plain of Tanzania. Olduvai, from which many important australopithecine fossils have been recovered, represents a geologic sequence that is almost continuous from the present to nearly 2 million years ago. Like our Grand Canyon, it was cut by river action. Because it was by the side of a lake, it was visited by many animals and was apparently a popular hominid camping ground.

that almost undoubtedly included some meat. The controversy became further complicated by the Leakeys' claim that another specimen, found at Olduvai, represented still another, more advanced species, *Homo habilis*, and that *Homo habilis* was the true antecedent of modern man. *Homo habilis* is now generally recognized, however, as being another form of *A. africanus*, probably the form transitional between *A. africanus* and modern hominids.

Thus, in short, there were at least three distinct groups: *A. robustus*, sometimes known as *Paranthropus*; *A. africanus*; and a form transitional between *A. africanus* and *Homo*, which is sometimes known as *Homo habilis*.

The oldest fossils, according to radioisotope dating methods, are the specimens found near Lake Omo in eastern Africa, representing remains of creatures that lived about four million years ago.

Part of the problem of classifying these hominids relates to the basic question of what constitutes a separate genus or species and to the even more complex question: What is man? Is he simply a bipedal primate? A toolmaker? Or does he have to be a certain size and have a certain brain capacity?

Another problem is that the fossil record of hominids is very scanty. At this stage of our knowledge, reconstructing human evolution is pretty much like working on a large and intricate jigsaw puzzle with most of the pieces missing. As a result of a recent surge of interest in this field of biology and numerous new digs and finds, especially in the area around the Olduvai Gorge in eastern Africa, much new data have come to light, but the interpretation of these data is not yet generally agreed upon. In other words, not only are a lot of pieces missing, but those we do have may be put together incorrectly.

MAN

The fifth major group of hominoids, the only modern hominids, is the genus *Homo*, of which there is only one modern species, *Homo sapiens*. Man is distinguished from the apes by his upright posture. (Bipedalism was gained at some cost. Members of this species suffer from backaches, rupture easily, and have far more trouble bearing young than any other animal.) Man is able to walk continuously and for long distances on two legs. This two-leggedness, or bipedalism, is probably the most important categorical difference between man and apes. Concomitantly, the pelvic bone is the most divergent skeletal structure between the two groups. The long, relatively narrow pelvis of the ape throws his spine forward, whereas man's hipbones are bent backward and downward, carrying the sacrum and spine into an erect position over his columnar legs. His lateral toes are short, and his big toe is in line with the others. His vertebral column makes an S curve balanced over his broad pelvis and his two fully extended, relatively straight legs.

There are also important differences in the skull: man's jaws are short, with a rounded dental arch, and his canines are usually no larger than his premolars. His brain is proportionately larger than that of other primates. Also man has a much larger and more convoluted cerebral cortex than the great apes. As a corollary, his skull is both larger and rounder. The normal range of cranial capacity among modern men is 1,000 to 2,000 cubic centimeters, with a mean of 1,300.

Like the chimpanzee, man is noisy, gregarious, and social. Unlike most other anthropoids (with the exception of the gibbon), he forms permanent or semi-

Adaptations to bipedalism. In man, as compared to the gorilla, the pelvis is shorter and more bowl-like. As a consequence, the legs move more efficiently, the pelvis is stabilized on the legs, and the knees and feet are close together.

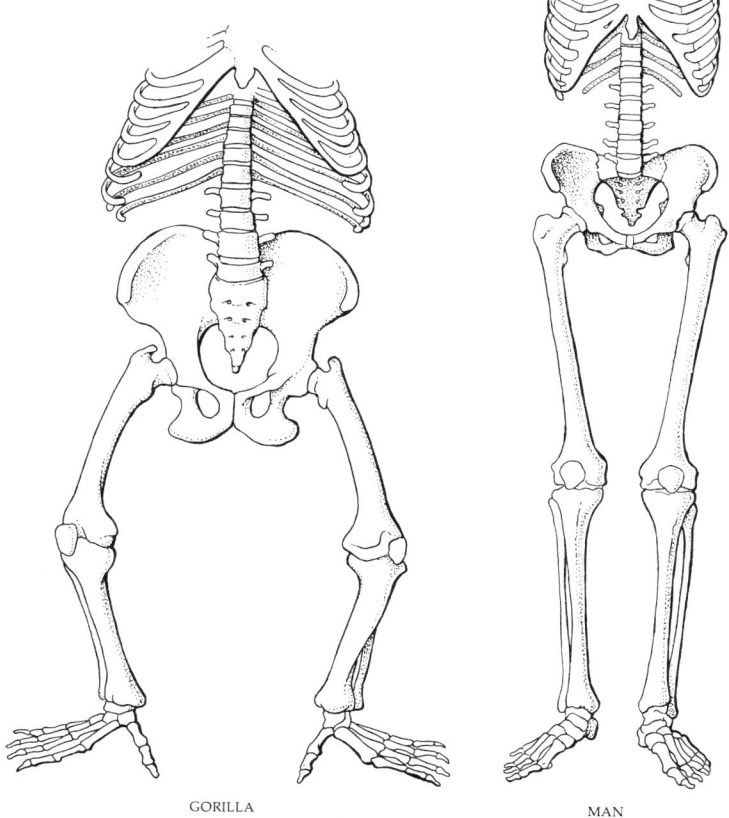

GORILLA   MAN

permanent pair bonds, and unlike other anthropoids, he regularly kills members of his own species. Unlike all other animals, with the possible but doubtful exception of certain cetaceans (whales and porpoises), he has a language and is capable of communicating abstract ideas. He both uses and makes tools.

How Did It Happen?

The faint traces left in the fossil rocks have enabled us to follow hominid footprints, though often very tentatively, over the past 14 million years. As is often the case, the unanswered questions are more interesting than those we have answered. For instance, what prompted one group of primitive apes to make the first fateful step to bipedalism? We know from pollen specimens that grasslands were spreading and forests contracting during the Miocene, the period during which *Ramapithecus* made its first appearance. The grasslands may have represented a new adaptive zone. The horse and other grazing animals evolved rapidly during this time, filling the new niches available to herbivores. Baboons also seem to have become terrestrial in this period. We do not know why one group of apes and not another moved into the grasslands. One possible explanation is that this group was driven out of the woodland by competition with other primates. The fact that *Ramapithecus* was smaller than

other contemporary apes is in keeping with this speculation. Perhaps they possessed some important preadaptation that we do not know about.

Standing upright on the ground would have greatly increased the range of vision of a ground-dwelling species with highly developed eyesight; other mammals, including bears, squirrels, and prairie dogs, often raise themselves on their two hindlegs to look around. Standing upright also, of course, frees the hands, both for toolmaking and, perhaps even more importantly at first, for carrying. One can make and use simple tools sitting as well as standing, and all the higher primates are very good sitters, but they cannot carry anything for long distances when traveling over the ground. Hence, they cannot go far from food or water. Ramapithecines, by freeing their hands for carrying, would have been able to extend their range of food searching. They could have also carried large sticks and other simple weapons to brandish at their enemies, as modern chimpanzees do; the possession of some kind of weapon would seem to have been almost essential for such a small and comparatively defenseless animal in open country. Use of weapons for defense correlates well with the reduction in the size of the canines among manlike fossil primates, as compared with baboons, for example, or gorillas, among which the males have large canines, which they display at their enemies. Early australopithecines apparently carried stones considerable distances; at one excavation site (Sterkfontein), among a number of animal bones, there are many rocks of a type that

48–23 *Bipedalism frees the upper extremities for a variety of activities. An adult female chimpanzee "fishes" for termites with a twig, while her infant sucks its opposable thumb.*

-24 *Although man's hands follow the basic primate pattern, his feet are very different from those of other primates. All primates except man have handlike feet with opposable first toes—in fact, the feet of many primates (for example, the tarsier) are better graspers than the hands. In man, as you can see, the first toe has lost its opposability and the other toes are very short, bringing all the joints at the base of the toes in line and making a ball to the foot. It is this ball that is the secret of man's ability to walk erect; it allows him to shift his weight from heel to ball and so stride forward. No other primate can do this, not even the gorilla, which has a larger heel and shorter side toes than those of the other apes.*

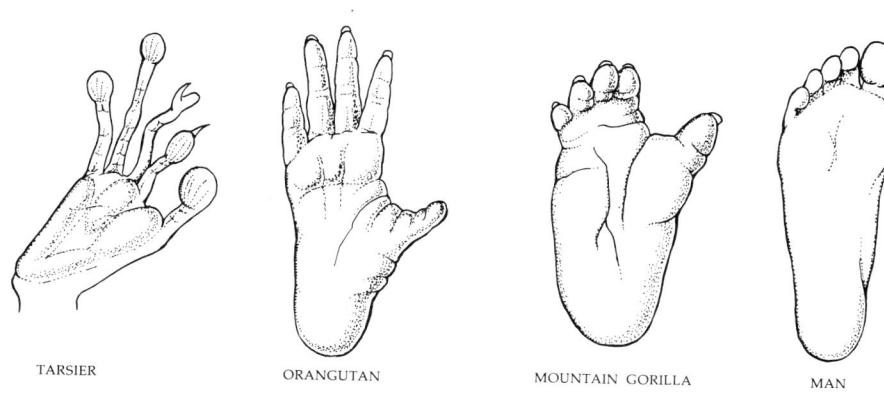

TARSIER   ORANGUTAN   MOUNTAIN GORILLA   MAN

must have come from a streamed a number of miles away. If these men-apes carried stones, it seems logical to believe they were in the habit of carrying other things as well. In any case, the advantages of bipedalism must have been very great for natural selection to have operated in its favor, because it is clear that the first bipedal primates must have moved slowly and inefficiently compared with the swift movement of other large animals either on the ground or in the trees.

Evolution of the Brain

As we noted previously, the earliest use of what archaeologists recognize as tools occurred at least more than two million years ago, and perhaps as long as three or four million years ago, when *Australopithecus* still had a brain no larger than that of an ape. Increasing dependence on simple tools, clearly, was one major cause of selection pressures on the evolution of the human hand, which differs from other higher primate hands, not so much in its capacity for fine manipulations, at which many other anthropoids are just as good, but in its capacity for a strong, precise grip. This grip is largely a consequence of man's large, more muscular, and completely opposable thumb.

By following this line of reasoning, we can explain at least hypothetically the major structural changes in man's limbs, pelvis, and spinal column in relation to the evolutionary pressures which brought them about. But how do we explain the most manlike change of all—the enormous increase in the size of that most mysterious organ, the brain?

In the course of hominid evolution, brain capacity increased greatly. It has doubled since the australopithecines, in the course of a mere 2 million years. The fossil record has no evidence of another evolutionary change as rapid as this. The horse, for example, was evolving extremely rapidly during this period, and insofar as it is possible to compare the forelimb of the horse and the brain of man, man's evolution was a hundred times more rapid.

Before we pursue this question any further, we should add that we do not really know the precise correlation between increasing brain size and increasing intelligence. Larger animals have large brains. As we saw in Chapter 39, in addition to the collection of processes we call thinking, the brain also controls all the musculature of the body and is a receiving center for the information

931   *The Primates*

from eyes, nose, and other sensory organs. As also noted, the brain of modern man varies greatly in capacity, from 1,000 to 2,000 cubic centimeters, and there is no clear correlation between its size and the intelligence of its possessor. Moreover, organization of the brain clearly plays an important part in its functional capacity, and we do not know anything about the organization of the brain of fossil men, and indeed not a great deal about the organization of our own brain. However, when we compare the intellectual capacities of animals of different species that have brains of different sizes in comparison to their weights, those with relatively larger brains are clearly the more intelligent. Therefore it is probably correct to assume that this tremendous change in brain size reflects a change in intelligence and that, during a relatively brief period in man's history, powerful evolutionary pressures were at work selecting individuals of slightly greater mental capacity over those of slightly less.

Before so much fossil evidence was available, it was popular to suppose that tool use progressed concurrently with the increase in brain size and each stimulated the other. This may be true, but, as we shall see in the following chapter, during the period of the most rapid change in brain size, the tools in use remained very simple. No really elaborate tools came into existence, according to the archeological evidence, until the appearance of Cro-Magnon man, sometime well after brain size had reached its peak. Some suggest that brain size increased rapidly at the time man was beginning to speak. Another suggestion takes note of the fact that hunting of large game began to occur during this period of brain increase. Such hunting activities would have required new skills and also, in particular, greatly increased social cooperation among the hunters, and this may have been the selection force for intelligence. In any case, modern man as he evolved is neither strong nor swift nor highly specialized for any ecological niche. He differs from all other animals chiefly by reason of his mind, the contents of his skull. And it is this difference that is responsible for most of the events that have taken place on the face of the planet in the brief period that he has inhabited it.

SUMMARY

The first mammals arose from primitive reptilian stock about 200 million years ago. One of the most crucial differences between reptiles and these early mammals is that the latter were homeothermic (warm-blooded) and so were more alert and active and, probably, nocturnal.

The extinction of the dinosaurs was followed by a rapid adaptive radiation of the mammals. The primates are the order of mammals that became adapted to arboreal life. Members of the primate order are generally less specialized anatomically than other mammals. Those adaptations they do have are related principally to arboreal life. Primates are characterized by five-digit extremities adapted for grasping with nails rather than claws, by freely movable limbs, and by relatively unspecialized dentition. They are dependent more upon vision than upon smell, and the higher primates all have stereoscopic vision with foveas for fine focus and cones for color vision. Another trend in primate evolution is pronounced reduction of the muzzle. Primates tend to sit and stand upright. Among the great apes, the forelimbs are longer proportionately than the hindlimbs so that the animals' bodies are semierect even when they

are on all fours. Because of this semierect posture, their heads are oriented forward. Arboreal living also increased selection pressures for increased maternal care of offspring and for intelligence.

The two principal groups of living primates are the prosimians and the anthropoids. Fossil prosimians were widespread and abundant during the beginnings of the Cenozoic era, some 65 to 40 million years ago. Modern prosimians include tree shrews (sometimes categorized as insectivores), lemurs, and tarsiers.

The anthropoids include the New World monkeys (the only primates other than man that inhabit the New World), the Old World monkeys, and the hominoids (apes and man). All of the New World monkeys are arboreal. Among the Old World monkeys, one group, which includes the macaques and baboons, is secondarily terrestrial.

The hominoids are larger than the monkeys and have proportionately larger brain cases. They are tailless. They are adapted for brachiation—moving through the trees arm over arm instead of on all fours (or fives) like the monkeys. Among modern forms, however, only the gibbons move primarily by brachiation. Early in hominoid evolution, in the Miocene epoch, the hominoids diverged into two lines, represented by *Pliopithecus* and *Dryopithecus*. *Pliopithecus* was ancestral to modern gibbons; *Dryopithecus* to orangutans, chimpanzees, and gorillas. *Ramapithecus*, believed to represent a fork in the evolutionary pathway from *Dryopithecus* to man, is known only from some jaw fragments dated 12 to 14 million years ago. A contemporary of *Dryopithecus*, *Ramapithecus* was smaller and more manlike in its dentition.

Gibbons are small (about three feet tall) and light in build with long arms. They apparently are the least intelligent of the great apes and are also the hairiest and the most agile brachiators. They move in small family groups and are territorial. Orangutans are larger than gibbons, more solitary, and strictly arboreal.

Chimpanzees and gorillas, which form closely related genera, are more adapted to terrestrial existence. Chimps spend much of their time on the ground, although they sleep in the trees and move through them easily by brachiation. Gorillas are almost entirely terrestrial, sleeping and feeding on the ground. Both chimps and gorillas can stand erect on two legs, but on the ground they usually move quadrupedally, on the soles of their feet and the knuckles of their hands. Both are mainly vegetarians, but the chimpanzees supplement their diets with meat. Chimpanzees also use simple tools. Members of both genera are social, living in groups that move through territories generally separate from the feeding ranges of other groups of the same species. The groups of chimpanzees are less well knit, however, and territories are not aggressively defended. Groups are socially structured, with dominance hierarchies. Gorilla groups are always led by an older male. Although basically similar in many characteristics, gorillas and chimpanzees differ in size, with chimpanzees averaging about three feet and 100 pounds compared to the six foot height and 350 to 450 pound weight of the adult male gorilla. They also differ in personality; chimpanzees are noisy, rowdy, gregarious, and curious, whereas gorillas are shy, apparently gentle (despite occasional fearsome threat displays), quieter, less curious, and more retiring.

The hominids include man (*Homo*) and the genus related to him, *Australo-pithecus*. *Australopithecus* is known from several hundred specimens, ranging in age from perhaps as long as four million years ago to as recently as a million.

Man is the only modern species of the genus *Homo*. He differs from the other hominoids by being bipedal. His hand, which has a larger thumb than that of any other primate, is better adapted for manipulation, and his feet, unlike those of other hominoids, are completely adapted to heel-to-toe walking. His jaws are short, his dental arch more rounded, and his canines smaller. His brain is larger and more rounded and so is his skull. His cranial capacity ranges from 1,000 to 2,000 cubic centimeters, with a mean of 1,300. He is social, like most other higher primates, and, unlike most other higher primates, often forms permanent or semipermanent pair bonds. He also makes and uses tools, eats meat, and has a language.

QUESTIONS

1. Define the following terms: primate, prosimian, hominid, hominoid, anthropoid.
2. Give some distinguishing features of each of these genera: *Dryopithecus, Tarsius, Australopithecus, Homo, Pan*.
3. Name five evolutionary trends among primates and name the probable selective value for each.

Chapter 49

The Emergence of Man

Modern man evolved in the Pleistocene (most recent—from *pleistos*, meaning "most"), which began some 2½ to 3 million years ago and ended 10,000 years ago. The Pleistocene was a period of extreme climatic changes that must have played a crucial role in the rapid evolution of the genus *Homo*.

In terms of the long history of the Earth, the drastic changes of the Pleistocene were unusual but not unique. During most of our planet's history, its climate has been warmer than it is at the present time. However, these long periods of milder temperatures have been interrupted periodically by Ice Ages, so called because they are characterized by glaciations, persistent accumulations of ice and snow. Such glaciations occur whenever the summers are not hot enough and long enough to melt ice that has accumulated during the winter. In many parts of the world, an alteration of only a few degrees in temperatures is enough to begin or end a glaciation.

49–1 *Glaciers are accumulations of snow and ice that flow across a land surface as a result of their own weight. This glacier is flowing into the sea. As the glacier moves forward, its edges melt; the rocks carried within it have become concentrated in the dark band at the right. This is part of the Greenland ice sheet, which has an area of about 1,726,000 square kilometers and at some places is more than 3 kilometers thick. Glaciers covered much of North America and Europe during the period of the evolution of the genus* Homo.

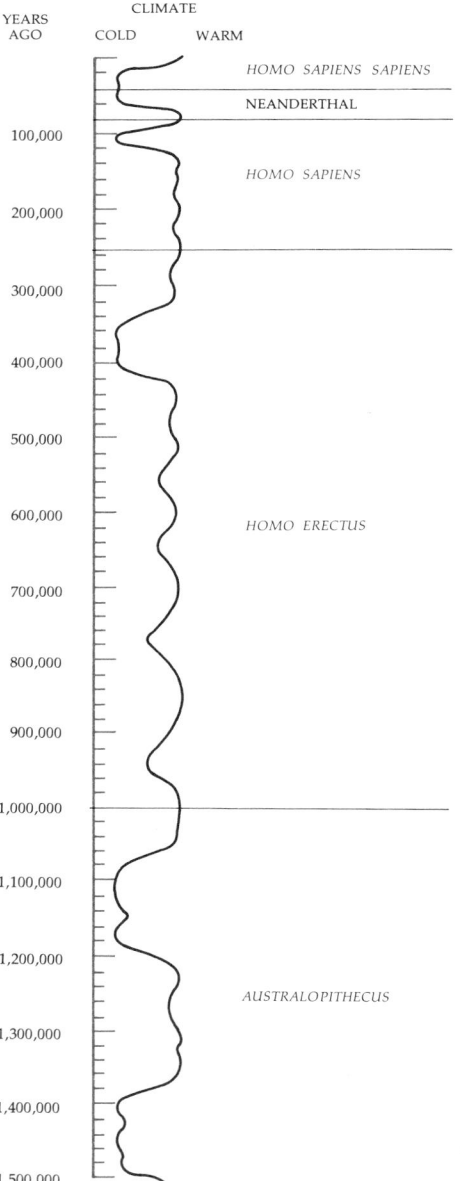

49–2 *A timetable of human evolution. The boundary lines are approximate and there was undoubtedly overlap among the various groups. Note the drastic climatic changes that accompanied the emergence of man.*

One Ice Age appears to have occurred at the beginning of the Paleozoic era some 600 million years ago. Another, marked by extensive glaciations in the Southern Hemisphere, closed the Paleozoic, some 250 million years ago. The reptiles evolved during this period, as did the conifers and possibly the angiosperms, as older forest types disappeared. A more recent, less severe cold, dry period occurred at the end of the Mesozoic, about 65 million years ago, and was perhaps a principal cause of the extinction of the dinosaurs and of the subsequent rise of the mammals. The most recent Ice Age has been marked by four extensive glaciations, which have covered large parts of North America, England, and northern Europe. Between the glaciations, there have been intervals—interglacials, as they are called—during which the climate has become warmer, although not, on the average, as warm as during the preceding period of the Earth's history. In each of these four glaciations, sheets of ice, as much as a mile thick in some regions, spread out locally from the poles, scraped their way over much of the continents—reaching as far south as the Ohio and Missouri rivers in North America and covering Scandinavia, most of Great Britain, northern Germany, and northern Russia—and then receded again. We are living at the end of the fourth glaciation, which began its retreat only some 10,000 years ago. During these periods of violent climatic changes, the fossil record shows that the populations of these regions were under extraordinary evolutionary pressures. Plant and animal populations had to move, to change, or to become extinct. In the interglacial periods, during which the average temperatures were sometimes warmer than those of today, the tropical forests and their inhabitants spread up through today's temperate zones. During the periods of glaciations, only animals of the northern tundra could survive in these same locations. Rhinoceroses, great herds of horses, large bears, and lions roamed Europe in the interglacial periods, and in North America, as the fossil record shows, there were camels and horses, saber-toothed cats, and great ground sloths, one species as large as an elephant. In the colder periods, reindeer ranged as far south as southern France, while during the warmer periods, the hippopotamus reached England. It was during this most recent period of glaciations and interglacials—the Pleistocene epoch—that hominids of the genus *Homo* evolved.

The reason for these large changes in temperature is one of the most controversial issues in modern science. They have been variously ascribed to changes in the Earth's orbit, variations in the Earth's angle of inclination toward the sun, migration of the magnetic poles, fluctuations in the solar energy, migration of the continents, higher elevations of the continental masses, and combinations of these and other causes. Also hotly (or coldly) debated is the question of whether this period of glaciations is over—perhaps for another 200 million years—or whether we are merely enjoying a brief interglacial before the ice sheets begin to creep toward the equator once more.

THE GENUS HOMO

At present, physical anthropologists recognize only two species of the genus *Homo, Homo sapiens,* the group to which we belong, and *Homo erectus,* which preceded *Homo sapiens* in time and is presumably his ancestor.

Homo erectus

Homo erectus is distinctly different from even the most advanced australopithecine. Members of this species had body skeletons much like our own and were about the same size as modern man. The bones of their legs indicate that they had a stride similar to our own. The chief differences between *Homo erectus* and modern man is in the skull. Skull specimens of *Homo erectus* are thick and massive, with a low forehead. Their jaws are large and chinless, with large teeth. Brain capacity ranges from 700 to 1,100 cubic centimeters, overlapping the capacity of modern man. There appears to be some correlation between the size of the brain and the age of the fossil, with the larger brains appearing in the more recent specimens.

Also, unlike the australopithecines, which have only been found in eastern and southern Africa, members of *Homo erectus* seem to have occupied a much larger range. In fact, the first fossil specimen, discovered in 1891, was found on the bank of the Solo River in central Java, and so became known as Java man. Fossils similar to Java man have been found not only along the Solo, which yielded a number of other fragments, but in China (Peking man), where more than three dozen partial specimens have been found, in eastern Africa (at Olduvai), in northern Africa (Algeria and Morocco), and in South Africa. There is also a jaw from Heidelberg and some skull fragments from Vértesszöllös, Hungary, that are believed to represent this species.

Along with these fossil bones, clear evidence has been found that members of this species made tools, hunted large animals, lived in caves, and had fire. We shall be discussing this evidence in subsequent pages.

Fossils of *Homo erectus* cover a period of about 700,000 years, from about 1 million to 300,000 years ago. For example, the group found in the Choukoutien caves in China—known as Peking man and dating from 300,000 to 400,000 years ago—lived there during a period of extreme cold (the latitude of Peking is about 40°, approximately the same as that of Philadelphia or Denver).

–3 *A reconstruction of the skull of* Homo erectus *from the Solo River, Java. The skull walls are thick and heavy, and the brow ridges are prominent. The protruding occipital crest was the point of attachment of strong, heavy neck muscles. The jaw was prognathous (protruding) and chinless.*

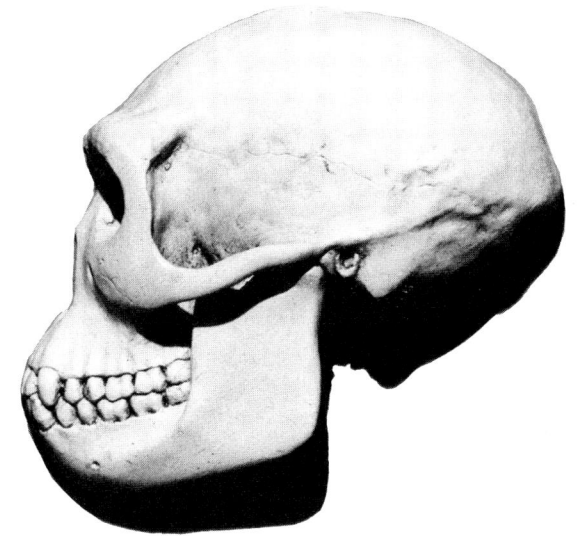

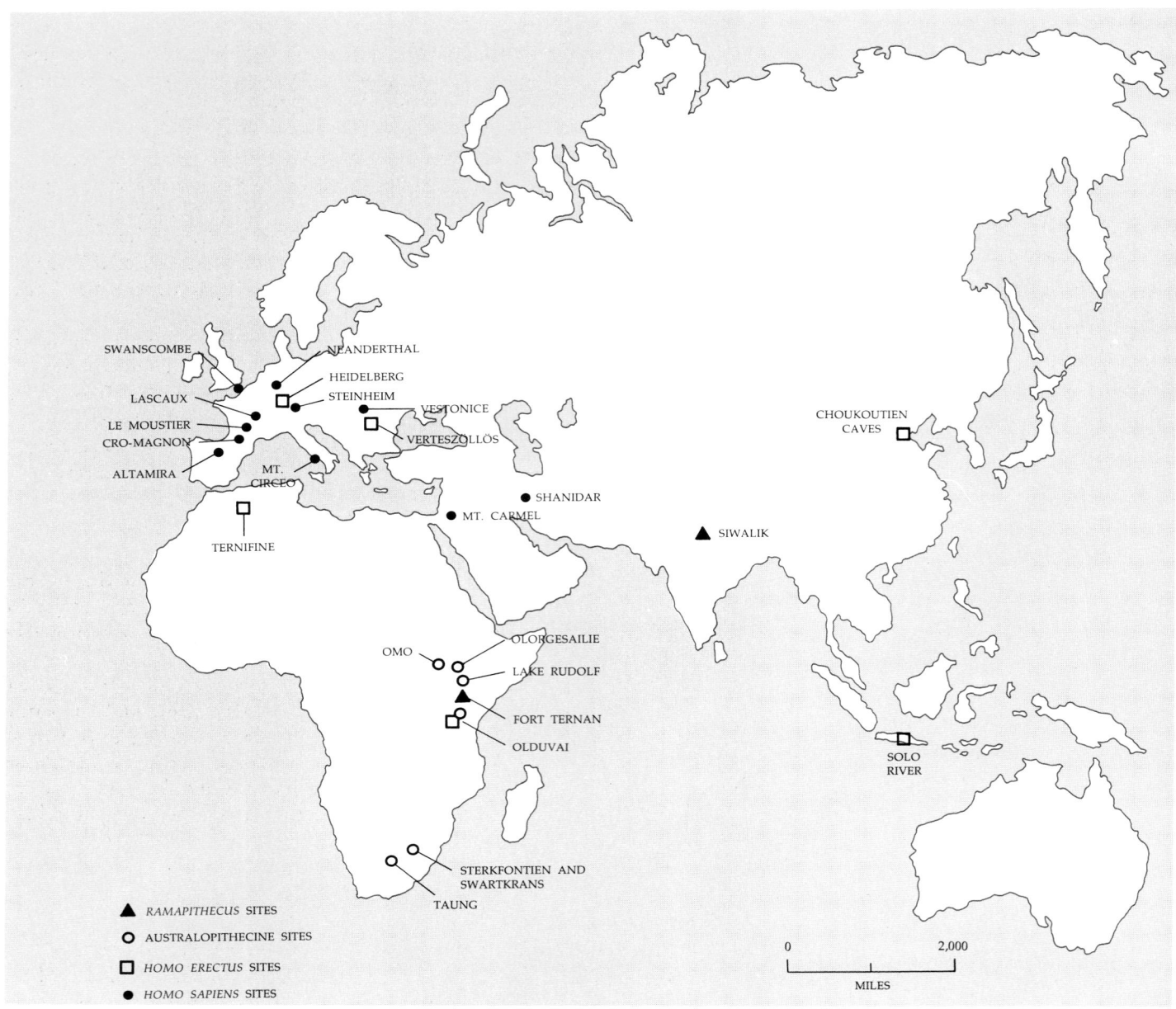

SWANSCOMBE

NEANDERTHAL

HEIDELBERG

LASCAUX

STEINHEIM

LE MOUSTIER

VESTONICE

CRO-MAGNON

VERTESZÖLLÖS

ALTAMIRA

MT.
CIRCEO

CHOUKOUTIEN
CAVES

SHANIDAR

MT. CARMEL

SIWALIK

TERNIFINE

OLORGESAILIE

OMO

LAKE RUDOLF

FORT TERNAN

SOLO
RIVER

OLDUVAI

STERKFONTIEN AND
SWARTKRANS

TAUNG

▲ *RAMAPITHECUS* SITES

○ AUSTRALOPITHECINE SITES

□ *HOMO ERECTUS* SITES

● *HOMO SAPIENS* SITES

0   2,000

MILES

49–4 *Map showing some sites of early man.*

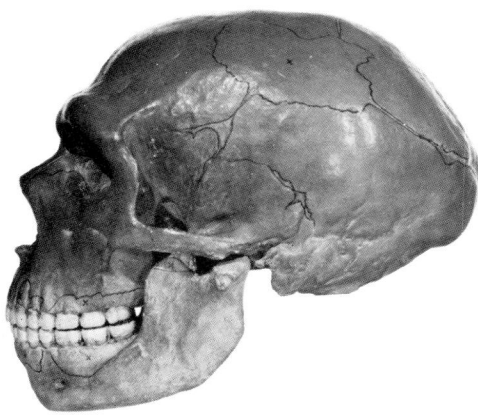

49–5 *The Neanderthal skull from La Chapelle-aux-Saints. It has a large cranial capacity and large face.*

Homo sapiens

The next group of fossil hominids are assigned to the species *Homo sapiens* ("knowing man"). These include the varieties known commonly as Neanderthal man and Cro-Magnon man, sometimes distinguished as *Homo sapiens neanderthales* and *Homo sapiens sapiens*. Cro-Magnon man, also often referred to as Upper Pleistocene man, is physically indistinguishable from modern man.

The earliest fossils that are classified as *Homo sapiens* come from Swanscombe in England and Steinheim in Germany. They consist only of some skull fragments, but these demonstrate clearly that the brains of these men were larger than those of *Homo erectus* and their skulls were less massive. They are dated at about 250,000 to 150,000 years ago, an interglacial period when Europe was warm, probably warmer than today.

Then we have another of those large and frustrating gaps in the fossil record. The next specimens that are known to us date from around 80,000 years ago to about 40,000 years ago, the period of the last glaciation. This period abounds with specimens of what we have come to call Neanderthal man.

Then, about 40,000 to 30,000 years ago, Neanderthal man became extinct, and modern man—sometimes known as *Homo sapiens sapiens*—appeared.

The Origins of Modern Man

Where did modern man come from? He clearly did not arise from the extreme Neanderthal variety of *Homo sapiens* previously present in Europe. He had a light skull, a high forehead, and a well-defined chin; suitably attired, he would be indistinguishable in a modern crowd.

One hypothesis is that Steinheim and Swanscombe man, then Neanderthal man, then *Homo sapiens sapiens* all came into Europe from the south or southeast (perhaps from around the Mediterranean where some suggestive fossil evidence has been found) and that all had ancestors in common. The classic Neanderthalers, perhaps because they were isolated by the return of the glaciers, represent an extreme variation. Modern men are the more generalized form.

What Became of the Neanderthalers?

What caused the sudden disappearance of the Neanderthalers? Perhaps they were exterminated in warfare—although there is no evidence of this. Perhaps they were simply unable to compete for food and living space with the better equipped Cro-Magnon type. Perhaps they intermarried, though there are no clear traces of intermediate forms. Some have suggested that modern men brought with them some disease to which they themselves were resistant and the Neanderthalers were not. In the course of later explorations, Europeans brought about the deaths of large numbers of previously isolated peoples by the accidental introduction of such diseases as measles and chicken pox, usually relatively harmless but often fatal in a previously unexposed population. In any case, soon after the appearance of *Homo sapiens sapiens*, there were no other hominids in Europe, and within the course of 10 to 20 thousand years, this new variety of primate was spread over the face of the Earth.

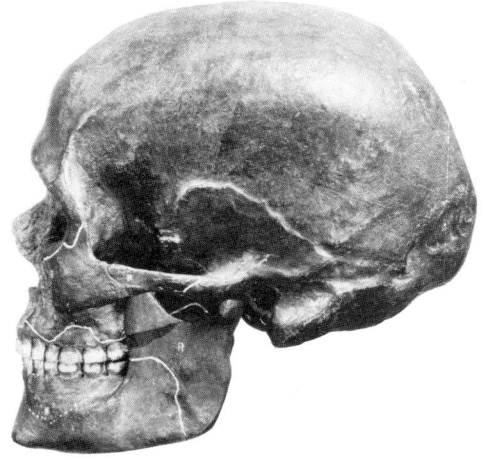

49–6 *The skull of Cro-Magnon man from the Dordogne region of southern France. This new subspecies (Homo sapiens sapiens) first appeared about 35,000 years ago. We are members of this subspecies.*

The Social Life of Early Man

What we know about how early man lived is largely pieced together from three sources of information, each with its own limitations. One of these sources is the other modern primates, which are now being studied intensively, almost for the first time, by scientists living near the animals in their own natural environments. (Some current observers also try to "explain" man on the basis of the behavior of such diverse animals as birds or fish, but, as we hope we have made clear, the behavior of even closely related animal groups is so disparate that even inferences drawn from other primates must be very cautious.)

The second source comprises the few groups of present-day people—such as the Bushmen and the pygmies of Africa, the Eskimos, the Australian aborigines, and some South American Indians—who are still, or were until recently, hunter-gatherers, using primitive tools.

The third source, and the only direct one, consists of the traces that early man has left behind—tools, food debris, hearths, homesites, works of art, and his own bones. In addition, pollen grains and fruit seeds preserved in the soil and at campsites provide evidence as to the countryside in which he lived and its climate.

Ecology of Early Man

As we saw in Section 7, the way an organism obtains its food supplies and the type of food it eats are primary factors in determining its ecological niche,

9–7 *Chimpanzee preparing to eat the fore-quarters of a young baboon.*

and, as a consequence, its identity as a species. Most large primates eat a varied diet of fruits, buds, shoots, roots, insects, eggs, and sometimes birds and small mammals. These are eaten as they are found, with nothing saved until later or shared in any organized way. There is no division of labor between the sexes. Infants are nursed for several years by species such as the chimpanzee, but they forage for themselves as soon as they are old enough. We have no reason to believe that the primates that were the early ancestors of man did not follow this general primate pattern.

Australopithecines, at least members of the species *Australopithecus africanus*, probably also foraged for their food in much the same way, but their diet also included more meat than the diet of modern nonhuman primates and, almost certainly, than their anthropoid contemporaries. At various australopithecine sites, fossil remains of these early man-apes have been found along with broken bones of small animals—frogs, lizards, rats, and mice—and of the young of larger animals, such as antelopes.

It used to be thought that one of the big differences between man and his fellow primates was his desire for meat. Recently, however, as we mentioned previously, field studies have shown that anthropoids (both chimpanzees and baboons) eat meat whenever they can get it and that it is clearly a highly prized dietary item, although not a staple food. So the transition to the meat-eating, omnivorous way of life characteristic of modern man was apparently not the result of some abrupt change in preman's nature. The reason that australopithecines ate more meat than other primates was probably simply that they were able to catch it more efficiently, perhaps because they used simple weapons to do so. Also they may have been more effective in forming hunting bands, which would indicate a more complex social organization.

The transition to a diet higher in proteins and richer in calories was important in human evolution because it reduced the amount of time spent in eating and so made other activities possible. Large herbivores—cows, horses, and gorillas, for instance—spend much of their waking hours feeding.

Members of the species we call *Homo erectus* became hunters on a different scale. In the Miocene, during which the hominids were evolving, not only were all the modern orders of mammals present but there were species that today are extinct. Many of these species were larger than the modern forms. For example, the biggest of all was a kind of rhinoceros 18 feet high at the shoulder and 25 feet long, the largest land mammal that has ever existed.

During various stages of the Pleistocene era, the epoch marked by the glaciations, there seem to have been far more grasslands than now exist. Some areas that are now deserts were tropical grasslands or savannas, and to the north, much land that is now forest was prairie or steppe, which was abundantly able to support large numbers of grazing and browsing animals. Man was a rare species then, greatly outnumbered by these large herbivores, and so would have had little difficulty finding game. And indeed, from sites as far apart as eastern Africa (Olduvai and Olorgesailie) and China (the Choukoutien caves, near Peking) and dating back from a million to half a million years ago, the bones of very large animals—elephants, rhinoceroses, antelopes, bears, hippopotamuses, and giant baboons—are found in the debris of human campsites. Here is clear evidence of hunting on a very large and well-organized scale.

49–8 *Pebble tools such as these have been found at Olduvai and other sites in eastern Africa in fossil strata of 2 million or more years ago. Australopithecines (presumably) took pebbles of lava and quartz and either struck off flakes in two directions at one end, making a somewhat pointed implement, or in a row on one side, making a chopper. They measure up to 4 inches in length and were probably used to prepare plant food and to butcher game. The flakes were also used apparently as scrapers.*

49–9 *Early hand axes, such as this, are stones trimmed on both sides of each edge into beak-shaped or pear-shaped implements. They came into use about 1 million years ago and are associated with early* Homo erectus. *They are characteristically flaked over most or all of their edges and usually measure from 4 to 6 inches long and about 2 inches thick. This one was made 400,000 years ago.*

We do not know how these early hunters killed animals so much larger and more ferocious than themselves. Stone implements found at the sites were, apparently, made to be held in the hand and used for chopping, cutting, or pounding—in other words, for butchering or preparing food rather than for killing game. There is some evidence that early man killed animals by stampeding them into marshes or over cliffs, perhaps with the help of fire to frighten them. American Indians slaughtered game in this way until very recent times. Certainly hundreds of thousands of years passed before men developed weapons as sophisticated as spears with stone or metal tips or bows and arrows. Anthropologists who have studied primitive tribes have been impressed by the fact that their weapons are not efficient and that hunters take little interest in perfecting them. These modern hunters emphasize patience, endurance, skill in stalking, and highly specialized knowledge of animals and terrain, and perhaps prehistoric man did also.

In any case, the hunting-foraging economy of early man was a determining factor in all other aspects of his physical and cultural evolution.

TOOLS

The making and using of tools is another highly distinctive feature of man, and one of the ways in which the genus *Homo* is usually defined. The observation by Goodall and others that chimpanzees make simple tools (such as leaf sponges) and keep them around for future use makes the distinction between man and his fellow primates a little less clear cut. However, it is still very likely that the use of tools by the australopithecines (and perhaps even by the ramapithecines) was a primary evolutionary force in achieving complete bipedalism (which freed the hands) and probably in the development of the brain.

Pebble Tools

The earliest tools of which we have any record are found in association with the remains of *Australopithecus africanus* at sites in eastern Africa. These stone tools are of the sort known as pebble tools, which is somewhat of a misnomer because they are actually, as you would imagine, not pebbles but rocks of a size that fit comfortably in a man's hand. These "pebbles" typically have flakes split off at one end to provide a chopping or scraping edge. Both the pebbles and the flakes were used as tools. These simple stone chopper tools and flakes were also used by early *Homo erectus* at sites in China and Southeast Asia occupied by these early men.

In Africa (but not in the Far East), presumably in the hands of *Homo erectus*, the pebble tool became a new and highly distinctive implement, the so-called hand ax. The hand ax is a stone that has been worked on all its surfaces to provide what appears to be a gripping surface and various combinations of cutting edges, sometimes with a more or less sharp point. Making such an implement requires both skill and time, as reported by anthropologists who have tried. Hand axes, therefore, were not always manufactured on the spot at the time they were needed but were obviously stockpiled for future use. In one of the butchering sites in East Africa, 459 hand axes were found, apparently discarded as they became dull through use.

10 *The thick, roughly shaped early hand ax was gradually improved into finer, better-proportioned instruments such as this, which is also almost 6 inches long, but only about ¾ inch thick with straight edges. Another characteristic tool in this tradition is the cleaver, which has a wide, straight cutting edge and is, experiment shows, particularly efficient for skinning and butchering carcasses.*

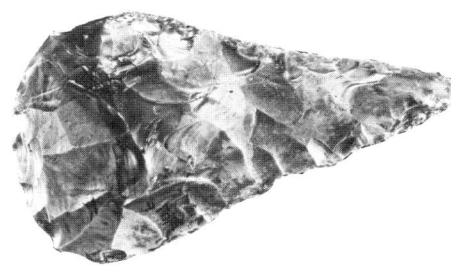

1 *Fossils of Neanderthal man are associated with a decrease in the use of core tools, such as the hand ax, and a predominance of flake tools. Multiple flakes were struck off from a central core and then retouched. They are known as points and scrapers, which also presumably describes their functions. They are usually 2 to 3 inches long. Some may have been used as points for spears or javelins, but, in general, it appears that they were hand-held. Spears, which were in use at that time, were of fire-hardened wood.*

FLAKE STRUCK FROM CORE

Hand axes were used extensively throughout Africa, India, and much of the Near East and Europe; tens of thousands have been found in the valleys of the Somme and Thames alone. In Asia, they were used only in the south and west; farther east, in China and Southeast Asia, men continued to use the pebble tools until a later date. In other words, the type of tool being used was not a reflection of the stage of physical evolution of the species but a consequence of cultural spread. This concept is confirmed by the close resemblance of one hand ax to another. Unlike the pebble tools, they were clearly fashioned according to a formal pattern, which indicates, of course, communication and exchange among the groups of men roaming over these vast territories at that time.

The Neanderthalers continued occasionally to use the hand ax but also developed a new type of tool, based on a new method of toolmaking. They first prepared a basic core, shaped like a disk or hemisphere, and then struck multiple flakes off its upper surface. This technique and the type of sharp, lightweight, versatile flakes it produces are so typical of the Neanderthal culture (often referred to as the Mousterian, from Le Moustier, a cave in southern France where many traces of Neanderthal man have been found) that wherever these cores and flakes are found, it is now possible to say with certainty that Neanderthal man was present. The form of many of the flakes is such that they would have been useful for scraping animal hides, suggesting the use of crude clothing, which is certainly in keeping with the cooler environment Neanderthal man inhabited.

The changes that took place in the stone tools during most of man's prehistory were very minor. Large improvements did not take place until modern man, *Homo sapiens sapiens*, appeared—between 40,000 and 30,000 years ago.

We do not know why Upper Pleistocene Cro-Magnon man was so much more proficient and prolific as an artisan than his immediate predecessors. As we mentioned previously, the brain of the Neanderthaler, on an average, was slightly larger than that of his successors—or of ours, for that matter. In this regard, it is worth noting that domesticated animals have brains about 20 percent smaller than those of the wild species from which they were derived. This difference has been interpreted by some observers as being related to the

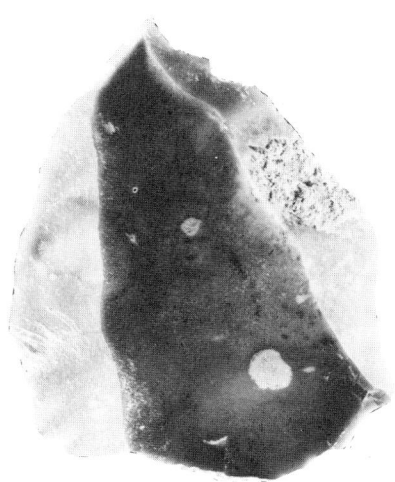

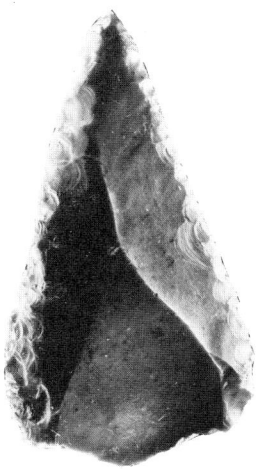

49–12 *Upper Pleistocene (Cro-Magnon) culture was characterized by a great increase in the types of specialized tools made from long and relatively thin flakes with parallel sides, called blades. These beautifully worked "laurel leaf" blades are often more than a foot long and ¼ inch thick.*

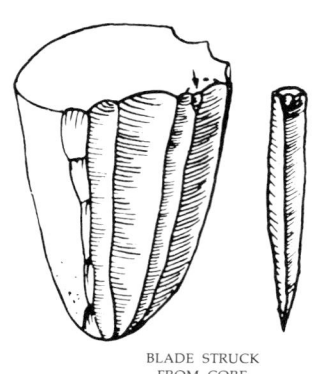

BLADE STRUCK
FROM CORE

enormously more complex environment in which a wild animal lives. Analogously, modern man may be a "domesticated" variety of his predecessors.

In any case, Upper Pleistocene man, when he first appeared in Europe, came bearing a new, quite different, and far better tool kit. His stone tools were essentially flakes—which, of course, had been in use for more than 2½ million years—but they were struck from a carefully prepared core with the aid of a punch, a tool made to make another tool. These flakes, usually referred to as blades, were smaller, flatter, and narrower, and most important, they could be and were shaped in a large variety of ways. They included, from the beginning, various scraping and piercing tools, flat-backed knives, awls, chisels, and a number of different engraving tools (burins). Using these tools to work other materials, especially bone and ivory, Upper Pleistocene man made a variety of projectile points, barbed points for spears and harpoons, fishing hooks, and—a remarkable and useful invention—needles. Thus, although Upper Pleistocene man lived much the same sort of existence as that of his forebears, he lived it with more possessions, more comfort, and more style.

SPEECH

Another categorical distinction between modern man and other animals is man's ability to speak. By speech, we mean the formulation and articulation of abstract ideas. Most higher animals can communicate the concept "I am hungry now." Only man can express "I was hungry yesterday." Yesterday is an abstraction.*

Man was clearly preadapted to speech simply by being a primate. No order of mammals is noisier. Social monkeys and apes hoot and chatter and bellow continuously, crying out with pleasure and alarm, threatening, calling, and just generally keeping in touch. The ability to communicate would have been a tremendous advantage to men hunting in groups ("Head off that hippopotamus!"), but we have no way of knowing whether they were able to speak. Chimpanzees, as Goodall observed, hunt cooperatively without formal speech.

When did speech evolve? Some students of human evolution argue that the repetition of tool forms over a long period among members of the *Homo erectus* species indicates that these individuals could talk to one another.

Some experts believe that Upper Pleistocene man was the first to really master speech, and that it was his power of speech that enabled him to supplant the Neanderthalers with such ease and also to develop the first complex culture. On the other hand, as we shall see, Neanderthal man buried his dead, sometimes with gifts of food and tools. Formal burials such as this suggest a concept of life after death, and it is difficult to understand how a society could hold such beliefs without some fairly sophisticated means of exchanging ideas among its members. Among the matters about which scientists know very little are the biological requirements for speech, both in the vocal apparatus and, more important, in the brain. What part of man's genetic endowment gives him this unique and powerful talent? When we know this, as someday we shall, it may be possible to speculate more fruitfully on the conversations of our ancestors.

* Some observers contend that some of the porpoises and dolphins speak to one another. We are deliberately avoiding this issue except to note that if these animals do speak, their capacity to do so evolved quite separately from man's.

HOME AND HEARTH

All higher primates except modern man are continuously on the move, and though they range back and forth over the same territory, they sleep in a different place every night. Any member of the group that cannot keep up is left behind. Early men were also nomads, as are most modern hunters, but there is evidence that they maintained base camps, some of which appear to have been fairly permanent, in the sense that they were returned to year after year. Hunting parties operated from these sites, leaving children and other members of the group behind and bringing game back with them. Apparently, if a very large animal was killed, the entire group would move to the butchering site for several days, consuming part of the kill on the spot and preparing the rest for carrying back to the base.

The australopithecines and the earliest men probably lived mostly in the open, although one australopithecine site is in the natural shelter of overhanging rocks. The campsites found in southeastern Africa are near streams or lakes, which would not only have provided water for the camper but would also have attracted game.

The Choukoutien caves, near Peking, inhabited 350,000 years ago, are among the earliest caves known to be used for human habitation. It is probably no coincidence that within these same caves is also one of the first traces of the human use of fire. Before men had fires, caves, which are also the homes of bears and other large carnivores, would have been too dangerous for habitation, but since fire made them available for human occupation, they have been used up to modern times. The caves near Shanidar, for example, in the remote mountains of Iraq have been occupied continuously—to the present day—for some 100,000 years.

3 *This large cave near the little village of Shanidar in northern Iraq has been continuously inhabited by man for more than 100,000 years. Nine Neanderthal skeletons have been found in the cave, including one who was, according to analysis of fossil pollen, buried on a bed of woody branches and June flowers gathered from the hillside.*

Prehistoric cave dwellers usually remained near the mouth of the cave, where there was light and where the smoke of the fire could escape. Often the caves had terraces outside them where, one could imagine, much of the toolmaking and other activities were carried on in good weather. Caves were the chief shelter of Neanderthalers, it would appear, and there are abundant traces of cave occupation during Upper Pleistocene times.

The earliest known examples of human architecture are the remains of crude huts built some 300,000 years ago on the shores of the Mediterranean, near what is now Nice. The huts were oval-shaped, from 26 to 49 feet long and from 13 to 20 feet wide; their outlines can be traced both from postholes and from lines of stones that apparently served to brace the walls. It appears that each year a hunting group would arrive at this area in the late spring (the time of year has been fixed by analysis of pollen in fossilized human feces), erect huts, hearths, and windscreens, and move on in the course of the summer. After the departure of the group, the simple structures would collapse. Wind would cover the living floor with sand, and rain would pack it down. When the hunters returned the following spring, they would construct the temporary dwelling again, often in exactly the same place. In one dune, there are 11 living floors, each constructed over the previous one; it is believed that these represent consecutive yearly visits, probably by the same group of people.

Man-made dwellings of far more recent times (about 20,000 years ago) have been found in Czechoslovakia, southern Russia, and Siberia, in vast caveless areas. Massive accumulations of mammoth bones and ivory have been discovered at these sites; the men who lived there were apparently primarily

49–14   *A contemporary of Neanderthal man and of his Upper Pleistocene successors, the woolly mammoth* (Mammuthus primigenius) *is an extinct elephant. About 10 feet at the withers, it is smaller than many living African and Indian ele-* *phants, but some males had tusks as long as 16 feet. Its short, stocky build, small ears and tail, woolly coat, and fat deposits are adaptations to environments such as the tundra and taiga, which were its habi-* *tat. Many specimens have been preserved frozen for thousands of years in permafrost soils in Siberia and Alaska. The mammoth was the first extinct mammal to be described as such (in 1695).*

mammoth hunters and are likely to have followed the great herds on their long seasonal migrations. Judging from the size of the living floors, some of these dwelling places were inhabited by as many as 90 or 100 people at a time. Perhaps a number of smaller groups would converge at such places at specific times of the year for hunting parties.

These dwellings were made of more substantial materials than the flimsy shelters found near Nice. Some, apparently winter dwellings, were made of earth and sunk slightly below ground; others, which resemble the summer homes of the Eskimos, were tentlike constructions of animal skins that presumably could be carried to new sites. However, like the shelters at Nice, most were either seasonal base camps (acres of animal bones have been found at some of these sites) or stopping-off places, way stations, on the perpetual journey characteristic of hunting men.

Fire

Man probably "stole" his first fires. Fires occur naturally as a consequence of lightning, in smoldering volcanoes, and, in some areas, from spontaneous combustion of gases. A fire, once captured, was probably tended carefully and carried from place to place.

In addition to making possible the use of caves for shelter, fires could be used for warmth, for fire-hardening wooden implements, and for cooking food. Although cooking undoubtedly made meat more palatable and certainly easier to chew, its biggest influence on man's diet was probably that it made it possible for him to eat certain vegetable products—in particular, seeds and grains—which otherwise cannot be digested by the human alimentary equipment.

The first known use of fire, as we mentioned before, occurred about 350,000 years ago; hearths have been found in both China and Hungary dating from that period. We do not know when man first learned to make fire, but it could have come about as a by-product of his stone toolmaking, which must, from time to time, have produced sparks. The first object interpreted as a fire-making tool dates from Upper Pleistocene times, about 30,000 years ago. It is a stone disk with a hole in its center, found with other objects in a grave.

The increasingly widespread use of fire for cooking—which may have originated as a happy accident—is reflected in a decrease in the size of the jaws and teeth of the human skull, a trend that is still continuing.

SOCIAL ORGANIZATION

The Tribe

Early man lived in groups. In this, he resembled almost all other higher primates. Tribal behavior appears to be a strong primate characteristic—the only known exceptions are the gibbons and orangutans, both of which live high in treetops and so have less need for the protection a group affords. (The only other mammals that live in groups containing more than one adult male are members of the dog family, such as wolves and jackals, that hunt in packs.) The chief function of grouping among primates—the reason apparently that group behavior was selected for—seems to be protection of its members, particularly of the females and the young. Male primates are characteristically considerably

-15 A Cro-Magnon firestone of iron pyrite, from a Belgium cave. The deep groove was worn in the stone by flint struck against it to produce sparks. The stone is shown here almost twice its natural size.

larger than the females, and they often have ferocious canines that are lacking in the females and that are used in fighting and threatening. When a baboon tribe moves across open country, for example, the females and juveniles keep to the center of the pack and are surrounded by adult males.

Primate groups typically number between 20 and 40 individuals, including infants. Group members are not held together by sexual ties between individuals —sexual relationships are totally promiscuous—but by affectionate relationships, which are particularly noticeable between mothers and offspring and between certain adult males. Friendships among group members are continually reinforced by mutual grooming.

On the basis of the species studied, the groups seemed to be organized internally into clear-cut social hierarchies—"pecking orders"—that, once established, serve to reduce intragroup fighting. Individuals recognize members of their own group and are suspicious and hostile toward other groups of the same species. However, fighting does not appear to be frequent between groups, probably because the members tend to stay within their home range or to retreat to it when confronted.

Based on the size of a single campsite (at Olduvai), it appears that bands of australopithecines probably contained three or four males plus females and children. Living areas of *Homo erectus* seem to be somewhat larger, perhaps big enough to accommodate 20 to 30 adults, which is about the size of some present-day hunting tribes. These little bands probably lived at relatively great distances from one another. Judging from contemporary hunting societies, several hundred square miles are needed to support a single hunting group.

At some distance from camps, other sites have been discovered at which game was butchered and presumably divided up among members of the hunting party. In other words, these early men not only hunted as a cooperative group but also shared their food. As we noted previously, such sharing does not really occur among other primates. Although it is popular to stress man's naturally aggressive and competitive nature, a capacity to get along equably with one's fellowmen was highly important for the survival of both the individual and the group during this long period of man's history, and the ties that bound man to his fellowman and held the group together must have been strong ones. Some psychologists attribute some of the dilemmas of human urban existence to the problem of man's trying to relate to large numbers of individuals when he is, by nature, attuned to close emotional involvements with a small group whose members are well known to him.

We have one fossil record of such ties, dating not from *Homo erectus*, however, but from Neanderthal times. In the Shanidar cave, archaeologists found a skeleton of a man who, because of a deformity obviously present since birth, had a withered right arm. The arm had been amputated below the elbow some years before the man's death. The teeth showed signs of unusual wear, suggesting that they had been used to compensate in some way for the lack of a right hand. And yet the man had lived to be about forty years old, a relatively advanced age in those days. Although he must have been an almost totally nonproductive member of the hunting group and completely dependent on it for his survival, he was obviously cared for throughout his lifetime, some 50,000 years ago.

The Family

Although man resembles other primates in his group behavior, he differs from most other primates in his formation of permanent male-female bonds. As we noted in Chapter 30, other female primates, and indeed all other female mammals, have periods of "heat," or estrus. Only during estrus periods, when pregnancy is possible, are they sexually receptive. Estrus does not occur in immature females, and mature females are usually either pregnant or nursing (which suppresses estrus). Thus sexual activity is relatively rare in these primate groups, although sex play occurs among the younger members of the tribe. (Apparently sexual behavior is learned, not innate, among higher primates; apes and monkeys raised in laboratories or zoos in isolation usually cannot be bred.) In these primate groups, the primary responsibility for child care devolves on the mother, who may receive help from other members of the group.

Most human societies, however, are based on some version of the nuclear family in which a single male has exclusive sexual rights to one or more females and in which both sexes cooperate in caring and providing for the young.

We do not know whether hominids grouped themselves first in family units and then the family units eventually banded together, or whether the tribe became subdivided into families. In either case, it is interesting to reflect upon what effects large-game hunting must have had on relationships between the sexes. To the extent that the tribe depended on elephant, antelope, and rhinoceros as its staple diet, the women and children were totally dependent on the men.

Dependency would also have increased because of the simultaneous growth in the size of the skull. The size of the brain at birth of a human infant today is only about one-fourth of its adult size; in other words, human beings are born at a very immature stage in their development. Delaying birth past nine months or so increases its difficulties; delivery of the head is much the most hazardous stage of human birth for both mother and child. Judging from the age at which eruption of the permanent teeth occurred, the young of *Homo erectus* matured at about the same rate as those of modern *Homo sapiens*. We guess that, on the basis of studies of today's hunter-gatherers, children in prehistoric families were spaced farther apart than modern children. This was due to prolonged nursing, which depresses fertility, high infant mortality, and, very likely, infanticide. Nevertheless, a female member of the tribe probably had immature children to take care of from the time she reached puberty until her death (which would probably have occurred by forty or forty-five years of age, at the latest).

Women played a very important role in the life of the tribe. Not only did they undoubtedly supervise the raising of the children but they also were probably responsible, as is the case in today's tribes, for the gathering of vegetable products and perhaps also of small game. These items probably made up most of the day-to-day menu, because they had no way of storing meat that we know of. But the woman's role must have been a completely different one from that of the man, and it has remained so until very recently, although now it is chiefly an anachronism from our hunter-gatherer past.

Among some primitive tribes, the headman or chief, often the most skilled and intelligent man of the group, has more wives than other males. If there was a headman in prehistoric tribes, and if he had more wives, he could clearly have

left many more offspring than the less skillful hunters. Moreover, because of his position in the group, his offspring would be more likely to survive in good health and to reproduce successfully. This could have been one method by which rapid selection for intelligence could have taken place.

Secondary Sexual Characteristics

One of the reasons we can be reasonably sure that pair-bonding has been in existence for a long time is that there are physical changes in the human female that must have occurred as a result of, or concomitant with, the evolution of bonding behavior. One of these changes is the suppression of estrus. The human female is the only female mammal continuously sexually receptive throughout her adult life. Although we often speak of our sexuality as "an animal instinct," and it clearly is, it is noteworthy that human males and females have a far more active sex life than most other animals.

Other physical changes related to the continuous sexual receptivity of the human female involve the secondary sex characteristics. Among other primates, the chief visible differences between males and females involve the larger size of the males and other male characteristics, such as large canines, associated with social dominance and defense. Special female characteristics—other than negative versions of the preceding male characteristics—are evident only when the female is in estrus. During this period, females of many species signal their sexual availability by swelling of the genital area and bright coloring of the skin surrounding the anus and the genitalia. By contrast, human females signal their continuous sexual receptivity by permanent signs, such as their enlarged mammae (other female primates provide adequate milk for their offspring but do not have protruding breasts), their distinctive distribution of hair and body fat, and their more delicate facial features, all of which are sexually attractive to the human male and apparently have been throughout much of history.

Incest Taboos and Exogamy

Because almost all peoples have laws or taboos against incest, it is believed that these strictures developed early in human culture. There are sound biological reasons for discouraging inbreeding, which tends to bring out genetic defects that would otherwise probably remain latent. It is unlikely, however, that prehistoric groups would have been aware of these effects. Incest taboos probably arose because intermarriage among fathers and sons and mothers and daughters would have disrupted the formation of nuclear families. There may also be some instinctive avoidance of inbreeding among primates. Young male chimpanzees, for example, will mate with other adult females in estrus but seem to avoid mating with their mothers, so perhaps there are more strictly biological factors as well as the social ones. Lower animals and plants have many devices to discourage self-fertilization and encourage outbreeding, and an outbreeding instinct among higher animals may be an extension of this. It is interesting to observe that young Israelis raised in the same kibbutz rarely marry each other, although there are no social strictures against it.

Seemingly related to incest taboos, but perhaps of quite separate origin, are tribal laws promoting exogamy—marrying outside the group. Such laws would

appear to have a highly practical basis. In the first place, a small band comprising only, for example, eight or ten families, might well not be able to provide a suitable mate for the right young man or woman at the right time; exogamy probably started for just this reason. Second, intermarriage to neighboring tribes promotes friendly relations on one's borders and also opens possibilities for cooperative activities when these would be useful—as in a large hunting expedition.

Among hunting people, the women usually leave their own tribe and go to live with the husband's tribe. This is a logical arrangement since much of the boy's preparation for manhood is gaining knowledge of the territory that will be his future hunting ground. Exogamy is important in man's cultural development because it clearly serves as a deliberate device for eliminating intraspecific fighting. In an analogous way, marriages were arranged among the children of various royal households until very close to the present time.

RITUAL, MAGIC, AND ART

Neanderthal man was, apparently, the first creature on Earth to bury its dead. Because burials often took place in the inhabited caves, a number of Neanderthal graves have been found. Often the bodies are gently flexed, as in natural sleep, but in some of them the limbs are pulled tightly to the body, suggesting that they may have been bound before burial. Such binding, when practiced by primitive tribes today, is done to keep the dead from rising to disturb the living. Some Neanderthal bodies were daubed with red ocher before burial, a custom also found among other primitive peoples. Many examples of burials with food and weapons—and one with spring flowers—have been uncovered. At es Skhul in Palestine, for example, where 10 bodies have been found in separate graves, one man was buried holding a haunch of meat. In the Russian cave of Teskik Tash, the body of a Neanderthal boy was surrounded by the horns of a Siberian mountain goat, their points stuck into the ground. Over another grave, also of a child, a large stone was placed, and on its undersurface are arrays of rough, cup-shaped pits, arranged in pairs. Their meaning is unknown.

Burials of Upper Pleistocene peoples, while generally similar to those of the Neanderthalers, were often much more elaborate. Some of the skeletons are richly decorated with headdresses and necklaces, and some are overlaid with shells, which appear to have been attached to garments.

One unusual and particularly interesting grave, about 20,000 years old, has been found at Vestonice in Czechoslovakia. The skeleton is that of a slightly built woman, 5 feet 3 inches tall, about forty years of age (old for that time). It was covered with red ocher and protected by two shoulderblades of a mammoth, one of which had a network of irregular lines scratched on its surface. Buried with her were stone tools. In one hand had been placed the paws and tail of an arctic fox (the swiftest and wiliest of animals), and in her other hand, its teeth. Clearly, she had been an important personage, the possessor of some special skill or wisdom or, most likely, of magical powers.

As we shall see, the cave paintings also bear witness to a culture rich in mystery and ceremony, but these we shall discuss under the heading of art, for such they clearly were.

Cannibalism

In the Choukoutien caves are numerous skulls of *Homo erectus*, which have been broken apart at the base in such a way as to suggest that the brains were removed for eating. Most animals do not eat members of their own species, and cannibalism among primitive peoples is almost always a ritual ceremony whose purpose is to acquire some of the powers or capacities of the dead person. Moreover, there are far more skulls at Choukoutien than there are other skeletal parts. For these reasons, these skulls from 350,000 years ago are often interpreted as evidence of ritual cannibalism.

This interpretation is supported to some extent by the discovery, in a cave in Italy at Mount Circeo, of the skull of a man who lived some 60,000 years ago, a Neanderthaler. This skull was surrounded by a ring of stone, and outside the ring of stone had been, apparently, a circle of bones of large mammals. The skull that was the center of all this attention had been broken open at the base in just the same way as the skulls in the distant Choukoutien caves.

Art

Unless we count the gravestone with its cup-marks or the circle of stones at Mount Circeo, the first known art appeared in Upper Pleistocene times, very late on the scale on which we have been viewing human history. During this period, art suddenly began to appear everywhere and in all its forms—sculpture, modeling, relief, engravings, drawings, and paintings. There are delicately worked tools and weapons, made of bone and mammoth ivory and often decorated with fine engravings, usually of game animals. There are small carved and modeled figures, some of animals, particularly game animals, and some of human beings, nearly all of which are women.

The most famous of these female figurines is a small statue of a female form, discovered near the beginning of this century, which has come to be known as the "Willendorf Venus." It was carved of limestone about 25,000 years ago. The figure is deliberately faceless; fat and fecund, with huge, sagging breasts, she is also probably pregnant. Numerous other "Venuses" have been found at sites from France to Siberia. (Upper Pleistocene art, or at least what remains of it, was confined almost exclusively to Europe.) There are variations among these "Venus" figures. Some are modeled of clay, and some are carved in relief on stone surfaces. Some are abbreviated, with two breasts hanging on a sticklike figure or sometimes just the breasts all by themselves, like a pair of discarded water wings. They are often found in the remains of human habitations, and they apparently performed some function related to fertility, childbirth, or nourishment. Or perhaps they were simply the guardians of hearth and household, the realm of the "female principle."

The art of Upper Pleistocene man reached its heights with the cave drawings and paintings of southern France and western Spain. The examples that remain to us, many surprisingly untouched by time, clearly form a part of a rich artistic tradition, integrated in its concepts and its vision, and enduring for a remarkably long time—at least 10,000 years.

What one usually notices first about the cave drawings is that they are almost entirely of animals. A census of the figures in the caves at Lascaux, France, reveals, for example, 10 ibexes, 7 bison, 1 musk ox, 1 reindeer, and 1

49–16   *The "Willendorf Venus," a small (11 centimeters) sculptured figurine, which may be a fertility or cult figure.*

49–17   *(Facing page) Bull from Lascaux, France. This wild ox is the aurochs,* Bos primigenius, *which became extinct in Europe during the seventeenth century. They were much larger than modern cattle, with the bulls often as much as 6 feet high at the shoulders.*

rhinoceros. All are game animals, although they are not represented in proportion to their consumption, judging from other evidence. In addition, there are six felines and one bear. There are no plants, unless one interprets a few scratchings as grasses. In Lascaux, as in the other caves, there are very few drawings of people, and those that do exist depict them as simple stick figures, often wearing animal heads or masks.

Paints were prepared from deposits of iron and manganese oxides, found in the soil, or from carbon blacks. These pigments were ground to a fine powder (special pestles and mortars have been found that were used for this purpose) and mixed with water. They were applied with fingers, with pointed instruments (perhaps quills), and with pads of moss or fur. It is thought that some of the colors were applied as a kind of spray, perhaps by blowing dry powder through a tube onto a wetted surface. Engraving was combined with drawing and painting to produce three-dimensional effects. The colors—reds, yellows, browns, and blacks—were shaded to suggest muscular contours.

In all cave art, the cave itself plays an important role. The placement of some of the drawings indicates that their form was suggested by the contours of the rock surface—which is incorporated as the swelling of a rump, for example—or by a natural marking, which becomes an eye or some other feature of the animal. More strikingly, the animals often appear to be moving into or out of tunnels, emerging from behind ledges, guarding entrances, or tumbling off precipices.

Cave art is almost never found near the mouths of caves, where daylight penetrates; it is almost always deep, and sometimes very deep, in the cave's interior. In fact, many of the paintings are in almost inaccessible places, even at the ends of tunnels, or on ceilings, where they are very difficult to see, and must have been even more difficult to execute. Some available surfaces are not used at all, while others nearby have been used over and over again. They must all have been painted and also viewed by the light of crude lamps or torches.

The meaning of these drawings and paintings has long been a matter of debate. Some of the animals are marked with darts or wounds (although very few appear to be seriously injured or dying). Such markings have led to the suggestion that the figures are examples of sympathetic magic, in which there is the notion that one can do harm to one's enemy by sticking needles in his image. The fact that many of the animals appear to be pregnant suggests that they may represent fertility, perhaps in some way related to the "Venuses." Many appear also to be in movement. Perhaps these animals, so vital to the hunters' welfare, were migratory in these areas, and they may have seemed to vanish at certain times in the year, mysteriously returning, heavy with young, in the springtime. This return of the animals might have been an event to be solicited or celebrated in much the same spirit as the rites of spring of more recent peoples or Easter are celebrated.

Cave art came to an end perhaps 8,000 or 10,000 years ago. Not only were the tools and pigments laid aside, but the sacred places—for such they seem to have been—were no longer visited. New forces were at work to mold the course of men's existences, and the end of this first great era in human art is our clearest line of demarcation between the old life of man the hunter and the new life that was to come.

9–18 (a) *Leaping cow and small horses at Lascaux, France. One of the most controversial questions concerning cave art is whether groupings such as this represent merely an attempt to cover all available space, as in graffitti on walls or subway cars, or whether these associations are meaningful. The illusion of movement is greatly enhanced by the patterns of light and shadow in the dark, narrow recesses of the cave. The lattice-like structure in front of the cow is variously interpreted as a trap or as some sort of shorthand notation, perhaps a fertility symbol. (b) In almost every cave, a small number of the animals show wounds, such as these seen in this bison from the Cave of Niaux in France. Although the animals are shown in realistic detail, the weapons and the hunters, if shown at all, are abstractions, perhaps to keep their identity a secret. (c) The cave paintings at Altamira, Spain, were the first-discovered examples of this artistic tradition, but are not as ancient as those at Lascaux. Most are bison. Although they are often more competently rendered than those at Lascaux, they seem to lack some of the strength and vitality. It has been suggested that this may be a consequence of using dead animals as models, a hypothesis supported by the fact that their feet don't seem to be touching the ground and that one often seems to be viewing them from slightly underneath, as the artist would if the animal were lying on its side.*

(a)

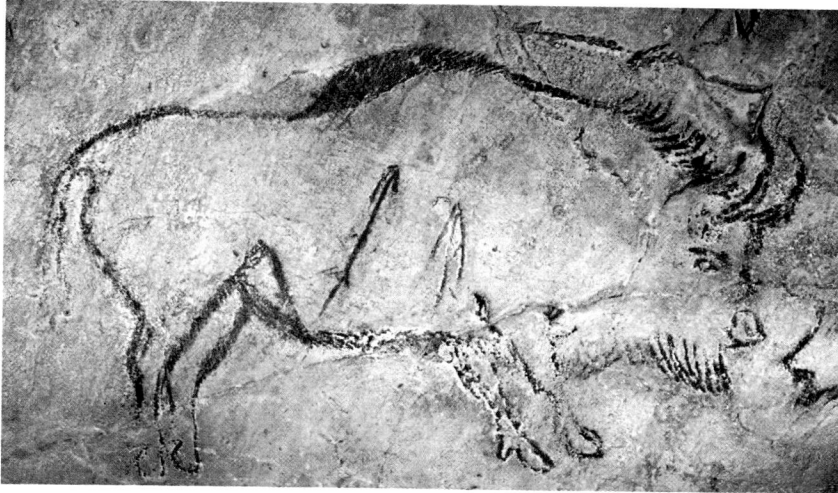

(b)

(c)

SUMMARY

Two species of the genus *Homo* are now widely recognized, *Homo erectus* and *Homo sapiens*. Fossil remains of *Homo erectus* have been found in Africa, Asia, and Europe. They date from about 1 million to 300,000 years ago. A few fossil fragments of *Homo sapiens*, dating from about 250,000 to 150,000 years ago (an interglacial), have been found in Europe. These are followed by numerous specimens of what has been called Neanderthal man, now recognized as a variety or subspecies of *Homo sapiens*. The fossil record of Neanderthal man dates from about 75,000 to about 30,000 years ago. The characteristics of the Neanderthal type may be related to the cold climate in which they lived or their genetic isolation or both (or neither).

Upper Pleistocene, or Cro-Magnon, man is, as far as can be told from fossils, indistinguishable from modern man. Upper Pleistocene man, *Homo sapiens sapiens*, appeared among the Neanderthalers about 30,000 to 40,000 years ago, during the most recent glaciation, and the Neanderthalers became extinct.

The cultural history of early man is tentatively reconstructed from three sources: observations of modern primates, studies of modern or recent hunter-gatherers, and analysis of fossil remains of early man, along with his artifacts, his food debris, and other evidence from his campsites and surroundings.

Early hominids—the australopithecines—appear to have been omnivorous, and to have eaten more meat (in the form of small game) than any present-day nonhuman primate. Australopithecines used simple "pebble" tools. They lived in small groups that may or may not have been subdivided into family units. They probably did not live in caves, and they do not appear to have had any sort of dwellings.

Individuals of the species *Homo erectus* lived in larger groups, numbering 20 to 30 adults. They were successful hunters of very large animals. Early members of the species used pebble tools; later, in Africa, the hand ax was developed. The widespread use of the hand ax throughout Africa, India, the Near East, and Europe (but not the Far East) indicates cultural exchanges between groups and suggests that members of this species possessed the capacity for speech.

By about 350,000 years ago some groups of men lived in caves and had fire, two developments probably related to each other. The extent and type of hunting engaged in by *Homo erectus* are evidence of division of labor between the sexes.

The Neanderthal variety of *Homo sapiens* lived in Europe during the last glacial advances. They had fire, inhabited caves, hunted large animals, and must have worn clothing, probably made from skins of prey animals. Neanderthalers continued to use hand axes but also made lightweight, versatile stone tools (Mousterian). They buried their dead, often with food and weapons. This practice indicates that the Neanderthalers had a culture that included rituals, religion, and speech.

Upper Pleistocene men, who replaced the Neanderthal variety, were nomads and hunters, like their predecessors. They had tools (blade tools) distinctly different, smaller, and far more varied than had been seen previously. Unlike their predecessors (as far as we know), they decorated themselves with ornaments. Their burials were elaborate. Their art, which covers a period of some

10,000 years, is evidence of a culture rich not only in beauty but in mystery, magic, and tradition. We are their direct descendents.

QUESTIONS

1. Define the following terms: hominoid, hominid, Upper Pleistocene, *Homo erectus*, *Homo sapiens sapiens*.
2. Man's aggressiveness toward his fellow man has been linked both by some scientists and some writers of popular treatises to his evolutionary history and origins. What evidence can you produce to support this point of view? To oppose it?

Chapter 50

The Ecology of Modern Man

The agricultural revolution, which occurred at the end of the Pleistocene, marked, for the great majority of mankind, the end of nomadic hunting as a way of life. It made way for the development of villages, towns, and finally cities, and for the industrial and scientific revolutions that were to follow. Indeed, more than any other occurrence in the history of mankind, it opened the floodgates for the inrush of civilization, with all its grandeurs and its discontents.

ORIGINS OF AGRICULTURE

The period immediately prior to the beginnings of agriculture was a period of adjustment to climatic change. The most recent of the glaciations began to retreat about 18,000 years ago, withdrawing slowly for about 10,000 years. As the glacier retreated, the plains of northern Europe and of North America, once cold grasslands or steppes, gave way to forest. The great herbivores that roamed these steppes retreated northward and eventually vanished; the woolly mammoth was last seen in Siberia about 12,000 years ago. While some animals became extinct, men adapted, as they had during the entire period of violent climatic changes that marked their evolutionary history. With the migratory animals gone, they shifted their attention to smaller game, such as deer, which tended to be resident in the same area throughout the year. The dog was domesticated during this period; it lived as a scavenger around human habitations, warned of the coming of strangers, and perhaps helped hunters to track game. Fishing became an important part of the economy at this time also; ponds and lakes were filling, and streams were rushing with waters from the melting glaciers. Hunters made canoes and paddles and seines and other fishing equipment. The water level in the seas rose several hundred feet. England became separated from the Continent, and the land bridge from northeast Siberia to the New World was inundated, as it had been in previous interglacials.

At the same time, around the Mediterranean, a change to a more arid climate appears to have taken place. Here also, as in the north, there is clear evidence of a more diversified hunting economy, directed more toward smaller animals, such as wild goats, with a greatly increased use of fish and, in particular, of shellfish. The hunting and gathering of small animals—in contrast to large migratory herbivores—would have resulted in a less nomadic existence for the hunters.

1 *A rock painting from Spain. Such paintings date from postglacial times, about
8000 to 3000 B.C. In contrast to the
earlier cave paintings, paintings such as
this were executed on sheltered rock surfaces, frequently show human figures,
and are small in scale. These paintings
typically show forest animals rather than
the great plains animals that stalked the
caves of Lascaux, France, and Altamira,
Spain. The deer are apparently being
driven toward the archers by beaters.*

THE TRANSITION

The earliest traces of agriculture are found in an area in the Near East generally known as the Fertile Crescent. Here were the raw materials required by an
agricultural economy: cereals, which are grasses with seeds capable of being
stored for long periods without serious deterioration, and herbivorous herd
animals, which can be readily domesticated. The grasses in the Fertile Crescent
were wild wheats and barley; they still grow wild in these foothills. The
animals were wild sheep and goats.

In late Paleolithic times, the people living in this area were hunter-gatherers
who set up temporary dwellings, often in caves or rock shelters. Then, about
11,000 years ago, new cultures appeared in this area; they were characterized
by implements associated with agriculture, such as flint sickle-blades, grinding
stones, or stone mortars and pestles. We do not know whether these peoples
cultivated crops. For instance, the sickle-blades may have been used for cutting
down rushes for baskets, or the mortars and pestles, for grinding pig-

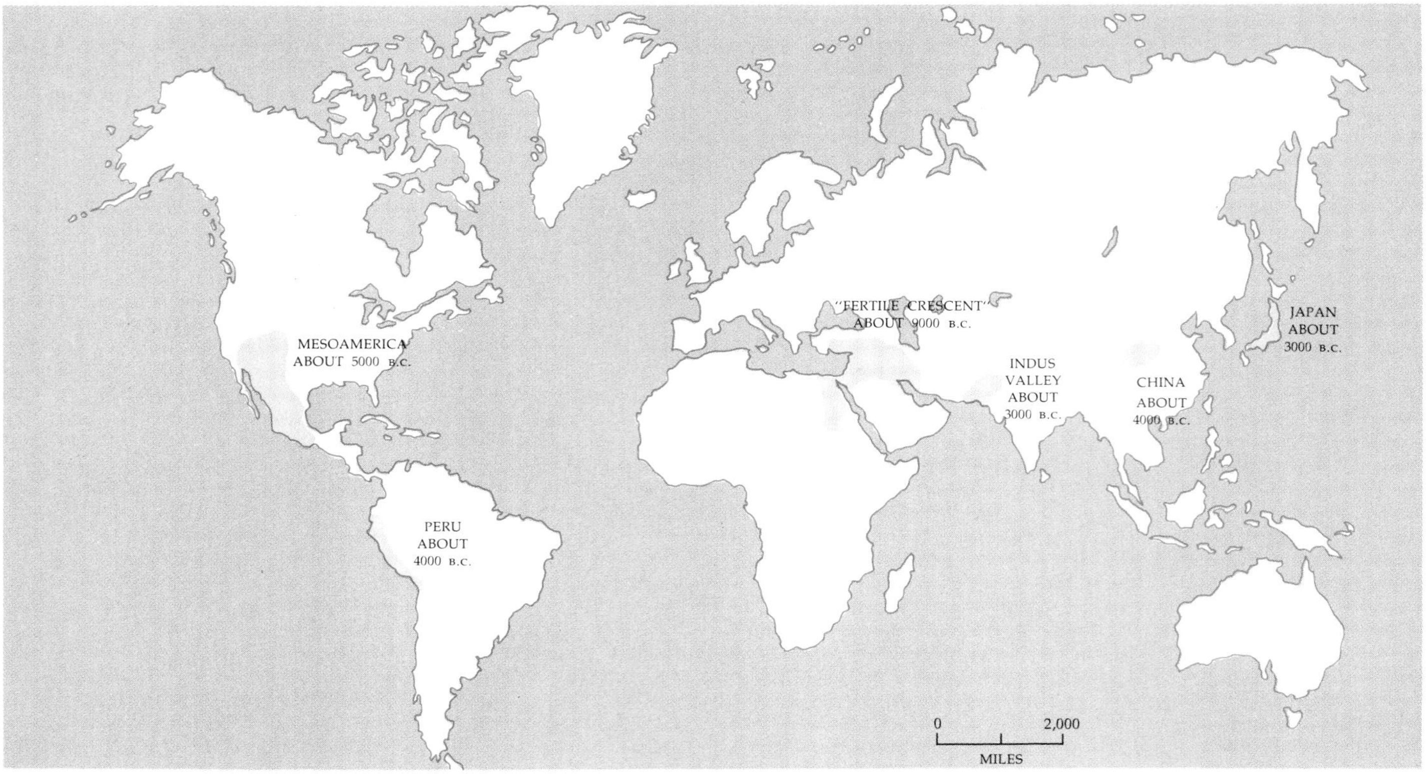

The map shows the following locations with labels:

- MESOAMERICA ABOUT 5000 B.C.
- PERU ABOUT 4000 B.C.
- "FERTILE CRESCENT" ABOUT 9000 B.C.
- INDUS VALLEY ABOUT 3000 B.C.
- CHINA ABOUT 4000 B.C.
- JAPAN ABOUT 3000 B.C.

0   2,000
MILES

50–2   *Origins of agriculture. Agriculture appears to have developed in a number of widely separated areas of the world at the same time.*

50–3   *Pottery probably had its origin in woven baskets, lined with clay so they would hold liquids, that were accidentally left by household fires. Not suitable for carrying long distances, pots are associated with a sedentary, food-storing way of existence. The earliest pots date from about 8,000 years ago.*

ments rather than grain. However, their widespread and quite sudden appearance indicates a change in how people were living and what they were doing. Furthermore, in a cave near Shanidar, dating back almost to this same period (more than 10,500 years ago), there is fairly good evidence for the presence of domesticated sheep, one of the first indications of domesticated animals kept for food.

Jarmo, one of the best studied of the early agricultural villages, was located in the Zagros Mountains toward the eastern end of the Fertile Crescent. Occupied for a few centuries about 8,500 years ago, Jarmo consisted of twenty-five houses constructed of packed mud, each with several small rectangular rooms containing built-in clay ovens and basins. Here at Jarmo is the first clear evidence of cultivated types of wheat and barley. The remains of domesticated goats and sheep were also found here. The food debris also contains wild game and wild plant products—such as acorns and pistachio nuts—in large enough quantities to indicate that they were important supplements to the diet. The village probably had about 150 inhabitants, including men, women, and children—about the size of villages in the area today.

Although we do not know exactly when or where it first happened, the first deliberate planting of seeds can be seen as the logical end of a simple series of events. The wild cereals are weeds, ecologically speaking; that is, they grow readily on open or disturbed areas, patches of bare land where there are few

Domesticated animals provided means of storing food. This rock painting from the Sahara shows herds of domesticated cattle from about 5,000 years ago. Their owners were probably nomads, as are the present herdsmen of the Sahara.

In both domesticated wheat, shown here, and the wild grasses from which modern wheat originated, the flowers, which are wind-pollinated, develop in clusters at the end of the stalk. In the wild grasses, as the fruits (grains) mature, the clusters break apart, dispersing the seed. In the domesticated grasses, the grains remain on the stalk and so can be more readily harvested. The great significance of grasses in the economy of man is that their fruits can be stored for a long period of time.

other plants to compete with them. Also, early man, judging from his campsites, was not very tidy. It is easy to envision how seeds might be spilled or discarded with garbage on open land around man's habitations. And the presence of permanent dwellings, such as those at Jericho (see page 962), certainly indicates that men were staying in one place long enough to recognize and reap their accidental harvest. From this it would have been an easy step to the deliberate saving of seeds and tending of crops.

Because men selected the seeds that they gathered and planted, they soon produced changes in the wild strains. In wild wheats, for instance, the stalk (rachis) on which the flower clusters grow and on which, eventually, the seeds develop, becomes brittle when the seeds mature and breaks off, scattering the seed. Among these wild plants, occasional mutants can be found in which the rachis is not brittle. It is these plants, at a disadvantage in the wild, that are more likely to be harvested by men and, as a consequence, more likely to be planted. Thus, just as men came to be dependent on their crop plants, plants —such as cereals that could not seed themselves—were produced that are dependent upon man.

JERICHO

Other evidence for a change in man's relation to nature is provided by the archaeological remains of the biblical town of Jericho, located at an oasis in the Valley of the Dead Sea. The earliest structure erected here, which dates back about 11,000 years ago, has been interpreted as a shrine visited by travelers. Overlying the shrine are the first remains of human habitations, and overlying this layer, dating from about 9,000 years ago, is a truly remarkable collection of one-room houses made of sun-dried bricks and shaped like beehives. The settlement covers about 10 acres, with buildings enough to accommodate from 2,000 to 3,000 people, and is surrounded by a stone wall 6 feet thick, with at least one tower containing an internal staircase 30 feet tall. Again, we are not sure that the inhabitants of this town were farmers. Some experts have expressed the opinion that only agriculture could have supported a community of this size. Others have suggested that it might have been a sort of trading center for the materials collected by outlying hunter-gatherers. In any case, this large town with its extraordinary walls is clear evidence that man had, apparently quite suddenly, begun to develop a very different sort of relation both with the land and with his neighbors.

(Left) Composed of flint-worked masonry, this 9,000-year-old tower was contained in the wall surrounding Jericho. The square opening leads into its hollow interior. (Right) Artifacts from the Jericho site: a dried-mud brick showing the brickmaker's thumbprints and a plaster-covered skull inset with cowrie shells, the significance of which is unknown.

-6 *Agriculture has resulted in marked changes in many food crops. (a) Fossil cobs of wild corn from the San Marcos Cave in the Tehuacan Valley of Mexico. The corn cobs, shown actual size, have been found among archaeological remains dating from 5200 to 3400 B.C. (b) Fossil* cobs *of early cultivated corn, from 3400 to 2300 B.C. These are much larger, probably as a result of irrigation and of weeding of competing plants. Like the wild cobs, these early cultivated ones are attached to the plant by only a fragile* stalk, *and each kernal is surrounded by bracts (chaff). (c) A fossil cob, about 2 to 3 centimeters long, from a cave in New Mexico compared with an ear of modern corn-belt corn (left) and large-seeded Peruvian flour corn (right).*

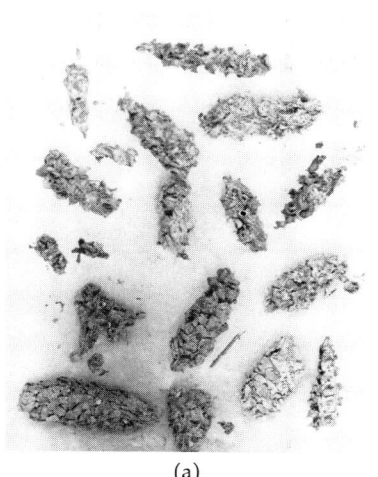 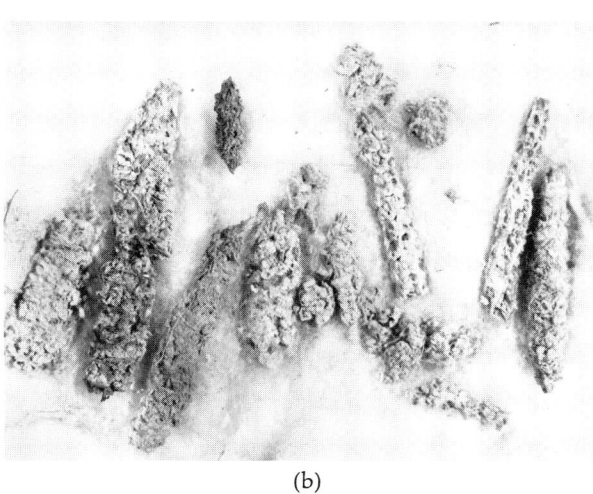 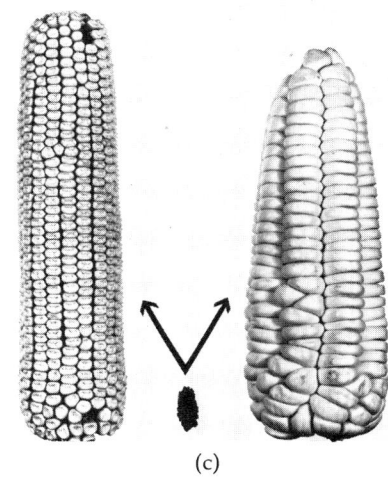

(a) (b) (c)

-7 *Teosinte, from Mexico, believed to be an ancestor of modern corn. Corn is one of the comparatively new major crops that came from the New World. Others are tobacco, tomatoes, and potatoes.*

Similar adaptations took place among animals as a result of selection of those with flat horns or no horns at all or of soft, woolly coats, both of which place the animals at a disadvantage in the wild but are useful to men.

By 8,100 years ago, agricultural communities were established in eastern Europe; by 7,000 years ago (about 5000 B.C.), agriculture had spread to the western Mediterranean and up the Danube into central Europe, and by 4000 B.C., to Britain. During this same period, agriculture originated separately in Central and South America, and perhaps slightly later in the Far East.

From the Old World sites, we have fossil imprints of cultivated wheat and barley, remains of domesticated goats, sheep, and cattle, and pottery vessels, stone bowls, and mortars. The farmers of the New World grew corn, pumpkins, squash, gourds, and cotton. The potato, the sweet potato, the peanut, and the tomato are also New World crops. Today, as you know, the vast majority of all the food we eat is the result of deliberate cultivation.

THE CAUSES OF THE AGRICULTURAL REVOLUTION

The question of why agriculture came about is a matter now under serious debate by prehistorians. In seeking any explanation for this economic revolution, it is important to remember that man had survived successfully as a hunter-gatherer for more than a million years. Based on analysis of food remains at prehistoric sites and also on observations of modern hunting peoples, the daily diet of the hunter-gatherer is probably of a generally higher nutritious

(a)

(b)

(c)

(d)

50–8   The Bushmen of the Kalahari Desert of
South Africa, who have been hunter-
gatherers for at least 11,000 years. (a)
Women and children beside a waterhole.
The groups, which average 15 to 25
members, move five or six times a year,
setting up a base camp near a waterhole.
The basic social unit is the family, and

marriages are very stable. The Bushmen
have no easily detectable protein or vitamin deficiencies, few chronic diseases,
low mortality, and a life span comparable to that of urban man. (b) A man
squeezing water from a large water root,
using the shell of a baoba fruit to catch
the water. (c) A young woman sharpen-

ing a digging stick. Women, who are the
chief gatherers of the plant products,
supply more than 50 percent of the food
and their social status in the group is
high. (d) Woman surrounded by mon-
gongo nuts. High in calories and protein
and drought-resistant because of their
hard shells, they are the staple diet of

(e)

quality than the diet of the majority of modern populations. Moreover, based again on studies of the Bushmen of South Africa, for example, and of the Australian aborigines, it is acquired by much less work. Among the Bushmen, for instance, the men hunt about three days a week, and the women, who gather nuts and other food and tend the campsites, are occupied with these tasks about five hours a day—a shorter work week for both sexes than that enjoyed by many college students and most writers of textbooks.

At this time, there are only suggestions as to why the change took place. As hunting became more diversified, it probably became possible for families to stay in one place longer, and a more sedentary existence may have brought about an increase in population. Such a population increase, coupled with a decrease in the extent of tropical grasslands because of increased aridity, might have placed populations under sufficient environmental pressures to result in new modes of existence. You will recall that a similar type of change during the Miocene is believed to account for the transition among certain primates from life in the trees to life on the ground.

(f)

(g)

(h)

the Bushmen. (e) *Men setting out from camp to hunt, using the same weapons and hunting the same game as they have from prehistoric times.* (f) *Man drinking from the shell of an ostrich egg. Bushmen make small holes in emptied egg shells, fill them with water, and bury them at strategic spots along their cus-* tomary routes. (g) *A hunter digging for edible roots on his way home from an unsuccessful hunt. Although women do most of the gathering, men share in this activity.* (h) *Hunter at water hole.*

50–9 *A pregnant Bushman woman carrying a child. Because their diet provides no soft food, mothers nurse their children, often until they are four or five years old. This long lactation period presumably has a contraceptive effect (as it does in other primates); the average length of time between pregnancies is four years, and the population remains stable in size.*

THE CONSEQUENCES OF THE AGRICULTURAL REVOLUTION

Whatever its causes, the change to agriculture had profound consequences. Populations were no longer nomadic. Thus they could store food not only in silos and granaries, but in the form of domesticated animals. In addition to food stores, other possessions could be accumulated to an extent far beyond that previously possible. Even land could be owned and accumulated and passed on by inheritance. Thus the world became divided into semipermanent groups of haves and have-nots, as it is today.

Because the efforts of a few could produce enough food for everyone, the communities became diversified. People became tradesmen, artisans, bankers, scholars, poets, all the rich mixture of which a modern community is composed. And these people could live much more densely than ever before. For hunting and food-gathering economies, 2 square miles, on the average, are required to provide enough for one family to eat.

One immediate and direct consequence of the agricultural revolution was an increase in populations. A striking characteristic of hunting groups is that they vigorously limit their numbers. A woman on the move cannot carry more than one infant along with her household baggage, minimal though that may be. When simple means of birth control—often just abstention—are not effective, she resorts to abortion or, more probably, infanticide. In addition, there is a high natural mortality, particularly among the very young, the very old, the ill, the disabled, and women at childbirth. As a result, populations dependent on hunting tend to remain small.

Once families became sedentary, there was no longer the same urgent need to limit the number of births, and probably, also, there was a decrease in the mortality rates.

About 25,000 years ago there were perhaps 3 million people. By the close of the Pleistocene era, some 10,000 years ago, the human population probably numbered a little more than 5 million, spread over the entire world. By 4000 B.C., about 6,000 years ago, the population had increased enormously, to more than 86 million, and by the time of Christ, it is estimated, there were 133 million people. In other words, the population increased by more than 25 times between 10,000 and 2,000 years ago.

By 1650, the world population had reached 500 million, many of them living in urban centers, and the development of science, technology, and industrialization had begun, bringing about further profound changes in the life of man and his relationship to nature.

LIFE VS. DEATH

One of the most important changes brought about in the last several hundred years is the reduction in the death rate, due partly to the discovery in the twentieth century of the new "wonder drugs" but even more to the earlier and less spectacular development of public health measures, such as clean water supplies, pasteurization of milk, and sanitary means for sewage disposal.

An infant born 300 years ago had little better than a fifty-fifty chance of surviving its first year, whereas in many countries today, infant mortality is less than 2 percent. (Incidentally, the United States ranks comparatively low

−10   *In many tropical countries, the death rate has fallen rapidly since 1940, resulting in a rapid growth of the population. The drop in death rate is the result of increased medical services, control of malaria by DDT, and the availability of new antibacterial drugs, especially the antibiotics. Note that the birthrate has also begun to fall. The area on the graph marked in color indicates population growth. The data shown here are from Ceylon.*

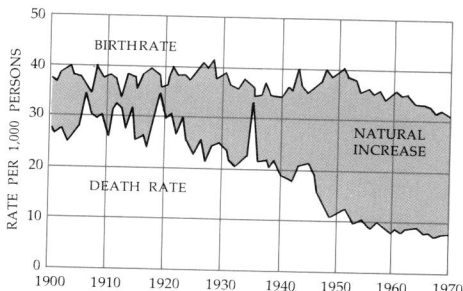

in its control of infant mortality: fifteenth among modern nations. This low figure is a reflection of the level of medical care received by the poor. The infant mortality among American nonwhites—28.8 per 1,000 live births—is nearly twice as high as for whites.)

Judging from the fossils, Neanderthal man, once he reached adulthood, might be expected to live to be about 35 years old. A man reaching adulthood in fourteenth century England had about the same life expectancy. In the United States the average life expectancy at birth, which has increased by more than 20 years since 1900, now has reached about 67 years for men and about 75 for women.

It would seem that these accomplishments, of which modern civilization is justly proud, have also resulted directly in what is considered by some observers to be the most pressing problem of our time—the explosive growth of the human population.

THE POPULATION EXPLOSION

In 1974, there were approximately 4 billion people on our planet. This is an almost incomprehensibly large figure. Imagine, for instance, that all the people in the world stood in one long single line. The line would reach 1¼ million miles. It could stretch to the moon and back twice or could circle the Earth more than 50 times at the equator.

Moreover, the rate of increase of this enormous population is unprecedented. The human family, according to the Population Reference Bureau, is presently gaining an average of 1,484,615 members a week, 212,088 a day, 8,837 an

11   *Growth of the human population. If reproduction rates in all countries could be brought down to replacement levels (about 2 children per family) by 1980, the population would still grow to 6.3 billion. If the replacement rate is reached by the year 2000 (which is still not very likely), the world population will be 8.2 billion by the year 2050.*

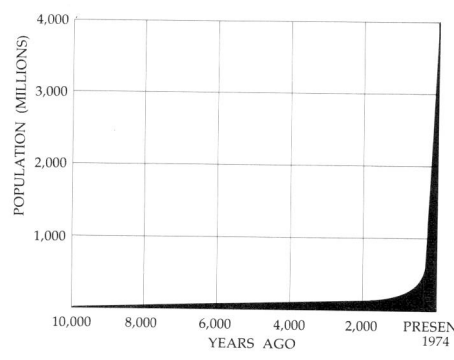

50–12   *Birthrates per 1,000 population for various geographic regions for 1972. The populations of the developed countries appear to be on the way to long-term stability, but they represent only about one-fourth of the world population. Thus, most of the growth of the human population expected in the next 25 years will be in the less developed countries.*

BIRTHS/1,000 POPULATION (1972)

- 0–20
- 20–30
- 30–40
- 40 AND OVER

hour, and 147 a minute. If it continues growing at its present rate of 2 percent a year, the world population will reach 5 billion in 1986 and 6 billion in 1995. Around 1650, when there were some 500 million people in the world, the doubling time of the population was approximately 1,000 years. By 1850, however, the population had already doubled, indicating a doubling time of 200 years. And by 1930, only 80 years later, there were 2 billion people. Now, 45 years later, it has doubled again. The present doubling time of the human population is estimated at approximately 35 years. It took all of man's history—perhaps a million years—for the human population to reach its first billion, which occurred shortly before 1850. At the present rate, by about 2000, well within your lifetimes, the population of our planet will be fast approaching 7 billion people.

Population is not increasing at a uniform rate in all parts of the world. In many of the countries of Latin America, Asia, and Africa—the so-called "less developed" countries—doubling times generally range from about 20 to 35 years. According to figures released by the Population Reference Bureau, some examples are: India and Turkey, 28; Nigeria and Guatemala, 27; Saudi Arabia and Brazil, 25; El Salvador, 22; and Pakistan, Mexico, and the Philippines, 21. For the "developed" countries, doubling times are from 50 to 250 years. For example, the population of the United States will double in 87 years, based on the 1973 growth rates; that of the United Kingdom will double in about 231 years; that of Italy in 99 years; France in 117 years; the Soviet Union in 70 years; and Japan in 58 years.

The growth of a population depends not only on its birthrate (the number of live births per 1,000 individuals) but also on the age distribution of its population (see Figure 50–13). Some authorities prefer to refer to fertility rates (the number of live births per 1,000 women of child-bearing age).

The unequal distribution of growth makes the problem more difficult. First, births are occurring at the greatest rates in precisely those areas where the new

50–13 *Charting populations by age and sex permits predictions about future growth rates. In India, for example, nearly 45 percent of the population is under fifteen years of age. Even if these young men and women limit their family size enough to produce only enough children to replace themselves (which means cutting the current birthrate in half), population growth will not level off until about the year 2040—and at a level of well over a billion. The population of Sweden, by contrast, will remain the same unless the birthrate is dramatically increased.*

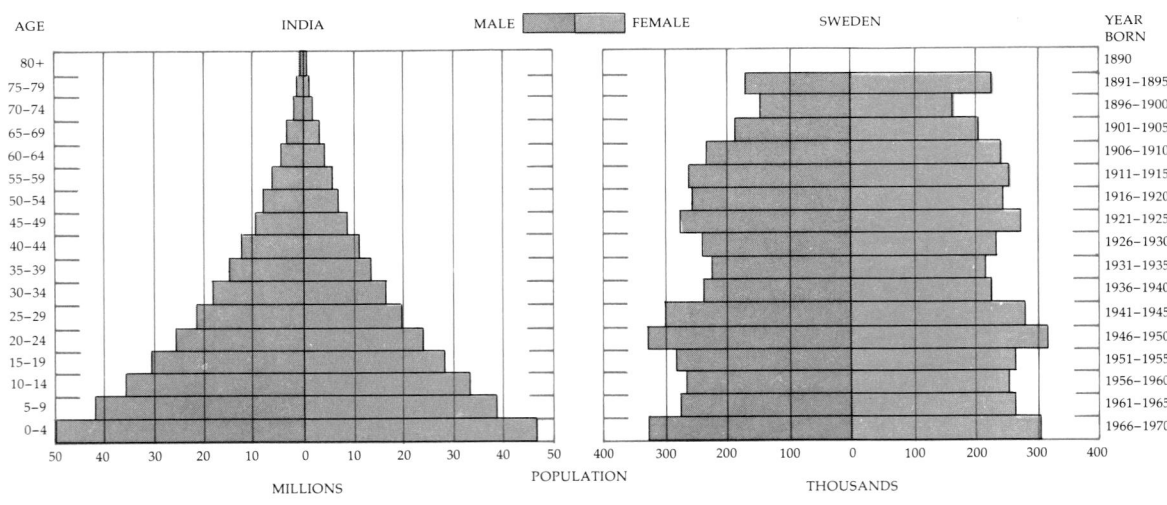

arrivals have the least chance of an adequate diet, good housing, schools, medical care, or future occupations. Moreover, because the affluent citizens of developed countries are not constantly reminded of the soaring population rate by the problems of hunger and crowding, so evident to the poor of other countries, they may feel that it is not their problem. Yet a child born to a middle-class American will consume, in his lifetime, a far greater amount of the limited resources of the world—more than twice the amount of food, for instance—than a child born in a less developed country. In the words of Jean Mayer, chairman of the National Council on Hunger and Malnutrition in the United States:

Rich people occupy much more space, consume more of each natural resource, disturb the ecology more, and create more land, air, water, chemical, thermal and radioactive pollution than poor people. So it can be argued that from many viewpoints it is even more urgent to control the numbers of the rich than it is to control the numbers of the poor.

POPULATION CONTROL

At some point, population growth must stop. It is simply a question of what stops it. One possibility is famine. Although there are reasons to believe that food production can be increased—a possibility that we shall discuss later in this chapter—it is clear that food production can never be increased as rapidly as the population is increasing at the present time. Famine could kill enough people, presumably, to keep the population in check. It could also, as any number of recent studies stress, leave indelible scars on any infants that survive in terms not only of physical deprivation but also of lowered intelligence and mental capabilities. Such a generation would be hard-pressed to find leaders to guide it out of the ruin that a great famine would leave behind.

Another possibility, of course, is war. There is already a sharp division between haves and have-nots. Among middle-aged citizens of the United States, one of the most pressing medical problems is obesity; at the same time half of the children of the world go to bed hungry. Yet until very recently the United States government was renting 20 million acres from our farmers so that they would not grow food on them, and a number of other countries, including Australia, Canada, and Argentina, have been involved in similar efforts to restrict production—at a time of worldwide hunger. The have nations, moreover, with their industrial economies and high consumer demands, are rapidly depleting the world's natural resources—minerals, timber, fuels—at the expense of the have-nots. All known reserves of natural gas, for example, will be entirely used up by 1985 at the present rate of consumption. The United States, with 6 percent of the world's population, accounts for almost 35 percent of the world's yearly consumption of fossil fuels.

As competition for these resources becomes more bitter and as the problems of hunger and crowding and unemployment become more pressing, war will become increasingly difficult to avoid. In a large-scale war, it is difficult to see how the use of nuclear weapons could be prevented. A nuclear war would, of course, solve the human population problem—perhaps permanently.

14 *"Excuse me, sir. I am prepared to make you a rather attractive offer for your square."*

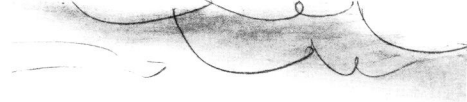

Birth-control Programs

A third possibility is a drastic reduction in the birthrate, which would limit the population to somewhere near its present size, that is, zero population growth. This solution also has its problems, but the problems, although difficult, are slight compared with those previously considered.

One of the problems is that an ideal method of birth control is not yet available. "The pill" (see page 536), which is the most reliable of all the methods except sterilization, carries with it certain hazards, although these hazards seem to be less than those associated with the pregnancies that it prevents. Also, "the pill" is relatively expensive and must be taken on schedule, which interjects problems of human carelessness. By far the most effective means of birth control and the safest, biologically speaking, is vasectomy of the male, a simple operation that prevents the sperm from reaching the seminal fluid. (See page 523.) Vasectomy is growing in popularity but is not yet widely accepted. There are three reasons apparently. One is a lack of information about the procedure; many men, according to a recent poll, believe, quite erroneously, that vasectomy interferes in some way with sexual performance or gratification. Second, even among men who do not plan to father additional children, there are some individuals to whom the idea of being sterile poses a psychological threat. Third—and this is the most realistic objection—there is at present no guarantee that a vasectomy can be reversed (although it often can be) and fertility restored. Therefore, many men wish to reserve the possibility of producing additional children in case they are divorced or their present children die. (Storage of frozen sperm in sperm banks has been suggested as a solution to this third objection.) Undoubtedly, better methods of birth control will eventually be found, but there may not be time to wait.

However, even if an ideal method of birth control were developed and made available on a worldwide basis, we have no reason to believe that any significant

50–15 *During 1974, the population of India grew by 13 million people, the grain harvest was seriously reduced by drought and lack of fertilizer, the country tested its first nuclear bomb, and expenditures for family planning were curtailed.*

reduction in population growth would immediately follow. Among many peoples of the world, including many Americans, opposition to birth control is deeply rooted in tradition, religious belief, or social custom. The domestic role assigned to women, with its consequent ideal of the "good mother," has contributed to this resistance since large families offer women status, an occupation, and a means of demonstrating success. And many families simply do not wish to limit the number of their children and resent the regulation of what seems to them to be a personal affair. There has been a tendency among middle-class Americans, and among the middle class of other Western nations, to regard family planning as primarily a way of controlling the numbers of the poor. In the words of a black delegate to a recent meeting on population control, "You've got Planned Parenthood ladies calling on us twice a day, but nobody is knocking on doors in suburbia and the rural areas."

Most importantly, there seems to be a correlation between economic deprivation and high birthrates. As we have seen, it is in the less developed countries—the countries that can least afford it—that the greatest rates of population growth are found today. In India, for example—a country in which large numbers of people are now starving—the annual rate of population increase is 2½ percent. According to the 1971 census, its population was 548 million. It could reach a billion between 1990 and 1995. Yet, according to statistics gathered in one of the family-planning programs now under way in India, most Indian women do not seek help in birth control until they have three or four children; about one-third of all sterilized Indian parents, in fact, have six living children. The desire for large families is deeply rooted in the Indian culture. Moreover, children are an economic advantage. They earn their cost before they leave home, and in Indian tradition, they provide security for their parents in old age. Indians know that there is strong probability that at least half of one's children will die before maturity and that there is a likelihood that all of them will. Two recent studies of life expectancy in India show that, with the high death rate among children, it is necessary for a mother to bear five children if the couple is to be 95 percent certain that one son will survive the father's sixty-fifth birthday.

Birthrates, Death Rates, and Social Security

Despite the fact that the most recent phase of the population increase is related to a reduction in death rates, particularly from infectious diseases, some experts believe that further reductions in death rates and a general increase in the standard of living will reduce the rate of population growth.

Barry Commoner of Washington University in St. Louis, who has long been in the front ranks of our environmental protectors, is of the opinion that population pressures can be significantly reduced by reducing death rates, particularly among infants. Wherever income rises and death rates drop below a critical point, birthrates begin to fall too, he points out. This was true historically in industrialized nations, and it is becoming true in some countries where living standards are just beginning to rise.

China offers an example of a country with a falling birthrate. The birthrate in rural Chinese compounds has now been reduced to 16 per 1,000 (about what it is in the United States). On being interviewed, both men and women stated

50–16 *Women in the United States are now having an average of 1.9 children each, slightly below the replacement level. However, because of the large number of young people in the country, there are still more births than deaths in a given year, and the birthrate must be held to the replacement level or lower for at least another generation before a stable population can be reached. About 50 million Americans will be added to the population in that time. The United States has only about one-third the population of India, but in ten days, the 200 million people of this country consume as much energy as India's 600 million consume in a year.*

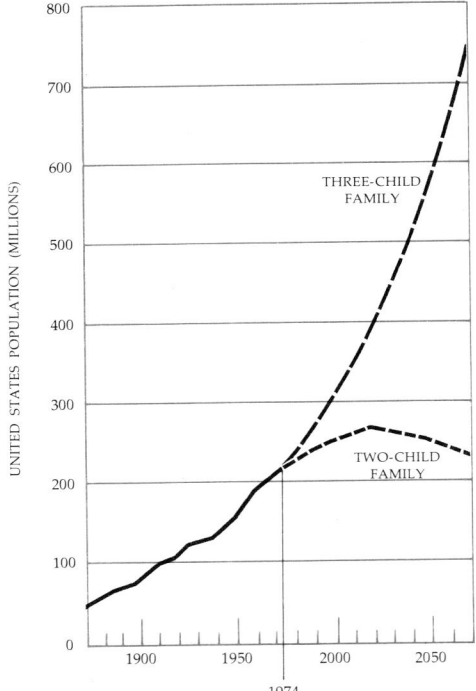

that they are voluntarily reducing the size of their families because their old-age security, which children used to provide, is now guaranteed and also because the women are actively employed outside the home.

According to two Western gynecologists who recently visited China, reduction of the birthrate is achieved, in part, by providing free contraceptives, abortions, and sterilization as part of the general health service program. A more important factor is late marriage combined with premarital chastity. Young people interviewed by the visitors said that premarital sex was illegal but they did not know the punishment for it because "it never happened." Some experts are skeptical of the reports from China, pointing to the lack of demographic data. However, the government has recently reported a total population of 695 million, which tends to confirm the success of the birth control program.

Japan, too, has a falling birthrate. In a period of about ten years, Japan cut its birthrate almost in half—from 30.2 per 1,000 in 1947 to 17.6 per 1,000 in 1959, and it is now being maintained at about this level. No new methods of birth control were introduced (abortion has been a major means of control), and there has been very little expenditure of government funds for public education or propaganda. However, the country is compact and highly literate.

A similar voluntary movement seems to be under way among young people in this nation. The U.S. birthrate was 23.7 per 1,000 in 1960, 17.3 in 1971, 15.8 in 1972, and 14.9 in 1973. The reason appears to be primarily that more young women are not marrying, are marrying later, or are marrying but nevertheless delaying having children. It is not yet clear whether this change in population growth is temporary or represents a major change in the life styles of young Americans. The 1973 ruling of the United States Supreme Court concerning the right of women to have abortions is an important reflection of changes in national sentiment and also a major development in population control.

Can we afford to become complacent? Leaders in the ecology movement such as Paul Ehrlich view the population explosion as the root cause of many, if not most, of our environmental problems. If educational programs, tax reforms (such as no longer granting deductions for unlimited numbers of children), and free medical treatment for prevention and termination of pregnancy do not begin to hold the population at its present level, Ehrlich and others believe that it may be necessary to pass laws prohibiting couples from having more than two children.

OUR HUNGRY PLANET

Of the world's 4 billion people, at least 2 billion are inadequately nourished. Malnutrition takes two forms. There may be a shortage of total calories, in which case the individual cannot meet his energy requirements and must draw on the reserve supplies of his own tissues until these, too, are exhausted. The second form of malnutrition, which may exist where the first type does not, involves a shortage of essential food elements, usually proteins and vitamins. Proteins are required for building and replenishing tissues and so are particularly important for children, who, unfortunately, are those most likely to be deprived of protein while receiving the dangerous comfort of enough calories. Examples include the "sugar babies" of the Caribbean, who are seldom

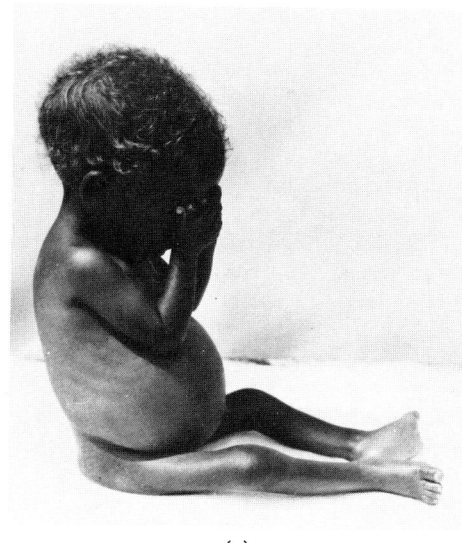
(a)

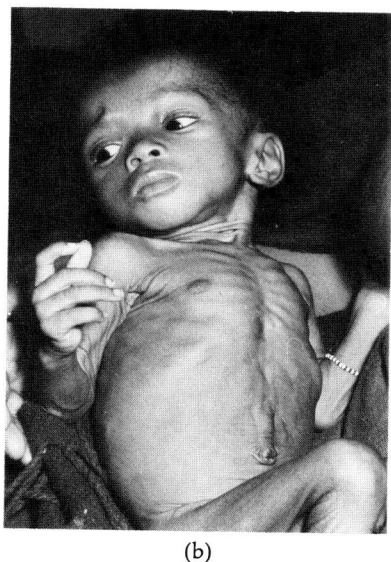
(b)

−17 (a) *A child with kwashiorkor, a West African word that means "the sickness a child develops when another child is born." The skin discolorations, swollen extremities, apathy and weakness are typical. The syndrome, which is the result of a deficiency in protein but not in total calories, usually develops after a child is weaned. Children with kwashiorkor can be found in the United States as well as in less developed countries. The swelling is caused by a deficit in blood proteins that results in osmotic movement of fluids out of the blood into the body tissues. (b) Child dying of starvation, caused by deficiency in total calories. Of the children being born today in the less developed countries, about 75 million will die from malnutrition or an associated disease.*

seen without their piece of sugar cane, and the children of our own cities whose diet is made up largely of candy bars and soft drinks—"empty calories," in the words of the nutritionists.

Protein shortages are more difficult to remedy than shortages in overall calories. Meat is a major source of protein in diets of the developed countries, as it seems to have been since man first became a hunter, but meat is now a luxury available in adequate amounts only to the relatively affluent. The reason for this is easily understood in terms of the losses at every trophic level which we discussed in Chapter 41. For example, 100,000 pounds of corn can produce, as a rough simplification, 10,000 pounds of beef, which, in turn, produces 1,000 pounds of humanity. The same 100,000 pounds of corn eaten as corn bread, tortillas, couscous, or polenta will provide 10,000 pounds of humanity.

Protein and Brain Development

The effects of protein deprivation are particularly serious in the young. As we saw in Chapter 31, the brain is proportionately extremely large in the embryo, and its growth continues to be very rapid during the period after birth. During the first three years of a child's life, his body grows to only 20 percent of its adult size, but the brain during this same period reaches 80 percent of its adult size. And, apparently, if it does not grow during this period, it cannot catch up later. More than half of the dry weight of the brain is protein. Studies of rats show that in infant rats who have been deprived of adequate protein, the brain size is smaller and so, often, is the cranium, the portion of the skull that holds the brain. There are indications that these same effects can be seen in undernourished children, and it is clear that intellectual function is impaired. For instance, in a study of slum children in Santiago, Chile, only 51 percent of malnourished children reached the normal range of mental development, as compared with 95 percent of children from the same background who received a supplemented diet.

The Reasons for Hunger

Famines have occurred in the past, but they have characteristically been short-term, localized events, usually the direct effects of particular natural or man-made disasters. This is the first time in human history that a large percentage of the population faces the prospect of never having enough to eat.

There are several reasons for the chronic malnutrition of our planet. The first is the population growth, previously discussed. However, although it is clear that food supply cannot keep up with the rapidly increasing demand—as we noted previously—the reasons for the present shortage of food are more complex and less obvious.

For one thing, unlike his hunter-gatherer predecessors, modern man no longer shares food. There is a large discrepancy in food consumption between the rich and the poor. It is comfortable to think that the world food shortage is a result of "their" population explosion so "they" must limit their population; however, the people of the countries that are the most crowded in population—the Netherlands, Great Britain, and Japan—enjoy a fairly high standard of living. It is the nations of Central and South America, with relatively low population densities, where the problems of malnutrition are most acute. A principal reason for the food shortages in these countries is that much of their food is exported to the rich nations. For example, the waters off the western coast of South America, as we noted in Chapter 40, abound with fish. In 1968, the United States imported 700,000 tons of this fish; it was used as feed for North American livestock. This amount would have provided enough protein for the entire population of Peru.

A third reason for the worldwide food shortage is one we have touched on previously: a significant number of people eat high on the food chain. In the

0–18 *Feedlots, where cattle are fattened for market by being fed grain. One of the principal problems in managing such an operation is waste disposal. Each steer excretes 64 pounds of manure per day, a total of 230 million pounds per year for a feedlot with 10,000 head of cattle.*

0–19 *The protein conversion ratio is simply the number of pounds of protein fed to livestock to produce 1 pound of protein for human consumption. Beef, as you can see, is by far the most wasteful protein source. (Actually cattle do not require cereal grains or other proteins; all the proteins they need can be provided by the cellulose-digesting bacteria in their rumen. High-protein food is given cattle to fatten or "finish" them before marketing; marbling the meat with fat improves its taste, although not its nutritional value, and enables cattle farmers to command a higher price for the livestock.)*

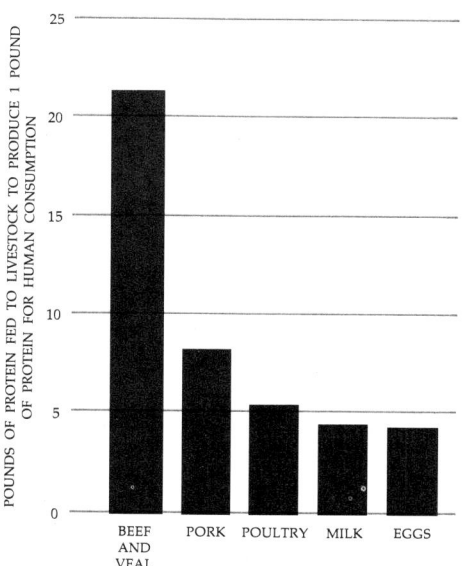

United States, for example, one-half of the harvested agricultural land is planted with feed crops. Livestock in this country are being fed about 20 million tons of protein per year, primarily from sources that could be used for human consumption: 89 percent of our corn crop, 87 percent of our oat crop, etc. Of this amount, only one-tenth—about 2 million tons—is retrieved for human consumption. The amount "lost"—18 million tons—is equivalent to 90 percent of the yearly world protein deficit—enough to provide 12 grams of protein a day for every person in the world.*

The Green Revolution

The most successful attempt to date to expand food production has involved the development by selective breeding of new strains of grain plants that produce higher yields. Under the sponsorship of the Rockefeller Foundation, several new strains of wheat have been developed in Mexico which, when heavily fertilized, can produce as much as twice the normal yield per acre. Moreover, these new strains have much less exacting photoperiod requirements than the older ones, so that two or even three crops can be grown in a single year. Using these new strains, Mexico has increased its wheat production sixfold in the last 20 years. The new strains were introduced into India more recently. Total Indian grain production went from 68.5 million metric tons in 1960–1961 to 92.8 million metric tons in 1970–1971. In the meantime, India's population increased from 440.3 million to 570 million—thus, the per capita grain production remained about constant. In the mid-1970s, as a result of crop failure coupled with the steady population growth, per capita production once more dropped to starvation levels.

* These statistics are from Frances Moore Lappé, *Diet for a Small Planet*, Ballantine Books, Inc., New York, 1971. According to her calculations, an adult of average size needs about 43 grams of protein a day.

50–20 (a) *Norman Borlaug, who was awarded the Nobel Peace Prize in 1970. Borlaug was the leader of a research project sponsored by The Rockefeller Foundation under which new strains of wheat were developed in Mexico. This new wheat, a cross between a native wheat and a Japanese dwarf wheat, has shorter, stiffer stalks, which are much less likely to topple even when heavily laden with grain.* (b) *Field workers weeding a rice plot at the International Rice Research Institute in the Philippines. The most recent strains are highly disease-resistant; however, they will grow only on irrigated land, which forms only about 30 percent of the cropland in Asia.*

(a)

(b)

A number of experts view the green revolution not only with skepticism but with alarm. One problem is that the genetic uniformity of these new strains makes them vulnerable to wholesale destruction on a scale never before possible by the "right" strain of pest or pathogen. A second is that they often must be cultivated in large, pure stands to prevent cross-pollination with older varieties. A third problem is that these new varieties are more productive only if grown in conjunction with optimum levels of irrigation water, chemical fertilizers, and pesticides.

Since 70 to 90 percent of Asian farm families have no irrigation water and no cash or credit for the purchase of chemical products, these new agricultural techniques tend to drive out the farmers with small land holdings. And indeed, the worldwide shortage of fertilizer (linked to the petroleum shortage) and the subsequent increase in price may make cultivation of these special varieties impossible in any less developed countries. (In the United States, we use as much fertilizer per year for nonagricultural purposes—such as golf courses, lawns, and ornamental gardens—as India can afford for all its needs.)

New Food Sources

A number of other new methods of increasing food production are being studied. Many of these seem promising, although none has yet proved itself on a mass basis. One possibility is to grow algae in large quantities as a source of animal fodder. Another is to grow microorganisms—probably bacteria or yeasts—on simple hydrocarbons, such as butane or methane, plus inorganic minerals as a source of food for domestic animals or humans. These organisms not only are good energy sources but are high in the essential amino acids. The problems are largely technological; methods are needed to grow the microorganisms on a large scale, to provide them with oxygen and prevent contamination, to harvest them, and to process them in some form acceptable for consumption. Present national interest in prolonged space flight may provide the impetus

Table 50–1 *Comparison of Average Yields of Hybrid and Nonhybrid Grains (In bushels per acre)*

	Hybrid grains		Nonhybrid grains	
	Sorghum	Corn	Wheat†	Soybeans
1930–34	12.5	22.6	13.4	14.2
1935–39	12.6	25.9*	13.0	18.1
1940–44	17.2	32.2*	17.1	18.1
1945–49	17.7	36.0	16.9	19.5
1950–54	19.4	39.4	17.3	20.3
1955–59	28.3*	48.7	22.2	22.6
1960–64	42.6*	62.5	25.2	24.0
1965–69	53.2	77.4	27.7	25.7
1970–73	55.9	87.3	32.4	27.5
Increase, 1970–73 compared with 1930–34	347%	286%	142%	94%

*Hybrids introduced on wide scale.
†Wheat hybrids now being tested will be introduced on a wide scale within the next five years.

20–21 *Results of irrigation. Beet fields in the Negev desert in Israel are irrigated principally with water delivered via pipeline from the freshwater Sea of Galilee.*

for the solution of some of these technical problems and also may make a diet of algae and bacteria seem more glamorous.

A particularly promising approach to increasing the world's harvests, and one which also needs a technological solution, is the desalination of water for the irrigation of the desert. About one-third of the land area of the world is made up of desert, and a surprisingly large proportion of this desert borders the ocean. Experiments with freshwater irrigation have proved that the desert, with its warm constant daytime temperatures and abundant sunlight with high light intensity, can be extremely productive in agriculture. Desalination of water is technically feasible but at present is very expensive in relation to the amount of water needed. It has been estimated that 200 gallons of water a day are required to grow the 2,500 calories needed by one person for one day. (Remember that more than 98 percent of the water taken up by a plant—corn, for instance—is transpired.) The solution may lie in the development of facilities using nuclear energy combined with means to reduce water requirements. The latter might include, for example, the selection of plants with a tolerance for sodium, so that slightly salty water could be used, and the construction of asphalt floors under the desert floors or plastic greenhouses over them to reduce water loss.

Plankton, which has been often mentioned as a potential food source, no longer seems so promising in this regard. In the open ocean, where the greatest total quantity of plankton is found, light is adequate only very near the surface of the water, and phosphorus and other minerals are scarce. As a consequence, the growth is very thin, which seems to make harvesting a technical impossibility. Also, removing large quantities of plankton from the ocean would further threaten the already waning supplies of fish, which depend directly or indirectly on this first trophic level.

A possibility which may or may not be practical but which is interesting to think about involves the importing of animals from Africa to fill unoccupied ecological niches in North America. According to this line of reasoning, these ecological niches have been vacant since the mysterious but apparently man-linked extinction of a number of large animals some 10,000 years ago. Some of these animals, such as one species of giant sloth and the American camel, were browsers that lived on tough leaves of plants of the chaparral, such as creosote bushes, yucca, and agave. Certain animals of the African savanna—the eland,

20–22 *Starving nomads and their emaciated cattle in the Upper Volta. Even with the return of normal rainfall in 1974, after six years of drought there is little chance that many of these people can replenish their herds and resume their ancient way of life.*

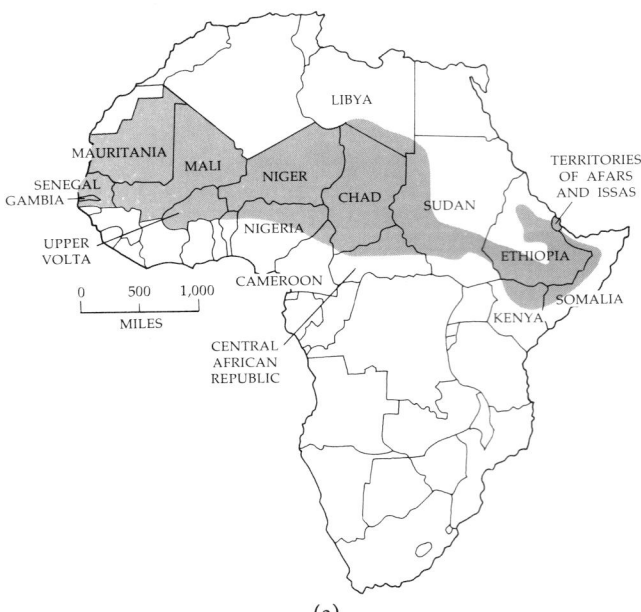

(a)

(b)

50-23 (a) *The problem of increasing food production and distribution has been exacerbated by a drought belt 4,000 miles long and 1,000 miles wide stretching across northern Africa. Some experts have attributed the six-year drought to a basic shift in weather patterns, perhaps caused by a gradual drop in the temperature of the Earth over the past few decades. Others assign equal blame to local political and economic changes which led to overgrazing in these regions and helped to destroy the already fragile ecosystem. (b) A starving Ethiopian child. Ethiopia is at the eastern end of the drought belt and did not receive the rains in 1974 that relieved the drought in other regions.*

for instance, which is domesticated easily—are able to subsist on these plants and might be used to convert them to protein for human consumption. There are 1 million square miles of such land in 11 western states and Mexico.

How much further we can increase grain yields by the introduction of new strains and whether or not any of the other means now being considered will be successful are simply not known at this time. The effect of all our present efforts may simply be to buy time—time in which, if the world is very lucky, a less temporary solution can be found.

POLLUTION

Another of the major threats posed by the cultural evolution of man is pollution. In one sense, the pollution problem is as old as life itself. All living creatures produce metabolic wastes, which are discarded into the air, the water, or the land. Oxygen itself was probably originally a toxic pollutant, the by-product or waste from early photosynthesizing organisms. When man was a wanderer, he simply moved to another campsite, and his excrement and debris were taken up by the land, broken down, and recycled through living systems. When people began to pack together in cities, waste disposal became a major problem. The solution of this problem, largely as a result of advances in the engineering of sanitation facilities—such as sewers—and of public health laws, brought about tremendous improvements in the livability of urban centers and striking decreases in mortality rates from infectious diseases. As a result of the efficient and largely silent and invisible systems, coupled with our general affluence, we have learned to throw things away in increasing numbers. The problem now is to find an "away" that can accommodate the growing mass of material to be discarded. "Away" no longer usually means an alley or the heads of our neighbors, as it did in medieval times, but it often means the river running past the next town or the offshore waters of our oceans.

(b)

(c)

(d)

(a)

—24 *Some major causes of air pollution:* (a)
motor vehicles , (b) *burning refuse,*
(c) *a steel plant, and* (d) *a power plant .*

50–25 *Abandoned cars under the Brooklyn Bridge in New York City.*

50–26 *Solid waste now exceeds 5 pounds per capita each day in the United States.*

The pollution problem imposed by our increasing numbers is made far more difficult by our advancing technology. To replace containers made of clay, wood, or skin, we have perfected the aluminum beer can, the disposable bottle, and the plastic bag, none of which decompose readily. In addition to our personal wastes, the industries that provide our necessities and luxuries spew their waste products into the air and waters at an ever increasing rate.

Although the composition of polluted air varies from city to city, it is usually composed of carbon monoxide, sulfur oxides, nitrogen oxides, hydrocarbons, and lead and other particulate substances. About 60 percent of the total pollution comes from internal combustion engines.

The pollution problem is the most soluble of our environmental crises—which does not necessarily mean that it will be quickly resolved. It must be made profitable to recycle inorganic materials, in the same way that it is profitable for the ecosystem to reuse its natural materials. Such recycling not only will prevent the accumulation of wastes but will retard the squandering of our resources—such as the thousands of acres of timberland turned into newsprint and other paper products each year and nonrenewable resources, such as metals, most of which are buried in our dumps or out at sea. Also, it must be made profitable for industry to control its wastes and the wastes of its finished products, such as the exhausts of automobiles. Or, to put it more harshly, it is necessary to make it unprofitable not to. These problems are largely matters of legislation; however, unfortunately, much of this legislation will be unpopular not only with producers but with many consumers.

Water pollution. (a) One of the victims of the Santa Barbara oil slick. The rise in the cost of fossil fuels is creating increasing pressure for extensions of off-shore drilling. (b) The Mississippi River (left) carrying waste and sewage sludge merges with the St. Croix River near Prescott, Wisconsin. (c) Estuaries and marshes, where fresh and salt water meet, not only are rich in crustaceans and shellfish but are the spawning grounds and shelter the young of many of the important food fish of the oceans. This polluted estuary is in Wilmington, Delaware. (d) Eutrophication—meaning, ironically, "good nourishment" —is a result of the pouring into lakes and ponds of excess phosphates and nitrates in the form of fertilizers, detergents, and sewage. The result is an explosive overgrowth of algae, which then die and are decomposed by bacteria whose respiratory activities consume so much oxygen that the fish and other aquatic animals can no longer survive.

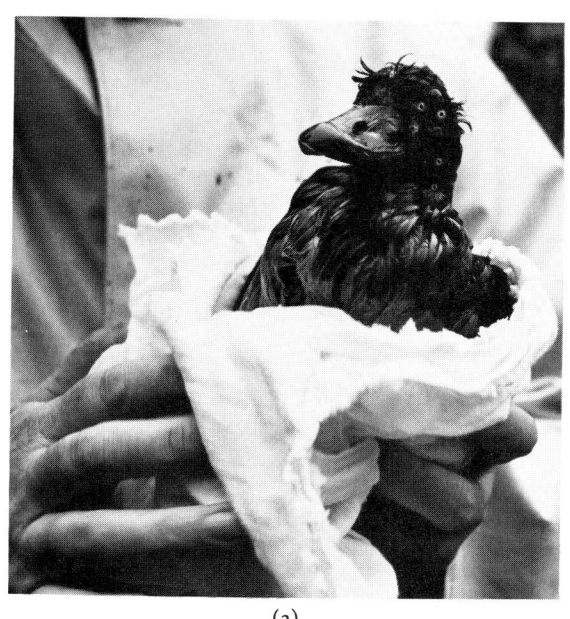

(a)

(b)

(c)

(d)

Insecticides

In addition to the pollutants that are by-products of our daily life and industry and that accidentally poison the living things in our environment, we have systematic programs for the development and use of materials with which we deliberately poison particular living organisms in our attempts to control our ecosystem. Chief among these poisons are insecticides.

DDT, the most widely used of the insecticides, was first introduced on a large scale during World War II and proved to be a spectacular success at curing body lice and, therefore, typhus. It soon became a major worldwide weapon against insect-borne diseases. Look again at the decline in the death rate in Ceylon (Figure 50–10). In 1945, it was 22 per 1,000. DDT was introduced in 1946, with the consequent rapid control of the malaria-carrying mosquitoes. The death rate dropped 34 percent between 1946 and 1947 and was down to 10 per 1,000 by 1954. By 1969, it was 8 per 1,000. Many of these lives were saved by DDT.

DDT is an efficient insect killer. In the nervous systems of both vertebrates and invertebrates, transmissions across synaptic junctions occur as the result of the discharge of certain chemicals, chief among which is acetylcholine. Acetylcholine, which acts swiftly to stimulate the adjacent neuron, is rapidly destroyed by an enzyme, cholinesterase. This enzymatic action is indispensable to the smooth functioning of the nervous system. DDT destroys cholinesterase in insects, causing their nervous system to run wild, and so kills them; the amount required is extremely small, far too little to have any immediate effect on other animals.

The problem with DDT is that it persists and it accumulates. Prior to World War II, most insecticides were organic compounds based on natural substances that have coevolved with the insects and that protect the plants from them. Nicotine compounds are an example. Such substances break down relatively quickly. DDT and other synthetic insecticides retain their activity far longer, however. And once farmers and gardeners came to recognize the increase in yields of crop plants, they began to apply it routinely, on fixed schedules, and DDT began to accumulate in the soil and water. A Long Island marsh that had been sprayed for 20 years to control mosquitoes was recently found, for example, to contain up to 32 pounds of DDT per acre. It is now estimated that more than a billion pounds of DDT is circulating in the biosphere at the present time; even if DDT and related insecticides were never to be used again—from this second onward—they would be with us for years to come.

Insecticides such as DDT, because they persist, accumulate along food chains. In 1949, 1954, and 1957, Clear Lake, California, was sprayed to control midges, a small, harmless insect that was an annoyance to vacationers. The pesticide used was DDD, a close chemical relation of DDT, less toxic but equally persistent. The amount present in the lake after each spraying was about 0.02 part per million, and after 2 weeks, in all three instances, no DDD was detectable in the water at all. Before 1950, Clear Lake had been a nesting ground for about 1,000 pairs of western grebes (ducklike birds that eat small fish and other aquatic organisms). From 1950 to 1961, no young were produced at all; in 1962, five years after the last spraying, one hatched. It was not until 1969 that they once more began to reproduce successfully. Studies undertaken to determine

50–28 *Concentration of DDT residues being passed along a simple food chain. As the biomass, or living material, is transferred from one link to the next in the chain, about half of it is usually consumed in respiration or is excreted; the remainder forms new biomass. The losses of DDT residue, however, are small in proportion to the respiration and excretion losses. Consequently, the concentration of DDT increases as the material passes along the chain, and high concentrations occur in the carnivores.*

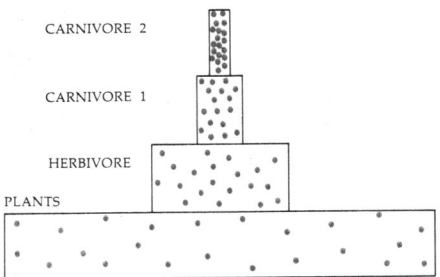

60–29 (a) *The female of the species* Anopheles *ingests blood (the male lives on nectar). If she draws blood from a person with malaria, she will probably pick up cells of* Plasmodium, *a parasitic protist that causes malaria. Worldwide use of DDT has virtually wiped out this species in many areas of human habitation and so eliminated malaria, a major cause of death in many tropical and subtropical countries. (b) An eaglet and an egg which will never hatch, photographed in a nest near the Muskegon River in Michigan. DDT causes a bird's liver to break down the hormones that mobilize calcium at the time of egg production, resulting in thin-shelled, fragile eggs, like the one shown here. Birds at the top of food chains, such as the osprey, peregrine falcon, and bald eagle, are principal victims. The effects of newer insecticides now being used in place of DDT have not yet been determined.*

the concentration of DDD along the various trophic levels revealed that the concentration in the plankton was about 250 times as much as that originally present in the water. The concentration in frogs was 2,000 times the original concentration; in sunfish, about 12,000 times; and in grebes, as high as 80,000 times. The reason that no DDD could be detected in the water immediately after spraying became all too clear: it had been taken up completely by the living organisms of the lake. In the meantime, by 1957, the midge and about 150 other species of insects had developed some immunity to the pesticide.

In birds, it is now known, DDT, DDD, and related pesticides collect in the liver and induce the production of enzymes that break down various hormones and so interfere with reproduction. A number of species of predatory birds, including the bald eagle, the peregrine falcon, and the brown pelican, are now under threat of extinction because of pesticides.

The effect of DDT on other animals is not yet known. Virtually all populations of vertebrate animals around the world now have appreciable amounts of DDT in their tissues. Even animals in the Arctic, where DDT has almost never been used, show traces of it, the result of airborne distribution. At present, human tissues contain, on the average, about 11 parts per million of DDT and related chemicals. (Seven parts per million is the statutory maximum for meat transported in interstate commerce.) DDT tends to concentrate in fat. Eventually, we may have accumulated so much DDT in our fat that it will be dangerous to go on a reducing diet that would release the stored DDT into our systems. It also accumulates in milk; milk samples taken from the breasts of American women have been found to contain 0.1 to 0.2 part per million, which

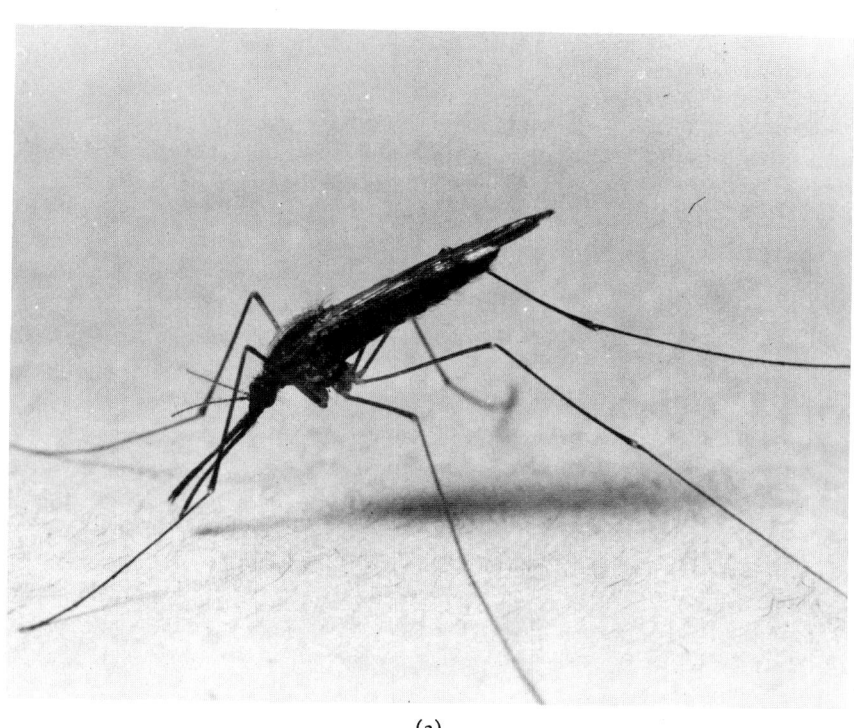

(a)

(b)

(a)

(b)

50–30 (a) *An alternative to DDT and other insecticides is the use of natural predators, such as the ladybug shown here eating its way through a feast of aphids.* (b) *Another alternative is to use synthetic pheromones, which, by simulating the chemical stimuli emanated by female insects at mating time, lure the males to their death. The huge feathery antennae of many male insects, such as the cecropia moth shown here, are capable of detecting these chemicals in very small amounts and at great distances.*

is several times more DDT than the federal government allows in dairy milk meant for human consumption.

Wide-scale routine applications of DDT have now been discontinued in most areas of the United States, and its use has been completely outlawed in some states. New means for controlling insect pests are being sought, though none as effective and inexpensive as DDT has yet been found. There is no agreement as to how much agricultural yields would be reduced if all insecticides were outlawed, as some ecologists recommend. Norman Borlaug warns, "If the use of pesticides in the U.S.A. were to be completely banned, crop losses would probably soar to 50 percent, and food prices would increase fourfold to fivefold. Who then would provide for the food needs of the low-income groups? Certainly not the privileged environmentalists."

Mercury

Mercury, a comparatively rare element, is used in chemical industries, in the manufacture of paints and paper, and in pesticides and fungicides. Inorganic mercury is relatively harmless; because of poor absorption in the digestive tract, a person could swallow a sizable amount without ill effects (absorption of inorganic mercury through the skin or inhalation of vapors, however, does produce toxic effects). On the other hand, organic compounds of mercury, particularly methyl mercury, are extremely toxic even when ingested, producing neurological damage and sometimes causing coma and death.

Some cases of mercury poisoning have been caused by eating grain intended for seed, and therefore treated with mercury-containing fungicides, or eating animals fed such grain. However, the greatest general hazard would appear to be from mercury pollution of waters. Bacteria and fungi in polluted waters convert

0–31 *Tomoko Uemura, a resident of Minamata, Japan, was poisoned by methylmercury prenatally. This photograph was taken when she was 16. She is being cared for by her mother who shows no apparent signs of poisoning because her fetus removed and concentrated the methylmercury from her bloodstream. The chemical plant responsible for polluting the bay has only recently been forced by government decree to assume responsibility for the mercury poisoning. The number of persons affected, estimated at as high as 10,000, is not known.*

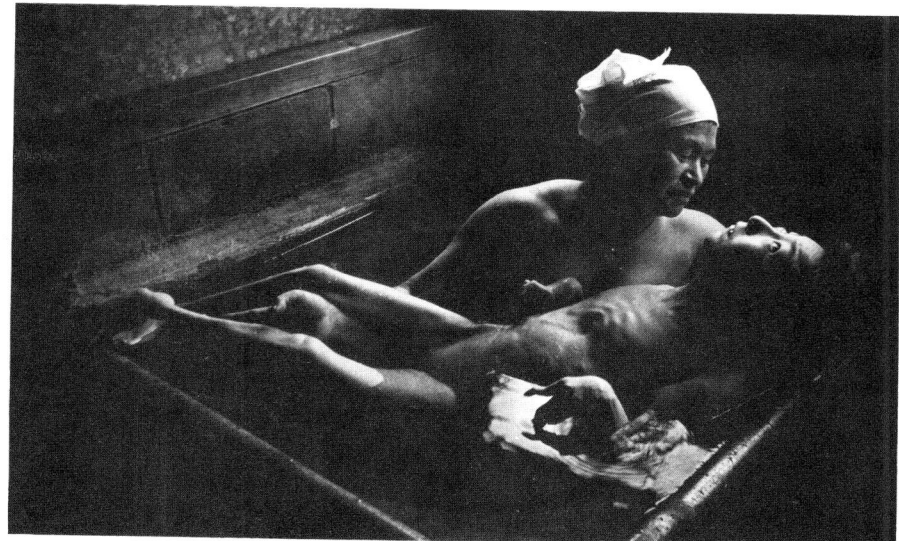

inorganic mercury to methylmercury. Moreover mercury, like DDT, works its way up the food chain.

In the 1950s, a group of fishermen and their families around Minamata Bay, Japan, were poisoned by eating mercury-contaminated fish. The source of the mercury was a nearby chemical plant that discharged waste into the fishing waters. Since that date, analyses of levels of mercury in water and in seafood from a variety of different sites in the United States and abroad have been carried out. Given the fact that there is, and always has been, a small amount of mercury present in the environment and that it is not known what concentration of mercury in human tissues, if any, can be regarded as safe, the present danger is difficult to evaluate. However, there is little controversy concerning the necessity of preventing recurrences of the tragedy of Minamata Bay.

Strontium-90

One more example. Strontium is an element closely related to calcium. Like calcium, it can be taken up by grasses through their foliage as well as through the soil, eaten by dairy cows, concentrated in milk, which is largely consumed by children, and deposited in growing bone tissues and teeth. Strontium-90, a radioactive form of strontium and a by-product of the testing of atomic weapons, follows exactly the same route. As a consequence of atomic tests in the 1950s, by 1959 the bones of children in North America and Europe averaged an estimated 2.6 micromicrocuries of strontium-90 per gram of bone calcium, compared with 0.4 micromicrocurie per gram of calcium in the bones of adults. This amount of radioactivity has not been proved dangerous, but exposure to radioactive elements is known to cause leukemia, bone cancers, and genetic abnormalities and generally to shorten the life span; and the minimum exposure that can produce such effects is not established. The half-life of strontium-90 is 28 years; in other words, it takes 28 years for half of the element to lose its radioactivity, so this exposure still continues.

Some years ago the World Health Organization instituted a program of spraying DDT widely on the tropical island of Borneo in an effort to control the mosquitoes. This program proved very effective in destroying mosquitoes. It also destroyed wasps, however. These wasps preyed on caterpillars, which, in turn, ate the thatched roofs of the peoples' houses. So as a direct consequence of the mosquito-control programs, the roofs collapsed inside the huts.

Meanwhile, a second program of spraying with DDT was initiated to kill houseflies. Up to that point, the houseflies had been controlled by small lizards—geckoes— which crawled over the ceilings and walls of the houses. When the flies were poisoned with DDT, the geckoes, which continued to eat them, also began to die. Falling to the floor of the huts, they were eaten by the house cats, which in turn began to die. Eventually, this led to an enormous increase in the rat population, which not only began to invade the houses and consume the food but also spread plague among the human population. Finally, a program of parachuting cats into remote Bornean villages was instituted in an effort to restore the ecological balance that was so badly disturbed by a seemingly reasonable program of spraying houseflies with DDT.

BORNEO

The channeling of strontium-90 along the calcium food chain might have been anticipated. Other results were less predictable. For example, in the Arctic, only slight radiation exposure from the atomic tests was expected because the amount of fallout that reaches the ground at the poles is much less than it is in the temperate parts of the United States. However, Eskimos in the Arctic were discovered to have amounts of radioactivity in their bodies that were much higher than those found in the inhabitants of temperate regions. The link in the chain was the lichens. Since they have no real roots, lichens absorb their minerals directly from the air, and so they absorbed a large amount of direct fallout, little of which had had time to decay and none of which was dissipated by absorption into the soil. In the winter, caribou live almost exclusively on lichens, and at the top of the food chain, Eskimos live largely on caribou.

The point of all these stories is, of course, that the world has now become, in many ways, a single ecosystem, that we are having profound effects on this ecosystem, and that we do not know what these effects or their consequences may be.

AGRICULTURE AND THE SOIL

In a natural ecosystem, as we have seen, organisms interact with one another and with their physical environment. As a result of these interactions, carbon, nitrogen, phosphates, and other materials are cycled and recycled through the ecosystem. In terrestrial ecosystems, one of the most important elements in this recycling, and certainly the most complex, is the soil. Natural soils contain a variety of minerals, which are taken up by plant and animal bodies and then returned in discarded leaves and branches, animal excreta, and decomposing

The prairie soils were once so bound together by the roots of grasses that they could not be cultivated until special plowing instruments were developed. But once the land was clear and the topsoil became depleted by intensive agriculture or overgrazing, the soil became vulnerable to wind erosion. Here an Oklahoma farmer and his children plod through a dust storm during the drought of the 1930s.

plant and animal bodies. In addition to these nonliving components, the soil contains a vast array of living creatures, a teeming underground world of animal life, of plant roots, of algae, fungi, bacteria, and protozoans. The insects, earthworms, and other invertebrates and the microorganisms that live in the soil decompose the organic material, converting it to a form in which it can be used by plants and also, as a consequence, turning the soil and aerating it.

When the land is turned to agriculture, the subtle relationships of this soil world are destroyed. The minerals are removed from the land when the crop plants are harvested or consumed by cattle and other herbivores. Insecticides and fungicides and herbicides kill the microorganisms of the soil. Plows rip up the roots that hold the soil together. One of the prime examples of the destructive consequences of agriculture is to be found in the tropical forests. The tropical rain forest is by far the richest, in terms of varieties of organisms, and the most luxuriant, in terms of density, of all biomes. Unlike the deciduous forests, which are typically carpeted with leaves and other debris, the floor of the tropical forest is bare; in this warm, damp climate, the pace of life is so swift that everything that falls to the ground that is not immediately devoured is quickly decomposed and reused. In fact, some recent studies of tropical soils indicate that perhaps there is no reservoir of minerals in the inorganic portion of the soil and that the only reservoir is the bodies of the soil microorganisms.

Because of the lush growth characteristics of the tropical forest, one might suppose that the areas of tropical forest would provide rich farmland, but this is not the case. Once the natural community of plants is removed and the soil plowed, the ground may become almost as hard as cement. This red, hard soil has a special name, "laterite"; it is useful for making bricks for building. The lifeless laterite soil will not hold any minerals; they are leached right through

When the rich flora of the tropical jungle (a) is slashed and burned for agriculture (b), the soil often turns hard and infertile. Mycorrhizae (associations between fungi and the roots of higher plants) have recently been found to play a crucial role in mineral cycling. In fact, current research indicates that in tropical soils, the chief reservoir of minerals is not in the soils themselves but in the soil fungi. Minerals are passed directly from the fungi to living root cells. In this way, few minerals leak into the soil, and leaching in undisturbed soils is at a minimum. In tilled tropical soils, on the other hand, the mycorrhizae are destroyed and the soils are rapidly depleted by leaching.

(a)

(b)

50–34 *In strip mining, the top, fertile layer of the soil is ripped off to get at the coal underneath. By 1974, 2.5 million acres of land had been strip-mined, and stripping was proceeding at a rate of 1,000 acres a week. In that year, the government finally passed legislation providing for restoration of stripped land, including land previously mined. Lobbyists for the coal industry contend that the new laws could reduce coal production by as much as one-third. The mine shown here is in southeast Ohio. A coke furnace is in the background.*

it with the tropical rainfall. For centuries, the farmers of the damp tropics have cleared patches of rain forest by slashing and burning, planted crops for two or three seasons until the ground hardened, and then moved on, leaving the wound in the forest to heal itself. Now, however, under the tremendous pressures of the population growth in these countries, more and more of the tropical forest is being destroyed and reduced to laterite. It is estimated that by the year 2000 almost none of the original tropical forest will remain. This is a double tragedy: most modern plant and animal life is believed to have had its beginnings in these forests, and yet we know little about them. Moreover, they are clearly the richest and most interesting of all biomes, and yet they have been little studied by ecologists and, once lost, can never be reconstructed. If they were contributing substantially toward feeding the hungry peoples of the world, one might be more easily reconciled to the loss, but because of our present lack of knowledge, they are not nor is it likely that they will be in the foreseeable future.

In the temperate zones, also, agriculture has left cruel scars on the land. When the roots and other organic materials that bind the soil together are ripped up by the plow, the land becomes much more vulnerable to the effects of the weather. Wind and water erode the soil, and the topsoil, which contains the humus, is lost. Minerals are leached from the soil and washed away, eventually to the sea.

Fertilizers

In order to combat the effects of the constant removal of nutrients from the soil, which occurs with agriculture, increasingly massive amounts of fertilizers must be applied. The modern use of fertilizers has often made possible spectacular increases in crop yield; one of the great advantages of the new dwarf strains of wheat and rice is their capacity to respond to large amounts of fertilizers. However, fertilizers have also produced some disastrous side effects.

One of the chief requirements of plants, as we have seen, is nitrogen, which must enter the plant in the form of an organic compound, almost always as a nitrate (NO_3^-). When nitrates are applied to crops in large amounts, plants may concentrate them in considerably greater quantities than are normal. Nitrates are relatively innocuous for humans, but nitrite (NO_2^-) compounds, always present in small amounts when nitrates are in solution, are not. Particularly in infants, high concentrations of nitrites interfere with the capacity of the red blood cells to carry oxygen, causing serious disease (called methemoglobinemia) and sometimes death. The evidence is mounting that some crops of commercially available spinach have such a high nitrite concentration as to be dangerous to infants if eaten in quantity. In several communities in the heavily farmed San Joaquin Valley of California, nitrites from fertilizers have reached the ground water and so are taken up again in wells for drinking purposes. The level of nitrites in the water is now so high that physicians in these communities routinely advise that infants drink bottled water until they are about six years old. And the use of nitrates has been steadily increasing. In the last 25 years, although the production of food in the United States has increased only in proportion to population growth, the production of nitrogen fertilizers is up 1,050 percent.

THE IMPOVERISHMENT OF THE ECOSYSTEM

How is it that our once sturdy little planet, on which life has persisted and even flourished for more than 3 billion years, has now become so fragile? The answer to this question can be found in a reexamination of the concept of mature and immature ecosystems (page 791). Any ecosystem under the "control" of man—and now man's influence is extending over the entire biosphere—is returned rapidly to an immature stage, a stage at which there are few species and therefore little complexity of interaction. A mature, complex ecosystem, with its complicated food webs, has many built-in checks and balances. Individual members of the plant and animal community may be sick or die, but the ecosystem itself is healthy and species tend to endure and in relatively stable numbers. Under agriculture, plants do not grow in complex communities, as they do in a forest, but in pure stands. A cornfield, for example, has little inherent stability. If not constantly guarded by man, it will be immediately overrun with insects and weeds. It is for this reason that insecticides and herbicides play such a large and, indeed, indispensable role in our modern life. The susceptibility of modern crops to predators and parasites was tragically illustrated by the great potato famine of Ireland, which was caused by a fungus infection. The famine of 1845–1847 was responsible for more than 1 million deaths from starvation and initiated large-scale emigration from Ireland to the United States; within a decade, the population of Ireland dropped from 8 million to 4 million. Virtually the entire Irish potato crop was wiped out in a single week in the summer of 1846. As we noted previously, a number of plant geneticists are warning that the new strains of wheat and rice, which promise to contribute so importantly toward feeding the growing populations, will be particularly susceptible, because of their genetic uniformity and widespread distribution, to such disasters.

-35 *Wheatfield, U.S.A.*

50–36 (a) *Some 360 animal species have become extinct since 1600, only half of them for natural reasons. The monkey-eating eagle shown here is only one of over 800 varieties of birds and mammals now considered endangered. (b) Due to pressures from the human population, polar bears are also considered endangered. Prized as trophies, they are now being hunted from the air. They are driven across the ice until exhausted and then shot by sportsmen. However, the large-scale extermination of plant and animal species that is now taking place is less the result of man's deliberate activities than of the changes in the natural environment as a consequence of his presence.*

(a)

(b)

The selection of certain plants and animals and the destruction of their enemies are part of a deliberate program designed for mankind's prosperity. At the same time, we are participating in accidental programs of selection and destruction. House mice, rats, fleas, houseflies, and cockroaches are remarkably well suited to living in man-made environments, and so such organisms are flourishing. On the other hand, many species are becoming extinct. Some, as we have mentioned, are being destroyed by DDT. Others, including some species of whales, crocodiles, alligators, and many of the large cats, are being hunted to death. Still other species are threatened with extinction because of disturbances of their customary feeding and breeding sites. Not only has the agricultural revolution changed the relative numbers of different species; it has caused the wholesale extinction of many animals and plants. Some of our concern about endangered species can be dismissed as sentimental or aesthetic; the world has learned to get along without the saber-toothed tiger and the dodo. However, there still remains the question of how many strands we can destroy in the web of life before the fabric of our own existence becomes threatened.

In short, we are creating a world that is more and more susceptible to damage, while at the same time, as a consequence of our increasing numbers and technology, we are amplifying our capacities to affect it. Although we can trace the biological roots of these phenomena, biology does not hold any ready solutions. There is an analogy here with the dilemma of twentieth-century physics. Through the gentle genius of scientists like Albert Einstein and Neils Bohr, physics created the knowledge that led to the atomic bomb, but physics could not tell us whether it should be detonated over Hiroshima. Biology has not created its equivalent of the atomic bomb but, like physics, it has created knowledge for which it cannot dictate the uses. Biology, for instance, can tell us the details of the reproductive process, and medical science, building on biological knowledge, can develop new means for limiting the birthrate, but biology cannot determine whether, by whom, or under what circumstances these new methods should be applied. Biology can tell us how an insecticide, amplified by its passage through the food chain, can strike a nesting tern, never for a moment its target, but in a world in which malaria and starvation are major causes of death, biology cannot decide the extent to which we should use insecticides to kill mosquitoes and to protect crop plants. Biology can tell us why animal protein is so much more expensive—in terms of human lives—than vegetable protein. Faced with the direct choice, few of us would rather eat a steak than share a potful of rice and beans with a group of hungry children. But we do not know how to share our excess with the millions of undernourished people.

In the words of Barry Commoner:

These are *value* judgments; they are determined not by scientific principle, but by the value we place on economic advantage and on human life. . . . These are matters of morality, of social and political judgments. In a democracy they belong not in the hands of "experts," but in the hands of the people and their elected representatives.*

* Barry Commoner, *The Closing Circle*, Alfred A. Knopf, New York, 1971.

0-37 *Ethiopian farmer preparing the ground with his traditional wooden plow. Although new agricultural techniques increase the yield, they add to the often-heavy burden of unemployment in developing countries.*

What are we, then, to do—not as experts, or biologists, or even students of biology, but just as ordinary individual human beings, inhabitants of a small planet?

We can always, of course, do nothing. Some contend that man is merely following his nature in his exploitation of his environment and of other living things, including his fellowman, and that this course will lead him inevitably to destruction. Others believe that things are so bad already that there is little hope in trying to do anything now; unfortunately, the gloomy portents of some authorities in the field, although they may not have had that intention, lend credence to this point of view. This attitude—a fashionable one in some circles —may, of course, be right. If more than 99 percent of all species are extinct, the odds are that *Homo sapiens* will follow this same route, and perhaps we should not be too sentimental about it, or too concerned about which other species may fill our ecological niche. Yet, although it may be the correct answer, few of us can live comfortably with it.

For those of us who do elect to take action, there are two kinds of action open to us. The first is individual action. We can try to limit our personal consumption and our personal waste—pick up our litter, clean up our automobile exhausts, learn to live with midges and gypsy moths and wash that is not whiter than white, limit the size of our families, and even try to change our personal diet so that there is a little left over for someone else. These actions may be important not only for their own sakes but because they serve to foster a new climate in which new ideas can take root.

The second type of action is political and social and—in the broadest sense— revolutionary. Nations—not just our own, but all nations—are dedicated, by the fact of their existence, to promoting the welfare of their own citizens, and within nations, governments find it necessary to support the interests of particular groups who play vital roles in the national economy. We are living in a country and at a time in which growth has been the proud criterion of progress for many decades: growth in population, in power consumption, in paved roads, in numbers of cars per family, and, above all, in gross national product, our total production of goods and services. If things are to change, it is necessary to adopt a new standard of values in which something much less measurable—such as quality of life not only for ourselves but for others—replaces our consumer's yardstick. Some authorities believe that this can take place only under conditions of much more rigid political, economic, and even psychological control. Others point to present changes—an apparent decrease in birthrates, the recent Supreme Court rulings concerning abortions, and widespread concern about pollution and conservation—and see in them evidence that Americans may be prepared to deal with these problems voluntarily and perhaps even effectively. It is too early to tell. Recent events in our history have shown, however, that governments do respond, although slowly, to massive change in the climate of public opinion. If we really do want a change, not only in our immediate environment but also in our political and social priorities, this desire will manifest itself in a large number of ways.

Finally, to return to biology, we should remember that man, although a member of the kingdom Animalia, is different in certain important respects from other animals. Because of these differences, it is largely irrelevant whether

man is naturally aggressive or passive, exploitative or generous. Unlike any of the other animals, man can make deliberate choices about his own course of behavior.

In the age of the dinosaurs, the earliest primates survived, it would appear, largely by their wits; now, if man is to survive the monsters of his own creating, he will have to do it, again, by the contents of his own skull. For within the mind of man—that complex collection of neurons and synapses—resides the uniquely human capacity to accumulate knowledge, to plan with foresight, and so to act with enlightened self-interest.

SUMMARY

The most important event in the cultural evolution of man was the advent of agriculture, about 10,000 years ago. Agriculture both ensured a more dependable food supply and permitted men to live together in communities, which hastened the spread of ideas and technology and, as a consequence, the pace of cultural change, and it enormously increased the growth of populations.

About 8000 B.C., when agriculture had its beginning, there were probably about 5 million people in the world. By 4000 B.C., the population had increased to 86 million; by the time of Christ, there were 133 million people in the world; and by 1650, there were 500 million.

In the last few hundred years, the population has been increasing at a faster and faster rate. By 1974, there were approximately 4 billion people. The doubling of the human population is now estimated at approximately 35 years, which means there will probably be about 7 billion people in the world by the close of the century. Because the capacity of the planet to support its human population is limited—although these limits have not been defined—attempts are under way to control the population by voluntary means rather than by the more ancient though highly effective means of war, pestilence, or famine. One obstacle to the reduction in births is the unavailability of an ideal contraceptive method. A much more stubborn obstacle is the resistance, for various religious, sociological, psychological reasons, of many individuals and groups to birth control. Many experts believe that only by improving health and living standards in the have-not countries can significant reductions in birthrate be achieved.

Of the some 4 billion people now present in the world, at least 2 billion are malnourished or undernourished. Malnutrition may involve either a shortage of total calories or a protein deficit, or both. Effects of protein deprivation are particularly serious in infants; protein shortages early in life permanently impair not only physical but mental development. Reasons for the present worldwide, chronic malnutrition of the human population include the growth of population, the growth in demand for high-protein foods, and the unequal distribution of world resources.

The development of new strains of wheat and rice has greatly increased food yields in Mexico, India, and Pakistan, and for a brief period, food production increased more rapidly than the birthrate in these countries. Yet despite this progress, there are more hungry people in the world today than ever before.

New methods of food production are being explored. These include the cul-

tivation on a large scale of algae, bacteria, and yeasts that can be used for animal fodder or processed for human consumption. Another approach involves the desalination of seawater for irrigation of the desert.

A major consequence of man's recent cultural evolution has been pollution of air, water, and land by industrial and agricultural waste products and by the by-products of human existence. Steps toward the solution of this problem are the development of means and facilities for recycling waste products and the imposition of strict regulations on both private individuals and industry.

In addition to accidental and careless pollution, in recent years we have deliberately added a number of new synthetic chemicals to the biosphere. Our experience with DDT, the most widely used and consequently the most studied of such chemicals, provides an example of the possible effects of this form of pollution. DDT is degraded very slowly and so persists, in contrast to natural products that are broken down rapidly under natural circumstances. Moreover, it accumulates along the food chain and so poses a particular threat to organisms of the higher trophic levels, such as birds of prey and man. Even if the use of DDT were discontinued immediately—which it will not be because of its importance in the control of malaria—its effects would persist in the ecosystem for at least another 20 years.

The extensive agriculture needed to grow food for the expanding human population has harsh effects upon the soil. The soils of tropical forests rapidly become infertile once the land is cleared for farming. They form a cementlike material known as laterite. Soils of the temperate zone can be kept fertile by massive applications of fertilizers. However, the nitrites which accumulate as a result of such fertilization programs are now also creating pollution problems in some heavily farmed areas.

In terms of its future effect on the living things on our planet, the most serious result of man's rise to ecological dominance may prove to be his simplification of the biosphere. Farming efficiency favors growing particular crops in large stands, and at the same time, as a result of agriculture, urbanization, and pollution, the number of species is being reduced. As a consequence, the biosphere is coming more and more to resemble an immature ecosystem, one in which there is little stability and in which the living components of the system are much more susceptible to the insults of the physical components of the environment and, undoubtedly, to man-made threats as well.

Evolution moves relentlessly in a single direction. We cannot regain a previous kind of relationship between man and nature any more than we can resurrect the dodo. But perhaps if we learn to use, increase, and respect our knowledge of our biosphere, to learn from our experiences, and to recognize that the future is already upon us, we may preserve some of what still remains.

QUESTIONS

1. Suppose the birthrate were to drop immediately to the so-called "replacement level"—two children per couple. Assuming that the great majority of children are born to women between 15 and 30 years of age, what would happen to the population of a country with a triangular population profile (Figure 50–13)? One with a barrel-shaped profile? One whose profile is an inverted pyramid?

2. In an editorial entitled "Food, Overpopulation, and Irresponsibility" that appeared recently in a scientific journal, the authors concluded: "Because it creates a vicious cycle that compounds human suffering at a high rate, the provision of food to the malnourished nations of the world that cannot, or will not, take very substantial measures to control their own reproductive rates is inhuman, immoral and irresponsible." What is your opinion?
3. Buckminster Fuller, the architect, has said, "Pollution is resources we are not harvesting." Give some examples to support this statement. Do you agree with his definition?

SUGGESTIONS FOR FURTHER READING

BAKER, HERBERT G.: *Plants and Civilization*, Wadsworth Publishing Company, Inc., Belmont, Calif., 1965.*

An introduction to the study of plants in relation to man, this book illustrates the profound influence of plants on man's economic, cultural, and political history.

BRESLER, JACK B. (ed.): *Human Ecology: Collected Readings*, Addison-Wesley Publishing Company, Inc., Reading, Mass., 1966.

A collection of articles, by experts in many different fields, on the wide variety of physical and biological factors that have influenced the evolution of man and that affect his present existence. The subjects range from biological clocks to insanity among lemmings (a cautionary tale) to the search for life on Mars.

BUETTNER-JANUSCH, JOHN: *Origins of Man*, John Wiley & Sons, Inc., New York, 1966.

A good introductory text by a physical anthropologist.

CAMPBELL, BERNARD: *Human Evolution*, Heinemann Educational Books, Ltd., London, 1967.

Campbell emphasizes the way that the special adaptations characterizing the hominids can be related to their environment, ecology, and culture.

CARSON, RACHEL: *Silent Spring*, Houghton Mifflin Company, Boston, 1962.*

This is the book that awakened the nation to the dangers of pesticides. Once seen as highly controversial, it is now regarded as a classic of modern ecology.

COMMONER, BARRY: *The Closing Circle*, Alfred A. Knopf, Inc., New York, 1971.

The author traces the relationship between the present ecological crisis and technological and social factors in our society. His title, The Closing Circle, *refers to his conviction that if we are to survive, "we must learn how to restore to nature the wealth we borrow from it." Commoner's book is criticized by Ehrlich and some other ecologists and demographers as presenting an oversimplified and too optimistic solution to environmental problems.*

EHRLICH, PAUL, and ANNE EHRLICH: *Population/Resources/Environment*, 2d ed., W. H. Freeman and Company, San Francisco, 1972.

Required, although not cheerful, reading for all concerned with the ecological crisis. The Ehrlichs present a wealth of useful data, even though you may not agree with all their conclusions.

HOWELLS, W. W.: *Mankind in the Making*, rev. ed., Doubleday & Company, Inc., Garden City, N.Y., 1967.

A clear and lively treatment of the story of human evolution, written for the general reader. The line drawings are charming.

* Available in paperback.

HOWELLS, WILLIAM: *Evolution of the Genus* Homo, Addison-Wesley Publishing Company, Inc., Reading, Mass., 1973.
A brief, up-to-date survey of human evolution.

JOLLY, ALISON: *The Evolution of Primate Behavior*, The Macmillan Company, New York, 1972.*
Few areas of modern biology are expanding as rapidly as studies of behavior of living primates. This is an excellent survey, covering the broad fields of ecology, society, and intelligence, from pottos to people.

LAPPÉ, FRANCES MOORE: *Diet for a Small Planet*, Ballantine Books, Inc., New York, 1971.*
Lappé clearly establishes the feasibility of eating "low on the food chain," thereby greatly increasing the availability of proteins to other human populations.

LE GROS CLARK, W. E.: *Antecedents of Man: An Introduction to the Evolution of the Primates*, Harper & Row, Publishers, Inc., New York, 1963.*
Le Gros Clark's emphasis is on the anatomy and physiology of the entire primate order.

PFEIFFER, JOHN E.: *The Emergence of Man*, 2d ed., Harper & Row, Publishers, Inc., New York, 1972.
Pfeiffer is one of the outstanding interpreters of complex scientific matters for the general public. This book is comprehensive, accurate, and thoroughly readable.

PILBEAM, DAVID: *The Ascent of Man: An Introduction to Human Evolution*, The Macmillan Company, New York, 1972.
A thorough survey of primate evolution, including evolution of the hominids.

SANDERS, N. K.: *Prehistoric Art in Europe*, Penguin Books, Ltd., Harmondsworth, England, 1968.
An illustrated history of art's first 30,000 years.

UCKO, PETER J., and ANDRÉE ROSENFELD: *Paleolithic Cave Art*, McGraw-Hill Book Company, New York, 1967.*
This book not only presents the cave art, in photos and drawings, but also examines its contents and context and discusses the various interpretations of it.

VAN LAWICK–GOODALL, JANE: *In the Shadow of Man*, Houghton Mifflin Company, Boston, 1971.
An absorbing personal account of eleven years spent observing the complex social organization of a single chimpanzee community in Tanzania.

* Available in paperback.

TEMPERATURE CONVERSION SCALE

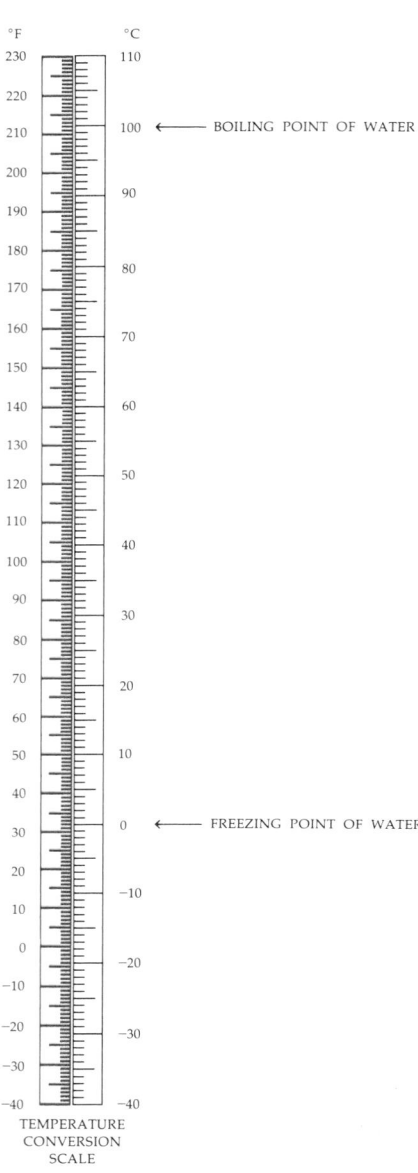

TEMPERATURE
CONVERSION
SCALE

FOR CONVERSION OF FAHRENHEIT TO CENTIGRADE,
THE FOLLOWING FORMULA CAN BE USED:
$$°C = \tfrac{5}{9}(°F - 32)$$

FOR CONVERSION OF CENTIGRADE TO FAHRENHEIT,
THE FOLLOWING FORMULA CAN BE USED:

$$°F = \tfrac{9}{5}°C + 32$$

APPENDIX B

METRIC TABLE

	Fundamental unit	Quantity	Numerical value	Symbol	English equivalent
Length	meter			m	39.4 inches
		kilometer	1,000 m	km	0.62137 miles
		centimeter	0.01 m	cm	0.3937 inches
		millimeter	0.001 m	mm	
		micrometer	0.000001 m	μm	
		nanometer	0.000000001 m	nm	
		angstrom	0.0000000001 m	Å	
Mass	gram			g	0.035 ounces
		kilogram	1,000 g	kg	2.2 pounds
		milligram	0.001 g	mg	
		microgram	0.000001 g	μg	
Time	second			sec	
		millisecond	0.001 sec	msec	
		microsecond	0.000001 sec	μsec	
Volume (solids)	cubic meter			m^3	1.308 cubic yards
		cubic centimeter	0.000001 m^3	cm^3	0.061 cubic inches
		cubic millimeter	0.000000001 m^3	mm^3	
Volume (liquids)	liter			l	1.06 quarts
		milliliter	0.001 liter	ml	
		microliter	0.000001 liter	μl	

APPENDIX C

CLASSIFICATION OF ORGANISMS

There are several ways to classify organisms. The one presented here is based closely upon that of Whittaker (in *Science*, **163**, 150–160, 1969). Organisms are divided into five major groups, or kingdoms: Monera, Protista, Fungi, Plantae, and Animalia.

The chief taxonomic divisions are kingdom, phylum, class, order, family, genus, species. The following classification includes all of the major phyla. Certain classes and orders, particularly those mentioned in this book, are also included, but the listing is far from complete. The number of species given for each group is the estimated number of living species described and named.

KINGDOM MONERA	Prokaryotes, i.e., cells which lack a nuclear envelope, chloroplasts and other plastids, mitochondria, and 9 + 2 flagella. Monera are unicellular but sometimes aggregate into filaments or other superficially multicellular bodies. Their predominant mode of nutrition is absorption, but some groups are photosynthetic or chemosynthetic. Reproduction is primarily asexual, by fission or budding, but conjugation occurs in some species.
PHYLUM	PHYLUM SCHIZOPHYTA: the bacteria. Unicellular; prokaryotic (lacking a nuclear membrane); reproduction usually asexual by cell division (fission); nutrition usually heterotrophic. About 3,000 species.
	PHYLUM CYANOPHYTA: blue-green algae. Unicellular or colonial; prokaryotic; chlorophyll, but no chloroplasts; nutrition usually autotrophic; reproduction by fission. Common on damp soil and rocks and in fresh and salt water. About 200 distinct, nonsymbiotic species.
KINGDOM PROTISTA	Eukaryotic unicellular organisms in which the cells are sometimes aggregated into filaments or other superficially multicellular bodies. Their modes of nutrition include ingestion, photosynthesis, and sometimes absorption. Reproduction both sexual (in some forms) and asexual. They move by 9 + 2 flagella or pseudopods or are nonmotile.
PHYLUM	PHYLUM PROTOZOA: microscopic, unicellular or simple colonial heterotrophic organisms; reproduction usually asexual by mitotic division; classified by type of locomotion. About 30,000 species.
CLASS	*Class Mastigophora:* protozoans with flagella, including a number of symbiotic forms such as *Trichonympha* and *Trypanosoma,* the cause of sleeping sickness.
	Class Sarcodina: protozoans with pseudopodia, such as amoebas. No stiffening pellicle outside cell walls.
	Class Ciliophora: protozoans with cilia, including *Paramecium* and *Stentor.*
	Class Sporozoa: parasitic protozoans; usually without locomotive organs during a major part of their life cycle. Includes *Plasmodium,* several species of which cause malaria.
PHYLUM	PHYLUM CHRYSOPHYTA: golden algae and diatoms. Autotrophic organisms with chlorophylls *a* and *c* and the accessory pigment fucoxanthin. Food stored as the carbohydrate leucosin or as large oil droplets. Cell walls consisting mainly of pectic compounds, sometimes heavily impregnated with siliceous materials. Some 6,000 to 10,000 living species.
CLASS	*Class Bacillariophyceae:* diatoms. Chrysophyta with double siliceous shells, the two halves of which fit together like a pillbox. They are sometimes motile by the secretion of mucilage fibrils along a specialized groove, the raphe. There are many extinct and 5,000 to 9,000 living species.
	Class Chrysophyceae: golden algae. A diverse group of organisms including flagellated, amoeboid, and nonmotile forms, some naked and others with a cell wall that may be ornamented with siliceous scales. At least 1,000 species.
PHYLUM	PHYLUM PYRROPHYTA: "fire" algae, sometimes called golden-brown algae. Autotrophic organisms with chlorophylls *a* and *c.* Food is stored as starch. Cell walls contain cellulose. The same phylum contains some 1,100 species, mostly biflagellated organisms, of which the great majority belong to the following class:

Class Dinophyceae: dinoflagellates. Pyrrophyta with lateral flagella, one of which beats in a groove that encircles the organism. They probably have no form of sexual reproduction and their mitosis is unlike that in any other organism. More than 1,000 species.

PHYLUM EUGLENOPHYTA: euglenoids. Autotrophic (or sometimes secondarily heterotrophic) organisms with chlorophylls *a* and *b*. They store food as paramylon, an unusual carbohydrate. Euglenoids usually have a single apical flagellum and a contractile vacuole. The flexible cell wall (pellicle) is rich in proteins. Sexual reproduction is unknown. Euglenoids occur mostly in fresh water. There are some 450 species.

PHYLUM GYMNOMYCOTA: the slime molds. Heterotrophic amoeboid organisms that mostly lack a cell wall but form sporangia at some stage in their life cycle. Predominant mode of nutrition is by ingestion. There are three classes:

Class Myxomycetes: plasmodial slime molds. Slime molds with multinucleate plasmodium which creeps along as a mass and eventually differentiates into sporangia, each of which is multinucleate and eventually gives rise to many spores. About 450 species.

Class Acrasiomycetes: cellular slime molds. Slime molds in which there are separate amoebas which eventually swarm together to form a mass but retain their identity within this mass, which eventually differentiates into a compound sporangium. Seven genera and about 26 species.

Class Protostelidomycetes: In this recently discovered group, the amoebas may remain separate or mass, but each one eventually differentiates into a simple stalked sporangium with one or two spores at its apex. Five genera and more than a dozen species.

KINGDOM FUNGI

Eukaryotic unicellular or multinucleate organisms in which the nuclei occur in a basically continuous mycelium; this mycelium becomes septate (partitioned off) in certain groups and at certain stages of the life cycle. They are heterotrophic, with nutrition by absorption. Reproductive cycles often include both sexual and asexual phases. Some 100,000 species of fungi have been named.

Class Oomycetes: Mostly aquatic fungi with motile cells characteristic of certain stages of the life cycle, their cell walls are composed of glucose polymers including cellulose. There are several hundred species.

Class Zygomycetes: terrestrial fungi, such as black bread mold, with the hyphae septate only during the formation of reproductive bodies; chitin predominant in the cell walls. The class includes several hundred species.

Class Ascomycetes: terrestrial and aquatic fungi, including *Neurospora*, powdery mildews, morels, and truffles. The hyphae are septate but the septa perforated; complete septa cut off the reproductive bodies, such as spores or gametangia. Chitin is predominant in the cell walls. Sexual reproduction involves the formation of a characteristic cell, the ascus, in which meiosis takes place and within which spores are formed. The hyphae in many ascomycetes are packed together into complex "fruiting bodies." Yeasts are unicellular ascomycetes that reproduce asexually by budding. About 30,000 species.

Classs Basidiomycetes: terrestrial fungi, including the mushrooms and toadstools, with the hyphae septate but the septa perforated; complete septa cut off reproductive bodies, such as spores or gametangia. Chitin is predominant in the cell walls. Sexual reproduction involves formation of basidia, in which meiosis takes place and on which the spores are borne. There are some 25,000 species.

Fungi Imperfecti: Mainly fungi with the characteristics of Ascomycetes but in which the sexual cycle has not been observed; a few probably belong to other classes. The Fungi

Imperfecti are classified by their asexual spore-bearing organs. There are some 25,000 species, including a *Penicillium*, the original source of penicillin, fungi which cause athlete's foot and other skin diseases, and many of the molds which give cheese, such as Roquefort and Camembert, their special flavor.

Lichens: The lichens are fungi (mostly Ascomycetes) symbiotic with unicellular algae. About 17,000 described species.

KINGDOM PLANTAE

Multicellular eukaryotes and related unicellular forms. Principal mode of nutrition by photosynthesis. Their photosynthetic pigment is chlorophyll *a*, with chlorophyll *b* and a number of carotenoids serving as accessory pigments. The cell walls contain cellulose. They are primarily nonmotile organisms attached to a substrate. There is considerable differentiation of organs and tissues in the more specialized forms. Their reproduction is primarily sexual with alternating gametophytic and sporophytic phases; the gametophyte phase has been progressively reduced in the course of evolution. The plants have clearly evolved on at least three occasions; the red algae and brown algae have an origin separate from the rest.

PHYLUM

PHYLUM RHODOPHYTA: red algae. Primarily marine plants characterized by the presence of chlorophyll *a* and pigments known as phycobilins. Their carbohydrate reserve is a special type of starch (floridian). No motile cells are present at any stage in the complex life cycle. The plant body is built up of closely packed filaments in a gelatinous matrix and is not differentiated into leaves, roots, and stem. It lacks specialized conducting cells. There are some 4,000 species.

PHYLUM PHAEOPHYTA: brown algae. Multicellular marine plants characterized by the presence of chlorophyll *a* and *c* and the pigment fucoxanthin. Their food reserve is a carbohydrate called laminarin. Motile cells are biflagellate, with one forward flagellum of the tinsel type and one trailing one of the whiplash type. A considerable amount of differentiation is found in some of the kelps, with specialized conducting cells for transporting photosynthate to dimly lighted regions of the plant present in some genera. There is however no differentiation into leaves, roots, and stem, as in the land plants. About 1,100 species.

PHYLUM CHLOROPHYTA: green algae. Unicellular or multicellular plants characterized by chlorophylls *a* and *b* and various carotenoids. The carbohydrate food reserve is starch. Motile cells have two whiplash flagella at the apical end. True multicellular genera do not exhibit complex patterns of differentiation. Multicellularity has arisen at least three times, and quite possibly more often. There are about 7,000 known species and possibly many more.

PHYLUM BRYOPHYTA: mosses, hornworts, and liverworts. Multicellular plants with the photosynthetic pigments and food reserves similar to those of the green algae. They have gametangia with a multicellular sterile jacket one cell layer thick. The sperm are biflagellate and motile. Gametophytes and sporophytes both exhibit complex multicellular patterns of development, but the conducting tissues are usually completely absent and not well differentiated when present. Most of the photosynthesis in these primarily terrestrial plants is carried out by the gametophyte, upon which the sporophyte is initially dependent. More than 23,500 species.

CLASS

Class Hepaticae: liverworts. The gametophytes are either thallose (not differentiated into roots, leaves, and stems) or leafy, and the sporophytes relatively simple in construction. About 9,000 species.

Class Antherocerotae: hornworts. The gametophytes are thallose. The sporophyte grows from a basal meristem for as long as conditions are favorable. Stomata are present on the sporophyte. About 100 species.

Class Musci: mosses. The gametophytes are leafy. Sporophytes have complex patterns of spore discharge. Stomata are present on the sporophyte. About 14,500 species.

PHYLUM	PHYLUM TRACHEOPHYTA: vascular plants. Terrestrial plants with complex differentiation of organs into leaves, roots, and stem. The only motile cells are the male gametes of some species, which are propelled by many cilia. The vascular plants have well-developed strands of conducting tissue for the transport of water and organic materials. The main trends of evolution in the vascular plants involve a progressive reduction in the gametophyte, which is green and free-living in ferns but heterotrophic and more or less enclosed by sporophytic tissue in the others; the loss of multicellular gametangia and motile sperm; and the evolution of the seed. The phylum includes the following subphyla with living representatives:

SUBPHYLUM

SUBPHYLUM LYCOPHYTINA: lycophytes. Homosporous and heterosporous vascular plants with microphylls; extremely diverse in appearance. All lycophytes have motile sperm. There are five genera and about 1,000 species.

SUBPHYLUM SPENOPHYTINA: horsetails. Homosporous vascular plants with jointed stems marked by conspicuous nodes and elevated siliceous ribs and sporangia borne in a strobilus at the apex of the stem. Leaves are scalelike. Sperm are motile. Although now thought to have evolved from a megaphyll, the leaves of the horsetails are structurally indistinguishable from microphylls. There is one genus, *Equisetum*, with about two dozen living species.

SUBPHYLUM PTEROPHYTINA: ferns, gymnosperms, and flowering plants. Although diverse, these groups possess in common the megaphyll, which in certain genera has become much reduced. About 260,000 species.

CLASS

Class Filicineae: the ferns. They are mostly homosporous although some are heterosporous. The gametophyte is more or less free-living and usually photosynthetic. Multicellular gametangia and free-swimming sperm are present. About 11,000 species.

Class Coniferinae: the conifers. Seed plants with active cambial growth and simple leaves, in which the ovules are not enclosed and the sperm are not flagellated. There are some 50 genera and about 550 species, the most familiar group of gymnosperms.

Class Cycadinae: cycads. Seed plants with sluggish cambial growth and pinnately compound, palmlike or fernlike leaves. The ovules are not enclosed. The sperm are flagellated and motile, but are carried to the vicinity of the ovule in a pollen tube. Cycads are gymnosperms. There are nine genera and about 100 species.

Class Ginkgoinae: ginkgo. Seed plants with active cambial growth and fan-shaped leaves with open dichotomous venation. The ovules are not enclosed and are fleshy at maturity. Sperm are carried to the vicinity of the ovule in a pollen tube, but are flagellated and motile. They are gymnosperms. There is one species only.

Class Angiospermae: flowering plants. Seed plants in which the ovules are enclosed in a carpel (in all but a very few genera), and the seeds at maturity are borne within fruits. They are extremely diverse vegetatively but characterized by the flower, which is basically insect-pollinated. Other modes of pollination, such as wind pollination, have been derived in a number of different lines. The gametophytes are much reduced, with the female gametophyte often consisting of only eight cells or nuclei at maturity. Double fertilization involving two of the three nuclei from the mature microgametophyte gives rise to the zygote and to the primary endosperm nucleus; the former becomes the embryo and the latter a special nutritive tissue, the endosperm. About 250,000 species.

SUBCLASS

Subclass Dicotyledonae: dicots. Flower parts are usually in fours or fives; leaf venation is usually netlike, pinnate, or palmate; there is true secondary growth with vascular cambium commonly present; there are two cotyledons; and the vascular bundles in the stem are in a ring. About 190,000 species.

Subclass Monocotyledonae: monocots. Flower parts are usually in threes, leaf venation is usually parallel, true secondary growth is not present, there is one cotyledon, and vascular bundles in the stem are scattered. About 60,000 species.

KINGDOM ANIMALIA			Eukaryotic multicellular organisms. Their principal mode of nutrition is by ingestion. Many animals are motile, and they generally lack the rigid cell walls characteristic of plants. Considerable cellular migration and reorganization of tissues often occurring during the course of embryology. Their reproduction is primarily sexual, with male and female diploid organisms producing haploid gametes which fuse to form the zygote. More than a million species have been described and the actual number may be close to 10 million.

PHYLUM

PHYLUM PORIFERA: sponges. Simple multicellular animals, largely marine, with stiff skeletons, and bodies perforated by many pores that admit water containing food particles. All have choanocytes, "collar cells." About 4,200 species.

PHYLUM COELENTERATA: coelenterates. Animals with radially symmetrical, "two-layered" bodies of a jellylike consistency. Reproduction is asexual or sexual. They are the only organisms with cnidoblasts, special stinging cells. All are aquatic and most are marine. About 11,000 species.

CLASS

Class *Hydrozoa: Hydra, Obelia,* and other *Hydra*-like animals. They are often colonial, and often have a regular alternation of asexual and sexual generations. The polyp form is dominant.

Class *Scyphozoa:* marine jellyfishes or "cup animals," including *Aurelia.* The medusa form is dominant. They have true muscle cells.

Class *Anthozoa:* sea anemones ("flower animals") and colonial corals. They have no medusa stage.

PHYLUM

PHYLUM CTENOPHORA: comb jellies and sea walnuts. They are free-swimming, often almost spherical animals. They are translucent, gelatinous, delicately colored, and often bioluminescent. They possess eight bands of cilia, for locomotion. About 80 species.

PHYLUM PLATYHELMINTHES: flatworms. Bilaterally symmetrical with three tissue layers. The gut has only one opening. They have no coelom or circulatory system. They have complex hermaphroditic reproductive systems and excrete by means of special (flame) cells. About 15,000 species.

CLASS

Class *Turbellaria:* planaria and other nonparasitic flatworms. They are ciliated, carnivorous, and have ocelli ("eyespots").

Class *Trematoda:* flukes. They are parasitic flatworms with digestive tracts.

Class *Cestoidea:* tapeworms. They are parasitic flatworms with no digestive tracts; they absorb nourishment through body surface.

PHYLUM

PHYLUM RHYNOCHOCOELA: proboscis, nemertine, or ribbon worms. They are nonparasitic, usually marine, and have a tubelike gut with mouth and anus, a protrusible proboscis armed with a hook for capturing prey, and simple circulatory and reproductive systems. About 600 species.

PHYLUM NEMATODA: roundworms. The phylum includes minute free-living forms, such as vinegar eels, and plant and animal parasites, such as hookworms. They are characterized by elongated, cylindrical, bilaterally symmetrical bodies. About 80,000 species.

PHYLUM ACANTHOCEPHALA: spiny-headed worms. They are parasitic worms with no digestive tract and a head armed with many recurved spines. About 300 species.

PHYLUM CHAETOGNATHA: arrow worms. Free-swimming planktonic marine worms, they have a coelom, a complete digestive tract, and a mouth with strong sickle-shaped hooks on each side. About 50 species.

PHYLUM NEMATOMORPHA: horsehair worms. They are extremely slender, brown or black worms up to 3 feet long. Adults are free-living, but the larvae are parasitic in insects. About 250 species.

PHYLUM ROTIFERA: microscopic, wormlike or spherical animals, "wheel animalcules." They have complete digestive tract, flame cells, and a circle of cilia on the head, the beating of which suggests a wheel; males are minute and either degenerate or unknown in many species. About 1,500 species.

PHYLUM GASTROTRICHA: These are microscopic, wormlike animals which move by longitudinal bands of cilia. About 140 species.

PHYLUM BRYOZOA: "moss" animals. Microscopic aquatic organisms, they are characterized by a U-shaped row of ciliated tentacles, with which they feed; they usually form fixed and branching colonies; they superficially resemble hydroid coelenterates but are much more complex; having anus and coelum, they retain larvae in special brood pouch. About 4,000 species.

PHYLUM BRACHIOPODA: lamp shells. Marine animals with two hard shells (one dorsal and one ventral), they superficially resemble clams. Fixed by a stalk or one shell in adult life, they obtain food by means of ciliated tentacles. About 260 living species; 3,000 extinct.

PHYLUM PHORONIDEA: sedentary, elongated, wormlike animals that secrete and live in a leathery tube. They have a U-shaped digestive tract and a ring of ciliated tentacles with which they feed. Marine. About 15 species.

PHYLUM ANNELIDA: ringed or segmented worms. They usually have a well-developed coelom, a one-way digestive tract, head, and circulatory system, nephridia, and well-defined nervous system. About 8,800 species.

CLASS

Class Archiannelida: small, simple, probably primitive, marine worms. About 35 species.

Class Polychaeta: mainly marine worms, such as *Nereis.* They have a distinct head with palps and tentacles and many-bristled lobose appendages. Parapodia are often brightly colored. About 4,000 species.

Class Oligochaeta: soil, freshwater, and marine annelids, including the earthworm (*Lumbricus*). They have scanty bristles and usually a poorly differentiated head. About 2,500 species.

Class Hirudinea: leeches. They have a posterior sucker and usually an anterior sucker surrounding the mouth. They are freshwater, marine, and terrestrial; either free-living or parasitic. About 300 species.

PHYLUM

PHYLUM MOLLUSCA: unsegmented animals, with a head, a mantle and a muscular foot, variously modified. They are mostly aquatic; soft-bodied, often with one or more hard shells, and a three-chambered heart. All the bivalves have a radula (rasplike organ used for scraping or marine drilling). About 110,000 species.

CLASS

Class Amphineura: chitons. The simplest type of mollusks, they have elongated body covered with mantle in which are embedded eight dorsal shell plates. About 700 species.

Class Pelecypoda: two-shelled mollusks, including clams, oysters, mussels, scallops. They usually have a hatchet-shaped foot and no distinct head. Generally sessile. About 15,000 species.

Class Scaphopoda: tooth or tusk shells. They are marine mollusks with a conical tubular shell. About 350 species.

Class Gastropoda: asymmetrical mollusks including snails, whelks, slugs. They usually have a spiral shell and a head with one or two pairs of tentacles. About 80,000 species.

Class Cephalopoda: octopus, squid, *Nautilus*. They are characterized by a "head-foot" with eight or ten arms or many tentacles, mouth with two horny jaws, and well-developed eyes and nervous system. The shell is external (*Nautilus*), internal (squid), or absent (octopus). All except *Nautilus* have ink glands. About 400 species.

PHYLUM

PHYLUM ARTHROPODA: The largest phylum in the animal kingdom, arthropods are segmented animals with paired jointed appendages, a hard jointed exoskeleton, a complete digestive tract, reduced coelom, no nephridia, a dorsal brain, and a ventral nerve cord with paired ganglia in each segment. About 765,257 species.

CLASS

Class Merostomata: horseshoe crabs. They are aquatic, with book gills. About 5 species

Class Crustacea: lobsters, crabs, crayfish, shrimps. Crustaceans are mostly aquatic, with two pairs of antennae, one pair of mandibles, and typically two pairs of maxillae. The thoracic segments have appendages, and the abdominal segments are with or without appendages. About 25,000 species.

Class Arachnida: spiders, mites, scorpions. Most members terrestrial, air-breathing; usually 4 or 5 pairs of legs; first pair of appendages used for grasping; have chelicerae (pincers or fangs) in place of jaws or antennae. About 30,000 species.

Class Onychophora: simple, terrestrial arthropods. All belong to one genus, *Peripatus*. They have many short unjointed pairs of legs. About 73 species.

Class Insecta: insects, including bees, ants, beetles, butterflies, fleas, lice, flies, etc. Most insects are terrestrial, and most breathe by means of trachea. They have one pair of antennae, three pairs of legs, and three distinct parts of the body (head, thorax, and abdomen). Most have 2 pairs of wings. About 700,000 species.

Class Chilopoda: centipedes. They have 15 to 173 trunk segments, each with one pair of jointed appendages. About 2,000 species.

Class Diplopoda: millipedes. They have an abdomen with 20 to 100 segments, each with two pairs of appendages. About 7,000 species.

PHYLUM

PHYLUM ECHINODERMATA: starfish and sea urchins. Echinoderms are radially symmetrical in adult stage, with a well-developed coelom, an endoskeleton of calcareous ossicles and spines, and a unique water vascular system. They have tube feet and are marine. About 6,000 species.

CLASS

Class Crinoidea: sea lilies and feather stars. Sessile animals, they often have a jointed stalk for attachment, and they have ten arms bearing many slender lateral branches. Most species are fossils.

Class Asteroidae: starfish. They have five to fifty arms, an oral surface directed downward, and two to four tube feet.

Class Ophiuroidea: brittle stars, serpent stars. They are greatly elongated, with highly flexible slender arms and rapid horizontal locomotion.

Class Echinoidea: sea urchins and sand dollars. Skeletal plates form rigid test which bears many movable spines.

Class Holothuroidea: sea cucumbers. They have a sausage-shaped or wormlike elongated body.

PHYLUM HEMICHORDATA: small group of wormlike marine animals, including the acorn, or proboscis, worms. They have a notochordlike structure in the head end, gill slits, and a solid nerve cord. About 91 species.

PHYLUM CHORDATA: animals having at some stage a notochord, pharyngeal gill slits, and a hollow nerve cord on the dorsal side. About 44,794 species.

SUBPHYLUM

SUBPHYLUM TUNICATA: tunicates or ascidians. Adults are saclike, usually sessile, often forming branching colonies. They feed by ciliary currents, have gill slits, a reduced nervous system, and no notochord. Larvae are active, with well-developed nervous system and notochord. They are marine. About 1,600 species.

SUBPHYLUM CEPHALOCHORDATA: lancelets. This small subphylum contains only *Amphioxus* and related forms. They are somewhat fishlike marine animals with a permanent notochord the whole length of the body, a nerve cord, a pharynx with gill slits, and no cartilage or bone. About 13 species.

SUBPHYLUM VERTEBRATA: the vertebrates. The most important subphylum of Chordata. In the vertebrates the notochord is replaced by cartilage or bone, forming the segmented vertebral column or backbone. A skull surrounds a well-developed brain. They usually have a tail. About 43,090 species.

CLASS

Class Agnatha: lampreys and hagfish. These are eel-like aquatic vertebrates without limbs, with a jawless sucking mouth, and no bones, scales, or fins.*

Class Chondrichthyes: sharks, rays, skates, and other cartilaginous fish. They have complicated copulatory organs, scales, and no air bladders. They are almost exclusively marine.*

Class Osteichthyes: the bony fish, including nearly all modern freshwater fish, such as sturgeon, trout, perch, anglerfish, lungfish, and some almost extinct groups. They usually have an air bladder or (rarely) a lung.*

Class Amphibia: salamanders, frogs, and toads. They usually breathe by gills in the larval stage and by lungs in the adult stage. They have incomplete double circulation and a usually naked skin. The limbs are legs. They were the first vertebrates to inhabit the land and are ancestors of the reptiles. Their eggs are unprotected by a shell and embryonic membranes. About 2,000 species.

Class Reptilia: turtles, lizards, snakes, crocodiles; includes extinct species such as the dinosaurs. Reptiles breathe by lungs and have incomplete double circulation. Their skin is usually covered with scales. The four limbs are legs (absent in snakes). They are cold-blooded. Most live and reproduce on land though some are aquatic. Their embryo is enclosed in an egg shell and has protective membranes. About 5,000 species.

Class Aves: birds. Birds are warm-blooded animals with complete double circulation and a skin covered with feathers. The forelimbs are wings. Their embryo is enclosed in an egg shell with protective membranes. Includes the extinct *Archaeopteryx*. About 8,590 species.

Class Mammalia: mammals. Mammals are warm-blooded animals with complete double circulation. Their skin is usually covered with hair. The young are nourished with milk secreted by the mother. They have four limbs, usually legs (forelimbs sometimes arms, wings, or fins), a diaphragm used in respiration, a lower jaw made up of a single pair of bones, 3 bones in each middle ear connecting eardrum and inner ear, and 7 vertebrae in the neck. About 4,500 species.

* The total number of species of fish is estimated to be about 23,000.

Subclass Prototheria: monotremes. These are the oviparous (egg-laying) mammals with imperfect temperature regulation. There are only two living species: the duckbill platypus and spiny anteater of Australia and New Guinea.

Subclass Metatheria: marsupials, including kangaroos, opossums, and others. Marsupials are viviparous mammals without a placenta (or with a poorly developed one); the young are born in an undeveloped state and are carried in an external pouch of the mother for some time after birth. They are found chiefly in Australia.

Subclass Eutheria: mammals with a well-developed placenta. This subclass comprises the great majority of living mammals. There are 12 principal orders of Eutheria:

Order

Insectivora: shrews, moles, hedgehogs, etc.

Edentata: toothless mammals—anteaters, sloths, armadillos, etc.

Rodentia: the rodents—rats, mice, squirrels, etc.

Artiodactyla: even-toed ungulates (hoofed mammals)—cattle, deer, camels, hippopotamuses, etc.

Perissodactyla: odd-toed ungulates—horses, zebras, rhinoceroses, etc.

Proboscidea: elephants

Lagomorpha: rabbits and hares

Sirenia: the manatee, dugong, and sea cows. Large aquatic mammals with the forelimbs finlike, the hind limbs absent.

Carnivora: carnivorous animals—cats, dogs, bears, weasels, seals, etc.

Cetacea: the whales, dolphins, and porpoises. Aquatic mammals with the forelimbs fins, the hind limbs absent.

Chiroptera: the bats. Aerial mammals with the forelimbs wings.

Primates: the lemurs, monkeys, apes, and man.

GEOLOGIC TIME SCALE

Years ago (millions)	Era	Period	Epoch	Life forms	Climates and major physical events
	CENOZOIC	Quaternary	Recent Pleistocene	Age of man. Planetary spread of *Homo sapiens;* extinction of many large mammals, including woolly mammoths. Deserts on large scale.	Fluctuating cold to mild. Four glacial advances and retreats (Ice Age); uplift of Sierra Nevada.
1½–7		Tertiary	Pliocene	Large carnivores. First known appearance of man-apes.	Cooler. Continued uplift and mountain building, with widespread extinction of many species.
7–26			Miocene	Whales, apes, grazing animals. Spread of grasslands as forests contract.	Moderate uplift of Rockies.
26–38			Oligocene	Large, browsing mammals. Apes appear.	Rise of Alps and Himalayas. Lands generally low. Volcanoes in Rockies area.
38–53			Eocene	Primitive horses, tiny camels, modern and giant types of birds.	Mild to very tropical. Many lakes in western North America.
53–65			Paleocene	First known primitive primates and carnivores.	Mild to cool. Wide, shallow continental seas largely disappear.
65–136	MESOZOIC	Cretaceous		Age of reptiles, extinction of dinosaurs. Marsupials, insectivores. Angiosperms become abundant.	Lands low and extensive. Last widespread oceans. Elevation of Rockies cuts off rain.
136–195		Jurassic		Dinosaurs' zenith. Flying reptiles, small mammals. Birds appear. Gymnosperms, especially cycads and ferns.	Mild. Continents low. Large areas in Europe covered by seas. Mountains rise from Alaska to Mexico. Breakup of Pangaea.
195–230		Triassic		First dinosaurs. Primitive mammals appear. Forests of gymnosperms and ferns.	Continents mountainous. Large areas arid. Eruptions in eastern North America. Appalachians uplifted and broken into basins.
230–280	PALEOZOIC	Permian		Reptiles evolve. Origin of conifers and possible origin of angiosperms; earlier forest types wane.	Extensive glaciation in Southern Hemisphere. Appalachians formed by end of Paleozoic; most of seas drain from continent.
280–345		Carboniferous Pennsylvanian Mississippian		Age of amphibians. First reptiles. Variety of insects. Sharks abundant. Great swamp forests of ferns, gymnosperms, and horsetails.	Warm. Lands low, covered by shallow seas or great coal swamps. Mountain building in eastern U.S., Texas, Colorado. Moist, equable climate, conditions like those in temperate or subtropical zones, little seasonal variation, root patterns indicate water plentiful.
345–395		Devonian		Age of fish. Amphibians appear. Shellfish abundant. Lungfish. Rise of land plants. Extinction of primitive vascular plants. Origin of modern subclasses of vascular plants.	Europe mountainous with arid basins. Mountains and volcanoes in eastern U.S. and Canada. Rest of North America low and flat. Sea covers most of land.
395–440		Silurian		First terrestrial plants. Rise of fish and reef-building corals. Shell-forming sea animals abundant. Modern groups of algae and fungi.	Mild. Continents generally flat. Mountain building in Europe. Again flooded.
440–500		Ordovician		First primitive fish. Invertebrates dominant.	Mild. Shallow seas, continents low; sea covers U.S. Limestone deposits; microscopic plant life thriving.
500–600		Cambrian		Age of marine invertebrates. Shell animals.	Mild. Extensive seas. Seas spill over continents.
Over 600	PRECAMBRIAN			Earliest known fossils.	Dry and cold to warm and moist. Planet cools. Formation of Earth's crust. Extensive mountain building. Shallow seas.

GLOSSARY

This list does not include units of measure, names of taxonomic groups, or geologic eras, which can be found in Appendixes B, C, and D, or terms that are used only once in the text and defined there.

A– [Gk. *an-*, not]: Prefix, negates the succeeding part of the word.

AB– [L., away, off]: Prefix, meaning "away from" or "off."

ABDOMEN: In vertebrates, that portion of the trunk containing visceral organs except for heart and lungs; in arthropods, the posterior portion of the body composed of a group of similar segments and containing the reproductive organs and posterior portion of the digestive tract.

ABSCISSION: In plants, the dropping of leaves, flowers, fruits, or stems at the end of a growing season, as the result of formation of an abscission layer, a layer of specialized cells, and the action of a hormone (abscisic acid).

ABSORPTION: The passage of water and dissolved substances into the cell or organism.

ABSORPTION SPECTRUM: The spectrum of light waves absorbed by a particular pigment.

ACETYLCHOLINE (a-**sea**-tell-**co**-leen): One of the known chemical transmitters of nervous impulses across synaptic junctions.

ACID [L. *acidus*, sour]: A substance that, on dissociation, releases hydrogen ions (H^+) but not hydroxyl ions (OH^-); having a pH of less than 7; the opposite of a base.

ACRO– [Gk. *akros*, highest, outermost]: Prefix, meaning "pertaining to the tip."

ACTIN: One of the two major proteins of muscle, makes up the thin filaments.

ACTION POTENTIAL: A rapid change in membrane potential resulting in transmission of a nerve impulse.

ACTION SPECTRUM: The spectrum of light waves that elicit a particular reaction.

ACTIVE SITE: That part of the surface of an enzyme molecule into which the substrate fits during a catalytic reaction.

ACTIVE TRANSPORT: The energy-expending process by which a cell moves a substance across the cell membrane from a point of lower concentration to a point of higher concentration, against the diffusion gradient.

AD– [L., toward, to]: Prefix, meaning "toward" or "to."

ADAPTATION [L. *adaptare*, to fit]: (1) The acquiring of characteristics by an organism (or group of organisms) which make it better suited to live and reproduce in its environment. (2) A peculiarity of structure, physiology, or behavior of an organism that especially aids in fitting the organism to its particular environment.

ADAPTIVE RADIATION: The evolution from a relatively primitive and unspecialized type of organism to several divergent forms specialized to fit numerous distinct and diverse ways of life, as occurred in Darwin's finches or the marsupials of Australia.

ADENOSINE TRIPHOSPHATE (ATP): The major source of usable energy in cell metabolism; composed of adenine, ribose, and three phosphate groups. On hydrolysis, ATP loses one phosphate and one hydrogen to become adenosine diphosphate (ADP), releasing energy in the process.

ADHESION [L. *adhaerere*, to stick to]: A sticking together of substances.

ADRENAL GLAND: A vertebrate endocrine gland. The cortex (outer surface) is the source of cortisol, aldosterone, and other steroid hormones; the medulla (inner core) secretes adrenaline.

ADRENALINE: A hormone produced by the medulla of the adrenal gland which increases the concentration of sugar in the blood, raises blood pressure and heartbeat rate, and increases muscular power and resistance to fatigue; also a chemical transmitter across synaptic junctions. Also called epinephrine.

AEROBIC [Gk. *aēr, aeros*, air + *bios*, life]: Requiring free oxygen for respiration.

AFFERENT [L. *ferre*, to bear]: Bringing inward to a central part, applied to nerves and blood vessels.

AGAR: A gelatinous substance derived from certain red algae which is used as a solidifying agent in the preparation of nutrient media for growing microorganisms.

ALDOSTERONE: A hormone produced by the adrenal cortex that stimulates the reabsorption of sodium from the kidney.

ALGA (**al**-gah): A photosynthetic organism lacking multicellular sex organs.

ALKALINE: Pertaining to substances which release hydroxyl (OH^-) ions in water; having a pH greater than 7; basic; opposite of acidic.

ALLANTOIS (al-**lan**-toe-iss) [Gk. *allantoeides*, sausage-shaped]: A saclike outgrowth from the ventral surface of the posterior part of the embryonic gut. It functions as an embryonic urinary bladder or as a respiratory extension of the hindgut in reptiles and birds, and in mammals it is modified to carry blood vessels to and from the placenta.

ALLELE (al-**eel**) (ALLELOMORPH) [Gk. *allelon*, of one another + -morphe, form]: One of the alternate forms of the same functional gene. Alleles occupy the same position (locus) on homologous chromosomes, and so are separated from each other at meiosis.

ALLOPOLYPLOID: A polyploid in which the different sets of chromosomes come from different species or widely different strains.

ALTERNATION OF GENERATIONS: A reproductive cycle in which a haploid ($1n$) phase, the gametophyte, gives rise to gametes which, after fusion to form a zygote, germinate to produce a diploid ($2n$)

phase, the sporophyte. Spores produced by meiotic division from the sporophyte give rise to new gametophytes, completing the cycle.

AMINO ACIDS (am-ee-no) [Gk. *Ammon*, referring to the Egyptian sun god, near whose temple ammonium salts were first prepared from camel dung]: Organic molecules containing nitrogen in the form of NH_2; the "building blocks" of protein molecules.

AMNION (am-neon) [Gk. dim. of *amnos*, lamb]: Inner, fluid-filled sac composed of a thin double membrane that surrounds the embryo in reptiles, birds, and mammals.

AMNIOTE EGG: An egg which is isolated from the environment by a more or less impervious shell during the period of its development and which is completely self-sufficient, requiring only oxygen from the outside.

AMOEBOID [Gk. *amoibe*, change]: Moving or eating by means of pseudopodia (temporary cytoplasmic protrusions from the cell body).

AMPHI— [Gk., on both sides]: Prefix, meaning "on both sides," "both," or "of both kinds."

AMPHIBIAN [Gk. *amphibios*, living a double life]: A class of vertebrates intermediate in many characteristics between fish and reptiles, which live part of the time in water and part on land.

AN— [Gk., not]: Prefix, equivalent to "a-", meaning "not"; used before a vowel or silent "h."

ANAEROBIC: Applied to cells (largely bacterial) that can live without free oxygen; obligate anaerobes cannot live in the presence of oxygen; facultative anaerobes can live with or without oxygen.

ANALOGOUS [Gk. *analogos*, proportionate]: Applied to structure similar in function but different in evolutionary origin, such as the wing of a bird and the wing of an insect.

ANAPHASE (anna-phase) [Gk. *ana*, up + *phasis*, form]: A stage in mitosis or meiosis in which the chromatids of each chromosome separate and move to opposite poles.

ANDRO— [Gk. *aner*, *andros*, man]: Prefix, meaning "man" or "male."

ANDROGEN: Male sex hormone.

ANGIOSPERM (an-jee-o-sperm) [Gk. *angion*, vessel + *sperma*, seed]: Literally, a seed borne in a vessel; thus, one of a group of plants whose seeds are borne within a matured ovary (fruit).

ANIMAL HEMISPHERE: That half of the egg or zygote in which there is less yolky material, which contains the nucleus, and which is most active in cell cleavage; contrasts with vegetal hemisphere.

ANIMAL POLE: Imaginary point on the surface of the egg or zygote which lies in the center of the animal hemisphere, opposite to the vegetal pole.

ANNUAL: A plant that completes its life cycle from seed germination to seed production, followed by death, in a single growing season.

ANTENNA: Long, paired sensory appendage on the head of many arthropods.

ANTERIOR [L. *ante*, before, toward, in front of]: The front end of an organism; in human anatomy, the ventral surface.

ANTHER [Gk. *anthos*, flower]: In plants, the pollen-bearing portion of a stamen.

ANTHROPO— [Gk. *anthropos*, man, human]: Prefix, meaning "man" or "human."

ANTHROPOID: A higher primate, includes monkeys, apes, and man.

ANTHROPOMORPHISM: Assignment of human characteristics, abilities, or feelings to nonhuman organisms.

ANTIBIOTIC: An organic compound formed and secreted by an organism that is toxic to organisms of other species.

ANTIBODY: A globular protein that is produced in the body fluids, in response to a foreign substance (antigen), with which it reacts specifically.

ANTIDIURETIC HORMONE (ADH): A hormone secreted by the hypothalamus that inhibits urine excretion by inducing the reabsorption of water from the nephrons of the kidneys, also called vasopressin.

ANTIGEN: A foreign substance, usually a protein or polysaccharide, that stimulates the formation of specific antibodies.

AORTA (a-ore-ta) [Gk. *airein*, to lift, heave]: The major artery in blood-circulating systems.

APICAL DOMINANCE: In plants, the influence of a terminal bud in suppressing the growth of lateral buds, sometimes the result of the release of the growth-regulating hormone auxin.

APICAL MERISTEM: The growing point, composed of meristematic tissue, at the tips of the root or stem in vascular plants.

ARBOREAL: Tree-dwelling.

ARCH—, ARCHEO— [Gk. *arche*, *archos*, beginning]: Prefix, meaning "first," "main," or "earliest."

ARCHEGONIUM, pl. ARCHEGONIA [Gk. *archegeonos*, first of a race]: In plants, multicellular egg-producing organ.

ARCHENTERON: The principal cavity within the early embryo (gastrula) of many animals; lined with endoderm, it opens to the outside by means of the blastopore and ultimately becomes the digestive tract.

ARTERY: A vessel carrying oxygenated blood from the heart to the tissues; it is usually thick-walled, elastic, and muscular.

ARTHROPOD [Gk. *arthron*, joint + *pous*, *podos*, foot]: An invertebrate animal with jointed appendages; a member of the phylum Arthropoda.

ASCUS, pl. ASCI (as-kus, as-i): A specialized cell, characteristic of the Ascomycetes, in which two haploid nuclei fuse to produce a zygote which immediately divides by meiosis; at maturity, an ascus contains ascospores.

ASEXUAL REPRODUCTION: Any reproductive process, such as fission or budding, that does not involve the union of gametes.

ATMOSPHERIC PRESSURE: The weight of the Earth's atmosphere over a unit area of the Earth's surface.

ATOM [Gk. *atomos*, indivisible]: The smallest unit into which a chemical element can be divided and still retain its characteristic properties.

ATOMIC NUCLEUS: The central core of an atom, containing protons and neutrons, around which electrons orbit.

ATOMIC NUMBER: The number of protons in the nucleus of an atom; equal to the number of electrons in the neutral atom.

ATOMIC WEIGHT: The average weight of an atom of an element relative to the weight of an atom of carbon (C^{12}), which is assigned the integral value of 12.

ATP: Abbreviation of adenosine triphosphate.

ATRIUM, pl. ATRIA (a-tree-um) [L., yard, court, hall]: A chamber of the heart that receives blood and passes it on to a ventricle.

AUTO— [Gk. *autos*, same, self]: Prefix, meaning "same" or "same self."

AUTONOMIC [Gk. *nomos*, law]: Self-controlling, independent of outside influences.

AUTONOMIC NERVOUS SYSTEM: A special system of motor nerves and ganglia in vertebrates that is not under voluntary control and that innervates the heart, glands, visceral organs, and smooth muscle. It is subdivided into the sympathetic and the parasympathetic nervous systems.

AUTOPOLYPLOID: A polyploid in which the chromosomes all come from the same source.

AUTOSOME: Any chromosome other than the sex chromosomes. Man has 22 pairs of autosomes and 1 pair of sex chromosomes.

AUTOTROPH [Gk. *trophos*, feeder]: An organism that is able to synthesize organic molecules from inorganic substances, in contrast to heterotroph. Most plants and some bacteria are autotrophs.

AUXIN [Gk. *auxein*, to increase]: One of a group of plant hormones with a variety of growth-regulating effects, including promotion of cell elongation.

AXIS: An imaginary line passing through a body or organ around which parts are symmetrically aligned.

AXON: The part of a neuron that carries impulses away from the cell body.

BACKCROSS: A cross between an individual that is heterozygous for a pair of alleles and a parent that is homozygous for the recessive alleles involved in the experiment; a test cross involving a homozygous recessive parent.

BACTERIOPHAGE [L. *bacterium* + Gk. *phagein*, to eat]: A virus that parasitizes a bacterial cell.

BACTERIUM [Gk. dim. of *baktron*, staff]: A small unicellular organism characterized by the absence of a formed nucleus; the genetic material may be dispersed throughout the cytoplasm.

BARK: All plant tissues outside of the cambium in a woody stem.

BASAL BODY: A cytoplasmic organelle that organizes cilia or flagella, identical in form and structure to a centriole.

BASAL METABOLISM: The metabolism of an organism when it is using just enough energy to maintain vital processes; measured as the amount of heat (in terms of quantity of oxygen consumed or carbon dioxide given off) produced by an animal at rest (but not asleep), determined at least 14 hours after eating and expressed as Calories per square meter of body surface (or per unit weight) per hour.

BASE: A substance that, on dissociation, releases hydroxyl (OH^-) ions but not hydrogen (H^+) ions; having a pH of more than 7; the opposite of an acid.

BASE-PAIRING RULE: The requirement that adenine must always pair with thymine (or uracil) and guanine with cytosine in a nucleic acid double helix.

BASIDIUM, pl. BASIDIA (ba-**sid**-i-um): A specialized reproductive cell of the Basidiomycetes, often club-shaped, in which nuclear fusion and meiosis occur.

BEHAVIORAL ISOLATING MECHANISMS: Modes of behavior, such as display rituals of courtship patterns, which serve to prevent mating between species.

BI— [L. *bis*, twice, double, two]: Prefix, meaning "two," "twice," or "having two points."

BIENNIAL [L. *annus*, year]: Occurring once in two years; a plant that requires two years to complete its reproductive cycle, with vegetative growth occurring in the first year and flowering, seed production, and death in the second.

BILATERAL SYMMETRY: An anatomical arrangement in which the right and left halves of an organism are approximate mirror images of each other.

BILE: A yellow secretion of the vertebrate liver, temporarily stored in the gallbladder and composed of organic salts that emulsify fats in the small intestine.

BINOMIAL SYSTEM: A system in which the name of an organism consists of two parts, the first designating the genus and the second designating the species.

BIO— [Gk. *bios*, life]: Prefix, meaning "pertaining to life."

BIOLOGICAL CLOCK: An unidentified internal factor (or factors) in plants and animals that governs the innate biological rhythms (growth and activity patterns) of the organism.

BIOMASS: Total weight of all organisms in a particular habitat or area.

BIOMES: A worldwide complex of communities, characterized by distinctive vegetation and climate; for example, the grassland areas collectively form the grassland biome, the tropical rain forests form the tropical rain forest biome, etc.

BIOSPHERE: The whole zone of air, land, and water at the surface of the Earth occupied by living things.

BIOSYNTHESIS: Formation of organic compounds from elements or simple compounds by living organisms.

BLADE: The broad, expanded part of a leaf or leaflike organ.

—BLAST, BLAST—, BLASTO— [Gk. *blastos,* sprout]: Suffix or prefix, meaning "pertaining to an embryo"; for example, osteoblast, an embryonic bone cell or a bone-forming cell.

BLASTODISC: Disklike area on the surface of a large yolky egg that undergoes cleavage and gives rise to the embryo.

BLASTOPORE: The opening connecting the archenteron of the gastrula stage of an embryo with the outside; represents the future mouth in some animals (protostomes), the future anus in others (deuterostomes).

BLASTULA: An animal embryo after cleavage and before gastrulation; usually consists of a hollow sphere the walls of which are composed of a single layer of cells.

BOND ENERGY: The energy required to break a bond.

BOWMAN'S CAPSULE: The bulbous unit of the nephron that encloses the glomerulus. It is the site of filtration of the renal fluid from the blood, the initial process in urine formation.

BRAINSTEM: The most posterior portion of the brain, includes medulla, pons, and midbrain.

BRONCHUS, pl. BRONCHI (**bronk**-us, **bronk**-eye) [Gk. *bronchos,* windpipe]: One of a pair of respiratory tubes branching into either lung at the lower end of the trachea; it subdivides into progressively finer passageways, the bronchioles, culminating in the alveoli.

BUD: (1) In plants, an embryonic shoot, including rudimentary leaves, often protected by specialized bud scales. (2) In animals, an asexually produced protuberance that develops into a new individual.

BUFFER: A substance that prevents appreciable changes of pH in solutions to which small amounts of acids or bases are added.

BULB: A modified bud with thickened leaves adapted for underground food storage.

BULK FLOW: The overall movement of a liquid induced by gravity, pressure, or an interplay of both.

CALORIE [L. *calor,* heat]: The amount of energy in the form of heat required to raise the temperature of 1 gram of water 1°C; in making metabolic measurements the kilocalorie (Calorie) is generally used. A Calorie is the amount of heat required to raise the temperature of 1 kilogram of water 1°C.

CALVIN CYCLE: The process by which carbon dioxide is reduced to carbohydrates during photosynthesis.

CAMBIUM [L. *cambiare,* to exchange]: See cork cambium, vascular cambium.

CAMERA EYE: An eye containing a lens system which concentrates light and focuses a precise image on a light-sensitive retina.

CAMOUFLAGE: Disguise resulting from color, pattern, shape, or a combination of these that tends to blend an organism with its surroundings and make it less visible to predator or prey.

CAPILLARY [L. *capillaris,* relating to hair]: A small, thin-walled blood vessel through which diffusion and filtration into the tissues occurs; connects arteries with veins.

CAPILLARY ACTION: The movement of water along a surface, against the action of gravity, resulting from the combined effect of water molecules cohering to each other and adhering to the molecules of the surface material.

CAPSULE (**kap**-sul): (1) A slimy layer around the cells of certain bacteria. (2) The sporangium of Bryophyta. (3) A dehiscent, dry fruit that develops from two or more carpels.

CARBOHYDRATE: An organic compound consisting of a chain of carbon atoms to which hydrogen and oxygen are attached in a 2:1 ratio; includes sugars, starch, glycogen, cellulose, etc.

CARNIVORE: An organism that obtains its food energy by eating animals.

CAROTENE [L. *carota,* carrot]: A yellow or orange pigment found in plants and some algae; converted into vitamin A in the vertebrate liver.

CAROTENOIDS: A class of pigments that includes the carotenes (yellows and oranges) and the xanthophylls (yellow); accessory pigments in photosynthesis.

CARPEL: A leaflike floral structure enclosing the ovule or ovules of the angiosperm, typically divided into ovary, style, and stigma; a flower may have one or more carpels, either single or fused.

CARTILAGE: The skeletal connective tissue of vertebrates; forms much of the skeleton of adult lower vertebrates and immature higher vertebrates.

CASPARIAN STRIP [after Robert Caspary, German botanist]: A thickened waxy strip that extends around and seals the walls of endodermal cells in plants, thus restricting the diffusion of solutes across the endodermis into the vascular tissues of the root.

CATALYST [Gk. *katalysis,* dissolution]: A substance that controls the rate of a chemical reaction but is not used up in the reaction; enzymes are catalysts.

CELL: The structural unit of protoplasm, composed of cytoplasm and one or more nuclei and surrounded by a membrane. In most plants, fungi, and bacteria there is a cell wall outside of the membrane.

CELL MEMBRANE: The outermost membrane of most animal cells.

CELL PLATE: A flattened structure that forms at the equator of the spindle in the dividing cells of land plants and a few green algae during early telophase; the predecessor of the middle lamella.

CELL WALL: A relatively rigid structure, produced by the cell and located outside the cell membrane in most plants, fungi, and bacteria; in plant cells, consists mostly of cellulose.

CELLULOSE: The chief constituent of the cell wall in all green plants; an insoluble complex carbohydrate formed of microfibrils of glucose molecules.

CENTRAL NERVOUS SYSTEM: In vertebrates, the brain and spinal cord; in invertebrates it usually consists of one or more cords of nervous tissue plus their associated ganglia.

CENTRIOLE (**sen**-tree-ole) [Gk. *kentron,* center]: A cytoplasmic organelle generally found in animal cells and in flagellated cells in other groups, usually outside of the nuclear membrane.

CENTROMERE (**sen**-tro-mere) [Gk. *kentron*, center + *meros*, a part]: See kinetochore.

CEPHALO– [Gk. *kephale*, head]: Prefix, meaning "head."

CEREBELLUM [L. dim. of *cerebrum*, brain]: An enlarged part of the dorsal side of the vertebrate brain; chief muscle-coordinating center.

CEREBRAL CORTEX: A layer of neurons (gray matter) forming the upper surface of the cerebrum, well developed only in mammals; the seat of conscious sensations and voluntary muscular activity.

CEREBRUM [L., brain]: The principal portion of the vertebrate brain, occupying the upper part of the cranium, consisting of two cerebral hemispheres united by the corpus callosum.

CHELICERA: First pair of appendages in arachnids; used for seizing and crushing prey.

CHEMICAL REACTION: A change of one or more substances into different substances by recombination of their constituent atoms into different kinds of molecules.

CHEMORECEPTORS: A cell or organ that detects substances according to their chemical structure; includes smell and taste receptors.

CHEMOTROPISM: The behavioral response of an organism to chemical stimulation; the turning to or away from a chemical stimulus.

CHIASMA, pl. CHIASMATA (**kye**-az-ma) [Gk. *chiasma*, a cross]: The X-shaped figure formed by the meeting of two nonsister chromatids of homologous chromosomes; the site of crossing over.

CHITIN (**kite**-n): A tough, resistant, nitrogen-containing polysaccharide present in the exoskeleton of arthropods, the epidermal cuticle or other surface structures of many other invertebrates, and the cell walls of certain fungi.

CHLORO– [Gk. *chlōros*, green]: Prefix, meaning "green."

CHLOROPHYLL: The green pigments of plant cells, necessary for photosynthesis.

CHLOROPLAST: A membrane-bound, chlorophyll-containing organelle in green plant cells; site of photosynthesis.

–CHORD, CHORDA– [L. *chorda*, cord, string]: Suffix or prefix, meaning "cord."

CHORDATE: Member of the animal phylum (Chordata) in which all members possess a notochord, dorsal nerve cord, and pharyngeal gill slits, at least at some stage of the life cycle.

CHORION (**core**-ee-on): The outermost embryonic membrane of reptiles, birds, and mammals; in placental mammals it contributes to the structure of the placenta.

CHROM– [Gk. *chrōma*, color]: Prefix, meaning "color."

CHROMATID (**crow**-ma-tid): One of the two daughter strands of a duplicated chromosome which are joined by a single centromere.

CHROMATIN (**crow**-ma-tin): The deeply staining nucleoprotein complex of the chromosomes.

CHROMOSOME [Gk. *soma*, body]: One of the bodies in the cell nucleus containing genes in a linear order; visualized as threads or rods of chromatin which appear in a contracted form during mitosis and meiosis.

CHROMOSOME MAP: A plan showing the relative position of the genes on the chromosome, determined chiefly by analysis of the relative frequency of crossing over between any two genes.

CILIUM, pl. CILIA (**silly**-um) [L. *cilium*, eyelash]: A short hairlike structure present on the surface of some cells, usually in large numbers and arranged in rows. Each cilium has a highly characteristic internal structure of two inner fibrils surrounded by nine pairs of outer fibrils.

CIRCADIAN RHYTHMS [L. *circa*, about + *dies*, day]: Regular rhythms of growth and activity that occur approximately on a 24-hour basis.

CLEAVAGE: The successive cell divisions of the fertilized egg to form the multicellular blastula.

CLIMAX COMMUNITY: Final or relatively stable community in a successional series.

CLINE [Gk. *klinein*, to lean]: A correlation between a series of gradual differences within a species with differences in climate or other geographical factors.

CLITORIS (**klit**-o-ris) [Gk. *kleitoris*, small hill]: A small erectile body at the anterior part of the vulva, homologous to the penis.

CLOACA [L., sewer]: The exit chamber from the digestive system; also may serve as exit for the reproductive system and urinary system.

CLONE [Gk. *klon*, twig]: A line of cells all of which have arisen from the same single cell by mitotic division; a population of individuals descended by asexual reproduction from a single ancestor.

CNIDOBLAST (**ni**-do-blast) [Gk. *knide*, nettle]: A stinging cell containing a nematocyst; characteristic of coelenterates.

COCHLEA [Gk. *kochlias*, snail]: Part of the inner ear of mammals; concerned with hearing.

CODON (**code**-on): Three adjacent nucleotides on a molecule of mRNA that form the code for a single amino acid.

–COEL, COELA–, COELO– [Gk., from *koilos*, hollow]: Suffix or prefix, meaning "cavity."

COELENTERON (see-**len**-t-ron): A digestive cavity with only one opening, characteristic of the phylum Coelenterata (jellyfish, hydra, corals, etc.).

COELOM (**seal**-um) [Gk. *koilos*, a hollow]: A body cavity formed between layers of mesoderm.

COENZYME: An organic molecule that plays an accessory role in enzyme-catalyzed processes, often by acting as a donor or acceptor of a substance involved in the reaction; NAD, NADP, and FAD are common coenzymes.

COHESION [L. *cohaerere*, to stick together]: The union or holding together of like molecules or like substances.

COHESION-TENSION THEORY: A theory accounting for upward water movement in plants. According to this theory, transpiration of a water molecule results in a negative (below 1 atmosphere) pressure in the leaf cells, inducing the entrance from the vascular tissue of another water molecule, which, because of the cohesive property of water, pulls with it a chain of water molecules extending up from the cells of the root tip.

COLEOPTILE (coal-ee-**op**-tile) [Gk. *koleon*, sheath + *ptilon*, feather]: A sheathlike structure covering the shoot of grass seedlings.

COLLAGEN [Gk. *kolla*, glue]: A fibrous protein material in bones, tendons, and other connective tissues.

COLON: The large intestine of vertebrates leading to the rectum or cloaca.

COLONY: A group of unicellular or multicellular organisms living together in close association.

COMMENSALISM [L. *com*, together + *mensa*, table]: The living together of two species in which one lives on or with the other without conferring either benefit or harm.

COMMUNITY: The organisms inhabiting a common environment and interacting with one another.

COMPETITION: Interaction between members of the same population or of two or more populations resulting from a greater demand for their supply of a mutually required resource.

COMPOUND [L. *componere*, to put together]: A combination of atoms in definite ratios, held together by chemical bonds; a substance containing only one kind of molecule, each molecule composed of two or more kinds of atoms.

COMPOUND EYE: In arthropods, a complex eye composed of a number of separate elements (ommatidia), each with light-sensitive cells and a refractory system that can form an image.

CONE: (1) In plants, the reproductive structure of a conifer. (2) In animals, a type of light-sensitive neuron in the vertebrate retina, concerned with the perception of color and with the most acute discrimination of detail.

CONJUGATION: The process in bacteria, protozoans, and certain algae and fungi by which genetic material is passed from one cell to another.

CONNECTIVE TISSUE: A type of tissue which lies between groups of nerves, glands, and muscle cells and beneath epithelial cells, in which the cells are irregularly distributed through a relatively large amount of intercellular material; includes bone, cartilage, blood, and lymph tissue.

CONVERGENT EVOLUTION [L. *convergere*, to turn together]: The independent development of similar structures in forms of life that are unrelated or only distantly related; often found in organisms living in similar environments, such as porpoises and sharks.

CORK: A secondary tissue produced by a cork cambium; made up of polygonal cells, nonliving at maturity, with walls infiltrated with suberin, a waxy or fatty material resistant to the passage of gases and water vapor.

CORK CAMBIUM: A lateral meristem producing cork in woody and some herbaceous plants; also called phellogen.

COROLLA (ko-**role**-a) [L. *corolla*, dim. of *corona*, wreath, crown]: Petals, collectively; usually the conspicuously colored flower parts.

CORTEX [L. *cortex*, bark]: (1) The outer layer. (2) In a stem or root, the primary tissue bounded externally by the epidermis and internally by the central cylinder of vascular tissue and consisting of parenchyma cells.

COTYLEDON (cottle-ee-don) [Gk. *kotyledon*, a cup-shaped hollow]: A leaflike structure of the embryo of a seed plant, concerned with digestion and storage of food.

COVALENT BOND: A chemical bond formed between atoms as a result of the sharing of a pair of electrons.

CROSS-FERTILIZATION: The mutual exchange of sperm between two hermaphroditic individuals and subsequent union of eggs and sperm, as in snails, earthworms.

CROSSING OVER: The exchange of corresponding segments of genetic material between chromatids of homologous chromosomes at meiosis.

CUTICLE (**ku**-tik-l) [L. *cuticula*, dim. of *cutis*, the skin]: Layer of waxy substance (cutin) on outer surface of plant cell walls.

CYCLOSIS (si-**klo**-sis) [Gk. *kyklosis*, circulation]: The circulation of protoplasm within a cell.

—CYTE, CYTO— [Gk. *kytos*, vessel, container]: Suffix or prefix, meaning "pertaining to cell."

CYTOCHROME: Heme-containing protein serving as an electron carrier in electron transport chains, involved in cellular respiration and photosynthesis.

CYTOKINESIS [Gk. *kinesis*, motion]: Division of the cytoplasm of a cell.

CYTOKININ [Gk. *kytos*, vessel + *kinesis*, motion]: One of a group of chemically related plant hormones that promote cell division, among other effects.

CYTOPLASM (**sight**-o-plazm): The living matter of a cell excluding the nucleus.

CYTOSTOME: In protozoans, the "mouth" opening.

DE— [L., away from, down, off]: Prefix, meaning "down," "away from," or "off"; for example, dehydration, "removal of water."

DECIDUOUS [L. *decidere*, to fall off]: Refers to plants that shed their leaves at a certain season.

DECOMPOSERS: Organisms (bacteria, fungi) in an ecosystem that convert dead organic material into plant nutrients.

DENDRITE [Gk. *dendron*, tree]: Nerve fiber, typically branched, that conducts impulses toward a nerve cell body.

DEOXYRIBONUCLEIC ACID (DNA) (dee-ox-y-rye-bo-new-**klee**-ick): The carrier of genetic information in cells, composed of two chains of phosphate, sugar molecules (deoxyribose), and purines and pyrimidines wound in a double helix; capable of self-replication as well as of determining RNA synthesis.

DERMIS [Gk. *derma*, skin]: The inner layer of the skin beneath the epidermis.

DEVELOPMENT: The progressive production of the phenotypic characteristics of an organism.

DI– [Gk. *dis*, twice, double, two]: Prefix, meaning "twice," "double"; like the Latin "bi-"; for example, dissect, "to cut in two."

DIAPHRAGM [Gk., from *diaphrassein*, to barricade]: A sheetlike muscle forming the partition between the abdominal and thoracic cavities.

DICOTYLEDON (**dye**-cottle-ee-don) [Gk. *kotyledon*, a cup-shaped hollow]: A subclass of angiosperms having two seed leaves or cotyledons among other distinguishing features; often abbreviated as dicot.

DIFFERENTIATION: The developmental process by which a relatively unspecialized cell or tissue undergoes a progressive change to become a more specialized cell or tissue.

DIFFUSION [L. *diffundere*, to pour out]: The movement of suspended or dissolved particles from a more concentrated to a less concentrated region as a result of the random movement of individual particles; the process tends to distribute them uniformly throughout a medium.

DIGESTION: The conversion of complex, usually insoluble foods into simple, usually soluble forms by means of enzymatic action.

DIKARYON (dye-**care**-ee-on) [Gk. *di*, two + *karyon*, kernel]: A cell or organism with paired but not fused nuclei derived from different parents. Found among fungi (Ascomycetes and Basidiomycetes).

DIOECIOUS (dye-**ee**-shuss) [Gk. *di*, two + *oikos*, house]: Having the male and female (or staminate and ovulate) elements on different individuals of the same species.

DIPLOID: The chromosome state in which each type of chromosome except for the sex chromosomes is represented twice ($2n$), in contrast to haploid ($1n$).

DOMINANT GENE: A gene that exerts its full phenotypic effect regardless of its allelic partner; a gene that masks the effect of its allele.

DORMANCY: A period during which growth ceases and is resumed only if certain requirements, as of temperature or day length, have been fulfilled.

DORSAL [L *dorsum*, back]: Pertaining to or situated near the back; opposite of ventral.

DOUBLE FERTILIZATION: The fusion of the egg and sperm (resulting in a $2n$ fertilized egg, the zygote) and the simultaneous fusion of the second male gamete with the polar nuclei (resulting in a $3n$ primary endosperm nucleus); a unique characteristic of all angiosperms.

DUODENUM (duo-**dee**-num) [L. *duodeni*, twelve each—from its length, about 12 fingers' breadth]: The upper portion of the small intestine in vertebrates, where food particles are broken down into molecules that can be absorbed by cells.

ECO– [Gk. *oikos*, house, home]: Prefix, meaning "house" or "home."

ECOLOGICAL DOMINANT: A species that, by virtue of size, number, or behavior, exerts a controlling influence on its environment and, as a result, determines what other kinds of organisms exist in that ecosystem.

ECOLOGICAL NICHE: The place occupied by a species in the community structure of which it is a part; the way in which an organism utilizes the resources of its ecosystem.

ECOLOGICAL SUCCESSION: The process by which a community goes through a number of temporary developmental stages, each characterized by a different species composition, before reaching a more stable condition, known as a climax community.

ECOLOGY [Gk. *logos*, a discourse]: The study of the interactions of organisms with their physical environment and with each other and of the results of such interactions.

ECOSYSTEM: All organisms in a community plus the associated environmental factors with which they interact.

ECTO– [Gk. *ektos*, outside]: Prefix, meaning "outside," "outer."

ECTODERM [Gk. *derma*, skin]: (1) The outermost layer of body tissue. (2) One of the three cell layers of the gastrula that give rise to all the tissues of the organism; the ectoderm gives rise to the epithelial tissue of the skin and sense organs, to the nerve cells, etc.

EFFECTOR: Cell, tissue, or organ (such as muscle or gland) capable of producing a response to stimuli.

EFFERENT [L. *ex*, out + *ferre*, to bear]: Carrying away from a center, applied to nerves and blood vessels.

EGG: A female gamete, or germ cell, which usually contains abundant cytoplasm and yolk; usually immotile, often larger than a male gamete.

ELECTRON: A subatomic particle with a negative electric charge equal in magnitude to the positive charge of the proton but with a mass $1/1,837$ that of the proton; normally orbits the atom's positively charged nucleus.

ELECTRON ACCEPTOR: Substance acting to receive electrons in an oxidation-reduction reaction.

ELECTRON CARRIER: A specialized protein, such as a cytochrome, that can gain and lose electrons reversibly and that functions to transfer electrons from organic nutrients to oxygen.

ELECTRON DONOR: Substance acting to donate electrons in an oxidation-reduction reaction.

ELEMENT: A substance composed of only one kind of atom; one of about 100 distinct natural or man-made types of matter which, singly or in combination, compose all materials of the universe.

EMBRYO [Gk. *en*, in + *bryein*, to swell]: The early developmental stage of an organism produced from a fertilized egg; a young organism before it emerges from the seed, egg, or the body of its mother.

ENDERGONIC: Energy-requiring, as in a chemical reaction; applied to an "uphill" process.

ENDO– [Gk. *endon*, within]: Prefix, meaning "within."

ENDOCRINE GLAND [Gk. *krinein*, to separate]: Ductless gland whose secretions (hormones) are released into the circulatory system; in vertebrates, includes pituitary, sex glands, adrenal, thyroid, and others.

ENDODERM [Gk. *derma*, skin]: (1) In animals, the innermost layer of body tissue. (2) One of the three cell layers of the gastrula that give rise to all the tissues of the organism; the endoderm gives rise to the epithelium that lines certain internal structures, such as most of the digestive tract and its outgrowths, most of the respiratory tract, and the urinary bladder. (3) In plants, a one-celled layer of specialized cells that lies between the cortex and the vascular tissues in young roots. The Casparian strip of the endodermis prevents diffusion of materials across the root.

ENDOMETRIUM: The glandular lining of the uterus in mammals; thickens in response to progesterone secretion during ovulation and is sloughed off in menstruation.

ENDOPLASMIC RETICULUM [L. *reticulum*, network]: An extensive system of double membranes present in most cells, dividing the cytoplasm into compartments and channels, often coated with ribosomes.

ENDOSPERM [Gk. *sperma*, seed]: In plants, a $3n$ tissue containing stored food that develops from the union of a male nucleus and the polar bodies of the egg; found only in angiosperms.

ENTROPY: The randomness or disorder of a system.

ENZYME: A protein molecule that regulates the rate of (catalyzes) a chemical reaction.

EPIDERMIS [Gk. *derma*, skin]: In plants and animals, the outermost layers of cells.

EPINEPHRINE: See adrenaline.

EPITHELIAL TISSUE [Gk. *thele*, nipple]: In animals, a type of tissue that covers a body or structure or lines a cavity; epithelial cells form one or more regular layers with little intercellular material.

EQUILIBRIUM: The state of a system in which no further net change is occurring.

ERYTHROCYTE (eh-**rith**-ro-site) [Gk. *erythros*, red + *kytos*, vessel]: Red blood cell, the carrier of hemoglobin.

ESTRUS [Gk. *oistros*, frenzy]: The mating period in female mammals, characterized by intensified sexual urge

EUKARYOTE (you-**car**-ry-oat) [Gk. *eu*, good + *karyon*, nut, kernel]: A cell having a membrane-bound nucleus, membrane-surrounded organelles, and multiple chromosomes in which the DNA is combined with special proteins; an organism composed of such cells.

EUSOCIAL: Applied to insect societies in which sterile individuals work on behalf of individuals involved in reproduction.

EVOLUTION [L. *e-*, out + *volvere*, to roll]: Any genetic change in organisms from generation to generation; the process by which, through a series of genetic changes or steps, any living organism or group of organisms acquires the characteristics that distinguish it from other organisms.

EXERGONIC: Energy-producing, as in a chemical reaction; applied to a "downhill" process.

EXOSKELETON: A skeleton covering the outside of the body; common in arthropods.

EXTRAEMBRYONIC MEMBRANES: Membranes formed of embryonic tissues that lie outside the embryo and are concerned with its protection and metabolism; include amnion, chorion, allantois, and yolk sac.

F_1 (FIRST FILIAL GENERATION): The offspring resulting from the crossing of plants or animals of the parental generation.

F_2 (SECOND FILIAL GENERATION): The offspring resulting from crossing members of the F_1 generation among themselves.

FALLOPIAN TUBES: In human females, the two reproductive tubes which carry the egg from the ovary to the uterus; oviducts.

FERTILIZATION: The fusion of two gametes, and so of their nuclei, to form a diploid or polyploid zygote.

FETUS [L. *fetus*, pregnant]: An unborn or unhatched vertebrate that has passed through the earliest developmental stages; a developing human from about the third month after conception until birth.

FIBRIL: Any minute, threadlike organelle within a cell.

FIBROUS PROTEIN: Insoluble structural protein, in which the polypeptide chain is extended or coiled along one dimension.

FILAMENT: (1) A chain of functionally separate cells. (2) In plants, the stalk of a stamen.

FISSION [L. *fissus*, split]: Asexual reproduction in cells or organisms by division of the body into two or more equal, or nearly equal, parts.

FITNESS: The relative ability to leave offspring.

FLAGELLUM, *pl.* FLAGELLA (fla-**jell**-um) [L. *flagellum*, whip]: A fine, long, threadlike structure, which protrudes from a cell body; it is longer than a cilium but has the same internal structure of nine pairs of microtubules encircling two central microtubules. Flagella are capable of vibratory motion and are used in locomotion and feeding.

FLOWER: The reproductive structure of angiosperms; a complete flower includes calyx, petals, stamens (male sex organs), and carpels (female sex organs), but some of these are absent in flowers of many species.

FOOD CHAIN, FOOD WEB: A sequence of organisms, including producers, herbivores, and carnivores, through which energy and materials move within an ecosystem.

FOSSIL [L. *fossilis*, dug up]: The remains of an organism, or direct evidence of its presence (such as tracks). May be an unaltered hard part (tooth or bone), a mold in a rock, petrifaction (wood or bone), unaltered or partially altered soft parts (a frozen mammoth).

FOSSIL FUELS: The remains of once-living organisms that are burned to release energy. Examples: coal, oil, and natural gas.

FREE ENERGY: That part of the total energy of a system that can do work under conditions of constant temperature and pressure.

FRUIT [L. *fructus*, fruit]: In plants, a matured, ripened ovary or group of ovaries and associated structures; formed from the ovule case of an angiosperm; contains the seeds.

FUNCTION: Characteristic role or action of any structure or process in the maintenance of normal metabolism or behavior of an animal.

GAMETE (**gam**-meet) [Gk., wife]: The mature functional haploid reproductive cell whose nucleus fuses with that of another gamete of an opposite sex (fertilization), with the resulting cell (zygote) developing into a new individual.

GAMETOPHYTE (gam-**meet**-o-fight): In plants having alternation of generations, the haploid (1n) gamete-producing generation.

GANGLION, *pl.* GANGLIA (**gang**-lee-on): Aggregate of nerve cell bodies.

GASTRULA [Gk. *gaster*, stomach]: An embryo in the process of gastrulation, the stage of development during which the blastula with its single layer of cells turns into a three-layered embryo, made up of ectoderm, mesoderm, and endoderm, often enclosing an archenteron.

GAUSE'S PRINCIPLE [after G. F. Gause, German geneticist]: The hypothesis that any two species with the same ecological requirements cannot coexist in the same locality and that occupation of that space will go to the species which is more efficient in utilizing the available resources.

GENE: A unit of heredity which is transmitted in the chromosome and which by interaction with internal and external environment controls the development of a trait; capable of self-replication. The sequence of nucleotides in a DNA molecule that dictates the amino acid sequence of a particular protein or the nucleotide sequence of an RNA molecule.

GENE FREQUENCY: The incidence (relative occurrence) of a particular allele in a population.

GENE POOL: All of the alleles of all of the genes in a population.

GENETIC CODE: The three-symbol system of base-pair sequences in DNA; referred to as a code because it determines the amino acid sequence in the enzymes and other protein components synthesized by the organism.

GENETIC ISOLATION: The inhibition of gene exchange between different species by morphological, behavioral, or physiological mechanisms.

GENOTYPE (**jean**-o-type): The genetic constitution, latent or expressed, of an organism, as contrasted with the phenotype; the sum total of all the genes present in an individual.

GENUS (**jean**-us): Inclusive group of related species.

GERM CELLS: Gametes or the direct antecedent cells of gametes.

GERMINATION: The resumption of growth by a spore or seed.

GIBBERELLINS (jibb-e-**rell**-ins) [Fr. *gibberella*, genus of fungi]: A group of chemically related plant growth hormones, whose most characteristic effect is stem elongation in dwarf plants and bolting.

GILL: The respiratory organ of aquatic animals, usually a thin-walled projection from some part of the external body surface or, in vertebrates, from some part of the digestive tract.

GLAND: A cell or organ producing one or more secretions which are discharged to the outside of the gland.

GLOBULAR PROTEIN: Protein in which the polypeptide chain is folded in three dimensions to form a globular shape.

GLOMERULUS (glom-**mare**-u-lus) [L. *glomus*, ball]: Cluster of capillaries enclosed by the Bowman's capsule; blood plasma minus large molecules filters through the walls of the glomerular capillaries into the renal tubules.

GLUCAGON: Hormone produced in pancreas that acts to raise concentration of blood sugar.

GLUCOSE: A six-carbon sugar ($C_6H_{12}O_6$); the most common monosaccharide in animals.

GLYCOGEN: A complex carbohydrate (polysaccharide); one of the main stored food substances of most animals and fungi; it is converted into glucose by hydrolysis.

GLYCOLYSIS (gly-**coll**-y-sis): The process by which the glucose molecule is changed anaerobically to two molecules of lactic acid with the liberation of a small amount of useful energy; reaction takes place in the cytoplasm.

GOLGI BODY (**goal**-jee): An organelle present in eukaryotic cells consisting of flat, disk-shaped sacs, tubules, and vesicles. It functions as a collecting and packaging center for substances the cell manufactures.

GONAD [Gk. *gone*, seed]: Gamete-producing organ of multicellular animals; ovary or testes.

GRANUM, *pl.* GRANA: A structure within the choloroplast, seen as a green granule with the light microscope and as a series of stacked thylakoids with the electron microscope. The grana contain chlorophylls and carotenoids and are the site of the light reactions of photosynthesis.

GROUND WATER: Water in the zone of saturation where all openings in rocks and soil are filled, the upper surface of which forms the water table.

HABITAT: The place normally occupied by a particular organism or population.

HALF-LIFE: The time required for disappearance or decay of one-half of a given substance.

HAPLOID [Gk. *haploos*, single]: Having only one of each type of chromosome (1n), in contrast to diploid (2n); characteristic of gametes and of the gametophyte generation in plants and of some microorganisms.

HARDY-WEINBERG LAW: The mathematical expression of the relationship between relative frequencies of two or more alleles in a population; it demonstrates that the frequencies of alleles in a gene pool are not changed by the process of sexual recombination.

HEME: The iron-porphyrin group of heme proteins.

HEMO–, HEMATO– [Gk. *haima*, blood]: Prefix, meaning "blood."

HEMOGLOBIN: The iron-containing protein in the blood that carries oxygen.

HEMOPHILIA: A hereditary disease in man characterized by failure of the blood to clot and excessive bleeding from even minor wounds.

HERBACEOUS (her-**bay**-shus): In plants, nonwoody.

HERBIVORE: An organism that eats plants or other photosynthetic organisms to obtain its food energy.

HEREDITY: The transfer of characteristics from parent to offspring by the transmission of genes from ancestor to descendent through the germ cells.

HERMAPHRODITE [Gk. *Hermes* and *Aphrodite*]: An organism possessing both male and female reproductive organs.

HETERO– [Gk. *heteros*, other, different]: Prefix, meaning "other" or "different."

HETEROSIS: Hybrid vigor, the superiority of the hybrid over either parent in any characteristic.

HETEROTROPH [Gk. *trophos*, feeder]: An organism that cannot manufacture organic compounds and so must feed on complex organic food materials that have originated in other plants and animals; in contrast to autotroph.

HETEROZYGOTE: A diploid organism that has two different alleles of the same gene.

HETEROZYGOTE SUPERIORITY: The greater fitness of a heterozygote as compared with the two homozygotes.

HIBERNATION: A period of dormancy and inactivity, varying in length depending on the organism and occurring in dry or cold seasons. During hibernation, metabolic processes are greatly slowed and, in mammals, body temperature may drop to just above freezing.

HOMEO–, HOMO–, HOMOLO– [Gk. *homos*, same, similar]: Prefix, meaning "similar" or "same."

HOMEOSTASIS (home-e-o-**stay**-sis) [Gk. *stasis*, standing]: The maintaining of a relatively stable internal physiological environment or equilibrium in an organism, population, or ecosystem.

HOMEOTHERM: Organism capable of maintaining a uniform body temperature independent of the environment; warm-blooded. Birds and mammals are homeotherms.

HOMINID: A member of the family of man, includes modern man and fossil man but not the apes.

HOMINOID: A hominid or one of the great apes.

HOMOLOGUES: Chromosomes that associate in pairs in the first stage of meiosis; each member of the pair is derived from a different parent.

HOMOLOGY [Gk. *homologia*, agreement]: Similarity in structure resulting from a common ancestry, regardless of function, such as the wing of a bird and the foreleg of a mammal.

HOMOZYGOTE: A diploid organism that has identical alleles of a particular gene.

HORMONE [Gk. *hormaein*, to excite]: A chemical substance secreted, usually in minute amounts, in one part of an organism and transported to another part of that organism where it has a specific effect.

HOST: An organism on or in which a parasite lives.

HYBRID: Offspring of two parents that differ in one or more heritable characters; offspring of two different varieties or of two different species.

HYDROGEN BOND: A weak molecular bond linking a hydrogen atom that is covalently bonded to another atom—usually oxygen, nitrogen, or fluorine—to the oxygen, nitrogen, or fluorine atom of another molecule.

HYDROLYSIS [Gk. *hydor*, water + *lysis*, loosening]: Splitting of one molecule into two by addition of H^+ and OH^- ions of water.

HYPER– [Gk., above, over]: Prefix, meaning "above" or "over."

HYPERTONIC [Gk. *hyper*, above + *tonos*, tension]: Having a concentration of solutes high enough to gain water across a selectively permeable membrane from another solution.

HYPHA [Gk. *hyphe*, web]: A single tubular filament of a fungus; the hyphae together make up the mycelium.

HYPO– [Gk., less than]: Prefix, meaning "under" or "less."

HYPOTHALAMUS [Gk. *thalamos*, inner room]: The floor and sides of the vertebrate brain just below the cerebral hemispheres; controls the autonomic nervous system and the pituitary gland and contains centers that regulate body temperature and appetite.

HYPOTHESIS [Gk. *hypo*–, under + *tithenai*, to put]: A temporary working explanation or supposition based on accumulated facts and suggesting some general principle or relation of cause and effect; a postulated solution to a scientific problem that must be tested by experimentation and, if not validated, discarded.

HYPOTONIC: Having a concentration of solutes low enough to lose water across a membrane to another solution.

IMPRINTING: A rapid and extremely narrow form of learning which occurs during a very short period in the early life of an organism, such as the following response in certain birds.

INBREEDING: The mating of individuals closely related genetically.

INDEPENDENT ASSORTMENT: See Mendel's second law.

INDUCTION [L. *inducere*, to induce]: The process in an animal embryo in which one tissue or body part causes the differentiation of another tissue or body part.

INSTINCT: A genetically determined pattern of behavior or response not based on the previous experience of the individual.

INSULIN: Hormone produced by the pancreas that lowers the concentration of sugar in the blood.

INTELLIGENCE: The capacity to profit by experience, based on analysis and association of ideas.

INTER– [L., between]: Prefix, meaning "between"; for example, intercellular, "between cells."

INTERPHASE: The stage between two mitotic or meiotic cycles.

INTRA— [L., within]: Prefix, meaning "within"; for example, intracellular, "within cells."

INVAGINATION [L. *in*, in + *vagina*, sheath]: The local infolding of a layer of tissue, especially in animal embryos, so as to form a depression or pocket opening to the outside.

ION (**eye**-on): An atom or molecule that has lost or gained one or more electrons. By this process, known as ionization, the atom becomes electrically charged.

IRRITABILITY: The ability to respond to stimuli or changes in environment; a general property of all living organisms.

ISO— [Gk. *isos*, equal]: Prefix, meaning "equal."

ISOGAMY: A type of sexual reproduction in algae and fungi in which the gametes are alike in size.

ISOLATING MECHANISMS: Mechanisms that prevent gene exchange between different species; may be morphological, behavioral, or physiological.

ISOTONIC: Having the same solute concentration as that of the substance on the other side of a selectively permeable membrane.

ISOTOPE [Gk. *topos*, place]: One of several possible forms of a chemical element, differing from other forms in the number of neutrons in the atomic nucleus and so in mass; some are unstable and emit radioactivity.

KARYOTYPE: The general appearance of the chromosomes with regard to number, size, and shape.

KERATIN [Gk. *keras*, horn]: One of a group of tough, fibrous proteins; a horny tissue formed by certain epidermal tissues, especially abundant in skin, claws, hair, feathers, and hooves.

KIDNEY: In vertebrates, the organ that regulates the balance of water and solutes and the excretion of nitrogen wastes in the form of urine.

KINETOCHORE: Region of constriction of chromosomes that holds sister chromatids together and to which the spindle fibers attach; also called the centromere.

LAMELLA (lah-**mell**-ah) [L. dim. of *lamina*, plate or leaf]: Layer, thin sheet.

LARVA [L., ghost]: An immature animal such as a caterpillar or tadpole that is morphologically very different from the adult.

LATERITE: A leached tropical soil with a high iron and aluminum concentration.

LEACHING: The removal of minerals and other elements from soil by the downward movement of water.

LEARNING: The process that produces adaptive change in individual behavior as the result of experience.

LEUKOCYTE [Gk. *leukos*, white + *kytos*, vessel]: White blood cell. Two of the principal types are neutrophiles, active in phagocytosis, and lymphocytes, involved in immune reactions.

LEUKOPLAST [Gk. *plastes*, molder]: In plants, a colorless cell organelle that serves as a starch repository; usually found in cells not exposed to light, such as roots and internal stem tissue.

LICHEN: Organism composed of a symbiotic alga and fungus.

LIFE CYCLE: The entire sequence of phases in the growth and development of any organism from time of zygote formation until gamete formation.

LIMITING FACTOR: Any environmental factor that determines whether or not an organism will grow in a particular environment.

LINKAGE: The tendency for certain genes to be inherited together owing to the fact that they are located on the same chromosome.

LIPID [Gk. *lipos*, fat]: One of a large variety of organic fat or fatlike compounds; includes fats, waxes, steroids, phospholipids, and carotenes.

LOCUS, *pl.* LOCI [L., place]: In genetics, the position of a gene on a chromosome.

LOOP OF HENLE [after F. G. J. Henle, German pathologist]: A hairpin-shaped portion of the kidney tubule system found in birds and mammals in which a hypertonic urine is formed by processes of osmosis and active transport.

LYMPH: Colorless fluid occurring in special lymph ducts, derived from blood by filtration through capillary walls.

—LYSIS, —LYTIC, —LYTE [Gk. *lysis*, a loosening]: Suffix, meaning "pertaining to dissolving."

LYSIS: Disintegration of cell.

LYSOGENIC BACTERIA (lye-so-**jenn**-ick): Bacteria carrying viruses (bacteriophages) which eventually break loose from the bacterial chromosome and set up an active cycle of infection, producing lysis in their bacterial hosts.

LYSOSOME: A membrane-surrounded organelle in which hydrolytic enzymes are segregated.

MACRO— [Gk. *makros*, large, long]: Prefix, meaning "large" or "long"; opposite of "micro-."

MACROMOLECULE: A molecule of very high molecular weight; refers specifically to proteins, nucleic acids, polysaccharides, and complexes of these.

MANTLE: (1) In mollusks, the outermost layer of the body wall or a soft extension of it; usually secretes a shell. (2) In geology, the region of Earth below the crust extending down to the molten outer core.

MARINE [L., from *mare*, the sea]: Living in salt water.

MARSUPIAL [Gk. *marsypos*, pouch, little bag]: A nonplacental mammal in which the female has a ventral pouch or folds surrounding the nipples; the premature young leave the uterus and crawl into the pouch where each one attaches itself by the mouth to a nipple until development is completed.

MASS-FLOW HYPOTHESIS: A hypothesis accounting for sap flow through the phloem system. According to this theory, the solution containing nutrient sugars is moved by pressure from the sieve tubes to the cells of the rest of the plant.

MECHANORECEPTOR: A sense cell or organ that receives mechanical stimuli such as those involved in touch, pressure, hearing, and balance.

MEDULLA (med-**dull**-a) [L. *medulla*, the innermost part]: (1) The inner as opposed to the outer part of an organ, as in the adrenal gland. (2) The most posterior region of the vertebrate brain; connects with the spinal cord.

MEDUSA: The free-swimming bell- or umbrella-shaped stage in the life cycle of many coelenterates; a jellyfish.

MEGA– [Gk. *megas*, great, large]: Prefix, meaning "large."

MEGASPORE: In plants, a haploid (1n) spore that develops into a female gametophyte.

MEIOSIS (my-o-sis) [Gk. *meioun*, to make smaller]: The two successive nuclear divisions in which the chromosome number is reduced from diploid (2n) to haploid (1n) and segregation and reassortment of the genes occur; gametes or spores may be produced as a result of meiosis.

MENDEL'S FIRST LAW: The factors for a pair of alternate characters are separate and only one may be carried in a particular gamete (genetic segregation).

MENDEL'S SECOND LAW: The inheritance of a pair of factors for one trait is independent of the simultaneous inheritance of factors for other traits, such factors "assorting independently" as though there were no other factors present (later modified by the discovery of linkage).

MENSTRUAL CYCLE [L. *mensis*, month]: In certain primates, the periodic discharge of blood and disintegrated uterine lining through the vagina.

MERI–, MERO–, –MER: Prefix or suffix, meaning "part" or "portion."

MERISTEM: The undifferentiated plant tissue from which new cells arise.

MESODERM [Gk. *derma*, skin]: One of the three primary cell layers in the early embryo that give rise to the tissues of the organism; the mesoderm gives rise to muscle, connective tissue, the circulatory system, and most of the excretory and reproductive systems.

MESSENGER RNA (mRNA): The RNA that carries genetic information from the gene to the ribosome, where it determines the order of the amino acids in the formation of a polypeptide.

METABOLISM [Gk. *metabole*, change]: The sum of all chemical reactions occurring within a living unit.

METAMORPHOSIS [Gk. *metamorphoun*, to transform]: Abrupt transition from larval to adult form, such as the transition from tadpole to adult frog.

METAPHASE [Gk. *meta*, middle + *phasis*, form]: The stage of mitosis or meiosis during which the chromosomes lie in the central plane of the spindle.

MICRO– [Gk. *mikros*, small]: Prefix, meaning "small."

MICROBE [Gk. *bios*, life]: A microscopic organism.

MICROFILAMENT: Small fibrous structure found in the cytoplasm of eukaryotic cells; appears to play a role in cellular and cytoplasmic motion.

MICRONUTRIENT: A mineral required in only minute amounts for plant growth, such as iron, chlorine, copper, manganese, zinc, molybdenum, and boron.

MICROSPORE: A spore which develops into a male gametophyte; in seed plants, it becomes a pollen grain.

MIDDLE LAMELLA: Distinct layer between adjacent cell walls, rich in pectic compounds, derived from the cell plate.

MIMICRY [Gk. *mimos*, mime]: The superficial resemblance in form, color, or behavior of certain organisms (mimics) to other more powerful or more protected ones (models), resulting in protection, concealment, or some other advantage for the mimic.

MINERAL: A naturally occurring chemical element or inorganic compound.

MITOCHONDRION, *pl.* MITOCHONDRIA: An organelle bound by a double membrane in which energy is captured in the form of ATP in the course of cellular respiration.

MITOSIS [Gk. *mitos*, thread]: Nuclear division characterized by exact chromosome duplication and the formation of two identical daughter cells; the means by which somatic cells divide.

MOLECULAR WEIGHT: The sum of the atomic weights of the constituent atoms in a molecule.

MOLECULE [L. *moles*, mass]: Smallest possible unit of a compound substance, consisting of two or more atoms.

MOLTING: Shedding of all or part of outer covering; in arthropods, periodic shedding of the exoskeleton to permit an increase in size.

MONO– [Gk. *monos*, single]: Prefix, meaning "one" or "single."

MONOCOTYLEDON [Gk. *kotyledon*, a cup-shaped hollow]: A subclass of angiosperms, characterized by a variety of features among which is the presence of a single seed leaf (cotyledon); abbreviated as monocot.

MONOECIOUS (mo-**nee**-shuss) [Gk. *monos*, single + *oikos*, house]: Hermaphroditic, having the anthers and carpels on the same individual but on different flowers.

MONOSACCHARIDE [Gk. *sakcharon*, sugar]: A simple sugar, such as five- and six-carbon sugars.

–MORPH, MORPHO– [Gk. *morphe*, form]: Suffix or prefix, meaning "form."

MORPHOGENESIS: The development of size, form, and other structural features of organisms.

MORPHOLOGY: The study of form and structure, at any level of organization.

MOTOR NEURON: Neuron that transmits nerve impulses from the central nervous system to skeletal muscle; efferent neuron.

MUSCLE FIBER: Muscle cell, a long, cylindrical, multinucleated cell containing numerous myofibrils.

MUTAGEN: An agent that increases the mutation rate.

MUTANT: A mutated gene or an organism carrying a gene that has undergone a mutation.

MUTATION [L., from *mutare*, change]: An inheritable change in the chromosomes; usually a change of a gene from one allelic form to another.

MYC–, MYCO– [Gk. *mykes*, fungus]: Prefix, meaning "pertaining to fungi."

MYCELIUM: The mass of hyphae forming the body of a fungus.

MYELIN SHEATH: A fatty material surrounding the axons of some nerve cells in vertebrates; made up of the membranes of Schwann cells.

MYO– [Gk. *mys*, muscle]: Prefix, meaning "muscle."

MYOFIBRIL: Contractile element of a muscle fiber, made up of thick and thin filaments arranged in sarcomeres.

MYOGLOBIN: An oxygen-binding, heme-containing globular protein found in muscles.

MYOSIN: One of the principal proteins in muscle; makes up the thick filaments.

MYX,– MYXO– [Gk. *myxa*, slime]: Prefix, meaning "slime."

NAD: Abbreviation of nicotinamide adenine dinucleotide, a coenzyme which functions as a hydrogen acceptor.

NATURAL SELECTION: The nonrandom reproduction of genotypes, resulting from the increased frequency of certain alleles in the population and the decreased frequency of others.

NECTAR [Gk. *nektar*, the drink of the gods]: A sugary fluid that attracts insects to plants.

NEMATOCYST [Gk. *nema*, *nematos*, thread + *kyst*, bladder]: A threadlike stinger, containing a poisonous or paralyzing substance, found in a cnidoblast of a coelenterate.

NEPHRIDIUM: A type of excretory organ found in many invertebrates.

NEPHRON [Gk. *nephros*, kidney]: A unit of the kidney structure in reptiles, birds, and mammals; a human kidney is composed of about a million nephrons.

NERVE: A group or bundle of nerve fibers with accompanying connective tissue.

NERVE FIBER: A filamentous process of a neuron; either dendrite or axon.

NERVE IMPULSE: A physiochemical excitation (excitement) transmitted along a nerve fiber from one part of an animal to another.

NERVOUS SYSTEM: All of the nerve cells of an animal; the receptor-conductor-effector system; in man, the nervous system consists of brain, spinal cord, and all nerves.

NEUR–, NEURO– [Gk. *neuron*, nerve]: Prefix, meaning "nerve."

NEURAL GROOVE: Dorsal longitudinal groove which forms in a vertebrate embryo; bordered by two neural folds; preceded by neural plate stage and followed by neural tube stage.

NEURAL PLATE: Thickened strip of ectoderm in early vertebrate embryos; forms along the dorsal side of the body and gives rise to the central nervous system.

NEURAL TUBE: Primitive hollow dorsal nervous system of the early vertebrate embryo formed by fusion of neural folds around neural groove.

NEUROHORMONE: A hormone secreted by a nerve ending; neurohormones include adrenaline, noradrenaline, and acetylcholine.

NEURON: A nerve cell, including the cell body, dendrites, and axon.

NEUROSECRETORY CELL: A neuron which produces one or more hormones.

NEUTRON (**new**-tron): An uncharged particle with a mass slightly greater than that of a proton. Found in the atomic nucleus of all elements except hydrogen, in which the nucleus consists of a single proton.

NICHE: See ecological niche.

NITRIFY: To convert organic nitrogen compounds into ammonium compounds, nitrates, and nitrites, as by nitrifying bacteria and fungi.

NITROGEN BASE: A nitrogen-containing molecule having basic properties (tendency to acquire an H atom); a purine or pyrimidine.

NITROGEN CYCLE: World-wide circulation and revitalization of nitrogen atoms, chiefly due to metabolic processes of plants and animals; plants take up inorganic nitrogen, convert it into organic compounds (chiefly proteins) which are assimilated into bodies of one or more animals; excretion and bacterial and fungal action on dead organisms return nitrogen atoms to inorganic state.

NITROGEN FIXATION: Incorporation of atmospheric nitrogen into inorganic nitrogen compounds available to green plants, a process that can be carried out only by certain microorganisms, or by higher plants in symbiotic association with microorganisms.

NODE [L. *nodus*, knot]: In plants, a joint of a stem; the place where branches and leaves are joined to stem.

NONDISJUNCTION: The failure of homologous chromosomes to separate during meiosis, resulting in one or more extra chromosomes in the cells of some offspring.

NOTOCHORD: A longitudinal, solid, elastic rodlike structure serving as the internal skeleton in the embryos of all chordates; in most adult chordates the notochord is replaced by a vertebral column which forms around (but not from) the notochord.

NUCLEAR ENVELOPE: The double membrane surrounding the nucleus within a cell.

NUCLEIC ACID: An organic acid consisting of nucleotides; the principal types are deoxyribonucleic acid (DNA) and ribonucleic acid (RNA).

NUCLEOLUS (new-**klee**-o-lus) [L. *nucleolus*, a small kernel]: A spherical body, containing DNA, RNA, and protein, present in nucleus of eukaryotic cells; site of production of ribosomal RNA.

NUCLEOTIDE: A single unit of nucleic acid composed of phosphate, a five-carbon sugar (either ribose or deoxyribose), and a purine or a pyrimidine.

NUCLEUS [L. *nucleus*, a kernal]: (1) The cellular structure that contains the genetic information in the form of DNA. In the eukaryotic cell, a specialized body bounded by a double membrane, containing the chromosomes. (2) The central part of an atom. (3) A group of nerve cell bodies in the central nervous system.

OCELLUS, pl. OCELLI [L. dim. of *oculus*, eye]: A simple light receptor common among invertebrates.

OLFACTORY [L. *olfacere*, to smell]: Pertaining to smell.

OLIGO– [Gk. *oligos*, few, small]: Prefix, meaning "few," "small"; for example, oligochaete, "having few bristles."

OMMATIDIUM, pl. OMMATIDIA [Gk. *ommos*, eye]: The single visual unit in the compound eye of arthropods; contains light-sensitive cells and a refractory system able to form an image.

OMNI– [L. *omnis*, all]: Prefix, meaning "all."

OMNIVOROUS: Eating "everything," for example, using both plants and animals as food.

ONTOGENY [Gk. *on*, being + *genesis*, origin]: The developmental history of an individual organism from zygote to maturity.

OO– [Gk. *oion*, egg]: Prefix, meaning "egg."

OOCYTE (**o**-us-sight) [Gk. *oion*, egg + *kytos*, vessel]: A cell that eventually gives rise by meiosis to an ovum.

OOGAMY: Kind of sexual reproduction in which one of the gametes (the egg) is large and nonmotile, while the other gamete (the sperm) is smaller and motile.

OPERATOR GENE: The gene in an operon which is affected by the repressor protein produced by the regulator gene for that particular operon and which, in turn, inhibits the activity of the other adjacent genes (the structural genes) in the operon.

OPERON: A group of adjacent genes whose functions are related to a particular synthetic activity and which are controlled by the repression of activation of a specific gene (the operator gene) in the group.

OPPOSABLE THUMB: Thumb that rotates at the joint so that the tip of the thumb can be placed opposite the tip of any one of the four fingers.

ORGAN [Gk. *organon*, tool]: A body part usually composed of several tissues grouped together in a structural and functional unit.

ORGANELLE: A formed body in the cytoplasm of a cell.

ORGANIC: Pertaining (1) to organisms or living things generally, or (2) to compounds formed by living organisms, or (3) to the chemistry of compounds containing carbon.

ORGANISM: Any individual living creature, either unicellular or multicellular.

ORGANIZER: Part of an embryo capable of inducing undifferentiated cells to follow a specific course of development; in particular, the dorsal lip of the blastopore in the amphibian.

OSMOSIS [Gk. *osmos*, impulse, thrust]: The movement of water between two solutions separated by a membrane that permits the free passage of water and that prevents or retards the passage of the solvent; the water tends to move from the side containing a lesser concentration of solute to the side containing a greater concentration.

OSMOTIC PRESSURE: Pressure generated by osmotic flow of water.

OV–, OVI– [L. *ovum*, egg]: Prefix, meaning "egg."

OVARY: (1) In animals, the egg-producing organ. (2) In flowering plants, the enlarged basal portion of a carpel or a fused carpel, containing the ovule or ovules; the ovary matures to become the fruit.

OVIDUCT [L. *ductus*, duct]: The tube serving to transport the eggs to the uterus or to the outside; Fallopian tubes (in humans).

OVULATION: Release of an animal egg from the ovary.

OVULE: In seed plants, the megasporangium, a structure composed of a protective outer coat, a tissue specialized for food storage, and a female gametophyte with egg cell; becomes a seed after fertilization.

OVUM, pl. OVA: The egg cell; female gamete.

OXIDATION: Loss of an electron by an atom. Oxidation and reduction (gain of an electron) take place simultaneously since an electron that is lost by one atom is accepted by another. Oxidation-reduction reactions are an important means of energy transfer within living systems.

OXIDATIVE PHOSPHORYLATION: Formation of ATP molecules from the energy of glycolusis.

PACEMAKER: Area of heart that initiates the heartbeat; located where the superior vena cava enters the right atrium; sinoatrial node.

PALEO– [Gk. *palaios*, old]: Prefix, meaning "old."

PALEONTOLOGY: The study of the life of past geologic times, principally by means of fossils.

PALISADE CELLS: In plants, columnar chloroplast-containing cells [part of the mesophyll ("middle leaf")].

PANCREAS (**pang**-kree-us) [Gk. *pan*, all + *kreas*, meat, flesh]: A small complex gland in vertebrates, located between the stomach and the duodenum, which produces digestive fluids and the hormones insulin and glucagon.

PARA– [Gk., from *para*, beside, akin to]: Prefix, meaning "akin to," "at the side of," or "beside."

PARASITE [Gk. *sitos*, food]: An organism that lives on or in an organism of a different species and derives nutrients from it.

PARASYMPATHETIC NERVOUS SYSTEM: A subdivision of the autonomic nervous system of vertebrates with centers located in the brain and in the most anterior part and most posterior parts of the spinal cord; stimulates digestion and has a generally inhibitory effect on other functions. Neurohormone: acetylcholine.

PARENCHYMA (pah-**renk**-ee-ma) [Gk. *en*, in + *chein*, to pour]: A plant tissue composed of living, thin-walled, randomly arranged cells with large vacuoles; usually photosynthetic or storage tissue.

PARTHENOGENESIS [Gk. *parthenos*, virgin]: The development of an egg without fertilization.

PECKING ORDER: The dominance hierarchy in poultry flocks in which each individual according to its rank can peck a number of subordinate fowl with impunity and must submit to pecking by a number of superiors without retaliation.

–PED, –PEDIA, PEDI– [L. *pes, pedis*, foot]: Suffix or prefix, meaning "foot"; like the Greek suffix "-pod"; for example, bipedal, "two-footed."

PELLICLE [L. dim. of *pellis*, skin]: A thin translucent envelope located outside of the cell membrane in many protozoans.

PENT–, PENTA– [Gk. *pente*, five]: Prefix, meaning "five"; for example, pentose, "five-carbon sugar."

PEPTIDE: Two or more amino acids linked together; molecules made up of a relatively small number of amino acids (2 to about 100) are called peptides while those formed of a larger number of amino acids are called polypeptides or proteins.

PEPTIDE BOND: The type of bond formed when two amino acid units are joined end to end; the acidic group (—COOH) of one amino acid is attached to the basic group (—NH$_2$) of the next, and a molecule of water (H$_2$O) is removed.

PERENNIAL [L. *per*, through + *annus*, year]: A plant which persists in whole or in part from year to year and flowers in more than one year.

PERI– [Gk., around]: Prefix, meaning "around"; for example, peristomial, "around the mouth"; peristalsis, "wavelike compression around a tubular organ," such as the gut.

PERITONEUM [Gk. *peritonos*, stretched over]: A mesodermal epithelial membrane lining the body cavity and forming the external covering of the visceral organs.

PERMEABLE [L. *permeare*, to pass through]: Penetrable, usually applied to membranes which let given substances pass through.

PETIOLE (**pet**-ee-ole): The stalk of a leaf.

pH: A symbol denoting the relative concentration of hydrogen ions in a solution; pH values range from 0 to 14; the lower the value the more acid a solution, that is, the more hydrogen ions it contains; pH 7 is neutral, less than 7 is acid, more than 7 is alkaline.

PHAGO–, –PHAGE [Gk. *phagein*, to eat]: Prefix or suffix, meaning "eating."

PHAGOCYTE (**fag**-o-sight): Any cell that engulfs foreign particles.

PHAGOCYTOSIS: Cell "eating"; the intake of solid particles by a cell, by flowing over and engulfing them; characteristic of amoebas, digestive cells of some invertebrates, and vertebrate white blood cells.

PHENOTYPE [Gk. *phainein*, to show]: The observable properties of an organism resulting from interaction between its genetic constitution (genotype) and the environment.

PHEROMONE (**fair**-o-moan): Substance secreted by an animal that influences the behavior or morphological development or both of other animals of the same species, such as sex attractants of moths, odor trail of ants.

PHLOEM (**flow**-em) [Gk. *phloos*, bark]: Vascular tissue that conducts sugars and other organic molecules from the leaves to other parts of the plant; composed of sieve cells (in gymnosperms) or sieve tubes and companion cells (in angiosperms), parenchyma, and fibers.

PHORO–, –PHORE [Gk. *-phoros*, bearing, carrying]: Prefix or suffix, meaning "bearing."

PHOTO– [Gk. *photos*, light]: Prefix, meaning "light."

PHOTON: Unit of light energy.

PHOTOPERIODISM: The response to relative day and night length, a mechanism by which organisms measure seasonal change.

PHOTOPHOSPHORYLATION: Formation of ATP molecules in the chloroplast using radiant energy.

PHOTORECEPTOR: A cell or organ capable of detecting light.

PHOTOSYNTHESIS [Gk. *syn*, together + *tithenai*, to place]: The conversion of light energy to chemical energy; the production of carbohydrate from carbon dioxide and water in the presence of chlorophyll, using light energy.

PHOTOTROPISM [Gk. *trope*, turning]: Movement in which the direction of the light is the determining factor, as the growth of a plant toward a light source; turning or bending response to light.

PHYCO– [Gk. *phykos*, seaweed]: Prefix, referring to aquatic plants; for example, phycomycete, "primitive (often aquatic) fungus."

PHYLLO–, –PHYLL [Gk. *phyllon*, leaf]: Prefix or suffix, meaning "leaf."

PHYLOGENY [Gk. *phylon*, race, tribe]: Evolutionary relationships among organisms; developmental history of a group of organisms.

PHYSIOLOGY [Gk. *physis*, nature + *logos*, a discourse]: The study of function in cells, organs, or entire organisms.

PHYTO–, –PHYTE [Gk. *phyton*, plant]: Prefix or suffix, meaning "plant."

PHYTOCHROME: A pigment found in green plants. It is a photoreceptor for red or far-red light and is involved with a number of developmental processes, such as flowering, dormancy, leaf formation, and seed germination.

PHYTOPLANKTON [Gk. *planktos*, wandering]: Aquatic free-floating microscopic photosynthetic organisms.

PIGMENT: A substance that absorbs light.

PINOCYTOSIS [Gk. *pinein*, to drink]: Cell "drinking"; the intake of fluid droplets by a cell.

PITUITARY [L. *pituita*, phlegm]: Endocrine gland in vertebrates; the anterior lobe (adenohypophysis) is the source of tropic hormones, growth hormone, and prolactin and is stimulated by neurosecretory cells in the hypothalamus; the posterior lobe stores and releases oxytocin and ADH produced by the hypothalamus.

PLACENTA [Gk. *plax*, a flat object]: A structure formed in part from the inner lining of the uterus and in part from the extra-embryonic membranes; develops in most species of mammals

and serves as the connection between the mother and the embryo during pregnancy through which exchanges occur between the blood of mother and of embryo.

PLANKTON [Gk. *planktos,* wandering]: Free-floating, mostly microscopic, aquatic organisms, both photosynthetic (phytoplankton) and heterotrophic (zooplankton).

PLANULA [L. dim. of *planus,* flat]: The ciliated, free-swimming larval type occurring in many coelenterates.

PLASMA–, PLASMO–, –PLAST [Gk. *plasma,* form, mold]: Prefix or suffix, meaning "formed" or "molded"; for example, protoplasm, "first-molded" (living matter); chloroplast, "green-formed" (body).

PLASMA: The clear, colorless fluid component of blood, containing dissolved salts and proteins; blood minus the corpuscles.

PLASMODESMATA [Gk. *plassein,* to mold + *desmos,* band, bond]: Minute cytoplasmic threads that extend through the pores in cell walls and connect the protoplasts of adjacent cells in higher plants.

PLASTID: A cytoplasmic, often pigmented organelle in plant cells (three types are leucoplasts, chromoplasts, and chloroplasts).

PLATELET (**plate**-let): In mammals, a minute, granular body suspended in the blood and involved in the formation of blood clots.

PLEIOTROPY (**plee**-o-trope-ee): The capacity of a gene to affect a number of different characteristics.

POLAR: Having parts or areas with opposed or contrasting properties, such as positive and negative charges, head and tail.

POLAR BODY: Minute nonfunctioning cell produced during meiotic divisions in egg cells; contains a nucleus but very little cytoplasm.

POLLEN [L., fine dust]: The male gametophytes of seed plants at the stage in which they are shed.

POLLINATION: The transfer of pollen from where it was formed (the anther) to a receptive surface (the stigma, in angiosperms) associated with an ovule of a flower.

POLY– [Gk. *polys,* many]: Prefix, meaning "many."

POLYGENIC INHERITANCE: The determination of a given characteristic, such as weight or height, by the complex interaction of many genes.

POLYMER: A large molecule composed of many identical molecular subunits.

POLYMORPHISM [Gk. *morphe,* form]: Occurrence together of two or more morphologically distinct forms of a population.

POLYP [Gk. *poly-,* many + *pous,* foot]: The sessile stage in the life cycle of coelenterates; a single attached individual of any colonial or solitary coelenterate.

POLYPEPTIDE: A molecule consisting of numerous amino acids linked together by peptide bonds.

POLYPLIODY [Gk. *ploos,* fold or times]: The possession of more than two complete sets of chromosomes per cell.

POLYSACCHARIDE: A carbohydrate composed of many joined monosaccharide units in a long chain; for example, glycogen, starch, cellulose.

POPULATION: Any group of individuals of one species; in genetic terms, an interbreeding group of organisms.

POST–, POSTERO– [L., behind, after]: Prefix, meaning "at," "near," or "toward the hind part"; opposite of "pre-," and "antero-."

POSTERIOR: Of or pertaining to the rear end; in man, the back of the body is said to be posterior.

PRE– [L., before, in front of]: Prefix, meaning "before," or "in front of"; opposite of "post-."

PREADAPTATION: The possession by an organism of a structure or function that develops new uses in the course of evolution.

PRIMARY GROWTH: In plants, growth originating in the apical meristem of the shoots and roots, as contrasted with secondary growth; results in increase in length.

PRIMARY STRUCTURE OF PROTEINS: The amino acid sequence of a protein and its inter- and intrachain disulfide bridges.

PRIMATE: A member of the order of mammals that includes man, apes, monkeys (anthropoids), and lower primates (prosimians).

PRIMITIVE [L., from *primus,* first]: Not specialized; at an early stage of evolution or development.

PRIMITIVE STREAK: The thickened dorsal longitudinal strip of ectoderm and mesoderm in early avian, reptilian, and mammalian embryos; equivalent to blastopore in other forms.

PROCAMBIUM: In plants, a primary meristematic tissue; gives rise to vascular tissues of primary plant body and to the vascular cambium.

PROGESTERONE: In mammals, steroid hormone produced by corpus luteum that helps prepare the uterus for reception of the ovum.

PROKARYOTE [L. *pro,* before + Gk. *karyon,* with nut, kernel]: A cell lacking a membrane-bound nucleus or membrane-bound organelles; a bacterium or a blue-green alga.

PROPHASE [Gk. *pro,* before + *phasis,* form]: An early stage in nuclear division, characterized by the condensing of the chromosomes and their movement to the equator of the spindle.

PROPRIOCEPTOR [L. *proprius,* one's own]: Internal sensory receptor that gives information to the brain regarding movements, position of body, or muscle stretch.

PROSIMIAN: A lower primate, includes lemurs, lorises, tarsiers, tree shrews, as well as many fossil forms.

PROSTAGLANDIN: One of a group of fatty acid hormones discovered in semen, now found to be present in many other tissues. Believed to play a role in fertilization.

PROSTATE GLAND: A mass of muscle and glandular tissue surrounding the base of the urethra in male mammals, through which the seminal vesicle passes and which secretes an alkaline fluid that has a stimulating effect on the sperm as they are released.

PROTEIN [Gk. *proteios,* primary]: A complex organic compound composed of one or more polypeptide chains, each made up of many (about 100 or more) amino acids joined by peptide bonds.

PROTO– [Gk. *protos,* first]: Prefix, meaning "first"; for example, Protozoa, "first animals."

PROTON: A subatomic particle with a single positive charge equal in magnitude to the charge of an electron and with a mass of 1; a component of every atomic nucleus.

PROTOPLASM [Gk. *plasma*, anything molded]: The living substance of all cells.

PROTOPLAST: A plant cell body, not including the cell wall.

PSEUDO– [Gk. *pseudes*, false]: Prefix, meaning "false."

PSEUDOPLASMODIUM: A multicellular mass of individual amoeboid cells, representing the aggregate phase in the cellular slime molds.

PSEUDOPOD [Gk. *pous*, pod-, foot]: A temporary cytoplasmic protrusion from an amoeboid cell which functions in locomotion or in feeding by phagocytosis.

PULMONARY ARTERY [L. *pulmonis*, lung]: In vertebrates, the artery carrying blood to the lungs.

PULMONARY VEIN: In vertebrates, a vein carrying oxygenated blood from the lungs to the left atrium, from which blood is pumped into the left ventricle and from there to the body tissues.

PUNNETT SQUARE: The checkerboard diagram used for analysis of gene segregation.

PUPA [L., girl, doll]: A developmental stage, nonfeeding, immotile, and sometimes encapsulated or in a cocoon, between the larval and adult phases in insects.

PURINE: A nitrogenous base such as adenine or guanine; one of the components of nucleic acids.

PYLORIC SPHINCTER: A ring of muscle at the junction of the stomach and duodenum which controls the entrance of the stomach contents into the duodenum.

PYRAMID OF ENERGY: Energy relationships between various feeding levels involved in a particular food chain; plants (at the base of the pyramid) represent the greatest amount of energy utilization, herbivores next, then primary carnivores, secondary carnivores, etc.

PYRIMIDINE: A nitrogenous base such as cytosine, thymine, or uracil; one of the components of nucleic acids.

QUEEN: The fertile, or fully developed, female of social bees, ants, and termites whose function is to lay eggs.

RADIAL SYMMETRY: The regular arrangement of parts around one longitudinal axis; any line drawn through this oral-aboral axis will divide similar halves; seen in coelenterates and adult echinoderms.

RADIATION [L., a spoke of a wheel, hence, a ray]: The movement of energy from one place to another.

RADIOACTIVE ISOTOPE: An isotope with an unstable nucleus that stabilizes itself by emitting radiation.

RECESSIVE GENE [L. *recedere*, to recede]: A gene whose phenotypic expression is masked by a dominant allele and so is manifest only in the homozygous condition. Heterozygotes involving recessives are phenotypically indistinguishable from dominant homozygotes.

RECOMBINATION: In genetics, the appearance of gene combinations in progeny which differ, as a result of the sexual process, from the combinations present in the parents.

REDUCTION [L. *reducere*, to lead back]: Gain of an electron by a compound; takes place simultaneously with oxidation (loss of an electron by an atom) since an electron that is lost by one atom is accepted by another.

REFLEX [L. *reflectere*, to bend back]: Unit of action of the nervous system involving a sensory neuron, often an interneuron or neurons, and one or more motor neurons.

REGULATOR GENE: A gene that influences whether or not a product, such as a polypeptide, is produced by another gene.

RENAL [L. *renes*, kidneys]: Pertaining to the kidney.

REPRESSOR: The substance produced by a regulator gene that represses protein formation.

RESOLVING POWER: Ability of a lens to distinguish two lines as separate.

RESPIRATION [L. *respirare*, to breathe]: (1) In organisms, the intake of oxygen and the liberation of carbon dioxide. (2) In cells, the oxidative breakdown and release of energy from fuel molecules.

RETICULAR FORMATION: A core of tissue which runs centrally through the entire brainstem; a weblike network of fibers and neurons; involved with consciousness.

RETICULUM [L. *reticulum*, network]: Any network of weblike structure, as in endoplasmic reticulum.

RETINA: The innermost nervous-tissue layer of the eyeball; contains several layers of neurons and light-receptor cells (rods and cones); receives the image formed by the lens and is connected to the brain by the optic nerve.

RHIZOID [Gk. *rhiza*, root]: Rootlike anchoring structure in nonvascular plants.

RHIZOME [Gk. *rhizoma*, mass of roots]: In plants, a horizontal underground stem, often enlarged for storage.

RIBONUCLEIC ACID (RNA) (rye-bo-new-**klee**-ick): A nucleic acid formed on chromosomal DNA and involved in protein synthesis; similar in composition to DNA, except that the pyrimidine uracil replaces thymine. RNA is the genetic material of many viruses.

RIBOSOME: A small organelle composed of protein and ribonucleic acid; the site of protein synthesis.

ROD: Light-sensitive nerve cell found in the vertebrate retina; sensitive to very dim light, responsible for "night vision."

ROOT: The descending axis of a plant, normally below ground and serving both to anchor the plant and to take up and conduct water and minerals.

SAPROPHYTE [Gk. *sapros*, rotten + *phyton*, plant]: An organism that secures its food directly from nonliving organic matter.

SARCOMERE: Functional and structural unit of contraction in striated muscle.

SECONDARY GROWTH: In plants, growth derived from secondary meristem, that is, the vascular cambium and the cork cambium; results in an increase in diameter and in the production of woody tissue.

SECONDARY SEX CHARACTERISTICS: External characteristics that distinguish between the two sexes but which have no direct role in reproduction, such as the rooster's comb.

SECRETION: Product of any cell, gland, or tissue which is released through the cell membrane.

SEED: A complex organ formed by the maturation of the ovule of seed plants following fertilization. In conifers, it consists of seed coat, embryo, and female gametophyte storage tissue. In angiosperms, it consists of seed coat and embryo; some angiosperm seeds also contain endosperm, a storage tissue.

SEGREGATION: The separation of two alleles into different gametes during meiosis; see Mendel's first law.

SELF-FERTILIZATION: The union of egg and sperm produced by a single hermaphroditic organism.

SELF-POLLINATION: The transfer of pollen from anther to stigma in the same flower or to another flower of the same plant, leading to self-fertilization.

SEMEN [L., seed]: Product of male reproductive system; includes sperm and the sperm-carrying fluid.

SENSILLUM, pl. SENSILLA: Sensory-receptor unit on body surface of arthropods; responsive to touch, smell, taste, and vibration (sound).

SENSORY NEURON: A neuron that carries impulses from a receptor to the central nervous system or central ganglion.

SENSORY RECEPTOR: A cell, tissue, or organ that detects internal or external stimuli.

SESSILE [L., from sedere, to sit]: Attached, not free to move about.

SEX CHROMOSOMES: Special sex-determining chromosomes, not occurring in identical numbers or shapes in both sexes.

SEX-LINKED CHARACTERISTIC: A genetic characteristic, such as color blindness, determined by a gene located either on the X or Y chromosome.

SEXUAL REPRODUCTION: Reproduction involving meiosis and syngamy (the union of gametes).

SHOOT: The aboveground portions, such as the stem and leaves, of a vascular plant.

SIEVE CELL: A long and slender sugar-conducting cell with relatively unspecialized sieve areas and with tapering end walls that lack sieve plates; found in the phloem of gymnosperms.

SIEVE TUBE: A vertical series of sugar-conducting cells (sieve-tube elements) of the phloem of angiosperms.

SMOOTH MUSCLE: Nonstriated muscle; lines the walls of internal organs and arteries and is under involuntary control.

SOCIAL DOMINANCE: A hierarchical pattern of physical domination of some members of a group by other members in a relatively orderly and long-lasting pattern.

SOLUTION: A mixture (usually liquid) in which one or more substances (the solute) are dispersed in the form of separate molecules or ions throughout the entire substance.

SOMATIC CELLS: The differentiated, usually diploid ($2n$) cells composing body tissues of multicellular plants and animals; all body cells except the germ cells.

SOMATIC NERVOUS SYSTEM: In vertebrates, the motor, sensory, and interneurons; the "voluntary" system, as contrasted with the "involuntary," or autonomic, nervous system.

SOMATO [Gk. soma, body]: Prefix, meaning "body," or "of the body."

SOMITE: One of the segments into which the body of many animals is divided, especially an incompletely developed embryonic segment.

SPECIALIZED: (1) Of organisms, having special adaptations to a particular habitat or mode of life. (2) Of cells, having particular functions.

SPECIES, pl. SPECIES [L., kind, sort]: A group of organisms that actually (or potentially) interbreed and are reproductively isolated from all other such groups.

SPECIES-SPECIFIC: Characteristic of (and limited to) a particular species.

SPECIFIC HEAT: The amount of heat (in calories) required to raise the temperature of 1 gram of a substance 1° C; the specific heat of water is 1 calorie.

SPECIFICITY: Uniqueness, as in proteins in a given organism and of enzymes in given reactions.

−SPERM, SPERME−, SPERMA−, SPERMATO− [Gk. sperma, seed]: Suffix or prefix meaning "seed."

SPERM [Gk. sperma, seed]: A mature male sex cell or gamete, usually motile and smaller than the female gamete.

SPERMATHECA: Receptacle for sperm storage, found in many types of female invertebrates.

SPERMATID: Each of four haploid cells resulting from the meiotic divisions of a spermatocyte, which become differentiated into a sperm cell.

SPERMATOCYTES: The diploid cells formed by the enlargement of the spermatogonia; give rise by meiotic division to the spermatids.

SPERMATOGENESIS [Gk. genesis, origin]: The process by which spermatogonia develop into sperm.

SPERMATOGONIA: The unspecialized diploid ($2n$) germ cell on the walls of the testes which by meiotic division become spermatocytes, then spermatids, then spermatozoa or sperm cells.

SPERMATOZOON, pl. SPERMATOZOA: A sperm cell.

SPINAL CORD: The part of the vertebrate central nervous system that consists of a thick longitudinal bundle of nerve fibers extending from the brain posteriorly along the dorsal side.

SPIRACLE [L. spirare, to breathe]: One of the external openings of the respiratory system in terrestrial arthropods.

SPORA–, SPORA– [Gk. *spora*, seed]: Prefix, meaning "seed."

SPORANGIOPHORE (spo-**ran**-ji-o-for) [Gk. *phore*, from *phorein*, go bear]: A branch bearing one or more sporangia.

SPORANGIUM, *pl.* SPORANGIA: A hollow unicellular or multicellular structure in which spores are produced.

SPORE: An asexual reproductive cell capable of developing into an adult without fusion with another cell; in contrast to a gamete.

SPOROPHYLL: Spore-bearing leaf; the carpels and stamens of flowers are modified sporophylls.

SPOROPHYTE: The spore-producing diploid ($2n$) phase in the life cycle of a plant having alternation of generations.

STAMEN [L., a thread]: The male organ of a flower which produces microspores or pollen; usually consists of a stalk, the filament, bearing an anther at its apex.

STARCH [M.E. *sterchen*, to stiffen]: A complex insoluble carbohydrate, the chief food-storage substance of plants, which is composed of several hundred glucose units ($C_6H_{12}O_6$); it is readily broken down enzymatically into these glucose units.

STATO– [Gk. *statos*, standing stationary, positioned]: Prefix meaning "stationary."

STATOCYST [Gk. *kystis*, sac]: An organ of balance, consisting of a vesicle containing granules of sand (statoliths) or some other material which stimulates sensory cells when the animal moves.

STEM: The part of the axis of vascular plants that is above ground, as well as anatomically similar portions below ground (such as rhizomes).

STEREOSCOPIC VISION [Gk. *stereos*, solid + *optikos*, pertaining to the eye]: Three-dimensional viewing of an object.

STIGMA: (1) The red eyespot of algae. (2) The region of a carpel serving as a receptive surface for pollen grains on which they germinate.

STIMULUS [L., goad, incentive]: Any internal or external change which influences the activity of an organism or of part of an organism.

–STOME, –STOMA, –STOMATA [Gk. *stoma*, mouth]: Suffix meaning "mouth"; for example, cytostome, "cell mouth."

STRIATED MUSCLE [L., from *striare*, to groove]: Skeletal, voluntary muscle and cardiac muscle.

STROMA: The granular substance that makes up the interior of the chloroplast in which the grana are dispersed.

STRUCTURAL GENE: One of the genes in an operon whose function is controlled by the operator and regulator genes.

SUB–, SUS– [L., under, below]: Prefix meaning "under" or "below"; for example, subepidermal, "underneath the epidermis."

SUBSTRATE [L. *substratus*, strewn under]: (1) The foundation to which an organism is attached. (2) A substance acted on by an enzyme.

SUCCESSION: In ecology, the slow, orderly progression of changes in species composition of a community during development of vegetation in any area.

SUCROSE: A common sugar; a disaccharide found in many plants.

SUGAR: Any monosaccharide or disaccharide.

SYM–, SYN– [Gk. *syn*, together, with]: Prefix meaning "together."

SYMBIOSIS [Gk. *bioun*, to live]: An intimate, protracted, and dependent relationship between two or more organisms of different species. Includes mutalism, in which the association is beneficial to both; commensalism, in which one benefits and the other is neither harmed or benefitted; and parasitism, in which one benefits and the other is harmed.

SYMPATHETIC NERVOUS SYSTEM: A subdivision of the autonomic nervous system, with centers in the midportion of the spinal cord; slows digestion, has a generally excitatory effect on other functions. Neurohormones: adrenaline or noradrenaline.

SYNAPSE [Gk. *synapsis*, a union]: The region of nerve impulse transfer between two neurons.

SYNAPSIS: The pairing of homologous chromosomes that occurs prior to the first meiotic division; crossing over occurs as a result of synapsis.

SYNGAMY (**sin**-gamy): The union of gametes in sexual reproduction; fertilization.

SYNTHESIS: The formation of a more complex substance from simpler ones.

SYSTEMATICS: Scientific study of the kinds and diversity of organisms and of the relationships between them.

TAXONOMY [Gk. *taxis*, arrange, put in order + *nomos*, law]: The science of the classification of organisms.

TELOPHASE [Gk. *telos*, end + *phasis*, form]: The last stage in mitosis and meiosis, during which the chromosomes become reorganized into two new nuclei.

TEMPLATE: A pattern or mold guiding the formation of a negative or complement. DNA replication is explained in terms of a template hypothesis.

TENTACLES [L. *tentare*, to touch]: Long, flexible protrusions located about the mouth in many invertebrates; usually prehensile or tactile.

TERRITORY: An area or space occupied by an individual or a group, trespassers into which are attacked (and usually defeated); may be the site of breeding, nesting, and food-gathering or any combination thereof.

TEST CROSS: A mating between a homozygous recessive individual and a dominant phenotype in order to determine the genetic constitution of the latter, that is, whether it is homozygous or heterozygous for a particular gene.

TESTIS, *pl.* TESTES [L., witness]: The male gamete-producing organ; also the source of male sex hormone.

TESTOSTERONE: A hormone secreted by the testes in higher vertebrates and stimulating the development and maintenance of male sex characteristics and the production of sperm; an androgen.

TETRAD [Gk. *tetras*, four]: In plants, a group of four spores formed by meiosis within a mother cell.

THALAMUS [Gk. *thalamos*, chamber]: A part of the vertebrate forebrain just posterior to the cerebrum; an important intermediary between all other parts of the nervous system and the cerebrum.

THALLUS: A simple plant body without true roots, leaves, or stems.

THEORY [Gk. *theorein*, to look at]: A generalization which is based on observation and experiments that were conducted to test the validity of a hypothesis and found to support the hypothesis.

THERMODYNAMICS [Gk. *therme*, heat + *dynamis*, power]: The study of energy, using heat as the most convenient form of measurement of energy. The first law of thermodynamics states that in all processes, the total energy of the universe remains constant. The second law states that the entropy, or degree of randomness, tends to increase.

THORAX: (1) In vertebrates, that portion of the trunk containing the heart and lungs. (2) In insects, the three leg-bearing segments between head and abdomen.

THYLAKOID: A saclike membranous structure in the chloroplast; stacks of thylakoids form the grana.

THYROID: An endocrine gland of vertebrates, located in the neck; source of an iodine-containing hormone (thyroxine) that increases the rate of oxidative processes (metabolic rate).

TISSUE [L. *texere*, to weave]: A group of similar cells organized into a structural and functional unit.

TRACHEA (**trake**-ee-a), pl. TRACHEAE [Gk. *tracheia*, rough]: An air-conducting tube, such as the windpipe of mammals and the breathing systems of insects.

TRACHEID (**tray**-key-idd) [Gk. *tracheia*, rough)]: An elongated, thick-walled conducting and supporting cell of xylem, characterized by tapering ends and pitted walls without true perforations. It is found in nearly all vascular plants and is dead at maturity.

TRANSCRIPTION: The enzymatic process, involving base pairing, by which the genetic information contained in DNA is used to specify a complementary sequence of bases in an RNA molecule.

TRANSDUCTION: The transfer of genetic material (DNA) from one bacterium to another by a lysogenic bacteriophage.

TRANSFER RNA (tRNA): The type of RNA that becomes attached to an amino acid and guides it to the correct position on the ribosome-mRNA complex for protein synthesis. There is at least one tRNA molecule for each amino acid. Each tRNA molecule is only about 80 nucleotides in length.

TRANSFORMATION: A genetic change produced by the incorporation into a cell of DNA from another cell.

TRANSLATION: The process by which the genetic information contained in a messenger RNA molecule dictates the sequence of amino acids in a protein molecule.

TRANSLOCATION: (1) In plants, the transport of the products of photosynthesis. (2) In genetics, the breaking off of a piece of a chromosome and its attachment to a nonhomologous chromosome.

TRANSPIRATION [L. *spirare*, to breathe]: In plants, the loss of water vapor from the stomata.

—TROPH, TROPHO— [Gk. *trophos*, feeder]: Suffix or prefix meaning "feeder," "feeding"; for example, autotrophic, "self-nourishing."

TROPHIC LEVEL: The position of a species in the food chain; a step in the movement of a biomass or energy through a system.

TROPIC [Gk. *trope*, a turning]: Pertaining to behavior or action brought about by specific stimuli, for example, phototropic ("light-oriented") motion, gonadotropic ("stimulating the gonads") hormone.

TUBER [L. *tuber*, bump, swelling]: A much-enlarged, short, fleshy underground stem, such as that of the potato.

TURGOR [L. *turgere*, to swell]: The rigid distention of a plant cell by its fluid contents.

UREA [Gk. *ouron*, urine]: An organic compound formed in the vertebrate liver from ammonia and carbon dioxide and excreted by the kidneys; the principal form of disposal of nitrogenous wastes in mammals.

URETER [Gk. from *ourein*, to urinate]: The tube carrying urine from the kidney to the cloaca (in reptiles and birds) or to the bladder (in amphibians and mammals).

URETHRA: The tube carrying urine from the bladder to the exterior in mammals.

URIC ACID: An insoluble nitrogenous waste product that is the principal excretory product of birds, reptiles, and insects.

URINE: The liquid waste filtered from the blood by the kidney and excreted by the bladder.

UTERUS [L., womb]: The muscular, expanded portion of the female reproductive tract modified for the storage of eggs or for housing and nourishing the developing embryo.

VACUOLE [L. *vacuus*, empty]: A space within a cell, bounded by a membrane and filled with water and solid materials in solution.

VAGINA [L. *vagina*, sheath]: The part of the female reproductive duct in mammals which receives the male penis during copulation.

VAGUS NERVE [L. *vagus*, wandering]: A nerve arising from the medulla of the vertebrate brain that innervates the visceral organs; carries parasympathetic fibers.

VALENCE [L. *valere*, to have power]: A measure of the bonding capacity of an atom, based on the number of electrons in its outer shell; equal to the charge which ions of that atom display in chemical reactions.

VAS DEFERENS (vass **deaf**-er-ens): In mammals, the tube carrying sperm from the testes to the urethra.

VASCULAR: Containing or concerning vessels that conduct fluid.

VASCULAR BUNDLE: In plants, a group of longitudinal supporting and conducting tissues (xylem and phloem).

VASCULAR CAMBIUM: A cylindrical sheath of meristematic cells which divide, producing secondary phloem and secondary xylem, but always with a cambial cell remaining.

VEIN: In plants, a vascular bundle forming a part of the framework of the conducting and supporting tissue of a leaf or other expanded organ. In animals, a blood vessel carrying blood from the tissues to the heart.

VENA CAVA (**vee**-na **cah**-va): A large vein which brings blood from the tissues to the right atrium of the heart. The superior vena cava collects blood from the forelimbs, head, and anterior or upper trunk; the inferior vena cava collects blood from the kidneys, the liver or lower gonads, and the general posterior body region.

VENTRAL [L. *venter*, belly]: Pertaining to the undersurface of an animal that moves on all fours; to the front surface of an animal that holds its body erect.

VENTRICLE [L. *ventriculus*, the stomach]: A chamber of the heart that receives blood from an atrium and pumps blood out of the heart.

VERTEBRAL COLUMN: The backbone; in nearly all vertebrates, it forms the supporting axis of the body and a protection for the spinal cord.

VESICLE [L. *vesicula*, a little bladder]: A small, intracellular membrane-bounded sac.

VESSEL: A tubelike structure of the xylem of angiosperms composed of dead cells (vessel elements) placed end to end. Its function is to conduct water and minerals from the soil.

VIABILITY: Ability to live.

VILLUS [L. *villus*, a tuft of hair]: In vertebrates, one of the minute fingerlike projections lining the small intestine that serve to increase the absorptive surface area of the intestine.

VIRUS [L., slimy, liquid, poison]: A submicroscopic noncellular particle, composed of a nucleic acid core and a protein shell; parasitic; reproduces only within a host cell.

VISCERA: The collective term for the internal organs of an animal.

VITAMIN [L. *vita*, life]: Any of a number of unrelated organic substances that cannot be synthesized by a particular organism and are essential in minute quantities for normal growth and function.

VULVA: External genitalia of the human female; includes the clitoris and the labia.

WATER TABLE: The upper limit of permanently saturated soil; the top layer of the ground water.

WILD TYPE: In genetics, the phenotype which is characteristic of the vast majority of individuals of a species in a natural environment.

WORKER: A member of the nonreproductive laboring caste in social insects.

XANTHOPHYLL [Gk. *xanthos*, yellow]: In a leaf, one of a group of yellow pigments; a member of the carotenoid group.

XYLEM [Gk. *xylon*, wood]: A complex vascular tissue through which most of the water and minerals are conducted from the roots to other parts of the plant; consists of tracheids or vessel elements, parenchyma cells, and fibers; constitutes the wood of trees and shrubs.

YOLK: The stored food material in egg cells that nourishes the embryo.

ZOOLOGY [Gk., from *zoe*, life + *logos*, a discourse]: The study of animals.

ZOOPLANKTON: A collective term for the nonphotosynthetic organisms present in plankton.

ZYGO– [Gk. *zygon*, yolk, pair]: Prefix, meaning "yolk" or "pair."

ZYGOTE (**zi**-got): The diploid ($2n$) cell resulting from the fusion of male and female gametes.

ILLUSTRATION ACKNOWLEDGMENTS

I–1 Craig MacFarland

I–2 Radio Times Hulton Picture Library

I–3 The American Museum of Natural History

I–4 K. Reichard, National Audubon Society

I–5 The American Museum of Natural History

I–6 M. P. L. Fogden, Bruce Coleman, Inc.

I–7 National Maritime Museum

I–9 Craig MacFarland

I–10 T. Davidson, National Audubon Society

I–11 The Bettmann Archive, Inc.

I–12 The Bettmann Archive, Inc.

1–1 © California Institute of Technology and the Carnegie Institution of Washington

1–2 NASA

1–4 © California Institute of Technology and the Carnegie Institution of Washington

1–5 (a) The Bettmann Archive, Inc.; (b) Yerkes Observatory, University of Chicago

Table 1–1 F. Bueche, *Physical Science*, Worth Publishers, Inc., New York, 1972

1–6 Photograph by S. Jonasson, from R. Anderson et al., *Science*, **148**, 1179–1190, 1965; copyright 1965 by The American Association for the Advancement of Science

1–7 C. Ponnamperuma

1–8 (a) E. S. Barghoorn; (b) M. I. Walker

1–9 George I. Schwartz

1–10 Eric V. Gravé

1–12 Eric V. Gravé

1–13 Jack Dermid

1–14 George I. Schwartz

1–15 D. P. Wilson

1–16 Craig MacFarland

1–17 (a) L. L. Rue, III, National Audubon Society; (b) Alan G. Nelson, National Audubon Society

2–1 John G. Torrey

Page 35 Lotte Jacobi

Page 37 Niels Bohr Library, Margrethe Bohr Collection, American Institute of Physics

Table 2–2 After H. J. Morowitz, *Energy Flow in Biology*, Academic Press, Inc., New York, 1968

Page 39 David S. Haglund

2–10 After E. J. DuPraw, *Cell and Molecular Biology*, Academic Press, Inc., New York, 1968

3–1 Grant M. Haist, National Audubon Society

3–2 (c) After James Sutcliffe, *Plants and Water*, St. Martin's Press, New York, 1968

3–4 Esso Research and Engineering Company

3–5 Larry Pringle, Photo Researchers, Inc.

3–6 Jack Dermid

4–1 DeBeers

Page 59 The Bettmann Archive, Inc.

Page 60 Thomas Eisner

4–11 (c) After Albert L. Lehninger, *Biochemistry*, Worth Publishers, Inc., New York, 1970

4–12 Myron C. Ledbetter

4–13 Keith R. Porter

4–16 The New York Times

Page 65 Culver Pictures, Inc.

5–5 Linus Pauling

5–7 After Lehninger, *op. cit.*

5–9 After Lehninger, *op. cit.*

5–10 After J. D. Watson, *Molecular Biology of The Gene*, 2d ed., W. A. Benjamin, Inc., Menlo Park, Calif., 1970

5–11 After Watson, *op. cit.*

5–15 Patricia Farnsworth

5–21 R. D. Preston

5–22 Larry West

5–24 After Lehninger, *op. cit.*

5–26 Keith R. Porter

5–27 B. E. Juniper

5–28 (a)–(i) After Lehninger, *op. cit.*; (j) Lee D. Simon

5–29 After Lehninger, *op. cit.*

5–30 (a) Keith R. Porter; (b) Emil Bernstein; (c) Jack Dermid

Page 90 (left) Rare Book Division, The New York Public Libary, Astor, Lenox, and Tilden Foundations; (right) from Arthur Hughes, *A History of Cytology*, Abelard-Schuman, London, 1959

6–2 J. Pickett-Heaps

6–3 J. D. Robertson

Pages 92–93 (a) Eric V. Gravé; (b) A. Jurand and G. G. Selman, *Anatomy of Paramecium aurelia*, St. Martin's Press, New York, 1969; (c) Eugene B. Small and Douglas S. Marszalek, *Science*, **163**, 1064–1065, 1969; copyright 1969 by The American Association for the Advancement of Science

6–4 (b) Myron C. Ledbetter

6–5 (a) D. Branton; (b) after DuPraw, *op. cit.*

6–6 Myron C. Ledbetter

6–7 Eldon H. Newcomb

6–8 Ursula Goodenough

6–10 (a) Jurand and Selman, *op. cit.*

6–11 R. D. Goldman and D. M. Knipe, "Functions of Cytoplasmic Fibers in Non-Muscle Cell Motility," *Cold Spring Harbor Symposium on Quantitative Biology*, **37**, 523–534, 1972

6–12 Eldon H. Newcomb

6–13 Jurand and Selman, *op. cit.*

6–14 Eric V. Gravé

6–15 A. V. Grimstone

6–16 After DuPraw, *op. cit.*

6–17 Etienne de Harven

7–1 W. G. Whaley, The University of Texas Electron Microscope Laboratory

7–3 P. Berger, National Audubon Society

7–5 After Lehninger, *op. cit.*

7–6 Thomas Eisner

7–10 Keith R. Porter

7–12 Gregory Antipa, University of Illinois

8–1 R. D. Estes

8–2 After Lehninger, *op. cit.*

8–3 Sun, Courtesy of DuPraw, *op. cit.*

8–9 The Bettmann Archive, Inc.

8–11 S. Pantovic, Photo Researchers, Inc.

8–12 Keith R. Porter

8–17 John Dominis, LIFE © 1971

Table 8–2 After Lehninger, *op. cit.*

9–1 Elihu Blotnick, BBM Associates

9–2 The Bettmann Archive, Inc.

9–3 The Bettmann Archive, Inc.

9–4 The Bettmann Archive, Inc.

9–5 N. Pfennig

9–6 L. K. Shumway

9–12 Daniel Branton

Page 153 Howard Towner

10–1 Kurt Hirschhorn, M.D., from *Birth Defects: Original Article Series*, vol. 4, no. 4, *Guide to Human Chromosome Defects* by Audrey Redding and Kurt Hirschhorn, M.D., The National Foundation

10–3 General Biological Supply

10–5 William Tai

10–6 Andrew S. Bajer

10–7 William Tai

10–8 (a) S. Inoué, "Polarization Optical Studies of the Mitotic Spindle. I. The Demonstration of Spindle Fibers in Living Cells," *Chromosome Bd.*, **5**, 487–500, Springer, 1953; (b) after B. John and K. R. Lewis, *Somatic Cell Division*, Oxford University Press, New York, 1972

10–9 M. Friedlander

10–10 Eric V. Gravé

10–11 James Cronshaw

11–1 E. J. DuPraw

11–2 The Bettmann Archive, Inc.

11–3 The Bettmann Archive, Inc.

11–4 The Bettmann Archive, Inc.

Page 173 Pasteur Institute and Rockefeller University Press

11–7 After Karl von Frisch, *Biology*, Harper & Row, Publishers, Inc., New York, 1964, translated by Jane Oppenheimer

12–1 (a) Tryggve Gustafson; (b) Everett Anderson

12–3 B. John

12–4 B. John

12–6 William Tai

12–7 Mary E. Clutter

12–8 Mary E. Clutter

12–9 B. John

12–10 Adapted from DuPraw, *op. cit.*

12–13 After John A. Moore, *Heredity and Development*, Oxford University Press, New York, 1963

12–14 (a) Victor A. McKusick; (b) Gary Laurish

13–1 C. G. G. J. van Steenis

13–2 Edward B. Gerard

13–6 (d) Photograph by F. B. Hutt, *Journal of Genetics*, **22**, 126, 1930

13–7 After E. D. Merrell, 1964

13–8 U.S. Public Health Service, U.S. HSM 73–1601, series II, no. 119, 1963–1965

13–9 Mary Eleanor Browning, National Audubon Society

14–5 B. John

14–7 After Watson, *op. cit.*

14–9 B. P. Kaufmann

14–10 After A. Srb., R. Owen, and R. Edgar, *General Genetics*, 2d ed., W. H. Freeman and Company, San Francisco, 1965

14–12 Photograph by The Bettmann Archive, Inc.

14–12 After I. Michael Lerner, *Heredity, Evolution and Society*, W. H. Freeman and Company, San Francisco, 1968

Page 224 Shirley Baty

15–1 The National Foundation – March of Dimes

15–2 Drawing by B. O. Dodge, from E. L. Tatum

15–6 After L. Pauling et al., "Sickle Cell Anemia, A Molecular Disease," *Science*, **110**, 543–548, 1949

15–7 After Derry D. Koob and William E. Boggs, *The Nature of Life*, Addison-Wesley Publishing Company, Reading, Mass., 1972

16–1 Lee D. Simon

16–2 After Watson, *op. cit.*

16–3 Lee D. Simon

16–4 From J. Cairns, G. S. Stent, and J. D. Watson (eds.), *Phage and the Origins of Molecular Biology*, Cold Spring Harbor Laboratory of Quantitative Biology, Cold Spring Harbor, N.Y., 1966

16–5 A. K. Kleinschmidt, D. Land, D. Jacherts, and R. K. Zahn, *Biochim. Biophys. Acta*, **61**, 857–864, 1962, Fig. 1

Table 16–1 After Erwin Chargaff, *Essays on Nucleic Acids*, 1963

16–6 From James D. Watson, *The Double Helix*, Atheneum, New York, 1968; copyright © 1968 by James D. Watson; reprinted by permission of the publisher

Page 245 From Watson, *op. cit.*, 1968

16–10 After Lehninger, *op. cit.*

16–11 After Lehninger, *op. cit.*

16–12 John Cairns

17–4 Ursula Goodenough

17–6 Hans Ris

17–8 O. L. Miller, Jr., and Barbara R. Beatty, Oak Ridge National Laboratory

17–9 O. L. Miller, Jr., Barbara A. Hamkalo, and C. A. Thomas, Jr., "Visualization of Bacterial Genes In Action," *Science*, **169**, 392–395, 1970; copyright © 1970 by the American Association for the Advancement of Science

17–11 After James D. Watson, *Molecular Biology of the Gene*, 2d ed., W. A. Benjamin, Inc., Menlo Park, Calif., 1970

17–12 After James D. Watson, *op. cit.*, 1970

17–14 Jack Griffith

18–1 E. J. DuPraw

18–2 George T. Rudkin

18–3 (a) J. Gall; (b) adapted from DuPraw, *op. cit.*

18–4 After DuPraw, *op. cit.*

18–5 Photograph by T. F. Anderson, from E. L. Wollman, F. Jacob, and W. Hayes, *Cold Spring Harbor Symposium on Quantitative Biology*, **21**, 141, 1950

18–10 Joseph L. Melnick, Baylor College of Medicine, Houston

19–1 Carl W. Rettenmeyer

19–2 University of Leiden Library, ms. Voss. Lat. Q.9, f.57

19–3 Fratelli Alinari

19–4 Burndy Library

19–5 (a) John S. Flannery, Bruce Coleman, Inc.; (b) Dale P. Hansen, National Audubon Society; (c) Jeanne White, National Audubon Society

19–8 (a), (b) Eric V. Gravé; (c) Michael P. Godomski, National Audubon Society; (d) Les Blacklock; (e) Alvin E. Staffan

19–9 (a), (c) Eric V. Gravé; (b) H. Forest

19–10 M. E. Edwards

19–14 Charles C. Brinton, Jr.

19–16 (a) Martin Dworkin; (b) Center for Disease Control; (c) Parke, Davis and Company

19–17 Peter Hirsch

19–18 R. S. Wolfe

19–19 M. W. Jennison

19–20 National Medical Audiovisual Center

19–21 D. Greenwood

19–23 M. Jost

19–24 (a) M. S. Fuller; (b) F. Palmer and E. Ordal

19–25 J. D. Almeida and A. F. Howatson, *Journal of Cell Biology*, **16**, 616, 1963

19–26 (a) Frederick A. Murphy; (b) R. C. Williams

19–27 Ny Carlsberg Glyptotek

20–1 Jeremy Pickett-Heaps

20–3 Thomas Uzzell and Christina Spolsky, "Mitochondria and Plastids as Endosymbionts: A Revival of Special Creation," *American Scientist*, **62**, 334–343, 1974

20–4 K. W. Jeon

20–5 (b) Eric V. Gravé

20–6 Eric V. Gravé

20–8 Eric V. Gravé

20–12 H. N. Guttmann

20–13 (a) Eric V. Gravé; (b) D. P. Wilson

20–14 Kent Cambridge Scientific Company

20–15 D. P. Wilson

20–16 D. Kubai and H. Ris

20–17 Larry West

20–18 J. T. Bonner

20–20 Jack Dermid

20–21 John S. Neiderhauser and William C. Cobb, "The Late Blight of Potatoes," *Scientific American*, May 1959

20–22 Alma W. Barksdale, *Mycologia*, **55**, 493–501, 1973

20–24 John H. Troughton

20–25 (a) Alvin E. Staffan; (b) John Shaw

20–26 C. Bracker

20–28 (a), (e) Larry West; (b) Jack Dermid; (c) Robert Carr; (d) Alvin E. Staffan

20–29 (a) Jack Dermid; (b), (d) Larry West; (c) Alvin E. Staffan

21–1 M. I. Walker

21–3 (a) D. P. Wilson; (b) B. J. W. Heath

21–4 (a) Don Longnecker; (b) D. P. Wilson; (c) William Randolph Taylor, *Plants of Bikini and Other Northern Marshall Islands*, University of Michigan Press, Ann Arbor, Mich., 1950; copyright 1950 by the University of Michigan

21–5 Ursula Goodenough

21–6 (a) Eric V. Gravé; (b) M. I. Walker; (c) George I. Schwartz

21–7 (a) Valerie Taylor, Ron Taylor Film Productions; (b) Oxford Scientific Films, Bruce Coleman, Inc.; (c) D. P. Wilson

21–14 John H. Troughton

21–15 Alvin E. Staffan

21–16 Turtox

21–17 E. S. Ross

21–19 (a) Bruce Coleman, Inc.; (b) John Shaw; (c) Robert Carr

21–20 © Field Museum of Natural History, painting by Charles R. Knight

21–21 (a) Harriet S. Gallant, Science Photographics, Ltd.; (b) Gene Ahrens, Bruce Coleman, Inc.

21–22 Charles Neidorf

21–24 S. Wilhelm, Bruce Coleman, Inc.

21–26 William M. Harlow

21–28 Photograph, Ward's Natural Science Establishment

21–29 Les Blacklock

21–33 (a) Jack Dermid; (b) Charles W. Mann, National Audubon Society; (c) L. L. Rue, III, National Audubon Society; (d) P. and S. Maslowski, Photo Researchers; (e) Robert C. Hermes, National Audubon Society

21–34 After T. Elliot Weir, C. Ralph Stocking, and Michael G. Barbour, *Botany*, 4th ed., John Wiley & Sons, Inc., New York, 1970

22–1 D. P. Wilson

22–2 Nat Fain

22–3 James McConnell

22–4 D. P. Wilson

22–6 Nat Fain

22–7 The American Museum of Natural History

22–11 After Ralph Buchsbaum and Lorus J. Milne, *The Lower Animals: Living Invertebrates of the World*, Doubleday & Company, Inc., Garden City, N.Y., 1962

22–13 Nat Fain

22–14 (a) D. P. Wilson; (b) G. Lower, National Audubon Society

22–15 Douglas Faulkner

22–18 Jack Dermid

22–20 Walter E. Harvey, National Audubon Society

22–23 After M. B. V. Roberts, *Biology: A Functional Approach*, The Ronald Press Company, New York 1971

22–25 British Museum (Natural History)

22–26 Maria Wimmer

22–27 David Pramer

22–28 After W. D. Russell-Hunter, *A Biology of Lower Invertebrates*, The Macmillan Company, New York, 1968

22–32 Alvin E. Staffan, National Audubon Society

22–33 Maria Wimmer

22–34 Hugh Spencer, National Audubon Society

22–35 (a) Jeanne White, National Audubon Society; (b), (c) Russ Kinne, Photo Researchers

22–38 After Roberts, *op. cit.*

22–39 Robert C. Hermes, National Audubon Society

22–40 Jack Dermid

22–42 (a) D. P. Wilson; (b) Robert C. Hermes, National Audubon Society; (c) Douglas Faulkner

22–43 D. P. Wilson

22–45 M. J. Wells and Heinemann Educational Books, Ltd.

23–1 E. S. Ross

23–2 Jack Dermid

23–3 Larry West

23–4 Edward O. Wilson et al., *Life on Earth*, Sinauer Associates, Inc., Stamford, Conn., 1973

23–5 Robert C. Hermes, National Audubon Society

23–9 E. S. Ross

23–10 (a) Jack Dermid; (b) Larry West

23–12 W. Clifford Healey, National Audubon Society

23–13 Larry West

23–16 (a), (b) Larry West; (c) E. S. Ross; (d) Jack Dermid

23–17 E. S. Ross

23–18 After Malcolm S. Gordon, *Animal Physiology: Principles and Adaptations*, 2d ed., The Macmillan Company, New York, 1972

23–19 After Roberts, *op. cit.*

23–20 Thomas Eisner

23–21 Larry West

23–24 E. S. Ross

23–25 Ross Hutchins

23–26 E. S. Ross

23–27 E. S. Ross

23–31 Donald H. Fritts

23–33 D. P. Wilson

23–34 The American Museum of Natural History

23–35 Mitchell Campbell, National Audubon Society

23–36 Norman Myers, Bruce Coleman, Inc.

23–37 Jack Dermid

23–38 Jack Dermid

23–39 Lynwood M. Chace, National Audubon Society

23–40 (a) The American Museum of Natural History; (b) G. Ronald Austing

23–41 After Cleveland P. Hickman and Cleveland P. Hickman, Jr., *Biology of Animals*, C. V. Mosby Company, St. Louis, 1972

23–42 The New York Zoological Society

23–44 Roberts Rugh and Landrum B. Shettles, M.D., *From Conception to Birth: The Drama of Life's Beginnings*, Harper & Row, Publishers, Inc., New York, 1971

24–1 Jack Dermid

24–6 C. D. K. Cook, *A Monographic Study of* Ranunculus *subgenus* Barrachium (*DC.*) *A. Gray*, **6**, 47–237, October 1966

24–7 Howard Towner

24–8 (a) Gary Laurish; (b) William Harlow

24–9 Photograph, Howard Towner

24–10 B. E. Juniper

24–11 From Myron C. Ledbetter and Keith R. Porter, *Introduction to the Fine Structures of Plant Cells*, Springer-Verlag, New York, 1970

24–12 From Ledbetter and Porter, *op. cit.*

24–14 (a) Ripon Microslides

24–15 Jack Dermid

24–16 William Tai

24–17 (a) William Harlow; (b), (c) Jack Dermid

24–18 William Harlow

24–19 From Richard Ketchum, *The Secret Life of the Forest*, American Heritage Press, New York, 1970

25–1 John H. Troughton

25–2 (b) B. E. Juniper

25–3 (a) J. Heslop-Harrison; (b) John H. Troughton

25–4 John H. Troughton

25–5 (a) Martin H. Zimmerman; (b) George A. Schaefers

25–7 Larry West

25–8 J. Heslop-Harrison

25–9 After M. Richardson, *Translocation in Plants*, St. Martin's Press, New York, 1968 (after Stout and Hoagland, 1939)

25–10 After Richardson, *op. cit.*

25–11 After A. C. Leopold, *Plant Growth and Development*, McGraw-Hill Book Company, New York, 1964 (after Kramer, 1937)

25–13 Alvin E. Staffan, National Audubon Society

25–14 J. Spurr, Bruce Coleman, Inc.

25–15 After Peter Ray, *The Living Plant*, 2d ed., Holt, Rinehart and Winston, Inc., New York, 1971

26–1 Jack Dermid

26–3 Jack Dermid

26–5 E. S. Ross

26–6 Jack Dermid

26–7 (a) Gene Ahrens, Bruce Coleman, Inc.; (b) E. S. Ross

26–9 (a), (c), (e), (g) E. S. Ross; (b) W. G. Harvey, National Audubon Society; (d) C. Allan Morgan; (f) L. West

26–10 Jack Dermid

26–11 (a), (b), (d) Patrick Echlin; (c) J. Heslop-Harrison

26–12 Myron C. Ledbetter

26–16 After Ray, *op. cit.*

26–17 Howard Towner

26–18 After Wilson et al., *op. cit.*

26–19 Peter Ray

26–20 Jack Dermid

26–23 L. J. Prater, U.S. Forest Service

27–1 Oscar Erpenstein, National Audubon Society

27–5 U.S. Department of Agriculture

27–8 J. P. Nitsch, *American Journal of Botany*, **37**, 3, 1950

27–10 S. H. Wittwer

27–11 J. E. Varner

27–12 After Wilson et al., *op. cit.*

27–14 H. R. Chen

27–18 Jack Dermid

27–20 After J. G. Torrey, *Development in Flowering Plants*, The Macmillan Company, New York, 1967

28–1 Marvin B. Winter

28–2 Hugh Spencer, National Audubon Society

28–3 After Aubrey W. Naylor, "The Control of Flowering," *Scientific American*, May 1952

28–5 From T. P. O'Brien and M. E. McCully, *Plant Structure and Function*, The Macmillan Company, New York, 1969

28–8 David Stetler

28–11 Jack Dermid

28–12 Jack Dermid

28–13 William Harlow

29–1 Ken Heyman

29–2 The American Museum of Natural History

29–5 Keith R. Porter

29–6 Keith R. Porter

29–7 Michael H. Ross and Edward J. Reith

29–9 Raymond C. Truex

30–1 Hugh Spencer, National Audubon Society

30–2 (a) Robert Carr; (b) J. MacFarland; (c) H. I. Fisher, National Audubon Society; (d) Mark Boulton, National Audubon Society

30–6 David M. Phillips

30–7 After Roberts, *op. cit.*

30–11 Man, Frank Skorski; lion, George Holton, Photo Researchers, Inc.; elephant seal, Al Lowry, Photo Researchers, Inc.; peacock, Jack Dermid

30–13 L. L. Rue, III, National Audubon Society

30–21 Carnegie Institution of Washington

Table 30–2 From J. E. Crouch and R. McClintic, *Human Anatomy and Physiology*, John Wiley & Sons, Inc., New York, 1971

31–1 Carnegie Institution of Washington

31–2 J. Smiles, *Biology of the Laboratory Mouse*, McGraw-Hill Book Company, New York, 1966; copyright 1966 by McGraw-Hill Book Company

31–3 Smiles, *op. cit.*

31–4 After Waddington, 1966

31–6 After Huettner, 1949

31–7 Tryggve Gustafson

31–8 L. M. Biedler

31–9 M. P. Olson

31–10 After Huettner, 1949

31–11 After Huettner, 1949

31–12 After Huettner, 1949

31–13 Photograph, John A. Moore

31–14 After Huettner, 1949

31–15 After Huettner, 1949

31–16 After Roberts Rugh, *Vertebrate Embryology*, Harcourt Brace and Company, New York, 1964

31–17 O. L. Miller and Barbara R. Beatty, Biology Division, Oak Ridge National Laboratory

31–19 After Waddington, 1966

31–20 After Grobstein, 1964

31–22 After Maurice Sussman, *Animal Growth and Development*, 2d ed., Prentice-Hall, Inc., Englewood Cliffs, N.J., 1964

31–23 After Charles W. Bodemer, *Modern Embryology*, Holt, Rinehart and Winston, Inc., New York, 1968

31–24 After Gustafson and Wolpert, 1963

31–25 After Torrey, 1962

31–26 After Sussman, *op. cit.*

31–27 After Torrey, 1962

31–29 Rugh and Shettles, *op. cit.*

31–30 Rugh and Shettles, *op. cit.*

31–31 Rugh and Shettles, *op. cit.*

31–32 After Lennart Nilsson and A. Ingelman-Sunberg, *A Child Is Born*, Delacourt Press, New York, 1965

31–33 After Nilsson and Ingelman-Sunberg, *op. cit.*

31–34 After Roberts, *op. cit.*

31–35 Carnegie Institution of Washington

31–36 Carnegie Institution of Washington

31–38 Rugh and Shettles, *op. cit.*

31–39 (a), (c) Rugh and Shettles, *op. cit.*; (b) H. Nishimura, from Rugh and Shettles, *op. cit.*; (d) E. Ludwig, from Rugh and Shettles, *op. cit.*

31–44 Rugh and Shettles, *op. cit.*

31–45 After Torrey, 1962

31–46 Rugh and Shettles, *op. cit.*

31–47 (a) After Swift, 1914; (b) after E. Witschi, *Development of Vertebrates*, W. B. Saunders Company, Philadelphia, 1956

31–48 After Witschi, *op. cit.*

Page 573 Courtesy of Walter Meagher

32–1 Mary E. Browning, National Audubon Society

32–3 United Press International

32–7 After N. J. Berrill, *Developmental Biology*, McGraw-Hill Book Company, New York, 1971

32–11 After Wilson et al., *op. cit.*

32–12 From William Bloom and Don W. Fawcett, *A Textbook of Histology*, 9th ed., W. B. Saunders Company, Philadelphia, 1968

Page 588 Rae Silver

32–16 Keith R. Porter

32–18 David H. Hubel

32–19 Keith R. Porter

32–20 Leon Root

32–22 Henri Cartier-Bresson, Magnum Photos

32–25 Flip Schulke, Black Star

32–27 (a) Keith R. Porter; (b) Hugh E. Huxley

32–28 After Keith R. Porter and Mary A. Bonneville, *Fine Structure of Cells and Tissues*, Lee and Febiger, Philadelphia, 1968

32–29 After Porter and Bonneville, *op. cit.*

32–30 R. Miledi

32–31 From Bloom and Fawcett, *op. cit.*

32–32 Keith R. Porter

33–1 Thomas L. Hayes and L. McDonald, *Experimental and Molecular Pathology*, vol. 10, no. 2, Academic Press, Inc., New York, 1969

33–6 Robert Rosenblum

33–7 After A. J. Vander, J. H. Sherman, and Dorothy S. Luciano, *Human Physiology*, McGraw-Hill Book Company, New York, 1969

33–8 After Roberts, *op. cit.*

33–9 Thomas Eisner

33–10 Bloom and Fawcett, *op. cit.*

33–12 After Roberts, *op. cit.*

Page 620 Parke-Davis Company

33–16 Pfizer, Inc.

33–17 Emil Bernstein, The Gillette Company Research Institute

33–19 Joseph Feldman

33–20 After Richard A. Lerner and Frank J. Dixon, "The Human Lymphocyte as an Experimental Animal," *Scientific American*, June 1973

33–21 After Lerner and Dixon, *Scientific American*, June 1973

33–22 Memorial Sloan–Kettering Cancer Center

34–1 Philippa Scott, Photo Researchers, Inc.

34–2 After Crouch and McClintic, *op. cit.*

34–4 After W. W. Howells, *Mankind in the Making*, rev. ed., Doubleday & Company, Inc., Garden City, N.Y., 1967

34–5 After Crouch and McClintic, *op. cit.*

34–6 I. Kaufman Arenberg, M.D., Department of Otolaryngology, Washington University School of Medicine, St. Louis, Missouri

34–9 Dr. Jeanne M. Riddle, Associate Professor of Pathology, Wayne State University School of Medicine

34–11 Beverly Farber, Johnson & Johnson

34–13 Bloom and Fawcett, *op. cit.*

Page 642 U.S. Department of Agriculture

34–14 Bloom and Fawcett, *op. cit.*

34–15 G. Decker

34–17 Keith R. Porter

34–18 The Bettmann Archive, Inc.

35–1 Alan Root

35–3 After Roberts, *op. cit.*

35–4 Thomas Eisner

35–6 J. and D. Bartlett, Bruce Coleman, Inc.

35–7 Photograph, Jack Dermid

35–8 Jack Dermid

35–10 (c) E. R. Wiebel

35–11 Keith R. Porter

Pages 658–659 Oscar Auerbach

35–13 After Knut Schmidt-Nielsen, *How Animals Work*, Cambridge University Press, New York, 1972

Page 663 Russ Kinne, Photo Researchers, Inc.

36–1 The American Museum of Natural History

36–2 Robert Chiarello

36–5 Russ Kinne, Photo Researchers, Inc.

36–6 From Lewis Carroll, *Alice's Adventures in Wonderland*, illustrated by John Tenniel

36–7 Bob Leatherman, National Audubon Society

36–8 After Crouch and McClintic, *op. cit.*

36–12 After Vander, Sherman, and Luciano, *op. cit.*

36–14 After Vander, Sherman, and Luciano, *op. cit.*

36–15 After Vander, Sherman, and Luciano, *op. cit.*

37–1 Jack Dermid

37–3 After Gordon, *op. cit.*

37–4 George Porter, National Audubon Society

37–5 Photograph, Allan D. Cruickshank, National Audubon Society; art, after Gordon, *op. cit.*

37–6 After Gordon, *op. cit.*

37–8 After Vander, Sherman, and Luciano, *op. cit.*

37–9 Spider, C. D.; Marvin B. Winter

37–10 After Schmidt-Nielsen, *op. cit.*

37–11 Bloom and Fawcett, *op. cit.*

37–12 (a) Nicholas Mrosovsky; (b) L. L. Rue, III, National Audubon Society; (c) S. C. Bisserot

37–13 Dominique Reger, UNESCO

Page 692 John H. Gerard

37–14 Phyllis McCutcheon, Photo Researchers, Inc.

37–15 Leon D. Harmon, Photo Researchers, Inc.

38–4 (a) Jeanne White, National Audubon Society; (b) Gordon Smith, National Audubon Society; (c) F. B. Grunzweig, Photo Researchers, Inc.; (d) A. W. Ambler, National Audubon Society

38–5 After R. L. Gregory, *Eye and Brain: The Psychology of Seeing*, McGraw-Hill Book Company, New York, 1966

38–6 After Gregory, *op. cit.*

38–8 E. R. Lewis

38–9 Keith R. Porter

38–11 T. Davidson, National Audubon Society

38–14 Jack Dermid

39–1 Arthur Jacques, Rhode Island Hospital

39–2 Jose M. R. Delgado

39–3 Aquarius Electronics, Mendocino, California; copyright 1971

39–4 W. R. Klemm, *Science, The Brain and Our Future*, Pegasus, New York, 1972

39–5 Photograph, Arthur Jacques, Rhode Island Hospital

39–7 D. H. Hubel

39–8 James Olds

39–15 Jerre Levy, Colwyn Trevarthen, and Roger W. Sperry, *Brain: The Journal of Neurology*, 1972

39–16 Laverne C. Johnson, "Are Stages of Sleep Related to Waking Behavior?", *American Scientist*, **61**, 3, 1973

39–17 Jerry Hecht, National Institute of Mental Health

39–18 E. R. Lewis

40–1 U.S. Forest Service

40–2 NASA

40–7 Photograph by Les Blacklock

40–8 Norman Myers, Bruce Coleman, Inc.

40–9 Russ Kinne, Photo Researchers, Inc.

40–10 (a) G. Harrison, Bruce Coleman, Inc.; (b) Bruce Coleman, Inc.; (c) A. Blank, Bruce Coleman, Inc.

40–11 Jack Dermid

40–12 Les Blacklock

40–13 Les Blacklock

40–14 (a)–(e) Alvin E. Staffan; (f) Jack Dermid

40–15 Les Blacklock

40–16 L. L. Rue, IV, National Audubon Society

40–18 (a) W. C. Denison; (b) Larry West

40–19 (a) G. Ahrens, Bruce Coleman, Inc.; (b) R. and J. Spurr, Bruce Coleman, Inc.

40–20 (a) Les Blacklock; (b) L. L. Rue, III, National Audubon Society; (c) Alvin E. Staffan

40–21 Willis Peterson

40–22 Les Blacklock

40–23 Patricia Caulfield

40–24 Jane Burton, Bruce Coleman, Inc.

40–25 Dennis Brokaw

40–26 Willis Peterson

40–27 (a) Allan D. Cruickshank, National Audubon Society; (b) B. Max Thompson, National Audubon Society; (c) N. Myers, Bruce Coleman, Inc.; (d) James Tallon

40–28 (a) J. and D. Bartlett, Bruce Coleman, Inc.; (b) Robert H. Wright, National Audubon Society; (c) Bill Ratcliffe; (d) Anthony Merieca, National Audubon Society; (e) Grace A. Thompson, National Audubon Society; (f) Willis Peterson; (g) H. L. Parent, National Audubon Society

40–29 L. R. Holdridge, *Science*, **105**, 367, 1947

40–30 John S. Flannery, Bruce Coleman, Inc.

40–32 Photograph by K. Gunnar, Bruce Coleman, Inc.

40–33 Les Blacklock

40–35 Les Blacklock

40–36 Les Blacklock

40–37 The American Museum of Natural History

40–40 D. P. Wilson

40–41 Allan Power, Bruce Coleman, Inc.

40–42 (a) Jeff Foote, Bruce Coleman, Inc.; (b) Charlie Ott, National Audubon Society; (c), (d) Jack Dermid

40–43 (a), (c), (d) Jack Dermid; (b) Marty Stouffer, National Audubon Society

41–1 Mark N. Boulton, National Audubon Society

41–3 (a) Alvin E. Staffan, National Audubon Society; (b) N. Myers, Bruce Coleman, Inc.; (c) James B. Faro

41–9 T. R. G. Gray

Page 777 Luoma, National Audubon Society

41–11 R. H. Wright, National Audubon Society

41–13 (a) The Nitragin Co.; (b) P. J. Dart

41–14 R. R. Hebert, R. D. Holsten, and R. W. F. Hardy, E. I. duPont de Nemours and Co., Inc.

41–17 G. E. Likens

41–18 (a) U.S. Forest Service; (b) Larry West

Page 787 Larry West

41–19 Larry West

41–21 Jack Dermid

41–22 (a) Jack Dermid; (b) U.S. Forest Service

41–23 Jack Dermid

42–1 Jeanne White, National Audubon Society

42–4 (a) Marineland of Florida; (b) C. I. R. H. Noailles; (c) A. D. Cruickshank, National Audubon Society; (d) Mark Boulton, National Audubon Society

42–8 Eric Hosking, National Audubon Society

42–10 Karl H. Maslowski, National Audubon Society

42–11 Redrawn from John L. Harper, *Symposia Society for Experimental Biology*, **15**, 25, 1961

42–12 C. H. Muller, Bulletin of the Torrey Botanical Club, **93**, 334, 1966

42–13 Jack Dermid

42–14 S. A. Wilde

42–15 E. S. Ross

42–16 Australian News and Information Service

42–17 John Markham

42–18 (a) Douglas Faulkner; (b) R. Mariscal, Bruce Coleman, Inc.; (c) C. Haagner, Bruce Coleman, Inc.; (d) N. Smythe, National Audubon Society

42–19 L. L. Rue, III, National Audubon Society

42–20 Biological Section Laboratory, Australian Department of Lands

42–21 George Pollak, *Science*, **176**, cover, 1972; copyright 1972 by the American Association for the Advancement of Science

42–22 (a) N. Myers, Bruce Coleman, Inc.; (b) James Simon, Bruce Coleman, Inc.; (c) Joy Spurr, Bruce Coleman, Inc.

42–23 (a) Karl Weidmann, National Audubon Society; (b) S. J. Kraseman, National Audubon Society; (c) Allan Power, National Audubon Society; (d) G. Ronald Austing

42–24 (a) Alvin E. Staffan; (b) E. S. Ross

42–25 (a), (b) E. S. Ross; (c) Jen and Des Bartlett, Bruce Coleman, Inc.; (d) Alan Blank, Bruce Coleman, Inc.; (e) G. R. Austing, Bruce Coleman, Inc.

42–26 (a) S. C. Bisserot, Bruce Coleman, Inc.; (b) E. S. Ross; (c) Alan Blank, Bruce Coleman, Inc.

42–27 (a) A. W. Ambler, National Audubon Society; (b) E. S. Ross

42–28 Thomas Eisner

42–29 E. S. Ross

42–30 Chester Karpiej

43–1 Masud Quraishy, Photo Researchers, Inc.

43–2 (a) Alexander M. Klots; (b) Larry West

43–3 J. Markham, Bruce Coleman, Inc.

43–4 (a) E. S. Ross; (b) Treat Davidson, National Audubon Society

43–5 E. S. Ross

43–6 E. S. Ross

43–7 E. S. Ross

43–8 Williams H. Gotwald, Jr.

43–9 Ross Hutchins

43–10 L. Biedler

43–11 (a) Norman Myers, Bruce Coleman, Inc.; (b) John Gerard, National Audubon Society

43–12 Charlie Ott, National Audubon Society

43–14 G. Silk

Page 833 Douglas Faulkner

43–15 T. W. Ransom

43–16 (a) Bill Gabriel, Photo Researchers, Inc.; (b) M. P. Kahl, Photo Researchers, Inc.

43–17 (a) R. R. Pawlaski, Bruce Coleman, Inc.; (b) S. Gaulin, Anthro/Photos; (c) L. L. Rue, III, Bruce Coleman, Inc.; (d) Douglas Faulkner

43–18 U.S. Fish and Wildlife Service

43–19 (a) Hermann Kacher; (b) Uzi Paz; (c) Willis Peterson

43–20 Patricia Caulfield

43–21 Patricia Caulfield

43–22 After L. David Mech, *The Wolf: The Ecology and Behavior of an Endangered Species*, Natural History Press, Garden City, N.Y., 1970

43–23 Russ Kinne, Photo Researchers, Inc.

43–24 Patricia Caulfield

43–25 After Mech, *op. cit.*

43–26 Patricia Caulfield

43–27 After Mech, *op. cit.*

43–28 L. David Mech

43–29 Toni Angermayer, Photo Researchers, Inc.

43–30 Norman Myers, Bruce Coleman, Inc.

43–31 Carleton Ray, Photo Researchers, Inc.

43–32 Irven deVore, Anthro/Photos

43–33 Irven deVore, Anthro/Photos

43–34 (a) T. McHugh, Photo Researchers, Inc.; (b) Irven deVore, Anthro/Photos

43–35 J. Popp, Anthro/Photos

43–36 J. Popp, Anthro/Photos

43–37 Irven deVore, Anthro/Photos

44–1 Baron Wolman

44–2 (a) Eric Hosking; (b) NBC

44–3 Philip Hyde

44–4 After J. Clausen and W. M. Hiesey, Carnegie Institution of Washington, publication 615, 1958

44–5 Evelyn M. Shafer

44–6 Mather and Harrison, *Heredity*, vol. 3, 1949

44–7 Lynwood M. Chace, National Audubon Society

44–8 After G. Ledyard Stebbins, *Processes of Organic Evolution*, Prentice-Hall, Inc., Englewood Cliffs, N.J., 1966

45–1 Herbert A. Fischler, Isaac Albert Research Institute of the Kingsbrook Jewish Medical Center

45–6 (b) Food and Agriculture Organization

45–7 Victor McKusick

46–1 British Museum (Natural History)

46–4 (a) Charles R. Knight, Field Museum of Natural History; (b), (c) Harold Pollock

46–5 (a) Ron Garrison, Zoological Society of San Diego; (b) L. L. Rue, III, Bruce Coleman, Inc.; (c) L. L. Rue, III, National Audubon Society

46–6 P. M. Sheppard

46–7 (a) Ross Hutchins; (b) Robert Mitchell; (c) Alan Blank, Bruce Coleman, Inc.; (d) Jeanne White, National Audubon Society

46–8 Jen and Des Bartlett, Bruce Coleman, Inc.

46–9 B. and C. Alexander, Bruce Coleman, Inc.

46–10 Daniel Janzen

46–11 M. W. F. Tweedie, Bruce Coleman, Inc.

46–12 Nebraska Game and Parks Commission

47–2 M. P. Harris

47–3 I. Eibl-Eibesfeldt, from *Ethology: Biology of Behavior*, Holt, Rinehart and Winston, Inc., New York, 1972

47–4 National Publicity Studios

47–5 After Edward O. Wilson and William H. Bossert, *A Primer of Population Biology*, Sinauer Associates, Inc., Stamford, Conn., 1971

47–6 Willis Peterson

47–7 John A. Moore

Page 898 John S. Shelton

47–8 Des Bartlett, Photo Researchers, Inc.

47–9 O. S. Pettingill, Jr., National Audubon Society

47–10 After Bruce Wallace and Adrian Srb, *Adaptation*, Prentice-Hall, Inc., Englewood Cliffs, N.J., 1964

47–11 After Wallace and Srb, *op. cit.*

47–13 A. W. Ambler, National Audubon Society

47–14 (a) J. and S. Brownlie, Bruce Coleman, Inc.; (b), (c), (d) Australian News and Information Bureau

47–15 (a) Charles Meyer, National Audubon Society; (b) Russ Kinne, Photo Researchers, Inc.

48–1 H. Albrecht, Bruce Coleman, Inc.

48–2 Robert C. Hermes, National Audubon Society

48–4 (a) H. Albrecht, Bruce Coleman, Inc.; (b) Des Bartlett, Bruce Coleman, Inc.

48–5 L. L. Rue, III, National Audubon Society

48–6 Mark N. Boulton, National Audubon Society

48–7 Zoological Society of San Diego

48–8 Zoological Society of San Diego

48–9 Zoological Society of San Diego

48–10 After W. W. Howells, *Mankind in the Making* (rev. ed.), Doubleday & Company, Inc., Garden City, N.Y., 1967

48–11 F. Erize, Bruce Coleman, Inc.

48–12 Des Bartlett, Bruce Coleman, Inc.

Page 919 C. R. Carpenter

48–14 Bruce Coleman, Inc.

48–15 Zoological Society of San Diego

48–16 H. Albrecht, Bruce Coleman, Inc.

48–17 George Holton, Photo Researchers, Inc.

48–19 The American Museum of Natural History

48–21 Robert F. Sisson; copyright, National Geographic Society

48–22 After Howells, *op. cit.*

48–23 T. W. Ransom

49–1 George Herben, Photo Researchers, Inc.

49–3 Tom McHugh, Photo Researchers, Inc.

49–5 American Museum of Natural History

49–6 American Museum of Natural History

49–7 Geza Teleki

49–8 The American Museum of Natural History

49–9 Lee Boltin

49–10 Lee Boltin

49–11 (a) From Jacques Bordaz, *Tools of The Old and New Stone Age*, The Natural History Press, Garden City, N.Y., 1970; (b) Lee Boltin

49–12 (a) J. Bordaz, *op. cit.*; (b) Peabody Museum, Harvard University, Cambridge

49–13 Ralph S. Solecki

49–14 The American Museum of Natural History

49–15 Patrimoine de l'Institut Royal des Sciences Naturelles de Belgique

49–16 Erich Lessing, Magnum Photos

49–17 Archives Photographique, Ministere des Affaires Culturelles Caisse Nationale des Monuments Historiques et des Sites

49–18 Niaux, Tom McHugh, Photo Researchers; Altamira, Rapho Guillumette; Lascaux, Art Reference Bureau

50–1 The American Museum of Natural History

50–3 The American Museum of Natural History

50–4 Erich Lessing, Magnum Photos

50–5 Irvin L. Oakes, Photo Researchers, Inc.

Page 962 Tower, Kathleen Kenyon; brick and skull, Ashmolean Museum

50–6 Paul C. Mangelsdorf, Richard S. MacNeish, and Walton C. Galinat, *Science*, **143**, 538–545, 1964; copyright 1964 by the American Association for the Advancement of Science

50–7 Allen Linn

50–8 (a), (c), (e) S. Trevor, Bruce Coleman, Inc.; (b), (d), (g) Irven deVore, Anthro/-Photos; (f) Norman Myers, Bruce Coleman, Inc.; (h) S. Washburn, Anthro/-Photos

50–9 S. Trevor, Bruce Coleman, Inc.

50–15 Nik Wheeler, SYGMA

50–16 The Chartmakers, Inc.; copyright © 1971, Time, Inc.

50–17 (a) Food and Agriculture Organization; (b) Pierre Pitet, Food and Agriculture Organization

50–18 Joe Munroe, Photo Researchers, Inc.

50–20 (a) Wide World Photos; (b) International Rice Research Institute

Table 50–1 U.S. Department of Agriculture

50–21 L. Goldman, Rapho Guillumette

50–22 Alain Nogues, SYGMA

50–23 (b) Woodfin Camp

50–24 (a), (d) United Press International; (b) National Air Pollution Control Administration; (c) Pennsylvania State Department of Health

50–25 Bernard P. Wolff, Photo Researchers, Inc.

50–26 J. P. Laffont, SYGMA

50–27 (a) Susan Landor; (b) Wide World Photos; (c), (d) Jack Dermid

50–29 (a) N. E. Beck, Jr., National Audubon Society; (b) Michigan Department of Natural Resources

50–30 (a) Jerome Wexler, National Audubon Society; (b) L. M. Chace, National Audubon Society

50–31 Eugene Smith, Magnum Photos

50–32 The Bettmann Archive, Inc.

50–33 Walter Dawn, National Audubon Society

50–34 Photo Researchers, Inc.

50–35 R. Gates, Frederic Lewis

50–36 (a) Leroy Simon, National Audubon Society; (b) Department of Indian Affairs and Northern Development

50–37 Food and Agriculture Organization

INDEX

Page numbers in italic type indicate references to illustrations, tables, and boxed discussions; boldface type is used to distinguish major text references from minor text references.

A horizon, 778
Abdominal cavity, *512*
Abscisic acid, 490
Abscission layer, and leaf drop, 443, *483–484*
Absorption
  of food molecules, 633, 639, 640, 641, **644**
  by plant roots, 438, 439, 451, 452, **456–457**
  of water through large intestine, 644
Absorption spectrum
  of carotenoids, 149
  of chlorophyll, 149
  measurement of, 148
Acacia, coevolution of, and ants, 882–883
Acetylcholine
  in nervous system, 592, 982
  as sleep inducer, 723
Acetyl CoA, in Krebs cycle, 132–133
Acid(s) (*see also specific acids*)
  defined, 53
  strength of, 54
Acid-base balance, in vertebrates, 669
Acidity
  defined, 52
  of gastric juice, 638, 641
  of human blood, 662
  measurement of, 53
  of soils, 778
Acquired characteristics, inheritance of (*see* Inheritance, of acquired characteristics)
Acrasiomycetes, 318, 319
Acromegaly, 579
Acrosome, 522
ACTH, 578, *583*, *585*
Actin, 601, 602

Actinosphaerium, *309*
Action potential, 590–591, 592
Action spectrum, *150*
Activation, of egg, 541–543
Activation energy, 44, 251
Active site
  in allosteric interaction, 77
  defined, 75
Active transport
  and cell membrane of axons, 590
  and countercurrent exchange, *678*
  defined, 116, 120
  of food molecules, 672
  and glomerular filtration, 676
  in plants, 450–451
  "revolving door" model of, *116*
Acupuncture, and perception of pain, 715
Adaptation(s) [*see also* Cline; Natural Selection; Polymorphism; Selective pressure(s)]
  of animals, *30*, *422*, *697*, *761–762*, *874*, *880*
    to extreme temperatures, 690–692, 753
    behavioral, 683, *813*
    as result of domestication, 963
  co- (*see* Coevolution)
  defined, 30, 878
  of humans, to bipedalism, *929*
  due to interspecific competition, 801, 802
  in macroevolution, 888–890
  to man-made selective pressures, 884–888
  due to natural selection, 874, 878
  of plants
    to cultivation, 961
    flowering, 462–463
    land, 441, 442, 443
  as property of living systems, **30–31**
  of protists, 306–307
Adaptive radiation, 902–904 (*see also* Ecological niche)
Adaptive zone, 902, 904
Addison's disease, 678–679
Adenine (*see also* Genetic code)
  in DNA, 234–235, *241*, 242, 245

  structure of, *66*
  as subunit of ATP, 66
Adenoids, 625
Adenosine diphosphate (*see* ADP)
Adenosine monophosphate (*see* AMP)
Adenosine triphosphate (*see* ATP)
Adenovirus, *302*
ADH
  function of, 580, *585*
  in loop of Henle, *678–679*
Adhesion, 49
ADP
  and ATP formation, 121, 126–135
  production of, **66**, 102
Adrenal cortex
  hormones of, 581–582, *585*
  pituitary control of, 578
Adrenal gland(s)
  effect of, on kidneys, 678–679
  in temperature regulation, 687
Adrenal medulla
  embryonic origin of, 583
  production of adrenaline by, 583, *585*, 592
Adrenaline
  chemical structure of, *583*
  effects of, 583, *585*, 613, 687
Adrenocorticotropic hormone, 578, *583*, *585*
Afferent neurons (*see* Sensory neurons)
Afferent vessel, in kidney, 674
Agglutination, and blood types, 197
Aggression (*see also* Social dominance; Territoriality)
  androgen responsible for, 527
  as inherited trait, *862*
  intraspecific, 838, 951
  in mating, 831–833
  and pecking order in chickens, 833
  territorial, 834–837
Aging (*see also* Life cycle)
  in animals, stopped by hibernation, 689
  in humans, and loss of hearing, 708
  in plants, 475–476
    effect of hormones on, 488, 490, 491
Agnatha, 416–417

and temperature regulation, 683
vision as stimulus for, 701–702
web-weaving, 400, 412
Belt, Thomas, 883
Beltian bodies, 883
Benthic zone, *762, 763*
Benzene, rings of, **59**
Bernard, Claude, 363
Berthold, A. A., 577–578
Beta cells, insulin production by, *584*
Beta-glucose, 81
Beta waves, 713
BFT, *712*
Bicarbonate, in human blood, 661–662
Biennials, 441–442
Bilateral symmetry, *373*
Bile, 641
Bilirubin, 641
Binomial system, 282–283
Biochemistry
    of cleavage, 550
    of embryonic development, 554
    of fermentation, 11
    of living organisms, 11–12
    of oogenesis, 550
    and study of muscles, 599
    and taxonomy, 284
Biofeedback training, *712*
Biogeochemical cycle(s), 776 [*see also*
    Carbon cycle; Mineral(s), cycling of]
Biological clock(s), 503–505, *579*
Biological time scale, *26*
Biology
    application of mathematics to, 175, *181*
    modern, and genetic mechanisms, 227
    molecular, use of bacteria in, 885, 886
Bioluminescence, 316
Biomass
    defined, 771–772
    versus productivity, 774, 791
    pyramid of, 774–775
Biomes
    defined, 736
    Holdridge system of classification of,
        756
    types of, 736–757 (*see also specific
        biomes*)
    and atmosphere, *737*
    defined, 18, 733–734
    entrance of energy into, *138*
    factors producing changes within,
        734–736
    as single ecosystem, 771
Biosynthesis
    of amino acids, 73, **227–231**, 296, 454,
        455, 644, 645
    of antibodies, 626
    of arginine, *228*
    of DNA, 265–266
    in embryonic cells, 531
    of enzymes, 260–263, 487, 586
    of fats, in mammals, 645
    of genes, 272
    of keratin, *516*
    of mRNA, 485, 487, 586
    of plant hormones, 481

of proteins, 250–263, *455*, 487, 527, 550
of RNA, 265–266, *455*
of vitamins, 644
of waxes, *84*
Bipedalism, 925, *929–931*
Birds
    adaptive radiation of, 902, 904
    cerebellum of, 716
    characteristics of, 920
    clutch size in, *875, 881*
    ear of, 706
    energy demands of, 688
    evolution of, 685
    gastrulation in, 548
    gizzards of, 634
    heart of, 611
    imprinting in, *831*
    interspecific competition in, 802
    mating behavior of, 832
    migratory, 749
    natural defense mechanisms of, 812
    photoperiodism in, 496
    predatory, threatened with extinction,
        983
    regulation of body temperature in,
        685, 687, **688**, 689
    reproduction in, 519, 524, **558–559**, 573,
        834–835, 983
    respiration in, 654, *655*
    ritual behavior in, 831, 832
    sex role reversal in, 834
    social parasitism of, 806
    taste in, 703
    territoriality in, 834–835, 837
    vision in, 698, 699, 702
    waste elimination in, 524, 668
Birth, human, 571–572, *579*
Birth control
    methods of, 536, *537,* 586, 970
    programs of, 970–971
Birth weight, human, 875
Birthrate (*see also* Natality)
    defined, 968
    reduction in, *967,* 970–972
Biston betularia, 884–885
Bivalvia, 386–387
Blade, of leaf, 332
Blagden, Charles, 682
Blastocoel, 543
Blastocyst, 561
Blastodisc, *546, 575*
Blastopore, 546
Blastula, 543, 546
Blending inheritance, theory of, 174–175,
    204, 207
Blindness
    color, *218*
    night, 700
Blind spot, of retina, *699*
Blood
    carbon dioxide in, 661–662
    circulation of, 609–621
    clotting of, 623–624
    oxygen in, 660–661
    pH of, 54
    volume of

in diving mammals, *663*
in man, *663*
Blood cells
    red, 78, 83, *91,* 196, **621,** 988
    white, 97, 117, **621–622**
Blood glucose, regulation of, 642, 644–645
Blood pressure, 615–616
    and cholesterol, 83
    defined, 615
Blood transfusions, *198*
Blood type(s), 197–199, 881
Blood vessels, 614–619
    cholesterol deposit on, 83
Body plan
    of animals, 377
    of land plants, *430*
Bohr, Niels, 36, 37
Bond (*see specific bonds*)
Bonding, of water molecules, 48
Bonner, James, 497–498
Bony palate, 911–912
Book lungs, 395, 396, 397, **400**
Borlaug, Norman, *976, 984*
Boron, plant use of, *455*
Botany (*see* Plants)
Botulism, cause of, *297*
Bowman's capsule, 674
Brachiation, 915, 920, 933
Brachydactylism, 196, *197*
Brain
    of arthropods, 396–397
    of cephalopods, 388, 390
    development of, *555, 564,* **973**
    of domesticated animals, 943–944
    evolution of, 715, **717–718,** 720, **931–932**
    human
        anatomy of, *714–720, 722*
        compared to brain of apes, 928
        evolution of, 931–932
        and learning, 725–726
        methods of studying, 712–714, 722
        relationship between eye and, *721*
        relationship between mind and, 711
        respiratory center of, 664–665
        and sleep, 723–725
    of mammals, 715, 717–718
    size of, and intelligence, 931–932, 943
    of vertebrates, 715, *717*
Brain waves, 712–713
Brainstem, of human brain, 714, 715
Breathing (*see also* Respiration)
    control of, 664–665
    in diving mammals, *663*
    mechanics of, 662
Breeding (*see also* Artificial selection;
        Crossbreeding)
    random, results of, *867, 869*
Briggs, R. W., 551
Bronchi, 656–657
Bronchioles, 656–657
Brower, Jane, 817
Brown fat, *688*
Brucellosis, 297
Bryophytes, 341, 342–343, 358
Büchner, Eduard, 11
Bud(s), growth of, 472–473, 482–483, 487

and temperature elevation in plants, 684
Flowering hormone, 488–489
Flukes, 376
Fly, touch receptors of, 408
Flying, energy requirements for, *655*
Follicles, ovarian, 531, 534–535
Follicle-stimulating hormone, 528, *529*, 534, 535, *585*
Following response, and imprinting, 831
Food (*see also* Digestion; Digestive system; Nutrition)
consumption of, inequality in, 974
oxidation of, in living systems, 59
production of, attempts to expand, 969, **975–978**
spoilage of by bacteria, 297
storage of
by diatoms, 314
in earthworms, 379
by Euglenophyta, 313
in plants, 332, 333, 343, 350, 435, 436, 437, 439, 486
Food chain
DDT in, 982–984
diatoms in, 314
eating "high" on, 973, 974–975
energy flow in, 771–775
mercury in, 984–985
strontium-90 in, 985, 986
trophic levels in, 771–775
Food poisoning, caused by bacteria, 297
Food supply
and energy expenditure, 688
as factor in ecological niches, 795, 796
and hibernation, 688, 689
limiting growth of populations, 800
in Malthus' theory, 8
Food web, *773*
Foraminifera, *309*
Forest(s)
climax, 790
recycling in ecosystem of, 784–785
succession in, and fire, 790
trophic levels of, *774, 775*
tropical, destructive effects of agriculture on, 987–988
types of, 738–748, 757
Fossil(s)
of algae, *24–25*
of bacteria, 24–25
of clam shell, *39*
earliest, 24–25
in evolutionary theory, 3
of hominoids, 920, 925, *927*, 928, 937, 939, 941, *943*
and identification of strata, 3
radioactive dating of, *39, 887*
of trilobites, *3*
Fossil record
and age of Earth, *887*
of climatic changes, 936
examples of adaptive radiation in, 902
macroevolution observed in, 888
Founder principle, 872–873
Fovea, 699, 915

FP (*see* Flavoproteins)
Franklin, Rosalind, *245*
Free energy (*see* Energy, free)
Frequency, of sound
detected by bats, 809
response of auditory cortex to, 718
response of basilar membrane to, 708
Frisch, Karl Von, 411–412
Frog(s)
adaptations of, *880*
development of, *542, 547*
leopard, subspeciation of, 896
mating timetable of, *900*
respiration in, 651, 652, *654*
tongue of, 635
visual stimulation of, 701–702
wood, speciation of, 896
From Fish to Philosopher (Smith), 667
Frontal lobes, of cerebral cortex, 720
Fructose
chemical characteristics of, 60–61
in glycolysis, 126–129
structure of, 60
Fruit (*see also* Angiosperms; Plants, flowering)
defined, 356
effect of auxin on, 483–484
formation of, 357, 467–468
ripening of, 490–491
Fruit fly (*see Drosophila*)
Fruiting bodies, 293
FSH, 528, *529*, 534, 535, *585*
Functional group, defined, 58
Fungi (*see also* Lichens)
chitin in cell walls of, 82
digestion in, 632
kingdom of, 286, *288*, **319–326**
in mineral cycling, *987*
reproduction in, 320–321
symbiosis of, with ants, 803–804
Fungi Imperfecti, 326
Fusion, of atomic nuclei, 17, *35*

G phase, 157–158
Galapagos finches, **7**, **891–894**, 899, 901, 902
Galapagos Islands, 7–8, 891
Galapagos tortoises, 7, 894, 895
Galileo, 21
Gametangia, 321, *322*
Gametes (*see also* Egg cell; Fertilization; Germ cell; Meiosis; Sex cell; Sperm cell)
defined, 174
DNA in, 240
plant, 465–466
size of, 337
of vertebrates, 518–519
Gametophyte, 338, 461, 465
Ganglia, 594
Garrod, Sir Archibald, 226–227
Gas exchange
across body surface of small animals, *650*
in plants, 433, **447–449**, 451 (*see also* Respiration, in plants; Water

movement, in plants)
in respiration, 660–662
Gastropoda, 387
Gastrulation, 543, 546–548
Gause, G. F., 795–796
Gemmules, in sponges, 366
Gene(s) (*see also* Chromosomes; DNA; Gene flow; Gene frequency; Gene pool; Genetics)
and amino acids, 227–231
chemistry of, 226, 232–235
and incomplete dominance, 204
interaction of, with environment, 207–209
lethal, 205
linkage groups of, 212–214
location of, 192–195, 210–225
and Mendel's experiments, 175–184
mutations of, 202–207
pleiotropy of, 205–206
and polygenic inheritance, 206–207
and proteins, 230–232
recessive, *212*, 870
recombination of, 864
regulation of activity of, **260–263**, 484, 487, 500
and regulation of enzyme biosynthesis, 226–231, 261–263
and structure of hemoglobin, 231
transferral of, 269
transcription of, *256, 258*
variations in, 863–864
Gene flow, 865, 896
Gene frequency, versus genotype frequency, 870
Gene pool
changes in, 858, 860
composition of, and Hardy-Weinberg law, 867, 869, 870, 871
defined, 857
Generation, spontaneous (*see* Spontaneous generation)
Genetic code, 250–264, 307
Genetic counseling, 220, 222–223, 274
Genetic defects, **196–197**, 218–223, 227, 231, 259, 271–272, 274
and Hardy-Weinberg law, 869, 870–871
produced by exposure of embryo to x-rays, 567
as result of inbreeding, 873
Genetic drift, defined, 872
Genetic engineering, 272–274
Genetic isolation
and geographic races, 896
through polyploidy, 900–901
as precondition for adaptive radiation, 902
provided by geographic barriers, 894–896
as requirement for speciation, 891, 894–896
as result of behavioral patterns, *899*
Genetic testing, use of amniotic fluid in, 558
Genetic uniformity, of selectively bred grains, 975, 989

Inducers of enzymes, 260–261, 263
Induction, embryonic, 553–554, 555
Infant mortality, 299, 966–967
Infection (*see also* Disease, bacterial; Disease, viral)
  bacterial
    controlled by "wonder drugs," 299
    of wounds, 297–298
  cycle of, in *E. coli*, 237–239
  defense mechanisms against, 622, 625–627
  fungus, as cause of Irish potato famine, 989
Inflammation
  as defense mechanism, 625, 626
  suppressed by glucocorticoids, 581
Influenza virus, 303
Information, genetic (*see also* Genetic code)
  and DNA, 240, 245–246
  and RNA, 250–254
  stored in living systems, 31
  transfer of, **237–239**, **268–272**, 291, 292, 306, 310
Informational macromolecules (*see* Macromolecules, informational)
Ingenhousz, Jan, 140, 141
Inheritance (*see also* Genetics; Heredity)
  of acquired characteristics, 5
  of behavior, *862*
  blending hypothesis of, 174–175
Inhibition
  chemical, in plants, 468, 479, 482, 490 (*see also* Hormones, plant)
  releasing of, in arthropods, 396–397
"Injury feigning," 812
Inorganic substances [*see* Biogeochemical cycle(s); Carbon cycle; Mineral(s), cycling of, in ecosystems]
Insecticide(s)
  alternatives to, *984*
  effect of, on nervous system, 593
  need for, 810
  as pollutants, 982–984
  resistance to, 886
Insectivores, 913
Insects (Insecta), 401–412 (*see also* Invertebrates)
  action of DDT against, 982, 984
  and adaptive radiation, 902
  body temperature of, 684
  chitin in shells of, 82
  chromosomes of, and RNA synthesis, 265
  evolution of, 462
  excretion of nitrogenous waste by, 668
  natural defense mechanisms of, 812, 814–*816*, *817*
  nonrandom mating in, 877
  photoperiodism in, 496
  plant defenses against, 809
  pseudocopulation of, *817*
  role of, in pollination, 460, **462–463**, *464*
  societies of, *820*, *821–829*
Instars, 403

Insulin, *77*, *583–584*, *585*, 642
Integration of bodily activities, 363, **577–608**
Intelligence
  and association cortex, in primates, 718
  and brain size, 931–932
  evolution of, and high constant temperature, 687
  human
    and protein intake during pregnancy, 569
    selection for, 950
  mammalian, 517
  and nature-nurture controversy, *862*
Interbreeding
  in definition of species, 283
  prevention of, between species, 896–897, 899–900
Intercostal muscles, 662
Interglacials, 936
Interneurons, 593
Interphase, 187
Interspecific competition (*see* Competition, interspecific)
Interstitial fluid, 672
Intestine(s), *640*, 644
  cells of, active transport into, 116
  role of bacteria in, 296
  origin of, 557–558
Intrauterine device, *537*
Inversions, chromosomal, *217*
Invertebrates, 362–364 (*see also specific invertebrates*)
  ecological importance of, 362
  phyla of, 364–393, 394–412 (*see also specific phyla*)
Iodine, and diet, 581
Ion(s)
  defined, 38, *42*, **46**
  hydrogen, and pH scale, 53
  regulation of concentration of, in body fluids, 582, 669
Ionic bonds, 41, *42*
Ionization, 41–42, 52
Iron, plant use of, *455*
Irrigation, of desert, 977
Islets of Langerhans, *584*, 642
Isogamy, *337*
Isolating mechanisms, 897, 899–900
Isotonic solution, 113
Isotopes
  defined, 34, 45
  of hydrogen, 36
  radioactive, 36, *39*
IUD, *537*

Janssens, F. A., 214
Janzen, David, 883
Jaundice, 641
Java man, 937
Jellyfish, 368, **370–371**, *391*
Jenner, Edward, 303
"Jet lag," 579
Jupiter, 19, *20*, 39
Jurassic period, *358*, *415*, 911

Kangaroo rat, 671, *672*
Karyotype(s), *220*, *221*, 222–223, 924
Kekulé, Friedrich, 59
Kelvin, Lord, *887*
Kelvin scale, *682*
Kennedy, Donald, 712
Kepler, Johannes, *21*
Keratin
  production of, by vertebrates, *85*
  role of, in human skin, *516*
Kettlewell, H. B. D., 884
Kidney
  evolution of, 670
  regulation of function of, 677–679
  structure of, 674–676
Kilocalorie, 44
Kinetic energy (*see* Energy, kinetic)
Kinetin, *487*, 488
Kinetochore, 159, 160–161
King, T. J., 551
Kingdom(s) (*see also specific kingdom*)
  classification of, *284–288*
Khorana, H. G., *260*
Koch, Robert, 299
Krebs cycle, 132–*133*
Kubai, Donna, 316
Kummer, Hans, 847–848
Kwashiorkor, *973*

Labia, 532
Labor, stages of, 571
Labyrinth, in fish, 706
Lactose, and milk drinking, 882
Lactic acid, *131*
Lactose
  structure of, 62
  in enzyme biosynthesis, 263
  as inducer, 260–261, 263
  polymorphism for digestion of, 882
  splitting of, *260*
"Ladder of life," 1, 5
Lake(s), 758–759, *760*
Lamarck, Jean Baptiste, 5, 866
Landsteiner, Karl, 197
Language (*see* Communication; Speech)
Laplace, P. S., 140–141
Large intestine, 644
Larvae, 403
Laryngitis, 656
Larynx, in humans, 656
Lashley, Karl, 725–726
Lateral-line system, of fish, 705–706
Lateral meristems, 473
Laterites, 739, 987–988
Latitude, and temperature, 736
Lavoisier, Antoine, 140–141
Law(s)
  of chance, applied to genetics, **181**, 867, 872
  of motion, *21*
  of probability, **181**, 867, 872
  of thermodynamics, 121, 123–124
Leaf (*see also* Photosynthesis; Stimuli, plant responses to)
  effect of cytokinin on, 487
Leaf(ves)

function of, 431, **433–434**
gas exchange in, 433, 434, *449*
growth of, 472
sessile, *431*
structure of, *352*, **433–434**, 446
variations in, *431–432*
venation in, *433*
wilting of, 502
Leaf drop, 443, 483–484
Leakey, Louis, 920, *927*
Leakey, Mary, 920, *927*
Learning
and association cortex, 718, 720
cerebral cortex as site of human, 722
"by cannibalism," 375
defined, 725
and imprinting, 831
mechanism of, 725–726
in octopus, 389–390
in planarians, 375
Learning period, 517 (*see also* Care of young)
Lederberg, Esther, 885, *886*
Lederberg, Joshua, 885, *886*
Leeches, 382
van Leeuwenhoek, Antony, 172, 174, 299
Legumes, symbiosis of, with bacteria, 803
Lehninger, Albert, 30–31
Lehrman, Daniel S., *587*
Lemmings, 749, *799*
Lemurs, 914, 916, *917*
Leukocytes, 621–622, 625
Leucoplasts, 98
Levene, P. A., 234–235
Lewontin, R. C., 878
LH
function of, 534, 535, 536, *585*
production of, 527–528, *529*
Lichens, **327–328**, *803*, 986
Liebig, Justus von, *11*, 783
Life [*see also* Biome(s); Living systems]
"code of," DNA as, 250 (*see also* DNA; Genetic code; RNA)
continuity of, 518 (*see also* Development; Growth; Reproduction)
defined, 26
earliest forms of, 24–25, *153*, 289
on Earth
reasons for, 19
requirements for, 733–734
form in, *27*
human, beginning of, 518
"language of," proteins as, 231, 250
origin of, 21–23
possibility of, on other planets, 19–21
raw materials for, 21
temperature limits for, 681
Life cycle(s)
of coelenterates, 371
of green algae, 334
of insects, 404–405
of *Obelia*, 369–370
of plants
annual, 441
biennial, 441–442

flowering, 465
land, 338
perennial, 442–443
and sexual reproduction, *196*
of Sporozoa, 312
vertebrate, 518
Life expectancy, 967, 971
Light
and air purification, 140, 141
effect of, on plants (*see* Photoperiodism)
fitness of, for living systems, 143–144
particle theory of, *143*
wave theory of, *143*
Light compensation level, defined, *758*
Light energy (*see* Energy, light)
Light reactions, of photosynthesis, 145, **151–152**, *155*
Light reception (*see* Vision)
Lignin, 94, 436
Limiting factors, of growth, 783–785, 798
Limnetic zone, 758, 759
Lind, James, 645
Linkage groups, 212–214
Linnaean system of classification, 282–283
Linnaeus, Carolus, 2, 282
Lipases, 640–641
Lipids, 33, 82–84
Lithosphere, 776
Littoral zone, 758–759, 767
Liver
cell division in, 158
functions of, 618, 642–*643*, **644–645**
Liverworts, 342–343, *358*
Living systems (*see also* Life)
adaptation of, *30*
biochemical unity of, 104, 230
chemistry of, 11–12, 33, 54
classification of, 281–286
common atoms in, 38–41
elements required by, 783
energy exchange in, **121–135**, *136* (*see also* Photosynthesis)
and entropy, 123–124
fitness of light for, 143–144
growth of, 157
levels of organization of, 86, 281
oxidation of food in, 59
pH values of, 54
properties of, **27–31**, 34, 88, 667
and reproduction, *29*
theoretical models of, 11
Lizards, 683, 810
Lobster, 398–399
Lockjaw, *293*, 297–298
Locomotion
in cephalopods, 388, 389
in *Chlamydomonas*, 334
in earthworms, 379
in fish, 654
in fungi, 320, 321
in *Gonium*, 335
in lobsters, 399
in medusa, 370
in nematodes, 377

in one-celled algae, 300, 313–314, 315, 318
in *Pandorina*, 335
in planarians, 373–374
in polychaetes, 382
in protozoans, **307**, 308, 309
in spiders, 399
in spirochetes, 293
in starfish, 383
in *Volvox*, 335
of white blood cells, 622
Locusts, 799
Loop of Henle, 676–678
Lorenz, Konrad, 831
Lucretius, 172
Lumen, 617
Lung(s) (*see also* Breathing; Respiration)
of birds, *655*
evolution of, 654
in fish, 418
in gas exchange, *650*
structure of, 651
Lung cancer, development of, *659*
Lungfish, **418**, 654, 902
Luria, Salvador, 237, *239*, 304
Luteinizing hormone (*see* LH)
Lycophytina, 344, *345*, *346*
Lyell, Charles, 6
Lymphatic system, 624–625
Lymphocytes, 622, 625, *626*, *629*
Lysenko, T. D., 5
Lysosomes, **97**, 98, *105*, 106, 117
Lysozyme, *72*, 291–292

Macaques, *918*, *919*
MacArthur, Robert, 796
Macroevolution, 888–890, 891
Macromolecules
informational [*see also* Enzymes; Hemoglobin; Hormone(s); Nucleic acid(s); Protein(s)]
structural, 69, 80–85
study of, to establish evolutionary relationships, *904*
Macronutrients, 455
Magnesium, plant use of, *455*
Magnification, versus resolving power, *92*
Mairan, Jean-Jacques de, 503
Malaria
cause of, *312*, 805
DDT used to control, 982
and sickling gene, 870–871
Malnutrition, 972–975
Malpighian tubules, 395–*396*
Malthus, Thomas, **8**
Mammalia, 420–422
Mammals (*see also* Human beings; Man; Vertebrates)
brain of, 715, 717–718
characteristics of, 420, 515, 517
cline of body size in, 878
coelom of, *512*
criteria for classification of, 911–912
ear of, 706
early, 911

Natural selection (*Continued*)
and macroevolution, 888–890
man-made pressures in, 884–888, 990
and myxoma virus in rabbits, 805
and pollination, 350, 462
polymorphism as result of, 879–882
and predation, 808, 809
requirements for, 874
and sexual reproduction, *194*
and size, 363
and social behavior, 821, 830
theory of, **8**, 857
and hypothesis of blending inheritance, 174–175
and man's role, 20
role of mutations in, 202–203, 240
types of, 875–876
for variability, 864
variations as raw material for, 859
Nature
balance of, *808*
man's changing relation to, *962*
Neanderthal man
brain of, 943
burial of dead by, 944, *945*, 948, 951
cannibalism of, 952
characteristics of, *940*
in evolution of modern man, 939
life expectancy of, 967
use of tools by, *943*
Negative feedback, in hormon regulation, 527, *534*, 582
Negative pressure, in plants, 452
Nematocysts, 366, *367*
Nematoda, 377
Nemertines, 376–377
Neocortex, evolutionary changes in, 717
Nephridia, 378–*379*, 380
Nephron, 674, *675*
Neptune, *20*
Nerve cord, 413
Nerve net, 369
Nervous system (*see also* Neurons)
action of DDT on, 982
of annelids, 378, *379*, **381**
of ants, *827*
of arthropods, 396–397, 408–409
of bivalves, 387
of gastropod, *387*
of *Hydra*, *368*
of medusa, 371
of octopus, *389*
of planarians, 375
of vertebrates
autonomic, *595*, **596**, *597*, *598*
and blood vessels, *598*
central, 594
development of, *548*, *550*, **554–555**, *564*
and muscle contraction, 603
parasympathetic, *596*, *597*, 605
peripheral, 594
sensory receptors of, 694–709
somatic, 595–596
sympathetic, 596, *597*, 605
"voluntary" versus "involuntary," 595

Net productivity, 772
Neural plate, in *549*
Neural tube, 548, *549*, 554–555
Neurobiology, learning and memory, 725–726
Neurons
in humans, 513, 515, *517*, 711
structure of, 588, *589*, *593*
types of, 593
Neurospora crassa, genetic research on, 227–230
Neutron, 34
Neutrophils, 622
Newton, Isaac, *21*, *142*
Nicotinamide, in NAD, *126*
Nitrates
applied to crops, 988
in formation of amino acids, 73
Nitrification, 779, 780
Nitrites
atomic structure of, 40, *41*
bacterial oxidation of, 296
as component of living systems, 33
excretion of, in mammals, *668*
in formation of amino acids, **73**
harmful effects of, 988
plant use of, *454*, 455
as raw material for life, 21
use of in fertilizers, 988
Nitrogen bases (*see also* Nucleotides)
in DNA, 234–235, 240, 242, 245
in nucleic acids, *79*
Nitrogen cycle, **779–780**, 785
role of bacteria in, 295–296
Nitrogen fixation, 780–781
by bacteria, 290
by blue-green algae, 301
Nocturnalism, 684–685, 690, 698, 699, 702, 911, 916, 917
Nomadism
of army ants, 826–827
of early man, 945, 966
of primates, 945
Nondisjunction, 221, 223
Notochord, 413, 548
Nuclear energy, 17, 35
Nuclear envelope, 94, *95*
Nuclear pores, 94, 95
Nuclei, in spinal cord, 594
Nucleic acid(s), 33, 69, **79**, 80, 159 (*see also* DNA; RNA)
Nucleolus, 95
during mitosis, 159, 160
in living systems, 33
role of, in genetic coding, *80*, 235, 241, *242*, *243*, 248, **250–259**
Nucleotides
structure of, 79, *80*, 234
as subunit of ATP, 66
as subunits of nucleic acids, 69, 79, 235
Nucleus (atomic), 34, *35*
Nucleus (of cell)
and cell size, 89
in ciliates, 309–310
function of, 95
in meiosis, 186–190

in mitosis, 160–163
structure of, 94–95
in various cells, *105*
"Nurse cells," of human egg cells, 117
Nutrients (*see also* Nutrition)
absorption of (*see* Absorption, of food molecules)
as limiting growth, 783, 784, 785
required by plants (*see* Micronutrients; Nutrition, in plants; *specific nutrients*)
in soil, 988 [*see also* Mineral(s)]
Nutrition (*see also* Autotrophs; Diet; Digestion; Food; Heterotrophs; Nutrients)
bacterial, 295–296
in blue-green algae, 301
in earthworms, 380
of embryo, 558, 562, 563
of fetus, and development of intelligence, 569
in fungi, 320, 321
in humans, 645, *646*
in lichens, 327
in plants, 447–457 (*see also* Food, storage of, in plants; Photosynthesis)
in ruminants, *642*
in vertebrates, 645
Nymphs, 403, 828

Obelia, 369–370
Oceans, 762–767
origin of life in, 22
temperature of, 51
trophic levels of, *774*
Ocelli, 371, *375*, 408
Octopus, 388, 389–390
compound eye of, 696
Odor (*see also* Smell)
as natural defense mechanism, 815–816
Odor
role of, in mating, 410–411
and territoriality, in wolves, 843
Olfactory bulbs, evolutionary changes in, 717
Olfactory epithelium, in man, 703–705
Olfactory stalks, 717
Oligocene epoch, *415*
Oligochaeta, 382
One-celled organisms
classification of, 286
cloning of, 273
lysosomal digestion in, 97
versus many-celled, 363
Oocytes, human, 530–*531*
Oogamy, *337*
Oogenesis, in humans, 530–531, 534–535
Oomycetes, 321, *322*, *326*
Operator gene, 261
Operon, 261–263
Optic chiasma, 722
Optic nerve, 698–699
Orangutans, 921–922
Order, 284
Ordovician period, *358*, *415*

Organ of Corti, 705, 707
Organelles (*see also* Mitochondria; *specific organelles*)
  autophagy of, 97
Organic chemistry, 57
Organic molecules, and origin of life, 21–22
Organism(s)
  characteristics of major groups of, *288*
  classification of, 281–286 (*see also specific groupings*)
  consequences of interactions of, 770
  defined, 362–363
  earliest forms of, **24–25**, *153*, 289
  ecological niches of, 795–796
  evolutionary relationships of, *286*
  integration of, 605 [*see also* Endocrine glands; Hormone(s); Muscle(s); Nervous system]
  interaction of, with environment, 733 (*see also* Biosphere; Ecology)
  number of, 796–800
  one-celled (*see* One-celled organisms)
  radial symmetry of, *366*
  requirements of, 577
Organization
  levels of biological, **10–12**, 281, 373
  as property of living systems, **27**, 34, 86
  social [*see also* Social behavior; Society(ies)]
    bases for vertebrate, 831–833
    of early man, 947–951
    principles of insect, 829
Organizer, 552
Organogenesis, 543, 554–559
Organ system, 513
Orgasm, 525, 532–533
Origin of Species, The (Darwin), 5, **9**
Oscula, *364*
Osmometer, 112
Osmosis
  in body fluids, 669, 672
  defined, 112, **120**
  in living organisms, 113–115
  measurement of pressure of, *112*
  movement of water in, 112–113
  in root entry, 452
  and turgor in plant cells, 114–115
  and water potential, 112, 113
Osteichthyes, 416, 417, 418
Osteoblasts, *514*
Osteoclasts, 584
Ova, 530 (*see also* Egg cells, human)
Ovary
  human, 530–531, *585*
  plant, *354*, 461
Overbeek, Johannes van, 487
Oviducts, 375
Ovists, 174
Ovulation, 530, 535
Ovules, plant, *354*, 461, 465–466, 467–468
Oxidation
  by bacteria, 295–296
  defined, 42
  of glucose

and respiration, 665
  stages of, 125–126 (*see also* Glycolysis; Respiration)
  in photosynthesis, 126
  of pyruvic acid molecule, 132
Oxidation-reduction reaction, 42, 59
Oxygen (*see also* Anaerobic pathway; Photosynthesis; Respiration)
  atmospheric, *153*, 649–650
  atomic structure of, 39, *40*
  as component of living systems, 33
  consumed by running animals, *648*
  diffusion of, through cell membrane, 111, 650
  as raw material for life, 21
  in respiration of organisms, 650–651
  transport of, in humans, 660–662
Oxygen cycle, *777*
Oxygen debt, 131
Oxytocin, 580, *585*
Ozone, *737*

Pacemaker, *612, 613*
Pain, and acupuncture, 715
Paleocene epoch, *415*, 916
Paleontology, 4
Paleozoic era
  evolution of land plants in, 349, *358*
  Ice Ages of, 936
  life forms of, *415*
Palisade cells, function of, 433
Pancreas
  hormones of, 583–584, *585*
  role of, in digestion, 642
Pancreatic duct, 640
Pandorina, *24*, 335–336
Pangaea, *898*
Panting, by dogs, *687*
Parallax, 698
Paramecium
  behavior of, *311*
  cell division in, 157
  cilia of, 101, *310*
  hypertonicity of, *113*
  and interspecific competition, 795
  lysosomal digestion in, 97, *310*
  multiple nuclei of, 89
  osmosis in, 114
  phagocytosis in, 117, *118–119*
  structure of, *92–93*, *310*
Paranthropus, 927–928
Parapod, 382
Parasitism
  bacterial, 295, 296
  defined, 802
  and disease, 804–805
  flukes, 376
  social, 806
  tapeworms, 376
Parasympathetic nervous system (*see* Nervous system, parasympathetic)
Parathyroid glands, 584, *585*
Parazoa, 364
Parenchyma cells, 433, *476*
Parthenogenesis, 542
Pasteur, Louis, *173*, 299

Pauling, Linus, 74, 231, 241
Pea plants, Mendel's experiments with, 175–180
Pecking order, 833, 948
Pectin, 484
Peking man, 937
Pelagic zone, *762, 763*
Pelecypoda, 386–387
Pellicle, of ciliates, 309
Pelvis, of man, 928, *929*
Penfield, Wilder, *719*
Penicillin, *292*, 299, 326
Penis, human, 524–525, 725
Pennsylvanian period, *358, 415*
Pepsin, 638
Peptide bond, 71, 73
Perennials, 442–443
Pericycle, 438
Peristalsis, 378, 633–634
Peritoneum, *634*, 638
Permafrost, 748
Permian period, 349, *358*, 415
Perspiration, 686 (*see also* Evaporation; Sweat; Sweat glands)
Petals, 460
Petiole, 434
pH (*see also* Acid-base balance)
  in body fluids, *669*
  in living systems, 54–55
  scale, *53, 54*
  of soils, 778, 788
Phaeophyta, 332, *339*
Phagocytosis, 117, *118*
Pharynx, 413, 636
Phenotype (*see also* Genetics; Genotype; Heredity)
  defined, 177, 859
  environmental effects on, 207–209
Phenylalanine
  genetic code for, 257
  and PKU, 227
Phenylketonuria (*see* PKU)
Pheromones, 410–411
  in army ants, 826–827
  as isolating mechanism, 899
  in mammals, 705
  and menstruation, 534
  of queen honeybee, 823, 824
Phloem
  defined, 344
  as florigen transporter, 489
  as site of translocation, 449, 450–451
  structure of, **436**, 437, 475
Phosphate group, structure of, *66*
Phospholipids, 82, *83*
Phosphorus
  as component of living systems, 33
  plant use of, *455*
  as pollutant, 783
Phosphorus cycle, *782*
Phosphorylation, 133, 151–152
Photochemical systems, of photosynthesis, 150–152
Photoelectric effect, *142*
Photoinduction, 488–489
Photomorphogenesis, 500

of water, *981*
Polychaeta, 382
Polydactylism, 873
Polyethylene, *59*
Polygenic inheritance, 206–207
Polymorphism
in clutch sizes of English swifts, 881
in color and banding in snails, *879*
in color of snow geese, 881, *885*
defined, 879
in human blood groups, 881
in lactose digestion by humans, 882
Polynomial system, 282
Polyoma virus, 271, 275
Polyp(s), 366, 368–*370*, 372
Polypeptide, *71*, 73
Polyploid cells, 186
Polyploidy, 186, 900–901
Polysaccharides
in blood, 197
structure of, 62–63, 81–82
Polysomes, *254*
Ponds, 761, *787*
Pons, of brainstem, 714, 715
Population(s) [*see also* Community(ies); Species; *other entries under* Population]
and agricultural revolution, 964, 966
and competitive exclusion, 795–796
defined, 795
interactions among, 800–802 (*see also* Parasitism; Symbiosis)
in Malthus' theory, 8
natural defenses of, 809–817
operation of founder principle in, 872–873
and predator-prey relationships, 806–809
size of
factors in, 796–800
and Hardy-Weinberg law, 872–873
survival curves of, 475
and territoriality, 837
variability in natural, 859–865
Population control, means of, 969–972
Population density, oscillations in, *800*
Population explosion, 299, 967–969, 971, 972
Population genetics, 857–873
defined, 857
and Hardy-Weinberg law, 866–873
Porifera, 364–366, *391*
Porphyrin ring, *78*
Porter, R. R., 626–627
Posterior, lobes of cerebral cortex, 720
Posture, of primates, 915–916
Potassium
plant use of, *455*
in propagation of nerve impulse, 590, 591
in stomatal activity, 448–449
Potato blight, *321*
Potential energy (*see* Energy, potential)
Power of Movement in Plants, The (Darwin), 480
Prairie, *775*, *779*

Prairie dog, 691, 750
Praying mantis, 396–397, 408, *409*
Preadaptation, 902
Precambrian era, *358*, *415*
Precipitation (*see* Rainfall, patterns of)
Predation
of arachnids, 399
and cephalization, 373
of cephalopods, 388
of coelenterates, 366
effect of, on populations, 800, 808
group life as protection against, 828
of lobsters, 399
natural defenses against, 809–817
of praying mantises, 408
of ribbon worms, 377
as selective pressure, 652
of spiders, 397
of wolves, 839, 840, **844–845**
Predator-prey relationships
between bats and moths, 808–809
in ecosystems, 806, 808
and introduction of new species, 808
parasitic diseases as, 804
Preevolutionists, 2
Pregnancy
hormones and, 536
test for, 536, 563
Prepuce, 533
Presentation, as gesture of submission, 848
Pressure
atmospheric, 649–650
hydrostatic, 109
osmotic, 112
Priestley, Joseph, 139–140, *141*
Primary succession, 786–788
Primates (*see also* Apes; Human beings; Man; *specific primates*)
adaptations of, 914–916
avoidance of inbreeding among, 950
behavior of (*see* Baboons; Social behavior)
brain of, 714, 717, 718, 719
care of young by, *916*
characteristics of, 422, *913–915*, *917*, *931*
communication among, 820, 944
evolution of, *912*, 913
phylogenetic tree of, *926*
sense of smell in, 702
sexual behavior of, 949, 950
social behavior of, 947–948
study of, and early man, 940
vision in, 702
Primitive streak, 548
Primordia, 472
Principles of Geology (Lyell), 6
Probability, laws of (*see* Laws, of chance: Laws, of probability)
Producers, in ecological pyramids, 773–775
Productivity, of ecosystems, 771, 774–775, 791
Profundal zone, 758, 759
Progesterone, *529*, 534, 536, *585*, *587*
structure of, *535*

Prokaryotic cells, 24, *105*, 289, *307* (*see also* Algae, blue-green; Bacteria; Monera; *specific prokaryotes*)
Prolactin, 578, 579, *580*, 587
Promoter gene, 261
Prophage, 270 (*see also* Virus)
Prophase
of meiosis, *186*, *187*, 189
of mitosis, 160–161
Proprioceptors, 408, 695
Prosimians, 916–917
Prostaglandins, 525, 586
Prostate gland, 525
Protein(s) [*see also* Amino acid(s); Enzymes; Hemoglobin]
biosynthesis of, 97, 250–263
in cell membrane, 82
changes in, and evolution, *904*, *924*, 925
conversion ratio of, *975*
and genes, 230–232, 238–239
influence of temperature on, 681–682
informational, 69–79
in nutrition, 645, 972–973, 975
and origin of life, 22
repressor, *263*
structural, 84–85
structure of, 69–79, 84–85
Prothrombin, 623
Protista, kingdom of, 286, *288*, 306–319 (*see also specific classes*)
Proton, 34
Protoplasm, defined, 91, 95
Protoplast, 94, 97
Protozoa, 306, 307–312
Provirus, 270 (*see also* Virus)
Pseudocoelom, 377, 378
Pseudocopulation, *817*
Pseudopod, *25*, 309
Pseudovirion, 272
Psilocybin, 324
Pterophytina, 344, *345*, 346
Puberty, environmental influences on, 208–209
Punnett squares, *178*, *181*, 867–868
Pupa, 403
Pupil, of eye, *696*
Purines, in DNA, 79, 234–235, 240, 242, 245
Pyramids, ecological, 773–775
Pyrimidines, in DNA, 79, 234–235, 240, 242, 245
Pyrrophyta, 315–317
Pyruvic acid, in cellular respiration, *126*–135

Quaternary period, *415*

Ray, John, 281–282
Reactions, chemical (*see* Chemical reactions)
Recessive traits, 176 (*see also* Mendel, Gregor)
"Red tides," 316
Redi, Francesco, *173*

Reduction, 42, 126 (*see also* Oxidation reduction)
Reductionism, 11
Redwoods, 747, *748*
Reflex arc, 595–596
Regulation, homeostatic (*see* Homeostatic regulation)
Regulator gene, 261
Regulatory systems, (*see* Endocrine glands; Hormones; Nervous system)
Relativity, theory of, *35*
REM sleep, 724–725
Replication (*see also* Cloning)
  of DNA, 240, 246–248, 259, 272, 275
    in bacteria, 290, 291
    in viruses, 302
  of *E. coli* chromosome, 268
  in *Euglena*, 314
  of proviruses, 270
  of RNA, 251, 275, 302
  self- (*see* Self-replication)
  viral, **237–239**, 240, 251
Repressor gene, 261, 263
Reproduction (*see also* Behavior, sexual; Copulation; Fertilization; Germination; Mating)
  in animals
    amphibians, 418
    birds, 519, 524, *587*, 834–835
    earthworms, 381
    Hydra, 369–370, *518*
    insects, 821–829
    mammals, 420–421, 517
    patterns of, 519
    planaria, *375*
    reptiles, 419, 519, 524
    salmon, *417*
    sponges, 366
    wolves, 840–841
  asexual
    cloning as method of, 272–274
    in *Neurospora*, 228
    in one-celled organisms, *194*
    in plants, 459
  in bacteria, 291
  in blue-green algae, 289, 300
  in colonial organisms, 335–336
  and definition of species, 283
  early ideas about, **172–174**, *574*
  in fungi, **320–321**, 322, 323, 324, 326
  human, 210–211, **520–538**, **540–541**, 586 (*see also* Development, human, embryonic; Genetics; Heredity, in man)
  in lichens, 328
  monoecious versus dioecious, 375
  in plants
    angiosperms, 354–357
    asexual, 459–460
    bryophytes, 342, **343**, 344
    Chlamydomonas, 337
    ferns, 346, 347
    gymnosperms, 349–352
    red algae, 333
    sexual, 459, 460–468
    Ulva, 338

potential for, 796, 798
as property of living systems, 28, *29*
in Protista, **306**, 308, 310, 313, 318
rate of, limitations on, 833, 837, 842–843
sexual
  dioecious, 461, 519
  energy of, *194*
  and life cycles, *196*
  meiosis in, *519*
  monoecious, 461
  in *Neurospora*, 228
  in plants, *194*, 459, **460–466**
  as source of variations, *194*, 859
  in vertebtrates, 518–519
  stability in process of, 857, 858–859
  vegetative, *194* (*see also* Reproduction, asexual)
  vertebrate, 518–519 (*see also* Reproduction, human)
Reptiles (Reptilia), **419**, 685
  cerebrum of, 717
  color changes in, 579
  desert, 753
  ear of, 706
  eggs of, 558–559
  evolution of, 936
  excretion in, 524, 558–559, 668
  regulation of temperature by, 683–684
  reproduction in, 519, 524, 548, **558–559**, 573
  respiration in, 654
  vision in, 699
  vomeronasal organ of, *703*
Resolving power, *92*, 699
Resources, natural, depletion of, 969
Respiration (*see also* Breathing; Gills)
  adrenaline as stimulus for, 583
  and air pressure, 649–650
  in animals
    air-breathing snails, 651
    amphibians, 654
    Amphioxus, 652
    arthropods, 395–396, *651*
    birds, 654
    as combustion, 140–141
    early studies of, 140–141
    fish, 418, 649, 654
    insects, 402
    mammals, *637*, 654, 662, *663*
    mollusks, 385, 387, 651
    reptiles, 651, 652, 654
    spiders, 400
    structures for, 651–654
    worms, **380**, 650–651
  and carbon cycle, 777
  cellular
    amino acids in, 668
    defined, 124–125
    effect of thyroxine on, 581
    and oxidation of glucose, 125–126, *135*, 644, 665
    in plants, 490–491
    role of diffusion in, 651
    stages of, 126–135
    use of oxygen in, 447

defined, 131, 648
and diffusion of carbon dioxide, 668
in eggs, 651
in embryo, 651
and energy, 648, 665
and energy flow in food chain, 772, 775
function of, 632
in humans, 656–665
means of, 650–651
photosynthesis as reverse of, 141
in plants, 447, 451, 490–491
Resting potential, of nerve membrane, 590
Retardation, mental (*see* Mental deficiency)
Reticular fibers, 513
Reticular system, *715*
Retina, *696*, 698–699
Retinal, 700
Rh factor, 198–199, 628
Rhabdom, 406–407
Rheumatic fever, 611
Rhizoids, 322
Rhizome, 346–347, 459
Rhizopus, 322
Rhodophyta, 333, *339*
Rhodopsin, *700*, 701
RhoGAM, 628
Rhynchocoela, 376–377, *391*
Rhythm method, 537
Ribbon worms, 376–377, *391*
Ribonucleic acid (*see* RNA)
Ribose, *66*
Ribosomal RNA, 95, 253–254
Ribosomes, 96, *98*, 105
Rickettsia, 293, 294, *298*
Ris, Hans, 316
Rituals, 831–833, 835–837, 838
Rivers, 761–762
RNA (*see also* Nucleic acid(s); *specific types of RNA*)
  biosynthesis of, 265–266
  in early development, 550, *551*
  and protein biosynthesis, 80, 254–256, 257–259, 260
  and regulation of gene activity, 260–263
  structure of, 79, *80*, 250
  types of, 250–254
  viral, 275, 302
RNA polymerase, *258*
Rocks, age of, *18*, 887
Rods, in retina, 699–701
Roots
  function of, 429, 438–439
  growth of, 451–452, 459, **470–471**, 482
  structure of, 438–439
  types of, *441*
Root cap, 470
Root hairs, *438*, 470–471
Root pressure, 453
Roundworms, 377
rRNA (*see* Ribosomal RNA)
Rubella, 567
Ruminants, digestion in, *642*

as component of living systems, 33
compartments, *671*, *672–673*
desalination of, 977
diffusion of, 111
heat of vaporization of, 51–52
and hydrogen bond, 48
imbibition of, 50
ionization of, 42, 52
molecular structure of, 39–40, 48, *52*
movement of (*see also* Active transport;
     Osmosis; Transport, in plants)
  in plants, 447, **451–452**, 457
  principles of, 109–115
nitrite concentration, 988
and origin of life, 21
in plants
  conservation of, 502
  and growth, 114
  movement of, 447, **451–452**, 457
as solvent for ions, *52*
specific heat of, 50–51
surface tension of, 49
temperature of, 50–51
Water balance
  in camels, 690
  in earthworms, 380
  evolution of systems of, 670–671
  in fish, 670, 671
  in insects, 402
  in kangaroo rat, 671, *673*, 677
  and perspiration, 686
  in planarians, 374
  in vertebrates, 667–679, 690
Water circulation, in oceans, 763

Water cycle, *760*
Water potential, 109–114
Water table, *760*
Watson, James D., 241–249
Wax(es), 84, 823
Weather, **735–736**, *737*, *978*
Weathering, and soil, 786
Wells, Martin, 390
Went, Frits W., 480–481
Wernicke's area, *720*
Whales, 662, *663*, 904
Wheat, 207, 208, *961*, *976*
White blood cells (*see* Blood cells, white;
     Leukocytes)
Whittaker, R. H., 809
Wiesel, Torsten N., *721*
"Wild-type," 211
Wilkins, Maurice, 241, 242, *245*
"Willendorf Venus," *952*
Wilting, 502
Wind currents, 735–736
Wöhler, Friedrich, *11*
Wolves
  evolution of, *841*
  mating in, 840–843
  packs of, 840–841
  as predators, 808, 809, 844–845
  social behavior among, 839–845
Women, role of
  and resistance to birth control, 971
  in tribal life, 949
Woodpecker finch, use of tools by, *893*
Woolly mammoth, *946*
Worker caste, 823–824, 828

Worms, **378–382**, *391*, 650–651 [*see also*
     Earthworm(s)]
Wrasses, *833*

X chromosomes
  abnormalities in number of, **223**
  and eye color, 212
  function of, 210–211
  and genetic information, 212, 218
  and hemophilia, 218–219, 221
  and vision, 218
Xanthophyll, 300
X-ray(s), *142*, 203, 567
X-ray diffraction, *245*
Xylem, 344, 433, 437, 475

Y chromosomes, 210–211, 212, 218, 223
Yeast, 130–131
Yolk, 544, 546
Yolk sac, 558, *562*
Young, care of (*see* Care of young)

Zebras, *897*
Zero population growth, 970
Zinc, *455*
Zinjanthropus, 927
Zona pellucida, *541*, 561
Zooflagellates, 308
Zygomycetes, 321–322, *326*
Zygospores, defined, 321, 322
Zygote [*see also* Egg cell(s)]
  in cleavage, 543
  defined, 466, 543
  formation of nucleus of, 541
  plant division of, 466